Handbook on

ENGINEERING PLASTICITY

Theoretical Plasticity & Applications of Plasticity Theory

Other Titles by the Author

Handbook on Mechanics of Inelastic Solids
(In 2 Volumes)
Volume 1: Plasticity, Creep and Viscous Deformation
Volume 2: Finite and Cyclic Deformation; Structural Applications
ISBN: 978-1-80061-206-8 (set)
ISBN: 978-1-80061-800-8 (vol. 1)
ISBN: 978-1-80061-801-5 (vol. 2)

Mechanics of Elastic Solids
ISBN: 978-1-78634-616-2

Mechanics of Solids and Structures
Second Edition
ISBN: 978-1-78326-395-0
ISBN: 978-1-78326-396-7 (pbk)

The Mechanics of Engineering Structures
ISBN: 978-1-78326-401-8
ISBN: 978-1-78326-408-7 (pbk)

Handbook on ENGINEERING PLASTICITY

Theoretical Plasticity & Applications of Plasticity Theory

David Rees D. Sc.

Brunel University of London, UK

World Scientific

NEW JERSEY • LONDON • SINGAPORE • BEIJING • SHANGHAI • HONG KONG • TAIPEI • CHENNAI • TOKYO

Published by

World Scientific Publishing Europe Ltd.
57 Shelton Street, Covent Garden, London WC2H 9HE
Head office: 5 Toh Tuck Link, Singapore 596224
USA office: 27 Warren Street, Suite 401-402, Hackensack, NJ 07601

Library of Congress Cataloging-in-Publication Data
Names: Rees, David W. A., 1947– author.
Title: Handbook of engineering plasticity : theoretical plasticity & applications of plasticity theory / David Rees, D.Sc., Brunel University London, UK.
Description: New Jersey : World Scientific, [2025] | Includes bibliographical references and index.
Identifiers: LCCN 2024044612 | ISBN 9781800619944 (hardcover) | ISBN 9781800619951 (ebook for institutions) | ISBN 9781800619968 (ebook for individuals)
Subjects: LCSH: Plasticity--Handbooks, manuals, etc. | Plastic analysis (Engineering)--Handbooks, manuals, etc. | Deformations (Mechanics)--Handbooks, manuals, etc.
Classification: LCC TA418.14 .R444 2025 | DDC 620.1/1233--dc23/eng/20241219
LC record available at https://lccn.loc.gov/2024044612

British Library Cataloguing-in-Publication Data
A catalogue record for this book is available from the British Library.

For any available supplementary material, please visit
https://www.worldscientific.com/worldscibooks/10.1142/Q0454#t=suppl

Desk Editor: Amanda Yun

Typeset by The Froebe Group/Manila Typesetting Company

CONTENTS

CHAPTER 2

STRAIN ANALYSIS

CHAPTER 3

YIELD CRITERIA

CHAPTER 4

NON-HARDENING PLASTICITY

CHAPTER 5

ELASTIC-PERFECT PLASTICITY

CHAPTER 6

SLIP LINE FIELDS

CHAPTER 7

LIMIT ANALYSIS

CHAPTER 8

CRYSTAL PLASTICITY

CHAPTER 9

THE FLOW CURVE

PART II: APPLICATIONS OF PLASTICITY THEORY

CHAPTER 10

PLASTICITY WITH HARDENING

CHAPTER 11

VISCOUS PLASTICITY

CHAPTER 12

ORTHOTROPIC PLASTICITY

CHAPTER 13

PLASTIC INSTABILITY

CHAPTER 14

STRESS WAVES IN BARS

CHAPTER 15

PRODUCTION PROCESSES

CHAPTER 16

APPLICATIONS OF FINITE ELEMENTS

PREFACE

This book brings together those elements of the mechanics of plasticity most pertinent to engineers. The presentation of the introductory material, the theoretical developments and the use of appropriate experimental data appear within a text of 16 chapters. A textbook style has been adopted in which worked examples and exercises illustrate the application of the theoretical material. The latter is provided with appropriate references to journals and other published sources. Therein, this book combines the reference material required of a researcher together with the detail in theory and application expected from a student. The topics chosen are primarily of interest to engineers as undergraduates, postgraduates and practitioners but they would also serve to capture a readership from among applied mathematicians, physicists and materials scientists. Much new work has been drawn from the research literature. Within this, there are few comparable texts with a similar breath in the subject range. The package of topics presented is intended to complement advanced monographs on the theory of plasticity. The unique blend of contents is intended to support syllabuses across a diversity of undergraduate courses including manufacturing, engineering mechanics, strength of materials and applied mathematics.

The first two chapters are concerned with the definitions of stress and strain analyses that underpin plasticity theory. Both the matrix and tensor notations are employed to emphasise their equivalence when describing constitutive relations, co-ordinate transformations, strain gradients and decompositions for both large and small deformations. Chapter 3 outlines the formulation of yield criteria and their experimental confirmation for different initial conditions of material: annealed, heat treated, rolled and extruded. Firstly, the identity between the yield function and a plastic potential is made to provide flow rules for the ideal plastic solids examined in chapters 4 and 5. Chapter 4 compares the predictions from the total and incremental theories of classical plasticity with experimental data. Differences between them have been attributed to a strain history dependence lying within non-radial loading paths. Chapter 5 compiles solutions to a number of elastic–perfect plastic structures. Ultimate loads, collapse mechanisms and residual stress are among the topics considered from the loading of non-hardening material beyond the initial yield point.

In chapter 6 it is shown how large scale plasticity in a number of forming processes can be described with slip line fields. For this an ideal, rigid-plastic, material is assumed. The theory identifies the stress states and velocities within a critical deformation zone under plane strain. The rolling Mohr's circle and hodograph constructions are particularly useful in providing a full slip line field for extrusion and forging of rectangular section. Alternative upper and lower bound analyses of the loads required for metal forming are given in chapter 7. These two bounding methods extend to circular section stock providing rapid approximations to forming within a useful load range.

Chapters 8–10 allow for material hardening behaviour and its influence upon practical plasticity problems. Firstly, in chapter 8, a description of hardening on a micro-scale is given. It is shown from the operating slip processes and their directions upon closely packed atomic planes, how there exists a yield criterion and a flow rule. There follows from this the concept of an initial and a subsequent yield surface, these being developed further in later chapters. The measurement and description of the flow curve (chapter 9) becomes an essential requirement when describing mathematically an observed, macro-plastic behaviour. Therein, chapter 10 outlines a simplified isotropic hardening model and, with its refinement, various models of kinematic hardening appear for when a description of a stress reversal is required. Chapter 11 introduces more recent developments within a time dependence of plasticity and an additional contribution to strain from viscous deformation.

In chapter 12 the theory of orthotropic plasticity for rolled sheet metals, extruded bars and tubes is given. Plastic instability of struts and plates appears in chapter 13 along with analyses of tensile necking. These enable predictions to plastic instability arising in structures and to a forming limit diagram for sheet metal forming.

A graphical analysis of the plasticity induced by longitudinal impact of bars is given in chapter 14. The plasticity arising from high impact stresses is shown to be carried by a stress wave which interacts with an elastic wave to distribute residual stress in the bar. Chapter 15 considers the control of plasticity arising in conventional production processes including: forging, extrusion, rolling and machining. Here, the detailed analyses of ram forces, roll torques and strain rates employ the principles of force equilibrium and strain compatibility. This approach recognises that there are alternatives to slip lines and bounding methods, all of which are complementary when applying plasticity theory in practice. Chapter 16 considers the application of finite elements (FE) to plasticity. The basic approach to defining a stiffness matrix for elasticity and plasticity is outlined. A number of FE applications are made to sheet metal forming in particular.

At the end of each chapter references are cited and where appropriate a bibliography is given for further reading. Exercises complete each chapter from which a selection of worked solutions are given in a separate appendix.

Thanks are due to the author's past teachers, students and conference organisers who kept him active in this area. In particular chapter 11 shows where the subject of plasticity has continued to develop. Fittingly, these days complex analyses are assisted by numerical methods. In this regard, the material presented here will serve to provide the essential mechanics required for any numerical implementation of a plasticity theory. One example of this appears in chapter 16 where the FE simulations of thin sheet metal forming, under a punch, in a die or by fluid pressure, are integrated with the critical combination of in-plane plastic strains that define the sheet's forming limit diagram (FLD).

D. Rees June 2024

LIST OF SYMBOLS

The intention within the various theoretical developments given in this book has been to define each new symbol where it first appears in the text. In this regard each chapter should be treated as self-contained in its symbol content. There are, however, certain symbols that re-appear consistently throughout the text, such as those representing force, stress and strain. These symbols are given in the following list along with others most commonly employed in plasticity theory.

α, β	curvilinear co-ordinates (slip lines)
α_{ij}	kinematic hardening translations
$\alpha_{ij}^{s}, \beta_{ij}^{s}$	Schmidt's orientation factors
β, φ	friction and shear angles
δ	rolling draft
ε, γ	normal and shear strains
$\dot{\varepsilon}, \dot{\gamma}$	normal and shear rates of strain
ε_{ij}^{P*}	micro-plastic strain tensor
$\bar{\varepsilon}^{P}$	equivalent plastic strain
σ, τ	direct and shear stress
$\sigma_1, \sigma_2, \sigma_3$	principal stresses
σ_m	mean or hydrostatic stress
σ_{ij}^{*}	micro-stress tensor
$\sigma_{ij}^{\prime}$	transformed stress
σ_t, σ_c	tensile and compressive strengths
$\bar{\sigma}$	equivalent stress
μ	friction coefficient
μ, ν	Lode's parameters
φ, λ	scalar multipliers
θ	angular twist
θ, α	die angles
ψ	hardening measure
ν	Poisson's ratio
ρ	density
λ	extension (stretch) ratio
χ, H, F	hardening functions
a, b, l, z	lengths
A	section or surface area
b, t	breadth and thickness
c	propagation velocity
C, T	torque
e_1, e_2, e_3	principal engineering strains
e_{ij}	distortions
ep	subscripts denoting elastic-plastic
E	superscript denoting elastic

E, G, K	elastic constants
f	yield function (plastic potential)
$F, G, H, L..$	anisotropy parameters
$L_1.., Q_1.., T_1..$	anisotropy parameters continued
F, P	force
H_{ij}, C_{ijkl}	orthotropic tensors
H_{ijkl}, H_{ijklmn}	orthotropic tensors continued
I, J	second moments of area
I_1, I_2, I_3	strain invariants
J_1, J_2, J_3	stress invariants
J_1', J_2', J_3'	stress deviator invariants
K	buckling coefficient
l, m, n	direction cosines
m, n	half-waves in buckling
M	bending moment
n	hardening exponent
p	pressure
P	superscript denoting plastic
Q	stress ratio
Q, S	shape and safety factors
$r\ \theta, z$	polar co-ordinates
r_1, r_2	incremental strain ratios (r-values)
R	extrusion ratio
R_1, R_2	radii of curvature
t_1, t_2	back tensions
u, v, w	displacements
U	strain energy
v, ω	linear and angular velocities
V	volume
W	work done
x, y, z	Cartesian co-ordinates
x_i	spacial co-ordinates
X_i	material co-ordinates
X, Z	equivalence coefficients
$Y (= \sigma_o), k$	tensile and shear yield stresses
z	Considére's subtangent
$\mathbf{\Omega}\ (= \omega_{ij})$	rotation tensor/matrix
$\mathbf{B}, \mathbf{C}, \mathbf{G}, \mathbf{L}$	deformation tensors
$\mathbf{E}\ (= \varepsilon_{ij})$	infinitesimal strain tensor/matrix
$\mathbf{F}, \mathbf{H}$	deformation gradients
$\mathbf{m}, \mathbf{n}, \mathbf{u}$	unit vectors
$\mathbf{M}\ (= l_{ij})$	rotation matrix
$\mathbf{S}$	nominal stress tensor
$\mathbf{T}\ (=\sigma_{ij})$	stress tensor/matrix
$\mathbf{T}'\ (=\sigma_{ij}')$	deviatoric stress tensor/matrix
$\mathbf{U}, \mathbf{V}$	stretch tensors

PART I

THEORETICAL PLASTICITY

CHAPTER 1

STRESS ANALYSIS

1.1 Introduction

Before proceeding to the study of flow in a deforming solid it is necessary to understand what is meant by the term stress. Various definitions of stress have been used so it is pertinent to begin with explanations as to how it arises and is quantified. Firstly, it is essential that the tensorial nature of stress is appreciated. It will be shown that stress is a symmetrical second order Cartesian tensor. Where deformation is small (infinitesimal) one can represent stress in both the tensor component and matrix notations. Stress is introduced for simple uniaxial and shear loadings. A combination of these loadings gives both normal and shear stress, these comprising two of the six independent components that are possible within a general stress tensor. The transformation properties of stress are to be examined following a rotation in the orthogonal co-ordinates chosen to define the stress state at a point. Alternative stress definitions are given when it becomes necessary to distinguish between the initial and current areas for large (finite) deformations. Finite deformation will affect the definition of stress because the initial and current areas can differ appreciably. The definition of stress becomes important when required to connect the stress and strain tensors within a constitutive relationship for both elastic and plastic deforming solids.

The following analyses will alternate between the engineering and mathematical co-ordinate notations listed in Table 1.1. This will enable the reader to interchange between notations in recognition of the equivalence between them.

Table 1.1 Symbol Equivalence in Engineering and Mathematical Notations

Quantity	Engineering Notation	Mathematical Notation
Material co-ordinates	x, y, z	x_1, x_2, x_3
Spacial co-ordinates	X, Y, Z	X_1, X_2, X_3
Material displacements	u, v, w	u_1, u_2, u_3
Spacial displacements	U, V, W	U_1, U_2, U_3
Unit co-ordinate vectors	$\mathbf{u}_x, \mathbf{u}_y, \mathbf{u}_z$	$\mathbf{u}_1, \mathbf{u}_2, \mathbf{u}_3$
Direction cosines	l, m, n	l_1, l_2, l_3
Unit normal equation	$\mathbf{u}_\mathrm{n} = l\mathbf{u}_\mathrm{x} + m\mathbf{u}_\mathrm{y} + n\mathbf{u}_\mathrm{z}$	$\mathbf{n} = l_1\,\mathbf{u}_1 + l_2\,\mathbf{u}_2 + l_3\,\mathbf{u}_3$
Unit normal column matrix	$\mathbf{u}_\mathrm{n} = \{l \;\; m \;\; n\}^\mathrm{T}$	$\mathbf{n} = \{l_1 \;\; l_2 \;\; l_3\}^\mathrm{T}$
Normal stress	$\sigma_x, \sigma_y, \sigma_z$	$\sigma_{11}, \sigma_{22}, \sigma_{33}$
Shear stress	$\tau_{xy}, \tau_{xz}, \tau_{yz}$	$\sigma_{12}, \sigma_{13}, \sigma_{23}$
Normal strain (see Ch. 2)	$\varepsilon_x, \varepsilon_y, \varepsilon_z$	$\varepsilon_{11}, \varepsilon_{22}, \varepsilon_{33}$
Shear strain (see Ch. 2)	$\tfrac{1}{2}\gamma_{xy}, \tfrac{1}{2}\gamma_{xz}, \tfrac{1}{2}\gamma_{yz}$	$\varepsilon_{12}, \varepsilon_{13}, \varepsilon_{23}$
Stresses on oblique plane	σ, τ	$\sigma_{11}', \sigma_{21}', \sigma_{31}'$

Note: A rotation matrix **M** employs the direction cosines in the above table for a co-ordinate transformation between Cartesian axes 1, 2 and 3, in each notation as follows:

$$\mathbf{M} = \begin{bmatrix} l_{11} & l_{12} & l_{13} \\ l_{21} & l_{22} & l_{23} \\ l_{31} & l_{32} & l_{33} \end{bmatrix} \equiv \begin{bmatrix} l_1 & m_1 & n_1 \\ l_2 & m_2 & n_2 \\ l_3 & m_3 & n_3 \end{bmatrix}$$

1.1.1 Direct Stress

Direct stress σ measures the intensity of a reaction to externally applied loading. In fact, σ refers to the internal force acting perpendicular to a unit of area within a material. For example, when a uniaxial external force is either tensile(+) or compressive(−), σ is simply

$$\sigma = \pm W / A \tag{1.1}$$

where W is the magnitude of the externally applied force and A is the original normal area (see Fig. 1.1a). The elastic reduction in a section area under stress is negligibly small and hence it is unnecessary to distinguish between initial and current areas within eq(1.1). Elasticity is clearly evident from the initial linear plot of stress versus strain in Fig. 1.1b.

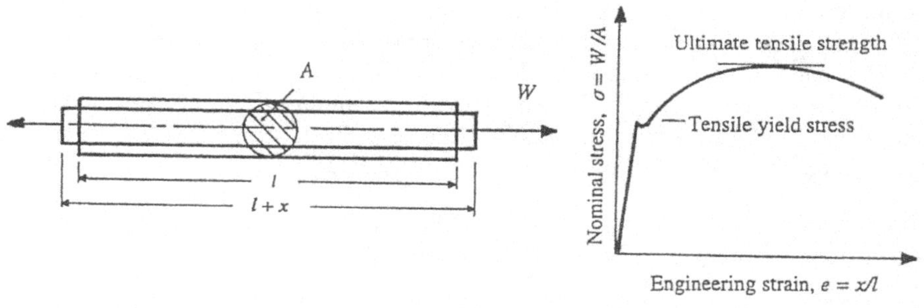

(a) (b)

Figure 1.1 Direct tensile stress showing elastic and plastic strain responses

Note: From Fig. 1.1a, the corresponding direct strain e is the amount by which the material extends per unit of its length as shown. For displacements under tension or compression, i.e. $\pm x$, occurring over a length l, the corresponding strains are

$$e = \pm x / l \tag{1.2}$$

This engineering definition of strain applies to small, elastic displacements. With larger deformations in the plastic range a true stress is calculated from the current area and plastic strains is calculated from referring the displacement to the current length. The true stress and true strain are developed further in this chapter and applied in most of the chapters involving plasticity that follow.

1.1.2 Shear Stress

Let an applied shear force F act tangentially to the top area A, as shown in Fig. 1.2a.

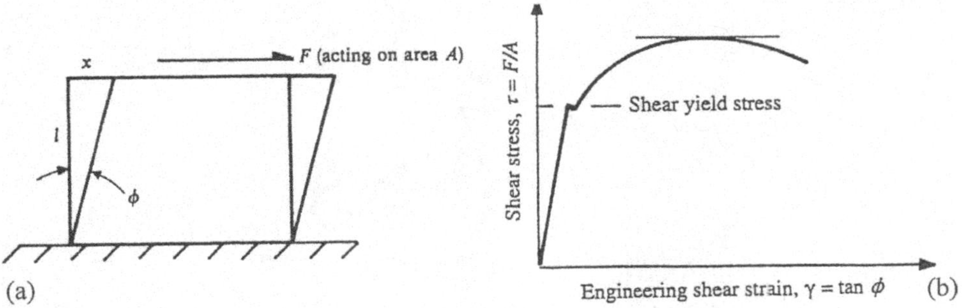

Figure 1.2 Shear distortion showing elastic and plastic strain responses

The shear stress intensity τ, sustained by the material as it maintains equilibrium with this force, is given by

$$\tau = F / A \tag{1.3}$$

The abscissa in the shear stress versus shear strain plot Fig. 1.2b refers to the angular distortion that a material suffers in shear. The shear strain is a non-dimensional measure of distortion and is defined in Fig. 1.2a as

$$\gamma = \tan \varphi = x / l \tag{1.4}$$

In eq(1.4) φ is the angular change in the right angle measured in radians. Within the elastic region the shear displacement x is small when it follows from eq(1.4) that, with a correspondingly small φ, the shear strain may be approximated as $\gamma \approx \varphi$ (rad).

The original area A in eq(1.3) will depend upon the mode of shear. For example, consider the two plates, in Fig. 1.3a joined with a single pin, subjected to tensile force F. Since the pin is placed in single shear, A refers to its cross-sectional area and F to the transverse shear force. In a double shear lap joint in Fig. 1.3b the effective area resisting F is doubled and so τ is halved.

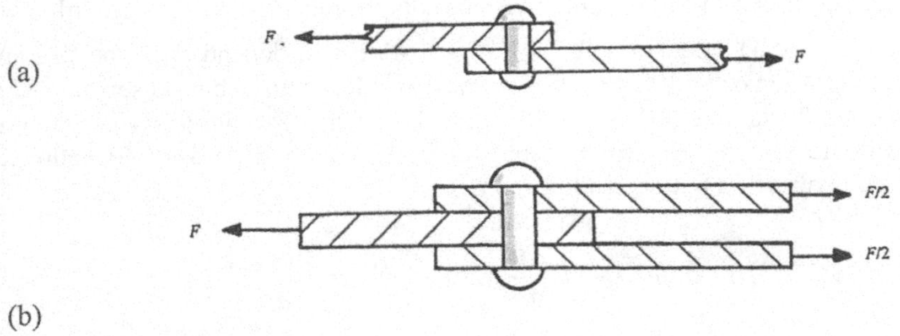

Figure 1.3 Pinned joints in single and double shear

1.2 Cauchy Definition of Stress

Consider an elemental area δa, on a plane B, that cuts through a loaded body in its deformed configuration (see Fig. 1.4).

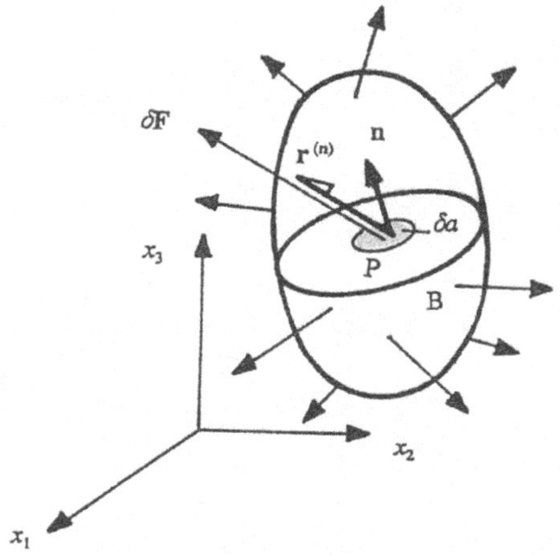

Figure 1.4 Force $\delta \mathbf{F}$ transmitted through area δa

Let a unit vector **n**, lying normal to δa at P, be directed outward from the positive side of B as shown. Due to the applied loading, an elemental resultant force vector $\delta \mathbf{F}$, acting in any direction on the positive side of δa, must also be transmitted to the negative side of B if the continuum is to remain in equilibrium. The traction acting across δa may be found from considering the lower half as a free body.

1.2.1 Stress Intensity

Let an average stress intensity, or traction vector $\mathbf{r}^{(\mathbf{n})}$, be the average force per unit area of δa, so that

$$\delta \mathbf{F} = \mathbf{r}^{(\mathbf{n})}\,\delta a \quad \text{or} \quad \delta F_i = r_i^{(\mathbf{n})}\,\delta a \tag{1.5a,b}$$

The alternative expression (1.5b) has employed the components $r_i^{(\mathbf{n})}$ of $\mathbf{r}^{(\mathbf{n})}$ in co-ordinates, x_i (where $i = 1, 2$ and 3). Equation (1.5a) shows that $\delta \mathbf{F}$ will depend upon the size and orientation of δa. The vector $\mathbf{r}^{(\mathbf{n})}$ emphasises this dependence upon the chosen area δa at P. For a given P, $\mathbf{r}^{(\mathbf{n})}$ is uniquely defined at the finite limit when δa tends to zero. This limit will further eliminate any moments of $\delta \mathbf{F}$ acting on δa. Thus, from eq(1.5a), the traction for any given normal direction **n**, through P, becomes

$$\mathbf{r}^{(\mathbf{n})} = \lim_{\delta a \to 0} \frac{\delta \mathbf{F}}{\delta a} = \frac{\mathrm{d}\mathbf{F}}{\mathrm{d}a} \quad \text{or} \quad r_i^{(\mathbf{n})} = \frac{\mathrm{d}F_i}{\mathrm{d}a} \tag{1.5c,d}$$

Equation (1.5d) reduces to the simple forms given in eqs(1.1) and (1.3) when a single force acts normal or parallel to a given surface. Where oblique forces act, the total stress vector $\mathbf{r}^{(\mathbf{n})}$ may be resolved into chosen co-ordinate directions, x_i. To define a general stress state completely, it is sufficient to resolve $\mathbf{r}^{(\mathbf{n})}$ into one normal and two shear stress components for the positive sides of orthogonal co-ordinate planes passing through point P. Such resolution reveals the tensor nature of stress since it follows that there will be nine traction components when three orthogonal planes are considered. To show this, let $\mathbf{n}_j$ ($j = 1, 2, 3$) be unit vectors in the direction of the co-ordinates x_i so that $\mathbf{r}^{(\mathbf{n}_1)}$, $\mathbf{r}^{(\mathbf{n}_2)}$, $\mathbf{r}^{(\mathbf{n}_3)}$ become the traction vectors on the three faces shown in Fig. 1.5.

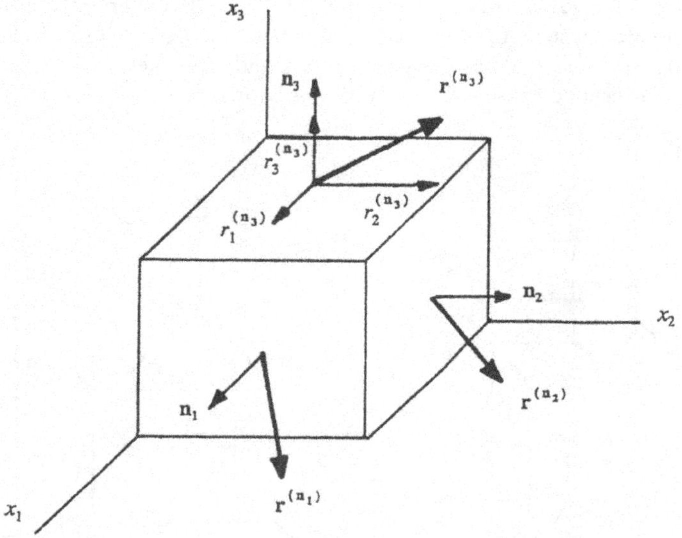

Figure 1.5 Traction across the three faces of a Cartesian element

The three traction vectors $\mathbf{r}^{(\mathbf{n}_j)}$ (in which $\mathbf{n}_j$ are also unit normals to the three orthogonal planes) may be written in terms of the scalar intercepts $r_i^{(\mathbf{n}_j)}$ that each vector makes with x_i as follows

$$\mathbf{r}^{(\mathbf{n}_1)} = r_1^{(\mathbf{n}_1)}\mathbf{n}_1 + r_2^{(\mathbf{n}_1)}\mathbf{n}_2 + r_3^{(\mathbf{n}_1)}\mathbf{n}_3 = r_i^{(\mathbf{n}_1)}\mathbf{n}_i$$
$$\mathbf{r}^{(\mathbf{n}_2)} = r_1^{(\mathbf{n}_2)}\mathbf{n}_1 + r_2^{(\mathbf{n}_2)}\mathbf{n}_2 + r_3^{(\mathbf{n}_2)}\mathbf{n}_3 = r_i^{(\mathbf{n}_2)}\mathbf{n}_i$$
$$\mathbf{r}^{(\mathbf{n}_3)} = r_1^{(\mathbf{n}_3)}\mathbf{n}_1 + r_2^{(\mathbf{n}_3)}\mathbf{n}_2 + r_3^{(\mathbf{n}_3)}\mathbf{n}_3 = r_i^{(\mathbf{n}_3)}\mathbf{n}_i$$

which, by the summation convention, may be contracted into a single equation

$$\mathbf{r}^{(\mathbf{n}_j)} = r_i^{(\mathbf{n}_j)}\mathbf{n}_i \tag{1.6a}$$

where $i, j = 1, 2$ and 3. The nine scalar components $r_i^{(\mathbf{n}_j)}$ form the components of a second order Cartesian stress tensor $\sigma_{ij} = r_i^{(\mathbf{n}_j)}$. Thus, the vector eq(1.6a) becomes

$$\mathbf{r}^{(\mathbf{n}_j)} = \sigma_{ij}\,\mathbf{n}_i \tag{1.6b}$$

Equation(1.6b) satisfies force equilibrium parallel to each co-ordinate directions. This equilibrium condition will appear later with the alternative engineering stress notation (see eqs(1.11a,b,c)). The Cauchy stress tensor **T**, with components σ_{ij} (where i, j = 1, 2, 3), is defined from eq(1.6b) when the co-ordinates x_i are referred to the deformed configuration.

1.2.2 General Stress State

Within a general three-dimensional stress state both normal and shear stresses components comprise the tensor components σ_{ij} within eq(1.6b). Two conventions are employed to distinguish between these components and to identify the directions in which they act. In the *engineering notation*, σ denotes normal stress and τ denotes shear stress. Let these appear within Cartesian co-ordinates x, y and z, as shown in Fig. 1.6a.

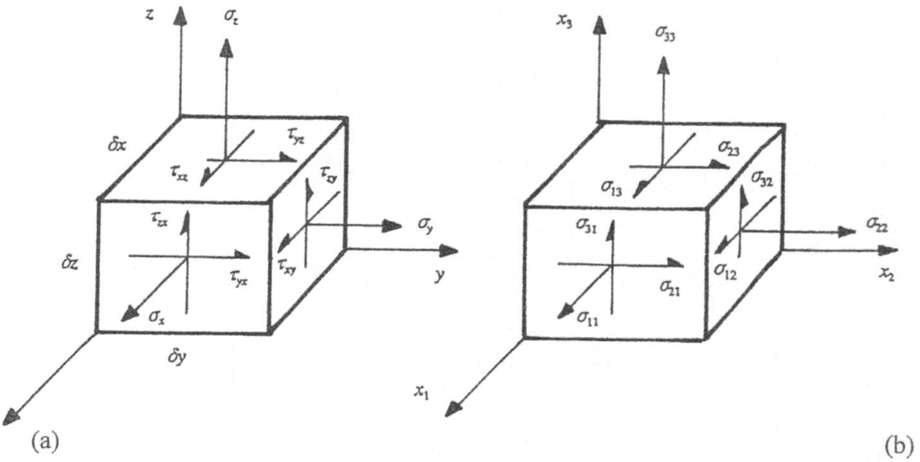

Figure 1.6 General stress states in (a) engineering and (b) mathematical notations

A single subscript on σ identifies the direction of the three normal stress components. The double subscript on τ distinguishes between the six shear components. The first subscript denotes the direction of the stress and the second the direction of the normal to the plane on which that stress acts, e.g. τ_{xy} is a shear stress aligned with the x–direction on the plane whose normal is aligned with the y–direction. (*Note*: some texts interchange these subscripts by writing the normal direction first). Only three shear stresses components are independent. The complementary nature of the shear stresses: $\tau_{xy} = \tau_{yx}$, $\tau_{xz} = \tau_{zx}$ and $\tau_{yz} = \tau_{zy}$, ensures that moments produced by the force resultants about any point are in equilibrium. To show this, take moments on four faces in the x–y plane about a point along the z-axis in Fig. 1.6a:

$$\sigma_y(\delta x)^2(\delta z)/2+\sigma_x(\delta y)^2(\delta z)/2+\tau_{xy}(\delta x)(\delta z)(\delta y) = \sigma_x(\delta y)^2(\delta z)/2+\sigma_y(\delta x)^2(\delta z)/2+\tau_{yx}(\delta x)(\delta y)(\delta z)$$

which leads to $\tau_{xy} = \tau_{yx}$. In Fig. 1.6b, an alternative Cartesian frame x_i (x_1, x_2 and x_3) is employed to identify the stress components according to the *mathematical tensor notation*. Here, the single symbol σ is used for both normal and shear stress components. They are distinguished with double subscripts referring to directions and planes as before. Thus, σ_{11} is a normal stress aligned with the 1–direction and the normal to its plane is also in the x_1–direction. Normal stresses will always appear with two similar subscripts in this notation.

Different subscripts denote shear stresses, e.g. σ_{12} is aligned with the x_1–direction but acts on the plane whose normal is aligned with the x_2–direction. Great care must be taken not to confuse generalised co-ordinates: x_1, x_2 and x_3 with the system of co-ordinates 1, 2 and 3 used in the following section to identify principal stresses. Since shear stress is absent in principal directions we may employ a single subscript 1, 2 or 3 with σ to identify principal stresses unambiguously.

1.2.3 Stress Tensor

It is seen that six independent scalar components of stress are required to define the general state of stress at a point. This identifies stress as a Cartesian tensor of second order. The components appear in the tensor notation as $\sigma_{ij} = \sigma_{ji}$ (where $i = j = 1$, 2 and 3). *Note*: A vector is a tensor of the first order as it is defined from the three scalar intercepts the vector makes with its co-ordinate axes. The following section shows that the scalar components of the stress tensor may be transformed for any given rotation in the co-ordinate axes. These tensor components are often expressed in the form of a symmetrical 3 × 3 matrix **T**. In full, the following two matrices of stress tensor components and their symbols σ_{ij} and **T** are all equivalent. In this and other chapters they are chosen according to reference co-ordinates: x, y and z in the engineering notation and 1, 2 and 3 in the tensor notation.

$$\sigma_{ij} = \begin{bmatrix} \sigma_x & \tau_{xy} & \tau_{xz} \\ \tau_{yx} & \sigma_y & \tau_{yz} \\ \tau_{zx} & \tau_{zy} & \sigma_z \end{bmatrix} \equiv \begin{bmatrix} \sigma_{11} & \sigma_{12} & \sigma_{13} \\ \sigma_{21} & \sigma_{22} & \sigma_{23} \\ \sigma_{31} & \sigma_{32} & \sigma_{33} \end{bmatrix} = \mathbf{T} \qquad (1.7\text{a,b})$$

For example, in engineering co-ordinates matrix eq(1.7a) shows the three independent direct stress components σ_x, σ_y and σ_z for the leading diagonal. Off diagonal shear stress components are symmetrical by virtue of complementary shear: $\tau_{xy} = \tau_{yx}$ etc.

1.3 Three-Dimensional Stress Analysis

Let an oblique triangular plane ABC in Fig. 1.7a cut through the stressed Cartesian element in Fig. 1.6a to produce a tetrahedron OABC. The six known independent stress components: $\sigma_x, \sigma_y, \sigma_z, \tau_{xy} = \tau_{yx}, \tau_{xz} = \tau_{zx}$ and $\tau_{yz} = \tau_{zy}$ now act on the back three triangular faces OAB, OBC and OAC in the negative co-ordinate directions.

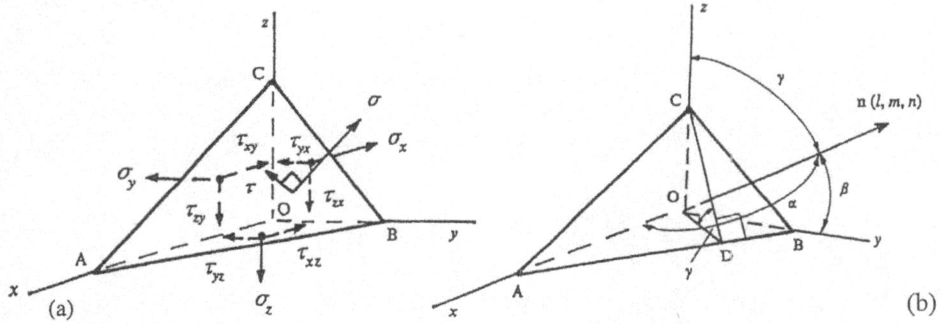

Figure 1.7 General stress state for a tetrahedron showing direction cosines for oblique plane ABC

Since the element must remain in equilibrium, the force resultants produced by the action of these stresses are equilibrated by a normal stress σ and a shear stress τ on the oblique plane ABC in Fig. 1.7a. The objective is to find, this stress state (σ, τ) in both magnitude and direction, by the methods of force resolution and tensor transformation.

1.3.1 Direction Cosines

It is first necessary to find the areas of each back face. Let the area ABC in Fig. 1.7b be unity. Construct a perpendicular CD to AB and join OD. A normal vector **n** to plane ABC is defined by direction cosines *l*, *m* and *n*, measured relative to *x*, *y* and *z* respectively as follows

$$l = \cos\alpha,\ m = \cos\beta \quad \text{and} \quad n = \cos\gamma \tag{1.8a,b,c}$$

Then, as Area ABC = ½AB × CD and Area OAB = ½AB × OD:

$$(\text{Area OAB})/(\text{Area ABC}) = \text{OD/CD} = \cos\gamma = n$$

Hence: Area OAB = n. Similarly: Area OBC = l and Area OAC = m. The direction cosines are not independent. Their relationship follows from the equation of vector **n**:

$$\mathbf{n} = n_x\,\mathbf{u}_x + n_y\,\mathbf{u}_y + n_z\,\mathbf{u}_z \tag{1.9a}$$

where $\mathbf{u}_x$, $\mathbf{u}_y$ and $\mathbf{u}_z$ are unit vectors and n_x , n_y and n_z are scalar intercepts with the co-ordinates x, y and z, as shown in Fig. 1.8a.

Figure 1.8 Scalar intercepts for (a) normal vector **n** and (b) unit normal $\mathbf{u}_n$

The unit vector $\mathbf{u}_n$, for the normal direction (see Fig. 1.8b), is found from dividing eq(1.9a) by the magnitude $|\mathbf{n}|$:

$$\mathbf{u}_n = \frac{n_x}{|\mathbf{n}|}\mathbf{u}_x + \frac{n_y}{|\mathbf{n}|}\mathbf{u}_y + \frac{n_z}{|\mathbf{n}|}\mathbf{u}_z \tag{1.9b}$$

Substituting from eqs(1.8a,b,c): $l = \cos\alpha = n_x/|\mathbf{n}|$, $m = \cos\beta = n_y/|\mathbf{n}|$ and $n = \cos\gamma = n_z/|\mathbf{n}|$, eq(1.9b) becomes

$$\mathbf{u}_n = l\,\mathbf{u}_x + m\,\mathbf{u}_y + n\,\mathbf{u}_z \tag{1.9c}$$

It follows that l, m and n are also the intercepts that the unit normal vector $\mathbf{u}_n$ makes with x, y and z (shown in Fig. 1.8b). Furthermore, since

$$(n_x)^2 + (n_y)^2 + (n_z)^2 = |\mathbf{n}|^2$$

$$\therefore \qquad \left(\frac{n_x}{|\mathbf{n}|}\right)^2 + \left(\frac{n_y}{|\mathbf{n}|}\right)^2 + \left(\frac{n_z}{|\mathbf{n}|}\right)^2 = 1$$

and the direction cosines obey the relationship:

$$l^2 + m^2 + n^2 = 1 \tag{1.10}$$

1.3.2 Force Resolution

(a) Magnitudes of σ and τ

Let σ and τ be the normal and shear stress components of the resultant force or traction vector **r**, acting upon plane ABC in Fig. 1.9a.

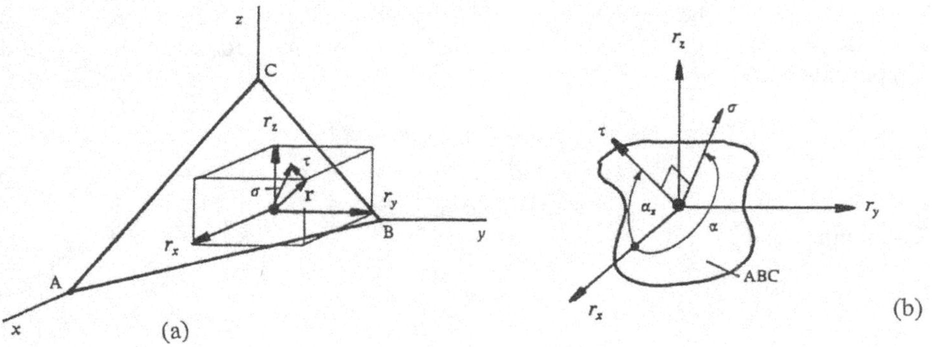

Figure 1.9 Stress state for the oblique plane ABC

The components of vector **r** are r_x, r_y and r_z as shown. Since **r** must equilibrate the forces due to stress components applied to the back faces (see Fig. 1.7a), it follows that

$$r_x = l\sigma_x + m\tau_{xy} + n\tau_{xz} \tag{1.11a}$$

$$r_y = m\sigma_y + n\tau_{yz} + l\tau_{yx} \tag{1.11b}$$

$$r_z = n\sigma_z + l\tau_{zx} + m\tau_{zy} \tag{1.11c}$$

Writing eqs(1.11a,b,c) in the contracted form: $r_i = \sigma_{ij}\, n_j$, it is seen that these become a re-statement of eq(1.6) in which $\sigma_{ij} = \sigma_{ji}$. Using the engineering notation, the corresponding matrix equation, **r** = **Tn**, gives

$$\begin{bmatrix} r_x \\ r_y \\ r_z \end{bmatrix} = \begin{bmatrix} \sigma_x & \tau_{xy} & \tau_{xz} \\ \tau_{yx} & \sigma_y & \tau_{yz} \\ \tau_{zx} & \tau_{zy} & \sigma_z \end{bmatrix} \begin{bmatrix} l \\ m \\ n \end{bmatrix}$$

Now as the area of ABC is unity, σ is the sum of the r_x, r_y and r_z force components resolved into the normal direction. This gives

$$\sigma = r_x \cos\alpha + r_y \cos\beta + r_z \cos\gamma = r_x l + r_y m + r_z n \qquad (1.12a)$$

From eqs(1.11a,b,c)

$$\sigma = \sigma_x l^2 + \sigma_y m^2 + \sigma_z n^2 + 2\,(lm\tau_{xy} + mn\tau_{yz} + ln\tau_{zx}) \qquad (1.12b)$$

The magnitude of the resultant force on ABC is expressed in two ways:

$$r^2 = r_x^2 + r_y^2 + r_z^2 = \sigma^2 + \tau^2$$

$$\therefore \tau^2 = r^2 - \sigma^2 = r_x^2 + r_y^2 + r_z^2 - \sigma^2 \qquad (1.12c)$$

and substituting eqs(1.11a–c) into (1.12c), τ can be found.

(b) Directions of σ and τ

Since σ lies parallel to **n**, the direction of σ is also defined by l, m and n for the plane ABC. The direction of τ in the plane ABC is defined by the directions: $l_s = \cos\alpha_s$, $m_s = \cos\beta_s$ and $n_s = \cos\gamma_s$ (see Fig. 1.9b). Because r_x, r_y and r_z are resultant forces of the x, y and z components of σ and τ, this gives

$$r_x = \sigma \cos\alpha + \tau \cos\alpha_s = l\sigma + l_s\tau$$
$$r_y = \sigma \cos\beta + \tau \cos\beta_s = m\sigma + m_s\tau$$
$$r_z = \sigma \cos\gamma + \tau \cos\gamma_s = n\sigma + n_s\tau$$

Re-arranging gives

$$l_s = (r_x - l\sigma)\,/\,\tau \qquad (1.13a)$$
$$m_s = (r_y - m\sigma)\,/\tau \qquad (1.13b)$$
$$n_s = (r_z - n\sigma)\,/\,\tau \qquad (1.13c)$$

Example 1.1 A stress resultant of 140 MPa makes respective angles of 43°, 75° and 50°53' with the x, y and z axes. Determine the normal and shear stresses, in magnitude and direction, on an oblique plane whose normal makes respective angles of 67°13', 30° and 71°34' with these axes.

Referring to Fig. 1.9a, $r = 140$ MPa is resolved into components for the x, y and z directions:

$$r_x = 140 \cos 43° = 102.39 \text{ MPa}$$
$$r_y = 140 \cos 75° = 36.24 \text{ MPa}$$
$$r_z = 140 \cos 50°53' = 88.33 \text{ MPa}$$

The normal stress is found from eq(1.12a), in which l, m and n are the direction cosines for the normal:

$$\sigma = r_x l + r_y m + r_z n = r_x \cos\alpha + r_y \cos\beta + r_z \cos\gamma$$
$$= 102.39 \cos 67°13' + 36.24 \cos 30° + 88.33 \cos 71°34' = 98.96 \text{ MPa}$$

Equation (1.12c) supplies the shear stress on this plane as

$$\tau = \sqrt{(r^2 - \sigma^2)} = \sqrt{(140^2 - 98.96^2)} = 99.03 \text{ MPa}$$

and eqs(1.13a,b,c) gives its direction cosines as

$$l_s = (r_x - l\sigma)/\tau = (102.39 - 98.96 \cos 67°13')/ 99.03 = 0.647 \quad (\alpha_s = 49°41')$$
$$m_s = (r_y - m\sigma)/\tau = (36.24 - 98.96 \cos 30°)/ 99.03 = - 0.500 \quad (\beta_s = 120°)$$
$$n_s = (r_z - n\sigma)/\tau = (88.33 - 98.96 \cos 71°34')/ 99.03 = 0.576 \quad (\gamma_s = 54°50')$$

from which a check shows: $l_s^2 + m_s^2 + n_s^2 = 1$.

1.3.3 Stress Transformations in Tensor and Matrix Notations

It is now shown that components σ and τ in eqs(1.12a,b) appear as components in a general tensor transformation law for stress. The general transformation law for a tensor follows from the dyadic product of two vectors. The equations for any pair of arbitrary vectors **a** and **b**, (see, for example, Fig. 1.10a) will appear in Cartesian co-ordinates: x_1, x_2 and x_3, as

$$\mathbf{a} = a_1\,\mathbf{u}_1 + a_2\,\mathbf{u}_2 + a_3\,\mathbf{u}_3 = a_i\,\mathbf{u}_i \qquad (1.14a)$$
$$\mathbf{b} = b_1\mathbf{u}_1 + b_2\mathbf{u}_2 + b_3\,\mathbf{u}_3 = b_i\,\mathbf{u}_i \qquad (1.14b)$$

where a_i and b_i are the scalar intercepts and $\mathbf{u}_i$ are unit co-ordinate vectors. Figure 1.10a shows each of these for the vector **a**.

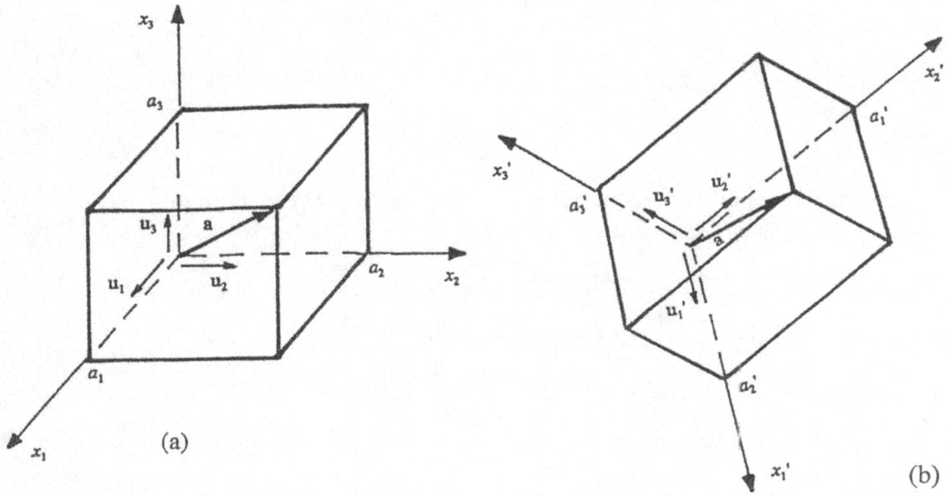

Figure 1.10 Components of a vector in co-ordinate frame x_1, x_2 and x_3 and x_1', x_2' and x_3'

Let the co-ordinate axes rotate about the origin to lie in the final orthogonal frame x_1', x_2' and x_3', as shown in Fig. 1.10b The equations of the stationary vectors **a** and **b** become

$$\mathbf{a} = a_1'\mathbf{u}_1' + a_2'\mathbf{u}_2' + a_3'\mathbf{u}_3' = a_j'\,\mathbf{u}_j' \qquad (1.15a)$$
$$\mathbf{b} = b_1'\mathbf{u}_1' + b_2'\mathbf{u}_2' + b_3'\mathbf{u}_3' = b_j'\,\mathbf{u}_j' \qquad (1.15b)$$

Consider the method for expressing this rotation. Equation (1.9c) showed that the components of a unit vector are the direction cosines. Thus, unit vectors $\mathbf{u}_1'$, $\mathbf{u}_2'$ and $\mathbf{u}_3'$ in the frame x_1', x_2' and x_3', may each be expressed in terms of unit vectors $\mathbf{u}_1$, $\mathbf{u}_2$ and $\mathbf{u}_3$, for the original frame x_1, x_2 and x_3, as follows:

$$\mathbf{u}_1' = l_{11}\,\mathbf{u}_1 + l_{12}\,\mathbf{u}_2 + l_{13}\,\mathbf{u}_3 \qquad (1.16a)$$
$$\mathbf{u}_2' = l_{21}\,\mathbf{u}_1 + l_{22}\,\mathbf{u}_2 + l_{23}\,\mathbf{u}_3 \qquad (1.16b)$$
$$\mathbf{u}_3' = l_{31}\,\mathbf{u}_1 + l_{32}\,\mathbf{u}_2 + l_{33}\,\mathbf{u}_3 \qquad (1.16c)$$

Using the summation convention eqs(1.16a,b,c) may be contracted to a single equation:

$$\mathbf{u}_i' = l_{ij}\,\mathbf{u}_j \qquad (1.16d)$$

To confirm this, sum eq(1.16d) over j = 1, 2 and 3 to give

$$\mathbf{u}_i' = l_{i1}\,\mathbf{u}_1 + l_{i2}\,\mathbf{u}_2 + l_{i3}\,\mathbf{u}_3$$

and substituting i = 1, 2 and 3 provides the three equations in (1.16a,b,c). The directions l_{ij} (i, j = 1, 2 and 3) in eq(1.16d), define each primed direction relative to the unprimed direction. That is: $l_{ij} = \cos(x_i', x_j)$. For example, $l_{11} = \cos(x_1', x_1)$, $l_{12} = \cos(x_1', x_2)$ and $l_{13} = \cos(x_1', x_3)$ define the directions of x_1' within the frame x_1, x_2 and x_3. It follows that the direction cosines l_{ij} are the components of the following rotation matrix $\mathbf{M}$:

$$\mathbf{M} = \begin{bmatrix} l_{11} & l_{12} & l_{13} \\ l_{21} & l_{22} & l_{23} \\ l_{31} & l_{32} & l_{33} \end{bmatrix} \qquad (1.17a)$$

An orthogonal property of this matrix is that $l_{ij}\,l_{kj} = \delta_{ik}$ or $\mathbf{MM}^{\mathrm{T}} = \mathbf{I}$. In full, this is

$$\mathbf{MM}^{\mathbf{T}} = \begin{bmatrix} l_{11} & l_{12} & l_{13} \\ l_{21} & l_{22} & l_{23} \\ l_{31} & l_{32} & l_{33} \end{bmatrix} \begin{bmatrix} l_{11} & l_{21} & l_{31} \\ l_{12} & l_{22} & l_{32} \\ l_{13} & l_{23} & l_{33} \end{bmatrix} \begin{bmatrix} 1 & 0 & 0 \\ 0 & 1 & 0 \\ 0 & 0 & 1 \end{bmatrix}$$

which contains the following relationships between cosines for each direction:

$$l_{11}^2 + l_{12}^2 + l_{13}^2 = 1 \;\; (\mathbf{u}_1' \bullet \mathbf{u}_1' = l_{1i}\,l_{1i} = \mathbf{u}_1{}^{\mathrm{T}}\mathbf{u}_1 = 1) \text{ for } x_1'$$
$$l_{21}^2 + l_{22}^2 + l_{23}^2 = 1 \;\; (\mathbf{u}_2' \bullet \mathbf{u}_2' = l_{2i}\,l_{2i} = \mathbf{u}_2{}^{\mathrm{T}}\mathbf{u}_2 = 1) \text{ for } x_2' \qquad (1.17b)$$
$$l_{31}^2 + l_{32}^2 + l_{33}^2 = 1 \;\; (\mathbf{u}_3' \bullet \mathbf{u}_3' = l_{3i}\,l_{3i} = \mathbf{u}_3{}^{\mathrm{T}}\mathbf{u}_3 = 1) \text{ for } x_3'$$

Additional relationships apply to pairs of orthogonal directions:

$$l_{11}\,l_{21} + l_{12}\,l_{22} + l_{13}\,l_{23} = 0 \;\; (\mathbf{u}_1' \bullet \mathbf{u}_2' = l_{1i}\,l_{2i} = \mathbf{u}_1{}^{\mathrm{T}}\mathbf{u}_2 = 0) \text{ for } x_1' \text{ and } x_2'$$
$$l_{21}\,l_{31} + l_{22}\,l_{32} + l_{23}\,l_{33} = 0 \;\; (\mathbf{u}_2' \bullet \mathbf{u}_3' = l_{2i}\,l_{3i} = \mathbf{u}_2{}^{\mathrm{T}}\mathbf{u}_3 = 0) \text{ for } x_2' \text{ and } x_3' \qquad (1.17c)$$
$$l_{11}\,l_{31} + l_{12}\,l_{32} + l_{13}\,l_{33} = 0 \;\; (\mathbf{u}_1' \bullet \mathbf{u}_3' = l_{1i}\,l_{3i} = \mathbf{u}_1{}^{\mathrm{T}}\mathbf{u}_3 = 0) \text{ for } x_1' \text{ and } x_3'$$

The abbreviated expressions in brackets show how equivalent equations appear in the respective notions of a direct tensor (i.e. the dot product), indicial tensor components and a matrix. Since $\mathbf{u}_1 = [\, l_{11} \;\; l_{12} \;\; l_{13}\,]^{\mathrm{T}}$ etc. denote column matrices it follows that a row matrix is formed from the transpose: $\mathbf{u}_1^{\mathrm{T}} = [l_{11} \;\; l_{12} \;\; l_{13}]$. Apart from the dot and cross products of vectors, the direct tensor notation will not be adopted further. Alternating between the tensor component and matrix notations of stress and strain is sufficient when expressing their transformation properties to follow.

Combining eqs(1.14) and (1.15), the vectors **a** and **b** may be expressed in both systems of co-ordinates as

$$\mathbf{a} = a_i\,\mathbf{u}_i = a_j'\,\mathbf{u}_j' = a_j'\,(l_{ji}\,\mathbf{u}_i)$$
$$\mathbf{b} = b_i\,\mathbf{u}_i = b_j'\,\mathbf{u}_j' = b_j'\,(l_{ji}\,\mathbf{u}_i)$$

from which the vector transformation laws follow:

$$a_i = a_j'\,l_{ji} = l_j\,a_j' \;\text{ and }\; b_i = b_j'\,l_{ji} = l_{ji}\,b_j \qquad (1.18\text{a,b})$$

In the matrix notation, eqs(1.18a,b) become

$$\mathbf{a} = \mathbf{M}^{\mathrm{T}}\,\mathbf{a}',\; \mathbf{b} = \mathbf{M}^{\mathrm{T}}\,\mathbf{b}' \qquad (1.18\text{c,d})$$

where **a**, **a**′, **b** and **b**′ are column matrices, e.g. $\mathbf{a} = [a_1 \; a_2 \; a_3]^{\mathrm{T}}$. To invert eqs(1.18a,b), multiply both sides by l_{ki}:

$$l_{ki}\,a_i = l_{ki}\,l_{ji}\,a_j' = \delta_{kj}\,a_j' = a_k'$$

where from eq(1.17b,c) $\delta_{ij} = 1$ for $k = j$ and $\delta_{kj} = 0$ for $k \neq j$. Reverting to i, j subscripts:

$$a_i' = l_{ij}\,a_j \;\text{ and, similarly, }\; b_i' = l_{ij}\,b_j \qquad (1.19\text{a,b})$$

Correspondingly, to invert the matrix eq(1.18c), pre-multiply both sides by $(\mathbf{M}^{\mathrm{T}})^{-1}$. This gives

$$(\mathbf{M}^{\mathrm{T}})^{-1}\,\mathbf{a} = (\mathbf{M}^{\mathrm{T}})^{-1}\,\mathbf{M}^{\mathrm{T}}\,\mathbf{a}' = \mathbf{I}\,\mathbf{a}' = \mathbf{a}' \qquad (1.19\text{c})$$

Since the inverse of the square matrix **M** will obey $\mathbf{M}\mathbf{M}^{-1} = \mathbf{I}$ and as $\mathbf{M}\mathbf{M}^{\mathrm{T}} = \mathbf{I}$, a further orthogonal property of the rotation matrix is that $\mathbf{M}^{-1} = \mathbf{M}^{\mathrm{T}}$. It then follows that

$$(\mathbf{M}^{\mathrm{T}})^{-1} = (\mathbf{M}^{-1})^{-1} = \mathbf{M}$$

and eq(1.19c) becomes

$$\mathbf{a}' = \mathbf{M}\mathbf{a} \;\text{ and, similarly, }\; \mathbf{b}' = \mathbf{M}\mathbf{b} \qquad (1.19\text{d,e})$$

A second-order Cartesian tensor may be formed from the dyadic product of two vectors. Note that this differs from the cross product which results in another vector lying normal to the plane containing the two vectors. The tensor or *dyadic product* of two vectors: **a** and **b**, is written as

$$\mathbf{a} \otimes \mathbf{b} = (a_i\,\mathbf{u}_i) \otimes (b_j\,\mathbf{u}_j) = a_i\,b_j\,(\mathbf{u}_i \otimes \mathbf{u}_j)$$

The tensor, so formed, appears as

$$\mathbf{K} = K_{ij}\,(\mathbf{u}_i \otimes \mathbf{u}_j)$$

where $K_{ij} = a_i\, b_j$ ($\mathbf{K} = \mathbf{ab}^{\mathrm{T}}$) are the components of a second order Cartesian tensor $\mathbf{K}$ for which the unit vectors (dyads) $\mathbf{u}_i$ and $\mathbf{u}_j$ appear in linear combination. The components K_{ij} may be referred to both sets of orthogonal axis x_i and x_i' through the vector transformation eqs(1.18a,b). These give

$$K_{ij} = a_i b_j = (\, l_{pi}\, a_p'\,)\,(\, l_{qj}\, b_q'\,)$$

and putting $K_{pq}' = a_p'\, b_q'$ leads to the transformation law

$$K_{ij} = l_{pi}\, l_{qj}\, K_{pq}' \tag{1.20a}$$

Equations(1.19a,b) provide the components of the inverse transformation matrix K_{ij}' as

$$K_{ij}' = a_i'\, b_j' = (l_{ip}\, a_p)(l_{jq}\, b_q)$$

Setting $K_{pq} = a_p b_q$, the general transformation law for any second order tensor is obtained:

$$K_{ij}' = l_{ip}\, l_{jq}\, K_{pq} \tag{1.21a}$$

In converting eqs(1.20a) and (1.21a) to matrix equations, similar subscripts must appear adjacent within each term to become consistent with matrix multiplication. That is

$$K_{ij} = l_{pi}\, l_{qj} K_{pq}' = l_{pi} K_{pq}'\, l_{qj} \quad \text{or} \quad \mathbf{K} = \mathbf{M}^{\mathrm{T}} \mathbf{K}'\, \mathbf{M} \tag{1.20b}$$

$$K_{ij}' = l_{ip}\, l_{jq} K_{pq} = l_{ip} K_{pq}\, l_{jq} \quad \text{or} \quad \mathbf{K}' = \mathbf{M\,K\,M}^{\mathrm{T}} \tag{1.21b}$$

Alternatively, direct matrix derivations are given by eqs(1.18c,d) as

$$\mathbf{K} = \mathbf{a\,b}^{\mathrm{T}} = (\mathbf{M}^{\mathrm{T}}\mathbf{a}')(\mathbf{M}^{\mathrm{T}}\mathbf{b}')^{\mathrm{T}} = \mathbf{M}^{\mathrm{T}}(\mathbf{a}'\,\mathbf{b}'^{\,\mathrm{T}})\,\mathbf{M} = \mathbf{M}^{\mathrm{T}}\mathbf{K}'\,\mathbf{M}$$

$$\mathbf{K}' = \mathbf{a}'\,\mathbf{b}'^{\,\mathrm{T}} = (\mathbf{M\,a})\,(\mathbf{M\,b})^{\mathrm{T}} = \mathbf{M}\,(\mathbf{a\,b}^{\mathrm{T}})\,\mathbf{M}^{\mathrm{T}} = \mathbf{M\,K\,M}^{\mathrm{T}}$$

Previously, it was recognised that the physical quantity called stress is a second order tensor. The stress tensor must therefore transform in the manner of eqs(1.20) and (1.21). Normally, it is required to transform the known components of the stress tensor σ_{pq} in axes x_1, x_2 and x_3 (Fig. 1.11a) to components σ_{ij}' in axes x_1', x_2' and x_3', as shown in Fig. 1.11b.

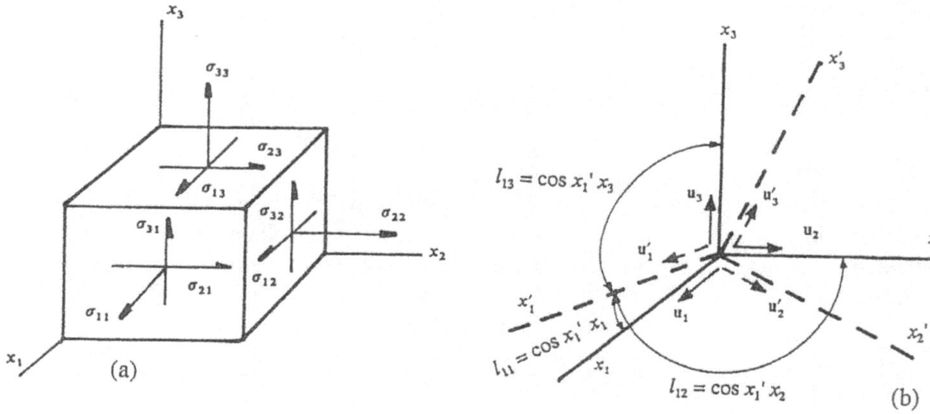

Figure 1.11 (a) Generalised stress components and (b) a rotation in orthogonal axes

It follows from eq(1.21a) that the law of transformation is

$$\sigma_{ij}' = l_{ip}\, l_{jq}\, \sigma_{pq} \quad \text{or} \quad \mathbf{T}' = \mathbf{M T M}^{\mathrm{T}} \tag{1.22a}$$

where $\mathbf{T} \equiv \sigma_{ij}$ and $\mathbf{T}' \equiv \sigma_{ij}'$ and $\mathbf{M} \equiv l_{ip}$. Writing the stress transformation law (1.22a) in full:

$$\begin{bmatrix} \sigma_{11}' & \sigma_{12}' & \sigma_{13}' \\ \sigma_{21}' & \sigma_{22}' & \sigma_{23}' \\ \sigma_{31}' & \sigma_{32}' & \sigma_{33}' \end{bmatrix} = \begin{bmatrix} l_{11} & l_{12} & l_{13} \\ l_{21} & l_{22} & l_{23} \\ l_{31} & l_{32} & l_{33} \end{bmatrix} \begin{bmatrix} \sigma_{11} & \sigma_{12} & \sigma_{13} \\ \sigma_{21} & \sigma_{22} & \sigma_{23} \\ \sigma_{31} & \sigma_{32} & \sigma_{33} \end{bmatrix} \begin{bmatrix} l_{11} & l_{21} & l_{31} \\ l_{12} & l_{22} & l_{32} \\ l_{13} & l_{23} & l_{33} \end{bmatrix} \tag{1.22b}$$

This gives one normal and two shear stresses for each of the three orthogonal planes in the x_i' ($i = 1, 2$ and 3) frame. In the analytical method, the stress state for a single oblique plane ABC (Fig. 1.7) was found. One can identify ABC with the plane lying normal to x_1' (say) with directions: l_{11}, l_{12} and l_{13}. The stress components for this plane (σ_{11}', σ_{21}' and σ_{31}'), are:

$$\begin{bmatrix} \sigma_{11}' \\ \sigma_{21}' \\ \sigma_{31}' \end{bmatrix} = \begin{bmatrix} l_{11} & l_{12} & l_{13} \end{bmatrix} \begin{bmatrix} \sigma_{11} & \sigma_{12} & \sigma_{13} \\ \sigma_{21} & \sigma_{22} & \sigma_{23} \\ \sigma_{31} & \sigma_{32} & \sigma_{33} \end{bmatrix} \begin{bmatrix} l_{11} \\ l_{12} \\ l_{13} \end{bmatrix} \tag{1.22c}$$

The normal and shear stress given in eqs(1.12b,c) are: $\sigma = \sigma_{11}'$ and $\tau = \sqrt{[(\sigma_{21}')^2 + (\sigma_{31}')^2]}$. Clearly, τ is the resultant shear stress acting on plane ABC and σ_{12}', σ_{13}' are its components aligned with the axes x_2' and x_3'.

Note: The prime placed upon stress in eqs(1.22a-c) refers to the normal and shear stress components for the transformed axes x_i'. They are not to be confused with deviatoric stresses σ_{ij}' and **T'**, shown in Fig. 3.3. The stress deviator has the hydrostatic part of the stress tensor removed, i.e. $\sigma_{ij}' = \sigma_{ij} - \tfrac{1}{3}\,\delta_{ij}\sigma_{kk}$, or, $\mathbf{T}' = \mathbf{T} - \tfrac{1}{3}\mathrm{tr}(\mathbf{IT})$ (see eqs(3.9a,b)) and retains the property of transformation, as in eq(1.22a).

1.4 Principal Stresses and Invariants

The three principal planes are orthogonal and free of shear stress. The three stresses normal to these planes are, by definition, principal stresses. Their magnitudes and orientation will now be derived from the known stress components for non-principal axes.

1.4.1 Magnitudes of Principal Stresses

For this let us employ the engineering notation, where the stress components shown in Fig. 1.6a correspond to the 3×3 matrix given in eq(1.7a). When the shear stress τ is absent for the plane ABC (l, m, n) in Fig. 1.7a then the normal stress σ becomes a principal stress. Force resolution in the x, y and z directions modifies eqs(1.11a–c) to:

$$r_x = l\,\sigma = l\sigma_x + m\tau_{xy} + n\tau_{xz}$$
$$r_y = m\sigma = m\sigma_y + n\tau_{yz} + l\tau_{yx}$$
$$r_z = n\sigma = n\sigma_z + l\tau_{zx} + m\tau_{zy}$$

That is

$$\begin{aligned} l\,(\sigma_x - \sigma) + m\tau_{xy} + n\tau_{xz} &= 0 \\ l\tau_{yx} + m\,(\sigma_y - \sigma) + n\tau_{yz} &= 0 \\ l\tau_{zx} + m\tau_{zy} + n\,(\sigma_z - \sigma) &= 0 \end{aligned} \tag{1.23a}$$

Writing the direction cosines in a column matrix $\mathbf{u} = [l\ m\ n]^{\mathrm{T}}$, the equivalent tensor component and matrix forms of eq(1.23a) will respectively appear as

$$(\sigma_{ij} - \sigma\,\delta_{ij})\,u_i = 0 \ \text{ or } \ (\mathbf{T} - \sigma\,\mathbf{I}\,)\,\mathbf{u} = 0 \tag{1.23b}$$

By Cramer's rule, the solution to σ is found from the determinant

$$\begin{vmatrix} (\sigma_x - \sigma) & \tau_{xy} & \tau_{xz} \\ \tau_{yx} & (\sigma_y - \sigma) & \tau_{yz} \\ \tau_{zx} & \tau_{zy} & (\sigma_z - \sigma) \end{vmatrix} = 0 \tag{1.23c}$$

Contracted forms of eq(1.23c) appear, in the two alternative notations, as

$$\det\,(\sigma_{ij} - \sigma\,\delta_{ij}) = 0 \ \text{ or } \ \det\,(\mathbf{T} - \sigma\,\mathbf{I}) = 0$$

Expanding eq(1.23c), leads to a cubic (or, characteristic) equation

$$(\sigma_x - \sigma)[(\sigma_y - \sigma)(\sigma_z - \sigma) - \tau_{yz}\tau_{zy}] - \tau_{xy}\,[\tau_{yx}(\sigma_z - \sigma) - \tau_{yz}\tau_{zx}] + \tau_{xz}\,[\tau_{yx}\tau_{zy} - \tau_{zx}\,(\sigma_y - \sigma)\,] = 0$$
$$\begin{aligned} &\sigma^3 - (\sigma_x + \sigma_y + \sigma_z)\sigma^2 + (\sigma_x\sigma_y + \sigma_y\sigma_z + \sigma_z\sigma_x - \tau_{xy}^2 - \tau_{yz}^2 - \tau_{zx}^2)\sigma \\ &- (\sigma_x\sigma_y\sigma_z + 2\tau_{xy}\tau_{yz}\tau_{zx} - \sigma_x\tau_{yz}^2 - \sigma_y\tau_{zx}^2 - \sigma_z\tau_{xy}^2) = 0 \end{aligned} \tag{1.24a}$$

The three roots (the eigen values) to eq(1.24a) give the principal stress magnitudes σ_1, σ_2 and σ_3. Equation (1.24a) is usually written as

$$\sigma^3 - J_1\sigma^2 + J_2\sigma - J_3 = 0 \tag{1.24b}$$

The principal stresses are unique for given stress tensor components. Therefore, coefficients J_1, J_2 and J_3 in eq(1.24b) are independent of the co-ordinate frame, x, y, z, in Fig. 1.6a, to which the components apply. Hence J_1, J_2 and J_3 are the *invariants* of the stress tensor σ_{ij}. Equation (1.24a) must include an orientation where x, y, z coincides with the principal stress directions 1, 2 and 3. Thus, invariants may be expressed either in terms of general stress components (subscripts x, y, z) or simplified with principal stresses (subscripts 1, 2, 3):

$$J_1 = \sigma_x + \sigma_y + \sigma_z = \sigma_1 + \sigma_2 + \sigma_3 = \sigma_{ii} = \mathrm{tr}\,\mathbf{T} \tag{1.25a}$$

$$\begin{aligned} J_2 &= \sigma_x\sigma_y + \sigma_y\sigma_z + \sigma_z\sigma_x - \tau_{xy}^2 - \tau_{yz}^2 - \tau_{zx}^2 = \sigma_1\sigma_2 + \sigma_2\sigma_3 + \sigma_1\sigma_3 \\ &= \tfrac{1}{2}(\sigma_{ii}\sigma_{jj} - \sigma_{ij}\sigma_{ji}) = \tfrac{1}{2}\,[\,(\mathrm{tr}\,\mathbf{T})^2 - \mathrm{tr}\,(\,\mathbf{T}^2)] \end{aligned} \tag{1.25b}$$

$$\begin{aligned} J_3 &= \sigma_x\sigma_y\sigma_z + 2\tau_{xy}\tau_{yz}\tau_{zx} - \sigma_x\tau_{yz}^2 - \sigma_y\tau_{zx}^2 - \sigma_z\tau_{xy}^2 = \sigma_1\sigma_2\sigma_3 \\ &= \det\,(\sigma_{ij}) = \det\,\mathbf{T} \end{aligned} \tag{1.25c}$$

Also given in eqs(1.25a–c) are the contracted tensor and matrix expressions. The former is to be employed with tensor subscripts $i, j = 1, 2, 3$. Repeated subscripts on a single symbol, or within a term, denote summation.

Where there are exact roots to eq(1.24a), the principal stresses are more conveniently found from expanding the determinant (1.23a) following substitution of the numerical values of the stress components. Otherwise, the major σ_1, intermediate σ_2 and minor σ_3 principal stresses ($\sigma_1 > \sigma_2 > \sigma_3$) must be found from the solution to the cubic eq(1.24a). The Cayley-Hamilton theorem states that a square matrix will satisfy its own characteristic equation. Here, for the 3 × 3 stress matrix, **T**, eq(1.24b) becomes

$$\mathbf{T}^3 - J_1\mathbf{T}^2 + J_2\mathbf{T} - J_3\,\mathbf{I} = 0 \tag{1.25d}$$

Substituting from eqs(1.25a–c), the theorem states that **T** must satisfy:

$$\mathbf{T}^3 - \mathbf{T}^2\mathrm{tr}\mathbf{T} + \tfrac{1}{2}\mathbf{T}\,[\,(\mathrm{tr}\mathbf{T})^2 - \mathrm{tr}\mathbf{T}^2\,] - (\det\mathbf{T})\,\mathbf{I} = 0 \tag{1.25e}$$

Taking the trace of eq(1.25d) gives an alternative expression for J_3

$$\mathrm{tr}\mathbf{T}^3 - J_1\,\mathrm{tr}\mathbf{T}^2 + J_2\,\mathrm{tr}\mathbf{T} - 3J_3 = 0$$

and substituting from eqs(1.25a and b) gives

$$J_3 = \frac{1}{6}\,(\,\mathrm{tr}\mathbf{T}\,)^3 - \frac{1}{2}\,\mathrm{tr}\mathbf{T}\;\mathrm{tr}\mathbf{T}^2 + \frac{1}{3}\,\mathrm{tr}\mathbf{T}^3$$

1.4.2 Principal Stress Directions

Let the direction cosines for σ_1 be l_1, m_1 and n_1 within a co-ordinate frame x, y, z. Substituting for the applied stresses into eq(1.23a), with $\sigma = \sigma_1$, leads to three simultaneous equations in l_1, m_1 and n_1. Only two of these are independent because of the relationship: $l_1^2 + m_1^2 + n_1^2 = 1$ (see eq(1.10)). A similar deduction can be made for further substitutions: $\sigma_2(l_2, m_2, n_2)$ and $\sigma_3(l_3, m_3, n_3)$ into eqs(1.23a). It follows from eq(1.9c) that the principal sets of direction cosines (l_1, m_1, n_1), (l_2, m_2, n_2) and (l_3, m_3, n_3) define the unit vectors aligned with the principal directions (see Fig. 1.12).

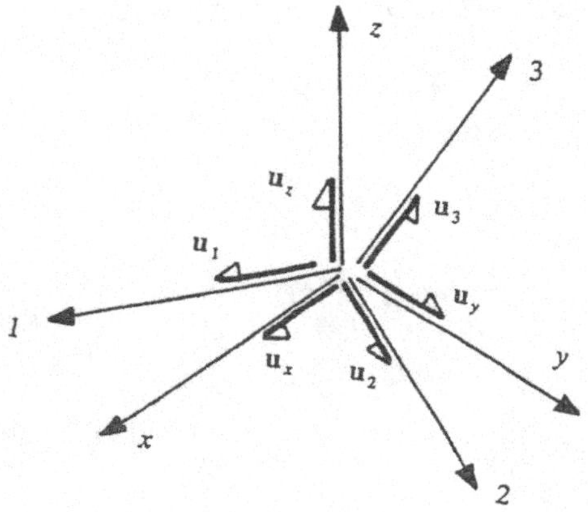

Figure 1.12 Principal directions

They are:

$$\mathbf{u}_1 = l_1\mathbf{u}_x + m_1\mathbf{u}_y + n_1\mathbf{u}_z \quad (1.26a)$$
$$\mathbf{u}_2 = l_2\mathbf{u}_x + m_2\mathbf{u}_y + n_2\mathbf{u}_z \quad (1.26b)$$
$$\mathbf{u}_3 = l_3\mathbf{u}_x + m_3\mathbf{u}_y + n_3\mathbf{u}_z \quad (1.26c)$$

The three unit vectors are orthogonal when their dot products are zero. For the 1 and 2 directions:

$$\mathbf{u}_1 \bullet \mathbf{u}_2 = (l_1\,\mathbf{u}_x + m_1\,\mathbf{u}_y + n_1\,\mathbf{u}_z) \bullet (l_2\,\mathbf{u}_x + m_2\,\mathbf{u}_y + \mathrm{n}_2\,\mathbf{u}_z) = 0$$

Now $\mathbf{u}_x \bullet \mathbf{u}_x = \mathbf{u}_y \bullet \mathbf{u}_y = \mathbf{u}_z \bullet \mathbf{u}_z = 1$ and $\mathbf{u}_x \bullet \mathbf{u}_y = \mathbf{u}_x \bullet \mathbf{u}_z = \mathbf{u}_y \bullet \mathbf{u}_z = 0$. The further dot products $\mathbf{u}_1 \bullet \mathbf{u}_3$ and $\mathbf{u}_2 \bullet \mathbf{u}_3$ determine the full orthogonality conditions:

$$l_1 l_2 + m_1 m_2 + n_1 n_2 = 0 \quad (1.27a)$$
$$l_2 l_3 + m_2 m_3 + n_2 n_3 = 0 \quad (1.27b)$$
$$l_1 l_3 + m_1 m_3 + n_1 n_3 = 0 \quad (1.27c)$$

The relationships (1.27a–c) are the only conditions that satisfy the simultaneous equations(1.26a–c), confirming that the principal stresses directions and their associated planes are orthogonal. In fact, $\mathbf{u}_\alpha$ ($\alpha = 1, 2, 3$) in eqs(1.26a,b,c) define the Eigen vectors for the characteristic eq(1.24b). Since its roots are the Eigen values σ_α ($\alpha = 1, 2, 3$), eq(1.23b) becomes

$$\sigma_{ij}\,l_{\alpha j} = \sigma_\alpha \delta_{ij}\,l_{\alpha j} = \sigma_\alpha\,l_{\alpha i} \quad \text{or} \quad (\mathbf{T} - \sigma_\alpha \mathbf{I}\,)\mathbf{u}_\alpha = 0 \quad (1.28a,b)$$

To show that these forms are identical to eqs(1.23a), put $\alpha = 1$ in eq(1.28a) and expand for $i = 1$ over $j = 1, 2$ and 3:

$$\sigma_{1j}\,l_{1j} = \sigma_1\,l_{11} \text{ (for } i = 1)$$
$$\sigma_{11}\,l_{11} + \sigma_{12}\,l_{12} + \sigma_{13}\,l_{13} = \sigma_1\,l_{11} \quad \text{(for } j = 1, 2, 3)$$
$$l_{11}\,(\sigma_{11} - \sigma_1) + l_{12}\,\sigma_{12} + l_{13}\,\sigma_{13} = 0$$

When converted to the engineering notation this equation becomes the first of eqs(1.23a), i.e. $l\,(\sigma_x - \sigma) + m\tau_{xy} + n\tau_{xz} = 0$. Post-multiplying eq(1.28a) by l_{pi}

$$\sigma_{ij}\,l_{\alpha j}\,l_{pi} = \sigma_\alpha\,l_{\alpha i}\,l_{pi} \quad (1.28c)$$

Then, for $\alpha = 1$ and $p = 2$ and $\alpha = 2$ and $p = 1$, eq(1.28c) gives

$$\sigma_{ij}\,l_{1j}\,l_{2i} = \sigma_1\,l_{1i}\,l_{2i}$$
$$\sigma_{ij}\,l_{2j}\,l_{1i} = \sigma_2\,l_{2i}\,l_{1i}$$

Subtracting these leads to

$$\sigma_{ij}\,l_{1j}\,l_{2i} - \sigma_{ij}\,l_{2j}\,l_{1i} = \sigma_1\,l_{1i}\,l_{2i} - \sigma_2\,l_{2i}\,l_{1i}$$
$$\sigma_{ij}\,l_{1j}\,l_{2i} - \sigma_{ij}\,l_{2i}\,l_{1j} = (\sigma_1 - \sigma_2)\,l_{1i}\,l_{2i}$$
$$(\sigma_{ij} - \sigma_{ji})\,l_{1j}\,l_{2i} = (\sigma_1 - \sigma_2)\,l_{1i}\,l_{2i}$$

but since $\sigma_{ij} = \sigma_{ji}$ and $\sigma_1 \neq \sigma_2$, it follows that

$$l_{1i}\,l_{2i} = l_{11}\,l_{21} + l_{12}\,l_{22} + l_{13}\,l_{23} = 0$$

In the engineering notation this is eq(1.27a): $l_1 l_2 + m_1 m_2 + n_1 n_2 = 0$, thereby confirming that directions 1 and 2 are orthogonal. Further pairs of substitutions ($\alpha = 1, p = 3; \alpha = 3, p = 1$) and ($\alpha = 2, p = 3; \alpha = 3, p = 2$) will confirm eqs(1.27b,c). Within the three column matrices $\mathbf{u}_1 = [l_{11}\, l_{12}\, l_{13}]^{\mathrm{T}}$, $\mathbf{u}_2 = [l_{21}\, l_{22}\, l_{23}]^{\mathrm{T}}$ and $\mathbf{u}_3 = [l_{31}\, l_{32}\, l_{33}]^{\mathrm{T}}$, which definite the 1, 2 and 3 directions respectively, the orthogonality conditions are written as

$$\mathbf{u}_1^{\mathrm{T}}\,\mathbf{u}_2 = \mathbf{u}_1\,\mathbf{u}_2^{\mathrm{T}} = 0$$
$$\mathbf{u}_1^{\mathrm{T}}\,\mathbf{u}_3 = \mathbf{u}_1\mathbf{u}_3^{\mathrm{T}} = 0$$
$$\mathbf{u}_2^{\mathrm{T}}\,\mathbf{u}_3 = \mathbf{u}_2\,\mathbf{u}_3^{\mathrm{T}} = 0$$

To show that a principal stress state exists when a rotation in the co-ordinates aligns them with the principal stress directions, eqs(1.22a) and (1.28b) are, respectively

$$\sigma_{i\alpha}' = l_{ip}\, l_{\alpha q}\, \sigma_{pq} \tag{1.29a}$$
$$\sigma_{pq}\, l_{\alpha q} = \sigma_\alpha\, l_{\alpha p} \tag{1.29b}$$

Pre-multiplying eq(1.29b) by l_{ip} and substituting from eq(1.29a) gives

$$l_{ip}\,\sigma_{pq}\,l_{\alpha q} = l_{ip}\,\sigma_\alpha\,l_{\alpha p}$$
$$\sigma_{i\alpha}' = l_{ip}\,l_{\alpha p}\sigma_\alpha = \delta_{i\alpha}\sigma_\alpha \quad \text{or} \quad \mathbf{T}' = \mathbf{I}\,\mathbf{t}$$

where **I** is the unit matrix and $\mathbf{t} = [\sigma_1 \;\; \sigma_2 \;\; \sigma_3]^{\mathrm{T}}$ are principal stresses.

Example 1.2 The stress components (in MPa) at a point within a loaded body are: $\sigma_x = 5$, $\sigma_y = 7$, $\sigma_z = 6$, $\tau_{xy} = 10$, $\tau_{xz} = 8$ and $\tau_{yz} = 12$. Find the magnitudes of the principal stresses and the maximum shear stress. Show that the principal stress directions are orthogonal.

It follows from eq(1.23c) that the principal stress's cubic (or characteristic equation) may be found either from (i) expanding the determinant:

$$\begin{vmatrix} 5-\sigma & 10 & 8 \\ 10 & 7-\sigma & 12 \\ 8 & 12 & 6-\sigma \end{vmatrix} = 0$$

or (ii) direct substitution into eq(1.24a). These give

$$\sigma^3 - 18\sigma^2 - 201\sigma - 362 = 0$$

from which the invariants in eq(1.24b) are identified as: $J_1 = 18$ MPa, $J_2 = -201$ (MPa)^2 and $J_3 = 362$ (MPa)^3. The roots to this cubic are identified with the principal stresses according to $\sigma_1 > \sigma_2 > \sigma_3$ as follows: $\sigma_1 = 26.2$, $\sigma_2 = -2.37$ and $\sigma_3 = -5.83$ MPa. The three direction cosines for the 1–direction (l_1, m_1, n_1) are found from substituting $\sigma_1 = 26.2$ MPa into eq(1.23a):

$$-21.2\, l_1 + 10m_1 + 8n_1 = 0$$
$$10\, l_1 + 19.2\, m_1 + 12n_1 = 0$$
$$8\, l_1 + 12\, m_1 - 20.2\, n_1 = 0$$

of which only two equations are independent. To solve for l_1, m_1 and n_1 from these equations let a vector $\mathbf{a} = a_1\mathbf{u}_1 + a_2\mathbf{u}_2 + a_3\mathbf{u}_3$ of arbitrary magnitude $|\mathbf{a}| = \sqrt{(a_1^2 + a_2^2 + a_3^2)}$ lie along the 1–direction. Setting $l_1 = a_1/|\mathbf{a}|$, $m_1 = a_2/|\mathbf{a}|$ and $n_1 = a_3/|\mathbf{a}|$ enables a_1 to be set to unity (say) from which $a_2 = 1.222$ and $a_3 = 1.122$. Hence $|\mathbf{a}| = 1.937$ and $l_1 = 0.516$, $m_1 = 0.631$ and $n_1 = 0.579$. Thus a unit vector (eq 1.26a) may be identified with the 1–direction as

$$\mathbf{u}_1 = 0.516\mathbf{u}_x + 0.631\mathbf{u}_y + 0.579\mathbf{u}_z$$

Similarly setting $\sigma_2 = -2.37$ MPa in eq(1.23a) leads to a unit vector for the 2–direction as

$$\mathbf{u}_2 = 0.815\mathbf{u}_x - 0.153\mathbf{u}_y - 0.560\mathbf{u}_z$$

Finally, setting $\sigma_3 = -5.83$ MPa in eq(1.23a), leads to a unit vector for the 3–direction as

$$\mathbf{u}_3 = 0.265\mathbf{u}_x - 0.761\mathbf{u}_y + 0.592\mathbf{u}_z$$

Taking the dot products of $\mathbf{u}_1$, $\mathbf{u}_2$ and $\mathbf{u}_3$ shows that $\mathbf{u}_1 \bullet \mathbf{u}_2 = \mathbf{u}_1 \bullet \mathbf{u}_3 = \mathbf{u}_2 \bullet \mathbf{u}_3 = 0$, so confirming that the directions 1, 2 and 3 are orthogonal.

1.4.3 Reductions to Plane Stress

Consider the non-zero plane stress components σ_x, σ_y and τ_{xy}, shown in Fig. 1.13a. Direction cosines $l = \cos\alpha$, $m = \cos\beta = \cos(90° - \alpha) = \sin\alpha$ and $n = 0$ define the direction normal to the oblique plane in x, y and z co-ordinates.

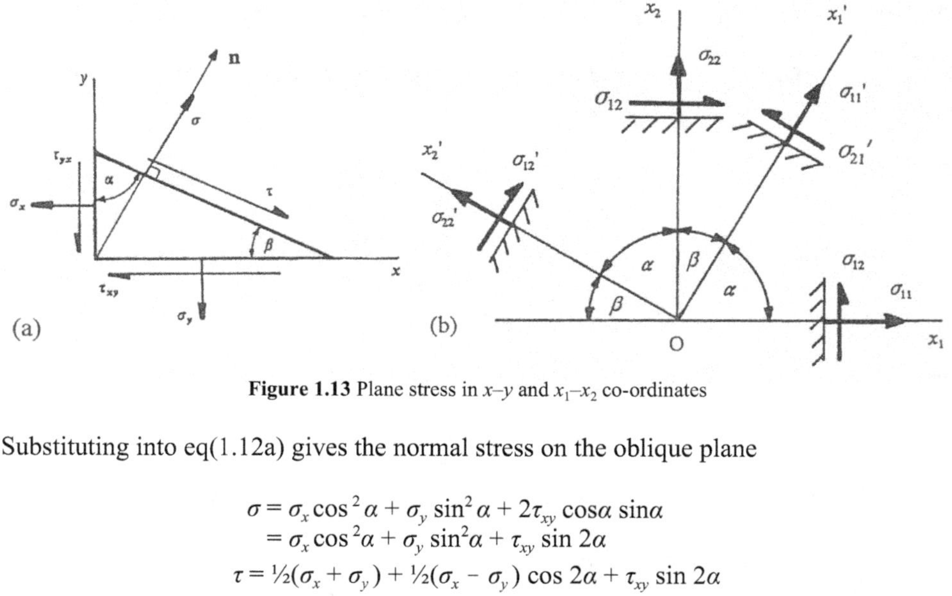

Figure 1.13 Plane stress in x–y and x_1–x_2 co-ordinates

Substituting into eq(1.12a) gives the normal stress on the oblique plane

$$\begin{aligned}\sigma &= \sigma_x \cos^2\alpha + \sigma_y \sin^2\alpha + 2\tau_{xy} \cos\alpha \sin\alpha \\ &= \sigma_x \cos^2\alpha + \sigma_y \sin^2\alpha + \tau_{xy} \sin 2\alpha \\ \tau &= \tfrac{1}{2}(\sigma_x + \sigma_y) + \tfrac{1}{2}(\sigma_x - \sigma_y)\cos 2\alpha + \tau_{xy} \sin 2\alpha\end{aligned}$$

and from eqs(1.11) and (1.12b), the components of the traction vector **r** are:

$$r_x = \sigma_x \cos\alpha + \tau_{xy} \sin\alpha,\ r_y = \sigma_y \sin\alpha + \tau_{yx} \cos\alpha \text{ and } r_z = 0$$

The shear stress on this plane is found from the substitution of r_x, r_y and r_z into eq(1.12b) as

$$\begin{aligned}\tau^2 &= (\sigma_x \cos\alpha + \tau_{xy}\sin\alpha)^2 + (\sigma_y \sin\alpha + \tau_{yx}\cos\alpha)^2 - (\sigma_x\cos^2\alpha + \sigma_y\sin^2\alpha + \tau_{xy}\sin 2\alpha)^2\\ &= (\sigma_x^2 - \sigma_x\sigma_y + \sigma_y^2)\sin^2\alpha\cos^2\alpha + \tau_{xy}\sin 2\alpha\,[\sigma_x(1 - 2\cos^2\alpha) + \sigma_y(1 - 2\sin^2\alpha)]\\ &\quad + \tau_{xy}^2(\sin^2\alpha + \cos^2\alpha - \sin^2 2\alpha)\\ &= \tfrac{1}{4}(\sigma_x - \sigma_y)^2\sin^2 2\alpha - \tau_{xy}(\sigma_x - \sigma_y)\sin 2\alpha\cos 2\alpha + \tau_{xy}^2\cos^2 2\alpha\\ &= [\,\tfrac{1}{2}(\sigma_x - \sigma_y)\sin 2\alpha - \tau_{xy}\cos 2\alpha\,]^2\end{aligned}$$

$$\therefore\ \tau = \tfrac{1}{2}(\sigma_x - \sigma_y)\sin 2\alpha - \tau_{xy}\cos 2\alpha$$

The respective matrix forms for this plane reduction are found from the transformation eq(1.22a): $\mathbf{T}' = \mathbf{MTM}^{\mathrm{T}}$ in which **T** and **M** are here 2 × 2 matrices:

$$\begin{bmatrix}\sigma_{11}' & \sigma_{12}'\\ \sigma_{21}' & \sigma_{22}'\end{bmatrix} = \begin{bmatrix}l_{11} & l_{12}\\ l_{21} & l_{22}\end{bmatrix}\begin{bmatrix}\sigma_{11} & \sigma_{12}\\ \sigma_{21} & \sigma_{22}\end{bmatrix}\begin{bmatrix}l_{11} & l_{21}\\ l_{12} & l_{22}\end{bmatrix}$$

where $l_{11} = \cos\alpha$, $l_{12} = \cos(90 - \alpha) = \sin\alpha$, $l_{21} = \cos(90 + \alpha) = -\sin\alpha$ and $l_{22} = \cos\alpha$. Figure 1.13b shows that the stress state $(\sigma_{11}', \sigma_{21}')$ applies to the x_1'–direction and thus the matrix equation reduces to

$$\begin{bmatrix}\sigma_{11}'\\ \sigma_{21}'\end{bmatrix} = \begin{bmatrix}l_{11} & l_{12}\\ l_{21} & l_{22}\end{bmatrix}\begin{bmatrix}\sigma_{11} & \sigma_{12}\\ \sigma_{21} & \sigma_{22}\end{bmatrix}\begin{bmatrix}l_{11}\\ l_{12}\end{bmatrix}$$

where the following associations are made with the engineering notation $\sigma_{11} = \sigma_x$, $\sigma_{22} = \sigma_y$ and $\sigma_{12} = \tau_{xy}$ in the co-ordinate directions and $\sigma_{11}' = \sigma$ and $\sigma_{21}' = \tau$ for the oblique plane.

In finding the plane principal stresses (σ_1 and σ_2), the invariants in eqs(1.25a–c) become:

$$J_1 = \sigma_x + \sigma_y,\quad J_2 = \sigma_x\sigma_y - \tau_{xy}^2 \ \text{ and }\ J_3 = 0$$

and the principal stress cubic eq(1.24b) reduces to a quadratic

$$\sigma^2 - (\sigma_x + \sigma_y)\sigma + (\sigma_x\sigma_y - \tau_{xy}^2) = 0$$

The solution gives the two principal stresses as roots

$$\sigma_{1,2} = \tfrac{1}{2}(\sigma_x + \sigma_y) \pm \tfrac{1}{2}\sqrt{[\,(\sigma_x - \sigma_y)^2 + 4\tau_{xy}^2\,]}$$

where the positive discriminant applies to σ_1.

1.5 Principal Stresses as Co-ordinates

Let the co-ordinate axes become aligned with the orthogonal principal directions 1, 2 and 3 so that the applied stresses are the principal stresses: $\sigma_1 > \sigma_2 > \sigma_3$, as shown in Fig. 1.14. Because there is no shear stress on faces ACO, ABO and BCO, the expressions for the normal and shear stresses (σ, τ) upon the oblique plane ABC are simplified.

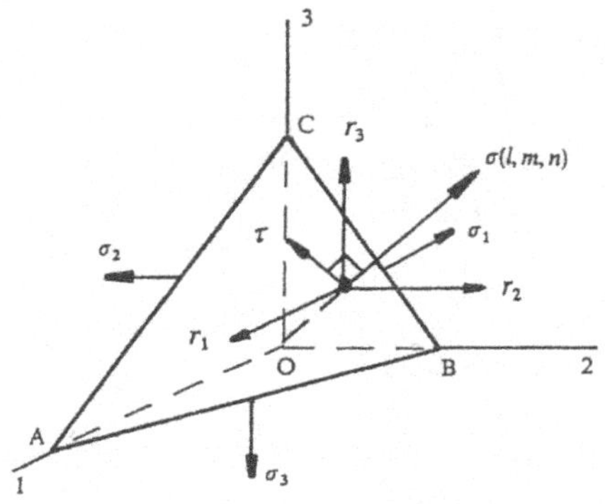

Figure 1.14 Oblique plane set in principal stress axes

Replacing x, y and z in eqs(1.11) and (1.12) with 1, 2 and 3 respectively and setting $\tau_{xy} = \tau_{yz} = \tau_{xz} = 0$ gives the following reduced forms

$$r_1 = l\sigma_1,\ \ r_2 = m\sigma_2,\ \ r_3 = n\sigma_3 \quad (1.30a,b,c)$$

$$\sigma = r_1 l + r_2 m + r_3 n$$
$$= \sigma_1 l^2 + \sigma_2 m^2 + \sigma_3 n^2 \quad (1.31a)$$

The traction across ABC can be written as

$$r^2 = r_1^{\,2} + r_2^{\,2} + r_3^{\,2} = \sigma^2 + \tau^2$$

from which

$$\tau^2 = r^2 - \sigma^2 = (r_1^{\,2} + r_2^{\,2} + r_3^{\,2}) - \sigma^2$$
$$= (l\sigma_1)^2 + (m\sigma_2)^2 + (n\sigma_3)^2 - (\sigma_1 l^2 + \sigma_2 m^2 + \sigma_3 n^2)^2$$
$$\therefore \tau = \surd[(l\sigma_1)^2 + (m\sigma_2)^2 + (n\sigma_3)^2 - (\sigma_1 l^2 + \sigma_2 m^2 + \sigma_3 n^2)^2] \quad (1.31b)$$

The direction cosines for τ are, from eqs(1.13)

$$l_s = (r_1 - l\sigma)/\tau = l(\sigma_1 - \sigma)/\tau \quad (1.32a)$$
$$m_s = (r_2 - m\sigma)/\tau = m(\sigma_2 - \sigma)/\tau \quad (1.32b)$$
$$n_s = (r_3 - n\sigma)/\tau = n(\sigma_3 - \sigma)/\tau \quad (1.32c)$$

In the mathematical notation, when x_1, x_2 and x_3 in Fig. 1.11a are aligned with the principal stress co-ordinates (1, 2 and 3), the transformation eq(1.22b) becomes

$$\begin{bmatrix} \sigma_{11}' & \sigma_{12}' & \sigma_{13}' \\ \sigma_{21}' & \sigma_{22}' & \sigma_{23}' \\ \sigma_{31}' & \sigma_{32}' & \sigma_{33}' \end{bmatrix} = \begin{bmatrix} l_{11} & l_{12} & l_{13} \\ l_{21} & l_{22} & l_{23} \\ l_{31} & l_{32} & l_{33} \end{bmatrix} \begin{bmatrix} \sigma_{11} & 0 & 0 \\ 0 & \sigma_{22} & 0 \\ 0 & 0 & \sigma_{33} \end{bmatrix} \begin{bmatrix} l_{11} & l_{21} & l_{31} \\ l_{12} & l_{22} & l_{32} \\ l_{13} & l_{23} & l_{33} \end{bmatrix}$$

This contains eqs(1.31a,b) for the normal and shear stresses on a single oblique plane ABC. Therein, direction cosines l_{11}, l_{12} and l_{13} define the normal to ABC gives its three primed stress components in which eqs(1.31a,b) re-appear asr $\sigma \equiv \sigma_{11}'$ and $\tau \equiv \surd\,[\,(\sigma_{21}')^2 + (\sigma_{31}')^2]$:

$$\begin{bmatrix} \sigma_{11}' \\ \sigma_{21}' \\ \sigma_{31}' \end{bmatrix} = \begin{bmatrix} l_{11} & l_{12} & l_{13} \end{bmatrix} \begin{bmatrix} \sigma_{11} & 0 & 0 \\ 0 & \sigma_{22} & 0 \\ 0 & 0 & \sigma_{33} \end{bmatrix} \begin{bmatrix} l_{11} \\ l_{12} \\ l_{13} \end{bmatrix}$$

1.5.1 Maximum Shear Stress

It can be shown from eq(1.31b) that maximum shear stresses act on planes inclined at 45° to two principal planes and are perpendicular to the remaining plane. For the 1–2 plane shown in Fig. 1.15, for example, the normal **n** to the 45° plane shown has directions $l = m = \cos 45° = 1/\surd 2$ and $n = \cos 90° = 0$.

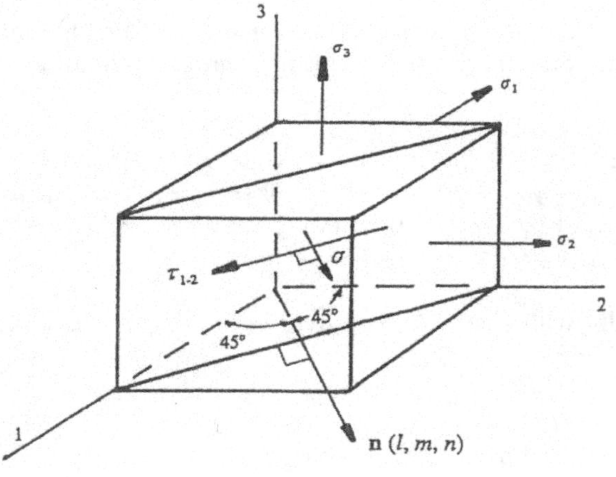

Figure 1.15 Maximum 45° shear plane

Substituting l, m and n into eq(1.31b) the maximum shear stress upon this plane is

$$(\tau_{1\text{-}2})^2 = \sigma_1^2/2 + \sigma_2^2/2 - (\sigma_1/2 + \sigma_2/2)^2 = \tfrac{1}{4}(\sigma_1^2 + \sigma_2^2 - 2\sigma_1\sigma_2) = \tfrac{1}{4}(\sigma_1 - \sigma_2)^2$$

$$\tau_{1\text{-}2} = \pm \tfrac{1}{2}(\sigma_1 - \sigma_2) \tag{1.33a}$$

Subscripts 1, 2 refer to the principal directions to which plane 1–2 is equally inclined at 45° and upon which a maximum shear stress $\tau_{1\text{-}2}$ acts. Similarly, for the plane 1–3 inclined at 45° to the 1 and 3 directions ($l = n = 1/\surd 2$ and $m = 0$), the maximum shear stress is

$$\tau_{1\text{-}3} = \pm \tfrac{1}{2}(\sigma_1 - \sigma_3) \tag{1.33b}$$

For the plane 2–3 inclined at 45° to the 2 and 3 directions, where $n = m = 1/\surd 2$ and $l = 0$, the maximum shear stress is

$$\tau_{2\text{-}3} = \pm \tfrac{1}{2}(\sigma_2 - \sigma_3) \tag{1.33c}$$

Alternating signs in eqs(1.33a–c) include the second complementary, maximum shear plane for each case. The greatest maximum shear stresses, for a principal stress system in which $\sigma_1 > \sigma_2 > \sigma_3$, is $\tau_{max} = \tau_{1\text{-}3}$. When the 45° shear planes are considered in all four quadrants they form a *rhombic dodecahedron*. The normal stress acting upon the planes of maximum shear stress are found from eq(1.31a). For example, with $l = m = 1/\surd 2$ and $n = 0$ for the 45° plane shown in Fig. 1.13, the normal stress is

$$\sigma = \tfrac{1}{2}(\sigma_1 + \sigma_2) \tag{1.33d}$$

1.5.2 Octahedral Plane

The octahedral plane is equally inclined to the principal directions. It follows from eq(1.10) that the direction cosines of its normal are $l = m = n = 1/\surd 3$ ($\alpha = \beta = \gamma = 54.8°$). Substituting these into eq(1.31a), gives the octahedral normal stress

$$\sigma_o = \sigma_1 l^2 + \sigma_2 m^2 + \sigma_3 n^2 = \sigma_1(1/\surd 3)^2 + \sigma_2(1/\surd 3)^2 + \sigma_3(1/\surd 3)^2$$
$$\therefore \sigma_o = \tfrac{1}{3}(\sigma_1 + \sigma_2 + \sigma_3) \tag{1.34a}$$

Since σ_o is the average of the principal stresses it is also called the mean or hydrostatic stress, σ_m. The octahedral shear stress is found from substituting $l = m = n = 1/\surd 3$ in eq(1.31b)

$$\tau_o^2 = (\sigma_1/\surd 3)^2 + (\sigma_2/\surd 3)^2 + (\sigma_3/\surd 3)^2 - [\sigma_1(1/\surd 3)^2 + \sigma_2(1/\surd 3)^2 + \sigma_3(1/\surd 3)^2]^2$$
$$= (\sigma_1 + \sigma_2 + \sigma_3)/3 - [(\sigma_1 + \sigma_2 + \sigma_3)/3]^2 = (2/9)(\sigma_1^2 + \sigma_2^2 + \sigma_3^2 - \sigma_1\sigma_2 - \sigma_1\sigma_3 - \sigma_2\sigma_3)$$
$$= (1/9)[(\sigma_1 - \sigma_2)^2 + (\sigma_2 - \sigma_3)^2 + (\sigma_1 - \sigma_3)^2]$$
$$\therefore \tau_o = \tfrac{1}{3}\surd[(\sigma_1 - \sigma_2)^2 + (\sigma_2 - \sigma_3)^2 + (\sigma_1 - \sigma_3)^2] \tag{1.34b}$$

Combining eq(1.34b) with eqs(1.33a,b,c) gives σ_o in terms of the three maximum shear stresses

$$\tau_o = (2/3)\surd(\tau_{1\text{-}2}^2 + \tau_{2\text{-}3}^2 + \tau_{1\text{-}3}^2) \tag{1.34c}$$

The direction of τ_o is found from eqs(1.32a,b,c) as

$$l_o = (\sigma_1 - \sigma_o)/(\surd 3\tau_o) \tag{1.35a}$$
$$m_o = (\sigma_2 - \sigma_o)/(\surd 3\tau_o) \tag{1.35b}$$
$$n_o = (\sigma_3 - \sigma_o)/(\surd 3\tau_o) \tag{1.35c}$$

When the eight octahedral planes in all four quadrants are considered they form the faces of the *regular octahedron*, shown in Fig. 1.16.

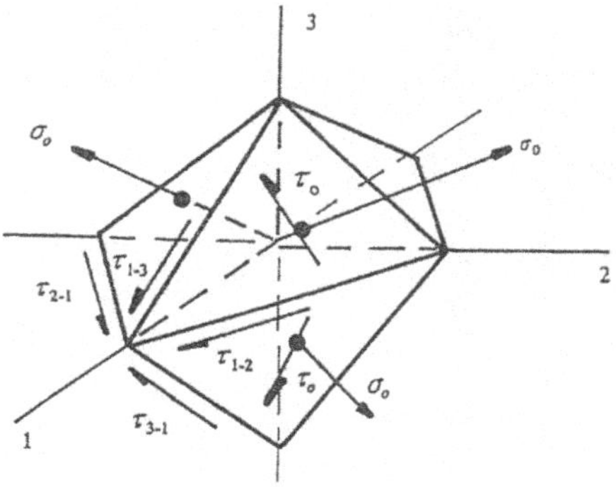

Figure 1.16 Octahedral planes

Here, σ_o and τ_o act on the eight planes while $\tau_{1\text{-}2}$, $\tau_{2\text{-}3}$ and $\tau_{1\text{-}3}$ act along their edges. The deformation that arises from a given principal stress state may be examined on an octahedral basis. Since σ_o acts with equal inclination and intensity it causes an elastic volume change δV (recoverable) provided by the bulk modulus K, i.e. $\delta V = \sigma_o V/K$. Superimposed upon this is the distortion produced by τ_o. Since eq(1.34b) shows that the magnitude of τ_o depends upon differences between the principal stresses, a critical value of τ_o will determine whether the deformation will be elastic or elastic–plastic. In chapter 3 it is shown that a yield criterion may be formulated on this basis.

Example 1.3 Given the principal stresses $\sigma_1 = 7.5$, $\sigma_2 = 3.1$ and $\sigma_3 = 1.4$ (MPa) find (i) the maximum shear stresses and their directions and (ii) the magnitude and direction of the normal and shear stresses for the octahedral plane. Confirm the answers using a Mohr's stress circle.

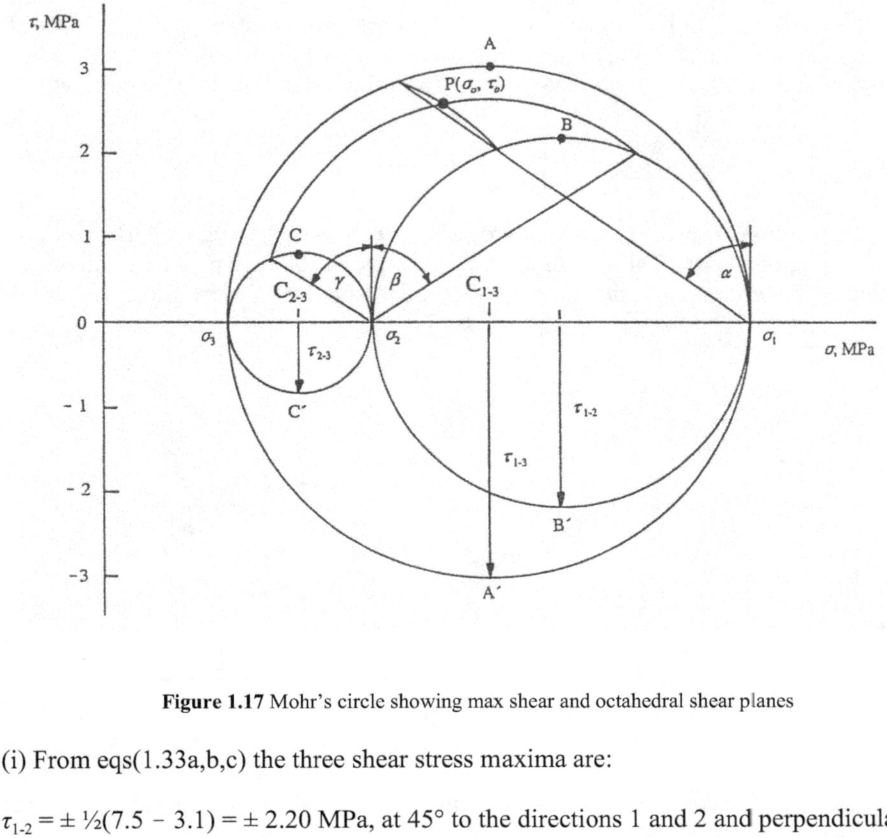

Figure 1.17 Mohr's circle showing max shear and octahedral shear planes

(i) From eqs(1.33a,b,c) the three shear stress maxima are:

$\tau_{1\text{-}2} = \pm\, ½(7.5 - 3.1) = \pm\, 2.20$ MPa, at 45° to the directions 1 and 2 and perpendicular to 3

$\tau_{2\text{-}3} = \pm\, ½(3.1 - 1.4) = \pm\, 0.85$ MPa, at 45° to the directions 2 and 3 and perpendicular to 1

$\tau_{1\text{-}3} = \pm\, ½(7.5 - 1.4) = \pm\, 3.05$ MPa, at 45° to the directions 1 and 3 and perpendicular to 2

(ii) From eq(1.34a) the normal stress on the octahedral plane is:

$\sigma_o = ⅓\,(7.5 + 3.1 + 1.4) = 4.0$ MPa equally inclined to directions 1, 2 and 3

The octahedral shear stress is, from eq(1.34b):

$$\tau_o = \tfrac{1}{3}\sqrt{[\,(7.5 - 3.1)^2 + (3.1 - 1.4\,)^2 + (7.5 - 1.4\,)^2]} = 2.57\ \text{MPa}$$

with direction cosines, from eq(1.35a,b,c):

$$l_o = (7.5 - 4.0\,) / (\sqrt{3} \times 2.57) = 0.786$$
$$m_o = (3.1 - 4.0) / (\sqrt{3} \times 2.57) = -\,0.202$$
$$n_o = (1.4 - 4.0) / (\sqrt{3} \times 2.57) = -\,0.584$$

Mohr's circle (see Fig. 1.17) confirms each answer. The maximum positive shear stresses are the vertical radii of the three circles at A, B and C. Complementary negative shear stresses are associated with the reflective points A′, B′ and C′. To locate the octahedral plane mark off lines at $\alpha = \beta = 54.7°$ ($l = m = n = 1/\sqrt{3}$) from σ_1, σ_2 and σ_3 as shown. With centres $C_{1\text{-}3}$ and $C_{2\text{-}3,}$ draw arcs to intersect at a point P whose co-ordinates are σ_o and τ_o. When P is co-incident with A, this locates a maximum shear plane where $\alpha = \beta = 45°$ and $\gamma = 90°$. Similarly, with P at B and C the orientation of two further shear planes are confirmed.

1.5.3 Reductions to Plane Principal Stress

The plane stress transformation equations are particular cases of the forgoing 3D equations. In Fig. 1.18a, applied principal stresses are: σ_1 and σ_2 and the oblique plane is defined with respective directions: $l = \cos\alpha$, $m = \cos(90° - \alpha) = \sin\alpha$ and $n = 0$ for directions 1, 2 and 3.

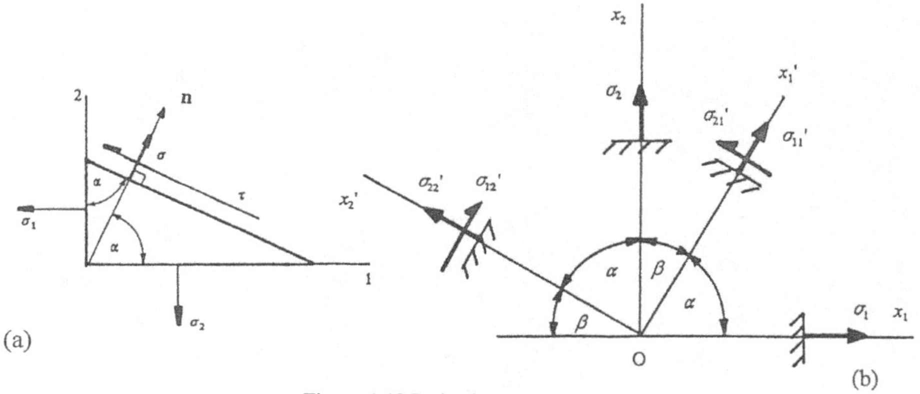

Figure 1.18 Reduction to plane stress

Substituting these into eq(1.31a,b) gives the normal and shear stresses for the oblique plane:

$$\sigma = \sigma_1 \cos^2\alpha + \sigma_2 \sin^2\alpha = \sigma_1 (1 + \cos 2\alpha)/2 + \sigma_2 (1 - \cos 2\alpha)/2$$
$$\sigma = \tfrac{1}{2}(\sigma_1 + \sigma_2) + \tfrac{1}{2}(\sigma_1 - \sigma_2)\cos 2\alpha$$

$$\tau^2 = \sigma_1^{\,2}\cos^2\alpha + \sigma_2^{\,2}\sin^2\alpha - (\sigma_1\cos^2\alpha + \sigma_2\sin^2\alpha)^2$$
$$= (\sigma_1 - \sigma_2)^2 \sin^2\alpha \cos^2\alpha$$
$$\tau = \tfrac{1}{2}(\sigma_1 - \sigma_2)\sin 2\alpha$$

Alternatively, these can be found from the transformation matrix $\mathbf{T}' = \mathbf{MTM}^{\mathrm{T}}$. In full, this takes the plane form

$$\begin{bmatrix} \sigma_{11}' & \sigma_{12}' \\ \sigma_{21}' & \sigma_{22}' \end{bmatrix} = \begin{bmatrix} l_{11} & l_{12} \\ l_{21} & l_{22} \end{bmatrix} \begin{bmatrix} \sigma_1 & 0 \\ 0 & \sigma_2 \end{bmatrix} \begin{bmatrix} l_{11} & l_{21} \\ l_{12} & l_{22} \end{bmatrix}$$

Referring to Fig. 1.18b, the direction cosines for the x_1–axis become: $l_{11} = \cos\alpha$ and $l_{12} = \cos(90 - \alpha) = \sin\alpha$ and for the x_2'–axis: $l_{21} = \cos(90 + \alpha) = -\sin\alpha$ and $l_{22} = \cos\alpha$. The required stress components σ_{11}' and σ_{21}' for the plane normal to x_1' are found from the reduced matrix multiplication:

$$\begin{bmatrix} \sigma_{11'} \\ \sigma_{21'} \end{bmatrix} = \begin{bmatrix} l_{11} & l_{12} \\ l_{21} & l_{22} \end{bmatrix} \begin{bmatrix} \sigma_1 & 0 \\ 0 & \sigma_2 \end{bmatrix} \begin{bmatrix} l_{11} \\ l_{12} \end{bmatrix}$$

where the correspondence is $\sigma_{11}' = \sigma$ and $\sigma_{12}' = \tau$. Full matrix multiplication will also supply those stress components σ_{12}', σ_{22}' on the plane normal to x_2', as shown in Fig. 1.18b.

1.6 Alternative Stress Definitions

The components σ_{ij}, of the Cauchy stress tensor $\mathbf{T}$, appearing in eq(1.6), refer to the elemental deformed area δa of surface on which the stress traction vector $\mathbf{r}^{(\mathbf{n})}$ acts. Thus, as the forgoing stress transformations are referred to the geometry of the deformed material, the stress components are said to be true stresses. Provided deformations are less than 1%, as with the elasticity of metallic materials, it is unnecessary to distinguish between the initial and current areas. A nominal stress, calculated from the original area, will give the Cauchy stress with acceptable accuracy. For example, in a tensile test on a metal, it will only become necessary to convert nominal stresses to true stresses within the reduced cross-sectional area when deformation advances well into the plastic range. However, it is unacceptable to calculate Cauchy elastic stress components from the original area where large elastic deformations arise in certain non-metals. When the elastic extension in an incompressible rubber reaches 300% the cross-sectional area will have diminished by 67%. This means that the nominal stress is only 1/3 of the true stress! Clearly, the choice between initial and current areas can become critical to defining stress properly. The Cauchy definition may be regarded as a true stress. Alternative definitions of nominal stress, given by Piola and Kirchhoff, employ the original area for convenience.

In Fig. 1.19 the resultant force $\delta\mathbf{F}$ is shown for both the reference and current configurations, X_i and x_i, respectively. The traction vectors $\mathbf{r}^{(\mathbf{N})}$ and $\mathbf{r}^{(\mathbf{n})}$ are associated with the vectors $\mathbf{N}$ and $\mathbf{n}$, lying normal to the reference and current areas δA and δa. The corresponding stress tensors are $\mathbf{T}$ (with Cauchy components σ_{ij} in x_i) and $\mathbf{S}$ (with nominal components S_{ij} in X_i). Since the force $\delta\mathbf{F}$ is common between the two configurations, it follows from eqs(1.5) and (1.6) that $\mathbf{r}^{(\mathbf{N})} = \mathbf{r}^{(\mathbf{n})}$, when:

$$\sigma_{ji} n_j \,\delta a = S_{ji} N_j \,\delta A \quad \text{or} \quad \mathbf{T}^{\mathrm{T}} \mathbf{n}\,\delta a = \mathbf{S}^{\mathrm{T}} \mathbf{N}\,\delta A \tag{1.36}$$

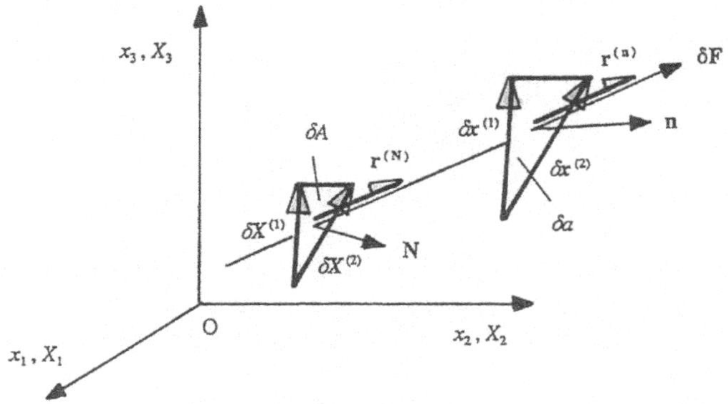

Figure 1.19 Traction in reference and current configurations

To express the relationship between **n**, **N**, δa and δA it is convenient to take, without loss of generality, infinitesimal triangular areas. Let vectors $\delta\mathbf{x}^{(1)}$ and $\delta\mathbf{x}^{(2)}$ lie along the sides of δa and vectors $\delta\mathbf{X}^{(1)}$ and $\delta\mathbf{X}^{(2)}$ lie along the sides of δA as shown. The areas appear within their cross products

$$n_i\,\delta a = \tfrac{1}{2}\varepsilon_{ijk}\,\delta x_j^{(1)}\,\delta x_k^{(2)} \quad \text{or} \quad \mathbf{n}\,\delta a = \tfrac{1}{2}\delta\,\mathbf{x}^{(1)} \times \delta\,\mathbf{x}^{(2)} \tag{1.37a}$$

$$N_i\,\delta A = \tfrac{1}{2}\varepsilon_{ijk}\,\delta X_j^{(1)}\,\delta X_k^{(2)} \quad \text{or} \quad \mathbf{N}\,\delta A = \tfrac{1}{2}\delta\mathbf{X}^{(1)} \times \delta\mathbf{X}^{(2)} \tag{1.37b}$$

where ε_{ijk} is the alternating unit tensor. The latter gives: (i) unity for clockwise permutations of differently valued indices, i.e. $\varepsilon_{123} = \varepsilon_{231} = \varepsilon_{312} = 1$, (ii) minus unity for anticlockwise differently valued indices, i.e. $\varepsilon_{132} = \varepsilon_{321} = \varepsilon_{213} = -1$ and (iii) zero when any two or more indices are the same, e.g. $\varepsilon_{112} = \varepsilon_{323} = 0$ etc. Now, since $\mathbf{x} = \mathbf{x}\,(\mathbf{X})$ or $x_i = x_i\,(X_j)$, it follows:

$$\delta x_i = (\partial x_i/\partial X_j)\,\delta X_j \quad \text{or} \quad \delta\mathbf{x} = (\partial\mathbf{x}/\partial\mathbf{X})\,\delta\mathbf{X} \tag{1.38}$$

Substituting eq(1.38) into eq(1.37a) and making appropriate changes to the indices

$$n_i\,\delta a = \tfrac{1}{2} \times \varepsilon_{ijk}\,(\partial x_j/\partial X_q)\,\delta X_q^{(1)}\,(\partial x_k/\partial X_r)\,\delta X_r^{(2)}$$

Multiplying both sides by $\partial x_i/\partial X_p$ leads to

$$n_i\,(\partial x_i/\partial X_p)\delta a = \tfrac{1}{2} \times \varepsilon_{ijk}\,(\partial x_i/\partial X_p)(\partial x_j/\partial X_q)(\partial x_k/\partial X_r)\delta X_q^{(1)}\,\delta X_r^{(2)} \tag{1.39}$$

Use is now made of the identity

$$\varepsilon_{pqr}\,\det\mathbf{F} = \varepsilon_{ijk}\,F_{ip}\,F_{jq}\,F_{kr} \tag{1.40a}$$

where: $F_{ip} = \partial x_i/\partial X_p$ etc, are the deformation gradients and $J = \det\mathbf{F} = |\partial x_i/\partial X_j|$ is the Jacobian determinant:

$$J = \begin{vmatrix} \partial x_1/\partial X_1 & \partial x_1/\partial X_2 & \partial x_1/\partial X_3 \\ \partial x_2/\partial X_1 & \partial x_2/\partial X_2 & \partial x_2/\partial X_3 \\ \partial x_3/\partial X_1 & \partial x_3/\partial X_2 & \partial x_3/\partial X_3 \end{vmatrix}$$

A simple physical interpretation of J is as a ratio between material densities (or volume elements) in the undeformed and deformed configurations. That is, $J = \rho_o/\rho = \mathrm{d}V/\mathrm{d}V_o$. Combining eqs(1.39) and (1.40):

$$n_i F_{ip}\,\delta a = \tfrac{1}{2}\varepsilon_{pqr}\,J\,\delta X_q^{(1)}\,\delta X_r^{(2)}$$

and substituting from eq(1.37b):

$$n_i F_{ip}\,\delta a = J\,N_p\,\delta A \quad \text{or} \quad \mathbf{n}^{\mathrm{T}}\mathbf{F}\,\delta a = J\,\mathbf{N}^{\mathrm{T}}\,\delta A$$

$$n_i\,\delta a = J\,(F_{pi}^{-1})\,N_p\,\delta A \quad \text{or} \quad \mathbf{n}^{\mathrm{T}}\,\delta a = J\,\mathbf{N}^{\mathrm{T}}\mathbf{F}^{-1}\,\delta A \tag{1.41}$$

The matrix form may further be expressed as

$$(\mathbf{n}^{\mathrm{T}})^{\mathrm{T}}\delta a = J\,(\mathbf{N}^{\mathrm{T}}\mathbf{F}^{-1})^{\mathrm{T}}\,\delta A$$

$$\mathbf{n}\,\delta a = J\,(\mathbf{F}^{-1})^{\mathrm{T}}\mathbf{N}\,\delta A$$

Substituting eq(1.41) into eq(1.36) gives

$$\sigma_{ji} J\,(F_{pj}^{-1}) N_p\,\delta A = S_{ji} N_j\,\delta A$$

$$\sigma_{ji} J\,(F_{pj}^{-1}) N_p\,\delta A = S_{pi} N_p\,\delta A$$

$$J\,\sigma_{ji}(F_{pj}^{-1}) = S_{pi} \quad \text{or} \quad J\,\mathbf{T}^{\mathrm{T}}(\mathbf{F}^{-1})^{\mathrm{T}} = J\,(\mathbf{F}^{-1}\mathbf{T})^{\mathrm{T}} = \mathbf{S}^{\mathrm{T}}$$

This transpose of the nominal stress tensor ($S_{ip} \equiv \mathbf{S}$) defines the first Piola-Kirchhoff stress tensor. It is seen that the components S_{pi} are related to the Cauchy stress components σ_{ji} and the deformation gradients as follows:

$$S_{pi} = (\det \mathbf{F})(\partial X_p/\partial x_j)\sigma_{ji} \quad \text{or} \quad \mathbf{S}^{\mathrm{T}} = (\det \mathbf{F})\,(\mathbf{F}^{-1}\mathbf{T})^{\mathrm{T}} \tag{1.42a}$$

Note, that the nominal stress tensor is

$$S_{ip} = (\det \mathbf{F})(\partial X_i/\partial x_j)\sigma_{jp} \quad \text{or} \quad \mathbf{S} = (\det \mathbf{F})\,\mathbf{F}^{-1}\mathbf{T} \tag{1.42b}$$

It follows from eqs(1.42a,b) that $S_{ip} \neq S_{pi}$ (or $\mathbf{S} \neq \mathbf{S}^{\mathrm{T}}$) so that this tensor is not symmetrical. A second, symmetric Piola-Kirchhoff stress tensor $\mathbf{S}'$ can be introduced within the relation

$$\mathbf{S} = \mathbf{S}'\,\mathbf{F}^{\mathrm{T}} \tag{1.43a}$$

Post-multiplying eq(1.43a) by $(\mathbf{F}^{\mathrm{T}})^{-1}$ gives

$$\mathbf{S}' = \mathbf{S}\,(\mathbf{F}^{\mathrm{T}})^{-1} \tag{1.43b}$$

and also

$$(\mathbf{S}^{\mathrm{T}}) = (\mathbf{S}'\,\mathbf{F}^{\mathrm{T}})^{\mathrm{T}} = \mathbf{F}\,\mathbf{S}'^{\mathrm{T}}$$

$$\mathbf{S}'^{\mathrm{T}} = \mathbf{F}^{-1}(\mathbf{S}^{\mathrm{T}}) \tag{1.43c}$$

Substituting eq(1.42b) into eqs(1.43b), it follows that

$$\mathbf{S}' = (\det \mathbf{F})\, \mathbf{F}^{-1} \mathbf{T}\, (\mathbf{F}^{\mathrm{T}})^{-1} \tag{1.44a}$$

from which the transpose is

$$\mathbf{S}'^{\mathrm{T}} = (\det \mathbf{F})\, \mathbf{F}^{-1} (\mathbf{F}^{-1} \mathbf{T})^{\mathrm{T}} = (\det \mathbf{F})\, \mathbf{F}^{-1} \mathbf{T}^{\mathrm{T}} (\mathbf{F}^{-1})^{\mathrm{T}} \tag{1.44b}$$

Equation (1.44b) can also be derived from substituting eq(1.42b) into eq(1.43c). With the following symmetries $(\mathbf{F}^{\mathrm{T}})^{-1} = (\mathbf{F}^{-1})^{\mathrm{T}}$ and $\mathbf{T} = \mathbf{T}^{\mathrm{T}}$, it follows from eqs(1.44a,b) that $\mathbf{S}' = \mathbf{S}'^{\mathrm{T}}$ but this tensor has no physical interpretation. Finally, the relationship between the two Piola-Kirchhoff stress tensors is found from eq(1.43a):

$$(\mathbf{S}^{\mathrm{T}}) = (\mathbf{S}'\mathbf{F}^{\mathrm{T}})^{\mathrm{T}} = \mathbf{F}\,\mathbf{S}'^{\mathrm{T}} = \mathbf{F}\,\mathbf{S}' \quad \text{or} \quad \mathbf{S}' = \mathbf{F}^{-1}(\mathbf{S}^{\mathrm{T}})$$

Now, from eq(1.44a), the tensor $\mathbf{S}'$, with components S_{ip}', is written in terms of the Cauchy stress tensor as

$$S_{ip}' = (\det \mathbf{F})\, (\partial X_i / \partial x_j)\, (\partial X_j / \partial x_i)\, \sigma_{jp} \tag{1.44c}$$

This form of nominal stress tensor is suitable for the formulation of certain finite constitutive relations. When the deformation gradients remain small during infinitesimal straining reductions to eqs(1.44a,b,c) show that the distinction between the three stress definitions disappears to give $\mathbf{T} = \mathbf{S}'$ and $\mathbf{S}^{\mathrm{T}} = \mathbf{I}\,\mathbf{S}'$.

Example 1.4 Express the Cauchy stress tensor **T** in terms of deformation gradients **F**, the density ratio ρ/ρ_o and (a) the nominal stress tensor **S**, (b) the first Piola-Kirchhoff stress tensor $\mathbf{S}^{\mathrm{T}}$ and (c) the second Piola-Kirchhoff stress tensor $\mathbf{S}'$.

(a) Writing eq(1.42a) as

$$\mathbf{S}^{\mathrm{T}} = J\,(\mathbf{F}^{-1}\mathbf{T})^{\mathrm{T}}$$

the transpose gives **S** as

$$\mathbf{S} = J\,\mathbf{F}^{-1}\mathbf{T}$$

Pre-multiplying by **F** gives the Cauchy stress **T** as

$$\mathbf{T} = (1/J)\mathbf{F}\,\mathbf{S} = (\rho/\rho_o)\,\mathbf{F}\,\mathbf{S} \quad \text{or} \quad \sigma_{ij} = (\rho/\rho_o)\, F_{ik} S_{kj} \tag{i}$$

(b) Again, from eq(1.42a):

$$\mathbf{S}^{\mathrm{T}} = J\,\mathbf{T}^{\mathrm{T}} (\mathbf{F}^{-1})^{\mathrm{T}}$$

Post-multiplying by $\mathbf{F}^{\mathrm{T}}$:

$$\mathbf{S}^{\mathrm{T}}\mathbf{F}^{\mathrm{T}} = J\,\mathbf{T}^{\mathrm{T}}(\mathbf{F}^{-1})^{\mathrm{T}}\mathbf{F}^{\mathrm{T}} = J\,\mathbf{T}^{\mathrm{T}}(\mathbf{F}\,\mathbf{F}^{-1})^{\mathrm{T}} = J\,\mathbf{T}^{\mathrm{T}}$$

from which

$$\mathbf{T}^{\mathrm{T}} = (1/J)\,\mathbf{S}^{\mathrm{T}}\mathbf{F}^{\mathrm{T}} = (\rho/\rho_o)\,\mathbf{S}^{\mathrm{T}}\mathbf{F}^{\mathrm{T}} = (\rho/\rho_o)\,(\mathbf{F}\,\mathbf{S})^{\mathrm{T}} \quad \text{or} \quad \sigma_{ji} = (\rho/\rho_o)\, S_{jk} F_{ki} \tag{ii}$$

(c) Writing eq(1.44a) as

$$\mathbf{S}' = J\,\mathbf{F}^{-1}\mathbf{T}\,(\mathbf{F}^{\mathrm{T}})^{-1}$$

Post-multiply by $\mathbf{F}^{\mathrm{T}}$:

$$\mathbf{S}'\mathbf{F}^{\mathrm{T}} = J\,\mathbf{F}^{-1}\mathbf{T}$$

Pre-multiply by **F**:

$$\mathbf{T} = (1/J)\,\mathbf{F}\,\mathbf{S}'\mathbf{F}^{\mathrm{T}} = (\rho/\rho_o)\,\mathbf{F}\,\mathbf{S}'\mathbf{F}^{\mathrm{T}} \quad \text{or} \quad \sigma_{ij} = (\rho/\rho_o)\, F_{ip} F_{jq} S_{pq}' \tag{iii}$$

Equations (i) and (ii) show that since $\mathbf{T} = \mathbf{T}^T$ then $\mathbf{FS} = (\mathbf{FS})^T$. That is, the matrix product **FS** is symmetrical. The Cauchy and Piola-Kirchhoff stress tensors, as linked within eqs(i)–(iii), are most often employed in finite deformation studies, depending upon whether stress is to be referred to the current or initial area.

Bibliography

Arridge R. G. C. *Mechanics of Polymers*, 1975, Clarendon, Oxford.
Arridge R. G. C. *An Introduction to Polymer Mechanics*, 1985, Taylor and Francis.
Backofen W. A. *Deformation Processing*, 1972, Addison-Wesley.
Barnes H. A., Hutton J. F. and Walters K. *An Introduction to Rheology*, Ser. 3, 1989, Elsevier.
Billington E. W. and Tate A. *The Physics of Deformation and Flow*, 1981, McGraw-Hill.
Billington E. W. *Introduction to the Mechanics and Physics of Solids*, 1986, Adam Hilger.
Boresi A. P. and Sidebottom O. M. *Advanced Mechanics of Materials*, 1985, Wiley and Son.
Calcote L. R. *Introduction to Continuum Mechanics*, 1968, van Nostrand Co. Inc.
Ford H. and Alexander J. M. *Advanced Mechanics of Materials*, 1963, Longmans.
Lubliner J. *Plasticity Theory*, 1990, Macmillan.
Malvern L. E. *Introduction to the Mechanics of a Continuous Medium*, 1969, Prentice-Hall.
Rees D. W. A. *Mechanics of Solids and Structures*, 2016, IC Press, World Scientific.
Rees D. W. A. *Mechanics of Elastic Solids*, 2019, IC Press, World Scientific.
Spencer A. J. M. *Continuum Mechanics*, 1980, Longman.
Tanner R. I. *Engineering Rheology*, 1985, Clarendon Press, Oxford.
Williams J. G. *Stress Analysis of Polymers*, 1980, Ellis Horwood.
Zyczkowski M. *Combined Loadings in the Theory of Plasticity*, 1981, PWN.

Exercises

1.1 Given the following stress tensor components (in MPa): $\sigma_{11} = 2$, $\sigma_{22} = 2$, $\sigma_{12} = 3$, $\sigma_{13} = 1$, $\sigma_{23} = 3$ and the first invariant $J_1 = 5$ MPa, determine σ_{33}, the principal stresses and the maximum shear stress.
See solution: 1, 6.523, 0.523, - 2.045, 4.284 (MPa)

1.2 Find the principal stresses for a stress tensor with the components: $\sigma_{11} = 1$, $\sigma_{22} = 4$, $\sigma_{33} = 1$, $\sigma_{12} = \sigma_{21} = 2$, $\sigma_{13} = \sigma_{31} = 3$ and $\sigma_{23} = \sigma_{32} = 6$ (MPa). Show that their directions are orthogonal.
Ans: $\sigma_1 = 10$, $\sigma_2 = 0$ and $\sigma_3 = -4$ (MPa)

1.3 The principal stresses: $\sigma_1 = 15.4$, $\sigma_2 = 12.65$ and $\sigma_3 = 6.8$ (MPa) act at a point. Determine the normal and resultant shear stresses acting upon an oblique plane whose normal is defined by the unit vector: $\mathbf{u}_n = 0.732\,\mathbf{u}_1 + 0.521\mathbf{u}_2 + 0.439\mathbf{u}_3$. What is the state of stress upon the octahedral plane? Check your answers graphically.
See solution: $\sigma = 13$, $\tau = 3.2$, $\tau_o = 11.6$, $\tau_o = 3.6$ (MPa)

1.4 The stress state at a point is described by the stress components: $\sigma_{11} = 14$, $\sigma_{22} = 10$, $\sigma_{33} = 35$, $\sigma_{12} = \sigma_{21} = 7$, $\sigma_{13} = \sigma_{31} = -7$ and $\sigma_{23} = \sigma_{32} = 0$ (MPa). Find the normal and resultant shear stresses acting upon a plane whose normal is defined by the direction cosines: $l = 2/\sqrt{14}$ $m = -1/\sqrt{14}$ and $n = 3/\sqrt{14}$.
See solution: $\sigma = 19.21$, $\tau = 14.95$ (MPa)

1.5 The stress state at a point is described by the stress components: $\sigma_{11} = 6$, $\sigma_{22} = 0$, $\sigma_{33} = 0$, $\sigma_{12} = \sigma_{21} = 2$, $\sigma_{13} = \sigma_{31} = 2$ and $\sigma_{23} = \sigma_{32} = 4$ (MPa). Find the principal stresses, the greatest shear stress and the state of stress upon the octahedral plane.
Ans: $\sigma_1 = 8$, $\sigma_2 = 2$ and $\sigma_3 = -4$, $\tau_o = 4.9$, $\sigma_o = 2$ (MPa)

1.6 Determine the invariants J_1, J_2 and J_3 for the stress tensor components: $\sigma_{11} = 6$, $\sigma_{22} = 6$, $\sigma_{33} = 8$, $\sigma_{12} = \sigma_{21} = -3$, $\sigma_{13} = \sigma_{23} = 0$ (MPa). Show that the same invariants values apply to the principal stresses for this system. Determine the octahedral shear stress and the maximum shear stress and the planes on which they act.
Ans: $J_1 = 20$, $J_2 = 123$, $J_3 = 216$, $\tau_o = 2.625$, $\tau_{max} = 3$ (MPa)

1.7 Transform the following non-zero components of a stress matrix: $\sigma_{11} = 15$, $\sigma_{22} = 5$, $\sigma_{33} = 20$, $\sigma_{12} = \sigma_{21} = -10$ and $\sigma_{13} = \sigma_{23} = 0$ (MPa) using the following non-zero components of a rotation matrix: $l_{11} = 0.6$, $l_{22} = 1$, $l_{33} = 0.6$, $l_{13} = -l_{31} = -0.8$.
See solution: $\sigma_{11}' = 18.2$, $\sigma_{22}' = 5$, $\sigma_{33}' = 16.8$, $\sigma_{12}' = \sigma_{21}' = -6$, $\sigma_{13}' = \sigma_{31}' = -2.4$, $\sigma_{23}' = \sigma_{32}' = -8$ (MPa)

1.8 What are the deviatoric stress components corresponding to each of the stress tensors given in exercises 1.1–1.4?
Ans: $\sigma_{11}' = ⅓$, $\sigma_{22}' = ⅓$, $\sigma_{33}' = -⅔$; $\sigma_{11}' = -1$, $\sigma_{22}' = 2$, $\sigma_{33}' = -1$; $\sigma_{11}' = 3.78$, $\sigma_{22}' = 1.03$, $\sigma_{33}' = -4.81$, $\sigma_{11}' = -5.67$, $\sigma_{22}' = -9.67$, $\sigma_{33}' = 15.34$ (MPa)

1.9 Show, from the general transformation law $\sigma_{ij}' = l_{ip} l_{jq} \sigma_{pq}$, that the relationship between the shear stress τ expression (1.12c) for any oblique plane is the resultant of its two shear traction components.

1.10 The following stress tensor components (in MPa) $\sigma_{11} = 1$, $\sigma_{12} = 5$, $\sigma_{13} = -5$, $\sigma_{22} = 0$, $\sigma_{23} = 0$ and $\sigma_{33} = -1$ apply to Cartesian axes x_1, x_2 and x_3. Determine the stress components for a rotation to new axes x_1', x_2' and x_3' defined by the respective orthogonal vectors:

$$\mathbf{a} = \mathbf{u}_1 + 2\mathbf{u}_2 + 3\mathbf{u}_3,\ \mathbf{b} = \mathbf{u}_1 + \mathbf{u}_2 - \mathbf{u}_3 \ \text{ and } \ \mathbf{c} = -5\mathbf{u}_1 + 4\mathbf{u}_2 - \mathbf{u}_3.$$

See solution: $\sigma_{11}' = -1.286$, $\sigma_{12}' = 1.389$, $\sigma_{13}' = 1.980$, $\sigma_{22}' = 6.667$, $\sigma_{23}' = -2.762$, $\sigma_{33}' = -5.381$ (MPa)

CHAPTER 2

STRAIN ANALYSIS

2.1 Introduction

The amount of stretch, compression and distortion within a deforming solid is defined in terms of its strain. Various definitions have been used so it is pertinent to begin with explanations as to how strain is defined. As with stress, it is essential that the tensor nature of strain is understood. Where deformation is small (infinitesimal) we examine: (i) the representations of strain in the tensor component and matrix notations and (ii) the manner in which distortion is decomposed into strain and rigid body rotation. It will be recognised from this that stress and strain are identical second rank Cartesian tensors with similar transformation properties. In fact, displacement derivatives will define the strain tensor components when metallic materials suffer infinitesimal elastic deformations. The strain at the limit of elasticity may be exceeded one hundred times by plastic deformation in a metal. Comparatively large elastic strains in non metals can arise from rapid rates of straining. Since the infinitesimal definition of strain is inappropriate under these conditions, alternative finite strain measures are discussed. Among these are the natural or logarithmic strain, the deformation gradient and the extension ratio.

In chapter 1, direct and shear strain components were shown to accompany stress for respective uniaxial and shear loadings (see Figs. 1.1 and 1.2). The two strains can co-exist in various combinations under combined loadings. In general, six independent strain components appear within the general strain tensor. As with stress, the following strain analyses will alternate between the engineering and mathematical notations (see table 1.1) to enable the reader to recognise their equivalence.

2.2 Infinitesimal Strain Tensor

A line element in an unloaded body may distort in a number of ways when forces are applied to it. In general, the element can translate, rotate and change its length. Length changes and angular distortions are normally associated with direct and shear strain components respectively. Rotations and translations, arising from rigid body motion, do not strain a body. When infinitesimal displacements of line elements arise from a combination of these motions, it becomes necessary to separate the physical deformation from rigid body motion.

2.2.1 Distortion and Rotation

Here it will be necessary to specify the deformation that arises from applying the six independent components of the stress tensor σ_{ij} to a body element. In the engineering notation, Fig. 1.6a identifies these components as σ_x, σ_y, σ_z, τ_{xy}, τ_{xz} and τ_{yz}. Two types of distortion arise: (i) direct strains ε_x, ε_y and ε_z from σ_x, σ_y and σ_z and (ii) angular distortions e_{xy}, e_{xz} and e_{yz} from τ_{xy}, τ_{xz} and τ_{yz}. Let one corner of the body element, $\delta x \times \delta y \times \delta z$, originally lie at the origin O in the x–y plane, as shown in Fig. 2.1a.

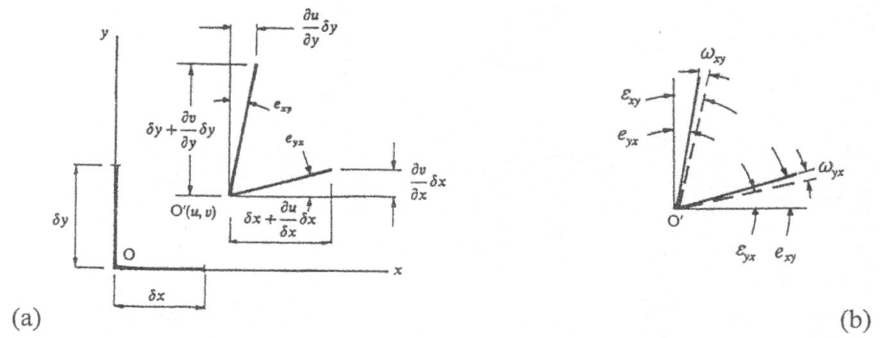

Figure 2.1 Distortion of a corner in the x–y plane

The angular change to the corner is composed of pure shear strains, due to τ_{xy}, τ_{xz} and τ_{yz}, with superimposed rigid body rotations. By subtracting rotations from the angular change, the shear strain associated with shape change is found. In Fig. 2.1b, the angular changes, e_{xy} and e_{yx}, to the corner are shown. With superimposed direct strains, ε_x and ε_y, due to σ_x and σ_y, the corner is also displaced from O to O' in the x and y directions. The respective displacements, u and v, are each functions of the Cartesian co-ordinates: $u = u(x, y, z)$ and $v = v(x, y, z)$. Thus, line elements δx and δy will change their lengths in proportion to the displacement gradients. The latter define the two direct strains in the x–y plane as

$$\varepsilon_x = \frac{\partial u}{\partial x} \quad \text{and} \quad \varepsilon_y = \frac{\partial v}{\partial y} \tag{2.1a,b}$$

In general, $\delta x \neq \delta y$ and $\varepsilon_x \neq \varepsilon_y$ and so the angular changes are unequal. That is, $e_{xy} \neq e_{yx}$ where $e_{xy} = \partial u/\partial y$ and $e_{yx} = \partial v/\partial x$. The engineering shear strain γ_{xy} is defined as the total angular change to the right angle. In Fig. 2.1b the distorted corner is rotated by ω_{xy} and ω_{yx} so that it becomes equally inclined to the x and y directions. This configuration defines two tensor shear strain components from γ_{xy} as

$$\gamma_{xy} = \frac{\partial u}{\partial y} + \frac{\partial v}{\partial x} \quad \text{and} \quad \varepsilon_{xy} = \varepsilon_{yx} = \frac{1}{2}\left(\frac{\partial u}{\partial y} + \frac{\partial v}{\partial x}\right) \tag{2.1c}$$

Their associated rotations are

$$\omega_{xy} = e_{xy} - \varepsilon_{xy} = \partial u/\partial y - ½(\partial u/\partial y + \partial v/\partial x) = ½(\partial u/\partial y - \partial v/\partial x)$$

$$\omega_{yx} = e_{yx} - \varepsilon_{yx} = \partial v/\partial x - ½(\partial u/\partial y + \partial v/\partial x) = -½(\partial u/\partial y - \partial v/\partial x)$$

which obey skew-symmetry: $\omega_{xy} = -\omega_{yx}$. Further strain-displacement relationships, similar in form to eqs(2.1a,b,c), will apply to normal and shear distortions in the x–z and y–z planes. These relationships involve a further displacement function: $w = w(x, y, z)$, for the z direction.

Consequently, the complete distortion of a volume element $\delta x \times \delta y \times \delta z$ may be expressed as the sum of corresponding strains and rotations in the matrix form

$$\begin{bmatrix} e_x & e_{xy} & e_{xz} \\ e_{yx} & e_y & e_{yz} \\ e_{zx} & e_{zy} & e_z \end{bmatrix} = \begin{bmatrix} \varepsilon_x & \varepsilon_{xy} & \varepsilon_{xz} \\ \varepsilon_{yx} & \varepsilon_y & \varepsilon_{yz} \\ \varepsilon_{zx} & \varepsilon_{zy} & \varepsilon_z \end{bmatrix} + \begin{bmatrix} \omega_x & \omega_{xy} & \omega_{xz} \\ \omega_{yx} & \omega_y & \omega_{yz} \\ \omega_{zx} & \omega_{zy} & \omega_z \end{bmatrix} \tag{2.2a}$$

where $\omega_x = \omega_y = \omega_z = 0$. In the tensor component notation, eq(2.2a) appears as

$$e_{ij} = \varepsilon_{ij} + \omega_{ij} \quad \text{(for } i \text{ and } j = 1, 2 \text{ and } 3) \tag{2.2b}$$

The strain and rotation components for each tensor in eq(2.2b) may be expressed in terms of their respective displacement gradients. Firstly, the angular changes appear as

$$e_{ij} = \begin{bmatrix} \frac{\partial u}{\partial x} & \frac{\partial u}{\partial y} & \frac{\partial u}{\partial z} \\ \frac{\partial v}{\partial x} & \frac{\partial v}{\partial y} & \frac{\partial v}{\partial z} \\ \frac{\partial w}{\partial x} & \frac{\partial w}{\partial y} & \frac{\partial w}{\partial z} \end{bmatrix}$$

Then, it follows that the strain matrix is

$$\varepsilon_{ij} = \begin{bmatrix} \frac{\partial u}{\partial x} & \frac{1}{2}\left(\frac{\partial u}{\partial y} + \frac{\partial v}{\partial x}\right) & \frac{1}{2}\left(\frac{\partial u}{\partial z} + \frac{\partial w}{\partial x}\right) \\ \frac{1}{2}\left(\frac{\partial u}{\partial y} + \frac{\partial v}{\partial x}\right) & \frac{\partial v}{\partial y} & \frac{1}{2}\left(\frac{\partial v}{\partial z} + \frac{\partial w}{\partial y}\right) \\ \frac{1}{2}\left(\frac{\partial u}{\partial z} + \frac{\partial w}{\partial x}\right) & \frac{1}{2}\left(\frac{\partial v}{\partial z} + \frac{\partial w}{\partial y}\right) & \frac{\partial w}{\partial z} \end{bmatrix}$$

and that the rotation matrix is

$$\omega_{ij} = \begin{bmatrix} 0 & \frac{1}{2}\left(\frac{\partial u}{\partial y} - \frac{\partial v}{\partial x}\right) & \frac{1}{2}\left(\frac{\partial u}{\partial z} - \frac{\partial w}{\partial x}\right) \\ -\frac{1}{2}\left(\frac{\partial u}{\partial y} - \frac{\partial v}{\partial x}\right) & 0 & \frac{1}{2}\left(\frac{\partial v}{\partial z} - \frac{\partial w}{\partial y}\right) \\ -\frac{1}{2}\left(\frac{\partial u}{\partial z} - \frac{\partial w}{\partial x}\right) & -\frac{1}{2}\left(\frac{\partial v}{\partial z} - \frac{\partial w}{\partial y}\right) & 0 \end{bmatrix}$$

All direct and shear strain components appear within a linear combination of displacement derivatives. This defines the *Euler infinitesimal strain tensor*:

$$\varepsilon_{ij} = \frac{1}{2}\left(\frac{\partial u_i}{\partial x_j} + \frac{\partial u_j}{\partial x_i}\right) \qquad \text{or} \qquad \mathbf{E} = ½\,(\mathbf{u}^{\mathrm{T}}\nabla + \nabla^{\mathrm{T}}\mathbf{u}) \tag{2.3}$$

The *Euler rotation tensor* is defined as

$$\omega_{ij} = \frac{1}{2}\left(\frac{\partial u_i}{\partial x_j} - \frac{\partial u_j}{\partial x_i}\right) \qquad \text{or} \qquad \mathbf{\Omega} = ½\,(\mathbf{u}^{\mathrm{T}}\nabla - \nabla^{\mathrm{T}}\mathbf{u}) \tag{2.4}$$

in which x_i are the spacial, or current, co-ordinates of the deformed body (see Fig. 1.19). When the corresponding engineering notation is required, the subscripts i and j on the symbols u and x are read as: $u_1 = u$, $u_2 = v$, $u_3 = w$, $x_1 = x$, $x_2 = y$ and $x_3 = z$. *Note*: Symbols $\mathbf{u}$ and ∇ in eqs(2.3) and (2.4) define column matrices of their components: $\mathbf{u} = [u_1\, u_2\, u_3]^{\mathrm{T}}$ and $\mathbf{V} = [\partial/\partial x_1 \;\; \partial/\partial x_2 \;\; \partial/\partial x_3]^{\mathrm{T}}$. Hence, $\mathbf{u}^{\mathrm{T}}$ and ∇^{T} define their row matrices. Equation (2.2b) shows that the 3 × 3 strain matrix **E** is symmetric (i.e. $\varepsilon_{ij} = \varepsilon_{ji}$) and that the 3 × 3 rotation matrix $\mathbf{\Omega}$ is skew symmetric (i.e. $\omega_{ij} = -\,\omega_{ji}$). Under applied principal stresses, when all the rotations $\omega_{ij} = 0$, the deformation is said to be *irrotational*.

2.2.2 Strain Transformations in Tensor and Matrix Notations

It may be deduced that the transformation properties of the symmetric 3 × 3 strain matrix ε_{ij} are identical to those of Cauchy stress, the latter being a symmetric 3 × 3 matrix ($\sigma_{ij} = \sigma_{ji}$). Thus, the transformation of an Euler strain tensor, following a rotation in the orthogonal axes from x_i to x_i', will obey a law equivalent in form to eq(1.22a):

$$\varepsilon_{ij}' = l_{ip}\, l_{jq}\, \varepsilon_{pq}, \quad \text{or} \quad \mathbf{E}' = \mathbf{MEM}^{\mathrm{T}} \tag{2.5a}$$

In full, eq(2.5a) becomes

$$\begin{bmatrix} \varepsilon_{11}' & \varepsilon_{12}' & \varepsilon_{13}' \\ \varepsilon_{21}' & \varepsilon_{22}' & \varepsilon_{23}' \\ \varepsilon_{31}' & \varepsilon_{32}' & \varepsilon_{33}' \end{bmatrix} = \begin{bmatrix} l_{11} & l_{12} & l_{13} \\ l_{21} & l_{22} & l_{23} \\ l_{31} & l_{32} & l_{33} \end{bmatrix} \begin{bmatrix} \varepsilon_{11} & \varepsilon_{12} & \varepsilon_{13} \\ \varepsilon_{21} & \varepsilon_{22} & \varepsilon_{23} \\ \varepsilon_{31} & \varepsilon_{32} & \varepsilon_{33} \end{bmatrix} \begin{bmatrix} l_{11} & l_{21} & l_{31} \\ l_{12} & l_{22} & l_{32} \\ l_{13} & l_{23} & l_{33} \end{bmatrix} \tag{2.5b}$$

It follows that there is a correspondence between the stress and strain transformation expressions. For this to be achieved, τ must be interchanged with the tensor definition of shear strain ($\gamma/2$). For example, to find the normal strain on an oblique plane in terms of engineering co-ordinate strains, eq(1.12b) is converted from stress to strain as follows:

$$\varepsilon = \varepsilon_x\, l^2 + \varepsilon_y\, m^2 + \varepsilon_z\, n^2 + 2\,(\, lm\varepsilon_{xy} + mn\varepsilon_{yz} + ln\varepsilon_{zx}\,) \tag{2.6a}$$

$$= \varepsilon_x\, l^2 + \varepsilon_y\, m^2 + \varepsilon_z\, n^2 + lm\gamma_{xy} + mn\gamma_{yz} + ln\gamma_{zx} \tag{2.6b}$$

Similarly, the principal strain cubic is deduced from eq(1.24b) as

$$\varepsilon^3 - I_1\,\varepsilon^2 + I_2\,\varepsilon - I_3 = 0 \tag{2.7a}$$

where the strain invariant are

$$I_1 = \varepsilon_x + \varepsilon_y + \varepsilon_z = \varepsilon_{ii} = \text{tr}\,\mathbf{E} \tag{2.7b}$$

$$I_2 = \varepsilon_x\varepsilon_y + \varepsilon_y\varepsilon_z + \varepsilon_z\varepsilon_x - \varepsilon_{xy}^{\ 2} - \varepsilon_{yz}^{\ 2} - \varepsilon_{zx}^{\ 2} = \tfrac{1}{2}\,(\varepsilon_{ii}\varepsilon_{jj} - \varepsilon_{ij}\varepsilon_{ji}) = \tfrac{1}{2}[(\text{tr}\,\mathbf{E})^2 - \text{tr}\,\mathbf{E}^2] \tag{2.7c}$$

$$\begin{aligned} I_3 &= \varepsilon_x\varepsilon_y\varepsilon_z + 2\varepsilon_{xy}\varepsilon_{zx}\varepsilon_{yz} - \varepsilon_x\varepsilon_{yz}^{\ 2} - \varepsilon_y\varepsilon_{zx}^{\ 2} - \varepsilon_z\varepsilon_{xy}^{\ 2} \\ &= \varepsilon_x\varepsilon_y\varepsilon_z + \tfrac{1}{4}\gamma_{xy}\gamma_{zx}\gamma_{yz} - \tfrac{1}{4}\varepsilon_x\gamma_{yz}^{\ 2} - \tfrac{1}{4}\varepsilon_y\gamma_{zx}^{\ 2} - \tfrac{1}{4}\varepsilon_z\gamma_{xy}^{\ 2} = \det(\varepsilon_{ij}) = \det\mathbf{E} \end{aligned} \tag{2.7d}$$

Because of the identical nature of stress and strain transformations, the directions of the principal stress and strain axes are coincident. It is not possible to convert the general shear stress expression in eq(1.12c) directly to shear strain without specifying the initial perpendicular directions to which the shear strain applies. Performing the matrix multiplication in eq(2.5b) for any one tensor shear strain (i.e. with $i \neq j$) provides a method for finding the shear strain between a pair of perpendicular lines. For example, with $i = 1$ and $j = 2$, eq(2.5b) becomes $\varepsilon_{12}' = l_{1p}\, l_{2q}\, \varepsilon_{pq}$. In the engineering notation this is equivalent to a reduced matrix multiplication:

$$\frac{\gamma}{2} = \begin{bmatrix} l_1 & m_1 & n_1 \end{bmatrix} \begin{bmatrix} \varepsilon_x & \varepsilon_{xy} & \varepsilon_{xz} \\ \varepsilon_{xy} & \varepsilon_y & \varepsilon_{yz} \\ \varepsilon_{xz} & \varepsilon_{yz} & \varepsilon_z \end{bmatrix} \begin{bmatrix} l_2 \\ m_2 \\ n_2 \end{bmatrix} \tag{2.8a}$$

$$= \begin{bmatrix} (l_1\,\varepsilon_x + m_1\,\varepsilon_{xy} + n_1\,\varepsilon_{xz}) & (l_1\,\varepsilon_{xy} + m_1\,\varepsilon_y + n_1\,\varepsilon_{yz}) & (l_1\,\varepsilon_{xz} + m_1\,\varepsilon_{yz} + n_1\,\varepsilon_z) \end{bmatrix} \begin{bmatrix} l_2 \\ m_2 \\ n_2 \end{bmatrix}$$

$$= (l_1\varepsilon_x + m_1\varepsilon_{xy} + n_1\varepsilon_{xz})l_2 + (l_1\varepsilon_{xy} + m_1\varepsilon_y + n_1\varepsilon_{yz})m_2 + (l_1\varepsilon_{xz} + m_1\varepsilon_{yz} + n_1\varepsilon_z)n_2$$

$$= l_1 l_2\varepsilon_x + m_1 m_2\varepsilon_y + n_1 n_2\varepsilon_z + (l_1 m_2 + l_2 m_1)\varepsilon_{xy} + (m_1 n_2 + m_2 n_1)\varepsilon_{yz} + (l_1 n_2 + l_2 n_1)\varepsilon_{xz} \tag{2.8b}$$

Equivalently, a generalised shear strain, eq(2.8b), appears within the tensor and matrix notations:

$$\tfrac{1}{2}\,\gamma = l_{1i}\,\varepsilon_{ij}\; l_{2j} = \mathbf{u}_1'^{\,\mathrm{T}}\;\mathbf{E}\;\mathbf{u}_2' \tag{2.8c}$$

where column matrices $\mathbf{u}_1' = [\, l_{11}\;\; l_{12}\;\; l_{13}\,]^{\mathrm{T}}$ and $\mathbf{u}_2' = [\, l_{21}\;\; l_{22}\;\; l_{23}\,]^{\mathrm{T}}$ express the direction cosines of unit vectors aligned with a given pair of perpendicular directions: x_1' and x_2'. It is possible to identify the correspondence between τ in the stress eq(1.12c) and $\gamma/2$ in eq(2.8b). Referring to the plane ABC in Fig. 1.7a, the perpendicular directions are those defined by the direction cosines: eqs(1.13a,b,c) for the shear stress τ and eqs(1.8a,b,c) for the normal stress σ. That is, in eq(2.8b) l_2, m_2 and n_2 are identified with l_s, m_s and n_s in eqs(1.13a–c) and l_1, m_1 and n_1 with l, m and n in eqs(1.8a–c). Equation (2.8b) will then supply the shear strain for original perpendicular directions: one parallel to the normal stress σ and the other aligned with the direction of the resultant shear stress, τ in Fig. 1.7a.

Example 2.1 The state of strain at a point is specified, in the engineering notation, by the following components of its strain tensor ($\times 10^{-4}$): $\varepsilon_x = 1$, $\varepsilon_y = 1$, $\varepsilon_z = 4$, $\varepsilon_{xy} = -3$, $\varepsilon_{xz} = \surd 2$ and $\varepsilon_{yz} = -\surd 2$. Determine: (a) the normal strain component in a direction defined by the unit vector: $\mathbf{u} = (1/2)\mathbf{u}_x - (1/2)\mathbf{u}_y + (1/\surd 2)\mathbf{u}_z$, (b) the shear strain between the normal and a perpendicular direction whose unit vector equation is: $\mathbf{u} = -(1/2)\mathbf{u}_x + (1/2)\mathbf{u}_y + (1/\surd 2)\mathbf{u}_z$, (c) the invariants and (d) the principal strains.

(a) Substituting $l = 1/2$, $m = -1/2$ and $n = 1/\surd 2$ together with the component strains into eq(2.6a) gives the normal strain

$$\varepsilon/10^{-4} = [(1/2)^2 \times 1] + [(-1/2)^2 \times 1] + [(1/\surd 2)^2 \times 4]$$
$$+ 2[(1/2)(-1/2)(-3) + (-1/2)(1/\surd 2)(-\surd 2) + (1/2)(1/\surd 2)(\surd 2)] = 6$$

(b) The engineering shear strain γ is found from eq(2.8b). The direction cosines: $l_1 = 1/2$, $m_1 = -1/2$ and $n_1 = 1/\surd 2$ apply to the normal direction and $l_2 = -1/2$, $m_2 = 1/2$ and $n_2 = 1/\surd 2$ apply to the perpendicular direction. (*Note*: $l_1 l_2 + m_1 m_2 + n_1 n_2 = 0$)

$$\gamma/2 \times 10^{-4} = [(1/2)(-1/2) \times 1] + [(-1/2)(1/2) \times 1] + [(1/\surd 2)(1/\surd 2) \times 4]$$
$$+ [(1/2)(1/2) + (-1/2)(-1/2)](-3) + [(-1/2)(1/\surd 2) + (1/2)(1/\surd 2)](-\surd 2)$$
$$+ [(1/2)(1/\surd 2) + (-1/2)(1/\surd 2)](\surd 2) = 0$$

This zero value for γ shows that there is no change in the right angle between the two perpendicular directions, i.e. they are principal directions and therefore the normal strain found in (a) is one of the principal strains.

(c) The invariant I_1, I_2 and I_3 are found from eqs(2.7b–d):

$$I_1/10^{-4} = (1 + 1 + 4)10^{-4} = 6$$
$$I_2/10^{-8} = [(1)(1) + (1)(4) + (4)(1) - (-3)2 - (-\surd 2)2 - (\surd 2)2] = -4$$
$$I_3/10^{-12} = \det(\varepsilon_{ij}) = 1(4 - 2) + 3(-12 + 2) + \surd 2(3/2 - \surd 2) = -24$$

(d) The invariant give a principal strain cubic, from eq(2.7a)

$$\varepsilon^3 - (6 \times 10^{-4})\varepsilon^2 - (4 \times 10^{-8})\varepsilon + (24 \times 10^{-12}) = 0$$

for which one root is 6×10^{-4}. The two remaining principal strains may be found from the coefficients a, b and c within the quadratic equation ($a\varepsilon^2 + b\varepsilon + c$) as follows

$$[\varepsilon - (6 \times 10^{-4})](a\varepsilon^2 + b\varepsilon + c) = \varepsilon^3 - (6 \times 10^{-4})\varepsilon^2 - (4 \times 10^{-8})\varepsilon + (24 \times 10^{-12}) = 0$$

Equating coefficients gives $a = 1$, $b = 0$ and $c = -4 \times 10^{-8}$. The quadratic equation simplifies to $\varepsilon^2 - 4 \times 10^{-8} = 0$, for which the roots are: $\varepsilon = \pm 2 \times 10^{-4}$. Thus, the principal strains are identified as: $\varepsilon_1 = 6 \times 10^{-4}$, $\varepsilon_2 = 2 \times 10^{-4}$ and $\varepsilon_3 = -2 \times 10^{-4}$.

2.2.3 Reduction to Plane Strain

Putting $l = \cos\alpha$, $m = \cos(90° - \alpha) = \sin\alpha$ and $n = \cos 90° = 0$ in eq(2.6b), leads to the normal strain in the direction of x' in Fig. 2.2a.

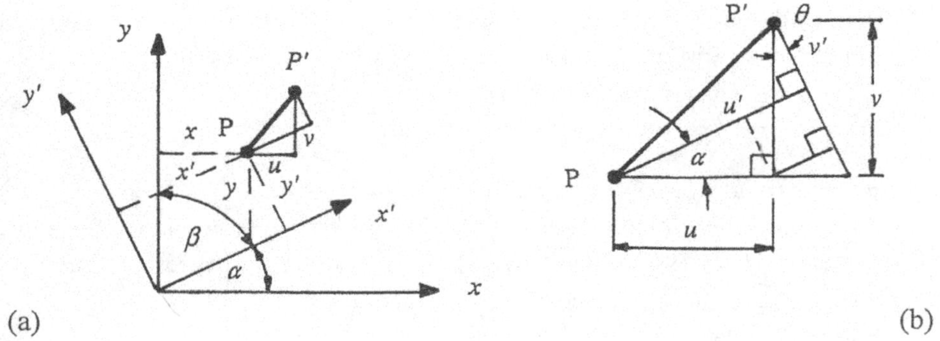

Figure 2.2 (a) Plane co-ordinate and (b) displacement transformations

$$\begin{aligned}\varepsilon_x' &= \varepsilon_x\, l^2 + \varepsilon_y\, m^2 + \varepsilon_z\, n^2 + \gamma_{xy}\, lm + \gamma_{yz}\, mn + \gamma_{zx}\, ln \\ &= \varepsilon_x \cos^2\alpha + \varepsilon_y \sin^2\alpha + \gamma_{xy} \sin\alpha \cos\alpha \\ &= ½(\varepsilon_x + \varepsilon_y) + ½(\varepsilon_x - \varepsilon_y) \cos 2\alpha + (½\gamma_{xy}) \sin 2\alpha \qquad (2.9a)\end{aligned}$$

The normal strain in the y'–direction is found for: $l = \cos(90 + \alpha) = -\sin\alpha$, $m = \cos\alpha$ and $n = 0$ in eq(2.9a). This gives

$$\begin{aligned}\varepsilon_y' &= \varepsilon_x \sin^2\alpha + \varepsilon_y \cos^2\alpha - \gamma_{xy} \sin\alpha \cos\alpha \\ &= ½(\varepsilon_x + \varepsilon_y) - ½(\varepsilon_x - \varepsilon_y) \cos 2\alpha - (½\gamma_{xy}) \sin 2\alpha \qquad (2.9b)\end{aligned}$$

The shear strain between the primed directions x' and y' is found from eq(2.8b). For the x'–direction, set: $l_1 = \cos\alpha$, $m_1 = \cos(90° - \alpha) = \sin\alpha$ and $n_1 = \cos 90° = 0$ and for the y'–direction, set: $l_2 = \cos(90 + \alpha) = -\sin\alpha$, $m_2 = \cos\alpha$ and $n_2 = 0$:

$$\begin{aligned}½\gamma_{xy}' &= -\varepsilon_x \sin\alpha \cos\alpha + \varepsilon_y \sin\alpha \cos\alpha + \gamma_{xy}(\cos^2\alpha - \sin^2\alpha) \\ &= -½\varepsilon_x \sin 2\alpha + ½\varepsilon_y \sin 2\alpha + ½\gamma_{xy} \cos 2\alpha \\ &= -½(\varepsilon_x - \varepsilon_y) \sin 2\alpha + ½\gamma_{xy} \cos 2\alpha \qquad (2.9c)\end{aligned}$$

The strains given in eqs(2.9a–c) may be confirmed from direct differentiation of the displacements according to the infinitesimal strain definitions (2.1a–c). Figure 2.2a provides the geometrical relationships between the co-ordinates:

$$\begin{aligned}x &= x' \cos\alpha - y' \sin\alpha \\ y &= x' \sin\alpha + y' \cos\alpha\end{aligned}$$

The displacements of a point P as it moves to P′ (see Fig. 2.2b), are given by

$$\begin{aligned}u' &= u \cos\alpha + v \sin\alpha \\ v' &= -u \sin\alpha + v \cos\alpha\end{aligned}$$

where u and v are the displacements of P′ along x and y and u' and v' are the displacements of P′ along x' and y', as shown. The normal strain and shear strains in the x'–y' plane are found from eqs(2.1a–c):

$$\begin{aligned}\varepsilon_x' &= \partial u'/\partial x' = (\partial u'/\partial x)(\partial x/\partial x') + (\partial u'/\partial y)(\partial y/\partial x')\\ &= [(\partial u/\partial x)(\partial u'/\partial u) + (\partial v/\partial x)(\partial u'/\partial v)](\partial x/\partial x') + [(\partial u/\partial y)(\partial u'/\partial u) + (\partial v/\partial y)(\partial u'/\partial v)](\partial y/\partial x')\\ &= [(\partial u/\partial x)\cos\alpha + (\partial v/\partial x)\sin\alpha]\cos\alpha + [(\partial u/\partial y)\cos\alpha + (\partial v/\partial y)\sin\alpha]\sin\alpha\\ &= \varepsilon_x \cos^2\alpha + \varepsilon_y \sin^2\alpha + \gamma_{xy}\sin\alpha\cos\alpha\\ &= \tfrac{1}{2}(\varepsilon_x + \varepsilon_y) + \tfrac{1}{2}(\varepsilon_x - \varepsilon_y)\cos 2\alpha + \tfrac{1}{2}\gamma_{xy}\sin 2\alpha\end{aligned}$$

$$\begin{aligned}\varepsilon_y' &= \partial v'/\partial y' = (\partial v'/\partial y)(\partial y/\partial y') + (\partial v'/\partial x)(\partial x/\partial y')\\ &= [(\partial u/\partial y)(\partial v'/\partial u) + (\partial v/\partial y)(\partial v'/\partial v)](\partial y/\partial y') + [(\partial u/\partial x)(\partial v'/\partial u) + (\partial v/\partial x)(\partial v'/\partial v)](\partial x/\partial y')\\ &= [-(\partial u/\partial y)\sin\alpha + (\partial v/\partial y)\cos\alpha]\cos\alpha + [-(\partial u/\partial x)\sin\alpha + (\partial v/\partial x)\cos\alpha](-\sin\alpha)\\ &= \varepsilon_x \sin^2\alpha + \varepsilon_y \cos^2\alpha - \gamma_{xy}\sin\alpha\cos\alpha\\ &= \tfrac{1}{2}(\varepsilon_x + \varepsilon_y) - \tfrac{1}{2}(\varepsilon_x - \varepsilon_y)\cos 2\alpha - \tfrac{1}{2}\gamma_{xy}\sin 2\alpha\end{aligned}$$

$$\begin{aligned}\gamma_{xy}' &= \partial v'/\partial x' + \partial u'/\partial y'\\ &= (\partial v'/\partial x)(\partial x/\partial x') + (\partial v'/\partial y)(\partial y/\partial x') + (\partial u'/\partial x)(\partial x/\partial y') + (\partial u'/\partial y)(\partial y/\partial y')\\ &= [(\partial u/\partial x)(\partial v'/\partial u) + (\partial v/\partial x)(\partial v'/\partial v)](\partial x/\partial x') + [(\partial u/\partial y)(\partial v'/\partial u) + (\partial v/\partial y)(\partial v'/\partial v)](\partial y/\partial x')\\ &\quad + [(\partial u/\partial x)(\partial u'/\partial u) + (\partial v/\partial x)(\partial u'/\partial v)](\partial x/\partial y') + [(\partial u/\partial y)(\partial u'/\partial u) + (\partial v/\partial y)(\partial u'/\partial v)](\partial y/\partial y')\\ &= [-(\partial u/\partial x)\sin\alpha + (\partial v/\partial x)\cos\alpha]\cos\alpha + [-(\partial u/\partial y)\sin\alpha + (\partial v/\partial y)\cos\alpha]\sin\alpha\\ &\quad + [(\partial u/\partial x)\cos\alpha + (\partial v/\partial x)\sin\alpha](-\sin\alpha) + [(\partial u/\partial y)\cos\alpha + (\partial v/\partial y)\sin\alpha]\cos\alpha\\ &= -2(\partial u/\partial x)\sin\alpha\cos\alpha + 2(\partial v/\partial y)\sin\alpha\cos\alpha + [(\partial v/\partial x) + (\partial u/\partial y)](\cos^2\alpha - \sin^2\alpha)\end{aligned}$$

$$\tfrac{1}{2}\gamma_{xy}' = -\tfrac{1}{2}(\varepsilon_x - \varepsilon_y)\sin 2\alpha + \tfrac{1}{2}\gamma_{xy}\cos 2\alpha$$

It is shown in § 2.4 that infinitesimal strain transformation equations refer to line elements in the current configuration (the deformed solid). However, with the small elastic strains involved here, there is negligible error in choosing line elements in the original reference configuration (the undeformed solid). Thus, the strain components ε_x, ε_y and γ_{xy} may be identified with the nominal (engineering) strains in which displacements are more conveniently referred to the initial unstrained geometry.

2.3 Large Strain Definitions

When strains exceed 1 %, one cannot employ the approximations used previously for deriving the infinitesimal strain–displacement relations. A number of large (finite) strain definitions are available and these are reviewed here. They can be applied wherever large deformations arise, say, in the stretching of rubbery materials and superplastic metallic alloys and for straining materials at very rapid rates under impact loading.

2.3.1 Natural (True or Logarithmic) Strain

The theory of metal plasticity employs an incremental strain definition. This was first employed by Ludwik [1] and Hencky [2], who referred the change in length δl of a given line element to its current length, l. This gives a natural incremental strain $\delta\varepsilon = \delta l/l$, for which the true, or logarithmic strain is characterised by either of the integrated quantities:

$$\begin{aligned}\varepsilon &= \int \delta l / l && (2.10a)\\ &= \ln (l / l_o) = \ln [(l_o + \Delta l) / l_o] \\ &= \ln (1 + \Delta l / l_o) = \ln (1 + e) && (2.10b)\\ &\approx e - e^2/2 + e^3/3 - ...\end{aligned}$$

where the nominal, or engineering strain $e = \delta l/l_o$, refers δl to its original length l_o. These two measures of strain, e and ε, are approximately equal for nominal strain values less than 2 %. The logarithmic strain provides the correct measure of the final strain when deformation takes place in a series of increments. For example, let an initial length l_o increase to a final length l_f, under an incremental tensile loading for which there are six intermediate lengths: l_1, l_2, l_3, l_4, l_5 and l_6. It follows from eq(2.10a) that the final, true strain $\varepsilon = \ln (l_f / l_o)$, will be given by the sum of the increments:

$$\begin{aligned}\varepsilon &= \ln (l_1 / l_o) + \ln (l_2 / l_1) + \ln (l_3 / l_2) + \ln (l_4 / l_3) + \ln (l_5 / l_4) + \ln (l_6 / l_5) + \ln (l_f / l_6)\\ &= \ln [(l_1 / l_o)(l_2 / l_1)(l_3 / l_2)(l_4 / l_3)(l_5 / l_4)(l_6 / l_5)(l_f / l_6)] = \ln (l_f / l_o)\end{aligned}$$

Here, as with all proportional loading paths, the final strain does not depend upon the intermediate strains, i.e. the same strain would be achieved if a single continuous loading were to produce l_f. In contrast, applying an engineering strain definition to each stage of the deformation, shows that successive strains are not additive under uniaxial loading. That is

$$(l_1 - l_o)/l_o + (l_2 - l_1)/l_1 + (l_3 - l_2)/l_2 + (l_4 - l_3)/l_3 + (l_5 - l_4)/l_4 + (l_6 - l_5)/l_5 \neq (l_f - l_o)/l_o$$

Therefore, a logarithmic strain measure will account for the influence of a strain path more reliably than engineering strain. This applies particularly when the load path changes direction under a non-proportional loading, giving a final strain that depends, correspondingly, upon the path of strain. Tensor transformation of normal strain is retained within this logarithmic definition but there is not a corresponding shear strain. Large shear strain is expressed as $\gamma = \tan\varphi = x/l$ (see Fig. 1.2) but this will only approximate to the engineering shear strain, i.e. $\gamma = \tan\varphi \approx \varphi$ rad, for distortion angles $\varphi < 10°$.

2.3.2 Extension Ratio

The extension ratio is a measure of large deformation, often used in rubber and polymer mechanics. It is, simply, the ratio between the final length l and the initial length l_o:

$$\lambda_{(\mathbf{u})} = l / l_o \qquad (2.11a)$$

where the unit vector **u** subscript denotes the original direction of l_o. Elastomeric fibres and polymers can stretch by multiples of their original lengths to give typical engineering strains of between 500 % and 600 %. The extension ratio bears a simple relationship to the engineering strain e. When l_o is aligned with a unit vector $\mathbf{u}_x$ in the x–co-ordinate direction:

$$e_x = (l - l_o)/l_o = \lambda_{(\mathbf{u}_x)} - 1 \qquad (2.11b)$$

If an incompressible (constant volume) deformation of a unit cube occurs without shear, eq(2.11b) supplies a relation between the principal extension ratios:

$$(1 + e_1)(1 + e_2)(1 + e_3) - 1 = 0 \quad \text{or} \quad \lambda_1 \lambda_2 \lambda_3 - 1 = 0 \qquad (2.12a,b)$$

Note: Only when strains are small can the left-hand side of eq(2.12a) be approximated to:

$$e_1 + e_2 + e_3 = 0 \quad \text{giving} \quad \lambda_1 + \lambda_2 + \lambda_3 = 3$$

2.3.3 Finite Homogenous Strains

(a) Direct Strain

Consider uni-directional straining of a line element PQ, aligned with the x–direction, as shown in Fig. 2.3. Let end points P and Q, with co-ordinates P(x, 0, 0) and Q($x + \delta x$, 0, 0), move to new positions P′ and Q′.

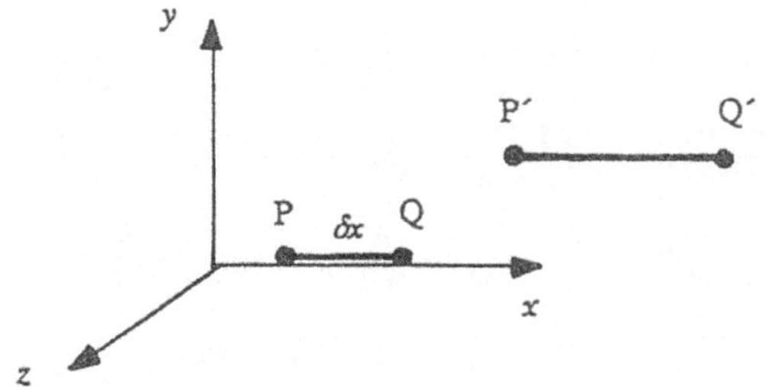

Figure 2.3 Finite extension of a line element aligned with the x–direction

Given that the displacements of point P to P′ are u, v and w in the x, y and z directions respectively, the co-ordinates of P′ become P′($x + u$, v, w). Now, these displacements are functions of the co-ordinates: $u = u(x, y, z)$, $v = v(x, y, z)$ and $w = w(x, y, z)$, and so it follows that the co-ordinates of point Q′ are

$$Q' \, [x + u + \delta\,(x + u), \;\; v + \delta v, \;\; w + \delta w\,]$$

and, since both δy and δz are zero in the unstrained configuration:

$$\delta\,(x + u) = \delta x + (\partial u/\partial x)\delta x, \; \delta v = (\partial v/\partial x)\delta x \;\; \text{and} \;\; \delta w = (\partial w/\partial x)\delta x \qquad (2.13a)$$

The finite normal strain for P′Q′ is expressed as

$$\{(P'Q')^2 - (PQ)^2\}/(PQ)^2 = \{[\delta x + (\partial u/\partial x)\delta x]^2 + [(\partial v/\partial x)\delta x]^2 + [(\partial w/\partial x)\delta x]^2 - (\delta x)^2\}/(\delta x)^2$$
$$= [1 + (\partial u/\partial x)]^2 + (\partial v/\partial x)^2 + (\partial w/\partial x)^2 - 1$$

$$\therefore \;\; 2\varepsilon_x{}^G = 2\{(\partial u/\partial x) + \tfrac{1}{2}\,[(\partial u/\partial x)^2 + (\partial v/\partial x)^2 + (\partial w/\partial x)^2\,]\} \qquad (2.13b)$$

The quantity { } in eq(2.13b) was proposed by Green and Adkins [3] as a finite strain measure. This is written as ε^G and bears the following relationship with the extension ratio $\lambda = \lambda_{(\mathbf{u}_x)} = P'Q'/PQ$

$$\lambda^2 - 1 = 2\varepsilon^G \quad \text{or} \quad \varepsilon^G = \tfrac{1}{2}\,(\lambda^2 - 1) \qquad (2.13c)$$

Also, ε^G relates to the engineering strain $e = P'Q'/PQ - 1$:

$$\begin{aligned} e &= \surd\{[1 + (\partial u/\partial x)]^2 + (\partial v/\partial x)^2 + (\partial w/\partial x)^2\} - 1 \\ &= \surd[1 + 2\{(\partial u/\partial x) + \tfrac{1}{2}[(\partial u/\partial x)^2 + (\partial v/\partial x)^2 + (\partial w/\partial x)^2]\}] - 1 \\ &= \surd(1 + 2\varepsilon^G) - 1 = \lambda - 1 \end{aligned} \quad (2.13d)$$

where

$$\varepsilon^G = \tfrac{1}{2}\,[(1 + e)^2 - 1] = e + \tfrac{1}{2}\,e^2 \quad (2.13e)$$

(b) Volumetric Strain

The strained volume V, accompanying finite deformation, appears in terms of the three principal extension ratios λ_1, λ_2 and λ_3 and the original volume V_o, as

$$(V / V_o)^2 = (\lambda_1 \lambda_2 \lambda_3)^2 \quad (2.14a)$$

Substituting eq(2.13c) into eq(2.14c)

$$(V / V_o)^2 = (1 + 2\varepsilon_1^G)(1 + 2\varepsilon_2^G)(1 + 2\varepsilon_3^G)$$

giving

$$\begin{aligned} \tfrac{1}{2}[\,(V / V_o)^2 - 1] &= (\varepsilon_1^G + \varepsilon_2^G + \varepsilon_3^G) + 2(\varepsilon_1^G\varepsilon_2^G + \varepsilon_1^G\varepsilon_3^G + \varepsilon_2^G\varepsilon_3^G) + 4\varepsilon_1^G\varepsilon_2^G\varepsilon_3^G \\ &= I_1^G + 2\,I_2^G + 4\,I_3^G \end{aligned} \quad (2.14b)$$

Note: I_1^G, I_2^G and I_3^G are the invariant of a principal *Green strain cubic*

$$(\varepsilon^G)^3 - I_1^G(\varepsilon^G)^2 + I_2^G\varepsilon^G - I_3^G = 0$$

which is identical in form to eq(2.7a).

A relationship between the three deformation measures is found from substituting the engineering volumetric strain $\Delta = (V - V_o) / V_o$, into eqs(2.14a,b):

$$\Delta + \Delta^2/2 = I_1^G + 2\,I_2^G + 4\,I_3^G = \tfrac{1}{2}\,[\,(\lambda_1\lambda_2\lambda_3)^2 - 1] \quad (2.14c)$$

Equation (2.14c) reduces to a simple approximation for infinitesimal straining

$$\Delta \approx e_1 + e_2 + e_3$$

(c) Shear Strain

Finite shear strain refers to an angular distortion similar to the infinitesimal shear strain eq(2.1c). Consider the shear deformation between two originally perpendicular line elements PQ and PR, lying in the X–Y plane (Fig. 2.4a). Finite shear deformation is identified with the cosine of the angle φ between P′Q′ and P′R′ in the deformed configuration (see Fig. 2.4b). Local co-ordinates x, y and z, with origin at P′, are aligned parallel to the original axes X, Y and Z, whose origin lies at point P. *Note*: Points R′ and Q′ do not remain in the x–y plane. Green's finite shear strain is defined as: $2\varepsilon_{xy}^G = \cos\varphi$, where $\cos\varphi$ is the dot product of unit vectors $\mathbf{n}_1$ and $\mathbf{n}_2$ aligned with P′Q′ and P′R′. The scalar intercepts made by $\mathbf{n}_1$ and $\mathbf{n}_2$ with x, y and z, are the direction cosines for each vector. Hence, using eq(2.13a)

$$\mathbf{n}_1 = (1 + \partial u/\partial x)\mathbf{u}_x + (\partial v/\partial x)\mathbf{u}_y + (\partial w/\partial x)\mathbf{u}_z \quad (2.15a)$$

$$\mathbf{n}_2 = (\partial u/\partial y)\mathbf{u}_x + (1 + \partial v/\partial y)\mathbf{u}_y + (\partial w/\partial y)\mathbf{u}_z \quad (2.15b)$$

where, respectively, $\mathbf{u}_x$, $\mathbf{u}_y$ and $\mathbf{u}_z$ are unit vectors in the x, y and z directions.

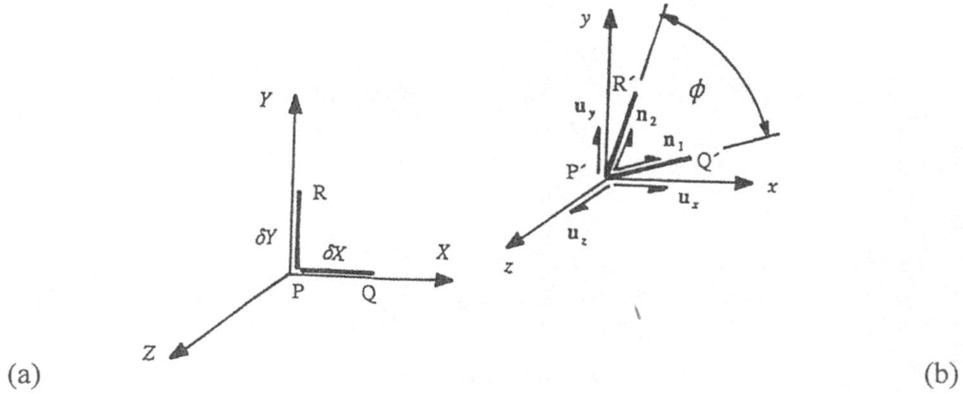

Figure 2.4 Finite shear deformation referred to an *x*, *y*, *z* frame

The finite shear strain becomes

$$\mathbf{n}_1 \bullet \mathbf{n}_2 = \cos\varphi = 2\varepsilon_{xy}{}^G \tag{2.16a}$$

Substituting eqs(2.15a,b) into eq(2.16a) gives

$$2\varepsilon_{xy}{}^G = \cos\varphi = (\partial u/\partial y)(1 + \partial u/\partial x) + (\partial v/\partial x)(1 + \partial v/\partial y) + (\partial w/\partial x)(\partial w/\partial y)$$
$$= \{\partial u/\partial y + \partial v/\partial x + (\partial u/\partial y)(\partial u/\partial x) + (\partial v/\partial x)(\partial v/\partial y) + (\partial w/\partial y)(\partial w/\partial x)\} \tag{2.16b}$$

In the small strain theory, product terms in eq(2.16b) are ignored. Taking the result with the identity $\cos\varphi = \sin(\pi/2 - \varphi) \simeq (\pi/2 - \varphi)$ rad, the infinitesimal shear strain is recovered:

$$\gamma_{xy} = 2\varepsilon_{xy} = (\pi/2 - \varphi) = \partial u/\partial y + \partial v/\partial x$$

In finite deformation, the engineering shear strain γ_{xy} is the angular change to the right angle, originally between PR and PQ. The relationship between the finite shear strain $2\varepsilon_{xy}{}^G$ and γ_{xy} follows from an alternative expression for the dot product of the vectors **P′R′** and **P′Q′**:

$$\mathbf{P'R'} \bullet \mathbf{P'Q'} = |\mathbf{P'R'}|\,|\mathbf{P'Q'}|\cos(\pi/2 - \gamma_{xy}) = |\mathbf{P'R'}|\,|\mathbf{P'Q'}|\sin\gamma_{xy}$$

$$\sin\gamma_{xy} = \frac{\mathbf{P'R'} \bullet \mathbf{P'Q'}}{\mathrm{PR}\,\lambda_{(\mathbf{u}_y)}\,\mathrm{PQ}\,\lambda_{(\mathbf{u}_x)}} = \frac{\mathbf{n}_1 \bullet \mathbf{n}_2}{\lambda_{(\mathbf{u}_x)}\lambda_{(\mathbf{u}_y)}}$$

where PR = |**P′R′**| and PQ = |**P′Q′**| under pure shear. Then, from eqs(2.13c) and (2.16a)

$$\sin\gamma_{xy} = \frac{2\,\varepsilon_{xy}^G}{\sqrt{[(1 + 2\,\varepsilon_x^G)(1 + 2\,\varepsilon_y^G)]}} \tag{2.16c}$$

and again, when the shear strains are small, $\sin\gamma_{xy} \approx \gamma_{xy} \approx 2\varepsilon_{xy}$.

(d) Combined Normal and Shear Strain

Figure 2.5 shows the line element PQ with unstrained co-ordinates P(x, y, z) and Q($x + \delta x$, $y + \delta y$, $z + \delta z$). In general, PQ will displace to P′Q′ with both extension and rotation in the manner shown.

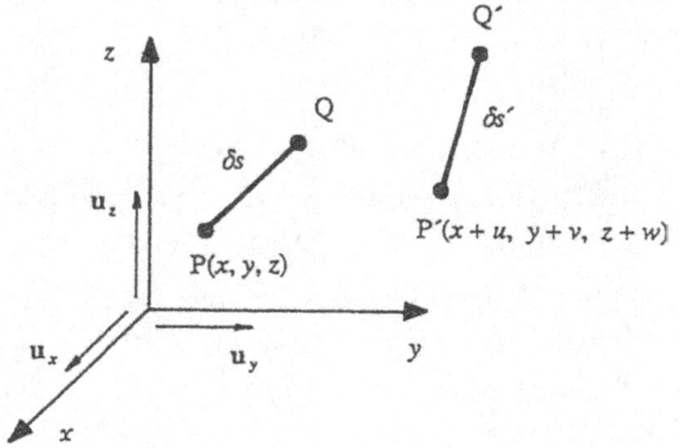

Figure 2.5 Finite extension and rotation of a line element PQ

Let the displacements of point P to P′ be u, v and w, so that the co-ordinates of P′ become P′($x + u$, $y + v$, $z + w$). The co-ordinates of Q′ become [$x + u + \delta(x + u)$, $y + v + \delta(y + v)$, $z + w + \delta(z + w)$], where

$$\delta(x + u) = \delta x + (\partial u/\partial x)\delta x + (\partial u/\partial y)\delta y + (\partial u/\partial z)\delta z \qquad (2.17a)$$
$$\delta(y + v) = \delta y + (\partial v/\partial x)\delta x + (\partial v/\partial y)\delta y + (\partial v/\delta z)\delta z \qquad (2.17b)$$
$$\delta(z + w) = \delta z + (\partial w/\partial x)\delta x + (\partial w/\partial y)\delta y + (\partial w/\partial z)\delta z \qquad (2.17c)$$

Set the infinitesimal lengths of PQ and P′Q′ in Fig. 2.5 to δs and $\delta s'$ respectively. The direction cosine l' for P′Q′ is found from dividing eqs(2.17a) by $\delta s'$. This gives

$$l' = \delta(x + u)/\delta s' = \delta x/\delta s' + (\partial u/\partial x)\delta x/\delta s' + (\partial u/\partial y)\delta y/\delta s' + (\partial u/\partial z)\delta z/\delta s'$$
$$= (\delta s/\delta s')[(\delta x/\delta s) + (\partial u/\partial x)(\delta x/\delta s) + (\partial u/\partial y)(\delta y/\delta s) + (\partial u/\partial z)(\delta z/\delta s)]$$

Substituting for the direction cosines of PQ in the undeformed configuration: $l = \delta x/\delta s$, $m = \delta y/\delta s$ and $n = \delta z/\delta s$, with $\lambda = \delta s'/\delta s$, gives

$$\lambda\, l' = (1 + \partial u/\partial x)l + (\partial u/\partial y)m + (\partial u/\partial z)n \qquad (2.18a)$$

Similarly, from eqs(2.17b,c), the two remaining direction cosines for P′Q′ appear as

$$\lambda\, m' = (\partial v/\partial x)l + (1 + \partial v/\partial y)m + (\partial v/\partial z)n \qquad (2.18b)$$
$$\lambda\, n' = (\partial w/\partial x)l + (\partial w/\partial y)m + (1 + \partial w/\partial z)n \qquad (2.18c)$$

Squaring and adding eqs(2.18a–c) and noting that $(l')^2 + (m')^2 + (n')^2 = 1$, leads to an expression in the extension ratio

$$\lambda^2 - 1 = 2\{l^2\varepsilon_x^G + m^2\varepsilon_y^G + n^2\varepsilon_z^G + 2\,(lm\varepsilon_{xy}^G + mn\varepsilon_{yz}^G + ln\varepsilon_{xz}^G)\}$$

from which Green's finite strain is defined as

$$\varepsilon^G = \tfrac{1}{2}(\lambda^2 - 1) = \{l^2\varepsilon_x^G + m^2\varepsilon_y^G + n^2\varepsilon_z^G + 2(lm\varepsilon_{xy}^G + mn\varepsilon_{yz}^G + ln\varepsilon_{xz}^G)\} \quad (2.19a)$$

Within the right hand side of eq(2.19a) are the components of ε^G, aligned with the co-ordinate axes x, y and z. These are of the form ε_x^G and $2\varepsilon_{xy}^G$, as in eqs(2.13b) and (2.16b). These are, in full:

$$\varepsilon_x^G = (\partial u/\partial x) + \tfrac{1}{2}[(\partial u/\partial x)^2 + (\partial v/\partial x)^2 + (\partial w/\partial x)^2] = \tfrac{1}{2}[(\lambda_x)^2 - 1] \quad (2.19b)$$

$$\varepsilon_y^G = (\partial v/\partial y) + \tfrac{1}{2}[(\partial u/\partial y)^2 + (\partial v/\partial y)^2 + (\partial w/\partial y)^2] = \tfrac{1}{2}[(\lambda_y)^2 - 1] \quad (2.19c)$$

$$\varepsilon_z^G = (\partial w/\partial z) + \tfrac{1}{2}[(\partial u/\partial z)^2 + (\partial v/\partial z)^2 + (\partial w/\partial z)^2] = \tfrac{1}{2}[(\lambda_z)^2 - 1] \quad (2.19d)$$

$$2\varepsilon_{xy}^G = \{\partial u/\partial y + \partial v/\partial x + (\partial u/\partial y)(\partial u/\partial x) + (\partial v/\partial x)(\partial v/\partial y) + (\partial w/\partial y)(\partial w/\partial x)\} = \lambda_{xy}^2 \quad (2.19e)$$

$$2\varepsilon_{xz}^G = \{\partial w/\partial x + \partial u/\partial z + (\partial u/\partial z)(\partial u/\partial x) + (\partial v/\partial x)(\partial v/\partial z) + (\partial w/\partial z)(\partial w/\partial x)\} = \lambda_{xz}^2 \quad (2.19f)$$

$$2\varepsilon_{yz}^G = \{\partial w/\partial y + \partial v/\partial z + (\partial u/\partial y)(\partial u/\partial z) + (\partial v/\partial z)(\partial v/\partial y) + (\partial w/\partial y)(\partial w/\partial z)\} = \lambda_{yz}^2 \quad (2.19g)$$

where eqs(2.19b and e) correspond to expressions (2.13b) and (2.16b), derived previously for finite uniaxial strain and finite shear strain. An alternative derivation to eq(2.19a) employs $2\varepsilon^G = [(P'Q')^2 - (PQ)^2] / (PQ)^2$. Substituting from eqs(2.17a–c), gives the finite normal strain measure ($2\varepsilon^G$) for P′Q′ as

$$\{[(x + u) + \delta x + (\partial u/\partial x)\delta x + (\partial u/\partial y)\delta y + (\partial u/\partial z)\delta z - (x + u)]^2$$
$$+ [(y + v) + \delta y + (\partial v/\partial x)\delta x + (\partial v/\partial y)\delta y + (\partial v/\partial z)\delta z - (y + v)]^2$$
$$+ [(z + w) + \delta z + (\partial w/\partial x)\delta x + (\partial w/\partial y)\delta y + (\partial w/\partial z)\delta z - (z + w)]^2\} \div [(\delta x)^2 + (\delta y)^2 + (\delta z)^2] - 1$$

It can be seen that the normal strain eq(2.19a) is identical to its infinitesimal counterpart eq(2.6a), previously associated with a tensor transformation for strain. Here it becomes apparent that finite strain components are defined to preserve a similar tensor transformation. That is: $\varepsilon_{ij}^{G\prime} = l_{ip}l_{jq}\varepsilon_{pq}^G$, where the prime refers to the rotated axes. *Note*: From eqs(2.19e–g), the *shear extension ratios* λ_{xy}, λ_{xz} and λ_{yz}, are related to Green's shear strain. Substituting eqs(2.19e–g) into eq(2.19a), gives $2\varepsilon^G$ in terms of extension ratios

$$(\lambda^2 - 1) = l^2(\lambda_x^2 - 1) + m^2(\lambda_y^2 - 1) + n^2(\lambda_z^2 - 1) + 2(lm\lambda_{xy}^2 + mn\lambda_{yz}^2 + ln\lambda_{xz}^2) \quad (2.20a)$$
$$= l^2\lambda_x^2 + m^2\lambda_y^2 + n^2\lambda_z^2 - (l^2 + m^2 + n^2) + 2(lm\lambda_{xy}^2 + mn\lambda_{yz}^2 + ln\lambda_{xz}^2)$$

from which

$$\lambda^2 = l^2\lambda_x^2 + m^2\lambda_y^2 + n^2\lambda_z^2 + 2(lm\lambda_{xy}^2 + mn\lambda_{yz}^2 + ln\lambda_{xz}^2) \quad (2.20b)$$

The right-hand side of eq(2.20b) is identical in form to the right-hand sides of eqs(2.6b) and (2.19a). It follows from this that the finite shear strain definition will also ensure that extension ratios conform to a transformation equation: $\lambda_{ij}^{2\prime} = l_{ip}l_{jq}\lambda_{pq}^2$.

Example 2.2 Convert the following matrix of Green's strain ε_{pq}^G to an appropriate matrix of extension ratios that will allow their transformation.

$$\varepsilon_{pq}^G = \begin{bmatrix} 1 & 3 & -2 \\ 3 & 1 & -2 \\ -2 & -2 & 6 \end{bmatrix}$$

Then apply the transformation $\lambda_{ij}^{2\prime} = l_{ip}\, l_{jq} \lambda_{pq}^{2}$, given rotations in the co-ordinates from x to x' and y to y'. Unit vectors, aligned with the directions x' and y' respectively, are given by

$$\mathbf{u}_x' = \sqrt{(2/3)}\mathbf{u}_x + (1/\sqrt{6})\mathbf{u}_y + (1/\sqrt{6})\mathbf{u}_z \quad \text{and} \quad \mathbf{u}_y' = (1/\sqrt{3})\mathbf{u}_x - (1/\sqrt{3})\mathbf{u}_y - (1/\sqrt{3})\mathbf{u}_z$$

Confirm the result from the corresponding transformation: $\varepsilon_{ij}^{G\prime} = l_{ip}\, l_{jq}\, \varepsilon_{pq}^{G}$.

Firstly, the unit vector $\mathbf{u}_z$, describing the rotation from z to z', can be found from the cross product:

$$\mathbf{u}_z = \mathbf{u}_x \times \mathbf{u}_y = \begin{bmatrix} \mathbf{u}_x & \mathbf{u}_y & \mathbf{u}_z \\ \sqrt{\dfrac{2}{3}} & \dfrac{1}{\sqrt{6}} & \dfrac{1}{\sqrt{6}} \\ \dfrac{1}{\sqrt{3}} & -\dfrac{1}{\sqrt{3}} & -\dfrac{1}{\sqrt{3}} \end{bmatrix} = \frac{1}{\sqrt{2}}\mathbf{u}_y - \frac{1}{\sqrt{2}}\mathbf{u}_z$$

Note: $\mathbf{u}_x' \bullet \mathbf{u}_y' = \mathbf{u}_x' \bullet \mathbf{u}_z' = \mathbf{u}_y' \bullet \mathbf{u}_z' = 0$. Converting ε_{pq}^{G} to λ_{pq}^{2} from eqs(2.19b–g), the required transformation $\lambda_{ij}^{2\prime} = l_{ip}\, l_{jq} \lambda_{pq}^{2}$ becomes

$$\lambda_{ij}^{2\prime} = \begin{bmatrix} \sqrt{(2/3)} & 1/\sqrt{6} & 1/\sqrt{6} \\ 1/\sqrt{3} & -1/\sqrt{3} & -1/\sqrt{3} \\ 0 & 1/\sqrt{2} & -1/\sqrt{2} \end{bmatrix} \begin{bmatrix} 3 & 6 & -4 \\ 6 & 3 & -4 \\ -4 & -4 & 13 \end{bmatrix} \begin{bmatrix} \sqrt{(2/3)} & 1/\sqrt{3} & 0 \\ 1/\sqrt{6} & -1/\sqrt{3} & 1/\sqrt{2} \\ 1/\sqrt{6} & -1/\sqrt{3} & -1/\sqrt{2} \end{bmatrix} = \begin{bmatrix} 4.667 & -0.943 & 2.888 \\ -0.943 & 2.333 & 8.166 \\ 2.888 & 8.166 & 12.002 \end{bmatrix}$$

The result may be checked from $\varepsilon_{ij}^{G\prime} = l_{ip}\, l_{jq}\, \varepsilon_{pq}^{G}$

$$\varepsilon_{ij}^{G\prime} = \begin{bmatrix} \sqrt{(2/3)} & 1/\sqrt{6} & 1/\sqrt{6} \\ 1/\sqrt{3} & -1/\sqrt{3} & -1/\sqrt{3} \\ 0 & 1/\sqrt{2} & -1/\sqrt{2} \end{bmatrix} \begin{bmatrix} 1 & 3 & -2 \\ 3 & 1 & -2 \\ -2 & -2 & 6 \end{bmatrix} \begin{bmatrix} \sqrt{(2/3)} & 1/\sqrt{3} & 0 \\ 1/\sqrt{6} & -1/\sqrt{3} & 1/\sqrt{2} \\ 1/\sqrt{6} & -1/\sqrt{3} & -1/\sqrt{2} \end{bmatrix} = \begin{bmatrix} 1.883 & -0.4715 & 1.444 \\ -0.4715 & 0.6666 & 4.083 \\ 1.444 & 4.083 & 5.501 \end{bmatrix}$$

from which $\lambda_{ij}^{2\prime}$ are found from eqs(2.19b–g). For example, $\lambda_x^{2\prime} = 1 + (2 \times 1.833) = 4.667$ and $\lambda_{xy}^{2\prime} = 2 \times (-0.4715) = -0.943$.

2.4 Finite Strain Tensors

In Fig. 2.6, two sets of independent Cartesian co-ordinates are superimposed. *Lagrangian material co-ordinates* $X_i = X_1, X_2$ and X_3, define the unstrained line element PQ in the material at time $t = 0$. *Eulerian spacial co-ordinates* $x_i = x_1, x_2$ and x_3 define the strained line element pq after time t.

The co-ordinates X_i and x_i allow two mappings of the deformation. In the first mapping the current position x_i is a function of the original position X_i:

$$x_i = x_i(X_i, t) = x_i(X_1, X_2, X_3, t) \quad \text{or} \quad \mathbf{x} = \mathbf{x}(\mathbf{X}, t) \tag{2.21a}$$

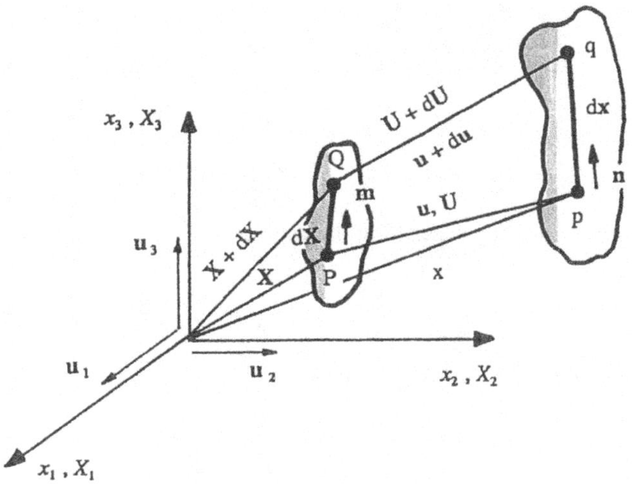

Figure 2.6 Finite deformation within material X_i and spacial x_i co-ordinates

The components $\mathrm{d}x_i$, of the current differential line element pq, are those of vector $\mathbf{dx}$:

$$\mathrm{d}x_i = (\partial x_i / \partial X_j)\, \mathrm{d}X_j \quad \text{or} \quad \mathbf{dx} = \mathbf{F}\, \mathbf{dX} \tag{2.21b}$$

where $\mathbf{F}$ is a matrix of *material deformation gradient* components:

$$F_{ij} = \partial x_i / \partial X_j \quad \text{or} \quad \mathbf{F} = \mathbf{x}\, \forall^{\mathrm{T}} \tag{2.21c}$$

where $\forall = [\partial/\partial X_1 \;\; \partial/\partial X_2 \;\; \partial/\partial X_3]^{\mathrm{T}}$. The second mapping retraces the original position from the current position. This appears in the function

$$X_i = X_i(x_i, t) = X_i(x_1, x_2, x_3, t) \quad \text{or} \quad \mathbf{X} = \mathbf{X}(\mathbf{x}, t) \tag{2.22a}$$

This inverse function will exist when the determinant of the Jacobian eq(1.40b) does not vanish. The components $\mathrm{d}X_i$ of the vector $\mathbf{dX}$, for the original differential line element PQ, are then

$$\mathrm{d}X_i = (\partial X_i / \partial x_j)\, \mathrm{d}x_j \quad \text{or} \quad \mathbf{dX} = \mathbf{H}\, \mathbf{dx} \tag{2.22b}$$

where $\mathbf{H} = \mathbf{F}^{-1}$, is the matrix of *spacial deformation gradients*

$$H_{ij} = \partial X_i / \partial x_j \quad \text{or} \quad \mathbf{H} = \mathbf{X}\, \nabla^{\mathrm{T}} \tag{2.22c}$$

where $\nabla = [\partial/\partial x_1 \;\; \partial/\partial x_2 \;\; \partial/\partial x_3]^{\mathrm{T}}$. The square of the current differential length pq is given as

$$\mathrm{d}x_i\, \mathrm{d}x_i = \delta_{ij}\, \mathrm{d}x_i\, \mathrm{d}x_j \quad \text{or} \quad |\mathbf{dx}|^2 = \mathbf{dx}^{\mathrm{T}}\, \mathbf{dx} \tag{2.23a}$$

Substituting eq(2.21b) into eq(2.23a) with k and l as dummy subscripts

$$\begin{aligned} |\mathbf{dx}|^2 &= \delta_{ik} (\partial x_i / \partial X_j)\, \mathrm{d}X_j (\partial x_k / \partial X_l)\, \mathrm{d}X_l = (\partial x_k / \partial X_j)\, \mathrm{d}X_j\, (\partial x_k / \partial X_l)\, \mathrm{d}X_l \\ &= (\partial x_k / \partial X_i)(\partial x_k / \partial X_j)\, \mathrm{d}X_i\, \mathrm{d}X_j = G_{ij}\, \mathrm{d}X_i\, \mathrm{d}X_j \end{aligned} \tag{2.23b}$$

or
$$|\mathbf{dx}|^2 = \mathbf{dX}^{\mathrm{T}} \mathbf{G}\, \mathbf{dX} \tag{2.23c}$$

The components G_{ij} of the right Cauchy-Green deformation tensor **G** are defined from eq(2.23b) as follows

$$G_{ij} = (\partial x_k / \partial X_i)(\partial x_k / \partial X_j) = F_{ki} F_{kj} \quad \text{or} \quad \mathbf{G} = \mathbf{F}^{\mathrm{T}} \mathbf{F} \tag{2.23d}$$

Now, the square of the current differential length PQ is

$$\mathrm{d}X_i\, \mathrm{d}X_i = \delta_{ij}\, \mathrm{d}X_i\, \mathrm{d}X_j \quad \text{or} \quad |\mathbf{dX}|^2 = \mathbf{dX}^{\mathrm{T}}\, \mathbf{dX} \tag{2.24a}$$

Substituting eq(2.22b) into eq(2.24a) leads to

$$|\mathbf{dX}|^2 = \delta_{ik} (\partial X_i / \partial x_j)\, \mathrm{d}x_j (\partial X_k / \partial x_l)\, \mathrm{d}x_l = (\partial X_k / \partial x_j)\, \mathrm{d}x_j (\partial X_k / \partial x_l)\, \mathrm{d}x_l$$
$$= (\partial X_k / \partial x_i)(\partial X_k / \partial x_j)\, \mathrm{d}x_i\, \mathrm{d}x_j = C_{ij}\, \mathrm{d}x_i\, \mathrm{d}x_j$$

or
$$|\mathbf{dX}|^2 = \mathbf{dx}^{\mathrm{T}}\, \mathbf{C}\, \mathbf{dx} \tag{2.24b}$$

where **C** is the Cauchy deformation tensor with components C_{ij}:

$$C_{ij} = (\partial X_k / \partial x_i)(\partial X_k / \partial x_j) = H_{ki} H_{kj} \quad \text{or} \quad \mathbf{C} = \mathbf{H}^{\mathrm{T}} \mathbf{H} \tag{2.24c}$$

The motion of point P to p may be expressed in terms of either the material or the spacial components of the respective displacement vectors, **U** or **u**, in Fig. 2.6. Working with the former material displacement vector **U**:

$$x_i = U_i + X_i \quad \text{or} \quad \mathbf{x} = \mathbf{U} + \mathbf{X} \tag{2.25a}$$

for which its derivative is

$$\partial x_i / \partial X_j = \partial U_i / \partial X_j + \partial X_i / \partial X_j \tag{2.25b}$$

Substituting from eq(2.21c) into eq(2.25b) gives

$$F_{ij} = \partial U_i / \partial X_j + \delta_{ij} \tag{2.25c}$$

where $\partial U_i / \partial X_j$ are the *material displacement gradients*. When working with the spacial displacement vector **u**:

$$x_i = u_i + X_i \quad \text{or} \quad \mathbf{x} = \mathbf{u} + \mathbf{X} \tag{2.26a}$$

the derivative is

$$\partial x_i / \partial x_j = \partial u_i / \partial x_j + \partial X_i / \partial x_j \tag{2.26b}$$

and substituting eq(2.22c) into eq(2.26b) gives

$$H_{ij} = \delta_{ij} - \partial u_i / \partial x_j \tag{2.26c}$$

where $\partial u_i / \partial x_j$ are the *spacial displacement gradients*.

Example 2.3 Find the components of **F, H, G** and **C** given that Eulerian components of a particle motion $x_i = x_i(X_i, t)$ are as follows: $x_1 = X_1 + X_3(\mathrm{e}^t - 1)$, $x_2 = X_2 + X_3(\mathrm{e}^t - \mathrm{e}^{-t})$ and $x_3 = X_3\, \mathrm{e}^t$, where e is the exponential constant.

Note: When time $t = 0$, $x_1 = X_1$, $x_2 = X_2$ and $x_3 = X_3$, the spacial x_i and material X_i co-ordinates coincide. The components of the deformation gradient are, from eq(2.21c)

$$F_{11} = \partial x_1/\partial X_1 = 1,\ F_{12} = \partial x_1/\partial X_2 = 0,\ F_{13} = \partial x_1/\partial X_3 = \mathrm{e}^{t} - 1$$
$$F_{21} = \partial x_2/\partial X_1 = 0,\ F_{22} = \partial x_2/\partial X_2 = 1,\ F_{23} = \partial x_2/\partial X_3 = \mathrm{e}^{t} - \mathrm{e}^{-t}$$
$$F_{31} = \partial x_3/\partial X_3 = 0,\ F_{32} = \partial x_3/\partial X_2 = 0 \text{ and } F_{33} = \partial x_3/\partial X_3 = \mathrm{e}^{t}$$

These define the material deformation gradient matrix:

$$\mathbf{F} = \begin{bmatrix} 1 & 0 & (\mathrm{e}^{t} - 1) \\ 0 & 1 & (\mathrm{e}^{t} - \mathrm{e}^{-t}) \\ 0 & 0 & \mathrm{e}^{t} \end{bmatrix}$$

Since $J = \det \mathbf{F} = \mathrm{e}^{t} \neq 0$, the inverse function $X_i = X_i(x_i, t)$ exists. Its components are

$$X_1 = x_1 - x_3(1 - \mathrm{e}^{-t}),\quad X_2 = x_2 - x_3(1 - \mathrm{e}^{-2t}) \text{ and } X_3 = x_3\mathrm{e}^{-t}$$

Hence, using eq(2.22c), the spacial deformation gradient matrix $\mathbf{H}$ is formed from its components H_{ij} as follows

$$H_{11} = \partial X_1/\partial x_1 = 1,\ H_{12} = \partial X_1/\partial x_2 = 0,\ H_{13} = \partial X_1/\partial x_3 = -(1 - \mathrm{e}^{-t})$$
$$H_{21} = \partial X_2/\partial x_1 = 0,\ H_{22} = \partial X_2/\partial x_2 = 1,\ H_{23} = \partial X_2/\partial x_3 = -(1 - \mathrm{e}^{-2t})$$
$$H_{31} = \partial X_3/\partial x_3 = 0,\ H_{32} = \partial x_3/\partial X_2 = 0,\ H_{33} = \partial x_3/\partial X_3 = \mathrm{e}^{-t}$$

$$\mathbf{H} = \begin{bmatrix} 1 & 0 & -(1 - \mathrm{e}^{-t}) \\ 0 & 1 & -(1 - \mathrm{e}^{-2t}) \\ 0 & 0 & \mathrm{e}^{-t} \end{bmatrix}$$

which may be confirmed from $\mathbf{H} = \mathbf{F}^{-1}$.

The right Cauchy-Green deformation tensor $\mathbf{G}$ is found from eq(2.23d) as

$$\mathbf{G} = \begin{bmatrix} 1 & 0 & 0 \\ 0 & 1 & 0 \\ (\mathrm{e}^{t} - 1) & (\mathrm{e}^{t} - \mathrm{e}^{-t}) & \mathrm{e}^{t} \end{bmatrix} \begin{bmatrix} 1 & 0 & (\mathrm{e}^{t} - 1) \\ 0 & 1 & (\mathrm{e}^{t} - \mathrm{e}^{-t}) \\ 0 & 0 & \mathrm{e}^{t} \end{bmatrix} = \begin{bmatrix} 1 & 0 & (\mathrm{e}^{t} - 1) \\ 0 & 1 & (\mathrm{e}^{t} - \mathrm{e}^{-t}) \\ (\mathrm{e}^{t} - 1) & (\mathrm{e}^{t} - \mathrm{e}^{-t}) & 3\mathrm{e}^{2t} - 2\mathrm{e}^{t} + \mathrm{e}^{-2t} - 1 \end{bmatrix}$$

The Cauchy deformation tensor $\mathbf{C}$ is found from $\mathbf{H}$ in eq(2.24c):

$$\mathbf{C} = \begin{bmatrix} 1 & 0 & 0 \\ 0 & 1 & 0 \\ -(\mathrm{e}^{t} - 1) & -(1 - \mathrm{e}^{-2t}) & \mathrm{e}^{-t} \end{bmatrix} \begin{bmatrix} 1 & 0 & -(1 - \mathrm{e}^{-t}) \\ 0 & 1 & -(1 - \mathrm{e}^{-2t}) \\ 0 & 0 & \mathrm{e}^{-t} \end{bmatrix} = \begin{bmatrix} 1 & 0 & -(1 - \mathrm{e}^{-t}) \\ 0 & 1 & -(1 - \mathrm{e}^{-2t}) \\ -(1 - \mathrm{e}^{-t}) & -(1 - \mathrm{e}^{-2t}) & 2 - 2\mathrm{e}^{-t} + \mathrm{e}^{-4t} \end{bmatrix}$$

Alternatively, **H** and **F** in eqs(2.25c) and (2.26c) can be found the displacement gradients. The material displacement components follow from eq(2.25a) as

$$U_1 = x_1 - X_1 = X_3(\mathrm{e}^t - 1),\ \ U_2 = x_2 - X_2 = X_3(\mathrm{e}^t - \mathrm{e}^{-t}) \ \text{and}\ \ U_3 = x_3 - X_3 = X_3(\mathrm{e}^t - 1)$$

These give the material displacement gradients:

$$\partial U_1/\partial X_1 = \partial U_1/\partial X_2 = \partial U_2/\partial X_1 = \partial U_2/\partial X_2 = \partial U_3/\partial X_1 = \partial U_3/\partial X_2 = 0$$

$$\partial U_1/\partial X_3 = (\mathrm{e}^t - 1),\ \partial U_2/\partial X_3 = (\mathrm{e}^t - \mathrm{e}^{-t}) \ \text{and}\ \partial U_3/\partial X_3 = (\mathrm{e}^t - 1)$$

when, from eq(2.25c):

$$\mathbf{F} = \begin{bmatrix} 0 & 0 & (\mathrm{e}^t - 1) \\ 0 & 0 & (\mathrm{e}^t - \mathrm{e}^{-t}) \\ 0 & 0 & (\mathrm{e}^t - 1) \end{bmatrix} + \begin{bmatrix} 1 & 0 & 0 \\ 0 & 1 & 0 \\ 0 & 0 & 1 \end{bmatrix} = \begin{bmatrix} 1 & 0 & (\mathrm{e}^t - 1) \\ 0 & 1 & (\mathrm{e}^t - \mathrm{e}^{-t}) \\ 0 & 0 & \mathrm{e}^t \end{bmatrix}$$

The spacial displacement components follow from eq(2.26a) as

$$u_1 = x_1 - X_1 = x_3(1 - \mathrm{e}^{-t}),\ u_2 = x_2 - X_2 = x_3(1 - \mathrm{e}^{-2t}) \ \text{and}\ \ u_3 = x_3 - X_3 = x_3(1 - \mathrm{e}^{-t})$$

These give the spacial displacement gradients:

$$\partial u_1/\partial x_1 = \partial u_1/\partial x_2 = \partial u_2/\partial x_1 = \partial u_2/\partial x_2 = \partial u_3/\partial x_1 = \partial u_3/\partial x_2 = 0$$

$$\partial u_1/\partial x_3 = (1 - \mathrm{e}^{-t}),\ \partial u_2/\partial x_3 = (1 - \mathrm{e}^{-2t}) \ \text{and}\ \partial u_3/\partial x_3 = (1 - \mathrm{e}^{-t})$$

when, from eq(2.26c)

$$\mathbf{H} = \begin{bmatrix} 1 & 0 & 0 \\ 0 & 1 & 0 \\ 0 & 0 & 1 \end{bmatrix} - \begin{bmatrix} 0 & 0 & (1 - \mathrm{e}^{-t}) \\ 0 & 0 & (1 - \mathrm{e}^{-2t}) \\ 0 & 0 & (1 - \mathrm{e}^{-t}) \end{bmatrix} = \begin{bmatrix} 1 & 0 & -(1 - \mathrm{e}^{-t}) \\ 0 & 1 & -(1 - \mathrm{e}^{-2t}) \\ 0 & 0 & \mathrm{e}^{-t} \end{bmatrix}$$

Note: In this example **F, H, G** and **C** show time and not co-ordinate dependency. In general, **F** and **G** will depend upon t and X_i while **H** and **C** will depend upon t and x_i.

2.4.1 Lagrange Finite Strain Tensor

Two finite strain tensors arise from taking the difference between the squares of the current and original differential lengths. They are the Lagrangian and Eulerian measures, which follow in separate descriptions.

To find the Lagrangian finite strain take the difference between eqs(2.23b) and (2.24a) and substitute from eq(2.25b):

$$
\begin{aligned}
|\mathbf{dx}|^2 - |\mathbf{dX}|^2 &= [\,(\partial x_k/\partial X_i)(\partial x_k/\partial X_j) - \delta_{ij}\,]\,\mathrm{d}X_i\,\mathrm{d}X_j \\
&= [\,(\partial U_k/\partial X_i + \delta_{ki})(\partial U_k/\partial X_j + \delta_{kj}) - \delta_{ij}\,]\,\mathrm{d}X_i\,\mathrm{d}X_j \\
&= [\,\partial U_j/\partial X_i + \partial U_i/\partial X_j + (\partial U_k/\partial X_i)(\partial U_k/\partial X_j)]\,\mathrm{d}X_i\,\mathrm{d}X_j \\
&= 2L_{ij}\,\mathrm{d}X_i\,\mathrm{d}X_j
\end{aligned} \tag{2.27a}
$$

L_{ij} is the *Lagrangian or Green finite strain tensor*, defined from eqs(2.23c) and (2.27a) as

$$L_{ij} = \tfrac{1}{2}[(\partial x_k/\partial X_i)(\partial x_k/\partial X_j) - \delta_{ij}] = \tfrac{1}{2}[\,\partial U_j/\partial X_i + \partial U_i/\partial X_j + (\partial U_k/\partial X_i)(\partial U_k/\partial X_j)] \tag{2.27b}$$

In symbolic notation:

$$\mathbf{L} = \tfrac{1}{2}(\mathbf{G} - \mathbf{I}) = \tfrac{1}{2}(\mathbf{F}^{\mathrm{T}}\mathbf{F} - \mathbf{I}) \tag{2.27c}$$

Note: The L_{ij} tensor components (2.27b) were those identified previously within eqs(2.19b–g) as Green's strains in the engineering notation. They correspond to a Lagrangian strain since this measure refers to the original material co-ordinates X_i.

2.4.2 Euler Finite Strain Tensor

Taking the difference between the squares of the current and original differential lengths, from eqs(2.23a) and (2.24b), gives

$$
\begin{aligned}
|\mathbf{dx}|^2 - |\mathbf{dX}|^2 &= [\delta_{ij} - (\partial X_k/\partial x_i)(\partial X_k/\partial x_j)]\mathrm{d}x_i\,\mathrm{d}x_j \\
&= [\delta_{ij} - (\delta_{ki} - \partial u_k/\partial x_i)(\delta_{kj} - \partial u_k/\partial x_j)]\mathrm{d}x_i\,\mathrm{d}x_j \\
&= [\partial u_j/\partial x_i + \partial u_i/\partial x_j - (\partial u_k/\partial x_i)(\partial u_k/\partial x_j)]\mathrm{d}x_i\,\mathrm{d}x_j \\
&= 2E_{ij}\,\mathrm{d}x_i\,\mathrm{d}x_j
\end{aligned} \tag{2.28a}
$$

E_{ij} is the *Eulerian or Almansi finite strain tensor* defined as:

$$E_{ij} = \tfrac{1}{2}[\delta_{ij} - (\partial X_k/\partial x_i)(\partial X_k/\partial x_i)] = \tfrac{1}{2}[\partial u_j/\partial x_i + \partial u_i/\partial x_j - (\partial u_k/\partial x_i)(\partial u_k/\partial x_j)] \tag{2.28b}$$

Symbolically:

$$\mathbf{E} = \tfrac{1}{2}(\mathbf{I} - \mathbf{C}) = \tfrac{1}{2}(\mathbf{I} - \mathbf{H}^{\mathrm{T}}\mathbf{H}) = \tfrac{1}{2}(\mathbf{I} - \mathbf{B}^{-1}) \tag{2.28c}$$

where $\mathbf{C} = \mathbf{B}^{-1}$ in which $\mathbf{B}$ is the *left Cauchy-Green deformation tensor*:

$$B_{ij} = (\partial x_i/\partial X_k)(\partial x_i/\partial X_k) \quad \text{or} \quad \mathbf{B} = \mathbf{F}\mathbf{F}^{\mathrm{T}} \tag{2.28d}$$

Note: The Eulerian tensor $\mathbf{E}$ refers to spacial co-ordinates x_i and is less useful when it is required to refer finite deformation to the original material co-ordinates X_i. However, in the case of infinitesimal straining the product term in eqs(2.27b) and (2.28b) may be neglected. This results in coincident Lagrangian and Eulerian infinitesimal strain tensors respectively. Since the reference configuration is unimportant to small strain measures, it is traditional to refer the Eulerian tensor in eq(2.3) to the material co-ordinates.

Example 2.4 Determine the Lagrangian, Eulerian and Cauchy finite strain tensors for the Eulerian description of motion $\mathbf{x} = \mathbf{x}(\mathbf{X}, t)$, given in example 2.3.

Using the matrices: **G**, **C** and **F** previously found, eq(2.27c) gives the Lagrangian strain tensor, **L** = ½(**G** − **I**):

$$\mathbf{L} = \frac{1}{2}\begin{bmatrix} 1 & 0 & (e^t - 1) \\ 0 & 1 & (e^t - e^{-t}) \\ (e^t - 1) & (e^t - e^{-t}) & (3e^{2t} - 2e^t + e^{-2t} - 1) \end{bmatrix} - \frac{1}{2}\begin{bmatrix} 1 & 0 & 0 \\ 0 & 1 & 0 \\ 0 & 0 & 1 \end{bmatrix} = \frac{1}{2}\begin{bmatrix} 0 & 0 & (e^t - 1) \\ 0 & 0 & (e^t - e^{-t}) \\ (e^t - 1) & (e^t - e^{-t}) & (3e^{2t} - 2e^t + e^{-2t} - 2) \end{bmatrix}$$

From eq(2.28c), the Eulerian tensor, **E** = ½(**I** − **C**), becomes

$$\mathbf{E} = \frac{1}{2}\begin{bmatrix} 1 & 0 & 0 \\ 0 & 1 & 0 \\ 0 & 0 & 1 \end{bmatrix} - \frac{1}{2}\begin{bmatrix} 1 & 0 & -(1 - e^{-t}) \\ 0 & 1 & -(1 - e^{-2t}) \\ -(1 - e^{-t}) & -(1 - e^{-2t}) & (2 - 2e^{-t} + e^{-4t}) \end{bmatrix} = \frac{1}{2}\begin{bmatrix} 0 & 0 & (1 - e^{-t}) \\ 0 & 0 & (1 - e^{-2t}) \\ (1 - e^{-t}) & (1 - e^{-2t}) & (2e^{-t} - e^{-4t} - 1) \end{bmatrix}$$

From eq(2.28d) the left Cauchy-Green deformation tensor is $\mathbf{B} = \mathbf{F}\mathbf{F}^{\mathrm{T}}$

$$\mathbf{B} = \begin{bmatrix} 1 & 0 & (e^t - 1) \\ 0 & 1 & (e^t - e^{-t}) \\ 0 & 0 & e^t \end{bmatrix}\begin{bmatrix} 1 & 0 & 0 \\ 0 & 1 & 0 \\ (e^t - 1) & (e^t - e^{-t}) & e^t \end{bmatrix} = \begin{bmatrix} 1 + (e^t - 1)^2 & (e^{-t} - 1)(e^t - e^{-t}) & e^t(e^t - 1) \\ (e^t - e^{-t})(e^t - 1) & 1 + (e^t - e^{-t})^2 & e^t(e^t - e^{-t}) \\ e^t(e^t - 1) & e^t(e^t - e^{-t}) & e^{2t} \end{bmatrix}$$

Each component of **L**, **E** and **B** can be evaluated for a given time, *t*.

2.4.3 Extension (or Stretch) Ratios

The infinitesimal definition of the extension ratio is $\lambda = |\mathrm{d}\mathbf{x}|/|\mathrm{d}\mathbf{X}|$. This may be expressed in terms of both the material and spacial co-ordinates. With unit vectors **m** and **n**, aligned with line elements PQ and pq respectively in Fig. 2.6, it follows that

$$\mathbf{m} = \mathrm{d}\mathbf{X} \,/\, |\mathrm{d}\mathbf{X}|, \quad \mathbf{n} = \mathrm{d}\mathbf{x} \,/\, |\mathrm{d}\mathbf{x}| \qquad (2.29\mathrm{a,b})$$

when, from eqs(2.23c) and (2.29a)

$$\lambda^2_{(\mathbf{m})} = |\mathrm{d}\mathbf{x}|^2 / |\mathrm{d}\mathbf{X}|^2 = G_{ij}\,(\,\mathrm{d}X_i / |\mathrm{d}\mathbf{X}|\,)\,(\,\mathrm{d}X_j / |\mathrm{d}\mathbf{X}|\,)$$

$$\lambda^2_{(\mathbf{m})} = G_{ij} m_i m_j \quad \text{or} \quad \lambda^2_{(\mathbf{m})} = \mathbf{m}^{\mathrm{T}}\mathbf{G}\mathbf{m} \qquad (2.30\mathrm{a})$$

where $\mathbf{m} = [m_1\ m_2\ m_3]^{\mathrm{T}}$ is a column matrix of the direction cosines m_i. From eq(2.27a)

$$|\mathrm{d}\mathbf{x}|^2 / |\mathrm{d}\mathbf{X}|^2 - 1 = 2L_{ij}\,(\,\mathrm{d}X_i / |\mathrm{d}\mathbf{X}|\,)\,(\,\mathrm{d}X_j / |\mathrm{d}\mathbf{X}|\,)$$

From eq(2.30a)

$$\lambda^2_{(\mathbf{m})} - 1 = 2L_{ij} m_i m_j \quad \text{or} \quad \lambda^2_{(\mathbf{m})} - 1 = \mathbf{m}^{\mathrm{T}}2\mathbf{L}\mathbf{m} \qquad (2.30\mathrm{b})$$

Combining eqs(2.30a,b) connects the material finite strain tensors **L** and **G** with the extension ratio

$$\lambda^2_{(\mathbf{m})} = \mathbf{m}^{\mathrm{T}}\, 2\mathbf{L}\mathbf{m} + 1 = \mathbf{m}^{\mathrm{T}}\, \mathbf{G}\mathbf{m} \tag{2.30c}$$

Also, from eqs(2.24c) and (2.29b)

$$1/\lambda^2_{(\mathbf{n})} = |\mathrm{d}\mathbf{X}|^2 / |\mathrm{d}\mathbf{x}|^2 = C_{ij}\, (\mathrm{d}x_i/|\mathrm{d}\mathbf{x}|)\, (\mathrm{d}x_j/|\mathrm{d}\mathbf{x}|)$$

$$1/\lambda^2_{(\mathbf{n})} = C_{ij} n_i n_j \quad \text{or} \quad 1/\lambda^2_{(\mathbf{n})} = \mathbf{n}^{\mathrm{T}}\mathbf{C}\mathbf{n} \tag{2.31a}$$

where $\mathbf{n} = [n_1\ n_2\ n_3]^{\mathrm{T}}$ is a column matrix of the direction cosines n_i. Then, from eq(2.28b)

$$1 - |\mathrm{d}\mathbf{X}|^2 / |\mathrm{d}\mathbf{x}|^2 = 2E_{ij}\, (\mathrm{d}x_i/|\mathrm{d}\mathbf{x}|)\, (\mathrm{d}x_j/|\mathrm{d}\mathbf{x}|)$$

From eq(2.29b)

$$1 - 1/\lambda^2_{(\mathbf{n})} = 2E_{ij} n_i n_j \quad \text{or} \quad 1 - 1/\lambda^2_{(\mathbf{n})} = \mathbf{n}^{\mathrm{T}}\, 2\mathbf{E}\mathbf{n} \tag{2.31b}$$

Combining eqs(2.31a,b) connects spacial finite strain tensors, **E** and **C**, to an extension ratio:

$$1/\lambda^2_{(\mathbf{n})} = 1 - \mathbf{n}^{\mathrm{T}}\, 2\mathbf{E}\mathbf{n} = \mathbf{n}^{\mathrm{T}}\mathbf{C}\mathbf{n} \tag{2.31c}$$

In general, $\lambda_{(\mathbf{m})} \neq \lambda_{(\mathbf{n})}$, from eqs(2.30c) and (2.31c), unless **n** remains in the same direction as **m**. One example where $\lambda_{(\mathbf{m})} = \lambda_{(\mathbf{n})}$ applies is to a stretch inclined at 45° to a pure shear deformation. This stretch occurs without rotation so that **m** and **n** remain parallel at 45°.

Referring to Fig. 2.7, the angular change between the two line elements PQ and PR is required when these elements both stretch and rotate to their deformed positions pq and pr. Figure 2.7 represents this deformation, within both material co-ordinates X_i and spacial co-ordinates x_i, in which the vectors $\mathrm{d}\mathbf{X}^{(1)}$ and $\mathrm{d}\mathbf{X}^{(2)}$ map into $\mathrm{d}\mathbf{x}^{(1)}$ and $\mathrm{d}\mathbf{x}^{(2)}$.

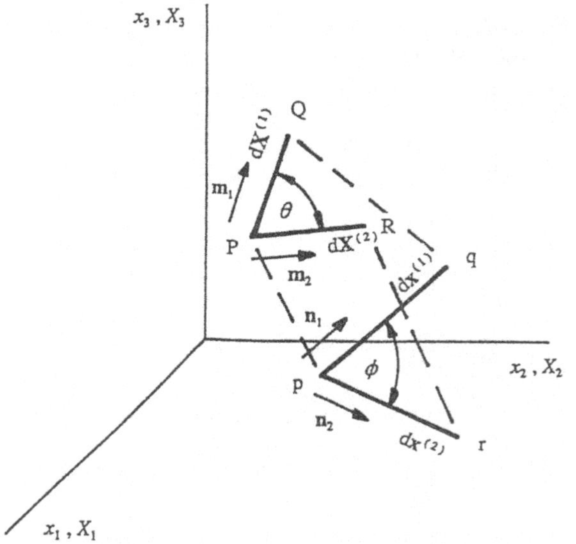

Figure 2.7 Shear deformation in material and spacial co-ordinates

From the dot product: $\mathbf{dx}^{(1)} \bullet \mathbf{dx}^{(2)} = |\mathbf{dx}^{(1)}|\,|\mathbf{dx}^{(2)}| \cos\varphi$:

$$\cos\varphi = \frac{\mathbf{dx}^{(1)}}{dx^{(1)}} \bullet \frac{\mathbf{dx}^{(2)}}{dx^{(2)}} = \frac{(\mathbf{dx}^{(1)})^{\mathrm{T}}}{dx^{(1)}} \frac{\mathbf{dx}^{(2)}}{dx^{(2)}} \tag{2.32a}$$

where $dx = |\mathbf{dx}|$. Substituting eqs(2.21b) and (2.23b) into eq(2.32a)

$$\cos\varphi = \frac{[\mathbf{F}\,\mathbf{dX}^{(1)}]^{\mathrm{T}}}{\sqrt{\mathbf{dX}^{(1)\mathrm{T}}\,\mathbf{G}\,\mathbf{dX}^{(1)}}} \times \frac{[\mathbf{F}\,\mathbf{dX}^{(2)}]}{\sqrt{\mathbf{dX}^{(2)\mathrm{T}}\,\mathbf{G}\,\mathbf{dX}^{(2)}}} \tag{2.32b}$$

The numerator and denominator of eq(2.32b) are divided by the product $dX^{(1)}\,dX^{(2)}$ to allow $\cos\varphi$ to be referred to unit vectors $\mathbf{m}_1 = \mathbf{dX}^{(1)}/dX^{(1)}$ and $\mathbf{m}_2 = \mathbf{dX}^{(2)}/dX^{(2)}$, aligned with the unstrained line elements PQ and PR:

$$\cos\varphi = \frac{(\mathbf{F}\,\mathbf{m}_1)^{\mathrm{T}}}{\sqrt{\mathbf{m}_1^{\mathrm{T}}\mathbf{G}\,\mathbf{m}_1}} \times \frac{(\mathbf{F}\,\mathbf{m}_2)}{\sqrt{\mathbf{m}_2^{\mathrm{T}}\mathbf{G}\,\mathbf{m}_2}} = \frac{\mathbf{m}_1^{\mathrm{T}}(\mathbf{F}^{\mathrm{T}}\mathbf{F})\mathbf{m}_2}{\sqrt{\mathbf{m}_1^{\mathrm{T}}\mathbf{G}\,\mathbf{m}_1}\,\sqrt{\mathbf{m}_2^{\mathrm{T}}\mathbf{G}\,\mathbf{m}_2}} \tag{2.32c}$$

Substituting eqs(2.23c) and (2.30a) into eq(2.32c) gives

$$\cos\varphi = \frac{\mathbf{m}_1^{\mathrm{T}}\,\mathbf{G}\,\mathbf{m}_2}{\lambda_{(\mathbf{m}_1)}\,\lambda_{(\mathbf{m}_2)}} \tag{2.32d}$$

The original included angle θ, between PQ and PR in Fig. 2.7, is found from the dot product in the spacial frame:

$$\mathbf{dX}^{(1)} \bullet \mathbf{dX}^{(2)} = |\mathbf{dX}^{(1)}|\,|\mathbf{dX}^{(2)}| \cos\theta$$

$$\cos\theta = \frac{\mathbf{dX}^{(1)}}{dX^{(1)}} \bullet \frac{\mathbf{dX}^{(2)}}{dX^{(2)}} = \frac{(\mathbf{dX}^{(1)})^{\mathrm{T}}}{dX^{(1)}} \frac{\mathbf{dX}^{(2)}}{dX^{(2)}} \tag{2.33a}$$

Substituting eqs(2.22b) and (2.24b) into eq(2.33a)

$$\cos\theta = \frac{[\mathbf{H}\,\mathbf{dx}^{(1)}]^{\mathrm{T}}}{\sqrt{\mathbf{dx}^{(1)\mathrm{T}}\,\mathbf{C}\,\mathbf{dx}^{(1)}}} \times \frac{[\mathbf{H}\,\mathbf{dx}^{(2)}]}{\sqrt{\mathbf{dx}^{(2)\mathrm{T}}\,\mathbf{C}\,\mathbf{dx}^{(2)}}} \tag{2.33b}$$

Dividing eq(2.33b) by the product $dx^{(1)}\,dx^{(2)}$ and introducing the unit vectors $\mathbf{n}_1 = \mathbf{dx}^{(1)}/dx^{(1)}$ and $\mathbf{n}_2 = \mathbf{dx}^{(2)}/dx^{(2)}$, aligned with strained line elements, gives

$$\cos\theta = \frac{(\mathbf{H}\,\mathbf{n}_1)^{\mathrm{T}}}{\sqrt{\mathbf{n}_1^{\mathrm{T}}\mathbf{C}\,\mathbf{n}_1}} \times \frac{(\mathbf{H}\,\mathbf{n}_2)}{\sqrt{\mathbf{n}_2^{\mathrm{T}}\mathbf{C}\,\mathbf{n}_2}} = \frac{\mathbf{n}_1^{\mathrm{T}}(\mathbf{H}^{\mathrm{T}}\mathbf{H})\,\mathbf{n}_2}{\sqrt{\mathbf{n}_1^{\mathrm{T}}\mathbf{C}\,\mathbf{n}_1}\,\sqrt{\mathbf{n}_2^{\mathrm{T}}\mathbf{C}\,\mathbf{n}_2}} \tag{2.33c}$$

Substituting eqs(2.24c) and (2.31a) into eq(2.33c) gives

$$\cos\theta = \lambda_{(\mathbf{n}_1)}\lambda_{(\mathbf{n}_2)}\left(\mathbf{n}_1^{\mathrm{T}}\,\mathbf{C}\,\mathbf{n}_2\right) \tag{2.33d}$$

which requires the spacial unit vectors $\mathbf{n}_1$ and $\mathbf{n}_2$ for the elements pq and pr.

Example 2.5 A unit square lies in the X_2–X_3 plane with one corner at the origin, as shown in Fig. 2.8a. Sketch the deformed shape when the Eulerian description of motion $\mathbf{x} = \mathbf{x}(\mathbf{X})$ at a given time is $x_1 = X_1 + BX_2X_3$, $x_2 = X_2 + BX_3^2$ and $x_3 = X_3 + BX_1^2$. Determine: (i) the extension ratio and the engineering strain for the diagonal joining the origin (0,0,0) to the corner Q(0,1,1) and (ii) the change in the right angle at the corner point Q when $B = 1$.

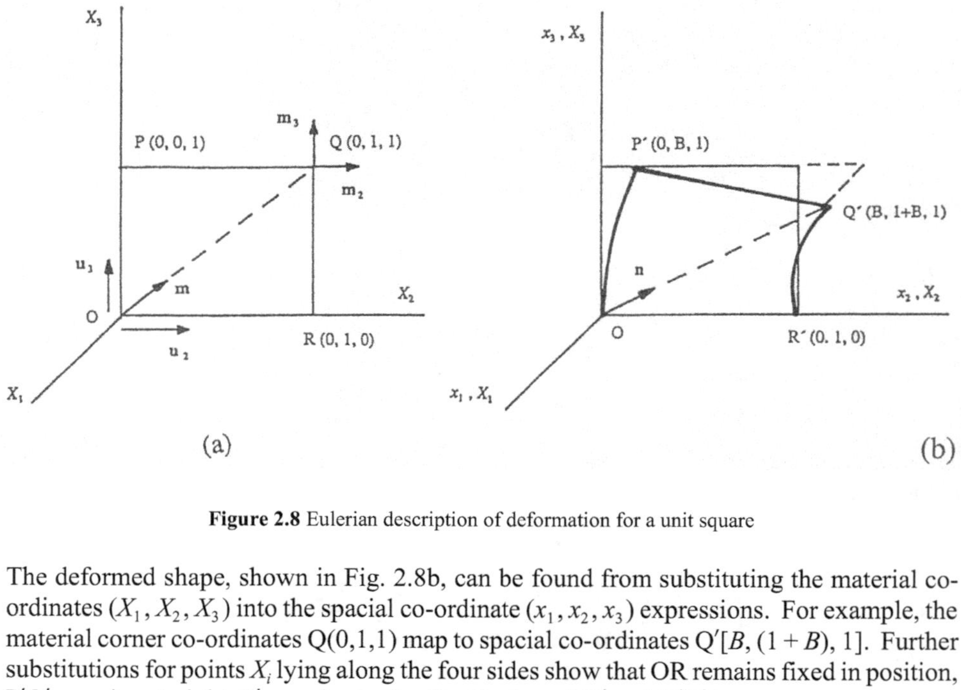

Figure 2.8 Eulerian description of deformation for a unit square

The deformed shape, shown in Fig. 2.8b, can be found from substituting the material co-ordinates (X_1, X_2, X_3) into the spacial co-ordinate (x_1, x_2, x_3) expressions. For example, the material corner co-ordinates Q(0,1,1) map to spacial co-ordinates Q′[B, $(1 + B)$, 1]. Further substitutions for points X_i lying along the four sides show that OR remains fixed in position, P′Q′ remains straight, P′ remains in the X_2, X_3 plane, OP′ and Q′R′ become curved.

Equation (2.30a) provides an extension ratio where the unit vector $\mathbf{m}$ is aligned with the unstrained diagonal OQ. That is: $\mathbf{m} = (1/\sqrt{2})\mathbf{u}_2 + (1/\sqrt{2})\mathbf{u}_3$, from which the column matrix is $\mathbf{m} = [0 \quad 1/\sqrt{2} \quad 1/\sqrt{2}]^{\mathrm{T}}$. Now, $\mathbf{F}$ is the matrix of derivates $\partial x_i/\partial X_j$. Hence, $\mathbf{G} = \mathbf{F}^{\mathrm{T}}\,\mathbf{F}$ is

$$\mathbf{G} = \begin{bmatrix} 1 & 0 & 2BX_1 \\ BX_3 & 1 & 0 \\ BX_2 & 2BX_3 & 1 \end{bmatrix}\begin{bmatrix} 1 & BX_3 & BX_2 \\ 0 & 1 & 2BX_3 \\ 2BX_1 & 0 & 1 \end{bmatrix} = \begin{bmatrix} [1 + (2BX_1)^2] & BX_3 & (BX_2 + 2BX_1) \\ BX_3 & [(BX_3)^2 + 1] & [(BX_2)(BX_3) + 2BX_3] \\ (BX_2 + 2BX_1) & [(BX_2)(BX_3) + 2BX_3] & [(BX_2)^2 + (2BX_3)^2 + 1] \end{bmatrix}$$

Substituting $\mathbf{m}$ and $\mathbf{G}$ into eq(2.30c), with $X_1 = 0$, $X_2 = X_3 = 1$, gives

$$\lambda^2_{(\mathbf{m})} = \begin{bmatrix} 0 & \frac{1}{\sqrt{2}} & \frac{1}{\sqrt{2}} \end{bmatrix} \begin{bmatrix} 1 & B & B \\ B & 1+B^2 & B(2+B) \\ B & B(2+B) & (1+5B^2) \end{bmatrix} \begin{bmatrix} 0 \\ \frac{1}{\sqrt{2}} \\ \frac{1}{\sqrt{2}} \end{bmatrix}$$

$$= \begin{bmatrix} \sqrt{2}B & \left(\frac{1}{\sqrt{2}} + \sqrt{2}B + \sqrt{2}B^2\right) & \left(\frac{1}{\sqrt{2}} + \sqrt{2}B + 3\sqrt{2}B^2\right) \end{bmatrix} \begin{bmatrix} 0 \\ \frac{1}{\sqrt{2}} \\ \frac{1}{\sqrt{2}} \end{bmatrix} = 1 + 2B + 4B^2$$

The engineering strain, is from eq(2.11b)

$$e_{(\mathrm{m})} = \lambda_{(\mathrm{m})} - 1 = \sqrt{(1 + 2B + 4B^2)} - 1$$

From the displacement of Q to Q′ shown, it follows that

$$\mathrm{OQ}' = \sqrt{[B^2 + (1 + B)^2 + 1^2]} = \sqrt{[2(1 + B + B^2)]}$$

and

$$\mathrm{OQ}'/\,\mathrm{OQ} = \sqrt{(1 + B + B^2)}$$

This is not the same as $\lambda_{(\mathbf{m})}$ because the deformation in OQ is non-linear. In fact, $\lambda_{(\mathbf{m})}$ would describe the ratio between the length of a curved trajectory OQ′ to its original length. Moreover, one should not expect the same stretch ratio $\lambda_{(\mathbf{n})}$ from eq(2.31c) since the directions of **n** and **m** differ.

Unit vector components, which describe the unstrained sides PQ and RQ, are respectively: $\mathbf{m}_2 = [0\ \ 1\ \ 0]^{\mathrm{T}}$ and $\mathbf{m}_3 = [0\ \ 0\ \ 1]^{\mathrm{T}}$. Thus, the numerator in eq(2.32d) becomes

$$\mathbf{m}_2^{\mathbf{T}}\mathbf{G}\,\mathbf{m}_3 = \begin{bmatrix} 0 & 1 & 0 \end{bmatrix} \begin{bmatrix} 1 & B & B \\ B & (1+B^2) & B(2+B) \\ B & B(2+B) & (1+5B^2) \end{bmatrix} \begin{bmatrix} 0 \\ 0 \\ 1 \end{bmatrix} = \begin{bmatrix} B & (1+B^2) & B(2+B) \end{bmatrix} \begin{bmatrix} 0 \\ 0 \\ 1 \end{bmatrix} = B(2+B)$$

In the denominator of eq(2.32d), the stretch ratios for PQ and RQ are, respectively

$$\lambda^2_{(\mathbf{m}_2)} = \mathbf{m}_2^{\mathrm{T}}\mathbf{G}\,\mathbf{m}_2 = \begin{bmatrix} 0 & 1 & 0 \end{bmatrix} \begin{bmatrix} 1 & B & B \\ B & (1+B^2) & B(2+B) \\ B & B(2+B) & (1+5B^2) \end{bmatrix} \begin{bmatrix} 0 \\ 1 \\ 0 \end{bmatrix} = \begin{bmatrix} B & (1+B^2) & B(2-B) \end{bmatrix} \begin{bmatrix} 0 \\ 1 \\ 0 \end{bmatrix} = 1 + B^2$$

$$\lambda^2_{(\mathbf{m}_3)} = \mathbf{m}_3^{\mathrm{T}}\mathbf{G}\,\mathbf{m}_3 = \begin{bmatrix} 0 & 0 & 1 \end{bmatrix} \begin{bmatrix} 1 & B & B \\ B & (1+B^2) & B(2+B) \\ B & B(2+B) & (1+5B^2) \end{bmatrix} \begin{bmatrix} 0 \\ 1 \\ 0 \end{bmatrix} = \begin{bmatrix} B & B(2+B) & (1+5B^2) \end{bmatrix} \begin{bmatrix} 0 \\ 0 \\ 1 \end{bmatrix} = 1 + 5B^2$$

Substituting into eq(2.32d):

$$\cos\varphi = \frac{B(2+B)}{\sqrt{(1+B^2)(1+5B^2)}}$$

which, for $B = 1$, gives $\cos\varphi = \sqrt{3}/2$. That is, $\varphi = 30°$ and so the angular change to the original right angle PQR is 60°.

2.5 Polar Decomposition

Equations (2.26a,b,c) show that an infinitesimal displacement gradient is composed of the sum of strain and rotation tensors. In the case of finite straining, a similar decomposition can be made to the deformation gradient **F**. Let an intermediate position vector $\mathbf{dx}'$ lie between the mapping of line element vector $\mathbf{dX}$ in its initial position to its current position $\mathbf{dx}$, in Fig. 2.6. The current position may be traced with two alternative local sequential mappings: (i) a stretch $\mathbf{dx}' = \mathbf{U}\,\mathbf{dX}$, followed by a rigid rotation $\mathbf{dx} = \mathbf{R}\,\mathbf{dx}'$, or, (ii) a rigid rotation $\mathbf{dx}' = \mathbf{R}\,\mathbf{dX}$, followed by a stretch $\mathbf{dx} = \mathbf{V}\mathbf{dx}'$. Thus, the final position $\mathbf{dx}$ is

$$\mathbf{dx} = (\mathbf{RU})\,\mathbf{dX} = (\mathbf{VR})\,\mathbf{dX} \tag{2.34a}$$

Comparing eqs(2.21b) and (2.34a), it follows that **F** is

$$\mathbf{F} = \mathbf{RU} = \mathbf{VR} \quad \text{or} \quad F_{ij} = \partial x_i / \partial X_j = R_{ik}U_{kj} = V_{ik}R_{kj} \tag{2.34b}$$

where **U** and **V** are the *right and left stretch tensors*, being positive, symmetric and possessing the same principal stretch values. Since $\det \mathbf{F} = \rho_o/\rho > 0$, **R** is an orthogonal rotation matrix, obeying

$$\mathbf{RR}^{\mathrm{T}} = \mathbf{R}^{\mathrm{T}}\mathbf{R} = \mathbf{I} \quad \text{or} \quad R_{ik}R_{jk} = R_{ki}R_{kj} = \delta_{ij} \tag{2.35a}$$

F, R, U and **V** will all depend upon position unless the deformation is homogenous, where $\mathbf{x} = \mathbf{FX}$ and the components of **F, R, U** and **V** become constants. Combining eqs(2.34b) and (2.35a) gives relationships between **U** and **V**

$$\mathbf{U} = \mathbf{R}^{\mathrm{T}}\mathbf{VR} \quad \text{and} \quad \mathbf{V} = \mathbf{RUR}^{\mathrm{T}} \tag{2.35b,c}$$

Now, from eqs(2.34b) and (2.23c), **G** and its components G_{ij} may be written in the two notations as

$$\begin{aligned} \mathbf{G} &= \mathbf{F}^{\mathrm{T}}\mathbf{F} = (\mathbf{RU})^{\mathrm{T}}(\mathbf{RU}) \\ &= \mathbf{U}^{\mathrm{T}}(\mathbf{R}^{\mathrm{T}}\mathbf{R})\mathbf{U} = \mathbf{U}^{\mathrm{T}}\mathbf{I}\,\mathbf{U} \end{aligned}$$

and as $\mathbf{U}^{\mathrm{T}} = \mathbf{U}$ and $\mathbf{IU} = \mathbf{U}$

$$\mathbf{G} = \mathbf{U}^2$$

$$\begin{aligned} G_{ij} &= F_{ik}{}^{\mathrm{T}}F_{kj} = F_{ki}F_{kj} \\ &= (R_{kp}U_{pi})(R_{kq}U_{qj}) \\ &= (R_{kp}R_{kq})U_{pi}U_{qj} \\ &= \delta_{pq}U_{pi}U_{qj} = U_{qi}U_{qj} \end{aligned} \tag{2.36}$$

Let $\mathbf{U}^*$ contain the principal stretches α_1, α_2 and α_3 of **U**. Substituting eq(2.36) into eq(2.30a) gives

$$\lambda^2_{(\mathbf{m})} = \mathbf{m}^{\mathrm{T}}(\mathbf{U}^{*2})\,\mathbf{m} \tag{2.37a}$$

Taking the principal stretch directions of $\mathbf{U}^*$ as co-ordinates, so that $\mathbf{m}_1$ is a unit vector aligned with the major principal stretch direction, eq(2.37a) gives

$$\lambda^2_{(\mathbf{m}_1)} = \begin{bmatrix} 1 & 0 & 0 \end{bmatrix} \begin{bmatrix} \alpha_1^2 & 0 & 0 \\ 0 & \alpha_2^2 & 0 \\ 0 & 0 & \alpha_3^2 \end{bmatrix} \begin{bmatrix} 1 \\ 0 \\ 0 \end{bmatrix} = \alpha_1^2 \tag{2.37b}$$

Thus, from eq(2.37b), $\lambda_{(\mathbf{m}_1)} = \alpha_1$. Similarly, when further unit vectors $\mathbf{m}_2 = [0\ \ 1\ \ 0]^{\mathrm{T}}$ and $\mathbf{m}_3 = [0\ \ 0\ \ 1]^{\mathrm{T}}$ are aligned with their respective principal directions: $\lambda_{(\mathbf{m}_2)} = \alpha_2$ and $\lambda_{(\mathbf{m}_3)} = \alpha_3$. This means that the stretch ratios for the principal directions are the principal values of **U**.

From eqs(2.34b) and (2.28d), **B** and its components B_{ij} are correspondingly:

$$\begin{aligned} \mathbf{B} &= \mathbf{F}\mathbf{F}^{\mathrm{T}} = (\mathbf{V}\mathbf{R})(\mathbf{V}\mathbf{R})^{\mathrm{T}} \\ &= \mathbf{V}(\mathbf{R}\mathbf{R}^{\mathrm{T}})\mathbf{V}^{\mathrm{T}} = \mathbf{V}\mathbf{I}\mathbf{V}^{\mathrm{T}} \end{aligned}$$

and as $\mathbf{V}^{\mathrm{T}} = \mathbf{V}$ and $\mathbf{V}\mathbf{I} = \mathbf{V}$:

$$\mathbf{B} = \mathbf{V}^2$$

$$\begin{aligned} B_{ij} &= F_{ik}F_{kj}^{\mathrm{T}} = F_{ik}F_{jk} \\ &= (V_{ip}R_{pk})(V_{jq}R_{qk}) \\ &= V_{ip}V_{jq}(R_{pk}R_{qk}) \\ &= V_{ip}V_{jq}\delta_{pq} = V_{iq}V_{jq} \end{aligned} \tag{2.38}$$

Let the principal directions of **U** be given by the three unit vectors $\mathbf{u}_i^*$ ($i = 1, 2, 3$), with their components in the X_i frame. They will satisfy an equation similar to eq(1.23b):

$$(\mathbf{U} - \alpha_i \mathbf{I})\,\mathbf{u}_i^* = 0 \tag{2.39a}$$

Since $\mathbf{R}^{\mathrm{T}}\mathbf{R}\mathbf{u}^* = \mathbf{I}\,\mathbf{u}^* = \mathbf{u}^*$, eq(2.39a) can be re-written as

$$\mathbf{R}\,(\mathbf{U} - \alpha_i\mathbf{I})(\mathbf{R}^{\mathrm{T}}\mathbf{R})\mathbf{u}_i^* = \left[\mathbf{R}\mathbf{U}\mathbf{R}^{\mathrm{T}} - \mathbf{R}\,(\alpha_i\mathbf{I})\mathbf{R}^{\mathrm{T}}\right](\mathbf{R}\mathbf{u}_i^*) = 0$$

Substituting from eq(2.35c)

$$(\mathbf{V} - \alpha_i\mathbf{I})(\mathbf{R}\mathbf{u}_i^*) = 0 \tag{2.39b}$$

When the principal directions of **V** are defined by the unit vectors $\mathbf{v}^*_i$ ($i = 1, 2, 3$) they will satisfy the equation

$$(\mathbf{V} - \beta_i\mathbf{I})\,\mathbf{v}_i^* = 0 \tag{2.39c}$$

Comparing eqs(2.39b,c) shows: (i) the principal values of **U** and **V** are the same since $\alpha_i = \beta_i$ and (ii) **R** connects the principal directions of **U** and **V** within: $\mathbf{v}_i^* = \mathbf{R}\mathbf{u}_i^*$ or $\mathbf{u}_i^* = \mathbf{R}^{\mathrm{T}}\,\mathbf{v}_i^*$. Equations (2.36) and (2.38) show that **U** and **V** provide an equivalent physical interpretation of **G** and **B** though **G** and **B** remain the more convenient measures of strain.

Principal stretch tensors $\mathbf{U}^*$ and $\mathbf{V}^*$ are calculated from the square roots of the Eigen values of **G** and **B** respectively. A 3 × 3 transformation matrix **M** of Eigen vector components (for **G** and **B**) will not, in general, equal the components of the rotation matrix **R** as the following example shows. **M** appears in tensor transformation laws for **U** and **V**, similar to eq(1.21b)

$$\mathbf{U}^* = \mathbf{MUM}^{\mathrm{T}} \text{ and } \mathbf{V}^* = \mathbf{MVM}^{\mathrm{T}} \tag{2.40a,b}$$

Reversing eqs(2.40a,b) gives (see eq 1.20b)

$$\mathbf{U} = \mathbf{M}^{\mathrm{T}}\mathbf{U}^*\mathbf{M} \text{ and } \mathbf{V} = \mathbf{M}^{\mathrm{T}}\mathbf{V}^*\mathbf{M} \tag{2.40c,d}$$

Alternatively, eqs(2.35b,c) allow the calculation of **U** from **V** and vice-versa, without having to find their principal values.

Example 2.6 Find **U**, **V** and **R** for a Euler description x_i ($i = 1, 2, 3$) of the following motion in Lagrange components X_j (j = 1, 2, 3): $x_1 = aX_1 + bX_2$, $x_2 = -aX_1 + bX_2$ and $x_3 = cX_3$ (where $a > b > c$). Show that the principal values of **U** and **V** are equal.

Having found the components $F_{ij} = \partial x_i / \partial X_j$, **G** is found from eq(2.23c) as

$$\mathbf{G} = \mathbf{F}^{\mathrm{T}}\mathbf{F} = \begin{bmatrix} a & -a & 0 \\ b & b & 0 \\ 0 & 0 & c \end{bmatrix} \begin{bmatrix} a & b & 0 \\ -a & b & 0 \\ 0 & 0 & c \end{bmatrix} = \begin{bmatrix} 2a^2 & 0 & 0 \\ 0 & 2b^2 & 0 \\ 0 & 0 & c^2 \end{bmatrix}$$

from which **U** is obtained:

$$\mathbf{U} = \sqrt{\mathbf{G}} = \begin{bmatrix} \sqrt{2}a & 0 & 0 \\ 0 & \sqrt{2}b & 0 \\ 0 & 0 & c \end{bmatrix} = \mathbf{U}^*$$

This result reveals that the directions of the principal, right stretches are aligned with the co-ordinates X_j. The rotation **R** then follows from eq(2.34b) as

$$\mathbf{R} = \mathbf{F}\,\mathbf{U}^{-1} = \begin{bmatrix} a & b & 0 \\ -a & b & 0 \\ 0 & 0 & c \end{bmatrix} \begin{bmatrix} \dfrac{1}{\sqrt{2}a} & 0 & 0 \\ 0 & \dfrac{1}{\sqrt{2}b} & 0 \\ 0 & 0 & 1/c \end{bmatrix} = \begin{bmatrix} \dfrac{1}{\sqrt{2}} & \dfrac{1}{\sqrt{2}} & 0 \\ -\dfrac{1}{\sqrt{2}} & \dfrac{1}{\sqrt{2}} & 0 \\ 0 & 0 & 1 \end{bmatrix}$$

These components are direction cosines corresponding to rotating X_1 and X_2 by 45° about the fixed X_3 axis. **B** is found from eq(2.28d) as

$$\mathbf{B} = \begin{bmatrix} a & b & 0 \\ -a & b & 0 \\ 0 & 0 & c \end{bmatrix} \begin{bmatrix} a & -a & 0 \\ b & b & 0 \\ 0 & 0 & c \end{bmatrix} = \begin{bmatrix} (a^2 + b^2) & (-a^2 + b^2) & 0 \\ (-a^2 + b^2) & (a^2 + b^2) & 0 \\ 0 & 0 & c^2 \end{bmatrix}$$

The stretch tensor **V** is more conveniently found from $\mathbf{V}^* = \sqrt{\mathbf{B}^*}$ when $\mathbf{B}^*$ is referred to its principal axes. The principal values of $\mathbf{B}^*$ (i.e. Eigen values) are the three real and positive roots β_i ($i = 1, 2, 3$) of the characteristic equation for **B**. They follow from the expansion to the determinant $(\mathbf{B} - \beta_i \mathbf{I}) = 0$:

$$\begin{vmatrix} (a^2+b^2)-\beta & (-a^2+b^2) & 0 \\ (-a^2+b^2) & (a^2+b^2)-\beta & 0 \\ 0 & 0 & c-\beta \end{vmatrix} = 0$$

which gives $\beta_1 = 2a^2$, $\beta_2 = 2b^2$ and $\beta_3 = c^2$. Hence, for principal axes:

$$\mathbf{B}^* = \begin{bmatrix} 2a^2 & 0 & 0 \\ 0 & 2b^2 & 0 \\ 0 & 0 & c^2 \end{bmatrix}, \quad \mathbf{V}^* = \sqrt{\mathbf{B}} = \begin{bmatrix} \sqrt{2}a & 0 & 0 \\ 0 & \sqrt{2}b & 0 \\ 0 & 0 & c \end{bmatrix}$$

showing that the principal values of **U** and **V** (i.e. the components of $\mathbf{U}^*$ and $\mathbf{V}^*$) are equal.

The Eigen vectors $\mathbf{v}_i^*$ ($i = 1, 2, 3$), which define orthogonal principal axes of $\mathbf{V}^*$, are found from the solution to:

$$(\mathbf{B} - \beta_i \mathbf{I})\,\mathbf{v}_i^* = 0 \qquad \text{(i)}$$

For $i = 1$, let $\mathbf{v}_1^* = l_1\mathbf{u}_1 + m_1\mathbf{u}_2 + n_1\mathbf{u}_3$ where $\mathbf{u}_1$, $\mathbf{u}_2$ and $\mathbf{u}_3$ are unit vectors aligned with the co-ordinates X_i (see Fig. 2.6) and l_1, m_1 and n_1 are the intercepts (direction cosines) with X_i. In eq(i) write vector components $\mathbf{v}_1^* = [l_1\ \ m_1\ \ n_1]^{\mathrm{T}}$ and substitute for matrix **B** to give:

$$(a^2 + b^2 - \beta_1)\, l_1 + (-a^2 + b^2)\, m_1 = 0 \qquad \text{(ii)}$$

$$(-a^2 + b^2)\, l_1 + (a^2 + b^2 - \beta_1)\, m_1 = 0 \qquad \text{(iii)}$$

$$(c^2 - \beta_1)\, n_1 = 0 \qquad \text{(iv)}$$

Substituting $\beta_1 = 2a^2$ in eqs (ii)–(iv) gives: $l_1 = 1/\sqrt{2}$, $m_1 = -1/\sqrt{2}$ and $n_1 = 0$ so that:

$$\mathbf{v}_1^* = (1/\sqrt{2})\mathbf{u}_1 - (1/\sqrt{2})\mathbf{u}_2$$

Similarly, for $i = 2$ and 3, the remaining Eigen vectors are:

$$\mathbf{v}_2^* = (1/\sqrt{2})\mathbf{u}_1 + (1/\sqrt{2})\mathbf{u}_2 \ \text{ and } \ \mathbf{v}_3^* = \mathbf{u}_3$$

in which $\mathbf{v}_1^* \bullet \mathbf{v}_2^* = \mathbf{v}_2^* \bullet \mathbf{v}_3^* = \mathbf{v}_3^* \bullet \mathbf{v}_1^* = 0$ confirms orthogonality in these vectors. Their components define the transformation matrix **M** in eq(2.40a–d) as follows:

$$\mathbf{V} = \begin{bmatrix} 1/\sqrt{2} & 1/\sqrt{2} & 0 \\ -1/\sqrt{2} & 1/\sqrt{2} & 0 \\ 0 & 0 & 1 \end{bmatrix} \begin{bmatrix} \sqrt{2}a & 0 & 0 \\ 0 & \sqrt{2}b & 0 \\ 0 & 0 & c \end{bmatrix} \begin{bmatrix} 1/\sqrt{2} & -1/\sqrt{2} & 0 \\ 1/\sqrt{2} & 1/\sqrt{2} & 0 \\ 0 & 0 & 1 \end{bmatrix} = \begin{bmatrix} (a+b)/\sqrt{2} & -(a-b)/\sqrt{2} & 0 \\ (a-b)/\sqrt{2} & (a+b)/\sqrt{2} & 0 \\ 0 & 0 & c \end{bmatrix}$$

Finally, this result may be checked from **F** = **VR**

$$\mathbf{F} = \begin{bmatrix} (a+b)/\sqrt{2} & -(a-b)/\sqrt{2} & 0 \\ (a-b)/\sqrt{2} & (a+b)/\sqrt{2} & 0 \\ 0 & 0 & c \end{bmatrix} \begin{bmatrix} 1/\sqrt{2} & 1/\sqrt{2} & 0 \\ -1/\sqrt{2} & 1/\sqrt{2} & 0 \\ 0 & 0 & 1 \end{bmatrix} = \begin{bmatrix} a & b & 0 \\ -a & b & 0 \\ 0 & 0 & c \end{bmatrix}$$

2.6 Alternative Strain Definitions

It is seen that there is no single fundamental definition of strain. Indeed, the expressions for the natural, Lagrangian, Eulerian strains and the extension ratio will all differ, even for a simple uniaxial, finite deformation. Thus, the flow behaviour of a material will depend upon the chosen definition of strain. The Lagrangian finite strain tensor, **L** in eq(2.27b,c), is more commonly used for referring finite deformation in a solid to its original shape.

For infinitesimal strains, the differential products in eqs(2.27b) and (2.28b) may be ignored. The Eulerian and Langrangian finite strain tensors **E** and **L**, will then reduce to eq(2.3), implying that there is negligible displacement between the material and spacial co-ordinates. Seth [4] related the three uniaxial strains to the extension ratio within a single formula. Given initial and final lengths l_o and l respectively, these strains are expressed from eqs(2.11b), (2.30c) and (2.31c) as

$$\varepsilon^{(a)} = (1/a)[1 - (l_o/l)^a] = (1/a)[1 - (1/\lambda)^a]$$

where $a = -1$, $a = -2$ and $a = +2$ according to the engineering definition and those of Lagrange and Euler respectively. Other values of a may lead to simpler constitutive relations, e.g. when a is chosen to provide linear stress–strain behaviour, but the corresponding stress definition becomes uncertain and the tensor nature of strain is lost.

The reader will find further discussion of strain tensors and their transformation properties in books on continuum mechanics [5–9].

References

1. Ludwik P. *Elemente der Technologischen Mechanik*, 1909, Springer, Berlin.
2. Hencky H. "Uber die form des elastizitatsgesetzes bei ideal elastischen stoffen", *Z.T.P.* 1928, **9**, 214-223.
3. Green A. E. and Adkins J. E. *Large Elastic Deformations*, 1960, Oxford University Press, London.
4. Seth B. R. "Generalised strain measure with applications to physical problems", Proc: *Second Order Effects*, 1964, IUTAM, Pergamon Press.
5. Farrashkhalaut M. and Miles J. P. *Tensor Methods for Engineers and Scientists*, 1990, Ellis-Horwood.
6. Lai W. M., Rubin D. and Krempl E. *Introduction to Continuum Mechanics*, Pergamon Engineering Series, Vol.17, 1978, Pergamon Press.
7. Mase G. *Continuum Mechanics*, 1970, Schaum Outline Series, McGraw-Hill.
8. Mase G. E. and Mase G. T. *Continuum Mechanics for Engineers*, 1992, CRC Press.
9. Spencer A. J. M. *Continuum Mechanics*, 1980, Longman.

Exercises

2.1 A displacement vector **u** is described by its three components u_i in spacial co-ordinates x_j as

$$u_1 = (k_1 x_1 - k_3 x_3)^2, \quad u_2 = (k_2 x_2 + k_3 x_3)^2 \text{ and } \quad u_3 = (k_1 x_1)(k_2 x_2)$$

Determine components of infinitesimal strain and rotation matrices, ε_{ij} and ω_{ij}, for point P(0, 1, 2), when $k_1 = k_2 = 1 \times 10^{-6}$ and $k_3 = 2 \times 10^{-6}$. What then is the change in the right angle between a pair of perpendicular lines passing through P with direction cosines: (1/√3, 1/√3, −1/√3) and (1/√14, 2/√14, 3/√14) with respect to x_1, x_2 and x_3?
See solution

2.2 The following infinitesimal, micro-strain components: $\varepsilon_{11} = 600$, $\varepsilon_{22} = \varepsilon_{33} = 0$, $\varepsilon_{12} = \varepsilon_{21} = 200$, $\varepsilon_{23} = \varepsilon_{32} = 400$ and $\varepsilon_{13} = \varepsilon_{31} = 200$ (all multiplied by 10^{-6}) define a strain tensor. Determine the magnitude and direction of the principal strains, the maximum and octahedral shear strains. What is the state of strain for a plane with normal direction cosines $\alpha = 0.53$, $\beta = 0.35$ and $\gamma = 0.77$?
Ans: 800, 200, − 400, 600, 9.8, 622 ($\times 10^{-6}$)

2.3 The nine components of a strain tensor (all multiplied by 10^{-6}) are: $\varepsilon_{11} = 100$, $\varepsilon_{22} = \varepsilon_{23} = 0$, $\varepsilon_{33} = -100$, $\varepsilon_{12} = \varepsilon_{21} = 500$ and $\varepsilon_{13} = \varepsilon_{31} = -500$ in x_1, x_2 and x_3 co-ordinates. Transform these to x_1', x_2' and x_3' co-ordinates given that the direction cosines for the x_1'–direction are: $l_{11} = 1/\surd 14$, $l_{12} = 2/\surd 14$, $l_{13} = 3/\surd 14$ and those for the x_2'–direction are: $l_{21} = 1/\surd 3$, $l_{22} = 1/\surd 3$, $l_{23} = -1/\surd 3$.
See solution: $\varepsilon_{11}' = -128.6$, $\varepsilon_{22}' = 666.7$, $\varepsilon_{33}' = 538.1$, $\varepsilon_{12}' = 138.9$, $\varepsilon_{13}' = 198$ and $\varepsilon_{23}' = -276.2$

2.4 Under plane-strain deformation, a line element increases in length by 17 % in the X_1–direction and by 22 % in the X_2–direction. What is the percentage increase in length and the rotation of a line element originally at 45° to the reference configuration X_1 and X_2?
Ans: 27 %, 2.4° to X_1

2.5 Given the three components of a spacial co-ordinate function $\mathbf{x} = \mathbf{x}(\mathbf{X}, t)$ are: $x_1 = A_1(X_1 + BX_2)$, $x_2 = A_2 X_2$ and $x_3 = A_3 X_3$, where A and B are constants at time t, determine: (a) the components of the deformation gradients **F** and $\mathbf{F}^{-1}$, (b) the left and right Cauchy-Green deformation tensors **G**, $\mathbf{G}^{-1}$, **B** and $\mathbf{B}^{-1}$ and (c) the Lagrangian and Eulerian strain tensors **L** and **E,** respectively.

2.6 The spacial co-ordinates x_i, in exercise 2.5, describe homogenous deformation at a given time t. Find the new direction and the stretch of a line element originally equally inclined to a reference configuration X_i ($i = 1$, 2 and 3) at time $t = 0$.

2.7 Given that the *spacial* components of motion, $\mathbf{x} = \mathbf{x}(\mathbf{X}, t)$ are:

$$x_1 = X_1 + X_2(\mathrm{e}^t - K),$$
$$x_2 = X_1(\mathrm{e}^{-t} - K) + X_2$$
$$x_3 = X_3,$$

show that the *material* components of motion $\mathbf{X} = \mathbf{X}(\mathbf{x}, t)$ are

$$X_1 = [-x_1 + x_2(\mathrm{e}^t - K)] / [K(K - \mathrm{e}^t - \mathrm{e}^{-t})]$$
$$X_2 = [x_1(\mathrm{e}^{-t} - K) - x_2] / [K(K - \mathrm{e}^t - \mathrm{e}^{-t})]$$
$$X_3 = x_3$$

See solution

2.8 The spacial position vector **x** has components: $x_1 = X_1$, $x_2 = X_2 + KX_3$ and $x_3 = X_3 + KX_2$, where K is a constant. Determine the components of the displacement vector $\mathbf{u} = \mathbf{x} - \mathbf{X}$ (see Fig. 2.6) with **x** and **X** referred to spacial and material co-ordinates respectively.
Ans: $u_1 = 0$, $u_2 = KX_3 = k(x_3 - Kx_2)/(1 - K^2)$ and $u_3 = KX_2 = k(x_2 - Kx_3)/(1 - K^2)$

2.9 Show that the components of **G** and **L** for x_i in exercise 2.8 are given as:

$$G_{ij} = \begin{bmatrix} 1 & 0 & 0 \\ 0 & 1+K^2 & 2K \\ 0 & 2K & 1+K^2 \end{bmatrix}, \quad L_{ij} = \begin{bmatrix} 0 & 0 & 0 \\ 0 & K^2 & 2K \\ 0 & 2K & K^2 \end{bmatrix}$$

Determine the magnitude and direction of the principal Lagrange strains for $K = 1$.

2.10 Given that the material components of the displacement vector **U** (see Fig. 2.6) are: $U_1 = X_1 X_3^2$, $U_2 = X_1^2 X_2$ and $U_3 = X_2^2 X_3$, determine the components of the material displacement gradient $\partial U_i / \partial X_j$ and the material deformation gradient $F_{ij} = \partial x_i / \partial X_j$. Hence, confirm the derivative $F_{ij} = \partial U_i / \partial X_j + \delta_{ij}$.
See solution

2.11 Calculate the stretch ratio for a line element vector aligned with the X_1–axis and another with direction cosines (0, 1/√2, 1/√2) w.r.t. X_i, following a shear deformation whose components x_i, are given in exercise 2.8. *Hint*: use eq(2.13c) with the corresponding right Cauchy-Green components G_{ij} from exercise 2.9.
Ans: 1, $(1 + K)$

2.12 Show from eq(2.28d) that the extension ratio $1 + K$, found in exercise 2.11, applies when inverting the components from x_i in $\mathbf{x} = \mathbf{x}(\mathbf{X})$ to X_i in $\mathbf{X} = \mathbf{X}(\mathbf{x})$ upon forming the inverse of the left Cauchy-Green components B_{ij}. Explain why this is.

2.13 Determine the extension ratios $\lambda_{\mathbf{X}}$ and $\lambda_{\mathbf{x}}$ from eqs(2.30a) and (2.31a) for a line element originally aligned with the X_2–axis, following a motion with x_i components given in exercise 2.8. Explain why these differ.
See solution

2.14 Determine, for the unit square shown in Fig. 2.8a, the extension ratio for the diagonal PR and the change in the angle OPQ, when it is subjected to the same Lagrangian description $\mathbf{x} = \mathbf{x}(\mathbf{X})$ of motion.
Hint: See text example 2.5.

2.15 Determine the deformation gradient **F** for a Lagrangian motion: $x_1 = X_1 + 2X_3$, $x_2 = X_2 - 2X_3$ and $x_3 = -2X_1 + 2X_2 + X_3$. Find **R**, **U** and **V** by polar decomposition and show that the principal values of **U** and **V** are equal.

CHAPTER 3

YIELD CRITERIA

3.1 Introduction

Experiment has shown that an assumption of initial isotropy is usually a good approximation when quoting a yield stress value for a polycrystalline material. One exception is where the yield stress becomes direction-dependent following a history of severe deformation processing, as with the cold-rolling of sheet metals. There are two approaches, microscopic and macroscopic, used in the analysis of this form of anisotropy. The first approach examines the manner in which various slip systems control anisotropy within a given structure. The second approach, which will be adopted here, quantifies the effect that anisotropy has upon the initial yield surface. Typically, the yield surface can display different initial yield stresses in tension and compression, so that it appears asymmetric relative to its stress co-ordinate axes. As with an annealed, isotropic material, it is convenient to express the inherent, or residual anisotropy, within the structure of a worked material, with an appropriate yield function. Suitable functions are given for conditions of isotropy, transverse isotropy and orthotropy. A second form of anisotropy, which accentuates yield surface asymmetry, is induced by subsequent straining of a hardening material beyond its initial yield point. This is a deformation-induced anisotropy, of which the *Bauschinger effect* is one manifestation. To describe this, the initial yield function requires a modification using a *hardening rule* (see chapter 10, § 10.3).

Similar functions are often employed to describe the initial *yield surface* in metals and the *failure surface* for brittle non-metals, despite their different responses to loading. In a non-hardening metal, the yield and fracture surfaces are coincident and a single function applies. The term *elastic, perfectly-plastic* is used to describe this. The material response is elastic within the surface's interior region. Plastic flow occurs under a constant yield stress for stress states lying upon the yield surface. In contrast, a purely brittle material remains elastic to the point of fracture, where the critical stress combination lies on its failure surface. However, as both yield and failure surfaces bound an elastic interior they appear similar mathematically.

A *flow rule* describes the manner in which plastic strain depends upon a given combination of stress. It is shown in chapter 4 that there are two types of flow rule for elastic-perfect plasticity. These rules differ between their incremental and total strain formulations: the former with and the latter without an association to the yield surface. The incremental strain approach will be outlined here for when hardening accompanies plasticity.

3.2 Isotropic Yielding Theory

In Fig. 1.1b it is seen that a ductile metallic material will begin to deform and harden plastically under a uniaxial stress when the yield stress is exceeded. In practice, the stress state is biaxial or triaxial and the question arises as to what magnitudes of combined stress will cause the onset of yielding? Those who first attempted to answer this question sought a suitable criterion based upon stress, strain or strain energy for the complex stress system that could be related to the corresponding quantity at the uniaxial yield point. For an isotropic material the critical value of the chosen parameter at yield is independent of the orientation of the stress system. All criteria may then be related to a constant uniaxial yield stress Y, which is most conveniently measured in a simple tension test (see Fig. 3.1a). Isotropic yield criteria are normally expressed in principal stress form. This form refers to the stress system σ_1, σ_2 and σ_3 in Fig. 3.1b, where, conventionally, $\sigma_1 > \sigma_2 > \sigma_3$. It will be seen that stress transformation equations, given in chapter 1, enable these yield criteria to be expressed in non principal form. Therein, any two or three-dimensional combination of applied direct and shear stress is admitted.

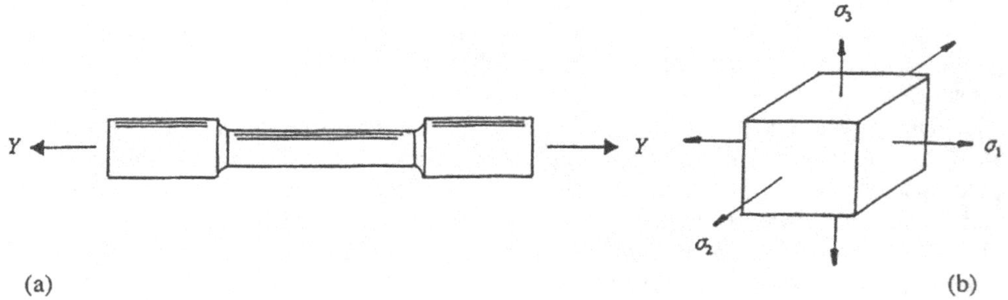

Figure 3.1 Uniaxial and triaxial principal stress states

A number of yield criteria have been proposed over the past two centuries [1], though it is now accepted that the criteria commonly attributed to von Mises [2] and Tresca [3] are most representative of initial yielding in isotropic, metallic materials. It is instructive to consider the derivation and verification of these two criteria using appropriate experimental data available in the literature.

3.2.1 Maximum Shear Stress

Attributed jointly to Tresca, Coulomb and Guest, the maximum shear stress theory assumes that yielding, under the principal stress system in Fig. 3.1b, begins when the maximum shear stress reaches a critical value. The latter is taken as the maximum shear stress k at the point of yielding under simple tension or compression. That is, $k = Y/2$, which acts along planes at 45° to the tensile stress axis, shown in Fig. 3.2a.

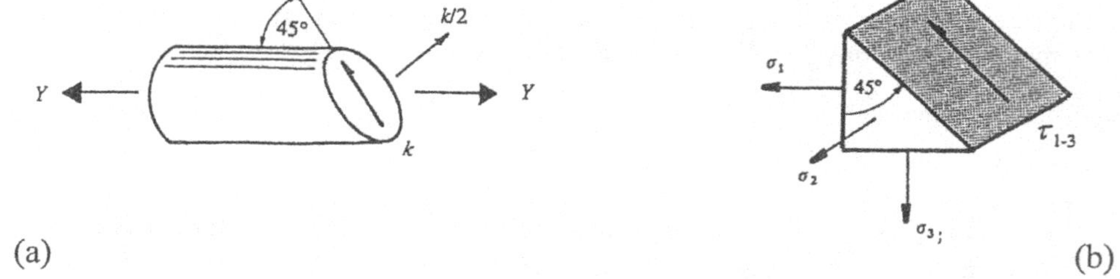

Figure 3.2 Maximum shear stress under uniaxial and triaxial stress states

Under a principal, triaxial stress state (Fig. 3.1b), the greatest shear stress is, from eq(1.33b), $\tau_{1\text{-}3} = \frac{1}{2}(\sigma_1 - \sigma_3)$. This acts along the plane inclined at 45° to the 1 and 3–directions, as shown in Fig. 3.2b. Equating $\tau_{1\text{-}3}$ to the shear stress $k = Y/2$, for the uniaxial case, leads to the Tresca yield criterion:

$$\sigma_1 - \sigma_3 = Y \tag{3.1a}$$

Numerical values of σ_1 and σ_3 must be substituted into the left-hand side of eq(3.1a) with signs denoting tension or compression, e.g. if $\sigma_1 = 30$, $\sigma_2 = -15$ and $\sigma_3 = -20$ MPa, then the left side value becomes: $30 - (-20) = 50$ MPa, which shows that the intermediate stress value is irrelevant in Tresca's criterion. Often, eq(3.1a) appears in a written form

Greatest Principal Stress − *Least Principal Stress* = *Tensile Yield Stress* (3.1b)

In the case of plane stress, where σ_x, σ_y and τ_{xy} are non-zero, the principal stresses have been derived in § 1.4:

$$\sigma_1, \sigma_2 = \tfrac{1}{2}(\sigma_x + \sigma_y) \pm \tfrac{1}{2}\sqrt{[(\sigma_x - \sigma_y)^2 + 4\tau_{xy}^2]} \text{ and } \sigma_3 = 0 \tag{3.2}$$

where σ_1 is tensile (greatest positive value) and σ_2 is compressive (least negative value). Setting $\sigma_y = 0$ in eq(3.2) and substituting into eq(3.1b) simplifies Tresca's criterion:

$$\sigma_x^2 + 4\tau_{xy}^2 = Y^2 \tag{3.3}$$

which describes the equation of a two-dimensional elliptical yield locus in axes of σ_x and τ_{xy}.

3.2.2 Shear Strain Energy

This yield criterion is used most commonly. It is named after Richard von Mises but the criterion is also associated with others: Maxwell, Huber and Hencky. Huber proposed that the total strain energy was composed of a dilation (volumetric) component and a distortion (shear) component. The former depends upon the mean, or hydrostatic, component of the applied stress and the latter upon the remaining reduced, or deviatoric, component of stress, as shown in Figs 3.3a–c.

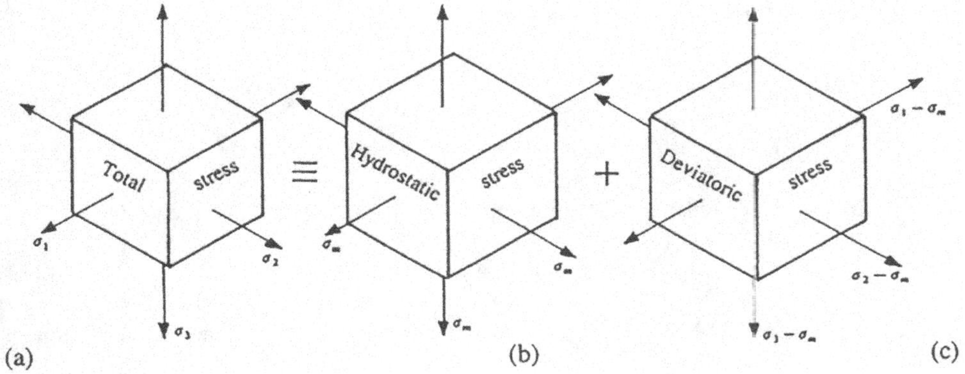

Figure 3.3 Components of a triaxial stress state

Maxwell assumed that hydrostatic stress played no part upon yielding. Therein, it was proposed that yielding occurred when the shear strain energy component of the total energy reached a critical value. This value is taken to be the shear strain energy at the tensile yield point. Alternative approaches, adopted by von Mises and Hencky, which lead to the same yield criterion, are outlined below. It will be shown in the following section that there is considerable experimental evidence supporting a von Mises criterion of yielding for ductile isotropic metals.

(a) Shear Strain Energy U_s

The total energy density for Fig. 3.3a is given by

$$U = \int_\varepsilon \sigma \, d\varepsilon = \int_\varepsilon (\sigma_1 \, d\varepsilon_1 + \sigma_2 \, d\varepsilon_2 + \sigma_3 \, d\varepsilon_3)$$

Substituting from the elastic constitutive relations:

$$\begin{aligned} \varepsilon_1 &= (1/E)[\sigma_1 - v(\sigma_2 + \sigma_3)] \\ \varepsilon_2 &= (1/E)[\sigma_2 - v(\sigma_1 + \sigma_3)] \\ \varepsilon_3 &= (1/E)[\sigma_3 - v(\sigma_1 + \sigma_2)] \end{aligned}$$

where E is Young's modulus and v is Poisson's ratio. This leads to

$$U = (\sigma_1^2 + \sigma_2^2 + \sigma_3^2)/(2E) - (v/E)(\sigma_1\sigma_2 + \sigma_2\sigma_3 + \sigma_1\sigma_3) \tag{3.4}$$

The volumetric strain energy density arises from the hydrostatic component of stress (see Fig. 3.3b):

$$U_v = \int_\varepsilon \sigma \, d\varepsilon = \int_\varepsilon \sigma_m \, d\varepsilon + \sigma_m \, d\varepsilon + \sigma_m \, d\varepsilon$$

Substituting $\varepsilon = \sigma_m/(3K)$ for the linear strain along an edge, where K is the bulk modulus:

$$\begin{aligned} U_v &= \sigma_m^2/(2K) = 3(1 - 2v)\sigma_m^2/(2E) \\ &= (1 - 2v)(\sigma_1 + \sigma_2 + \sigma_3)^2/(6E) \end{aligned} \tag{3.5}$$

where $\sigma_m = \sigma_{kk}/3$ with a relationship between the three elastic constants: $E = 3K(1 - 2v)$. Subtracting eq(3.5) from eq(3.4) leads to the shear strain energy U_S associated with the reduced (or deviatoric) stress shown in Fig. 3.3c:

$$\begin{aligned} U_S &= U - U_v \\ &= (\sigma_1^2 + \sigma_2^2 + \sigma_3^2)/(2E) - v(\sigma_1\sigma_2 + \sigma_2\sigma_3 + \sigma_1\sigma_3)/E - (1 - 2v)(\sigma_1 + \sigma_2 + \sigma_3)^2/(6E) \\ &= (1 + v)[(\sigma_1 - \sigma_2)^2 + (\sigma_2 - \sigma_3)^2 + (\sigma_1 - \sigma_3)^2]/(6E) \end{aligned} \tag{3.6a}$$

The value of U_S at the tensile yield point applies to: $\sigma_1 = Y$, $\sigma_2 = \sigma_3 = 0$ in eq(3.6a)

$$U_S = (1 + v)Y^2/(3E) \tag{3.6b}$$

Equating (3.6a) and (3.6b) provides the principal stress form of the von Mises criterion:

$$(\sigma_1 - \sigma_2)^2 + (\sigma_2 - \sigma_3)^2 + (\sigma_3 - \sigma_1)^2 = 2Y^2 \tag{3.7a}$$

If one principal stress, say σ_3, is zero the biaxial stress form of eq(3.7a) becomes

$$\sigma_1^{\,2} - \sigma_1\sigma_2 + \sigma_2^{\,2} = Y^2 \tag{3.7b}$$

Substitutions from eqs(3.2) for σ_1 (tensile) and σ_2 (compressive) will provide a general plane stress form in terms of σ_x, σ_y and τ_{xy}. Setting $\sigma_y = 0$ within this reveals a common plane stress form of the Mises criterion:

$$\sigma_x^{\,2} + 3\tau_{xy}^{\,2} = Y^2 \tag{3.8}$$

Subscripts x and y are often omitted from eq(3.8) to account for yielding under a given combination of direct and shear stress, set within either Cartesian or polar co-ordinates.

(b) Octahedral Shear Stress τ_o

This approach proposes that yielding, under the triaxial stress system in Fig. 3.1b, commences when τ_o in eq(1.34b) reaches its critical value at the tensile yield point, for which $\sigma_1 = Y$ and $\sigma_2 = \sigma_3 = 0$. This gives

$$\tfrac{1}{3}\surd[(\sigma_1 - \sigma_2)^2 + (\sigma_2 - \sigma_3)^2 + (\sigma_3 - \sigma_1)^2] = \tfrac{1}{3}\surd\,(2Y^2)$$

which again leads to eq(3.7a). This is because the normal stress for the octahedral planes, i.e. σ_o in eq(1.34a), is numerically equal to the mean stress $\sigma_m = ⅓\sigma_{kk}$, which produces dilatation only. Thus, the distortion which occurs under τ_o, leads to yielding under a critical combination of principal stress differences. Clearly, energy considerations are not required when formulating the yield criterion in this equivalent manner.

3.2.3 Deviatoric Stress Invariants

Because the mean or hydrostatic stress plays no part in yielding, it follows that the yield criterion should be a function of the deviatoric, or reduced, stress in Fig. 3.3c. This shows that the deviatoric stress tensor σ_{ij}' is the remaining part of the absolute tensor σ_{ij} after the mean or hydrostatic stress σ_m has been subtracted. This reduction applies only to the normal stress components σ_{11}, σ_{22} and σ_{33}, since shear stresses σ_{12}, σ_{13} and σ_{23} will cause no dilatation. The unit matrix **I**, equivalent to Kronecker's delta δ_{ij}, ensures a correct reduction in each notation:

$$\sigma_{ij}' = \sigma_{ij} - \delta_{ij}\sigma_m \quad \text{or} \quad \mathbf{T}' = \mathbf{T} - \frac{1}{3}\,\mathrm{tr}\,(\mathbf{T}) \times \mathbf{I} \tag{3.9a,b}$$

where δ_{ij} is unity for $i = j$ and zero for $i \neq j$. For example, with $i = 1$ and $j = 1$, 2 and 3, eq(3.9a) gives

$$\sigma_{11}' = \sigma_{11} - \sigma_m = \sigma_{11} - ⅓\,(\sigma_{11} + \sigma_{22} + \sigma_{33})$$
$$\sigma_{12}' = \sigma_{12} \ \text{(no change)}$$

With further reductions, it follows that the resulting deviatoric stress matrix $\mathbf{T}'$, in eq(3.9b), is composed of σ_{11}', σ_{22}' and σ_{33}' and the original shear stresses σ_{12}, σ_{13} and σ_{23}:

$$\begin{bmatrix} \sigma_{11}' & \sigma_{12} & \sigma_{13} \\ \sigma_{21} & \sigma_{22}' & \sigma_{23} \\ \sigma_{31} & \sigma_{32} & \sigma_{33}' \end{bmatrix} = \begin{bmatrix} \sigma_{11} & \sigma_{12} & \sigma_{13} \\ \sigma_{21} & \sigma_{22} & \sigma_{23} \\ \sigma_{31} & \sigma_{32} & \sigma_{33} \end{bmatrix} - \frac{\sigma_{11} + \sigma_{22} + \sigma_{33}}{3}\begin{bmatrix} 1 & 0 & 0 \\ 0 & 1 & 0 \\ 0 & 0 & 1 \end{bmatrix} \tag{3.9c}$$

When σ_{ij}' is expressed in the form $f(\sigma_{ij}')$ = constant, this function is known as a yield criterion, predicting the onset of plasticity under a critical deviatoric stress state. Moreover, as yielding is a property of the material itself f must be independent of the co-ordinates used to define σ_{ij}'. Yielding must, therefore, be a function of the deviatoric stress invariants: J_1', J_2' and J_3' since each remains unaltered by co-ordinate transformation. Deviatoric invariants are obtained by subtracting σ_m from the absolute stress invariants, J_1, J_2 and J_3, given in eq(1.25a–c). In terms of principal stresses the deviatoric invariants become

$$\begin{aligned} J_1' &= \sigma_1' + \sigma_2' + \sigma_3' = (\sigma_1 - \sigma_m) + (\sigma_2 - \sigma_m) + (\sigma_3 - \sigma_m) \\ &= (\sigma_1 + \sigma_2 + \sigma_3) - 3\sigma_m = 0 \end{aligned} \tag{3.10a}$$

$$\begin{aligned} -J_2' &= \sigma_1'^2 + \sigma_2'^2 + \sigma_3'^2 = (\sigma_1 - \sigma_m)^2 + (\sigma_2 - \sigma_m)^2 + (\sigma_3 - \sigma_m)^2 \\ &= ⅓\,[(\sigma_1\sigma_2 + \sigma_2\sigma_3 + \sigma_1\sigma_3) - (\sigma_1^2 + \sigma_2^2 + \sigma_3^2)] \\ &= -\,1/6\,[(\sigma_1 - \sigma_2)^2 + (\sigma_2 - \sigma_3)^2 + (\sigma_1 - \sigma_3)^2] \end{aligned} \tag{3.10b}$$

$$\begin{aligned} J_3' &= \sigma_1'\sigma_2'\sigma_3' = (\sigma_1 - \sigma_m)(\sigma_2 - \sigma_m)(\sigma_3 - \sigma_m) \\ &= \sigma_1\sigma_2\sigma_3 - \sigma_m(\sigma_1\sigma_2 + \sigma_2\sigma_3 + \sigma_1\sigma_3) + 2\sigma_m^3 \\ &= \sigma_1\sigma_2\sigma_3 - ⅓(\sigma_1 + \sigma_2 + \sigma_3)(\sigma_1\sigma_2 + \sigma_2\sigma_3 + \sigma_1\sigma_3) + 2/27(\sigma_1 + \sigma_2 + \sigma_3)^3 \\ &= (⅓)^4\,[(2\sigma_1 - \sigma_2 - \sigma_3)^3 + (2\sigma_2 - \sigma_1 - \sigma_3)^3 + (2\sigma_3 - \sigma_1 - \sigma_2)^3] \end{aligned} \tag{3.10c}$$

The sign of the J_2' expression (3.10b) is made positive within its corresponding deviatoric stress cubic as follows:

$$(\sigma')^3 - J_2'(\sigma') - J_3' = 0$$

Note: For a general stress deviator σ_{ij}', the invariants in eqs(3.10a,b,c) appear in the tensor and matrix notations as

$$J_1' = \sigma_{ii}' = \mathrm{tr}\,\mathbf{T}' = 0 \tag{3.11a}$$

$$J_2' = ½\sigma_{ij}'\,\sigma_{ji}' = ½\,\mathrm{tr}\,(\mathbf{T}')^2 \tag{3.11b}$$

$$J_3' = ⅓\,\sigma_{ij}'\,\sigma_{jk}'\,\sigma_{ki}' = ⅓\,\mathrm{tr}\,(\mathbf{T}')^3 \tag{3.11c}$$

Yielding begins when a function of the two, non-zero, deviatoric invariants J_2' and J_3' attains a critical constant value C:

$$f(J_2', J_3') = C \tag{3.12}$$

C is normally defined from reducing eq(3.12) to yielding in simple tension or torsion. For example, when J_3' is omitted from the function in eq(3.12), f is simply equated to J_2' to give the von Mises yield function:

$$J_2' = k^2 \tag{3.13a}$$

where k is the shear yield stress. Setting $\sigma_1 = Y$ (the tensile yield stress) with $\sigma_2 = \sigma_3 = 0$ in eq(3.10b) shows $k^2 = Y^2/3$, when from eqs(3.10b) and (3.13a):

$$[(\sigma_1 - \sigma_2)^2 + (\sigma_2 - \sigma_3)^2 + (\sigma_1 - \sigma_3)^2]/\,6 = Y^2/3 \tag{3.13b}$$

which gives again the yield criterion in eq(3.8). This derivation, in which the second invariant of deviatoric stress attains a critical value at yield, describes the yield criterion normally associated with von Mises.

3.3 Experimental Verification

3.3.1 Yield Point Definition

Where a material displays a sharp yield point, as for low carbon steel in Fig. 3.4a, the division between the elastic and elastic–plastic regions is clearly defined and a yield stress is evident.

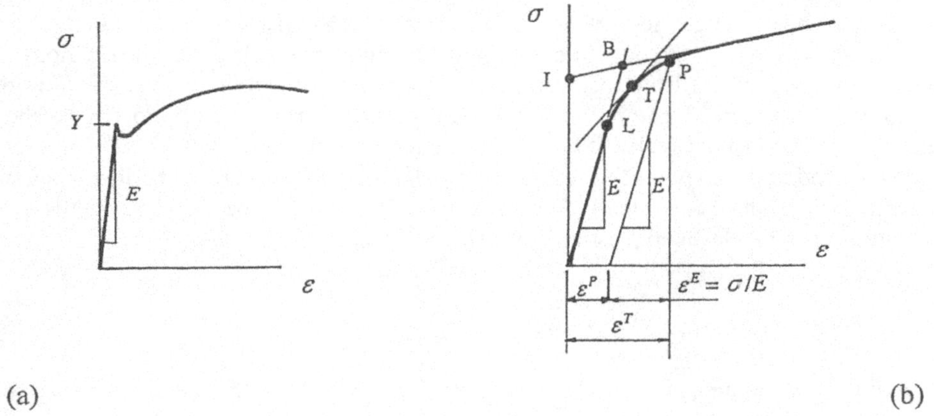

Figure 3.4 Definitions of yielding

Key: B-Back extrapolation, I-Intercept stress, L-Limit of Proportionality, P-Proof stress, T-Tangent stress

However, there is doubt as to what the initial yield stress is for the many metallic materials that display a gradual transition between their elastic and elastic–plastic regions. Various definitions of yield have been employed to overcome such uncertainty in the determination of yield stress. The most commonly used definition is the *proof stress*, i.e. the stress at P in Fig. 3.4b. The proof stress corresponds to a plastic strain ε^P, offset by a given small amount, usually taken from the range: 0.001 % – 0.01 %. Under combined stresses, a similar magnitude of equivalent plastic strain value is chosen to determine the amount that the component strains are to be offset. This method is often employed to determine the combined yield stresses necessary to construct a yield locus. One difficulty arising with this is that the following equivalent plastic strain expression pre-supposes a Mises yield surface:

$$\bar{\varepsilon}^P = \sqrt{\frac{2}{3}\left(\varepsilon_{ij}^P \varepsilon_{ij}^P\right)}$$

If the yield points, so determined, do not lie on a Mises locus the implication is that another $\bar{\varepsilon}^P$ definition is required, corresponding to a locus that would contain the yield points. Unfortunately, there is no way of knowing in advance what the true yield function can be. All that can be said is that the von Mises offset strain method checks the validity of the Mises function. Many investigations on the shape of the yield surface, reviewed in [4], have attempted to avoid this problem by employing low $\bar{\varepsilon}^P$ Mises strain values. This ensures that yield stresses, found from offsetting strains by other definitions of $\bar{\varepsilon}^P$, would all lie within a narrow range, just beyond the limit of elastic proportionality, L in Fig. 3.4b.

An alternative, *back-extrapolated* yield point, B in Fig. 3.4b, is much simpler to locate but its application requires large excursions into the plastic range. Since B is the junction in the bi-linear approximation to a stress–strain curve, it estimates the yield stress at large plastic strain. The extrapolation is suited to materials, e.g. alloy steels, that harden in an approximately linear manner. It is less applicable to annealed materials with well-rounded plastic regions, e.g. copper and aluminium. The figure shows that the proof stress P, found from increasing the offset strain, will approach the back-extrapolation stress at point B. Here the determination of both P and B involves considerable plastic strain, which demands a new testpiece for every experimental stress probe. Otherwise, with repeated probing upon a single testpiece, the plastic strains accumulate to confuse the material's initial condition.

Of all the yield stress definitions available, that which derives its value from the *limit of proportionality* L, is the only one with physical significance. It divides the regions of elastic lattice distortion from inelastic slip. It is because the stress value at the limit of proportionality is sensitive to individual judgement that alternative definitions have been employed. Mincho and Findley [5] have discussed instances in Fig. 3.4b, where the stress intercept I and a tangent point T, of predetermined slope, have defined yield. More recently, point L has been found consistently from the stepped change in temperature that occurs with the transition from elastic to plastic deformation [6].

3.3.2 Comparisons Between Yield Loci, Plane Stress

Experimental determination of yield loci are usually conducted within the simplest of two-dimensional stress states. These allow comparisons to be made between experimental loci and those of Tresca, von Mises and other theoretical predictions. An ideal comparison should employ a metallic material with a well-defined yield point, such as low carbon steel, as in Fig. 3.4a. It is also instructive to examine the influence of yield point definition for other materials with less distinct yield points. For these, initial yield stresses can be determined at the limit of proportionality by the proof strain and extrapolation methods. The comments made previously upon this issue are pertinent to the following comparisons.

(a) The σ_1 versus σ_2 plane

A principal, biaxial stress state is achieved in the wall of a thin-walled tube when it is subjected to combined internal pressure and axial load. The radial stress σ_3 may be ignored provided the diameter to thickness ratio of the tube exceeds 15. Yield loci have either been determined from proportional or step-wise loading paths, these producing tensile stresses lying within the first (positive) quadrant. In the remaining quadrants of the yield locus, compressive buckling can precede yielding, particularly with yield definitions involving larger amounts of plastic strain.

Table 3.1 summarises the test conditions for published data [7–12] used in the construction of Fig. 3.5. The materials investigated had received a processing method, involving heat treatment, to attain an initially isotropic condition. There was an obvious yield point (y. p.) for En 24 steel [7]. In the remaining investigations [8–12], the yield stress was identified with the limit of proportionality (l. p.) or with the proof stress at the indicated offset strain value ($\bar{\varepsilon}^P$). All stress paths were radial, with one exception [9], in which it was shown that it was possible to determine the full yield locus in alloy steels from applying a sequence of non-radial stress probes to a single testpiece. Here the choice of yield stress at the l. p. minimised the accumulation of plastic strain arising from the repeated probing. To avoid strain history effects from radial probing well into the plastic range, a new testpiece should be employed for each stress ratios (σ_2/σ_1). This also applies when defining a yield point at a large offset strain or by back extrapolation.

Table 3.1 Experimental yield point investigations in σ_1, σ_2 space

Material	Heat Treatment	Yield Definition	Symbol	Reference
En 24 carbon steel	annealed	y. p.	×	7
SAE 1045 carbon steel	hot rolled	$\bar{\varepsilon}^P = 0.7$ %	■	8
2¼% Cr, 1% Mo steel	stress relieved	l. p.	○	9
X-60 alloy steel	normalised	l. p.	∇	9
304 stainless steel	stress relieved	l. p.	Δ	9
306 stainless steel	solution treated	l. p.	□	9
M-63 brass	annealed	l. p.	+	10
14S-T4 Al alloy	hot rolled	$\bar{\varepsilon}^P = 0.2$ %	□	11
Ni-Cr-Mo steel	annealed	$\bar{\varepsilon}^P = 0.2$ %	●	12

The theoretical loci, shown in Fig. 3.5, are plotted with their axes normalised by the tensile yield stress Y.

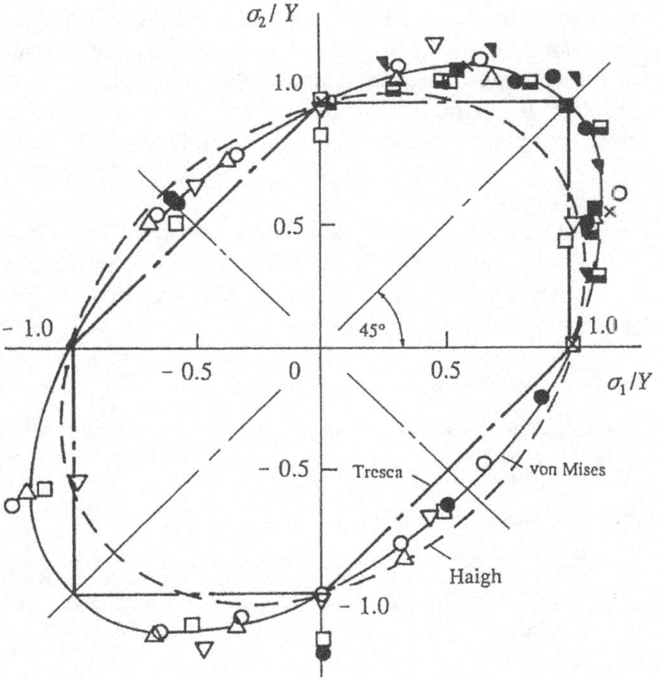

Figure 3.5 Tresca, von Mises and Haigh yield loci in σ_1, σ_2 space

The von Mises prediction, eq(3.7b), becomes

$$(\sigma_1 / Y)^2 - (\sigma_1 / Y)(\sigma_2 / Y) + (\sigma_2 / Y)^2 = 1$$

which defines an ellipse with a 45° orientation to its stress axes. The Tresca locus is found from applying eq(3.1b) to each quadrant. For example, in quadrant 1, where $0 \le \sigma_1 /Y < 1$, the greatest principal stress is $\sigma_2/Y = 1$ (constant) and the least is $\sigma_3/Y = 0$. Also, in this quadrant, where $0 \le \sigma_2/Y < 1$, the greatest principal stress is $\sigma_1/Y = 1$ and the least value is $\sigma_3/Y = 0$. The respective yield criteria, $\sigma_2/Y = 1$ and $\sigma_1/Y = 1$, thus describe the horizontal and vertical sides of the Tresca hexagon as shown. In quadrant 2, σ_2/Y is greatest in tension while σ_1/Y is least in compression. Hence, the left sloping side of the hexagon conforms to $\sigma_2/Y - \sigma_1/Y = 1$. The completed Tresca hexagon is inscribed within the Mises ellipse. The Haigh total energy yield criterion, as derived from eq(3.4), is also shown in this figure. The yield locus associated with this criterion has an elliptical equation involving Poisson's ratio:

$$(\sigma_1/Y)^2 - 2\nu\,(\sigma_1/Y)(\sigma_2/Y) + (\sigma_2/Y)^2 = 1$$

but this is not often used since it assumes an influence of hydrostatic stress upon yield. Overall, the superimposed experimental data in Fig. 3.5 lie closer to the von Mises prediction and appear independent of the test conditions. The conclusion to be drawn from this is that an initially isotropic von Mises yield condition applies to ductile polycrystalline materials. Most data lies outside the hexagon, confirming that a Tresca prediction of yielding for these materials is conservative. It is shown later that where the yield points for an individual material lie between or outside the two loci they may be represented by the function in eq(3.12). The uncertainties in yield point determination may however cast doubt on the need for an initial yield function to fit the test data precisely. This author's view is that a von Mises function is adequate provided the material has been heat-treated to an approximately isotropic condition. An examination of plastic strain paths under proportional loading may provide a more definitive test of an appropriate yield function, given that normality rule between the plastic strain path and the yield locus is accepted.

(b) The σ versus τ plane

The experiments, detailed in Table 3.2, were all conducted on thin-walled tubes, subjected to torsion combined with either a circumferential tension [7] or an axial tension/compression [13–19]. The respective Tresca and Mises yield criteria are given in eqs(3.3) and (3.8). The total energy theory (Haigh) follows from eq(3.4) as

$$(\sigma/Y)^2 + 2\,(1 + \nu)(\tau/Y)^2 = 1$$

Here the subscripts x and y are dropped when applying yield criteria to this simple, combined stress state, i.e. σ and τ.

Table 3.2 Experimental yield point investigations in σ, τ space

Material	Heat Treatment	Yield Definition	Symbol	Reference
En 24 carbon steel	annealed	y. p.	×	7
En 25 carbon steel	annealed	y. p.	○	13
19S Al alloy	stress relieved	l. p.	▽	14
1100-0 Al	annealed	l. p.	■	15
PA6 Al	as-received	$\bar{\varepsilon}^P = 50\mu\varepsilon$	+	16
Brass	as-received	$\bar{\varepsilon}^P = 200\mu\varepsilon$	⊾	17
Ti-50A Ti-alloy	stress relieved	$\bar{\varepsilon}^P = 20\mu\varepsilon$	●	18
Copper (99.8%)	annealed	b. e.	△	19
Aluminium (99.7%)	annealed	b. e.	□	19

These tension-torsion combinations include the original experiments of Taylor and Quinney [19] who, in 1931, determined the yield stress by applying the backward extrapolation method (b. e.). With the exception of the stepped-stress probes, employed by Ivey [14] and Phillips and Tang [15], all other investigations, reported in Table 3.2, employed radial loading. Ellyin and Grass [18] applied multiple, radial probes, each emanating from the stress origin, to a single testpiece. The final probe, which duplicated their initial probe, revealed little difference between their yield stresses. Consequently, the effect of accumulated plastic strain, arising from intermediate probing, was avoided by employing a small, offset strain ($\bar{\varepsilon}^P = 20\mu\varepsilon$) to define yield.

The comparison between theory and experiment in Fig. 3.6 reveals again that most of the experimental data in quadrants 2 and 4 lie closer to the yield locus of Mises than to Tresca, irrespective of the chosen yield definition. There is an obvious difference between the von Mises and Tresca predictions to the shear yield stress k. That is, their intercepts upon the semi-minor axis are predicted as: $k = Y/\sqrt{3}$ and $Y/2$ respectively.

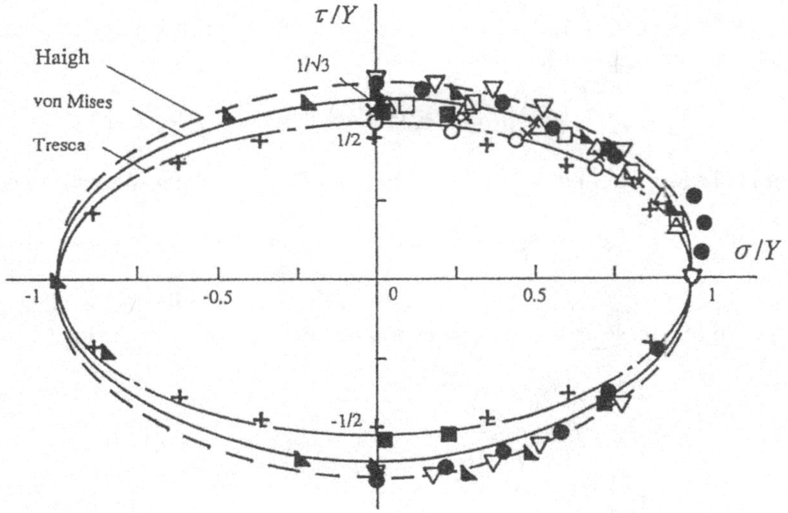

Figure 3.6 Tresca and Mises loci in σ, τ space

Figure 3.6 confirms that a Mises initial yield condition applies to ductile, isotropic materials under combined σ, τ stress states. This observation was first made by Taylor and Quinney [19], from their own results for copper and aluminium. Similar comments, made previously on the influence of the definition of yielding in Table 3.1, again apply to Table 3.2. Despite the uncertainty in yield point determination, it appears, from Fig. 3.6, that Tresca's yield locus will provide a conservative estimate of yield by any definition.

Note: Hot-rolled [8,11] and as-received material [16,17], given in Tables 3.1 and 3.2, did not receive any further heat treatment prior to test. It is possible that their departure from a Mises initial condition is due to the influence of an initial residual anisotropy. This condition is confirmed when initial tensile yield stresses for orthogonal directions are dissimilar, but they were not reported in these investigations.

There are many practical instances where direct and shear stress states are combined. They arise from loading a shaft axially in tension or compression in combination with a transverse shear force or axial torsion. Combining a bending moment with a torque produces a similar stress state. A combination of direct stress and shear stress exists at points in the cross-section of a transversely loaded beam, away from its neutral axis. Equations (3.3) and (3.8) apply to yielding under each combination of a single direct stress and a single shear stress arising from these superimposed loadings.

3.3.3 *Influence of the Third Invariant*

(a) Symmetrical Functions
The fact that some initial yield points for isotropic material do not lie on the Mises loci, in Figs 3.5 and 3.6, suggests that initial yielding may conform to a more general function containing both deviatoric invariants. Firstly, note that Tresca's yield criterion may be expressed as the following complex function [20] of these invariants

$$f = 4J'^3_2 - 27J'^2_3 - 36k^2J'^2_2 + 96k^4J'_2 = 64\,k^6 \qquad (3.14a)$$

where k, the shear yield stress, bears a simple relationship to the tensile yield stress Y, i.e. $k = Y/2$. Substituting eqs(3.10b,c) into (3.14a) and factorising gives

$$[(\sigma_1' - \sigma_2')^2 - 4k^2][(\sigma_2' - \sigma_3')^2 - 4k^2][(\sigma_1' - \sigma_3')^2 - 4k^2] = 0 \qquad (3.14b)$$

The familiar Tresca criterion follows from the last of the following solutions to eq(3.14b):

$$(\sigma_1 - \sigma_2) = 2k, \qquad (\sigma_2 - \sigma_3) = 2k \qquad \text{and} \qquad (\sigma_1 - \sigma_3) = 2k \qquad (3.14c)$$

in which $(\sigma_1 - \sigma_3) = (\sigma_1' - \sigma_3')$. There are many other isotropic functions containing both invariants. Among these are the following homogenous stress functions for f in eq(3.12)

$$f = J'^3_2 - c\,J'^2_3 = k^6 \qquad (3.15a)$$
$$f = J'_2 - b\,(J'_3/J'_2)^2 = k^2 \qquad (3.15b)$$
$$f = (J'_2)^{3/2} - d\,J'_3 = k^3 \qquad (3.15c)$$

where c, b and d are material constants and k is the shear yield stress. Unlike eq(3.14a), it is now possible to select a value for the constants in eqs(3.15a–c) to fit the initial yield behaviour of most metallic materials [21]. Of these three yield functions, the most well-known is eq(3.15a), proposed by Drucker [22]. This is a homogenous function in stress to the sixth degree, for which c must lie in the range $-27/8 \le c \le 9/4$, to ensure convexity of the yield surface. These limits are shown in Fig. 3.7 where bounding loci, corresponding to the positive and negative limiting values $c_{crit} = 9/4$ and $-27/8$, ensure a closed surface [23]. It is seen that corners begin to appear at the negative limit for c_{crit}.

Figure 3.7 also demonstrates how closely the data given for stainless steel [9] can be represented by taking $c = -2$ in eq(3.15a). The corresponding locus f_o represents the data better than would be found from either Mises or Tresca. The following chapter will show, by including J_3' in a homogenous yield function, that the associated stress versus plastic strain relations become somewhat cumbersome. Consequently, the simplified relations associated with the Mises yield function $f = J_2'$ are often preferred.

The subsequent experimental yield locus f, shown in Fig. 3.7, is almost a rigid translation of the initial yield locus f_o. Locus f applies to a plastic pre-stress P reached under a radial path having a constant stress ratio $R = \sigma_2/\sigma_1$.

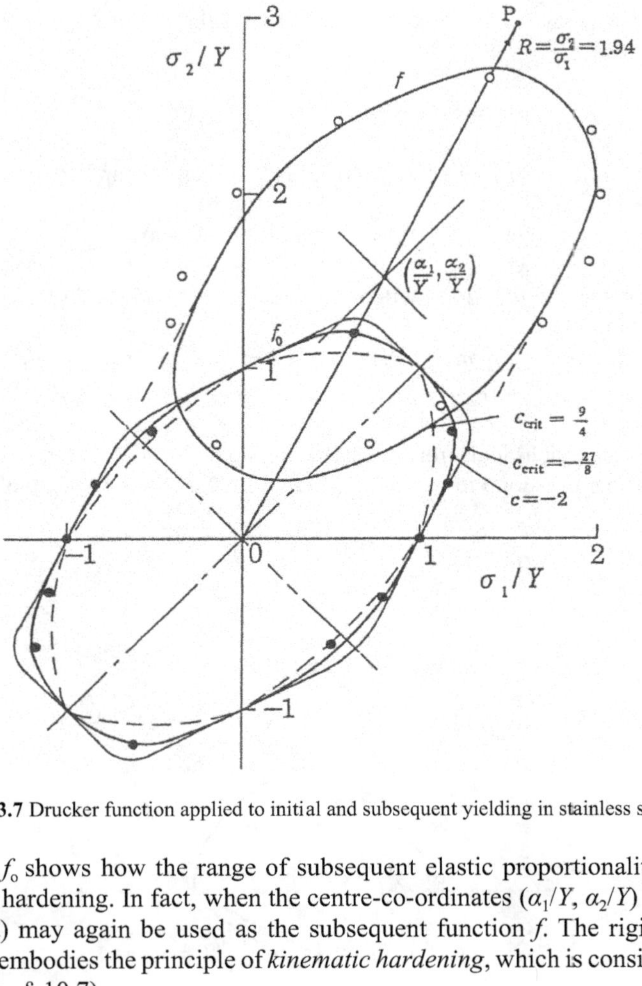

Figure 3.7 Drucker function applied to initial and subsequent yielding in stainless steel

Comparing f to f_o shows how the range of subsequent elastic proportionality have been altered by work hardening. In fact, when the centre-co-ordinates (α_1/Y, α_2/Y) re-define the origin, eq(3.15a) may again be used as the subsequent function f. The rigid translation between f_o and f embodies the principle of *kinematic hardening*, which is considered further in chapter 10 (see § 10.7).

(b) Non-Symmetrical Functions

Consider a yield function as a linear combination of the two deviatoric invariants, J_2' and J_3', in the following form [21]

$$f = J'^{\,n}_2 + (p/\sigma_t)^{(2n-1)} J'_3 \tag{3.16}$$

in which n and p are constants. The tensile yield stress σ_t is employed within the denominator of the second term in eq(3.16) to ensure homogeneity in stress. Functions of this form account for so-called *second order* effects arising from incompressible plasticity [24]. Such effects include: (i) non-linear plastic strain paths from radial loading [21], (ii) the accumulation of axial strain under pure torsion [25] and (iii) a difference between the initial tensile and compressive (σ_c) yield stresses. Explicit forms of eq(3.16) are found from substituting J_2' and J_3' from eqs(3.10b,c). For example, taking $n = 1$ in eq(3.16), within a principal biaxial stress state (i.e. $\sigma_3 = 0$), gives an initial yield function

$$f = \tfrac{1}{3}(\sigma_1^2 - \sigma_1\sigma_2 + \sigma_2^2) + (\tfrac{1}{3})^4\,[(2\sigma_1 - \sigma_2)^3 + (2\sigma_2 - \sigma_1)^3 - (\sigma_1 + \sigma_2)^3] \tag{3.17}$$

The effect of (iii) above is revealed from respective substitutions $\sigma_1 = \sigma_t$ and $\sigma_1 = -\sigma_c$, each with $\sigma_2 = 0$. This gives

$$f = \sigma_t^2 (1/3 + 2p/27) = \sigma_c^2/3 - 2p\sigma_c^3/(27\sigma_t) \tag{3.18a}$$

Employing the ratio $\rho = \sigma_c/\sigma_t$ in eq(3.18a) leads to a relationship between p and ρ :

$$(2p/9)\rho^3 - \rho^2 + (2p/9 + 1) = 0 \tag{3.18b}$$

One may normalise the function f with σ_t from equating (3.17) and (3.18a). This gives

$$(\sigma_1/\sigma_t)^2 - (\sigma_1/\sigma_t)(\sigma_2/\sigma_t) + (\sigma_2/\sigma_t)^2 + (p/27)[(2\sigma_1/\sigma_t - \sigma_2/\sigma_t)^3 + (2\sigma_2/\sigma_t - \sigma_1/\sigma_t)^3 - (\sigma_1/\sigma_t + \sigma_2/\sigma_t)^3] = 1 + 2p/9 \tag{3.19}$$

Figure 3.8 presents yield loci, from eq(3.19), with $\rho = ½, ⅔, 1, 2$ and 3. Equation (3.18b) shows that these correspond to $p = -3, -27/14, 0, 3/2$ and 9/7 respectively.

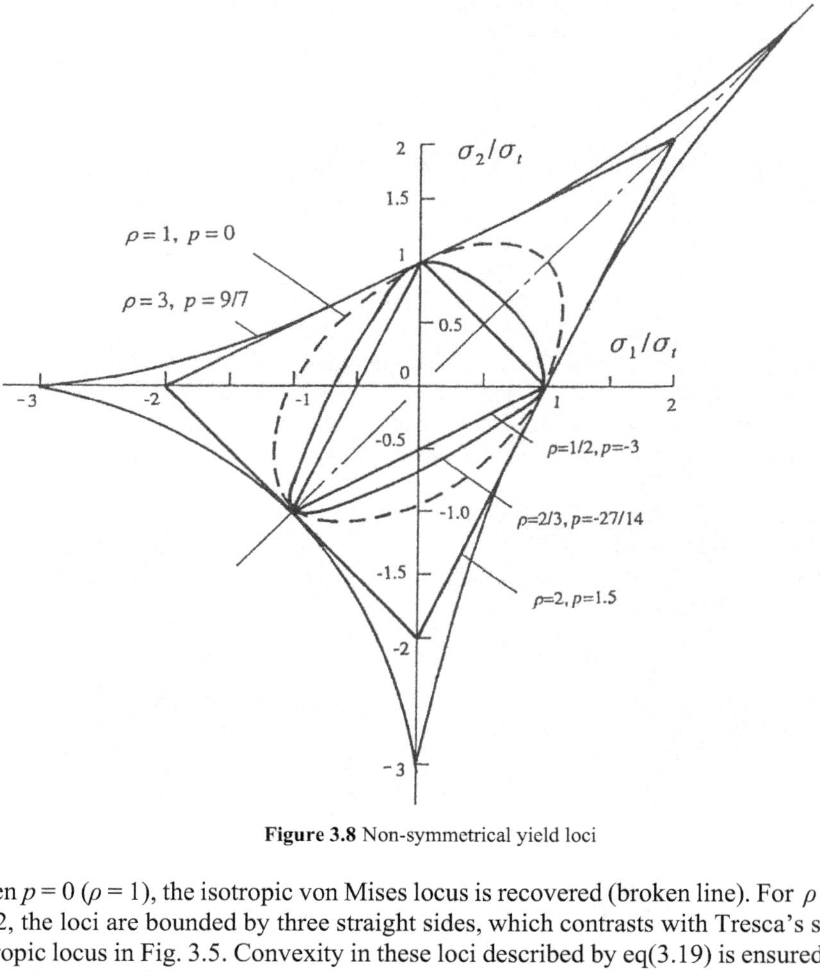

Figure 3.8 Non-symmetrical yield loci

When $p = 0$ ($\rho = 1$), the isotropic von Mises locus is recovered (broken line). For $\rho = ½$ and $\rho = 2$, the loci are bounded by three straight sides, which contrasts with Tresca's six-sided isotropic locus in Fig. 3.5. Convexity in these loci described by eq(3.19) is ensured when p lies in the range $-3 < p < 3/2$. This means that eq(3.18a) is restricted to where one uniaxial yield stress is not more than 50 % of the other. For example, taking an intermediate value $p = -27/14$ at $\rho = 2/3$, then σ_t exceeds σ_c by 33 %. Because stress deviator invariants appear in eq(3.16), the function preserves plastic incompressibility if it is required to account for such a large difference in a ductile metal.

Uncertainty in yield point determination can cloud the influence that J_3' has upon the initial yield stress. It is also possible that initial anisotropy has a greater influence than J_3', in promoting deviations from the Mises and Tresca loci, shown in Figs 3.5 and 3.6. This is more likely for the non-heat treated materials in Table 3.2. It will, therefore, be appropriate to consider later (see § 3.4) a number of anisotropic yield functions.

3.3.4 Superimposed Hydrostatic Stress

It has been seen that when a yield function is formulated from the stress deviator invariants it implies that initial yielding is unaffected by the magnitude of hydrostatic stress. This simplifying assumption has enabled the development of various isotropic yield functions for polycrystalline materials. The literature reveals experimental evidence in support of this assumption. High magnitudes of hydrostatic pressure, up to 3 kbar, when superimposed upon torsion [26–28], tension [29, 30] and compression [31], have not altered significantly the yield stresses for mild steel, copper, aluminium and brass. To show this theoretically, we assume a yield function of the general isotropic form $f(J_2', J_3')$. With incremental changes to J_2' and J_3', the change in f becomes:

$$df = [(\partial f/\partial J_2')\, dJ_2' + (\partial f/\partial J_3')\, dJ_3'] \tag{3.20a}$$

Since f, J_2' and J_3' are functions of σ_{ij}', the total differential is

$$df/d\sigma_{ij}' = [(\partial f/\partial J_2')(\partial J_2'/\partial \sigma_{ij}') + (\partial f/\partial J_3')(\partial J_3'/\partial \sigma_{ij}')] \tag{3.20b}$$

Under torsion, with $J_2' = \tau^2$ and $J_3' = 0$:

$$df/d\tau = (\partial f/\partial J_2') \times 2\tau$$

Under tension, the second invariant is

$$J_2' = \tfrac{1}{2}(\sigma'^2_{11} + \sigma'^2_{22} + \sigma'^2_{33})$$

so that: $\partial J_2'/\partial \sigma'_{11} = \sigma'_{11}$, $\partial J_2'/\partial \sigma'_{22} = \sigma'_{22}$ and $\partial J_2'/\partial \sigma'_{33} = \sigma'_{33}$. The third invariant is

$$J_3' = \tfrac{1}{3}(\sigma'^3_{11} + \sigma'^3_{22} + \sigma'^3_{33})$$

from which $\partial J_3'/\partial \sigma'_{11} = \sigma'^2_{11}$, $\partial J_3'/\partial \sigma'_{22} = \sigma'^2_{22}$ and $\partial J_3'/\partial \sigma'_{33} = \sigma'^2_{33}$. Substituting into eq(3.20b) gives:

$$df/d\sigma'_{11} = [(\partial f/\partial J_2')\, \sigma'_{11} + (\partial f/\partial J_3')\, \sigma'^2_{11}] \tag{3.21a}$$
$$df/d\sigma'_{22} = [(\partial f/\partial J_2')\, \sigma'_{22} + (\partial f/\partial J_3')\, \sigma'^2_{22}] \tag{3.21b}$$
$$df/d\sigma'_{33} = [(\partial f/\partial J_2')\, \sigma'_{33} + (\partial f/\partial J_3')\, \sigma'^2_{33}] \tag{3.21c}$$

It is apparent from eqs(3.21a–c) that a change $df > 0$, depends upon deviatoric stress and the changes to the stress deviators. Thus, the flow stress remains unaffected by a superimposed mean, or hydrostatic, stress σ_m of any intensity. This can also be shown from the total differential, with respect to σ_{kk}:

$$df/d\sigma_{kk} = [(\partial f/\partial J_2')(\partial J_2'/\partial\sigma_{kk}) + (\partial f/\partial J_3')(\partial J_3'/\partial\sigma_{kk})] = 0$$

where, for a hydrostatic pressure $-p$ imposed upon σ_{ij}:

$$\sigma_{kk} = 3\sigma_m = [(\sigma_{11} - p) + (\sigma_{22} - p) + (\sigma_{33} - p)]$$

It follows that plastic flow is associated with the partial derivatives of f with respect to the deviatoric stress. However, since σ_{ii} has been removed from within the deviatoric invariant function f, plasticity will depend directly upon the partial derivatives $\partial f/\partial\sigma_{ij}$. Thus, incremental plastic strains appear within an *associated flow rule*:

$$d\varepsilon_{ij}^P = d\lambda\frac{\partial f}{\partial\sigma_{ij}} \tag{3.22}$$

where $d\lambda$ is a scalar multiplier, which derives from the unique, equivalent stress–strain relation for a given material (see § 4.2.2, chapter 4). *Note*: A superimposed hydrostatic pressure can increase the strain to fracture by inhibiting void formation prior to onset of fracture [32]. In the absence of a superimposed pressure, the mean component of the applied stress, $\sigma_m = ⅓(\sigma_{11} + \sigma_{22} + \sigma_{33})$, being comparatively low, does not influence the yield stress. Few experimental data are available that show an influence of mean stress upon yielding of anisotropic polycrystals. When unaffected by σ_m, an appropriate yield function and its associated flow rule should ensure that mean stress does not influence anisotropic yielding and that subsequent plastic flow remains incompressible.

Under certain conditions, an allowance for the influence of superimposed hydrostatic pressure may be required. For example, an upper shear yield stress $k = 200$ MPa, for mild steel decreased by 5–6 % under $p = 3$ kbar [26]. For pre-strained brass, with k in the range 110–190 MPa, variations to k between −6 % and 3 % were observed with superimposed pressure between 1 and 4 kbar [30]. To account for this, Hu [27] separated the influence of σ_m upon yielding by including $J_1 = \sigma_{kk}$ in the yield function $f(J_1, J_2', J_3')$. Functions of this kind are also required where the magnitude of σ_m is critical to the failure in brittle, non-metals and to plastic flow in porous compacts. The following section develops one such failure criterion for cast iron.

3.3.5 Influence of the First Invariant

Equation (3.16) does not allow the mean, or hydrostatic, stress to promote a difference between σ_t and σ_c. Here, a yield, or fracture, function should contain the mean stress. Conveniently, σ_m is proportional to the first invariant of absolute stress: $J_1 = \sigma_{ii} = 3\sigma_m$. Thus, for brittle materials, which are influenced by hydrostatic stress, J_1 is combined with J_2' to formulate the function. The latter will define a fracture surface in the absence of plasticity. A notable example is cast iron, for which a fracture surface follows from a simple linear sum of J_1' and J_2':

$$J_2' + \alpha J_1 = c \tag{3.23a}$$

Substituting from eqs(3.10a,b)

$$(1/6)[(\sigma_1 - \sigma_2)^2 + (\sigma_2 - \sigma_3)^2 + (\sigma_3 - \sigma_1)^2] + \alpha(\sigma_1 + \sigma_2 + \sigma_3) = c \tag{3.23b}$$

Constant c is found from substituting the uniaxial condition $(\sigma_1, 0, 0)$ in eq(3.23b):

$$(1/6)(2\sigma_1^{\,2}) + \alpha\,\sigma_1 = c \tag{3.24a}$$

Since $\sigma_1 = \sigma_t$ and $\sigma_1 = -\sigma_c$ are the roots of eq(3.24a), it follows that

$$(\sigma_1 - \sigma_t)(\sigma_1 + \sigma_c) = \sigma_1^{\,2} + 3\alpha\,\sigma_1 - 3c$$
$$\sigma_1^{\,2} + \sigma_1(\sigma_c - \sigma_t) - \sigma_c\sigma_t = \sigma_1^{\,2} + 3\alpha\,\sigma_1 - 3c \tag{3.24b}$$

Coefficients in eq(3.24b) give $\alpha = (\sigma_c - \sigma_t)/3$ and $c = \sigma_c\sigma_t/3$. Substituting into eq(3.23b)

$$[(\sigma_1 - \sigma_2)^2 + (\sigma_2 - \sigma_3)^2 + (\sigma_3 - \sigma_1)^2\,] + 2\sigma_t(\sigma_c/\sigma_t - 1)(\sigma_1 + \sigma_2 + \sigma_3) = 2\sigma_c\sigma_t \tag{3.25a}$$

Putting $\rho = \sigma_c/\sigma_t$ in eq(3.25a), leads to the Stassi fracture criterion [33]

$$[(\sigma_1 - \sigma_2)^2 + (\sigma_2 - \sigma_3)^2 + (\sigma_3 - \sigma_1)^2] + 2\sigma_t(\rho - 1)(\sigma_1 + \sigma_2 + \sigma_3) = 2\rho\sigma_t^{\,2} \tag{3.25b}$$

Setting $\sigma_3 = 0$ in eq(3.25b), provides a normalised, biaxial criterion used to predict the Stassi fracture loci shown in Fig. 3.9.

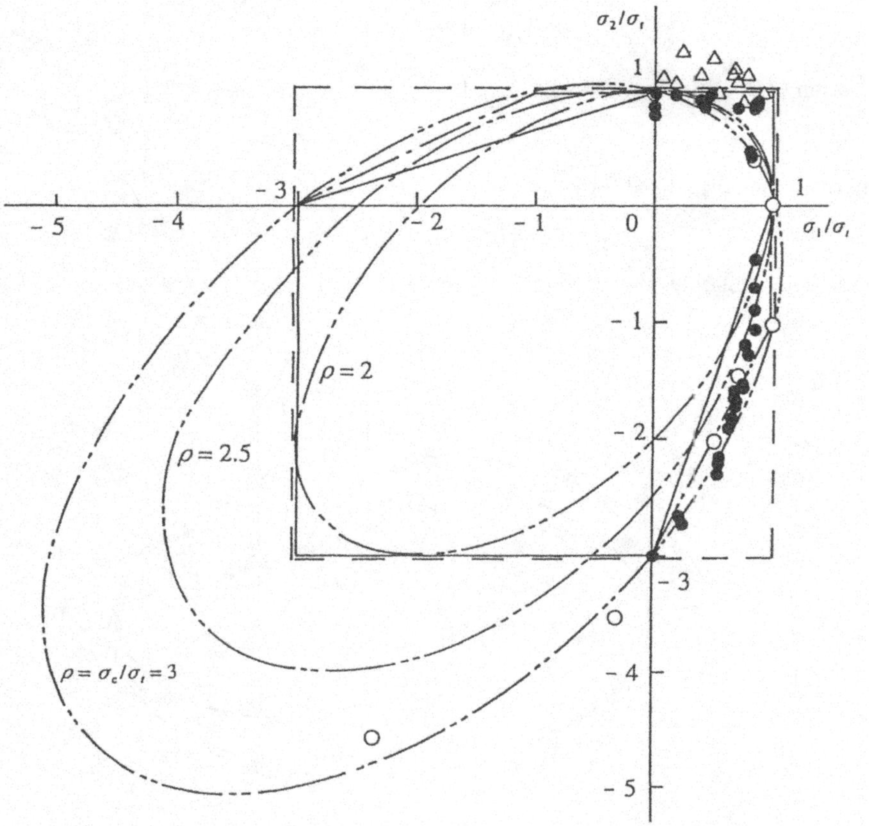

Figure 3.9 Brittle fracture criteria applied to cast iron

Key: ——— — —— Stassi; ———— Coulomb-Mohr; — — — Rankine; ——— — ——— Modified-Mohr

That is:

$$(\sigma_1/\sigma_t)^2 - (\sigma_1/\sigma_t)(\sigma_2/\sigma_t) + (\sigma_2/\sigma_t)^2 + (\rho - 1)(\sigma_1/\sigma_t + \sigma_2/\sigma_t) = \rho \tag{3.25c}$$

for which $\rho = \sigma_c/\sigma_t = 2$, 2.5 and 3 when principal stresses are normalised with σ_t. Stassi did not employ invariants for his derivation of eq(3.25b) but it is clear that hydrostatic stress accounts for the difference between tensile and compressive strengths in this material. In Fig. 3.9, a comparison is made between various strength criteria, including eq(3.25c), and experimental data for grey cast iron. A detailed description of the alternative predictions: Rankine, Coulomb, Mohr etc., has been given elsewhere [1]. It appears, from Fig. 3.9, that experiment is unable to discriminate between the various predictions since test data falls mainly within quadrants 1 and 4. This is largely due to the difficulty in acquiring reliable data under biaxial compression. Under biaxial tension, the low-stress combinations at failure are not inclined to any one theory. However, few data lying within quadrant 3 in Fig. 3.9 suggest that Stassi's criterion will predict enhanced compressive strengths most reliably by using the appropriate ρ–value.

When the derivation is extended to include shear stress, J_1 remains unchanged but additional shear stress σ_{12}, σ_{23} and σ_{13} appear within J_2':

$$(\sigma_{11} - \sigma_{22})^2 + (\sigma_{22} - \sigma_{33})^2 + (\sigma_{11} - \sigma_{33})^2 + 6(\sigma_{12}^2 + \sigma_{13}^2 + \sigma_{23}^2) + 2\sigma_t(\rho - 1)(\sigma_{11} + \sigma_{22} + \sigma_{33}) = 2\rho\,\sigma_t^2 \tag{3.26a}$$

Equation (3.26a) defines a locus in a reduced space σ_{11}, σ_{12}, as

$$\sigma_{11}^2 + 3\sigma_{12}^2 + \sigma_t(\rho - 1)\sigma_{11} = \rho\,\sigma_t^2 \tag{3.26b}$$

for which the normalised form is

$$(\sigma_{11}/\sigma_t)^2 + 3(\sigma_{12}/\sigma_t)^2 + (\rho - 1)(\sigma_{11}/\sigma_t) = \rho \tag{3.26c}$$

Figure 3.10 illustrates the fracture loci from eq(3.26c) for $\rho = 1.25$, 2, 2.5 and 3.

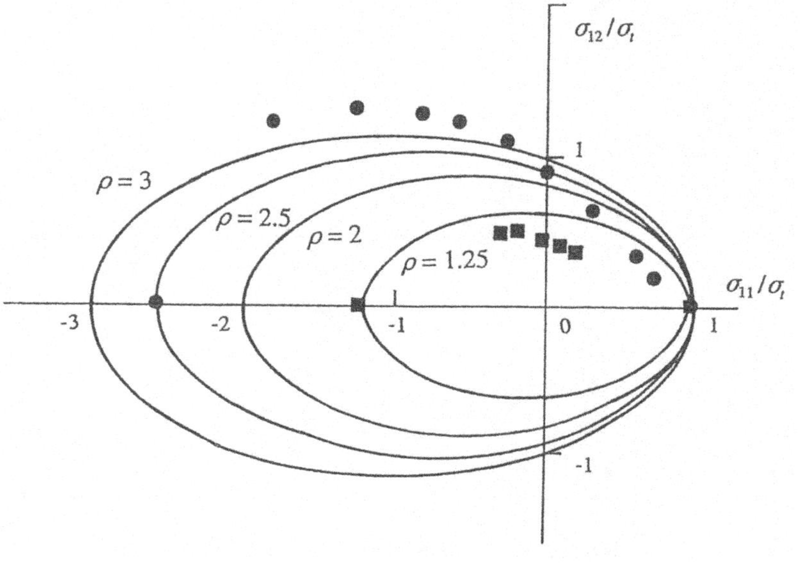

Figure 3.10 Failure loci in σ_{11}, σ_{12} space (Key: ● cast iron, ■ mmc)

Comparisons are made in Fig. 3.10 with this author's data for cast iron (5 % ferrite, 6–12 mm graphite flakes) and a metal matrix composite (17 % volume of 3μ, Si-C particles in a 2124 Cu-Al metal matrix). Dissimilar, uniaxial strengths in each material and the combined stress fracture values are predictable within the range of ρ shown.

It is seen that the major advantage of an invariant function lies in its reduction to any given stress state. When $\sigma_c = \sigma_t$, i.e. $\rho = 1$, the von Mises function is recovered and hydrostatic stress plays no part in failure.

3.4 Anisotropic Yielding Theory

In metallic materials, plastic-anisotropy is associated with a directional variation in yield stress. None of the forgoing isotropic functions can display similar behaviour. Also, with the exception of eq(3.16), they show yield stresses of equal magnitude irrespective of the sense of stress. For example, from the symmetry of Figs 3.5–3.7, the magnitude of the uniaxial tensile and compressive yield stresses are equal and the positive and negative shear yield stresses are also equal. This assumption is reasonably consistent with the observed, initial yield behaviour of annealed, or stress-relieved, metals and alloys. In the absence of heat treatment, the orthotropic form of initial anisotropy in polycrystals is most common, arising from the processing method, e.g. rolling, drawing and extrusion. An orthotropic material has different initial yield stresses for axes lying parallel and normal to the direction in which it has been worked. Cold-rolled sheet and extruded bar will often continue to display distinctly different tensile stress–strain behaviour along these orthotropic axes, well into the plastic range. This is apparent from a divergence in these stress-strain curves when overlaid.

3.4.1 Initial Yielding

In his pivotal paper of 1948, Hill [34] generalised the von Mises isotropic yield criterion to account for anisotropic yielding in an orthotropic material. Hill's yield function describes a yield surface whose stress axes align with the three principal axes of orthotropy in the material. To establish the yield condition an applied stress σ_{ij} must, therefore, be resolved along the orthotropic axes. With the latter as x, y and z, the yield function is written as

$$2f(\sigma_{ij}) = F(\sigma_{yy} - \sigma_{zz})^2 + G(\sigma_{xx} - \sigma_{zz})^2 + H(\sigma_{xx} - \sigma_{yy})^2 + 2L\tau_{yz}^{\ 2} + 2M\tau_{xz}^{\ 2} + 2N\tau_{xy}^{\ 2} = 1 \qquad (3.27)$$

where F, G, H, L, M and N are six coefficients that characterise an orthotropic symmetry in the yield stresses. In fact, eq(3.27) is a restricted form of a general quadratic yield function given later by Edelman and Drucker [35]:

$$f(\sigma_{ij}) = \tfrac{1}{2}\, C_{ijkl}\, \sigma_{ij}\sigma_{kl} \qquad (3.28)$$

The fourth rank tensor C_{ijkl} contains 81 components, similar to that used for defining a strain energy function for anisotropic elasticity [36]. The number of independent components reduces to 21 with imposed symmetry conditions: (i) $\sigma_{ij} = \sigma_{ji}$ in the Cauchy stress tensor, giving $C_{ijkl} = C_{jikl} = C_{ijlk}$ and (ii) coincidence between the axes of stress and orthotropy, which gives $C_{ijkl} = C_{klij}$. With these reduced coefficients, the function can account for an influence of hydrostatic stress upon anisotropic yielding. When a condition of incompressibility is imposed upon eq(3.28), the number of independent coefficients is reduced further to 15. The corresponding relation is found from substituting eq(3.28) into the flow rule, eq(3.22). The incremental plastic strains become

$$\delta\varepsilon_{ij}^{P} = \delta\lambda\, \partial f(\sigma_{ij})/\partial\sigma_{ij} = \delta\lambda \times \tfrac{1}{2} C_{ijkl}\,\sigma_{kl}$$

Multiplying by δ_{ij}

$$\delta\varepsilon_{ij}^{P}\,\delta_{ij} = \delta\lambda \times \tfrac{1}{2} C_{ijkl}\,\sigma_{kl}\,\delta_{ij}$$

The left-hand side is $\delta\varepsilon_{ii}^{P} = 0$, when it follows that $C_{iikl} = 0$. This condition applies to normal stress components, for which $k = l$, so that

$$C_{ii11} = C_{ii22} = C_{ii33} = 0$$

Alternatively, the 15 independent components will appear directly from the expansion of a Mises generalised quadratic potential, written in stress deviator form:

$$f(\sigma_{ij}') = \tfrac{1}{2} H_{ijkl}\,\sigma_{ij}'\,\sigma_{kl}' = 1 \tag{3.29}$$

In general, the elastic–plastic deformation of an anisotropic solid is characterised by two fourth-order tensors with 21 elastic components [36] and 15 plastic components. These tensors will ensure elastic compressibility and plastic incompressibility, in common with metals. With further reductions to the number of independent components H_{ijkl}, eq(3.29) reveals particular material symmetries. Among these are: orthotropy (six coefficients), transverse isotropy (three coefficients) and a cubic form of anisotropy characterised by two coefficients [37]. The first of these corresponds to Hill's eq(3.27). The remaining two symmetries follow from eq(3.27) by setting: (i) $G = H$ with $M = N$ and (ii) $F = G = H$ with $L = M = N$ respectively. Wider application of the general forms of eqs(3.28) and (3.29) have been demonstrated in the absence of such symmetry [38, 39]. Further applications and developments to Hill's quadratic function have been made under particular conditions of plane stress and plane strain [40–42]. Alternative yield criteria [43–50] for anisotropic metals are summarised in Table 3.3 for various stress states (general, principal and plane).

Table 3.3 Yield criteria for various initially anisotropic conditions in metals

ID	Reference	Anisotropic Yield Function
(a)	[43]	$f = C_{11}\sigma_{11}^2 + C_{12}\sigma_{11}\sigma_{22} + C_{13}\sigma_{11}\sigma_{12} + C_{22}\sigma_{22}^2 + C_{23}\sigma_{22}\sigma_{12} + C_{33}\sigma_{12}^2 = 1$
(b)	[44]	$f = 3/2[(\sigma_1/\sigma_{1y})^2 - (\sigma_1/\sigma_{1y})(\sigma_2/\sigma_{2y}) + (\sigma_2/\sigma_{2y})^2]^3$ $- \tfrac{1}{8}(\sigma_1/\sigma_{1y} + \sigma_2/\sigma_{2y})^2(\sigma_1/\sigma_{1y} - 2\sigma_2/\sigma_{2y})^2(\sigma_2/\sigma_{2y} - 2\sigma_1/\sigma_{1y})^2 = 1$
(c)	[45]	$f = J_2'^2(J_2' + A_{ijkl}\sigma_{ij}'\sigma_{kl}') - CJ_3'^2 = k^6$
(d)	[46]	$f = f(\sigma_2 - \sigma_3)^m + g(\sigma_1 - \sigma_3)^m + h(\sigma_1 - \sigma_2)^m$ $+ a(2\sigma_1 - \sigma_2 - \sigma_3)^m + b(2\sigma_2 - \sigma_3 - \sigma_1)^m + c(2\sigma_3 - \sigma_1 - \sigma_2)^m = \sigma^m$
(e)	[47]	$f = 3\sigma_x^3 - 6\sigma_x^2\sigma_y - 6\sigma_x\sigma_y^2 + 4\sigma_y^3 + (4\sigma_x + 21\sigma_y)\tau_{xy}^2$
(f)	[48]	$f = \sum_{ijk} A_{ijk}\sigma_x^i\sigma_y^j\tau_{xy}^{2k}$, for $i + j + 2k \le 4$
(g)	[49]	$f = a_1(\sigma_x - \sigma_y)^2 + a_2(\sigma_x - \sigma_z)^2 + a_3(\sigma_y - \sigma_z)^2 + a_4\tau_{yz}^2 + a_5\tau_{xy}^2 + a_6\tau_{xz}^2$ $+ a_7(\sigma_x - \sigma_y) + a_8(\sigma_x - \sigma_z) + a_9(\sigma_y - \sigma_z) + a_{10}\tau_{yz} + a_{11}\tau_{xy} + a_{12}\tau_{xz} = 1$
(h)	[50]	$f = A\sigma_1^2 + B\sigma_2^2 + C\sigma_3^2 - D\sigma_1\sigma_2 - E\sigma_2\sigma_3 - F\sigma_1\sigma_3 + L\sigma_1 + M\sigma_2 + N\sigma_3 = 1$

In the plane, quadratic stress function (a) in Table 3.3, the cross-product terms in $\sigma\tau$ introduce two additional constants (Hill has a total of four). This form was used to predict the initial yield stresses in anisotropic sheet metal. The function (b) is a normalised form of Drucker's eq(3.11a) for a principal, biaxial stress state. The stresses, σ_1 and σ_2, are divided by their respective yield stresses, σ_{1y} and σ_{2y}, which differ between the 1 and 2–directions. In eq(c), Drucker's isotropic function is modified with an orthotropic quadratic term. This form accounted for deviations from a Mises isotropic condition due to the combined influences of J_3' and initial anisotropy. A non-quadratic function (d) was proposed by Hill [46] for principal stresses aligned with a material's orthotropic axes. The constants σ^m, a, b, c, f, g and h are all positive. It can be seen how this function has modified the two isotropic, deviatoric invariants:

$$J_2' = (1/6)[(\sigma_1 - \sigma_2)^2 + (\sigma_2 - \sigma_3)^2 + (\sigma_3 - \sigma_1)^2]$$
$$J_3' = (1/3)^4[(2\sigma_1 - \sigma_2 - \sigma_3)^3 + (2\sigma_2 - \sigma_3 - \sigma_1)^3 + (2\sigma_3 - \sigma_1 - \sigma_2)^3]$$
$$= (1/3)^3(2\sigma_1 - \sigma_2 - \sigma_3)(2\sigma_2 - \sigma_3 - \sigma_1)(2\sigma_3 - \sigma_1 - \sigma_2)$$

It is also apparent from the first equation how Hill's orthotropic yield function (3.21) modifies J_2'. Using a non-integer exponent for eq(d) in the range $1 \le m \le 2$ extends its application beyond the quadratic form. In particular, by taking $a = b = 0$ and $f = g = 0$, this equation has been shown to predict the plastic strain ratios arising from applying uniaxial and equi-biaxial tension to rolled, polycrystalline sheets with planar-isotropy [51, 52]. Interestingly, in 1935, Bailey [53] predicted multiaxial, secondary creep rates observed in steam piping at 850°C. The anisotropic flow potential, which was given later by Davis [54], becomes similar to eq(d) when a, b and c are set to zero. *Note*: It is permissible to compare the creep potential to a yield function in this way when creep rates identify with time-dependent plastic strains. Hill [46] showed that the terms containing the constants f, g, h and m, when taken alone, cannot admit all forms of anisotropy. For example, the four or more ears, which have appeared from cupping rolled sheet metal, are unpredictable. The quadratic function, eq(3.27), can provide for two and, at most, four ears in certain cases. An account of more ears requires a higher order polynomial yield function. Bourne and Hill [47] showed that with five anisotropy coefficients, their cubic stress function (e) matched the six ears observed in brass. A quartic function (f), proposed by Gotoh [48], represents up to a maximum of eight ears observed in soft aluminium and its alloy. The explicit, plane stress, form of eq(f) is

$$f = A_1\sigma_x^4 + A_2\sigma_x^3\sigma_y + A_3\sigma_x^2\sigma_y^2 + A_4\sigma_x\sigma_y^3 + A_5\sigma_y^4 + (A_6\sigma_x^2 + A_7\sigma_x\sigma_y + A_8\sigma_y^2)\tau_{xy}^2 + A_9\tau_{xy}^4$$

where eight coefficients $A_2 \ldots A_9$, together with $A_1 = 1$, provide a description of eight ears. A further term: $A_0(\sigma_x + \sigma_y)^2$, is added where an account of compressibility is required. These coefficients are found from the yield stresses under uniaxial tension and either through-thickness plane strain compression, or in-plane equi-biaxial tension. Further consideration of the yield criteria employed for rolled sheet metals is given in chapter 12 (see § 12.8).

3.4.2 The Bauschinger Effect

Quadratic stress yield functions, in eqs(3.27) and (3.28), assume that the tensile and compressive yield stresses are equal for each principal direction in the material. This also applies to all other homogenous stress functions in Table 3.3. It should be noted that Hill had originally intended his 1948 function for backward-extrapolated yield points, which do show similar stresses in tension and compression. The unequal tensile and compressive yield stresses which appear within non-homogenous yield functions are consistent with other yield point definitions. The term *Bauschinger effect* is adopted here to denote this form of asymmetry in the initial yield surface, though the effect is more often associated with a subsequent yield surface (as with *f* in Fig. 3.7). A simple account of different positive and negative yield stress values appears with linear stress terms in the yield function. Two such functions are given in Table 3.3: eq(g) in plane principal space and eq(h) in general stress space. Stassi-D'Alia [49] examined both forms after placing restrictions on certain constants. One can arrive at a more general formulation of linear plus quadratic stress terms when yielding remains independent of a superimposed hydrostatic stress. Equation (3.29) is modified to

$$f(\sigma_{ij}') = H_{ij}\sigma_{ij}' + \tfrac{1}{2}H_{ijkl}\,\sigma_{ij}'\,\sigma_{kl}' = 1 \qquad (3.30)$$

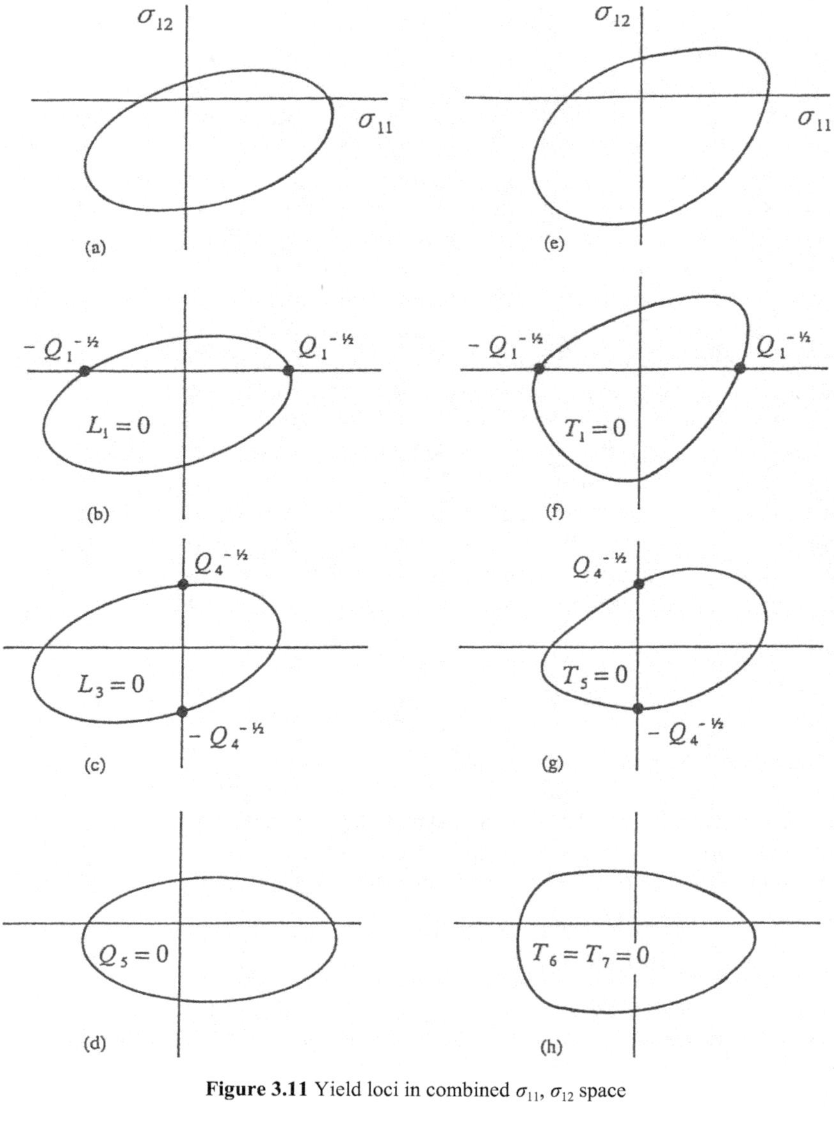

Figure 3.11 Yield loci in combined σ_{11}, σ_{12} space

When the 1 and 2–axes of orthotropy coincide with the axes of a plane stress state, σ_{11}, σ_{22} and σ_{12}, eq(3.30) expands into

$$f = L_1\sigma_{11} + L_2\sigma_{22} + L_3\sigma_{12} + Q_1\sigma_{11}^2 + Q_2\sigma_{22}^2 + Q_3\sigma_{11}\sigma_{22} + Q_4\sigma_{12}^2 + Q_5\sigma_{11}\sigma_{12} + Q_5\sigma_{22}\sigma_{12} = 1 \tag{3.31a}$$

The coefficients L and Q are derived from H_{ij} and H_{ijkl} using plastic incompressibility [55]. In a reduced stress space, σ_{11} and σ_{12}, eq(3.30) becomes

$$f = L_1\sigma_{11} + L_3\sigma_{12} + Q_1\sigma_{11}^2 + Q_4\sigma_{12}^2 + Q_5\sigma_{11}\sigma_{12} = 1 \tag{3.31b}$$

Figures 3.11a–d show four cases of anisotropy provided by eq(3.31b). The coefficients L and Q provide equal and unequal yield stress intercepts as shown. When tensile and compressive yield stresses differ, they become the roots to $Q_1\sigma_{11}^2 + L_1\sigma_{11} - 1 = 0$, in Figs 3.11a,c,d. When forward and reversed shear yield stresses differ, they become the roots to $Q_4\sigma_{12}^2 + L_3\sigma_{12} - 1 = 0$, in Figs 3.11a,b,d. The product term $Q_5\sigma_{11}\sigma_{12}$, in eq(3.31b), provides an inclination to the yield locus, as shown in Figs 3.11a,b,c. A simple check for the inclination is whether axial strain is produced from torsion. This requires an examination of the direction of the outward normal at a point of intersection between the yield locus and the axis σ_{12}. An inclined normal has a component of axial strain aligned with the axis σ_{11}, in addition to a shear strain component, aligned with the axis σ_{12}. This effect is due to anisotropy and is not the second-order effect, previously associated with J_3'. Where no axial strain is observed, Q_5 is set to zero, resulting in the locus of Fig. 3.11d.

When shear stress is absent along a material's orthotropic directions 1 and 2, eq(3.31a) shows that the principal, biaxial stresses σ_1 and σ_2 appear in non dimensional form:

$$f = L_1(\sigma_1/\sigma_{1t}) + L_2(\sigma_2/\sigma_{1t}) + Q_1(\sigma_1/\sigma_{1t})^2 + Q_2(\sigma_2/\sigma_{1t})^2 + Q_3(\sigma_1/\sigma_{1t})(\sigma_2/\sigma_{1t}) = 1 \tag{3.31c}$$

in which σ_{1t} is the tensile yield stress in the 1–direction. Applications of eq(3.31c) have shown good agreement with initial yield loci for anisotropic sheets of Ti-Al alloy and zircaloy [55]. An example of the former material [56] is shown in Fig. 3.12.

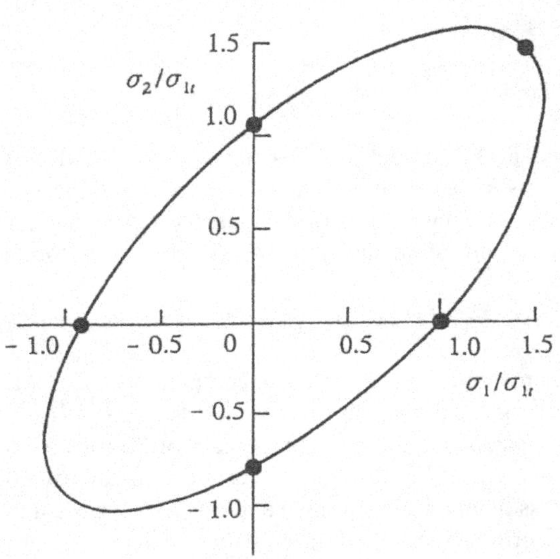

Figure 3.12 Initial yield locus for anisotropic Ti-Al alloy

It is apparent that, within this elliptical function, the five yield points may be fitted exactly with an equal number of coefficients in L and Q. Comparisons of this kind are best made at the limit of proportionality since an equivalent, anisotropic, offset strain cannot be defined. Moreover, if yield were to be so defined, anisotropy, initially present in the yield locus, would diminish with increasing offset strain.

To recover an initial Mises condition from eq(3.30), identify H_{ij} and H_{ijkl} with the corresponding isotropic tensors:

$$H_{ij} = \delta_{ij} \tag{3.32}$$

and, for the general fourth-order isotropic tensor [57]:

$$H_{ijkl} = I_{ijkl} = \lambda\delta_{ij}\delta_{kl} + \mu\delta_{ik}\delta_{jl} + \nu\,\delta_{il}\delta_{kj} \tag{3.33a}$$

Substituting eqs(3.32) and (3.33a) into eq(3.30) gives

$$\begin{aligned} f(\sigma_{ij}') &= \delta_{ij}\sigma_{ij}' + \tfrac{1}{2}\,(\lambda\delta_{ij}\delta_{kl} + \mu\delta_{ik}\delta_{jl} + \nu\,\delta_{il}\delta_{kj})\,\sigma_{ij}'\,\sigma_{kl}' \\ &= \sigma_{ii}' + \tfrac{1}{2}\,(\lambda\sigma_{ii}'\,\sigma_{kk}' + \mu\sigma_{kj}'\,\sigma_{kj}' + \nu\,\sigma_{lj}'\,\sigma_{jl}') \\ &= \tfrac{1}{2}\,(\,\mu\,\sigma_{kj}'\,\sigma_{kj}' + \nu\,\sigma_{lj}'\,\sigma_{lj}'\,) \end{aligned}$$

where $\sigma_{ii}' = \sigma_{kk}' = 0$. Putting $\mu = \nu = \tfrac{1}{2}$ leads to the von Mises form

$$f(\sigma_{ij}') = \tfrac{1}{2}\,\sigma_{ij}'\,\sigma_{ij}'$$

It follows that, by setting $\lambda = 0$ and $\mu = \nu = \tfrac{1}{2}$ in eq(3.33a), a reduced form of fourth order, isotropic tensor is suited to stress deviators:

$$I_{ijkl} = \tfrac{1}{2}\,(\delta_{ik}\,\delta_{jl} + \delta_{il}\,\delta_{jk}) \tag{3.33b}$$

since this will lead directly to the Mises condition. *Note*: Equation (3.30) describes an ellipsoidal yield surface and will not account for any distortion that may be present initially.

3.4.3 Distorted Yield Loci

Experiments described in Tables 3.3 and 3.4 have often revealed distorted yield loci. Distortion, which appears to be a dominant feature in certain alloys, can be modelled when cubic stress terms appear in the yield function. Consider, for example, an anisotropic yield function formed from the sum of quadratic and cubic stress deviator terms [58]:

$$f(\sigma_{ij}') = \tfrac{1}{2}\,H_{ijkl}\sigma_{ij}'\,\sigma_{kl}' + \tfrac{1}{3}\,H_{ijklmn}\,\sigma_{ij}'\,\sigma_{kl}'\,\sigma_{mn}' = 1 \tag{3.34}$$

Restrictions upon the number of sixth-order tensor components H_{ijklmn} follow from symmetry in the stress tensor: $\sigma_{ij}' = \sigma_{ji}'$, $\sigma_{kj}' = \sigma_{jk}'$ and $\sigma_{mn}' = \sigma_{nm}'$. This gives $H_{ijklmn} = H_{jiklmn} = H_{ijlkmn} = H_{ijklnm}$. With the axes of stress coincident with material orthotropy axes, then furthermore: $H_{ijklmn} = H_{klmnij} = H_{mnijkl} = H_{ijmnkl} = H_{mnklij} = H_{klijmn}$. Moreover, a condition of incompressibility is implied in using stress deviators in eq(3.34) and this reduces the number of coefficients to 35. Hill's modified orthotropic function, eq(d) in Table 3.3, is one form of eq(3.34), where $m = 2$ in the first line of eq(d) and $m = 3$ in the second line. The reduction in eq(3.34) to a direct stress combined with a shear stress becomes:

$$f = Q_1 \sigma_{11}^{\ 2} + Q_4 \sigma_{12}^{\ 2} + Q_5 \sigma_{11} \sigma_{12} + T_1 \sigma_{11}^{\ 3} + T_5 \sigma_{12}^{\ 3} + T_6 \sigma_{11} \sigma_{12}^{\ 2} + T_7 \sigma_{11}^{\ 2} \sigma_{12} = 1 \tag{3.35a}$$

Figures 3.11e–h apply to eq(3.35a). These contrast with the quadratic predictions (Fig. 3.11a–d) for four similar cases of anisotropy described previously. Clearly distortion is now a consistent feature within each case. Equation (3.34) can be written in a non dimensional form when principal, biaxial stress directions are aligned with orthotropic axes 1 and 2:

$$Q_1 (\sigma_1/\sigma_{1t})^2 + Q_2 (\sigma_2/\sigma_{1t})^2 + Q_3 (\sigma_1/\sigma_{1t})(\sigma_2/\sigma_{1t})$$
$$+ T_1 (\sigma_1/\sigma_{1t})^3 + T_2 (\sigma_2/\sigma_{1t})^3 + T_3 (\sigma_1/\sigma_{1t})^2 (\sigma_2/\sigma_{1t}) + T_4 (\sigma_1/\sigma_{1t})(\sigma_2/\sigma_{1t})^2 = 1 \tag{3.35b}$$

The coefficients Q and T in eqs(3.35a,b) are derived from H_{ijkl} and H_{ijklmn} in [48]. Equations (3.30) and (3.35b) are applied to an initial yield locus for extruded magnesium in Fig. 3.13.

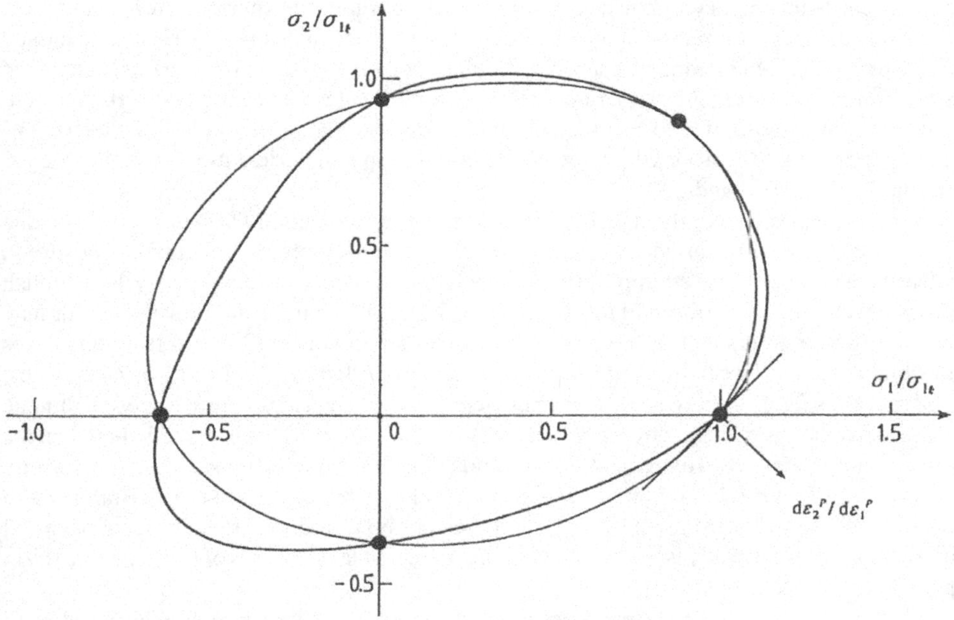

Figure 3.13 Yield locus for an orthotropic magnesium extrusion

The five constants, L and Q in eq(3.30), are determined directly from the five experimental yield points shown. The determination of seven constants in eq(3.35b) requires an additional yield point (interpolated) and a strain vector direction as indicated. It is seen that both predictions represent the pronounced 'Bauschinger effect' along each axis of orthotropy. Both quadratic and cubic yield functions appear to describe the measured initial yield points equally well. However, eq(3.35b) provides the better account of distortion appearing in a locus connecting these yield points. Such distortion is linked to the direction of the plastic strain increment vector through the normality rule, i.e. this vector is aligned with the direction of the exterior normal to the yield locus. It appears from Fig. 3.13 that additional yield points would be required to test an initial distortion prediction more precisely. In general, the literature reveals far stronger evidence for a strain-induced form of distortion within a subsequent yield locus than there is for a distortion within the initial yield surface.

When it is required to recover a condition of isotropy from eq(3.34), H_{ijkl} is defined from eq(3.33b) and H_{ijklmn} appears as a general sixth order isotropic tensor [23]:

$$H_{ijklmn} = I_{ijklmn} = a\delta_{ij}\delta_{kl}\delta_{mn} + b(\delta_{ij}\delta_{km}\delta_{ln} + \delta_{ij}\delta_{kn}\delta_{ml} + \delta_{kl}\delta_{im}\delta_{jn} + \delta_{kl}\delta_{in}\delta_{jm} + \delta_{mn}\delta_{ik}\delta_{jl} + \delta_{mn}\delta_{il}\delta_{jk})$$
$$+ c(\delta_{ik}\delta_{lm}\delta_{nj} + \delta_{ik}\delta_{jm}\delta_{nl} + \delta_{il}\delta_{jm}\delta_{nk} + \delta_{il}\delta_{jn}\delta_{mk} + \delta_{im}\delta_{jk}\delta_{ln} + \delta_{im}\delta_{jl}\delta_{kn} + \delta_{in}\delta_{jk}\delta_{lm} + \delta_{in}\delta_{jl}\delta_{km}) \quad (3.36)$$

Substituting from eq(3.36) reduces the second term in eq(3.34) to the third deviatoric invariant: $J_3' = \frac{1}{3}\sigma_{ij}'\,\sigma_{jk}'\,\sigma_{ki}'$, when $a = b = 0$ and $c = \frac{1}{8}$. It follows that the first two terms in eq(3.36) are unnecessary for the recovery of this isotropic stress deviator.

3.5 Choice of Yield Function

It has been seen that the classical, von Mises and Tresca theories of yielding for metals may each be formulated as a function of the stress deviator invariants. Such macroscopic predictions to the initial yield surface provide for all possible stress combinations. However, these do not offer information about the micro-structural mechanisms of yielding, discussed in chapter 8. In consideration of subsequent yielding, where plastic strain paths, calculated from $\varepsilon^P = \varepsilon^T - \sigma/E$ (see Fig. 1.1), remain linear, this indicates that a simple rule of *isotropic hardening* may be applied for loading within the plastic range. This employs the concept of an expanding yield surface which retains its initial shape and orientation, to be developed further in chapters 9 and 10.

It has been seen that the departure from the isotropic function $f(J_2')$ may either be due to the influence of J_3' in an isotropic material, or, to the presence of initial anisotropy. Clearly, when selecting an appropriate yield function, checks are necessary to establish precisely the initial condition of the material. Anisotropic yielding and flow behaviour may be identified with distinct differences in the stress–strain curves obtained from testpieces machined from different directions in the material. For rolled sheets, the off-axis tensile test will reveal differences between directional yield stresses when anisotropy is present, but the more usual measure of anisotropy in sheet metal is the r–value. This is the ratio between the plastic components of an incremental width and thickness strain in a tension test. Anisotropy is revealed when $r \neq 1$. The forming of sheet metal is enhanced when $r > 1$. Bramley and Mellor [59] showed, from eq(3.27), how it was possible to describe the effects of initial anisotropy in transversely isotropic sheets using the constant gradients of linear plastic strain paths (see § 12.5, chapter 12).

Alternatively, when isotropy is assumed in a tension test the material may be taken to conform to the general isotropic function $f(J_2', J_3')$. Substituting this for f into eq(3.22) gives the plastic strain increment tensor

$$d\varepsilon_{ij}^P = d\lambda\, [\, (\partial f/\partial J_2')(\partial J_2'/\partial\sigma_{ij}) + (\partial f/\partial J_3')(\partial J_3'/\partial\sigma_{ij})] \quad (3.37)$$

If σ_1 is the non-zero tensile stress, it follows from eqs(3.10b,c) that $\partial J_2'/\partial\sigma_1 = 2\sigma_1/3$, $\partial J_2'/\partial\sigma_2 = -\,\sigma_1/3$, $\partial J_3'/\partial\sigma_1 = 2\sigma_1^2/9$ and $\partial J_3'/\partial\sigma_2 = -\,\sigma_1^2/9$. Then, from eq(3.37), the axial and lateral plastic strain increments become

$$d\varepsilon_1^P = d\lambda\, [(\partial f/\partial J_2')(2\sigma_1/3) + (\partial f/\partial J_3')(2\sigma_1^2/9)] \quad (3.38a)$$
$$d\varepsilon_2^P = d\lambda\, [(\partial f/\partial J_2')(-\sigma_1/3) + (\partial f/\partial J_3')(-\sigma_1^2/9)] \quad (3.38b)$$

Dividing eqs(3.38a,b) gives the constant ratio $d\varepsilon_2^P/d\varepsilon_1^P = -\frac{1}{2}$, irrespective of the yield function $f(J_2', J_3')$. That is, the gradient of the lateral versus axial plastic strain plot remains linear, with a gradient of $-\frac{1}{2}$. Figures 3.14a–e compares this isotropic prediction with experimental plastic strain paths for aluminium, steel, copper and brass.

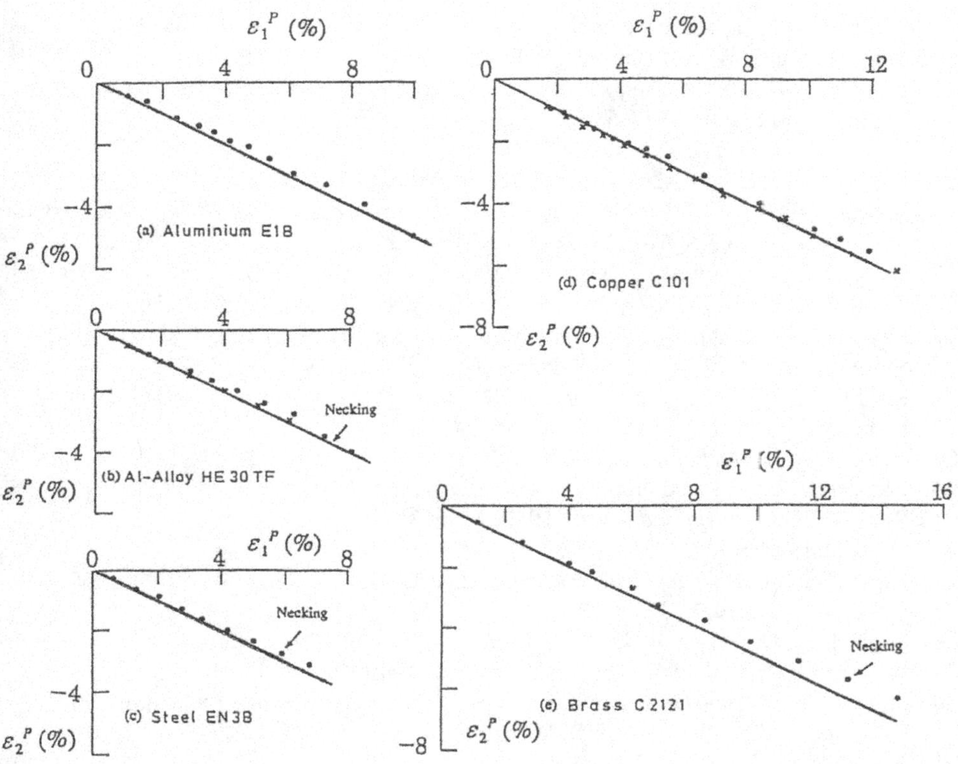

Figure 3.14 Lateral versus axial plastic strain paths under tension

The agreement found between the gradient of each plot and the theoretical gradients value of − ½ confirms that testpieces machined longitudinally from extruded bars become almost isotropic following heat treatment. In contrast, the inter-stage anneals employed for rolling sheet metal may not leave the material in an isotropic condition. It will be seen later that it is desirable to retain anisotropy arising from rolling when it serves to enhance forming.

References

1. Rees D. W. A. *The Mechanics of Solids and Structures*, 2nd Ed, 2016, IC Press, U.K.
2. von Mises R. *Nachr. Ges. Wiss*. Gottingen, 1913, 582. (also *ZAMM* 1928, **8**, 161)
3. Tresca H. "Memoire sur l'ecoulement des corps solides", *Memoirs Par Divers Savants*, Paris, 1968, **18**, 733 and 1972, **20**, 75.
4. Ikegami K. J. *Soc. Mat. Sci*. 1975, **24**(261), 491 and **24**(263), 709 (see also BISI translation 14420, The Metals Soc., Sept 1976)
5. Michno M. J. and Findley W. N. *Int. J. Non-Linear Mech*., 1976, **11**, 59.
6. Banabic D. et al, Proceedings: *4th ESAFORM Conf. on Material Forming*, (Ed. Habraken A. M.) April 2001, Vol. 1, p. 297, University of Liége, Belgium.
7. Shahabi S. N. and Shelton A. J. *Mech. Eng. Sci*., 1975, **17**, 82.
8. Johnson K. R. and Sidebottom O. M. *Expl. Mech*., 1972, **12**, 264.

9. Moreton D. N., Moffat D. G. and Parkinson D. B. *J. Strain Anal.*, 1981, **16**, 127.
10. Miastkowski J. and Szczepinski W. *Int. J. Solids and Structures*, 1965, **1**, 189.
11. Marin J., Hu L. W. and Hamburg J. F. *Proc. A. S. M.*, 1953, **45**, 686.
12. Lessels J. M. and MacGregor, C. W. *Jl Franklin Inst.*, 1940, **230**, 163.
13. Rogan J. and Shelton A. *J. Strain Anal.*, 1969, **4,** 127.
14. Ivey H. J. *J. Mech. Eng. Sci.*, 1961, **3**, 15.
15. Phillips A. and Tang J-L. *Int. J. Solids and Struct.*, 1972, **8**, 463.
16. Kowalewski Z. L., Dietrich L. and Socha G. Proceedings: *Third Int. Congress on Thermal Stress*, 1999, 181–184, University of Kracow.
17. Shiratori E., Ikegami K. and Kaneko K. *Trans. Japan Soc. Mech. Engrs.*, 1973, **39**, 458.
18. Ellyin F. and Grass J-P. *Trans. Can. Soc. for Mech. Engrs.*, 1975, **3**, 156.
19. Taylor G. I. and Quinney, H. *Phil. Trans. Roy. Soc. A*, 1931, **230**, 323.
20. Shrivastava H. P., Mroz, Z. and Dubey, R. N. *ZAMM*, 1973, **53**, 625.
21. Rees D. W. A. *Proc. R. Soc. Lond.*, 1982, **A383**, 333.
22. Drucker D. C. *J. Appl. Mech.*, 1949, **16**, 349.
23. Betten J. *Acta Mechanica*, 1976, **25**, 79.
24. Freudenthall A. M. and Gou P. F. *Acta Mechanica*, 1969, **8**, 34.
25. Rees D. W. A. *J. Applied Mech.*, 1982, **49**, 663.
26. Crossland B. *Proc. I. Mech. E.*, 1954, **168**, 935.
27. Hu L. W. Proc: *Naval Structures*, (Eds Lee E. H. and Symonds P. S.) 1960, 924, Pergamon, Oxford, U. K.
28. Pugh H. Ll. D. and Green D. *Proc. I. Mech. E.*, 1964, **179**, 1.
29. Ratner S. I. *Zeitschrift fur Technische Physik*, 1949, **19**, 408.
30. Fung P. K., Burns D. J. and Lind N. C. Proc: *Foundation of Plasticity*, (Ed. A. Sawczuk) 1973, 287, Noordhoff, Warsaw.
31. Ros M. and Eichinger A. *Eidgenoess, Materialpruef. Versuchanstalt Ind. Bauw.*, 1929, **34**, 1 Gewerbe, Zurich.
32. Brandes M. in: *The Mechanical Behaviour of Materials Under Pressure*, (Ed. Pugh H. Ll.) 1971, Ch. 6, 236, Applied Science, London.
33. Stassi-D'Alia F. *Meccanica*, 1967, **2**, 178.
34. Hill R. *Proc. Roy. Soc.*, 1948, **A193**, 281.
35. Edelman F. and Drucker D. C. *J. Franklin Inst.*, 1951, **251**, 581.
36. Hearmon R. F. S. *Applied Anisotropic Elasticity*, 1961, Clarendon Press, Oxford.
37. Olsak W. and Urbanowski W. *Arch. of Mech.*, 1956, **8**, 671.
38. Sobotka, Z. *ZAMM*, 1969, **49**, 25.
39. Troost A. and Betten J. *Mech. Res. Comms.*, 1974, **1**, 73.
40. Fava F. *J. Annals of the C.I.R.P.*, 1967, **15**, 411.
41. Hu L. W. *J. Appl. Mech.*, 1956, **23**, 444.
42. Hazlett T. H., Robinson A. T. and Dorn J. E. *Trans. ASME*, 1950, **42**, 1326.
43. Jones S. E. and Gillis P. P. *Met. Trans. A*, 1984, **15A**, 129.
44. Hu L. W. and Marin J. *J. Appl. Mech.*, 1955, **22**, 77.
45. Takeda T. and Nasu Y. *J. Strain Analysis*, 1991, **26**, 47.
46. Hill R. *Math. Proc. Camb. Phil. Soc.*, 1979, **85**, 179.
47. Bourne L. and Hill R. *Phil. Mag.*, 1950, **41**, 671.
48. Gotoh M. *Int. J. Mech. Sci.*, 1977, **19**, 505.
49. Stassi-D'Alia F. *Meccanica*, 1969, **4**, 349.
50. Harvey S. J., Adkin P. and Jeans P. J. *Fatigue of Eng. Mats. and Struct.*, 1983, **6**, 89.
51. Dodd B. *Int. J. Mech. Sci.*, 1984, **12**, 587.
52. Parmar A. and Mellor P. B. *Int. J. Mech. Sci.*, 1978, **20**, 385.

53. Bailey R. W. *Proc. I. Mech. E.*, 1935, **131**, 131.
54. Davis E. A. *Trans. ASME*, 1961, **28E**, 310.
55. Rees D. W. A. *Acta Mechanica*, 1982, **43**, 223.
56. Lee D. and Backofen W. A. *Trans. Met. Soc. AIME*, 1966, **236**, 1077.
57. Spencer A. J. M. *Continuum Mechanics*, 1980, Longman, London.
58. Rees D. W. A. *Acta Mechanica*, 1984, **52**, 15.
59. Mellor P. B. in: *Mechanics of Solids*, (Eds Hopkins H. G. and Sewell M. J.) 1982, 383, Pergamon Press, Oxford, U.K.

Exercises

3.1 Principal stresses (in MPa) $\sigma_1 = 250$, $\sigma_2 = 25$ and $\sigma_3 = 10$ are applied to a material whose uniaxial yield strength is 300 MPa. Find the safety factors imposed by the von Mises and Tresca criteria of yielding. Show that principal stress combinations: $\sigma_1 \geq \sigma_2 \geq \sigma_3$ are equally safe by von Mises and Tresca when they conform to the condition: $(\sigma_1 - \sigma_2)(\sigma_2 - \sigma_3) = 0$. In how many ways could the applied stresses be altered to satisfy this condition?
Ans: 1.288, 1.25

3.2 Principal stresses (in MPa) $\sigma_1 = 200$, $\sigma_2 = 100$ and $\sigma_3 = -100$ exist within a material whose uniaxial yield strength is known to be 400 MPa. Find the safety factors imposed by the von Mises and Tresca criteria of yielding. Explain geometrically the difference between these safety factors. By changing σ_2 find the principal stress combinations for which the two safety factors are equal.
Ans: 1.51, 1.33

3.3 A steel pressure vessel having internal and external radii 170 mm and 175 mm is required to contain a fluid at a gauge pressure of 80 bar. Determine the safety factors that apply based upon the Tresca and von Mises criteria of yielding given a tensile yield strength of 275 MPa. What internal pressure would cause the vessel to yield by von Mises? If this pressure is exceeded by a small amount find the ratio of the principal plastic strain increments between the axial and circumferential directions ($\delta\varepsilon_z^P/\delta\varepsilon_\theta^P$) and between the radial and circumferential directions ($\delta\varepsilon_r^P/\delta\varepsilon_\theta^P$). Take the stresses in these directions to be given by: $\sigma_z = pr/2t$, $\sigma_\theta = pr/t$ and $\sigma_r = 0$, where r is the internal radius, t is the wall thickness and p is the gauge pressure.
Ans: 1.01, 1.167; 93.4 bar; 0, −1

3.4 A thin walled steel cylindrical pressure vessel with hemispherical ends is required to contain a fluid at a gauge pressure of 50 bar. The cylinder has internal and external radii 170 mm and 175 mm. Each hemisphere has and internal radius of 170 mm but the outer radius is chosen to ensure compatibility in hoop strain at each welded join. Show that ratio between spherical and cylindrical thicknesses is 3/7 given a Poisson's ratio of ¼ for steel. Determine the safety factors that apply to the vessel's body and ends based upon the Tresca and von Mises criteria of yielding for a tensile yield strength of 275 MPa. Is the vessel safe? The hoop, meridional and radial stresses for the thin walled ends are given by: $\sigma_\theta = pr/2t$, $\sigma_\varphi = pr/2t$ and $\sigma_r = 0$. Refer to exercise 3.3 for the cylinder stresses.
Ans: 1.618, 1.867; 1.39, 1.39 (Yes)

3.5 Find the internal pressure to cause yielding of a thin walled cylindrical vessel of inner diameter 300 mm and wall thickness 5 mm when radial stress is (a) ignored and (b) taken to equal the internal pressure. Compare predictions for (a) and (b) according to the Tresca and von Mises criteria of yielding with a material uniaxial yield stress of 300 MPa. Refer to exercise 3.4 for the cylinder's membrane stresses.
Ans: (a) 100, 115.5; (b) 103.5, 119.4 (bar)

3.6 Show that the constants *c*, *b* and *d* in eqs(3.15a–c) are defined from the relationship between Y and k as follows:

(a) $Y = k\,(1/27 - 4c/9^3\,)^{1/6}$,

(b) $Y = 9k\,/\,(27 - 4b)^{½}$

(c) $Y = k\,/\,(1/3^{3/2} - 2d/27)^{1/3}$.

See solution

3.7 Normalise the Stassi fracture criterion in eq(3.25b) with the compressive fracture stress σ_c and plot the family of loci in σ_1, σ_2 space for $\rho = \sigma_c / \sigma_t = 2$, 3 and ∞.

3.8 Construct the family of yield loci in σ, τ space from the unsymmetrical yield function:

$$f = J_2' + (p\,/\,Y)\,J_3'$$

within the range $1.5 \leq p \leq -3$.

3.9 Construct a family of yield loci for the unsymmetrical yield function (3.16) for $n = 2$. Determine the range of *p*–values that will ensure a closed yield surface.

3.10 Establish the right-hand sides of the expressions (c) and (d) given in Table 3.3 in terms of the known yield stress σ_{1y} along the principal 1–axis of orthotropy.

3.11 Compare the yield functions (a)–(h), listed in Table 3.3, when they are reduced to a principal biaxial stress space, i.e. for $\sigma_3 = 0$ and $\tau_{xy} = \tau_{xz} = \tau_{yz} = 0$, as appropriate.

3.12 Derive from eq(3.22) the principal, plastic constitutive relations corresponding to each of the isotropic yield functions given in eqs(3.15a–c).

3.13 The coefficients L_1, L_2, Q_1, Q_2 and Q_3 in eq(3.31c) can account for four cases in which the tensile and compressive yield stresses in the 1 and 2–directions may be the same or different under a principal biaxial stress state σ_1, σ_2. Show each case with a plot similar to Fig. 3.11a–d.

3.14 Show how the coefficients T_1, T_2, T_3, T_4, Q_1, Q_2 and Q_3 in eq(3.35b) can account for several conditions of anisotropy under a principal biaxial stress state σ_1, σ_2. Present the locus for each case with plots similar to Figs 3.11e–h.

3.15 Reduce eq(3.27) to criteria of yielding under a principal, biaxial stress state (σ_1, σ_2) for material: (a) orthotropy and (b) transverse isotropy, where yield stresses are the same in the 2– and 3–directions.
See solution

3.16 Show that Hill's yield criterion (3.27) reduces to a plane stress form:

$$(G + H)\sigma_x^{\,2} - 2H\,\sigma_x\,\sigma_x + (H + F)\sigma_x + 2N\tau_{xy}^{\,2} = 1$$

when plane stresses σ_x, σ_y and τ_{xy} are applied within the plane of a thin sheet and aligned with the orthotropic axes *x* and *y* (axis *z* is through the thickness). Use this equation to show that the variation in yield stress Y with orientation θ to x within the plane of the sheet, is given by:

$$Y(\theta) = \frac{1}{\sqrt{\left[F\sin^2\theta + G\cos^2\theta + H + \frac{1}{4}(2N - F - G - 4H)\sin^2 2\theta\right]}}$$

3.17 If, in exercise 3.16, the yield stresses in the three orthogonal directions x, y and z are denoted by X, Y and Z respectively and P is the shear yield stress in the x–y plane of the sheet, show that the coefficients F, G, H and N are given by:

$$F = \frac{1}{2}\left(\frac{1}{Y^2} + \frac{1}{Z^2} + \frac{1}{X^2}\right),$$

$$G = \frac{1}{2}\left(\frac{1}{Z^2} + \frac{1}{X^2} + \frac{1}{Y^2}\right),$$

$$H = \frac{1}{2}\left(\frac{1}{X^2} + \frac{1}{Y^2} + \frac{1}{Z^2}\right),$$

$$N = \frac{1}{2P^2}$$

3.18 Using the orthotropic yield function in eq(3.27), examine how you would predict the onset of yielding in a rolled sheet metal when a direct stress σ is combined with a shear stress τ in each of the following cases:
(a) when both σ and τ are aligned with the in-plane orthotropic axes x and y
(b) when they are inclined at θ to x and y, as shown in Fig. 3.15.

Note: The presence of complementary shear applies to each case.

See solution: (a) $\sigma_x^2 + \dfrac{2N}{H + G}\tau_{xy}^2 = X^2$ where X is the yield stress in the x–direction.

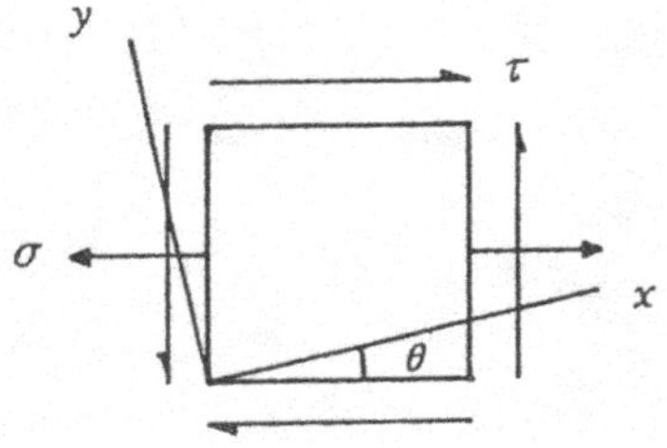

Figure 3.15

3.19 Adopting Hill's orthotropic function, examine how you would predict yielding from eq(3.27) in an orthotropic sheet under in-plane biaxial stresses σ_1 and σ_2 for each of the following cases:
(a) when σ_1 and σ_2 are aligned with the in-plane orthotropic directions x and y respectively
(b) when σ_1 and σ_2 are inclined at θ to x and y, as shown in Fig. 3.16.

Note: Axis z is radial, normal to the thickness.

Ans: (i) $\sigma_x^2 - \dfrac{2H}{H + G}\sigma_x\sigma_y + \dfrac{H + F}{H + G}\sigma_y^2 = X^2$ where X is the yield stress in the x–direction.

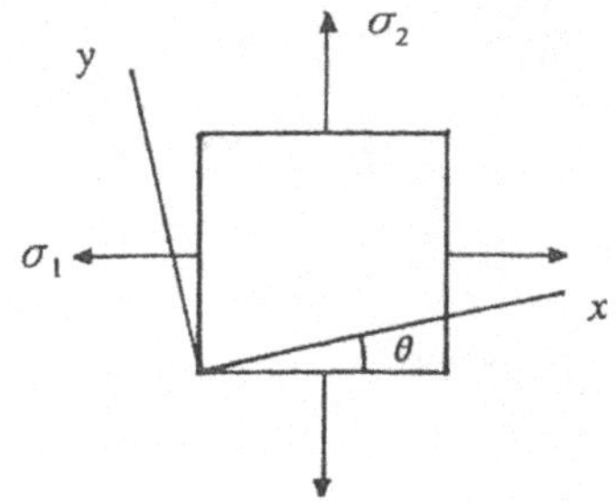

Figure 3.16

3.20 A cast iron column has ultimate tensile and compressive strengths of 300 and 900 MPa respectively. The column is required to carry both axial compression and torsion. Find: (i) the critical combination of axial and shear stress at failure when applied in the ratio 4/1and (ii) when the compressive stress is increased in the presence of a constant shear stress of 75 MPa.

3.21 Construct in the manner of Fig. 3.8 a family of unsymmetrical yield loci in non-dimensional co-ordinates $x = \sigma/Y$ and $y = \tau/Y$. Take $n = 1$ in eq(3.16) with $1.5 \leq p \leq -3$.

CHAPTER 4

NON-HARDENING PLASTICITY

4.1 Introduction

Non-hardening theories of plasticity are very useful for providing solutions to practical problems with static indeterminancy, i.e. where satisfying force equilibrium conditions at yield is insufficient. Solutions are reached from combining the yield condition with equations of equilibrium with and strain compatibility. Here it is convenient to distinguish between uniform and non-uniform stress states. Two, classical multiaxial theories are applied which are referred to as Hencky-Ilyushin and Prandtl-Reuss. Though formulated differently, each theory can provide the total strain arising from practical, non-uniform, stress states. The question arises as to which theory is the more reliable? To answer this, comparisons are made between predictions to the deformation in solid and hollow bars of circular section, subjected to combinations of axial load and torque and experimental results.

4.2 Classical Theories of Plasticity

Two classical theories of plasticity are the total strain (or deformation) theory of Hencky [1], Illyushin [2] (also Nadai [3]) and the incremental (or flow) theory of Prandtl-Reuss [4, 5]. It is instructive to examine the manner in which each general theory accounts for the plasticity of a non-hardening material subjected to both uniform and non-uniform stress states. It will be seen that, although each theory is constructed from different assumptions, they do provide comparable solutions to the load-deformation behaviour and the internal stress distributions under simple loading paths. Firstly, an outline of each theory is given.

4.2.1 Hencky-Ilyushin

In the formulation of the deformation theory of non-hardening plasticity, total strains are used. The strains that arise from the application of normal stresses, within the plastic range, are the sum of deviatoric and volumetric strain components. The latter is associated with the mean normal stress. Shear deformation occurs independently under the application of shear stress. In this way the total strain tensor must appear, in the respective tensor subscript and matrix notations, as

$$\varepsilon_{ij}^{\,t} = \varphi\, \sigma_{ij}' + (\delta_{ij}\,\sigma_{kk}) \,/\, (9K) \quad \text{or} \quad \mathbf{E} = \varphi\, \mathbf{T}\,' + (\mathrm{tr}\,\mathbf{T})\,\mathbf{I} \,/\, (9K) \qquad (4.1\text{a,b})$$

where the scalar $\varphi > 0$ during loading and $\varphi = 0$ with unloading. The use of the Kronecker delta δ_{ij} in eq(4.1a) ensures that the final volumetric strain term does not contribute to the components of shear strain, since $\delta_{ij} = 0$ for $i \neq j$. The absolute stress appears with the substitution of the deviatoric (or reduced) stress expression $\sigma_{ij}' = \sigma_{ij} - \delta_{ij}\sigma_{kk}/3$, into eq(4.1a)

$$\varepsilon_{ij}^{\,t} = \varphi\,\sigma_{ij} + [(1 - 2\nu)/E - \varphi\,]\delta_{ij}\sigma_{kk}/3 \tag{4.2a}$$

or

$$\mathbf{E} = \varphi\,\mathbf{T} + [1/(3K) - \varphi\,]\,(\mathrm{tr}\,\mathbf{T})\,\mathbf{I}\,/\,3 \tag{4.2b}$$

where $E = 3K(1 - 2\nu)$ connects the three elastic constants E, K and ν. Equation (4.2a) indicates a considerable simplification from assuming $\nu = ½$, i.e. for an elastically incompressible material. This provides a good approximation to the total strain tensor:

$$\varepsilon_{ij}^{\,t} = \varphi\,\sigma_{ij} - \varphi\,\delta_{ij}\,\sigma_{kk}/\,3 \tag{4.2c}$$

This total strain theory has been used mostly for the determination of limit loads in structures of elastically–perfectly plastic material [6, 7]. Such loading involves combinations of tension, torsion, bending, pressure and shear force. The theory provides for the resulting stress and strain distributions of structures in both the elastic–plastic and fully–plastic conditions [6]. Equations (4.2a,b) are not linked to a particular yield function and therefore may be combined with any suitable yield function, including those of Tresca and von Mises. The plastic strain supplied by eq(4.2b) is the earliest example of a *non associated flow rule.*

4.2.2 Prandtl-Reuss

In the derivation of the flow theory of Prandtl-Reuss, the total incremental strain tensor $\mathrm{d}\varepsilon_{ij}^{\,t}$ is the sum of elastic and plastic incremental strain tensors. This gives

$$\mathrm{d}\varepsilon_{ij}^{\,t} = \mathrm{d}\varepsilon_{ij}^{\,P} + \mathrm{d}\varepsilon_{ij}^{\,e} = \mathrm{d}\lambda\,\sigma_{ij}' + \left[\frac{(1-2\nu)\,\delta_{ij}\,\mathrm{d}\sigma_{kk}}{3E} + \frac{\mathrm{d}\sigma_{ij}'}{2G}\right] \tag{4.3a}$$

In matrix notation:

$$\mathrm{d}\mathbf{E} = \mathrm{d}\lambda\,\mathbf{T}' + \frac{(1-2\nu)}{3E}\,(\mathrm{tr}\,\mathrm{d}\mathbf{T})\,\mathbf{I} + \frac{1}{2G}\,\mathrm{d}\mathbf{T}' \tag{4.3b}$$

Components of the plastic strain tensor $\mathrm{d}\varepsilon_{ij}^{\,P}$ are identified with the flow rule of Levy-Mises. That is, f in eq(3.22) is associated with the von Mises criterion of yielding $f = ½\sigma_{ij}'\sigma_{ij}'$, in which the stress deviator σ_{ij}' ensures plastic incompressibility. The incremental, recoverable, elastic strain $\mathrm{d}\varepsilon_{ij}^{\,e}$ appears in two terms within [ ] in eq(4.3a). This tensor component form of Hooke's law reveals that elastic strain is the sum of hydrostatic and deviatoric stress contributions to elasticity. There is no contribution from the former under shear stress, since $\delta_{ij} = 0$. Under normal stress, where $\delta_{ij} = 1$, the hydrostatic component is small without a superimposed pressure. Here, an assumption of elastic incompressibility ($\nu = ½$) will reduce eq(4.3a) to an acceptable approximation in a sum of plastic and elastic strain components:

$$\mathrm{d}\varepsilon_{ij}^{\,t} = \mathrm{d}\lambda\,\sigma_{ij}' + \mathrm{d}\sigma_{ij}'\,/\,(2G) \tag{4.3c}$$

In eqs(4.3a,c), the scalar multiplier $\mathrm{d}\lambda$, characterises the changing nature of the plastic hardening behaviour for the material. Normally $\delta\lambda$ is defined from uniaxial tension where the equivalent stress and the equivalent, incremental plastic strain are connected from within the first term in eq(4.3c)

$$\mathrm{d}\bar{\varepsilon}^{P} = \mathrm{d}\lambda\left(\bar{\sigma} - \tfrac{1}{3}\bar{\sigma}\right) \quad \Rightarrow \quad \mathrm{d}\lambda = \frac{3\,\mathrm{d}\bar{\varepsilon}^{P}}{2\bar{\sigma}}$$

in which $\bar{\sigma} = \bar{\sigma}(\bar{\varepsilon}^P)$. In the case of non-hardening (perfectly–plastic) material $\bar{\sigma} = Y =$ constant. *Note*: It is then possible to integrate eq(4.3a) to give the components of a total strain tensor as

$$
\begin{aligned}
\varepsilon_{ij}^{\,t} &= \int_0^{\bar{\varepsilon}^P} \frac{3\sigma_{ij}'}{2Y}\,\mathrm{d}\bar{\varepsilon}^P + \int_0^{\sigma_{kk}} \frac{(1-2\nu)}{3E}\delta_{ij}\,\mathrm{d}\sigma_{kk} + \int_0^{\sigma_{ij}'} \frac{1}{2G}\,\mathrm{d}\sigma_{ij}' \\
&= \frac{3\bar{\varepsilon}^P}{2Y}\sigma_{ij}' + \frac{(1-2\nu)}{3E}\delta_{ij}\sigma_{kk} + \frac{1}{2G}\sigma_{ij}'
\end{aligned} \tag{4.3d}
$$

The integration assumes that the components of the stress deviator tensor σ_{ij}' in eq(4.3d) increase: (a) proportionately or (b) in a stepwise manner. This allows stress components to be separated within the integral. The following section § 4.2.3 shows how a loading (b) applies to a combination of a single shear stress and a single direct stress.

4.2.3 Theoretical Comparison

One can examine further differences in formulation between the compressible forms of the deformation and flow theories from making a theoretical comparison between eqs(4.1a) and (4.3a). Take, for example, a stress system composed of tension σ_{11} combined with one shear stress component σ_{12}. The Hencky eq(4.1a) gives the total direct and shear strains as

$$
\begin{aligned}
\varepsilon_{11}^{\,t} &= \varphi\,\sigma_{11}' + \delta_{11}\sigma_{11}/(9K) \\
&= \sigma_{11}[\,2\varphi/3 + 1/(9K)\,]
\end{aligned} \tag{4.4a}
$$

$$
\begin{aligned}
\varepsilon_{22}^{\,t} = \varepsilon_{33}^{\,t} &= -\,\varphi\,\sigma_{11}/3 + (1-2\nu)\sigma_{11}/(3E) \\
&= \sigma_{11}[-\,\varphi/3 + 1/(9K)\,] = \varepsilon_{33}^{\,t}
\end{aligned} \tag{4.4b}
$$

$$
\begin{aligned}
\varepsilon_{12}^{\,t} &= \varphi\,\sigma_{12} + \delta_{12}\sigma_{12}/(9K) \\
\gamma_{12}^{\,t} &= 2\varepsilon_{12}^{\,t} = 2\varphi\,\sigma_{12}
\end{aligned} \tag{4.4c}
$$

where $\delta_{11} = 1$ and $\delta_{12} = 0$. From the Prandtl-Reuss eq(4.3d), the total strains are

$$
\begin{aligned}
\varepsilon_{11}^{t} &= \int_0^{\bar{\varepsilon}^P} \frac{3\,\mathrm{d}\bar{\varepsilon}^P}{2Y}\left(\frac{2\sigma_{11}}{3}\right) + \int_0^{\sigma_{11}}\left[\frac{(1-2\nu)}{3E} + \frac{1}{3G}\right]\mathrm{d}\sigma_{11} \\
&= \frac{\bar{\varepsilon}^P}{Y}\sigma_{11} + \left[\frac{(1-2\nu)}{3E} + \frac{1}{3G}\right]\sigma_{11}
\end{aligned} \tag{4.5a}
$$

$$
\begin{aligned}
\varepsilon_{22}^{\,t} &= \int_0^{\bar{\varepsilon}^P} \frac{3\,\mathrm{d}\bar{\varepsilon}^P}{2Y}\sigma_{22}' + \int_0^{\sigma_{11}} \frac{(1-2\nu)}{3E}\delta_{22}\,\mathrm{d}\sigma_{11} + \int_0^{\sigma_{22}'} \frac{1}{2G}\,\mathrm{d}\sigma_{22}' \\
&= -\int_0^{\bar{\varepsilon}^P} \frac{\mathrm{d}\bar{\varepsilon}^P}{2Y}\sigma_{11} + \int_0^{\sigma_{11}}\left[\frac{(1-2\nu)}{3E} - \frac{1}{6G}\right]\mathrm{d}\sigma_{11} \\
&= -\frac{\bar{\varepsilon}^P}{2Y}\sigma_{11} + \left[\frac{(1-2\nu)}{3E} - \frac{1}{6G}\right]\sigma_{11} = \varepsilon_{33}^{\,t}
\end{aligned} \tag{4.5b}
$$

$$\varepsilon_{12}^{\,t} = \int_0^{\bar{\varepsilon}^P} \frac{3\,\mathrm{d}\bar{\varepsilon}^P}{2Y}\,\sigma_{12} + \int_0^{\sigma_{11}} \frac{(1-2\nu)}{3E}\,\delta_{12}\,\mathrm{d}\sigma_{11} + \int_0^{\sigma_{12}} \frac{1}{2G}\,\mathrm{d}\sigma_{12}$$

$$= \frac{3\bar{\varepsilon}^P}{2Y}\,\sigma_{12} + \frac{1}{2G}\,\sigma_{12}$$

$$\mathrm{d}\gamma_{12}^{\,t} = 2\varepsilon_{12}^{\,t} = \frac{3\bar{\varepsilon}^P}{Y}\,\sigma_{12} + \frac{1}{G}\,\sigma_{12} \tag{4.5c}$$

It will be seen that total strain predictions differ except for when the stress increases proportionately, i.e. σ_{12}/σ_{11} = constant. The corresponding total strain ratio is, from the Hencky-Ilyushin eqs(4.4a,c):

$$\frac{\gamma_{12}^{\,t}}{\varepsilon_{11}^{\,t}} = \left(\frac{2\varphi}{2\varphi/3 + 1/(9K)}\right)\frac{\sigma_{12}}{\sigma_{11}} \tag{4.6a}$$

and, from the Prandtl-Reuss eqs(4.5a,c), this ratio becomes

$$\frac{\gamma_{12}^{\,t}}{\varepsilon_{11}^{\,t}} = \left(\frac{3\bar{\varepsilon}^P/Y + 1/G}{\bar{\varepsilon}^P/Y + 1/(9K) + 1/(3G)}\right)\frac{\sigma_{12}}{\sigma_{11}} \tag{4.6b}$$

The total strain ratios in eqs(4.6a,b) become equal when φ identifies with $(3\bar{\varepsilon}^P)/(2Y)$ given that the reciprocals of the elastic constants are negligibly small. The ratios in eqs(4.6a,b) are exactly equal when the deviatoric stress contribution to plasticity is separated from within each ratio. That is, both give $\gamma_{12}^{\,P}/\varepsilon_{11}^{\,P} = 3(\sigma_{12}/\sigma_{11})$. In general, the plastic response from the deformation and flow theories of plasticity agree under proportional (radial) loading conditions.

Note: Unlike the flow theory of Prandtl-Reuss, Hencky's eq(4.2a) does not separate total strain into elastic and plastic components. Nor does Hencky depend upon the strain path as it provides the final total strain under a given stress condition. Consequently, Hencky's load limit predictions may involve error for certain non-proportional stress paths in which the history of deformation is influential. In contrast, the flow theory traces the path through an incremental strain summation. Theoretical uncertainty raised here is examined next from making further comparisons with experiment. Also, a consideration is given to the influence of the elastic component of strain upon compressibility and total strain.

4.3 Application of Theory to Uniform Stress States

Near-uniform stress states are achieved from the application of various loadings to thin-walled tubes and plates. Experiments that combine tension, torsion and internal pressure in thin-walled tubes have often been used to provide data for the appraisal of a plasticity theory. The nature of hardening assumed should reflect that in a given material. Here, we examine variations in stress with strain from two theories of non-hardening plasticity without a simplifying assumption of incompressible elasticity. A comparison is made with the results from four different experiments in which one component of total strain is held constant.

4.3.1 Combined Tension-Torsion with Constant Shear Strain

Hohensemer [8] conducted the original experiment in which a thin-walled tube of non-hardening, low-carbon steel was extended longitudinally while the shear strain γ_o at yield was held constant. This gave

$$\gamma_o = k/G = 2Y(1+\nu)/(\sqrt{3}E) = \text{constant} \tag{4.7}$$

where $k = Y/\sqrt{3}$ and $E = 2G(1+\nu)$. It is required to determine the variations in the ensuing axial σ and shear τ stresses with total axial strain ε^t, according to the two theories of plasticity. The von Mises yield criterion, eq(3.8), is applied where the shear stress through the wall of the tube is taken to be uniform, and the influence of radial stress and hoop stress are small enough to be neglected. Stress subscripts will be omitted for simplicity when we identify $\sigma = \sigma_{11}$ and $\tau = \sigma_{12}$. Equation (3.8) is written, for $S = \sigma/Y$ and $T = \tau/Y$, as

$$\sigma^2 + 3\tau^2 = Y^2 \quad \Rightarrow \quad S^2 + 3T^2 = 1 \tag{4.8a,b}$$

(a) Hencky-Ilyushin

The deformation theory (4.2a) provides the following total axial and shear strain components under the stress combination σ, τ :

$$\varepsilon^t = \frac{2}{3}\varphi\sigma + \frac{(1-2\nu)}{3E}\sigma \tag{4.9a}$$

$$\gamma^t = 2\varphi\,\tau \tag{4.9b}$$

Setting $\gamma^t = \gamma_o$ = constant in eq(4.9b) defines a scalar multiplier for the ensuing deformation as: $\varphi = \gamma_o/(2\tau)$. Substituting into eq(4.9a) and expressing τ from within eq(4.8a) leads to

$$\varepsilon^t = \frac{\sigma\gamma_o}{\sqrt{3(Y^2-\sigma^2)}} + \frac{(1-2\nu)\sigma}{3E} \tag{4.10a}$$

Substituting from eq(4.7) gives the normalised form of eq(4.10a), which is to be used with the normalised yield criterion, eq(4.8b).

$$\frac{\varepsilon^t E}{Y} = \frac{S}{3}\left[(1-2\nu) + \frac{2(1+\nu)}{\sqrt{1-S^2}}\right] \tag{4.10b}$$

(b) Prandtl-Reuss

The application of the flow theory (4.3a) supplies total incremental strains, each as the sum of elastic and plastic components:

$$d\varepsilon^t = \frac{2\,d\lambda\,\sigma}{3} + \frac{d\sigma}{E} \tag{4.11a}$$

$$d\gamma^t = 2\,d\lambda\,\tau + \frac{d\tau}{G} \tag{4.11b}$$

Setting $d\gamma^t = 0$ in eq(4.11b) gives $d\lambda = -\, d\tau/(2\tau G)$. Substituting this into eq(4.11a):

$$d\varepsilon^t = -\frac{\sigma\, d\tau}{3\tau G} + \frac{d\sigma}{E} \tag{4.12a}$$

where from eq(4.8a), $d\tau = -\sigma\, d\sigma/(3\tau)$. Substituting this into eq(4.12a) gives

$$d\varepsilon^t = \frac{\sigma^2\, d\sigma}{9\tau^2 G} + \frac{d\sigma}{E} = \frac{\sigma^2 d\sigma}{3G(Y^2 - \sigma^2)} + \frac{d\sigma}{E} \tag{4.12b}$$

Integrating eq(4.12b) by partial fractions

$$\varepsilon^t = \frac{1}{3G}\int_0^\sigma \left(\frac{Y/2}{Y+\sigma} + \frac{Y/2}{Y-\sigma} - 1 \right) d\sigma + \frac{1}{E}\int_0^\sigma d\sigma$$

$$= \frac{Y}{6G} \ln\left(\frac{Y+\sigma}{Y-\sigma} \right) - \frac{\sigma}{3G} + \frac{\sigma}{E} \tag{4.12c}$$

Taken with $E = 2G(1 + \nu)$, eq(4.12c) reduces to the normalised form

$$\frac{\varepsilon^t E}{Y} = \frac{(1-2\nu)S}{3} + \frac{(1+\nu)}{3} \ln\left(\frac{1+S}{1-S} \right) \tag{4.12d}$$

Figure 4.1a–c compares the predictions from eqs(4.10b) and (4.12d) with Hohensemer's experimental results.

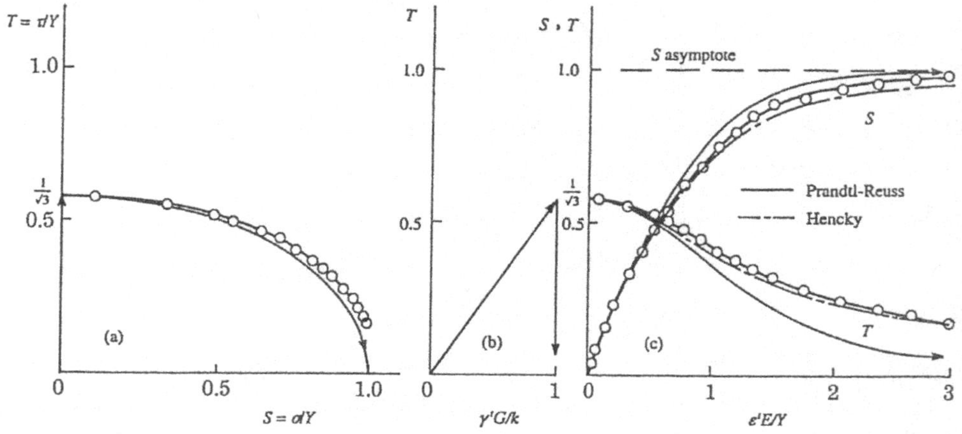

Figure 4.1 Shear yield strain constrained under increasing tension

Both theories appear to be in reasonable agreement with observed experimental trends. Figure 4.1c shows that the total axial strain increases with S initially at a gradient of E. Thereafter, $\varepsilon^t E/Y$ becomes infinite as S approaches unity asymptotically. Also, $T = \tau/Y$

approaches zero simultaneously as γ^t becomes a permanent strain. The fact that Hencky appears to be more representative of this data led early workers [9, 10] to favour a total strain formulation. Later, however, the view held generally became that Prandtl-Reuss provides a more reliable theory in the presence of stress gradients and non-radial loading. The deviation found between the flow theory and this experiment is likely to be due to the occurrence of some hardening and a disparity with the von Mises yield criterion for the steel in question. The Hencky theory retains more than an historical interest since, as we see here and later, for other uniform stress states, it is easier to apply and provides perfectly acceptable predictions.

The Hencky eq(4.10a) can also be applied where an elastic shear strain $\gamma_o = \tau_o/G$, in which $\tau_o < k$ (k is the shear yield stress) is held constant under increasing tension. According to Prandtl-Reuss, the material first yields under a stress state σ_o, τ_o, according to

$$\sigma_o^{\,2} + 3\tau_o^{\,2} = Y^2 \tag{4.13}$$

With Y constant and $\sigma > \sigma_o$, it follows from eq(4.13) that $\tau < \tau_o$. The ensuing, total axial strain again follows from eq(4.12c) but with modified integration limits

$$\varepsilon^t = \frac{1}{3G}\int_{\sigma_o}^{\sigma}\left(\frac{Y/2}{Y+\sigma} + \frac{Y/2}{Y-\sigma} - 1\right)\mathrm{d}\sigma + \frac{1}{E}\int_{\sigma_o}^{\sigma}\mathrm{d}\sigma \tag{4.14a}$$

integrating eq(4.14a):

$$\varepsilon^t = \frac{Y}{6G}\ln\left[\frac{(Y+\sigma)(Y-\sigma_o)}{(Y+\sigma_o)(Y-\sigma)}\right] + \left(\frac{1}{E} - \frac{1}{3G}\right)(\sigma - \sigma_o) \tag{4.14b}$$

Equation (4.14b) relates normalised stress to total strain in the manner of Figs 4.2a–c.

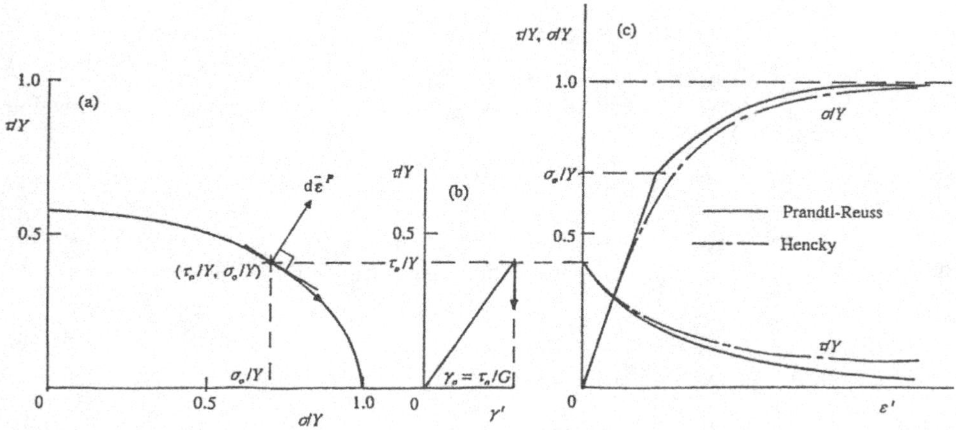

Figure 4.2 Elastic shear strain constrained under increasing tension

Initial loading to σ_o is elastic with slope E. Thereafter, with $\sigma > \sigma_o$, σ and τ vary with ε^t in a similar asymptotic manner to Fig. 4.1c. As the stress state follows the boundary of the yield locus, the plastic strain increment vector $d\bar{\varepsilon}^P$ aligns with its exterior normal (Fig. 4.2a). The two components of this vector identify the axial and plastic shear strains. Figure 4.2b shows that restraining the initial elastic shear strain results in its gradual conversion to plastic strain.

(c) Other Theories

The incompressible forms of Prandtl-Reuss and Hencky (or Nadai) are found from putting $\nu = ½$. In Hohensemer's experiment, for example, substituting $\nu = ½$ in eqs(4.10a) and (4.12d) reduces them to simplified forms, originally given by Prager [11] and Neuber [12]:

$$S = \tanh(\varepsilon E/Y) \text{ and } S = (\varepsilon E/Y)/[1 + (E\varepsilon/Y)] \tag{4.15a,b}$$

These predictions to the S variations would be less satisfactory than those of compressible theories shown in Fig. 4.1. Betten [13] generalised eqs(4.15a,b) to become

$$S = [\tanh(\varepsilon E/Y)^n]^{1/n} \text{ and } S = (\varepsilon E/Y)/[1 + (E\varepsilon/Y)^n]^{1/n} \tag{4.16a,b}$$

Thus, with an integer value $n > 1$, eqs(4.16a,b) ensured an improved fit to this data. However, it is doubtful whether Hencky's incompressible theory could allow a similar modification with its application to other reduced stress states.

4.3.2 Combined Tension-Torsion with Constant Axial Strain

Consider a thin tube subjected to constrained deformation in which the tensile yield strain is held constant under increasing torsion. The following analyses provide predictions to the growth in shear strain.

(a) Hencky Ilyushin

The axial strain at yield is

$$\varepsilon_o = \frac{Y}{E} = \frac{\sqrt{3}k}{2G(1+\nu)} \tag{4.17a}$$

Putting $\varepsilon^t = \varepsilon_o$ = constant, in eq(4.9a), defines the scalar multiplier for the ensuing deformation as

$$\varphi = \frac{3\varepsilon_o}{2\sigma} - \frac{(1-2\nu)}{2E} \tag{4.17b}$$

Substituting eq(4.17b) into eq(4.9b) and eliminating σ from eq(4.8a) leads to

$$\gamma^t = \frac{3\tau\varepsilon_o}{\sqrt{Y^2 - 3\tau^2}} - \frac{(1-2\nu)\tau}{E} \tag{4.18a}$$

Employing eq(4.17a), and setting $Y = \sqrt{3}k = E\varepsilon_o$, gives a normalised form of eq(4.18a)

$$\frac{\gamma^t G}{\sqrt{3}k} = \frac{3T}{2(1+\nu)\sqrt{1-3T^2}} - \frac{(1-2\nu)T}{2(1+\nu)} \tag{4.18b}$$

(b) Prandtl-Reuss

Substitute $\mathrm{d}\varepsilon^{t} = 0$ in eq(4.11a) to give the scalar multiplier

$$\mathrm{d}\lambda = -\frac{3\,\mathrm{d}\sigma}{2E\sigma} \tag{4.19a}$$

Then, from eq(4.8a)

$$\frac{\mathrm{d}\sigma}{\sigma} = -\frac{3\tau\mathrm{d}\tau}{\sigma^2} = \frac{3\tau\mathrm{d}\tau}{(3\tau^2 - Y^2)} \tag{4.19b}$$

Substituting into eq(4.11b)

$$\mathrm{d}\gamma^{t} = \frac{-9\tau^2\,\mathrm{d}\tau}{E(3\tau^2 - Y^2)} + \frac{\mathrm{d}\tau}{G} \tag{4.20a}$$

Integrating by partial fractions

$$\gamma^{t} = \frac{3}{E}\int_0^{\tau}\left(-1 + \frac{Y/2}{Y - \sqrt{3}\tau} + \frac{Y/2}{Y + \sqrt{3}\tau}\right)\mathrm{d}\tau + \int_0^{\tau}\frac{\mathrm{d}\tau}{G}$$

$$= \frac{3}{E}\left[-\tau + \frac{Y}{\sqrt{3}}\ln\left(\frac{Y + \sqrt{3}\tau}{Y - \sqrt{3}\tau}\right)\right] + \frac{\tau}{G} \tag{4.20b}$$

The corresponding non-dimensional form of eq(4.20b) is

$$\frac{G\gamma^{t}}{\sqrt{3}k} = \frac{\sqrt{3}}{4(1+\nu)}\ln\left(\frac{1 + \sqrt{3}T}{1 - \sqrt{3}T}\right) - \frac{(1-2\nu)T}{2(1+\nu)} \tag{4.20c}$$

The stress variations, predicted from eqs(4.18a) and (4.20b), are shown in Figs 4.3a–c.

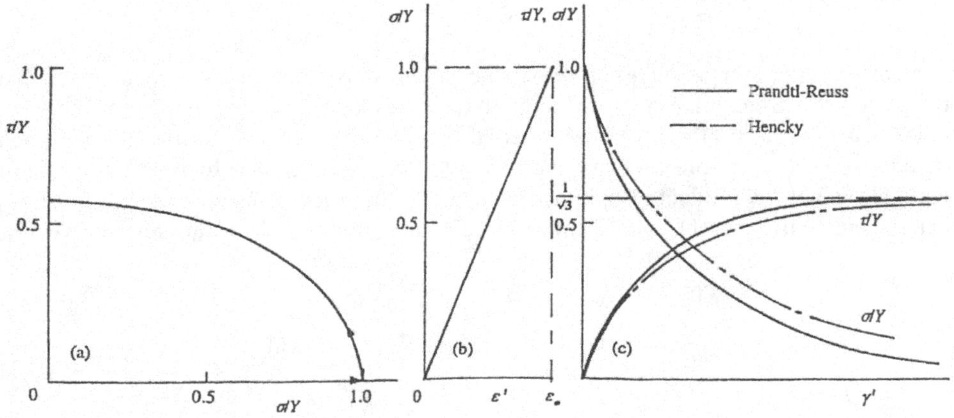

Figure 4.3 Tensile yield strain constrained under increasing torsion

In this case, Fig. 4.3b shows that the axial strain becomes permanent as the corresponding stress reduces to zero.

When an initially elastic, axial strain $\varepsilon_o^{\,e}$ is held constant under increasing torsion, the lower limit of the integral in eq(4.20b) is altered to τ_o. This provides the total shear strain for the ensuing deformation under $\tau > \tau_o$ as

$$\gamma^{\,t} = \frac{3}{E}\int_{\tau_o}^{\tau}\left(-1 + \frac{Y/2}{Y - \sqrt{3}\tau} + \frac{Y/2}{Y + \sqrt{3}\tau}\right)\mathrm{d}\tau + \int_{\tau_o}^{\tau}\frac{\mathrm{d}\tau}{G}$$

$$= \frac{3}{E}\left(-(\tau - \tau_o) + \frac{Y}{2\sqrt{3}}\ln\left[\frac{(Y + \sqrt{3}\tau)(Y - \sqrt{3}\tau_o)}{(Y - \sqrt{3}\tau)(Y + \sqrt{3}\tau_o)}\right]\right) + \frac{(\tau - \tau_o)}{G}$$

$$= \left(\frac{1}{G} - \frac{3}{E}\right)(\tau - \tau_o) + \frac{\sqrt{3}Y}{2E}\ln\left[\frac{(Y+\sqrt{3}\tau)(Y-\sqrt{3}\tau_o)}{(Y-\sqrt{3}\tau_o)(Y+\sqrt{3}\tau_o)}\right] \tag{4.21}$$

Equation (4.21) will also apply to a stress path in which an initially elastic, axial stress σ_o, is allowed to alter under an increasing shear stress. This is because σ_o can only alter in a non-hardening material by maintaining the total axial strain constant. That is, $\mathrm{d}\varepsilon^{t} = 0$ for the ensuing deformation, when from eq(4.11a), the axial plastic strain will alter according to:

$$\mathrm{d}\varepsilon^{P} = (2\mathrm{d}\lambda/3)\sigma = -\mathrm{d}\sigma/E \tag{4.22a}$$

Integration of eq(4.22a), reveals a conversion from initial elastic strain $\varepsilon_o^{\,e} = \sigma_o/E$, to plastic strain in the form

$$\varepsilon^{P} = (\sigma_o - \sigma)/E \tag{4.22b}$$

where $\sigma_o > \sigma$. Clearly, $\varepsilon^{P} = \varepsilon_o^{\,e}$ when σ in eq(4.22b) falls to zero. The incremental plastic shear strain component $\mathrm{d}\gamma^{P}$ is associated with the first term in eq(4.11b). Dividing this by eq(4.22a) gives the ratio between the incremental plastic strains as

$$\frac{\mathrm{d}\gamma^{P}}{\mathrm{d}\varepsilon^{P}} = \frac{3\tau}{\sigma} \tag{4.23a}$$

All tension-torsion stress paths conform to the relationship given in eq(4.23a), showing that the plastic strain increment vector is aligned with the exterior normal to the yield surface. One can also show this from eq(4.8a), where the gradient of a tangent to this surface is given by $\mathrm{d}\sigma/\mathrm{d}\tau = -\sigma/(3\tau)$. It follows that the gradient to the normal equals the right-hand side of eq(4.23a) since the product of these gradients is -1. In order to separate the plastic strain component from eq(4.21), eqs(4.22a) and (4.23a) are combined with eq(4.8a) to give

$$\mathrm{d}\gamma^{P} = \frac{-\sqrt{3}}{E}\left[\frac{\sqrt{Y^2 - \sigma^2}\,\mathrm{d}\sigma}{\sigma}\right]$$

$$\gamma^{P} = \frac{-\sqrt{3}Y}{E}\int_0^{S}\frac{\sqrt{1 - S^2}\,\mathrm{d}S}{S}$$

Making a trigonometric substitution, $S = \sin\theta$, leads to

$$\gamma^P = \frac{\sqrt{3}Y}{E}\left[\left(\frac{1}{S} + \sqrt{\sin^2 S - 1}\right)\sin^{-1}S + \sqrt{1 - S^2}\right] \tag{4.23b}$$

The difference $\gamma^t - \gamma^P$, found from eqs(4.21) and (4.23b), is identified with the elastic component of the total shear strain.

4.3.3 Neutral Loading under Internal Pressure Combined with Torsion

Cylindrical co-ordinates will be employed for the plane stress state arising in the wall of a tube under combined torsion and internal pressure. That is, the non-zero stress components are written as σ_θ, σ_z and $\tau_{z\theta}$. The small magnitude of radial stress can be ignored when the wall is thin. Gill [14] conducted a neutral loading experiment on a tube of non-hardening, pre-strained, 70/30 brass. A total shear pre-strain of 0.44 % was sufficient to ensure that the material did not harden further when combined stress components were adjusted to follow the contour of the yield locus. The latter is derived from eq(3.13a) as

$$\sigma_z^2 - \sigma_z\sigma_\theta + \sigma_\theta^2 + 3\tau_{z\theta}^{\,2} = Y^2 \tag{4.24a}$$

Note: Under internal pressure, where $\sigma_\theta = R\sigma_z$ $(R = 2.2)$, two independent stress components are required to define the yield criterion. Thus, eq(4.24a) appears in terms of σ_z and $\tau_{z\theta}$ as

$$Q\,\sigma_z^2 + 3\tau_{z\theta}^{\,2} = Y^2 \tag{4.24b}$$

where $Q = 1 - R + R^2$. It follows from the analysis of Hohensemer's experiment in § 4.3.1, that 'neutral loading' applies to a non-hardening material when shear yield strain for the ensuing deformation is constrained:

$$\gamma_o^{\,t} = k/G = 2Y(1 + \nu)/(\sqrt{3}E) = \text{constant} \tag{4.25}$$

The following classical theoretical solutions apply with $\gamma_o^{\,t}$ constant.

(a) Hencky-Ilyushin

The total strains are, from eq(4.2a)

$$\begin{aligned}\varepsilon_z^{\,t} &= \varphi\,\sigma_z + [(1 - 2\nu)/E - \varphi](\sigma_z + \sigma_\theta)/3\\ &= \sigma_z\{\varphi + 1/3(1 + R)[(1 - 2\nu)/E - \varphi]\}\end{aligned} \tag{4.26a}$$

$$\begin{aligned}\varepsilon_\theta^{\,t} &= \varphi\,\sigma_\theta + [(1 - 2\nu)/E - \varphi](\sigma_z + \sigma_\theta)/3\\ &= \sigma_z\{R\varphi + 1/3(1 + R)[(1 - 2\varphi)/E - \varphi]\}\end{aligned} \tag{4.26b}$$

$$\gamma_{z\theta}^{\,t} = 2\varphi\,\tau_{z\theta} \tag{4.26c}$$

where $R = \sigma_\theta/\sigma_z$. Setting $\gamma_{z\theta}^{\,t} = \gamma_o^{\,t} =$ constant in eq(4.26c) gives the scalar $\varphi = \gamma_o^{\,t}/(2\tau_{z\theta})$. Substituting φ into eqs(4.26a,b) and eliminating $\tau_{z\theta}$ from eq(4.24b) gives the total axial and circumferential strains respectively:

$$\varepsilon_z^t = \frac{(2-R)\,\gamma_o^t\,\sigma_z}{2\sqrt{3\,(Y^2 - Q\,\sigma_z^2)}} + \frac{(1+R)(1-2\nu)\sigma_z}{3E} \tag{4.27a}$$

$$\varepsilon_\theta^t = \frac{(2R-1)\,\gamma_o^t\,\sigma_z}{2\sqrt{3\,(Y^2 - Q\,\sigma_z^2)}} + \frac{(1+R)(1-2\nu)\sigma_z}{3E} \tag{4.27b}$$

From eq(4.25) and setting $S_z = \sigma_z / Y$, the non-dimensional forms of eqs(4.27a,b) become

$$\frac{\varepsilon_z^t E}{Y} = \frac{(2-R)(1+\nu)S_z}{3\sqrt{(1 - Q\,S_z^2)}} + \frac{(1+R)(1-2\nu)S_z}{3} \tag{4.27c}$$

$$\frac{\varepsilon_\theta^t E}{Y} = \frac{(2R-1)(1+\nu)S_z}{3\sqrt{(1 - Q\,S_z^2)}} + \frac{(1+R)(1-2\nu)S_z}{3} \tag{4.27d}$$

(b) Prandtl-Reuss

The total incremental strains are, from eq(4.3b)

$$\mathrm{d}\varepsilon_z{}^t = (2\,\mathrm{d}\lambda/3)(\sigma_z - \sigma_\theta/2) + (1/E)(\mathrm{d}\sigma_z - \nu\,\mathrm{d}\sigma_\theta) \tag{4.28a}$$

$$\mathrm{d}\varepsilon_\theta{}^t = (2\,\mathrm{d}\lambda/3)(\sigma_\theta - \sigma_z/2) + (1/E)(\mathrm{d}\sigma_\theta - \nu\,\mathrm{d}\sigma_z) \tag{4.28b}$$

$$\mathrm{d}\gamma_{z\theta}{}^t = 2\,\mathrm{d}\lambda\tau_{z\theta} + \mathrm{d}\tau_{z\theta}/G \tag{4.28c}$$

and with $\mathrm{d}\gamma_{z\theta}{}^t = 0$ in eq(4.28c):

$$\mathrm{d}\lambda = -\,\mathrm{d}\tau_{z\theta} / (2\tau_{z\theta}G) \tag{4.29a}$$

Differentiating eq(4.24b)

$$3\tau_{z\theta}\,\mathrm{d}\tau_{z\theta} = -\,Q\,\sigma_z\,\mathrm{d}\sigma_z \tag{4.29b}$$

Substituting eqs(4.29a,b) into eqs(4.28a,b,c) provides the total, axial and circumferential strain increments

$$\mathrm{d}\varepsilon_z^t = \frac{Q(2-R)}{6G} \times \frac{\sigma_z^2\,\mathrm{d}\sigma_z}{(Y^2 - Q\sigma_z^2)} + \frac{(1-R\nu)\,\mathrm{d}\sigma_z}{E} \tag{4.30a}$$

$$\mathrm{d}\varepsilon_\theta^t = \frac{Q(2R-1)}{6G} \times \frac{\sigma_z^2\,\mathrm{d}\sigma_z}{(Y^2 - Q\sigma_z^2)} + \frac{(R-\nu)\,\mathrm{d}\sigma_z}{E} \tag{4.30b}$$

Integrating eqs(4.30a,b) by partial fractions leads to non-dimensional total strains:

$$\frac{E\,\varepsilon_z^t}{Y} = \frac{(1-2\nu)(1+R)S_z}{3} + \frac{(1+\nu)(2-R)}{6\sqrt{Q}}\ln\left(\frac{1+\sqrt{Q}S_z}{1-\sqrt{Q}S_z}\right) \tag{4.30c}$$

$$\frac{E\,\varepsilon_\theta^t}{Y} = \frac{(1-2\nu)(1+R)S_z}{3} + \frac{(1+\nu)(2R-1)}{6\sqrt{Q}}\ln\left(\frac{1+\sqrt{Q}S_z}{1-\sqrt{Q}S_z}\right) \tag{4.30d}$$

The predictions from eqs(4.27c,d) and (4.30c,d) are compared with Gill's results in Figs 4.4a–d, for $\nu = 1/3$.

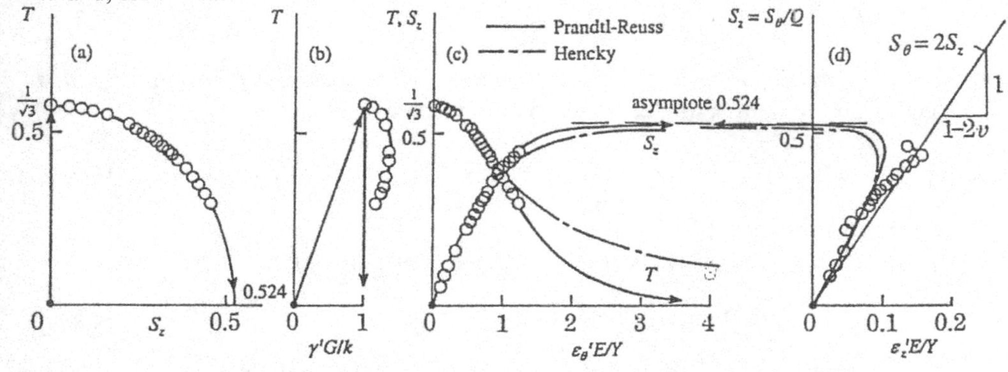

Figure 4.4 Neutral loading under combined internal pressure and axial torsion

It is seen from Fig. 4.4a that the relation between the stress components is in good agreement with the yield function, eq(4.24b). Figure 4.4b confirms, approximately, that the shear strain remains constrained. Both predictions, given in Figs 4.4c,d, show that $T\,(=\tau/Y)$ diminishes to zero as $S_z\,(=\sigma_z/Y)$ becomes asymptotic to $1/\sqrt{Q}\,(=0.524)$. Correspondingly, the predicted strains, $\varepsilon_\theta{}^t$ and $\varepsilon_z{}^t$, become infinite in tension and compression respectively. Because the experiment was terminated well before large values of these strains were achieved, it does not reveal which theory is the better. It is seen that the axial strain is extremely sensitive to the stress ratio. With $\sigma_\theta = R\sigma_z$ $(R = 2)$, both eqs(4.27c) and (4.30c) give an elastic axial strain response according to

$$\varepsilon_z{}^e E/Y = (1-2\nu)S_z \tag{4.31}$$

Equation (4.31) defines the line of slope 3 in Fig. 4.4d, which is apparently in closer agreement with the experimental data. Gill was surprised to find plastic deformation from his 'neutral' path but one would expect this path to result in plasticity for a non-hardening material. When a neutral path coincides with the initial yield condition in a hardening material, i.e. it follows the boundary of an initial yield locus, the stress state will remain at the yield point without producing plasticity.

4.3.4 Internal Pressure Combined with Axial Tension

When a thin walled, closed tube is pressurized, the radial stress is negligible compared to the axial and circumferential stresses $(\sigma_z, \sigma_\theta)$. This allows a comparison to be made between the total and incremental theories within a biaxial, principal stress space. In Schlafer and

Sidebottom's experiments [15], the axial and circumferential strain was constrained in separate experiments on a thin-walled cylinder of non-hardening, annealed SAE 1035 steel. In their first experiment, the cylinder was pre-strained to the uniaxial yield strain $\varepsilon_o{}^t$, under circumferential tension, then extended under axial tension with the pre-strain held constant. The von Mises yield criterion, eq(4.24a), reduces to a common form for to the two theories

$$\sigma_z{}^2 - \sigma_z \sigma_\theta + \sigma_\theta{}^2 = Y^2 \tag{4.32a}$$

The normalised form of eq(4.32a) is

$$S_z{}^2 - S_z S_\theta + S_\theta{}^2 = 1 \tag{4.32b}$$

where $S_z = \sigma_z / Y$ and $S_\theta = \sigma_\theta / Y$. To eliminate one stress component, say σ_θ, within the first quadrant of stress σ_θ, σ_z, eq(4.32a) gives, with a positive discriminant:

$$\sigma_\theta = \tfrac{1}{2}\, \sigma_z \pm \tfrac{1}{2} \sqrt{(4Y^2 - 3\sigma_z{}^2)} \tag{4.32c}$$

(a) Hencky Ilyushin

Using cylindrical co-ordinates, the total strains are, from eq(4.2a)

$$\varepsilon_\theta{}^t = \varphi\, \sigma_\theta + [(1 - 2\nu)/E - \varphi](\sigma_z + \sigma_\theta)/3 = \varepsilon_o{}^t \tag{4.33a}$$

$$\varepsilon_z{}^t = \varphi\, \sigma_z + [(1 - 2\nu)/E - \varphi](\sigma_z + \sigma_\theta)/3 \tag{4.33b}$$

Equation (4.33a) provides the scalar φ as

$$\varphi = \frac{3E\varepsilon_o^t - (1 - 2\nu)(\sigma_z + \sigma_\theta)}{E(2\sigma_\theta - \sigma_z)} \tag{4.34}$$

Substituting eq(4.34) into eq(4.33b) leads to

$$\varepsilon_z^t = \frac{\varepsilon_o^t (2\sigma_z - \sigma_\theta)}{(2\sigma_\theta - \sigma_z)} + \frac{3(1 - 2\nu)(\sigma_z^2 - \sigma_\theta^2)}{3E(2\sigma_\theta - \sigma_z)} \tag{4.35a}$$

One may apply eq(4.35a) to the deformation ensuing from the yield strain $\varepsilon_o{}^t = Y/E$, under stress components connected through the yield criterion. The following normalised form of eq(4.35a) is found when σ_θ is eliminated from eq(4.32c)

$$\frac{E\varepsilon_z^t}{Y} = \frac{3\sigma_z/2 - 2\sqrt{Y^2 - 3\sigma_z^2/4} + \left[(1 - 2\nu)/Y\right]\left(Y^2 + \sigma_z\sqrt{Y^2 - 3\sigma_z^2/4} - 3\sigma_z^2/2\right)}{2\sqrt{Y^2 - 3\sigma_z^2/4}}$$

$$= \frac{3S_z/2 - 2\sqrt{1 - 3S_z^2/4} + \left(1 - 2\nu\right)\left(1 + S_z\sqrt{1 - 3S_z^2/4} - 3S_z^2/2\right)}{2\sqrt{1 - 3S_z^2/4}} \tag{4.35b}$$

where ν = 0.285 and E/Y = 1345.3. The variation in $E\varepsilon_z^t/Y$ with S_θ follows from the simultaneous eqs(4.32b) and (4.35b). Substituting $S_z = 0$ in eq(4.35b) gives the origin of normalised strain as $E\varepsilon_z^t/Y = -\,0.785$ (see Fig. 4.5).

(b) Prandtl-Reuss

For the circumferential pre-straining, the volume of the cylinder material remains constant:

$$\mathrm{d}\varepsilon_r^{\,P} + \mathrm{d}\varepsilon_z^{\,P} + \mathrm{d}\varepsilon_\theta^{\,P} = 0$$

$$\therefore\ \mathrm{d}\varepsilon_r^{\,P} = \mathrm{d}\varepsilon_z^{\,P} = -\,\tfrac{1}{2}\,\mathrm{d}\varepsilon_\theta^{\,P} \tag{4.36}$$

Now, from eqs(4.3a) and (4.36), $\mathrm{d}\varepsilon_z^{\,t}$ bears the following relationship to $\mathrm{d}\varepsilon_\theta^{\,P}$

$$\mathrm{d}\varepsilon_z^{\,t} = \mathrm{d}\varepsilon_z^{\,P} + \mathrm{d}\varepsilon_z^{\,e} = -\,\tfrac{1}{2}\,\mathrm{d}\varepsilon_\theta^{\,P} - \nu\,\mathrm{d}\sigma_\theta/E \tag{4.37a}$$

Integrating eq(4.37a), with limits of σ_θ from 0 to Y, defines a normalised axial strain origin:

$$\frac{E\,\varepsilon_z^{\,t}}{Y} = -\,\frac{E\,\varepsilon_\theta^{\,P}}{2\,Y} - \nu \tag{4.37b}$$

That is, with $\varepsilon_\theta^{\,P} = Y/E$ and $\nu = 0.285$, eq(4.37b) gives $E\varepsilon_z^t/Y = -\,0.785$, which agrees with Hencky. Under constrained deformation, the total incremental strains are found from the Prandtl-Reuss eq(4.3b)

$$\mathrm{d}\varepsilon_z^{\,t} = (2\,\mathrm{d}\lambda/3)(\sigma_z - \sigma_\theta/2) + (1/E)(\mathrm{d}\sigma_z - \nu\,\mathrm{d}\sigma_\theta) \tag{4.38a}$$

$$\mathrm{d}\varepsilon_\theta^{\,t} = (2\,\mathrm{d}\lambda/3)(\sigma_\theta - \sigma_z/2) + (1/E)(\mathrm{d}\sigma_\theta - \nu\,\mathrm{d}\sigma_z) \tag{4.38b}$$

where σ_z is the total axial stress. Stress variations are found from setting $\mathrm{d}\varepsilon_\theta^{\,t} = 0$ in eq(4.38b). This gives

$$\mathrm{d}\lambda = -\,3(\mathrm{d}\sigma_\theta - \nu\,\mathrm{d}\sigma_z)/[E(2\sigma_\theta - \sigma_z)] \tag{4.39a}$$

where from eq(4.32a)

$$\mathrm{d}\sigma_\theta = -\,(2\sigma_z - \sigma_\theta)\mathrm{d}\sigma_z/(2\sigma_\theta - \sigma_z) \tag{4.39b}$$

Substituting eqs(4.39a,b) into eq(4.38a) leads to

$$\mathrm{d}\varepsilon_z^{\,t} = \frac{[2Y^2(2-\nu) + (2\nu - 1)(3\sigma_z\sigma_\theta - Y^2)]\,\mathrm{d}\sigma_z}{E(2\sigma_\theta - \sigma_z)^2} \tag{4.40a}$$

Substituting eq(4.32c) into eq(4.40a) and integrating provides the total axial strain

$$\varepsilon_z^{\,t} = \frac{Y^2(5-4\nu)}{E}\int_0^{\sigma_z}\frac{\mathrm{d}\sigma_z}{\left(4Y^2 - 3\sigma_z^2\right)} + \frac{3(2\nu-1)}{2E}\int_0^{\sigma_z}\frac{\sigma_z^2\,\mathrm{d}\sigma_z}{\left(4Y^2 - 3\sigma_z^2\right)} \pm \frac{3(2\nu-1)}{2E}\int_0^{\sigma_z}\frac{\sigma_z\,\mathrm{d}\sigma_z}{\sqrt{4Y^2 - 3\sigma_z^2}}$$

$$= \frac{Y(5-4\nu)}{4\sqrt{3}E}\ln\left(\frac{Y+\sqrt{3}\sigma_z/2}{Y-\sqrt{3}\sigma_z/2}\right) + \frac{Y(2\nu-1)}{2\sqrt{3}E}\left[\ln\left(\frac{Y+\sqrt{3}\sigma_z/2}{Y-\sqrt{3}\sigma_z/2}\right) - \frac{\sqrt{3}\sigma_z}{Y}\right] + \frac{(2\nu-1)}{E}\left[Y - \sqrt{Y^2 - \left(\frac{\sqrt{3}\sigma_z}{2}\right)^2}\,\right] \tag{4.40b}$$

$$= \frac{\sqrt{3}Y}{4E}\ln\left(\frac{Y+\sqrt{3}\sigma_z/2}{Y-\sqrt{3}\sigma_z/2}\right) + \frac{(2\nu-1)}{E}\left[Y - \frac{\sigma_z}{2} - Y\sqrt{1-\left(\frac{\sqrt{3}\sigma_z}{2Y}\right)^2}\,\right]$$

The corresponding non-dimensional form of eq(4.40b) is

$$\frac{E\varepsilon_z^t}{Y} = \frac{\sqrt{3}}{4}\ln\left(\frac{1+\sqrt{3}S_z/2}{1-\sqrt{3}S_z/2}\right) - (1-2\nu)\left(1 - S_z/2 - \sqrt{1-3S_z^2/4}\right) \quad (4.40c)$$

Figure 4.5a–c shows the Hencky and Prandtl-Reuss predictions, from eqs(4.35b) and (4.40c) respectively. These are consistent with the observation of Schlafer and Sidebottom, in that S_z increases continuously while S_θ increases initially before it decreases.

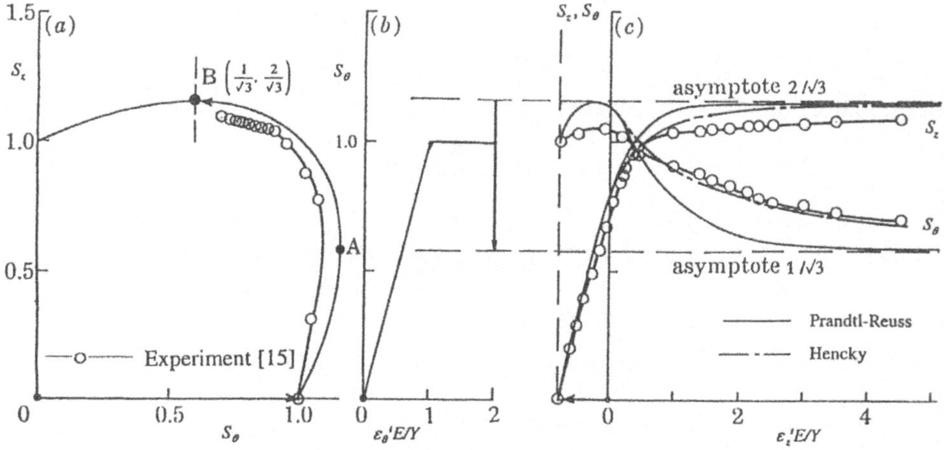

Figure 4.5 Circumferential yield strain constrained under axial tension

Figure 4.5c shows that the rapid onset of infinite axial strain, observed for asymptotic stress values of 2/√3 and 1/√3, is closer to the Hencky prediction. Prandtl-Reuss predictions approach each asymptote for strain less than was observed. The theories converge for large axial strains up to five times the yield value. The deviation from the Mises yield locus, in Fig. 4.5a, could be attributed to possible influences of the third invariant or initial residual anisotropy following annealing. Also, there may have been experimental difficulty in fully constraining the circumferential pre-strain, as it should appear from Fig. 4.5b.

Figure 4.6a–c shows that similar conclusions apply to the magnitude of the circumferential strain in a second test, on the same material, when an axial pre-strain was held constant. The deformation under increasing circumferential tensile stress may be deduced from replacing θ with z in eqs(4.35b) and (4.40c). This gives the Hencky and Prandtl-Reuss prediction respectively:

$$\frac{E\varepsilon_\theta^t}{Y} = \frac{3S_\theta/2 - 2\sqrt{1-3S_\theta^2/4} + (1-2\nu)\left(1 + S_\theta\sqrt{1-3S_\theta^2/4} - 3S_\theta^2/2\right)}{2\sqrt{1-3S_\theta^2/4}}$$

$$\frac{E\varepsilon_\theta^t}{Y} = \frac{\sqrt{3}}{4}\ln\left(\frac{1+\sqrt{3}S_\theta/2}{1-\sqrt{3}S_\theta/2}\right) - (1-2\nu)\left(1 - \frac{S_\theta}{2} - \sqrt{1-\frac{3S_\theta^2}{4}}\right)$$

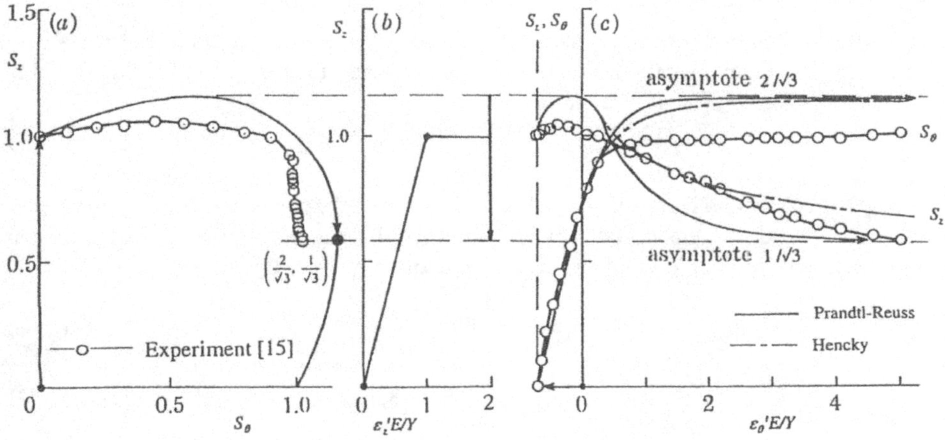

Figure 4.6 Axial yield strain constrained under circumferential tension

The corresponding origin in Fig. 4.6c is found from interchanging θ with z in eq(4.37b):

$$E\varepsilon_\theta^{\,t} / Y = -\ (E\varepsilon_z^{\,P}) / (2Y) - \nu = -\ 0.785$$

where $\varepsilon_z^{\,P} = Y/E$. The theoretical asymptotes in Figs 4.5c and 4.6c limit the stresses with greater accuracy than has been found from a Tresca yield criterion [15]. However, in practice, before the infinite theoretical strains $\varepsilon_z^{\,t}$ and $\varepsilon_\theta^{\,t}$ in Figs 4.5c and 4.6c could be reached, a thin-walled tube would become unstable from local necking and bulging.

Selected solutions to similar problems under uniform stress states for hardening materials will be considered later in chapter 10 (see § 10.6).

4.4 Application of Theory to Non-Uniform Stress States

This section examines the ability with which the two, classical theories of plasticity can provide solutions to elastic–perfectly plastic deformation under non-uniform stress states. For the purpose of making an experimental appraisal, results will be used for solid circular bars, subjected to different combinations of axial tension and torsion. It will be seen that, though constructed differently, the two theories supply comparable solutions to the deformation response and to the internal stress distributions from loading bars under relatively simple loading paths. The Hencky-Ilyushin theory supplies closed-form equations that may be solved by trial. The Prandtl-Reuss equations, on the other hand, require numerical solutions, unless, as will be seen, the stresses are uniformly distributed. A further assumption of elastic incompressibility (i.e. $\nu = ½$), which lead to eqs(4.2b) and (4.3c), will simplify the application of each theory without incurring unacceptable error. Applying the 'incompressible' theory to a given stress state provides the loading corresponding to elastic–plastic and fully–plastic bars. These solutions also employ a criterion of yielding with the force and torque equilibrium equations.

In the case of a tensile force P, combined with a torque C, the equilibrium equations for a fully-plastic, solid bar of outer radius r_o, are

$$P = 2\pi \int_0^{r_o} \sigma \times r\,\mathrm{d}r, \qquad C = 2\pi \int_0^{r_o} \tau \times r^2\,\mathrm{d}r \tag{4.41a,b}$$

For a thick-walled tube, the inner radius r_i is the lower limit of integration in eqs(4.41a,b). It is convenient to employ the following non-dimensional forms of P and C for a solid bar

$$n = \frac{P}{P_y} = \frac{P}{\pi Y r_o^2} \quad \text{and} \quad m = \frac{C}{C_y} = \frac{2\surd 3\, C}{\pi Y r_o^3} \qquad (4.42a,b)$$

where P_y and C_y, found from eqs(4.41a,b), apply to when the bar first becomes plastic under the separate action of tension and torsion. In a similar manner, the combined strains are normalised with respect to yield strains $\varepsilon_y = Y/E$ and $\gamma_y = Y/(\surd 3G)$:

$$e = \varepsilon^t/\varepsilon_y = \varepsilon^t E/Y \qquad (4.43a)$$

$$g = \gamma^t/\gamma_y = \surd 3G\gamma^t/Y = \surd 3G\,(\chi\, r_o)/Y \qquad (4.43b)$$

where $\chi = \theta/l$ is the twist/unit length giving χr_o as the outer diameter shear strain. The following loading paths facilitate comparison between theory with experiment.

4.4.1 Radial Loading

Let P and C increase proportionately (i.e. n/m = constant) from zero in a solid circular bar. The theoretical analyses each employ the von Mises yield criterion (4.8).

(a) Hencky-Ilyushin

The total strain components are again expressed from eqs(4.9a,b) assuming that any radial and hoop stress arising from P and C, are negligibly small. Eliminating φ between eqs(4.8) and (4.9), the normalised axial stress $S = \sigma/Y$, for the plastic region, is found from the solution to

$$(1-2\nu)^2 S^4 - 6(1-2\nu)eS^3 + [9e^2 - (1-2\nu)^2 + 4\,(1+\nu)2g^2\rho^2]S^2 + 6(1-2\nu)eS - 9e^2 = 0 \qquad (4.44)$$

Equation (4.44) applies to given values of $\rho = r/r_o$, e, and g from eqs(4.43a,b). The corresponding normalised shear stress $T' = \tau/k$, is found from eqs(4.8a) and (4.9b)

$$T' = \frac{2g\rho\,(1+\nu)}{\sqrt{[3e - S(1-2\nu)]^2 + 4(1+\nu)^2 g^2\rho^2}} \qquad (4.45)$$

Note: In eq(4.8b) $T = T'/\surd 3$. Now, consider the elastic–plastic bar, where the plastic zone has penetrated from the outside radius to a radius r_{ep} in the cross-section. This interface radius is normalised $\rho_{ep} = r_{ep}/r_o$ so that, for the elastic core $0 \le \rho \le \rho_{ep}$, the normalised stresses are simply

$$S = e, \quad T' = g\rho \qquad (4.46a,b)$$

The equilibrium equations (4.41a,b) for P and C involve contributions from the elastic and plastic zones. The axial force is

$$P = 2\pi \int_0^{r_{ep}} \sigma\, r\, \mathrm{d}r + 2 \int_{r_{ep}}^{r_o} \sigma\, r\, \mathrm{d}r \qquad (4.47a)$$

where, from eqs(4.42), (4.43) and (4.46), the non-dimensional form becomes

$$n = 2\int_0^{\rho_{ep}} S(\rho)\,\rho\,\mathrm{d}\rho + 2\int_{\rho_{ep}}^{1} S(\rho)\,\rho\,\mathrm{d}\rho = e\rho_{ep}^{\,2} + 2\int_{\rho_{ep}}^{1} S(\rho)\,\rho\,\mathrm{d}\rho \tag{4.47b}$$

Similarly, the torque C is normalised to give m:

$$C = 2\pi\int_0^{r_{ep}} \tau\, r^2\,\mathrm{d}r + 2\pi\int_{r_{ep}}^{r_o} \tau\, r^2\,\mathrm{d}r \tag{4.48a}$$

$$m = 4\int_0^{\rho_{ep}} T'(\rho)\,\rho^2\,\mathrm{d}\rho + 4\int_{\rho_{ep}}^{1} T'(\rho)\,\rho^2\,\mathrm{d}\rho = g\,\rho_{ep}^{\,4} + 4\int_{\rho_{ep}}^{1} T'(\rho)\,\rho^2\,d\rho \tag{4.48b}$$

The second integral in eqs(4.47b) and (4.48b) may be evaluated numerically, e.g. by Simpson's rule. Here $S(\rho)$ and $T'(\rho)$ are the plastic zone stresses supplied by eqs(4.44) and (4.45). The condition that S and T' are singular on the elastic–plastic boundary is ensured from equating either (4.45) and (4.46b) or (4.44) and (4.46a). This gives

$$e^2 + g^2\rho_{ep}^{\,2} = 1 \tag{4.49}$$

When $\rho_{ep} = 1$, the cylinder is fully elastic (f.e.) and eqs(4.47)–(4.49) supply relationships between the applied loading:

$$n = e \text{ and } m = g \tag{4.50a}$$

$$n^2 + m^2 = 1 \tag{4.50b}$$

For a given plastic penetration ρ_{ep}, the solution to $S(\rho)$, $T'(\rho)$, n and m is found as follows. Firstly, assume a value for e and from eq(4.49), find g. At a given normalised radius ρ, where $\rho_{ep} \le \rho \le 1$ in the plastic zone, S and T' are found from eqs(4.44) and (4.45). A value for n/m follows from the integration of eqs(4.47b) and (4.48b). If this does not equal the true applied ratio: $n/m = (P/C)[r_o/(2\surd 3)]$, found from eqs(4.42a,b), then another e value must be assumed and the procedure repeated until agreement is found. This method will supply the stress distributions rapidly when programmed. The original limit load theory [6] was simplified on setting $\nu = ½$, from assuming elastic incompressibility. Therein, eqs(4.44) and (4.45) provide the following closed solutions within the plastic region: $\rho_{ep} \le \rho \le 1$, for an elastic–plastic cylinder:

$$S = \frac{e}{\sqrt{e^2 + g^2\rho^2}}\,, \qquad T' = \frac{g\rho}{\sqrt{e^2 + g^2\rho^2}} \tag{4.51a,b}$$

Since eqs(4.46a,b) remain valid for the elastic core, $0 \le \rho \le \rho_{ep}$, their substitutions, together with those of eqs(4.51a,b), into (4.47b) and (4.48b), lead to

$$n = e\rho_{ep}^{\,2} + (2e/g^2)(e^2 + g^2)^{1/2} - (2e/g^2)(e^2 + g^2\rho_{ep}^{\,2})^{1/2} \tag{4.52a}$$

$$m = g\rho_{ep}^{\,4} + (4/g)(e^2 + g^2)^{1/2} - [8/(3g^3)](e^2 + g^2)^{3/2} - (4\rho_{ep}^{\,2}/g)(e^2 + g^2\rho_{ep}^{\,2})^{1/2} + [8/(3g^3)](e^2 + g^2\rho_{ep}^{\,2})^{3/2} \tag{4.52b}$$

With a given n/m value, e and g are found from the simultaneous solution to eqs(4.49) and (4.52a,b). Take, for example, radial loading of a solid circular bar under $n/m = 0.8$ to an elastic–plastic interface radius $\rho_{ep} = 0.4$. Then from eq(4.49): $e^2 = 1 - 0.16g^2$. Substituting this into eqs(4.52a,b) leads to the following expression in g:

$$0.0204\,g + (3.18/g)(1 + 0.84g^2)^{1/2} - (2.12/g^3)(1 + 0.84g^2)^{3/2} - 0.509/g + 2.12/g^3$$
$$= 0.16\,(1 - 0.16g^2)^{1/2} + (2/g^2)(1 - 0.16g^2)^{1/2}(1 + 0.84g^2)^{1/2} - (2/g^2)(1 - 0.16g^2)^{1/2}$$

A trial solution gives $g = 1.19$ and hence $e = 0.879$, $m = 0.926$ and $n = 0.741$. Equations (4.46a,b) supply the elastic zone stress state: $S = e = 0.879$ and $T' = g\rho = 1.19\rho$. Equations (4.51a,b) supply the plastic zone stress state:

$$S = 1/(1 + 1.835\rho^2)^{1/2} \quad \text{and} \quad T' = 1.355\rho/(1 + 1.835\rho^2)^{1/2}$$

In the fully plastic (f. p.) case, where $\rho_{ep} = 0$, eq(4.49) gives $e = 1$. Equations (4.52a,b) reduce to

$$n = (2e/g^2)(e^2 + g^2)^{1/2} - 2e^2/g^2 \tag{4.53a}$$
$$m = (4/g)(e^2 + g^2)^{1/2} - [8/(3g^3)](e^2 + g^2)^{3/2} + 8e^3/(3g^3) \tag{4.53b}$$

Equations(4.53a,b) provide the following expression in g for the ratio $n/m = 0.8$.

$$(3.18/g)(1 + g^2)^{1/2} - (2.12/g^3)(1 + g^2)^{3/2} + 2.12/g^3 = (2/g^2)(1 + g^2)^{1/2} - 2/g^2$$

giving $g = 1.38$, $m = 0.93$ and $n = 0.744$. The fully plastic stresses follow from eqs(4.51a,b):

$$S = 1/(1 + 1.896\rho^2)^{1/2} \quad \text{and} \quad T' = 1.377\rho/(1 + 1.896\rho^2)^{1/2}$$

The strains corresponding to each ρ_{ep} value may be found from eqs(4.43a,b), given Y and E for the material. With increasing strain beyond full plasticity, i.e. $e > 1$ and $g > 1.38$, eqs(4.53a,b) show that n and m remain constant. Zyczkowski [6] has shown that when e and g are eliminated between eqs(4.53a,b), three relations between m and n apply

$$m = (2/3)(1 - n)^{1/2}(2 + n) \tag{4.54a}$$
$$m = (2/3)(2 - 3n)/(1 - n)^{3/2} \tag{4.54b}$$
$$m = (2/3)(1 + n)^{1/2}(2 - n) \tag{4.54c}$$

It is seen from Fig. 4.7 that only eq(4.54a) expresses the true interaction between m and n at the f. p. load-limit carrying capacity.

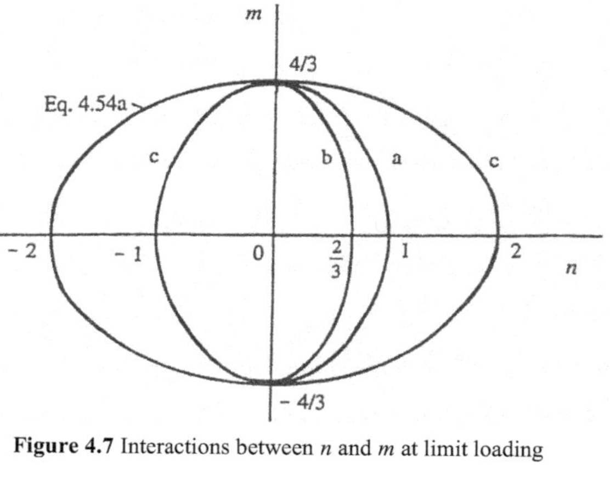

Figure 4.7 Interactions between n and m at limit loading

With $n/m = 0.8$, eq(4.54a) is satisfied by $m = 0.93$ and $n = 0.744$. The inner curve b must be rejected because it does not give $n = 1$ under a fully-plastic tensile load. Figures 4.8a,b give the normalised stress distributions with incompressible elasticity. These are found from eqs(4.51a,b), for 0.2 intervals in ρ_{ep} from the fully elastic (f. e.) to the fully plastic (f. p.) condition.

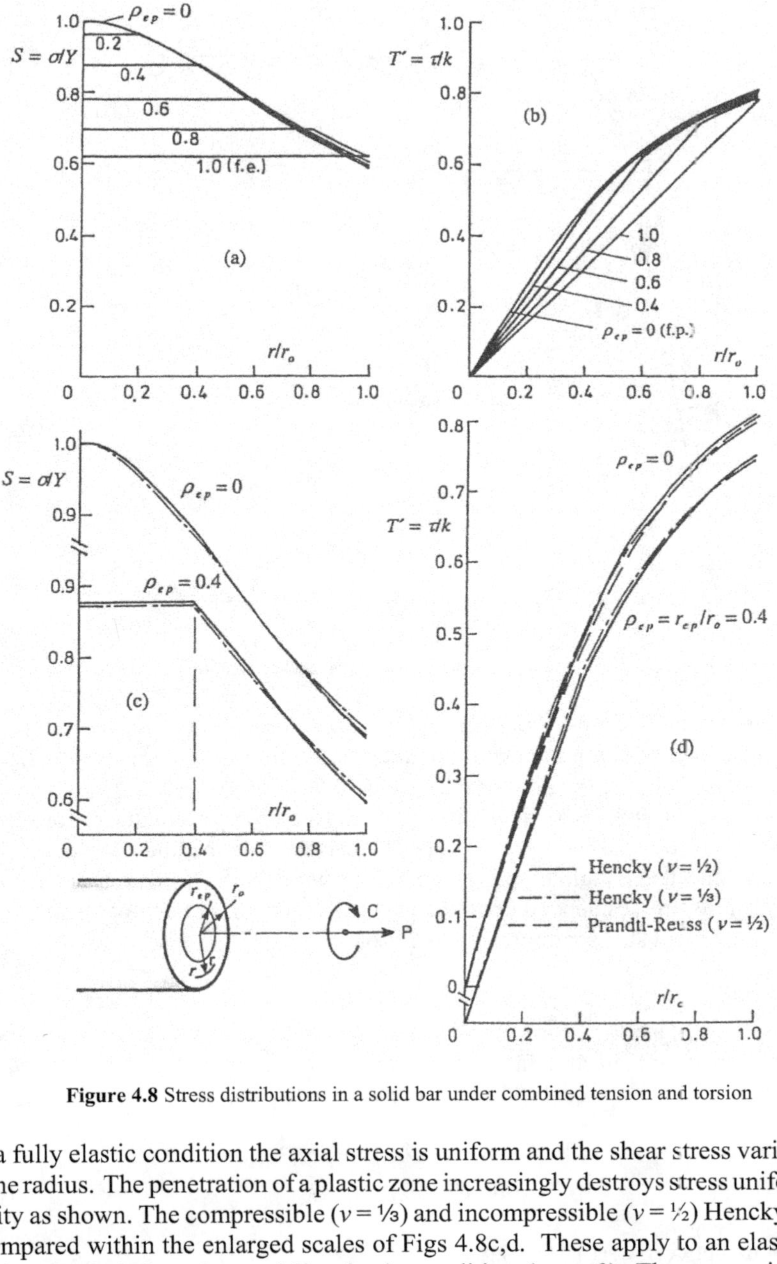

Figure 4.8 Stress distributions in a solid bar under combined tension and torsion

With a fully elastic condition the axial stress is uniform and the shear stress varies linearly with the radius. The penetration of a plastic zone increasingly destroys stress uniformity and linearity as shown. The compressible ($\nu = ⅓$) and incompressible ($\nu = ½$) Hencky solutions are compared within the enlarged scales of Figs 4.8c,d. These apply to an elastic–plastic condition ($\rho_{ep} = 0.4$) and to a fully plastic condition ($\rho_{ep} = 0$). The comparison shows negligible error in the stress magnitudes for $\nu = ½$. If follows that the incompressible theory provides realistic distributions of stress as the bar deforms from an initially elastic condition, through the elastic–plastic region, to attain full plasticity.

The strains increase rapidly to infinity under the limit loading in the manner of Fig. 4.9. Experimental data [16] is compared with the plot between n and m, predicted from eqs(4.42a,b) and with a plot between the outer diameter axial and shear strain, ε and γ, from eqs(4.43a,b).

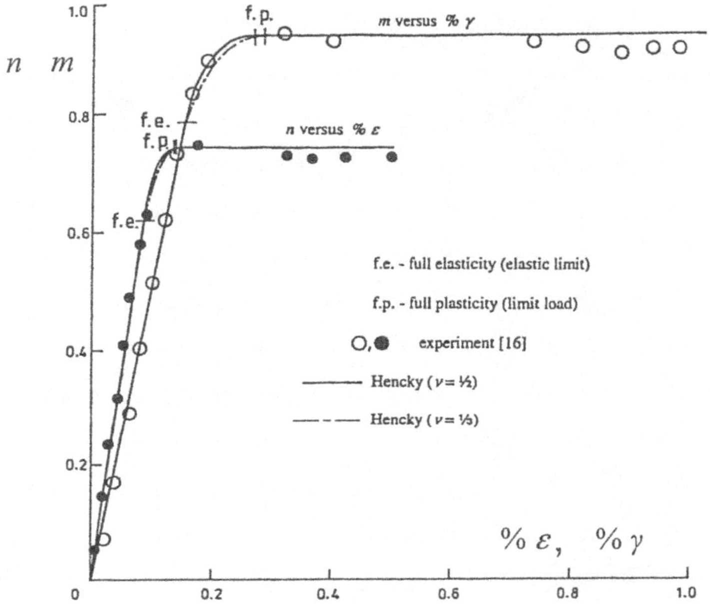

Figure 4.9 Load-deformation behaviour in a solid bar under combined tension-torsion

Clearly, limit loads from the incompressible, total-strain theory are accurate. Slight differences observed between total strain predictions, with and without incompressibility, are confined only to the region between the f. e. and f. p. conditions.

Strictly, a triaxial stress state will arise from the Hencky-Ilyushin theory for a tension-torsion member of material with $\nu \neq ½$ [17]. The perturbation method of solution showed uniform radial and tangential compressive stresses within the elastic zone of a partially plastic bar. These stresses depend upon the penetration depth, but do not amount to more than 10 % of the yield stress within the plastic zone.

(b) Prandtl-Reuss

Gaydon [18] derived a closed-form, non-hardening, Prandtl-Reuss solution to this problem from assuming that $\nu = ½$ in eq(4.3a). Within the plastic region ($\rho_{ep} \le \rho \le \rho_o$), the total axial and shear strain increments, eqs(4.11a,b), were written as

$$d\varepsilon^t = dl/l = (2d\lambda/3)\sigma + d\sigma/(3G) \qquad (4.55a)$$

$$d\gamma^t = r\, d\theta/l = 2d\lambda\tau + d\tau/G \qquad (4.55b)$$

Eliminating $d\lambda$ between eqs(4.55a,b) gives

$$\frac{1}{l} \times \frac{\mathrm{d}l}{\mathrm{d}\theta} = \frac{\sigma}{3\tau} \times \frac{r}{l} - \left(\frac{\sigma}{3G\tau}\right)\frac{\mathrm{d}\tau}{\mathrm{d}\theta} + \frac{1}{3G} \times \frac{\mathrm{d}\sigma}{\mathrm{d}\theta} \tag{4.56a}$$

Since the stresses must satisfy the von Mises yield criterion (4.8a), σ in eq(4.56a) may be eliminated, to give

$$\frac{3}{l} \times \frac{\mathrm{d}l}{\mathrm{d}\theta} = \frac{r}{\tau l}\sqrt{Y^2 - 3\tau^2} - \frac{Y^2}{G\tau\sqrt{Y^2 - 3\tau^2}} \times \frac{\mathrm{d}\tau}{\mathrm{d}\theta} \tag{4.56b}$$

With the substitutions $T' = \tau/k = \sqrt{3}\tau/Y$ and $g = \sqrt{3}(G/Y)(r_o/l_o)\theta$, where l_o is the initial length of the bar, eq(4.56b) becomes

$$\frac{\mathrm{d}T'}{\mathrm{d}g} = \frac{rl_o}{r_o l}\left(1 - T'^2\right) - \frac{\alpha T'}{\sqrt{3}}\sqrt{1 - T'^2} \tag{4.56c}$$

In eq(4.56c), $\alpha = (\sqrt{3}G/Yl)(\mathrm{d}l/\mathrm{d}g)$ is a variable which integrates to

$$\int \alpha\, \mathrm{d}g = (\sqrt{3}G/Y)\ln(l/l_o) \tag{4.57a}$$

$$l/l_o = \mathrm{e}^{\frac{Y}{\sqrt{3}G}\int \alpha\, \mathrm{d}g} \tag{4.57b}$$

The axial strain is defined as

$$\varepsilon^{t} = \ln(l/l_o) \tag{4.58a}$$

for which the normalised form is

$$e = \varepsilon^{t}/(Y/E) = \sqrt{3}\int \alpha\, \mathrm{d}g \tag{4.58b}$$

where $E = 3G$. Now, $S = e$ and $T' = g\rho$ for the elastic core: $0 \le \rho \le \rho_{ep}$. Then, from eq(4.8), $S^2 + T'^2 = 1$ at the elastic–plastic interface. Substituting from eqs(4.49) and (4.58) gives

$$3\left(\int_0^g \alpha\, \mathrm{d}g\right)^2 + (\rho_{ep}\, g)^2 = 1 \tag{4.59}$$

The integral term ( ) in eq(4.59) must take the constant value e at full elasticity ($\rho_{ep} = 1$). This condition is given by $e^2 + g^2 = 1$, where $e = n$ and $g = m$. Under radial loading, e is found from combining eqs(4.42a,b) and (4.50a,b) to give

$$[P/(\pi Y r_o^2)]^2 + [2\sqrt{3}C/(\pi Y r_o^3)]^2 = 1 \tag{4.60a}$$

$$e = n = P/(\pi Y r_o^2) = [1 + (12/r_o^2)(C/P)^2]^{-1/2} \tag{4.60b}$$

As the normalised load and torque are again found from eqs(4.47a) and (4.48a), the ratio $n/m = (Pr_o)/(2\sqrt{3}C)$ is a constant, since P/C = constant. The three equations (4.56c), (4.59) and (4.60b) are then solved by trial from assuming an initial value for α.

In the case of a fully plastic bar, putting $\mathrm{d}T'/\mathrm{d}g = 0$ and $l/l_o \approx 1$ in eq(4.56c) leads to

$$T'(\rho) = \rho/(\rho^2 + \alpha^2/3)^{1/2} \tag{4.61a}$$

$$S(\rho) = (1 - T'^2)^{1/2} = \alpha/(\alpha^2 + 3\rho^2)^{1/2} \tag{4.61b}$$

Also, with $\rho_{ep} = 0$, there follows from eqs(4.47b) and (4.48b)

$$\frac{n}{m} = \frac{Pr_o}{2\surd 3C} = \frac{S(\rho)\rho\,\mathrm{d}\rho}{2[T'(\rho)\rho^2\mathrm{d}\rho]} \tag{4.62a}$$

Substituting eqs(4.61a,b) into eq(4.62a) leads to

$$\frac{Pr_o}{C} = \frac{\int_0^1 \alpha\rho\,\mathrm{d}\rho/(\alpha^2+3\rho^2)^{1/2}}{\int_0^1 \rho^3\,\mathrm{d}\rho/(\alpha^2+3\rho^2)^{1/2}} \tag{4.62b}$$

from which α may be solved. In the radial loading test employed by Shammamy and Sidebottom [16] the test conditions were $P/C = 0.374$ mm^{-1} (9.5 in^{-1}) and $r_o = 7.37$ mm (0.29 in). The properties of the non-hardening, SAE 1035 steel used were: $Y = 283.3$ MPa (41.1×10^3 lbf/in^2) and $E = 207$ GPa (30×10^6 lbf/in^2). We find by trial that eq(4.62b) is satisfied by $\alpha = 1.261$. The stresses are then found from eqs(4.61a,b). The comparison between the Hencky and Prandtl-Reuss stress distributions at full plasticity in Fig. 4.8 reveals that these are almost coincident. A further check can be made on the assumption that $l / l_o \approx 1$ at full plasticity, from eq(4.57b). Correspondingly, when $\rho_{ep} = 0$, eq(4.59) gives $\int \alpha dg = 0.5773$. Substituting this, together with Y and $E = 3G$, into eq(4.57b) gives $l / l_o = 1.0014$. The assumption is valid because, when the bar first becomes fully plastic, the strains are of an elastic order. It may be deduced from Fig. 4.8d that distributions of stress, found from the 'incompressible' Prandtl-Reuss theory, are again realistic as the bar deforms in the elastic–plastic region.

Sved and Brooks [19] have given the governing equations for a Prandtl-Reuss solution to this problem when accounting for compressible elasticity. When $\nu \neq \frac{1}{2}$, it becomes necessary to admit the tangential and radial stresses in addition to the axial and shear stresses. They showed that this leads to the requirement for a numerical solution to a system of four, simultaneous, partial differential equations. The additional stresses, so found, were of similar small magnitude to those found from the compressible Hencky theory [17]. Tangential and radial stresses are absent when $\nu = \frac{1}{2}$ but, as has been seen, a different ν does not alter the more dominant axial and shear stresses by more than a few percent.

4.4.2 Constrained Deformation

Consider a deformation path where the tensile yield strain ε_y is held constant while increasing the shear strain γ, within a thick-walled tube of non-hardening material. The theories predict the following variations in both P and C with γ.

(a) Hencky-Ilyushin

According to eqs(4.11a,b), when $\nu = \frac{1}{2}$ and $\varepsilon_y = Y/E$, the stresses during the ensuing deformation are given by

$$\sigma = 3\varepsilon_y/(2\varphi), \quad \tau = \gamma/(2\varphi) \tag{4.63a,b}$$

Substituting eqs(4.63a,b) into eq(4.8a) supplies the scalar multiplier φ

$$\varphi = \surd(9\varepsilon_y^2 + 3\gamma^2)/(2Y) \tag{4.64}$$

Equations (4.63a,b) and (4.64) provide the following dimensionless stress components

$$S = (1 + \rho^2 g^2)^{-1/2} \quad \text{and} \quad T' = \rho g\,(1 + \rho^2 g^2)^{-1/2} \tag{4.65a,b}$$

where

$$S^2 + T'^2 = 1 \tag{4.65c}$$

Note: Here $T' = \tau/k$, which is not to be confused with $T = \tau/Y$, used for the uniform stress analysis given in § 4.3. The relationship $E = 3G$ is implied with assumed incompressibility. Since the tube is fully plastic, n and m follow from the corresponding forms of eqs(4.41a,b) and (4.42a,b):

$$n = \frac{2}{(1 - D^2)}\int_0^1 \rho S\,\mathrm{d}\rho\,, \qquad m = \frac{4}{(1 - D^4)}\int_0^1 \rho^2 T'\,\mathrm{d}\rho \tag{4.66a,b}$$

where $D = r_i/r_o < 1$, is the tube diameter ratio. Combining eqs(4.65a,b) and (4.66a,b) leads to the integrated forms for n and m

$$n = \frac{2\left[(1 + g^2)^{\frac{1}{2}} - (1 + D^2 g^2)^{\frac{1}{2}}\right]}{g^2(1 - D^2)} \tag{4.67a}$$

$$m = \frac{4\left[(1 + g^2)^{\frac{1}{2}} - \dfrac{2}{3g^2}(1 + g^2)^{\frac{3}{2}} - D^2(1 + D^2 g^2)^{\frac{1}{2}} + \dfrac{2}{3g^2}(1 + D^2 g^2)^{\frac{3}{2}}\right]}{g(1 - D^4)} \tag{4.67b}$$

With a given value of g for the ensuing deformation, the distribution of stress through the wall of the cylinder is supplied by eq(4.65a,b). Equations (4.67a,b) express the loading necessary to achieve that stress state.

(b) Prandtl-Reuss

The total, incremental strains are given by the Prandtl-Reuss eqs(4.11a,b). Since $\mathrm{d}\varepsilon^{t} = 0$, it follows from eqs(4.8a) and (4.11a) that the scalar $\mathrm{d}\lambda$ is

$$\mathrm{d}\lambda = -3\mathrm{d}\sigma/(2E\sigma) = 9\tau\,\mathrm{d}\tau/(Y^2 - 3\tau^2) \tag{4.68}$$

Substituting eq(4.68) into eq(4.11b) and integrating between 0 and τ leads to a γ^t prediction, given by Smith and Sidebottom [20]:

$$\gamma^t = r\chi = (2\nu - 1)\tau/E + (\sqrt{3}Y/E)\tanh^{-1}(\sqrt{3}\tau/Y) \tag{4.69a}$$

Employing the present non-dimensional parameters, eq(4.69a) becomes

$$(1 + \nu)g\rho = (\nu - \tfrac{1}{2})T' + \tfrac{3}{4}\ln\left[(1 + T')/(1 - T')\right] \tag{4.69b}$$

Thus, T' can be found from eq(4.69b) for a given g over the range $D \le \rho \le 1$. S follows from eq(4.65c) and n and m from the numerical integration of eqs(4.66a,b). The incompressible solution is found from putting $\nu = \tfrac{1}{2}$, giving a relationship between the elastic constants as $1/E = 1/(9K) + 1/(3G)$. Then, eqs(4.65c) and (4.69b) provide non-dimensional stresses in closed form

$$S = \text{sech}\,(\rho g\,), \quad T' = \tanh\,(\rho g\,) \tag{4.70a,b}$$

Equations (4.70a,b) originally appeared in the work of Gaydon [18], Prager and Hodge [21] and Nadai [22]. Gaydon also solved the case of loading where the axial strain was initially elastic. Combining eqs(4.66) and (4.70), the corresponding loading parameters become:

$$n = \frac{2}{(1 - D^2)} \int_D^1 \rho\,\text{sech}(\rho g)\,\text{d}\rho\;, \quad m = \frac{4}{(1 - D^4)} \int_D^1 \rho^2 \tanh(\rho g)\,\text{d}\rho \tag{4.71a,b}$$

The few published results, which enable comparison with this theory, are again due to Shammamy and Sidebottom [16]. These apply to the same grade of non-hardening SAE 1035 steel in a tube with D = 0.93. Figures 4.10a,b compare the theoretical stress distributions and, in Fig. 4.10c, a comparison is made between experiment and the predictions from eqs(4.66a,b), (4.67a,b) and (4.71a,b).

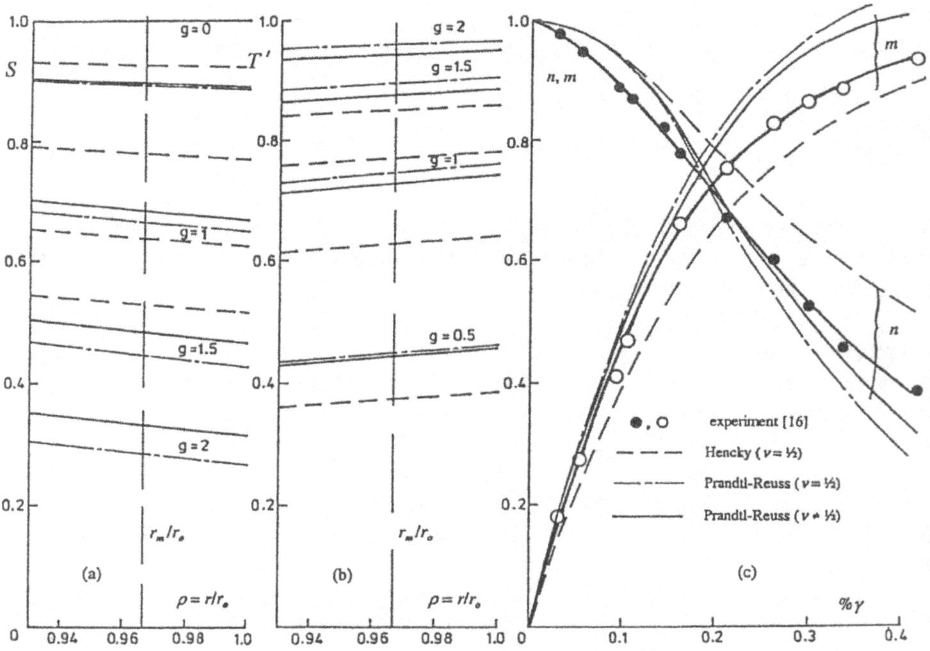

Figure 4.10 Constrained deformation in a thick-walled tube under combined tension and torsion

It is seen from Fig. 4.10c that deviations between the three predictions increase with increasing shear deformation. The greatest deviation in the stress distributions, for a given g value in Figs 4.10a,b, occurs between the incremental and total strain theories for $\nu = ½$. This applies further in respect of loading parameters n and m in (c), although experiment is unable to distinguish clearly between the various theoretical predictions. Again, the $\nu = ½$ assumption leads to acceptable simplified solutions. A similar test, conducted on a solid bar of non-hardening material, would be more conclusive for making a comparative study of this nature. Because it is evident from Figs 4.10a,b that the wall stress gradients for this tube are not severe, a thin-walled theory would approximate its singular wall stress value to the indicated mid-wall radius.

As the material in Fig. 4.10 is non-hardening, the theory need not account for the influence of the initial 0.2 % plastic loading strain. In fact, a general feature of the Hencky theory is that it applies only to the current stress state, i.e. it does not refer to how that state was achieved. This deficiency is corrected with an incremental strain summation and explains why the Prandtl-Reuss theory is generally believed to be more representative of path dependent plasticity under non-proportional loading.

Sved and Brooks [19] applied the compressible Prandtl-Reuss solution to the converse problem in which the twist was held constant at its yield value, in a solid bar, under increasing axial strain. They showed that radial and hoop stresses also arose with maximum values in compression at the bar centre. These were never greater than 5 % of the shear yield stress, even at full plasticity with low Poisson's ratio: $v \approx 0.1$. Their work confirmed that the variation in v is of sufficiently small influence, justifying Gaydon's[17] application of an incompressible Prandtl-Reuss theory.

4.4.3 Stepped Loading

Here an elastic torque C_o is held constant under an increasing axial force P (Fig. 4.11a). This gives $m < 1$, and with $v = ½$, it is first necessary to find n from eq(4.50a) for the fully-elastic condition ($\rho_{ep} = 1$). For increasing n within the elastic–plastic region ($0 \le \rho_{ep} \le 1$), eqs(4.46a,b), (4.49), (4.51a,b) and (4.52a,b) are applied to determine e, g and n for a given $\rho_{ep} = r_{ep}/r_o$ and m.

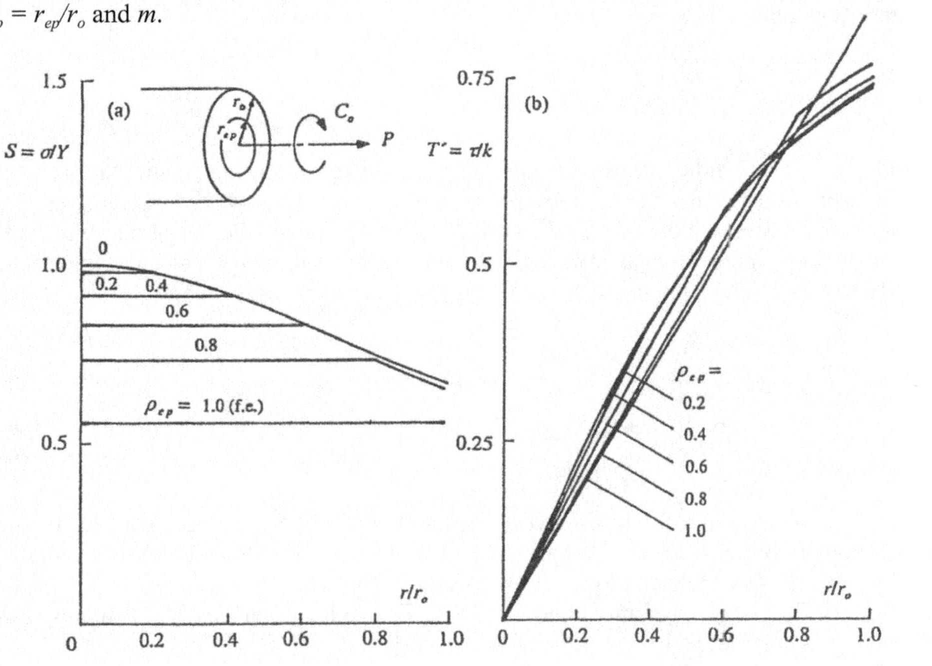

Figure 4.11 Normalised stress distributions for a stepped loading of increasing P under constant C_o

The following procedure applies to the simplified 'incompressible' Hencky theory. Say one wishes to establish the stress distributions for $m = 0.83$ = constant, when the force P has produced: (i) a fully elastic bar ($\rho_{ep} = 1$) and (ii) an elastic-plastic bar with $\rho_{ep} = 0.6$. For (i), it follows from eqs(4.46a,b) that $T' = m\rho = 0.83\rho$ (i.e. linear in ρ) in which $0 \le \rho \le 1$ and $S = n = 0.56$. For condition (ii), eq(4.49) gives $e^2 = 1 - 0.36g^2$. Substituting into eq(4.52b):

$$0.83 = 0.13g + (4/g)(1 + 0.64g^2)^{1/2} - [8/(3g^3)](1 + 0.64g^2)^{3/2} - 1.44/g + 8/(3g^3) \qquad (4.72)$$

A trial solution to eq(4.72) gives $g = 0.934$ and $e = 0.828$. Substituting the appropriate e, g and ρ_{ep} values into eq(4.52a) provide $n = 0.77$, the non-dimensional measure of the increased force necessary for partial yielding of the section. Within the elastic zone ($0 \le \rho \le 0.6$), the stresses are found from eqs(4.46a,b) to be: $S = e = 0.828$ and $T' = g\rho = 0.934\rho$. In the plastic zone ($0.6 \le \rho \le 1$), eqs(4.51a,b) give

$$S = 1/(1 + 1.272\rho^2)^{1/2} \quad \text{and} \quad T' = 1.128\rho/(1 + 1.272\rho^2)^{1/2} \qquad (4.73\text{a,b})$$

The normalised stresses, from eqs(4.73a,b), are distributed according to Figs 4.11a and b. Other distributions shown are calculated in a similar manner for interface radii: $\rho_{ep} = 0$ (i.e. fully plastic), 0.2, 0.4, 0.8 and 1.0 (i.e. fully elastic). For each ρ_{ep} value, a zero shear stress occurs at the bar centre, corresponding to maximum axial stress. Note that the elastic and elastic-plastic stresses must agree at the interface radius. Experiments [18] show that even in mildly hardening materials, n and m continue to increase beyond their fully plastic values. The rule of *isotropic hardening* explained in chapter 10 (see § 10.3) can account for this by allowing the yield surface to expand. This rule has enabled a corresponding extension to the Hencky theory to account for the greater load carrying capacity which most metallic materials display [23].

4.4.4 Residual Stress Distributions

Self-equilibrium residual stresses result from unloading an elastic–plastic bar. They are found from subtracting elastic stresses, S_e and T_e', that recover with unloading, from elastic–plastic stresses that arise when the loading is applied. For example, with radial loading of incompressible elastic material in Fig. 4.8, the Hencky eqs(4.46a,b), (4.50a,b) and (4.51a,b) supply the following normalised, residual stress expressions

$$S_R = S - S_e \qquad (4.74\text{a})$$
$$= (1 + g^2\rho^2/e^2)^{-1/2} - n \qquad (4.74\text{b})$$

$$T_R' = T' - T_e' \qquad (4.75\text{a})$$
$$= (g/e)\rho\,(1 + g^2\rho^2/e^2)^{-1/2} - \rho m \qquad (4.75\text{b})$$

where e, g, m and n are found from eqs(4.49) and (4.52a,b) for a given n/m and ρ_{ep}. The distributions, found from eqs(4.74b) and (4.75b), are shown in the upper two diagrams of Fig. 4.12a. These correspond to an elastic–plastic cylinder with $\rho_{ep} = 0.6$ and a fully plastic cylinder (i.e. $\rho_{ep} = 0$). The latter condition increases the magnitude of the residual stresses following radial loading of a non-hardening material.

The residual stress distributions, S_R and T_R', resulting from the release of a non-radial loading in Fig. 4.10 need separate consideration. When eqs(4.74a) and (4.75a) are applied to the constrained test conditions, S_R and T_R' are distributed in the manner of Fig. 4.12b. Clearly, for constrained deformation, the residuals show a greater dependency upon the particular theory chosen to determine S and T'. That is, differences between the Hencky and

Prandtl-Reuss predictions, shown here for $g = 2$, are a consequence of variations between the theoretical loading stresses given in Figs. 4.10a,b. In Fig. 4.12c, the residuals are shown for $\rho_{ep} = 0$ and 0.6, based upon the Hencky applied stress distributions (see Fig. 4.11) under stepped loading. These residuals are similar to those following the release of a radial loading (Fig. 4.12a) with comparable elastic–plastic interface radii. It has long been known that residual stresses can have the beneficial effect of enhancing fatigue life. Here, it becomes important to establish the residual stresses with reasonable accuracy, particularly in the case of non-radial loading, where large errors can arise by over simplifying the solution to the resulting stress distribution [24, 25].

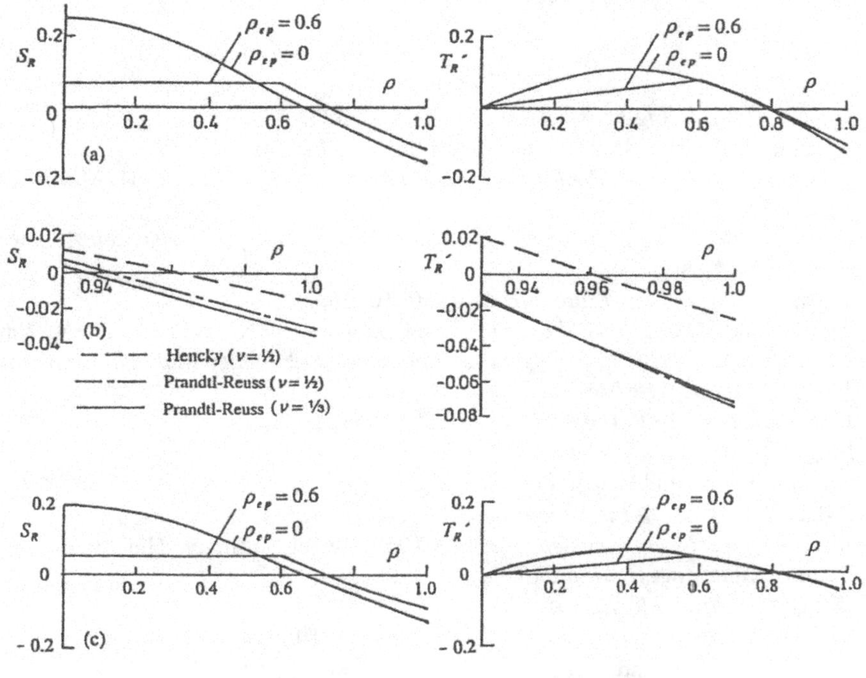

Figure 4.12 Residual stress resulting from various loadings

4.5 Concluding Remarks

Choosing between the two classical descriptions of deformation arising from a combined stress state will depend upon the manner in which stress is distributed. When the latter is uniform, there is little difference between predictions from the Hencky deformation theory and the Prandtl-Reuss flow theory. In fact, results of experiments conducted under non-radial loading paths, reviewed in this chapter, appear to agree better with simpler, closed-form Hencky predictions to deformation. In the analysis of the deformation behaviour of structures under non-uniform stress states, experiment shows that the effects of history are insignificant with proportional loadings from zero stress. The Hencky total-strain deformation theory will represent, within closed solutions of acceptable accuracy, the distribution of stress and the load-deformation response with the transition from elastic to fully plastic behaviour. Deviations arise between the total and incremental theories under non-proportional loadings, including stepped-stress paths or paths in which one component of strain is constrained. Although the Hencky theory appears to remain conservative in its prediction of strain, the generally held view, partly substantiated here, is that the incremental Prandtl-Reuss theory (with elastic compressibility) provides more realistic predictions whenever history effects are present. However, it is known now that errors do arise from the Prandtl-Reuss theory. In particular it was seen in chapter 3 that the latter theory does not account for directional plasticity arising in anisotropic material and softening behaviour following stress reversals (see § 3.3 and 3.4). Another main deficiency of the said theory has been recognised where plastic strain increments are written as strain rates $\delta\varepsilon^P \rightarrow \delta\varepsilon^P/\delta t$. Time-dependent straining is manifested in a strain rate dependency of material when loading is rapid and in its accumulation of creep strain when loading over a prolonged periods. Chapter 11 considers these influences upon the theory in greater detail.

References

1. Hencky H. *Zeit. Angew. Math. Mech.*, 1924, **4**, 323.
2. Ilyushin A. A. *Prik. Matem. Mekh.*, AN SSSR, 1946, **10**, 347.
3. Nadai A. *Plasticity: A Mechanics of the Plastic State of Matter*, 1931, McGraw-Hill, New York.
4. Prandtl L. *Proc. 1st Int. Congr. Appl. Mech. Delft*, 1924, p. 43 (also *Zeit Angew Math Mech* 1928, **8**, 85.)
5. Reuss A. *Zeit. Angew. Math. Mech.*, 1930, **10**, 266.
6. Zyczkowski M. *Combined Loadings in the Theory of Plasticity*, 1981, PWN, Warsaw.
7. Massonnet C. *Plasticity in Structural Engineering, Fundamentals and Applications*, 1979, Ch. 1, Springer Verlag.
8. Hohensemer K. *Zeit. Angew. Math. Mech.*, 1951, **11**, 515.
9. Prager W. *J. Appl. Phys.*, 1944, **15**, 65.
10. Hohensemer K. and Prager W. *Zeit. Angew. Math. Mech.*, 1932, **12**, 1.
11. Prager W. Proc.: *5th Int. Congr. Appl. Mech.*, 1938, S.234, Cambridge.
12. Neuber H. *Kerbspannungslehre, 2 Aufl*, 1958, Berlin Gottingen, Heidelberg.
13. Betten J. *Zeit. Angew. Math. Mech.*, 1975, **55**, 119.
14. Gill S. S. *J. Appl. Mech.*, 1956, **23**, 497.
15. Schlafer J. L. and Sidebottom O. M. *Expl. Mech.*, 1969, **9**, 500.
16. Shammamy M. R. and Sidebottom O. M. *Expl. Mech.*, 1967, **7**, 497.
17. Zyczkowski M. *Rozpr. Inz.*, Polska Akademia, 1955, **3**, 285.
18. Gaydon F. A. *Quart. J. Mech. and Appl. Math.*, 1952, **5**, 29.
19. Sved G. and Brooks D. S. *Acta Techn. Hung.*, 1965, **50**, 337.
20. Smith J. O. and Sidebottom O. M. *Inelastic Behaviour of Load-Carrying Members*, 1965, J. Wiley and Son, New York.
21. Prager W. and Hodge P. *Theory of Perfectly Plastic Solids*, 1951, J. Wiley and Son, New York.
22. Nadai A. *Theory of Flow and Fracture of Solids*, 1950, McGraw-Hill, London.
23. Rees D. W. A. *Applied Solid Mechanics*, (Eds Tooth A. S. and Spence J.) 1988, Chapter 17, 321–346 Elsevier.
24. Hoskin B. C. *Austral. J. Appl. Sci.*, 1961, **12**, 255.
25. Mii H. J. *Japan Soc. Appl. Mech.*, 1950, **3**, 195, Tokyo, .

Exercises

Uniform Stress States

4.1 A thin walled cylinder is stressed to the point of yield under an axial-shear stress ratio $\sigma/\tau = 2$. The shear strain is then held constant while the cylinder is subjected to increasing tension. Derive expressions for the subsequent variation in axial strain with the axial and shear stresses (σ, τ) according to (a) Hencky and (b) Prandtl-Reuss theory. Compare, graphically showing the normalised stress asymptotes σ/Y and τ/Y. Take $G = 77$ GPa, $\nu = 0.3$ and $Y = 230$ MPa.
See solution

4.2 A thin walled cylinder is stressed to the point of yield under an axial-shear stress ratio $\sigma/\tau = 2$. The axial strain is then held constant while the cylinder is subjected to increasing torsion. Derive expressions for the subsequent variation in shear strain with the axial and shear stresses (σ, τ) according to the Hencky and Prandtl-Reuss theories. Compare, graphically showing the normalised stress asymptotes σ/Y and τ/Y. Take $G = 77$ GPa, $\nu = 0.3$ and $Y = 230$ MPa.

4.3 The internal pressure in a closed-end, thin walled tube is increased to produce yielding of the tube material under a stress ratio $\sigma_\theta/\sigma_z = 2$. The circumferential strain is then held constant while the tube is progressively twisted. Show, according to the Prandtl-Reuss theory, that total shear strain γ^t varies with that $T' = \tau/k$ according to the following equation. Explain why axial plastic strain absent when the radial stress is ignored?

$$\gamma^t = \frac{k(2 - \nu)}{2G(1 + \nu)} \ln\left(\frac{1 + T'}{1 - T'}\right) - \frac{(1 - 2\nu)T'}{(1 + \nu)}$$

4.4 In a thin-walled tube of non-hardening material the axial strain at yield is held constant. Show, with progressively increasing circumferential tension, that the Prandtl-Reuss prediction to the ensuing total hoop strain ε_θ^P is given by the following equation, where $S_\theta = \sigma_\theta/E$. Hence, show that the normalised stress asymptotes are $\sigma_\theta/Y = 2/\surd 3$ and $\sigma_z/Y = 1/\surd 3$.
See solution

$$\varepsilon_\theta^t = \frac{\surd 3\, Y}{4E} \ln\left(\frac{1 + \surd 3 S_\theta/2}{1 - \surd 3 S_\theta/2}\right) + \frac{Y(2\nu - 1)}{E}\left[\left(1 - \frac{3S_\theta}{2}\right) - \sqrt{1 - \frac{3S_\theta^2}{4}}\right]$$

4.5 A thin-walled tube is twisted under torsion to the point of yield. The total shear strain is then held constant while the tube is subjected to an increasing internal pressure for which $\sigma_\theta = 2\sigma_z$. Show, for the ensuing deformation, that the respective total circumferential and axial strains are given by the respective equations where $S_z = \sigma_z/Y$. Plot S_z versus $\varepsilon_\theta{}'E/Y$ and $\varepsilon_z{}'E/Y$ showing the S_z asymptotes.

$$\varepsilon_\theta^t = \frac{Y}{E}\left[(1 - 2\nu)S_z + \frac{(1 + \nu)}{2\surd 2} \ln\left(\frac{1 + \surd 2 S_z}{1 - \surd 2 S_z}\right)\right]$$

$$\varepsilon_z^t = \frac{Y}{E}(1 - 2\nu)S_z$$

4.6 A non-hardening, thin-walled cylinder, with inner and outer radii 7.1 and 7.6 mm respectively, is strained to the tensile yield point. Thereafter, the cylinder is subjected to an increasing torque whilst the simultaneous tension is adjusted to maintain the tensile yield strain Y/E constant. Compare predictions to the deformation that ensues from the Hencky incompressible theory ($\nu = ½$) and the Prandtl-Reuss theory with $\nu = ¼$. Take $E = 207$ GPa and $Y = 283$ MPa.

Non-Uniform Stress States

4.7 A solid cylindrical bar, 14.75 mm diameter, is made from a non-hardening steel with a yield stress of 283 MPa. Given that the torque C and the axial load P increase proportionately in the ratio $P/C = 0.375$ mm^{-1}, find P and C: (a) at the limit of elastic deformation, (b) at the elastic–plastic mean radius and (c) for a fully plastic condition. Assume elastic incompressibility with $E = 207$ GPa.
See solution: (a) 80.5 Nm, 30.1 kN; (b) 94.12 Nm, 35.3 kN; (c) 95.7 Nm, 35.8 kN

4.8 Plot for the cylinder in exercise 4.7, the distribution of tensile and shear stress with radius for each of the conditions (a), (b) and (c). Assuming elastic unloading, plot the distribution of residual stress corresponding to (b) and (c). Take $E = 207$ GPa and assume elastic incompressibility.

4.9 A solid cylindrical bar, 12.8 mm diameter, is made from a non-hardening steel with a yield stress of 305 MPa. The bar is subjected to a constant torque of 60.7 Nm with a steadily increasing axial force. Calculate the value of the force corresponding to: (a) the limit of elasticity, (b) an elastic–plastic interface coincident with the mean radius and (c) full plasticity. Take $E = 207$ GPa and $\nu = ½$.
See solution: (a) 21.5 kN, (b) 30.8 kN, (c) 31.5 kN

4.10 Plot the distributions of axial and shear stress with radius for the cylinder of exercise 4.9 in each of (a), (b) and (c). Determine which condition (ii) and (iii) has the greatest extent of the residual compression following unloading.
See solution

4.11 Apply the Prandtl-Reuss solution, with compressible elasticity, to the deformation in a solid bar subjected to torque and axial load which increase proportionately within the plastic range.

4.12 Derive expressions for the normalised stress components $S = \sigma_z / Y$ and $T = \tau / k$ from the incompressible Prandtl-Reuss theory when the shear strain in a solid circular bar is increased while the elastic axial strain is held constant.

4.13 A solid circular bar supports a constant, initially elastic–plastic torque while being subjected to an increasing tension. Determine the Hencky stress distributions for suitable stages of the ensuing deformation to full plasticity.

4.14 Derive the Prandtl-Reuss solutions, for $\nu = ½$, to the stress distributions in a solid circular bar under increasing axial force combined with a constant, initially elastic torque.

4.15 Establish the stress distributions from exercise 4.14 for $m = 0.83$, when the force has produced a fully elastic bar ($\rho_{ep} = 1$) and an elastic-plastic bar with $\rho_{ep} = 0.6$. Compare graphically with the Hencky solution (see Fig. 4.11) for the same conditions.

4.15 Establish the governing equations from the Prandtl-Reuss theory, with compressible elasticity, for the deformation which ensues when an initially elastic torque C_o is held constant under an increasing force P in a solid bar.

CHAPTER 5

ELASTIC-PERFECT PLASTICITY

5.1 Introduction

With ideal, elastic–perfectly plastic behaviour, elastic strain accompanies plastic strain in the absence of hardening. This simplification will often provide closed-form solutions to the stress distribution in loaded structures, revealing a clear distinction between zones of elasticity and plasticity. Using this model, the non-hardening elastic-plastic behaviour of beams in bending and solid shafts under torsion are examined. A yield criterion is introduced when the stress state is multi-axial. This is illustrated in the determination of principal stress distributions for axially symmetric structures, including a thick-walled pressurized cylinder, with different end conditions, an annular disc and a rotating disc. All solutions to the plastic-zone stresses employ the equilibrium condition and the yield criterion to ensure stress continuity at the elastic-plastic interface. Both elastic and plastic zone stress distributions must satisfy the conditions of internal force equilibrium and strain compatibility. From applying these, the limiting plastic loads may be found for beams, torsion bars, thick cylinders and discs. Modifications to such load limit predictions are discussed for when the material within these structures is strain hardening.

5.2 Elastic-Plastic Bending of Beams

Let an applied moment M_E lead to yielding within the most highly stressed surface furthest from the neutral axis (NA). As all other interior positions in the cross-section remain elastic, the beam is then in its fully elastic condition. As the moment is increased beyond M_E, the cross-section becomes partially plastic as successive interior layers, approaching the NA, reach the yield stress Y of the beam material. This state constitutes an elastic–plastic beam under an applied moment M_{ep}. When the plastic zone has penetrated through the whole cross-section on the tensile and compressive sides of the NA, this condition will determine the ultimate plastic moment M_P a given beam can withstand. A beam of given cross-section collapses under an M_P which depends upon that section. It is assumed that in deforming the beam material behaves in an elastic–perfect plastic manner. Thus, no increase in stress beyond Y occurs during the inward plastic penetration from the outer surface layers.

Consider a beam with a rectangular section, breadth b, depth d, under a positive (hogging) bending moment M (see Fig. 5.1a). Initially, the material of the beam is stressed elastically along its length by its curvature, giving tension above and compression below the unstressed neutral axis. As M increases, the yield stress is reached along the top and bottom edges of the section. Then plastic penetration begins, resulting in an elastic–plastic section, prior to the beam collapsing under the fully plastic under M_P. It is required to derive the bending moment at each stage of deformation. This analysis employs the stress distributions involved in the transition from elastic to fully plastic behaviour, as shown in Figs 5.1b–d.

5.2.1 Fully Elastic Section

Under M_E in Fig. 5.1b, the outer fibres are stressed to the yield point but the interior cross-section remains elastic. Therefore, the engineer's theory of bending applies to the outer fibres where $\sigma = Y$ for $y = d/2$.

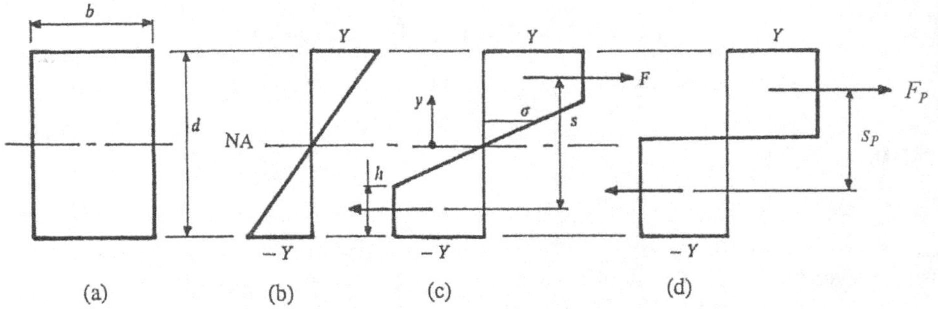

Figure 5.1 Penetration of a plastic zone in a rectangular section beam

This theory gives the fully elastic moment expression for initial yielding as

$$M_E = \frac{YI}{y} = \frac{Y(bd^3/12)}{(d/2)} = \frac{bd^2Y}{6} \tag{5.1}$$

with a radius of curvature:

$$R_E = \frac{Ey}{\sigma} = \frac{Ed}{2Y} \tag{5.2}$$

5.2.2 Elastic–Plastic Section

Under an intermediate elastic–plastic moment M_{ep} in Fig. 5.1c, the plastic zone has penetrated by the amount h, from the top and bottom surfaces. The cross-section's moment of resistance is composed of elastic and plastic components, such that:

$$M_{ep} = M_e + M_p = YI_e/y + Fs$$

where $s = d - h$ is the distance between the net force $F = Ybh$, in each plastic region. Dimension $y = (d - 2h)/2$ defines the distance of the elastic–plastic interface from the NA. The second moment of area for the inner elastic core is

$$I_e = b(d - 2h)^3/12$$

Then M_{ep} may be expressed in a non-dimensional form as follows:

$$M_{ep} = \frac{2Yb(d - 2h)^3}{12(d - 2h)} + (Ybh)(d - h)$$

$$= \frac{bd^2Y}{6}\left[1 + \frac{2h}{d}\left(1 - \frac{h}{d}\right)\right] \tag{5.3a}$$

$$\therefore \quad \frac{M_{ep}}{M_E} = 1 + \frac{2h}{d}\left(1 - \frac{h}{d}\right) \tag{5.3b}$$

The radius of curvature R_{ep} of the NA, within an elastic–plastic section, is found from applying the elastic bending theory to the inner elastic core:

$$\frac{M_e}{I_e} = \frac{E}{R_{ep}} = \frac{Y}{y}$$

and with $y = (d - 2h)/2$, this gives

$$R_{ep} = \frac{Ey}{Y} = \frac{E(d - 2h)}{2Y} \tag{5.4a}$$

or

$$R_{ep} = \frac{EI_e}{M_e} = \frac{EI_e\,y}{YI_e} = \frac{E(d - 2h)}{2Y} \tag{5.4b}$$

5.2.3 Full Plasticity

Under the fully–plastic moment M_P in Fig. 5.1d, the plastic zone has penetrated to half depth both the tensile and compressive sides of the NA. The resistive moment of the fully-plastic section becomes

$$M_P = F_P \times s_P = \frac{bdY}{2} \times \frac{d}{2} = \frac{bd^2Y}{4} \tag{5.5}$$

(a) Shape Factor

The shape factor Q is defined as the ratio between the ultimate and initial yield moments

$$Q = \frac{M_P}{M_E} > 1 \tag{5.6a}$$

Substituting eqs(5.1) and (5.5) into eq(5.6a), the shape factor becomes

$$Q = \frac{M_P}{M_E} = \frac{bd^2Y/4}{bd^2Y/6} = \frac{3}{2} \tag{5.6b}$$

Equation (5.6b) shows that the constant Q value depends solely upon the beam cross-section, i.e. Q is independent of the applied loading.

(b) Load Factor

This is defined as the ratio between the corresponding collapse load W_P and the safe elastic working load W

$$F = W_P / W \tag{5.7}$$

Hence F will depend upon the section, the applied loading and the manner of its support. The mechanism of collapse in a beam involves the formation of the minimum number of plastic hinges. That number will depend upon the manner of the support while the geometry of the hinge depends upon the whether the loading is concentrated or distributed.

5.2.4 Residual Bending Stress

When an elastic-plastic beam is fully unloaded from the moment M_{ep}, a state of residual stress σ_R will remain in the cross-section following recovery of elastic bending stresses σ_E. It is assumed that purely elastic unloading occurs from the elastic–plastic stress state σ that prevails under M_{ep}. This gives

$$\sigma_R = \sigma - \sigma_E \tag{5.8}$$

For the elastic–plastic rectangular section in Fig. 5.1c, the elastic stress recovers over the whole depth $-d/2 \le y \le d/2$, following unloading from M_{ep}. Elastic bending theory gives the recovered elastic stress when written as

$$\sigma_E = \frac{M_{ep}\,y}{I} = \frac{M_{ep}}{M_E} \times \frac{M_E\,y}{I}$$

where $I = bd^3/12$. Substituting from eqs(5.1) and (5.3b)

$$\begin{aligned}\sigma_E &= \frac{M_{ep}}{M_E} \times \frac{2Yy}{d}\\ &= \frac{2Y}{d}\left[1 + \frac{2h}{d}\left(1 - \frac{h}{d}\right)\right]y\end{aligned} \tag{5.9}$$

Within the plastic zone $\sigma = Y$, for y lying in the outer regions $\pm(d/2 - h) \le y \le \pm d/2$. The residual stresses follow from eqs(5.85) and (5.9) as

$$\sigma_R = Y - \frac{2Y}{d}\left[1 + \frac{2h}{d}\left(1 - \frac{h}{d}\right)\right]y \tag{5.10}$$

Within the elastic zone, y lies in the inner regions: $0 \le y \le \pm(d/2 - h)$. Bending theory supplies the applied stress as

$$\sigma = \frac{M_e\,y}{I} = \frac{Yb(d - 2h)^2 y}{6} \times \frac{12}{b(d - 2h)^3} = \frac{2Yy}{(d - 2h)} \tag{5.11}$$

which can be checked readily from similar triangles in Fig. 5.1c. The residual stress for this zone is found from subtracting eq(5.9) from eq(5.11):

$$\begin{aligned}\sigma_R &= \frac{2Yy}{(d - 2h)} - \frac{2Y}{d}\left[1 + \frac{2h}{d}\left(1 - \frac{h}{d}\right)\right]y\\ &= Yy\left\{\frac{2}{(d - 2h)} - \frac{2}{d}\left[1 + \frac{2h}{d}\left(1 - \frac{h}{d}\right)\right]\right\}\end{aligned} \tag{5.12}$$

Equations (5.10) and (5.12) show that σ_R varies linearly with y within each zone. Figure 5.2a illustrates σ_R corresponding to an elastic–plastic condition. In Fig. 5.2b, σ_R for the fully plastic condition is given where, with $h = d/2$ in eq(5.10), $\sigma_R = Y(1 - 3y/d)$. This shows that the maximum residual stress possible, $\sigma_R = Y$, occurs at the beam centre where $y = 0$. At the top and bottom edges, where $y = \pm\, d/2$, $\sigma_R = \pm\, Y/2$.

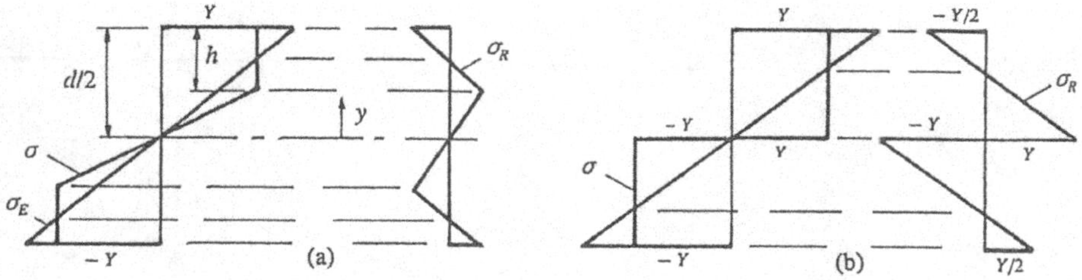

Figure 5.2 Residual stress distributions

Residual stresses have an important influence upon fatigue strength of a structure, where it is desirable that they serve to prolong life. Here, the sense of the residual stress must oppose the mean stress within a superimposed fatigue cycle arising from external loading.

5.2.5 Residual Strain and Curvature

The elastic bending theory also provides the relationship between the longitudinal bending strain ε and the radius of curvature R of the neutral axis:

$$\varepsilon = \sigma / E = y / R \tag{5.13a}$$

where y is the distance from the NA to any point within the elastic core. When the outer fibres reach the yield stress Y, eq(5.2) gives the neutral axis radius of curvature for the fully elastic beam. Therefore, the longitudinal strain at depth position y is as:

$$\varepsilon_E = y / R_E = (2Yy) / (Ed) \tag{5.13b}$$

Now, from eqs(5.2a) and (5.4a), the normalised curvature of an elastic–plastic section under M_{ep} is defined from its elastic core:

$$\frac{R_{ep}}{R_E} = \frac{E(d - 2h)}{2Y} \times \frac{2Y}{Ed} = 1 - \frac{2h}{d} \tag{5.14a}$$

Equation (5.14a) shows that the ratio between these elastic curvatures diminishes from unity to zero as h penetrates to the full depth $d/2$. Inverting the ratio in eq(5.14a) defines the normalised elastic–plastic strain when, from eq(5.13a)

$$\frac{\varepsilon_{ep}}{\varepsilon_E} = \frac{y/R_{ep}}{y/R_E} = \frac{R_E}{R_{ep}} = \frac{1}{1 - 2h/d} \tag{5.14b}$$

Equation (5.14b) applies to all y within the elastic core $0 \le y \le (d/2 - h)$. Figure 5.3 shows that the strain ratio $\varepsilon_{ep}/\varepsilon_E$ becomes infinite as the ratio M_{ep}/M_E approaches its asymptote 3/2.

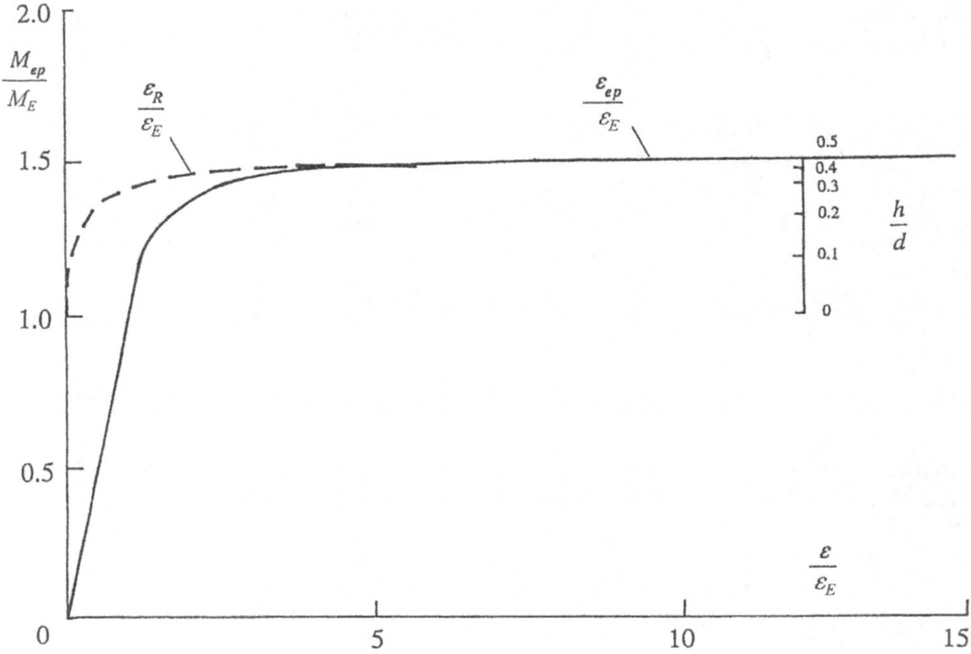

Figure 5.3 Relationship between longitudinal strain and the elastic–plastic moment

Following the removal of M_{ep}, the residual strain is found from

$$\varepsilon_R = \varepsilon_{ep} - \varepsilon_E \tag{5.15a}$$

where the elastic recovered strain is $\varepsilon_E = (M_{ep}y)/(IE)$. Normalising eq(5.15a) with the elastic core strain ε_E from eq(5.13b) and substituting from eq(5.1), leads to

$$\frac{\varepsilon_R}{\varepsilon_E} = \frac{\varepsilon_{ep}}{\varepsilon_E} - \frac{M_{ep}\, y}{IE\,\varepsilon_E} = \frac{\varepsilon_{ep}}{\varepsilon_E} - \frac{M_{ep}}{M_E} \tag{5.15b}$$

where from eq(5.1) $M_E = 2IY/d$. Substituting eqs(5.3b) and (5.14b) into eq(5.15b) gives the residual strain:

$$\frac{\varepsilon_R}{\varepsilon_E} = \frac{1}{(1 - 2h/d)} - \left[1 + \frac{2h}{d}\left(1 - \frac{h}{d}\right)\right] = \frac{2(h/d)^2(3 - 2h/d)}{(1 - 2h/d)} \tag{5.16a,b}$$

Note: Here ε_E is defined by eq(5.13b), for which $0 \le y \le (d/2 - h)$. Equation (5.16b) shows that $\varepsilon_R/\varepsilon_E$ increases from zero, at $h/d = 0$, to infinity when $h/d = 0.5$ as M_P is reached (see Fig. 5.3). Comparing eqs(5.12) and (5.16a) shows that residual stress and strains in the elastic core obey a simple elastic relation $\varepsilon_R = \sigma_R/E$. The same applies to residual strain within the outer plastic zones in non-hardening material, i.e. where σ_R is given by eq(5.10) for $\pm(d/2-h) \leq y \leq \pm d/2$. It follows that the residual strain distribution is identical to Fig. 5.2b when a scaling factor E is applied. This fact implies that originally plane cross-sections of the rectangular section beam remain plane. Because eq(5.16b) applies to the elastic core, $0 \leq y \leq (d/2-h)$, which exists under M_{ep}, it will also provide a residual curvature R_R corresponding to a given penetration depth

$$R_R = \frac{y}{\varepsilon_R} = \frac{y}{\varepsilon_E} \times \frac{\varepsilon_E}{\varepsilon_R} \tag{5.17a}$$

Substituting eqs(5.13b) and (5.16b) into eq(5.17a) gives

$$R_R = R_E\left(\frac{\varepsilon_E}{\varepsilon_R}\right) = \frac{Ed}{2Y}\left(\frac{\varepsilon_E}{\varepsilon_R}\right) = \frac{Ed^3(1 - 2h/d)}{4Yh^2(3 - 2h/d)} \tag{5.17b}$$

Note: In eqs(5.16a,b) and (5.17b) the strain ratios are non-dimensional at a given y but individually strains are distributed linearly within the elastic core, i.e. for $0 \leq y \leq d/2$.

5.2.6 Axi-Symmetric Sections

The results that apply to a rectangular cross-section given above may be extended to other cross sections built from rectangles and symmetrical about the neutral axis. Consider, for example, the two aluminium alloy extruded sections shown in Figs 5.4a,b that were loaded as beams to collapse under four point loading.

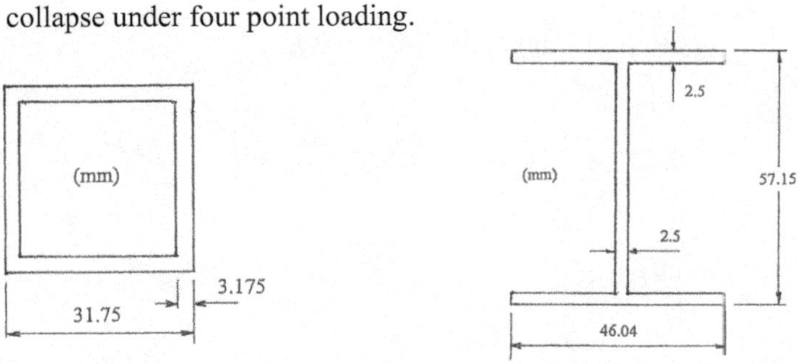

(a) (b)

Figure 5.4 Square tube and I section geometry

Concentrated, equal loads $W/2$, kN were applied through two rollers ≈ ⅓ m apart to the top surface of each beam with length 0.57 m supported on rollers. With l as the distance between one load point and a reaction roller, the applied moment becomes: $M = W/2 \times l$. For the box section, $M = W/2 \times (0.57-0.33)/2 = 0.06W$ kNm. For the I-section, $M = W/2 \times (0.57 - 0.30)/2 = 0.0675W$ kNm. Each moment is constant within the separation of the loading rollers. Loads W were applied in 0.5 kN increments to the point of collapse and the respective deflections Δ_1 and Δ_2 at the beam centre and beneath one load roller were measured in mm. The plots between W and Δ_1 and Δ_2 given in Figs 5.5a,b serve to establish the fully elastic and fully plastic (at collapse) loads for each beam. Comparisons are made with the following theory in which the corresponding moments depend solely upon the section dimensions and the yield stress $Y = 300$ MPa.

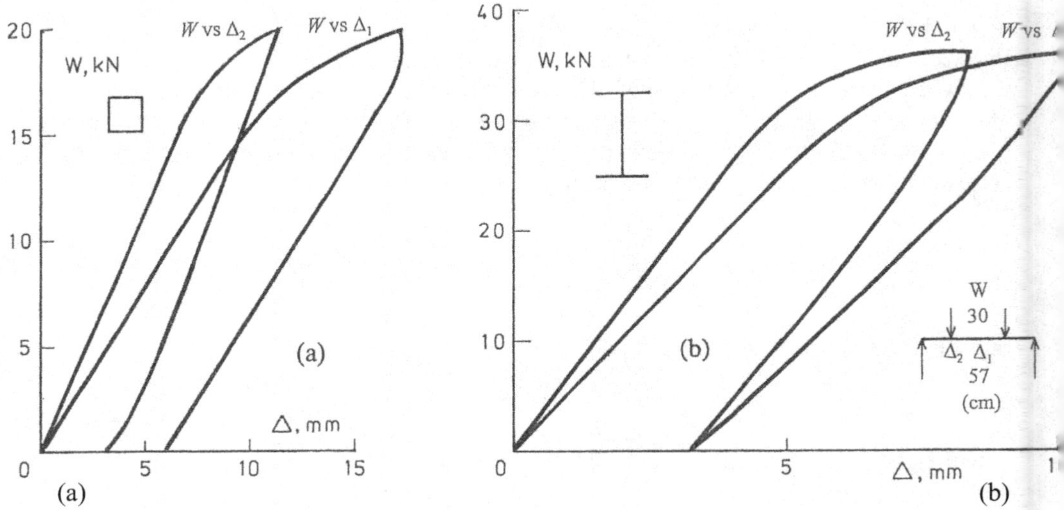

Figure 5.5 Load–deflection plots for tube and I-section

(a) Square Box Section

$$M_E = \frac{IY}{y} = \frac{(a_1^4 - a_2^4)}{12} \times \frac{2Y}{a_1}$$

$$= \frac{(31.75^4 - 25.4^4) \times 300}{6 \times 31.75 \times 10^6} = 0.945 \text{ kNm}$$

$$\therefore W_E = M_E/0.06 = 15.75 \text{ kN}$$

$$M_P = \frac{a_1^2}{2} \times Y \times \frac{a_1}{2} - \frac{a_2^2}{2} \times Y \times \frac{a_2}{2} = \frac{(a_1^3 - a_2^3)Y}{4}$$

$$= \frac{(31.75^3 - 25.4^3) \times 300}{4 \times 10^6} = 1.171 \text{ kNm}$$

$$\therefore W_P = M_P/0.06 = 19.52 \text{ kN}$$

which are to be compared with experimental loads: $W_E = 15$ kN and $W_P = 20$ kN.

(b) I Section

$$M_E = \frac{IY}{y} = \frac{(b_1 d_1^3 - b_2 d_2^3)}{12} \times \frac{2Y}{d_1}$$

$$= \frac{(46.04 \times 57.15^3 - 43.54 \times 52.15^3) \times 300}{6 \times 57.15 \times 10^6} = 2.12 \text{ kNm}$$

$$\therefore W_E = M_E/0.0675 = 31.41 \text{ kN}$$

$$M_P = \left(\frac{b_1 d_1}{2} \times Y \times \frac{d_1}{2}\right) - \left(\frac{b_2 d_2}{2} \times Y \times \frac{d_2}{2}\right) = \frac{(b_1 d_1^2 - b_2 d_2^2)Y}{4}$$

$$= \frac{(46.04 \times 57.15^2 - 43.54 \times 52.14^2) \times 300}{4 \times 10^6} = 2.40 \text{ kNm}$$

$$\therefore W_P = M_P/0.0675 = 35.6 \text{ kN}$$

which are to be compared with experimental loads: $W_E = 27.5$ kN and $W_P = 35.5$ kN. Hence, the predictions made in (a) and (b) above to the 'global' yielding and collapse loading for these sections are acceptable given that any hardening of this alloy and local buckling of the section walls and the flange had little influence on global behaviour.

5.2.7 Non-Symmetric Section

In the T-section of Fig. 5.6a, the unstressed, neutral axis does not lie at the central depth but passes originally through the centroid of its area. Example 5.1 below shows that yielding under M_E will commence at the web bottom since this is the more highly stressed surface. As the bending moment is increased to M_{ep}, the plastic zone penetrates inwards from the bottom surface, as shown in Fig. 5.6b. During penetration, the horizontal tensile and compressive forces, T and C respectively, remain balanced, i.e. $T = C$. For this to occur, the stress re-distributes so that Fig. 5.6b accommodates a shift in the NA to the position y_{ep} shown, giving $M_{ep} = Cs = Ts$. With increasing M_{ep}, the section stress continues to redistribute, maintaining $C = T$, until full plasticity is reached under M_P in Fig. 5.6c. The NA then divides the section area equally.

5.2.8 Influence of Hardening

It is possible to account for the effect of linear hardening upon the moment carrying capacity of the section. The shaded areas in Fig. 5.6c represent this increased capacity when the gradient of the stress–strain curve $K = d\sigma/d\varepsilon^P$ is identified with the gradient of the stress distribution [1]. This is illustrated in the following example.

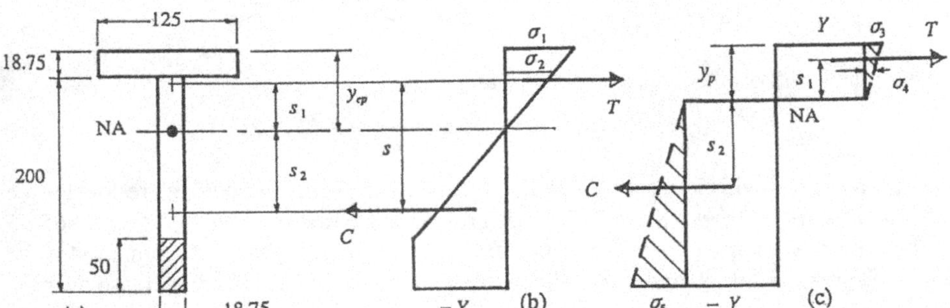

Figure 5.6 Elastic–plastic bending stress in a T-section

Example 5.1 In the T-section of Fig. 5.6a, yielding has occurred 50 mm from the web bottom. If the yield stress is constant at $Y = 278$ MPa, find the applied moment, the stress at the flange top and the position of the neutral axis. Estimate the fully plastic moment when (i) the beam material is non-hardening and (ii) when it hardens according to the linear law $\sigma = Y + K\varepsilon^P$, in which $K = 1/10$.

In Fig. 5.6b, the condition $T = C$ gives

$$(125 \times 18.75)(\sigma_1 + \sigma_2)/2 + (y_{ep} - 18.75)18.75\sigma_2/2 = (18.75 \times 50)Y + (168.75 - y_{ep})18.75Y/2 \quad \text{(i)}$$

where the stress ordinates appear in terms of y_{ep}. From the geometry of Fig. 5.6b:

$$\sigma_1/Y = y_{ep}/(168.75 - y_{ep}) \text{ and } \sigma_2/Y = (y_{ep} - 18.75)/(168.75 - y_{ep}) \quad \text{(ii, iii)}$$

Substituting eqs(ii) and (iii) into eq(i) leads to an equation which locates the position of the neutral axis at $y_{ep} = 72.84$ mm. Equation (ii) then gives the elastic stress at the flange top as $\sigma_1 = 211.1$ MPa. The forces $T = C = 510.6$ kN follow from the left- and right-hand sides of eq(i). These act at the respective centroids for the areas above and below the neutral axis. That is: $s_1 = \sum(A_i y_i)/A = 52.465$ mm and $s_2 = 72.955$ mm. The moment M_{ep} is found from

$$M_{ep} = Cs = C(s_1 + s_2) = 510.6\,(52.465 + 72.955)10^{-3} = 64.04 \text{ kNm}$$

When the fully plastic moment is reached (see Fig. 5.6c), the condition $T = C$ becomes

$$[(125 \times 18.75) + (y_p - 18.75) \times 18.75]Y = (218.75 - y_p) \times 18.75Y \quad \text{(iv)}$$

The neutral axis now divides the section into two equal areas to give $y_p = 56.25$ mm. The centroids of these areas are: $s_1 = \sum(A_i y_i)/A = 40.385$ mm and $s_2 = 81.25$ mm. Equation (iv) gives $T = C = 847.04$ kN and M_P is found from

$$M_P = Cs = (218.75 - y_p)18.75Y(s_1 + s_2)$$
$$= 162.5 \times 18.75 \times 278\,(40.385 + 81.25)10^{-6} = 103.03 \text{ kNm}$$

To account for hardening, the force within each shaded area in Fig. 5.6c must be added to eq(iv). This modifies the position of the neutral axis to y_p'. Using the condition $T' = C'$:

$$[(125 \times 18.75) + (y_p' - 18.75)18.75]Y + (y_p' - 18.75)\sigma_4/2 + (18.75 \times 125)(\sigma_3 + \sigma_4)/2$$
$$= (218.75 - y_p')18.75Y + (218.75 - y_p')18.75\sigma_5/2 \quad \text{(v)}$$

in which the stress ordinates (see Fig. 5.6c) are:

$$\sigma_3 = Ky_p', \quad \sigma_4 = K(y_p' - 18.75) \text{ and } \sigma_5 = K(218.75 - y_p') \quad \text{(vi-viii)}$$

Substituting eqs(vi–viii) into eq(v), with $K = 1/10$, leads to an equation which re-locates the position of the neutral axis to $y_p' = 57.38$ mm. From this, we find from eq(v) $T' = C' = 865.55$ kN. The centroid for the upper area is modified to $s_1' = \sum(A_i y_i)/A = 41.23$ mm and for the lower web area $s_2 = 80.69$ mm. Hence the collapse moment shows a slight increase over the non-hardening value:

$$M_P' = C'(s_1' + s_2') = 865.55\,(41.23 + 80.68)10^{-3} = 105.52 \text{ kNm}$$

5.3 Beam Collapse

Collapse of a beam will depend upon the formation of the minimum number of plastic hinges necessary to form a mechanism. The behaviour of each beam depends upon the manner of its loading and support, as will be shown here for common case examples.

5.3.1 Simply Supported – Central Load

Consider a simply supported beam that collapses from the formation of a single plastic hinge beneath a central load as shown in Fig. 5.7a.

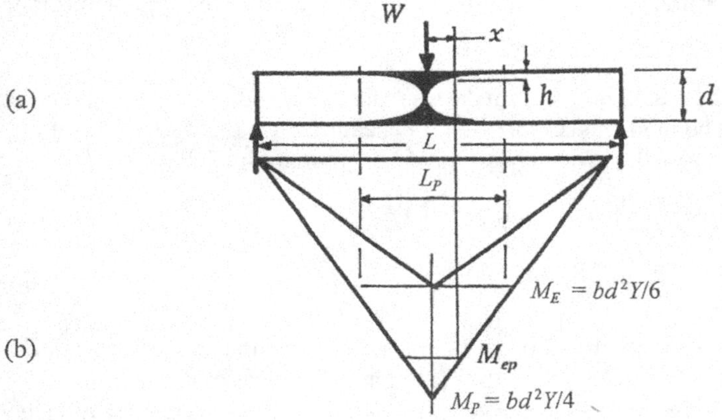

Figure 5.7 Plastic collapse of a simply-supported beam

(a) Loading
The elastic safety factor S is

$$S = \sigma_a / \sigma_w \tag{5.18a}$$

where σ_a is the allowable working stress and σ_w is the safe working stress. Putting $\sigma_a = Y$ in eq(5.18a) gives:

$$\sigma_w = Y/S \tag{5.18b}$$

Now, from bending theory, the corresponding safe working moment is

$$M = \frac{\sigma_w I}{y} = \frac{YI}{Sy} \tag{5.19}$$

The maximum bending moment $M = WL/4$ occurs beneath the load. It follows from eq(5.19) that the safe working load is

$$W = \frac{4M}{L} = \frac{4YI}{LSy} = \frac{2Ybd^2}{3LS} \tag{5.20}$$

for which the fully elastic load $W = W_E$ applies to $S = 1$. Substituting from eq(5.5) gives the fully plastic (ultimate) load for this beam as

$$W_P = \frac{4M_P}{L} = \frac{bd^2Y}{L} \tag{5.21}$$

Employing eqs(5.6), (5.7), (5.20) and (5.21), the load factor F may be combined with the shape factor Q and the safety factor S to reveal a simple general form

$$F = \frac{(bd^2Y/L)}{(2Ybd^2)/(3LS)} = QS \tag{5.22}$$

On taking $S = 1$ in eq(5.20), the safe load becomes the initial yield load. Here, equal load and shape factors, $F = Q = 3/2$, apply to a simply supported beam with rectangular section.

(b) Plastic Hinge

The plastic hinge length L_p is determined from the ultimate moment diagram at the point of collapse (see Fig. 5.7). The moments are: $M = M_P$, beneath the load and $M = M_E$, at the extremities of the hinge. From the geometry of the M–diagram in Fig. 5.2b:

$$2\,(M_P - M_E)/\,L_p = 2\,M_P/L \tag{5.23a}$$

giving

$$L_p = L\,(1 - M_E/M_P) = L\,(1 - 1/Q) \tag{5.23b}$$

Substituting $Q = 3/2$ from eq(5.6b) into eq(5.23a) shows that the hinge length of a rectangular cross-section is $L_p = L/3$. That is, the central hinge extends over one third of the beam's length. For other axi-symmetric beam sections, Q in eq(5.23b) is a numerical value defining the cross-section. For example, it can be shown that $Q = 1.7$ for a solid circular section and $Q = 2$ for a square section when oriented with one diagonal as the horizontal axis of symmetry [2].

The shape of the hinge follows from similar triangle geometry in Fig. 5.7b. Also, by employing eqs(5.3b) and (5.6b) there follows, for the rectangular section

$$\frac{x}{L} = \frac{1}{2}\left(1 - \frac{M_{ep}}{M_P}\right)$$

$$= \frac{1}{2}\left\{1 - \frac{2}{3}\left[1 + \frac{2h}{d} - 2\left(\frac{h}{d}\right)^2\right]\right\}$$

$$\frac{x}{L/2} = \frac{1}{3}\left(1 - \frac{h}{d/2}\right)^2 \tag{5.23c}$$

Equation (5.23c) shows full penetration $h = d/2$ for $x = 0$ at the centre and $x = L_P/2 = L/6$ for $h = 0$. Within these the hinge has a parabolic profile in normalised axes $\dfrac{x}{L/2}$ versus $\dfrac{h}{d/2}$ as shown. The gradient of the elastic load versus deflection plot follows from the central deflection expression for this beam $W/\Delta = 48EI/L^3$. Beyond W_E there is a loss of elastic section with increasing depth penetrations h of plasticity inwards from the top and bottom surfaces at the centre position. Hence I is modified to $b(d - 2h)^3/12$ for the central elastic rectangular section wherein the stiffness W/Δ diminishes non-linearly from full elasticity under W_E to zero at full plasticity under W_P, consistent with observation.

5.3.2 Simply Supported – Distributed Load

Figure 5.8 shows the moment diagram that applies to the fully plastic condition in which the central moment is M_P.

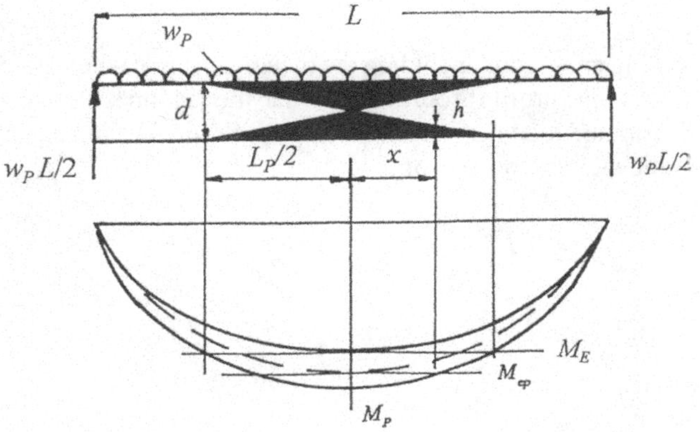

Figure 5.8 Hinge geometry at collapse in a simply supported beam with udl

To determine the hinge geometry, the elastic–plastic moment, distance x right of centre, is found from the two moments for the length of beam $L/2 - x$ to the right of x, i.e. hogging due to the distributed loading (positive) and sagging due to RH support reaction (negative). Respectively:

$$M_{ep} = \frac{w_P}{2}\left(\frac{L}{2} - x\right)\left(\frac{L}{2} - x\right) - \frac{w_P L}{2}\left(\frac{L}{2} - x\right)$$

$$= -\frac{w_P}{2}\left(\frac{L^2}{4} - x^2\right) = M_P\left[1 - \left(\frac{2x}{L}\right)^2\right] \quad (5.24a)$$

where $M_P = -wL^2/8$. Dividing both sides of eq(5.24a) by M_P and employing M_{ep} for a rectangular section from eq(5.3b)

$$\frac{M_{ep}}{M_P} = \frac{M_E}{M_P} \times \frac{M_{ep}}{M_E}$$

$$1 - \left(\frac{2x}{L}\right)^2 = \frac{2}{3}\left[1 + 2\left(\frac{h}{d}\right) - 2\left(\frac{h}{d}\right)^2\right]$$

$$\frac{x}{L/2} = \frac{1}{\surd 3}\left(1 - \frac{h}{d/2}\right) \quad (5.24b)$$

Here a linear hinge profile appears in normalised axes $\dfrac{x}{L/2}$ versus $\dfrac{h}{d/2}$ in Fig. 5.8a. Equation (5.24b) shows a hinge half length: $x = L_P/2 = L/(2\surd 3)$ for $h = 0$ at its ends where the moment is fully elastic. Thus, the hinge extends by this amount either side of the centre with full length $L_P = L/\surd 3$. Alternatively, set $M_{ep} = M_E$ in eq(5.24a) for $x = L_P/2$:

$$M_E = -\frac{w_P L^2}{8}\left[1 - \left(\frac{L_P}{L}\right)^2\right] = M_P\left[1 - \left(\frac{L_P}{L}\right)^2\right]$$

$$\therefore \frac{L_P}{L} = \left(1 - \frac{M_E}{M_P}\right)^{\frac{1}{2}} = \left(1 - \frac{1}{Q}\right)^{\frac{1}{2}}$$

which applies to any uniform section including the rectangle above when $Q = 3/2$. However, it would appear that the hinge profile is only linear when the section is square or rectangular. For the latter, the distributed loading w_P to cause collapse follows from equating its plastic moment to maximum bending moment for the beam:

$$M_P = \frac{bd^2 Y}{4} = \frac{w_P L^2}{8} \qquad \therefore w_P = \frac{2bd^2 Y}{L^2}$$

5.3.3 Encastré – Concentrated Load

More than one hinge is necessary for the failure mechanism to include the fixed supports. For the encastré beam in Fig. 5.9a, the collapse mechanism requires that M_P is reached beneath the off-centre load and at each fixed-end simultaneously.

(a)

(b)

(c)

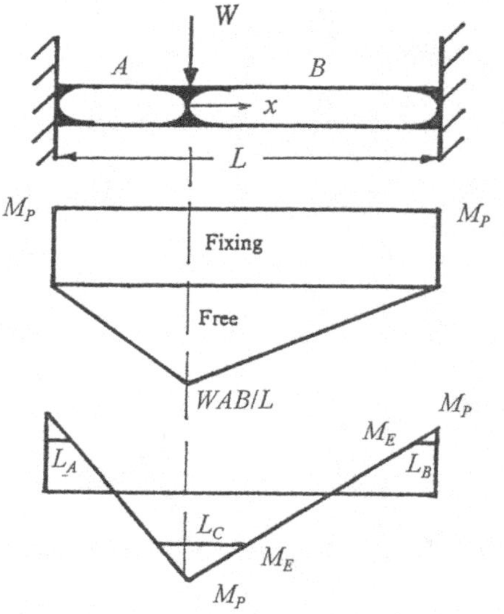

Figure 5.9 Collapse of an encastré beam

Equating the hogging fixing moment at one end to the net sagging moment at the centre in Fig. 5.9b

$$M_P = -(M_P - W_P AB/L) \tag{5.25a}$$

Thus, the collapse load is $W_P = 2M_P L/AB$ and the plastic hinge lengths L_A, L_B and L_C follow from the geometry of similar triangles within Fig. 5.9c

$$\frac{M_P - M_E}{L_A} = \frac{2M_P}{A}$$

$$L_A = \frac{A}{2}\left(1 - \frac{M_E}{M_P}\right) = \frac{A}{2}\left(1 - \frac{1}{Q}\right) \tag{5.25b}$$

$$\frac{M_P - M_E}{L_B} = \frac{2M_P}{B}$$

$$L_B = \frac{B}{2}\left(1 - \frac{M_E}{M_P}\right) = \frac{B}{2}\left(1 - \frac{1}{Q}\right) \tag{5.25c}$$

$$L_C = (A + B)\left(1 - \frac{M_E}{M_P}\right) = \frac{L}{2}\left(1 - \frac{1}{Q}\right) \tag{5.25d}$$

where Q in eqs(5.25b–d) depends upon the beam cross-section. For example, with a rectangular section (breadth b, depth d): $Q = 3/2$ and $L_A = A/6$, $L_B = B/6$ and $L_C = L/6$.

These geometrical solutions to hinge lengths and its profile may be determined from the moment equation at position x from the centre under the collapse condition:

$$M_x = M_P - \frac{WA}{L}(B - x)$$

$$\frac{M_x}{M_P} = 1 - \frac{2L}{WAB} \times \frac{WA}{L}(B - x) = -1 + \frac{2x}{B} \tag{5.26a}$$

which, for $0 \le x \le B$, confirms $M_x = M_P$ for $x = 0$ and $x = B$. Within the hinge region, where $0 \le x \le B/6$, eq(5.26a) applies for $M_x = -M_{ep}$, giving:

$$\frac{M_{ep}}{M_P} = -1 + \frac{2x}{B} = -\frac{2}{3}\left[1 + \frac{2h}{d}\left(1 - \frac{h}{d}\right)\right]$$

$$\therefore \quad \frac{2x}{B} = \frac{1}{3}\left(1 - \frac{2h}{d}\right)^2 \tag{5.26b}$$

Equation (5.26b) reveals a parabolic hinge symmetry lying between the fully elastic and plastic cross-sections at its end and centre: $x = B/6$ for $h = 0$ and $x = 0$ for $h/d = ½$ respectively.

With the off-centre loading, the fully elastic load W_Y rests with the greatest of three elastic moments being the first to reach the fully elastic moment M_E. These are respectively: for left end fixing, beneath the load and at the right end fixing [3]:

Moments: $$M = \frac{WAB^2}{L^2};\quad \frac{-2WA^2B^2}{L^3};\quad \frac{WA^2B}{L^2} \tag{5.27a–c}$$

The greatest M, eq(5.27a), applies to the left end when $B > A$ and $L > 2A$. Hence, with yield stress Y at that end's top and bottom surfaces of rectangular section, breadth $b \times$ depth d:

Left End: $$M_E = \frac{W_E AB^2}{L^2} = \frac{bd^2Y}{6} \quad \therefore\ W_E = \frac{bd^2Y}{6} \times \frac{L^2}{AB^2}$$

With increasing loading plasticity penetrates inwards from the left end surfaces as the central and right end moments approach their respective M_E, at which the corresponding loads are:

Load Position: $$|M_E| = \frac{2W_E A^2B^2}{L^3} = \frac{bd^2Y}{6} \quad \therefore\ W_E = \frac{bd^2Y}{6} \times \frac{L^3}{2A^2B^2}$$

Right End: $$M_E = \frac{W_E A^2B}{L^2} = \frac{bd^2Y}{6} \quad \therefore\ W_E = \frac{bd^2Y}{6} \times \frac{L^2}{A^2B}$$

These show that the elastic moments at the right end fixing and at the load position will lag the moment at the left end fixing until M_P is reached there first. Increasing loading thereafter develops hinges at each other position until M_P is reached successively for the collapse mechanism to occur, as illustrated in the following example.

Example 5.2 Determine the sequence of loads necessary for a fully plastic, triple hinge formation that leads to the collapse of an encastré beam with load offset from the centre with: $A/L = 1/4$ and $B/L = 3/4$ (see Fig. 5.9a). Under which load will the beam collapse?

For the three moments in eq(5.27a–c), the respective collapse loads for fully formed plastic hinges at the LH fixing, beneath the load and at the RH fixing are respectively:

$$W_P = M_P \times \frac{L^2}{AB^2}; \qquad W_P = M_P \times \frac{L^3}{2A^2B^2}; \qquad W_P = M_P \times \frac{L^2}{A^2B};$$

where $M_P = bd^2Y/4$. Taking the ratio between the first and the second loads and again between the first and the thirds load gives $2A/L$ and A/B respectively. For the A/L and B/L ratios given it can be seen that $2A/L = 1/2$ and $A/B = 1/3$. This means that collapse loads, coincident with fully formed hinges beneath the load and at the RH fixing, for this beam are respectively two and three times the collapse load for the LH fixing. Hence the third load is required to be reached in order to collapse this beam.

In example 5.2, the respective collapse loads occur from fully formed hinges at each position ($h/d = ½$). It is unlikely that the gradual collapse sequence implied by this would occur in practice. In fact, the central and end penetrations may not develop fully should their remaining central elastic sections be unable to bear the load. Then, an instantaneous collapse occurs by redistribution of the moments M_P. The load at which this occurs is the lesser of the solutions applied to centre and right end for which (say) the depth of penetration lies at 90 % of the section depth (i.e. $h/d = 0.45$). Respectively:

$$\frac{M_{ep}}{M_P} = \frac{2WA^2B^2}{L^3} \times \frac{4}{bd^2Y} = \frac{2}{3}\left[1 + 2\left(\frac{h}{d}\right) - 2\left(\frac{h}{d}\right)^2\right]$$

$$\frac{M_{ep}}{M_P} = \frac{WA^2B}{L^2} \times \frac{4}{bd^2Y} = \frac{2}{3}\left[1 + 2\left(\frac{h}{d}\right) - 2\left(\frac{h}{d}\right)^2\right]$$

5.3.4 Encastré – Distributed Load

Figures 5.10a–c combine the free and fixing moment diagrams for this beam. With the distributed loading w extending the full length L, the separated diagrams in Fig. 5.10b have the same area. Elastic ends moments are $wL^2/12$ and the central net moment (free – fixing) $= -wL^2/24$ [3]. A fully elastic condition applies when the end moments allow the yield stress Y to be reached at the top and bottom surfaces of the beam's rectangular cross-section (breadth $b \times$ depth d). From this condition the fully elastic loading w_E/unit length is found:

$$M_E = \frac{w_E L^2}{12} = \frac{bd^2 Y}{6}, \qquad \therefore\ w_E = \frac{2bd^2 Y}{L^2}$$

The collapse condition involves a redistribution of the end and central moments. When, with increasing distributed loading, all three moments attain the plastic moment M_P, they enable the collapse from fully formed hinges at the three positions (as for Fig. 5.10c).

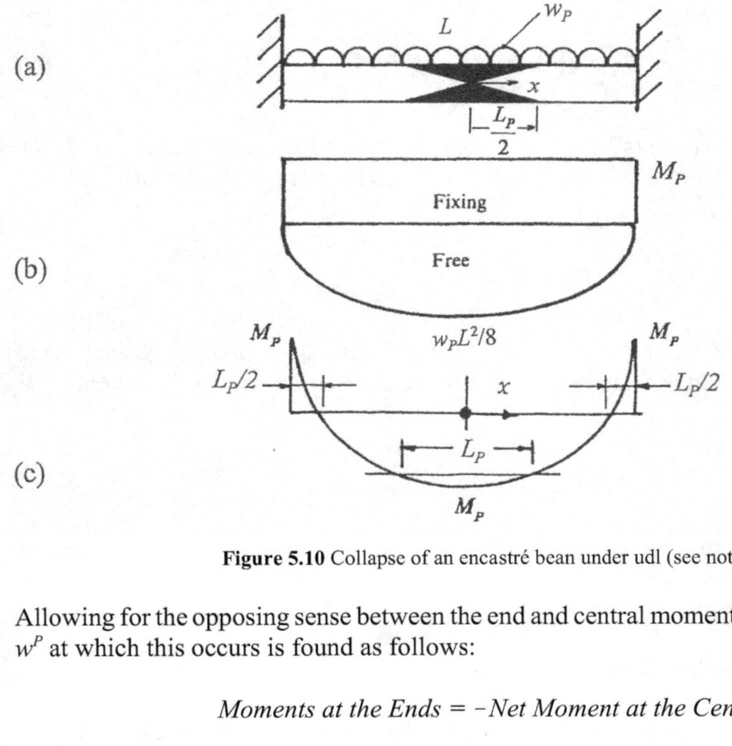

Figure 5.10 Collapse of an encastré bean under udl (see notes)

Allowing for the opposing sense between the end and central moments, the collapse loading w^P at which this occurs is found as follows:

Moments at the Ends = –Net Moment at the Centre

$$M_P = -(M_P - w_P L^2/8), \qquad \therefore\ M_P = w_P L^2/16$$

$$M_P = \frac{w_P L^2}{16} = \frac{bd^2 Y}{4}, \qquad \therefore\ w_P = \frac{4bd^2 Y}{L^2}$$

Figure 5.10c shows hinges extending for length L_P at the centre and $L_P/2$ at each end. To find what proportion of the beam length L each hinge occupies and the hinge profiles, moments are taken at position x within the half hinge length right of centre. There the moment is elastic–plastic given that plasticity has penetrated inwards to depth h from the outer surfaces:

$$M_x = M_P + \frac{w_P}{2}\left(\frac{L}{2} - x\right)^2 - \frac{w_P L}{2}\left(\frac{L}{2} - x\right)$$

$$\frac{M_x}{M_P} = 1 + \left(\frac{w_P}{2} \times \frac{16}{w_P L^2}\right)\left(\frac{L}{2} - x\right)^2 - \left(\frac{w_P L}{2} \times \frac{16}{w_P L^2}\right)\left(\frac{L}{2} - x\right)$$

$$= 1 + 2\left(1 - \frac{x}{L/2}\right)^2 - 4\left(1 - \frac{x}{L/2}\right) = -1 + 2\left(\frac{x}{L/2}\right)^2 \qquad (5.28a)$$

Equation (5.28a) applies to two regions: the full half length ($0 \le x \le L/2$) and the central hinge half length ($0 \le x \le L_P/2$). The first region confirms the collapse condition that $M_x = -M_P$ for $x = 0$ and $M_x = M_P$ for $x = L/2$. For the second hinge region, $M_x = M_{ep,}$ within which moments are negative. Substituting from eq(5.3a):

$$\frac{M_{ep}}{M_P} = -\frac{2}{3}\left[1 + \frac{h}{d/2}\left(1 - \frac{1}{2} \times \frac{h}{d/2}\right)\right] = -1 + 2\left(\frac{x}{L/2}\right)^2 \qquad (5.28b)$$

This shows that: (i) a full penetration $h = d/2$ applies to the centre $x = 0$ where $M_{ep} = M_P$ and (ii) zero penetration $h = 0$ applies to the hinge end $x = L_P/2$, where a fully elastic (negative) moment prevails for $M_{ep} = M_E$:

$$\frac{M_E}{M_P} = -\frac{2}{3} = -1 + 2\left(\frac{L_P}{L}\right)^2$$

consistent with the ratio $|M_E/M_P| = 2/3$ from eq (5.6b). The linear relationship between x and h found from eq(5.28b) defines a straight-sided hinge geometry. In normalised form:

$$\frac{x}{L/2} = \frac{1}{\sqrt{6}}\left(1 - \frac{h}{d/2}\right)$$

this showing: $x = 0$ when $h = d/2$ and when $h = 0$, $x = L_P/2 = L/(2\sqrt{6})$, giving the central hinge length as $L_P = L/\sqrt{6}$.

5.4 Elastic–Plastic Torsion of Bars

Elastic–plastic torsion theory of a circular shaft is similar to elastic–plastic bending of beams in that three conditions apply to the shaft: fully elastic, elastic–plastic and fully plastic. The corresponding torques are derived for both solid and hollow circular section shafts.

5.4.1 Solid Circular Section

Consider a solid circular section of outer radius r_o subjected to an increasing torque T. Provided the cross-section is elastic, torsion theory will describe the twist, shear stress and strain. This includes the case where the outer fibres have reached the shear yield stress k. Here, an initial yield torque T_E for the limiting, fully–elastic section (subscript E) is

$$T_E = \frac{Jk}{r_o} = \frac{\pi r_o^4 k}{2r_o} = \frac{\pi r_o^3 k}{2} \tag{5.29}$$

Increasing the torque beyond T_E results in the penetration of a plastic zone inwards from the outer radius. This produces an elastic–plastic section, shown in Fig. 5.11a, where an elastic core is surrounded by a plastic annulus, interfaced at a common elastic-plastic radius r_{ep}.

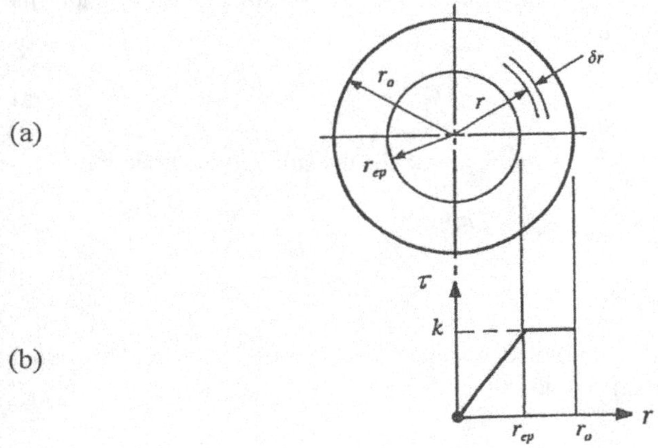

Figure 5.11 Elastic–plastic bar in torsion

Since the material does not strain-harden (see Fig. 5.7b) k is constant for $r_{ep} \le r \le r_o$. It follows that there is a linear variation in elastic shear stress τ within the core, $0 \le r \le r_{ep}$:

$$\tau = k\,(r / r_{ep}) \tag{5.30}$$

The corresponding elastic–plastic torque T_{ep} is found from the equilibrium condition:

$$T_{ep} = 2\pi \int_{r_{ep}}^{r_o} k r^2 \mathrm{d}r + 2\pi \int_0^{r_{ep}} \tau r^2 \mathrm{d}r \tag{5.31a}$$

$$= 2\pi k \int_{r_{ep}}^{r_o} r^2 \mathrm{d}r + \frac{2\pi k}{r_{ep}} \int_0^{r_{ep}} r^3 \mathrm{d}r$$

$$= \frac{2\pi k}{3} \left| r^3 \right|_{r_{ep}}^{r_o} + \frac{2\pi k}{4 r_{ep}} \left| r^4 \right|_0^{r_{ep}}$$

$$= \frac{\pi k r_o^3}{2} \left\{ \frac{4}{3} \left[1 - \left(\frac{r_{ep}}{r_o} \right)^3 \right] + \left(\frac{r_{ep}}{r_o} \right)^3 \right\} \tag{5.31b}$$

Dividing eq(5.31b) by T_E from eq(5.29) gives the normalised torque:

$$\frac{T_{ep}}{T_E} = \frac{4}{3} \left[1 - \frac{1}{4} \left(\frac{r_{ep}}{r_o} \right)^3 \right] \tag{5.32}$$

The angle of twist θ_{ep} is found from applying elastic torsion theory to the core. At the elastic–plastic interface, where $\tau = k$ for $r = r_{ep}$ (Fig. 5.11b) and for a shaft of length l:

$$\theta_{ep} = \frac{kl}{Gr_{ep}} = \frac{T_e l}{GJ_e} \tag{5.33}$$

in which $J_e = \pi r_{ep}^4/2$ and T_e is the torque carried by the elastic core, i.e. the second integral in eq(5.31a). The elastic torque contribution T_e is not to be confused with the fully elastic torque in eq(5.29). The section becomes fully plastic through to the bar centre under its ultimate torque T_P. Putting $r_{ep} = 0$ in eq(5.31b) gives

$$T_P = (2/3)(\pi r_o^3 k) \tag{5.34a}$$

and substituting from eq(5.29) defines the *shape factor* for the solid circular section

$$T_P / T_E = 4/3 \tag{5.34b}$$

(a) Residual Shear Stress

When the elastic–plastic bar is fully unloaded, elastic stresses recover to leave a residual shear stress distribution τ_R. This distribution is found from

$$\tau_R = \tau - \tau_E \tag{5.35}$$

where τ is the shear stress in either zone under the elastic–plastic torque T_{ep}. Torsion theory supplies the elastic shear stress τ_E that recovers upon removal of T_{ep}:

$$\tau_E = \frac{T_{ep} r}{J} \tag{5.36a}$$

where from eq(5.29)

$$\frac{T_{ep}}{J} = \frac{T_{ep}}{T_E} \times \frac{T_E}{J} = \frac{T_{ep}}{T_E} \times \frac{k}{r_o} \tag{5.36b}$$

Equations (5.32), (5.35) and (5.36b) provide the residual shear stress distribution τ_R in the plastic zone, $r_{ep} \le r \le r_o$ (see Figs 5.12a,b). There, $\tau = k$ and $J = \pi r_o^4/2$, which give:

$$\tau_R = k - \frac{T_{ep} r}{J} = k - \left(\frac{T_{ep}}{T_E}\right)\left(\frac{kr}{r_o}\right)$$

$$= k\left\{1 - \frac{4}{3}\left(\frac{r}{r_o}\right)\left[1 - \frac{1}{4}\left(\frac{r_{ep}}{r_o}\right)^3\right]\right\} \tag{5.37a}$$

and, from eqs(5.30), (5.32), (5.35) and (5.36b), the residual shear stress distribution within the elastic zone, $0 \le r \le r_{ep}$, is

$$\tau_R = k\left(\frac{r}{r_{ep}}\right) - \frac{4k}{3}\left(\frac{r}{r_o}\right)\left[1 - \frac{1}{4}\left(\frac{r_{ep}}{r_o}\right)^3\right]$$

$$= k\left(\frac{r}{r_{ep}}\right)\left\{1 - \frac{4}{3}\left(\frac{r_{ep}}{r_o}\right)\left[1 - \frac{1}{4}\left(\frac{r_{ep}}{r_o}\right)^3\right]\right\} \tag{5.37b}$$

The τ, τ_E and τ_R distributions are represented graphically in Fig. 5.12b. The residual stresses supplied by eqs(5.37a,b) are simply the difference between stress ordinates τ and τ_E within each zone, their signs determined from eq(5.35). It is seen that τ_R has its two largest values at the outer and interface radii, the greater value depending upon the depth of penetration.

(a)

(b)

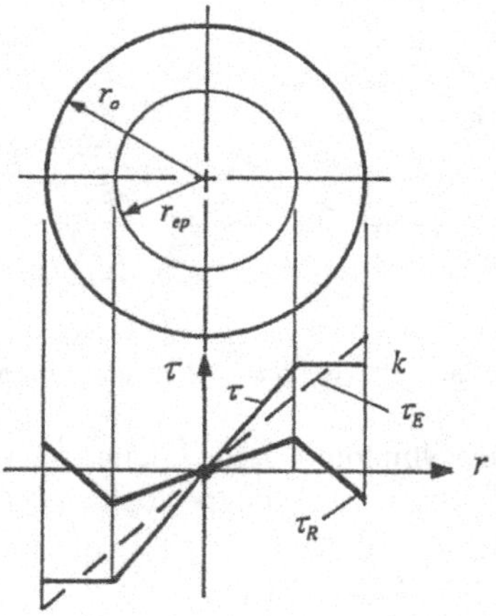

Figure 5.12 Residual stress in a partially-plastic solid bar

(b) Residual Angular Twist

When the bar's outer fibres reach the shear yield stress k, the fully-elastic angular twist θ_E is supplied by torsion theory:

$$\theta_E = \frac{kl}{Gr_o} = \frac{T_E l}{JG} \tag{5.38}$$

Combining eqs(5.33) and (5.38) gives a twist ratio for the elastic–plastic bar

$$\frac{\theta_{ep}}{\theta_E} = \frac{(kl)/(Gr_{ep})}{(kl)/(Gr_o)} = \frac{r_o}{r_{ep}} \tag{5.39}$$

The following normalised, torque–twist relationship is found from combining eq(5.32) with eq(5.39)

$$\frac{T_{ep}}{T_E} = \frac{4}{3}\left[1 - \frac{1}{4}\left(\frac{\theta_E}{\theta_{ep}}\right)^3\right] \tag{5.40}$$

The graph of eq(5.40), given in Fig. 5.13, shows that the twist ratio becomes infinite as the normalised torque increases beyond unity to its ultimate value of 4/3.

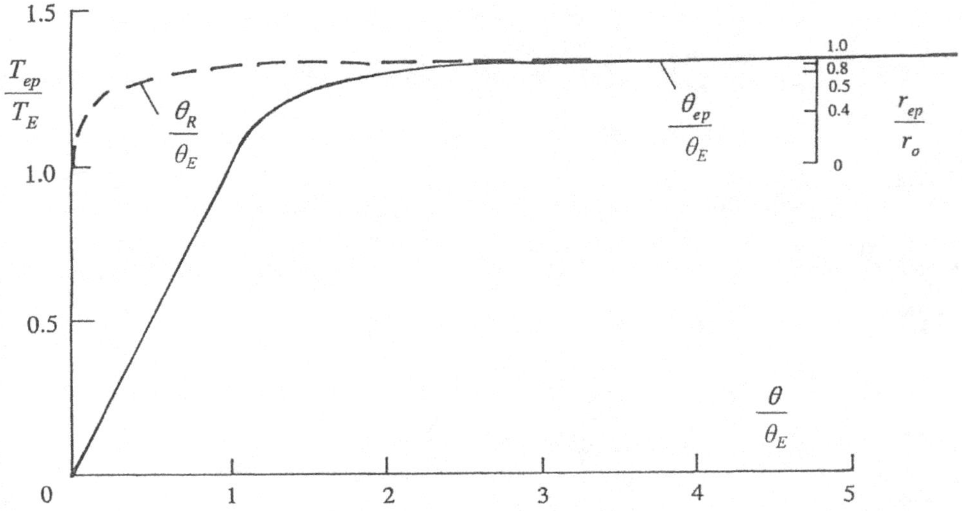

Figure 5.13 Dependence of twist upon applied torque in an elastic–plastic bar

Since the ordinate T_P/T_E is the asymptote 4/3, the implication is that T_P could not be attained in practice if the material was truly elastic–perfectly plastic. However, most metallic materials do strain-harden to some degree, which increases the torque carrying capacity above T_P.

Following unloading from an intermediate elastic–plastic torque ($T_E < T_{ep} < T_P$), the state of residual twist θ_R is found from

$$\theta_R = \theta_{ep} - \theta_E \qquad (5.41a)$$

where θ_E is the angular twist which recovers elastically from the removal of T_{ep}:

$$\theta_E = (T_{ep}\, l)/(GJ) \qquad (5.41b)$$

Substituting eq(5.41b), together with eq(5.33), into eq(5.41a) gives

$$\theta_R = \frac{kl}{Gr_{ep}} - \frac{T_{ep} l}{GJ} \qquad (5.41c)$$

Normalising θ_R with θ_E from eq(5.38):

$$\frac{\theta_R}{\theta_E} = \frac{(kl)/(Gr_{ep})}{(kl)/(Gr_o)} - \frac{(T_{ep} l)/(GJ)}{(kl)/(Gr_o)} \qquad (5.42a)$$

Substituting eqs(5.29), (5.39) and (5.40) into eq(5.42a) gives this twist ratio:

$$\frac{\theta_R}{\theta_E} = \frac{r_o}{r_{ep}} - \frac{4}{3}\left[1 - \frac{1}{4}\left(\frac{r_{ep}}{r_o}\right)^3\right] \tag{5.42b}$$

$$= \frac{\theta_{ep}}{\theta_E} - \frac{4}{3}\left[1 - \frac{1}{4}\left(\frac{\theta_E}{\theta_{ep}}\right)^3\right] \tag{5.42c}$$

Equations (5.40) and (5.42c) establish a dependence between T_{ep}/T_E, θ_{ep}/θ_E and θ_R/θ_E, shown in Fig. 5.13. A residual, angular twist exists following unloading from a torque which exceeds T_E. The ratio θ_R/θ_E increases to infinity under T_P. When $T \le T_E$, elastic torsion theory describes the torque versus twist line of unit slope, since $T/T_E = \theta/\theta_E = (Tr_o)/(Jk)$.

5.4.2 Hollow Circular Section

Figure 5.14a shows a partially yielded hollow cylinder, with inner and outer radii r_i and r_o respectively. The plastic zone has penetrated to intermediate radius r_{ep} under a torque T_{ep}. The distribution of shear stress is given in Fig. 5.14b. Again, an elastic–perfectly plastic material is assumed in which a plastic zone penetrates under a constant shear yield stress k.

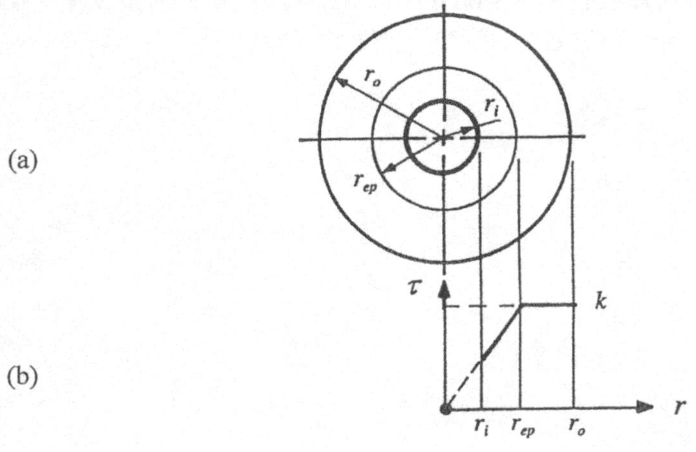

Figure 5.14 Elastic–plastic hollow cylinder under torsion

Torque equilibrium expresses the sum of elastic and plastic torques from each zone

$$T_{ep} = T_e + T_p = \frac{2\pi k}{r_{ep}}\int_{r_i}^{r_{ep}} r^3\,\mathrm{d}r + 2\pi k\int_{r_{ep}}^{r_o} r^2\,\mathrm{d}r \tag{5.43a}$$

$$= \frac{\pi k}{2r_{ep}}\left(r_{ep}^4 - r_i^4\right) + \frac{2\pi k}{3}\left(r_o^3 - r_{ep}^3\right) \tag{5.43b}$$

Setting $r_{ep} = r_o$ in eq(5.43b) gives the initial yield torque T_E as

$$T_E = \frac{\pi k}{2r_o}(r_o^4 - r_i^4) \tag{5.44}$$

Putting $r_{ep} = r_i$ in eq(5.43b) gives the ultimate plastic torque T_P. When T_P is normalised with T_E, from eq(5.44):

$$\frac{T_P}{T_E} = \frac{4\,[1 - (r_i/r_o)^3\,]}{3\,[1 - (r_i/r_o)^4\,]} \tag{5.45}$$

Dividing eqs(5.43b) and (5.44) normalises the elastic–plastic torque T_{ep} for interface radii in the range $r_i \le r_{ep} \le r_o$

$$\frac{T_{ep}}{T_E} = \frac{4r_o(r_o^3 - r_{ep}^3)}{3\,(r_o^4 - r_i^4)} + \frac{r_o(r_{ep}^4 - r_i^4)}{r_{ep}(r_o^4 - r_i^4)}$$

$$= \frac{\frac{4}{3}\left[1 - \frac{1}{4}\left(\frac{r_{ep}}{r_o}\right)^3\right] - \frac{r_i}{r_{ep}}\left(\frac{r_i}{r_o}\right)^3}{1 - \left(\frac{r_i}{r_o}\right)^4} \tag{5.46}$$

Equation (5.46) reduces to eq(5.32) with $r_i = 0$ for a solid cylinder. The angular twist θ_e is found from the inner elastic core, $r_i \le r \le r_{ep}$, where the elastic torsion theory applies:

$$\frac{T_e}{J_e} = \frac{G\theta}{l} = \frac{k}{r_{ep}} \tag{5.47}$$

The elastic twist is, from eq(5.47)

$$\theta_e = \frac{T_e l}{GJ_e} = \frac{kl}{Gr_{ep}} \tag{5.48a}$$

where J_e applies to the elastic annulus

$$J_e = \frac{\pi}{2}\left(r_{ep}^4 - r_i^4\right) \tag{5.48b}$$

and T_e is supplied from the first term in eq(5.43b)

$$T_e = \frac{\pi k}{2r_{ep}}(r_{ep}^4 - r_i^4) \tag{5.48c}$$

A residual angle of twist θ_R will remain when T_{ep} is released. Since the recovered twist θ_E is assumed to be elastic, eq(5.41b) again applies with $J = \pi\,(r_o^4 - r_i^4)/2$ for the full cross-section. Substituting eq(5.43b) into eq(5.41b)

$$\theta_R = \frac{kl}{Gr_{ep}} - \frac{\pi k l r_o^3}{2JG}\left\{\frac{4}{3}\left[1 - \frac{1}{4}\left(\frac{r_{ep}}{r_o}\right)^3\right] - \left(\frac{r_i}{r_{ep}}\right)\left(\frac{r_i}{r_o}\right)^3\right\} \tag{5.49a}$$

which is constant for a given r_{ep}. Normalising eq(5.49a) with the twist at the yield point for the outer radius, $\theta_E = kl/Gr_o$, gives

$$\frac{\theta_R}{\theta_E} = \frac{r_o}{r_{ep}} - \frac{\pi r_o^4}{2J}\left\{\frac{4}{3}\left[1 - \frac{1}{4}\left(\frac{r_{ep}}{r_o}\right)^3\right] - \left(\frac{r_i}{r_{ep}}\right)\left(\frac{r_i}{r_o}\right)^3\right\} \tag{5.49b}$$

Putting $r_i = 0$ in eq(5.49b), with $J = \pi r_o^4/2$, leads to the residual twist for a solid shaft, given previously in eq(5.42b).

The residual shear stress in the outer, plastic zone, $r_{ep} \le r \le r_o$, is found from substituting eqs(5.36a) and (5.43b) into eq(5.35)

$$\tau_R = \tau - \frac{T_{ep}\,r}{J} = k - k\left(\frac{T_{ep}}{T_E}\right)\left(\frac{r}{r_o}\right)$$

$$= k\left[1 - \left(\frac{T_{ep}}{T_E}\right)\left(\frac{r}{r_o}\right)\right]$$

$$= k\left\{1 - \frac{4r}{3}\left[\frac{(r_o^3 - r_{ep}^3)}{(r_o^4 - r_i^4)} + \frac{3(r_{ep}^4 - r_i^4)}{4r_{ep}(r_o^4 - r_i^4)}\right]\right\} \tag{5.50a}$$

The residual stress in the inner, elastic zone $r_i \le r \le r_{ep}$, is found from substituting eqs(5.30), (5.36b) and (5.43b) into eq(5.35)

$$\tau_R = \tau - \frac{T_{ep}r}{J} = k\left(\frac{r}{r_{ep}}\right) - k\left(\frac{T_{ep}}{T_E}\right)\left(\frac{r}{r_o}\right)$$

$$= k\left[\frac{r}{r_{ep}} - \left(\frac{T_{ep}}{T_E}\right)\left(\frac{r}{r_o}\right)\right]$$

$$= k\left\{\frac{r}{r_{ep}} - \frac{4r}{3}\left[\frac{(r_o^3 - r_{ep}^3)}{(r_o^4 - r_i^4)} + \frac{3\,(r_{ep}^4 - r_i^4)}{4\,r_{ep}(r_o^4 - r_i^4)}\right]\right\} \tag{5.50b}$$

Alternatively, one could employ eq(5.46) for T_{ep}/T_E within the second line of each derivation of eqs(5.50a,b). It is seen that when $r_i = 0$ in eqs(5.50a,b), they reduce to the solid bar residual stresses, given previously in eqs(5.37a,b).

5.4.3 Sandhill Analogy

A simple analogy can be employed to find the fully plastic (ultimate) torque of solid, non-circular sections [4]. To show this, compare the expression for the volume V of a cone of sand that would rest on the end of an upright circular bar (see Table 5.1) with eq(5.34a). This shows that $T_P = 2V$, where $k = h/r_o$ defines the sloping side of the cone. The ultimate torque T_P for the rectangular and equilateral triangular sections can be derived in a similar manner, i.e. from knowing the volume of a prism of sand each section would support and that each side gradient is k.

Table 5.1 Fully plastic torques T_P for solid sections

Section	V	Slope	$T_P = 2V$
Circle (radius r_o)	$\pi r_o^2 h/3$	$k = h/r_o$	$2\pi r_o^2 h/3 = 2\pi r_o^3 k/3$
Rectangle ($b \times t$)	$(th/6)(3b - t)$	$k = 2h/t$	$kt^2(3b - t)/6$
Triangle (side a)	$a^2 h/(4\sqrt{3})$	$k = 2\sqrt{3}h/a$	$ka^3/12$

With I, T and U sections under torsion, we take the sum of T_P for each rectangular web and flange. The T_P expression for a rectangle will also apply to thin-walled, curved sections, where b is the perimeter length and t is the constant wall thickness. For example, $b = 2\pi r_m$ in a split, thin-walled circular tube, of mean wall radius r_m, giving $T_P \approx \pi k t^2 r_m$. For closed tubes the fourth column in Table 5.1 applies to the difference in T_P for the inner and outer 'solid' sections.

5.5 Closed-End Thick-Walled Cylinder

Thick-walled cylinders may be employed safely even when an internal pressure induces a partially plastic state within its wall. This pressure is intermediate to those pressures associated with full elasticity and full plasticity. As with the elastic–perfectly plastic beam and the torsion bar, the non-hardening cylinder is also a statically determinate problem. That is, the critical pressures may be found from combining an equilibrium equation with a yield criterion. These are sufficient to ensure that a condition of strain compatibility exists at the elastic–plastic interface.

5.5.1 Initial Yielding

Let a thick-walled cylinder of non-hardening material, with inner and outer radii r_i and r_o respectively, be subjected to a steadily increasing internal (gauge) pressure. The external pressure remains atmospheric at zero gauge pressure. Under elastic conditions, a principal, triaxial stress state is distributed through the wall. This state consists of radial, hoop and axial stresses σ_r, σ_θ and σ_z respectively, these being expressed from the Lamé theory [4] as

$$\sigma_r = \frac{p_i r_i^2 (1 - r_o^2/r^2)}{(r_o^2 - r_i^2)} = \frac{p_i (1 - r_o^2/r^2)}{(K^2 - 1)} \tag{5.51a}$$

$$\sigma_\theta = \frac{p_i r_i^2 (1 + r_o^2/r^2)}{(r_o^2 - r_i^2)} = \frac{p_i (1 + r_o^2/r^2)}{(K^2 - 1)} \tag{5.51b}$$

$$\sigma_z = \frac{(\sigma_r + \sigma_\theta)}{2} = \frac{p_i}{(K^2 - 1)} \tag{5.51c}$$

where $K = r_o/r_i$. As the pressure increases, the bore fibres are the first to reach the yield stress Y. Let the material conform to the von Mises yield criterion. Equation (3.7a) appears in its principal, polar co-ordinate form

$$(\sigma_\theta - \sigma_z)^2 + (\sigma_\theta - \sigma_r)^2 + (\sigma_r - \sigma_z)^2 = 2Y^2 \tag{5.52}$$

Substituting eqs(5.51a–c) into eq(5.52), with $r = r_i$, gives the initial yield pressure

$$p_Y = \frac{Y(K^2 - 1)}{\sqrt{3}K^2} \tag{5.53}$$

In Tresca's prediction to the yield pressure, eq(3.1b) is written as

$$\sigma_\theta - \sigma_r = Y \tag{5.54}$$

From eqs(5.51a,b) and (5.54), with $r = r_i$

$$p_Y = \frac{Y(K^2 - 1)}{2K^2} \tag{5.55}$$

Comparing eqs(5.53) and (5.55), the cylinder first yields under Tresca's pressure. The von Mises internal pressure is increased by a further 15 % before yielding commences.

5.5.2 Elastic–Plastic Cylinder

Here it is required to determine the internal pressure which penetrates a plastic zone to a radial depth r_{ep}, where $r_i \le r_{ep} \le r_o$. Since the outer annulus $r_{ep} \le r \le r_o$ is elastic, the radial and hoop stresses are again supplied by the Lamé eqs(5.51a,b,c). The boundary conditions are: (i) at the interface radius r_{ep}, $\sigma_r = -p_{ep}$ and (ii) at outer radius r_o, $\sigma_r = 0$. These give

$$\sigma_r = \frac{p_{ep} r_{ep}^2 (1 - r_o^2/r^2)}{(r_o^2 - r_{ep}^2)} \tag{5.56a}$$

$$\sigma_\theta = \frac{p_{ep} r_{ep}^2 (1 + r_o^2/r^2)}{(r_o^2 - r_{ep}^2)} \tag{5.56b}$$

$$\sigma_z = \frac{\sigma_\theta + \sigma_r}{2} = \frac{p_{ep} r_{ep}^2}{(r_o^2 - r_{ep}^2)} \tag{5.56c}$$

At the interface radius, the material is at its initial yield point. Substituting eqs(5.56a–c) into eq(5.52), with $r = r_{ep}$, leads to the interface pressure

$$p_{ep} = \frac{Y}{\sqrt{3}}\left(1 - \frac{r_{ep}^2}{r_o^2}\right) \tag{5.57}$$

Substituting eq(5.57) into eqs(5.56a,b,c), provides the triaxial stresses in the elastic zone

$$\sigma_r = \frac{Y}{\sqrt{3}}\left(\frac{r_{ep}}{r_o}\right)^2\left(1 - \frac{r_o^2}{r^2}\right) \tag{5.58a}$$

$$\sigma_\theta = \frac{Y}{\sqrt{3}}\left(\frac{r_{ep}}{r_o}\right)^2\left(1 + \frac{r_o^2}{r^2}\right) \tag{5.58b}$$

$$\sigma_z = \frac{Y}{\sqrt{3}}\left(\frac{r_{ep}}{r_o}\right)^2 \tag{5.58c}$$

A radial equilibrium equation applies to the axial symmetry [2]:

$$\sigma_\theta - \sigma_r = r\,\frac{d\sigma_r}{dr} \tag{5.59}$$

When the LHS of eq(5.59) is constant it becomes easy to apply. Within the plastic annulus $r_i \le r \le r_{ep}$, axial plastic strain is ignored when applying internal pressure to a closed-end cylinder. Substituting $d\varepsilon_z^P = 0$ within the first term of eq(4.3c) leads to

$$d\varepsilon_z^P = 0 = d\lambda\,[\sigma_z - \tfrac{1}{2}(\sigma_r + \sigma_\theta)]$$

$$\therefore\quad \sigma_z = \tfrac{1}{2}(\sigma_r + \sigma_\theta) \tag{5.60}$$

Equation (5.60) holds consistently with the axial stress in elastic region, given by eq(5.56c). Substituting eq(5.60) into eq(5.52), the von Mises criterion, is simply

$$\sigma_\theta - \sigma_r = 2Y/\sqrt{3} \tag{5.61}$$

Combining eqs(5.59) and (5.61) allows integration

$$\sigma_r = \frac{2Y}{\sqrt{3}}\ln r + A \tag{5.62a}$$

in which A is found from the condition that $\sigma_r = -\,p_{ep}$ for $r = r_{ep}$. Substituting from eq(5.57)

$$A = -\,p_{ep} - \frac{2Y}{\sqrt{3}}\ln r_{ep} = -\frac{Y}{\sqrt{3}}\left(1 - \frac{r_{ep}^2}{r_o^2}\right) - \frac{2Y}{\sqrt{3}}\ln r_{ep} \tag{5.62b}$$

Combining eqs(5.62a,b) with eqs(5.60) and (5.61) leads to expressions for the triaxial stress state in the plastic zone

$$\sigma_r = \frac{2Y}{\surd 3}\ln\left(\frac{r}{r_{ep}}\right) - \frac{Y}{\surd 3}\left(1 - \frac{r_{ep}^2}{r_o^2}\right) \tag{5.63a}$$

$$\sigma_\theta = \frac{2Y}{\surd 3}\ln\left(\frac{r}{r_{ep}}\right) + \frac{Y}{\surd 3}\left(1 + \frac{r_{ep}^2}{r_o^2}\right) \tag{5.63b}$$

$$\sigma_z = \frac{2Y}{\surd 3}\ln\left(\frac{r}{r_{ep}}\right) + \frac{Y}{\surd 3}\left(\frac{r_{ep}^2}{r_o^2}\right) \tag{5.63c}$$

Now as $\sigma_r = -\,p_i$ for $r = r_i$, eq(5.63a) gives the corresponding internal pressure

$$p_i = \frac{2Y}{\surd 3}\ln\left(\frac{r_{ep}}{r_i}\right) + \frac{Y}{\surd 3}\left(1 - \frac{r_{ep}^2}{r_o^2}\right) \tag{5.64a}$$

The pressure p_{ult}, required to produce a fully plastic cylinder, is found from putting $r_{ep} = r_o$ in eq(5.64a):

$$p_{ult} = \frac{2Y}{\surd 3}\ln\left(\frac{r_o}{r_i}\right) \tag{5.64b}$$

which depends upon the yield stress of the cylinder material and its radius ratio.

Figure 5.15a shows that the hoop stress is greatest in tension at the interface radius. At the bore, the magnitude of the maximum, compressive radial stress equals the internally applied pressure. Also, the axial stress averages the radial and hoop stresses across both zones.

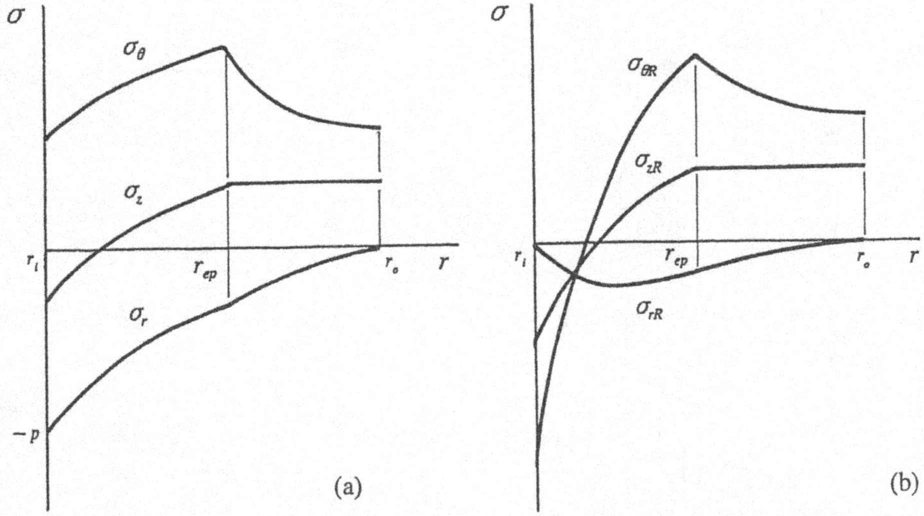

Figure 5.15 (a) Elastic–plastic section and (b) residual stress distributions within cylinder wall

Crossland et al. [5, 6] showed that eq(5.64b) was in reasonable agreement with their burst pressures for pressure vessel steel cylinders, when Y was identified with the upper yield point. However, an empirical modification to eq(5.64a) was made for when material within an elastic–plastic cylinder displayed both upper and lower yield points. They used

$$p_i = \frac{Y_2}{\sqrt{3}} \ln\left(\frac{r_{ep}}{r_i}\right)^2 + \frac{Y_1}{\sqrt{3}}\left[1 - \left(\frac{r_{ep}}{r_o}\right)^2\right] \tag{5.64c}$$

where $(Y_1/\sqrt{3})$ and $(Y_2/\sqrt{3})$ were identified with the respective upper and lower shear yield stresses in a torsion test. The elastic and plastic eqs(5.58a–c) and (5.63a–c) respectively give radial, hoop and axial stress under pressure p_i. They are represented graphically in Fig. 5.15a for when the interface radius coincides with the mean wall radius, i.e. $r_{ep} = \frac{1}{2}(r_i + r_o)$.

5.5.3 Residual Stress

A triaxial state of residual stress will remain distributed throughout the wall of the cylinder when an elastic–plastic pressure is released. That is, when the bore pressure p_i in eq(5.64a) is released, elastic stresses σ_E recover to leave a residual stress distribution σ_R. The latter is found from applying $\sigma_R = \sigma - \sigma_E$, where σ are the elastic-plastic, applied stresses appearing in eqs(5.58a–c) and (5.63a–c). An elastic recovery occurs over the whole section $r_i \le r \le r_o$, upon releasing p_i. Thus, σ_E assumes the usual Lamé forms, given in eq(5.51a–c). Subtracting σ_E from the corresponding stresses σ under pressure in each zone leads to the required residuals. For the inner plastic zone, where $r_i \le r \le r_o$

$$\sigma_{\theta R} = \sigma_\theta - \sigma_{\theta E}$$

$$= \frac{2Y}{\sqrt{3}} \ln\left(\frac{r}{r_{ep}}\right) + \frac{Y}{\sqrt{3}}\left(1 + \frac{r_{ep}^2}{r_o^2}\right) - \frac{p_i r_i^2 (1 + r_o^2/r^2)}{(r_o^2 - r_i^2)} \tag{5.65a}$$

$$\sigma_{rR} = \sigma_r - \sigma_{rE}$$

$$= \frac{2Y}{\sqrt{3}} \ln\left(\frac{r}{r_{ep}}\right) - \frac{Y}{\sqrt{3}}\left(1 - \frac{r_{ep}^2}{r_o^2}\right) - \frac{p_i r_i^2 (1 - r_o^2/r^2)}{(r_o^2 - r_i^2)} \tag{5.65b}$$

$$\sigma_{zR} = \sigma_z - \sigma_{zE}$$

$$= \frac{2Y}{\sqrt{3}} \ln\left(\frac{r}{r_{ep}}\right) + \frac{Y}{\sqrt{3}}\left(\frac{r_{ep}^2}{r_o^2}\right) - \frac{p_i r_i^2}{(r_o^2 - r_i^2)} \tag{5.65c}$$

For the outer elastic zone, where $r_{ep} \le r \le r_o$

$$\sigma_{\theta R} = \sigma_\theta - \sigma_{\theta E}$$

$$= \frac{Y}{\sqrt{3}}\left(\frac{r_{ep}}{r_o}\right)^2\left(1 + \frac{r_o^2}{r^2}\right) - \frac{p_i r_i^2 (1 + r_o^2/r^2)}{(r_o^2 - r_i^2)} \tag{5.65d}$$

$$\sigma_{rR} = \sigma_r - \sigma_{rE}$$
$$= \frac{Y}{\surd 3}\left(\frac{r_{ep}}{r_o}\right)^2\left(1 - \frac{r_o^2}{r^2}\right) - \frac{p_i r_i^2(1 - r_o^2/r^2)}{(r_o^2 - r_i^2)} \tag{5.65e}$$

$$\sigma_{zR} = \tfrac{1}{2}(\sigma_{\theta R} + \sigma_{rR})$$
$$= \frac{Y}{\surd 3}\left(\frac{r_{ep}}{r_o}\right)^2 - \frac{p_i r_i^2}{(r_o^2 - r_i^2)} \tag{5.65f}$$

where p_i is given by eq(5.64a). Equations (5.65a–f) are distributed in the manner of Fig. 5.15b. Compressive stresses left at the bore are particularly beneficial to improving fatigue strength. They oppose and reduce the magnitudes of peak, cyclic tensile stresses due to fluctuating, internal service pressure. The process of pre-pressurising thick cylinders to produce compressive residual stresses is known as *autofrettage*. It has long been used for strengthening cylindrical pressure vessels and cannon tubing against fatigue failure [7].

A simpler, alternative analysis employs the Tresca yield criterion in eq(5.54) instead of the von Mises eq(5.61). Since no assumption needs to be made about axial strain, eq(5.50) is readily combined with the equilibrium eq(5.59) to give the radial stress $\sigma_r = Y \ln r + C$. It follows from eq(5.62a) that one can arrive at the Tresca solutions from multiplying, with the constant factor $\surd 3/2$, all the forgoing (von Mises) pressures and stresses, including the residuals in eqs(5.65a–f).

5.6 Open End Thick Walled Cylinder/Disc

In an open-end, thick-walled cylinder, bore pistons produce and contain an internal fluid pressure. When the force exerted by each piston upon the fluid is reacted externally, e.g. by hydraulic jacks, the cylinder wall remains unstressed axially, giving $\sigma_z = 0$. The latter also applies under a radial pressure expansion of a hole in a disc. Here, the z–dimension is small but the diameter ratio is comparable with that of a thick cylinder, i.e. a plane stress condition applies. In both the disc and the open-ended cylinder the axial stress σ_z remains the intermediate (i.e. absent) stress $\sigma_\theta > \sigma_z > \sigma_r$. It follows, therefore, that Tresca's hoop and radial stresses in a closed-end cylinder will also apply to a cylinder with open-ends. Again, these are found from multiplying the σ_θ and σ_r eqs(5.58a,b), (5.63a,b) and (5.65a,b,d,e) by the factor $\surd 3/2$. The von Mises yield criterion is more difficult to combine with the equilibrium condition for an open-end cylinder. Nadai's parametric approach [1] is particularly useful for providing a solution to the radial and hoop stresses in an elastic–plastic, open-end cylinder and a thin disc of non-hardening material. In contrast to Tresca's common solution, it will now be shown that a von Mises, open-end solution differs from its closed-end solution, given above.

5.6.1 Initial Yielding

With $\sigma_z = 0$ for open ends, the von Mises criterion, eq(5.52), reduces to its biaxial form:

$$\sigma_\theta^{\,2} - \sigma_\theta\sigma_r + \sigma_r^{\,2} = Y^2 \tag{5.66a}$$

With an initial yielding of the bore fibres, we put $r = r_i$ and $p_i = p_Y$ in the Lamé elastic stress eqs(5.51a,b) to give the bore stresses σ_θ and σ_r. Substituting these into eq(5.66a) provides the condition for yielding within the bore

$$p_Y^2[(K^2+1)^2 + (K^2-1)(K^2+1) + (K^2-1)^2]/(K^2-1)^2 = Y^2 \tag{5.66b}$$

where $K = r_o/r_i$. Re-arranging eq(5.66b) supplies the yield pressure as

$$p_Y = \frac{Y(K^2-1)}{\sqrt{(3K^4+1)}} \tag{5.67}$$

Clearly, eq(5.67) differs from the closed-end yield pressure in eq(5.53).

5.6.2 Fully Plastic Section

The bore pressure is required that will produce full plasticity in a thin disc or open cylinder of non-hardening material. For at any radius in the wall, the von Mises yield criterion eq(5.66a) and the radial equilibrium eq(5.59) apply. Making the following substitutions

$$\sigma = \frac{1}{\sqrt{2}}(\sigma_\theta + \sigma_r), \quad \sigma' = \frac{1}{\sqrt{2}}(\sigma_\theta - \sigma_r) \tag{5.68a,b}$$

in eq(5.66a) reduces it to

$$\frac{\sigma^2}{a^2} + \frac{\sigma'^2}{b^2} = 1 \tag{5.69}$$

where $a = \sqrt{2}Y$ and $b = \sqrt{(2/3)}Y$. Equation (5.69) describes the ellipse, shown in Fig. 5.16.

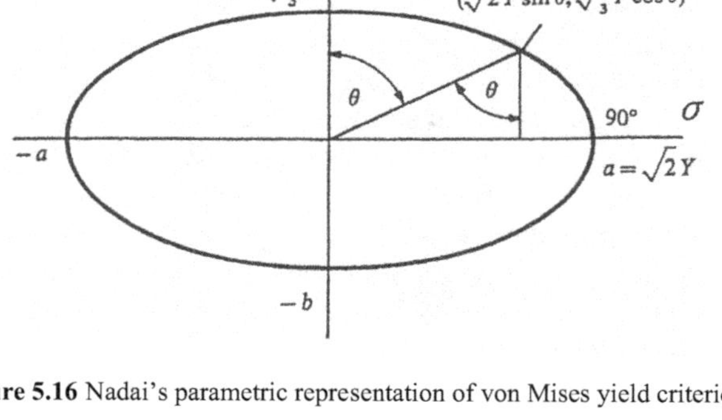

Figure 5.16 Nadai's parametric representation of von Mises yield criterion

The respective lengths, a and b, of the semi-major and -minor ellipse axes, re-appear within Nadai's parameters [1]

$$\sigma = a\sin\theta = \sqrt{2}Y\sin\theta \tag{5.70a}$$

$$\sigma' = b\cos\theta = \sqrt{(2/3)}\,Y\cos\theta \tag{5.70b}$$

Solving eqs(5.68a,b) for σ_θ and σ_r and combining with eqs(5.70a,b)

$$\sigma_\theta = \frac{1}{\sqrt{2}}(\sigma + \sigma') = Y\left[\sin\theta + \frac{1}{\sqrt{3}}\cos\theta\right] = \frac{2Y}{\sqrt{3}}\sin\left(\theta + \frac{\pi}{6}\right) \quad (5.71a)$$

$$\sigma_r = \frac{1}{\sqrt{2}}(\sigma - \sigma') = Y\left[\sin\theta - \frac{1}{\sqrt{3}}\cos\theta\right] = \frac{2Y}{\sqrt{3}}\sin\left(\theta - \frac{\pi}{6}\right) \quad (5.71b)$$

Substituting eqs(5.71a,b) into eq(5.59) leads to a differential equation

$$r\frac{d}{dr}\left[\sin\left(\theta - \frac{\pi}{6}\right)\right] = \left[\sin\left(\theta + \frac{\pi}{6}\right) - \sin\left(\theta - \frac{\pi}{6}\right)\right] = \cos\theta$$

Putting $y = \cos\theta$ enables the variables to be separated into the form

$$\frac{dy}{\sqrt{1 - y^2}} + \frac{dy}{\sqrt{3}} = -\frac{2}{\sqrt{3}}\frac{dr}{r}$$

Integrating leads to:

$$\frac{A^2}{r^2} = y\,e^{-\sqrt{3}\cos^{-1}y} = \cos\theta \times e^{-\sqrt{3}\theta} \quad (5.72a)$$

where A is an integration constant, found from the condition that $\sigma_r = 0$ for $r = r_o$. Correspondingly, eq(5.71b) gives $\theta_o = \pi/6$ and eq(5.72a) becomes

$$A^2 = \frac{\sqrt{3}}{2}r_o^2\,e^{-\sqrt{3}\pi/6}$$

$$\frac{r_o^2}{r^2} = \frac{2}{\sqrt{3}}\cos\theta\;e^{\sqrt{3}(\pi/6 - \theta)} \quad (5.72b)$$

Equation (5.72b) enables θ to be found at any radius $r_i \le r \le r_o$, when σ_θ and σ_r will follow from eqs(5.71a,b). For example, with $r_o = 62.5$ and $r_i = 25$ mm, the stress distributions follow the dashed lines in Fig. 5.17.

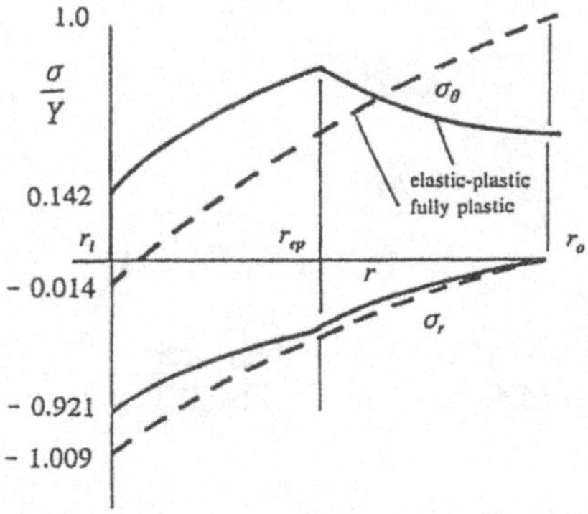

Figure 5.17 Yielding in a thin pressurised disc

The internal pressure to produce full plasticity identifies with the radial stress at the bore: $\sigma_r = -p_i$ for $r = r_i$ in eq(5.71b). This gives

$$p_i = \frac{2Y}{\surd 3} \sin\left(\frac{\pi}{6} - \theta_i\right) \tag{5.72c}$$

where θ_i at r_i follows from eq(5.72b)

$$\left(\frac{r_o}{r_i}\right)^2 = \frac{2}{\surd 3} \cos\theta_i \; e^{\surd 3(\pi/6 - \theta_i)}$$

The radial stress remains compressive for valid values of θ in the range: $-\pi/3 < \theta < \pi/6$. For example, with $r_o/r_i = 2.5$, the disc becomes fully plastic (the lower dashed line in Fig. 5.17), when from eq(5.72b)

$$\left(\frac{62.5}{25}\right)^2 = \frac{2}{\surd 3} \cos\theta_i \; e^{\surd 3(\pi/6 - \theta_i)}$$

A trial solution gives $\theta_i = -30.9° = -0.539^c$. The corresponding normalised, internal pressure is, from eq(5.72c)

$$\frac{p_i}{Y} = \frac{2}{\surd 3} \sin\left(\frac{\pi}{6} + 0.539\right) = 1.009$$

Figure 5.17 shows that the maximum radial compressive stress is $\sigma_r/Y = -p_i/Y = -1.009$ within the bore, where the hoop stress is $\sigma_\theta/Y = -0.014$. Substituting these stresses into the yield criterion (5.66a), confirms that the bore is at the point of yield. The maximum hoop stress is tensile for $r = r_o$. Since $\sigma_r = 0$ for $r = r_o$, eq(5.66a) confirms a maximum value $\sigma_\theta = Y$ (i.e. $\sigma_\theta/Y = 1$) for a fully plastic disc.

5.6.3 Elastic–Plastic Section

We can now determine the internal pressure p_i to yield the disc to any intermediate radius r_{ep}, where $r_i \le r_{ep} \le r_o$ (see Fig. 5.17). The Lamé stress in the outer elastic zone $r_{ep} \le r \le r_o$ follows from eqs(5.51a,b)

$$\sigma_\theta = \frac{p_{ep}\, r_{ep}^2 \left(1 + r_o^2/r^2\right)}{\left(r_o^2 - r_{ep}^2\right)} \tag{5.73a}$$

$$\sigma_r = \frac{p_{ep}\, r_{ep}^2 \left(1 - r_o^2/r^2\right)}{\left(r_o^2 - r_{ep}^2\right)} \tag{5.73b}$$

with $\sigma_z = 0$. These are shown as the continuous lines in Fig. 5.17. The interface yield pressure p_{ep} is found from eq(5.67), where $K = r_o/r_{ep}$. This gives

$$p_{ep} = \frac{Y[(r_o/r_{ep})^2 - 1]}{\sqrt{3(r_o/r_{ep})^4 + 1}} \tag{5.74}$$

The stresses within the inner, plastic zone $r_i \le r \le r_{ep}$, are again supplied by eqs(5.71a,b). The radial and hoop stress plastic distributions must each match the corresponding elastic values at the elastic–plastic, interface radius. The condition that σ_r is common to both zones at $r = r_{ep}$ is used to re-determine the constant A in eq(5.68a) for an elastic–plastic disc. That is, from eqs(5.71b) and (5.74), $\sigma_r = -p_{ep}$ for $\theta = \theta_{ep}$:

$$\frac{Y[1 - (r_o/r_{ep})^2]}{\sqrt{3(r_o/r_{ep})^4 + 1}} = \frac{2Y}{\sqrt{3}}\sin\left(\theta_{ep} - \frac{\pi}{6}\right)$$

$$\frac{1 - (r_o/r_{ep})^2}{\sqrt{(r_o/r_{ep})^4 + 1/3}} = 2\sin\left(\theta_{ep} - \frac{\pi}{6}\right) \tag{5.75}$$

where θ_{ep} must be found by trial. It follows from eq(5.72a) that

$$A^2 = r_{ep}^2 \cos\theta_{ep}\, \mathrm{e}^{-\sqrt{3}\theta_{ep}}$$

$$\frac{r_{ep}^2}{r^2} = \frac{\cos\theta}{\cos\theta_{ep}}\, \mathrm{e}^{\sqrt{3}(\theta_{ep} - \theta)} \tag{5.76}$$

Equation (5.76) enables θ to be found at any plastic radius $r_i \le r \le r_{ep}$. In particular, the solution $\theta = \theta_i$ for $r = r_i$, when substituted into eq(5.72c), supplies the internal pressure p_i to produce an elastic–plastic disc. For example, with r_i = 25 mm and r_o = 62.5 mm, the pressure to yield the disc to its mean radius: $r_{ep} = ½(25 + 62.5) = 43.75$ mm, follows:

$$r_{ep} / r_i = 43.75 / 25 = 1.75$$

$$r_o / r_{ep} = (r_o / r_i) \times (r_i / r_{ep}) = 2.5 / 1.75 = 1.4286$$

Substituting into eq(5.75)

$$-0.4908 = 2\sin(\theta_{ep} - \pi/6)$$

A trial solution gives $\theta_{ep} = 15.8° = 0.2758^c$, which is a constant for a given depth of penetration. From eq(5.76), with $r = r_i$ and $\theta = \theta_i$

$$1.75^2 = 1.0393\cos\theta_i \times \exp[\sqrt{3}(0.2758 - \theta_i)]$$

Here a trial solution yields $\theta_i \simeq -22.9° = -0.3997^c$. Hence, from eq(5.72c)

$$p_i/Y = (2/\sqrt{3})\sin(\pi/6 + 0.3997) = 0.921$$

and from eq(5.71a), $\sigma_\theta/Y = 0.142$. The stress distribution within the plastic zone in Fig. 5.17 is found for further solutions to θ from eq(5.76) at given radii r within the range $r_i \le r \le r_{ep}$. Equations (5.73a,b) and (5.74) supply the elastic zone stress distributions for $r_{ep} \le r \le r_o$.

5.6.4 Residual Stress

When the elastic–plastic pressure p_i in eq(5.72c) is released, it is assumed that elastic stress σ_E is recovered for all radii according to the Lamé eqs(5.51a,b). Applying $\sigma_R = \sigma - \sigma_E$ to the plastic zone: $r_i \le r \le r_{ep}$, eqs(5.51a,b) are subtracted from eqs(5.71a,b) to give the hoop and radial, residual stresses as:

$$\sigma_{\theta R} = \frac{2Y}{\sqrt{3}} \sin\left(\theta + \frac{\pi}{6}\right) - \frac{p_i(1 + r_o^2/r^2)}{r_o^2/r_i^2 - 1} \tag{5.77a}$$

$$\sigma_{rR} = \frac{2Y}{\sqrt{3}} \sin\left(\theta - \frac{\pi}{6}\right) - \frac{p_i(1 - r_o^2/r^2)}{r_o^2/r_i^2 - 1} \tag{5.77b}$$

In the elastic zone: $r_{ep} \le r \le r_o$, the applied stresses σ are given by eqs(5.73a,b) so that the residuals within this zone become

$$\sigma_{\theta R} = \frac{p_{ep}(1 + r_o^2/r^2)}{r_o^2/r_{ep}^2 - 1} - \frac{p_i(1 + r_o^2/r^2)}{r_o^2/r_i^2 - 1} \tag{5.77c}$$

$$\sigma_{rR} = \frac{p_{ep}(1 - r_o^2/r^2)}{r_o^2/r_{ep}^2 - 1} - \frac{p_i(1 - r_o^2/r^2)}{r_o^2/r_i^2 - 1} \tag{5.77d}$$

where p_{ep} is given by eq(5.74). According to eqs(5.73a–d), the distributions in residual stress $\sigma_{\theta R}$ and σ_{rR} are similar to those of $\sigma_{\theta R}$ and σ_{rR} in Fig. 5.15b for a closed-end cylinder. The maximum residual hoop stress is compressive within the bore fibres. *Note*: It is not permissible for $\sigma_{\theta R}$ to exceed Y for $r = r_i$ in a cylinder of non-hardening material. If eq(5.77a) predicts $\sigma_{\theta R} > Y$, the implication is that reversed yielding of the bore has occurred, so invalidating any assumption of an elastic recovery following pressure release. The experimental determination of residual stresses, by the hole boring technique [7], confirms that this assumption will overestimate the compressive residual hoop stresses in the bore fibres whenever reversed yielding occurs.

5.7 Rotating Disc

Here the angular velocities of a spinning disc are required to: (i) initiate yielding, (ii) penetrate a plastic zone into the wall and (iii) produce full plasticity. Consider a uniformly thin, solid disc of outer radius r_o and let the respective velocities be ω_Y, ω_{ep} and ω_P. Firstly, a Tresca yield criterion is employed to simplify each solution. Thereafter, Nadai's parameters define the plane-stress state providing solutions for a similar disc obeying the von Mises yield criterion.

5.7.1 Initial Yielding

The elastic, radial and hoop stresses σ_r, σ_θ, at a radius $r \le r_o$ in a solid disc, rotating at a speed ω rad/s, are [1]

$$\sigma_r = \frac{\rho\omega^2}{8}(3 + \nu)(r_o^2 - r^2) \tag{5.78a}$$

$$\sigma_\theta = \frac{\rho\omega^2}{8}\left[(3 + \nu)r_o^2 - (1 + 3\nu)r^2\right] \tag{5.78b}$$

where ρ is the density of the disc material. Equations (5.78a,b) are distributed in the manner shown in Fig. 5.18a.

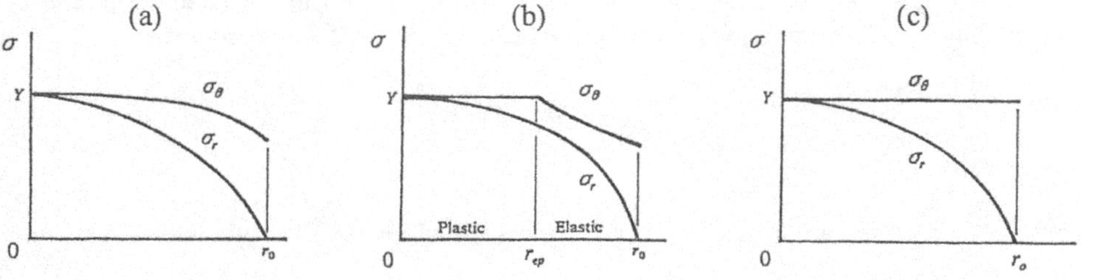

Figure 5.18 Radial and hoop stress distributions

Since both σ_θ and σ_r are tensile, the Tresca criterion (5.54) employs the magnitude of the greater tensile stress together with the least stress (i.e. zero axial stress σ_z). This gives

$$\sigma_\theta = Y \tag{5.79}$$

Clearly, yielding will commence at the centre where σ_θ is a maximum. Substituting eq(5.78b) into (5.79) for $r = 0$ supplies the speed ω_Y to initiate yielding

$$\frac{1}{8}\rho\omega_Y^2(3 + \nu)r_o^2 = Y$$

$$\therefore\ \omega_Y^2 = \frac{8Y}{\rho(3 + \nu)r_o^2} \tag{5.80}$$

5.7.2 Elastic–Plastic Disc

The equilibrium eq(5.59) for an elastic–plastic disc is modified for the centrifugal force within the inner plastic core $0 \le r \le r_{ep}$. This gives

$$\sigma_r + r\frac{d\sigma_r}{dr} - \sigma_\theta + \rho\omega^2 r^2 = 0 \tag{5.81}$$

Combining eq(5.79) with eq(5.81) provides a solution to σ_r as follows

$$\frac{d}{dr}(r\sigma_r) = Y - \rho\omega_{ep}^2 r^2$$

$$r\sigma_r = Yr - \frac{\rho\omega_{ep}^2 r^3}{3} + A$$

$$\sigma_r = Y - \frac{\rho\omega_{ep}^2 r^2}{3} + \frac{A}{r} \tag{5.82a}$$

where ω_{ep} is the speed to produce partial yielding, yet to be determined. Figure 5.18b shows $\sigma_\theta = \sigma_r = Y$ for $r = 0$, wherein the integration constant $A = 0$ in eq(5.82a). Thus, at the elastic–plastic radius $r = r_{ep}$, from within the plastic zone, eq(5.82a) becomes

$$\sigma_r = Y - \frac{\rho\omega_{ep}^2 r_{ep}^2}{3} \tag{5.82b}$$

For the outer, elastic annulus $r_{ep} \le r \le r_o$, Lamé's equations are modified with a centrifugal stress term [1]

$$\sigma_r = a - \frac{b}{r^2} - (3 + \nu)\frac{\rho\omega_{ep}^2 r^2}{8} \tag{5.83a}$$

$$\sigma_\theta = a + \frac{b}{r^2} - (1 + 3\nu)\frac{\rho\omega_{ep}^2 r^2}{8} \tag{5.83b}$$

Note: Equations(5.83a,b) reduce to eqs(5.78a,b) for an elastic disc when the constants $b = 0$ and $a = (3 + \nu)\rho\omega^2 r_o^2/8$, apply to the respective conditions: (i) that the stresses cannot be infinite at the disc centre and (ii) $\sigma_r = 0$ for $r = r_o$.

For an elastic–plastic disc, the following two boundary conditions enable the constants a and b in eqs(5.83a,b) to be found: (i) $\sigma_r = 0$ for $r = r_o$

$$0 = a - \frac{b}{r_o^2} - (3 + \nu)\frac{\rho\omega_{ep}^2 r_o^2}{8}$$

and (ii) $\sigma_\theta = Y$ at $r = r_{ep}$

$$Y = a + \frac{b}{r_{ep}^2} - (1 + 3\nu)\frac{\rho\omega_{ep}^2 r_{ep}^2}{8}$$

Solving these simultaneously for a and b:

$$a = \left\{ Y - \frac{\rho\omega_{ep}^2}{8}\left[(3 + \nu)r_o^2 - (1 + 3\nu)r_{ep}^2\right]\right\}\frac{r_{ep}^2}{(r_o^2 + r_{ep}^2)} + (3 + \nu)\frac{\rho\omega_{ep}^2 r_o^2}{8} \tag{5 .84a}$$

$$b = \left\{ Y - \frac{\rho\omega_{ep}^2}{8}\left[(3 + \nu)r_o^2 - (1 + 3\nu)r_{ep}^2\right]\right\}\frac{r_o^2 r_{ep}^2}{(r_o^2 + r_{ep}^2)} \tag{5.84b}$$

The angular speed ω_{ep} is found from the condition that σ_r is common to both zones at the interface radius r_{ep}. Within the elastic annulus, eq(5.83a) gives, for $r = r_{ep}$

$$\sigma_r = a - \frac{b}{r_{ep}^2} - (3 + \nu)\frac{\rho \omega_{ep}^2 r_{ep}^2}{8} \tag{5.85a}$$

Substituting a and b from eqs(5.84a,b) into eq(5.85a) leads to

$$\sigma_r = (3 + \nu)\frac{\rho\omega_{ep}^2}{8}(r_o^2 - r_{ep}^2) - \frac{Y(r_o^2 - r_{ep}^2)}{(r_o^2 + r_{ep}^2)}$$

$$+ \frac{\rho\omega_{ep}^2}{8}\left[(3 + \nu)r_o^2 - (1 + 3\nu)r_{ep}^2\right]\frac{(r_o^2 - r_{ep}^2)}{(r_o^2 + r_{ep}^2)} \tag{5.85b}$$

Equating (5.78b) and (5.81b)

$$2Yr_o^2 = \frac{\rho\omega_{ep}^2}{8}\left\{(3 + \nu)(r_o^2 - r_{ep}^2)(2r_o^2 + r_{ep}^2) + r_{ep}^2\left[\frac{8}{3}(r_o^2 + r_{ep}^2) - (1 + 3\nu)(r_o^2 - r_{ep}^2)\right]\right\}$$

giving the speed to penetrate plasticity to radius r_{ep} for an elastic–plastic solid disc

$$\omega_{ep}^2 = \frac{16\,Y}{\rho r_o^2\left\{(3+\nu)\left(1 - \frac{r_{ep}^2}{r_o^2}\right)\left(2 + \frac{r_{ep}^2}{r_o^2}\right) + \left(\frac{r_{ep}^2}{r_o^2}\right)\left[\frac{8}{3}\left(1 + \frac{r_{ep}^2}{r_o^2}\right) - (1 + 3\nu)\left(1 - \frac{r_{ep}^2}{r_o^2}\right)\right]\right\}} \tag{5.86}$$

Equation (5.86) shows a speed $\omega_{ep}^2 = (2.671Y)/(\rho r_o^2)$, when the elastic–plastic interface radius coincides with the mean radius $r_{ep}/r_o = ½$, in a steel disc with $\nu = ¼$. Dividing by ω_Y from eq(5.80) gives $\omega_{ep}/\omega_Y = 1.042$. This shows that only a 4.2 % increase in the initial yield speed is required to produce this amount of plastic penetration.

5.7.3 Fully Plastic Disc

When the disc becomes fully plastic, eq(5.82a) applies to $0 \le r \le r_o$. Again, $\sigma_r = Y$ for $r = 0$, giving $A = 0$. Also, because $\sigma_r = 0$ for $r = r_o$, eq(5.82b) gives

$$0 = Y - \frac{\rho\,\omega_P^2\,r_o^2}{3}$$

Hence the fully plastic speed ω_P is

$$\omega_P^2 = \frac{3\,Y}{\rho r_o^2} \tag{5.87}$$

Dividing eq(5.87) with ω_Y, from eq(5.80) gives the angular speed ratio

$$\frac{\omega_P}{\omega_Y} = \sqrt{\frac{3(3 + \nu)}{8}}$$

This gives $\omega_P/\omega_Y = 1.104$ for $\nu = ¼$, showing that a 10.4% increase in the initial yield speed is required to attain full plasticity in a solid disc. Distributions in σ_r and σ_θ for the fully plastic disc are shown in Fig. 5.18c. Clearly, the use of the Tresca criterion, where $\sigma_\theta = Y$ is constant throughout the region of plasticity, has simplified this solution.

5.7.4 Tresca versus von Mises

A numerical, von Mises solution to the stress distribution within a non-hardening, elastic–plastic, solid disc is also possible. The elastic zone stresses are again given by eq(5.83a,b) for the same boundary condition: $\sigma_r = 0$ for $r = r_o$. Moreover, eqs(5.83a,b) must satisfy the von Mises yield condition at the interface radius, i.e σ_θ and σ_r must equal the corresponding plastic zone stresses at r_{ep}. Within the inner plastic core, these two stresses must satisfy the von Mises yield criterion and the equilibrium equation simultaneously:

$$\sigma_\theta^2 - \sigma_\theta \sigma_r + \sigma_r^2 = Y^2 \tag{5.88a}$$

$$\sigma_r + r\frac{d\sigma_r}{dr} = \sigma_\theta - \rho\omega^2 r^2 \tag{5.88b}$$

Using Nadai's eqs(5.71a,b) to separate σ_θ and σ_r in eq(5.88a) and then substituting these into eq(5.88b) leads to a first order differential equation in r and θ [8]:

$$\frac{d\theta}{dr} = \frac{2Y\cos\theta - \sqrt{3}\rho\omega^2 r^2}{Yr(\sqrt{3}\cos\theta + \sin\theta)} \tag{5.89}$$

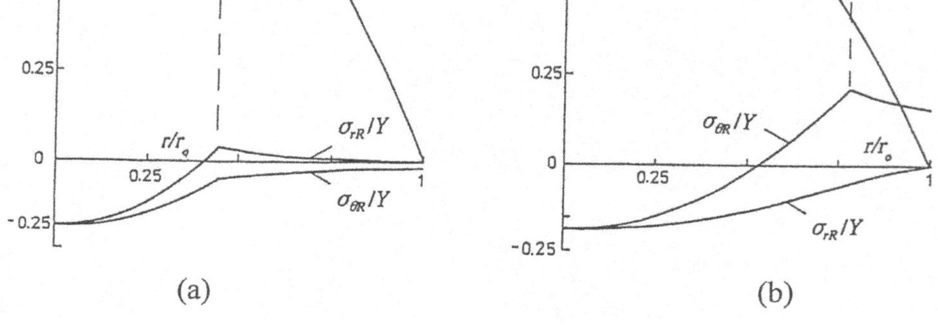

Figure 5.19 (a) von Mises and (b) Tresca stresses for solid disc at similar speeds

One may solve eq(5.89) using a Runga-Kutta method, taking a single starting value for θ, depending upon whether the disc is solid or hollow (i.e. $\theta_o = \pi/2$ or $\pi/6$ respectively). The von Mises, solid disc solution shows that both σ_r and σ_θ must remain very nearly equal to Y in this zone, in contrast to the Tresca solution (see Figs 5.19a,b).

The hollow disc solution reveals a similar σ_r distributions in Figs 5.20a,b but the von Mises σ_θ is not constant, increasing above Y in the plastic zone interior $r_i < r < r_{ep}$.

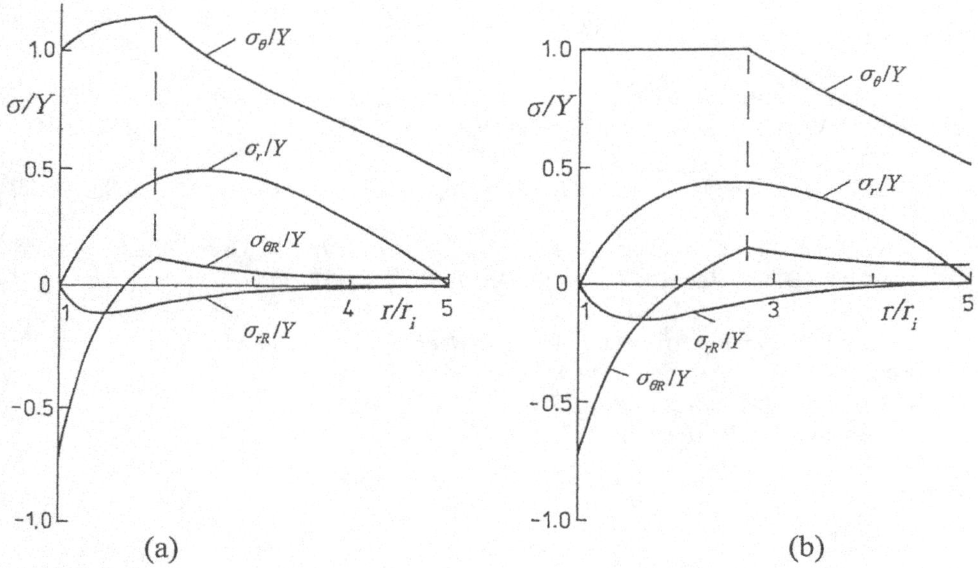

Figure 5.20 (a) von Mises, (b) Tresca stress distributions for a hollow rotating disc

The greater differences arising in σ_θ and σ_r, from applying different yield criteria to a solid disc, are reflected in their residual stress estimates, $\sigma_{\theta R}$ and σ_{rR}, as shown. For example, when applying $\sigma_R = \sigma - \sigma_E$ to each zone in Fig. 5.19b, Tresca's residual stress follow from eqs(5.78a,b), (5.82a) and (5.83a,b) as:

$0 \le r \le r_{ep}$

$$\sigma_{\theta R} = Y - \frac{\rho\omega_{ep}^2}{8}\left[(3+\nu)r_o^2 - (1+3\nu)r^2\right] \tag{5.90a}$$

$$\sigma_{rR} = Y - \frac{\rho\omega_{ep}^2 r^2}{3} - \frac{\rho\omega_{ep}^2}{8}(3+\nu)(r_o^2 - r^2) \tag{5.90b}$$

$r_{ep} \le r \le r_o$

$$\sigma_{\theta R} = a + \frac{b}{r^2} - (1+3\nu)\frac{\rho\omega_{ep}^2 r^2}{8} - \frac{\rho\omega_{ep}^2}{8}\left[(3+\nu)r_o^2 - (1+3\nu)r^2\right] \tag{5.90c}$$

$$\sigma_{rR} = a - \frac{b}{r^2} - (3+\nu)\frac{\rho\omega_{ep}^2 r^2}{8} - \frac{\rho\omega_{ep}^2}{8}(3+\nu)(r_o^2 - r^2) \tag{5.90d}$$

where a, b and ω_{ep} are given by eqs(5.84a,b) and (5.86). Equations (5.90a–d) were applied to a solid steel disc with $Y = 310$ MPa, $\nu = 0.28$, $\rho = 7750$ kg/m^3 and $r_o = 250$ mm, following a speed $N = 13000$ rev/min. Figures 5.20a,b show that Tresca's residuals are marginally greater, arising from an increased penetration, compared to von Mises residuals [8]. The latter contrast with the conservative nature of Tresca, which predicts a greater spread of plasticity at similar rotational speeds. *Note*: Original solutions to the elastic–plastic rotating disc problem were given by Hodge [9] and Heyman [10].

References

1. Nadai A. *Theory of Flow and Fracture of Solids*, 1950, McGraw-Hill, London.
2. Rees D. W. A. *Mechanics of Solids and Structures*, 2nd Ed, 2016, ICP, World Scientific, London.
3. Rees D. W. A. *The Mechanics of Engineering Structures*, 2015, ICP, World Scientific, London
4. Timoshenko S. and Goodier J. N. *Theory of Elasticity*, 1951, McGraw-Hill, New York.
5. Crossland B. and Bones J. A. *Proc. I. Mech. E.*, 1958, **172**, 777.
6. Crossland B. *Proc. I. Mech. E.*, 1954, **168**, 935.
7. Franklin G. J. and Morrison J. L. *Proc. I. Mech. E.*, 1960, **174**, 947.
8. Rees D. W. A. *Zeit. Angew. Math. Mech.*, 1999, **79**, 281.
9. Hodge P. G. *Jl. Applied Mech.*, 1955, **24**, 311.
10. Heyman J. *Proc. I. Mech. E.*, 1958, **172**, 531.

Exercises

Elastic–Plastic Beams

5.1 A steel beam has a uniform rectangular cross section breadth $b = 50$ mm and depth $d = 100$ mm and a tensile yield strength of $Y = 320$ MPa. Find the bending moments required to initiate yielding within this section and that which would cause the beam to collapse. Explain how the ratio between these moments becomes a design parameter for the beam section. Assuming an elastic–perfectly plastic behaviour of the steel, what intermediate bending moment would be required to penetrate plasticity inward from top and bottom surfaces to a depth of 25 mm? Find for this intermediate moment: (a) the corresponding central concentrated load for a 2.5 m length of simply supported beam, (b) the central length of beam over which plasticity extends within a plastic hinge and (c) the normalised residual stress distribution (w.r.t. Y) that remains in the cross section upon removal of the load.
Ans: 26.6 kNm, 40 kNm, 36.6 kNm, (a) 58.67 kN, (b) 0.682 m, (c) ± 0.375, ± 0.3125
See solution

5.2 A steel beam has a uniform rectangular cross section breadth 15 mm and depth 35 mm. A length of 3 m is simply supported when supporting a central concentrated load. The strength and stiffness of the steel are given as: $Y = 300$ Mpa and $E = 210$ GPa. Assuming elastic–perfect plasticity, find the moment and the load required to initiate yielding within this section and those which would cause the beam to collapse. What intermediate bending moment would be required to penetrate plasticity inward from top and bottom surfaces to occupy 20 % of the central section area? Under what load and beam curvature does this condition apply? Find for this intermediate moment: (a) the central length of beam over which plasticity extends within a plastic hinge and (b) the residual stress distribution in MPa that remains in the cross section upon its removal. Over what central length of beam does plasticity extend for the collapse condition?
Ans: 918.75 Nm, 1225 N; 1378.125 Nm, 1837.5 N; 1084.13 Nm, 1445.5 N, 5.315 m; (a) 0.458 m, (b) ± 54 MPa, ± 16.8 MPa; 1 m.

5.3 A 3 m long steel beam has a uniform rectangular section breath 15 mm and depth 35 mm. The beam supports with fixed ends a concentrated load 1 m from its left end. Find the fully plastic moment and the collapse load and the spread of plasticity within each hinge length.
Ans: 1378.13 Nm, 4.134 kN; left 1/6 m, centre ½ m, right 1/3 m

5.4 A simply supported beam of length L, with a rectangular section breadth b and depth d, is to carry a uniformly distributed load w/unit length and a central concentrated load W. Derive expressions that provide the ultimate moment, the collapse loading and the plastic hinge length, given that the yield stress Y is constant (a) when w acts alone upon and the beam and (b) when W and w act together.
See solution

5.5 Determine the collapse load and the length of the plastic hinge when a simply supported beam of length l, with rectangular section $b \times d$, carries a single concentrated vertical load W, applied: (a) at the centre and (b) at a position that divides the length l into p and q where $p < q$.
Ans: (a) $M_P = (bd^2Y)/l$, $l/3$; (b) $M_P = (bd^2lY)/(4pq)$, $(p/3 + q/3)$

5.6 Show that the collapse moment for a ⊥ section, made from equal rectangles each of length a and thickness t, is given by $M_P = Yah(a + h)/2$.

5.7 Examine the manner in which a cantilever of length l will collapse when carrying a uniformly distributed load w/unit length, with a prop to prevent deflection at its free end. Show that the fully elastic loading is $w_E = 8M_E/l^2$ and the collapse loading is $w_P = 11.73M_P/l^2$. Hence find the ratio between these loads for a rectangular section $b \times d$. *Hint*: The prop reaction is given by $3wl/8$.

5.8 A 40 mm thick T-section with flange and web lengths 280 mm and 160 mm receives an elastic–plastic bending moment for which the neutral axis becomes coincident with the underside of the flange as the yield stress is reached within its top surface. Find the moment and the depth of plastic penetration in the web and distribution of residual stress when the moment is released.
Ans: 168 kNm, 120 mm, 130.5 MPa flange top, 131.9 MPa web bottom.

Elastic–Perfectly Plastic Torsion Bars

5.9 Find the residual stress distribution following the removal of a torque that has penetrated plasticity to a mean radius of a solid shaft with radius 25 mm. What is the angular twist per unit length of this shaft in rad/m before and after this torque is released? Take $k = 150$ MPa, $G = 80$ GPa.
Ans: $\tau_R = 53.1$ MPa at $r_{ep} = 12.5$ mm, $\tau_R = -43.8$ MPa at $r_o = 25$mm; $\theta/L = 0.15\,^c$/m, $0.053\,^c$/m

5.10 A solid circular shaft 100 mm diameter is subjected to an applied torque of 30 kNm. Show that the shaft is in an elastic–plastic condition given that its shear yield stress is 125 MPa. Find the radial depth of plastic penetration and the angular twist over a 3 m length for a shear modulus of 82 GPa. Find the distribution of residual stress that remains in the section when the applied torque is removed.
See solution

5.11 A bar of diameter d and length l is bored to diameter $d/2$ over half its length. If the outer diameter of the solid shaft reaches its yield point under an applied torque, show that the diameter of the elastic–plastic interface d_{ep}, within the hollow shaft section, may be found from the solution to the quartic equation: $d_{ep}^4 - d^3 d_{ep} + (3/16)\, d^4 = 0$. Show that the ratio between the angular twists for the hollow and solid shafts is given by d/d_{ep}.
See solution

5.12 A tapered solid plug is subjected to a fully elastic torque at its larger end radius R_o. Show that the smaller end radius $r_o = 0.918R_o$ if plasticity is to penetrate to its mean radius? Given $R_o = 50$ mm and a 5° taper find the plug length and the angular twist. Take G = 82 GPa.
Ans: 46.86 mm, 1.09°

5.13 Derive and plot the normalised torque-twist relationship for a hollow bar. Show the accumulation of normalised residual strain on this plot in a similar manner to that given in Fig. 5.9 for a solid bar.

5.14 What value of torque is required for an elastic-plastic interface to lie at the mean radius in a tube of inner and outer diameters 25 and 100 mm respectively? Determine the residual stress distribution and the residual twist when this torque is subsequently removed. Take $k = 230$ MPa and $G = 78$ GPa.

Elastic-Perfectly Plastic Cylinders and Discs

5.15 What is the maximum, limiting diameter ratio of an annular disc beyond which it is not possible to achieve a full spread of plasticity when radial pressure is applied to the inner diameter. At what pressure does this occur?
See solution: 2.963, $2Y/\sqrt{3}$

5.16 An annular disc, with inner and outer radii 25 mm and 62.5 mm respectively, is machined from an alloy steel with a yield stress 500 MPa. Determine the internal pressures necessary to: (a) initiate yielding, (b) produce a fully plastic disc and (c) produce partial plasticity to the mean radius.
Ans: 2.415 kbar, 5.045 kbar, 4.584 kbar

5.17 Compare the residual stress distributions in open-end and closed-end thick walled cylinders resulting from an autofrettage pressure sufficient to penetrate an elastic-plastic interface to coincide with the mean wall radius. Employ a von Mises yield criterion and normalise the stresses with the yield stress Y for a cylinder of diameter ratio 3.

5.18 Show that the speeds necessary to initiate yielding within solid and hollow discs are each independent of the yield criterion and are respectively:
See solution

$$\omega_Y^2 = \frac{8Y}{(3+\nu)\rho r_o^2}, \qquad \omega_Y^2 = \frac{4Y}{\rho\left[r_i^2(1-\nu) + r_o^2(3+\nu)\right]}$$

5.19 Using the Nadai's parameter approach, determine numerically the von Mises stress distributions in an elastic-plastic, hollow disc, $r_i = 50$ mm and $r_o = 250$ mm, rotating at a speed of 11000 rev/min. Derive from these the residual stresses distribution for when the disc is brought to rest. The following conditions apply to the disc material: $Y = 310$ MPa, $\nu = 0.28$, and $\rho = 7750$ kg/m^3. Compare and contrast each distribution with the corresponding Tresca solution.
Ans: similar to Figs 5.20a,b

CHAPTER 6

SLIP LINE FIELDS

6.1 Introduction

Rigid plasticity theory ignores elastic strain and assumes that plastic deformation occurs without hardening under a constant yield stress. This material model is often justified for the description of a forming process involving large plastic strains. When extruding carbon steel, for example, the maximum elastic and plastic strains are typically of contrasting magnitudes: 0.1 % and 50 % respectively. In general, a plastic-rigid model is an appropriate choice for the analyses of large scale deformation processes. These processes may be framed within co-ordinates of plane strain and axial symmetry, depending upon the shape that is produced. In this chapter it is shown how the slip-line field (SLF) method can represent the flow behaviour of plane strain processes with a plastic-rigid material model. Slip-lines can be constructed only where plane strain conditions are upheld. Consequently, this method is restricted to processes that include: extrusion of rectangular stock, indentation under a narrow, parallel die, rolling of thin sheet and orthogonal machining. Where plane strain conditions do not exist, as with the extrusion of circular bar, alternative analyses are required. Among these are the limit and slab analyses. A limit analysis provides simple, upper and lower bound approximations to the forming forces. A slab analysis is based upon the force equilibrium of a deforming element. In principle, these techniques place no restriction upon the deformation mode and are employed most often given that they lead to convenient closed solutions. In particular, analyses are made in chapters 7 and 15 of the plastic collapse of beams and circular plates under lateral loadings, hot and cold forging, rolling, wire and strip drawing.

Note: In the absence of elasticity and hardening, SLF theory takes no account of the effects of creep, strain rate, temperature generation and thermal gradients. A more serious effect is likely to arise when plane cross sections distort out of their original planes. This can occur under extrusion for the very severe deformation conditions in the die region. Its extent has been examined from the distortion that occurs to an orthogonal grid [1].

6.2 Slip Line Field Theory

A plane strain condition refers to a process where strain in a direction perpendicular to the plane of deformation is zero or a constant. This applies to rolling, extrusion and indentation processes, where lateral flow is negligible. The slip-line field will approximate the forming loads required for each process. In an extrusion and indentation process, the boundary conditions for the contacting surfaces may vary from a constant coefficient of friction lying between zero and a sticking value.

Presented here is a graphical interpretation of equations, due to Hencky [2] and Geiringer [3], which govern plane strain processes. Three diagrams are: (i) the physical plane, defining the geometry of the process, (ii) the stress plane, showing the variation in normal stress on a slip-line and (iii) the hodograph of velocity streamlines.

6.2.1 *Yield and Flow*

When $f = J_2'$, given by eq(3.10b), is substituted into eq(3.22) it results in the *Levy-Mises flow rule*. Partial differentiation gives the principal plastic strain increments:

$$d\varepsilon_1 = \tfrac{2}{3}\, d\lambda\, [\sigma_1 - \tfrac{1}{2}(\sigma_2 + \sigma_3)] \tag{6.1a}$$
$$d\varepsilon_2 = \tfrac{2}{3}\, d\lambda\, [\sigma_2 - \tfrac{1}{2}(\sigma_3 + \sigma_1)] \tag{6.1b}$$
$$d\varepsilon_3 = \tfrac{2}{3}\, d\lambda\, [\sigma_3 - \tfrac{1}{2}(\sigma_1 + \sigma_2)] \tag{6.1c}$$

In eqs(6.1a,b,c), the strain superscript P is omitted since $d\varepsilon$ become total strain increments in the absence of elasticity for the rigid material assumed. Here, 1 and 3 to lie in the plane of deformation, with $d\varepsilon_2 = 0$ for plane strain in a perpendicular direction. From eq(6.1b):

$$\sigma_2 = \tfrac{1}{2}(\sigma_1 + \sigma_3) \tag{6.2}$$

Thus, σ_2 is the intermediate stress $\sigma_1 \geq \sigma_2 \geq \sigma_3$, giving Tresca's yield criterion (3.1a) as

$$\sigma_1 - \sigma_3 = 2k \tag{6.3}$$

where k is the constant shear yield stress. Equation (6.3) remains unchanged when eq(6.2) is substituted into the von Mises criterion (eq 3.7a). However, the relation between k and Y differ between the two criteria. For Tresca $k = Y/2$ and for von Mises $k = Y/\surd 3$. With plane-strain yielding under general 2D stress components: σ_x, σ_y and τ_{xy}, the in-plane principal stresses are

$$\sigma_{1,3} = \tfrac{1}{2}(\sigma_x + \sigma_y) \pm \tfrac{1}{2}\surd[(\sigma_x - \sigma_y)^2 + 4\tau_{xy}^2] \tag{6.4a}$$

Substituting eq(6.4a) into eq(6.3), the Tresca yield condition becomes

$$\tfrac{1}{2}(\sigma_1 - \sigma_3) = \tfrac{1}{2}\surd[(\sigma_x - \sigma_y)^2 + 4\tau_{xy}^2] = k \tag{6.4b}$$

Equation (6.4b) is the limiting radius of a Mohr's circle (see Fig. 6.1a) describing a general stress state at the point of yield.

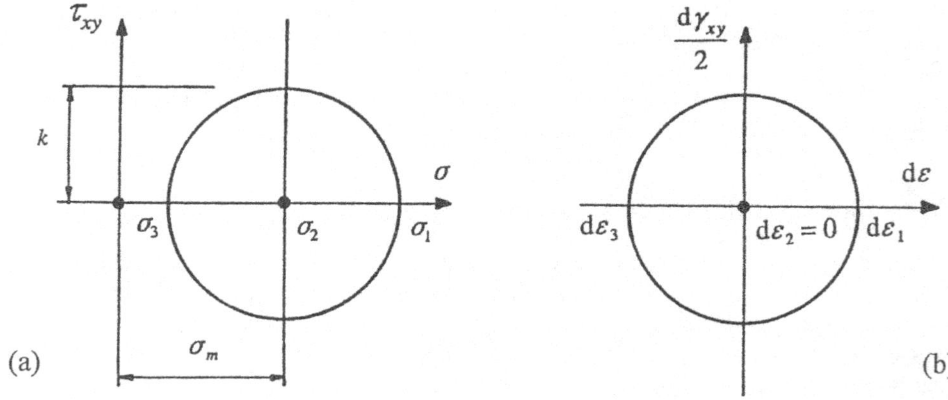

Figure 6.1 Mohr's circles of stress and strain

The centre of Mohr's stress circle follows from eqs(6.2) and (6.4a):

$$\tfrac{1}{2}(\sigma_1 + \sigma_3) = \sigma_2 = \tfrac{1}{2}(\sigma_x + \sigma_y) \tag{6.4c}$$

The circle centre identifies the *mean (or hydrostatic) stress* $\sigma_m = ⅓\, \sigma_{kk}$:

$$\sigma_m = ⅓\,(\sigma_1 + \sigma_2 + \sigma_3)$$
$$= ⅓\,[\sigma_1 + ½(\sigma_1 + \sigma_3) + \sigma_3] = ½\,(\sigma_1 + \sigma_3)$$

which equals the intermediate stress σ_2 in eq(6.2). Now, as the volume of material is conserved $dV/V \simeq d\varepsilon_{kk} = 0$ when, from eqs(6.1a,c), $d\varepsilon_1 = -d\varepsilon_3$. Hence the relationship between $d\varepsilon_1$ and $d\varepsilon_3$ and the component strains $d\varepsilon_x$, $d\varepsilon_y$ and $d\gamma_{xy}$, where also $d\varepsilon_x = -\, d\varepsilon_y$, is given by

$$d\varepsilon_{1,3} = \frac{1}{2}(d\varepsilon_x + d\varepsilon_y) \pm \frac{1}{2}\sqrt{(d\varepsilon_x - d\varepsilon_y)^2 + (d\gamma_{xy})^2}$$

The radius of the strain circle (see Fig. 6.1b) becomes

$$\frac{1}{2}(d\varepsilon_1 - d\varepsilon_3) = \pm\frac{1}{2}\sqrt{(d\varepsilon_x - d\varepsilon_y)^2 + (d\gamma_{xy})^2} \qquad (6.5a)$$

with centre co-ordinate

$$\frac{1}{2}(d\varepsilon_1 + d\varepsilon_3) = d\varepsilon_2 = \frac{1}{2}(d\varepsilon_x + d\varepsilon_y) = 0 \qquad (6.5b)$$

It follows, from eqs(6.4a) and (6.5a), that the stress and strain circles are geometrically similar: (i) for the shear strain ordinate $½d\gamma_{xy}$, given in Fig. 6.1b and (ii) when the mean component of normal stress is subtracted from the stress abscissa in Fig. 6.1a. This does not effect the radius $½[\sigma_1 - \sigma_m) - (\sigma_3 - \sigma_m)]$ of the stress circle, given in eq(6.4a).

The Mohr's circles describe the stress and strain states at a given point. The top and bottom of a vertical diameter represent two maximum shear planes, lying at right angles to each other. Upon these planes of maximum shear, an association is made between: (i) the maximum shear stress aligned with the direction of these planes and (ii) a hydrostatic stress lying normally to the planes, for which there is no direct strain. In the theory, the maximum shear stress τ_{max} has attained the shear yield stress k so that maximum shear planes become known as *slip lines* or *shear lines*. The stress state along a slip line will vary from point to point since the hydrostatic stress σ (subscript m omitted) is a variable. This variation can be observed within a slip-line field construction, shown in Fig. 6.2.

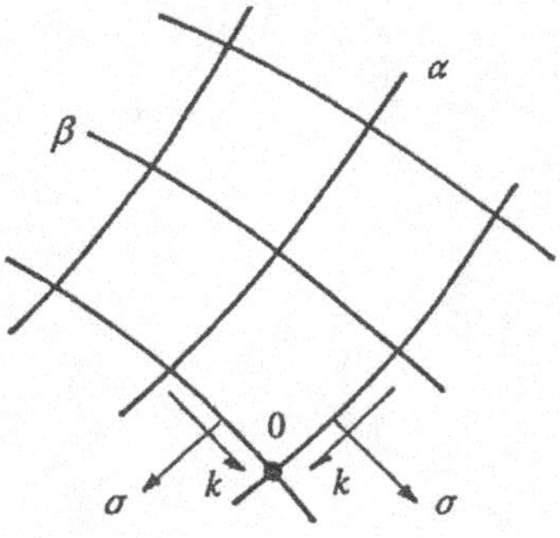

Figure 6.2 A slip line field

The SLF is an orthogonal, curvilinear network of α and β lines. The convention for constructing Mohr's circle for an intersecting pair of α, β lines is that tensile mean stress is positive and a clockwise shear stress is positive, pointing to the right of the slip line normal.

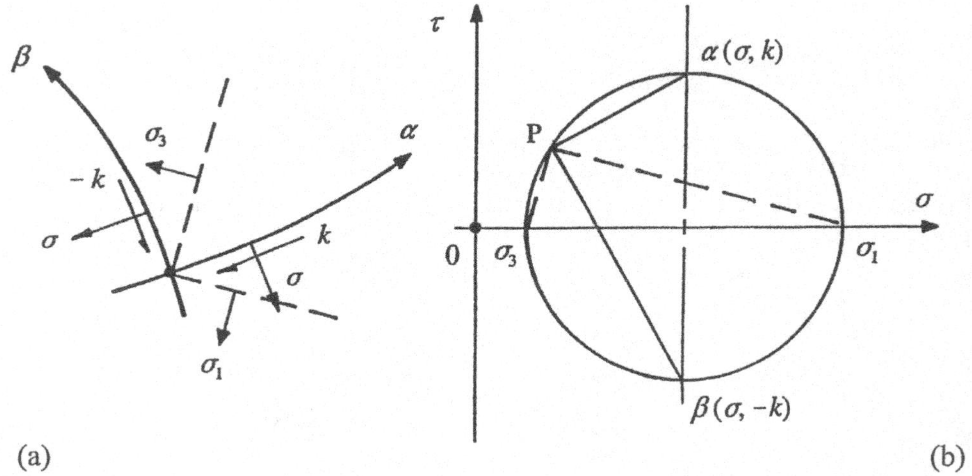

Figure 6.3 Stress state for a pair of intersecting slip lines

It follows from Fig. 6.3a that α–lines are all associated with a positive shear stress ($+k$) and the β–lines with a negative shear stress ($-k$). The corresponding Mohr's circle construction is given in Fig. 6.3b. Firstly, the co-ordinates: α (σ, k), β (σ, $-k$) are located on the circle according to our convention. The circle is drawn with points α and β at opposite ends of a vertical diameter. The directions of α and β are projected through points α and β to intersect the circle in a common pole point, P. Actually, one such projection is sufficient to locate the pole. It follows that the planes on which principal stresses σ_1 and σ_3 act are parallel to the broken lines joining P to σ_1 and P to σ_3. Transferring principal planes to Fig. 6.3a shows that the direction of the major principal stress σ_1 lies between α and β.

6.2.2 Hencky's Theorems

A unique relationship will hold between σ and k for each line. This is found from applying force equilibrium to Cartesian axes x and y, inclined at φ to any pair of slip lines α and β in Fig. 6.4.

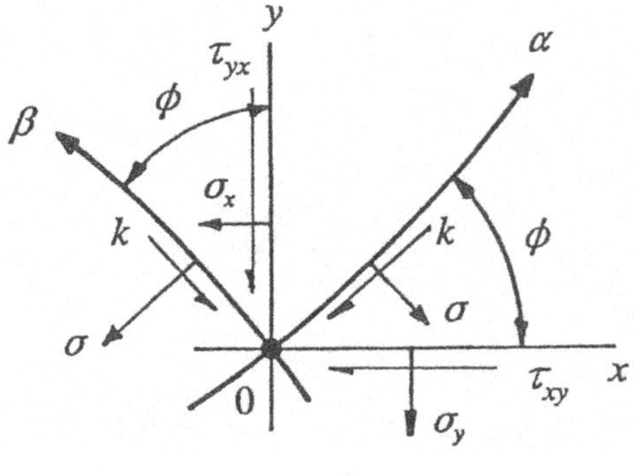

Figure 6.4 Equilibrium in Cartesian axes

The plane equilibrium conditions are referred to x and y axes:

$$\frac{\partial \sigma_x}{\partial x} + \frac{\partial \tau_{xy}}{\partial y} = 0$$
$$\frac{\partial \sigma_y}{\partial y} + \frac{\partial \tau_{yx}}{\partial x} = 0 \qquad (6.6a,b)$$

The Cartesian stress components σ_x, σ_y and τ_{xy} are derived from σ and k by inverting the stress transformation law in eq(1.22a). Firstly, note that α and β become the primed axes along which the known stress components (σ, k) appear in a 2×2 matrix $\mathbf{T}'$. The inverse transformation from $\mathbf{T}'$ to $\mathbf{T}$ in Cartesian axes x, y requires pre- and post-multiplication by $\mathbf{M}^{\mathrm{T}}$ and $\mathbf{M}$ respectively as follows:

$$\mathbf{M}^{\mathrm{T}}\mathbf{T}'\mathbf{M} = \mathbf{M}^{\mathrm{T}}(\mathbf{M}\mathbf{T}\mathbf{M}^{\mathrm{T}})\mathbf{M} = (\mathbf{M}^{\mathrm{T}}\mathbf{M})\mathbf{T}(\mathbf{M}^{\mathrm{T}}\mathbf{M})$$

Given that the orthogonal rotation matrix $\mathbf{M}$ obeys $\mathbf{I} = \mathbf{M}^{\mathrm{T}}\mathbf{M} = \mathbf{M}\mathbf{M}^{\mathrm{T}}$, then:

$$\mathbf{T} = \mathbf{M}^{\mathrm{T}}\mathbf{T}'\mathbf{M}$$

In full:

$$\begin{bmatrix} \sigma_x & \tau_{xy} \\ \tau_{yx} & \sigma_y \end{bmatrix} = \begin{bmatrix} \cos\varphi & -\sin\varphi \\ \sin\varphi & \cos\varphi \end{bmatrix} \begin{bmatrix} \sigma & k \\ k & \sigma \end{bmatrix} \begin{bmatrix} \cos\varphi & \sin\varphi \\ -\sin\varphi & \cos\varphi \end{bmatrix}$$

in which the components of $\mathbf{M}$ are: $l_{11} = \cos(\angle\alpha 0x) = \cos\varphi$, $l_{12} = \cos(\angle\alpha 0y) = \cos(90 - \varphi) = \sin\varphi$, $l_{21} = \cos(\angle\beta 0x) = \cos(90 + \varphi) = -\sin\varphi$ and $l_{22} = \cos(\angle\beta 0y) = \cos\varphi$. The matrix multiplication gives

$$\sigma_x = \sigma - k\sin 2\varphi \qquad (6.7a)$$
$$\sigma_y = \sigma + k\sin 2\varphi \qquad (6.7b)$$
$$\tau_{xy} = \tau_{yx} = k\cos 2\varphi \qquad (6.7c)$$

Substituting eqs(6.7a–c) into eqs(6.6a,b) provides two relationships

$$\frac{\partial}{\partial x}(\sigma - k\sin 2\varphi) + \frac{\partial}{\partial y}(k\cos 2\varphi) = 0$$
$$\frac{\partial}{\partial y}(\sigma + k\sin 2\varphi) + \frac{\partial}{\partial x}(k\cos 2\varphi) = 0$$

In the limit of these equations $\sin 2\varphi \to 0$ and $\cos 2\varphi \to 1$ as $\varphi \to 0$. Then, for the α–line

$$\frac{\partial \sigma}{\partial x} - 2k\frac{\partial \varphi}{\partial x} = 0 \qquad (6.8a)$$

and for the β–line

$$\frac{\partial \sigma}{\partial y} + 2k\frac{\partial \varphi}{\partial y} = 0 \qquad (6.8b)$$

Integrating eqs(6.8a,b), leads to the Hencky equations for any pair of orthogonal slip lines

$$\sigma - 2k\varphi = c_1 \quad \text{and} \quad \sigma + 2k\varphi = c_2$$

where c_1 and c_2 are constants. More usually, σ is replaced by the mean compressive stress ($\sigma = -p$) acting upon α and β to give:

$$p + 2k\varphi = c_1 \quad \text{and} \quad p - 2k\varphi = c_2 \tag{6.9a,b}$$

where a counter clockwise φ is positive. Equations (6.9a,b) lead to Hencky's *first theorem,* which states that when two α–slip lines are cut by β–slip lines (see Fig. 6.5), then the angle subtended by tangents to the α–line at the intersection points (A and B) and (P and Q) is constant along the length of α.

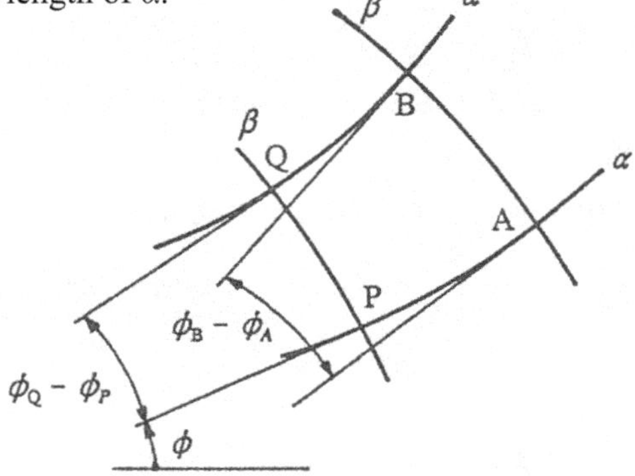

Figure 6.5 Hencky's theorem

That is, $(\varphi_Q - \varphi_P) = (\varphi_B - \varphi_A)$. The proof uses eqs(6.9a,b) with a fixed horizontal datum for φ as shown. Along α, eq(6.9a) gives

$$p_Q + 2k\varphi_Q = c_1 \quad \text{and} \quad p_B + 2k\varphi_B = c_1$$
$$p_P + 2k\varphi_P = c_1 \quad \text{and} \quad p_A + 2k\varphi_A = c_1$$

Along β, eq(6.9b) gives

$$p_Q - 2k\varphi_Q = c_2 \quad \text{and} \quad p_B - 2k\varphi_B = c_2$$
$$p_P - 2k\varphi_P = c_2 \quad \text{and} \quad p_A - 2k\varphi_A = c_2.$$

Now

$$\begin{aligned} p_Q - p_A &= (p_Q - p_B)_\alpha + (p_B - p_A)_\beta \\ &= 2k(\varphi_B - \varphi_Q) + 2k(\varphi_B - \varphi_A) \end{aligned} \tag{6.10a}$$

and

$$\begin{aligned} p_Q - p_A &= (p_Q - p_P)_\beta + (p_P - p_A)_\alpha \\ &= 2k(\varphi_Q - \varphi_P) + 2k(\varphi_A - \varphi_P) \end{aligned} \tag{6.10b}$$

Equating (6.10a,b):

$$2k(\varphi_B - \varphi_Q) + 2k(\varphi_B - \varphi_A) = 2k(\varphi_Q - \varphi_P) + 2k(\varphi_A - \varphi_P)$$

$$\therefore \varphi_Q - \varphi_P = \varphi_B - \varphi_A$$

so confirming a constant angle of intersection.

In his *second theorem*, Hencky related the radius of curvature of a slip line to the distance along it. This has been found useful for numerical solutions to slip line fields as discussed by Hill [4], Prager and Hodge [5] and Chakrabarty [6]. These are not employed here. Instead, Prager's geometrical interpretation [7] of this theorem is employed. This enables a convenient graphical construction, as outlined in the following section.

6.2.3 Stress Plane

To follow Prager's geometrical construction of Hencky's eqs (6.9a,b), consider firstly the variation in stress between points 1 and 2 along an α–line in Fig. 6.6a.

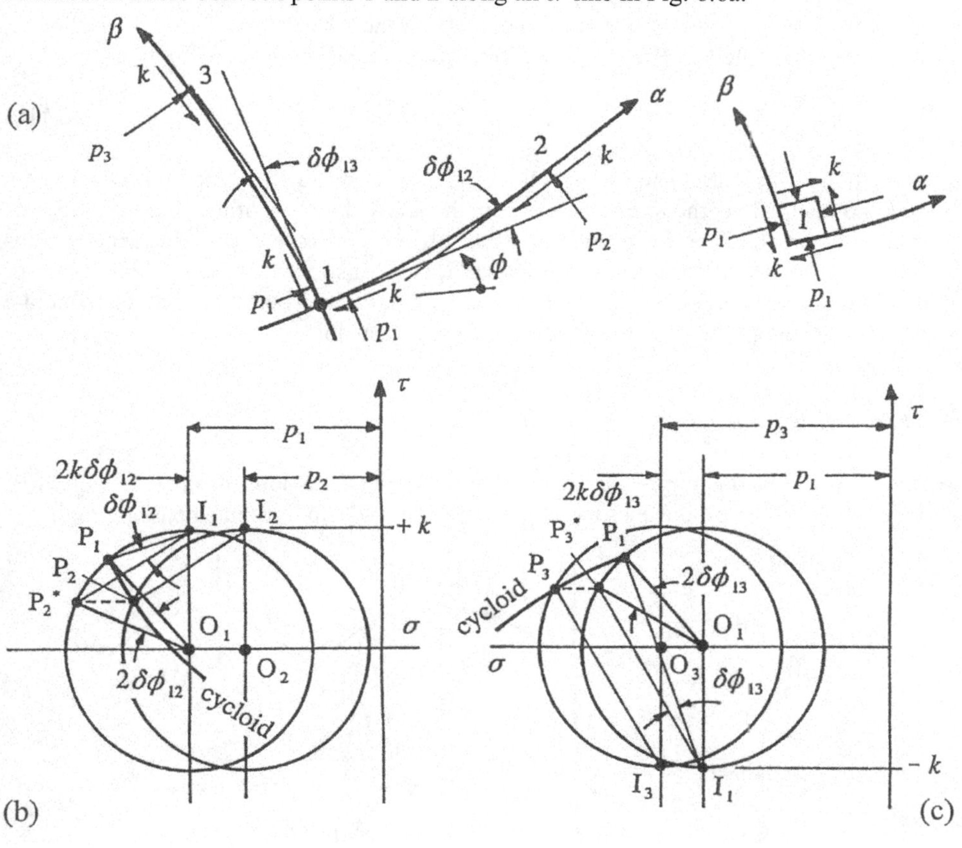

Figure 6.6 Prager's geometrical interpretation

If there is no variation in p, the constant stress state $(-p_1, k)$ appears as a point I_1 on a fixed circle with centre O_1. The pole point P_1 for this circle is the intersection point, found from projecting the direction of the plane on which $-p$ and k act, through I_1 (see Fig. 6.6b). One could also have arrived at P_1 from the constant stress state $(-p, -k)$ along the β–line, for which I_1 would have lain at the opposite end of the vertical diameter. Now let the mean pressure vary along each slip line: from point 1 to 2 along α and from 1 to 3 along β (as in Fig. 6.6a). Let the stress state at point 2 on the α–line be $(-p_2, k)$, where $p_1 \geq p_2$. This state locates a point I_2 on another circle, with centre O_2 and pole P_2. If the positive, anti-clockwise (ACW) difference between the slopes of the tangents at points 1 and 2 is $\delta\varphi_{12}$, then the pressure difference is found from eq(6.9a) to be

$$p_2 - p_1 = -\ 2k\delta\varphi_{12} \tag{6.11a}$$

confirming that p_2 is less than p_1. As $\delta\varphi_{12} = P_1 I_1 P_2^*$ in Fig. 6.6b, this locates the image, P_2^* of P_2, on the original circle such that the angle subtended at the centre is $P_1 O_1 P_2^* = 2\delta\varphi_{12}$.

The image point P_2^* reveals that a translation from P_1 to P_2 could be obtained from rolling the original circle without slip along the ordinate $\tau = +k$. The geometrical interpretation of eq(6.11a) is that the pole will trace a cycloid $P_1 P_2 \ldots$ by rolling in the same direction as the changing slope of the α–line (ACW). A similar conclusion may be drawn from stress state $(p_3, -k)$ at point 3 on the β–line (Fig. 6.6a). Here the circle construction, in Fig. 6.6c, locates I_3 and the pole P_3. The pressure difference is found, from eq(6.9b), to be

$$p_3 - p_1 = 2k\delta\varphi_{12} \tag{6.11b}$$

showing that p_3 is greater than p_1. A cycloid $P_1 P_3 \ldots$ is generated by the pole as this circle rolls without slip along the ordinate $\tau = -k$, in the ACW direction of the changing slope of the β–line. Thus, slip lines in the physical plane will map into a system of congruent cycloids in Mohr's stress plane. In each of Figs 6.6b,c, I_1 is the instantaneous centre of rotation of point P_1 and $I_1 P_1$ is parallel to the respective α– and β–lines. Therefore, each element of a slip line is orthogonal to its corresponding element of cycloid.

6.2.4 Stress Discontinuities

Where, in Fig. 6.7a, the normal and shear stress σ_N and τ for a point A on an irregular free surface are known to produce a state of yielding, the α, β slip lines must pass through A.

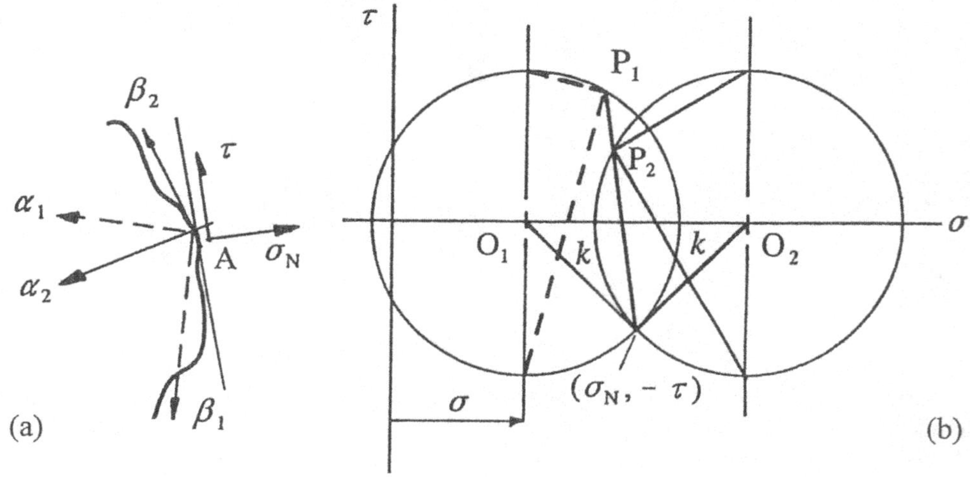

Figure 6.7 Strong and weak circles

However, the stress plane construction in Fig. 6.7b shows that α and β are not defined uniquely. Marking the radius k from the stress point $(\sigma_N, -\tau)$ in (b) locates the centres, O_1 and O_2, of two circles with respective poles P_1 and P_2. Joining the pole to points on the vertical diameter of each circle shows that there are two sets of possible slip lines, one pair (α_1, β_1), being weak (broken lines), the other pair (α_2, β_2) being strong (continuous lines). The latter contains the larger hydrostatic stress σ at O_2. Recall that the α–line is associated with $+k$ and the β - line with $-k$. Thus, in the physical plane, the two α–lines are drawn parallel to the lines from P_1 and P_2 to the top of each circle. The two β–lines are parallel to the lines joining P_1 and P_2 to the bottom of each circle. It will be seen that it is usually possible to judge which of the two solutions is the appropriate one for a particular problem.

Next, consider the curved boundary between two plastic zones in which the adjacent material experiences different hoop tensions, i.e. $\sigma_x \neq \sigma_x'$ in Fig. 6.8a. Equilibrium is maintained through a continuity in both radial stress σ_y and in shear stress ($\tau_{xy} = \tau_{yx}$) across the boundary.

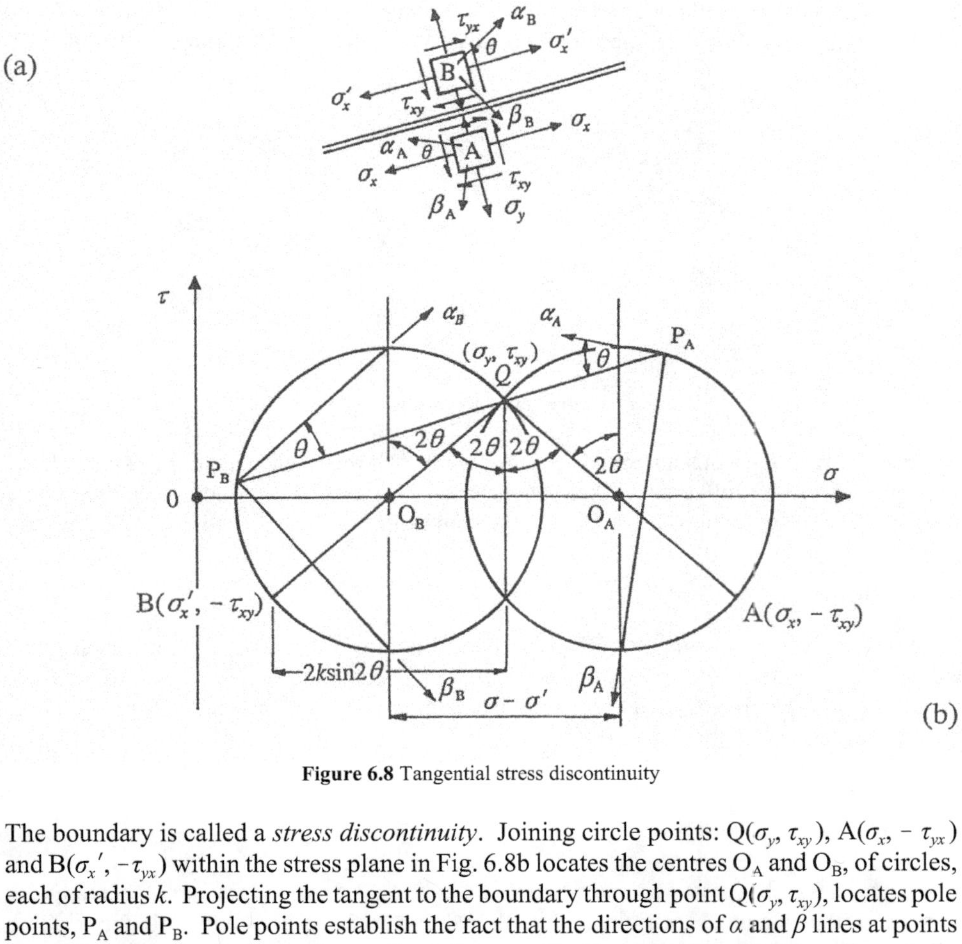

Figure 6.8 Tangential stress discontinuity

The boundary is called a *stress discontinuity*. Joining circle points: $Q(\sigma_y, \tau_{xy})$, $A(\sigma_x, -\tau_{yx})$ and $B(\sigma_x', -\tau_{yx})$ within the stress plane in Fig. 6.8b locates the centres O_A and O_B, of circles, each of radius k. Projecting the tangent to the boundary through point $Q(\sigma_y, \tau_{xy})$, locates pole points, P_A and P_B. Pole points establish the fact that the directions of α and β lines at points A and B in the physical plane are reflected across the discontinuity. That is, adjacent slip lines α_A and α_B for maximum positive shear stress lie with an equal inclination θ to the discontinuity. Also, the orthogonal slip lines β_A and β_B for the maximum negative shear stress will lie with an equal inclination $\theta + 90°$ to the discontinuity (see Fig. 6.8a). Generally, a stress discontinuity can coincide neither with a slip-line nor, as it will be shown, with a velocity discontinuity. The magnitude of the stress discontinuity is, from the circles

$$\sigma_x - \sigma_x' = 4k \sin 2\theta$$

where θ is the inclination of the reflection α_A to α_B across the discontinuity. When $\theta = 0°$, there is no discontinuity and the circles coincide. When $\theta = 45°$, the maximum difference arises, $\sigma_x - \sigma_x' = 4k$, as the circles become tangential at point Q lying on the σ axis. Here, the greatest difference between the mean stress components is $\sigma - \sigma' = 2k$. Figure 6.8b refers to material that has yielded on either side of the discontinuity band while the interior of the band may remain elastic. As the band is penetrated from both sides σ_x' and σ_x will change continuously until they equalise. This must be so if circles, with diminishing radii less than k, pass through the common point Q, to ensure σ_y and τ_{xy} remain unchanged.

Where there is an axis of symmetry (a centre line) in the physical plane then this axis and another axis perpendicular to it will co-incide with the principal stress planes. Planes of maximum shear stress, i.e. the slip lines, lie at 45° to the principal planes (see Fig. 6.9).

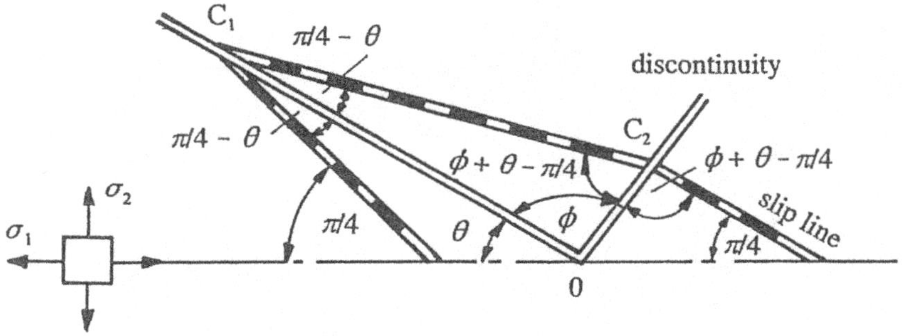

Figure 6.9 Axis of symmetry

Figure 6.9 also shows that a stress discontinuity must be reflected at an axis of symmetry if slip lines are to be reflected at their intersections with discontinuities. For the simple configuration of reflections at C_1 and C_2, the sum of included angles within OC_1C_2 gives

$$(\pi/4 - \theta) + \varphi + (\varphi + \theta - \pi/4) = \pi$$

from which $\varphi = 90°$, showing that OC_1 is perpendicular to OC_2. The full picture displays the mirror image of these slip lines and discontinuities beneath the horizontal symmetry axis. *Note*: It is possible for OC_1 and OC_2 to subtend angles of $\pi/4$ and $3\pi/4$ at the symmetry axis and so co-incide with the slip lines [8].

6.2.5 Geiringer's Equations

Shear strain, which is at its maximum between slip lines, is a radian measure of their angular distortion (the change to the right angle). A hydrostatic stress does not strain the slip line permanently. A further pair of compatibility relationships will ensure that both α and β slip lines are inextensible (see Fig. 6.10).

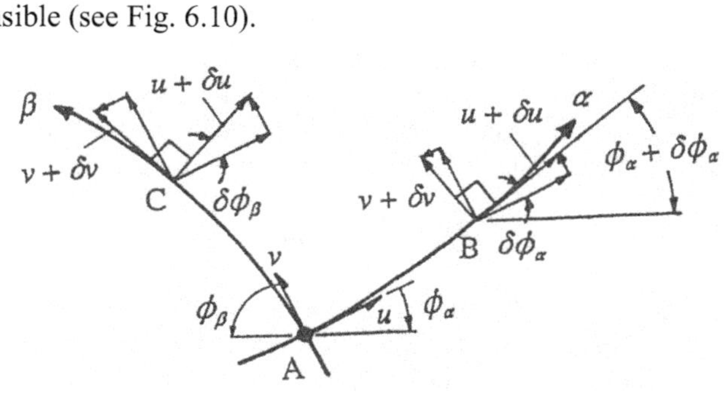

Figure 6.10 Velocity variation

Let u and v be the respective velocities in the directions of α and β at point A. These velocities increase by δu and δv at point B along the α–line as shown. Since the velocities $u + \delta u$ and $v + \delta v$ at B have rotated by $\delta\varphi_\alpha$ they must be resolved in the directions of u and v at A. The fact that the α–line does not change its length is expressed in

$$(u + \delta u)\cos\delta\varphi_\alpha - (v + \delta v)\sin\delta\varphi_\alpha = u$$

Similarly at point C, inextensibility in the β–line is stated as

$$(v + \delta v)\cos\delta\varphi_\beta + (u + \delta u)\sin\delta\varphi_\beta = v$$

As $\delta\varphi \to 0$, $\cos\delta\varphi \to 1$ and $\sin\delta\varphi \to \delta\varphi$. Ignoring the products of infinitesimals leads to the *Geiringer equations* [3].

$$\frac{\partial u}{\partial \varphi_\alpha} - v = 0$$
$$\frac{\partial v}{\partial \varphi_\beta} + u = 0 \qquad (6.12a,b)$$

Note: Tangential velocities, u and v, when normalised by a given length become strain rates.

6.2.6 Hodograph (Velocity Plane)

The hodograph is a graphical construction of the tangential velocities u and v which satisfy eqs(6.12a,b). Let resultant velocities, v_A and v_B, apply to respective points A and B along an α–line (see Fig. 6.11a). It is seen in Fig. 6.11b that their projections onto AB must be equal if the initially straight length of AB is not to change.

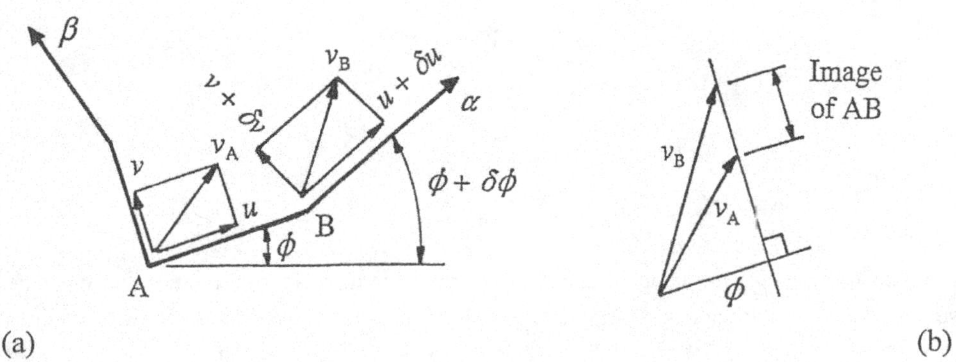

Figure 6.11 Hodograph for a portion AB of a straight slip line

The line joining v_A to v_B in Fig. 6.11b is the orthogonal image of the slip line between A and B in the physical plane (Fig. 6.11a). Because elements in the physical plane and stress plane are orthogonal, it follows that elements in the hodograph and the stress plane will be parallel. Where the α–slip line is curved between A and B (see Fig. 6.12a), it may be deduced that the curve joining v_A to v_B in Fig. 6.12b becomes the orthogonal image of the slip line.

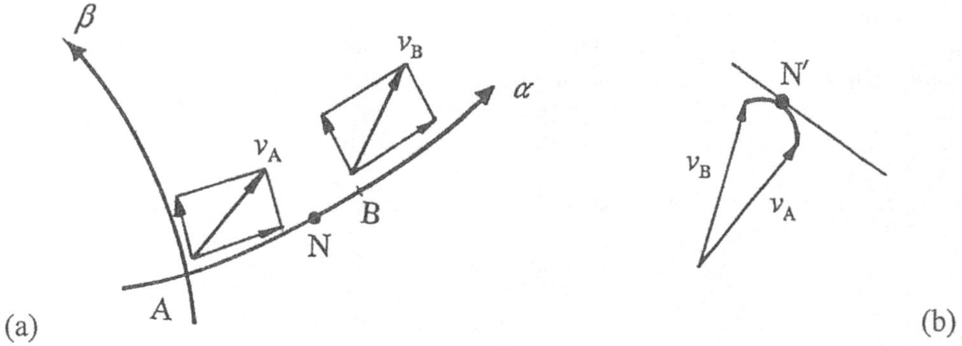

Figure 6.12 Hodograph for a curved slip line

The construction in which the tangent at point N in Fig. 6.12a is orthogonal to the tangent at point N′ in Fig. 6.12b ensures that the initially curved length of AB does not change.

6.2.7 Velocity Discontinuities

A velocity discontinuity can arise when, in Fig. 6.13, the tangential velocities on either side of the discontinuity are unequal, i.e. $u \neq u'$.

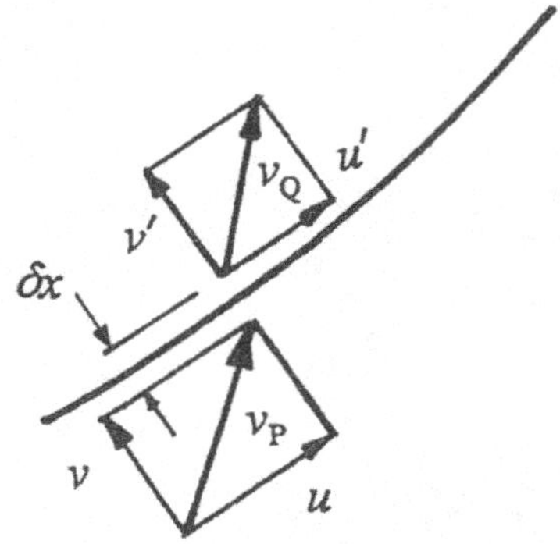

Figure 6.13 Tangential velocity discontinuity

Let the variation $\delta u = u - u'$ occur in a distance δx that straddles the discontinuity. It can be seen that as $\delta x \to 0$, the shear strain rate $\delta u/\delta x \to \infty$. One may conclude from this that lines of velocity discontinuity and slip lines are coincident. Applying Geiringer's eq(6.12a,b) to regions adjacent to the discontinuity gives

$$\mathrm{d}u = v\,\mathrm{d}\varphi \quad \text{and} \quad \mathrm{d}u' = v'\,\mathrm{d}\varphi \tag{6.13a,b}$$

Normal velocities must remain equal, i.e. $v = v'$, if the material is to deform without volume change. Equations (6.13a,b) give

$$\mathrm{d}u = \mathrm{d}u' \quad \text{or} \quad u - u' = \text{constant} \tag{6.14}$$

indicating that the velocity discontinuity has a constant magnitude. Since $u \neq u'$, the resultant velocities, v_P and v_Q in Fig. 6.13a, will differ in both magnitude and direction as the slip line is crossed. Where u and u' vary along the slip line, the magnitudes and directions of the resultant velocities are not constant. Because $v = v'$, eq(6.14) implies that the vector sum of v_P and v_Q will be constant at all points along the slip line/velocity discontinuity. This is seen readily within a hodograph construction (Fig. 6.14b) for a length of α–slip line. Here the resultant velocities are v_{P1}, v_{Q1} at point 1 and v_{P2}, v_{Q2} at point 2, as shown in Fig. 6.14a.

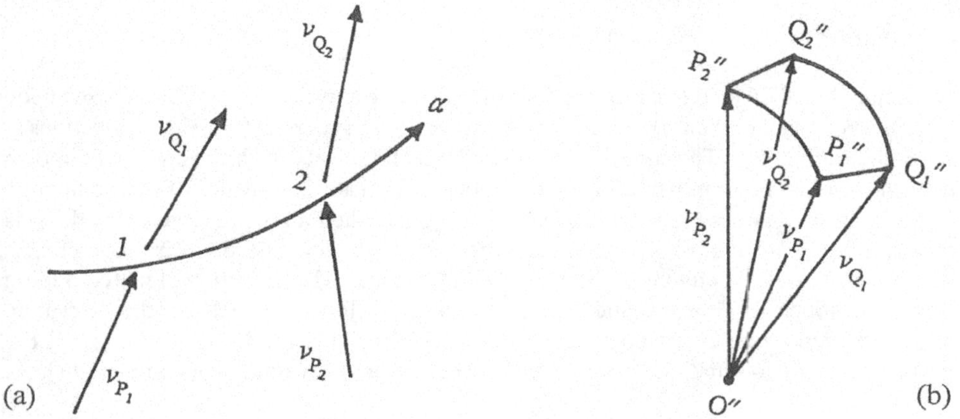

Figure 6.14 Equal velocity discontinuities

When constructing the hodograph in Fig. 6.14b, the velocities v_P and v_Q, are drawn through a common origin O″, parallel to velocity vectors in the physical plane. The magnitudes of the velocities are determined from two requirements:

(i) that $P_1''Q_1''$ and $P_2''Q_2''$ are equal and parallel to the respective tangents at points 1 and 2. Their magnitudes represent the jump in the tangential u component of velocity from crossing the α line at points 1 and 2 (see eq 6.14),

(ii) that the inextensibility of the α line requires $P_1''P_2''$ and $Q_1''Q_2''$ to be orthogonal images of the slip line between points 1 and 2.

It is possible for u, as well as v, on one side of a slip line to be uniform when the material moves as a rigid body. The resultant velocity is then constant, as shown by v_P in Fig. 6.15a.

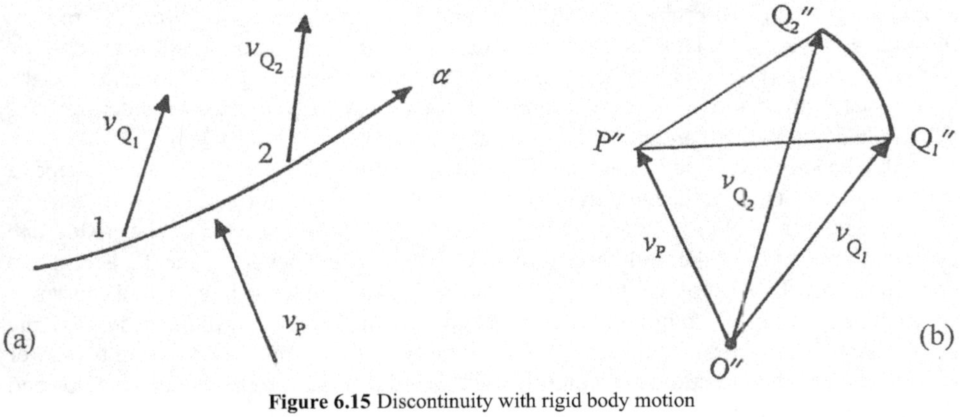

Figure 6.15 Discontinuity with rigid body motion

Here u' varies between points 1 and 2, on the other side of the slip line. The resultant velocities v_{Q1} and v_{Q2}, will again differ in magnitude and in direction as shown. The simplified hodograph (see Fig. 6.15b) shows that $Q_1''Q_2''$ is the orthogonal image of slip line 12 and $P''Q_1'' = P''Q_2''$ is the jump in the tangential velocity at 1 and 2. This hodograph reveals that when v_P is constant then v_Q must vary. It is not possible, therefore, for the material on both sides of the slip line to move as rigid bodies between points 1 and 2.

6.2.8 Slip Line Field Construction

The general method employed for the determination of a slip-line field (SLF) is based upon what is known about their 45° intersections with axes of symmetry and their reflections at stress discontinuities. The stress plane and hodograph follow from knowing that their own elements are parallel but perpendicular to those in the physical plane. If the hodograph closes then it is a possible solution, but is valid only when a positive rate of working is confirmed. Finally, it must be checked that the stress state within dead metal zones remains elastic. In this way, a unique solution can be established. The distortion of material lying within the slipped region is found from integrating slip-line velocities and the deforming forces are found from integrating the slip-line stress distributions. In constructing a SLF it is also necessary to satisfy the following three conditions prevailing at a boundary [1].

(a) To determine the SLF for interior yielding given surface stress components.

In the physical plane of Fig. 6.16a, the normal and shear stresses stress are known for points 1 and 2 on a curved boundary. It is possible to construct the stress plane in Fig. 6.16b where stress co-ordinates locate points 1′ and 2′ on circles each of radius k. Pole points P_1 and P_2 are found from the projections of tangents to the boundary at 1 and 2. The directions of the α– and β–lines lie parallel to lines connecting P_1 and P_2 to $\pm k$. In this case, the α–line is taken from circle 1 and the β–line from circle 2. The α–line may be extended from the cycloid generated by point P_1 as circle 1 rolls along $\tau = +k$ in Fig. 6.16b. Similarly, P_2 traces a cycloid when circle 2 is rolled along $\tau = -k$. Cycloid intersection occurs at point a′. The corresponding intersection point a in the physical plane is constructed from the tangents at a′ and $P_{1,2}$ with adjustment for equal lengths aside the change in slip line direction (Fig. 6.16a). In the complete field, both the α– and β–lines intersect at the boundary points 1 and 2. The remaining slip-lines (broken lines) connect pole points to points of maximum shear. At their intersections, the α– and β–lines are subjected to the same hydrostatic stress. The *strong* solution to this problem would correspond to the reflection of circles of radii k, through points 1 and 2. This corresponds to an α–line with the greatest possible hydrostatic stress at point 1. It is possible to select the appropriate solution to a given problem from the known deformation mode. *Note*: Where surface tractions are given, these are interpreted as the resultant stresses σ_{R1} and σ_{R2} in Fig. 6.16a.

A given absolute velocity v of boundary material implies that points 1, 2 and c_A all move with velocity v. Thus, in the hodograph (Fig. 6.16c) the corresponding points, 1″ 2″ and c_A'', show a velocity v relative to a stationary origin, O″. The remainder of the hodograph construction employs the orthogonal relation between elements of the SLF and the corresponding elements of the hodograph. The additional points, a, b, c_B, d and e, assume the lower positions shown in Fig. 6.16a. This ensures that, when a slip line is crossed, the proper identity is made with a velocity discontinuity. For example, in crossing the lower β–line edb, equal magnitude discontinuities are found from joining points 2″e″, c_B''d″ and a″b″ in the hodograph.

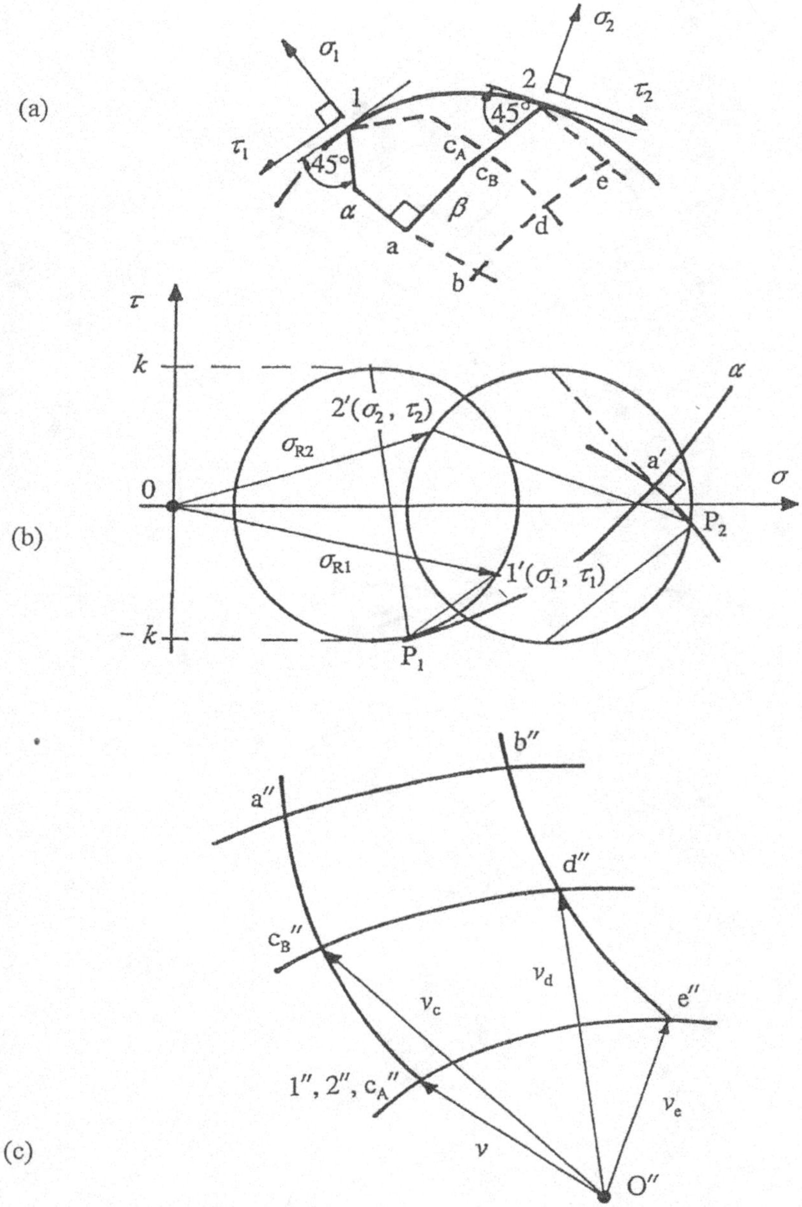

Figure 6.16 Known boundary stresses

In crossing the lower α–slip line ab, the corresponding, equal-magnitude, velocity discontinuities are: a″c_B″ and b″d″. Absolute velocities, v_c, v_d and v_e, are found from joining the respective points, c″, d″ and e″, to the origin O″.

(b) To determine the SLF between two, bounding, intersecting α, β - slip lines.

Firstly, establish the circle for the stress state at the intersection point 1 in Fig. 6.17a. As in (a) the pole position 1' in Fig. 6.17b follows from projecting tangents to α and β, at 1.

(a)

(b)

(c)

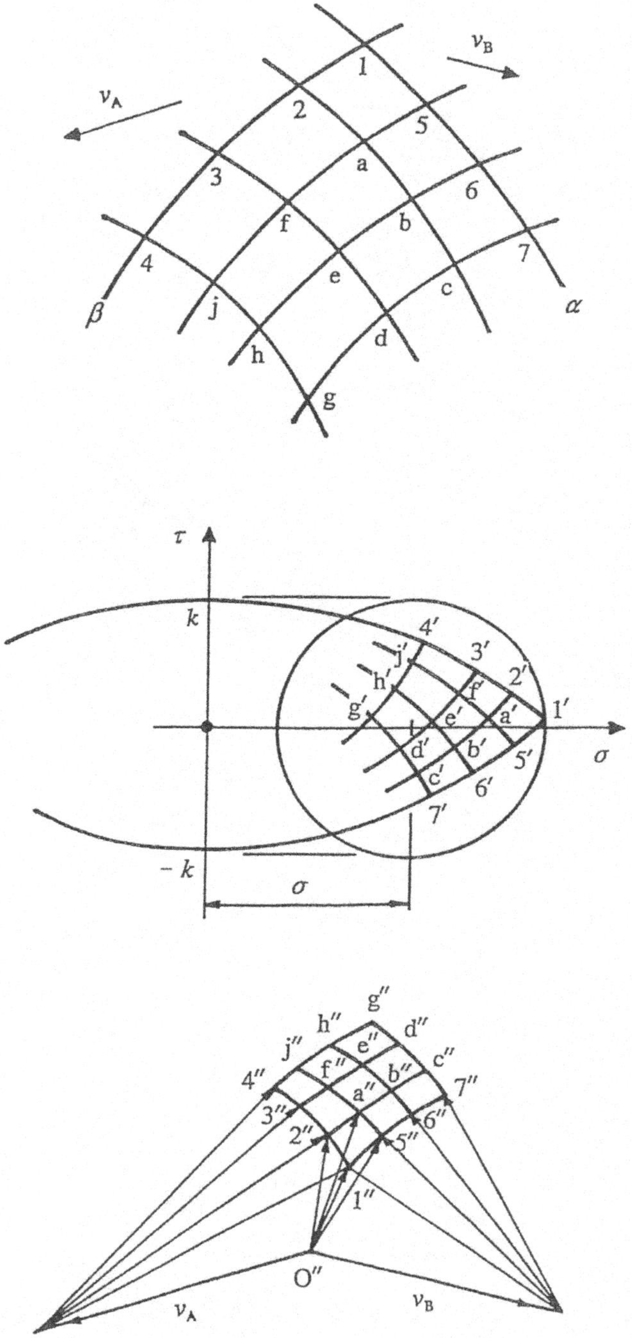

Figure 6.17 Bounding intersecting slip lines

Correspondence must apply between the normals at additional points 2, 3, 4, 5, 6 and 7 in the physical plane and the tangents to these points in the stress plane. The tangents to the slip lines rotate clockwise (CW) for the α–line and anti-clockwise (ACW) for the β–line. It follows that the circle must roll ACW along $\tau = -k$ through β–points 2′, 3′ and 4′ and CW along $\tau = +k$ through α–points 5′, 6′ and 7′. Interior cycloid intersections at a′, b′, c′, d′, e′, f′, g′, h′ and j′ are found from reversing the direction of rotation at points 2′, 3′, 4′, 5′, 6′ and 7′ about the ordinate $\tau = -k$. Tangents to the orthogonal network of cycloids are then used to construct the SLF by extending slip lines, e.g. point a is found from points 2 and 5, b from a and 6, c from b and 7 etc.

Where rigid body velocities v_A and v_B, exterior to the boundary, are known and α– and β–lines are discontinuities, it is possible to construct the hodograph for material enclosed by the slip lines. In the physical plane, the SLF network above is positioned within lower regions where velocities are unknown. The absolute velocities v_A and v_B are first drawn in Fig. 6.17c. Point 1″ is determined from the intersection between tangents to α and β at 1. Points 2″, 3″ and 4″ follow from the direction of the tangents to points 2, 3 and 4, their magnitudes being equal. Points 5″, 6″ and 7″ are established similarly. Point a″ is then derived from 2″ and 5″ using the tangents at corresponding points in the physical plane. The construction for the remaining intersection points within the hodograph is similar to that used for extending slip lines. This is because the interior SLF must be an orthogonal image of the hodograph. The resulting hodograph becomes a composite of the two discontinuities discussed separately in Figs 6.14 and 6.15. For example, the construction in Fig. 6.14 applies to crossing an interior slip line 5afj. Thus, in Fig. 6.17c, constant magnitude velocity discontinuities are found from joining 1″5″, 2″a″, 3″f″ and 4″j″. The absolute velocities on adjacent sides of this slip line are found from joining points 1″, 5″, 2″, a″, 3″, f″, 4″ and j″ to the origin 0″. Figure 6.15 applies to the absolute velocities in crossing the outer α–line at points 1, 5, 6 and 7, i.e. their magnitudes are found from joining 1″, 5″, 6″ and 7″ to O″ in the hodograph. Fig. 6.17c shows that equal tangential velocity discontinuities along α appear as a fan of vectors originating from the end of v_B.

(c) To construct the SLF where a given slip line meets an axis of symmetry.

In Fig. 6.18a, an inclination of 45° exists between the given β–line and the symmetry axis, the latter being a principal stress direction. The circle in Fig. 6.18b applies to the stress state (σ, $-k$) at point 1. The pole 1′ is located from the tangent direction to β at point 1. As the circle rolls CW along $\tau = -k$, the pole traces the portion of cycloid 1′, 2′, 3′ as shown. For the intermediate positions 2′ and 3′, the circle is rolled ACW along $\tau = k$ to trace portions of orthogonal cycloids which meet the axis of symmetry at a′ and c′. Lastly, a circle with pole a′ is rolled CW along $\tau = -k$ to give the image of the β–line a′ b′. Tangents to β at points 2′ and a′ in the stress plane locate the positions of 2 and a in Fig. 6.18a. The remaining points 3, b and c are found from the tangents at points 3′, b′ and c′ in a similar manner.

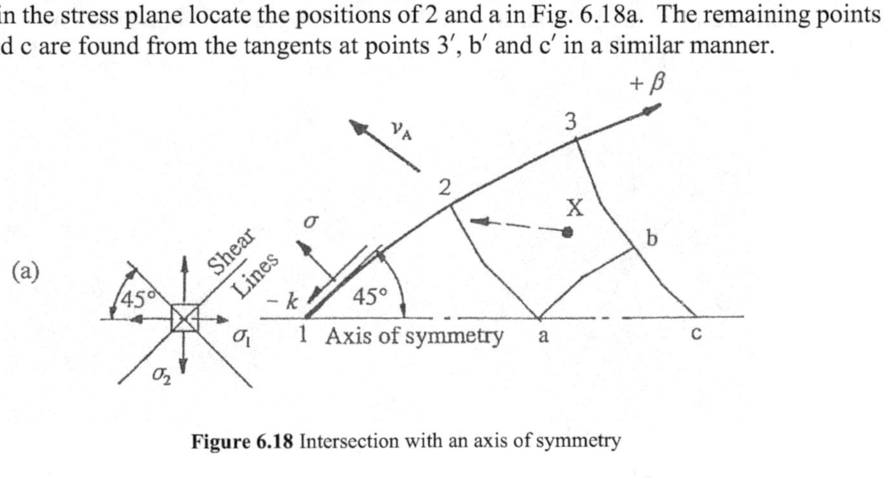

Figure 6.18 Intersection with an axis of symmetry

(b)

(c)

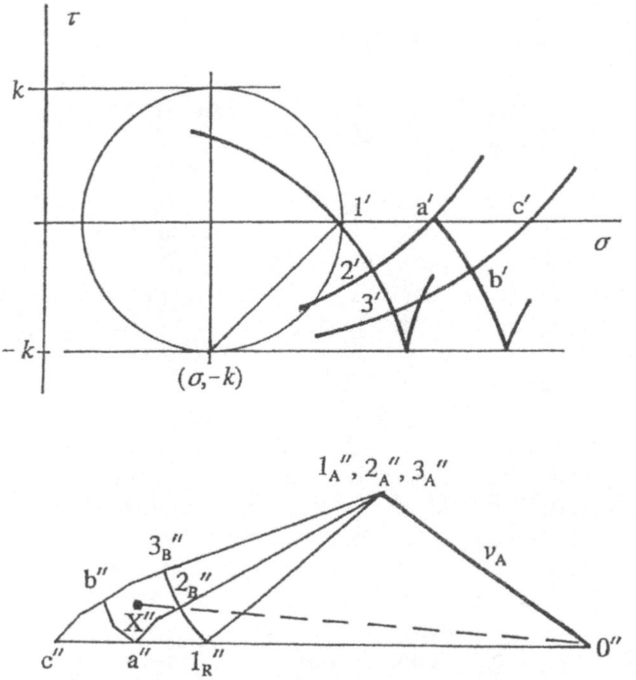

Figure 6.18 Intersection with an axis of symmetry (contd)

With a rigid body velocity v_A, exterior to the β–line in Fig. 6.18a, a fan of discontinuities ($1_A''\ 1_R''$, $2_A''\ 2_B''$ and $3_A''\ 3_B''$) in Fig. 6.18c originates from the end of v_A ($1_A''$, $2_A''$, $3_A''$). Subscripts refer to positions above, right and below the given point. The hodograph is extended to points a″, b″ and c″ employing parallel elements from the stress plane as shown. The absolute velocity of any arbitary point X is found from joining the corresponding point X″ in the hodograph to its origin O″.

6.2.9 *Resultant Force and Pressure on a Curved Slip Line*

The stress plane supplies stress states for points along a curved slip line. Where horizontal forming forces are required, the stress distribution is integrated by one of two methods that follow.

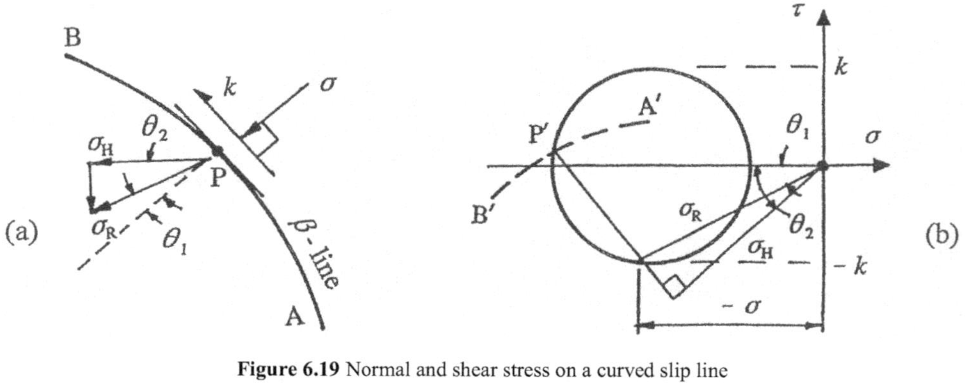

Figure 6.19 Normal and shear stress on a curved slip line

(a) Resultant Stress

Consider a point P on a β slip-line AB in Fig. 6.19a. In the stress plane, the corresponding circle (Fig. 6.19b) has its pole at P′. The stress state at P ($-\sigma$, $-k$) in (a) follows from projecting the tangent to P through P' in Fig. 6.19b. This intersects the circle with these co-ordinates ($-\sigma$, $-k$), confirming AB as a β–line. Let the resultant stress (force on unit area) at P, i.e. $\sigma_R = \surd(\sigma^2 + k^2)$, have a horizontal component σ_H. We may regard σ_R and σ_H as forces/unit length of slip line when the thickness is unity. The construction for σ_R and σ_H shown applies to a given cycloidal image of the slip line within the stress plane. Because the cycloid is the locus of pole points, it is only necessary to connect the pole either to $+k$, for an α–line, or to $-k$, for a β–line, on the corresponding circle. The length of the perpendicular, dropped from the stress origin to the extension of these lines (see Fig. 6.19b), is the required component of horizontal stress σ_H. Taking other points between A′ and B′, we can repeat this construction to establish the variation of σ_H over AB, shown in Fig. 6.20.

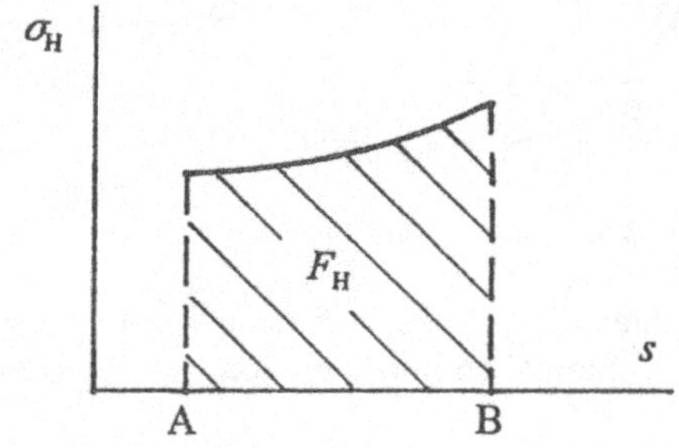

Figure 6.20 Resultant horizontal force on a slip line

It follows that the net horizontal force on a slip line is

$$F_H = \int_{s_A}^{s_B} \sigma_H \, \mathrm{d}s \tag{6.15}$$

which is the area beneath Fig. 6.20, when drawn on the true, base length s of AB.

(b) Cartesian Stress

In Fig. 6.21a, the portion AB of β–slip line is approximated with a series of vertical and horizontal planes to reveal Cartesian stress components σ_x, $\tau_{xy} = \tau_{xy}$ as shown.

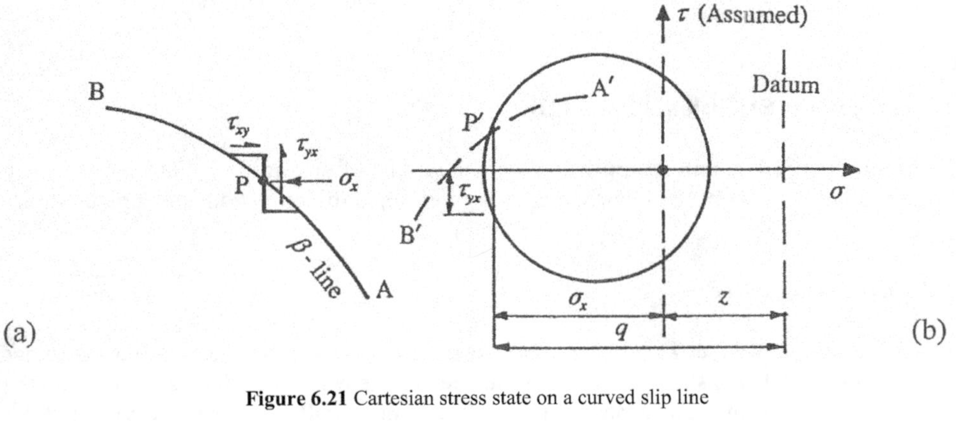

Figure 6.21 Cartesian stress state on a curved slip line

The plane stress state P(σ_x, τ_{yx}) upon the vertical plane lies at the intersection between a vertical line through pole P′ and the circle in Fig. 6.21b. Repeating this construction for a number of poles between A' and B' will establish the variations in σ_x and τ_{xy} with the projected lengths of AB, i.e. a and b in directions x and y respectively (see Fig. 6.22).

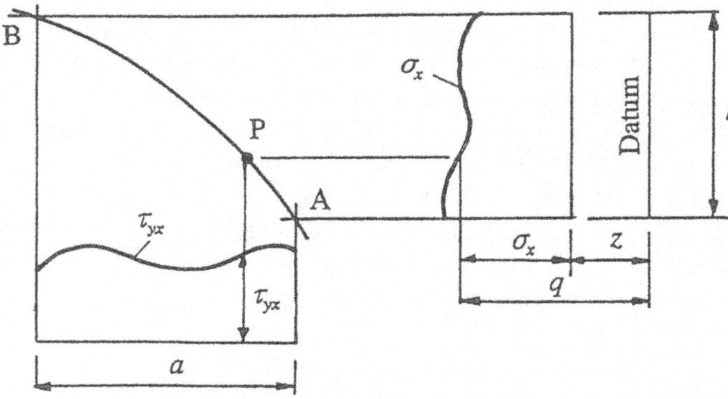

Figure 6.22 Stress component variations

Now τ_{xy}, upon the horizontal plane (Fig. 6.21a), is complementary to τ_{yx} and acts in opposition to σ_x. The net horizontal compressive force is found from

$$F_H = \int_0^b \sigma_x \, dy - \int_0^a \tau_{xy} \, dx \tag{6.16a}$$

Equation(6.16a) expresses the difference between the respective enclosed areas in Fig. 6.22. When the origin of stress is unknown, but F_H is known, one may work from an assumed origin for τ. Let $q = \sigma_x + z$ locate be the true datum for τ, so modifying eq(6.16a) to

$$F_H = \int_0^b \sigma_x \, dy + \int_0^b z \, dy - \int_0^a \tau_{xy} \, dx \tag{6.16b}$$

When $F_H = 0$ in eq(6.16b), the separation z between the true and assumed τ axes becomes

$$z = \frac{1}{b}\left[\int_0^a \tau_{xy} \, dx - \int_0^b \sigma_x \, dy\right] \tag{6.16c}$$

6.3 Frictionless Extrusion, Parallel Dies

The following SLF solutions apply to a rectangular die of unspecified width for the given height reduction ratio $R = H/h$. Force analyses given apply to a unit width under plane strain.

6.3.1 Extrusion Ratio R = 2

In a lubricated vessel (see Fig. 6.23a) there are no frictional forces parallel to the container walls and so interfacial shear stress is absent. Hence, the walls are principal planes and the slip-lines, i.e. maximum shear planes, meet the container walls (points 0 and 5) at 45°.

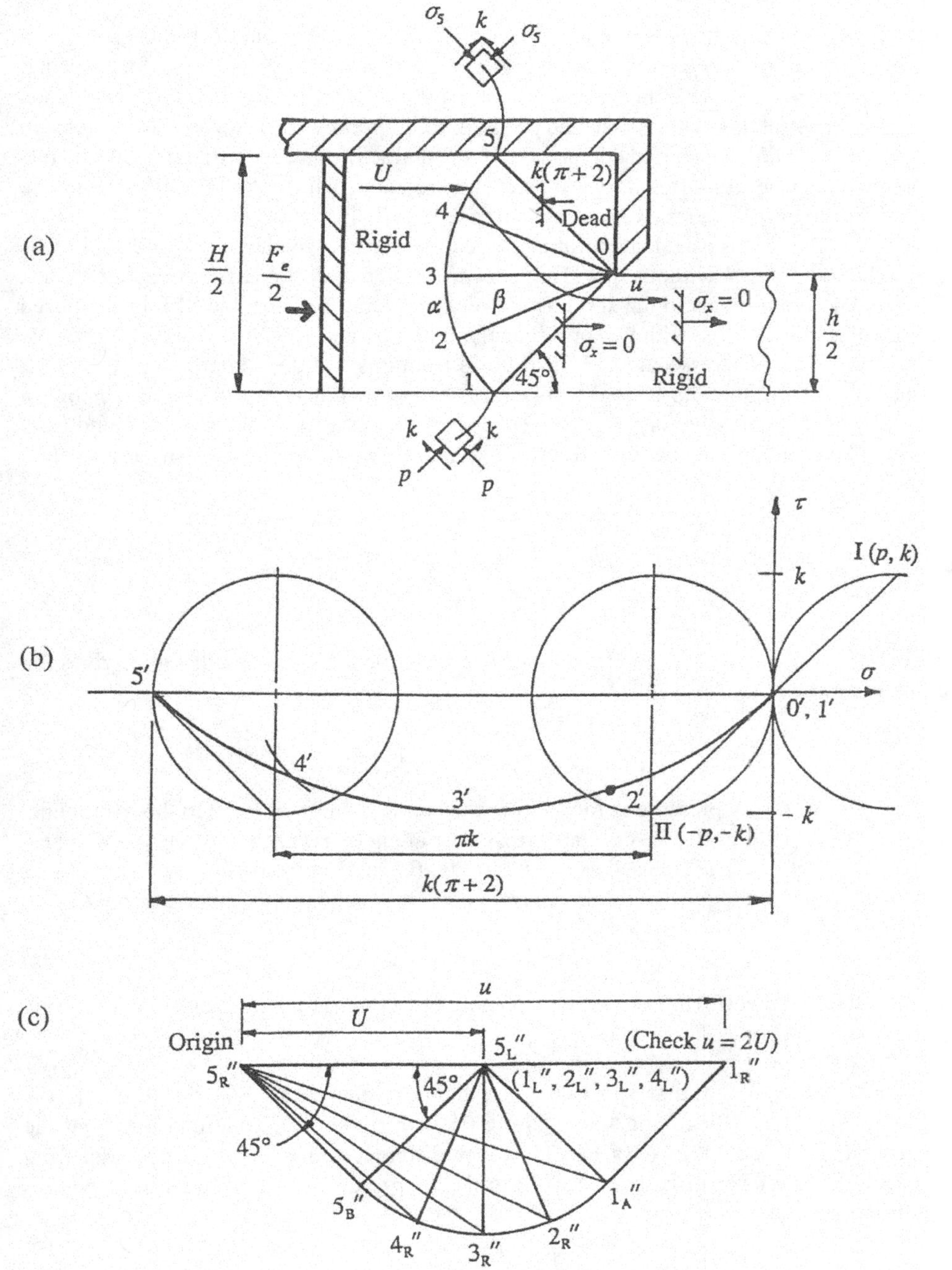

Figure 6.23 Frictionless extrusion $R = 2$

The slip line, joining points 0 and 5 in the physical plane (see Fig. 6.23a), is assumed to be straight. Another slip line 01 must be orthogonal to 05 at 0 and meet the centre-line at 45°. The problem is then to construct the slip line connecting points 1 to 5. This begin with the fact that the major principal stress σ_x, normal to the free-end of the extruded billet, is zero.

This determines that the pole point $0'1'$ is coincident with the origin of the stress plane in Fig. 6.23b. In order to construct the initial circle, the complete stress state for a second plane is required. Because the shear stress along 01 is k, two circles, I and II, apply when the mean stress for this plane is assumed tensile and compressive respectively. Mean tension is impossible under the applied compressive force and so the weaker circle II applies to this problem. The true stress state for 01 is found by projecting a line parallel to 01 through the pole. This gives $p = -k$, $\tau = -k$, so establishing 01 as a β–line. It follows that 12345 will be an α–line. This line is established from the cycloid traced out by the pole as the circle rolls along the ordinate $+ k$ without slip. Tangents drawn to this cycloid at intermediate points, $2'$, $3'$ and $4'$, are normals to the slip line at points 2, 3 and 5. Thus, the SLF is composed simply of a single α–line with two orthogonal β–lines, 01 and 05. *Note*: Point $5'$ is the pole of the circle for the stress state at 5. It follows from this that the compressive stress on a vertical plane through 5 is $k(\pi + 2)$. Since this acts perpendicular to the front face of the container, the die pressure is $p_d = k(\pi + 2)$. The mean extrusion pressure p_e is found from equating horizontal forces above and below the centre line. With unit width and an extrusion ratio $R = H/h = 2$:

$$p_e H = p_d (H - h)$$

$$p_e H = k(\pi + 2)(H - h)$$

$$p_e = k(1 + \pi/2) \tag{6.17a}$$

Multiplying p_e in eq(6.17a) by the ram area (or multiplying p_d by the die area) gives the extrusion force F_e/unit area

$$F_e = p_e H = p_d (H - h)$$

$$F_e / H = k(1 + \pi/2) \tag{6.17b}$$

Alternatively, the die pressure is found from the application of eq(6.9a) to the α–slip line, 15. Because $\sigma_x = 0$ for 01, the constant stress state along 01 is $(-p, -k)$ where $p = k$. Thus, an element at point 1, is stressed as shown in Fig. 6.23a. Let the datum for φ (increasing CW) be the tangent to α at 1. It follows from eq(6.9a) that at point 5, where $\varphi = -\pi/2$:

$$\sigma_5 + 2k(-\pi/2) = c_1 \tag{6.18a}$$

The constant c_1 in eq(6.18a) is found from applying eq(6.9a) to point 1 on α, where $\varphi = 0$:

$$p + 2k(0) = c_1 = k \tag{6.18b}$$

It is seen from eq(6.18a) that $\sigma_5 = k(1 + \pi)$, so that an element at point 5 in Fig. 6.23a is stressed as shown. Resolving forces along 05 within the dead zone confirms that the horizontal die pressure is $p_d = k(\pi + 2)$. However, different estimates to p_e and F_e arise from equating the rate of external plastic work to the energy dissipated within a homogenous deformation zone:

$$\dot{W} = \int p_e \dot{V} = \left[\int Y \,\mathrm{d}\varepsilon \right] \dot{V} \tag{6.19}$$

where $Y = 2k$ is the constant, plane strain yield stress, assumed within a deforming zone under an incremental strain $\mathrm{d}\varepsilon = \mathrm{d}l/l$. Dividing eq(6.19) by the extruded volume/unit time $\dot{V}$ gives the extrusion pressure as

$$p_e = \int_\varepsilon Y \,\mathrm{d}\varepsilon = Y \int_l \mathrm{d}l/l = Y \ln(l_2/l_1) = 2k \ln(l_2/l_1) \tag{6.20a}$$

In preserving volume, the ratio between final and initial lengths is $l_2/l_1 = A_1/A_2 = H/h = R$. Then, from eq(6.20a)

$$p_e = F_e/H = 2k\ \ln R \tag{6.20b}$$

It follows that the limiting reduction occurs when $p_e = 2k$, giving $R = 2.72$. With $R = 2$, eq(6.20b) gives $p_e/2k = 0.693$, which is considerably less than the value 1.286 found from the SLF (eq 6.17a). As the work formula gives the minimum possible work for a forming process, the amount by which it underestimates the actual forming force may be attributed to the effects of frictional work and the redundant shear work.

In the construction of the hodograph in Fig. 6.23c, let the velocity within the container be U and that of the extruded material be u. An incompressible material obeys: $U(H \times 1) = u\,(h \times 1)$, so that $u = 2U$ for $H = 2h$. The origin, $5_R''$, refers to the stationary material within the dead-metal zone to the right of point 5 in Fig. 6.23a. To the left of point 5, the material in the body of the container has velocity U which appears as the horizontal vector joining points $5_R''$ and $5_L''$ in the hodograph. Crossing the slip lines from each of these regions to a point directly beneath point 5, establishes the vectors, $5_R''5_B''$ and $5_L''5_B''$. Their directions are parallel to the slip lines at point 5. Leftward velocity points: $1_L''$, $2_L''$, $3_L''$ and $4_L''$, must be coincident with $5_L''$, since all move horizontally with velocity U. A fan of vectors: $4_L''4_R''$, $3_L''3_R''$, $2_L''2_R''$ and $1_L''1_A''$, emanate from the singular point $5_L''$ in directions parallel to the tangents at points 4, 3, 2 and 1 in the physical plane. In order to ensure inextensibility of slip-line 1 2 3 4 5, the curve $5_B''\,4_R''\,3_R''\,2_R''\,1_A''$ becomes the parallel image of points $5'\ 4'\ 3'\ 2'\ 1'$ in the stress plane. Similarly, the remainder of the hodograph $1_L''1_R''1_A''$ is constructed for velocities to the left, right and above point 1 in the physical plane. Knowing that vector $5_R''1_R''$ is the horizontal billet velocity u, the hodograph must terminate at the closing point $1_R''$ where $u = 3U$. Within the hodograph (see Fig. 6.23c), there are identified velocity discontinuities associated with rigid body motion adjacent to a slip line (refer to Fig. 6.15):

(i) $5_R''5_B''$, along 05, with stationary dead-metal to the right,

(ii) $1_A''1_R''$, along 01, with a rigid-body velocity u to the right and

(iii) the fan of constant magnitude vectors $1_L''1_A''$, $2_L''2_R''$, $3_L''3_R''$ etc, at points 1, 2, 3 etc. along the α–line when, to its left, the rigid body motion is U.

It follows that the slip lines are also velocity discontinuities bounding the deformation zone. Resultant velocities at points 1, 2, 3, 4 and 5 on the right side of α differ. Their magnitudes and directions are given by the vectors: $5_R''1_A''$, $5_R''2_R''$, $5_R''3_R''$ and $5_R''4_R''$. Their directions enable the motion of particles to be traced from container to billet in the physical plane as shown. The solution applies only to an extrusion ratio of 2. Greater or lesser ratios need separate consideration as follows.

6.3.2 Extrusion for 2 < R < 3

In the SLF construction given in Fig. 6.24a, the weaker starting circle has its pole $0'1'$ at the origin of the stress plane in Fig. 6.24b. This corresponds to the known β–slip line, 01 as before. It is seen, however, that end point $5'$ for cycloid $1'\ 2'\ 3'\ 4'\ 5'$, which establishes an orthogonal α–line, cannot be fixed at the container wall without an extension to the field. The requirement is that another α–line, originating at point 6 and the extension to the β–line from 5 to 9, must both meet the wall at 45°.

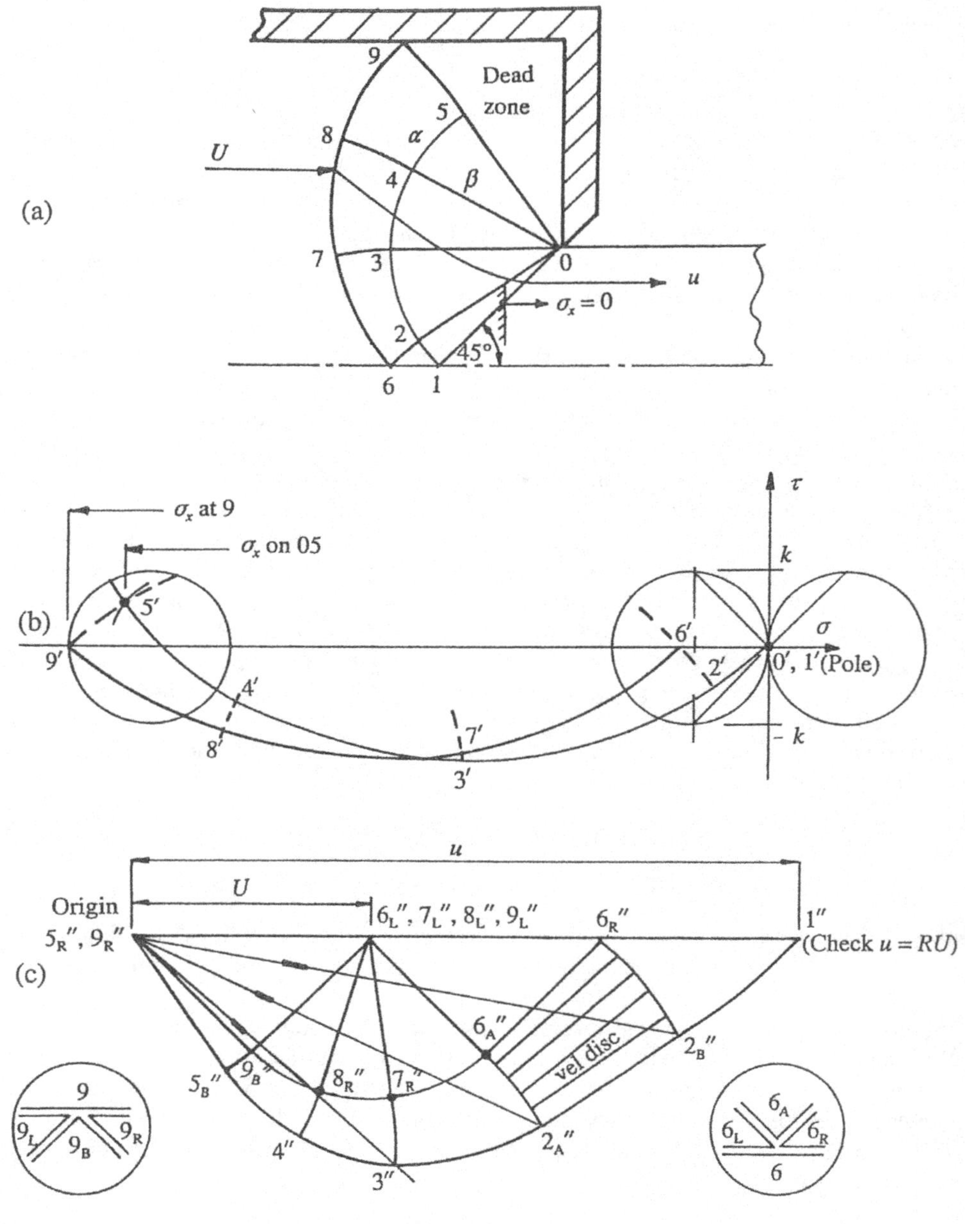

Figure 6.24 Frictionless extrusion for $2 < R < 3$

Thus, in the stress plane, both 6′ and 9′ lie on the $\tau = 0$ axis. Point 6′ is derived from an orthogonal cycloid originating from the selected point, 2′. Cycloid 2′6′ is found from the rotation of a circle with pole point 2′ and radius k, which rolls along $\tau = -k$. Point 6 is found graphically from extending 02 to meet the centre-line with a 45° inclination. This employs equal line elements whose directions are the normals to the corresponding cycloid at points 2′ and 6′. A similar construction is used to establish the position of point 7 from the extension to the slip lines at points 3 and 6. This employs normals to points 2′, 3′, 6′ and 7′ in the stress plane. The construction is equivalent to replacing the extended portion of the slip-line with circular arcs. In repeating this construction, point 8 is obtained from the extensions to slip lines at points 4 and 7 and finally, point 9 from the extensions at points 5 and 8. If 9 does not fall on the inner wall of the container, the position of point 2 must be adjusted and the construction repeated until this is achieved. *Note*: The arc 1345 and radii 01, 03 and 04, together with the extensions 37 and 48, are construction lines and not part of the final SLF solution. That is, the final SLF comprises an orthogonal network with a single α–line 6789 and two β–lines, 026 and 059, containing the deformation zone.

The construction of the hodograph in Fig. 6.24c for the α–line 6789 is similar to that given previously. Again, there is rigid body motion U to its left. Therefore, the coincident points, $6_L''$, $7_L''$, $8_L''$, $9_L''$, all move with velocity U relative to zero velocity in the dead metal zone $5_R''$. The fan of velocity discontinuity vectors, emanating from these coincident points, are equal in magnitude and lie parallel to tangents at points 9, 8, 7 and 6 along the slip line. They terminate at points $9_B''$, $8_R''$, $7_R''$ and $6_A''$, giving an orthogonal image of this slip line. Point $6_R''$ is found from drawing $6_A''6_R''$, parallel to the tangent at 6. The extension to the hodograph is achieved using a construction similar to that for extending slip-lines. It begins with establishing point $5_B''$ from $5_R''$ and $9_B''$. To achieve this, the velocity directions are drawn orthogonal to the tangents at points 5 and 9 in the physical plane. Point 4″ is then found from $5_B''$ and $8_R''$, employing construction lines drawn perpendicular to the tangents at 5 and 8. *Note*: A single velocity applies at 4 and 3 since no slip lines are crossed. However, in crossing slip line 026 it is again necessary to distinguish between a point above ($2_A''$) and below ($2_B''$) point 2, where a velocity discontinuity arises. With points taken above ($6_A''$) and to the right ($6_R''$) of point 6, the hodograph shows that the magnitude of this discontinuity is constant at all points between 2 and 6 in the manner of Figs 6.14a,b. Here again all slip lines in Fig. 6.24a are velocity discontinuities. Having arrived at the destination point 1″ in Fig. 6.24c, the magnitude of u must be checked from the fact that the final velocity $u = RU$. The absolute velocity vectors, originating from stationary points $5_R''$ and $9_R''$, enable the path of a particle through the deformation zone to be shown as a streamline.

To find the net horizontal force F_H from eq(6.16b), the variations in σ_x and τ_{xy} are required either along 059 or 6789. The circles, with pole points 0′, 5′ and 9′ in Fig. 6.24b, show that σ_x varies from zero at point 0 to the indicated values at points 5 and 9. The corresponding shear stress τ_{xy} varies from zero at point 0, through a maximum of $\tau = +k$ between 0 and 5, falling to zero at point 9. Clearly, it would be necessary to select more intermediate values when applying eq(6.16b) to solve this problem.

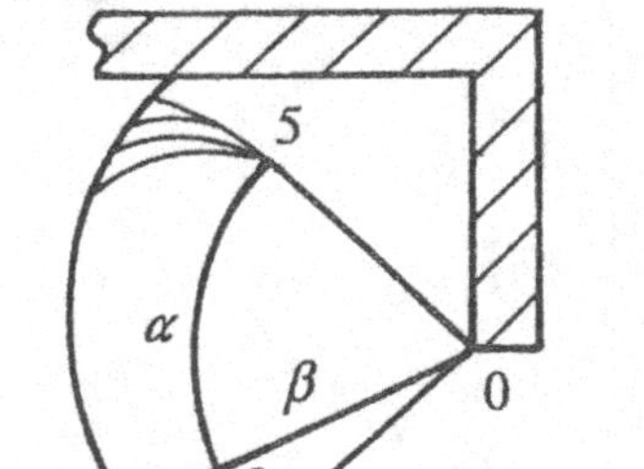

Figure 6.25 Alternative SLF for $2 < R < 3$

There is an alternative solution [1] for this extrusion ratio. In this, the single slip line 59 in Fig. 6.24a is replaced with a fan of slip lines, shown in Fig. 6.25. The envelope originates from 5 permitting a variation in the velocity discontinuity from the dead-metal zone across the fan. For this to be possible, the discontinuities cannot be coincident with each of the interior slip lines.

6.3.3 Extrusion for 1 < R' < 2

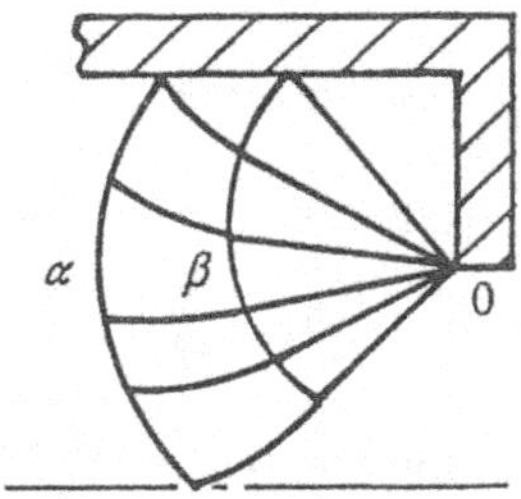

Figure 6.26 Frictionless extrusion, $1 < R < 2$

Figure 6.26 shows the SLF for an extrusion ratio $R' = H/h$, less than 2 [4]. This appears as an inversion of the SLF for $2 < R < 3$ about the singular point 0 in Fig. 6.24a. For an exact inversion to apply, the % increase in the extrusion ratio above 2 must equal the % decrease below 2. In Fig. 6.24 this gives $R' + R = 4$, so if $R = 2.4$, then $R' = 1.6$. Under the latter ratio the stress plane and the hodograph also become lateral inversions of Figs 6.24b,c about their origins [1]. However, where the indicated β–line is partly curved for $R' < 2$, the circle is not stationary and the true stress origin is unknown. Here, an origin for stress is assumed for completing the construction. The true origin is found thereafter from eq(6.16c), where the total horizontal force is zero to the right of this β–line. The origin for the inverted hodograph becomes, simply, the checkpoint 1″ in Fig. 6.24c.

6.3.4 Extrusion for R = 3

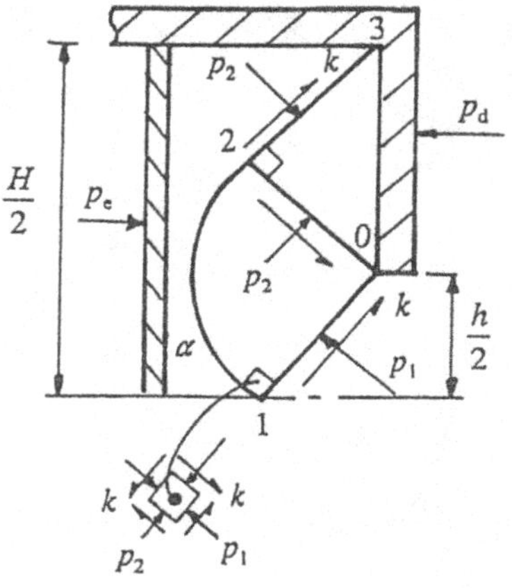

Figure 6.27 Frictionless extrusion for $R = 3$

Lee [9] and Green [10] proposed a SLF for $R = 3$, given in Fig. 6.27. In the absence of a dead-zone, Lee [9] showed that this field provided the minimum possible extrusion force. Moreover, alternative solutions, discussed in the next section, were found to violate the yield criterion when applied to $R = 3$. As with $R = 2$, eqs(6.9a,b) may be applied to the SLF for $R = 3$. Again $\sigma_x = 0$ for 01 and the constant stress state along 01 is (p_1, k) where $p_1 = k$. Thus, an element at point 1 is stressed as shown. To find p_2, let the datum for φ be the tangent to α–line at 1. Applying eq(6.9a) to point 2, where $\varphi = -\pi/2$:

$$p_2 + 2k(-\pi/2) = c_1 \tag{6.21a}$$

in which $c_1 = k$ for $\varphi = 0$. Equation (6.21a) gives

$$p_2 = k(1 + \pi) \tag{6.21b}$$

so that element 023 is stressed as shown. Resolving forces gives the die pressure as

$$p_d = k(2 + \pi) \tag{6.22a}$$

from which the extrusion pressure is

$$p_e = \left(1 - \frac{1}{R}\right) p_d = \frac{2p_d}{3} \tag{6.22b}$$

Combining eqs(6.22a and b) gives $p_e/2k = 1.714$.

6.3.5 *Extrusion for $R > 3$*

The SLF, given in Fig. 6.28a, will apply in the absence of a dead-metal zone. It is an extension to the SLF for $R = 3$ discussed previously.

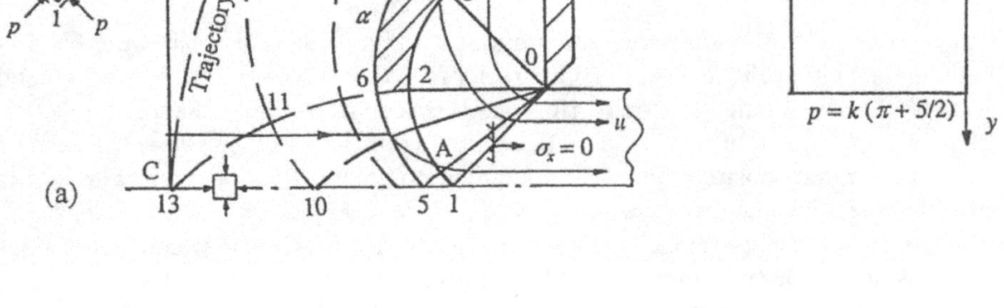

Figure 6.28 Frictionless extrusion $R > 3$

(b)

(c)

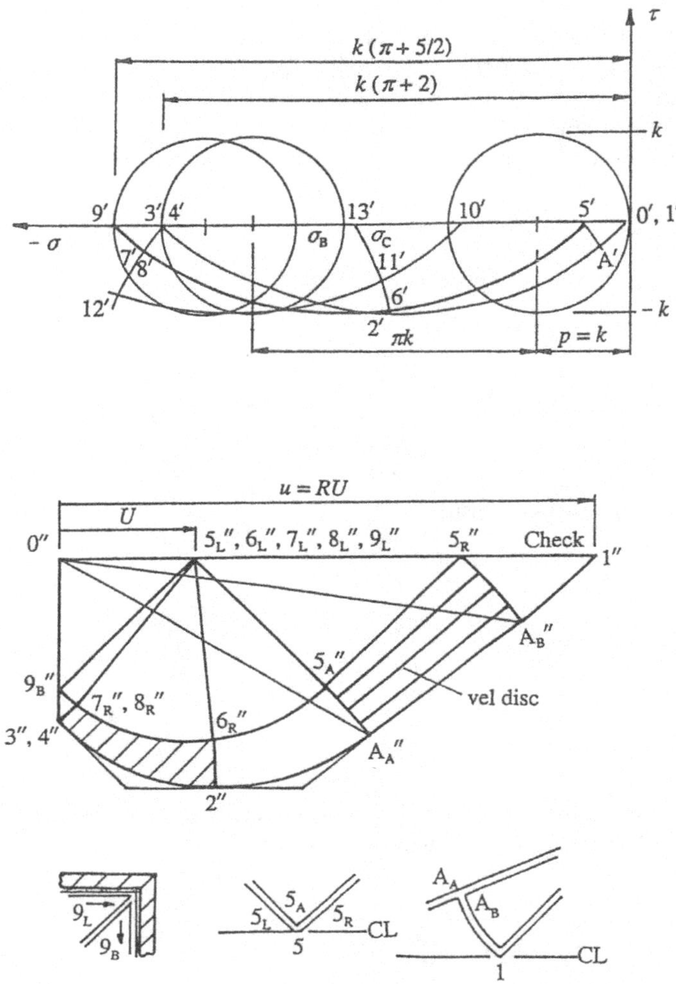

Figure 6.28 Frictionless extrusion $R > 3$ (continued)

It has been shown [11] that the extrusion pressure for Fig. 6.28a is less than would be found from extending either Fig. 6.24 or 6.25 to greater R. Moreover, when these were extended to give $R = 3$, they violated the yield criterion. It is seen from Fig. 6.28a that Lee's α–line 1234 and two β–lines, 01 and 03, would fit exactly within a vessel for $H/h = 3$. These lines are now used in the construction. To extend them to greater R values, select an arbitrary point A on the α–line and identify A′ in the stress plane of Fig. 6.28b. Next, we locate 5′ on the $\tau = 0$ axis from the orthogonal cycloid and thus its position 5 on the centre line. For the correct position of 5, end points 9′ and 9 will give zero shear stress at the corner of the container. Normals, drawn at the intersection points 2′3′4′ and 5′6′7′8′, within the orthogonal network of cycloids, extend the slip-line field in the usual way. The hodograph origin 0″ in

(c) refers to any point on the stationary vessel. Thus, vector $0''\ 9_B''$ is identified with the vertically downward motion of material beneath point 9 relative to the top corner point. Coincident points, $5_L''$, $6_L''$, $7_L''$, $8_L''$ and $9_L''$, in the region to the left of the α–line, move with rigid body velocity U. Velocity discontinuities arise in crossing this α–line, their magnitudes being given by velocity vectors: $9_L''9_B''$, $8_L''8_R''(\equiv 7_L''7_R'')$, $6_L''6_R''$ and $5_L''\ 5_A''$. Their terminal points are joined by the orthogonal image of the α–line. The point $5_R''$ is then located from points $5_A''$ and $5_L''$. Tangents drawn at intersection points $2'3'4'$ and $5'6'7'8'$, within the orthogonal network of cycloids in the stress plane, are used to extend the hodograph. This gives points $3''(\equiv 4'')$, $2''$, A_A'' and A_B''. A fan of constant-magnitude, velocity discontinuities applies to all points between 5 and A in the manner of Fig. 6.14. Finally, tangents projected through points A_B'' and $5_R''$ locate the checkpoint, $1''$, for which $u = RU$. Absolute velocity vectors, with common origin at $0''$, may be transferred to their appropriate slip lines to trace the motion of a particle through the deformation zone. Three such streamlines are illustrated as horizontals (→) initially in Fig. 6.28a.

The mean extrusion pressure is found from the known normal stresses at points 0, 4 and 9 along the die face. The stress plane shows that these are principal stresses (i.e. $\tau = 0$ along 049). Their magnitudes are $k(\pi + 2)$ = constant along 04, increasing to a maximum value of $k(\pi + 5/2)$ at point 9. Assuming the linear variation in normal stress between 4 and 9 (see Fig. 6.28a) enables the mean extrusion pressure p_e to be estimated from

$$p_e H = k(H - h)(\pi + 2) + \frac{k}{2}\left[H - (R - 1)h\right]\left[\left(\pi + \frac{5}{2}\right) - (\pi + 2)\right]$$

$$p_e = k\left(1 - \frac{1}{R}\right)(\pi + 2) + \frac{k}{2}\left[1 - \frac{R - 1}{R}\right]\left[\left(\pi + \frac{5}{2}\right) - (\pi + 2)\right] \qquad (6.23)$$

where $R = H/h$. Taking, for example, $R = 4$ in eq(6.23) gives $p_e = 3.92k$. The total force/unit width of die becomes $F_e = Hp_e$.

In the test for a complete solution, two conditions must be satisfied. Firstly, there is the requirement for positive plastic work: $\int \sigma_{ij}\,d\varepsilon_{ij} > 0$. This will involve a consistency check between relative directions in the hodograph and the deformation within a constant region of stress, e.g. the region 2376 bounded by slip lines in Fig. 6.28a. The pole positions in the stress plane reveal that the shear yield stresses $\pm k$ act in the directions shown in Fig. 6.29.

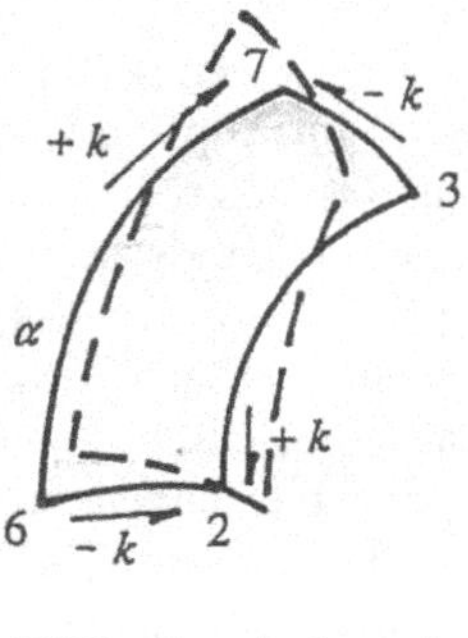

Figure 6.29 Positive plastic work check

The compressive hydrostatic stress acting on the α, β boundaries causes no deformation, the latter being due entirely to the shear stress k. The shear distortion is that indicated with broken lines. The element of the hodograph in Fig. 6.28c confirms the inward direction of point 3 relative to 6 along the line connecting 3 to 6. It follows that the product of a compressive stress and strain rate in the 36 direction will produce the required rate of positive plastic work. Four examples of inadmissible elements of hodograph are given in Figs 6.30a–d for the element of SLF in Fig. 6.29.

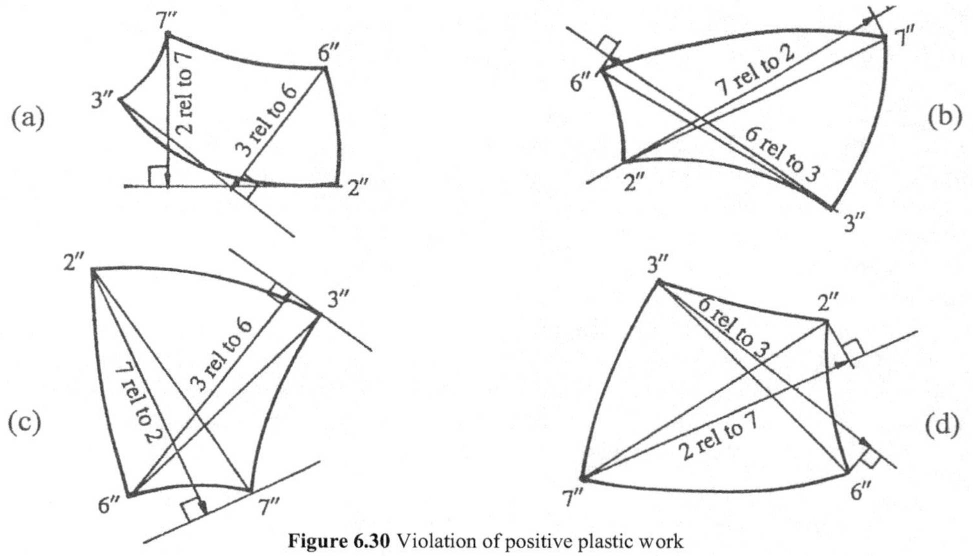

Figure 6.30 Violation of positive plastic work

These could arise when there is more than one hodograph fulfilling the required conditions. However, since these show relative motions between points 2, 3, 6 and 7 that are inconsistent with the known deformation from Fig. 6.29, they must be rejected in favour of the unique solution supplying positive plastic work.

Secondly, a check is made that the yield criterion is nowhere violated within the rigid material. This means that a rigid-plastic zone should not deform under the stresses imposed around its boundary. Bishop [12] showed that a slip-line field is complete where it can be extended into the rigid zone with pseudo slip-lines that become coincident with a stress free boundary. The approach has been applied to frictionless extrusion with $R = 2$ [1] and $R = 3$ [11]. In applying Bishop's method to $R > 3$ in Fig. 6.28a, it must ensure that the material to the right of 01 is non-deforming. The initial circle shows that the principal stresses for all points along 01 are $\sigma_y = 2k$ and $\sigma_x = 0$, acting in the vertical and horizontal directions respectively. The material to the right of the slip line suffers less vertical stress and therefore, the application of eq(6.3) shows that this material has not yielded. In the rigid region to the left of 56789 an orthogonal network of cycloids are constructed within the stress plane to enable the extension of pseudo slip lines (broken lines). Their end-points are located on a principal stress trajectory 9–12–13 which lies at 45° to α and β and is tangential to a stress free point 9 on the container wall (Fig. 6.28a). The principal stresses, normal to the trajectory, are found from projecting the tangent to the trajectory in (a) through the corresponding pole point in (b). In this manner σ_B and σ_C act normal to the trajectory at points B and C as shown. Equilibrium requires that σ_B and σ_C act with equal magnitudes in vertical and horizontal directions. Since absent shear stress is a consequence of frictionless conditions, σ_B and σ_C are transmitted into the remaining material to align with the container wall. The greatest principal stress difference $\sigma_B - \sigma_C$, applies to the top left corner. The stress plane reveals that this difference is less than $2k$ and hence the yield criterion is not violated anywhere.

6.4 Frictionless Extrusion, Inclined Dies

The extrusion pressure can be reduced by spreading the extrusion force over a larger area but there is a practical limit to increasing the width of the ram. However, the die angle may be inclined to achieve the same effect.

6.4.1 Extrusion for $R = 2$

Taking a die angle of $\theta = 30°$, the simple field in Fig. 6.31a applies for $R = H/h = 2$. This is constructed from the requirement that slip lines meet the die face and the axis of symmetry at 45°.

(a)

(b)

(c)

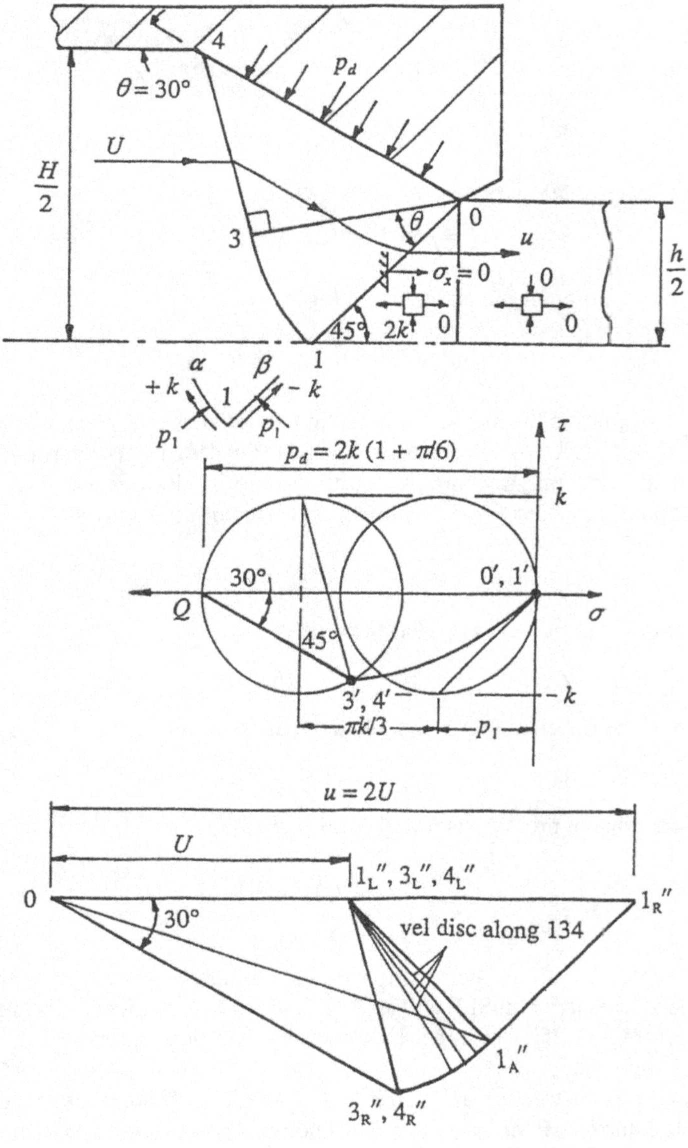

Figure 6.31 Frictionless extrusion through inclined dies with $R = 2$

The initial circle in Fig. 6.31b applies to the β–line 01, with its pole position coincident with the origin, so giving $\sigma_x = 0$. In rolling this circle along the ordinate $\tau = +k$, the variation in mean stress is established for that portion 13 of the α–line. The remainder 34 of this slip line must remain straight if it is to meet the die face at 45°, with the circle being stationary with co-incident pole points 3′ and 4′. Projecting a line parallel to the die face through pole 4′ gives the stress state at Q where the normal stress is identified with a constant die pressure:

$$p_d = 2k\,(1 + \pi/6) \tag{6.24a}$$

The extrusion pressure p_e follows from equating horizontal forces/unit width. This gives

$$p_e H = \frac{p_d (H - h)\sin\theta}{\sin\theta} = 2k\left(1 + \frac{\pi}{6}\right)(H - h)$$

$$p_e = 2k(1 + \pi/6)(1 - 1/R) = k(1 + \pi/6) \tag{6.24b}$$

from which the horizontal extrusion force/unit width is

$$F_e = p_e \times H = H\,k\,(1 + \pi/6)$$

In general, for the particular die geometry in Fig. 6.31a:

$$R = H/h = 1 + 2\sin\theta \tag{6.25}$$

Equation (6.25) applies to the maximum reduction ratio that is possible with an inclined die [1]. For example, $R = 2$ for $\theta = 30°$ and $R = 2.414$ for $\theta = 45°$. The corresponding extrusion pressure is found from applying eq(6.9a) to the α–line 134. Referring to Fig. 6.31a, $p_1 = k$ at point 1. At point 3, where $\varphi = -\theta$, substitution into eq(6.9a) shows

$$p_3 + 2k\,(-\theta) = c_1 \tag{6.26a}$$

where $c_1 = k$ for $\varphi = 0$. Equation (6.26a) then gives

$$p_3 = k\,(1 + 2\theta) \tag{6.26b}$$

Resolving forces for the element 034 supplies the die face pressure

$$p_d = 2k\,(1 + \theta) \tag{6.27a}$$

As before, the extrusion pressure is found from eq(6.24b)

$$\frac{p_e}{2k} = (1 + \theta)\left(1 - \frac{1}{R}\right) \tag{6.27b}$$

Equations (6.25) and (6.27b) supply an optimum die angle associated with the lowest extrusion pressure. A different SLF will apply to a given R–value when θ is greater than the limiting value from eq(6.25). This is because the redundant work increases as frictional work decreases with increasing θ.

The hodograph construction in Fig. 6.31c shows rigid body horizontal velocity U to the left of points 1, 3 and 4. Material to the right of points 3 and 4 moves in the direction of the die face. Points $1_A''$, $3_R''$ and $4_R''$ are located from projecting the tangent directions at 1, 3 and 4 through points $1_L''$, $3_L''$ and $4_L''$. The vectors $3_L''3_R''$ and $1_L''1_R''$, identify velocity discontinuities of the type in Fig. 6.15, in which the end points $3_R''(\equiv 4_R'')$ and $1_A''$ lie on a curve that is the orthogonal image of 13. Finally, the horizontal velocity u of a point to the right of 1 ($1_R''$) must give $u = 2U$. The particle flow direction, shown in Fig. 6.31a, is found from the directions of the absolute velocities on adjacent sides of the slip lines it crosses.

6.4.2 Extrusion for R > 2

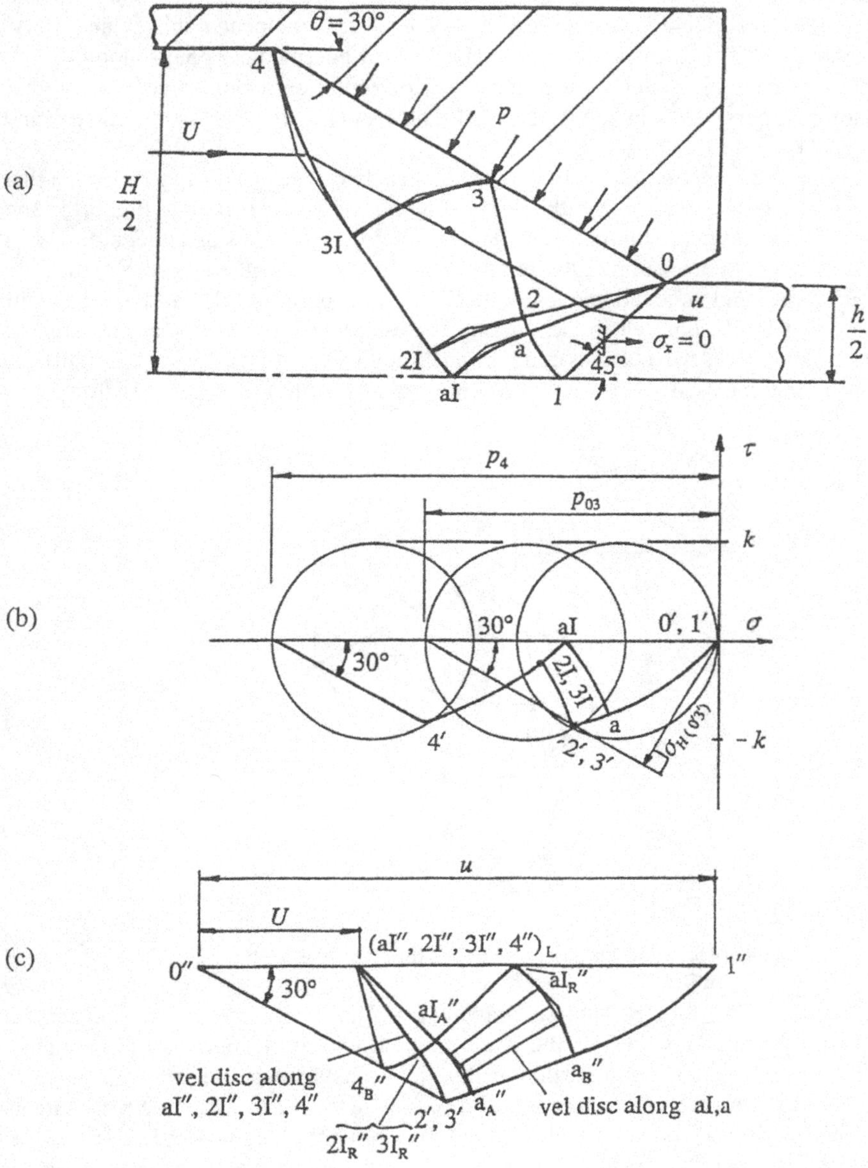

Figure 6.32 Frictionless extrusion through inclined dies with $R > 2$

The limiting solution, discussed in the previous section, may be employed to extend slip-line fields for extrusion ratios that are either greater or less than the ratio given by eq(6.25). Take the case of $R > 2$, with $\theta = 30°$, shown in Fig. 6.32a. A point a, between 1 and 2, must be found such that when the field is extended by the usual method, the point 4 lies in the container corner. *Note*: A similarity appears between this slip-line field and that for square dies with $R \geq 3$ in Fig. 6.28a. The stress plane in Fig. 6.32b shows that the normal pressure p along the inclined die is constant at p_{03} along 03 but increases from 3 to a maximum value p_4 at 4. One method for obtaining the extrusion pressure is to integrate σ_x and τ_{xy} along the die face in the manner of eq(6.16b). Alternatively, eq(6.15) could be used with σ_H typically as indicated for 0′3′. However, the stress plane can only supply stresses at points 0′, 3′ and 4′ since 034 is not a slip line. To improve accuracy, a mesh refinement would be necessary, giving further intersections between pseudo slip-lines at intermediate points along 03 and 34. For this reason it is better to determine the horizontal extrusion force F_H by either method along the curved α–slip line: aI, 2I, 3I, 4. The mean extrusion pressure is then found from $p_e = F_H /$ ram area.

The hodograph construction in Fig. 6.32c reveals two types of discontinuity: (i) that associated with rigid body motion to the left of slip line: 4, 3I, 2I, aI (refer to Fig. 6.15) and (ii) the fan of constant magnitude velocity vectors for adjacent points on either side of aI,a (refer to Fig. 6.14). For example, the relative velocity is constant for points above and below point a (a_A'' and a_B'') and above and to the right of point aI (aI_A'' and aI_R''). Their directions vary with the normals to the slip line at these points. It is seen that the path of a streamline now involves two changes of direction through the deformation zone. Finally the checkpoint 1″ may be used to ensure $U = Ru$ in obedience with incompressible flow.

6.4.3 Extrusion for $R < 2$

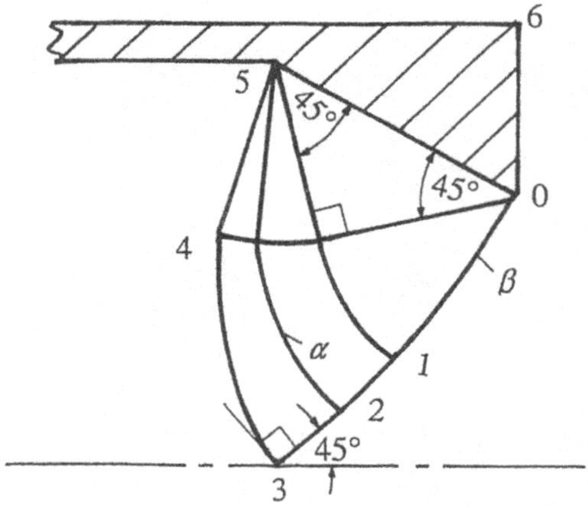

Figure 6.33 Frictionless extrusion through inclined dies with $R < 2$

Figure 6.33 gives a SLF for $R < 2$ in strip drawing [4]. Here, it is necessary to find point 1 on the β–line by trial. This enables the β–extension 23 to reach the axis of symmetry. Intersections at points 2 and 3 with orthogonal α–lines must meet at the top corner point 5 of the inclined die. As the α–lines cannot all meet the container at 45°, the stress state at point 4 is discontinuous. Because the circle is not stationary along points 1, 2 and 3, the stress plane is constructed for an assumed origin [1]. The true origin is found from applying eq(6.16c) in which the total horizontal force is zero for material to the right of the β–line.

6.5 Extrusion with Friction, Parallel Dies

These solutions apply where the coefficient of friction is sufficiently high to induce frictional shear forces which yield the material in contact with the container walls. It follows that slip lines will meet the wall tangentially and normally within the deformation zone. The slip line field solution for a reduction ratio of 2 is given in Fig. 6.34. The requirement is that the α–line BD meets the axis of symmetry at 45° and the container wall at 90°. To achieve this, the position of 3, i.e. point 3′ in the stress plane (Fig. 6.34b), is varied until the pole position D′ lies on the $\tau = -k$ ordinate. A line drawn perpendicular to the wall through D′ confirms that BD is an α–line associated with $\tau = +k$.

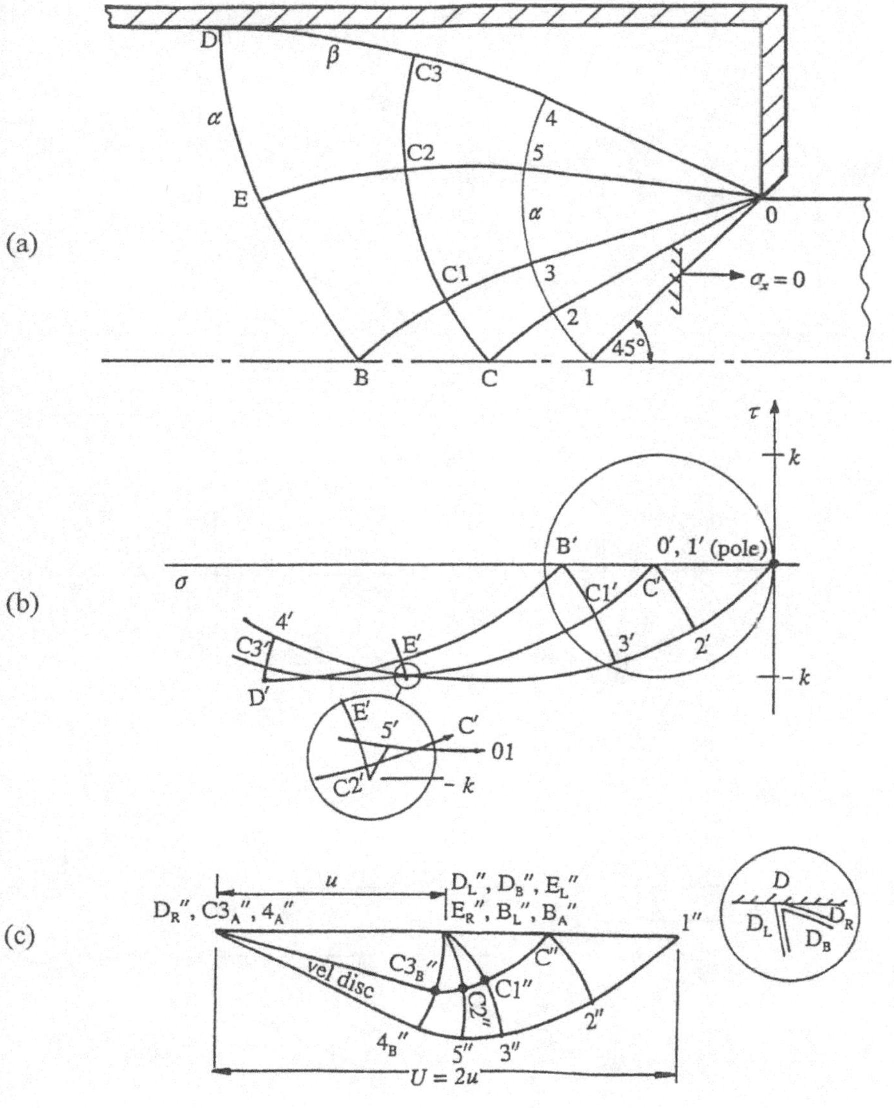

Figure 6.34 Extrusion with friction, $R = 2$

The tangential β–line at D is determined from the orthogonal cycloid at D′ which terminates at point 4′. Intermediate slip lines are constructed from point C lying between 1 and B and from 5 between 3 and 4. Points to the right of D, and above C3 and 4, lie within the dead metal zone and, therefore, lie at the origin for the hodograph in Fig. 6.34c. Points beneath D, C3 and 4 are associated with the fan of velocity discontinuities. No discontinuity arises between points to the left and right of D, E and B, since these all move horizontally with a common velocity u. The checkpoint 1″ is found from extending the hodograph, with directions parallel to those in the stress plane, to give the final velocity $U = 2u$.

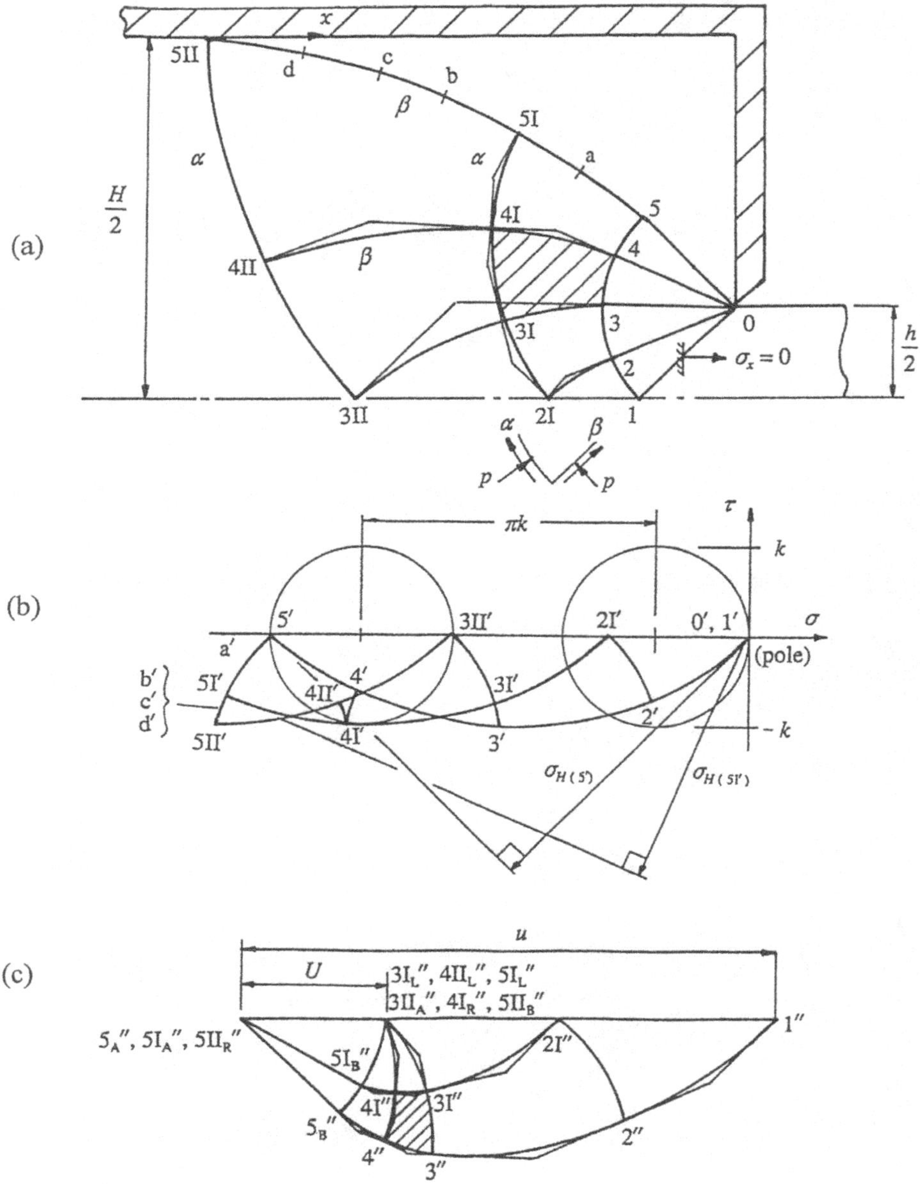

Figure 6.35 Extrusion with friction for $R = 4$

The solutions for different extrusion ratios and sticking friction are less distinct than those previously considered without friction. For example, the SLF solution for $R = 4$ (see Fig. 6.35a) is similar to that for $R = 2$ and, therefore, one need not detail the corresponding constructions for Figs 6.35b,c again.

6.5.1 Net Extrusion Force and Mean Pressure

The additional construction lines in Fig. 6.35b enable the determination of the total extrusion force/unit width by the method outlined in Fig. 6.19. This applies to the β–line 0, 5, 5I, 5II taking intermediate points a, b, c and d, to bound the dead metal zone. The corresponding poles, located on the orthogonal image in the stress plane, enable the centre and the ordinate position $\tau = -k$ to be located within each circle. Connecting the pole of each circle to its ordinate reveals the lengths of the perpendicular projections σ_H from the stress origin. These are expressed as a ratio of the circle's semi-circumferential length πk and plotted against the true length s of the β–line in Fig. 6.36.

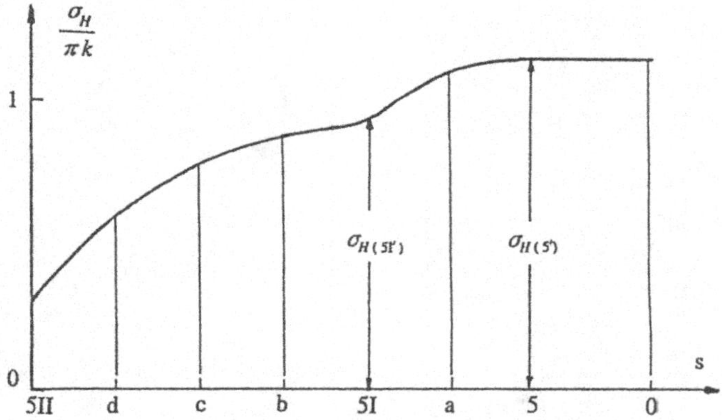

Figure 6.36 Variation in resultant horizontal stress along 0, 5, 5I, 5II

The area beneath the curve in Fig. 6.36 gives the horizontal force/unit width upon the upper rectangular die face. Applying the trapezoidal rule, this area is $6.165\pi k$, which gives a net horizontal force $F_H = 12.33\pi k$ and a mean die pressure/unit width:

$$p_d = \frac{F_H}{(H - h)} = \frac{4.11\pi k}{h}$$

The mean extrusion pressure p_e is found from

$$Hp_e = (H - h)p_d$$

$$p_e = 0.75 p_d = \frac{3.083\pi k}{h}$$

Finally, a check is made that the rate of plastic work is positive. The shaded region within the hodograph in Fig. 6.35c shows that the motion of point 4I relative to 3 is consistent with the directions of k for the corresponding shaded region given in the stress plane (Fig. 6.35a).

6.6 Notched Bar in Tension

Deformation in a notched bar is in pseudo-steady state in which the bar size but not its shape changes with time. Fig. 6.37a shows one half of a symmetrical, double-notched tension bar. It will be shown that the semi-circular notch retains its shape when the through-thickness strain is zero under a plane strain condition.

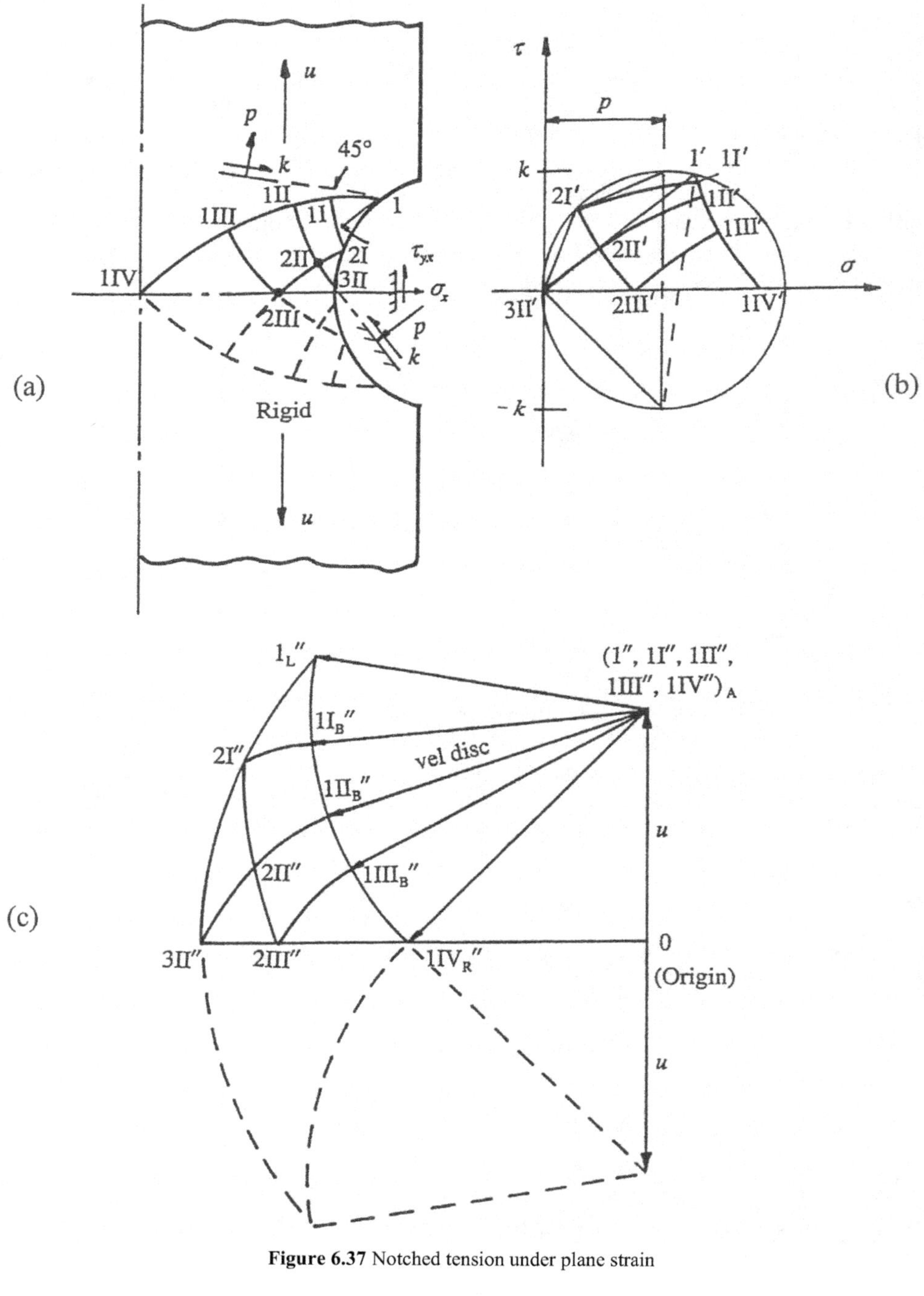

Figure 6.37 Notched tension under plane strain

The symmetry of the problem reduces the physical plane to ¼ of the bar only. Since both σ_x and τ_{yx} are zero at the notch root 3II, the circle in Fig. 6.37b must pass through a stress origin coincident with its pole point 3II′. The circle applies to the strong solution where the mean stress p is tensile. This circle will also represent the zero radial and tangential surface stress states for points 1 and 2I but with the different pole positions 1′ and 2I′ indicated. They are found from projecting tangents at 1 and 2I through the stress origin. The cycloid pattern shown originates from these pole positions when the circle rolls along $\tau = \pm k$. Points 2III′ and 1IV′ along the circle axis are associated with zero shear stress. Physically, they lie on the horizontal axis of symmetry, as this is a principal plane. The β–line: 3II, 2II, 1II, and the α–line: 2I, 2II, 2III, are derived from the corresponding cycloids in the stress plane. The problem lies in locating point 1. The requirement is that the α–line: 1, 1I, 1II, 1III, 1IV, is to terminate at the bar centre. Point 1II′ in the stress plane is determined by trial until this is achieved. We see that this is a particular type of boundary problem in which the surface tractions are zero. The applied force may be estimated from a stress distribution based upon the major principal (vertical) stresses at points 3II, 2III and 1IV.

In constructing the hodograph (Fig. 6.37c), the material outside the deformation zone moves with rigid body velocity u. This fixes co-incident points: 1″, 1I″, 1II″, 1III″, 1IV″, above the bounding α–line. Within the deformation zone, the magnitude of the velocity discontinuities: $1''1_L''$, $1I''1I_B''$, $1II''1II_B''$, $1III''1III_B''$ and $1IV''1IV_R''$, are fixed since a point to the right of 1IV (i.e. $1IV_R''$) must also move horizontally relative to the origin. Remaining elements in the hodograph are established from orthogonality with the physical plane. It is seen that the deformed notch retains a semi-circular shape as points on its surface move radially inwards with uniform absolute velocity. That is, points $1_L''$, 2I″ and 3II″ lie on an arc with centre at the origin of the hodograph.

6.7 Die Indentation

With flat die indentation of a plate (see Fig. 6.38), the requirement is to develop a SLF for a given impression in material beneath the die. This leads to the die pressure and applied force F. Plasticity is localised for small b/t ratios, i.e. for narrow impressions in thick plates, but spreads through the thickness as b/t increases with wider impressions in thinner sheet. Solutions are available for full friction, frictionless indentation and with an intermediate Coulomb friction lying between the contacting surfaces.

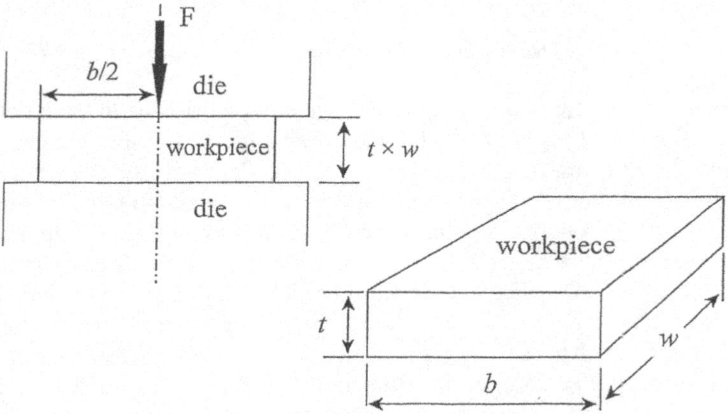

Figure 6.38 Flat die indentation of a plate

The Coulomb shear stress is the product of the frictional coefficient and the pressure normal to the die surface. Here, slip lines can meet the die surface at varying angles, in contrast to the respective tangential and 45° directions, with and without friction [8]. The SLF solutions for a blunt wedge indentor is surrounded by a dead metal zone similar to the SLF for flat die indentation of a thick plate. Thus flat die indentation is regarded as a limiting case of the wedge indentor where a plate of finite thickness t is compressed along a vertical axis between two flat platens each of breadth b (Fig. 6.38). Symmetry restricts the analysis necessary to ¼ of the deformation zone. Plane strain conditions apply when the plate width w is large compared to the die breadth b. This means that plastic strain in the width direction is absent, this promoting an elastic constraint in each undeformed region adjacent to the deformation zone. The following geometries are considered from the interactions that arise between opposing slip line fields. The distinction between dies without friction (smooth) and with friction (rough) will need to be made for b/t ratios greater than unity.

6.7.1 Indentation for $b/t < 0.115$

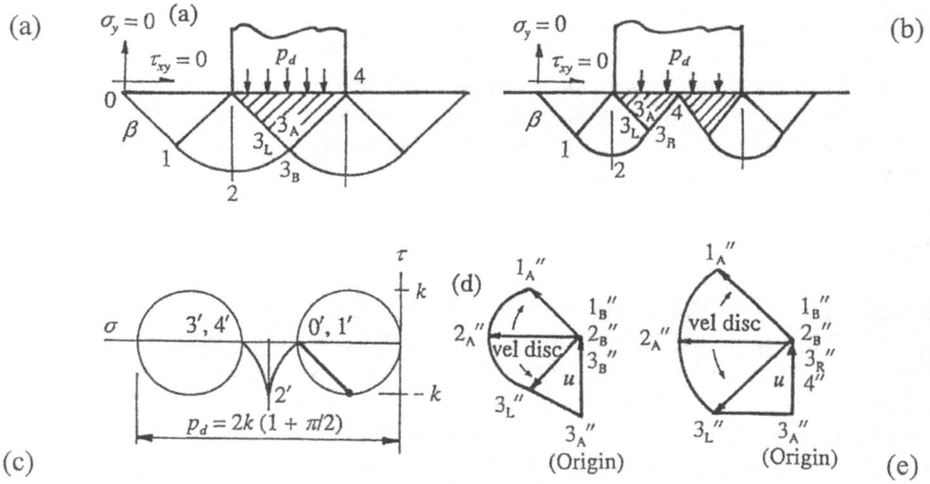

Figure 6.39 Flat die indentation for a thick plate

Two alternative SLF's for this b/t range (see Figs 6.39a,b) are due to the pioneering work of Prandtl [13] and Hill [4] respectively. Experiment has shown that Prandtl's SLF in (a) is more applicable to rough dies where there is less tendency for the dead zone (shaded) to slide on its contacting surface. However, should frictional forces along the narrow die face be zero, then surface roughness is unimportant. Hill's solution in (b) applies strictly to frictionless conditions in which slip lines intersect the die centre and edges at 45°. The same stress plane (Fig. 6.39c) applies to both slip-line fields in Figs 6.39a,b. The initial, weak circle applies to the zero stress state (σ_y, τ_{xy}) at point 0 on the free surface. Projecting the horizontal plane through the stress origin locates the pole at $0'1'$. Thus 01 is a β–line, which is extended to points 2 and 3 by rolling the circle along $\tau = -k$ to trace the cycloid image points $2'$ and $3'$. Beyond 3, the circle remains stationary with β adjacent to the dead zone. That is, pole points $3'$ and $4'$ coincide, corresponding to a straight line 34, bounding the dead zone. Projecting a horizontal line, parallel to the die surface, through pole 3′4′ intersects this circle to give the stress normal to this surface. This is the die pressure

$$p_d = 2k\,(1 + \pi/2) \tag{6.28}$$

Prandtl and Hill's hodographs (Figs 6.39d,e respectively) show the downward velocity u of rigid wedges (points $3_A''$) relative to each origin. The latter are the coincident points $1_B''$, $2_B''$, $3_{B,R}''$ and $4''$ for stationary material. A fan of velocity discontinuities appears where the outer β–line is crossed from stationary to moving material (points: $1_A''$, $2_A''$, $3_L''$). A further discontinuity $3_A''3_L''$ arises as the wedge is crossed from a point above and to the left of 3.

Both SLF solutions apply to regions beneath each die surface that do not penetrate to the horizontal axis of symmetry. In fact, the upper field would not be altered by the removal of the lower die, i.e. with the plate bottom resting upon a flat surface (and vice versa).

6.7.2 Indentation for 0.115 < b/t < 1

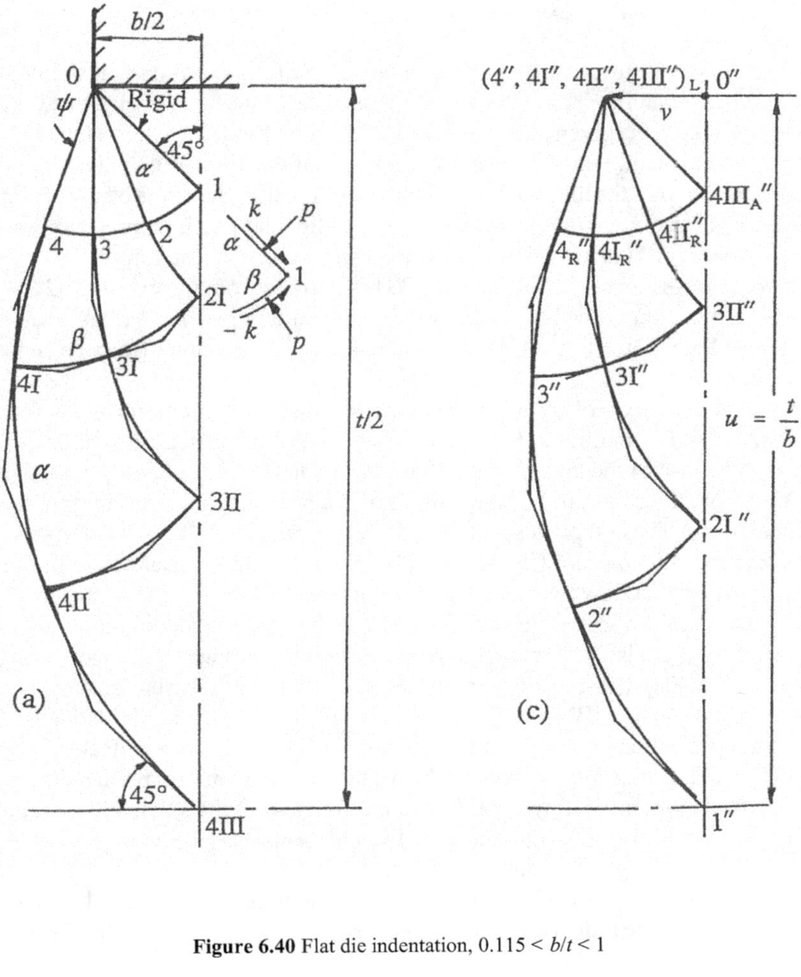

Figure 6.40 Flat die indentation, 0.115 < b/t < 1

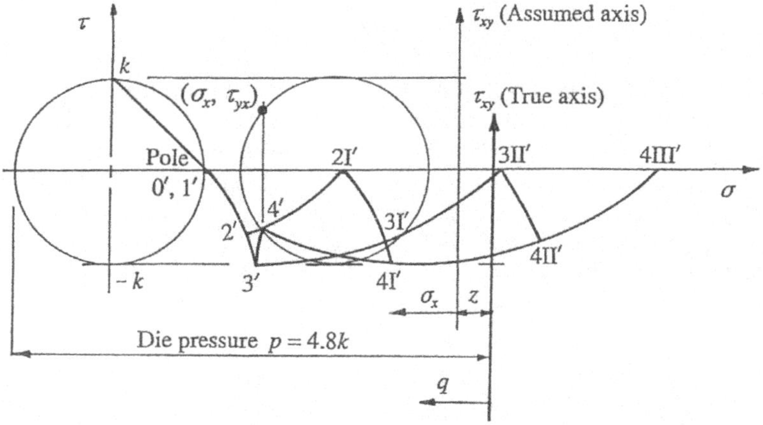

(b)

Figure 6.40 Flat die indentation, $0.115 < b/t < 1$ (continued)

Hill [14] derived a ratio $b/t = 0.115$, for opposing SLFs to just meet at the centre line. Where $b/t > 0.115$, an interaction between opposing SLF modifies their shape. A typical example for $b/t = 0.14$ is given in Figs 6.40a–c. As with Prandtl's solution (see Fig. 6.39a), the wedge of dead metal lying beneath the full width of the die in Fig. 6.40a, implies that it is unimportant whether the die face is rough or smooth. Since the stress origin for Fig. 6.40b is not known, a stationary circle for the straight slip line 01, is chosen arbitrarily (left position). Assuming this an α–line, the stress state will lie at the top of the circle, from which the pole 0′1′ is located. The β–line 1234 is derived from the cycloid 1′2′3′4′, traced by this pole as the circle rolls in a clockwise direction along $\tau = -k$. Within the network of orthogonal cycloids point 4′ is selected in Fig. 6.40b to locate point 4III at the centre of the deformation zone.

The lateral velocity v of rigid material exterior to the deformation zone is found from an incompressibility condition: $b/t = v/u$, where u is the vertical die velocity. In the hodograph (Fig. 6.40c) the coincidence of leftward points: $4_L''$, $4I_L''$, $4II_L''$ and $4III_L''$ is set at $v = 1$ unit from the origin 0″. The elements of the hodograph are orthogonal to those of the physical plane. These show discontinuities: $4_L''4_R''$, $4I_L''4I_R''$, $4II_L''4II_R''$ and $4III_L''4III_A''$, from crossing the outer α–line. Checkpoint 1″ reveals that the material within the dead zone moves with the correct downward velocity: $u = tv/b = 7.14v$.

The position of the stress origin is found from the second method outlined in § 6.2.9. The requirement there is that the total horizontal force $F_H = 0$ when applied to the outer, α slip line in Fig. 6.40a. Cartesian normal σ_x and shear stress τ_{xy} distributions are established from the points 0′, 4′, 4I′, 4II′ and 4III′ in the stress plane. These vary with the vertical and the horizontal projections of the slip line as shown in Fig. 6.41. *Note*: Shear stress τ_{yx} is the full projection of a slip line that curves back on itself. For example, at point 4 the stress state (σ_x, τ_{yx}) is found from projecting a vertical line through 4′ in the stress plane. This shows that the sense of σ_x and τ_{xy} ($= \tau_{yx}$) oppose and so eq(6.16b) may be applied as in Figs 6.21 and 6.22. Here, q in Fig. 6.41 must be replaced with: (i) $\sigma_x + z$, for σ_x compressive in the region $c \le y \le b$ and (ii) $-(\sigma_x - z)$, for σ_x tensile in the region $0 \le y \le c$. Using ordinates in multiples of the circle radius k, the integrals in eq(6.16c) are evaluated by the trapezoidal rule. Setting $F_H = 0$ in eq(6.16b) determines z and the position of the true stress origin. From this, the left point in the stress plane gives a normalised die pressure: $p_d/(2k) = 2.4$.

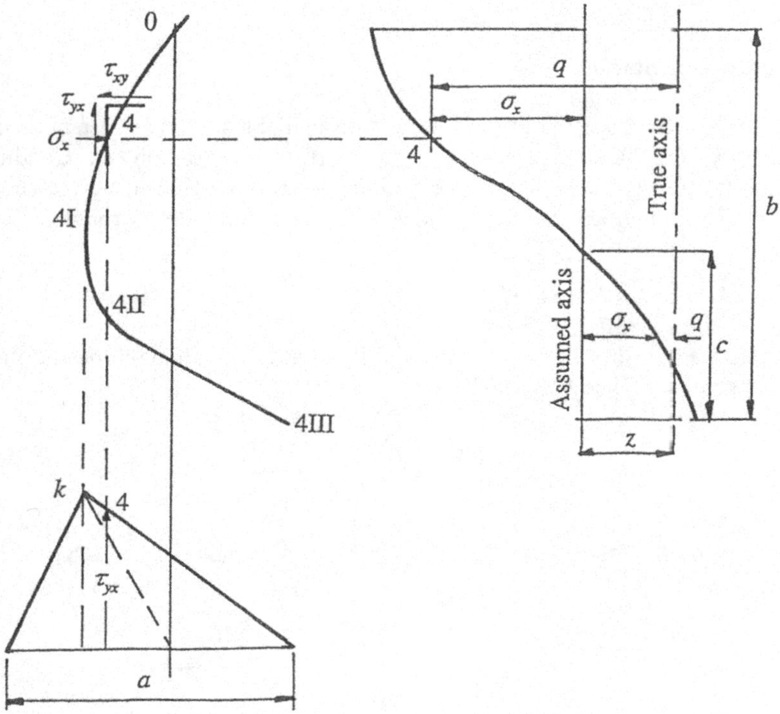

Figure 6.41 Variation in Cartesian stress components along α slip line

6.7.3 Indentation for b/t = 1

When $b/t = 1$, the SLF (Fig. 6.42a) consists of a single wedge of dead metal beneath the die face. As before, this implies the field is equally applicable to both rough and smooth dies.

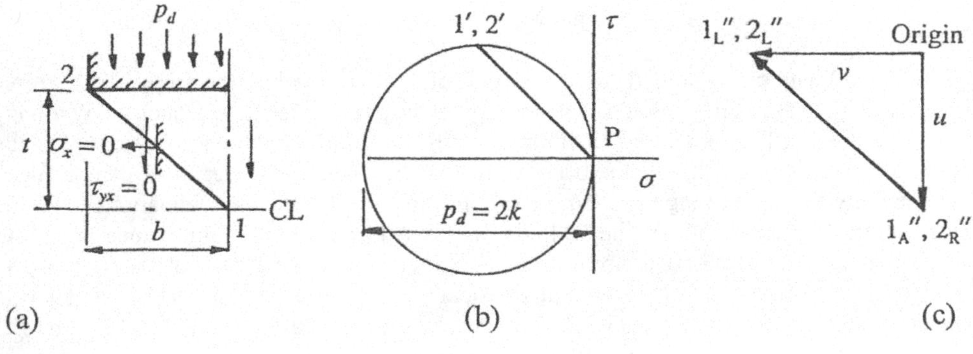

Figure 6.42 Flat die indentation, $b/t = 1$

With the pole P, at the origin of the stress plane in Fig. 6.42b, the constant die pressure is: $p_d/(2k) = 1$. In the hodograph (Fig. 6.42c), the lateral and vertical velocities are the same, i.e. $u = v$ and these are connected by the common velocity discontinuity: $1_A''1_L''$, $2_R''2_L''$.

6.8 Rough Die Indentation

In the case of partially rough dies, SLF solutions are available for: (i) a constant shear stress at the boundary, less than k, and (ii) a constant coefficient of friction [1]. Condition (ii) allows the ratio between the shear force and the normal reaction to be matched more realistically. The following deals with an ideal case having completely rough dies.

6.8.1 Indentation for 1 < b/t < 3.64

A detailed analysis [8] has shown that the slip line field in Fig. 6.43a does not meet the boundary tangentially when $b/t < 3.64$.

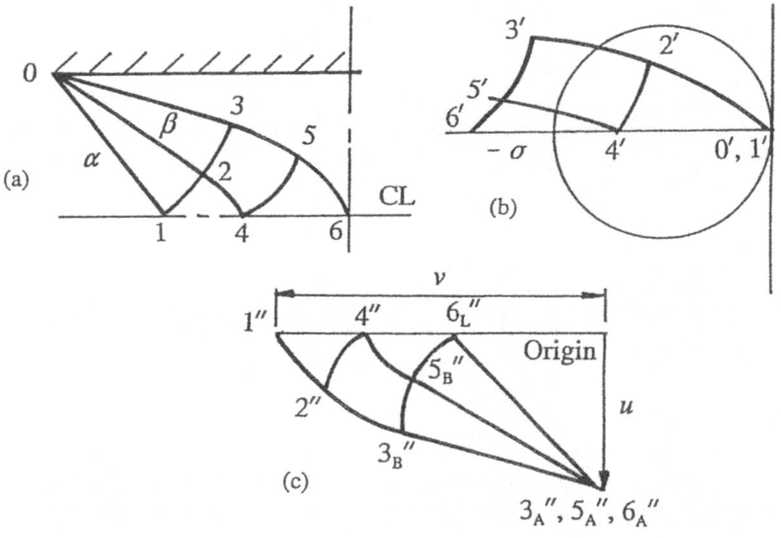

Figure 6.43 Rough die indentation, $1 < b/t < 3.64$

The 45° inclination of α–line 01 locates the pole of the weak circle at the origin of the stress plane in Fig. 6.43b. The cycloid 1′2′3′ is traced by the pole as the circle rotates ACW along $\tau = -k$ corresponding to β–line 123. Point 3 is adjusted until point 6 on α–line 356, which bounds the rigid zone, lies at the bar centre. This line can only meet the indentor edge tangentially for the limiting ratio: $b/t = 3.64$. For lower b/t, the edge inclinations of 0356 and another, intermediate α–line 024, lie between 0 and 45°. Slip lines meet axes of symmetry at 45° since these are principal stress directions. Elements of the hodograph in Fig. 6.43c are orthogonal to those in the physical plane (see Fig. 6.43a). The latter shows an array of discontinuity points, $3_B''$, $5_B''$ and $6_L''$, below the wedge. Points $3_A''$, $5_A''$ and $6_A''$ move rigidly downwards with velocity u. Checkpoint 1″ confirms that the lateral velocity is $v = (b/t)u$.

6.8.2 *Indentation for b/t > 3.64*

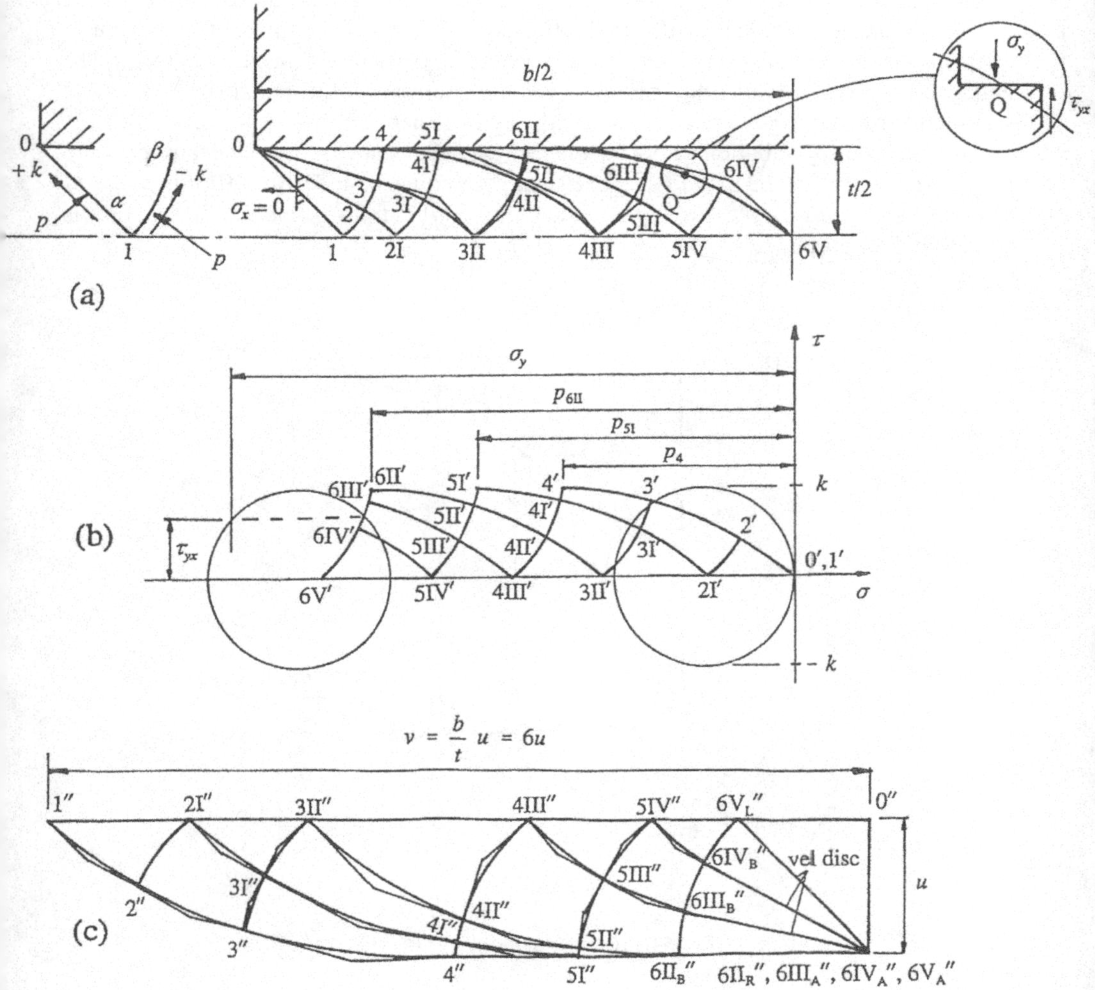

Figure 6.44 Rough die indentation $b/t = 6$

With $b/t > 3.64$ under full frictional conditions, the slip-lines will meet a rough die surface tangentially to maintain a constant shear stress k at all points of contact. For example, a SLF solution for $b/t = 6$ is given in Fig. 6.44a. For the stress plane (Fig. 6.44b), the initial weak circle, with pole $0'1'$ at the origin, applies to the stress state along α–line 01. The image of the β–line 1234 is a cycloid $1'2'3'4'$, traced by the pole as this circle rolls along $\tau = -k$. The α–lines that intersect with points 2, 3 and 4 are constructed in the usual way. Point 3 must be adjusted so that point 6V lies at the bar centre. This shows that α–line: 6II . . . 6V, contains the rigid wedge over half the die breadth. This SLF solution may be scaled to match any intermediate b/t ratio between 3.64 and 6.

Within the hodograph (Fig. 6.44c), the coincident points $6\text{II}_R''$, $6\text{III}_A''$, $6\text{IV}_A''$, and $6\text{V}_A''$, in the wedge zone, all move downwards with die velocity u. Crossing this α–line to points $6\text{II}_B''$ $6\text{III}_B''$, $6\text{IV}_B''$ and $6\text{V}_L''$, shows that this slip line is also a velocity discontinuity. The constant velocity is fixed by the horizontal motion of point $6V_L''$. The remaining elements of the hodograph are constructed from the orthogonal SLF elements, given the horizontal motion of points, 1″, 2I″, 3II″, 4III″, and 5IV″, along the axis of symmetry. Checkpoint 1″ confirms $v = 6u$ for rigid material to the left of 01, under constant volume deformation. The hodograph also reveals how the horizontal velocity increases from zero along any vertical section, i.e. by locating the intersections with slip lines between Figs 6.44a,c.

The pressure distribution across the die face is not constant. A mean indentor pressure p_m can be found from the magnitudes of σ_y and τ_{yx} at points 0, 4, 5I, 6II, 6III, 6IV and 6V. These are provided by the stress plane (see Fig. 6.44b), from which the distribution is derived in Fig. 6.45.

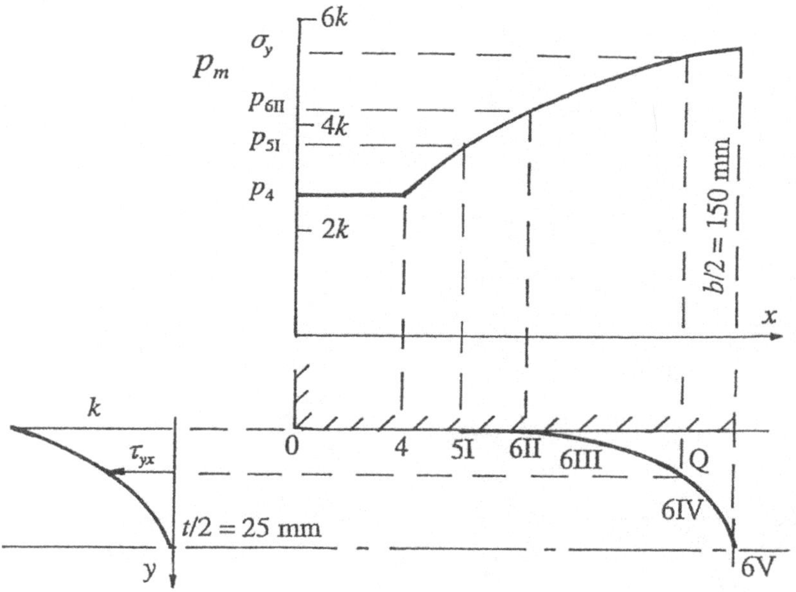

Figure 6.45 Cartesian stress distributions on die face

Using a similar derivation to eq(6.16a), the net compressive vertical force, per unit width, is given by

$$F_V = p_m b = \int \sigma_y \, dx - \int \tau_{yx} dy \tag{6.29}$$

in which the stress component directions are opposed, as illustrated for point Q in Fig. 6.44a. Applying eq(6.29), with $b = 300$ mm, $t = 50$ mm and $k = 150$ MPa, leads to: $F_V = 1400k$ per mm of width. Then, with k and p_m in MPa: $p_m/(2k) = 700/300 = 2.34$.

The forgoing solutions to each indentor enable a plot (see Fig. 6.46) of normalised pressure $p/(2k)$ versus b/t for the rough die indentation ($2k$ is the diameter of Mohr's circle). It is seen that a minimum appears in Figure 6.46 for the co-ordinates (1, 1) and pressure is constant for $b/t < 0.115$. Hill [4] calculated the pressure across the die face for b/t ratios between 0 and 1. Experiments by Watts and Ford [15] confirmed Hill's theoretical prediction. With the exception of some integral ratios b/t, mean pressures must be employed for $b/t > 1$.

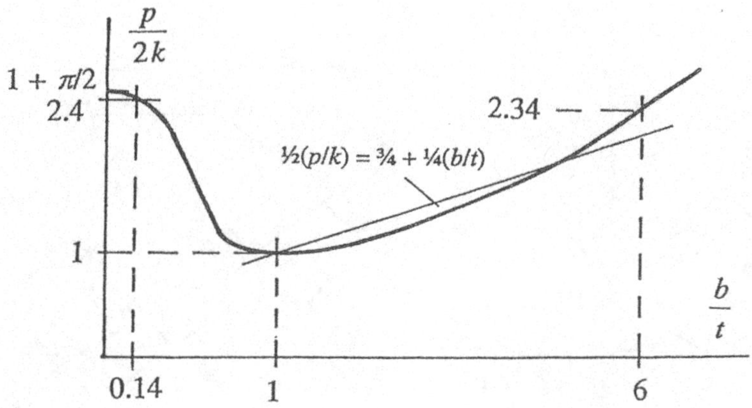

Figure 6.46 Pressure variation with *b*/*t* ratio for rough dies

Hill [16] estimated these from the following linear relation which is seen to be in approximate agreement with the ratio $p/(2k) = 2.34$, derived above for $b/t = 6$.

$$\frac{p}{2k} = \frac{3}{4} + \frac{b}{4t}$$

6.9 Smooth Die Indentation

Complex, slip-line fields for perfectly lubricated (smooth) dies, with $1 > b/t > 2$, were developed by Green [17]. Because slight differences arose for the intermediate critical ratio $b/t = \sqrt{2}$, a further subdivision of the b/t was found to be necessary.

6.9.1 Indentation for $1 < b/t < \sqrt{2}$

The SLF, given in Fig. 6.47a, shows a single dead-metal zone lying at the centre of the die surface. Elsewhere, slip-lines intersect with the die face and both axes cf symmetry at 45°.

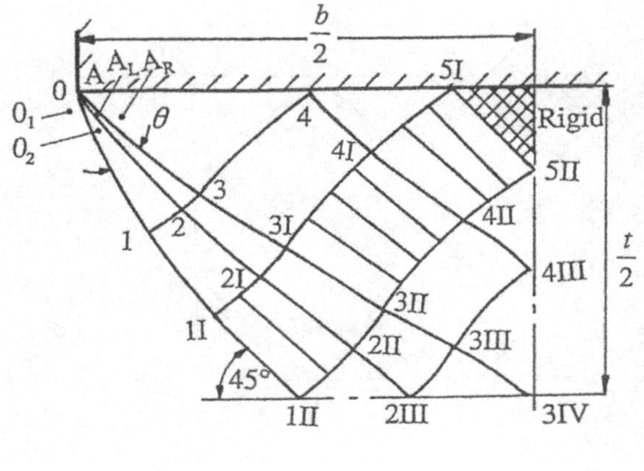

(a)

Figure 6.47 Smooth die indentation, $1 < b/t < \sqrt{2}$

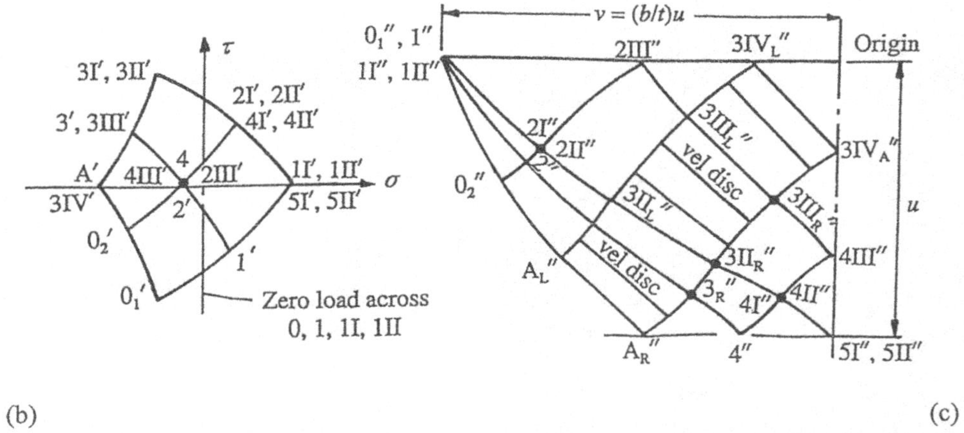

Figure 6.47 Smooth die indentation, $1 < b/t < \sqrt{2}$ (continued)

The angle θ must be adjusted in Fig. 6.47a, along with the shape of the α–line: 0, 1, 1I, 1II, in order to locate point 3IV correctly on the vertical centre line. In Fig. 6.47b, the τ–axis is chosen to remove horizontal loading and, in Fig. 6.47c, the velocity of all leftward points $0_1''$, $1''$, $1I''$ and $1II''$ must conform to volume deformation.

6.9.2 Indentation for $\sqrt{2} < b/t < 2$

Within this b/t range, a lesser die pressure applies to a SLF with two dead-metal zones (see Fig. 6.48a). This also involves a more complex arrangement of velocity discontinuities as seen in Fig. 6.48c. Again, θ and the shape of the α–line 0, 1, 1I, 1II in Fig. 6.48a, must be adjusted to locate points 6II correctly at the centre. The stress origin is fixed from the condition that the net horizontal force is zero across this slip line. Outside the deformation zone the incompressible material flows laterally with velocity $v = (b/t)u$.

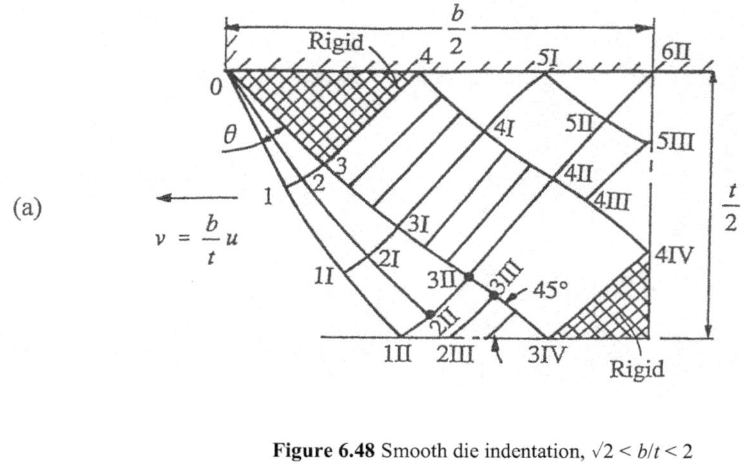

Figure 6.48 Smooth die indentation, $\sqrt{2} < b/t < 2$

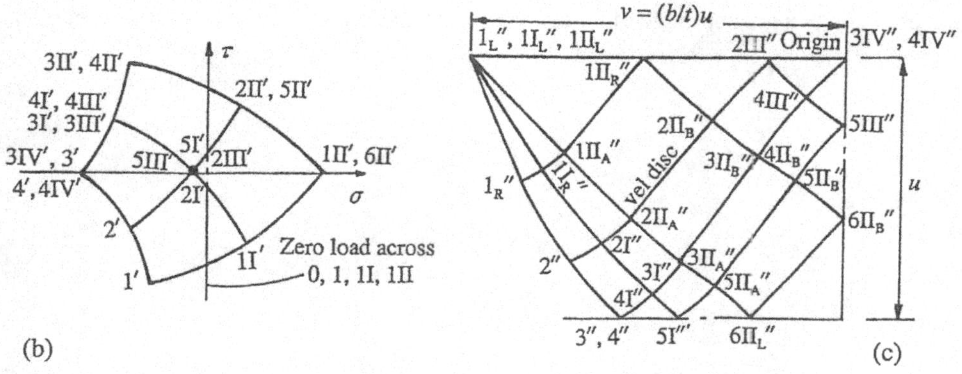

Figure 6.48 Smooth die indentation, $\sqrt{2} < b/t < 2$ (continued)

6.9.3 Indentation for $b/t \geq 2$

Complete SLF solutions, available for smooth dies with integral ratios $b/t = 2, 3, 4$ etc, show rigid, block slip. For example, a simple SLF applies to $b/t = 2$, as shown in Fig. 6.49a.

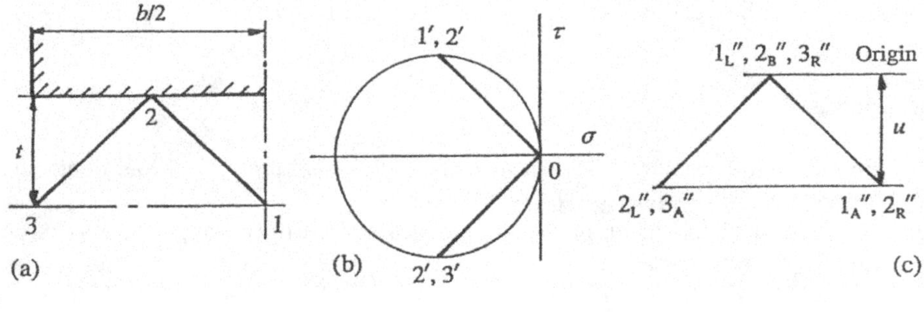

Figure 6.49 Smooth die indentation, $b/t = 2$

The weak, compressive circle in Fig. 6.49b will have its pole at the origin because $\sigma_x = 0$ in all horizontal directions. It follows that the mean stress on slip-line is constant. That is, slip-lines remain straight, with every stress state represented within a single Mohr's circle. The maximum positive shear stress occurs at points 1′ and 2′, so identifying the α–line 12. Similarly, the β–line 23 is derived from points 2′, 3′ in the stress plane. The dead-zone, above point 1 and to the right of 2, moves downwards with rigid velocity, u. This fixes points $1_A''$ and $2_R''$ within the hodograph (Fig. 6.49c). Velocity discontinuities coincide with slip lines, showing that the material points $1_L''$, $2_B''$ and $3_R''$ move leftward rigidly. Consideration of the horizontal and vertical components of velocity at points 2 and 3 reveals that the shape of elements within slip lines are retained with progressive deformation. With higher b/t ratios, a similar, extended pattern of slip lines and velocity discontinuities apply. These extensions introduce further 45° elements of alternating slope to reach the die edge.

Watts and Ford [15] recommended that b/t ratios within the range $2 \leq b/t \leq 4$ be employed for the plane strain compression test. Here t is the current thickness and the sheet width w is taken from the range $5 \leq w/b \leq 12$ (see Fig. 6.38). Their experiments confirmed the theoretical variation in $p/(2k)$ with b/t for frictionless compression in non-hardening material. Over a greater range of b/t the pressure variation is cyclic, as shown in Fig. 6.50.

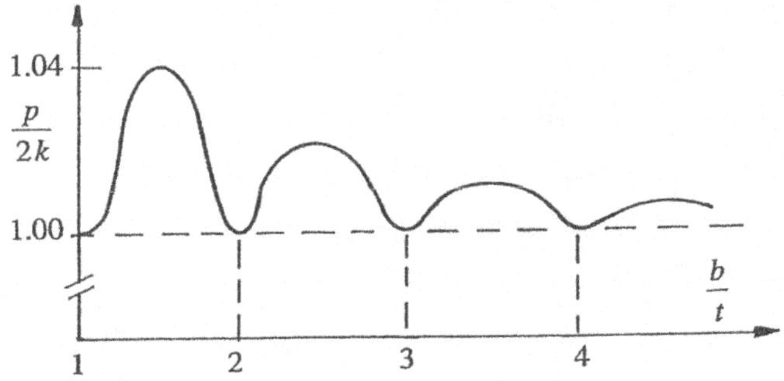

Figure 6.50 Cyclic variation in smooth die pressure with b/t ratio

Minimum values, $p/(2k) = 1$, occur for each integral ratio. With $b/t < 1$, the $p/(2k)$ variation will coincide with that found for rough dies (see Fig. 6.46). Testing within a limited b/t range allows for natural strain up to 200 % (see exercise 6.11). This is a far greater strain than can be reached in uniaxial compression upon a cylindrical testpiece where barrelling and buckling impede the uniform strain required (see § 9.6).

Finally, the reader may wish to be challenged by noting that SLFs for smooth indentations with fractional b/t ratios beyond 2 are not known.

References

1. Ford H. and Alexander J. M. *Advanced Mechanics of Materials*, 1963, Longmans.
2. Hencky H. *Zeit. Angew. Math. Mech.*, 1923, **3**, 241.
3. Geiringer H. "Fondements mathematiques de la theorie des corps plastiques isotropes", *Mem. Sci. Math.*, 1937, 86.
4. Hill R. *The Mathematical Theory of Plasticity*, 1950, Clarendon Press, Oxford.
5. Prager W. and Hodge P. G. *The Theory of Perfectly Plastic Solids*, 1951, Wiley, New York.
6. Chakrabarty J. *Theory of Plasticity*, 3rd Ed, 2006, Elsevier Butterworth-Heinmann.
7. Prager W. "A geometrical discussion of the slip-line field in plane plastic flow", *Trans. Roy. Inst. of Tech.*, 1953, 65, Stockholm.
8. Ellis F. *Jl. Strain Anal.*, 1967, **2**, 52.
9. Lee E. H. "On stress discontinuities in plane plastic flow", Proc: *3rd Symposium in Applied Math*, 1950, 213, McGraw-Hill.
10. Green W. A. *Jl. Mech. Phys. Solids*, 1962, **10**, 225.
11. Alexander J. M. *Quart. Appl. Maths.*, 1961, **19**, 32.
12. Bishop J. F. W. *Jl. Mech. Phys. Solids*, 1953, **2**, 43.
13. Prandtl L. *Nachr. Ges. Wiss. Gottingen*, 1920, 74.
14. Hill R. *Jl. Iron and Steel Inst.*, 1947, **156**, 513.
15. Watts A. B. and Ford H. *Proc. I. Mech. E.*, 1952–3, **1**B, 448.
16. Hill R. In: *Strength of Materials*, 1950, HMSO, **6**(1), 87.
17. Green A. P. *Phil. Mag.*, 1951, **42**(7), 900.

Exercises

6.1 Construct extended, pseudo slip lines to show that, within the rigid body regions for $R = 2$ extrusion (see Fig. 6.23a), the yield criterion is nowhere violated.

6.2 Show that the slip line field and hodograph solutions for an extrusion ratio $R = 3$ in Fig. 6.27 produces a consistent positive rate of plastic work.

6.3 Determine the horizontal extrusion force F_H from the application of eq(6.16a) to the α–slip line aI, 2I, 3I, I4 in Fig. 6.32. Take the extrusion ratio $R = 3.3$ for a die inclination $\theta = 30°$.

6.4 In Fig. 6.51 the magnitude and direction of respective velocities v_A and v_B for points 1 and 2 on an α–line, together with v_C and v_D for points 1 and 3 on a β–line, are known. Construct the hodograph for the region enclosed within α and β.

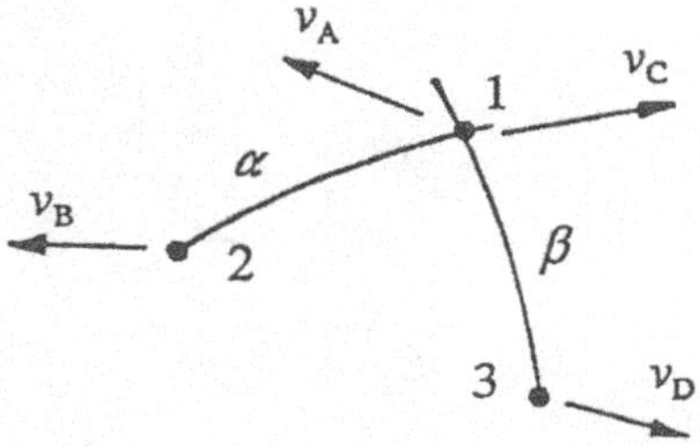

Figure 6.51

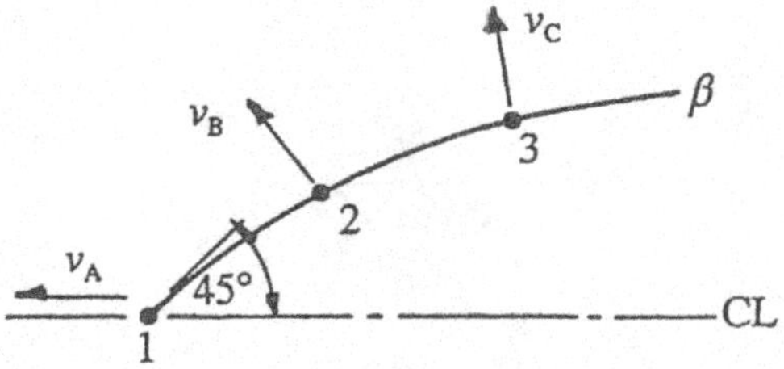

Figure 6.52

6.5 The magnitude and direction of absolute velocities v_A, v_B and v_C at points 1, 2 and 3 on a β–line are known. Construct the hodograph for the region enclosed within this β–line and an axis of symmetry (C_L) when the intersection is as shown in Fig. 6.52.

6.6 The absolute velocities at two points 1 and 2 on a free boundary in Fig. 6.53 are known. Construct the hodograph for the lower, interior region.

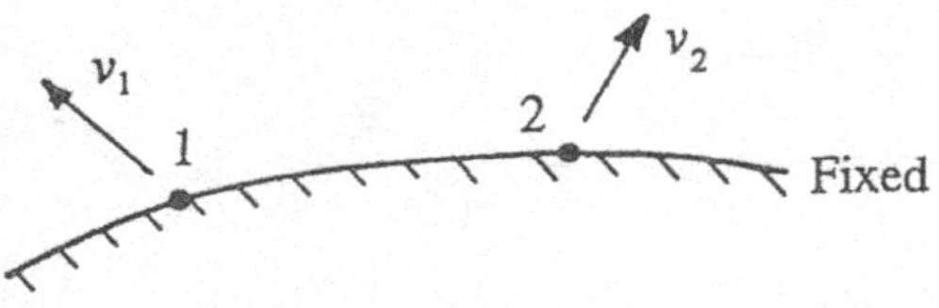

Figure 6.53

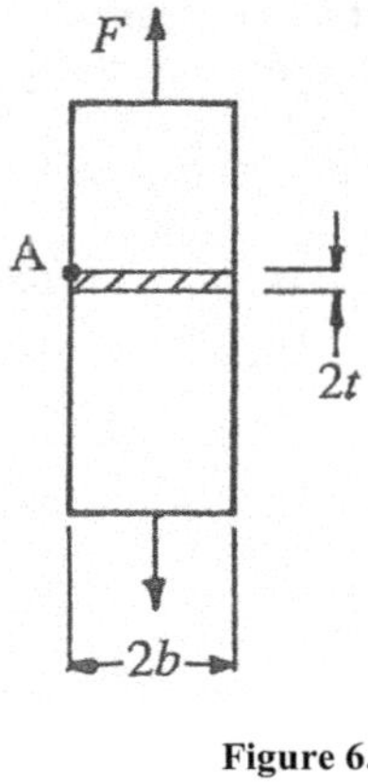

Figure 6.54

6.7 Figure 6.54 shows two rigid blocks connected with a layer of solder $2b \times 2t \times w$, where the width w is large compared to b and t. When carrying a tensile force F/unit width, the solder flows plastically under a shear yield stress k. Sketch the slip line field and hodograph and show that when $b/t = 11.5$, $\sigma = k(2.5\pi + 1)$ at point A, where the slip line bounding the central dead zone reaches the surface of the block. Estimate F and the mean normal stress along the interface for this condition.

6.8 In the side extrusion container in Fig. 6.55, the ram exerts a vertical force P to extrude horizontally a rectangular strip of thickness h. Assuming frictionless plane-strain conditions, determine appropriate upper bounds and associated hodographs for the extrusion pressure/unit width of container. How would these be modified by base friction?

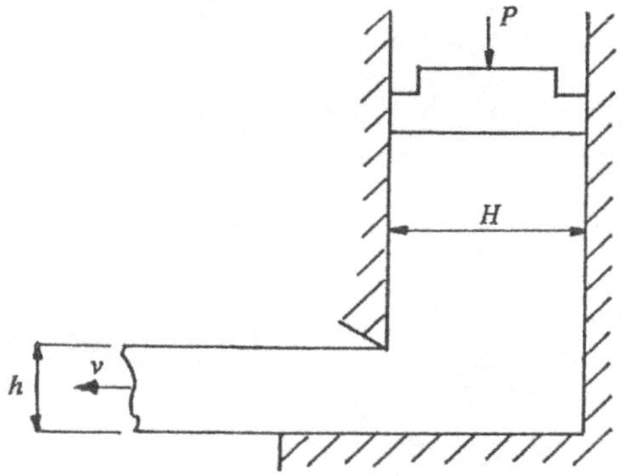

Figure 6.55

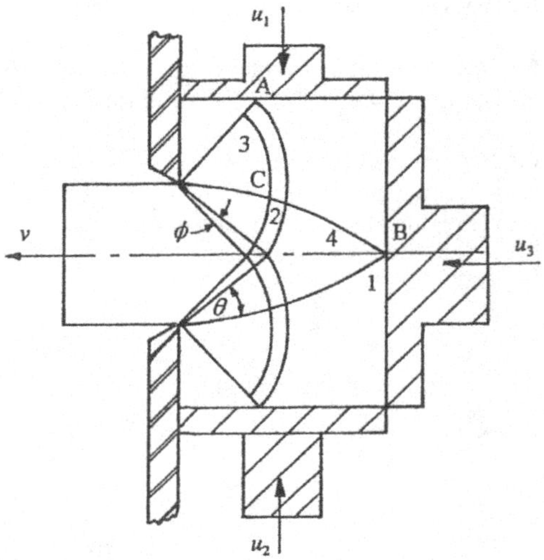

Figure 6.56a

6.9 Figures 6.56a and 6.56b give the general SLF and hodograph solutions respectively for frictionless extrusion under three rams. The inward speeds u_1 and u_2 of the lateral rams are equal. Determine (a) the correct angle of intersection of the slip lines with the rams, (b) the α– and β–lines for which θ and φ re-appear within the hodograph as shown, (c) velocity vectors within the hodograph corresponding to the absolute velocities at points 1, 2, 3 and 4 and the pressure exerted by each ram at A and B. How would you modify the SLF in Fig. 6.55a with the presence of friction along each lateral ram face and the hodograph in Fig. 6.56b, when $u_1 \neq u_2$ with $u_3 = 0$?

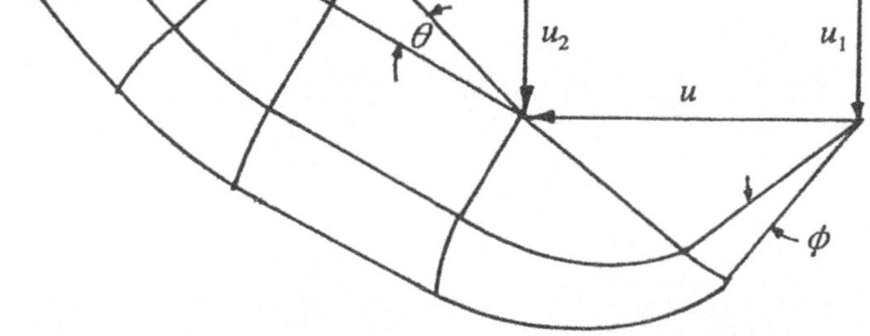

Figure 6.56b Hodograph for extrusion by three rams

6.10 A quarter section of a wide plate with a circular arc notch is shown in Fig. 6.57a. The plate is a rigid perfectly plastic material and is subjected to a tensile force P along the vertical axis. Construct the SLF for when the plate yields initially. Find P from a summation of forces around the plastic rigid boundary interpolating as necessary between cycloids for the stress plane given in Fig. 6.57b. Find the position of the notch boundary for a 6 mm upward displacement (a stretch) of the quadrant shown, given that the SLF remains unaltered.

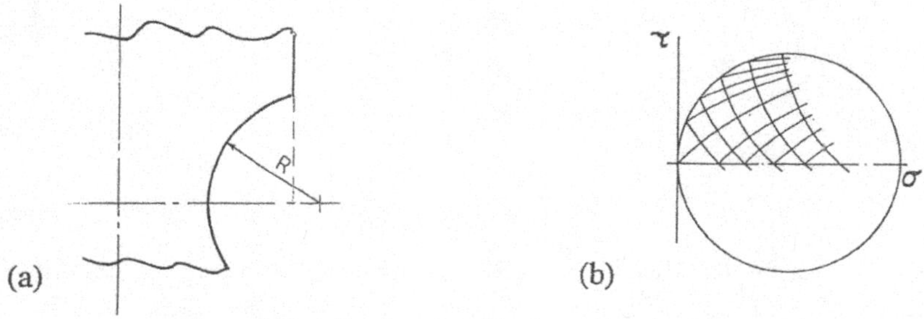

Figure 6.57a,b Tension of a wide plate with a circular arc notch

6.11 Watts and Ford [15] recommended that b/t ratios within the range $2 \le b/t \le 4$ be employed for the plane strain compression test. Here t is the current thickness and the sheet width w is taken from the range $5 \le w/b \le 12$ (see Fig. 6.38).

The following data were obtained for compression of 5754 Al-alloy strip at a rate of 3 mm/min in which the diminishing thickness t was measured under each load. A strip width of 30.4 mm and initial thickness 2.48 mm, cut at 45° to the roll, maintained the second equality. An interchange of dies with diminishing b range: 4, 3.2, 2.5 and 1.28 mm die changes maintained the first inequality. Assuming frictionless compression with lubrication of a hardening material, employ these data to produce the equivalent stress versus integrated plastic strain for the material. Check the validity of the data with a suitable fit from the Hollomon, Ludwik and Swift laws.

$b = 4$ mm, $t_o = 2.48$ mm

W, kN	23.8	27.8	30.5	32.5	34.5	36.0	37.5	38.6
t, mm	2.45	2.40	2.35	2.28	2.20	2.12	2.02	1.92

$b = 3.2$ mm, $t_o = 1.92$ mm

W, kN	29.7	31.2	32.0	32.6	33.0	33.5	34.0
t, mm	1.89	1.82	1.73	1.65	1.55	1.47	1.38

$b = 2.5$ mm, $t_o = 1.38$ mm

W, kN	26.5	27.2	27.5	27.7	28.0	28.4	28.7
t, mm	1.35	1.27	1.18	1.10	1.00	0.91	0.80

$b = 1.28$ mm, $t_o = 0.8$ mm

W, kN	15.0	15.1	15.2	15.3	15.5	15.7	16.1	16.7	17.4	18.2	20.5
t, mm	0.743	0.685	0.614	0.549	0.477	0.413	0.351	0.291	0.235	0.187	0.140

Ans: *See solution*

6.12 A slip line field is proposed for frictionless sheet drawing through a tapered die is shown in Fig. 6.58 along with a portion of the stress plane. Draw a hodograph for this solution and check whether it fulfils the boundary conditions on velocity. Determine the pressure upon the die face and the drawing force/unit width. Show that a stress discontinuity must exist at the exit from the die.

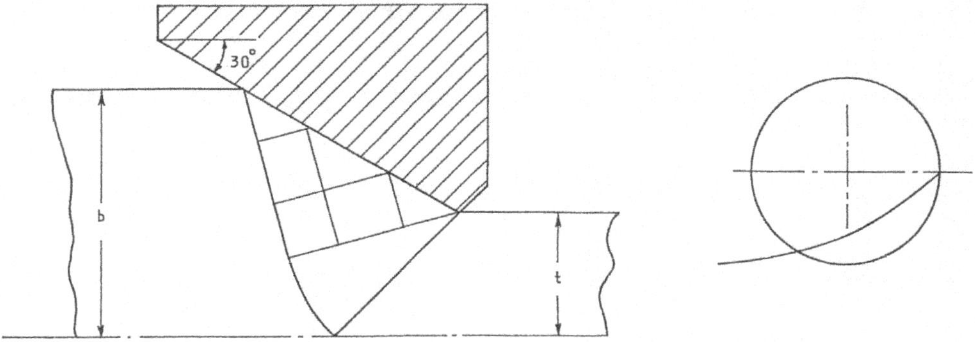

Figure 6.58 SLF for frictionless sheet drawing through a tapered die

6.13 A slip line field is proposed for frictionless extrusion through an inclined die as shown in Fig. 6.59 along with a portion of the stress plane. Given an extrusion ratio $H/h = (1 + \sqrt{3})/1$, draw a hodograph for this solution and check that it fulfils the boundary conditions on velocity. Determine the pressure upon the die face given by this solution considering the stresses acting upon the boundary slip line at entry.

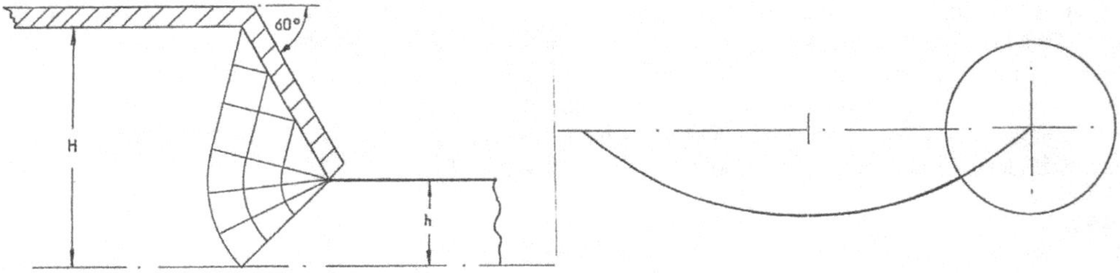

Figure 6.59 SLF for frictionless extrusion through an inclined die

6.14 Figure 6.60a shows a suggested slip line field for frictionless plane strain indentation of a block of thickness t by a flat die of width w. Figure 6.59b shows the corresponding stress plane. Construct the hodograph and show that the SLF is a possible solution. Determine the position of the origin of the stress plane and hence the pressure between the die and block.

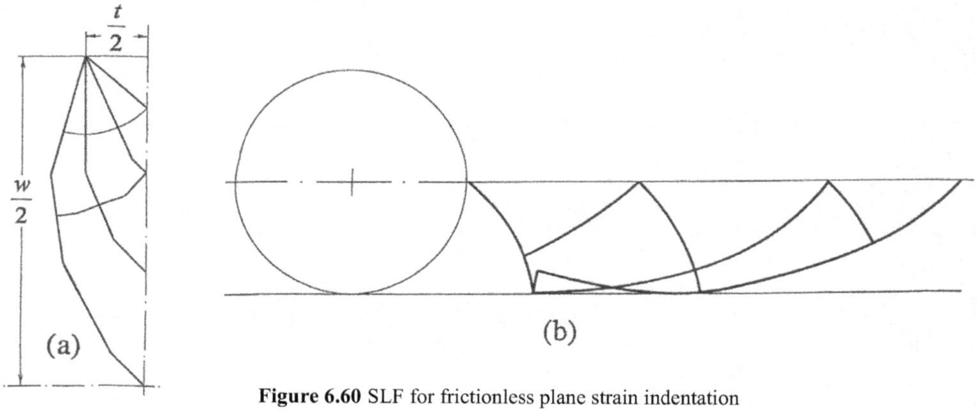

Figure 6.60 SLF for frictionless plane strain indentation

CHAPTER 7

LIMIT ANALYSIS

7.1 Introduction

The plastic behaviour of a structure or a process involving plastic-rigid material may often be represented in lower- and upper-bounds. Two principles are employed to obtain these bounds: (i) that a statically admissible stress field is in equilibrium, satisfying the forces applied to the boundary and (ii) that a kinematically admissible velocity field is compatible with internal strains, satisfying displacements imposed at the boundary. Principle (i) provides a lower-bound solution, where the admissible stress field will not exceeds the yield stress anywhere. Principle (ii) is associated with an upper-bound solution, which provides the loading associated with the admissible displacement field. Since the work dissipated by the upper bound loading is usually more than that done by the actual forces, the true solution will lie between the bounds. The latter impose limits to plastic behaviour but in certain structures, as for the beams that follow, no difference exists between them.

7.2 Upper and Lower Bounds for Beams

In consideration of the limit loading for a simply-supported beam in Fig. 7.1a, the lower bound load W_L is supplied from the moment diagram (Fig. 7.1b). In the beam cross-section beneath W there exists an admissible stress field. The collapse moment is given by [1].

$$M_P = \frac{abW}{a + b}$$

A lower bound load follows:

$$W_L \leq \frac{M_P(a + b)}{ab} \tag{7.1a}$$

The beam displacement diagram, in Fig. 7.1c, shows the work done by an upper-bound load. Normally, this exceeds the energy dissipated with the rotation θ of the hinge. This gives

$$W_U\delta \geq M_P\theta$$

Substituting $\theta = \delta/a + \delta/b$:

$$W_U \geq \frac{M_P(a + b)}{ab} \tag{7.1b}$$

In general, the true loading W lies in the range $W_L < W < W_U$, but eqs(7.1a, b) show no distinction between the bounds in this case. If the ends of the beam in Fig. 7.1a were both fixed, collapse would occur from the formation of three plastic hinges: one at each end and one beneath W. The condition for collapse is that equal, fully–plastic moments M_P are attained simultaneously within each hinge.

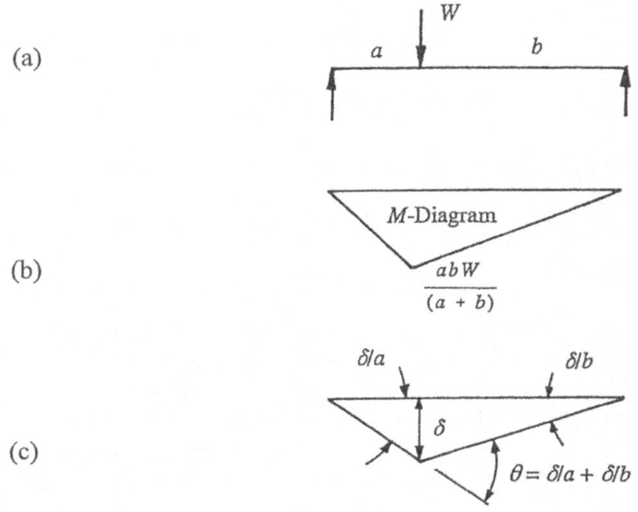

Figure 7.1 Collapse of a simply supported beam

Their magnitudes are each 100 % greater than M_P for a simply supported beam. This gives, from modifying eq(7.1a)

$$W_L = \frac{2M_P(a + b)}{a\,b} \tag{7.2a}$$

The deflected shape of the encastre beam is similar to Fig. 7.1c. The work of collapse is given by

$$W_U\delta = M_P\theta + \left(M_P \times \frac{\delta}{a}\right) + \left(M_P \times \frac{\delta}{b}\right)$$

$$W_U = \frac{2\,M_P(a + b)}{a\,b} \tag{7.2b}$$

Again, there is no distinction between the bounds in eqs(7.2a,b). *Note*: For both beam fixings, the shape of the cross-section is not specified when deriving the lower-bound collapse load W_L. Generally, this is true for a beam with a uniform cross-section.

To take an example of a non-uniform section, consider a plane, vee-notched beam subjected to a uniform moment M/unit width (see Fig. 7.2a). A lower bound moment M_L is found from the fully plastic admissible stress field in the weakest section. This gives

$$M_L = F_L d = Y\left(1 \times \frac{a}{2}\right)\frac{a}{2} = 0.25\,Ya^2 \tag{7.3a}$$

With an upper bound collapse under a single plastic hinge (see Fig. 7.2b), the energy dissipated is $2M_U\delta\theta$. Here M_U follows from the shear yield stress $k = Y/2$, acting around the plastic zone radius r.

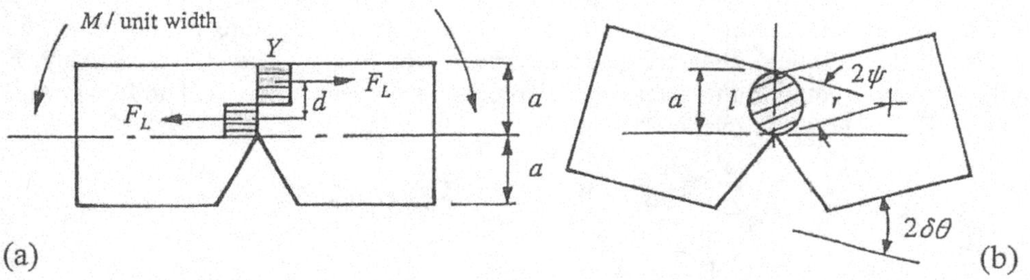

Figure 7.2 Plane vee-notched beam

This gives

$$M_U = klr = \frac{Ylr}{2} \tag{7.3b}$$

where $l = 2\psi r$ is the arc length. From eq(7.3b), collapse under a single plastic hinge is assumed to occur when the hinge geometry product lr is at its minimum:

$$lr = 2\psi r^2 = 2\psi\left(\frac{a^2}{4\sin^2\psi}\right) = \frac{a^2\psi}{2\sin^2\psi}$$

$$\frac{\mathrm{d}}{\mathrm{d}\psi}(lr) = \frac{a^2}{2\sin^2\psi} - \frac{a^2\psi\cos\psi}{\sin^3\psi} = 0$$

$$\tan\psi = 2\psi$$

That is, $\psi = 1.166$ rad and $lr = 0.69a^2$. Substituting into eq(7.3b) gives

$$M_U = 0.345Ya^2 \tag{7.3c}$$

The true collapse moment M lies within the range bounded by eqs(7.3a,c):

$$0.25Ya^2 < M < 0.345Ya^2$$

A slip-line field solution gives $0.315Ya^2$. Generally, it is found that SLF solutions will be closer to upper-bound estimates.

7.3 Collapse of Structures

7.3.1 Principal of Virtual Work

A simpler method to obtain collapse loading of beams and plane frames is to treat collapse as a rigid mechanism under an equilibrium system of co-planar, non-concurrent forces F_i where $i = 1, 2 \ldots N$. Using vectors, the force and moment equilibrium conditions are

$$\sum_{i=1}^{N} \mathbf{F}_i = 0 \quad \text{and} \quad \sum_{i=1}^{N} \mathbf{M}_i = 0 \tag{7.4a,b}$$

where the moments $M_i = |\mathbf{M}_i|$ of forces must sum to zero at any arbitrary point. Since the displacements and rotations that develop in a collapsing structure do not alter the equilibrium eqs(7.4a,b), they are taken as *virtual*, written as Δ^v and θ^v, respectively. It follows that the equilibrium system of *real* forces and moments forces will do zero *virtual* work. That is, the scalar relationships apply

$$\sum_{i=1}^{N} F_i \Delta_i^v = 0 \quad \text{and} \quad \sum_{i=1}^{N} M_i \theta_i^v = 0 \qquad (7.5a,b)$$

Combining eqs(7.5a,b) leads to a useful form

$$\sum_{i=1}^{N} F_i \Delta_i^v = \sum_{i=1}^{N} M_i \theta_i^v = 0 \qquad (7.6)$$

In eq(7.6), θ_i^v refers to the rotation at each hinge under a collapse moment and Δ_i^v to the deflections beneath the corresponding forces. To apply eq(7.6), a collapse geometry must be assumed. Where a number of collapse mechanisms are possible, the true collapse load is that which minimises eq(7.6).

7.3.2 Beams

The first beam structure in (a) below reveals that it is possible for a non hardening material to require a rising load versus deflection diagram from a fully elastic condition through to plastic collapse. Specific examples (b) and (c) that follow apply PVW at collapse to reveal the likely collapse mode.

(a) Propped Cantilever - Central Load

The transition from a fully elastic cantilever in Fig. 7.3a to plastic collapse appears within the ordinates of the respective bending moment diagrams given in Fig. 7.3b. The net moment diagram splits into two components: hogging (positive) due to W and sagging (negative) due to P as shown in Fig. 7.3b.

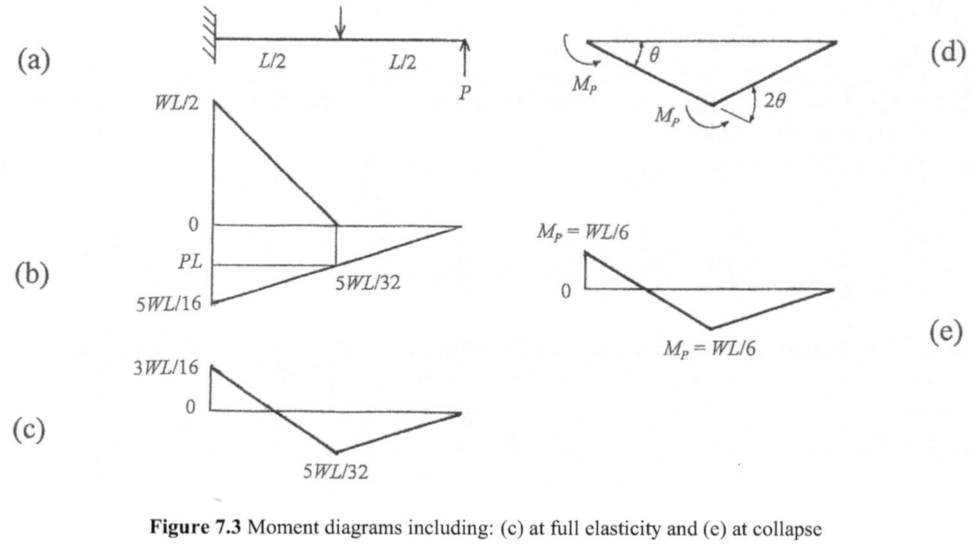

Figure 7.3 Moment diagrams including: (c) at full elasticity and (e) at collapse

Referring to Fig. 7.3b, with zero deflection at the prop, Mohr's deflection theorem reveals the prop reaction P to fully elastic loading, i.e. when the yield stress is reached.

$$EI\delta_P = 0 = \text{Moment of area of } M\text{-Diagram from A to } P \text{ about } P$$

$$\therefore \text{Moment of hogging diagram} - \text{Moment of sagging diagram} = 0$$

$$(WL/2)(L/4)[L/2 + (2/3 \times L/2)] - (PL \times L/2 \times 2L/3) = 0$$

$$5WL^3/48 - PL^3/3 = 0 \quad \therefore \; P = 5W/16$$

Hence in Fig. 7.3c the net moment diagrams appears showing:

$$\text{Net fixing moment: } M = WL/2 - P = WL/2 - 5W/16 = 3WL/16$$

$$\text{Net central moment: } M = PL/2 = 5W/16 \times L/2 = 5WL/32$$

The central deflection follows from Mohr's theorem:

$$EI\delta_W = \text{Moment of area of } M\text{-Diagram from A to } W \text{ about } W$$

$$= [WL/2 \times 2/3(L/2) \times L/4] - \{(5WL/32 \times L/2 \times L/4) + [5WL/32 \times L/4 \times 2/3(L/2)]\}$$

$$= WL^3/24 - (5WL^3/256 + 5WL^3/384)$$

$$\therefore \; \delta_W = \frac{7WL^3}{768EI} \tag{7.7a}$$

Thus, the central deflection δ_W is linear with the load W. To attain a fully elastic fixing moment the load is:

$$W_E = \frac{16M_E}{3L} \tag{7.7b}$$

where M_E depends upon the yield stress Y and the properties of the beam cross-section area with respect to its neutral axis: depth y and second moment of area I (i.e. $M_E = YI/y = a^3Y/6$). For example, consider a 750 mm long, 8 mm square section, propped cantilever in low carbon steel. Equations (7.7a,b) provide the central loading gradient and yield load, respectively:

$$\frac{W}{\delta_W} = \frac{768EI}{7L^3} = \frac{768 \times (200 \times 10^3) \times (8^4/12)}{7 \times 750^3} = 17.75 \, \frac{\text{N}}{\text{mm}}$$

$$W_E = \frac{16YI}{3yL} = \frac{8a^3Y}{9L} = \frac{8 \times 8^3 \times 200}{9 \times 750} = 121.4 \, \text{N}$$

which were confirmed experimentally [2]. The fully plastic collapse load is found from the moment diagram in which the fixing and central moments each equal M_P by stress redistribution for the loading beyond W_E. Figure 7.3d shows the collapse mechanism for the cantilever in which PVW applies:

$$M_P\theta^v + M_P(2\theta^v) = W_P\Delta^v$$

and with $\Delta^v = L\theta^v/2$, equal moments apply to the end and centre (see Fig. 7.3e):

$$M_P = W_P L/6$$

Taking $M_P = a^3Y/4$, the collapse load W_P is

$$W_P = \frac{6a^3Y}{4L} = \frac{6 \times 8^3 \times 200}{4 \times 750} = 204.8 \text{ N}$$

confirming the observation that the load increases well beyond W_E with re-distribution under a constant yield stress for an elastic–perfectly plastic material.

(b) Propped Cantilever - Distributed Loading
For a propped cantilever in Fig. 7.4a, the mechanism of collapse shown in Fig. 7.4b is assumed. The uniformly distributed collapse load w_P and the position z_o of the second hinge are to be determined.

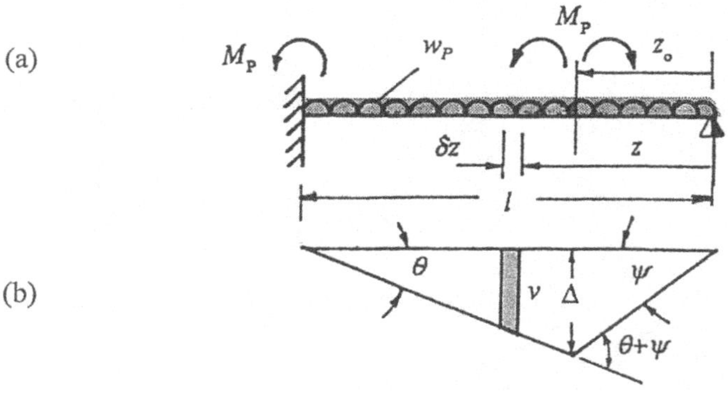

Figure 7.4 Collapse mechanism in a propped cantilever with distributed loading

Applying eq(7.6) and omitting the 'virtual' superscript v for simplicity:

$$M_P\theta + M_P(\theta + \psi) = w_P \int_0^l v\,dz \tag{7.8a}$$

where v is the deflection at position z and wdz is the elemental force at this position. The integral in eq(7.8a) is the enclosed area

$$\int_0^l v\,dz = \frac{l\Delta}{2} \tag{7.8b}$$

The rotation ψ is found from

$$\Delta = (l - z_o)\theta = z_o\psi \tag{7.8c}$$

which gives

$$\psi = \frac{(l - z_o)\theta}{z_o} \tag{7.8d}$$

Substituting eqs(7.8b,c,d) into eq(7.8a):

$$w_P = \frac{2M_P(l + z_o)}{l z_o(l - z_o)} \tag{7.9a}$$

To find the z_o value which minimises eq(7.9a), set $dw_P/dz_o = 0$. This gives

$$z_o^2 + 2z_o l - l^2 = 0 \tag{7.9b}$$

for which $z_o = (\sqrt{2} - 1)l = 0.4141l$. Then, from eqs(7.9a,b), $w_P = 11.66M_P/l^2$.

(c) Continuous Beam

Figure 7.5a shows a stepped, two-bay beam in which the plastic collapse moment for the second bay is twice that of the first. The beam is fixed at one end and rests upon two simple supports. To determine the collapse load W, all possible modes of failure (see Figs 7.5b–d) must be considered.

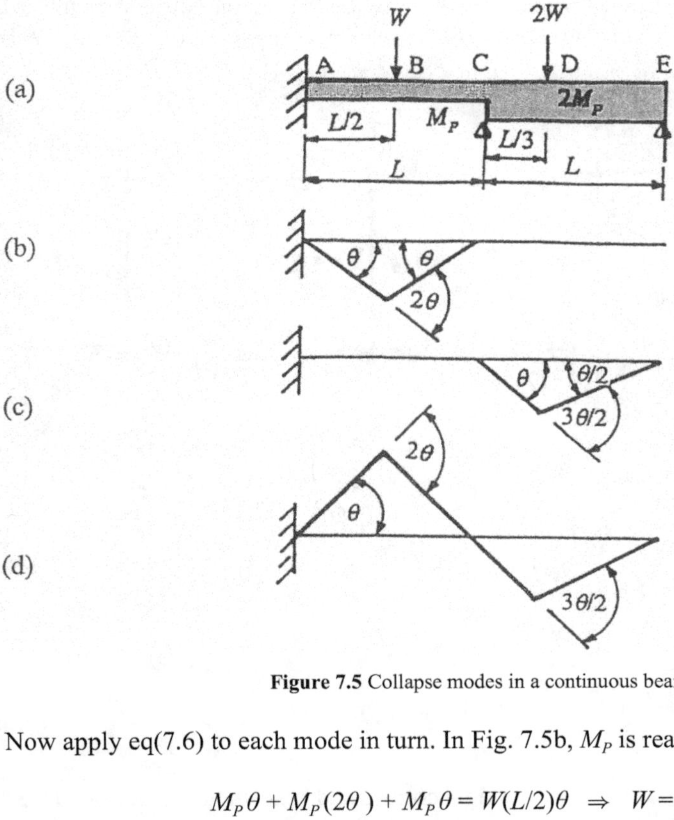

Figure 7.5 Collapse modes in a continuous beam

Now apply eq(7.6) to each mode in turn. In Fig. 7.5b, M_P is reached at A, B and C:

$$M_P\theta + M_P(2\theta) + M_P\theta = W(L/2)\theta \;\Rightarrow\; W = 8M_P/L \tag{7.10a}$$

In Fig. 7.5c, M_P is reached in a hinge to the left of C and $2M_P$ must be reached for collapse at D. The rotation at E is not accompanied by a moment (free-end). This gives

$$M_P\theta + 2M_P(3\theta/2) = 2W(L/3)\theta \Rightarrow W = 6M_P/L \tag{7.10b}$$

In Fig. 7.5d, collapse moments of M_P, M_P and $2M_P$ are reached at A, B and D respectively:

$$M_P\theta + M_P(2\theta) + 2M_P(3\theta/2) = -W(L/2)\theta + 2W(L/3)\theta \Rightarrow W = 36M_P/L \tag{7.10c}$$

The least load from eqs(7.10a–c) is the collapse condition: $W = 6M_P/L$.

7.3.3 Portal Frames

The analysis of collapse in a portal frame involves the application of eq(7.6) to the most obvious failure modes [3]. Firstly, the application of eq(7.6) is demonstrated in an example. There then follows a broader discussion upon weight to strength requirements in portal frame design.

Example 7.1 Determine the collapse loading for the portal frame shown in Fig.7.6a from a consideration of (b) beam collapse, (c) stanchion side sway and (d) combined collapse of beam and stanchions. The beam vertical load W is twice that of the left corner inward horizontal load. The moment carrying capacity of the beam M_P is increased by 20 % for each stanchion.

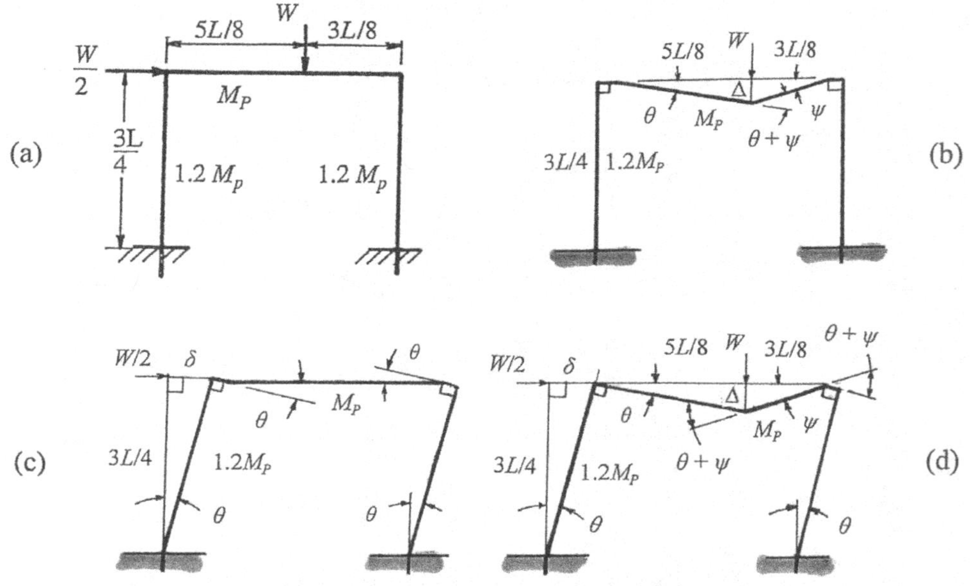

Figure 7.6 Modes of failure in a portal frame

(b) Beam Collapse

Hinges form on the beam side of the left and right corners and beneath the vertical load (see Fig. 7.6b). The virtual rotations at each hinge are respectively θ^v, ψ^v and $\theta^v + \psi^v$ radian and the vertical beam deflection is Δ^v. These bear the following relationships:

$$\Delta^v = \frac{5L\theta^v}{8} = \frac{3L\psi^v}{8}, \quad \therefore \ \psi^v = \frac{5\theta^v}{3} \qquad \text{(i, ii)}$$

Applying PVW, eq(7.6):

$$M_P\theta^v + M_P(\theta + \psi)^v + M_P\psi^v = W\Delta^v$$

From eqs(i) and (ii):

$$M_P\theta^v + M_P(\theta + 5\theta/3)^v + (M_P \times 5\theta^v/3) = W \times 5L\theta^v/8$$

$$16M_P\theta^v/3 = 5WL\theta^v/8$$

$$\therefore \qquad W = 128M_P/15L = 8.5\dot{3}\,M_P/L$$

(c) Stanchion Side Sway

Hinges form in Fig. 7.6c at each stanchion base and on the beam side of each corners, each with rotation θ^v, to give a top horizontal displacement δ^v:

$$\delta^v = 3L/4 \times \theta^v \qquad \text{(iii)}$$

Applying eq(7.6):

$$1.2M_P\theta^v + M_P\theta^v + M_P\theta^v + 1.2M_P\theta^v = W\delta^v/2$$

where from eq(iii):

$$4.4M_P\theta^v = 3WL\theta^v/8$$

$$W = 176M_P/15L = 11.7\dot{3}\,M_P/L$$

(d) Combined Collapse of Beam and Stanchions

A minimum number of hinges form at each stanchion base, on the beam side of the right corner and beneath the vertical load (Fig. 7.6d). The rotation at each stanchion base hinge is θ^v and of the two beam hinges is $\theta^v + \psi^v$ radian. The vertical beam deflection is Δ^v and the horizontal side sway is δ^v. These bear the following relationships:

$$\Delta^v = \frac{5L\theta^v}{8} = \frac{3L\psi^v}{8}, \quad \psi^v = \frac{5\theta^v}{3}, \quad \delta^v = \frac{3L\theta^v}{4} \qquad \text{(iv–vi)}$$

Applying eq(7.6):

$$2\left[1.2M_P\theta^v + M_P(\theta + \psi)^v\right] = W\delta^v/2 + W\Delta^v$$

From eqs(iv)–(vi):

$$2\left[1.2M_P\theta^v + M_P(\theta + 5\theta/3)^v\right] = (W/2 \times 3L\theta^v/4) + (W \times 5L\theta^v/8)$$

$$116M_P\theta^v/15 = WL\theta^v$$

$$W = 116M_P/15L = 7.7\dot{3}\,M_P/L$$

which is the lowest failure load predicted among (b)–(d).

The frame given in example 7.1 is of good design with the collapse loads all being relatively close, which may not apply to other portal frame designs. For example, let the frame ABCDE in Fig. 7.7a carry both and horizontal and vertical loading at B and C. When supports at A and E are hinged they carry no moments. Consequently, collapse can occur in the horizontal beam (Fig. 7.7b), the stanchions may sway (Fig. 7.7c) and in Fig. 7.7d a combination of collapse and sway applies.

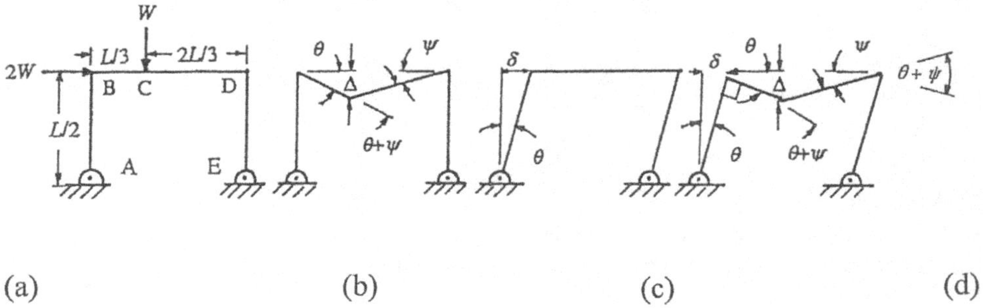

Figure 7.7 Collapse modes in a portal frame

Corner collapse will occur in the limb with the lower collapse moment where different sections exist between the limbs. Assuming that the stanchions have a 50 % greater moment capacity than the beam, all the hinges will form on the beam side of the corners under M_P, as shown. Failure from each mechanism occurs with hinges at: (b) B, C and D, (c) B and D and (d) C and D. Applying eq(7.6) to each mode in Figs 7.5 b, c and d, respectively:

$$M_P\theta + M_P(\theta + \psi) + M_P\psi = W\Delta \quad (7.11a)$$

$$2M_P\theta = 2W\delta \quad (7.11b)$$

$$M_P(\theta + \psi) + M_P(\theta + \psi) = W\Delta + 2W\delta \quad (7.11c)$$

where superscript v is dropped. Substituting $\psi = \theta/2$, $\delta = L\theta/2$ and $\Delta = L\theta/3$ into eqs(7.11a,b,c) leads to the respective collapse load $W = 9M_P/L$, $2M_P/L$ and $9M_P/4L$, the lowest showing that a sway collapse would occur. *Note*: These calculations show that the collapse moment ($1.5M_P$) for each stanchion is not reached. It would be more economical on material to employ a uniform cross section. With a single collapse moment M_P for the whole structure hinges form in the top corners and eqs(7.11a–c) again apply.

With encastré base fixings in Fig. 7.7a, further hinges must form at A and E under the side-sway and combined collapse modes (e.g. see Figs 7.8d,f). Equations (7.11b,c) are modified:

$$M_P\theta + M_P(\theta + \psi) + M_P\psi = W\Delta \quad (7.12a)$$

$$2M_P\theta + 2(3M_P/2)\theta = 2W\delta \quad (7.12b)$$

$$M_P(\theta + \psi) + M_P(\theta + \psi) + 2\,(3M_P/2)\theta = W\Delta + 2W\delta \quad (7.12c)$$

The respective collapse loads are $W = 9M_P/L$, $5M_P/L$ and $9M_P/(2L)$, indicating that a combined mode of failure occurs. The structure can be designed more economically when both sway and combined mode failures occur simultaneously. Let M_P' ($< M_P$) be the modified collapse moment for the stanchions. With stanchion hinges appearing in the corresponding failure modes, shown in Figs 7.8c,e and g, eqs(7.11a,b,c) are modified:

$$M_P'\,\theta + M_P'(\theta+\psi) + M_P'\psi = W\Delta \tag{7.13a}$$

$$2M_P'\theta + 2M_P'\theta = 2W\delta \tag{7.13b}$$

$$2M_P'\theta + M_P(\theta+\psi) + M_P'(\theta+\psi) = W\Delta + 2W\delta \tag{7.13c}$$

In eq(7.13c), the respective terms on the left hand side apply to hinges formed at each stanchion base with rotation θ, a hinge under the beam load with rotation $\theta+\psi$ and a hinge at the top of the right stanchion with rotation $\theta+\psi$. *Note:* Under a combined collapse mode there is no hinge formed at the top of the left stanchion. The corner right angle is preserved. Equation (7.13b) shows $M_P' = WL/4$. Substituting this into eq(7.13c) gives $M_P = 11WL/36$. Therefore $W = (36M_P)/(11L)$ with $M_P' = 9M_P/11$, which confirms that the given stanchions are over-designed at a moment capacity of $3M_P/2$. Should the modes in eqs(7.13a,c) be made to coincide then $M_P' = WL/9$ and $M_P = 17WL/27$ giving $M_P' = 3M_P/17$, i.e. a much reduced collapse moment for the stanchions. In a fully optimised design it is possible to arrange that all three modes occur together under a given loading. This serves to minimise weight by the method in the section that follows.

7.3.4 Minimum Weight Solution

PVW is employed to design a beam or a frame with minimum weight. To do this, even the most unlikely failure mode has to be accounted for. In the portal frame of Fig.7.8a there are in fact six modes (b–g) as shown. With the deflections δ, rotations θ and collapse moments M_P indicated, the application of virtual work to each mode results in a corresponding system of non-dimensional moment equations:

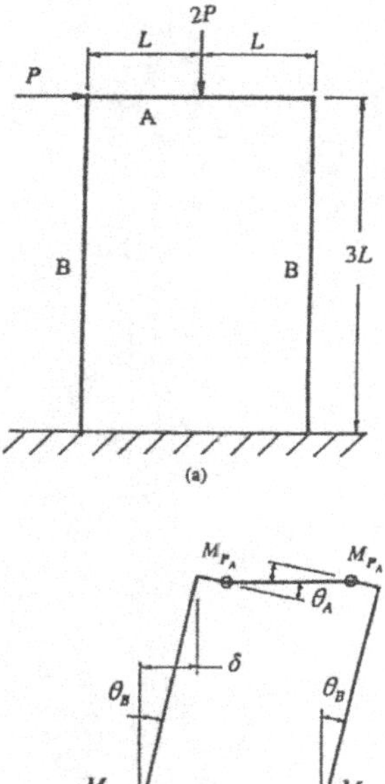

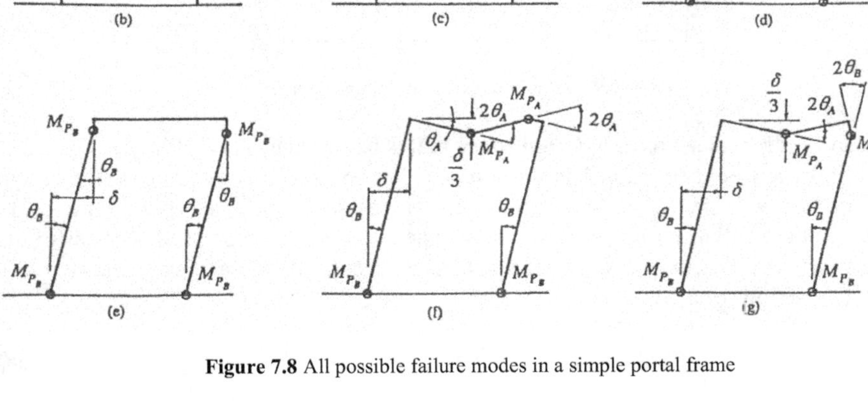

Figure 7.8 All possible failure modes in a simple portal frame

Fig. 7.8b: $2P\delta = 4M_{P_A}\theta_A \quad \Rightarrow \quad \dfrac{M_{P_A}}{PL} \geq 0.5$ (7.14a)

where $\theta_A = \delta/L$

Fig. 7.8c: $2P\delta = 2M_{P_A}\theta_A + 2M_{P_B}\theta_B \quad \Rightarrow \quad \dfrac{M_{P_A}}{PL} + \dfrac{M_{P_B}}{PL} \geq 1.0$ (7.14b)

where $\theta_A = \theta_B = \delta/L$

Fig. 7.8d: $P\delta = 2M_{P_A}\theta_A + 2M_{P_B}\theta_B \quad \Rightarrow \quad \dfrac{M_{P_A}}{PL} + \dfrac{M_{P_B}}{PL} \geq \dfrac{3}{2}$ (7.14c)

where $\theta_A = \theta_B = \delta/(3L)$

Fig. 7.6e: $P\delta = 4M_{P_B}\theta_B \quad \Rightarrow \quad \dfrac{M_{P_B}}{PL} \geq \dfrac{3}{4}$ (7.14d)

where $\theta_B = \delta/(3L)$

Fig. 7.8f: $P\delta + 2P\dfrac{\delta}{3} = 4M_{P_A}\theta_A + 2M_{P_B}\theta_B \quad \Rightarrow \quad 2\dfrac{M_{P_A}}{PL} + \dfrac{M_{P_B}}{PL} \geq \dfrac{5}{2}$ (7.14e)

where $\theta_A = \theta_B = \delta/(3L)$

Fig. 7.8g: $P\delta + 2P\dfrac{\delta}{3} = 2M_{P_A}\theta_A + 4M_{P_B}\theta_B \quad \Rightarrow \quad \dfrac{M_{P_A}}{PL} + 2\dfrac{M_{P_B}}{PL} \geq \dfrac{5}{2}$ (7.14f)

where $\theta_A = \theta_B = \delta/(3L)$.

Each of eqs(7.14a–f) plots as a straight line within the axes given in Fig. 7.9.

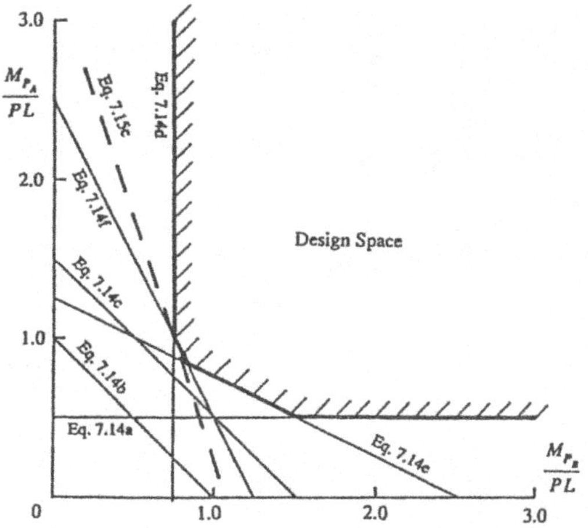

Figure 7.9 Design space for portal frame collapse

The inequalities reveal a working *design space* that is bounded by eqs(7.14a, d, e and f), as shown. If now a weight function X is superimposed upon Fig. 7.9, one point on the design space boundary will provide a safe, minimum weight solution. The weight/unit length depends upon the collapse (ultimate) moment. The latter can be expressed in terms of the yield stress and the dimensions of the section area, as in eq(5.5). From this one may express the weight function for a whole structure [4]:

$$X = \sum M_P L \tag{7.15a}$$

and for the portal frame in Fig. 7.8a, eq(7.15a) becomes

$$X = (M_{P_A} \times 2L) + 2(M_{P_B} \times 3L) \tag{7.15b}$$

In non-dimensional form:

$$\frac{X}{PL^2} = 2\frac{M_{P_A}}{PL} + 6\frac{M_{P_B}}{PL} \tag{7.15c}$$

The right-hand side of eq(7.15c) describes a straight line of gradient −3, as shown. A minimum value of $X/(PL^2)$ is chosen to lie at that corner of the design space in Fig. 7.9 where the collapse modes in Figs 7.8e,g occur simultaneously. This requires a simultaneous solution to the corresponding eqs(7.14d,f), which gives the minimum weight conditions:

$$\frac{M_{P_A}}{PL} = 1\ , \quad \frac{M_{P_B}}{PL} = \frac{3}{4} \quad \text{and} \quad \frac{X}{PL^2} = 6.5$$

Commercial sections will not conform exactly to these conditions but the method allows their selection to lie within the design space, close to the minimum weight point.

7.4 Die Indentation

7.4.1 Lower Bounds

In order to estimate a maximum lower-bound, fields of all possible stress discontinuities are used. Consider applying this approach to flat die indentation of a large plate (see Fig. 6.38). The simplest possible configuration of stress discontinuities is that given in Fig. 7.10a.

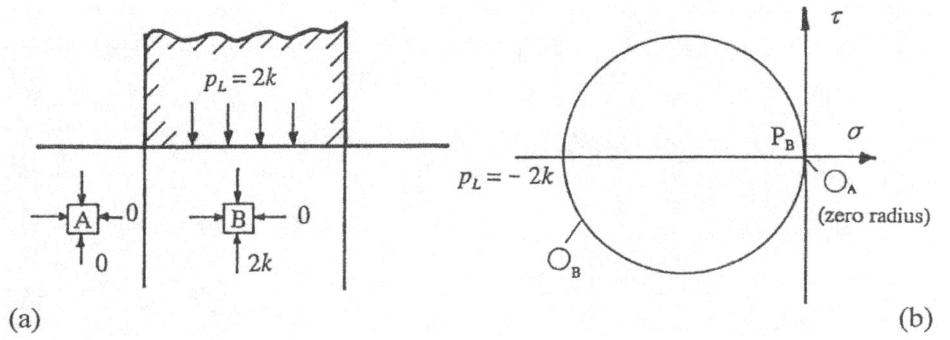

Figure 7.10 Simple stress discontinuity pattern for flat die indentation

Two constant stress regions A and B are defined in Fig. 7.10a by the given states: A is unstressed and B satisfies the plane strain yield criterion $\sigma_1 - \sigma_3 = 2k$. In the stress plane (see Fig. 7.10b), point A defines the origin and B becomes a circle $\bigcirc_B$ with pole P_B at the origin. Clearly, the constant die pressure $p_L = 2k$ is far less than that found previously from the SLF solution, eq(6.28): $p_d = 2k(1 + \pi/2) = 5.142k$.

To raise the lower bound p_L, the stress states may be altered to those in Fig. 7.11a.

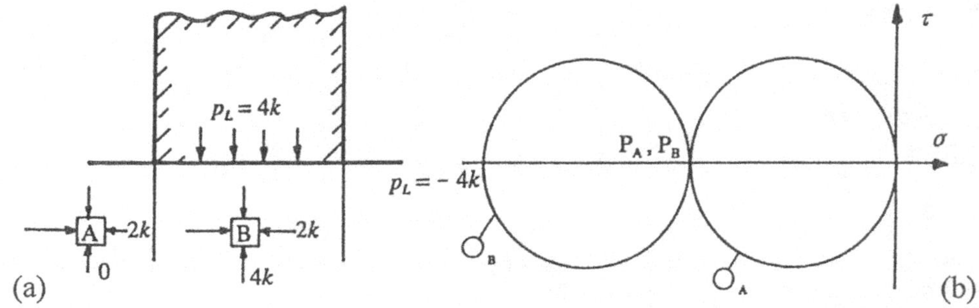

Figure 7.11 Stress discontinuities for flat die indentation

Both regions A and B now satisfy the yield criterion. The stress plane (Fig. 7.11b) consists of two circles ○$_A$ and ○$_B$, each of radius k, with common poles P_A and P_B. Projecting the horizontal plane of the die face through P_B intersects ○$_B$ to give a die pressure $p_L = 4k$, which is twice that found from Fig. 7.10b.

Now examine whether the two alternative patterns of stress discontinuities on either side of the vertical centre line in Fig. 7.12a are admissible.

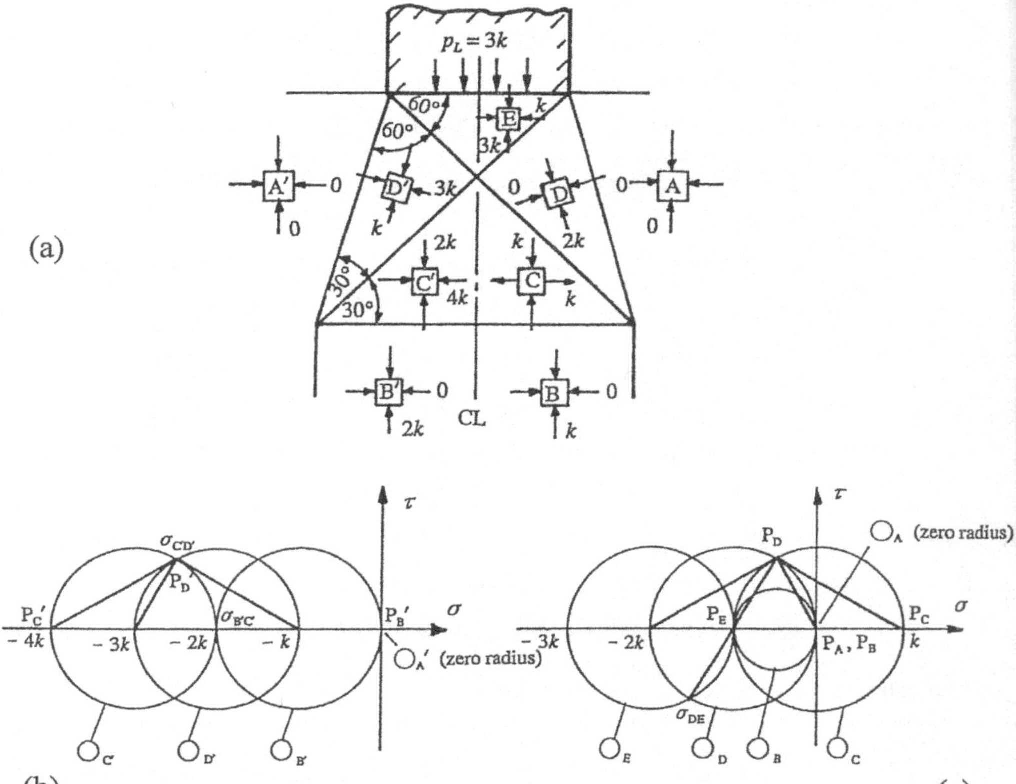

Figure 7.12 Complex stress discontinuities for flat die indentation

First work in an anticlockwise direction, to the left of the centre-line, starting with a stress free condition for region A′. Circles $\bigcirc_B'$ and $\bigcirc_C'$, for yielding within regions B′ and C′, are shown in Fig. 7.12b, with their respective pole points P_B' and P_C'. The pole P_D' for $\bigcirc_D'$ is found from projecting plane C′D′ through P_C'. Circle$\bigcirc_D'$ gives the principal stresses $\sigma_1 = -k$ and $\sigma_3 = -3k$, within region D′, orientated as shown. However, because there is no continuity in the stress normal to plane D′A′ when returning to A′ from region D′, this patten of discontinuities is inadmissible.

Starting again with a stress free condition for region A in Fig. 7.12a, work clockwise across regions B, C, D and E, to the right of the vertical centre-line. This shows elastically stressed material in region B, implying that $\sigma_1 = +k$, $\sigma_3 = -k$ within the yielded region C. Circle $\bigcirc_C$ has its centre at the stress origin in Fig. 7.12c. Projecting plane CD through pole P_C establishes pole P_D for $\bigcirc_D$. Projecting planes parallel and perpendicular to DA through P_D intersects $\bigcirc_D$ to give a principal stress state $\sigma_1 = 0$, $\sigma_3 = -2k$, within region D. Since state D maintains continuity with zero stress for region A, one can now determine the stress state within region E. Projecting direction DE through P_D gives a normal stress σ_{DE} within $\bigcirc_D$, that must also apply to region E. This establishes the pole P_E for $\bigcirc_E$ and a principal stress state within E: $\sigma_1 = -k$, $\sigma_3 = -3k$, as shown. The latter defines the normal die pressure as $p_L = 3k$. In searching for an even greater lower bound, another discontinuous stress field is considered in Fig. 7.13a.

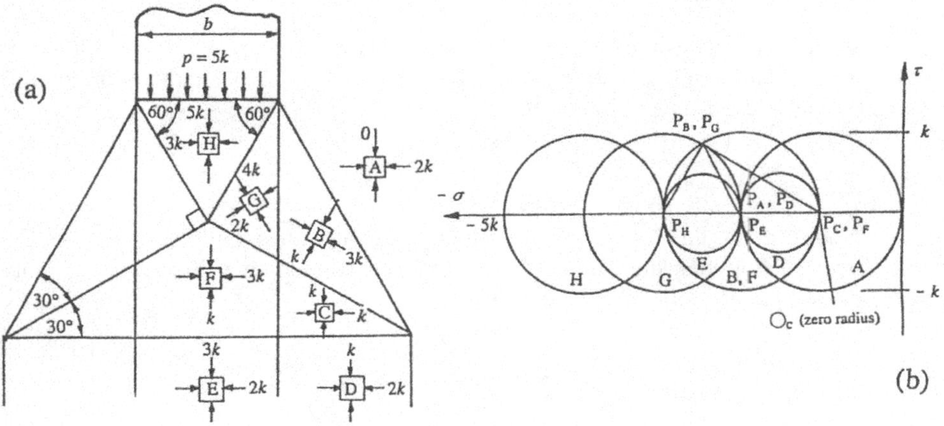

Figure 7.13 Further complex stress discontinuities for flat die indentation

The solution begins with the stress state $\sigma_1 = 0$ and $\sigma_3 = -2k$, within region A. Following the construction of $\bigcirc_A$ in Fig. 7.13b, the pole point P_B and $\bigcirc_B$ are determined. This gives $\sigma_1 = -k$ and $\sigma_3 = -3k$ for region B. Locating the position of pole P_C reveals that $\sigma_3 = -k$ within region C. It can be seen from route ADC that poles P_A and P_D are co-incident and regions C and D are elastically stressed. That is, the radii of circles $\bigcirc_C$ and $\bigcirc_D$, being zero and $k/2$ respectively, are less than k. Pole P_E is located from the continuous stress state from D to E. Similarly P_F is found from continuity in C to F. The radii of circles $\bigcirc_E$ and $\bigcirc_F$ are $k/2$ and k, respectively. Pole P_G follows from P_F and $\bigcirc_G$ gives: $\sigma_1 = -4k$, $\sigma_3 = -2k$. Pole P_H follows from P_G when, finally, $\bigcirc_H$ reveals a die pressure of $p_L = 5k$.

Alternative patterns of stress discontinuities, e.g. that given in Fig. 7.14a, have not served to raise the die pressure above $p_L = 5k$, indicating that this is the true lower-bound solution. Here, the stress plane (see Fig. 7.14b) shows elastic material in regions D and G and yielded material elsewhere, when starting with the assumed state shown in region F.

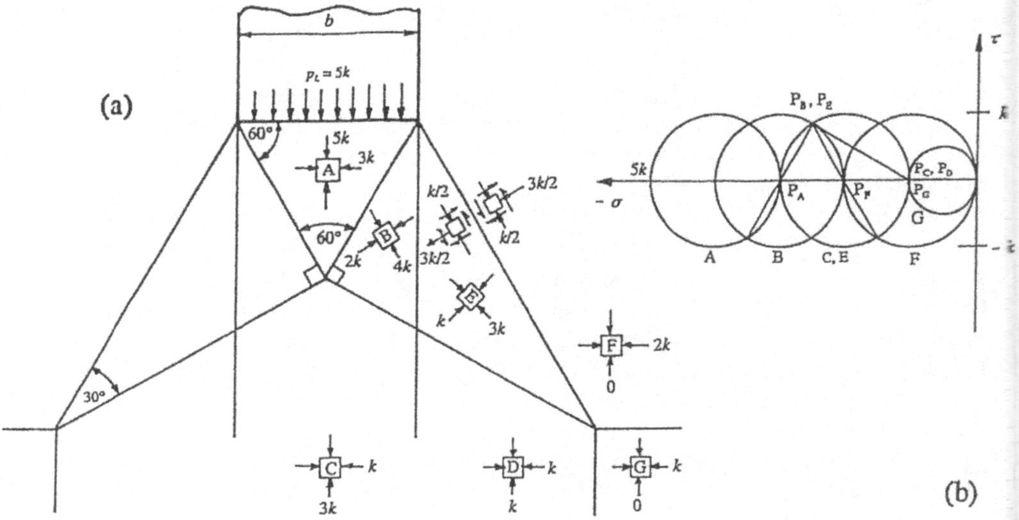

Figure 7.14 Alternative stress discontinuity pattern for flat die indentation

Other principal stress states shown follow from the corresponding circles after locating their pole points in the order of P_F, P_E, P_B P_C, P_D and P_G. A similar pattern of stress discontinuities appears later as a lower bound for extrusion.

7.4.2 Upper Bounds

Upper bound solutions employ patterns of velocity discontinuities. Let the element ABCD in Fig. 7.15 be distorted in shear to A′B′C′D′ after it has crossed the velocity discontinuity YY. The discontinuity could also be a slip-line.

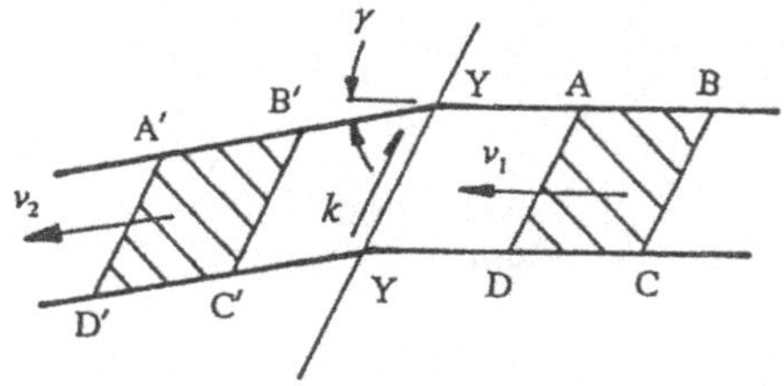

Figure 7.15 Shearing at a velocity discontinuity

The absolute velocities on either side of YY differ in magnitude by $v_2 - v_1$ and in direction by γ, as shown. The rate of the shear work of distortion per unit volume is

$$\dot{w} = k\dot{\gamma} = k\,(v_2 - v_1)/\mathrm{AB}$$

from which the total work rate for a unit thickness (of the die) is

$$\dot{W} = \dot{w}\,V = k\,(v_2 - v_1)(\mathrm{AD} \times \mathrm{AB} \times 1)\,/\,\mathrm{AB}$$

$$= k\,(v_2 - v_1)\,\mathrm{AD} = k \times \mathrm{YY} \times v_{\mathrm{YY}} \qquad (7.16)$$

where $v_{YY} = v_2 - v_1$ is the magnitude of the velocity discontinuity and AD = YY is the length of the discontinuity line.

Consider two possible upper bound mechanisms for die indentation in Figs 7.16a,b.

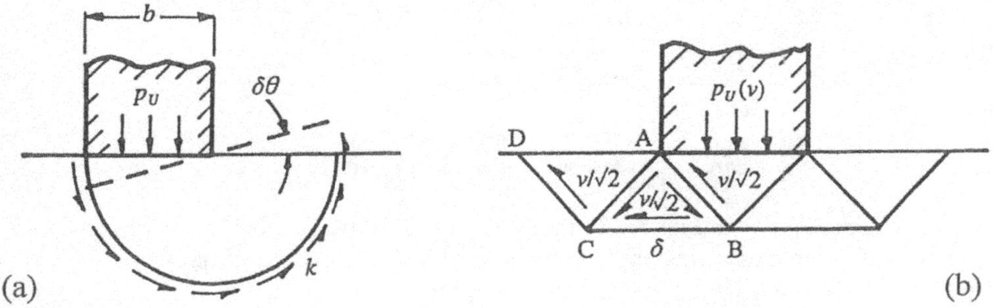

Figure 7.16 Upper bounds for flat die indentation

A simplified rigid rotation $\delta\theta$ of the semi-circular zone of material beneath the die is assumed in Fig. 7.16a. Here the average external work done by p_U is equated to the work produced by shear yielding on radius b. When rotation $\delta\theta$ occurs in a time δt, the tangential velocity discontinuity is $v = b\delta\theta/\delta t = b\dot{\theta}$ and eq(7.16) gives, for a die of unit thickness,

$$p_U(b \times 1)(b \times \dot{\theta}/2) = k(\pi b \times 1)(b\dot{\theta}) \tag{7.17a}$$

from which $p_U = 2\pi k = 6.28k$, to a first approximation. The true, upper bound will reduce p_U to its minimum value. For example, consider the rigid motion of five triangular blocks within the SLF in Fig. 7.16b. When the die velocity is v downwards, the resolved velocity discontinuities are as shown. The rate of external work by p_U is equated to the net shear work rate, produced by shear forces acting along the slip lines AB, BC, AC and CD. Equation (7.16) gives, for mirrored slip on both sides of the die, with breadth b and unit thickness,

$$p_U(b \times 1)v = 2k\left\{\left[\frac{b}{\sqrt{2}}\left(\frac{v}{\sqrt{2}} + \frac{v}{\sqrt{2}}\right)\right]_{AB} + (b\,v)_{BC} + \left[\left(\frac{b}{\sqrt{2}}\right)\left(\frac{v}{\sqrt{2}}\right)\right]_{AC} + \left[\left(\frac{b}{\sqrt{2}}\right)\left(\frac{v}{\sqrt{2}}\right)\right]_{CD}\right\} \tag{7.17b}$$

from which $p_U = 6k$. Any further attempt to reduce the upper bound estimate of p_U should aim for the die pressure given by the SLF solution, i.e. $p_d = 5.14k$ from eq(6.28).

7.5 Extrusion

7.5.1 Lower Bounds

A simple, lower-bound estimate for extrusion consists of the two stress discontinuities given in Fig. 7.17a.

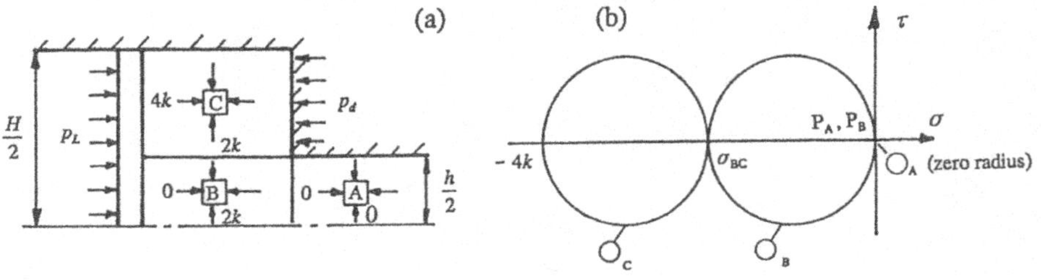

Figure 7.17 Possible stress discontinuity pattern for extrusion

Using the stress plane in Fig. 7.17b, the stress states for the yielded regions B and C are found from the unstressed material in region A. This shows that the die pressure is $p_d = 4k$, from which the lower bound ram pressure p_L becomes

$$4k\,(H - h) \times 1 = (H \times 1)\,p_L$$

$$\frac{p_L}{2k} = 2\left(1 - \frac{1}{R}\right) \tag{7.18}$$

where $R = H/h$. Alexander [5] improved upon eq(7.18) by raising p_L for $R > 3$. The stress discontinuity pattern (Fig. 7.18a) is similar to that given in Fig. 7.14a for flat die indentation. The lettering of uniform regions of stress, A, B, C, D, E, F and G, correspond between these two figures.

(a)

(b)

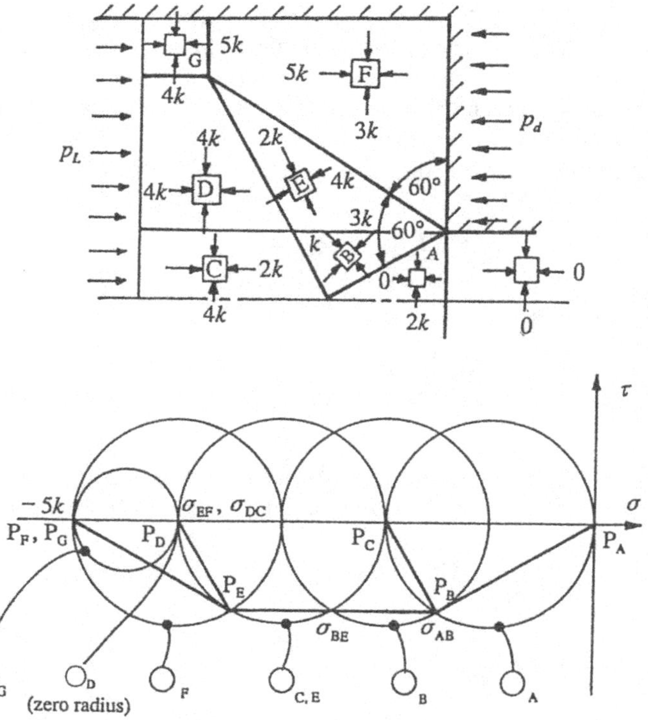

Figure 7.18 Complex stress discontinuity for plane extrusion

However, different stress states apply to each region. Those in Fig.7.18a follow from the stress plane in Fig. 7.18b. It is seen that if the normal stress is to remain continuous between adjacent regions, then $\bigcirc_D$ and $\bigcirc_G$ must have radii of zero and $k/2$ respectively. Within region F, circle $\bigcirc_F$ gives $p_d = 5k$, from which:

$$\frac{p_L}{2k} = 2.5\left(1 - \frac{1}{R}\right) \tag{7.19}$$

7.5.2 Upper Bounds

Upper bound solutions are more useful to manufacturing process design than lower bounds. They are not prone to large overestimations because in practice materials will harden to some degree. A good upper bound solution requires an accurate estimate of the true deformation zone. When slip-line fields are available they may be employed to guide the correct zone shape. Upper-bounding for limit loading is always possible even when SLFs are not available. The procedure then followed is to examine a number of possible velocity discontinuity patterns and select the one which gives the lowest, upper-bound estimate.

Figure 7.19a gives a simple velocity discontinuity pattern for smooth $R = 2$ extrusion.

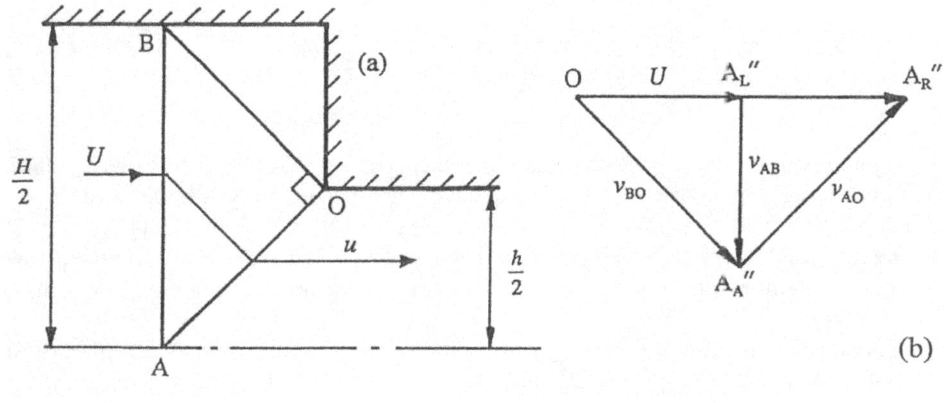

Figure 7.19 Simple velocity discontinuity pattern for $R = 2$ extrusion

The straight line BA replaces the semi-circular arc 1, 2 . . 5 of the α–slip line in Fig. 6.23a. Applying eq(7.16) to the corresponding hodograph in Fig. 7.19b, gives

$$W = p_U(H \times 1)U = k\,[v_{AB}\,AB + v_{BO}\,BO + v_{AO}\,AO\,]$$

$$= 2k\left[U\frac{H}{2} + \sqrt{2}U\frac{H}{2\sqrt{2}} + \sqrt{2}U\frac{H}{2\sqrt{2}}\right] = 3kUH$$

$$\frac{p_U}{2k} = 1.5$$

which compares with $p_e/(2k) = 1.285$, the SLF solution from eq(6.17a).

A further pattern of velocity discontinuities for $R = 3$ is given in Fig. 7.20a. Here, the α–slip line's circular arc 12 in Fig. 6.27 has been replaced with a straight line AB. Correspondingly, there results a different hodograph (see Fig. 7.20b), which requires the application of eq(7.16) for unit width across the full container depth to find the upper bound.

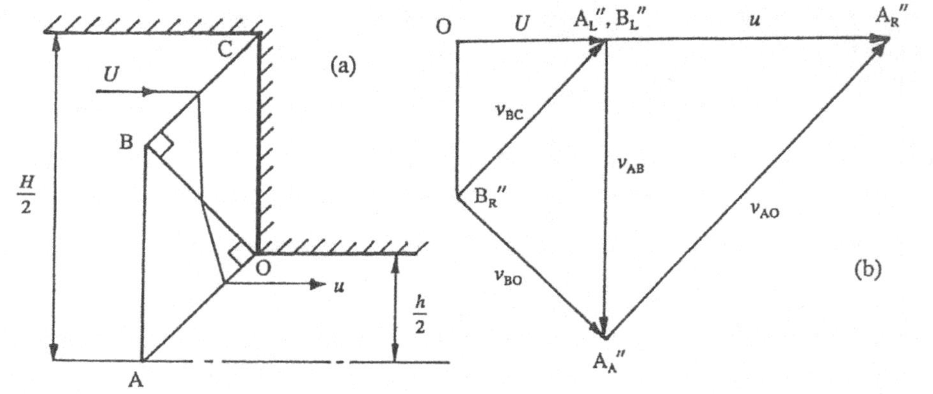

Figure 7.20 Velocity discontinuity pattern for $R = 3$ extrusion

This gives:

$$W = p_U(H \times 1)U = k\,[v_{AB}\,AB + v_{BO}\,BO + v_{AO}\,AO + v_{BC}\,BC\,]$$

$$= k\left[\left(2U \times \frac{2H}{3}\right) + \left(\sqrt{2}U \times \frac{\sqrt{2}}{3}H\right) + \left(2\sqrt{2}U \times \frac{\sqrt{2}}{3}H\right) + \left(\sqrt{2}U \times \frac{\sqrt{2}}{3}H\right)\right] = 4kUH$$

$$\therefore \quad \frac{p_U}{2k} = 2$$

which compares with $p_e/2k = 1.71$ from SLF theory in eq(6.22a,b). If there were sticking friction along the inner die wall, an additional velocity discontinuity would arise along OC. Adding $v_{OC}\,OC = U \times 2H/3$ to W, given above, leads to $W = 14kUH/3$ and $p_U/(2k) = 7/3$, which is a 16.7 % increase over the frictionless case. With both wall and die friction present, one could obtain an upper bound solution from simplifying the corresponding SLF solutions given in Figs 6.34 and 6.35.

Contrast the solution for $R = 3$ given above with that for an alternative pattern of four velocity discontinuities AB, AD, DB and BC shown in Fig. 7.21a. Region AED is the dead metal zone in which DE = DC.

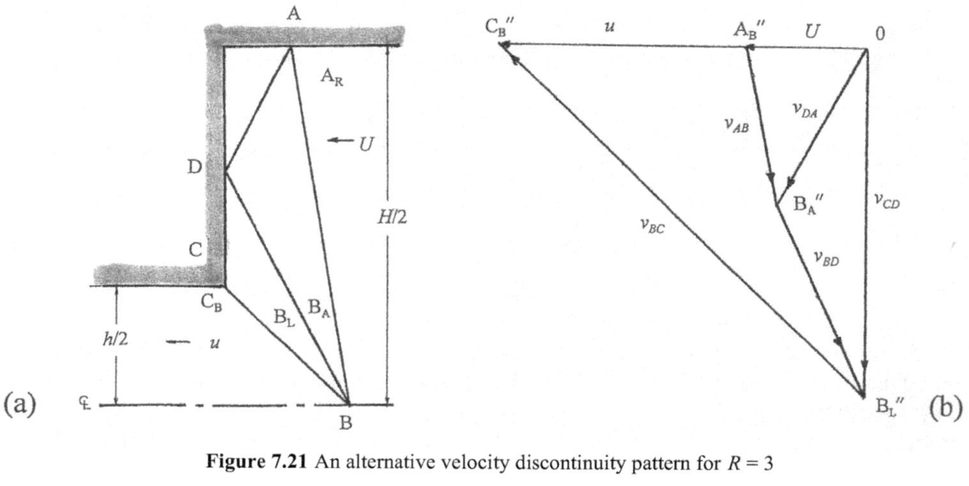

Figure 7.21 An alternative velocity discontinuity pattern for $R = 3$

With a 45° inclination for BC, the geometry shows a 30° inclination of AD to the vertical wall. Other lengths are scaled from Fig.7.21a as follows:

$$AB = 1.01H/2,\ AD = 0.39H/2,\ BD = 0.75H/2 \text{ and } BC = 0.49H/2$$

The block sliding velocities are scaled from the velocity diagram in Fig. 7.21b to give

$$v_{AB} = 1.4U,\ v_{AD} = 1.62U,\ v_{BD} = 1.8U \text{ and } v_{BC} = 4.29U$$

from which the ram pressure/unit width follows from the work rate as

$$W = p_U(H \times 1)U = 2k\,[v_{\rm AB}(\rm AB \times 1) + v_{\rm AD}(\rm AD \times 1) + v_{\rm DB}(\rm DB \times 1) + v_{\rm BC}(\rm BC \times 1)]$$

in which

$$W = p_U(H \times 1)U = 2k\,[(1.4U \times 1.01H/2) + (1.62U \times 0.39H/2)$$

$$+ (1.80U \times 0.75H/2) + (4.29U \times 0.49H/2)] = 5.434kUH$$

$$\therefore p_U/2k = 2.717$$

Comparing with Fig. 7.20, this velocity field does not provide the lowest upper bound.

Johnson [6] proposed an alternative pattern of velocity discontinuities in Fig. 7.22a. This provides an upper-bound solution for frictionless extrusion ratios R in a medium range from 3 to 10.

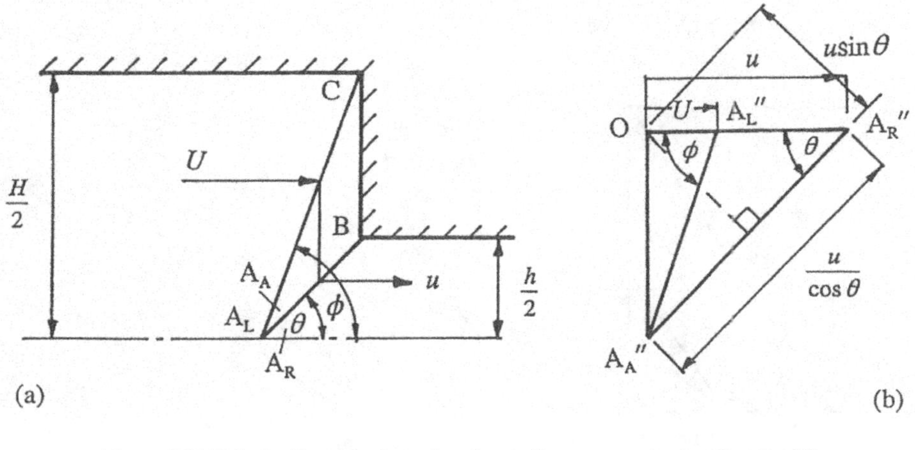

Figure 7.22 Velocity discontinuity pattern for medium range extrusion ($3 < R < 10$)

The geometry of the physical plane and hodograph in Figs 7.22a,b shows an extrusion ratio:

$$R = \frac{H}{h} = \frac{\tan\varphi}{\tan\theta} = \frac{u}{U} \tag{7.20}$$

The rate of external work is $p_U(H \times 1)U$. Working with a unit volume of material extruded in unit time, the specific external work w is simply the extrusion pressure p_U. Let $v_{\rm N}$ and $v_{\rm T}$ be the normal and tangential velocities to AB and AC. Equating w to the distortion energy:

$$w = p_U = k\left(\gamma_{AB} + \gamma_{AC}\right)$$

$$p_U = k\left[\left(\frac{v_T}{v_N}\right)_{AB} + \left(\frac{v_T}{v_N}\right)_{AC}\right]$$

$$= k\left(\frac{1}{\cos\theta\ \sin\theta} + \frac{1}{\cos\varphi\ \sin\varphi}\right)$$

$$= k\left[\frac{(1 + \tan^2\theta)}{\tan\theta} + \frac{(1 + \tan^2\varphi)}{\tan\varphi}\right] \tag{7.21a}$$

The same result would be found from eq(7.16). Substituting eq(7.20) into eq(7.21a)

$$\frac{p_U}{k} = \frac{(1 + R)(1 + R\tan^2\theta)}{R\tan\theta} \tag{7.21b}$$

which has a minimum for $\mathrm{d}(p_U/k) = 0$. That is, $R\tan^2\theta = 1$ or $\tan\varphi = 1/\sqrt{R}$. Substituting into eq(7.21b), for the upper bound:

$$\frac{p_U}{2k} = \frac{1 + R}{\sqrt{R}} \tag{7.21c}$$

which is known to overestimate p_U at higher R. Johnson [6] reduced p_U for extrusion ratios exceeding 10 with the alternative velocity discontinuity pattern shown in Fig. 7.23a.

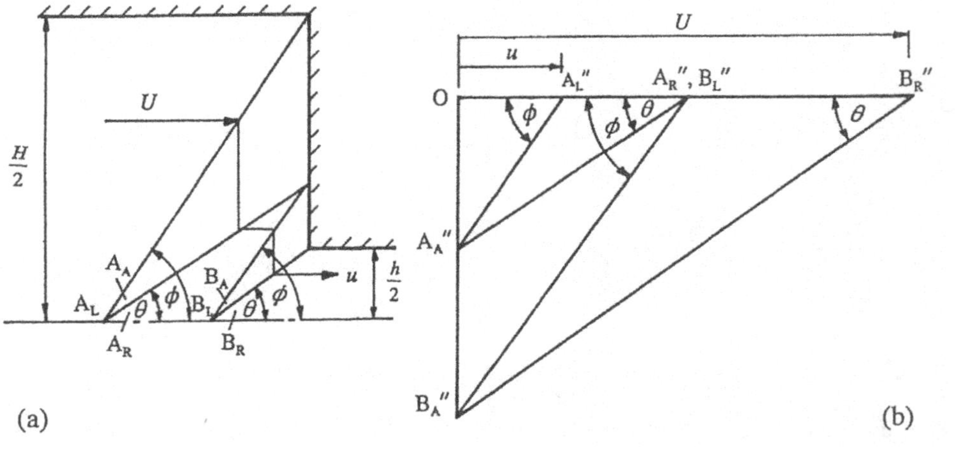

Figure 7.23 Velocity discontinuity pattern for higher R extrusion ratios

For the geometry given in Fig. 7.23a, it follows that

$$R = \frac{H}{h} = \frac{\tan^2\varphi}{\tan^2\theta} = \frac{u}{U} \tag{7.22}$$

The hodograph (Fig. 7.23b) shows that the dissipation rate of shear strain energy will be twice that for Fig. 7.20b. That is:

$$w = p_U = 2k\left[\frac{(1 + \tan^2\theta)}{\tan\theta} + \frac{(1 + \tan^2\varphi)}{\tan\varphi}\right] \quad (7.23a)$$

Substituting eq(7.22) into eq(7.23a):

$$\frac{p_U}{2k} = \frac{(1 + \sqrt{R})(1 + \sqrt{R}\tan^2\theta)}{\sqrt{R}\tan\theta} \quad (7.23b)$$

Equation (7.23b) is a minimum for $\sqrt{R}\tan^2\theta = 1$. This gives $\tan^2\theta = 1/\sqrt{R}$, so that

$$\left(\frac{p_U}{2k}\right)^2 = \frac{4(1 + \sqrt{R})^2}{\sqrt{R}} \quad (7.23c)$$

Table 7.1 lists $p/(2k)$ values from the upper and lower-bound estimates given in eqs(7.18), (7.19), (7.21c) and (7.23c).

Table 7.1 Upper and lower bound pressures for given extrusion ratios

	$p_L/2k$		$p_U/2k$	
R	$2(1 - 1/R)$	$2.5(1 - 1/R)$	$(1 + R)/\sqrt{R}$	$[4(1 + \sqrt{R})^2/\sqrt{R}]^{1/2}$
2	1.000	1.250	2.12	4.06
3	1.333	1.666	2.31	4.16
9	1.778	2.222	3.33	4.62
25	1.920	2.400	5.20	5.37
36	1.944	2.430	6.17	5.70

For a given extrusion ratio, the highest lower-bound and the lowest upper-bound are selected from Table 7.1. The extrusion pressure is expected to lie between these two bounds, as exact solutions show. *Note*: The p_U's in the final column are only appropriate for $R > 10$ and that p_L estimates converge with increasing R.

7.6 Further Upper Bounds

Though the velocity discontinuity provision for an upper bound solution is the most convenient method for the assessment of forming forces it does not guarantee accuracy when compared to the slip line field. A number of attempts with discontinuity patterns deemed possible are required before the lowest upper bound is found to be acceptable. Therein, the following are examples where corresponding SLF fields would also need to be investigated.

7.6.1 Extrusion with Tapered Die

Assuming that the sloping die surface is frictionless and that plane strain applies to a plastic rigid material, Fig. 7.24a shows one pattern of velocity discontinuities for an extrusion ratio $H/h = 4$ (i.e. a 75 % reduction to H). Here, estimates are required of: (i) ram pressure and (ii) the temperature distribution across the pattern, given the shear yield stress k, the specific heat under constant volume C_v and the density ρ of the rigid plastic material under extrusion.

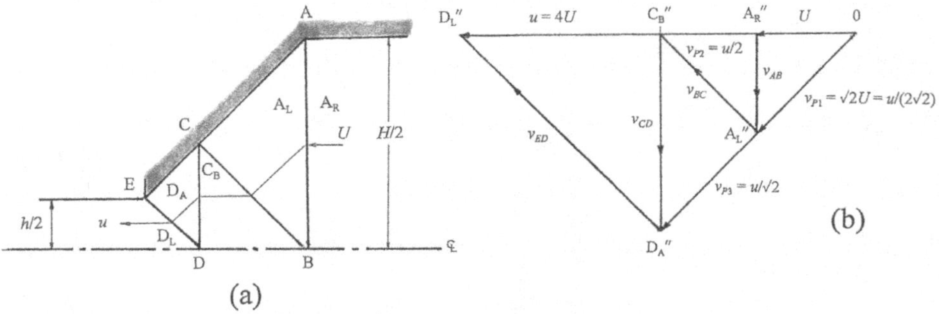

Figure 7.24 Velocity discontinuity pattern within sloping die for $H/h = 4$

With the slope of the wall container at 45° and $H/h = 4$ the geometry of this field reveals the length of each discontinuity when $h = 1$:

$$AB = H/2,\ BC = H/(2\sqrt{2}),\ CD = H/4 \text{ and } DE = H/(4\sqrt{2})$$

The hodograph given in Fig. 7.24b provides the following velocity of particles lying upon each discontinuity:

$$v_{AB} = U,\ v_{BC} = \sqrt{2}U,\ v_{CD} = 2U,\ v_{DE} = 2\sqrt{2}U$$

These are equal to the relative velocities between adjacent blocks along the discontinuity in the absence of friction. Otherwise, kinetic friction (with coefficient μ_k) reduces the sliding speed below that of the particle velocity. Quantitatively, the reduction factor depends upon an opposing friction force F, proportional to the normal force R upon the discontinuity. The friction force F opposes the tangential force T along the discontinuity. Both R and T arise from resolving the ram's horizontal force $H \times 1 \times p$ from its pressure p. The net force $T - F$ determines the acceleration a of each block of mass m from which the sliding velocity v_s follows when starting from rest. The corresponding equations are proposed for an estimate of v_s:

$$F = \mu_k R$$

$$T - F = ma$$

$$v_s = at$$

where t is the particle velocity time of travel for each discontinuity and $m = \rho A$ per unit width where A are the triangular block areas EDC, CBD etc. In the absence of friction the solutions are simplified in which the ram's upper bound pressure p_u is based solely upon the work rate equality per unit width:

$$p_U(H \times 1)U = 2k\,[v_{AB}(AB \times 1) + v_{BC}(BC \times 1) + v_{CD}(CD \times 1) + v_{DE}(DE \times 1)]$$
$$= 2k[(U \times H/2) + (\sqrt{2}U \times H/(2\sqrt{2})) + (2U \times H/4) + (2\sqrt{2}U \times H/(4\sqrt{2})] = 4kHU$$
$$\therefore \quad p_U/2k = 4$$

The temperature change experienced by a particle that crosses a discontinuity requires that its direction is traced from within the hodograph. The path shown in Fig. 7.24a is based upon the direction of motion of the particle within each block relative to the ground point O. For example, within block ABC, any point A_L'' left of A in the hodograph defines the vector $v_{p1,}$ which lies at 45° to O, its magnitude given by $|v_{p1}| = \sqrt{2}U = u/(2\sqrt{2})$. Similarly, for block BCD, any point C_B'' beneath C in the hodograph defines a horizontal vector v_{p2} relative to 0 with magnitude $|v_{p2}| = 2U = u/2$. The path of the particle from entry to exit is as shown in Fig. 7.24a. Table 7.2 gives the transverse velocities across each discontinuity.

Table 7.2 Particle transverse and sliding velocities relative to each discontinuity

	AB	BC	CD	DE
Transverse v_P	$u/4$	$u/(2\sqrt{2})$	$u/2$	$u/\sqrt{2}$
Sliding v	$u/4$	$u/(2\sqrt{2})$	$u/2$	$u/\sqrt{2}$

Table 7.2 shows that the magnitude of each transverse velocity equals that of block sliding along the respective discontinuity. The temperature change ΔT accompanying each velocity traversal v_P follows from the power relationship/zone area (kJ/m^3 × m/s = kW/m^2):

$$m\,C_V \Delta T\, v_P / V = kv$$

$$\rho C_V \Delta T = k \times \frac{v}{v_P} \quad \Rightarrow \quad \Delta T = \frac{k}{\rho C_V} \times \frac{v}{v_P}$$

and with a temperature T within the entry zone, the passage of the particle through each zone is accompanied by a temperature change with absolute temperature, as in Table 7.3:

Table 7.3 Temperature distribution within discontinuity field

Zone	v/v_P	ΔT	T
Entry	1	–	T
ABC	1	$k/\rho C_V$	$T + k/\rho C_V$
BCD	1	$k/\rho C_V$	$T + 2k/\rho C_V$
CDE	1	$k/\rho C_V$	$T + 3k/\rho C_V$
Exit	1	$k/\rho C_V$	$T + 4\,k/\rho C_V$

7.6.2 Machining

One velocity discontinuity field that is possible for an edge machining process (shaping) is shown in Fig. 7.25a. The perpendicular rake face is assumed to be frictionless for a rigid plastic material in which a plane strain condition applies to the chip with an exit velocity v.

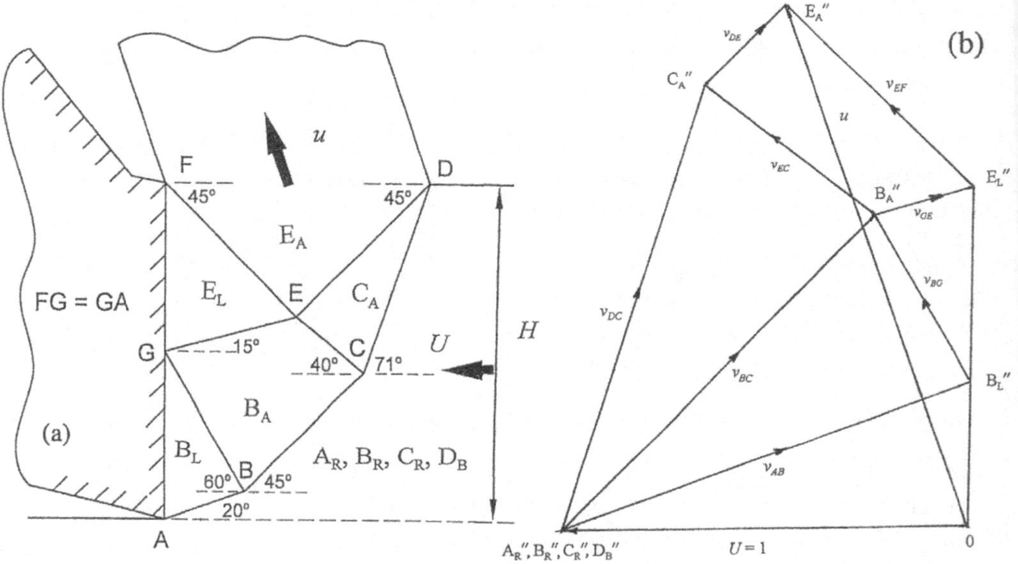

Figure 7.25 Velocity discontinuities for edge machining

Taking a unit velocity U between the tool and the workpiece, Figure 7.25b shows the hodograph, from which the block sliding velocities within the field are:

$$v_{AB} = 1.06\,,\ v_{BC} = 1.09\,,\ v_{CD} = 1.16,\ v_{DE} = 0.27,\ v_{EF} = 0.64,$$

$$v_{EG} = 0.24,\ v_{CE} = 0.52 \text{ and } v_{BG} = 0.47$$

With a unit height H for the vertical rake face AGF, i.e. the depth of cut, the length of each discontinuity is scaled as follows:

$$\text{AB} = 0.25,\ \text{BC} = 0.5,\ \text{CD} = 0.6,\ \text{DE} = 0.55,\ \text{EF} = 0.55,$$

$$\text{EG} = 0.4,\ \text{CE} = 0.25 \text{ and BG} = 0.475$$

For a cutting speed U, the normal force F acting upon the rake face is found from the power balance equation/unit width:

$$FU = k[\text{AB}v_{AB} + \text{BC}v_{BC} + \text{CD}v_{CD} + \text{DE}v_{DE} + \text{EF}v_{EF} + \text{EG}v_{EG} + \text{CE}v_{CE} + BGv_{BG}] \times U$$

$$= kU[(0.25 \times 1.06) + (0.5 \times 1.09) + (0.6 \times 1.16) + (0.55 \times 0.27) + (0.55 \times 0.64) + (0.4 \times 0.24) + (0.25 \times 0.52) + (0.475 \times 0.47)] = 2.456kU$$

which gives $F = 2.456k$ N for $U = 1$ m/s. The hodograph shows a chip velocity $u = 1.36$ m/s, which is required to preserve the volume of material undergoing block shear. Under plane strain machining, a unit width of material in contact with the rake face is discharged into the chip without volume change. Therefore, the volume preservation is expressed simply as $UH = uh$, where $h/H = U/u = 1/1.36 = 0.735$ is the chip depth lying perpendicular to the direction of u in good agreement with the geometry of Fig 7.25a. In fact, whatever velocity discontinuity pattern is attached to the rake face, its geometry must ensure $h = 0.735$ with U and H unitised with appropriate units in this case.

7.6.3 Tube Extrusion

The extrusion of tube from a solid circular billet employs an axially symmetric tapered punch and die, as shown in section within Fig. 7.26a. The billet of radius H is rammed at speed U into a tapered aperture between the stationary punch and die to extrude tube of wall of thickness h at an exit velocity u.

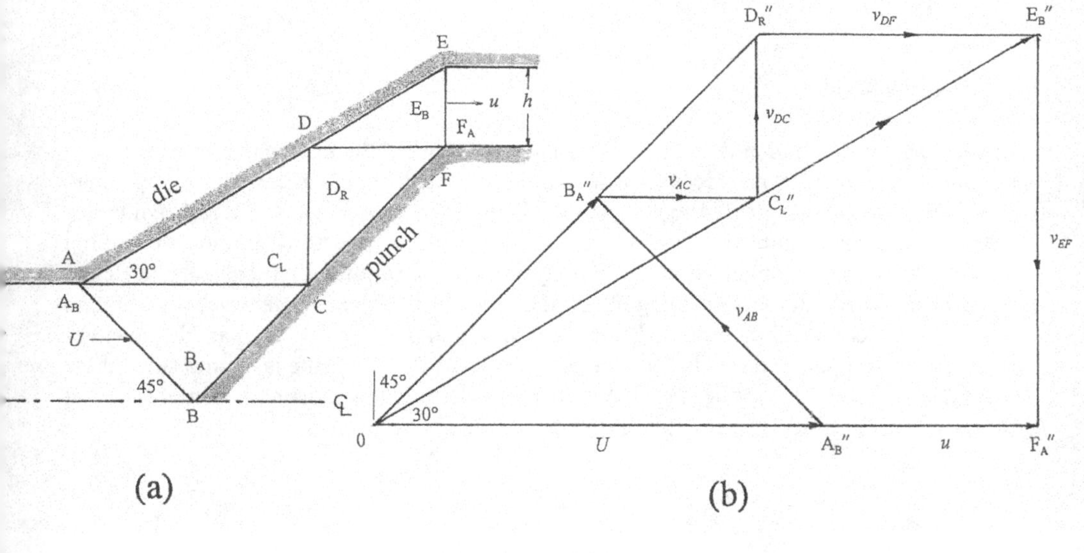

Figure 7.26 Tube extrusion within stationary punch and die

Figure 7.26a shows the five velocity discontinuities within the concentric taper by which the tube forms from block shear. The extrusion conditions assumed are that the die and punch surfaces are frictionless for plastic rigid material under plane strain. It is required to find the extrusion ratio H/h and the ram pressure from the geometry within the field of discontinuities shown. Figure 7.26b shows the corresponding velocity diagram (hodograph) in which the direction of the resultant velocity within each block is applied to the diagram. That is, blocks ABC and CDF move parallel to the punch surface while blocks ACD and DEF move parallel to the die surface. The horizontal entry and exit velocities and the two sloping wall directions contain the five discontinuities with the aligned velocities given below. The lengths of the discontinuities are calculated from Fig. 7.26a as

$$AB = \sqrt{2}H,\ AC = 2H,\ DC = 1.155H,\ DF = 1.155H \text{ and } EF = 0.667H$$

Note: With EF = h, the die geometry corresponds to the extrusion ratio $H/h = 1.5$.

The magnitudes of the velocities aligned with these discontinuities are scaled from Fig. 7.26b as follows:

$$v_{AB} = U/\sqrt{2},\ v_{AC} = 0.366U,\ v_{CD} = 0.366\text{U},\ v_{DF} = 0.634\text{U} \text{ and } v_{EF} = 0.866U$$

Axi-symmetric plane strain implies that the sectional view upon the vertical diameter in Fig. 7.26a may be taken with a unit width to estimate the work rate provided by the punch and its dissipation within shear deformation at the discontinuities. This simplification provides the ram pressure p_U, as with any rectilinear extrusion:

$$p_U(H \times 1)U = k\,[v_{AB}(AB \times 1) + v_{AC}(AC \times 1) + v_{CD}(CD \times 1) + v_{DF}(DF \times 1) + v_{EF}(EF \times 1)]$$

$$= k\left[\left(\frac{U}{\sqrt{2}} \times \sqrt{2H}\right) + (0.366U \times 2H) + (0.366U \times 1.155H) + (0.866U \times 0.667H)\right] = 3.465\,kHU$$

$$\therefore p_U/2k = 1.733$$

7.7 Strip Rolling

In the simplest two-high mill in Fig. 7.27a, a plate of original thickness h_1 is rolled to a final strip thickness h_2 by two rolls of radius R. Normally, the thickness at entry is not more than $h_1 = 5$ mm, giving a typical strip width to thickness ratio $w/h_1 \approx 10$, with a roll radius $R \approx 100h_1$. There are a number of approaches that have been used to determine the rolling torque. The elementary plane-strain, slab analysis technique allows for a constant coefficient of friction μ between the deforming material and the rolls. Because plane sections remain plane, the directions of the frictional forces $\mu p(R\delta\varphi)$ oppose at positions, A and B, on either side of a neutral plane (dashed vertical line in Fig. 7.21a). This plane is neutral since there is no relative movement between roll and strip. In addition, a homogenous compression of the strip thickness is assumed under a constant yield stress.

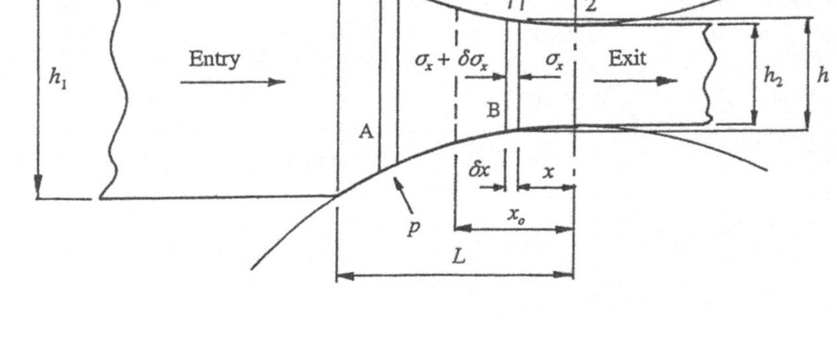

(b)

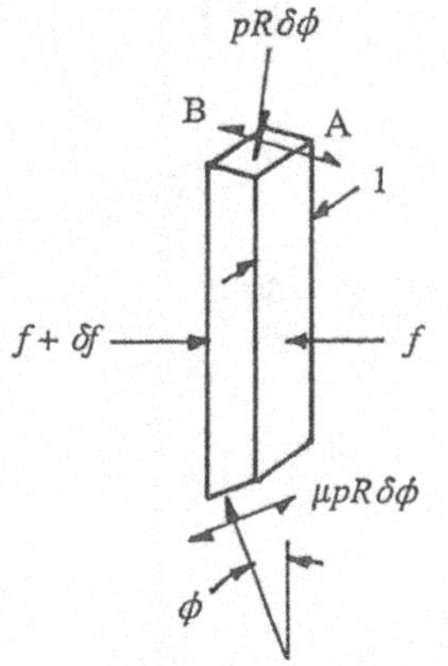

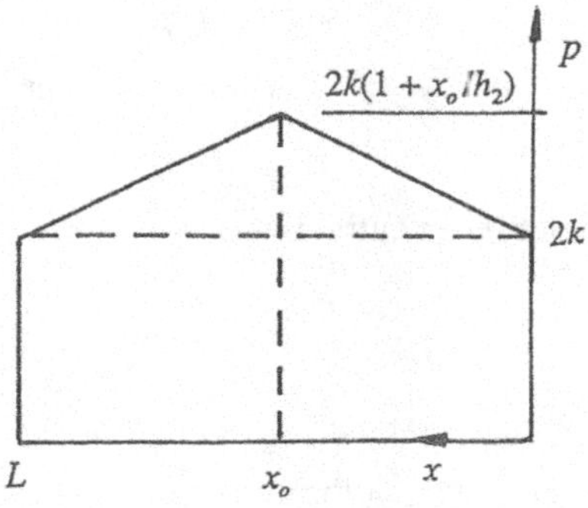

(c)

Figure 7.27 Slab analysis for strip rolling of thin, rectangular section

The elastic deformation of the strip is normally ignored, but more exact theories [1, 7, 8] attempt to account for this. Moreover, the roll itself will elastically deform. This becomes important in cold rolling sheet where, under a high radial pressure p, an elastic flattening of the roll increases its radius (see chapter 14). Here, as with the hot rolling of strip, these elastic effects can be ignored since the radial pressure is far less. In Fig. 7.27b, the slab forces acting at position B, within the arc of contact, are shown. For a slab at position A the friction forces are reversed as shown. Where σ_x is uniform the horizontal force is $f = \sigma_x h$/unit width. Force equilibrium equations in the x–direction for each slab at A and B are:

$$(f + \delta f) + 2\mu pR\delta\varphi \cos\varphi = f + 2pR\delta\varphi \sin\varphi \quad \text{at A}$$

$$(f + \delta f) = f + 2\mu pR\delta\varphi \cos\varphi + 2pR\delta\varphi \sin\varphi \quad \text{at B}$$

Setting $k = \mu p$ for sticking friction leads to the combined form

$$\delta f = 2pR\delta\varphi \sin\varphi \pm 2kR\delta\varphi \cos\varphi$$

$$\frac{\mathrm{d}}{\mathrm{d}\varphi}(\sigma_x h) = 2(pR\,\delta\varphi \sin\varphi \pm kR\,\delta\varphi \cos\varphi) \tag{7.24a}$$

Further equations are required to relate the variables σ_x, p and h before a solution from this elementary theory can be found. The simplest possible relation follows from the assumption that the roll arcs may be replaced with flat platens. It follows that $h = h_2$ is constant, $\varphi = 0$ and $R\delta\varphi = \delta x$. Equation (7.24a) becomes

$$h_2 \frac{\mathrm{d}\sigma_x}{\mathrm{d}x} = \pm\, 2k \tag{7.24b}$$

where the origin for x lies on the roll's vertical centre-line. Integrating eq(7.24b) for the exit side (−) leads to

$$\sigma_x = \frac{2kx}{h_2} + C_2 \tag{7.25a}$$

Now, $\sigma_x = 0$ for $x = 0$ and therefore $C_2 = 0$, which reduces eq(7.25a) to

$$\sigma_x = \frac{2kx}{h_2} \quad \text{for} \quad x_o \le x \le 0 \tag{7.25b}$$

It follows that at the neutral plane, where $x = x_o$, eq(7.25b) becomes

$$\sigma_x = \frac{2kx_o}{h_2} \tag{7.25c}$$

Integrating eq(7.24b) for the entry side (+) gives

$$\sigma_x = -\frac{2kx}{h_2} + C_1 \tag{7.26a}$$

where C_1 is found from knowing σ_x on the exit side of the neutral plane. Substituting eq(7.25c) into eq(7.26a), gives

$$\frac{2kx_o}{h_2} = -\frac{2kx_o}{h_2} + C_1 \;\Rightarrow\; C_1 = \frac{4kx_o}{h_2}$$

and, hence, eq(7.26a) becomes

$$\sigma_x = \left(\frac{2k}{h_2}\right)(2x_o - x) \quad \text{for} \quad L \le x \le x_o \tag{7.26b}$$

Equations (7.25b) and (7.26b) may now be combined with the plane strain yield criterion, when σ_x and p are taken to be principal stresses. This gives

$$p - \sigma_x = 2k$$

The roll pressures become

$$\begin{aligned} p &= 2k\left(1 + x/h_2\right) && \text{for} \quad 0 \le x \le x_o \\ p &= 2k\left[1 + (2x_o - x)/h_2\right] && \text{for} \quad L \le x \le x_o \end{aligned} \tag{7.27a,b}$$

which are distributed in Fig. 7.21c as a 'friction hill'. The area enclosed beneath the hill gives the total roll force/unit width as

$$P = 2kL\left(1 + \frac{x_o}{2h_2}\right) \tag{7.28}$$

When P is taken to act vertically through the centre of the arc of contact, the torque exerted about the centre line of each roll is simply $PL/2$. The length L of the contact arc is found from the geometry of Fig. 7.21a:

$$\tan\varphi_1 = (h_1 - h_2)/L \quad \text{and} \quad \varphi_1\,(\text{rad}) \approx L/R$$

The two equations may be combined for small φ to give L as

$$L = \sqrt{R(h_1 - h_2)} \tag{7.29}$$

Thus, from eqs(7.28) and (7.29), the total roll torque (with 2 rolls) becomes

$$T = P\sqrt{R(h_1 - h_2)}$$
$$= 2k\left[1 + x_o/(2h_2)\right] \times \sqrt{R(h_1 - h_2)} \tag{7.30a}$$

Equation (7.30a) will provide the drive power $T\omega$, where ω is the roll's angular velocity. Experiment [8] has shown that $x_o \approx L/2$. A factor of 0.8 replaces unity in eq(7.30a) for a similar analysis based upon inclined roll platens. These modifications give

$$T = 2k\left[0.8 + \frac{\sqrt{R(h_1 - h_2)}}{4h_2}\right] \times \sqrt{R(h_1 - h_2)} \tag{7.30b}$$

Taking, for example, $R = 100$ mm, $h_1 = 10$ mm and $h_2 = 5$ mm in eq(7.30b), gives $T/2k = 42.9$ for T in Nm and k in MPa.

Using their SLF solution as a guide, Johnson and Mellor [9] derived an upper-bound solution to strip rolling based upon the velocity discontinuities given in Fig. 7.28a. It was shown that two discontinuities, composed of the circular arcs AB and BC, were sufficient to ensure that material in contact with the roll rotated with it. Rigid body motion prevails on the entry and exit sides. To construct the discontinuities, an arc BC is drawn tangentially to the roll at exit (where $k = \mu p$) which intersects the strip's principal stress axis (CL) at 45°. Similarly, BC is tangential at A and meets CL at 45°.

(a)

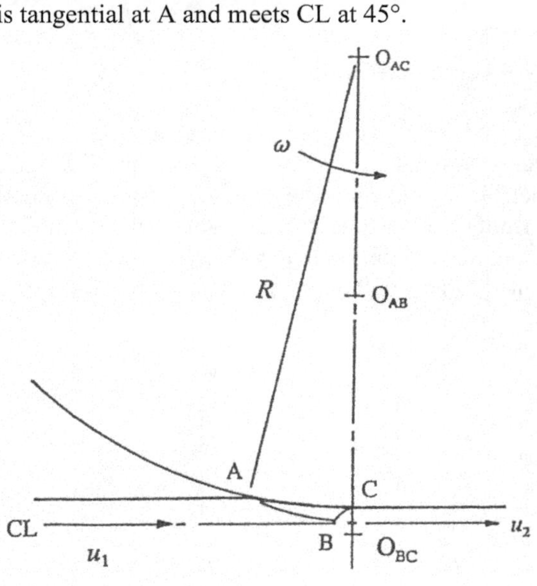

(b)

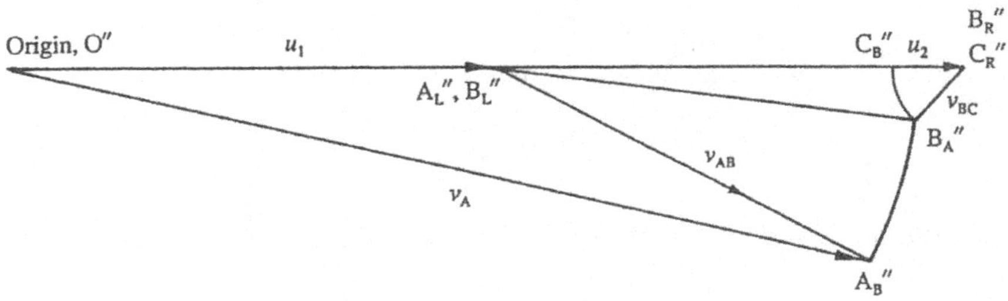

Figure 7.28 Upper bound for strip rolling

The chord of arc BC is inclined at 22.5° to the horizontal at point C and its perpendicular bisector locates the centre position $\bigcirc_{BC}$. Similarly, bisecting the chord AB locates the centre, $\bigcirc_{AB}$, of arc AB.

The corresponding hodograph is given in Fig. 7.22b. Material to the left of A and B (i.e. A_L'' and B_L'') has a rigid horizontal velocity u_1. The peripheral velocity ωR of material in contact with the roll at point A fixes point A_B'' in a tangential direction from the origin, O″. Discontinuities connect A_L'' to A_B'' and B_L'' to B_A'' as shown. The horizontal peripheral velocity ωR for material at point C fixes point C_B''. Finally, discontinuities connect C_B'' to C_R'' and B_A'' to B_R'', as shown. The arcs A_B'' B_A'' and B_A'' C_B'' are orthogonal images of arcs AB and BC and enclose a fan of discontinuities for all positions along each arc. The checkpoints B_R'' and C_R'' give $u_2 = (h_1/h_2)u_1$ when the spread in breadth is neglected. To enable a comparison between this upper bound and the solution $T/(2k) = 42.9$ from eq(7.30b), the hodograph is constructed for $R = 100$ mm, $h_1 = 10$ mm and $h_2 = 5$ mm. Applying eq(7.16) gives the total rate of working as

$$W = T\omega = 2k\,[(AB \times v_{AB}) + (BC \times v_{BC})]$$

where $\omega = v_A/R$. Substituting for ω and the two discontinuities in terms of u_1, from the scale of the hodograph, leads to a comparable upper bound value $T/(2k) = 48.7$ (same units).

7.8 Transverse Loading of Circular Plates

Up to now, all examples have been plane-strain in nature. This chapter concludes with a bounding solution to an axially symmetric problem. It is required to find the loading for plastic collapse of circular plates under various lateral loadings. An upper-bound to the collapse loading follows from the kinematically admissible velocity field [9].

In Fig. 7.29a, an annular plate of thickness t, with an encastré support around its outer radius r_o, carries a distributed force f/unit circumferential length around its inner radius r_i.

(a)

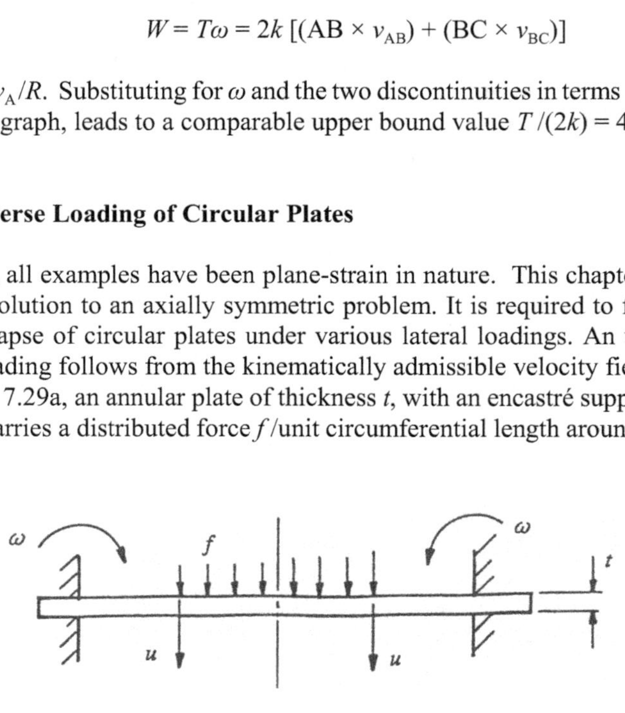

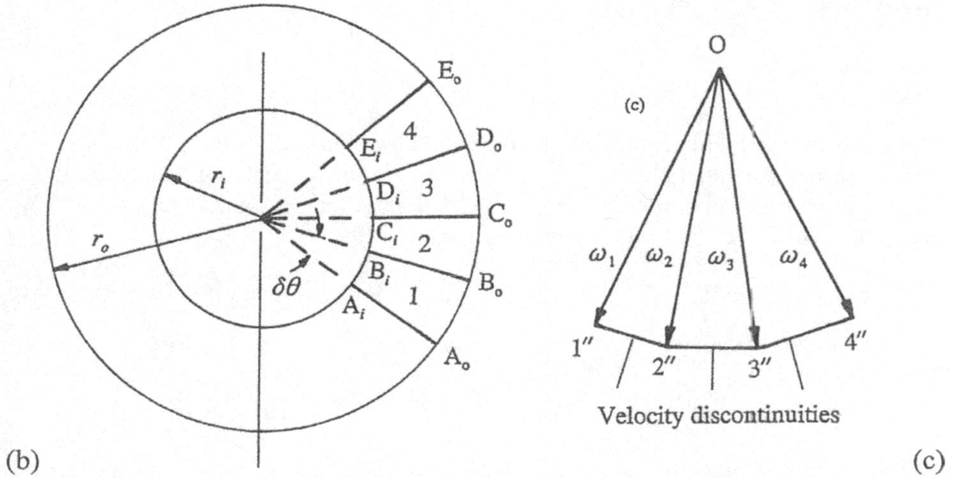

Figure 7.29 Annular, outer encastre plate with inner circumferential loading

One mechanism of plastic collapse involves the formation of radial and circumferential plastic hinges which bound elemental sectors 1, 2, 3, 4 etc., as shown in Fig. 7.29b. Plastic hinges A_oB_o and B_iB_o, are formed between sectors 1 and 2. Simultaneously, the hinges B_oC_o and C_iC_o are formed between sectors 2 and 3, etc. As the inner radius r_i descends vertically with velocity u, the sectors rotate rigidly about their respective plastic hinges. That is: sector 1 rotates about A_oB_o, sector 2 rotates about B_oC_o, etc, in an anti-clockwise sense at a constant angular velocity $\omega = u/(r_o - r_i)$. The hodograph in Fig. 7.29c represents these angular velocities as vectors ω_1, ω_2 etc, by applying the right-hand screw rule to each hinge rotation. The hodograph displays velocity discontinuities: 1″2″, 2″3″, 3″4″ etc, across the respective radial hinges: B_iB_o, C_iC_o, D_iD_o, etc. In the limiting case, where $\delta\theta \to 0$, the hodograph becomes a circle of radius ω. At the point of collapse, the rate of external work done $\dot{W}$ is given by

$$\dot{W} = (2\pi r_i)f_U \times u = (2\pi r_i)f_U \times (r_o - r_i)\omega \tag{7.31a}$$

The rate of internal energy dissipation is

$$\dot{U} = 2\pi r_o \omega M_U + 2\pi(r_o - r_i)\omega M_U \tag{7.31b}$$

In the first term of eq(7.31b), M_U is the upper-bound collapse moment/unit length of the circumferential hinge. In the second term, M_U is expressed per unit length of the radial hinge. The collapse mechanism requires equal radial and circumferential moments. Equating $\dot{W}$ and $\dot{U}$, from eqs(7.31a,b), gives

$$(2\pi r_i)f_U \times (r_o - r_i)\omega = 2\pi\omega M_U(2r_o - r_i)$$

which gives the collapse force f_U/unit of inner circumference as

$$f_U = \frac{(2r_o - r_i)}{r_i(r_o - r_i)} M_U \tag{7.32a}$$

The total collapse force becomes

$$F_U = 2\pi r_i f_U = 2\pi \frac{(2r_o - r_i)}{(r_o - r_i)} M_U \tag{7.32b}$$

With $r_i = 0$ for a solid plate, the central concentrated collapse force is $F_U = 4\pi M_U$. If the outer radius of the annular plate in Fig. 7.29 were to rest upon simple supports, the collapse mechanism is due solely to the formation of radial hinges under the limiting moment M_U. From eq(7.31b), the energy dissipation simplifies to

$$\dot{U} = 2\pi(r_o - r_i)\omega M_U \tag{7.32c}$$

Equating (7.32c) to eq(7.31a) gives

$$(2\pi r_i)f_U \times (r_o - r_i)\omega = 2\pi(r_o - r_i)\omega M_U$$

from which $f_U = M_U/r_i$ and $F_U = 2\pi r_i f_U = 2\pi M_U$. This total collapse load also applies to an annular plate, simply supported at its inner radius r_i, with a limiting line load $f_U = M_U/r_o$ applied around its outer radius r_o.

Next, consider a plate where the inner radius r_i is clamped and the outer radius r_o supports the line load f_U as in Fig. 7.30.

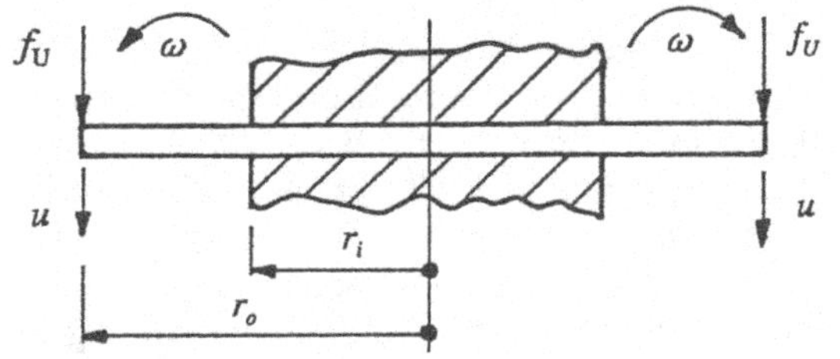

Figure 7.30 Annular plate clamped at inner radius with outer rim circumferential loading

The plate collapses when circumferential and radial plastic hinges form at r_i simultaneously under the limiting moment M_U. The previous boundary conditions are reversed so that one may substitute r_i for r_o in eqs(7.31a,b) to give

$$2\pi r_o f_U u = 2\pi r_i \omega M_U + 2\pi(r_o - r_i)\omega M_U$$

$$2\pi r_o f_U \times (r_o - r_i)\omega = 2\pi \omega M_U r_o \tag{7.33}$$

This leads to the limiting collapse loading:

$$f_U = \frac{M_U}{(r_o - r_i)} \quad \text{and} \quad F_U = 2\pi r_o f_U = \frac{2\pi r_o M_U}{(r_o - r_i)}$$

where the clamping radius $r_i \to 0$, the total collapse load $F_U \to 2\pi M_u$.

Now, let an annular plate be clamped around its outer edge and carry, instead of line loading, a uniform normal pressure p over its top surface area, as shown in Fig. 7.31.

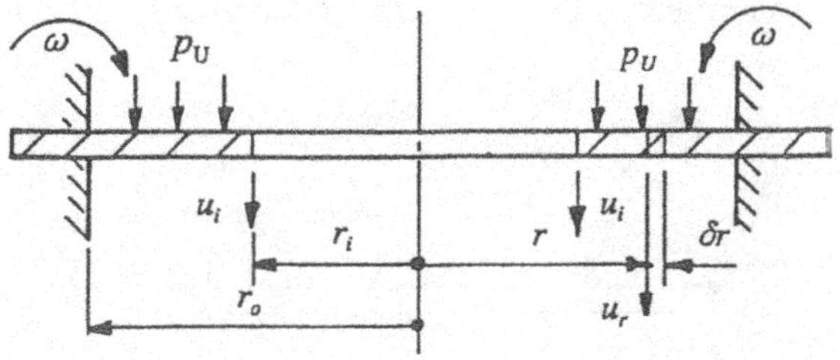

Figure 7.31 Annular plate with outer, encastre fixing and uniform pressure loading

In formulating the external work rate $\dot{W}$, note that the descending velocities u_r are required at all radii: $r_i \le r \le r_o$. Assuming a linear variation in velocity between u_i for $r = r_i$ and zero velocity for $r = r_o$ gives

$$u_r = u_i(r_o - r)/(r_o - r_i) \tag{7.34}$$

At collapse is the external work rate:

$$\dot{W} = \int_r (2\pi r\,\mathrm{d}r)p_U \times u_r \tag{7.35a}$$

Substituting from eq(7.34) gives

$$\dot{W} = 2\pi p_U u_i/(r_o - r_i) \int_{r_i}^{r_o} r(r_o - r)\mathrm{d}r$$

$$= \frac{\pi}{3} p_U\,\omega\,(r_o - r_i)^2(r_o + 2r_i) \tag{7.35b}$$

Equating $\dot{W}$ from eq(7.35b) to $\dot{U}$ from eq(7.31b) gives

$$\frac{1}{3}\pi p_U \omega(r_o - r_i)^2(r_o + 2r_i) = 2\pi\omega M_U(2r_o - r_i)$$

leading to an upper pressure bound

$$p_U = \frac{6\,M_U(2r_o - r_i)}{(r_o - r_i)^2(r_o + 2r_i)} \tag{7.36}$$

For a solid plate $r_i = 0$, eq(7.36) reduces to $p_U = 12M_U/r_o^2$. If the plate in Fig. 7.31 were simply supported around its outer radius, the simplified energy dissipation equation (7.32c) again applies. Equating this to the work rate in eq(7.35b) gives an upper bound pressure:

$$\frac{1}{3}\pi p_U \omega(r_o - r_i)^2(r_o + 2r_i) = 2\pi(r_o - r_i)\omega M_U$$

$$p_U = \frac{6\,M_U}{(r_o - r_i)(r_o + 2r_i)}$$

For a solid, simply supported plate, take $r_i = 0$, to give $p_U = 6M_U/r_o^2$.

Finally, for the plate in Fig. 7.32, the inner radius r_i is clamped while the remaining surface area is subjected to its uniform collapse pressure p_U, as shown.

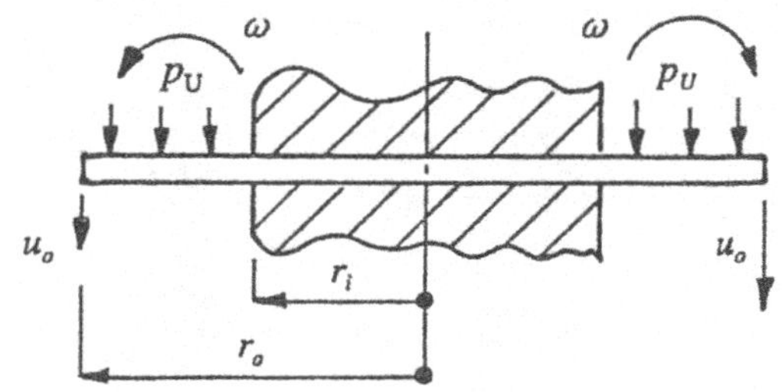

Figure 7.32 Annular plate with inner encastre fixing under uniform pressure loading

This loading configuration reverses the boundary conditions given for Fig. 7.31. Hence, the descent velocity u_r at any radius $r_i \le r \le r_o$ becomes $u_r = u_o(r - r_i)/(r_o - r_i)$. Substituting into eq(7.35a), the external work rate becomes

$$\begin{aligned} \dot{W} &= 2\pi p_U u_o/(r_o - r_i) \int_{r_i}^{r_o} r(r - r_i)\mathrm{d}r \\ &= \frac{\pi}{3} p_U \omega (r_o - r_i)^2 (r_i + 2r_o) \end{aligned} \qquad (7.37)$$

The energy dissipation rate is identical for the plate in Fig. 7.30. Equation (7.33) gives

$$\frac{1}{3}\pi p_U \omega (r_o - r_i)^2 (r_i + 2r_o) = 2\pi r_i \omega M_U + 2\pi (r_o - r_i)\omega M_U$$

$$\frac{1}{3}\pi p_U \omega (r_o - r_i)^2 (r_i + 2r_o) = 2\pi \omega M_U r_o$$

The upper bound pressure is

$$p_U = \frac{6 M_U r_o}{(r_o - r_i)^2 (r_i + 2r_o)} \qquad (7.38)$$

As the clamping radius $r_i \to 0$, eq(7.38) shows that the collapse pressure approaches that for a solid plate, i.e. $p_U \to 3M_U / r_o^2$.

7.9 Concluding Remarks

This introduction to limit analysis has demonstrated that an assumption of plastic-rigidity facilitates bounding solutions to the collapse condition in many load-bearing structures. The simplest, upper-bound solution is particularly useful when it is required to obtain, with ease and certainty, the maximum forming forces required. There are many further applications of these techniques, e.g. to forging and machining. These appear in more specialised texts [1, 7, 9–12], to which the interested reader should refer.

References

1. Ford H. and Alexander J. M. *Advanced Mechanics of Materials*, 1963, Longman.
2. Geirenger H. "Fondements mathematiques de la theorie des corps plastiques isotropes", *Mem. Sci. Math.*, 1937, 86.
3. Rees, D. W. A. *Mechanics of Inelastic Solids*, 2021, in preparation, IC Press, London.
4. Neal B. G. *Plastic Methods of Structural Analysis*, 1964, Chapman and Hall.
5. Alexander J. M. *Quart. Appl. Math*., 1961, **19**, 32.
6. Johnson W. *Proc. I. Mech. E.*, 1959, **173**, 61.
7. Prager W. and Hodge P. G. *The Theory of Perfectly Plastic Solids*, 1951, Wiley, N.Y.
8. Orowan E. *Proc. I..Mech. E.*, 1943, **150**, 140.
9. Johnson W. and Mellor P. B. *Engineering Plasticity*, 1983, Ellis-Horwood.
10. Alexander J. M. and Brewer R. C. *Manufacturing Properties of Materials*, 1963, van Nostrand Co. Ltd.
11. Calladine C. R. *Plasticity for Engineers*, 2000, Horwood Pub. Ltd.
12. Chakrabarty J. *Theory of Plasticity*, 3rd Ed, 2006, Butterworth-Heinemann.

Exercises

7.1 Obtain upper and lower bounds for a beam with the triangular cut out, base $2a$, height a, shown in Fig. 7.33, when the beam supports a uniform bending moment M
See solution: $M_L = ka^2$, $M_U = 1.38ka^2$

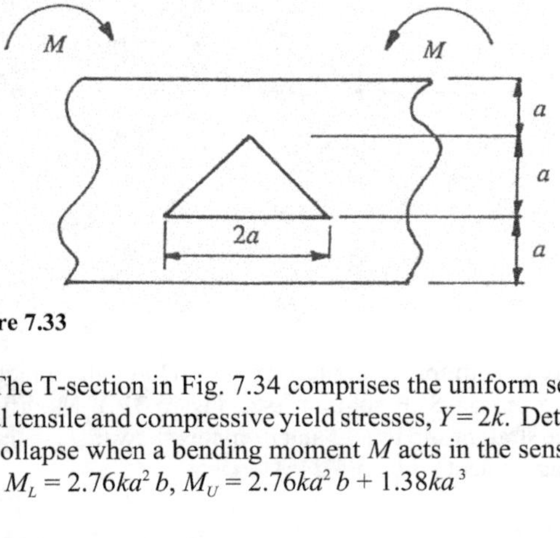

Figure 7.33

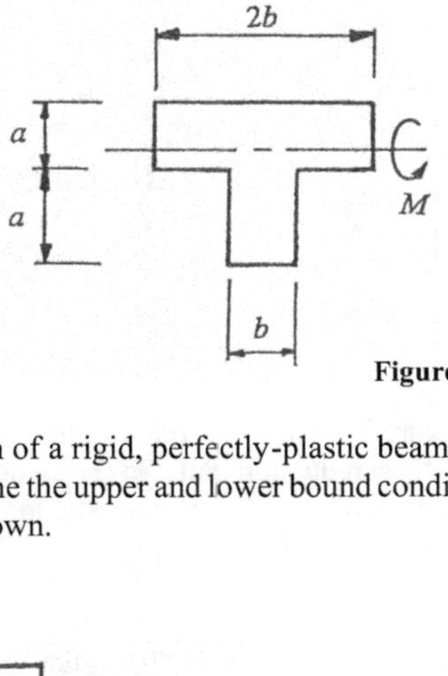

Figure 7.34

7.2 The T-section in Fig. 7.34 comprises the uniform section of a rigid, perfectly-plastic beam with equal tensile and compressive yield stresses, $Y = 2k$. Determine the upper and lower bound conditions for collapse when a bending moment M acts in the sense shown.
Ans: $M_L = 2.76ka^2 b$, $M_U = 2.76ka^2 b + 1.38ka^3$

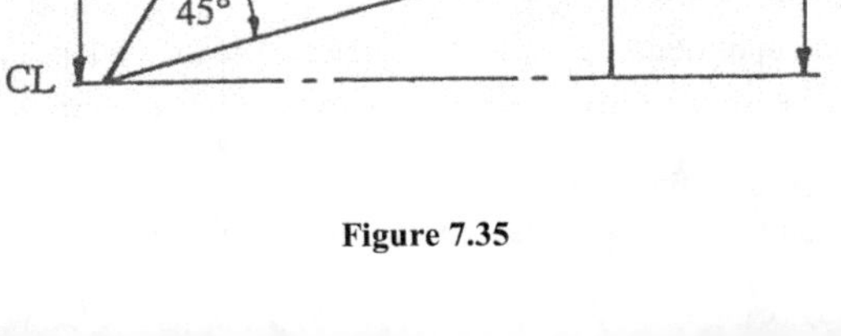

Figure 7.35

7.3 The stress discontinuity pattern in Fig. 7.35 may be used to obtain a lower bound solution to the problem of frictionless plane strain extrusion through a square die. Construct the stress plane and show that the extrusion pressure is

$$\frac{p}{2k} = 2\left[1 - \frac{2R}{(2R - 1)^2 + 1}\right]$$

where $R = H/h$. Compare the value of $p/2k$ for $R = 3$ with that from the slip-line field (see Fig. 6.27). *See solution*

7.4 Fig. 7.36 shows ¼ of a testpiece in which the ratio of the notched to un-notched breadths is 2/3. Determine the axial force P required to initiate yield. What is the new position of the notch boundary after the rigid region moves away from the horizontal centre line by 6.5 mm?

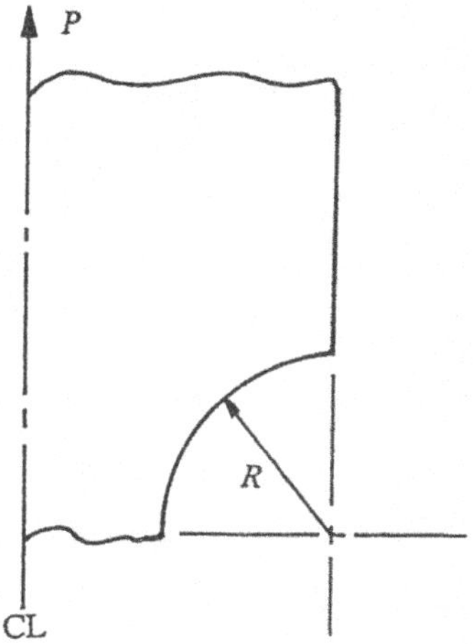

Figure 7.36

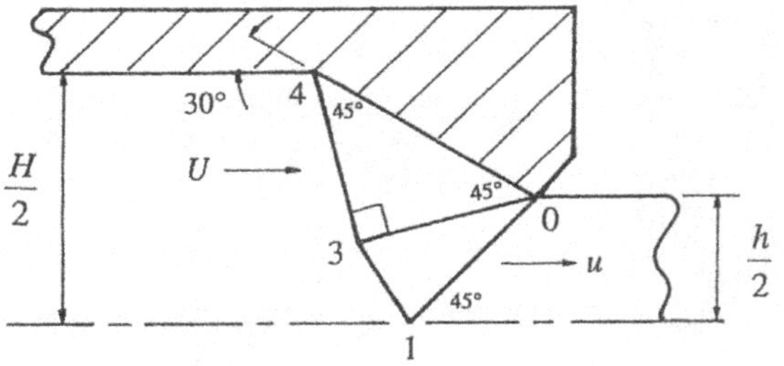

Figure 7.37

7.5 Figure 7.37 shows a velocity discontinuity pattern for frictionless extrusion through an inclined die with a reduction ratio of 2, as derived from an SLF solution (see Fig. 6.31a). Modify the hodograph in Fig. 6.31b to correspond with the straight slip line 13 and from this show that the upper-bound extrusion pressure is $p_U/2k = 0.76$, in agreement with eq(6.24b).

7.6 A square section beam of side a and length L is built in at one end and simply supported at the other end. A vertical load F is applied at a distance x from the built-in end. When the collapse moment is M, show that F and the position in the length for which F is a minimum become

See solution

$$\frac{F}{M} = \frac{1}{(L - x)} + \frac{2}{x}, \quad \text{where } x = \frac{\surd 2L}{(1 + \surd 2)}$$

7.7 A thin, flat triangular plate is fixed in position around its perimeter whilst carrying a normal force F at its centroid. Find an upper bound solution to F in terms of the collapse moment M_U.

7.8 A thin, flat triangular plate is fixed in position along each side whilst carrying a normal pressure p. Find an upper-bound solution to p in terms of the collapse moment M_U.

7.9 Determine the collapse loading for each of the three failure modes (for beam, stanchions and both) possible in the portal frame given in Fig. 7.38. Express each answer in terms of the collapse moment M_P and length L for the beam. What is implied when the collapse loads between agree?
Ans: $W = 3.5M_P/L$, $W = 2M_P/L$, $W = 2M_P/L$

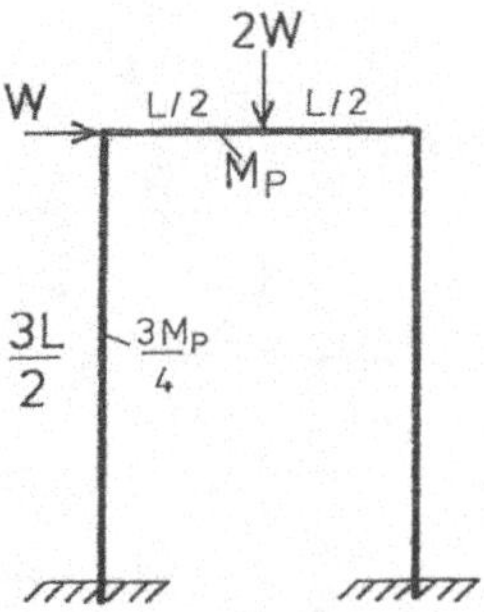

Figure 7.38

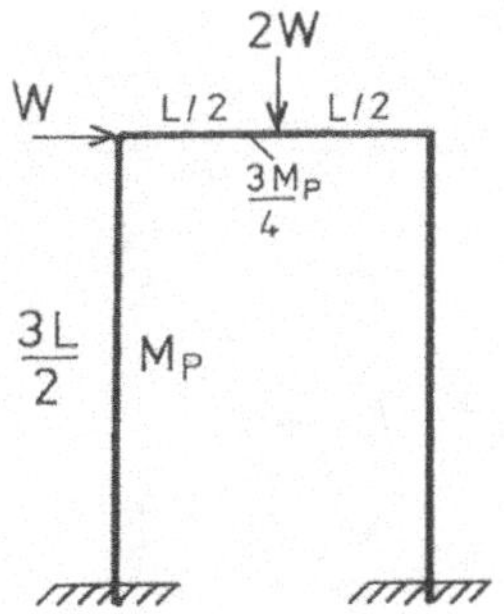

Figure 7.39

7.10 The moment carrying capacities for the beam and stanchions in Fig. 7.39 interchange those of Fig. 7.38 without changing their lengths or the loading applied. Examine the influence upon the collapse loadings and establish which frame is closer to an optimal design.
Ans: $W = 3M_P/L$, $W = 7M_P/3L$, $W = 2M_P/L$

7.11 Determine the collapse loading for each of the three separate failure modes: for the beam, the stanchions and both beam and stanchion in the portal frame given in Fig. 7.40. Express each answer in terms of the variables M_P and L. If the base fixings are replaced by pinned connections what then are the collapse loads from each failure mode? What is the multiplying factor x in modifying the moment capacity for the beam to xM_P if separate beam collapse and stanchion sway modes are to occur in the fixed base frame under the similar loading?
Ans: $W = 11.25M_P/L$, $W = 4M_P/L$, $W = 5.7M_P/L$; $W = 11.25M_P/L$, $W = 2M_P/L$, $W = 4.5M_P/L$, $x = 2/7$

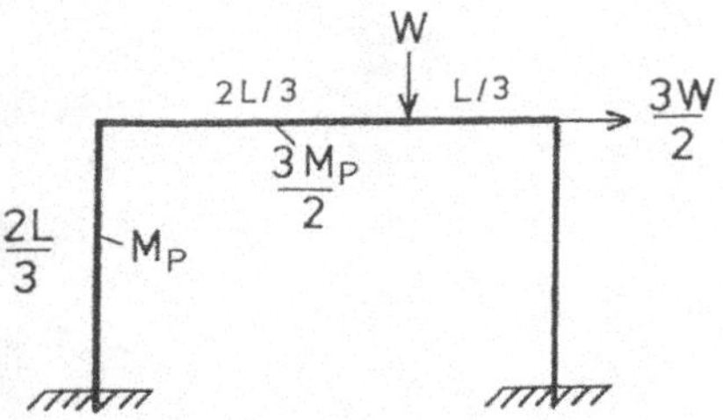

Figure 7.40

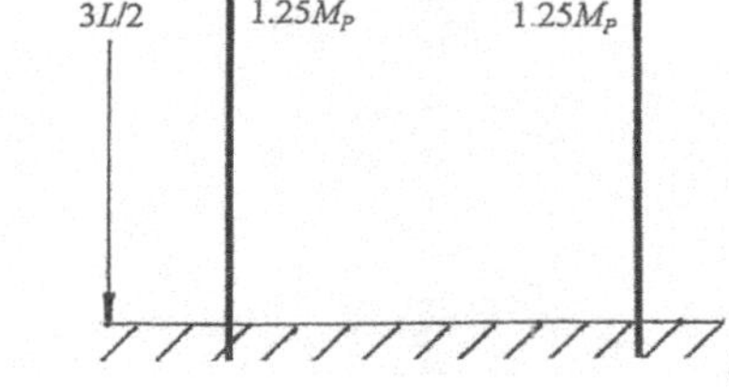

Figure 7.41

7.12 Determine the collapse loading for each of the three failure modes (for beam, stanchions and both) possible in the portal frame given in Fig. 7.41. Express each answer in terms of the collapse moment M_P and length L for the beam. How are the collapse loads altered when the frame is pinned at the base of each fixed stanchion?
Ans: $W = 32M_P/3L$, $W = 6M_P/L$, $W = 7M_P/L$; $W = 32M_P/3L$, $W = 8M_P/3L$, $W = 16M_P/3L$

7.13 Derive, from energy considerations, expressions for the collapse load of the fixed-base portal frame in Fig. 7.42 under: (a) a central vertical load $W/2$, (b) a horizontal load W and (c) a combination of the loads in (a) and (b).
Note: The moment carrying capacity for the left and right stanchions differ, being $1.5M_P$ and $0.5M_P$ respectively.
Ans: $15M_P/L$, $7M_P/L$, $39M_P/11L$

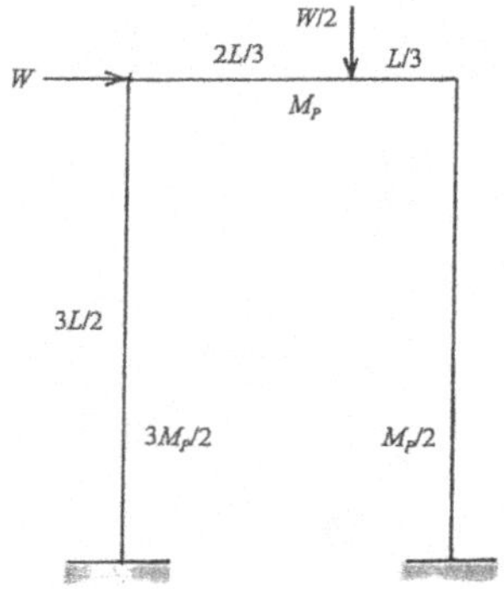

Figure 7.42

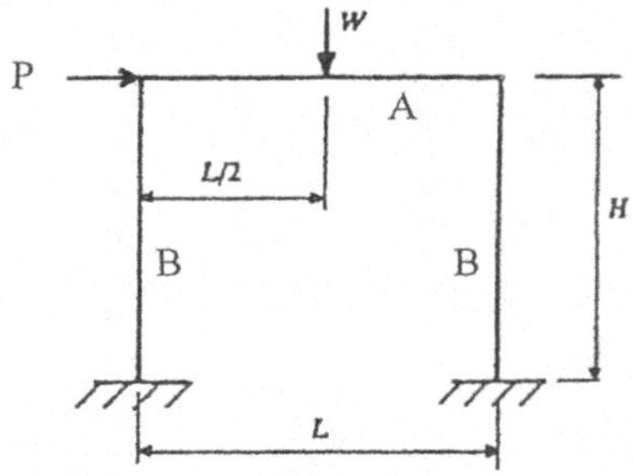

Figure 7.43

7.14 Derive, from energy considerations, expressions for the collapse load of the fixed-base portal frame in Fig. 7.43 under: (a) a central vertical load W, (b) a horizontal load P and (c) combined loads W and P. The frame has uniform section of moment carrying capacity M_P. Show graphically for (c) the manner on which the collapse condition varies with the ratios $\alpha = P/W$ and $\beta = H/L$ when $\beta =$ constant and for $\beta = \alpha/2$, $\beta = \alpha$ and $\beta = 2\alpha$.
See solution: $W = 8M_P/L$, $W = 4M_P/H$, $M_P/WL = (1 + 2\alpha\beta)/12$

CHAPTER 8

CRYSTAL PLASTICITY

8.1 Introduction

Much work has been done to increase our understanding of the microstructural mechanisms responsible for plastic flow. It is not the purpose of this chapter to detail all possible dislocation mechanisms by which atomic slip can occur as these have been adequately documented elsewhere [1, 2]. Instead, the discussion is limited to a consideration of those aspects of the micro-mechanics of deformation that lend support to the key elements employed in continuum theories of plasticity. Important among these, as following chapters will show, are the physical interpretation of work hardening, the Bauschinger effect, the shape of the initial and subsequent yield surfaces, the associated flow rule and the manner in which strain within a grain contributes to the observed strain for a polycrystal. The plastic response of single crystals and polycrystals to a tensile stress involves slip with hardening. Anisotropic or directional hardening occurs in a single crystal. A polycrystal with many randomly oriented grains has an aggregated isotropic structure that hardens anisotopically under a plastic strain imposed directionally. However, an aggregated isotropic hardening behaviour is often assumed. The two behaviours influence the subsequent yield surface and its manner of representation within respective isotropic and kinematic hardening models.

External parameters that influence plastic flow include the environmental conditions and the nature of the loading. Internal, micro-structural parameters involve the orientation of the slip plane, lattice imperfections, alloying and impurity elements as follows:

1. Particular slip systems operating within close-packed planes in close-packed directions, that enable individual crystals to deform plastically.
2. The behaviour of slip mechanisms in the presence of existing defects in real lattices. These include: (i) point defects–interstitial and vacancies, (ii) line defects–dislocations, (iii) plane defects–grain boundaries and (iv) volume defects–cracks and atomic stacking faults.
3. The movement, multiplication and interaction between dislocations and the extent to which these are impeded by other defects acting as obstacles.

The important point to note is that such defects present in metallic crystals facilitate the process of block slip to occur between adjacent planes of atoms at considerably reduced stress levels. It has been observed [1] that the shear stress required for slip is four orders of magnitude lower than a theoretical cohesive strength, calculated from $G/(2\pi)$, in which G is the shear modulus. On a micro-scale, slip is the major contributor to low temperature plasticity in metals. Other lesser influences such as twinning, void growth and micro-shear banding can contribute to plasticity under certain conditions. For example, void growth occurs in dispersion hardened alloys at higher temperatures and twinning arises in a hexagonal close-packed lattice to enhance the plasticity from slip on its single slip plane. This chapter also surveys the metallurgy of heat treatment and hardening mechanisms within aluminium alloys and steels given their importance in structural design.

8.2 Resolved Shear Stress and Strain

8.2.1 Simple Tension

In order to understand the nature of hardening in a polycrystalline aggregate it is instructive, firstly, to analyse the stress in a single, cylindrical crystal of metal, loaded in tension along its axis (see Fig. 8.1a).

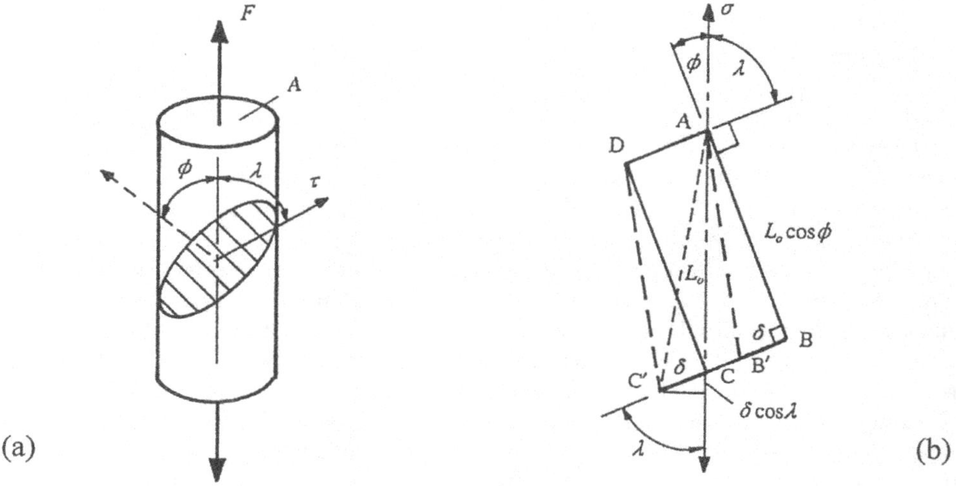

Figure 8.1 Single crystal under tension

Let a tensile force F act normal to section area A of the crystal to induce a nominal tensile stress $\sigma = F/A$. The normal to the inclined slip plane and the slip direction make respective angles φ and λ with the tensile axis as shown. The resolved shear stress τ is aligned with the slip direction lying in the slip plane. The magnitude of τ is found from dividing that component of F, aligned with the slip direction, by the area of the slip plane. The latter is an ellipse with an area $A/\cos\varphi$. Thus:

$$\tau = \frac{F\cos\lambda}{A/\cos\varphi} = \sigma\cos\varphi\,\cos\lambda \qquad (8.1)$$

It follows from eq(8.1) that τ is zero for the $\varphi = 0°$ and $90°$ principal directions, the former being associated with a cleavage fracture plane. In contrast, τ will reach its maximum value of $\sigma/2$ for a $\varphi = 45°$ slip plane. The magnitude of this resolved shear stress that initiates slip upon a 45° plane is called the *critical resolved shear stress* τ_{cr}. Planes with different orientation will slip once τ_{cr} is attained. The corresponding *resolved shear strain*, or glide strain γ, is a more appropriate measure of deformation to use than percentage elongation or nominal engineering strain. An often-used approximation to γ is applied when the axial strain ε_z is small. Thus, in Fig. 8.1b, ABCD is a rectangular element lying between two parallel slip planes, separated by length L_o along the tensile axis z. Let δ equal the amount of slip, BB′ and CC′, along the lower slip plane. Within the deformed element AB′C′D, the geometry shows that AC has extended by the amount $\delta\cos\lambda$. Hence, the axial strain is

$$\varepsilon_z = \frac{\delta}{L_o}\cos\lambda \qquad (8.2a)$$

Now AB = $L_o \cos\varphi$ is the perpendicular distance between the original slip planes. AB will also rotate towards the tensile axis during deformation. When the rotation is small the glide strain may be defined simply in terms of the angular change to the right angle ∠ DAB. That is, γ becomes the tangent of the angle ∠ BAB′:

$$\gamma = \tan(\mathrm{BAB}') \simeq \frac{\delta}{L_o \cos\varphi} \quad (\mathrm{rad}) \tag{8.2b}$$

Combining eqs(8.2a,b) provides the glide strain in terms of the measured axial strain:

$$\gamma \simeq \frac{\varepsilon_z}{\cos\varphi \, \cos\lambda} \tag{8.2c}$$

Honeycombe [2] employed alternative measures of γ for large deformation. These involve the geometry of a normal to the slip plane AB that rotates to AB′, i.e. towards the tensile z–axis with increasing strain. Given final orientations φ_f and λ_f of a slip plane, the resolved shear strain becomes

$$\gamma = \frac{\cos\lambda_f}{\sin(90 - \varphi_f)} - \frac{\cos\lambda}{\sin(90 - \varphi)} \tag{8.2d}$$

where φ and λ define the initial orientations of this plane (Fig. 8.1a). Alternatively, when the final, nominal axial strain ε_{zf} is measured, γ follows as

$$\gamma = \frac{1}{\sin(90 - \varphi)}\left[\sqrt{(1 + \varepsilon_{zf})^2 - \sin^2\lambda} - \cos\lambda\right] \tag{8.2e}$$

Equations (8.1) and (8.2c–e) define shear stress and strain for a slip plane. Plots between τ and γ reveal that three distinct regions of hardening accompany slip within a single crystal.

8.2.2 General Stress State

Now, instead of an applied tension, let a general stress tensor σ_{ij} be defined in arbitrary axes x_i ($i = 1, 2$ and 3). Firstly, σ_{ij} has to be transformed to give the shear stress upon the active slip plane. This identifies with the shear stress component σ_{12}' in eq(1.22b), written here as σ_{12}^s, where axes x_1^s and x_2^s lie parallel and normal to an active slip plane s in Fig. 8.2a.

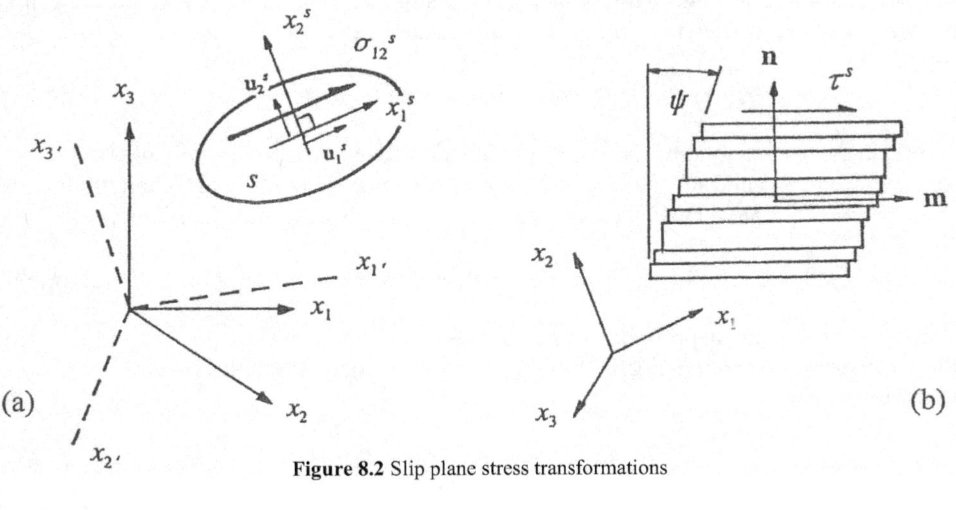

Figure 8.2 Slip plane stress transformations

Equation (1.22a) gives this shear stress component directly when $i = 1$ and $j = 2$:

$$\sigma_{12}^{\ s} = l_{1p}^{\ s}\, l_{2q}^{\ s}\, \sigma_{pq} = (l_{1i}^{\ s}\, l_{2j}^{\ s})\, \sigma_{ij} = \alpha_{ij}^{\ s}\, \sigma_{ij} \tag{8.3a}$$

where the *Schmidt orientation factor* $\alpha_{ij}^{\ s} = l_{1i}^{\ s}\, l_{2j}^{\ s}$ employs the symmetry of the stress tensor $\sigma_{ij} = \sigma_{ji}$. The Schmidt factor employs a set of direction cosines $l_{1i}^{\ s}$ for axis $x_1^{\ s}$ and another set $l_{2j}^{\ s}$ for axis $x_2^{\ s}$, with respect to the reference stress axes x_i (see Fig. 8.2a).

It is more convenient to align reference axes with the cubic axes x_i' ($i = 1, 2,$ and 3) of a unit cell, given that the slip plane and slip direction are defined with Miller's indices (see § 8.3). Referring to Fig. 8.2a, two stress transformations become necessary to find $\sigma_{12}^{\ s}$: From eq(1.22a), transform firstly from arbitrary stress axes x_i to reference (cube) axes x_i':

$$\sigma_{ij}' = l_{ip}\, l_{jq}\, \sigma_{pq} \quad \text{or} \quad \mathbf{T}' = \mathbf{L}\mathbf{T}\mathbf{L}^{\mathrm{T}} \tag{8.3b}$$

Then, transform from reference axes x_i' to the slip plane:

$$\sigma_{12}^{\ s} = l_{1p}^{\ s}\, l_{2q}^{\ s}\, \sigma_{pq}'$$

These two transformations are combined as follows:

$$\begin{aligned}
\sigma_{12}^{\ s} &= l_{1p}^{\ s}\, l_{2q}^{\ s}\, \sigma_{pq}' = l_{1p}^{\ s}\, l_{2q}^{\ s}\, (l_{pr} \quad l_{qs}\, \sigma_{rs}) \\
&= (l_{1p}^{\ s}\, l_{pr}\;\; l_{2q}^{\ s}\, l_{qs})\, \sigma_{rs} = (l_{1p}^{\ s}\, l_{pi}\;\; l_{2q}^{\ s}\, l_{qj})\, \sigma_{ij} \\
&= \alpha_{pq}^{\ s}\, (l_{pi}\, l_{qj})\sigma_{ij}
\end{aligned} \tag{8.3c}$$

in which the Schmidt factor $\alpha_{pq}^{\ s} = l_{1p}^{\ s}\, l_{2q}^{\ s}$ again appears. The matrix form of eq(8.3a) employs column matrices of direction cosines for each unit vector, i.e. $\mathbf{u}_1^{\ s} = [l_{11}^{\ s}\ \ l_{12}^{\ s}\ \ l_{13}^{\ s}]^{\mathrm{T}}$ and $\mathbf{u}_2^{\ s} = [l_{21}^{\ s}\ \ l_{22}^{\ s}\ \ l_{23}^{\ s}]^{\mathrm{T}}$ (see Fig. 8.2a). Within eq(8.3c), the transformation matrix $\mathbf{L}$ between axes x_i and x_i' in eq(8.3b) must also appear. Equations(8.3a,c) become, respectively,

$$\sigma_{12}^{\ s} = (\mathbf{u}_1^{\ s})^{\mathrm{T}}\, \mathbf{T}'\, \mathbf{u}_2^{\ s} \quad \text{and} \quad \sigma_{12}^{\ s} = (\mathbf{u}_1^{\ s})^{\mathrm{T}}\, (\mathbf{L}\ \mathbf{T}\ \mathbf{L}^{\mathrm{T}})\mathbf{u}_2^{\ s} \tag{8.4a,b}$$

where $\mathbf{T} \equiv \sigma_{ij}$ is the Cauchy stress matrix (1.17). For simplicity, $\sigma_{12}^{\ s}$ is replaced with τ^s and unit vectors **n** and **m** align with the normal and slip directions respectively for the cell's reference axes x_i, as shown in Fig. 8.2b. Equations (8.3a) and (8.4a) become

$$\tau^s = m_i\, n_j\, \sigma_{ij} = \alpha_{ij}^{\ s}\, \sigma_{ij} \equiv \mathbf{m}^{\mathrm{T}}\mathbf{T}\mathbf{n} \tag{8.5a}$$

where, in the matrix notation, m_i and n_j are direction cosine components of column vectors $\mathbf{m} = [m_1\ \ m_2\ \ m_3]^{\mathrm{T}}$ and $\mathbf{n} = [n_1\ \ n_2\ \ n_3]^{\mathrm{T}}$ in cube axes x_i. According to Schmidt's law, it follows from eq(8.5a) that slip occurs when the resolved shear stress attains a critical value:

$$\tau^s = \alpha_{ij}^{\ s}\, \sigma_{ij} = \tau_{\mathrm{cr}}^{\ s} \tag{8.5b}$$

In chapter 1 the identical properties of stress and strain transformations were revealed. Here the shear strain $\gamma^s = \tan\psi$ in Fig. 8.2b may be transformed to the cubic axes x_i by an inverse transformation:

$$\varepsilon_{ij}^{\ P} = \tfrac{1}{2}\,(n_i\, m_j + m_i\, n_j)^s\, \gamma^s = \alpha_{ij}^{\ s}\, \gamma^s \tag{8.6a}$$

where $\alpha_{ij}{}^s = ½\,(n_i m_j + m_i n_j)^s = ½\,(\mathbf{nm}^T + \mathbf{mn}^T)^s$. The slip plane orientation will be altered by the rotation component of the shear distortion. Figure 8.3a shows plane shear distortions e_{12} and e_{21}, from which the shear strain $\gamma = \varepsilon_{12}{}^P + \varepsilon_{21}{}^P$ is derived. These are accompanied by plastic rotation (spin) $\omega_{12}{}^P$ and $\omega_{21}{}^P$ of the crystal body about the 3–axis (see Fig. 8.3b).

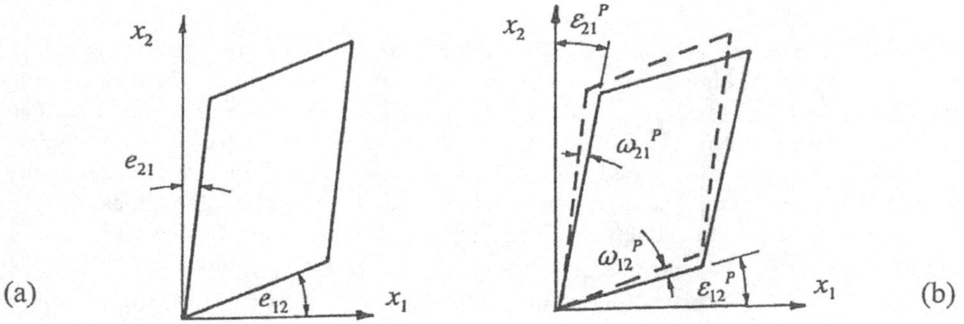

Figure 8.3 Decomposition of shear distortions

The decomposition shown conforms to eq(2.2b). In general, the 3D rotations are found from dyadic products of the unit vectors **m** and **n** as

$$\omega_{ij}{}^P = ½\,(n_i m_j - m_i n_j)^s\,\gamma^s = \beta_{ij}{}^s\,\gamma^s \tag{8.6b}$$

where $\beta_{ij}{}^s = ½\,(n_i m_j - m_i n_j)^s \equiv ½\,(\mathbf{nm}^T - \mathbf{mn}^T)^s$.

Where multiple slip systems, $s = 1, 2, 3 \ldots N$, exist within a single crystal, they contribute to the overall observed strain in the cubic reference frame. Their summations reveal the incremental plastic strain components from eq(8.6a), as follows:

$$\mathrm{d}\varepsilon_{ij}^P = \sum_{s=1}^{N} \alpha_{ij}^s\,\mathrm{d}\gamma^s = \frac{1}{2}\sum_{s=1}^{N}\left(n_i m_j + m_i n_j\right)^s \mathrm{d}\gamma^s \tag{8.7}$$

For example, with $N = 5$ slip systems, eq(8.7) shows

$$\begin{aligned}
\mathrm{d}\varepsilon_{11}{}^P &= (n_1 m_1)^1\,\mathrm{d}\gamma^1 + (n_1 m_1)^2\,\mathrm{d}\gamma^2 + \ldots\ldots + (n_1 m_1)^5\,\mathrm{d}\gamma^5\\
\mathrm{d}\varepsilon_{22}{}^P &= (n_2 m_2)^1\,\mathrm{d}\gamma^1 + (n_2 m_2)^2\,\mathrm{d}\gamma^2 + \ldots\ldots + (n_2 m_2)^5\,\mathrm{d}\gamma^5\\
\mathrm{d}\varepsilon_{33}{}^P &= -\,(\mathrm{d}\varepsilon_{22}{}^P + \mathrm{d}\varepsilon_{22}{}^P)\\
\mathrm{d}\varepsilon_{12}{}^P = \mathrm{d}\varepsilon_{21}{}^P &= (n_1 m_2 + m_1 n_2)^1\,\mathrm{d}\gamma^1 + (n_1 m_2 + m_1 n_2)^2\,\mathrm{d}\gamma^2 + \ldots\ldots + (n_1 m_2 + m_1 n_2)^5\,\mathrm{d}\gamma^5\\
\mathrm{d}\varepsilon_{13}{}^P = \mathrm{d}\varepsilon_{31}{}^P &= (n_1 m_3 + m_1 n_3)^1\,\mathrm{d}\gamma^1 + (n_1 m_3 + m_1 n_3)^2\,\mathrm{d}\gamma^2 + \ldots\ldots + (n_1 m_3 + s_1 m_3)^5\,\mathrm{d}\gamma^5\\
\mathrm{d}\varepsilon_{23}{}^P = \mathrm{d}\varepsilon_{32}{}^P &= (n_2 m_3 + m_2 n_3)^1\,\mathrm{d}\gamma^1 + (n_2 m_3 + m_2 n_3)^2\,\mathrm{d}\gamma^2 + \ldots\ldots + (n_2 m_3 + m_2 n_3)^5\,\mathrm{d}\gamma^5
\end{aligned}$$

in which plastic incompressibility $\mathrm{d}\varepsilon_{ii}{}^P = 0$ and strain symmetry $\mathrm{d}\varepsilon_{ij}{}^P = \mathrm{d}\varepsilon_{ji}{}^P$ apply. Mathematically, the 5 independent strain components may be taken to arise from 5 independent slip systems. In a real crystal lattice, more or less slip systems may operate depending upon its atomic arrangement.

Finally, the incremental work done by the stress tensor must equal the work done in shearing the slip plane. To show this, eqs(8.5a) and (8.6a) express the specific work done by the applied stress and the incremental plastic strain tensors:

$$\sigma_{ij} \times \mathrm{d}\varepsilon_{ij}{}^P = \tau^s\,(\alpha_{ij}{}^s)^{-1} \times \alpha_{ij}{}^s\,\mathrm{d}\gamma^s = \tau^s \times \mathrm{d}\gamma^s \tag{8.8}$$

in which $m_i n_j = \frac{1}{2}(n_i m_j + m_i n_j) = \alpha_{ij}^s$. It is seen that the final term in eq (8.8) is the specific work done upon the slip plane, i.e. the product of its resolved shear stress and shear strain.

8.3 Lattice Slip Systems

Schmidt [3] showed experimentally that τ_{cr} is almost constant for a pure metal. This means that for a crystal of high purity, τ_{cr} is independent of the orientation of the slip plane. Equation (8.1) shows that it is possible to vary σ, φ and λ and so maintain τ_{cr} constant. The active slip plane will depend upon the atomic packing within a unit cell of a metallic lattice.

Three *Miller indices*, are used to identify a slip plane. The indices are derived from the intercepts between the plane and orthogonal cube axes *x*, *y* and *z*. Reciprocals of the intercepts are then taken and any resulting fractions cleared. Thus, for the shaded plane (110), shown in Fig. 8.4a, the intercepts are 1, 1 and ∞, for which their reciprocals become 1, 1 and 0. Brackets are placed around the Miller indices as follows: ( ) to denote all planes parallel to this and { } for all planes of a similar type.

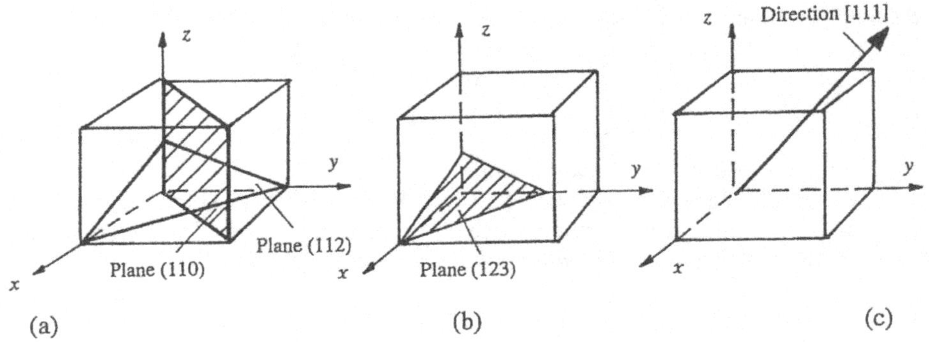

Figure 8.4 Slip planes and directions

Further examples of planes (112) and (123) are shown in Figs 8.4a,b. Three further Miller indices are used to denote a slip direction within a plane. Firstly, the direction vector must be translated to pass through the origin of *x*, *y* and *z* in the sense required. Direction indices become integer co-ordinates for the vector's tip. For example, Fig. 8.4c shows a [111] slip direction. Here, brackets [ ] denote all parallel directions and ⟨ ⟩ denote all directions of that type. Metals such as Fe, Mo and W, with body centred cubic (b.c.c.) cells, have no obvious close-packed planes. It is found that plastic flow may activate one or more of twelve slip planes, each with slightly different atomic packing density. Of these, three planes in particular, {110}, {112} and {123} (see Fig. 8.4a,b), activate slip most commonly in ⟨111⟩ close-packed directions. With four ⟨111⟩ slip directions and twelve slip planes per b.c.c. cell, there are a maximum of 48 possible slip systems. However, many of these systems are extremely sensitive to temperature and impurity and may not operate. Consequently, a b.c.c. structure can be ductile at high temperature and brittle at low temperature. It follows that the critical resolved shear stress τ_{cr}, required to initiate b.c.c. yielding, is extremely structure sensitive.

Slip in f.c.c.and h.c.p lattices is more distinct than in b.c.c. The (111) plane in Fig. 8.5a has three slip directions: $[10\bar{1}]$, $[0\bar{1}1]$ and $[\bar{1}10]$, where a bar denotes a negative intercept.

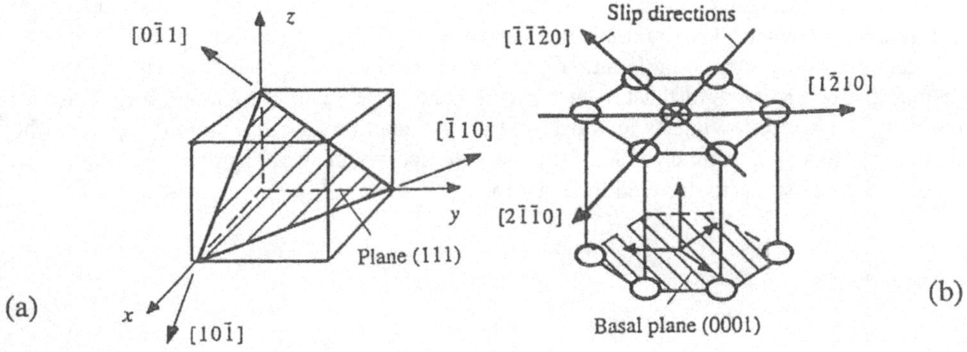

Figure 8.5 Slip planes for f.c.c.and h.c.p. cells

The square brackets shown embrace parallel directions to each of the three shown, for which one could contain them all as $\langle 10\bar{1}\rangle$, indicating directions of the $10\bar{1}$ type. In fact, Fig. 8.5a applies to a face-centred cubic cell (f.c.c.), typical of Al, Cu, Ag, Au and Pt. Their slip occurs on close-packed octahedral $\{111\}$ planes, in close-packed $\langle 10\bar{1}\rangle$ directions. Now, as there are four $\{111\}$ planes within a single f.c.c. cell and three slip directions, it follows that there are twelve slip systems by which an f.c.c. metal can deform. These many possibilities for slip to occur account for the ductile nature of a f.c.c. metal.

Planes and directions in a hexagonal close-packed (h.c.p.) structure are specified with four *Miller-Bravais indices*. A given plane is identified from the reciprocals of its intercepts with four axes originating from the basal plane, shown shaded in Fig. 8.5b. Three axes lie in this plane at 120° and a fourth axis lies perpendicular to it. Directions are again specified with their lowest integer co-ordinates. Slip within a h.c.p. cell can only occur on the close-packed basal plane (0001) in close-packed $\langle 2\bar{1}\bar{1}0\rangle$ directions. The latter includes the three directions shown: $[2\bar{1}\bar{1}0]$, $[\bar{1}210]$ and $[\bar{1}\bar{1}20]$. Since there is only one basal plane, containing three slip directions per cell (see Fig. 8.5b), the number of slip systems is restricted to three. As a consequence, h.c.p. metals (Be, Cd, Mg, Ti, Zn and Zr) are less ductile. However, h.c.p metals can also deform by twinning, in which a mirror of homogenous shear occurs across a band of atomic planes, as shown in Fig. 8.6.

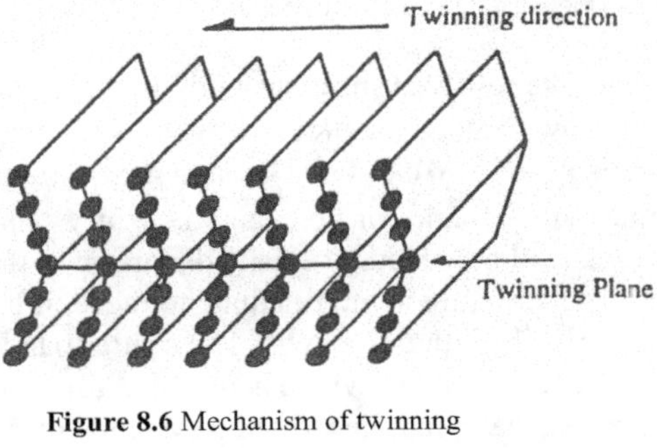

Figure 8.6 Mechanism of twinning

For h.c.p: $\{10\bar{1}2\}$ defines the twinning planes and $\langle 10\bar{1}\bar{1}\rangle$ defines the twinning directions. Deformation twinning is less common in a b.c.c. lattice, being restricted to respective planes and directions: $\{112\}$ and $\langle 111\rangle$. Twinning is rarely seen in a deforming f.c.c. lattice, though it has been observed for {111} planes and in ⟨112⟩ directions with low-temperature deformation of copper. The twinning mechanism requires that a critical shear stress displaces adjacent planes by an amount less than the atomic spacing. These sum to yield large shear displacements on layers furthest from the twin plane. The effect of twinning is to rotate the basal plane back following its forward rotation to accommodate slip upon this plane. The reversal produces a geometric softening of a lattice which would otherwise require an increased shear stress to produce further slip.

8.4 Hardening

8.4.1 Single Crystals

Figures 8.7a and b shows the extent to which the critical resolved shear stress τ_{cr} for various cells is dependent upon temperature and purity.

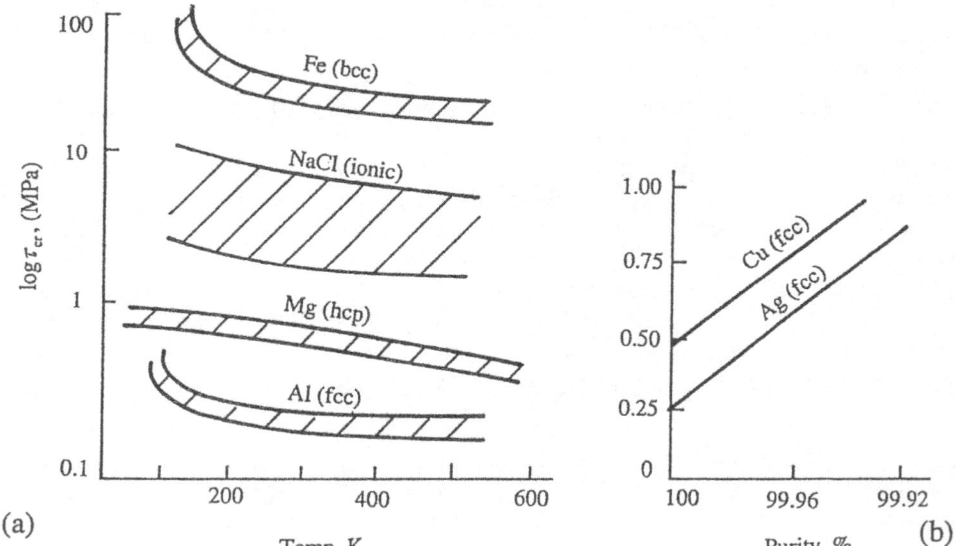

Figure 8.7 Dependence of τ_{cr} upon temperature and purity

Figure 8.7a shows that the resolved shear strength of a b.c.c. metal is far greater than both f.c.c. and h.c.p. metals over a wide temperature range. At room temperature, a typical value for a b.c.c. structure is $\tau_{cr} = 50$ MPa. For f.c.c and h.c.p. structures, $\tau_{cr} \simeq 1$ MPa. Despite the even lower values of τ_{cr}, found for f.c.c. and h.c.p. crystals of super purity (Fig. 8.7b), the initial yield point is still preceded by elastic distortion in these structures. The strength of an ionic NaCl crystal structure is intermediate to those of h.c.p and b.c.c. These results apply only to single crystals with these cells. The scatter band in Fig. 8.7a is due to small variations in purity, the sensitivity of which is revealed for two f.c.c. metals in Fig. 8.7b. It is important to emphasise that τ_{cr} applies to a slip system which is the first to activate under an applied tensile stress. Equation (8.1) identifies this with the slip system that is the first to reach the critical value of the Schmidt factor: $m = \tau_{cr}/\sigma = \cos\varphi \cos\lambda$. With the large number of slip systems available in b.c.c. and f.c.c cells, it is possible for two or more systems to operate together as τ_{cr} is attained simultaneously. This is apparent from measurements of the resolved shear stress required to maintain subsequent plastic deformation. For example, Fig. 8.8a shows that two types of behaviour, both A and B, are possible for a deforming f.c.c single crystal.

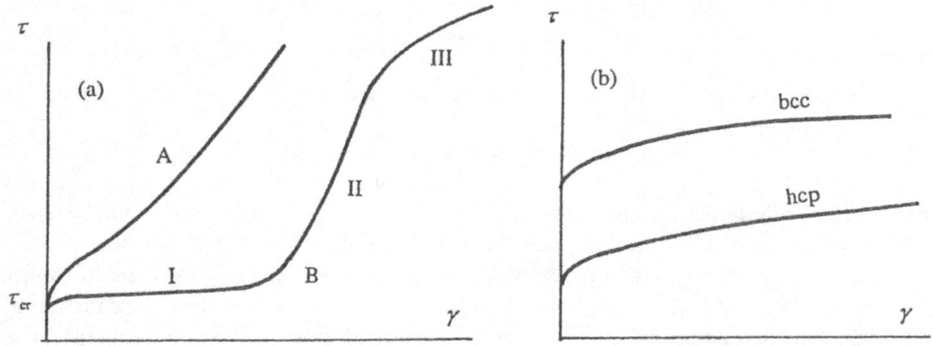

Figure 8.8 Hardening in a single cells: (a) f.c.c., (b) b.c.c and h.c p.

Plasticity appears within both curves at a similar τ_{cr} value, typically 1–10 MPa, depending upon metal purity. Subsequent flow behaviour is controlled by the number of active slip systems. For curve A in Fig. 8.8a, several slip systems operate simultaneously, leading to a high rate of work hardening $d\tau/d\gamma$. For curve B, a single slip system operates during stage I at a very low, constant rate of work hardening. That rate and duration of this linear stage depend strongly upon the Schmidt factor m. The occurrence of multiple slip on intersecting slip planes promotes stage II, where the rate of hardening has increased dramatically to attain a constant $d\tau/d\gamma \approx G/300$. This is followed by stage III in which $d\tau/d\gamma$ falls with increasing resolved shear strain γ. The resolved shear stress τ that initiates stage III depends upon the temperature, purity and the rate of straining, similar to the critical τ_{cr} to commence stage I. Stage III is also associated with thermally activated dynamic recovery. That recovery requires a sufficient density of dislocations to be developed within stage II before it can occur. Recovery processes, climb and cross slip, then become active, reducing the dislocation density and thereby softening the material.

A three-stage curve may not result from continued slip within other cells. For a h.c.p. cell, slip is confined to the basal plane (see Fig. 8.5b) and so it will remain in stage I, as shown in Fig. 8.8b. However, it is possible for dislocations to interact between the basal {0001} and other {1211} planes in certain h.c.p. metals. Dislocation interaction presents sessile barriers to slip and can result in a three-stage curve, similar to that for a f.c.c. cell. Moreover, with twinning present in a h.c.p. cell, there will result a discontinuous (serrated) stress–strain curve. Twinning can also arise in a b.c.c. cell but subsequent slip normally remains in stage I under a low rate of work hardening. The latter is maintained by each of the many slip systems for this cell operating sequentially as they are activated under τ_{cr} by an increasing applied stress. The b.c.c. stage I stress levels are higher than those for h.c.p. (see Fig. 8.8b) and may be high enough to promote a brittle fracture at low temperatures.

8.4.2 Polycrystals

Figure 8.9 shows that the slip directions within individual grains of a polycrystal are randomly oriented to the axis of tensile stress σ.

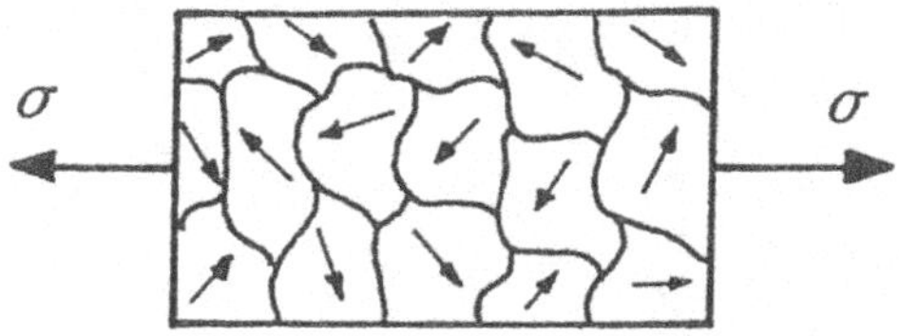

Figure 8.9 Grains in a polycrystalline material

Slip induced plasticity within a polycrystal is similar to that for a single crystal but, because the grain orientations differ, the Schmidt factor will be high for some grains and low for others. Therefore, before gross yielding by slip can occur, the applied stress must be raised to produce yield in grains with initially low Schmidt factors. In addition, there are constraints imposed by surrounding grains upon the slip within an individual grain. Consequently, the flow stresses in polycrystals are substantially higher than those for single crystals. For example, in annealed copper the respective yield stresses are 3.5 and 0.65 **MPa**. It is often assumed that polycrystalline flow behaviour is an average of f.c.c. stages I – III (Fig. 8.8a) when they co-exist simultaneously within many differently oriented grains. Increasing the stress beyond the initial yield value activates further slip planes because the increased slip produced in one grain must be relieved by slip in its neighbours. Without the relief necessary to maintain continuity of slip, cracking would occur. The grain boundary presents a barrier to slip which effectively raises both the initial and subsequent flow stresses. The initial yield stress Y increases with decreasing grain size d according to Petch's relation [4]

$$Y = A + B/\sqrt{d} \tag{8.9}$$

where A and B are constants. Also, the rate of macro-hardening $d\sigma/d\varepsilon^P$ is greater for a crystal with finer grains than one with coarse grains. A varying grain orientation prevents a homogeneity of slip throughout an aggregate. As the slip differential within neighbouring grains is transmitted through the grain boundary, the stress and strain within individual grains becomes uneven. The strain is greater at the grain boundary than at its centre but can become more uniform with increasing strain in smaller grains. The linear stage I hardening seen in a single crystal is never observed in a fine-grained f.c.c. polycrystal. Consequently, a parabolic flow curve follows from its initial yield point. Here this curve becomes an average of simultaneous stage II and stage III hardening as the applied stress can never distinguish between each stage. A mixed stage II and III hardening within individual grains and the barriers to slip presented by grain boundaries resist plastic flow. Stage I hardening in h.c.p and b.c.c. depend upon grain orientation relative to the axis of an applied stress. The result is that stage I is more advanced in aligned grains but an aggregate behaviour appears across all grains. These results indicate that when any polycrystalline material is loaded into the plastic region it becomes harder to deform. In overcoming these barriers the terms *work-hardening or strain-hardening* are used for the plasticity that occurs under a continuously increasing flow stress. In fact, the parabolic flow curve revealed by tensile testing shows that stress σ increases as the 'rate' of hardening $d\sigma/d\varepsilon$ diminishes with increasing strain ε. Allowing for a known hardening behaviour is key to plasticity theory (see chapters 9 and 10). This becomes important at the higher rates of hardening under low strain. The idealisations of elastic–perfectly plasticity and plastic rigidity, adopted in chapters 5 and 6, are simplifications which apply at best to large strains where rates of hardening are low.

8.4.3 Solution Hardening

Other types of hardening to be considered in plasticity theory will depend upon the material composition. The strength within alloys rests upon the manner in which the different atoms interact to resist plasticity. Solution hardening relies upon an elastic interaction between dislocation motion and impurity atoms. The latter fall into two categories, substitution and interstitial, which result in positive and negative volumetric distortions depending upon the relative sizes of the impurity and host atoms. The elastic strains created around impurity atoms and around dislocations interact within their hydrostatic and shear components. Substitution atoms are attracted to a hydrostatic field and interstitials atoms to a shear strain field. Impurity atom distributions can prevent the motion of dislocation directly by pining them and also by raising the lattice friction to inhibit slip on close packed planes. Such 'grown-in' dislocations are re-activated under the yield stress along with an increase in dislocation density, typically by two orders, at the highest rate of work hardening. The Peierls stress, the consistent measure of lattice friction upon a slip plane, is raised intermittently by the presence of solute atoms, which control the flow stress.

8.4.4 Precipitation Hardening

Alloying atoms distribute themselves within the hot lattice as a solid solution or by the formation of an intermediate phase. The respective equilibrium diagrams for two element alloys show completely soluble atoms in one and insoluble atoms in the other. Precipitation hardening refers to partial solubility with a fall in solid solubility with equilibrium cooling. The loss in solubility results in the distribution of a second phase as the solute atoms precipitate through the lattice. If cooling is too rapid the precipitation is inhibited and the resulting *supersaturated solid solution* requires gradual re-heating. Such controlled precipitation from this *age hardening treatment* is adopted for wrought aluminium–1.6 % copper, to which many plasticity studies apply. The two-stage heat treatment of this alloy involves a quench followed by an anneal. Solution treatment at 600°C dissolves inter-metallic θ-compounds ($CuAl_2$) resulting in a single phase α–solid solution. With an equilibrium cooling to room temperature, precipitation of solute atoms occurs, distributing θ coarsely throughout the lattice. A finer, more coherent precipitation in a distribution with Cu/Al atom matching is found from annealing following a rapid quench to room temperature. The quench suppresses precipitation, giving a super saturated solid solution. By re-heating that solution in a temperature range 100°C–300°C, precipitation is controlled. The precipitates then form as a finer dispersal of discs of pure copper with plates and rods of alloy compounds into *Guinier-Preston zones*. Lattice strain, arising from the 12 % difference between the size of atoms within this dispersal, interacts with dislocation to enhance hardening in the manner described above in § 8.4.3. The increase in yield stress depends upon the volume fraction of the precipitate and the interaction energy. With these being mutually exclusive, a dislocation shearing and looping of precipitates accounts for the greater degree of hardening observed. The optimum condition depends upon the size and spacing of precipitates, these depending upon the particular combination of annealing time and temperature, in which a raised temperature reduces the time required. With an over-aged condition, arising from increased time and temperature, the precipitates attain an equilibrium condition. The latter appear as non-coherent clusters, embedded almost independently of the surrounding solid solution. The non-optimum micro-structure is less effective in resisting dislocation motion and therefore in raising strength.

8.4.5 Recovery and Recrystallisation

Plasticity of metals raises the dislocation density which in turn increases its hardness. In the present context the metal behaviour is more often referred to as work or strain hardening. This appears, macroscopically, as an increase in flow stress required to increase plastic strain by slip on close packed planes. Associated with the dense, tangled dislocation arrangement that presents a barrier to slip is a sufficient amount of stored energy to initiate its release by recovery and recrystallisation processes.

(a) Recovery
Stage III recovery of an f.c.c. single crystal in Fig. 8.8a refers to cross slip of dislocations by thermal activation under the application of shear stress. In the absence of stress recovery, processes that occur during annealing involve a non conservative dislocation motion known as climb. The higher the temperature, the higher the rate of recovery where climb of dislocations occurs: (i) with opposite sign to annihilate one another or (ii) with similar sign to polygonise within the formation of stable walls. Both climb-driven events draw upon the stored energy of the pre-deformed structure so reducing that store and making the metal softer. As an example of climb mechanism (i), the leading edge dislocations on parallel slip plane climb towards each other and annihilate mutually. Thereby, the dislocation density is reduced within pile ups that had contributed previously to strain hardening. Strain is preserved within mechanism (ii) but again the polygonised rearrangements soften by absorbing the stored energy necessary for their formation. The walls of edge dislocations in a polygonised structure produce low angle tilt boundaries between parallel slip planes.

(b) Recrystallisation
Metals annealed at near melt temperature grow new strain-free grains to replace those heavily deformed by pre-deformation. The energy within the latter contributes to the driving force necessary for nucleation. Therein, the energy is supplied to a small volume and activated by the annealing temperature. Thus nucleation is made easier with higher temperatures. Many dislocations disappear at grain boundaries and voids or escape at the surface. The temperature also assists the climb of partial dislocations when these are widely spaced, re-arranging them with lower energy. Although these mechanisms are similar to those of recovery, it is unlikely that the two softening processes work together. It is the degree of cold work and the annealing temperature which control the activation of one process or another. Generally, recrystallisation of grains favours a high purity metal. Recovery at lower temperatures, which does not alter the micro-structure in this way, is more likely to occur with a high dislocation density.

8.5 Steels

The structure and properties of plain carbon steels is relevant to many plasticity studies with this alloy of small amounts of carbon in iron. There are many conditions of steel available in service: annealed, normalised, hardened and tempered. Carbon contents up to 1.5 % include the low, medium and high carbon steel compositions. These lie within a wide spread of strength and ductility associated with different phases that form with an equilibrium cooling from above 700°C in the austenite region, i.e. a solid solution in which all carbon is absorbed within f.c.c. iron. The slow cooling results in the allotropic change in iron from f.c.c. to b.c.c. For particular C-compositions up to 0.8 %, two phases appear, as in Fig. 8.10.

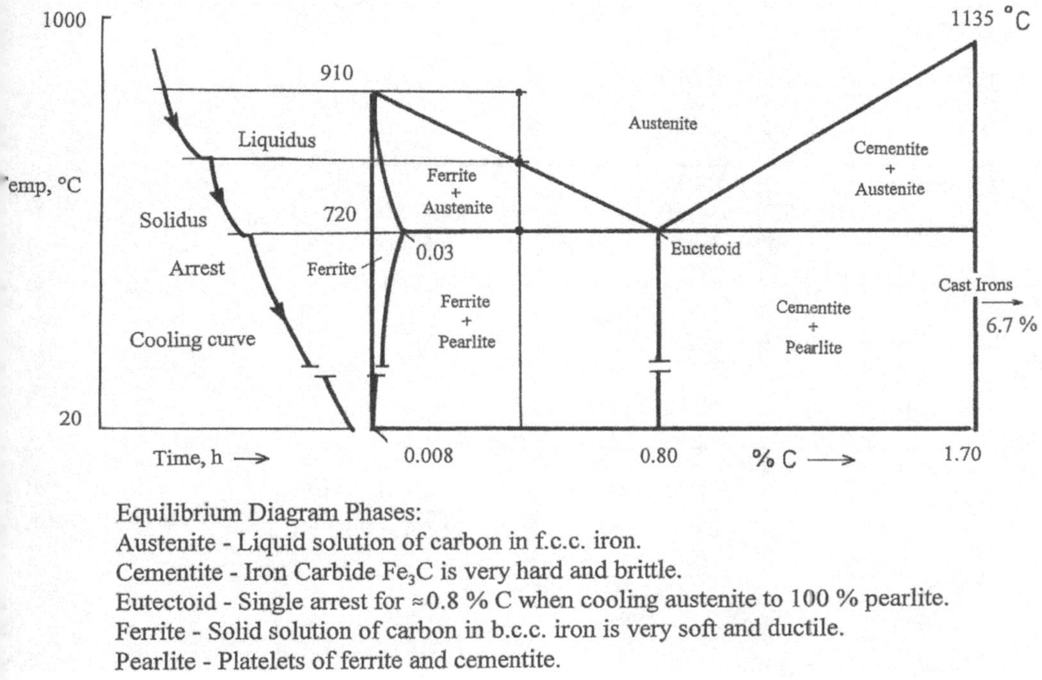

Figure 8.10 Equilibrium cooling of steels

8.5.1 Equilibrium Phases

Ferrite is a ductile solid solution of carbon in body centred iron with maximum solubility of 0.03 % at a lower critical temperature of 723°C. *Pearlite* is a eutectoid decomposition of f.c.c austenite into alternate plates of ductile ferrite and brittle Fe_3C carbide (*cementite*) with a thickness ratio of 7/1. The reduced ductility of these mixed phases is compensated by an increase in strength. When cooling the eutectoid composition (0.8 % carbon) and allowing time at 723°C, all the austenite transforms into pearlite, giving maximum strength with minimum ductility. In the slow cooling of carbon compositions beyond 0.8 %, a carbide precipitation occurs at austenite grain boundaries before the eutectoid transformation of remaining austenite to pearlite occurs. This retains the boundary network for pearlite where the cementite phase remains to embrittle the steel with some loss in strength. As a measure of the scale of the three phases across the carbon range in Fig. 8.10, a 1.2 % hypereutectoid composition shows 1/7 the amount of cementite at boundaries as there is ferrite at boundaries for a 0.4 % hypoeutectoid steel. The following example shows how to calculate the relative amounts of each phase for both steels and their individual C-compositions.

Example 8.1 Use the 'lever rule' to calculate the relative amounts of ferrite, pearlite and cementite phases for 0.4 % and 1.2 % C-steels and the C-compositions within each phase.

The said rule works along the horizontal eutectoid line at 723°C and the cooling lines that separate the decomposition of austenite into the two phases at each composition. Refer symbols α and γ to solid solutions of ferrite and austenite respectively and Fe_3C to the compound that is cementite. *Note*: From Fig. 8.10, the intercepts upon the eutectoid line occur for C-compositions 0.03, 0.8 and 6.7 %. Then, with the two C-compositions:

For a 0.4% C-steel:

(a) Ferrite/Pearlite 52/48 = 1.083

Phase (≈730°C):	α	γ
% C-content	0.03	≈0.8
Lever Rule:	$\frac{(0.8-0.4)100}{0.8-0.03} = 52\ \%$	$\frac{(0.4-0.03)100}{0.8-0.03} = 48\ \%$
Decomp^{n} (<723°C):	52 % Ferrite	48 % Pearlite

(b) Ferrite/Cementite: 94.5/5.5 = 17.2

Phase (<723°C)	α	Fe_3C
% C-content	0.03	6.7
Lever Rule:	$\frac{(6.7-0.4)100}{6.7-0.03} = 94.5\ \%$	$\frac{(0.4-0.03)100}{6.7-0.03} = 5.5\ \%$

For a 1.2% C-steel:

(a) Pearlite/Cementite 93/7 = 13.3

Phase (≈730°C):	γ	Fe_3C
% C-content	0.8	6.7
Lever Rule:	$\frac{(6.7-1.2)100}{6.7-0.8} = 93\ \%$	$\frac{(1.2-0.8)100}{6.7-0.8} = 7\ \%$
Decomp^{n} (<723°C):	93 % Pearlite	7 % Cementite

(b) Ferrite/Cementite 82.5/17.5 = 4.7

Phase (<723°C)	α	Fe_3C
% C-content	0.03	6.7
Lever Rule:	$\frac{(6.7-1.2)100}{6.7-0.03} = 82.5\ \%$	$\frac{(1.2-0.03)100}{6.7-0.03} = 17.5\ \%$

8.5.2 Annealing and Normalising

These two treatments refer to an equilibrium cooling at which two phases, as calculated in example 8.1 above, will form from decomposing austenite. Both treatments may be applied to fully formed components to relieve residual stress, providing a more uniform micro-structure with improved properties. Annealing requires furnace cooling from just above the temperature 723°C in the case of hyper-eutectoid steels to avoid carbide embrittlement at grain boundaries. Normalising employs faster air cooling to refine pearlite and increase its strength over that of annealing with some loss of ductility.

8.5.3 Hardening of C-Steel

Very low carbon sheet steel may be precipitation hardened across the solvus line that bounds ferrite. With maximum solubility of 0.03% carbon at 723°C, the aging process following solution treatment has deleterious effects with the formation of unsightly Luder's lines if time is allowed for natural ageing. Therefore, sheets are formed in the as solution treated condition to avoid Luder's lines and the diffusion of impurity atoms to dislocation sites that would result otherwise in strain ageing. The latter refers to a raising of the upper yield point in mild steel intermittently when ageing periods are allowed between plastic straining.

More important to steel production generally across its carbon range is the eutectoid decomposition associated with hardening and tempering. Interstitial carbon within f.c.c. iron, identified as the *austenite* phase of an equilibrium cooling diagram, exists only above 723°C. The carbon distribution is homogenous and therefore the transformation of austenite to pearlite involves diffusion. The faster cooling rate followed in the heat treatment of steels depresses the transformation temperature to well below 723°C at its completion (≈700°C). This transformation is faster than when more time is allowed for it and consequently a finer pearlite appears with reduced spacing between many more carbide plates. An even finer pearlite resulting from fast cooling was re-named *troostite*. A decreasing rate of carbon diffusion applies to cooling from 600°C, where other phases form in place of pearlite. In the lower temperature range 200–400°C, *bainite* forms as a fine dispersion of carbide flakes in ferrite, this being harder than pearlite. With very rapid quenching from above 723°C, neither pearlite nor bainite can form. Instead a *martensite* phase appears, which is an unstable supersaturated solid solution of carbon in b.c.c. iron. Austenite converts to this form instantly at a single temperature without involving diffusion. That temperature, which lies below 300°C, will depend upon the carbon content (and other alloying elements). The transformation from a f.c.c. austenite to its metastable b.c.c. form must accommodate the five extra carbon atoms. Consequently, the b.c.c. cell is distorted to a body centred tetragonal cell. The degree of shear distortion depends upon C-content, but a ≈4 % increase in volume occurs with extensions of one atomic radius between neighbouring atoms for this cell. The shear strain field within an aggregate of tetragonal cells interacts with and locks dislocations to create solid solution hardening. Martensite is associated with reduced grain size and an increase in dislocation density which can also raise the flow stress marginally. In itself the martensite metastable phase is strong but, being extremely hard, it is very brittle and therefore of little practical use aside from glass cutting. Hence, a further heat treatment is necessary following quenching. Re-heating martensite shows a loss of hardness and improved ductility, providing the best engineering properties of all steel micro-structures. This tempering process releases carbon from the supersaturated solution to precipitate as iron carbide compounds. Typically, on holding a temper at 150°C for 1 hour, a very fine precipitate first appears as coherent particles. These accompany a loss of the b.c.c. cell's tetragonal distortion and its extreme hardness. Such loss is compensated by particle hardening but improvements in strength and ductility are lost progressively if fine particles are allowed to coarsen at extended time for a 150 °C temper. A rise in temperature between 200 and 400°C coarsens particles more rapidly and introduces a blue brittleness where carbide forms at grain boundaries by the diffusion of carbon. In general, tempers are always below 723°C to allow the formation of stable ferrite and cementite that would appear with an equilibrium cooling from the austenite phase. Hypereutectoid steels can retain austenite when quenched from an austenitizing temperature much in excess of 723°C. Hardness is impaired by the retention of austenite and here, beneficially, a higher temper recovers hardness with a conversion of remaining austenite to the bainite phase. A linear expansion of ≈0.025 mm/mm occurs for each percentage of retained austenite that transforms. The correct pre-quench temperature for high carbon steels should lie just above the lower critical range where undissolved cementite accompanies austenite at the eutectoid composition. Then the quench produces thin martensite plates in twelve orientations of the 'habit' plane within the austenite grain from which it forms. A higher temper of 300°C for medium to high C-steels ensures a useful combination of strength and ductility where toughness is required,

8.5.4 Alloy Steels

The higher strengths of alloy steels are due to the manner in which specific elements affect constitution, the properties of ferrite, the decomposition of austenite and the temper reaction. Most alloy steels contain less than 5 % alloying elements. Low 0.15 % carbon ferritic steels may contain 0.75 % Mn, 0.5 % Cu, 0.6 % Ni, 0.25 % Mo and 0.1 % P. These delay the ferrite and pearlite phase transformations to a lower temperature, giving an increased strength of ≈50 % from finer phases formed with a stronger ferrite solid solution. Copper improves corrosion resistance, manganese lowers the ductile-brittle transition temperature and Ni and Mo improve hardenability, i.e. they ensure more uniformity in hardness from a rapid quench. Consequently, low carbon martensite steels temper to a strength of 700 MPa, three times the strength of hot rolled steel. In *maraged* steel a large nickel addition 18–25 % to a very low carbon steel (< 0.1 %), in addition to Al, Co, Mo and Ti, lowers the temperature at which austenite becomes unstable. The martensite is formed by natural cooling and then subsequently aged by precipitation of the alloy compounds Ni_3Ti and Ni_3Mo. Left in a soft condition following cooling, the alloy is worked easily to the shape required before being strengthened further by ageing.

Medium carbon alloy steels in the tempered condition have similar properties across the alloying range. That range is chosen primarily to improve hardenability by lowering the cooling rate and retarding phase formation. For similar reasons, *martempering* employs a hot oil quench to reduce temperature gradients and the risk of stress gradient cracking. Compared to plane carbon steel, both strength and ductility are also raised by Ni, Cr and Mo, typically to 1550 MPa and 12 %. Toughness is raised by alloying 0.3–0.45 % C-steels. Furthermore, the alloying delays softening when tempering, which allow higher tempers to relieve residual stress and avoid the temper brittle range. Further additions of Ni (2.5 %), Cr (1 %), and Mo (0.5 %) can raise tempered strength to very high levels (2000 MPa) but with the risk of failure from having a low fracture toughness, i.e. they are susceptible to crack propagation from inclusions, poor surface finish and stress concentrations. These steels permit *austforming*, a hot forming and rolling when partially quenched to 500°C. A further quench produces martensite from worked austenite with high dislocation density. A final 'return temper' of 500°C produces a very high strength 2750 MPa alloy steel.

High carbon (hypereutectoid) steels involve pearlite in combination with the cementite phase in a spheroidal form from an equilibrium cooling. These two phases are suppressed with rapid cooling from the austenite-cementite region of the iron-carbon equilibrium diagram in Fig. 8.10 with C-content from 0.8–1 %. With the austenite close to the 0.8 % C eutectoid composition, cooling from just above the lower critical temperature 723°C results in fine grained martensite with no retained austenite. A low temper raises toughness and impairs harness only slightly, properties that ensure respective impact and wear resistance in tooling applications where ductility is less important. There are very many types of tool steels listed in specialist texts [5, 6].

8.6 Yield Surface

8.6.1 Isotropic Aggregate

It has been seen that twelve slip systems are possible in a face-centred cubic lattice. When slip is the only source of plastic strain, each slip system identifies a pair of complementary yield planes $f(\sigma_{ij}) = 0$ that bound an elastic region (Fig. 8.11a). A rule of flow shows that $d\varepsilon_{ij}^{P}$ lies normal to each yield plane for a slip system in which the resolved shear stress is critical. When a number of slip systems operate together in cubic lattices, a yield polyhedron will apply to each grain within a randomly oriented aggregate of grains. Lin [7] and Kocks [8] derived plane sections of this surface for cubic metals. An example of a f.c.c. polyhedral yield surface is given in Fig. 8.11b, in which the stress axes are σ_{11}, σ_{12} and σ_{13}.

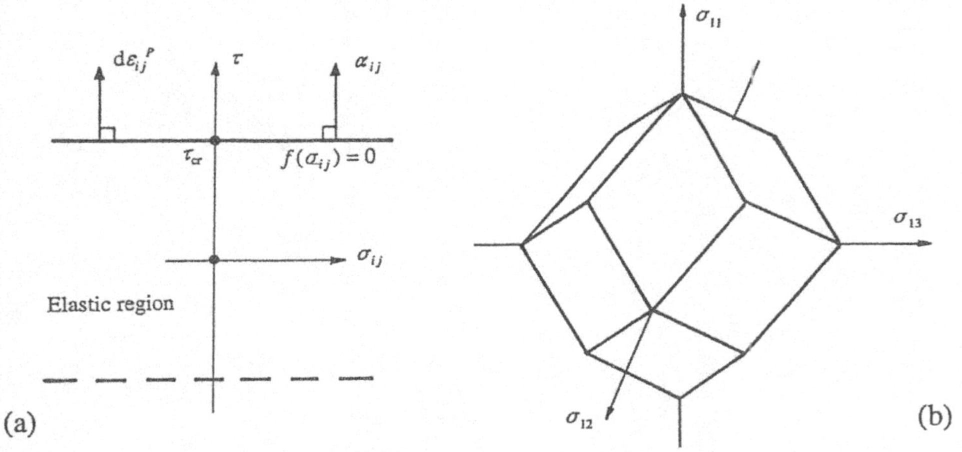

Figure 8.11 Yield planes for slip in an f.c.c. grain

When the stress axes coincide with the f.c.c. cell axes, the direction cosines for the vectors **n** and **m** in eq(8.5a) equate directly to the Miller indices for the slip plane and slip direction for each of 12 slip systems. The yield planes meet to form edges and corners of the yield polyhedron. Two slip systems operate together along the edges of the yield surface and four or more systems operate simultaneously at its corners. In an ideal polycrystal, Tresca's maximum shear stress criterion of yielding applies as the critical shear stress is attained on the 24 planes in each grain simultaneously. Thus, Tresca's yield criterion applies to an ideal aggregate of crystals, each with similar elastic moduli and critical resolved shear stresses for the active slip planes. This gives a *lower bound* solution. In a real f.c.c. crystal, where these properties may vary between grains, a homogenous yielding does not occur. It has been seen that the constraint exerted upon a grain by its surrounding matrix raises the critical resolved stress. Raising the applied stress enables slip along the most favourably oriented planes to spread into neighbouring grains. The yield stress associated with the initial spread of plasticity will lie on a three-dimensional yield surface framed in the applied stress axes. This smooth convex surface is bounded by the yield planes for slip within all individual grains of an aggregate [9] to give an *upper bound* solution.

Yield loci under plane stress states are constructed from a generalised Schmidt's law, eq(8.5b). A lower-bound yield locus is found when this law is applied to a minimum number of active slip systems within a single grain of a polycrystal. The grain must be taken in isolation under its biaxial stress state. Correspondingly, the sum of the separate shear strains attains a least value [10]. More realistically, the micro-mechanics of yielding in cubic metals should involve a restricted crystallographic slip within an aggregate of grains. The constraints presented to slip in a single grain, from its boundary with differently oriented neighbouring grains, will raise the yield stress. In their account of restricted slip, Bishop and Hill [11] showed that an isotropic, upper-bound yield locus may be established from the *maximum work principal*. This assumes that inhomogenous, simultaneous slip is the only mechanism of spreading plasticity in a polycrystalline solid. To examine the work done in deforming a unit volume, let $(\sigma_{ij})_A$ be an initially elastic, *macroscopic* stress tensor. This stress corresponds to point A lying within the yield surface *f* in Fig. 8.12a. The elastic energy stored $W_A^{\,e}$ is identified with an enclosed triangular area beneath component stress–strain plots, typified by Fig. 8.12b.

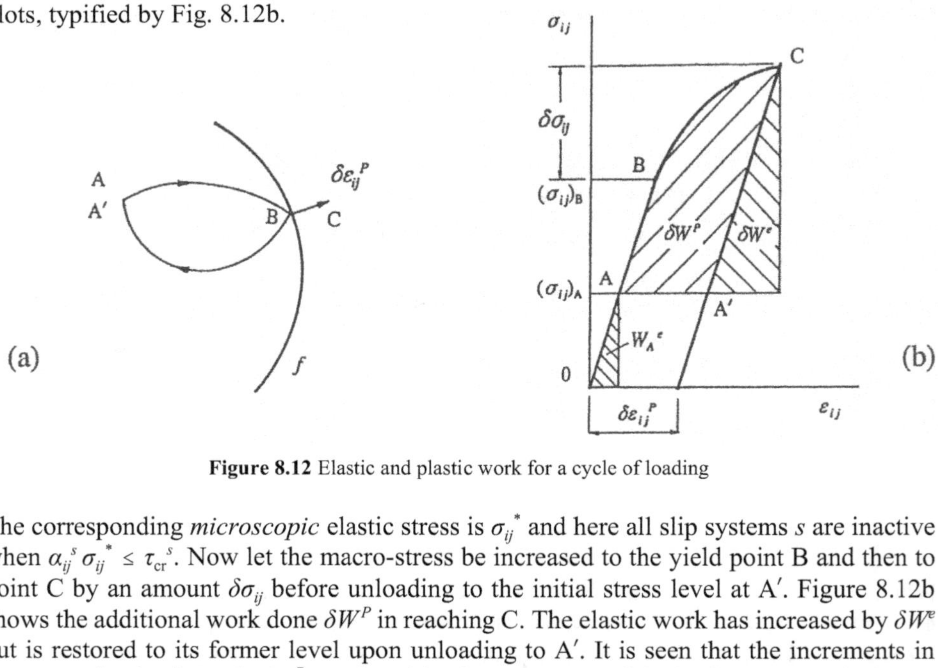

Figure 8.12 Elastic and plastic work for a cycle of loading

The corresponding *microscopic* elastic stress is $\sigma_{ij}^{\,*}$ and here all slip systems *s* are inactive when $\alpha_{ij}^{\,s}\,\sigma_{ij}^{\,*} \le \tau_{cr}^{\,s}$. Now let the macro-stress be increased to the yield point B and then to point C by an amount $\delta\sigma_{ij}$ before unloading to the initial stress level at A′. Figure 8.12b shows the additional work done δW^P in reaching C. The elastic work has increased by δW^e but is restored to its former level upon unloading to A′. It is seen that the increments in macroscopic plastic strain $\delta\varepsilon_{ij}^{\,P}$, produced by $\delta\sigma_{ij}$, are responsible for the permanent change to the stored energy. The incremental plastic work δW^P for loading cycle ABCA′ may be written as

$$[(\sigma_{ij})_B - (\sigma_{ij})_A]\delta\varepsilon_{ij}^{\,P} + \delta\sigma_{ij}\,\delta\varepsilon_{ij}^{\,P} > 0 \tag{8.10a}$$

When point A lies on the yield surface, then $(\sigma_{ij})_B = (\sigma_{ij})_A$ and eq(8.10a) simplifies to

$$\delta\sigma_{ij}\,\delta\varepsilon_{ij}^{\,P} > 0 \tag{8.10b}$$

Correspondingly, the microscopic stress distribution is altered by $\delta\sigma_{ij}^{\,*}$ and an incremental, plastic micro-strain $\delta\varepsilon_{ij}^{\,P\,*}$ arises from the slip $\delta\gamma^s$ in systems where $\alpha_{ij}^{\,s}\,\sigma_{ij}^{\,*} > \tau_{cr}^{\,s}$. Here:

$$\delta\sigma_{ij}^{\,*}\,\delta\varepsilon_{ij}^{\,P\,*} > 0 \tag{8.10c}$$

The *maximum external work principal* combines eqs(8.10b,c) within volume *V*, as

$$\delta\sigma_{ij}\,\delta\varepsilon_{ij}^{P} = \int_V \mathrm{d}\sigma_{ij}^{*}\,\mathrm{d}\varepsilon_{ij}^{P\,*}\,\mathrm{d}V \ge 0 \tag{8.11a}$$

where the equality means that no work is done when $\delta\sigma_{ij}$ refers to an unloading or a neutral loading around the yield surface. Within eqs(8.10a,b), the macro-plastic strain increment tensor becomes a volume average of the micro-plastic strain increment tensor:

$$\delta\varepsilon_{ij}^{P} = \int_V \mathrm{d}\varepsilon_{ij}^{P*}\,\mathrm{d}V \tag{8.11b}$$

Taylor [10] employed eq(8.8) within a *minimum shear principle*:

$$\sigma_{ij}\,\mathrm{d}\varepsilon_{ij}^{P} \le \tau^{s}\,\mathrm{d}\gamma^{s} \tag{8.12a}$$

Equation (8.12a) is integrated for a plastic aggregate of volume V

$$\int_V \sigma_{ij}\,\mathrm{d}\varepsilon_{ij}^{P}\,\mathrm{d}V \le \int_V \left(\sum_{s=1}^{N} \tau^{s}\,|\,\mathrm{d}\gamma^{s}|\right) \mathrm{d}V \tag{8.12b}$$

where stress and strain distributions may vary continuously whilst satisfying both equilibrium and compatibility. The exact calculation of the yield surface, based upon eq(8.12b), is complex. Certain simplifications [12] will produce upper and lower-bound yield surfaces to satisfy either equilibrium or compatibility, but not both. For example, compatibility is satisfied in Taylor's lower bound with the assumption of uniform strain within the grains equal to the macroscopic strain (elastic strain is neglected). Employing eq(8.7) for N slip systems gives the uniform strain increments as

$$\mathrm{d}\varepsilon_{ij}^{P*} = \mathrm{d}\varepsilon_{ij}^{P} = \sum_{s=1}^{N} \alpha_{ij}^{s}\,\mathrm{d}\gamma^{s} \tag{8.13a}$$

but the condition of equilibrium is violated at grain boundaries. Conversely, when $\sigma_{ij} = \sigma_{ij}^{*}$ is uniform in eq(8.12b), one may write from eq(8.5b):

$$\sigma_{ij}^{*} = \sigma_{ij} = (\alpha_{ij}^{s})^{-1}\,\tau^{s} \tag{8.13b}$$

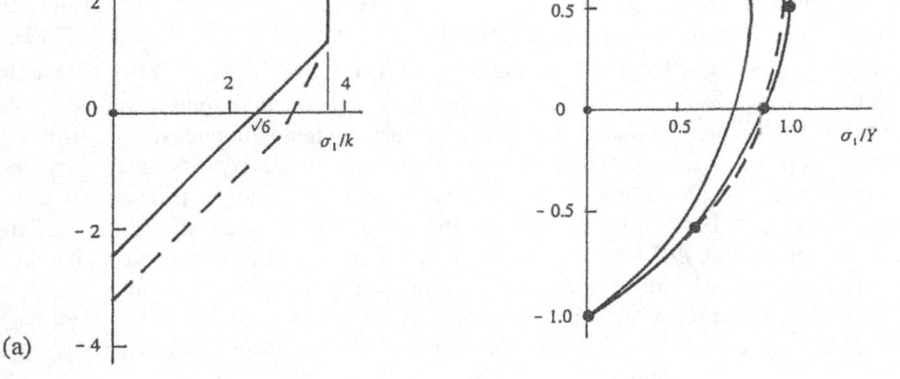

Figure 8.13 Bounding loci for an (a) isotropic (b) anisotropic texture

Here, as an upper bound, equilibrium is achieved for a uniform distribution of microscopic stress throughout the grains, but compatibility is violated. Equations (8.13a,b) apply to an isotropic aggregate which hardens uniformly, i.e. it shows no Bauschinger effect. It may be deduced from this that the yield surface lies between the upper and lower bounds. For example, Fig. 8.13a shows these two bounds for an ideal cubic lattice [13]. This restricted the slip under a principal, biaxial stress state to the (111) plane in a $[11\bar{2}]$ direction.

Further bounding yield loci may be derived for different operating slip systems. When extended from an f.c.c. grain to a polycrystal, the bounding is equivalent to taking its plane principal yield locus to lie between the von Mises and Tresca criteria. This was confirmed experimentally (see § 3.3.2) in the continuum theory of yielding for a polycrystalline aggregate, where both criteria identify with functions of the stress deviator invariants. Again, it is implied here that an aggregate of small, randomly orientated grains is isotropic macroscopically, i.e. where the flow stress is independent of the material direction.

8.6.2 Texture

The plastic flow behaviour of single crystals is anisotropic in that it depends upon the relative orientations of the slip plane and the stress axis. While the yield stress of an aggregate of equi-axed grains is usually direction independent, that for non equi-axed grains is not. For example, anisotropic yielding is found where the grains of a polycrystal exhibit a preferred orientation, as in a heavily rolled sheet metal. The degree of anisotropy may approach that for a single crystal when the directionality is severe. Anisotropy arising from cold rolling non-ferrous sheets cannot be entirely removed by a recrystallisation anneal. Indeed, the slip and any twinning sites from prior-working become thermally activated at the annealing temperature. The likely influences of crystallographic anisotropy is to distort, translate and rotate the yield surface. Such features are associated with the anisotropic hardening that is known to arise from plastic pre-straining (see § 8.9). A wrought, fibrous structure is associated with an alignment of impurities and voids along the forging direction. A hot forged structure is less likely to have been hardened anisotropically than a cold-rolled structure but the former may display anisotropy in stress and strain at fracture.

Upper and lower-bound estimates of yielding in textured f.c.c aggregates of copper and aluminium were made by Althoff and Wincierz [14]. In their analysis of textures, a bound was taken to be the average of a number of yield loci. Each locus was found from an individual texture component. Their method is similar to placing bounds upon yielding within individual grains of an isotropic solid. Figure 8.13b illustrates the result of compounding the residual textures arising from the recrystallisation of a dominant (112), $[11\bar{1}]$ texture in an extruded aluminium tube. The experimental data suggests that the yield locus, defined by 0.1 % offset plastic strain, lies between the upper and lower-bounds. Figures 8.13a,b show that the bounds for equi-axed and textured structures do not display: (i) the smooth curved boundary of the measured yield loci and (ii) any difference between yield stresses in tension and compression. Observations (i) and (ii) imply that strain within a grain is not distributed uniformly. Strain uniformity is assumed when maximising external work for an upper bound. Piehler and Backofen [13, 15] employed an alternative technique for bounding yield loci in rolled sheets with cubic lattices. Here a deformation texture, representative of planar isotropy, was established from rotating an ideal texture about the sheet normal. Figures 8.14a–e illustrate the construction of lower-bound yield loci in normalised stress axes by applying Schmidt slip to each stage of this rotation. Lower bounds with corners again result from their technique. In Fig. 8.14e, the average lower-bound locus for the full rotation is compared with an upper bound from applying the maximum work principle [15]. It is seen that the region between the bounds, within which the true locus is expected to lie, can accommodate a limited amount of unsymmetrical distortion and a difference between initial yield stresses. Kocks [9] gave computer simulations of rotating textures providing upper-bound yield loci for rolled and pre-strained polycrystals. These loci were found to be irregularly shaped with flattened sides and pointed vertices.

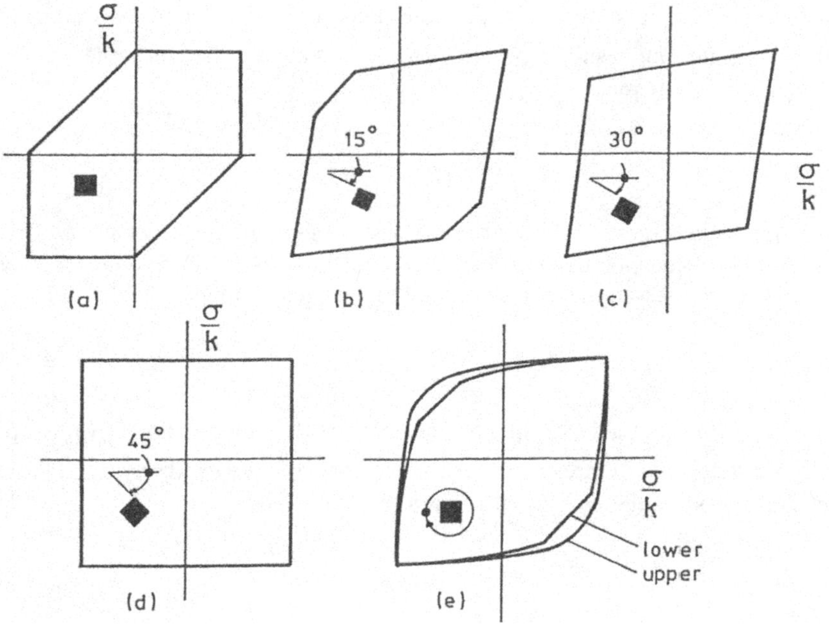

Figure 8.14 Lower bound yield loci for a cubic texture

Good agreement has been found between theoretical and experimental yield loci for textures under biaxial stress states [16–19]. Some typical observations on the initial yield behaviour of rolled sheet metal can be seen in chapter 12 (see § 12.8). In the continuum plasticity theory, initial anisotropy appears within a suitable yield function containing the macroscopic stress and plastic strain history. Many yield functions have been employed (see a–h in table 3.3) to represent irregular shaped yield surfaces in polycrystalline aggregates. Here, tensor formulations admit all macro-stress and plastic strain states. An *associated flow rule* identifies each yield function with a plastic potential for which stress derivatives equate to an aggregated stress–strain (constitutive) relation. This classical macroscopic approach to yielding is clearly more convenient than bounding with Schmidt's active micro-slip systems. Realistically, though, that the macro derives from the micro is unquestionable.

8.7 The Flow Rule

The decomposition of the total strain into elastic and inelastic components is a common feature in the classical theory of plasticity. The theory assumes that elasticity is due to lattice distortion and that slip on crystallographic planes accounts for all inelastic deformation. The latter is taken to be quasi-static for low rates of straining, so ignoring the effect of a discrete dislocation substructure. Strictly, the dynamics of dislocation motion renders the inelastic strain component as inherently time-dependent or visco-plastic [20]. To admit that strain, it becomes necessary to consider a similar decomposition in the total rate of deformation. Verification of the plastic potential, employed in classical theory, follows from a consideration of crystallographic slip and the motion of discrete dislocation lines [21]. When the strains are small, grain rotation can be ignored so that slip alone is responsible for the macroscopic plastic strain ε_{ij}^{P}. The latter is taken as the surplus $\varepsilon_{ij} - \varepsilon_{ij}^{e}$, where ε_{ij} defines a homogenous strain field and ε_{ij}^{e} is the contribution to elastic lattice distortion. When inelastic strain rates $\dot{\varepsilon}_{ij}^{P}$ result from a general stress state σ_{ij}, the work rate from V_s, slip systems within volume V, is expressed from re-writing eq(8.12a) as

$$\sigma_{ij}\,\dot{\varepsilon}_{ij}^{P} = \frac{1}{V}\int_{V_s} \tau^s \dot{\gamma}^s \,\mathrm{d}V \tag{8.14}$$

In eq(8.14), τ^s is the resolved shear stress when each slip system shears at a rate $\dot{\gamma}^s$. Let f be the force on a dislocation line L and L_d refer to all such lines within volume V. The work rate may also be expressed in the positive rate of change of slipped area:

$$\sigma_{ij}\,\dot{\varepsilon}_{ij}^{P} = \frac{1}{V}\int_{L} f\, v \,\mathrm{d}L \tag{8.15}$$

where v is the velocity of a dislocation line segment dL. Let the function $\dot{\varepsilon}_{ij}^{P}(\sigma_{ij})$ apply to a given slipped state where, correspondingly, shear strain $\gamma = \dot{\gamma}(\tau)$ and $v = v(f)$. Following an infinitesimal stress change $\mathrm{d}\sigma_{ij}$, eqs(8.14) and (8.15) become

$$\mathrm{d}\sigma_{ij}\,\dot{\varepsilon}_{ij}^{P} = \frac{1}{V}\int_{V_s} (\dot{\gamma}^s \,\mathrm{d}\tau^s)\,\mathrm{d}V \tag{8.16a}$$

$$\mathrm{d}\sigma_{ij}\,\dot{\varepsilon}_{ij}^{P} = \frac{1}{V}\int_{L_d} (v\,\mathrm{d}f)\,\mathrm{d}L \tag{8.16b}$$

As both integrals, eqs(8.16a,b), are identical, they may be replaced with a potential function $f(\sigma_{ij})$, whose partial derivatives in σ_{ij} supply strain rate components, $\dot{\varepsilon}_{ij}^{P}$. Introducing a scalar multiplier $\dot{\lambda}$, for stress to strain conversion, leads to a flow rule for strain rates:

$$\dot{\varepsilon}_{ij}^{P} = \dot{\lambda}\,\frac{\partial f(\sigma_{ij})}{\partial \sigma_{ij}} \quad \text{or} \quad \dot{\mathbf{E}}^{P} = \dot{\lambda}\,\frac{\partial f(\boldsymbol{T})}{\partial \boldsymbol{T}} \tag{8.17a,b}$$

Equation (8.17a) implies normality of $\dot{\varepsilon}_{ij}^{P}$ to a convex yield surface. It follows that the yield function $f(\sigma_{ij})$ in eq(8.17a) may be identified with either right-hand side in eqs(8.16a,b):

$$f(\sigma_{ij}) = \frac{1}{V\dot{\lambda}}\int_{V_s}\left[\int_{\tau^s} \dot{\gamma}^s \,\mathrm{d}\tau^s\right]\mathrm{d}V = \frac{1}{V\dot{\lambda}}\int_{L_d}\left[\int_{f} v\,\mathrm{d}f\right]\mathrm{d}L \tag{8.18}$$

Geometrically, the yield function $f(\sigma_{ij})$ in eqs(8.17a) and (8.18) follows from micro-slip considerations alone simply by re-arranging Schmidt's law (8.5b) as

$$f(\sigma_{ij}, \tau_{cr}^{\ s}) = \alpha_{ij}^{\ s}\sigma_{ij} - \tau_{cr}^{\ s} = 0 \tag{8.19}$$

for when the resolved shear stress τ_{cr}^{s} is allowed to vary within each slip system. Again, normality is implied from combining eq(8.6a) with eq(8.19):

$$\mathrm{d}\varepsilon_{ij}^{P} = \alpha_{ij}^{s}\,\mathrm{d}\gamma^{s} = \mathrm{d}\gamma^{s}\frac{\partial f}{\partial \sigma_{ij}} \tag{8.20}$$

Comparison between eqs(8.17a) and (8.20) identifies $\dot{\lambda}$ with the rate of slip $\mathrm{d}\gamma^{s}/\mathrm{d}t$. In the continuum theory, the *plastic potential* function f in eq(8.18) is identified with a yield function describing a convex yield surface. Convexity is implied within the maximum work principle through eqs(8.10a,b) (see Drucker's postulate § 10.2.3). Either of eqs(8.17a,b) are is known as the *associated flow rule*, of which normality between f and $\dot{\varepsilon}_{ij}^{P}$ is a consequence. When quasi-static, eq(8.20) shows there is normality between f and $\mathrm{d}\varepsilon_{ij}^{P}$ for $\mathrm{d}\lambda = |\mathrm{d}\gamma^{s}|$. For compressible materials, this association can be too restrictive. Then f and the constitutive relation between stress and plastic strain are disconnected in a *non-associated flow rule*.

8.8 Micro- to Macro-Plasticity

The macro-plastic strain follows from the volume average of the microscopic strains. The latter may be expressed in the microscopic parameters, either from continuum slip or dislocation motion. They are, respectively,

$$\varepsilon_{ij}^{P} = \frac{1}{V}\int_{V_s} \tfrac{1}{2}(n_i m_j + m_i n_j)\gamma\,\mathrm{d}V \quad \text{or} \quad \mathbf{E}^{P} = \frac{1}{V}\int_{V_s} \tfrac{1}{2}(\mathbf{n}\,\mathbf{m}^{\mathrm{T}} + \mathbf{m}\,\mathbf{n}^{\mathrm{T}})\gamma\,\mathrm{d}V \tag{8.21a}$$

$$\varepsilon_{ij}^{P} = \frac{1}{A}\int_{A_s} \tfrac{1}{2}(n_i b_j + b_i n_j)\,\mathrm{d}A \quad \text{or} \quad \mathbf{E}^{P} = \frac{1}{A}\int_{A_s} \tfrac{1}{2}(\mathbf{n}\,\mathbf{b}^{\mathrm{T}} + \mathbf{b}\,\mathbf{n}^{\mathrm{T}})\,\mathrm{d}A \tag{8.21b}$$

Equation (8.21a) follows from eq(8.6a), in which slip systems are referred to the active volume V_s. In eq(8.21b), A_s is the collective area of active slip planes, **n** is the outward normal vector to a slip plane, **m** defines the slip vector and **b** is Burger's vector, with magnitude $|\mathbf{b}| = \gamma V/A_s$. The shear strain rate in eq(8.16a) is defined as

$$\dot{\gamma}^{s} = |\boldsymbol{b}|\,\rho\,\langle v\rangle = \frac{\gamma\,V\rho\,\langle v\rangle}{A_s}$$

where $\rho = L_d/V$ is the mobile dislocation density and $\langle v\rangle$ is the mean dislocation velocity.

The compatibility condition for a polycrystal requires that the deformation in each grain is transmitted to adjoining grains across their grain boundaries. Consequently, the internal stress within an individual grain will depend upon the applied macro-stress and the micro-stress imposed from neighbouring grains. A mathematical averaging technique is used to simplify the evolution from slip within a single grain to the stress–strain curve for a polycrystalline aggregate. Taylor [10] was the first to employ this technique in 1928 for an aluminium polycrystal composed of fine, randomly orientated, equi-axed grains with typical diameters 10^{-4} m. He assumed that at least 5 of the 12 slip systems would activate to accommodate plasticity and preserve volume under tension. Slip systems were activated by the least amount of work when axial strain was uniformly dispersed throughout the grains.

Normally, numerical integration is necessary when evaluating eq(8.21a) to give the strain in a representative volume element. However, a closed-solution arises from integrating slip within a hemispherical volume element composed of ideal grains, e.g. as in a single-phase crystalline solid. Grains of similar orientation within the element have parallel slip planes. Euler angles α, β and φ determine the orientation of the stress axes x_i relative to the normal **n** and direction **m** for an element of slip plane with solid angle dΩ, as shown in Fig. 8.15a. The number of grains with similar orientation is proportional to the product (d$\Omega \times$ dφ) and therefore these grains contribute to slip γ by an amount dγ = (d$\Omega \times$ dφ)γ. The plastic strain ε_{ij}^{P} in stress axes x_i (i = 1, 2, 3) then follows from eq(8.7). This gives the strain arising from all active, parallel slip systems within the unit hemisphere [22, 23]:

$$\varepsilon_{ij}^{P} = \frac{1}{2}\int_{\Omega} \mathrm{d}\Omega \int_{-\pi/2}^{\pi/2} (m_i n_j + n_i m_j)\,\gamma\,\mathrm{d}\varphi \tag{8.22a}$$

where, in spherical co-ordinates (r, α, β) and with r = 1:

$$\mathrm{d}\Omega = (r \sin\alpha\,\mathrm{d}\beta \times r\,\mathrm{d}\alpha) = \sin\alpha\,\mathrm{d}\alpha\,\mathrm{d}\beta \tag{8.22b}$$

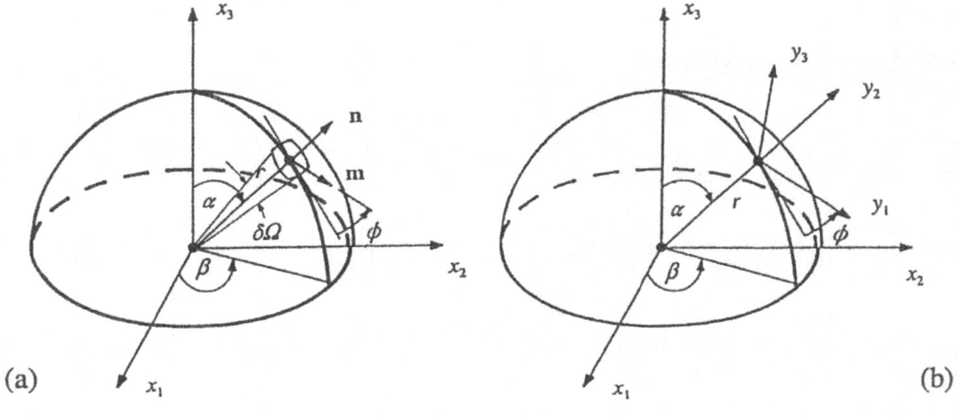

Figure 8.15 Euler angles and loading axes for a hemi-spherical crystal

By setting $\gamma = \gamma(\tau) = \sum c_n (\tau/\tau_{cr} - 1)^n$ as an n-term polynomial in eq(8.22a), it is seen how macroscopic plastic strain accumulates with slip when $\tau \geq \tau_{cr}$. This effect of this simplified approach is to distribute strain uniformly. Alternatively, it may be more realistic to identify the macroscopic strain in a polycrystal with an aggregate of all its non-uniform granular strains. To do this, let y_i (i = 1, 2, 3) be the orientation of a point on the slip plane within a single spherical grain (see Fig. 8.15b). Transforming slip from all active planes to axes y_i gives a microscopic strain tensor ε_{ij}^{P*} for that grain. Again, using spherical co-ordinates (r, α, β) and eq(8.22b), one may refer ε_{ij}^{P*} to stress axes x_i. This gives the macroscopic strain tensor for the grain [19]:

$$\varepsilon_{ij}^{P} = \frac{1}{4\pi^2}\int_H\!\!\int \sin\alpha\,\mathrm{d}\alpha\,\mathrm{d}\beta \int_{-\pi}^{\pi} \varepsilon_{ij}^{P*}\,\mathrm{d}\varphi \tag{8.23a}$$

where eq(8.7) applies to ε_{ij}^{P*}, providing tensor strain components in the crystal axes. Integration applies over a hemi-sphere H for all grain orientations φ. The macroscopic stress becomes a simple volume average of the microscopic stresses in all spherical grains.

$$\sigma_{ij} = \frac{1}{V}\int_V \sigma_{ij}^{*}\,\mathrm{d}V \tag{8.23b}$$

This method shows how the macro stress–strain behaviour evolves from micro-slip. However, many have recognised the need for a micro–macro evolution to account for the restricted, heterogenous deformation within grains. Here, the models of Eshelby [24] and Kroner [25] are useful. The underlying principal is that of amassing all grains of identical orientation into a single spherical or ellipsoidal grain. The 'grain', with orientation α, lies within an infinite isotropic matrix, as shown in Fig. 8.16a. When the 'composite' structure is stressed under a remote macroscopic stress σ_{ij}, the deformation differs between the spherical inclusion and the surrounding matrix. This results in an internal stress σ_{ij}^* within the grain. When matrix and grain are analysed separately under σ_{ij} (Figs 8.16b,c) an internal stress difference $\sigma_{ij} - \sigma_{ij}^*$ must exist to maintain compatibility, i.e. the grain must fit the hole. Eshelby [24] examined an elastic interaction between the anisotropic deformation produced from the uniform internal stress σ_{ij}^* in the grain (* refers to the grain) and the homogenous deformation assumed for the surrounding matrix. Kroner [25] extended this approach to include a plastic interaction between grain and matrix. Here, the integral in eq(8.23a) supplies the plastic strain in the anisotropic, spherical grain. Equilibrium and compatibility conditions must apply to the stress σ_{ij} and plastic strain ε_{ij}^P within a randomly orientated, isotropic matrix and stress σ_{ij}^* and strain ε_{ij}^{P*} within the anisotropic grain.

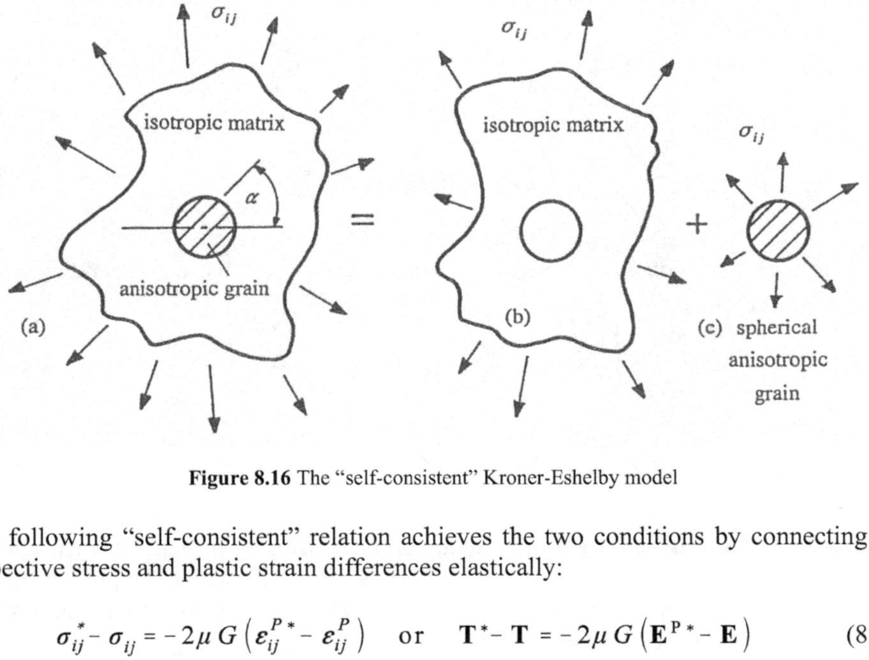

Figure 8.16 The "self-consistent" Kroner-Eshelby model

The following "self-consistent" relation achieves the two conditions by connecting the respective stress and plastic strain differences elastically:

$$\sigma_{ij}^* - \sigma_{ij} = -2\mu G\left(\varepsilon_{ij}^{P*} - \varepsilon_{ij}^{P}\right) \quad \text{or} \quad \mathbf{T}^* - \mathbf{T} = -2\mu G\left(\mathbf{E}^{P*} - \mathbf{E}\right) \tag{8.24}$$

where $\mu = (7 - 5\nu)/[15(1 - \nu)]$, G and ν are the usual elastic constants. The plastic strain difference embodies plastic incompressibility to ensure that the interaction between grain and matrix is independent of hydrostatic stress. Equation (8.24) may be applied in turn to all spherical grains of varying orientation and finally averaged to give the matrix stress and strain $\varepsilon_{ij}^P = (\varepsilon_{ij}^{P*})_{av}$ and $\sigma_{ij} = (\sigma_{ij}^*)_{av}$. However, eq(8.24) is known to over-estimate the stress difference $\sigma_{ij}^* - \sigma_{ij}$. The coefficient μ was altered by Kroner [25] to account for the non-uniform plastic strain in the matrix but, in retaining an elastic interaction, eq(8.24) still exaggerated the grain's internal stress, σ_{ij}^*. Hill [26] replaced the coefficient μG in eq(8.24) with a tensor parameter that minimised structural stiffness as the spread of plasticity "softened" the interaction. Berveiller and Zaoui [27] showed that an approximate method for accommodating a plastic interaction with stress relaxation between grain and matrix was to replace $2\mu G$ with $2\alpha\mu G$ in eq(8.24), where

$$\alpha = \left[1 + (3G\bar{\varepsilon}^P)/(2\bar{\sigma})\right]^{-1} \tag{8.25}$$

The use of macroscopic equivalent stress $\bar{\sigma}$ and plastic strain $\bar{\varepsilon}^P$ within eq(8.25) enabled α to be determined under a uniaxial stress. In order to include an instantaneous response from the heterogenous plastic volume V, they [28] added a further 'Hill-term' to eq(8.24):

$$\sigma_{ij}^* = \sigma_{ij} - 2\alpha\mu G\left(\varepsilon_{ij}^{P*} - \varepsilon_{ij}^P\right) + \frac{2\alpha\mu G}{V}\sum N_{ijkl}\left(\varepsilon_{ij}^{P*} - \varepsilon_{ij}^P\right) \tag{8.26}$$

Here, α is defined in eq(8.25) and N_{ijkl} is a fourth order interaction tensor of tangent modulus.

It is now possible to devise a method by which either of eqs(8.24) and (8.26) are applied to many crystalline slip systems, $s = 1, 2, 3 \ldots N$, within a single grain. Firstly, eq(8.6a) expresses the total strain in the grain as

$$\varepsilon_{ij}^{P*} = \frac{1}{2}\sum_{s=1}^{N}\left(n_i m_j + n_j m_i\right)^s \gamma^s = \sum_{s=1}^{N}\alpha_{ij}^s \gamma^s \tag{8.27a}$$

or

$$\mathbf{E}^{P*} = \frac{1}{2}\sum_{s=1}^{N}[\mathbf{n}\,\mathbf{m}^T + \mathbf{m}\,\mathbf{n}^T]^s \gamma^s = \sum_{k=1}^{N}\mathbf{A}^s \gamma^s \tag{8.27b}$$

in which there is no summation over s. Equation (8.6b) gives the corresponding plastic rotation in the grain:

$$\omega_{ij}^{P*} = \frac{1}{2}\sum_{s=1}^{N}\left(n_i m_j - n_j m_i\right)^s \gamma^s = \sum_{s=1}^{N}\beta_{ij}^s \gamma^s \tag{8.28a}$$

or

$$\mathbf{\Omega}^{P*} = \frac{1}{2}\sum_{s=1}^{N}[\mathbf{n}\,\mathbf{m}^T - \mathbf{m}\,\mathbf{n}^T]^s \gamma^s = \sum_{s=1}^{N}\mathbf{B}^s \gamma^s \tag{8.28b}$$

where $\mathbf{A}^s \equiv \alpha_{ij}^s$ and $\mathbf{B}^s \equiv \beta_{ij}^s$ are the orientation factors for a given slip system s. Let a number of grains, $q = 1, 2, 3 \ldots M$, each with volume V_q, lie within the polycrystalline volume V. The macroscopic plastic strain and rigid rotation of the matrix are found from the volume averages of eqs(8.27a) and (8.28a):

$$\varepsilon_{ij}^P = \frac{1}{V}\int_V \varepsilon_{ij}^{P*}\,\mathrm{d}V = \frac{1}{V}\sum_{q=1}^{M}\sum_{s=1}^{N} V_q \alpha_{ij}^{qs}\gamma^{qs} \tag{8.29a}$$

$$\omega_{ij}^P = \frac{1}{V}\int_V \omega_{ij}^{P*}\,\mathrm{d}V = \frac{1}{V}\sum_{q=1}^{M}\sum_{s=1}^{N} V_q \beta_{ij}^{qs}\gamma^{qs} \tag{8.29b}$$

Equation (8.5b) shows for a given slip system s, within a single grain q, that the critical resolved shear stress τ_{cr}^s may be expressed in terms of the uniform stress field σ_{ij}^* within that grain as

$$\tau_{cr}^{s} = \sigma_{ij}^{*} n_i^s m_j^s = \sigma_{ij}^{*} \alpha_{ij}^{s} \tag{8.30a}$$

or

$$\tau_{cr}^{s} = \mathrm{tr}\,[\mathbf{T}^{*}\,\mathbf{n}^{s}\,(\mathbf{m}^{s})^{\mathrm{T}}] = \mathrm{tr}\,(\mathbf{T}^{*}\mathbf{A}^{s}) \tag{8.30b}$$

where $\alpha_{ij}^{s} = n_i^s m_j^s$, or $\mathbf{A}^s = \mathbf{n}^s(\mathbf{m}^s)^{\mathrm{T}}$. The macroscopic stress σ_{ij}, for M grains in volume V, derives from the volume average eq(8.23b):

$$\sigma_{ij} = \frac{1}{V}\sum_{q=1}^{M} \sigma_{ij}^{*} V_q \tag{8.31}$$

Each resolved shear stress in a f.c.c. lattice depends upon all possible $N = 24$ slip systems:

$$\tau_{cr}^{s} = h^{s}(\gamma^1, \gamma^2, \gamma^3 \ldots\ldots \gamma^{24}) \tag{8.32}$$

in which the hardening function h^s depends upon the Burgers vector and the dislocation density. The solution to the macro stress–strain behaviour combines a suitable micro-constitutive hardening law with the volume average, eq (8.31). Assume that a known σ_{ij} is applied and ε_{ij}^{P} is to be calculated. Firstly, σ_{ij}^{*} is estimated from eq(8.31). Then, τ_{cr}^{s} is found from eq(8.30a) for each slip system within a grain. It follows that the amounts of slip γ^s in eq(8.32) depend upon the chosen hardening law. Taylor [10] employed a linear isotropic hardening law:

$$\tau_{cr}^{s} = \tau_o + \sum_{k=1}^{N} h^{sk}\gamma^{k} \tag{8.33}$$

where τ_o is the shear stress necessary to initiate slip. Equation (8.33) allows the constant value of tangent modulus $h^{sk} = \partial\tau_{cr}^{s} / \partial\gamma^{k}$ to differ between slip systems. Other hardening laws have been employed. For example, Brown [29] employed a non-linear relation for each slip system in which c and n are empirical constants:

$$\tau_{cr}^{s} = c\,(\gamma^{s})^{n} \tag{8.34}$$

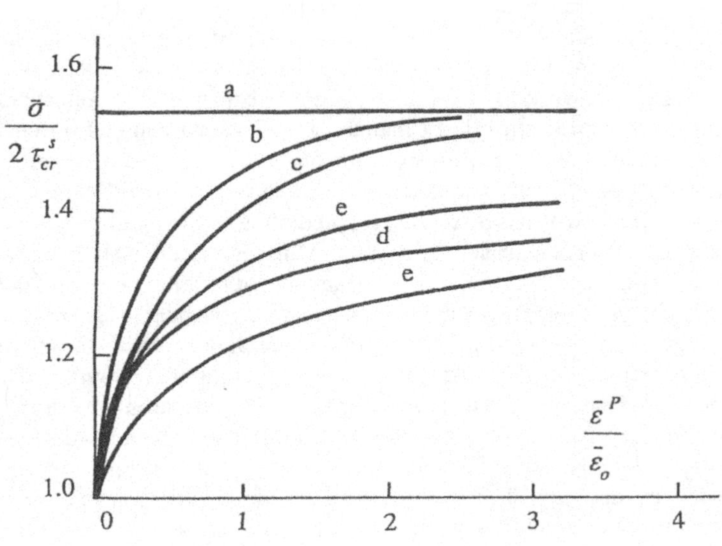

Figure 8.17 Theories of polycrystalline hardening ($\bar{\varepsilon}_o = Y/E$)

The micro– and macro–plastic strains follow from the application of eqs(8.27a) and (8.29a), respectively. Finally, eq(8.24) enables σ_{ij}^* to be re-calculated within an iteration procedure. Berveiller and Zaoui [27] and Hutchinson [30] compared uniaxial, stress–strain curves from this method by making various assumptions. Firstly, with ideal, elastic–perfectly plastic f.c.c. grains, $h^{sk} = 0$ in eq(8.33). Further assumptions, listed below, were followed by these and others in producing the respective curves a–d in Fig. 8.17:

(a) Taylor [10]: Large, uniform strain satisfying compatibility (trivially), elastic strain and work hardening ignored. Equilibrium condition for grain boundary stress is violated.
(b) Lin [7]: Uniform strain satisfying compatibility (trivially) but not restricted to large strain. Includes elastic strain component and plastic hardening. Equilibrium is violated.
(c) Eshelby [24] and Kroner [25]: Both equilibrium and compatibility satisfied. Uniform strain only within grains of identical orientations. Inhomogenous strain distribution elsewhere with an elastic interaction between grains.
(d) Hill [26]: Same as for (c) but with a rigorous analysis of the inelastic interaction between grains given.
(e) Berveiller and Zaoui [27]: Same as for (d) but with a less rigorous analysis of the inelastic interaction between grains. Consequently, application of the theory is simplified.

Method (d) is believed to be the most exact though it requires the greatest amount of computation. It was adopted by Weng [31, 32] to account for hardening in a two-phase alloy and the distortion found in its subsequent yield locus. With α defined from eq(8.25), it is seen that curve *e* is a reasonable approximation to curve *d* and can be found with less effort. The account of heterogenous deformation within the surrounding matrix (see curve *e*) has reproduced the observed non-linear, uniaxial stress-strain for a polycrystal to $\varepsilon^P \simeq 0.1$ %. Further modifications to the Eshelby-Kroner self-consistent relation, eq(8.24), were made to account for texture development and a Bauschinger effect in the grain [33].

Other methods of treating interactions between grains adapt further the ideas of Taylor et al. [10, 34] in preference to following a model of self-consistency [7, 35]. Taylor neglected elastic strains and equated the strain in each grain to an average macroscopic strain. Figure 8.17 shows that Taylor's curve *a* is realistic only at large strain, where individual grains do not work-harden further. A more realistic prediction, curve *b*, applies to grains that behave in an elastic–plastic manner, with linear hardening from eq(8.33). Both the Taylor and Lin predictions overestimate the observed response since the condition of equilibrium is violated at grain boundaries. Nonetheless, the simplifications afforded by linear hardening explain why it is often used for predicting polycrystalline deformation from single crystal plasticity. To improve the accuracy, a further refinement [36] has incorporated a non-linear hardening law to account for interaction and rotation between grains.

In practice, alloys are employed more often than metals to bear load. This is particularly so in hostile, high-temperature, corrosive environments. Understanding the micro-structural processes that constitute the observed strain is far more complex. Alloying raises the yield stress and the rate of hardening. The reason for this is that the stress necessary for a stage III dynamic recovery is increased as alloying atoms resist and impede the motion of dislocations. Additionally, as the Petch eq(8.9) shows, the yield stress is raised if the grain size is reduced by alloying. Theories of solid solution hardening are generally more successful than those of precipitation hardening [1, 2] in explaining these effects. This is largely because the dislocation interactions in solid solutions are more clearly defined. Raising temperature and/or reducing the rate of deformation assists with the processes of climb and cross slip of pinned dislocations by thermal activation. These reduce the yield stress and the hardening rate. Furthermore, when precipitation hardened alloys operate at higher temperatures, it becomes necessary to account for other diffusion processes: grain coarsening, phase changes, micro-voids and crack formation [37]. It is apparent therefore why the global macro theory has advanced plasticity theory further than of taking into account all the specific micro detail.

8.9 Subsequent Yield Surface

Qualitatively, the theory discussed in the previous section predicts a parabolic stress–strain curve for a polycrystal correctly. A further validation of this theory test would lie in its ability to describe the subsequent yield surface for a work-hardened material. For example, the assumption of plastic rigidity, employed by Taylor [10], implies that the initial yield surface applies to all subsequent work-hardened states. This oversimplification suggests that the initial distribution of grain orientations is not altered by plasticity. Clearly, this model ignores those influences of deformation texture upon the yield surface, as shown in Fig. 8.14.

8.9.1 Isotropic Hardening

Firstly, consider Taylor's linear hardening eq(8.33) modification. This gives equal increases in the resolved shear stress for both active and non-active (latent) slip planes. Thus, where the hardening behaviour neither depends upon the sense nor the direction of slip, a polycrystal will display an initial isotropy. Such behaviour is evident with equal yield stress in tension and compression of a fully annealed material, i.e. one whose structure is free from the history of previous deformation. With an isotropic hardening rule imposed under combined stress states, the initial yield surface is understood to expand uniformly, retaining its shape and orientation. Figure 8.18a illustrates this behaviour from within a principal biaxial stress space of σ_1 versus σ_2, in which hardening is non-linear.

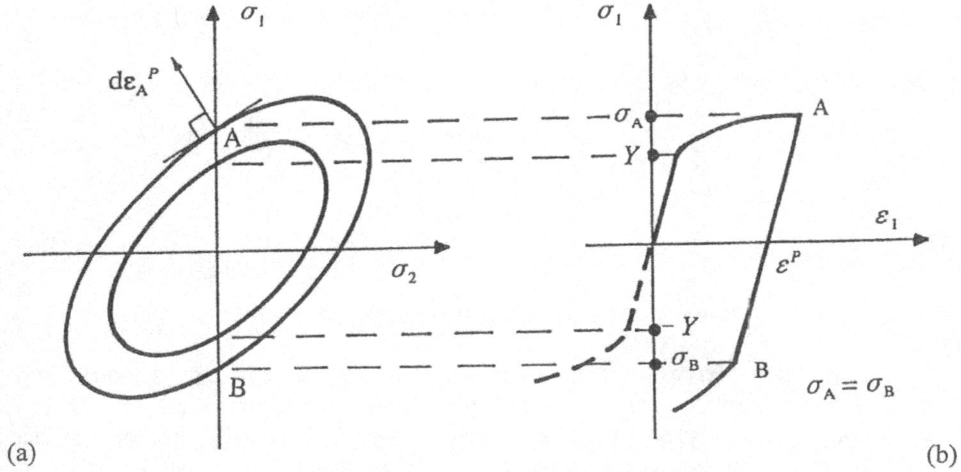

Figure 8.18 The rule of isotropic hardening

Figure 8.18b has projected the expanded yield locus into a stress-strain response for simple tension. The rule shows that when the direction of deformation is reversed upon reaching the work-hardened state A, yielding re-commences when the stress at B equals that at A. The major criticism of isotropic hardening is that it does not show a *Bauschinger effect*. The latter states $\sigma_B < \sigma_A$, which is always observed in stress reversal experiments. One should only expect the subsequent yield surface to be an inflation of the initial yield surface for the backward-extrapolation definition of yield (see § 3.3.1), employed by Taylor and Quinney [34]. Their yield point is detected from the low hardening rate pertaining to large plastic strain. In contrast, alternative definitions of yield employ little or no plastic strain. The identification of the forward and reversed yield stresses σ_A and σ_B at proportional limits (as shown) or from small offset proof strain will involve a Bauschinger effect under stress reversal. However, one may adopt the isotropic hardening rule for outward loading paths where there is no stress reversal. Figure 8.18a shows how combining a uniform expansion with the *normality rule* connects σ_A to an incremental plastic strain vector $\boldsymbol{\delta\varepsilon}_A{}^P$ at the load point. Chapter 10 shows where the rule of isotropic hardening can be combined with the flow rule to match certain experimental data from outward loading.

8.9.2 The Minimum Surface

It has been seen that the slip theory of Batdorf et al. [22, 23, 38] employs active slip systems within individual crystals comprising the polycrystalline aggregate. With only one independent slip system per crystal, this theory predicts a subsequent yield surface to be the minimum surface enclosing the prestress point and the initial yield surface. When the latter is defined by von Mises, it projects as a circle on the deviatoric plane (see Fig. 9.1). The circle is the cross-section of a cylinder with an axis of hydrostatic stress equally inclined to the axes of principal, deviatoric stress σ_1', σ_2' and σ_3', as shown in Fig. 8.19a.

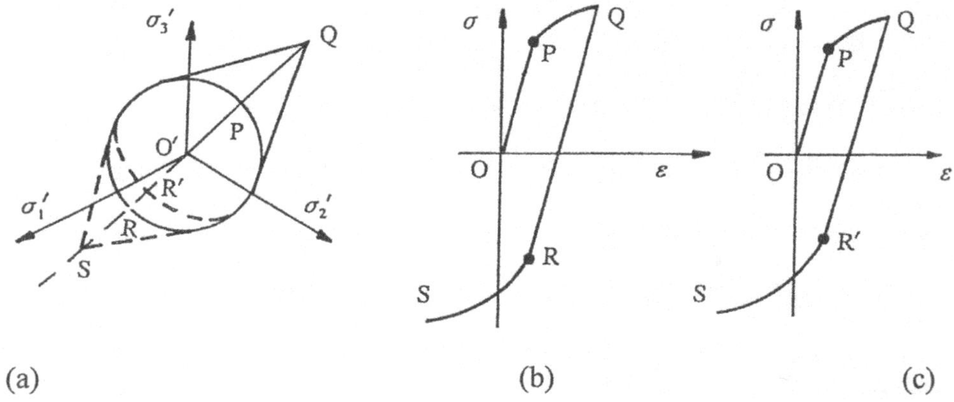

Figure 8.19 Minimum surface showing Bauschinger effect

Once the stress path pierces the initial surface at P, a subsequent surface is formed from projecting tangents from the initial surface to pass through the current stress point Q. This construction leaves the rear portion of the original surface unchanged. The minimum surface reveals that an isotropic material becomes anisotropic. It is seen, from the associated stress-plastic strain response (Fig. 8.19b) how a minimum surface displays a Bauschinger effect: $\sigma_R < \sigma_Q$. Figure 8.19c shows a realistic Bauschinger effect, where reversed yielding is expected at R′. Correspondingly, the rear of the minimum surface is modified [39, 40], as indicated with a dashed line in Fig. 8.19a.

Much experimental work was conducted in the decade 1950–60 to confirm whether the subsequent yield surface displayed the corner or pointed vertex at the load point, predicted by slip theory. When the findings from this work were reviewed objectively [41], they appeared to be inconclusive. The majority of investigations confirmed a local distortion of the subsequent yield surface. A blunt nose, not a corner, appears in the region where the surface was originally pierced by the stress vector. It can then be said that a unique normal still applies for stress points on all subsequent surfaces, e.g. at point 1 on the subsequent surface f_1, in Fig. 8.20a.

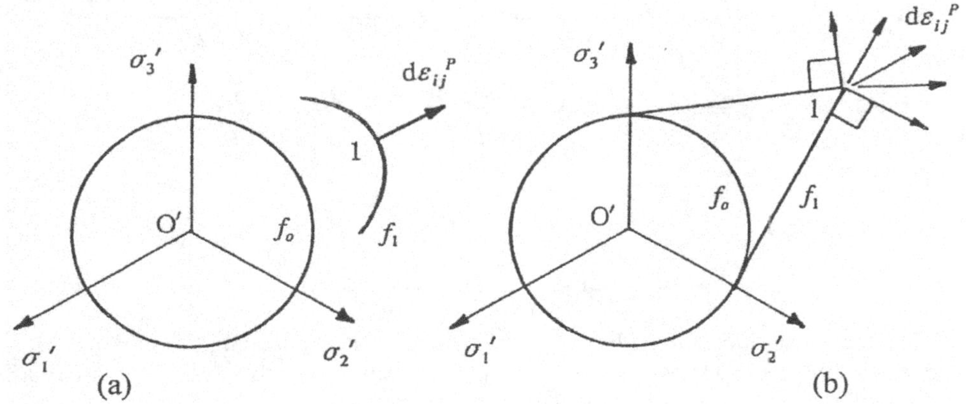

Figure 8.20 Surface with smooth contour and vertex point

The slip theory was 'validated' indirectly from experiments [42, 43] that demonstrated a fan of normals present at stress point 1 in Fig. 8.20b. The fan implies an undefined direction of the plastic strain increment vector at the load point, which would be predicted in the presence of a corner in the yield surface. Corners may be indeed be present for yielding in Tresca-type single crystals [7]. However, when the many slip systems are activated within the grains of a polycrystal, the aggregated yielding behaviour does provide, acceptably, the unique normal at the load point, given by macro-plasticity theory.

8.9.3 Kinematic Hardening

When the applied stress is removed from a plastically deformed polycrystal, an instantaneous elastic recovery occurs but a residual stress distribution will remain in self-equilibrium. Microscopically, an internal stress builds as the increasing dislocation density of forward deformation becomes less mobile. Dislocations entangle and are impeded by obstacles lying in active slip planes. A pile-up of dislocations promotes a back-stress which resists further slip under the resolved stress in these slip planes. Many of the crystals remain permanently slipped following an unloading of the aggregate from the plastic region. The remaining heterogenous distribution of residual stress carries a sense in opposition to the applied stress, as proposed originally by Kadashevitch and Novozhiliov [44]. Consider a reversal in the direction of uniaxial deformation, say at point A in Fig. 8.21a. The stress required to initiate reversed yielding at B is lowered because of the additive contribution from an internal stress of the same sense. Thus, the Bauschinger effect, $\sigma_B < \sigma_A$, can be explained from the existence of a residual micro-stress [45, 46].

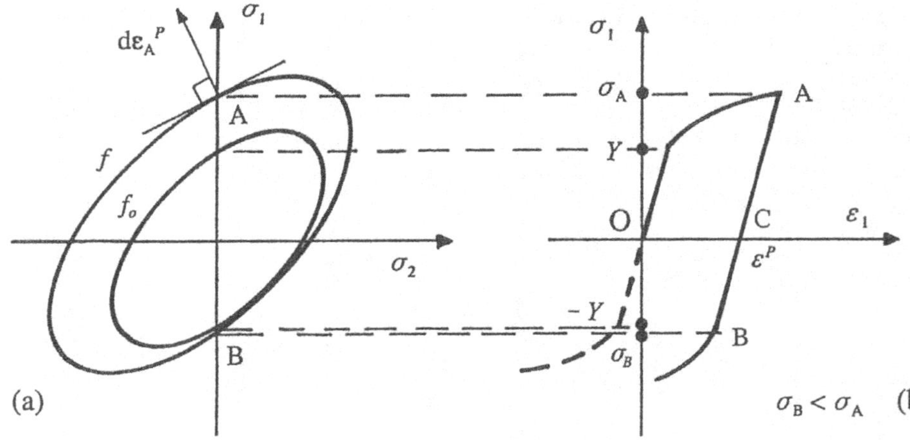

Figure 8.21 The rule of combined hardening

Next, consider uniaxial loading OA, unloading AC and re-loading CA in Fig. 8.21b. The sense of the internal stress, arising from a plastic loading–unloading cycle, opposes that of the applied stress upon re-loading. This explains, qualitatively, why a forward yielding will not occur until the stress level again reaches that at point A. The increase in the forward yield stress (i.e. $\sigma_A > Y$) is associated with the plastic pre-strain, ε^P. Clearly, if new yield points A and B are to lie on the subsequent yield surface f, then the initial yield surface f_o must combine an expansion with a translation from its origin, as shown in Fig. 8.21a.

The original *kinematic hardening rule* of Prager [47] refers to a simplified, rigid translation of the yield surface in the absence of an expansion (see Fig. 8.22a). Figure 8.22b shows that linear hardening is implied by this, i.e. the initial elastic range $2Y$ remains unaltered by plastic strain so that $\sigma_A - \sigma_B = 2Y$, where Y is the initial yield stress. The amount of hardening is measured by an increase in the flow stress: $\sigma_A - Y$. The amount of softening is measured by a decrease in flow stress: $Y - \sigma_B$. It follows that each change to Y will have the same magnitude for a given ε^P.

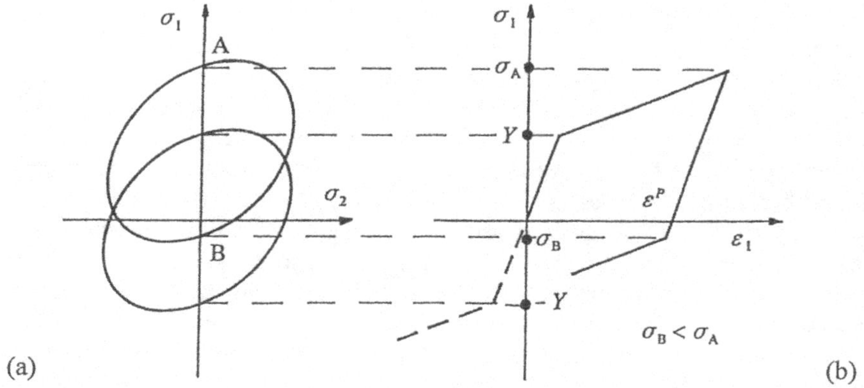

Figure 8.22 The rule of kinematic hardening

These observations reveal an equivalence between the internal stress and the centre of a translated yield surface. Centre co-ordinates a_{ij} define a strain history dependent translation. In the Kroner-Eshelby model, a_{ij} becomes the difference between the micro and macro stress tensors. Re-arranging eq(8.24) gives an incremental translation corresponding to each instantaneous stress change, i.e. when the time interval $\delta t \to 0$:

$$\delta a_{ij} = \delta(\sigma_{ij}^{*} - \sigma_{ij}) = 2\mu G\, \delta(\varepsilon_{ij}^{P} - \varepsilon_{ij}^{P*})$$

$$\mathrm{d}a_{ij} = \mathrm{d}\sigma_{ij}^{*} - \mathrm{d}\sigma_{ij} = 2\mu G\, (\mathrm{d}\varepsilon_{ij}^{P} - \mathrm{d}\varepsilon_{ij}^{P*})$$

or, in matrix form:

$$\mathrm{d}\mathbf{A} = \mathrm{d}\mathbf{T}^{*} - \mathrm{d}\mathbf{T} = 2\mu G\, (\mathrm{d}\mathbf{E}^{\mathrm{P}} - \mathrm{d}\mathbf{E}^{\mathrm{P*}})$$

Note: In these equations stress and strain rates apply to time-dependent plasticity. The theoretical predictions, curves *c*, *d* and *e* in Fig. 8.17, employ this model. Hence, by reversing the direction of deformation, each curve predicts softening behaviour and therefore can be expected to describe the Bauschinger effect realistically. Here, a dislocation pile-up is the major contributor to the development of the micro-stress, which is more often called the back (or internal) stress arising from this source.

Experiment shows that it is more appropriate to model the plasticity of metals by combining the rules of isotropic and kinematic hardening, as in Fig. 8.21. For example, Brown [48] described the translation for pre-strained aluminium with non-linear hardening, eq(8.34). Subsequent flow potentials for increasing times under stress (i.e. for cold creep) were concentric von Mises ellipsoids with a common centre a_{ij}. Experimental investigations have shown that both the initial and subsequent yield surface for a polycrystal are sensitive to the chosen definition of yield (see § 3.3.1). Certainly, where a linear region of elasticity exists in plastically pre-strained material, this is evidence [49] for a translation in the subsequent yield surface. The rigid kinematic translation shown in Fig. 8.22 is an extreme idealisation. Observed behaviour involving load reversal suggests that combined hardening including non linear translation should be the preferred model of plasticity (see exercises 8.15–8.17).

8.10 Concluding Remarks

This chapter has considered micro-mechanical descriptions of yield, flow and hardening behaviour. These are the key elements employed in the development of continuum theories of plasticity. Much of micro-mechanics involves the study of slip in single crystals and polycrystals for metals. It appears from the literature that additional sources of plasticity arise in metal alloys. This is not surprising since the essential purpose of alloying is to strengthen by inhibiting slip. Aluminium alloy and steel have provided examples here of what the real micro-structural behaviour involves. For example, in order to accommodate a loading strain internally, the identification of those slip planes which remain active would have to be considered along with the manner of crack and void formation that limit strain. Despite these additional processes, the elements of continuum plasticity theory remains largely unaltered. In its use of macro-stress and strain across a polycrystalline aggregate, the continuum theory does not differentiate between metals and alloys specifically. By employing an appropriate yield function and a universal flow rule, it supplies the macroscopic plastic strain within the bulk volume of any polycrystal. It will be seen that this can be achieved with the empirical representation of hardening behaviour, derived from uniaxial stress–strain curves. The latter depend upon the service conditions, among which the strain rate, test temperature, pre-strain and heat treatment are important. The most recent developments to plasticity theory recognise that all inelastic deformation is time dependent to a degree (see chapter 11). Practically, this means that stress depends upon the rate of straining and strain upon time under load. There is a range of rates and temperatures where plastic strain may be taken as quasi static or time independent. The character of that strain is quite unlike a true elastic strain which is instantaneous, independent of time and recoverable. Elastic strain is identified as lattice distortion that contributes to total strain by a far lesser amount usually than that attributed to a slipped and dislocated atomic arrangement of that lattice. Therein lies the residual stress, the plastic strain and void damage that current macro-models of plasticity attempt to replicate.

References

1. Cottrell A. *Theory of Crystal Dislocations*, 1964, Blackie & Son, London.
2. Honeycombe R. W. K. *The Plastic Deformation of Metals*, 1985, Arnold.
3. Schmidt E. and Boas W. *Plasticity of Crystals*, 1950, Hughes, London.
4. Petch N. J. *Jl. Iron and Steel Inst.*, 1953, **173**, 25.
5. Guy A. G. *Physical Metallurgy for Engineers*, 2nd Ed, Addison-Wesley, 1966.
6. *Metals Handbook: Vol. 1 Properties and Selection of Metals*, 8th Ed, 1961, ASM.
7. Lin T. F. *Advances in Applied Mechanics*, (ed. Chia-ShunY.) 1971, 255, Academic Press, New York.
8. Kocks U. F. *Metallurgical Trans.*, 1970, **1**, 1121.
9. Kocks U. F. *Unified Constitutive Equations for Creep and Plasticity*, (ed. Miller A. K.) 1987, **1**, Elsevier.
10. Taylor G. I. *Jl. Inst. Metals*, 1938, **62**, 307.
11. Bishop J. F. W. and Hill R. *Phil. Mag.*, 1951, **42**, 418.
12. Bishop J. F. W. and Hill R. *Phil. Mag.*, 1951, **42**, 1298.
13. Backofen W. A. *Deformation Processing*, 1972, Addison-Wesley.
14. Althoff J. and Wincierz P. *Z. Metallkde.*, 1972, **63**, 623.
15. Piehler H. R. and Backofen W. A. *Textures in Research and Fracture* (eds Grewen J. and Wassermann G.) 1969, 436, Springer-Verlag.
16. Grzesik D. *Z. Metallkde.*, 1972, **63**, 618.
17. Grzesik D. *Mechanische Anisotropie*, (ed. Stuwe H. P.) 1974, 231, Springer-Verlag.
18. Matucha K. H. and Wincierz P. ibid., 201.
19. Avery D. H., Hosford W. F. and Backofen W. A. *Trans. AIME*, 1965, **233**, 71.
20. Johnson W. G. and Gilman J. J. *Jl. Appl. Phy.*, 1959, **30**, 129.
21. Rice J. R. *Trans. ASME, Jl. Appl. Mech.*, 1970, **37**, 728.
22. Batdorf B. and Budianski B. 'A mathematical theory of plasticity based upon the concept of slip', *NACA Tech. Note* 1871, April 1949.
23. Batdorf S. B. and Budiansky B. *Jl Appl Mech*, 1954, **21**, 323.
24. Eshelby J. D. *Progress in Solid Mechanics*, 1961, **2**, 87, North Holland.
25. Kroner E. *Acta Metallurgica*, 1961, **9**, 155.
26. Hill R. *Jl. Mech. Phys. Solids*, 1965, **13**, 89.
27. Berveiller M. and Zaoui A. *Jl. Mech. Phys. Solids*, 1979, **26**, 325.
28. Berveiller M. and Zaoui A. *Trans. ASME, Jl. Eng. Mat. and Tech.*, 1984, **106**, 295.
29. Brown G. M. *Jl. Mech. Phys. Solids*, 1970, **18**, 367.

30. Hutchinson J. W. *Proc. Roy. Soc.*, 1970, **A319**, 247.
31. Weng G. J. *Int. Jl. Plasticity*, 1985, **3**, 275.
32. Weng G. J. *Int. Jl. Plasticity*, 1987, **3**, 315.
33. Tokuda M., Kratochvil J. and Ohashi Y. *Phys. Stat. Sol. (a)*, 1981, **68**, 629.
34. Taylor G. I. and Quinney H. *Phil. Trans. Roy. Soc.*, 1931, **A203**, 323.
35. Lin T. H. *Jl. Mech. Phys. Solids*, 1957, **5**, 143.
36. Havner K. S. *Int. Jl. Plasticity*, 1985, **1**, 111.
37. Argon A. S. *Constitutive Equations in Plasticity*, 1975, MIT Press.
38. Budiansky B. and Wu T. T. Proc.: *4th U.S. Nat. Cong. Applied Mechanics*, ASME, New York, 1963, **2**, 1175, Pergamom.
39. Como M. and D'Agostino S. *Meccanica*, 1969, **4**, 146.
40. Como M. and D'Agostino S. *Archives of Mechanics*, 1973, **25**, 685.
41. Michno M. J. and Findlay W. N. *Int. Jl. Non-Linear Mech.*, 1976, **11**, 59.
42. Naghdi P. M., Rowley J. C. and Beadle C. W. *Jl. Appl. Mech.*, 1955, **22**, 416.
43. Phillips A. and Gray G. A. *Trans. ASME, Jl. Basic Eng.*, 1961, **83**, 275.
44. Kadashevitch I. I. and Novozhilov V. V. *Prikl. Mat. Mekh.*, 1958, **22**, 78.
45. Como M. and D'Agostino S. *Meccanica*, 1969, **4**, 146.
46. Kafka V. *ZAMM*, 1968, **48**, 265.
47. Prager, W. *Jl. Appl. Mech.*, 1956, 23, 483
48. Brown G. M. *Jl. Mech. Phys. Solids*, 1970, **18**, 383.
49. Ikegami K. Translation *BISI* 14420, 1976, The Metals Society.

Exercises

8.1 Apply eq(8.1) to show that when a tensile stress σ is applied to a single crystal, the critical resolved shear stress τ_{cr} is at a maximum for $\varphi = \lambda = 45°$. Plot the variation in τ_{cr}/σ when the slip plane is inclined at various angles to the stress axis.
See solution

8.2 Sketch isometric views of: (a) a cube with the portion above the (111) plane removed and (b) a hexagonal prism with the portion above plane (01 12) removed, indicating the [1120] direction.

8.3 Show that eqs(8.11a,b) appear equivalently in matrix notation as, respectively,

$$\mathrm{tr}\,(\mathbf{T}\,\delta\mathbf{E}^{\mathrm{P}}) = \int_V \mathrm{tr}\,(\mathbf{T}^*\,\mathrm{d}\mathbf{E}^{\mathrm{P}*})\,\mathrm{d}V$$

$$\delta\mathbf{E}^{\mathrm{P}} = \int_V \mathrm{d}\mathbf{E}^{\mathrm{P}*}\,\mathrm{d}V$$

in which $\mathbf{T}$ is Cauchy stress matrix and $\delta\mathbf{E}^{\mathrm{P}}$ is the incremental plastic strain matrix, given in eqs (1.46) and (2.5b), respectively.
See solution

8.4 Show that observed values τ_{cr} = 6, 4 and 20 ($\times 10^6$ dynes/cm^2) for silver, aluminium and iron, respectively, are approximately four orders of magnitude less than their theoretical values given from $\tau_{cr} = G/(2\pi)$. Respective elastic constants (Young's modulus and Poisson's ratio) for the three metals are: E = 70 GPa, ν = 0.37; E = 71 GPa, ν = 0.34 and E = 206 GPa, ν = 0.29.
See solution

8.5 List the particular point, line, plane and volume defects most responsible for the large discrepancy referred to in exercise 8.4.
See solution

8.6 Examine the application of a micro-stress volume average, eq(8.23a), to a spherical f.c.c. grain when the distribution of plastic strain ε_{ij}^{P*} is according to eq(8.8).

8.7 Examine what modifications to eq(8.23a) would be required to accommodate an ellipsoidal grain.

8.8 Estimate the % increase in a unit volume at ambient temperature following the transformation of austenite to martensite for 0.4 % and 0.8 % C-steels. Assume that: (i) a 4 % increase in volume occurs with extensions of one atomic radius between neighbouring atoms for this cell and (ii) that the amount of martensite present is equivalent to the decomposition of austenite to pearlite under slow cooling for each C-composition.

8.9 Sketch the phases involved in cooling those steels with the compositions given in exercise 8.8 under equilibrium conditions at temperatures just above (730°C) and just below (720°C) the lower critical temperature (723°C).
Hint: see text example 8.1.

8.10 What are the three requirements of steel that it can be hardened by tempering?

8.11 List the specific properties more accurately than the description of martensite as a supersaturated solution of carbon in b.c.c. ferrite.

8.12 Use the lever rule to estimate the amount of ferrite and cementite that apply in the slow cooling of 0.8 % C-steel of eutectoid composition from a temperature of 800°C to 723°C. Then show that the pearlite formed following the transformation of austenite is 88 % ferrite and 12 % cementite.

8.13 Given the critical resolved shear stress for aluminium is 250 kPa calculate the tensile stress required to cause yielding when the tensile axis is [001].

8.14 A zinc single crystal is oriented with the normal to the basal plane inclined at 60° to the tensile axis. Three slip systems within the crystal make angles of 38°, 45° and 84° with the tensile axis. Calculate the critical resolve shear stress when plasticity first appears at a tensile stress of 2.25 MPa.

8.15 The following function is proposed as a combined hardening model for the subsequent yielding of polycrystalline solid. Identify, with the respective symbols given, the inherent anisotropy present in that solid and the anisotropy induced by plastic straining

$$f(\sigma_{ij}\, \varepsilon_{ij}^{P}) = C_{ijkl}(\sigma_{ij} - \alpha_{ij})(\sigma_{kl} - \alpha_{kl}) = \chi(\psi)$$

8.16 What explicit forms of α_{ij}, χ, ψ could apply to the function given in exercise 8.15 when the solid remains incompressible under plastic deformation?

8.17 Reduce the function given in exercise 8.15 to a principal biaxial stress combination in the case of an initially isotropic condition where subsequent incompressible plasticity conforms to Hodge's combined hardening. Plot the initial yield locus and the subsequent yield locus for a stress path coincident with a one principal axis.
Hint: Refer to Fig. 3.7 as a guide.

CHAPTER 9

THE FLOW CURVE

9.1 Introduction

In the earlier chapters, ideal materials were assumed to flow under a constant yield stress. Some metals do approximate to this behaviour but most will require an increase in stress to progress plastic flow. The hardening rate $d\sigma/d\varepsilon^P$ describes this increase in stress with increasing plastic strain. This rate is at its greatest immediately beyond the yield point, where there is a rapid increase in dislocation density. Thus, it is in the lower strain region of plasticity where an account of hardening is made, usually under a monotonic loading. If plastic slip processes are to continue, the stress must be raised further to overcome the barriers presented by a dislocated structure. The rate of hardening will, however, diminish to zero at large strain, where the energy of the structure assists with slip by activating additional climb and cross slip recovery mechanisms.

Theories of plasticity account for hardening by employing various empirical descriptions of uniaxial or torsional plastic flow behaviour. The latter is found from adopting standard laboratory tests, the choice between them depending upon the range of plastic strain required. Testing is normally performed under increasing loading so that the description derived applies to forward deformation. Tension and compression tests are useful when limited to the small total strains that apply, say, to a plasticity theory for an elastic–plastic structure. The appearance of necking in the tension test and barrelling and buckling under compression prevents each test from attaining large uniaxial strain values. To overcome these experimental difficulties, a modified uniaxial compression test, plane strain and bulge testing have been employed. These enable the flow behaviour to be described at much larger strains, typical of those reached in metal forming processes.

In this chapter, these standard tests are described and their data are employed to demonstrate empirical descriptions of flow curves. The latter appear in axes of equivalent stress versus equivalent plastic strain. A definition of equivalence is required to correlate flow curves under different stress states in material with an initial conditions known to be isotropic and anisotropic. Particular attention is paid to annealed, isotropic materials obeying the von Mises and Drucker yield conditions. The distinction between the hypotheses of strain and work hardening in isotropic material is also examined.

9.2 Equivalence in Plasticity

Consider an initially isotropic material whose yield surface is described by a function of the stress deviator invariants $f(J_2', J_3')$. The von Mises function [1] $f = J_2'$ is by far the most common form since it relies on the second invariant alone. However, the determination of one constant, say, in the Drucker function [2] $f = J'^2_2 - cJ'^2_3$, accommodates a wider range of isotropic behaviour. Here it will be shown that equivalence expressions for stress and plastic strain rest with the choice between a work and strain hardening hypothesis.

9.2.1 Equivalent Stress

(a) von Mises

When the isotropic yield function f is equated to the second invariant of deviatoric stress, this implies that hydrostatic stress does not contribute to plastic deformation:

$$f = J_2' = \frac{1}{2}\sigma_{ij}'\sigma_{ij}' \tag{9.1a}$$

Yielding begins under a given stress state when J_2' attains a critical value. The latter is determined from uniaxial yield stress Y as $J_2' = Y^2/3$. We then have

$$J_2' = \frac{1}{3}Y^2 = \frac{1}{2}\sigma_{ij}'\sigma_{ij}' \tag{9.1b}$$

from which the initial yield stress is

$$Y = \sqrt{\frac{3}{2}\sigma_{ij}'\sigma_{ij}'} \tag{9.2a}$$

Thus, Y may be regarded as an effective, or equivalent, yield stress for a general stress state. When this yield stress is exceeded, an equivalent flow stress $\bar{\sigma}$ ($\bar{\sigma} > Y$) for the plasticity that ensues is defined correspondingly:

$$\bar{\sigma} = \sqrt{\frac{3}{2}\sigma_{ij}'\sigma_{ij}'} \tag{9.2b}$$

Written in its fully expanded form, eq(9.2b) becomes

$$\bar{\sigma} = \frac{1}{\sqrt{2}}\sqrt{(\sigma_{11} - \sigma_{22})^2 + (\sigma_{22} - \sigma_{33})^2 + (\sigma_{11} - \sigma_{33})^2 + 6(\sigma_{12}^2 + \sigma_{23}^2 + \sigma_{13}^2)} \tag{9.2c}$$

For a principal stress system, the shear stress components σ_{12}, σ_{13} and σ_{23} are absent and eq(9.2c) reduces to

$$\bar{\sigma} = \frac{1}{\sqrt{2}}\sqrt{(\sigma_1 - \sigma_2)^2 + (\sigma_2 - \sigma_3)^2 + (\sigma_1 - \sigma_3)^2} \tag{9.2d}$$

Equations (9.2c,d) are the equivalent stresses used most commonly. Equation (9.2b) provides an alternative $\bar{\sigma}$ expression for principal deviatoric stress space (see Fig. 9.1).

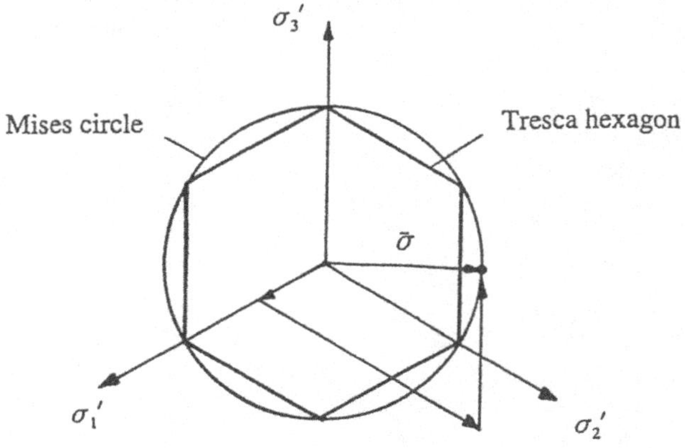

Figure 9.1 The deviatoric plane

Geometrically, eq(9.2b) shows that $\bar{\sigma}$ describes the radius of a circle upon the deviatoric plane. A projection factor of 3/2 allows σ_i' to appear in full, as shown in Fig. 9.1. That is:

$$\bar{\sigma} = \sqrt{\frac{3}{2}\left(\sigma_1'^2 + \sigma_2'^2 + \sigma_3'^2\right)} \tag{9.2e}$$

where $\sigma_1' = \sigma_1 - ⅓ (\sigma_1 + \sigma_2 + \sigma_3)$ etc and $\sigma_1' + \sigma_2' + \sigma_3' = 0$. This circle is the cross-section of the uniform, von Mises cylinder whose axis is equally inclined to orthogonal axes of principal deviatoric stress. Tresca's cylinder, found from eq(3.14c), appears as a hexagonal section inscribed within the von Mises circle upon the deviatoric plane. Since the axes of both cylinders are coincident with the axis of hydrostatic stress, the latter does not affect the radius of the cylinder cross-section, i.e. the equivalent flow stress. However, as $\bar{\sigma}$ increases with plastic hardening, the initial yield surface is taken to expand concentrically by the isotropic hardening rule to contain the current equivalent stress point.

(b) Drucker [3]
Here the yield function is of the form

$$f(J_2', J_3') = J_2'^3 - J_3'^2 \tag{9.3a}$$

where, for a uniaxial equivalent stress, the second and third stress deviator invariants become

$$J_2' = \frac{1}{2}\sigma_{ij}'\sigma_{ij}' = \frac{1}{3}\bar{\sigma}^2 \quad \text{and} \quad J_3' = \frac{1}{3}\sigma_{ij}'\sigma_{jk}'\sigma_{ki}' = \frac{2}{27}\bar{\sigma}^3 \tag{9.3b,c}$$

Combining eqs(9.3a–c), Drucker's function gives

$$f = J_2'^3 - cJ_3'^2 = \left(\frac{\bar{\sigma}^2}{3}\right)^3 - c\left(\frac{2\bar{\sigma}^3}{27}\right)^2$$

from which the equivalent stress is

$$\bar{\sigma} = \left[\frac{J_2'^3 - cJ_3'^2}{(1/3)^3 - c(2/27)^2}\right]^{\frac{1}{6}} \tag{9.4a}$$

$$= \left[\frac{\frac{27}{8}\left(\sigma_{ij}'\sigma_{ij}'\right)^3 - 3c\left(\sigma_{ij}'\sigma_{jk}'\sigma_{ki}'\right)^2}{1 - 4c/27}\right]^{\frac{1}{6}} \tag{9.4b}$$

The general, expanded forms of eq(9.4a,b) are cumbersome but they can be reduced to give $\bar{\sigma}$ for any stress subspace without difficulty. For yielding under torsion, the numerator in eq(9.4a) is simply the shear yield stress $(k^6)^{1/6} = k$. This enables the constant c to be found from the initial tensile yield stress $\bar{\sigma} = Y$, as

$$Y = \frac{k}{\left[(1/3)^3 - c(2/27)^2\right]^{1/6}} \tag{9.4c}$$

It will be shown later in this chapter that by including J_3' in the yield function, f can account for a greater range of experimental observations.

9.2.2 Equivalent Plastic Strain

Equivalent plastic strain expressions may be deduced from the corresponding equivalent stresses. A non-rigorous approach [3] exploits the fact that second order tensors of deviatoric stress and plastic strain increment have identical symmetrical:

$$\mathbf{T}' = \begin{bmatrix} \sigma_{11}' & \sigma_{12}' & \sigma_{13}' \\ \sigma_{21}' & \sigma_{22}' & \sigma_{23}' \\ \sigma_{31}' & \sigma_{32}' & \sigma_{33}' \end{bmatrix}, \qquad \mathrm{d}\mathbf{E}^P = \begin{bmatrix} \mathrm{d}\varepsilon_{11}^P & \mathrm{d}\varepsilon_{12}^P & \mathrm{d}\varepsilon_{13}^P \\ \mathrm{d}\varepsilon_{21}^P & \mathrm{d}\varepsilon_{22}^P & \mathrm{d}\varepsilon_{23}^P \\ \mathrm{d}\varepsilon_{31}^P & \mathrm{d}\varepsilon_{32}^P & \mathrm{d}\varepsilon_{33}^P \end{bmatrix}$$

Correspondence in symmetry between the matrix components gives $\sigma_{ii}' = 0$ (tr $\mathbf{T}' = 0$) and $\mathrm{d}\varepsilon_{ii}^P = 0$ (tr $\mathrm{d}\mathbf{E}^P = 0$). Provided the tensor shear strains are used, i.e: $\mathrm{d}\varepsilon_{12}^P = \mathrm{d}\varepsilon_{21}^P = ½\,\mathrm{d}\gamma_{12}^P$, $\mathrm{d}\varepsilon_{13}^P = \mathrm{d}\varepsilon_{31}^P = ½\,\mathrm{d}\gamma_{13}^P$ and $\mathrm{d}\varepsilon_{23}^P = \mathrm{d}\varepsilon_{32}^P = ½\,\mathrm{d}\gamma_{23}^P$, then the three invariants of deviatoric stress (J_1', J_2', J_3') are identical in their form to those of plastic strain (I_1, I_2, I_3). That is:

$$J_1 = \sigma_{ii}' = \sigma_{11}' + \sigma_{22}' + \sigma_{33}' = 0, \qquad I_1 = \mathrm{d}\varepsilon_{ii}^P = \mathrm{d}\varepsilon_{11}^P + \mathrm{d}\varepsilon_{22}^P + \mathrm{d}\varepsilon_{33}^P = 0$$

$$J_2' = \frac{1}{2}\,\sigma_{ij}'\,\sigma_{ij}', \qquad I_2 = \frac{1}{2}\,\mathrm{d}\varepsilon_{ij}^P\,\mathrm{d}\varepsilon_{ij}^P \tag{9.5a,b}$$

$$J_3' = \frac{1}{3}\,\sigma_{ij}'\,\sigma_{jk}'\,\sigma_{ki}', \qquad I_3 = \frac{1}{3}\,\mathrm{d}\varepsilon_{ij}^P\,\mathrm{d}\varepsilon_{jk}^P\,\mathrm{d}\varepsilon_{ki}^P$$

By employing the similarity between eqs(9.5a,b), the equivalent strain can be deduced directly from the equivalent stress that applies to a given yield function.

(a) von Mises

Let $\mathrm{d}\varepsilon_{11}^P$, $\mathrm{d}\varepsilon_{22}^P$, $\mathrm{d}\varepsilon_{33}^P$, $\mathrm{d}\varepsilon_{12}^P$, $\mathrm{d}\varepsilon_{13}^P$ and $\mathrm{d}\varepsilon_{23}^P$ be the incremental plastic strains arising from a change in a general deviatoric stress state from σ_{ij}' to $\sigma_{ij}' + \mathrm{d}\sigma_{ij}'$. The equivalent plastic strain increment expression is identified with the corresponding change to I_2. Let this change be equivalent to the change in I_2 for uniaxial tension within its three principal plastic strain increments, $\mathrm{d}\varepsilon_1^P$, $-½\mathrm{d}\varepsilon_1^P$ and $-½\mathrm{d}\varepsilon_1^P$, for an isotropic solid. Substituting the three strain components into eq(9.5b) leads to $I_2 = ¾(\mathrm{d}\varepsilon_1^P)^2 = ¾(\mathrm{d}\bar{\varepsilon}^P)^2$, from which

$$\mathrm{d}\bar{\varepsilon}^P = \sqrt{\frac{4I_2}{3}} = \sqrt{\frac{2}{3}\left(\mathrm{d}\varepsilon_{ij}^P\,\mathrm{d}\varepsilon_{ij}^P\right)} = \sqrt{\frac{2}{3}\,\mathrm{tr}\,(\mathrm{d}\mathbf{E}^P)^2} \tag{9.6a}$$

Equation (9.6a) expands into its general six component form

$$\mathrm{d}\bar{\varepsilon}^P = \frac{\sqrt{2}}{3}\Big\{(\mathrm{d}\varepsilon_{11}^P - \mathrm{d}\varepsilon_{22}^P)^2 + (\mathrm{d}\varepsilon_{22}^P - \mathrm{d}\varepsilon_{33}^P)^2 + (\mathrm{d}\varepsilon_{33}^P - \mathrm{d}\varepsilon_{11}^P)^2 + 6\left[\left(½\,\mathrm{d}\gamma_{12}^P\right)^2 + \left(½\mathrm{d}\gamma_{23}^P\right)^2 + \left(½\mathrm{d}\gamma_{31}^P\right)^2\right]\Big\}^{\frac{1}{2}} \tag{9.6b}$$

where $d\varepsilon_{12}^P = ½d\gamma_{12}^P$ etc. The $d\bar{\varepsilon}^P$ expression most widely used employs principal plastic strain increments. That is, for $d\gamma_{12}^P = d\gamma_{13}^P = d\gamma_{23}^P = 0$ in eq(9.6b), which gives

$$d\bar{\varepsilon}^P = \frac{\sqrt{2}}{3}\sqrt{\left(d\varepsilon_1^P - d\varepsilon_2^P\right)^2 + \left(d\varepsilon_2^P - d\varepsilon_3^P\right)^2 + \left(d\varepsilon_3^P - d\varepsilon_1^P\right)^2} \tag{9.6c}$$

An alternative principal strain form follows directly from eq(9.6a):

$$d\bar{\varepsilon}^P = \sqrt{\frac{2}{3}\left[\left(d\varepsilon_1^P\right)^2 + \left(d\varepsilon_2^P\right)^2 + \left(d\varepsilon_3^P\right)^2\right]} = \sqrt{\frac{2}{3}\,\mathrm{tr}\left(d\mathbf{E}^P\right)^2} \tag{9.6d}$$

Note: The incompressibility condition $d\varepsilon_{ii}^P = 0$ connects eq(9.6c) to eq(9.6d).

A difference in the numerical factor arises between the $\bar{\sigma}$ and $d\bar{\varepsilon}^P$ expressions because a uniaxial stress does not produce uniaxial plastic strain. Lateral strains, $d\varepsilon_2^P$ and $d\varepsilon_3^P$, accompany the axial strain $d\varepsilon_1^P$, in which $d\varepsilon_2^P = d\varepsilon_3^P = -½d\varepsilon_1^P$. Substituting this relationship into eqs(9.6c,d), it may be checked that these lead correctly to $d\bar{\varepsilon}^P = d\varepsilon_1^P$, i.e. they ensure that the axial strain becomes the equivalent strain under a uniaxial stress.

The octahedral shear stress and strain increment have also been used as equivalent measures for multi-axial plasticity [4]. The shear stress is defined from eq(1.34b) as

$$\tau_o = \frac{1}{3}\sqrt{(\sigma_1 - \sigma_2)^2 + (\sigma_2 - \sigma_3)^2 + (\sigma_3 - \sigma_1)^2}$$

The octahedral shear strain has a similar form within the interchange between tensor components of stress and strain: $\sigma_1 \to d\varepsilon_1^P$, $\sigma_2 \to d\varepsilon_2^P$, $\sigma_3 \to d\varepsilon_3^P$ and ital $\tau_o \to d\gamma_o^P/2$

$$d\gamma_o^P = \frac{2}{3}\sqrt{\left(d\varepsilon_1^P - d\varepsilon_2^P\right)^2 + \left(d\varepsilon_2^P - d\varepsilon_3^P\right)^2 + \left(d\varepsilon_3^P - d\varepsilon_1^P\right)^2}$$

The reduction to a uniaxial deformation employs the respective substitutions $(\sigma_1, 0, 0)$ and $(d\varepsilon_1^P, -½\, d\varepsilon_1^P, -½\, d\varepsilon_1^P)$. These give $\tau_o = \sqrt{2}\sigma_1/3$ and $d\gamma_o^P = \sqrt{2}d\varepsilon_1^P$, which reveal the general connection between the equivalent and octahedral stress and strain measures:

$$\bar{\sigma} = \frac{3}{\sqrt{2}}\tau_o \qquad \text{and} \qquad d\bar{\varepsilon}^P = \frac{d\gamma_o^P}{\sqrt{2}}$$

(b) Drucker

To correspond with the use of J_3' in defining $\bar{\sigma}$, I_3 is used to define $d\bar{\varepsilon}^P$. Thus, a strain equivalence is sought between changes to the invariant expression $I_2^3 - cI_3^2$ arising from the general and uniaxial systems. The latter has two non-zero invariants, $I_2 = ¾\,(d\varepsilon_1^P)^2$ and $I_3 = ¼\,(d\varepsilon_1^P)^3$, giving

$$I_2^3 - cI_3^2 = \left[\frac{3}{4}\left(d\varepsilon_1^P\right)^2\right]^3 - c\left[\frac{1}{4}\left(d\varepsilon_1^P\right)^3\right]^2$$

Since the axial strain is the equivalent strain:

$$\left(d\varepsilon_1^P\right)^6 = \left(d\bar{\varepsilon}^P\right)^6 = \frac{\left(I_2^3 - cI_3^2\right)}{\left[(3/4)^3 - c(1/4)^2\right]}$$

Substituting for I_2 and I_3 from eq(9.5b), in the general case:

$$\mathrm{d}\bar{\varepsilon}^P = \left[\frac{2(\mathrm{d}\varepsilon_{ij}^P \mathrm{d}\varepsilon_{ij}^P)^3 - (16c/9)(\mathrm{d}\varepsilon_{ij}^P \mathrm{d}\varepsilon_{jk}^P \mathrm{d}\varepsilon_{ki}^P)^2}{27/4 - c} \right]^{\frac{1}{6}} \tag{9.7a}$$

The principal strain form follows from the expansion to eq(9.7a):

$$\mathrm{d}\bar{\varepsilon}^P = \left\{ \frac{2\left[(\mathrm{d}\varepsilon_1^P)^2 + (\mathrm{d}\varepsilon_2^P)^2 + (\mathrm{d}\varepsilon_3^P)^2\right]^3 - (16c/9)\left[(\mathrm{d}\varepsilon_1^P)^3 + (\mathrm{d}\varepsilon_2^P)^3 + (\mathrm{d}\varepsilon_3^P)^3\right]^2}{27/4 - c} \right\}^{\frac{1}{6}} \tag{9.7b}$$

or

$$\mathrm{d}\bar{\varepsilon}^P = \left\{ \frac{2\left[\mathrm{tr}(\mathrm{d}\mathbf{E}^P)^2\right]^3 - (16c/9)\left[\mathrm{tr}(\mathrm{d}\mathbf{E}^P)^3\right]^2}{27/4 - c} \right\}^{\frac{1}{6}} \tag{9.7c}$$

Equivalent strain expressions, corresponding to other isotropic yield functions $f(J_2', J_3')$, may be found in a similar manner. However, literature reviews [5, 6] show an almost exclusive use of von Mises $\mathrm{d}\bar{\varepsilon}^P$ forms, eqs(9.6a–d).

It has been observed [3] that constant c appearing in eqs(9.4a–c) may be sufficient to correlate flow stress data without it re-appearing in the $\mathrm{d}\bar{\varepsilon}^P$ expressions (9.7a–c). However, equivalence is required in plastic strain as well as in stress. For example, the determination of c from eq(9.4c) will only ensure that the initial yield points in tension and torsion assume the same value of equivalent stress. As the material hardens subsequently, there is no guarantee that their equivalent stresses would agree for similar amounts of von Mises equivalent plastic strain, where in torsion $\bar{\varepsilon}^P = \gamma^P/\sqrt{3}$, from eq(9.6b). Consistently, one might seek an agreement in the flow behaviour between the two tests, with $\bar{\varepsilon}^P$ defined from eq(9.7b). That is, substituting $\mathrm{d}\varepsilon_1^P = \frac{1}{2}\mathrm{d}\gamma^P$, $\mathrm{d}\varepsilon_2^P = -\frac{1}{2}\mathrm{d}\gamma^P$ and $\mathrm{d}\varepsilon_3^P = 0$ in eq(9.7b) and then integrating gives $\bar{\varepsilon}^P = \gamma^P/(27 - 4c)^{1/6}$. Experiment can show which approach is preferable but, firstly, it is necessary to examine in more detail the experimental errors that can arise between equivalent flow data. These normally apply to tension, torsion and compression tests. However, alternative tests may be used to extend the range of plastic strain, e.g. plane strain compression and balanced biaxial tension (bulge test).

9.3 Uniaxial Tests

When metallic materials harden with increasing plastic strain, the flow stress increases. Normally, a suitable empirical account of this behaviour is derived from stress–strain curves under uniaxial tension or compression.

9.3.1 Simple Tension

The engineering stress and strain expressions employ the original testpiece area and length. Errors will arise in stress and strain calculations when large changes to testpiece dimensions occur within the plastic range. For example, the ultimate tensile strength of a metal will not be a true fracture stress if no account is made of the reduction in testpiece area that occurs during necking. Moreover, the stress in a material will actually continue to rise, even though the load falls, as the neck develops before final fracture. It is possible to correct the nominal tensile stress–strain curve to reflect this behaviour. The true stress σ and natural strain increment $\delta\varepsilon$ are defined as follows, where A and l are the current section area and gauge length

$$\sigma = W/A \quad \text{and} \quad \delta\varepsilon = \delta l/l \tag{9.8a,b}$$

(a) True Stress and Strain

Unlike elasticity, the plastic deformation that is produced by stressing beyond the elastic limit occurs without volume change. Thus, if A_o and l_o are the original area and length, respectively, the nominal (engineering) stress and strain are

$$\sigma_o = W/A_o \quad \text{and} \quad \varepsilon_o = (l - l_o)/l_o \tag{9.9a,b}$$

The incompressibility (constant volume) condition gives

$$A\,l = A_o l_o \quad \Rightarrow \quad A = \frac{A_o l_o}{l} = \frac{A_o}{1 + \varepsilon_o} \tag{9.10a,b}$$

Substituting eq(9.10b) into eqs(9.8a) leads to the true stress

$$\sigma = \frac{W}{A_o}(1 + \varepsilon_o) = \sigma_o(1 + \varepsilon_o) \tag{9.11}$$

Integrating eq(9.8b) provides a measure of the true, or logarithmic, total strain between the original and current limits of length as

$$\varepsilon = \ln\left(\frac{l}{l_o}\right) = \ln(1 + \varepsilon_o) \tag{9.12a}$$

Removing the elastic strain component $\varepsilon^E = \sigma/E$ from eq(9.12a) leaves the plastic component of strain ε^P

$$\varepsilon^P = \varepsilon - \varepsilon^E = \ln(1 + \varepsilon_o) - \frac{\sigma}{E} \tag{9.12b}$$

Equation (9.11) modifies stress in a nominal curve $\sigma_o = \sigma_o(\varepsilon_o)$, in the manner of Fig. 9.2a. In Figs 9.2b,c, eqs(9.12a,b) have also modified the strain axes to natural (true) strain and natural plastic strain, respectively.

(a)

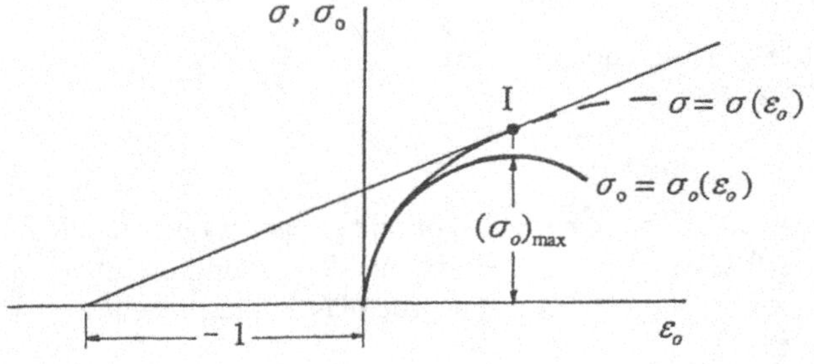

Figure 9.2 Hardening in nominal and true stress–strain axes

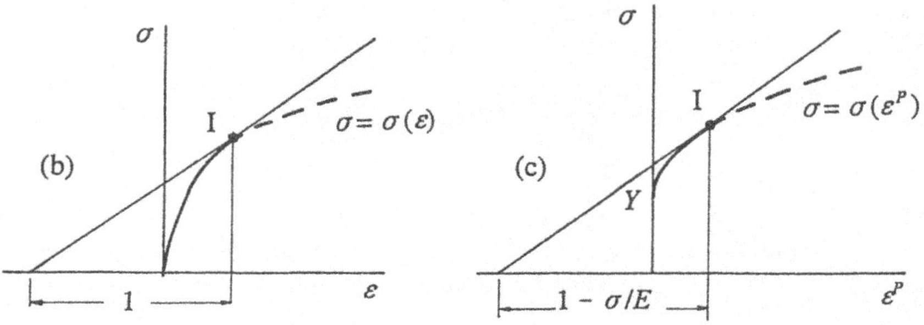

Figure 9.2 Hardening in true stress–strain axes (continued)

Note: It has been shown previously in § 9.2.2 that the axial plastic strain ε^P in tension (Fig. 9.2c) is the equivalent plastic strain.

(b) Necking

One limitation of a tension test is that the range of uniform strain is limited by the formation of a neck. In Considere's construction [7], the true stress at the point of instability corresponds to the maximum ordinate for σ_o in Fig. 9.2a. This condition is expressed as

$$\frac{d\sigma_o}{d\varepsilon_o} = 0 \tag{9.13a}$$

Substituting from eq(9.11) and differentiating the quotient leads to

$$\frac{d\sigma_o}{d\varepsilon_o} = (1 + \varepsilon_o)\frac{d\sigma}{d\varepsilon_o} - \sigma = 0 \quad \Rightarrow \quad \frac{d\sigma}{d\varepsilon_o} = \frac{\sigma}{(1 + \varepsilon_o)} \tag{9.13b}$$

Equation (9.13b) shows that the tangent to a curve with an ordinate of true stress intersects the engineering strain axis at $\varepsilon_o = -1$ (see Fig. 9.2a).. Now, from eq(9.12a):

$$\frac{d\varepsilon}{d\varepsilon_o} = \frac{1}{(1 + \varepsilon_o)} \tag{9.13c}$$

Converting the instability condition (9.13b) to true strain:

$$\frac{d\sigma}{d\varepsilon} = \frac{d\varepsilon_o}{d\varepsilon} \times \frac{d\sigma}{d\varepsilon_o} \tag{9.14a}$$

and substituting eqs(9.13b,c) into eq(9.14a) gives

$$\frac{d\sigma}{d\varepsilon} = \sigma \tag{9.14b}$$

Equation (9.14b) shows that a subtangent value of unity appears on a true, total strain axis in Fig. 9.2b. *Note*: If eq(9.12b) had been used for this derivation instead of eq(9.12a), the instability condition in axes of true stress versus true plastic strain becomes

$$\frac{d\sigma}{d\varepsilon^P} = \frac{\sigma}{(1 - \sigma/E)} \tag{9.14c}$$

Here, the flow curve $\sigma = \sigma(\varepsilon^P)$ commences at the yield stress Y, not at the origin (see Fig. 9.2c). Since there is no appreciable difference between eqs(9.14b,c), a unit sub-tangent is applied to locate the point of instability in these axes.

True stress and true strain are adequate for describing the uniform state of plastic strain in a tensile testpiece up to the start of necking. However, the critical, uniaxial strain is often less than 20 %, even in ductile materials. Continued straining is complicated by the non-uniform, triaxial stress state within the neck. Bridgeman [8] analysed the triaxial state of stress arising in tensile necking of a solid cylinder. He expressed the dependence of radial σ_r and tangential σ_θ stress upon the applied axial stress σ_z. Setting $\sigma_r = \sigma_\theta$ enabled an equivalent stress to be derived for the triaxial stress state shown in Fig. 9.3.

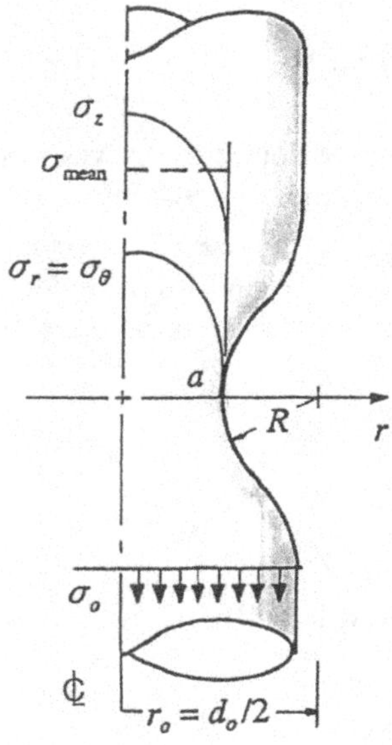

Figure 9.3 Stress state for a tensile neck

According to the von Mises criterion (9.2a), yielding commences in proportion to the root-mean-square of these stresses. From this, Bridgeman found the radial and axial stresses at the narrowest section, with neck diameter $2a$ and radius of curvature R (see Fig. 9.3):

$$\sigma_r = \sigma_\theta = \bar{\sigma} \ln\left(\frac{a^2 + 2aR - r^2}{2aR}\right) \tag{9.15a}$$

$$\sigma_z = \bar{\sigma}\left[1 + \ln\left(\frac{a^2 + 2aR - r^2}{2aR}\right)\right] \tag{9.15b}$$

where $0 \le r \le a$ is an intermediate radius within this section and $\bar{\sigma}$ is the true equivalent stress in the neck. The distribution of σ_r and σ_z is shown in Fig. 9.3. The mean axial stress σ_m in the neck and the axial load W are in equilibrium:

$$W = \pi a^2 \sigma_m = 2\pi \int_0^a r\sigma_z \, \mathrm{d}r$$

Substituting from eq(9.15b) gives

$$\pi a^2 \sigma_m = 2\pi \bar{\sigma} \int_0^a \left[r + r \ln \left(\frac{a^2 + 2aR - r^2}{2ar} \right) \right] \mathrm{d}r$$

$$= 2\pi\bar{\sigma}\left\{ \left(aR + a^2/2 \right) \left[\ln (a^2 + 2aR) - \ln (2aR) \right] \right\}$$

$$= \pi \bar{\sigma} \left[(a^2 + 2aR) \ln \left(1 + \frac{a}{2R} \right) \right]$$

$$\frac{\bar{\sigma}}{\sigma_m} = \frac{1}{(1 + 2R/a) \ln \left[1 + a/(2R) \right]} = \frac{1}{(1 + 1/X) \ln (1 + X)} \qquad (9.16)$$

where $X = a/(2R)$. Bridgeman derived correction curves from eq(9.16) in which a/r_o was plotted against $\bar{\sigma}/\sigma_m$ for a given material (e.g. see Fig. 9.22). Hence, if the smallest neck diameter $d = 2a$ is measured during an incremental loading to W, beyond the point of instability, the true equivalent stress for each load increment may be calculated.

The true strain can also be estimated from the neck diameter d using eq(9.12a) and the constant volume condition $\pi d_o^2 l_o/4 = \pi d^2 l/4$:

$$\varepsilon = \ln \left(\frac{l}{l_o} \right) = 2 \ln \left(\frac{d_o}{d} \right) \qquad (9.17)$$

The dashed lines in Figs 9.2b,c show how Bridgeman's necking-phase corrections extend each true stress–strain curve.

9.3.2 Simple Compression

Larger strains can be achieved from compression tests on short cylinders (see Fig. 9.4a) but unless the ends are well lubricated, a non-uniform deformation arises. Barrelling at the testpiece centre (Fig. 9.4b) is caused by cones of undeformed (elastic) material penetrating inwards from each end, as shown.

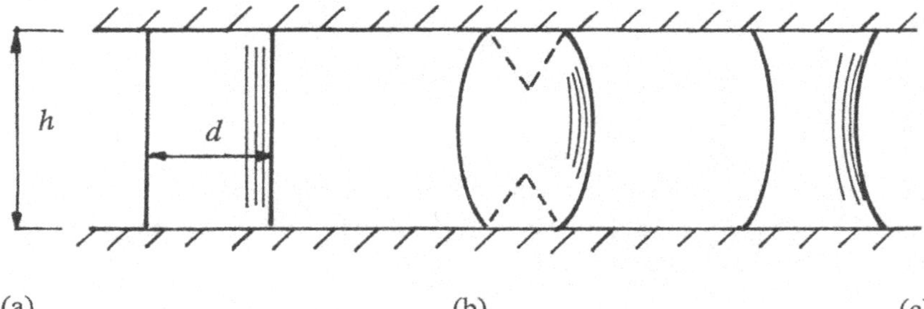

Figure 9.4 Uniform deformation, barrelling and 'bollarding' under compression

The end diameter, in contact with the platens, forms the base of each cone. It follows that any transverse cross-section cutting the cone has an inner elastic core surrounded by an elastic–plastic annulus. When the apex of each cone has not penetrated to half the depth, there exists a fully plastic central region at the barrel's maximum diameter. Within the central region, the stress state is triaxial, being composed of equal radial and circumferential stresses, in addition to the axial compression. It has been shown [9] that an effective stress $\bar{\sigma}$ may be related to the mean axial pressure p in the following form

$$p = \left(1 + \frac{\mu d_o}{3 h_o}\right) \bar{\sigma} \tag{9.18}$$

where d_o and h_o are the cylinder's original diameter and height, respectively. Equation (9.18) shows that a uniaxial stress state $p = \sigma_e$ is achieved from compression only when the friction coefficient $\mu = 0$.

(a) Lubrication

Two methods ensure good lubrication of a short cylinder during compression: (i) to entrap molybdenum disulphide (MoS_2) paste within concentric grooves machined into its ends and (ii) to insert thin sheets of polytetraflouroethylene (ptfe) over the contact areas. The latter is particularly suitable for eliminating barrelling. In fact, with a ptfe thickness of 0.15 mm, the reverse behaviour of central waisting or "bollarding" (see Fig. 9.4c) can occur [10]. This behaviour may be minimised with the use of an optimum ptfe thickness, between 0.05 and 0.15 mm, depending upon testpiece height. To attain large strains in the central region, the method is employed best with fresh sheets applied to the ends between a repeated loading-unloading test sequence.

(b) True Stress and Strain

For a homogoneous compression (Fig. 9.4a), the true strain may be calculated from either expression in eq(9.17), depending upon whether the current height h or diameter d of the cylinder is measured. A minus sign is normally avoided by inverting the height ratio to give the natural compressive strain ε in magnitude

$$|\varepsilon| = \ln\left(h_o / h\right) \tag{9.19a}$$

where $h_o > h$. Similarly, the engineering strain is defined positively:

$$\varepsilon_o = (h_o - h)/h_o \tag{9.19b}$$

Percentage deformation is another measure of compressive strain. It is derived from eq(9.19b) as $\% D = 100\varepsilon_o$. Combining eqs(9.19a,b) leads to

$$|\varepsilon| = \ln\left(\frac{1}{1 - \varepsilon_o}\right) \tag{9.20}$$

The true compressive stress is positive when eq(9.11) is written as

$$\sigma = \frac{|W|}{A} = \frac{|W|}{A_o}(1 - \varepsilon_o) \tag{9.21}$$

where current area $A > A_o$ and $|W|$ is the magnitude of an increasing compressive force. From eqs(9.20) and (9.21), the plastic component of strain is positive when

$$\varepsilon^P = \varepsilon - \varepsilon^E = \ln\left(\frac{1}{1-\varepsilon_o}\right) - \frac{\sigma}{E}$$

$$= \ln\left(\frac{1}{1-\varepsilon_o}\right) - \frac{|W|(1-\varepsilon_o)}{A_o E} \tag{9.22a}$$

Equations (9.21) and (9.22a) allow a true stress versus natural plastic strain curve to be derived from a monotonic loading in compression, when h and W are the measured quantities. Given A_o and h_o are constants, Young's modulus E is derived from the elastic region as

$$E = \frac{\sigma_o}{\varepsilon_o} = \frac{|W|h_o}{A_o(h_o - h)} \tag{9.22b}$$

An alternative to eq(9.22a) applies when the elastic strain is allowed to recover with an unloading from compression. Given h_p as the deformed height, the natural plastic strain is simply

$$\varepsilon^P = \ln\left(\frac{h_o}{h_p}\right) \tag{9.22c}$$

9.4 Torsion Tests

9.4.1 Thin-Walled Tubes

Torsion tests are conducted on thin-walled cylindrical tubes when it is required to minimise the wall shear stress gradient. The theory of torsion [11] gives the mean shear stress as

$$\tau = Tr_m/J \tag{9.23a}$$

where $r_m = \frac{1}{4}(d_i + d_o)$ is the mean wall radius, shown in Fig. 9.5, and J is the polar second moment of area

$$J = \pi(d_o^4 - d_i^4)/32$$

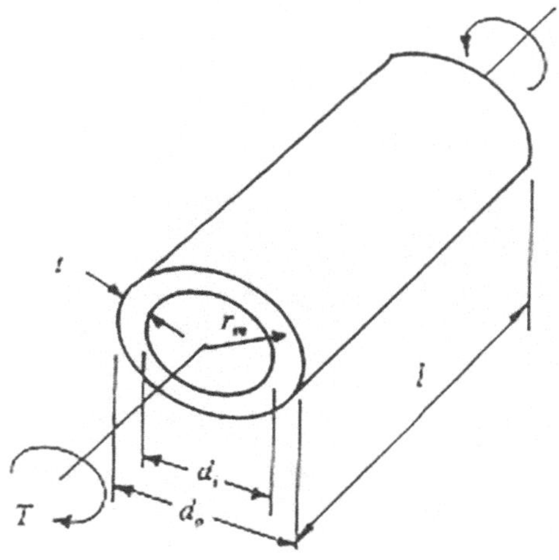

Figure 9.5 Thin-walled tube under torsion

Alternatively, where the wall thickness $t = ½(d_o - d_i)$ is small compared to the mean diameter $d_m = 2r_m$, the Bredt-Batho theory of torsion [11] gives the mean shear stress as

$$\tau = \frac{T}{2At} \tag{9.23b}$$

where $A = \pi r_m^2$ is the area enclosed by the mean wall circle. The engineering shear strain is also referred to the mean radius as

$$\gamma = \frac{r_m \theta}{l} \tag{9.24a}$$

where θ is the angular twist (in radians) over a fixed length l. The plastic component of shear strain is

$$\gamma^P = \gamma - \gamma^E = \gamma - \frac{\tau}{G} \tag{9.24b}$$

Unlike a tension, the torsion of an isotropic tube does not alter its length or cross-section and so a conversion to true stress and strain does not arise. However, large angular twist requires that shear strain be calculated from eq(1.4): $\gamma = \tan\varphi$, where $\varphi = r_m\theta/l$, since the small-strain approximation $\gamma \simeq \varphi$ becomes invalid. The shearing occurs in the z–θ plane (≡1–2) plane in Fig. 9.5, where z (≡1) is axial and θ (≡2) is circumferential. Hence, the non-zero stress and strain components are $\tau \equiv \sigma_{12}$ and $\gamma^P \equiv \varepsilon_{12}^P + \varepsilon_{21}^P = 2\varepsilon_{12}^P$ within eqs(9.2c) and (9.6c). This leads to an equivalent stress and equivalent plastic strain for torsion

$$\bar{\sigma} = \sqrt{3}\tau \quad \text{and} \quad \bar{\varepsilon}^P = \gamma^P/\sqrt{3} \tag{9.25a,b}$$

9.4.2 Solid Bar

Thin-walled tubes are prone to torsional buckling well before fracture, thus limiting the range of shear strain. It is possible to achieve far greater strain from a torsion test on a solid cylindrical bar of hardening material. Nadai's account [12] of the shear stress gradient enables the stress–strain curve for the outer diameter (i.e. τ_o versus γ_o) to be derived from the torque versus unit twist curve (T versus θ/l) in Fig. 9.6.

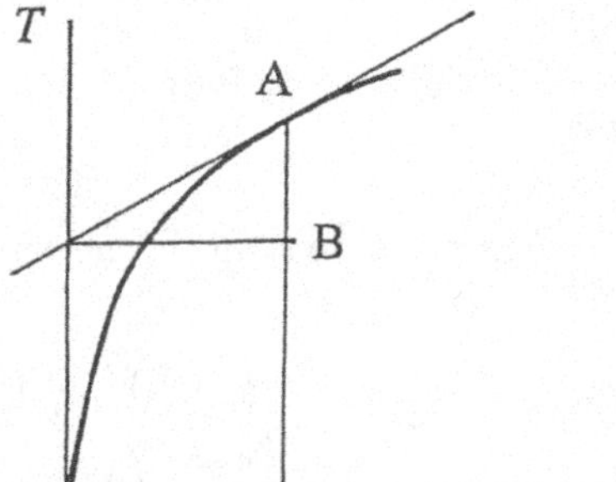

Figure 9.6 Nadai's construction

A solid bar, with outer radius r_o, is in equilibrium with the applied torque T when

$$T = 2\pi \int_0^{r_o} \tau\, r^2\, dr \tag{9.26a}$$

Substituting $\gamma = r\text{æ}$ for $\text{æ} = \theta / l$ = constant into (9.26a) gives

$$T = \frac{2\pi}{\text{æ}^3}\int_0^{\gamma_o} \tau(\gamma)\gamma^2\,\mathrm{d}\gamma \tag{9.26b}$$

where the function $\tau = \tau(\gamma)$ applies at radius r. Differentiating eq(9.26b) with respect to γ_o

$$\frac{1}{2\pi}\frac{\mathrm{d}}{\mathrm{d}\gamma_o}(T\text{æ}^3) = \tau_o(\gamma_o)\gamma_o^2$$

Since $\gamma_o = r_o\text{æ}$, it follows that

$$\frac{1}{2\pi}\frac{\mathrm{d}}{\mathrm{d}\text{æ}}(T\text{æ}^3) = \tau_o(\gamma_o)r_o^3\text{æ}^2$$

giving

$$\tau_o(\gamma_o) = \frac{1}{2\pi r_o^3\text{æ}^2}\times\frac{\mathrm{d}}{\mathrm{d}\text{æ}}(T\text{æ}^3) = \frac{1}{2\pi r_o^3}\left(3T + \text{æ}\frac{\mathrm{d}T}{\mathrm{d}\text{æ}}\right) \tag{9.27a}$$

The geometrical interpretation of eq(9.27a), is given by the Nadai construction in Fig. 9.6, in which

$$\tau_o = \frac{1}{2\pi r_o^3}(3\times\mathrm{AC} + \mathrm{AB}) \tag{9.27b}$$

Nadai's method will establish hardening behaviour at the outer diameter. This takes the form of an equivalent stress versus equivalent plastic strain plot with conversions from τ_o and γ_o to $\bar{\sigma}$ and $\bar{\varepsilon}^P$ from eqs(9.25a,b). For example, assume that the torque-unit twist diagram for a solid bar can be expressed in the form $T = B\text{æ} + A\text{æ}^n$. The linear term represents the elastic line. With $n < 1$, the parabolic term dominates in the plastic range.

The slope to the T versus æ curve is $\mathrm{d}T/\mathrm{d}\text{æ} = B + nA\text{æ}^{n-1}$. Substituting into eq(9.27a):

$$\begin{aligned}\tau_o &= \frac{1}{2\pi r_o^3}\left[3(B\text{æ} + A\text{æ}^n) + \text{æ}(B + nA\text{æ}^{n-1})\right]\\ &= \frac{1}{2\pi r_o^3}\left[A\text{æ}^n(3+n) + 4B\text{æ}\right]\end{aligned}$$

The corresponding shear stress–strain relationship for the outer radius r_o is found from setting $\text{æ} = \gamma_o/r_o$:

$$\begin{aligned}\tau_o &= \frac{2B}{\pi r_o^4}\left[\gamma_o + (3+n)\left(\frac{Ar_o}{4B}\right)\left(\frac{\gamma_o}{r_o}\right)^n\right]\\ &= \frac{B}{J}\left[\gamma_o + \frac{A(3+n)r_o^{1-n}\gamma_o^n}{4B}\right]\end{aligned}$$

where $J = \pi r_o^4/2$. Finally, the equivalent stress versus equivalent plastic strain relationship is found from setting $\bar{\sigma} = \sqrt{3}\tau_o$ and $\bar{\varepsilon}^P = \gamma_o/\sqrt{3} - \tau_o/G$, where $\gamma_o = \text{æ}r_o$.

9.5 Uniaxial and Torsional Equivalence

The equivalent stress and plastic strain, defined in § 9.3 and § 9.4, are employed to correlate hardening behaviour of an isotropic material under various stress states. For example, with perfect lubrication of an isotropic material under compression, the equivalent stress–strain curve ($\bar{\sigma}$ versus $\int \mathrm{d}\bar{\varepsilon}^P$) would expect to coincide with the same for tension and torsion. Later in § 9.9, explicit forms of the strain and work hardening functions H and F are examined within their respective hypotheses:

$$\bar{\sigma} = H\left(\int_{\bar{\varepsilon}^P} \mathrm{d}\bar{\varepsilon}^P\right) = F\left(\int_{\bar{\varepsilon}^P} \bar{\sigma}\,\mathrm{d}\bar{\varepsilon}^P\right)$$

i.e. $\bar{\sigma}$ is a function of equivalent plastic strain or work. The following tests enable the respective functions H and F to determined empirically.

9.5.1 Tension and Compression of Carbon Steel

Consider, firstly, second invariant (von Mises) correlations between tension and compression test data. Figure 9.7 applies to solid circular cylinders of 0.17 % carbon steel with a common diameter of 9.75 mm.

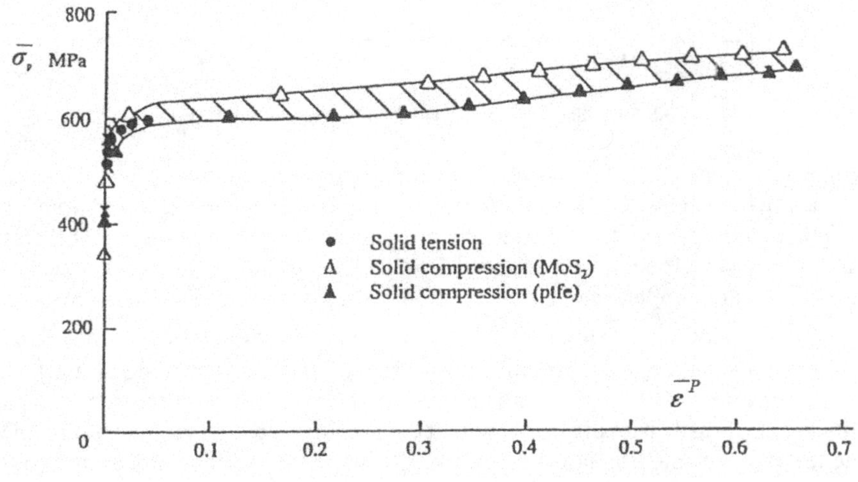

Figure 9.7 Steel flow curves from tension and compression tests on solid bars

Following the application of eqs(9.11), (9.12b), (9.21) and (9.22b) to convert nominal to true stress–strain data, two obvious points arise from Fig. 9.7. Firstly, the tension test upon a solid testpiece (•) is limited to a maximum logarithmic strain, $\varepsilon^P \approx 5$ %, by the formation of a neck. Secondly, large compressive strains to 70 % appear to show (▲) an irregular dependence upon stress. Normally, a smooth, convex curve is observed for a hardening material. Here, MoS_2 was used for a lubricant in producing the smoother upper curve (Δ) seen in Fig. 9.7. For the lower curve, the testpiece was compressed between ptfe sheets to improve lubrication. Both compression tests were unloaded repeatedly to allow measurement of permanent heights and re-lubrication of the platens. Despite the precautions taken, the compressive data for solid cylinders is unlikely to be purely uniaxial.

9.5.2 Tension and Torsion

(a) Mild Steel

Figure 9.8 shows that an equivalence correlation can be found between torsion and tension up to the point of tensile necking. The data applies to similar, tubular mild steel (En3B) testpieces used in each test: 16.5 mm outer diameter and 12.5 mm inner diameter.

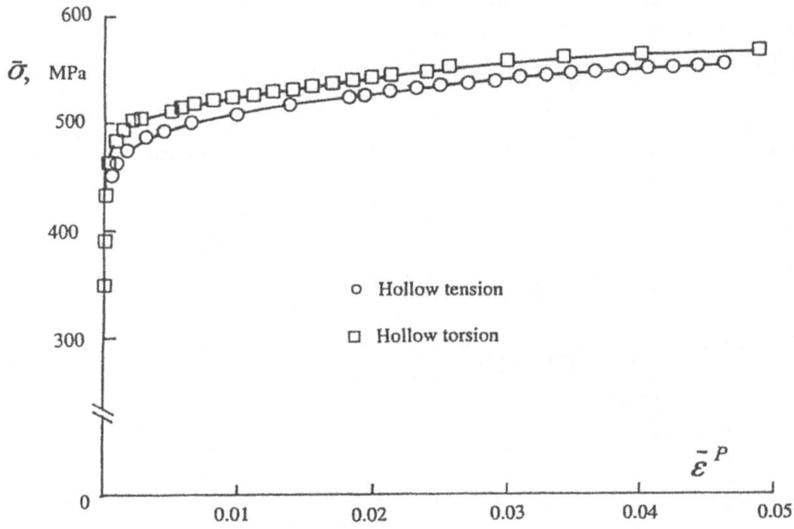

Figure 9.8 Equivalence between tension and torsion for En3B steel

The mean wall formulae (9.23a) and (9.24a,b) were applied to the torsion testpiece for r_m = 7.25 mm, where the angular twist θ was measured over a gauge length l = 50 mm. The same length was employed for the attachment of the tensile extensometer. The elastic modulus in tension and torsion were found as E = 207 GPa and G = 81.6 GPa, respectively. The flow curves are similar, geometrically showing, as expected, a diminishing rate of hardening $\mathrm{d}\bar{\sigma}/\mathrm{d}\bar{\varepsilon}^P$ with increasing strain. A greater equivalent strain of 23 % was achieved from torsion before the testpiece succumbed to buckling. The critical buckling strain greatly exceeds the scale of Fig. 9.8, this being chosen to accommodate a far lesser 5 % tensile instability strain. Tensile necking was apparent though this would be far more pronounced in a solid testpiece. Some triaxiality in stress exists in the tube wall but not at its two free surfaces. Consequently, the stress gradients are far less severe than for a solid section.

(b) Black Mild Steel

A further equivalence correlation between tension and torsion of black mild steel is given in Fig. 9.9. The range of strain has been extended by the respective Bridgeman and Nadai methods as described in § 9.3 and § 9.4. Here, eqs(9.16) and (9.27a) were applied to solid testpiece dimensions: d_o = 12.7 mm for tension and r_o = 12 mm, l = 76.2 mm for torsion. It is seen that Bridgeman's method results in a considerable extension to the tensile strain (○). In fact, the equivalent strain accommodates tensile instability and is comparable with that found from Nadai's torsion (■), i.e. each method extends $\bar{\varepsilon}^P$ to a maximum of 60–70 %. For an additional torsion test (□), conducted on a thin-walled cylinder with r_m = 10.8 mm and t = 2.56 mm, equivalent strain is limited to 25 %.

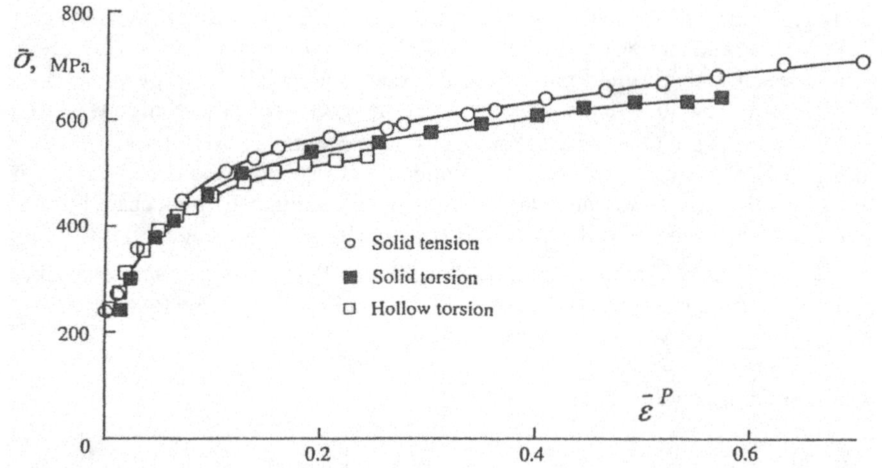

Figure 9.9 Equivalence between tension and torsion for black mild steel

Having applied eqs(9.25a,b), the plot of $\bar{\sigma}$ versus $d\bar{\varepsilon}^P$ reveals that the maximum $\bar{\varepsilon}^P$ is curtailed to 25 % by the onset of torsional instability. This is consistent with the maximum strain found for the En3B tube in Fig. 9.8. A reasonable correlation appears between the three curves though the tensile curve lies consistently above those of torsion over the full range of $\bar{\varepsilon}^P$.

(c) Pure Aluminium

For an isotropic von Mises material, it is expected that second invariants J_2' and I_2 will correlate hardening behaviour under all stress states, i.e. they provide a single curve of $\bar{\sigma}$ versus $\int d\bar{\varepsilon}^P$. In practice, data points from different tests may lie within a narrow scatter band in axes of $\bar{\sigma}$ versus $\bar{\varepsilon}^P$. This suggests that the equivalence may also depend upon the third invariant of stress and plastic strain. For example, Fig. 9.10a shows that eqs(9.25a,b) do not provide an equivalence between tension and torsion for annealed, E1A aluminium cylinders of 99.8 % commercial purity. Figure 9.10b shows an improved correlation where stress equivalence is based upon eq(9.4c). Here, $c = 2\frac{1}{4}$ lies at the lower limit of convexity ($2\frac{1}{4} \le c \le 3\frac{3}{8}$) for a Drucker yield surface.

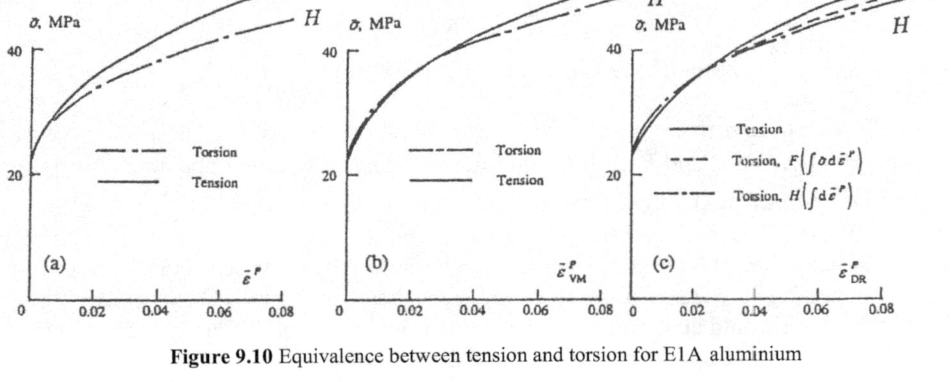

Figure 9.10 Equivalence between tension and torsion for E1A aluminium

Since this material does not show a well-defined yield point, c is calculated from applying eq(9.4c) to tensile and torsional proof stress (Y and k) at equivalent von Mises offset strains. In the absence of a sharp yield point, it is far more convenient to employ a von Mises equivalent offset strain to give the $\bar{\sigma}$ required but the equivalence correlation is of a mixed kind, i.e. for a stress ordinate by Drucker versus plastic strain by von Mises.

In Fig. 9.10c, the strain equivalence is altered corresponding to eq(9.7b). The work hardening hypothesis (F) correlates satisfactorily but the strain hardening chain line (H) for torsion has not improved its overlay with tension, indicating that a change to both definitions $\bar{\sigma}$ and $\bar{\varepsilon}^P$, as in Fig. 9.10b, is unnecessary (see also § 9.9). *Note*: It is possible that the curves in Fig 9.10c can be made to coincide for another c value, found by trial.

9.6 Modified Compression Tests

9.6.1 Cook and Larke

Longer cylinders placed under uniaxial compression are more likely to buckle than to barrel. In order to avoid bucking, Cooke and Larke [13] proposed an experimental method of extrapolating to the true stress in a cylinder of infinite length. Their method required the compression of four cylinders with equal initial diameters d_o and initial heights h_o providing their chosen ratios $d_o/h_o = ½$, 1, 2 and 3.

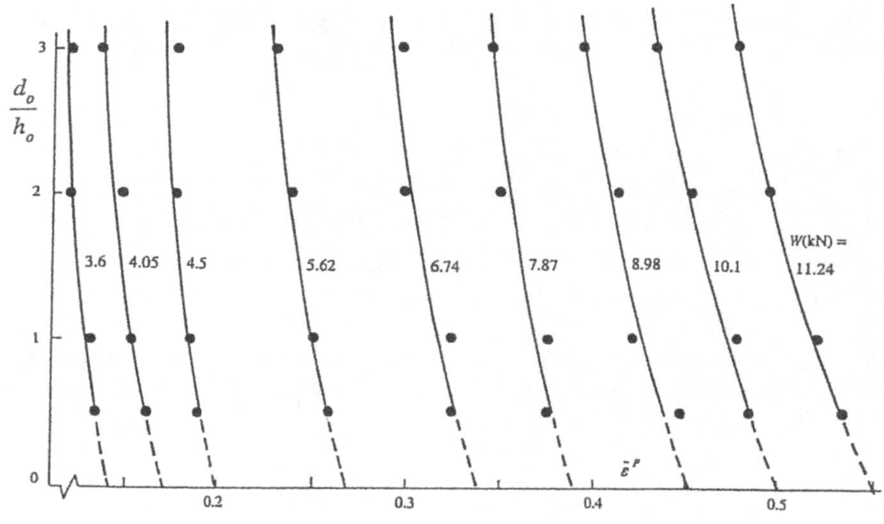

Figure 9.11 Extrapolation to infinite length in compression

Originally, the stress $\bar{\sigma}$ was extrapolated to $d_o/h_o = 0$ for equal percentage height reductions to each cylinder. Watts and Ford [14] modified this test with an extrapolation that gave $\bar{\varepsilon}^P$ for $d_o/h_o = 0$ from applying a series of equal loads to cylinders with similar d_o/h_o ratios (see Fig. 9.11). This requires the calculation of plastic strain from permanent height reductions. Therefore, it is advisable to unload from each of the increased load levels applied to each cylinder. The true stress and strain are calculated from eqs(9.21) and (9.22b). The plastic strain that would arise from applying similar loading to an infinitely long cylinder, i.e. $d_o/h_o = 0$, follows from the extrapolation (dashed lines).

Figure 9.11 applies this method to 99.9 % pure, annealed copper (B.S. 2870, C101). Testpieces were compressed in a sub-press between the platens of a 500 kN test machine. Watts and Ford's convenient experimental modification enables height reductions to be measured precisely and allows re-lubrication between loads. The results from this test also reveal the effect of *d*/*h* on the flow curve, as shown in Fig. 9.12.

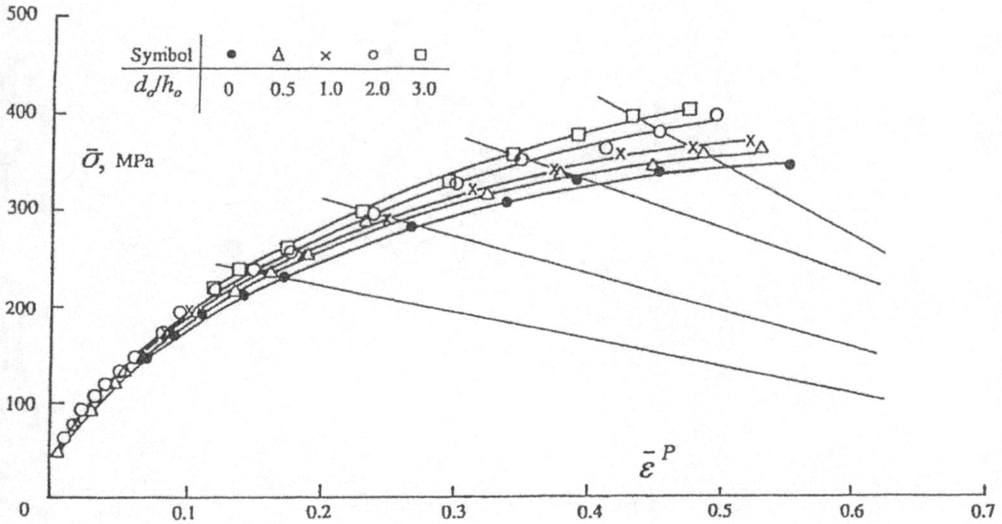

Figure 9.12 Effect of cylinder height upon compressive flow

The true, lower limiting, flow curve applies to a cylinder of infinite height. As the cylinder height is reduced, so the flow behaviour diverges from the true compressive behaviour. For this reason, compressive curves in Fig. 9.7 would be expected to lie above those from tension. In Fig. 9.12, co-ordinates $(\bar{\sigma}, \bar{\varepsilon}^P)$ under constant load lie on the straight lines shown. The equation of each line is 'equivalent' to eq(9.21):

$$\bar{\sigma} = \frac{|W|}{A_o}\left(1 - \bar{\varepsilon}^P\right)$$

where A_o is the section's original area. Since W is constant, each line connects a common point of unity on the plastic strain axis to the ordinate $\bar{\sigma} = |W|/A_o$. In the following section, the extrapolated curve for $d_o/h_o = 0$ (see Fig. 9.12) will be compared to the flow curve obtained from a plane-strain compression test on this material.

9.6.2 Plane Strain Compression

In the plane-strain compression test (see Fig. 9.13), strain in the direction of the width w is prevented by the unstressed material on either side of the die. When the die contact faces are polished and lubricated, Fig. 6.50 shows that the following conditions avoid a widely fluctuating die pressure to give a repeatable flow behaviour [15]:

$$2 \le \frac{b}{t} \le 4 \quad \text{and} \quad 5 \le \frac{w}{b} \le 12 \qquad (9.28a,b)$$

The force F may be applied in increments to produce thickness reductions between 1 and 2 %, though they may be as large as 10–15 %. Here, a die change to smaller breadth b will avoid violating eqs(9.28a,b) following large amounts of deformation. The variable to be measured is simply the current thickness t, corresponding to an unloading from a given F.

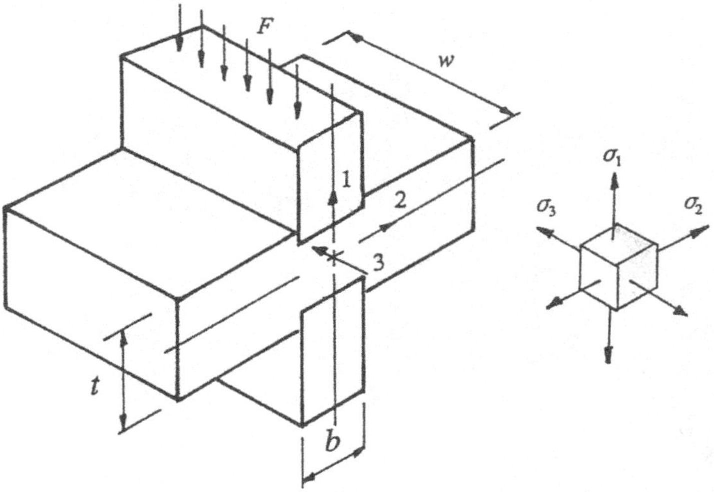

Figure 9.13 Plane strain compression test

(a) Theory

Equivalent stress and strain expressions follow from the Levy-Mises flow rule, i.e. by setting $f = J_2'$ in eq(3.22). For a point beneath the die, this rule gives the principal, plastic strain increments for the thickness, length and width directions 1, 2 and 3, respectively:

$$d\varepsilon_1{}^P = \tfrac{2}{3}\, d\lambda\, [\sigma_1 - \tfrac{1}{2}(\sigma_2 + \sigma_3)] \tag{9.29a}$$

$$d\varepsilon_2{}^P = \tfrac{2}{3}\, d\lambda\, [\sigma_2 - \tfrac{1}{2}(\sigma_3 + \sigma_1)] \tag{9.29b}$$

$$d\varepsilon_3{}^P = \tfrac{2}{3}\, d\lambda\, [\sigma_3 - \tfrac{1}{2}(\sigma_1 + \sigma_2)] \tag{9.29c}$$

The co-ordinates are shown in Fig. 9.13. Given $d\varepsilon_3^P = 0$ for plane strain and with no longitudinal constraint, then $\sigma_2 = 0$. From eq(9.29c), $\sigma_3 = \tfrac{1}{2}\sigma_1$, where the compressive stress is $\sigma_1 = -F/(bw)$. Substituting into eq(9.2d) gives the equivalent stress

$$\begin{aligned}\bar{\sigma} &= \frac{1}{\sqrt{2}}\sqrt{(\sigma_1 - 0)^2 + (0 - \sigma_1/2)^2 + (\sigma_1 - \sigma_1/2)^2} \\ &= \frac{\sqrt{3}}{2}\sigma_1 = -\frac{\sqrt{3}}{2}\left(\frac{F}{bw}\right)\end{aligned} \tag{9.30}$$

Similar substitutions into eqs(9.29a,b) result in principal, plastic strain increments

$$d\varepsilon_1^P = \frac{d\lambda}{2}\sigma_1, \quad d\varepsilon_2^P = -\frac{d\lambda}{2}\sigma_1, \quad \therefore\ d\varepsilon_2^P = -\, d\varepsilon_1^P \tag{9.31a,b,c}$$

Substituting eq(9.31c) into eq(9.6d) gives the equivalent plastic strain increment

$$d\bar{\varepsilon}^P = \sqrt{\frac{2}{3}\left[\left(d\varepsilon_1^P\right)^2 + \left(-d\varepsilon_1^P\right)^2 + (0)^2\right]}$$

$$= \frac{2}{\surd 3}\, d\varepsilon_1^P = \frac{2}{\surd 3}\frac{dt}{t}$$

Integrating for the total plastic strain:

$$\bar{\varepsilon}^P = \frac{2}{\surd 3}\ln\left(\frac{t}{t_o}\right) = -\frac{2}{\surd 3}\ln\left(\frac{t_o}{t}\right) \qquad (9.32b)$$

The minus signs in eqs(9.30) and (9.32b) are dropped in the plot of $\bar{\sigma}$ versus $\bar{\varepsilon}^P$. This is consistent with having taken a negative square root in the $\bar{\sigma}$ and $\bar{\varepsilon}^P$ expressions.

(b) Test Results

A plane-strain compression test was conducted on annealed, pure copper plate of initial dimensions $w = 50$ mm and $t_o = 3.65$ mm. Die breadths $b = 6.37$, 4.12 and 3.19 mm were selected to ensure $b/t < 4$ throughout deformation to a final, 84 % reduction in thickness. Loads were applied with MoS_2 lubricant applied to each die face. To prevent misalignment, the dies were guided vertically in a jig. The permanently deformed thickness t was measured following unloading from each of the rising loads beyond 10 kN as indicated in Table 9.1. This table also shows the permanent thickness t under each load and their conversions into equivalent stress and strain from eqs(9.30b) and (9.32b).

Table 9.1 Plane strain compression of annealed copper

$b = 6.37$ mm, $t_o = 3.65$ mm, $\bar{\varepsilon}_o^P = 0$

F, kN	10	25	40	50	60	70	80	90	100
t, mm	3.625	3.585	3.53	3.475	3.42	3.35	3.26	3.15	3.015
$\bar{\sigma}$, MPa	27.4	68.6	109.7	137.1	164.5	191.94	219.4	246.8	274.2
$\bar{\varepsilon}^P$	0.008	0.0203	0.0385	0.0563	0.0739	0.0995	0.131	0.170	0.220

continued

F, kN	110	120	125	130	135	140	145
t, mm	2.815	2.575	2.40	2.18	2.01	1.82	1.65
$\bar{\sigma}$, MPa	301.6	329.1	342.8	356.5	370.2	383.9	397.6
$\bar{\varepsilon}^P$	0.301	0.403	0.484	0.595	0.689	0.804	0.917

$b = 4.12$ mm, $t_o = 1.65$ mm, $\bar{\varepsilon}_o^P = 0.917$

F, kN	97.5	100	102.5	105	107.5
t, mm	1.50	1.35	1.24	1.14	1.05
$\bar{\sigma}$, MPa	413.4	424.0	434.6	445.2	455.8
$\bar{\varepsilon}^P$	1.027	1.149	1.247	1.344	1.439

$b = 3.19$ mm, $t_o = 1.05$ mm, $\bar{\varepsilon}_o^P = 1.439$

F, kN	82.5	85	87.5	92.5	98.75
t, mm	0.94	0.87	0.81	0.69	0.58
$\bar{\sigma}$, MPa	451.7	465.4	479.1	506.5	540.7
$\bar{\varepsilon}^P$	1.567	1.656	1.738	1.924	2.124

The equivalent flow curve, as derived from eqs(9.30) and (9.32b), is shown in Fig. 9.14.

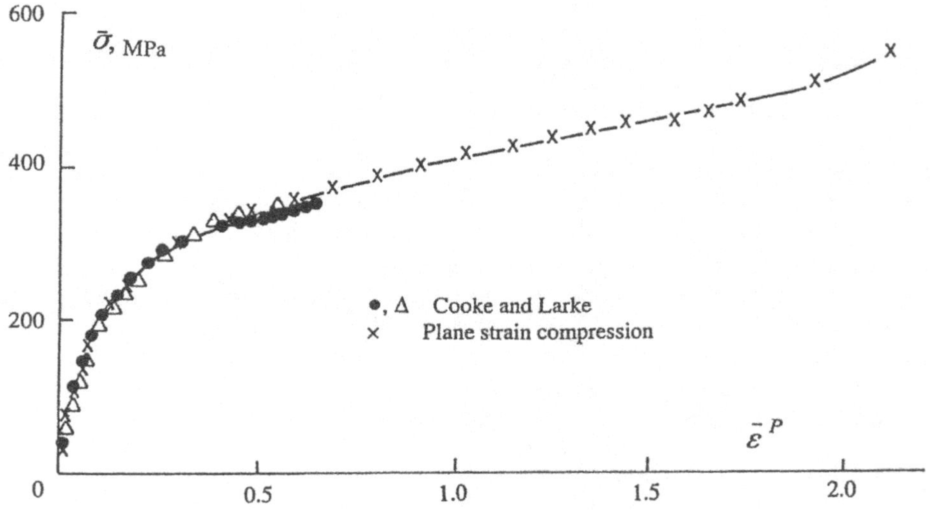

Figure 9.14 Compressive flow under plane strain and from C & L extrapolation

9.6.3 Equivalence in Compression Tests

It is seen that a very large equivalent plastic strain, beyond 200 %, accumulates from the plane-strain compression test with die interchange. Figure 9.14 shows that this magnitude of strain is far more than can be achieved from the maximum extrapolated strain ≃ 60 %, obtained from the Cook and Larke method. It is seen that within a comparable range of strain, the flow behaviour for annealed copper, derived by these two methods, is almost coincident.

Results from the two Cook and Larke experiments (•, ▵) in Fig. 9.14 were obtained from different test machines but this has not altered the material's flow behaviour as expected from a true material property. The flow behaviour of copper is, however, very sensitive to its initial condition. In the absence of annealing as-received copper, Fig. 9.15 shows that the flow stresses are higher while the rate of hardening is initially lower. Beyond 30 % strain, all flow curves begin to converge as each structure attains a comparable amount of slip and twinning. Figure 9.15 shows further that there is reasonable consistency in plane strain and direct compression when the as-received copper is lubricated with MoS_2 and ptfe. *Note*: Plane strain provides a smooth curve over a wider strain range without die interaction. The latter is evident when a concave curvature appears in the flow curve (see Fig. 9.14) .

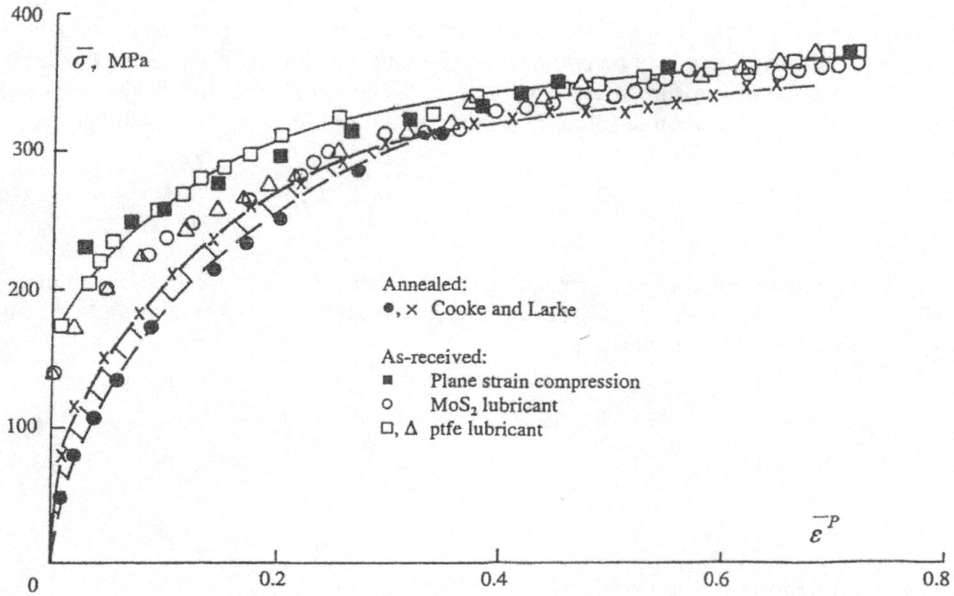

Figure 9.15 Superimposed compressive flow curves for copper

9.7 Bulge Test

Moderately large plastic strains can be achieved from a balanced biaxial tension. In practice, this condition prevails at the pole in a bulge test. Hydraulic pressure is applied to the underside of a thin circular plate that is clamped and sealed around its periphery. Alternatively, the sheet is forced by the pressure into an open circular die to form the bulge. The pole flow behaviour is calculated from the current pole thickness t and deflection h at the centre of the bulge, where the pole radius is R, as shown in Fig. 9.16.

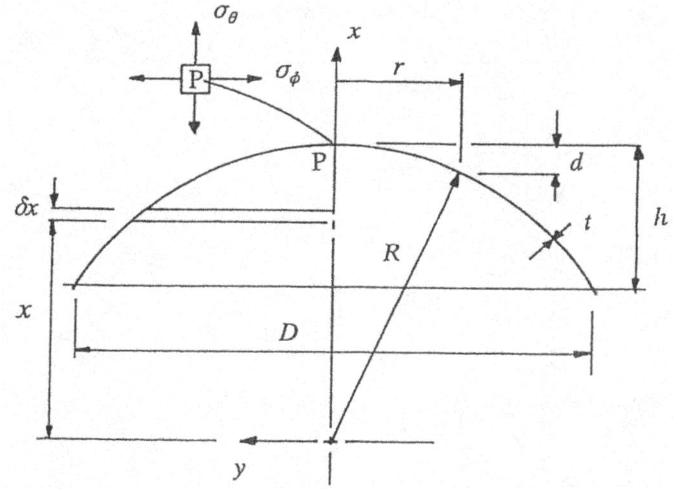

Figure 9.16 Geometry of the bulge test

9.7.1 Theory

Provided the plate is thin, compared to its diameter D, giving $D/t > 100$, the stresses due to bending and shear force may be ignored. Neglecting also any radial stress σ_r, this leaves only the membrane pressure stresses σ_θ and σ_φ. At the pole P, the meridional stress σ_φ is found from a consideration of force equilibrium in the x–direction when the bulge section is semi-circular. This gives

$$2\pi R t \sigma_\varphi = p\pi R^2 \qquad \therefore \ \sigma_\varphi = \frac{pR}{2t} \tag{9.33a}$$

Similarly, equilibrium for the y–direction gives the hoop stress $\sigma_\theta = pR/2t$, showing that an equi-biaxial stress state $\sigma_\theta = \sigma_\varphi$ exists everywhere. Substituting eq(9.33a) into eq(9.2d), with $\sigma_r = 0$, leads to an equivalent stress

$$\bar{\sigma}^2 = \sigma_\theta^2 - \sigma_\theta \sigma_\varphi + \sigma_\varphi^2$$

$$\bar{\sigma} = \sigma_\varphi = \frac{pR}{2t} \tag{9.33b}$$

where R and t are current pole values. The Levy-Mises flow rule $\mathrm{d}\varepsilon_{ij}^P = \mathrm{d}\lambda \sigma_{ij}'$ gives the three plastic strain increments as

$$\mathrm{d}\varepsilon_\theta^P = \tfrac{2}{3}\,\mathrm{d}\lambda\,[\sigma_\theta - \tfrac{1}{2}(\sigma_\varphi + \sigma_r)] = \tfrac{1}{3}\,\mathrm{d}\lambda\,\sigma_\theta \tag{9.34a}$$

$$\mathrm{d}\varepsilon_\varphi^P = \tfrac{2}{3}\,\mathrm{d}\lambda\,[\sigma_\varphi - \tfrac{1}{2}(\sigma_r + \sigma_\theta)] = \tfrac{1}{3}\,\mathrm{d}\lambda\,\sigma_\theta \tag{9.34b}$$

$$\mathrm{d}\varepsilon_r^P = \tfrac{2}{3}\,\mathrm{d}\lambda\,[\sigma_r - \tfrac{1}{2}(\sigma_\theta + \sigma_\varphi)] = -\tfrac{2}{3}\,\mathrm{d}\lambda\,\sigma_\theta \tag{9.34c}$$

and from eqs(9.34a–c):

$$\mathrm{d}\varepsilon_\theta^P = \mathrm{d}\varepsilon_\varphi^P \quad \text{and} \quad \mathrm{d}\varepsilon_r^P = -\,2\,\mathrm{d}\varepsilon_\theta^P \tag{9.35a,b}$$

which conform to the constant volume condition $\mathrm{d}\varepsilon_{kk}^P = 0$. Substituting eqs(9.35a,b) into eq(9.6d) gives the equivalent plastic strain increment

$$\begin{aligned}\mathrm{d}\bar{\varepsilon}^P &= \sqrt{\frac{2}{3}\left[(\mathrm{d}\varepsilon_\theta^P)^2 + (\mathrm{d}\varepsilon_\varphi^P)^2 + (\mathrm{d}\varepsilon_r^P)^2\right]} \\ &= \sqrt{\frac{2}{3}\left[(-\tfrac{1}{2}\,\mathrm{d}\varepsilon_r^P)^2 + (-\tfrac{1}{2}\,\mathrm{d}\varepsilon_r^P)^2 + (\mathrm{d}\varepsilon_r^P)^2\right]} = \mathrm{d}\varepsilon_r^P\end{aligned} \tag{9.36a}$$

The radial strain is defined as

$$\varepsilon_r^P = \int_{t_o}^{t} \frac{\mathrm{d}t}{t} = \ln\left(\frac{t}{t_o}\right) = -\ln\left(\frac{t_o}{t}\right) \tag{9.36b}$$

where t_o is the initial sheet thickness. From eqs(9.36a,b):

$$\bar{\varepsilon}^P = -\ln\left(\frac{t_o}{t}\right) \tag{9.36c}$$

The negative sign may be omitted when the negative root of eq(9.36a) is taken. It follows from eqs(9.33b) and (9.36c) that to construct the plot between $\bar{\sigma}$ and $\int d\bar{\varepsilon}^P$, the current values of the variables p, R and t are required. The radius R is calculated from the geometry in Fig. 9.16:

$$R^2 = r^2 + (R - d)^2 \;\Rightarrow\; R = \frac{(r^2 + d^2)}{2d} \tag{9.37a}$$

In early bulge tests [16, 17], r and d in eq(9.37a) were determined from dimensional changes to concentric circles scribed on the undeformed plate. The pole radius R was found by graphical extrapolation to $r \simeq 0$. Equation (9.35b) shows that $dt/t = -2dr/r$, indicating that thickness could also be calculated from this extrapolation:

$$\int_{r_o}^{r} \frac{dr}{r} = -\frac{1}{2}\int_{t_o}^{t} \frac{dt}{t} \quad\Rightarrow\quad t = \frac{t_o}{(r/r_o)^2} \tag{9.37b}$$

The left-hand integral in eq(9.37b) becomes the hoop strain when extrapolated to $r_o = 0$.

9.7.2 Instrumented Test

In an instrumented bulge test [18], transducers are incorporated within a spherometer to measure t at the pole and h above a horizontal plane of diameter D (Fig. 9.16). In addition to a pressure transducer, a flow meter, placed within the pressure line, provides the volume of pressurising fluid beneath the dome. This volume can be expressed in terms of the spherical dome radius R. Revolving the elemental strip in Fig. 9.16 about the x–axis gives $\delta V = \pi y^2 \delta x$. Substituting $y^2 = R^2 - x^2$ and integrating between limits of x gives

$$\begin{aligned} V &= \pi \int_{R-h}^{R} (R^2 - x^2)\, dx = \pi \left| R^2 x - \frac{x^3}{3} \right|_{R-h}^{R} \\ &= \pi h^2 \left(R - \frac{h}{3} \right) \quad\Rightarrow\quad R = \frac{V}{\pi h^2} + \frac{h}{3} \end{aligned} \tag{9.38a}$$

Equating R between eqs(9.38a) to (9.37a), with $r = D/2$ and $d = h$, leads to

$$\frac{V}{\pi h^2} + \frac{h}{3} = \frac{[(D/2)^2 + h^2]}{2h} \tag{9.38b}$$

Equation (9.38b) gives h as one root to the cubic equation

$$h^3 + \frac{3}{4} D h - \frac{6V}{\pi} = 0 \tag{9.38c}$$

Equation (9.38c) shows that with V measurement it is possible theoretically to dispense with pole height h but this is not advisable when deviations from a spherical dome occur, particularly in the region of the clamped rim. Such deviation leads to greater error in the calculation of h than of R. This is particularly true for bulging orthotropic sheet metals when eqs(9.33b) and (9.36a) appear with a common multiplication factor:

$$\bar{\sigma} = K\sigma_\varphi \quad \text{and} \quad d\bar{\varepsilon}^P = d\varepsilon_r^P / K$$

where K depends upon the anisotropy coefficients r_1 and r_2 for the sheet (see § 11.5.2):

$$K^2 = \frac{3(r_1 + r_2)}{2(r_1 + r_1 r_2 + r_2)}$$

9.7.3 Equivalence with Torsion

Figure 9.17 compares equivalent flow curves for a bulge test and a tubular torsion test for annealed brass of B. S. 2874 composition: 40 % Zn, 57 % Cu, 3 % Pb.

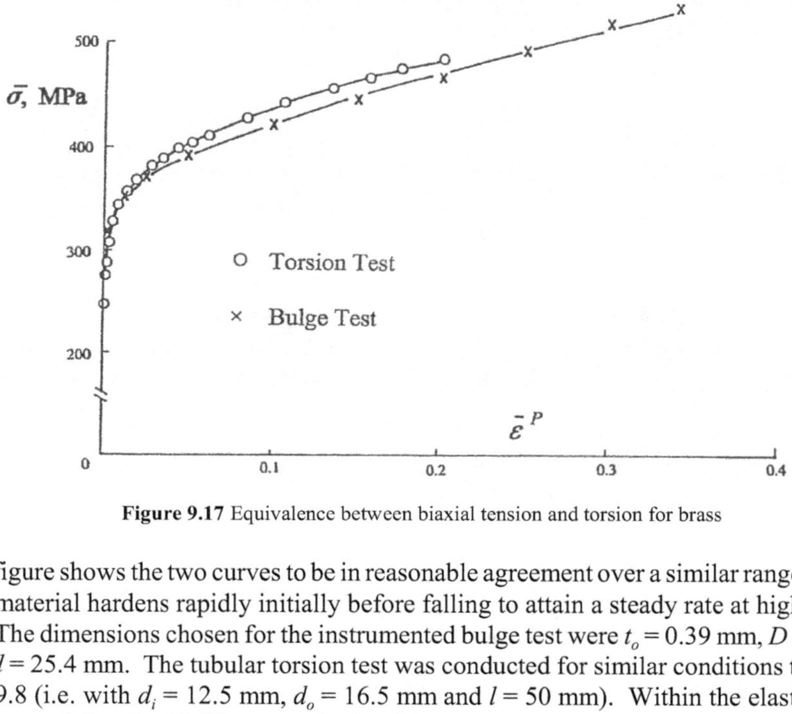

Figure 9.17 Equivalence between biaxial tension and torsion for brass

The figure shows the two curves to be in reasonable agreement over a similar range of strain. The material hardens rapidly initially before falling to attain a steady rate at higher strain.

The dimensions chosen for the instrumented bulge test were $t_o = 0.39$ mm, $D = 100$ mm and $d = 25.4$ mm. The tubular torsion test was conducted for similar conditions to those in Fig. 9.8 (i.e. with $d_i = 12.5$ mm, $d_o = 16.5$ mm and $l = 50$ mm). Within the elastic region, a shear modulus $G = 39$ GPa was calculated from the torque and twist measurements made. The elasticity modulus, E = 102 GPa, was calculated from $E = 2G(1 + \nu)$, assuming Poisson's ratio $\nu = ⅓$. Under continuous loading, E and G are used to subtract the elastic components of strain from the total strain in each test. In torsion, eq(9.25b) is applied to give the equivalent plastic strain. The strain at the point of torsional failure was $\bar{\varepsilon}^P \simeq 20$ %, corresponding to 140° twist. *Note*: This strain is consistent with a tensile elongation of 20–30 % for this material. The equivalent plastic strain under rising pressure in a bulge test modifies eq(9.36c) to

$$\bar{\varepsilon}^P = -\left[\ln\left(\frac{t_o}{t}\right) - \frac{\bar{\sigma}}{E}\right]$$

where $\bar{\sigma}$ is defined in eq(9.33b). Strictly, eq(9.36c) applies when t is measured following a pressure release from the plastic range. Here, however, t is sensed under an increasing pressure, and so elastic strain is included. The application of eq(9.36c) to estimate $\bar{\varepsilon}^P$ would involve negligible error due to elastic strain $\bar{\sigma}/E$. The advantage of the bulge test is that a greater fracture strain is reached compared to torsion. For example, in Fig. 9.17, the bulge would extend the strain to 64 %.

9.8 Equations to the Flow Curve

Most empirical representations [9, 19] of the uniaxial flow curve obey the strain hardening function: $\sigma = H(\int d\varepsilon^P)$. The corresponding non-dimensional forms of the functions H commonly used are shown in Figs 9.18a–f. Here, σ_o is the yield stress and ε_o is a constant.

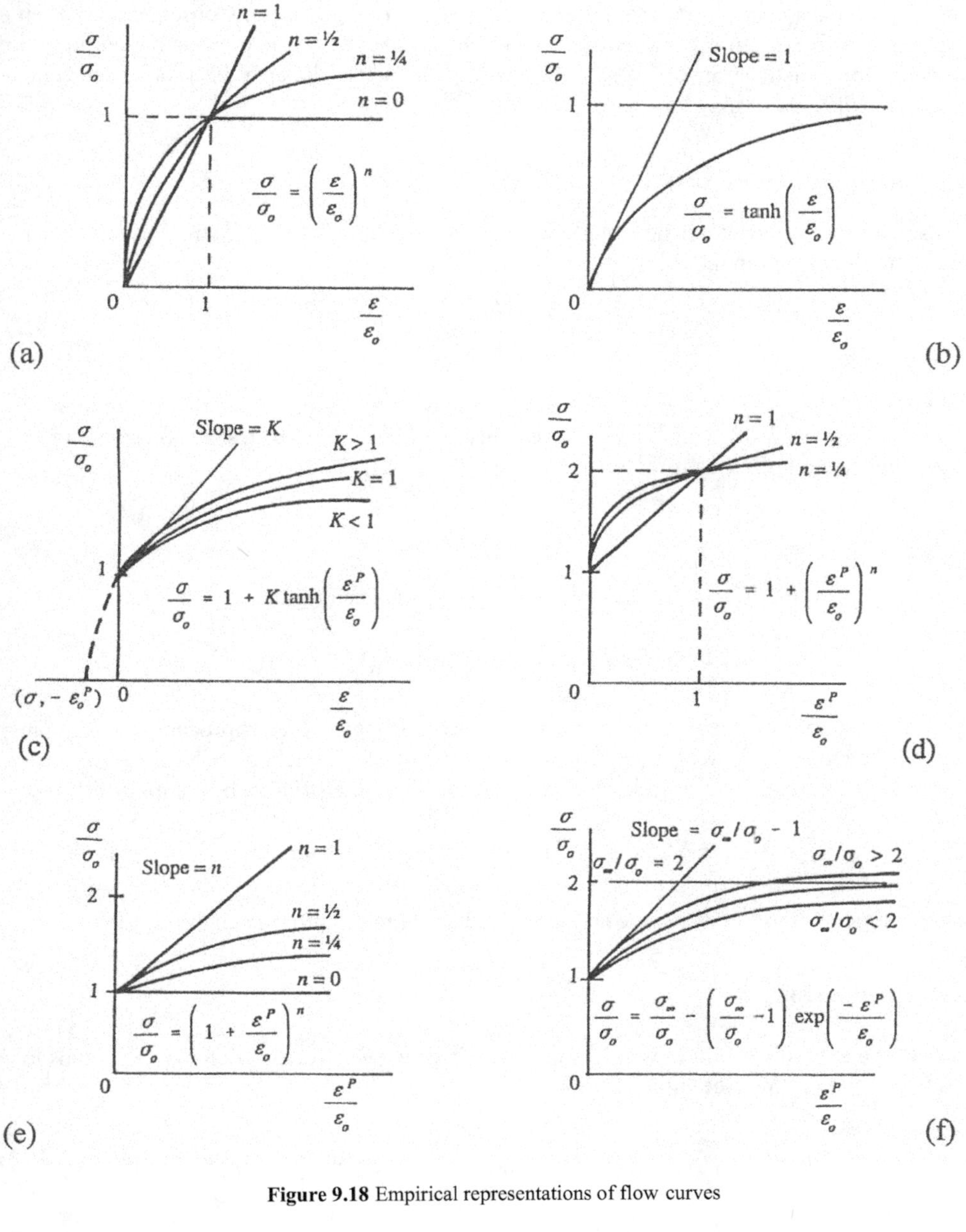

Figure 9.18 Empirical representations of flow curves

9.8.1 Hollomon [20]

The simplest, most popular function is the Hollomon power law (see Fig. 9.18a):

$$\sigma = A\,\varepsilon^{n} \quad \text{or} \quad \frac{\sigma}{\sigma_o} = \left(\frac{\varepsilon}{\varepsilon_o}\right)^{n} \tag{9.39a,b}$$

where A ($= \sigma_o/\varepsilon_o^{\,n}$) is a strength coefficient and n is a strain hardening exponent, which lies in the range $0.1 \le n \le 0.55$ for metals. Because eqs(9.39a,b) pass through the origin, strictly, they should not be applied to any curve with an intercept on the stress axis. When the curve's abscissa is plastic strain ε^P, this intercept becomes the yield stress. Consequently, when eqs(9.39a,b) are to represent a plastic flow curve $\sigma = H(\int d\varepsilon^P)$, the greatest deviation will arise for low plastic strains $\varepsilon^P < 0.2$ %. For $\varepsilon^P > 5$ %, where the error becomes negligible, then ε in eq(9.39a,b) may become ε^P reliably.

9.8.2 Ramberg-Osgood [21]

This well-known law describes total strain with the addition of elastic strain to eq(9.39a). It is usually written in the following form:

$$\varepsilon^{t} = \frac{\sigma}{E} + \left(\frac{\sigma}{E_o}\right)^{m} \tag{9.40}$$

where E is Young's modulus and E_o and m are plastic constants. As with any continuous function, H in eq(9.40) is unable to represent discontinuous flow curves, e.g. one with upper and lower yield points, as for a low carbon steel.

9.8.3 Prager [22]

Prager's hyperbolic flow curve is shown in Fig. 9.18b. The original expressions

$$\sigma = Y\tanh\left(E\varepsilon/Y\right) \quad \text{or} \quad \sigma/\sigma_o = \tanh\left(\varepsilon/\varepsilon_o\right) \tag{9.41a,b}$$

describe a curve originating from the origin, with slope E, that approaches rapidly an asymptote $\sigma/\sigma_o = 1$. As with eqs(9.39a,b), no yield point is identified. Where a hyperbolic expression originates at the yield stress ($\sigma = \sigma_o$ for $\varepsilon^P = 0$), eqs(9.41a,b) are modified:

$$\sigma = Y + K\tanh\left(E\,\varepsilon/Y\right) \quad \text{or} \quad \sigma/\sigma_o = 1 + K\tanh\left(\varepsilon/\varepsilon_o\right) \tag{9.42a,b}$$

where, as shown in Fig. 9.18c, K may be varied as shown to provide an asymptote.

9.8.4 Ludwik [23]

The simplest modification to eqs(9.39a,b), which give $\sigma = \sigma_o$ (or $\sigma = Y$) for $\varepsilon^P = 0$, leads to Ludwik's power law (see Fig. 9.18d):

$$\sigma = Y + A\left(\varepsilon^{P}\right)^{n} \quad \text{or} \quad \frac{\sigma}{\sigma_o} = 1 + \left(\frac{\varepsilon^{P}}{\varepsilon_o}\right)^{n} \tag{9.43a,b}$$

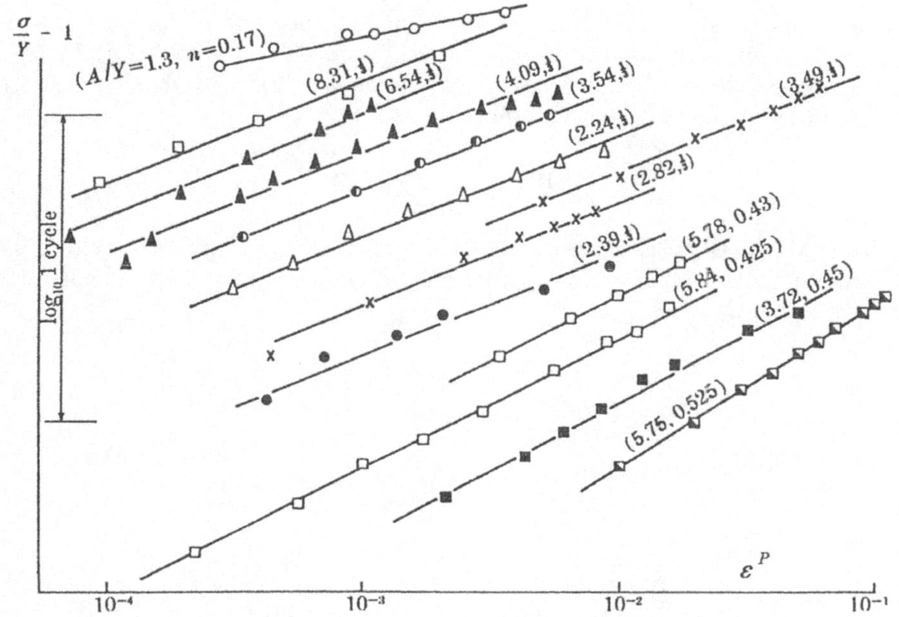

Figure 9.19 Application of the Ludwik law

Key: O, Δ steels; □,O, Δ, ● Al -alloys; × brass; ■ Ti-alloy, □ Mg-alloy (Points separated for clarity)

Equations (9.43a,b) displays parabolic strain hardening in which $\sigma > Y$ for $n < 1$. A and n are empirical constants, similar to those in Hollomon's law. It will be shown in chapter 10 that eq(9.43a) is particularly useful to the theory of plasticity, allowing incremental strain expressions to be integrated for the majority of loading paths. In particular, when $n = ⅓$, closed solutions apply. Ludwik's logarithmic plot (see Fig. 9.19), identifies n and A in eq(9.43a) with the slope and intercept, respectively, of the resulting straight line. Figure 9.19 shows that the flow behaviour of many engineering alloys in tension and torsion conforms to eq(9.43a) for plastic strains up to 10 % [24]. Some initial deviation from the indicated constants is found because of a large increase in the rate of hardening $d\sigma/d\varepsilon^P$ as $\varepsilon^P \to 0$. When $\varepsilon^P = 0$, eq(9.43a) gives $d\sigma/d\varepsilon^P = \infty$, in contrast to finite rates observed. Revised constants A and n can improve accuracy for a Ludwik application to a small strain range.

9.8.5 Swift [25]

Swift's power function, illustrated in Fig. 9.18e, represents hardening within three constants

$$\sigma = K\left(A + \varepsilon^P\right)^n \quad \text{or} \quad \frac{\sigma}{\sigma_o} = \left(1 + \frac{\varepsilon^P}{\varepsilon_o}\right)^n \qquad (9.44a,b)$$

In eq(9.44a), constants A, K and n bear a relationship to the yield stress: $Y = KA^n$. The finite hardening rate $d\sigma/d\varepsilon^P = nKA^{n-1}$ for $\varepsilon^P = 0$ is preferable to the infinite rate given by eq(9.43a). However, to separate the constants, one further point is required, either within a mid-range of strain with co-ordinates (σ, ε^P) or at the negative strain intercept (0, $-\varepsilon_o$). Alternatively, the normalised eq(9.44b) gives $\sigma/\sigma_o = 1$ and $\varepsilon^P = 0$ with $d(\sigma/\sigma_o)/d(\varepsilon^P/\varepsilon_o) = n$, this allowing n to be separated from ε_o. For example, half-hard aluminium has been represented [9] with the following form of eq(9.44b):

$$\sigma/105 = \left(1 + \varepsilon^P/0.222\right)^{0.25}$$

with σ in MPa over a plastic strain range $0 \le \varepsilon^P \le 0.4$. In applying eq(9.14c) to eq(9.44b) at the point of instability in a tension test, the true ultimate stress and plastic strain σ_u and ε_u^P are related to n and ε_o as follows:

$$\frac{\sigma_u}{\sigma_o} = \left(\frac{n}{\varepsilon_o}\right)^n \left(1 - \frac{\sigma_u}{E}\right)^n \approx \left(\frac{n}{\varepsilon_o}\right)^n \tag{9.44c}$$

$$\frac{\varepsilon_u^P}{\varepsilon_o} = \frac{n}{\varepsilon_o}\left(1 - \frac{\sigma_u}{E}\right) - 1 \approx \frac{n}{\varepsilon_o} - 1 \tag{9.44d}$$

Sometimes, binomial expansion approximations to eq(9.44b) are useful. For $\varepsilon^P/\varepsilon_o < 1$:

$$\frac{\sigma}{\sigma_o} = 1 + n\left(\frac{\varepsilon}{\varepsilon_o}\right) + \frac{n}{2}(n-1)\left(\frac{\varepsilon}{\varepsilon_o}\right)^2 + \frac{n}{6}(n-1)(n-2)\left(\frac{\varepsilon}{\varepsilon_o}\right)^3 + \ldots$$

and for $\varepsilon^P/\varepsilon_o > 1$:

$$\frac{\sigma}{\sigma_o} = \left(\frac{\varepsilon}{\varepsilon_o}\right)^n \left[1 + n\left(\frac{\varepsilon_o}{\varepsilon}\right) + \frac{n}{2}(n-1)\left(\frac{\varepsilon_o}{\varepsilon}\right)^2 + \frac{n}{6}(n-1)(n-2)\left(\frac{\varepsilon_o}{\varepsilon}\right)^3 + \ldots\right]$$

9.8.6 Vocé [26]

Finally, Vocé's exponential law (see Fig. 9.18f) is adapted as a strain hardening function:

$$\sigma = A + (Y - A)\,e^{-B\varepsilon^P} \quad \text{or} \quad \frac{\sigma}{\sigma_o} = \frac{\sigma_\infty}{\sigma_o} - \left(\frac{\sigma_\infty}{\sigma_o} - 1\right) e^{-\frac{\varepsilon^P}{\varepsilon_o}} \tag{9.45a,b}$$

Equation (9.45a) is consistent with $\sigma = Y$ for $\varepsilon^P = 0$, where there is a finite hardening rate: $d\sigma/d\varepsilon^P = B(Y - A)$. Constant A is determined from matching eq(9.45a) to a stress chosen at large strain. In eq(9.45b), the initial hardening rate is

$$\frac{d(\sigma/\sigma_o)}{d(\varepsilon^P/\varepsilon_o)} = \frac{\sigma_\infty}{\sigma_o} - 1$$

Should the curve be matched to the ultimate strength point (σ_u, ε_u^P), eqs(9.14c) and (9.45b) are combined. This provides the following relationships between σ_∞ and ε_o at the point of tensile instability:

$$\frac{\sigma_u}{\sigma_o} = \frac{\sigma_\infty}{\sigma_o}\left(\frac{1 - \sigma_u/E}{1 - \sigma_u/E + \varepsilon_o}\right) \approx \frac{\sigma_\infty}{\sigma_o(1 + \varepsilon_o)} \tag{9.45c}$$

$$e^{\varepsilon_u^P/\varepsilon_o} = \frac{(\sigma_\infty - \sigma_o)(1 - \sigma_u/E + \varepsilon_o)}{\varepsilon_o\sigma_\infty} \approx \frac{(\sigma_\infty - \sigma_o)(1 + \varepsilon_o)}{\varepsilon_o\sigma_\infty} \tag{9.45d}$$

The non-dimensional eqs(9.39b)–(9.45b) provide a more rational mathematical basis of hardening function. In these, σ_o is the yield stress and n, ε_o, K and σ_∞ are material constants. The graphical summary in Figs 9.18a–f gives the initial rate of hardening and shows trends appropriate to the choice of n, K and σ_∞/σ_o. Of the various functions mentioned here, the Hollomon, Ludwik and Swift forms are most often used in plasticity theory.

9.9 Strain and Work Hardening Hypotheses

In the rule of isotropic hardening, a flow potential f, appearing as a scalar function of the stress deviator invariants, is equated to a hardening function χ:

$$f(J_2', J_3') = \chi(\textstyle\int \mathrm{d}\psi) \tag{9.46}$$

where χ is an increasing function of a suitable scalar measure ψ of hardening. It was shown previously how to derive $\bar{\sigma}$ for a Drucker function $f(J_2', J_3')$ in eq(9.4b). The corresponding definition of $\mathrm{d}\bar{\varepsilon}^P$ will depend upon the hardening hypothesis used, e.g. strain hardening is given in eqs(9.7a–c). Using the work hypothesis ensures that the flow rule and equivalence relationships are associated with the same yield function but it is more difficult to derive an equivalent plastic strain [3]. However, this is simplified given the yield function in a reduced stress space. Take, for example, the Drucker yield function $f = J_2'^3 - cJ_3'^2$ for torsion. The equivalent flow stress follows from eq(9.4c) when $\bar{\sigma}$ replaces Y:

$$\bar{\sigma} = \frac{\tau}{\left[(1/3)^3 - 4c/9^3\right]^{1/6}}$$

Applying the work hypothesis

$$\bar{\sigma}\,\mathrm{d}\bar{\varepsilon}^P = \sigma_{ij}\,\mathrm{d}\varepsilon_{ij}^P = \tau\,\mathrm{d}\gamma^P$$

$$\mathrm{d}\bar{\varepsilon}^P = \frac{\tau\,\mathrm{d}\gamma^P}{\bar{\sigma}} = \left(\frac{1}{3^3} - \frac{4c}{9^3}\right)^{\frac{1}{6}}\mathrm{d}\gamma^P$$

which differs from $\mathrm{d}\bar{\varepsilon}^P = \mathrm{d}\gamma^P/(27 - 4c)^{1/6}$ for a strain hardening hypothesis (see § 9.2.2(b)). Consequently, the flow curves correlate differently between the two hypotheses, as illustrated for tension and torsion in Fig. 9.10c. Recall that the chain line H is found from the strain hardening hypothesis, with $\bar{\sigma}$ and $\bar{\varepsilon}^P$ derived from similar forms of the stress and strain invariant functions. The broken line F applies to the new position of the equivalent torsional flow curve when $\bar{\varepsilon}^P$ is defined by the work hypothesis. The latter provides a better correlation with tension for the same constant c in the yield function. One may attribute the deviation between F and H to the influence of J_3' within the chosen hardening hypothesis. It is only when J_3' is absent in the yield function, giving $f = J_2'$, are the two hardening hypotheses indistinguishable as shown below.

9.9.1 Strain Hardening Function H

Let f in eq(9.46) be equated to the second invariant of deviatoric stress

$$f = J_2' = \frac{1}{2}\sigma_{ij}'\sigma_{ij}' = \frac{1}{2}\mathrm{tr}\,\mathbf{T}'^2$$

and $\mathrm{d}\psi$ be the increment in the second invariant of plastic strain

$$\mathrm{d}\psi = \mathrm{d}I_2 = \frac{1}{2}\mathrm{d}\varepsilon_{ij}^P\,\mathrm{d}\varepsilon_{ij}^P = \frac{1}{2}\mathrm{tr}(\mathrm{d}\mathbf{E}^P)^2$$

These give one form of eq(9.46) as

$$J_2' = \chi\left(\int \mathrm{d}I_2\right) \tag{9.47a}$$

Moreover, with $J_2' = ⅓(\bar{\sigma})^2$ and $\mathrm{d}I_2 = ¾(\mathrm{d}\bar{\varepsilon}^P)^2$, eq(9.47a) becomes

$$\frac{\bar{\sigma}^2}{3} = \frac{3}{4}\chi\left[\int (\mathrm{d}\bar{\varepsilon}^P)^2\right] \tag{9.47b}$$

Equation (9.47b) shows that the current equivalent stress depends upon equivalent plastic strain. For convenience, strain hardening is expressed directly with $\psi = \int \mathrm{d}\bar{\varepsilon}^P$. Taken again with $f = J_2'$, the hypothesis of strain hardening [27] is normally written:

$$\bar{\sigma} = H\left(\int \mathrm{d}\bar{\varepsilon}^P\right) \tag{9.48}$$

where, from eqs(9.2) and (9.6)

$$\bar{\sigma} = \sqrt{\frac{3}{2}\sigma_{ij}'\sigma_{ij}'} = \sqrt{\frac{3}{2}\left[\,\mathrm{tr}\,(\mathbf{T}')^2\,\right]} \tag{9.49a}$$

$$\mathrm{d}\bar{\varepsilon}^P = \sqrt{\frac{2}{3}\mathrm{d}\varepsilon_{ij}^P\,\mathrm{d}\varepsilon_{ij}^P} = \sqrt{\frac{2}{3}\left[\,\mathrm{tr}\,(\mathrm{d}\mathbf{E}^P)^2\,\right]} \tag{9.49b}$$

The numerical factors in eqs(9.49a,b) enable the correct reduction to a uniaxial stress state. That is, $\bar{\varepsilon}^P$ and $\bar{\sigma}$ become the axial plastic strain and stress in a tension test. This test will establish which of the empirical formulae, discussed in the previous section, is the most suitable hardening function H in eq(9.48). The function χ which connects alternative $\bar{\sigma}$ and $\mathrm{d}\bar{\varepsilon}^P$

definitions, corresponding to $f(J_2', J_3')$ in eq(9.46), may be also be appraised in this way. These alternative definitions become necessary when $f(J_2', J_3')$ provides the required equivalence between tests. For an isotropic solid, $\bar{\sigma}$ and $\mathrm{d}\bar{\varepsilon}^P$ are derived from the invariants of the tensors of deviatoric stress and incremental plastic strain. For a Drucker function, the equivalent stress and strain, re-defined by eqs(9.4b) and (9.7a), apply to eq(9.48). In the determination of H, no further connection exists between $\bar{\sigma}$ and $\mathrm{d}\bar{\varepsilon}^P$ other than the similarity between the two invariants functions from which they derive.

9.9.2 Work Hardening Function F

Equating ψ to the specific work of plastic deformation, W^P/unit volume, gives

$$\psi = \int \mathrm{d}W^P = \int \sigma_{ij}\mathrm{d}\varepsilon_{ij}^P = \int \mathrm{tr}(\mathbf{T}\,\mathrm{d}\mathbf{E}^P) \tag{9.50}$$

Taking eq(9.50) with a von Mises potential $f = J_2' = \bar{\sigma}^2/3$, the hardening function in eq(9.46), becomes

$$\frac{\bar{\sigma}^2}{3} = \chi\left(\int \mathrm{d}W^P\right)$$

This hypothesis of work hardening [26] is normally given as

$$\bar{\sigma} = F\left(\int \mathrm{d}W^P\right) \tag{9.51a}$$

Equation (9.51a) shows that the current equivalent stress is a function of the specific work of plastic deformation. The hardening function F may be established from a uniaxial test where the argument $\int \mathrm{d}W^P$ equals the area beneath the stress versus plastic strain curve. To show this, apply eq(9.50) to a uniaxial stress:

$$W^P = \int \sigma_1 \mathrm{d}\varepsilon_1^P = \int \bar{\sigma}\,\mathrm{d}\bar{\varepsilon}^P \tag{9.51b}$$

In general, the increment of plastic work $\mathrm{d}W^P$ in eq(9.51a) may be expressed in absolute or deviatoric stress forms. For an absolute stress σ_{ij}:

$$\begin{aligned}\delta W^P &= \sigma_{ij}\delta\varepsilon_{ij}^P = \mathrm{tr}(\mathbf{T}\,\delta\mathbf{E}^P) \\ &= \sigma_{11}\delta\varepsilon_{11}^P + \sigma_{22}\delta\varepsilon_{22}^P + \sigma_{33}\delta\varepsilon_{33}^P + 2\,(\sigma_{12}\delta\varepsilon_{12}^P + \sigma_{13}\delta\varepsilon_{13}^P + \sigma_{23}\delta\varepsilon_{23}^P)\end{aligned} \tag{9.52a}$$

Substituting eq(3.9a) into eq(9.52a) leads to the deviatoric form

$$\begin{aligned}\delta W^P &= (\sigma_{ij}' + \tfrac{1}{3}\,\delta_{ij}\sigma_{kk})\delta\varepsilon_{ij}^P = \sigma_{ij}'\delta\varepsilon_{ij}^P + \tfrac{1}{3}\,\sigma_{kk}\delta\varepsilon_{ii}^P \\ &= \sigma_{ij}'\,\delta\varepsilon_{ij}^P = \mathrm{tr}(\mathbf{T}'\delta\mathbf{E}^P) \\ &= \sigma_{11}'\delta\varepsilon_{11}^P + \sigma_{22}'\delta\varepsilon_{22}^P + \sigma_{33}'\delta\varepsilon_{33}^P + 2(\sigma_{12}\delta\varepsilon_{12}^P + \sigma_{13}\delta\varepsilon_{13}^P + \sigma_{23}\delta\varepsilon_{23}^P) \\ &= \sigma_{11}'\delta\varepsilon_{11}^P + \sigma_{22}'\delta\varepsilon_{22}^P + (\sigma_{11}' + \sigma_{22}')(\delta\varepsilon_{11}^P + \delta\varepsilon_{22}^P) + 2(\sigma_{12}\delta\varepsilon_{12}^P + \sigma_{13}\delta\varepsilon_{13}^P + \sigma_{23}\delta\varepsilon_{23}^P)\end{aligned} \tag{9.52b}$$

which has employed $\delta\varepsilon_{ii}^{P} = \mathrm{tr}(\delta\mathbf{E}^{P}) = 0$ and $\sigma_{ii}' = \mathrm{tr}\mathbf{T}' = 0$. Thus, the equivalent stress $\bar{\sigma}$ and plastic strain increment $\mathrm{d}\bar{\varepsilon}^{P}$ identify three forms of δW^{P} within eq(9.51b):

$$\delta W^{P} = \bar{\sigma}\delta\bar{\varepsilon}^{P} = \sigma_{ij}\delta\varepsilon_{ij}^{P} = \sigma_{ij}'\delta\varepsilon_{ij}^{P} \tag{9.52c}$$

From eqs(9.49a,b), equivalent matrix forms of eq(9.52c) are

$$\begin{aligned}\delta W^{P} &= \sqrt{\left[\mathrm{tr}\,(\mathbf{T}')^{2}\right]\left[\mathrm{tr}\left(\delta\mathbf{E}^{P}\right)^{2}\right]}\\ &= \mathrm{tr}\left(\mathbf{T}\delta\mathbf{E}^{P}\right) = \mathrm{tr}\left(\mathbf{T}'\delta\mathbf{E}^{P}\right)\end{aligned} \tag{9.52d}$$

In the absence of a flow rule, the selection of $\delta\bar{\varepsilon}^{P}$ in eq(9.52c) can be arbitrary. While $\bar{\sigma}$ depends upon the chosen yield function, $\delta\bar{\varepsilon}^{P}$ derives from a similar function in the plastic strain invariants, as was shown for the strain hardening hypothesis. When the flow rule is associated with the yield function, then $\bar{\sigma}$ and $\delta\bar{\varepsilon}^{P}$ in eq(9.52c) are connected. For example, with a von Mises yield function, $\bar{\sigma}$ is defined from eq(9.49a). The associated flow rule is $\mathrm{d}\varepsilon_{ij}^{P} = \mathrm{d}\lambda\sigma_{ij}'$. Substituting these into eq(9.52c) gives

$$\begin{aligned}\bar{\sigma}\delta\bar{\varepsilon}^{P} &= \sqrt{(\sigma_{ij}'\delta\varepsilon_{ij}^{P})(\sigma_{ij}'\delta\varepsilon_{ij}^{P})} = \sqrt{(\sigma_{ij}'\delta\lambda\,\sigma_{ij}')(\delta\varepsilon_{ij}^{P}\,\delta\varepsilon_{ij}^{P})/\delta\lambda}\\ &= \sqrt{(\sigma_{ij}'\sigma_{ij}')(\delta\varepsilon_{ij}^{P}\,\delta\varepsilon_{ij}^{P})} = \sqrt{\tfrac{2}{3}\bar{\sigma}^{2}(\delta\varepsilon_{ij}^{P}\,\delta\varepsilon_{ij}^{P})}\\ \delta\bar{\varepsilon}^{P} &= \sqrt{\tfrac{2}{3}(\delta\varepsilon_{ij}^{P}\,\delta\varepsilon_{ij}^{P})} = \sqrt{\tfrac{2}{3}\left[\mathrm{tr}(\delta\mathbf{E}^{P})^{2}\right]}\end{aligned}$$

The fact that $\mathrm{d}\bar{\varepsilon}^{P}$ is identical to eq(9.49b) confirms that a von Mises material is independent of the hardening hypothesis. Thus, a plot of $\bar{\sigma}$ versus $\int\bar{\sigma}\,\mathrm{d}\bar{\varepsilon}^{P}$ will not correlate flow curves any better than a plot of $\bar{\sigma}$ versus $\int\mathrm{d}\bar{\varepsilon}^{P}$. However, equivalence plots for other materials obeying $f(J_2', J_3')$ will depend upon the hardening hypothesis. Consequently, this yield function provides a choice between an abscissa of equivalent plastic strain or specific plastic work.

9.9.3 Comparisons Between H and F

The respective strain and work hardening functions H and F, appearing in eqs(9.48) and (9.51a), are found by changing the abscissae within the following standard tests (H to the left, F to the right).

(a) Tension

$$\bar{\sigma} \text{ versus } \int\left(\frac{\mathrm{d}l}{l} - \frac{\mathrm{d}\bar{\sigma}}{E}\right) \qquad\qquad \bar{\sigma} \text{ versus } \int\bar{\sigma}\left(\frac{\mathrm{d}l}{l} - \frac{\mathrm{d}\bar{\sigma}}{E}\right)$$

$$\bar{\sigma} \text{ versus } \int \frac{dl}{l} - \frac{\bar{\sigma}}{E} \qquad \bar{\sigma} \text{ versus } \int \bar{\sigma}\frac{dl}{l} - \frac{\bar{\sigma}^2}{2E} \tag{9.53a,b}$$

(b) Uniaxial Compression

A similar expression to (a) above applies to lubricated compression under continuous loading.

$$\bar{\sigma} \text{ versus } \int \frac{dh}{h} - \frac{\bar{\sigma}}{E} \qquad \bar{\sigma} \text{ versus } \int \bar{\sigma}\frac{dh}{h} - \frac{\bar{\sigma}^2}{2E} \tag{9.54a,b}$$

(c) Torsion

$$\sqrt{3}\tau \text{ versus } \int \frac{d\gamma^P}{\sqrt{3}} \qquad \sqrt{3}\tau \text{ versus } \int \sqrt{3}\tau \times \frac{d\gamma^P}{\sqrt{3}}$$

$$\sqrt{3}\tau \text{ versus } \int \left(r\frac{d\theta}{l} - \frac{d\tau}{G} \right) \qquad \sqrt{3}\tau \text{ versus } \int \tau\left(\frac{r\,d\theta}{l} - \frac{d\tau}{G} \right)$$

$$\sqrt{3}\tau \text{ versus } \int r\frac{d\theta}{l} - \frac{\tau}{G} \qquad \sqrt{3}\tau \text{ versus } \int \tau\left(\frac{r\,d\theta}{l} \right) - \frac{\tau^2}{2G} \tag{9.55a,b}$$

(d) Bulge Test (Equi-Biaxial Tension)

$$\bar{\sigma} \text{ versus } \int \left(d\varepsilon_r^{\,t} - d\varepsilon_r^{\,e} \right) \qquad \bar{\sigma} \text{ versus } \int \bar{\sigma}\left(d\varepsilon_r^{\,t} - d\varepsilon_r^{\,e} \right) \tag{9.56a,b}$$

where, in eqs(9.56a,b), the total and elastic radial strain increments are

$$d\varepsilon_r^{\,t} = \frac{dt}{t} \text{ and } d\varepsilon_r^{e} = -\frac{\nu}{E}(d\sigma_\theta + d\sigma_\varphi) = -\left(\frac{2\nu}{E} \right) d\bar{\sigma} \tag{9.56c,d}$$

Substituting eqs(9.56c,d) into eqs(9.56a,b) leads to

$$\bar{\sigma} \text{ versus } \int \left(\frac{dt}{t} + 2\nu\frac{d\bar{\sigma}}{E} \right) \qquad \bar{\sigma} \text{ versus } \int \bar{\sigma}\left(\frac{dt}{t} + 2\nu\frac{d\bar{\sigma}}{E} \right)$$

$$\bar{\sigma} \text{ versus } \int \frac{dt}{t} + \frac{2\nu\bar{\sigma}}{E} \qquad \bar{\sigma} \text{ versus } \int \bar{\sigma}\frac{dt}{t} + \frac{\nu\bar{\sigma}^2}{E} \tag{9.56e,f}$$

Note: Increments:dl, dh, dθ and dt in eqs(9.53)–(9.56) denote the total dimensional changes arising from elastic–plastic straining. In interrupted tests, plastic lengths l, h and t are measured under no-load conditions. The first integral term then defines plastic strain and plastic work directly, a technique followed next in modifying compression tests.

(e) Cook and Larke

Under no-load measurements of the cylinder height, the equivalence plots are simplified to

$$\bar{\sigma} \text{ versus } \int \frac{\mathrm{d}h}{h} \qquad\qquad \bar{\sigma} \text{ versus } \int \bar{\sigma}\,\frac{\mathrm{d}h}{h} \tag{9.57a,b}$$

(f) Plane Strain Compression

For no-load, permanent thickness measurements, the equivalence relation is derived as

$$\bar{\sigma} \text{ versus } \frac{2}{\sqrt{3}} \int \frac{\mathrm{d}t}{t} \qquad\qquad \bar{\sigma} \text{ versus } \frac{2}{\sqrt{3}} \int \bar{\sigma}\,\frac{\mathrm{d}t}{t} \tag{9.58a,b}$$

The calculation of the equivalent stress $\bar{\sigma}$ within eqs (9.57) and (9.58) has been given with the separate description of each test in § 9.6.

9.9.4 Relationships Between F and H

The functions H and F can be found empirically from each of the plots described above. Alternatively, F can be derived from some strain hardening functions H.

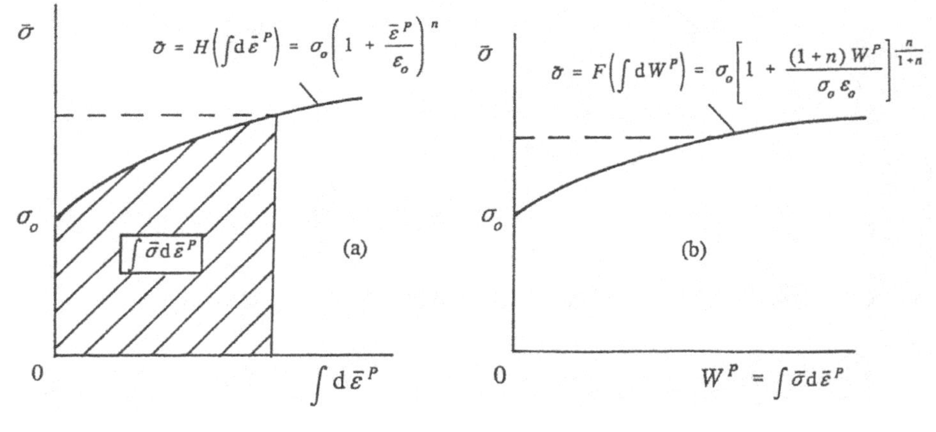

Figure 9.20 Arguments for the hypotheses of strain and work hardening

For example, in Fig. 9.20a H is identified with the Swift eq(9.44b). The argument of plastic work in eq(9.53b) becomes

$$W^P = \int \bar{\sigma}\,\mathrm{d}\bar{\varepsilon}^P = \int_0^{\bar{\varepsilon}^P} \sigma_o\left(1 + \frac{\bar{\varepsilon}^P}{\varepsilon_o}\right)^n \mathrm{d}\bar{\varepsilon}^P$$

$$= \frac{\sigma_o\,\varepsilon_o}{n+1}\left|\left(1 + \frac{\bar{\varepsilon}^P}{\varepsilon_o}\right)^{n+1}\right|_0^{\bar{\varepsilon}^P} = \frac{\sigma_o\,\varepsilon_o}{n+1}\left[\left(\frac{\bar{\sigma}}{\sigma_o}\right)^{\frac{n+1}{n}} - 1\right]$$

Transposing for $\bar{\sigma}$, determines the function F as

$$\bar{\sigma} = F(W^P) = \sigma_o \left[1 + \frac{(1+n)\, W^P}{\sigma_o\, \varepsilon_o} \right]^{\frac{n}{1+n}}$$

The explicit relationship between F and H is illustrated graphically in Fig. 9.20b.

On the other hand, with a function H, defined from Ludwik's eq(9.43b), the argument of plastic work becomes

$$W^P = \int_0^{\bar{\varepsilon}^P} \sigma_o \left[1 + \left(\frac{\bar{\varepsilon}^P}{\varepsilon_o} \right)^n \right] \mathrm{d}\bar{\varepsilon}^P$$

$$= \sigma_o\, \bar{\varepsilon}^P \left[1 + \frac{1}{n+1} \left(\frac{\bar{\varepsilon}^P}{\varepsilon_o} \right)^n \right] = \sigma_o\, \varepsilon_o \left(\frac{\bar{\sigma}}{\sigma_o} - 1 \right)^{\frac{1}{n}} \left[1 + \frac{1}{n+1} \left(\frac{\bar{\sigma}}{\sigma_o} - 1 \right) \right]$$

Here it can be seen that H does not supply a closed solution for F. In chapter 10 (see § 10.3.1), either F or H is used to define an isotropic hardening rule.

9.10 Concluding Remarks

Experiments with a von Mises material (i.e. $f = J_2'$) cannot distinguish between the work and strain hardening hypotheses. Hence, the strain hardening hypothesis is preferred for simplicity, since it identifies directly with the common hardening expressions of Hollomon, Swift, Ludwik etc. The choice between these expressions depends upon the material and the range of strain. A number of standard tests are available to establish an appropriate expression experimentally with the minimum of error. For isotropic materials obeying a more general yield function $f(J_2', J_3')$, it has been shown that it is possible to distinguish between the two hardening hypotheses, this resulting in different equivalent plastic strain expressions $\mathrm{d}\bar{\varepsilon}^P$. The latter follows from strain hardening arbitrarily but work hardening is more exacting as $\mathrm{d}\bar{\varepsilon}^P$ must satisfy the incremental plastic work expression $\bar{\sigma}\,\mathrm{d}\bar{\varepsilon}^P = \sigma_{ij}\,\mathrm{d}\varepsilon_{ij}^P$.

References

1. von Mises R. *Natchr. Ger. Wiss.*, 1913, 582, Gottingen.
2. Drucker D. *Quart. Appl. Math.*, 1950, **7**, 411.
3. White G. N. and Drucker D. C. *Jl. Appl. Phys.*, 1950, **21**, 1013.
4. Johnson A. E. *Complex Stress, Creep, Relaxation and Fracture of Metallic Alloys*, 1962, HMSO, Edinburgh.
5. Ikegami K. *Brit. Ind. & Sci. Int. Trans. Ser.* (The Metals Society), 1976, **1442**.
6. Michno M. J. and Findley W. N. *Acta Mech.*, 1973, **18**, 163.
7. Considere A. *Ann. Ponts et Chaussees*, 1885, **9**, 574.
8. Bridgeman P. W. *Studies in Large Plastic Flow and Fracture*, 1952, McGraw-Hill, New York.
9. Johnson W. and Mellor P. B. *Engineering Plasticity*, 1983, Ellis-Horwood.
10. Hsu T. C. *ASME*, 1967, W.A. / Met.11

11. Bredt R. *Zeitschrift des Vereines Deutscher Ingenieure*, 1896, **40**, 785, 813.
12. Nadai A. *Theory of Flow and Fracture of Solids*, 1950, McGraw-Hill.
13. Cook M. and Larke E. C. *Jl. Inst. Metals,* 1945, **71**, 371.
14. Watts A. B. and Ford H. *Proc. I. Mech. E.*, 1952, **1B**, 449.
15. Watts A. B. and Ford H. *Proc. I. Mech. E.*, 1955, **169**, 1141.
16. Brown W. F. and Sachs G. *Trans. ASME*, 1948, **70**, 241.
17. Mellor P. *Jl. Mech. Phys. Solids*, 1956, **5**, 41.
18. Easterbrook K. and Grieve R. *Jl. Mech. Work. Tech.*, 1985, **11**, 229.
19. Zyczkowski M. *Combined Loadings in the Theory of Plasticity*, 1981, PWN, Warsaw.
20. Hollomon J. H. *Trans Amer Inst Mech Eng*, 1945, **162**, 269.
21 Ramberg W. and Osgood W. *Nat. Adv. Comm. for Aero.*, 1943, TN902.
22. Prager W. *Proc. 5th ICAM Conference*, 1938, 234.
23. Ludwik P. *Elements der Technologischen Mechanik*, 1909, Springer, Berlin.
24. Rees D. W. A. *Proc. Roy. Soc.*, 1987, A**410**, 443.
25. Swift H. W. *Jl. Mech. Phys. Solids*, 1952, **1**, 1.
26. Vocé E. *Jl. Inst. Metals*, 1948, **74**, 537.
27. Bland D. R. *Proc. 9th Int. Conf. Appl. Mech.*, 1957, **8**, 45.

Exercises

9.1 Confirm the stress and strain predictions, defining the point of tensile instability, as given by the Swift eqs(9.44c,d).
See solution

9.2 Confirm the stress and strain predictions, defining the point of tensile instability, as given by the Vocé eqs(9.45c,d).

9.3 Examine whether it is possible to derive the function F in the work hardening hypothesis for the flow curves described by Hollomon, modified Prager and Vocé in Figs 9.19a, c and f respectively.

9.4 Is the strain hardening hypothesis, which defines $\bar{\varepsilon}^P$ in eqs(9.7c), consistent with a meaningful work increment expression? Examine this from a consideration of deformation under pure shear.

9.5 The following data applies to the hydrostatic bulging of 0.8 mm thick steel sheet through a circular die aperture 180 mm diameter:

p, bar	1.45	3.67	12.09	19.62	24.9	32.46	39.11	
	41.77	46.64	52.39	59.04	63.47	68.79	74.55	76.0
h, mm	0.15	3.00	8.85	12.9	16.35	20.55	24.6	
	25.95	28.95	32.55	36.9	40.65	46.05	53.7	58.35
t, mm	0.8	0.794	0.769	0.744	0.72	0.689	0.664	
	0.651	0.629	0.601	0.564	0.492	0.486	0.402	0.372

Construct the equivalent stress versus equivalent strain curve based upon the current pole heights and thickness values given.

9.6 The following data applies to a plane strain compression test upon an annealed aluminium strip with initial dimensions: breadth 50 mm and thickness 6 mm and a die of width 6 mm:

F, kN	4.98	7.48	9.97	12.46	14.95	17.44	19.94	22.43	24.92
	27.41	29.90	34.89	39.87	44.86	49.84	54.82	59.81	64.79
Δt, mm	0.30	0.32	0.36	0.40	0.58	0.75	0.90	1.11	1.39
	1.78	2.29	3.15	3.75	4.25	4.50	4.75	5.35	5.75

The thickness change Δt applies to an unloaded condition following the removal of each applied force level *F*. Plot the equivalent stress versus equivalent plastic strain curve and determine whether a Hollomon law can be applied to describe the curve.
See solution

9.7 The following torque-twist data applies to a torsion test upon a solid bar of mild steel, diameter 20.27 mm and gauge length 203.2 mm. Use the Nadai analysis to obtain a true, equivalent stress versus strain plot for the material.

T, Nm	28.25	45.21	75.73	96.08	118.7	144.7	168.0	191.0	198.9
	237.4	260.0	274.7	282.6	276.9	282.6	282.6	282.6	288.2
	288.2	291.1	295.0	301.8	310.8	327.8	340.22	350.4	361.7
	373.0	382.0	401.3	418.2	435.2	439.1	452 1	491.7	514.3
	542.5	553.9	582.1	587.8	594.5	605.8	616.0	626.2	640.9
	649.9	678.2	689.5	700.8					
θ °	0.2	0.4	0.6	0.8	1.0	1.2	1.4	1.6	1.8
	2.0	2.2	2.4	2.6	2.8	3.0	3.2	3.4	5.0
	17.5	20	22	25	30	35	40	45	50
	55	60	70	80	90	100	110	140	160
	200	220	250	270	300	330	360	420	480
	540	600	660	720					

See solution

9.8 The following force-extension data applies to tensile loading upon a cylindrical bar of mild steel, diameter 11.43 mm and gauge length 57.15 mm. Construct both the nominal and true stress-strain curves for the material, superimposed upon the same axes.

W, kN	9.47	13.71	17.54	21.93	25.42	29.61	33.19	36.08	38.68
	40.87	44.16	45.85	47.05	47.75	48.35	49.44	50.34	50.84
	51.43	52.33	53.33	53.93	54.63	55.32	56.02	57.02	58.01
	58.81	59.41	60.01	60.51	61.30	62.0	62.15	62.3	62.5
	62.7	62.8	62.8	62.8	62.5	62.2	61.3	44.36	

x, mm	0.0254	0.0508	0.0762	0.102	0.127	0.152	0.178	0.203	0.229
	0.254	0.305	0.356	0.406	0.457	0.508	0.635	0.762	0.889
	1.016	1.27	1.524	1.78	2.03	2.29	2.54	3.05	3.56
	4.06	4.57	5.08	5.59	6.604	7.62	8.128	8.636	9.144
	9.652	10.16	10.67	11.176	11.684	11.938	12.192	12.7	

9.9 The following force-extension data applies to loading followed by unloading in a tension test upon a cylindrical bar of mild steel, diameter 11.43 mm and gauge length 50.8 mm.

Loading:

W, kN	2.59	5.78	7.88	9.77	11.86	13.95	16.05	18.04	20.23
	22.53	26.62	30.60	33.89	36.78	39.08	40.77	42.26	43.46
	44.46	45.45	46.15	47.45	49.64	50.04	50.64	51.04	51.44
	51.73	52.03	52.23	52.53					

x, mm	0.0051	0.0102	0.0152	0.0203	0.0254	0.0305	0.0356	0.0406	0.0457
	0.051	0.061	0.071	0.081	0.091	0.102	0.112	0.122	0.132
	0.142	0.152	0.178	0.203	0.254	0.279	0.305	0.330	0.356
	0.381	0.406	0.432	0.457					

Unloading:

W, kN	41.07	32.09	22.73	13.46	5.085	0
x, mm	0.432	0.406	0.381	0.356	0.330	0.315

Determine the initial modulus of elasticity for the material and the proof stress for offset engineering strains of 0.1 % and 0.5 %. What does the modulus become when unloading from the plastic range? *See solution*

9.10 The following data applies to a tensile test upon a 12.7 mm diameter mild steel testpiece in which the diameter of the neck was recorded through to fracture. Calculate and plot the true stress versus the natural strain in the neck when based upon the Bridgeman correction curve given in Fig. 9.21.

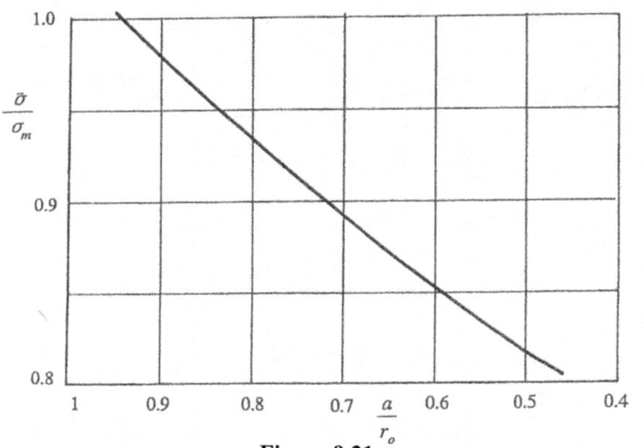

Figure 9.21

W, kN	34.27	37.83	44.5	46.73	51.18	53.4	56.29	57.85	58.43
	59.36	59.54	59.72	59.72	59.81	59.63	59.63	59.45	59.27
	58.74	58.3	57.94	57.85	56.96	56.52	55.63	55.18	54.29
	53.85	53.13	52.51	51.18	49.84	49.75	48.45	46.73	46.28

d, mm	12.61	12.57	12.496	12.446	12.357	12.268	12.116	12.014	11.938
	11.836	11.735	11.659	11.532	11.465	11.354	11.227	11.151	11.049
	10.846	10.719	10.592	10.363	10.236	10.058	9.931	9.804	9.728
	9.55	9.347	9.246	8.992	8.915	8.814	8.636	8.484	8.433

See solution

9.11 The following height reduction data apply to Cooke and Larke compression tests upon three solid copper cylinders, each with a common initial diameter $d_o = 18$ mm and initial heights h_o of 9, 18 and 36 mm as follows.

$\underline{h = 9 \text{ mm}}$

Δh, mm	0.03	0.14	0.30	0.50	0.73	1.00	1.26	1.53	1.83
	2.13	2.45	2.73	3.0	3.24	3.46	3.75	3.97	4.17
	4.37	4.53	4.81	5.0	5.22	(for $h_o = 9$ mm)			

$\underline{h = 18 \text{ mm}}$

Δh, mm	0.08	0.34	0.72	1.15	1.65	2.20	2.82	3.46	4.11
	4.77	5.37	5.95	6.54	7.03	7.53	8.01	8.43	8.86
	9.22	9.70	10.0	10.31	10.77	(for $h_o = 18$ mm)			

$\underline{h = 36 \text{ mm}}$

Δh, mm	0.05	0.44	1.15	2.02	3.04	4.14	5.44	6.72	8.12
	9.47	10.74	12.04	13.29	14.49	15.57	16.59	17.57	18.39
	19.14	20 29	20.82	21.34	22.29	(for $h_o = 36$ mm)			

Given that the following loads correspond to the increasing Δh values in each cylinder, construct the true stress versus natural strain curve for the material when based upon an extrapolation to a cylinder with $d_o/h_o = 0$.

W (kN)	10	20	30	40	50	60	70	80	90
	100	110	120	130	140	150	160	170	180
	190	200	220	230	250	(for all h_o values)			

See solution

9.12 The following load versus extension data apply to a tensile test upon a ductile, thin sheet steel. Testpiece dimensions were width 9.75 mm, thickness 0.2 mm and gauge length 50 mm. Calculate the true stress and natural plastic strains and show that these conform to a Hollomon law when applied: (a) to total strain and (b) plastic strain for strains greater than 1.5 %. Compare Hollomon's (b) analysis with a Ludwik law fit to these data. Take E = 180 GPa and Y = 256 MPa.

x, mm	0.2	0.4	0.6	0.8	1.0	1.2	1.4	1.6	1.8
W, kN	0.499	0.527	0.545	0.564	0.581	0.598	0.606	0.618	0.625

continued:

x, mm	2.0	2.4	2.8	3.2	3.6	4.0	4.4	4.8	5.2
W, kN	0.636	0.648	0.660	0.676	0.687	0.696	0.701	0.715	0.713

Ans: Hollomon: (a) $\sigma = 597(\varepsilon)^{0.171}$ for $\varepsilon > 0.015$

(b) $\sigma = 595(\varepsilon^P)^{0.168}$ for $\varepsilon^P > 0.015$

Ludwik: $\sigma = 256 + 1210(\varepsilon^P)^{0.86}$ for $\varepsilon^P > 0.015$

PART II

APPLICATIONS OF PLASTICITY THEORY

CHAPTER 10

PLASTICITY WITH HARDENING

10.1 Introduction

When the initial yield stress is exceeded, a subsequent yield surface is required to express the current flow stress for a strain hardened material. This means that it becomes necessary to account for the effect of plastic strain on the yield surface. In this chapter we restrict our attention to such accounts provided by the rules of isotropic and kinematic hardening. Thus, the initial yield surface may expand uniformly or translate rigidly to contain the current stress point by each respective rule. It appears that subsequent yield functions are mostly based upon initial von Mises and Tresca conditions for mathematical simplicity. For example, when the von Mises function $f = J_2'$ is employed within the simpler rule of isotropic hardening, it leads to the flow rule of Levy-Mises. Alternative constitutive relations apply from adopting different initial yield functions or the deformation theory instead of the flow theory. Important among these are the influences of J_3', initial anisotropy and path dependence upon plasticity. It will be shown, correspondingly, how alternative relations apply based upon: (i) an initial function $f(J_2', J_3')$, (ii) the orthotropic function, discussed in the following chapter and (iii) Hencky's deformation theory, extended with hardening.

It is also appropriate to compare the elements of the two hardening rules with experimental data. Therein, limitations of their usefulness can be stated. In particular, it will appear that a more versatile model of plasticity combines these two rules of hardening. These lead to a combined hardening concept with identifiable sources within lattice slip.

10.2 Subsequent Yield Surface

Any theory of isothermal plasticity should apply to its irreversible behaviour. Geometrically, the theory should: (i) match the shape of both the initial and subsequent boundaries between the elastic and plastic regions where they exist and (ii) provide the magnitude and path of plastic strain. Classical theory expresses plasticity in the following manner and couples this with Hookean elasticity.

10.2.1 Loading Conditions

The *subsequent yield surface* is expressed as a *loading function* of stress σ_{ij}, plastic strain ε_{ij}^{P} and a hardening parameter χ:

$$f(\sigma_{ij}, \varepsilon_{ij}^{P}, \chi) = 0 \tag{10.1}$$

where it has been seen previously that χ is either a function of plastic strain or work (see § 9.9, eq(9.46)). The total differential of f in eq(10.1) is

$$\mathrm{d}f = \left(\frac{\partial f}{\partial \sigma_{ij}}\right)\mathrm{d}\sigma_{ij} + \left(\frac{\partial f}{\partial \varepsilon_{ij}^{P}}\right)\mathrm{d}\varepsilon_{ij}^{P} + \left(\frac{\partial f}{\partial \chi}\right)\mathrm{d}\chi = 0 \tag{10.2a}$$

Where all changes occur in time dt, the rate form of eq(10.2a) becomes

$$\frac{\mathrm{d}f}{\mathrm{d}t} = \left(\frac{\partial f}{\partial \sigma_{ij}}\right)\left(\frac{\mathrm{d}\sigma_{ij}}{\mathrm{d}t}\right) + \left(\frac{\partial f}{\partial \varepsilon_{ij}^{P}}\right)\left(\frac{\mathrm{d}\varepsilon_{ij}^{P}}{\mathrm{d}t}\right) + \left(\frac{\partial f}{\partial \chi}\right)\left(\frac{\mathrm{d}\chi}{\mathrm{d}t}\right) = 0 \tag{10.2b}$$

where $\mathrm{d}\varepsilon_{ij}^{P}$ and $\mathrm{d}\chi$ depend upon the direction of the incremental stress vector $\mathrm{d}\sigma_{ij}$ from a point P on the surface $f = 0$. Three *loading conditions* arise, as shown in Fig. 10.1.

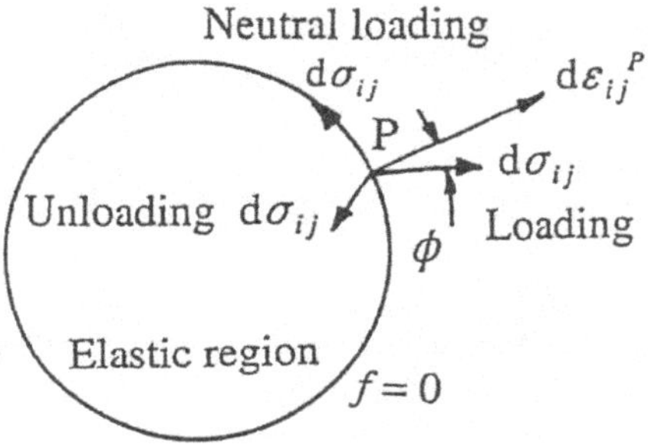

Figure 10.1 The three conditions of loading

They are loading, unloading and neutral loading. Each condition is associated with the change, or rate of change, to f produced by the corresponding stress change. Writing the change to f as $\mathrm{d}f_\sigma$, it follows from eqs(10.2a,b) that

$$\mathrm{d}f_\sigma = \frac{\partial f}{\partial \sigma_{ij}}\mathrm{d}\sigma_{ij} \quad \text{and} \quad \frac{\mathrm{d}f_\sigma}{\mathrm{d}t} = \frac{\partial f}{\partial \sigma_{ij}} \times \frac{\mathrm{d}\sigma_{ij}}{\mathrm{d}t} \tag{10.3a,b}$$

For an elastic–plastic loading we have $\mathrm{d}f_\sigma > 0$ so that $\mathrm{d}\varepsilon_{ij}^{P} > 0$ and $\mathrm{d}\chi > 0$. With an elastic unloading from the plastic region, $\mathrm{d}f_\sigma < 0$ and $\mathrm{d}\varepsilon_{ij}^{P} = \mathrm{d}\chi = 0$. Neutral loading refers to a change of stress state upon the yield surface in the absence of plastic strain, i.e. $\mathrm{d}f_\sigma = 0$ and $\mathrm{d}\varepsilon_{ij}^{P} = \mathrm{d}\chi = 0$. *Note*: A change to the elastic strain results from each of the three loadings.

10.2.2 Associated Flow Rule

Hill [1] and Drucker [2] were among the first to employ a plastic potential function for a plastic loading condition. The potential describes a closed, convex surface in strain space whose outward normal defines the direction of the plastic strain increment vector [3]. It is convenient to let this surface lie in stress space, where it differs from the plastic potential in elastic dimension only. It may then be taken that the incremental vector of plastic loading strain lies normal to the stress surface $g(\sigma_{ij})$. This normality condition is expressed in

$$\mathrm{d}\varepsilon_{ij}^{P} = \mathrm{d}\lambda \frac{\partial g(\sigma_{ij})}{\partial \sigma_{ij}} \tag{10.4a}$$

The elastic conversion between the strain and stress spaces identifies $g(\sigma_{ij})$ with the yield function f. The flow rule is then 'associated' with the yield function and eq(10.4a) becomes

$$d\varepsilon_{ij}^{P} = d\lambda \frac{\partial f}{\partial \sigma_{ij}} \tag{10.4b}$$

The positive scalar $d\lambda$ is defined from combining eqs(10.2a) and (10.4b) as

$$d\lambda = \frac{-(\partial f/\partial \sigma_{ij})d\sigma_{ij} - (\partial f/\partial \chi)d\chi}{(\partial f/\partial \varepsilon_{mn})(\partial f/\partial \sigma_{mn})} \tag{10.5}$$

Explicit forms of $d\lambda$ follow from eq(10.5) when f is based upon the von Mises yield function and an isotropic hardening rule:

$$f = \frac{1}{2}\sigma_{ij}'\sigma_{ij}' - \chi(\psi) = 0 \tag{10.6}$$

where $\psi = \int d\bar{\varepsilon}^{P}$ or $\psi = \int dW^{P}$. It is then straightforward to derive $d\lambda$ from the reduction of eqs(10.5) and (10.6) to simple tension. It will be seen that this allows χ to be identified with functions H and F from the strain and work hypotheses as required (see § 10.3).

10.2.3 Drucker's Postulate

Consider a unit volume of material in elastic equilibrium under a stress state σ_{ij}^{*} at point A in Fig. 10.2a. Drucker [2] considered the net change to the elastic energy stored when some external action, force moment, torque etc, applied a loading cycle. The sequence of loading produces a new stress state σ_{ij} at a point B on the yield surface, then pierces the surface with an infinitesimal stress change $d\sigma_{ij}$ before restoring it to σ_{ij}^{*}. This loading sequence produces a cyclic response ABCA′ within each stress–strain component plot (Fig. 10.2b).

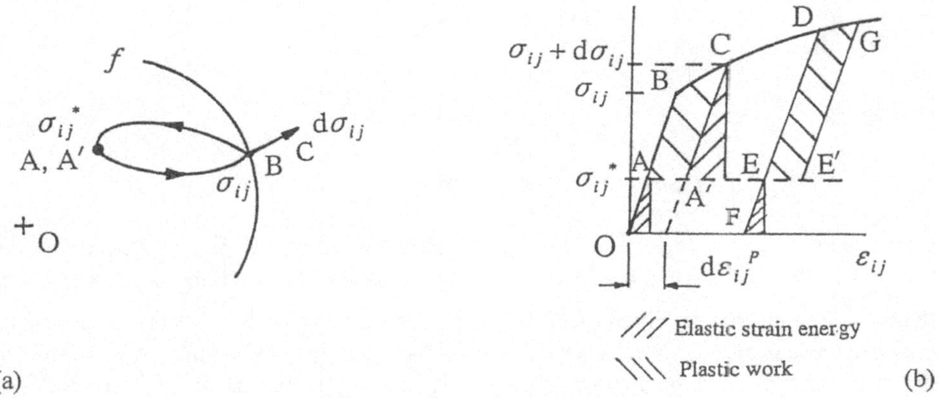

Figure 10.2 Drucker's postulate

It can be seen from Fig. 10.2b that the incremental plastic strain $d\varepsilon_{ij}^{P}$ under $d\sigma_{ij}$ is solely responsible for a permanent change to the stored energy. It was shown previously for a similar cycle (see Fig. 8.12) that elastic energy recovers, leaving the plastic work done as

$$dW^{P} = (\sigma_{ij} - \sigma_{ij}^{*})\, d\varepsilon_{ij}^{P} + d\sigma_{ij}\, d\varepsilon_{ij}^{P} > 0 \tag{10.7}$$

At its extremes, σ_{ij}^* may either define an unstressed state at the origin O ($\sigma_{ij}^* = 0$) or at the yield point B ($\sigma_{ij}^* = \sigma_{ij}$). The inequality (10.7) applies to the loading condition in Fig. 10.1. When $d\sigma_{ij}$ corresponds to unloading or neutral loading from B, then $dW^P = 0$. Splitting eq(10.7), it follows that two conditions apply to a loading from B, written in component and equivalent forms:

$$\left(\sigma_{ij} - \sigma_{ij}^*\right) d\varepsilon_{ij}^P > 0 \quad \text{or} \quad \left(\bar{\sigma} - \bar{\sigma}^*\right) d\overline{\varepsilon}^P \geq 0 \qquad (10.8a,b)$$

$$d\sigma_{ij} d\varepsilon_{ij}^P > 0 \quad \text{or} \quad d\bar{\sigma}\, d\overline{\varepsilon}^P \geq 0 \qquad (10.8c,d)$$

Drucker's postulate also applies to any subsequent yield surface where σ_{ij} identifies a new yield condition, e.g. point D instead of B in Fig. 10.2b. Thus, when $d\sigma_{ij}$ 'loads' from D to point G, the postulate will apply to the cycle EDGE′ in which the enclosed area (shaded) is the plastic work done by $\sigma_{ij} - \sigma_{ij}^*$. The difference now is that under σ_{ij}^* the initial store of recoverable elastic energy at point E (small triangle) is accompanied by the irrecoverable work of previous inelastic deformation (i.e. area ABCDE).

It is convenient to visualise each loading condition (10.8b,d) with scalar products between equivalent stress $\bar{\boldsymbol{\sigma}}$ and equivalent plastic strain increment $\mathbf{d}\overline{\boldsymbol{\varepsilon}}^P$ vectors. With a von Mises definition of $\bar{\sigma}$, the initial yield surface f appears as a circle of radius O′B on the deviatoric plane in Fig. 10.3a.

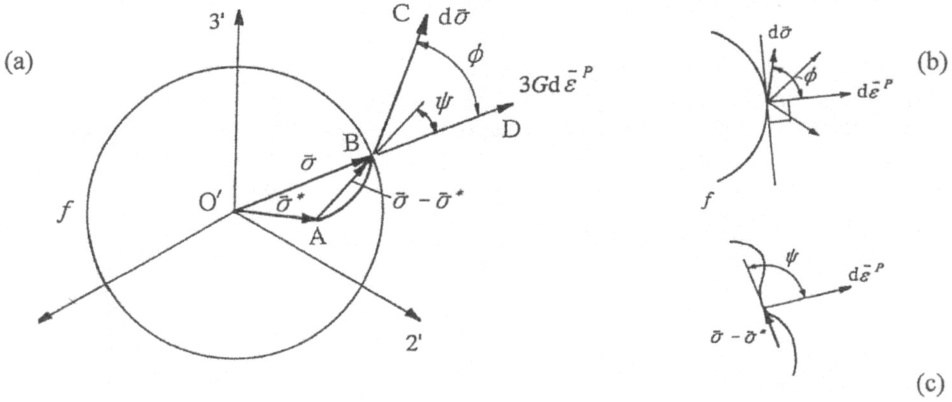

Figure 10.3 Drucker's postulate on the deviatoric plane

Within eq(10.8b), $\bar{\boldsymbol{\sigma}}^*$ and $\left(\bar{\boldsymbol{\sigma}} - \bar{\boldsymbol{\sigma}}^*\right)$ appear as respective vectors **O′A** and **AB** on this plane. Vector **BC** represents the change to the initial yield stress $\mathbf{d}\bar{\boldsymbol{\sigma}}$, this being accompanied by an equivalent plastic strain increment $\mathbf{d}\overline{\boldsymbol{\varepsilon}}^P$. The latter may be scaled (× $3G$) to stress vector **BD** on the deviatoric plane. To show this, firstly, the principal incremental deviatoric plastic strains are converted to dimensions of deviatoric stress with the constant multiplier $2G$ [4]. Then, the magnitude |**BD**| becomes the sum of 'stress components' $2Gd\varepsilon_1'^P$, $2Gd\varepsilon_2'^P$ and $2Gd\varepsilon_3'^P$:

$$|\mathbf{BD}| = 2G \times \sqrt{\frac{3}{2}\left[(d\varepsilon_1'^P)^2 + (d\varepsilon_2'^P)^2 + (d\varepsilon_3'^P)^2\right]} = 2G \times \frac{3}{2}\, d\overline{\varepsilon}^P \qquad (10.9a)$$

where eq(9.6d) and $d\varepsilon_{ij}^P = d\varepsilon_{ij}'^P$ have been used with a projection factor $\sqrt{(3/2)}$ from orthogonal deviatoric axes 1′, 2′ and 3′, each equally inclined at 60° to the deviatoric plane.

Equation (10.9a) shows that |**BD**| is proportional to $\mathrm{d}\overline{\varepsilon}^P$ and extends the $\overline{\sigma}$ vector **O'B**. The incremental stress and strain vectors form the scalar product:

$$\mathbf{BC} \bullet \mathbf{BD} = |\mathbf{BC}|\,|\mathbf{BD}| \cos\varphi = 3G\,(\overline{\sigma}\,\mathrm{d}\overline{\varepsilon}^P)\cos\varphi \tag{10.9b}$$

It follows from eq(10.8d) and (10.9b) that the included angle φ between $\mathrm{d}\overline{\sigma}$ and $\mathrm{d}\overline{\varepsilon}^P$ must be less than 90°. Hence, for loading, it may be deduced [2, 5, 6] that $\mathrm{d}\overline{\varepsilon}^P$ remains in the position of an exterior normal, independently of the direction of $\mathrm{d}\overline{\sigma}$ in Fig. 10.3b. Furthermore, eq(10.8b) shows an included angle $\psi < 90°$ between vectors **AB** and **BD** in Fig. 10.3a. The implication is that the yield surface *f* must always be convex. Figure 10.3c shows how it is possible for $\psi > 90°$ only when a surface is concave.

10.3 Isotropic Hardening

In the rule of isotropic hardening, the initial yield surface expands uniformly in stress space as a material is subjected to elastic–plastic deformation under an outward loading. Thus, all subsequent yield surfaces inflate the shape and orientation of the initial surface. Figure 10.4 illustrates this rule within two common plane stress states: (a) tension-torsion and (b) principal biaxial stress.

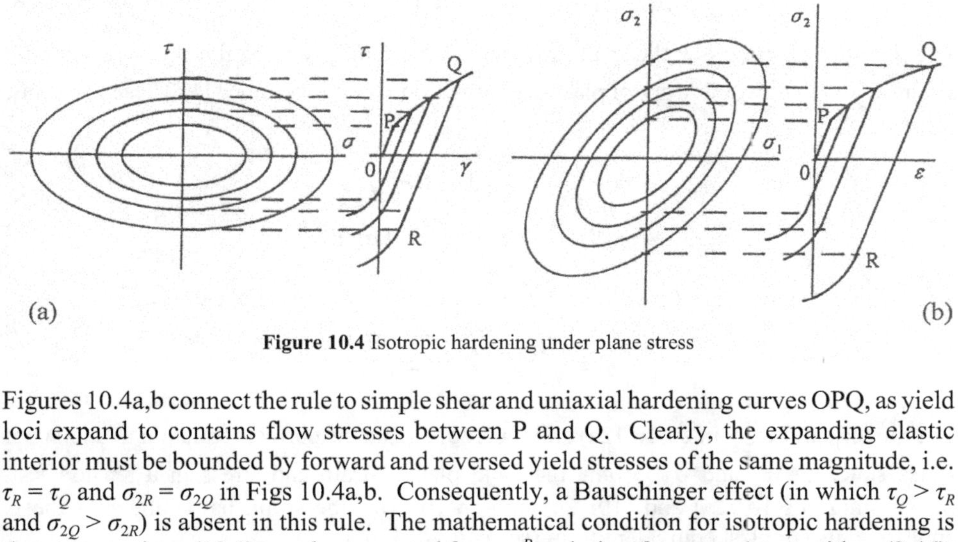

Figure 10.4 Isotropic hardening under plane stress

Figures 10.4a,b connect the rule to simple shear and uniaxial hardening curves OPQ, as yield loci expand to contains flow stresses between P and Q. Clearly, the expanding elastic interior must be bounded by forward and reversed yield stresses of the same magnitude, i.e. $\tau_R = \tau_Q$ and $\sigma_{2R} = \sigma_{2Q}$ in Figs 10.4a,b. Consequently, a Bauschinger effect (in which $\tau_Q > \tau_R$ and $\sigma_{2Q} > \sigma_{2R}$) is absent in this rule. The mathematical condition for isotropic hardening is that stress σ_{ij} in eq(10.1) may be separated from $\varepsilon_{ij}^{\ P}$ and χ in a form consistent with eq(9.46). Hence, the shape of the yield (or loading) surface depends solely upon σ_{ij}, while its size will depend upon χ. This is expressed as

$$f\,(\sigma_{ij}) = \chi\,(\psi) \tag{10.10a}$$

where χ is a hardening function of a suitable scalar quantity ψ of plastic strain or plastic work that increases monotonically under increasing σ_{ij}. Then the change to *f* in eq(10.10a), corresponding to a change in stress $\mathrm{d}\sigma_{ij}$, is given by eq(10.3a):

$$\mathrm{d}f = \frac{\partial f}{\partial \sigma_{ij}}\,\mathrm{d}\sigma_{ij} \tag{10.10b}$$

which depends upon the three conditions of loading given in § 10.2.1 as follows:

(i) $\mathrm{d}f > 0$, $\mathrm{d}f = \mathrm{d}\chi$ and $\mathrm{d}\varepsilon_{ij}{}^P > 0$ for elastic–plastic *loading*

(ii) $\mathrm{d}f < 0$, $\mathrm{d}\chi = 0$ and $\mathrm{d}\varepsilon_{ij}{}^P = 0$ for elastic *unloading*

(iii) $\mathrm{d}f = \mathrm{d}\chi = 0$ and $\mathrm{d}\varepsilon_{ij}{}^P = 0$ for *neutral loading*

No further hardening can occur as a material is either unloaded or is subjected to a neutral stress change. *Note*: In the case of a rigid plastic material, or one that has reached saturation hardening, it is possible for $\mathrm{d}\varepsilon_{ij}{}^P > 0$ under the condition $\mathrm{d}f = 0$.

The specific forms of eq(10.10a) that follow depend upon the function f and the argument ψ. The initial condition of the material governs the choice of yield function f, while ψ will depend upon the hypothesis of hardening.

10.3.1 Levy-Mises

The isotropic function f in eq(10.10a) is most often identified with the second invariant of deviatoric stress $J_2' = \frac{1}{2}\,\sigma_{ij}'\sigma_{ij}'$, so implying that hydrostatic stress does not contribute to plastic deformation. Equation (10.10a) becomes

$$J_2' = \chi(\psi) \quad \Rightarrow \quad \frac{1}{2}\sigma_{ij}'\sigma_{ij}' = \chi(\psi) \tag{10.11a,b}$$

The previous chapter took the scalar measure of hardening ψ to be the equivalent plastic strain $\bar{\varepsilon}^P$ or, the specific work of plastic deformation W^P/unit volume [1]. These give ψ as, respectively,

$$\begin{aligned} \psi &= \bar{\varepsilon}^P = \frac{2}{3}\int \left(\mathrm{d}\varepsilon_{ij}^P\,\mathrm{d}\varepsilon_{ij}^P\right)^{\frac{1}{2}} \\ \psi &= W^P = \int \sigma_{ij}\,\mathrm{d}\varepsilon_{ij}^P = \int \sigma_{ij}'\,\mathrm{d}\varepsilon_{ij}^P \end{aligned} \tag{10.12a,b}$$

The left-hand side of eq(10.11b) may be identified with an equivalent stress $\bar{\sigma}$:

$$\bar{\sigma} = \sqrt{3\,J_2'} = \sqrt{\frac{3}{2}\sigma_{ij}'\sigma_{ij}'} \tag{10.12c}$$

The numerical factors in eq(10.12a) and (10.12c) enable the correct reduction to a uniaxial stress state, i.e. $\bar{\varepsilon}^P$ and $\bar{\sigma}$ become the axial plastic strain and stress in a tension test. Substituting eqs(10.12a) and (10.12c) into eq(10.11b), the strain hardening hypothesis appears in its simplest equivalent form (see eq(9.48))

$$\bar{\sigma} = H\left(\int \mathrm{d}\bar{\varepsilon}^P\right) \tag{10.13a}$$

It follows that the hardening function H in eq(10.13a) may be established from simple tension. The most useful explicit forms of eq(10.13a) are those given by Hollomon [7], Ludwik [8] and Swift [9] (see § 9.8). Alternatively, substituting eqs(10.12b,c) into eq(10.11b) defines the work hardening hypothesis (see eq(9.51a)):

$$\bar{\sigma} = F\left(\int \mathrm{d}W^P\right) \tag{10.13b}$$

Equation (10.13b) is less convenient because function F must be established from the area beneath a curve of stress versus plastic strain. Equations (10.11b) and (10.12c) provide the relationship between the hardening functions in eqs(10.13a,b) as

$$\chi(\psi) = \frac{\bar{\sigma}^2}{3} \quad \text{where} \quad \bar{\sigma} = H\left(\int \mathrm{d}\bar{\varepsilon}^P\right) = F\left(\int \mathrm{d}W^P\right) \tag{10.13c}$$

Taking both stress deviator invariants $J_2' = \frac{1}{2}\sigma_{ij}'\sigma_{ij}'$ and $J_3' = \frac{1}{3}\sigma_{ij}'\sigma_{jk}'\sigma_{ki}'$ into function $f(\sigma_{ij})$ in eq(10.10a) will account for most experimental observations on isotropic yielding [10]. *Note*: The previous chapter showed that including J_3' required a re-definition of $\bar{\sigma}$ and $\bar{\varepsilon}^P$ according to the choice of yield function and hardening hypothesis.

The single most attractive feature of isotopic hardening over alternative hardening rules is its mathematical simplicity. The latter imposes limitations, as will be seen, but for most outward loading paths, the rule of isotropic hardening can predict plastic strains acceptably. Here, the incremental theory of plasticity combines the rule of isotropic hardening with the flow rule, eq(10.4b). Originally, Levy [11] and von Mises [12] did this by associating the plastic potential function $g(\sigma_{ij})$ in eq(10.4a) with a von Mises 'flow potential' $f(\sigma_{ij}) = J_2' = \frac{1}{2}\sigma_{ij}'\sigma_{ij}'$. That is, $f(\sigma_{ij})$ is the subsequent yield function given by isotropic hardening. This leads to a linear constitutive relation between the incremental plastic strain tensor $\mathrm{d}\varepsilon_{ij}^P$ and the current deviatoric stress tensor σ_{ij}':

$$\mathrm{d}\varepsilon_{ij}^P = \mathrm{d}\lambda\,\sigma_{ij}' \tag{10.14}$$

where $\sigma_{ij}' = \sigma_{ij} - \frac{1}{3}\delta_{ij}\sigma_{kk}$. Kronecker's delta, in which $\delta_{ij} = 1$ for $i = j$ and $\delta_{ij} = 0$ for $i \neq j$, ensures that normal stress components are reduced by their mean (or hydrostatic) stress value. The incremental scalar $\mathrm{d}\lambda$ is a factor of proportionality, linking the current equivalent stress to an increment of equivalent plastic strain. When referred to the co-ordinates ($\bar{\sigma}$, $\int \mathrm{d}\bar{\varepsilon}^P$) of a uniaxial flow curve, the mean stress $\sigma_m = \frac{1}{3}\bar{\sigma}$, giving $\sigma_1' = \bar{\sigma} - \frac{1}{3}\bar{\sigma} = \frac{2}{3}\bar{\sigma}$. Also, the axial tensile strain $\mathrm{d}\varepsilon_1^P = \mathrm{d}\bar{\varepsilon}^P$. Equation (10.14) gives $\mathrm{d}\varepsilon_1^P = \mathrm{d}\lambda\sigma_1'$, from which

$$\mathrm{d}\lambda = \frac{3\,\mathrm{d}\bar{\varepsilon}^P}{2\bar{\sigma}} \tag{10.15}$$

Equation (10.15) shows that $\mathrm{d}\lambda$ is inversely proportional to the incremental plastic modulus, i.e. proportional to the incremental compliance $\mathrm{d}\bar{\varepsilon}^P/\bar{\sigma}$. For non-linear hardening, it follows from the strain and work hardening hypotheses, eqs(10.13a,b), that

$$H' = \frac{\mathrm{d}\bar{\sigma}}{\mathrm{d}\bar{\varepsilon}^P} \tag{10.16a}$$

$$F' = \frac{\mathrm{d}\bar{\sigma}}{\mathrm{d}W^P} = \frac{\mathrm{d}\bar{\sigma}/\mathrm{d}\bar{\varepsilon}^P}{\mathrm{d}W^P/\mathrm{d}\bar{\varepsilon}^P} = \frac{\mathrm{d}\bar{\sigma}/\mathrm{d}\bar{\varepsilon}^P}{\bar{\sigma}} \tag{10.16b}$$

Correspondingly, the compliance may be expressed, from eqs(10.16a,b), in one of two forms:

$$\frac{\mathrm{d}\bar{\varepsilon}^P}{\bar{\sigma}} = \frac{\mathrm{d}\bar{\sigma}}{\bar{\sigma}H'} \quad \text{or} \quad \frac{\mathrm{d}\bar{\varepsilon}^P}{\bar{\sigma}} = \frac{\mathrm{d}\bar{\sigma}}{F'\bar{\sigma}^2} \tag{10.17a,b}$$

Combining eqs(10.15) and (10.17a,b) shows how $\mathrm{d}\lambda$ changes monotonically with an increasing $\bar{\sigma}$ during loading:

$$\mathrm{d}\lambda = \frac{3\,\mathrm{d}\bar{\sigma}}{2H'\bar{\sigma}} \quad \text{or} \quad \mathrm{d}\lambda = \frac{3\,\mathrm{d}\bar{\sigma}}{2F'\bar{\sigma}^2} \tag{10.18a,b}$$

Note: Eqs(10.18a,b) also derive from eq(10.5) provided f is expressed in an isotropic hardening form of eq(10.6). For a principal triaxial stress system σ_1, σ_2 and σ_3, eq(10.14) expands to give principal, plastic strain increments:

$$\mathrm{d}\varepsilon_1^P = \frac{\mathrm{d}\bar{\varepsilon}^P}{\bar{\sigma}}\left[\sigma_1 - \frac{1}{2}(\sigma_2 + \sigma_3)\right] \tag{10.19a}$$

$$\mathrm{d}\varepsilon_2^P = \frac{\mathrm{d}\bar{\varepsilon}^P}{\bar{\sigma}}\left[\sigma_2 - \frac{1}{2}(\sigma_3 + \sigma_1)\right] \tag{10.19b}$$

$$\mathrm{d}\varepsilon_3^P = \frac{\mathrm{d}\bar{\varepsilon}^P}{\bar{\sigma}}\left[\sigma_3 - \frac{1}{2}(\sigma_1 + \sigma_2)\right] \tag{10.19c}$$

where $\mathrm{d}\bar{\varepsilon}^P/\bar{\sigma}$ is given by either of eqs(10.17a,b). Numerical integration of eqs(10.19a,b,c) is usually required except in the case of uniform biaxial and triaxial stress, where components increase proportionately or follow a single step during loading. Equations (10.19a–c) are often compared with triaxial elastic relations. For the latter, ν replaces ½ for compressible elasticity and $\mathrm{d}\bar{\varepsilon}^P/\bar{\sigma}$ is replaced by an elastic compliance $1/E$. More important is a recognition of the essential difference between constitutive relations for non-linear plasticity and linear elasticity.

Example 10.1 Derive the strain and work hardening derivatives F' and H', respectively, for the non-dimensional form of the Ludwik hardening law, eq(9.43b).

Firstly, H' in eq(10.16a) follows from

$$H' = \sigma_o \times \frac{\mathrm{d}}{\mathrm{d}\bar{\varepsilon}^P}\left[1 + \left(\frac{\bar{\varepsilon}^P}{\varepsilon_o}\right)^n\right]$$

$$H' = \frac{\mathrm{d}\bar{\sigma}}{\mathrm{d}\bar{\varepsilon}^P} = n\left(\frac{\sigma_o}{\varepsilon_o}\right)\left(\frac{\bar{\sigma}}{\sigma_o} - 1\right)^{\frac{n-1}{n}} \tag{10.16c}$$

The specific plastic work as an argument in eq(10.13b) becomes

$$w^P = \sigma_o \int_0^{\bar{\varepsilon}^P}\left[1 + \left(\frac{\bar{\varepsilon}^P}{\varepsilon_o}\right)^n\right]\mathrm{d}\bar{\varepsilon}^P = \sigma_o\varepsilon_o\left(\frac{\bar{\sigma}}{\sigma_o} - 1\right)^{\frac{1}{n}}\left[1 + \frac{1}{(n+1)}\left(\frac{\bar{\sigma}}{\sigma_o} - 1\right)\right]$$

in which the derivative required is found from an inverted derivative:

$$F' = \frac{\mathrm{d}\bar{\sigma}}{\mathrm{d}w^P} = \frac{1}{(\mathrm{d}w^P/\mathrm{d}\bar{\sigma})}$$

where
$$\frac{\mathrm{d}w^P}{\mathrm{d}\bar{\sigma}} = \frac{2\varepsilon_o}{n+1}\left(\frac{\bar{\sigma}}{\sigma_o} - 1\right)^{\frac{1}{n}} + \sigma_o\varepsilon_o$$

10.3.2 Drucker's Potential

Employing the Drucker 'loading function' [13] as a flow potential, eq(10.10a) becomes

$$f(\sigma_{ij}) = J'^{3}_{2} - cJ'^{2}_{3} = \chi(\psi) \tag{10.20}$$

Now, for any isotropic function $f(J_2', J_3')$, the associated flow rule becomes

$$\mathrm{d}\varepsilon^P_{ij} = \mathrm{d}\lambda \frac{\partial f}{\partial \sigma'_{ij}} = \mathrm{d}\lambda \left[\left(\frac{\partial f}{\partial J_2'}\right)\left(\frac{\partial J_2'}{\partial \sigma'_{ij}}\right) + \left(\frac{\partial f}{\partial J_3'}\right)\left(\frac{\partial J_3'}{\partial \sigma'_{ij}}\right) \right]$$

$$= \mathrm{d}\lambda \left[\left(\frac{\partial f}{\partial J_2'}\right)\sigma'_{ij} + \left(\frac{\partial f}{\partial J_3'}\right)\left(\sigma'_{ik}\sigma'_{kj} - \frac{2}{3}J_2'\delta_{ij}\right) \right]$$

$$= \mathrm{d}\lambda \left[\left(\frac{\partial f}{\partial J_2'}\right)\sigma'_{ij} + \left(\frac{\partial f}{\partial J_3'}\right)t'_{ij} \right] \tag{10.21a}$$

where $t'_{ij} = \partial J_3'/\partial \sigma'_{ij} = \sigma'_{ik}\,\sigma'_{kj} - ⅔\, J_2'\delta_{ij}$. Substituting $f(\sigma_{ij})$ from eq(10.20) into eq(10.21a) gives Drucker's flow rule

$$\mathrm{d}\varepsilon^P_{ij} = \mathrm{d}\lambda\,(3J'^{2}_{2}\sigma'_{ij} - 2cJ'_3\,t'_{ij}) \tag{10.21b}$$

Principal plastic strain increments $\mathrm{d}\varepsilon_1^P$, $\mathrm{d}\varepsilon_2^P$ and $\mathrm{d}\varepsilon_3^P$ follow from eq(10.21b) with the following substitutions:

$$J_2' = ½\,(\sigma'^2_1 + \sigma'^2_2 + \sigma'^2_3), \quad J_3' = ⅓\,(\sigma'^3_1 + \sigma'^3_2 + \sigma'^3_3)$$

$$t_1' = \sigma'^2_1 - ⅔\,J_2' = \sigma'^2_1 - ⅓\,(\sigma'^2_1 + \sigma'^2_2 + \sigma'^2_3)$$

$$= ⅓\,(2\sigma'^2_1 - \sigma'^2_2 - \sigma'^2_3)$$

and similarly

$$t_2' = ⅓\,(2\sigma'^2_2 - \sigma'^2_1 - \sigma'^2_3)$$

$$t_3' = ⅓\,(2\sigma'^2_3 - \sigma'^2_1 - \sigma'^2_2)$$

Converting to absolute stress components

$$\sigma_1' = \sigma_1 - ⅓\,(\sigma_1 + \sigma_2 + \sigma_3)$$

$$\sigma_2' = \sigma_2 - ⅓\,(\sigma_1 + \sigma_2 + \sigma_3)$$

$$\sigma_3' = \sigma_3 - ⅓\,(\sigma_1 + \sigma_2 + \sigma_3)$$

giving

$$J_2' = [(\sigma_1 - \sigma_2)^2 + (\sigma_1 - \sigma_3)^2 + (\sigma_2 - \sigma_3)^2]\,/\,6$$

$$J_3' = [(2\sigma_1 - \sigma_2 - \sigma_3)^3 + (2\sigma_2 - \sigma_1 - \sigma_3)^3 + (2\sigma_3 - \sigma_1 - \sigma_2)^3]\,/\,81$$

$$t_1' = [(\sigma_1 - \sigma_2)^2 + (\sigma_1 - \sigma_3)^2 - 2(\sigma_2 - \sigma_3)^2]\,/\,9$$

$$t_2' = [(\sigma_1 - \sigma_2)^2 + (\sigma_2 - \sigma_3)^2 - 2(\sigma_1 - \sigma_3)^2]\,/\,9$$

$$t_3' = [(\sigma_1 - \sigma_3)^2 + (\sigma_2 - \sigma_3)^2 - 2(\sigma_1 - \sigma_2)^2]\,/\,9$$

The scalar $\mathrm{d}\lambda$ is again determined from the given yield function by setting $\sigma_2 = \sigma_3 = 0$ for a uniaxial stress. Equation (10.21b) leads to

$$\mathrm{d}\varepsilon^P_1 = \frac{2}{9}\,\mathrm{d}\lambda\,\sigma_1^5\left(1 - \frac{4c}{27}\right) \tag{10.22a}$$

so that
$$\mathrm{d}\lambda = \frac{9\,\mathrm{d}\bar{\varepsilon}^P}{2\bar{\sigma}^5(1 - 4c/27)} \tag{10.22b}$$

where the equivalent stress is given by eq(9.4a). The equivalent strain is defined either by the strain hardening hypothesis, eq(9.7c), or from the work hardening hypothesis, eq(9.52c). According to each hypothesis, $\mathrm{d}\lambda$ follows from eqs(10.16a,b) and (10.22b), respectively:

$$\mathrm{d}\lambda = \frac{9\,\mathrm{d}\bar{\sigma}}{2\bar{\sigma}^5 H'(1 - 4c/27)} \quad \text{and} \quad \mathrm{d}\lambda = \frac{9\,\mathrm{d}\bar{\sigma}}{2\bar{\sigma}^6 F'(1 - 4c/27)} \tag{10.23a,b}$$

The derivatives $H' = \mathrm{d}H/\mathrm{d}\bar{\varepsilon}^P$ and $F' = \mathrm{d}F/\mathrm{d}w^P$ eliminate strain in their arguments so allowing eqs (10.23a,b) to express the monotonic dependence of $\mathrm{d}\lambda$ upon $\bar{\sigma}$. *Note*: Eq(10.20) provides a uniaxial relationship with χ taken as each hardening function H and F:

$$\bar{\sigma}^6\left(\frac{1}{27} - \frac{4c}{9^3}\right) = \chi(\psi) \quad \text{where:} \quad \psi = \int \mathrm{d}\bar{\varepsilon}^P \quad \text{or} \quad \psi = \int \mathrm{d}w^P \tag{10.23c}$$

10.4 Verification of Levy-Mises and Drucker Flow Rules

Many workers have conducted radial outward loadings upon thin-walled cylinders by combining tension, torque and internal pressure. The linear plastic strain paths observed in annealed material have confirmed eq(10.14). That is, for a given uniform stress state within the wall, flow theory predicts a constant ratio between the plastic strain increment components under a proportional loading (e.g. see Fig. 10.5).

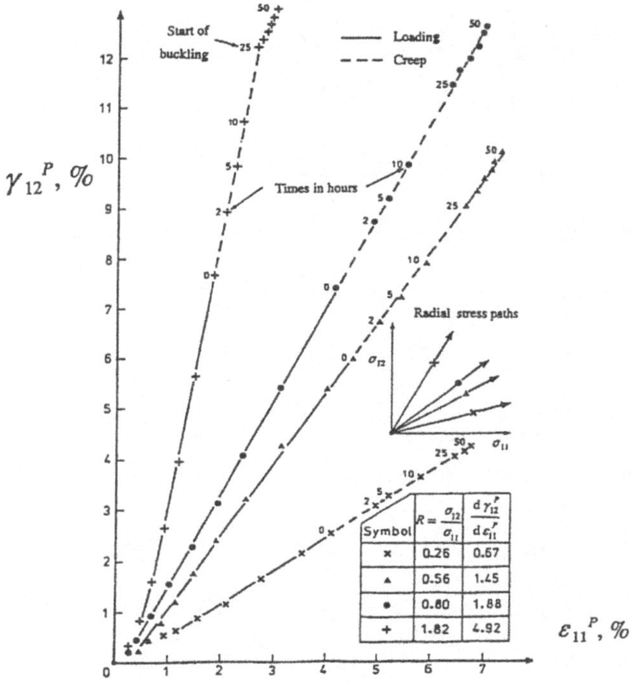

Figure 10.5 Linear plastic strain paths under radial tension-torsion

10.4.1 Combined Tension-Torsion

The linear plastic strain paths shown in Fig. 10.5 apply to radial loading under combined tension-torsion of a 99.8% commercially pure, annealed aluminium (E1A). Four constant stress ratios were employed: $R = \sigma_{12}/\sigma_{11} = 0.26$, 0.56, 0.80 and 1.82 [14]. In the absence of torsional buckling, Fig. 10.5 also shows that when constant, combined loads are maintained for a period of 50 hours, the strain paths continue to extend in time without changing their gradients. The occurrence of room-temperature creep conforms to earlier observations for a similar purity 1100 aluminium [15, 16]. The creep behaviour suggests that a strain rate form of eqs(10.19a,b,c) applies when the essential slip mechanisms of low-temperature deformation are common to plasticity and creep. This may not apply to higher temperatures, where creep deformation involves thermally activated climb and cross-slip of dislocations, together with grain coarsening and cavitation. Linearity has often been seen in many plastic strain paths arising from tension–torsion loading [15–20]. Their gradients, including those from Fig. 10.5, have been used to construct Fig. 10.6.

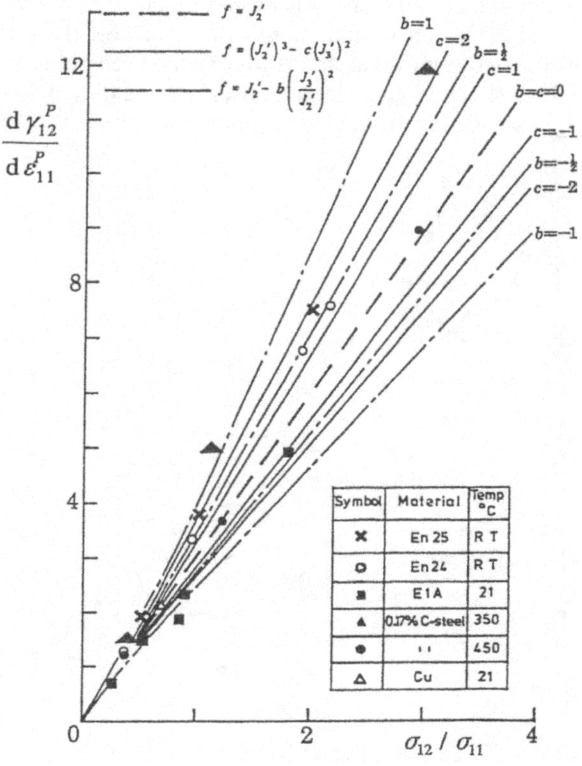

Figure. 10.6 Stress-dependent plastic strain increment ratio

These data apply to radial loadings in En steel [17, 18], low C-steels at elevated temperature [18, 19] and commercially pure copper [20]. Predictions from eq(10.14) are simply

$$d\gamma_{12}^P / d\varepsilon_{11}^P = 3\sigma_{12} / \sigma_{11} = 3R \tag{10.24a}$$

$$d\varepsilon_{22}^P / d\varepsilon_{11}^P = -1/2 \tag{10.24b}$$

Equation (10.24a) gives the broken line in Fig. 10.6, this being representative of J_2' materials only. Improved predictions are found [10, 21] for three homogenous flow potentials involving both invariants $f(J_2', J_3')$. For example, eq(10.21b) gives plastic strain increment ratios associated with Drucker's flow potential:

$$\frac{d\gamma_{12}^P}{d\varepsilon_{11}^P} = \frac{3R(R^2 + 1/3)^2 - (2Rc/3^2)(R^2 + 2/3^2)}{(R^2 + 1/3)^2 - c(2/3^3 + R^2/3^2)(R^2 + 2/3^2)} \tag{10.25a}$$

$$\frac{d\varepsilon_{22}^P}{d\varepsilon_{11}^P} = \frac{-(R^2 + 1/3)^2 - (2c/3^2)(R^2 + 2/3^2)(R^2 - 1/3)}{2(R^2 + 1/3)^2 - 2c(2/3^3 + R^2/3^2)(R^2 + 2/3^2)} \tag{10.25b}$$

It appears from Fig. 10.6, that eqs(10.25a,b) represent the data for certain steels better than $f = J_2'$ can. Heat treatment rendered these alloys with a uniform, isotropic grain size which permits the selection of a suitable alternative isotropic function $f(J_2', J_3')$. Figure 10.7 presents a more convincing demonstration of the J_3' influence. This is provided by experiments where both diametral and axial strains were found in combined tension–torsion experiments upon En 25 steel [22], copper, aluminium and two mild steels [23]. These show that the strain ratio $d\varepsilon_{22}^P/d\varepsilon_{11}^P$ differs from the $-\frac{1}{2}$ prediction from eq(10.24b).

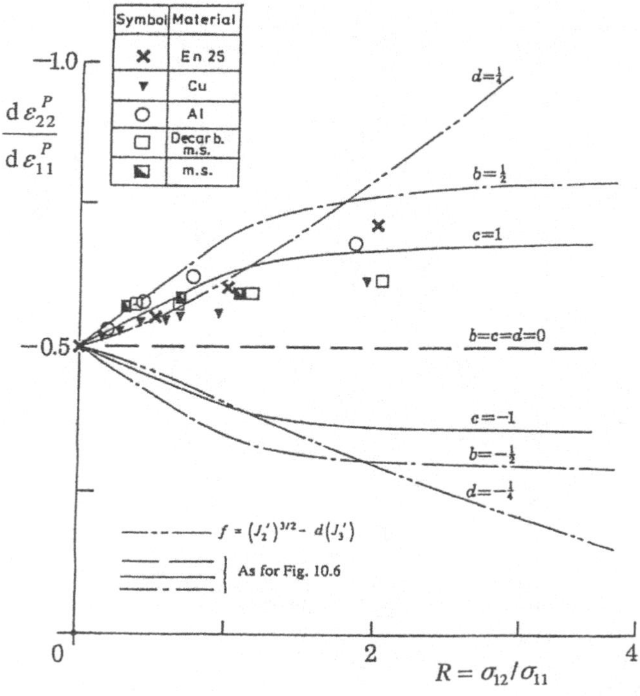

Figure 10.7 Departure from $f = J_2'$ theory

It is seen that the effect that stress ratio R has on decreasing the strain ratio can be predicted from eq(10.25b) and two further homogenous yield functions in J_2' and J_3', as indicated.

10.4.2 Lode Parameters

An alternative check on eq(10.14) is provided by Lode's [24] respective stress and strain parameters μ and ν:

$$\nu = \frac{2\mathrm{d}\varepsilon_3^P - \mathrm{d}\varepsilon_1^P - \mathrm{d}\varepsilon_2^P}{\mathrm{d}\varepsilon_1^P - \mathrm{d}\varepsilon_2^P} = \frac{2\mathrm{d}\lambda\sigma_3' - \mathrm{d}\lambda\sigma_1' - \mathrm{d}\lambda\sigma_2'}{\mathrm{d}\lambda\sigma_1' - \mathrm{d}\lambda\sigma_2'} = \frac{2\sigma_3 - \sigma_1 - \sigma_2}{\sigma_1 - \sigma_2} = \mu \tag{10.26}$$

An observed equality between μ and ν for a given radial path will validate eq(10.14). Substituting the principal plastic strains and stresses under tension-torsion (σ_{11}, σ_{12}) into eq(10.26) gives Lode's strain parameter:

$$\nu = -\frac{\mathrm{d}\varepsilon_{11}^P}{\sqrt{(\mathrm{d}\varepsilon_{11}^P)^2 + (2/3)^2(\mathrm{d}\gamma_{12}^P)^2}} = -\frac{1}{\sqrt{1 + (2/3)^2(\mathrm{d}\gamma_{12}^P/\mathrm{d}\varepsilon_{11}^P)^2}} \tag{10.27a}$$

Lode's stress parameter becomes

$$\mu = -\frac{\sigma_{11}}{\sqrt{\sigma_{11}^2 + 4\sigma_{12}^2}} = -\frac{1}{\sqrt{1 + 4(\sigma_{12}/\sigma_{11})^2}} \tag{10.27b}$$

Employing the observed plastic strain gradient $\mathrm{d}\gamma_{12}^P/\mathrm{d}\varepsilon_{11}^P$ for each stress ratio σ_{12}/σ_{11} within eqs(10.27a,b) has enabled the construction of Fig. 10.8.

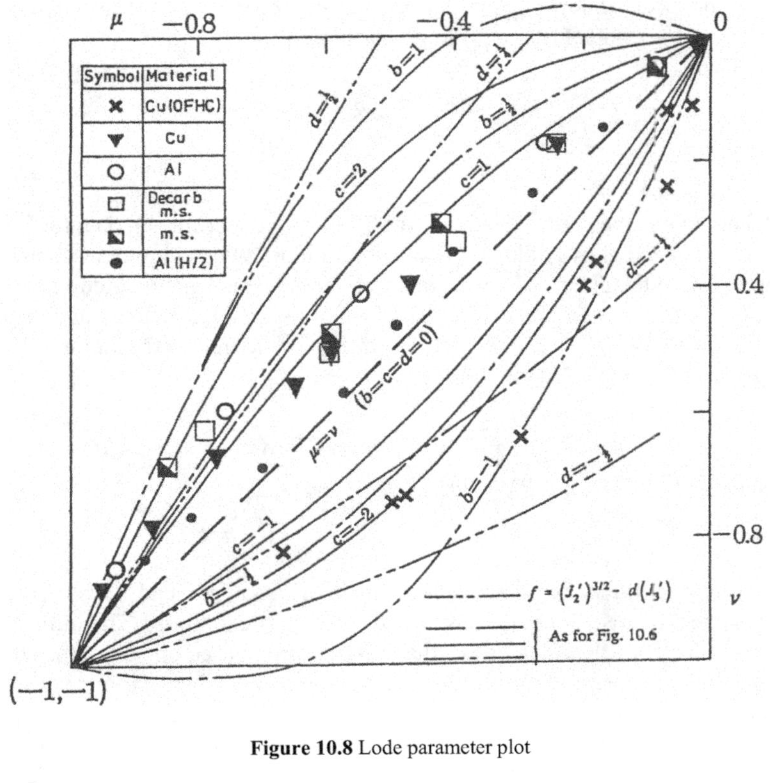

Figure 10.8 Lode parameter plot

This Lode parameter plot applies to steels [22, 23], aluminium [23, 25] and copper [23, 26]. It also reproduces the μ, ν data from pioneering experiments performed by Taylor and Quinney in 1931 [23]. They were the first to reveal a systematic deviation from $\mu = \nu$ in each material. The precise reason for the deviation remains uncertain. It may be due to a residual anisotropy in these materials because tests were not conducted to establish beyond doubt a condition of initial isotropy. However, each material in Fig. 10.8 was annealed so, when isotropy is assumed, it may imply that any deviation from $\mu = \nu$ is due to the influence of the third invariant J_3'. For example, using Drucker's yield function in eq(10.20), it can be shown [10] that the non-linear relationship between μ and ν is given by

$$\nu = \mu \left[\frac{3^2(\mu^2 + 3)^2 - 2^2 c(\mu^2 - 3)(\mu^2 - 3^2)}{3^3(\mu^2 + 3)^2 + 2^3 c\mu^2(\mu^2 - 3^2)} \right] \tag{10.28}$$

Figure 10.8 shows that eq(10.28) and a further homogenous function $f(J_2', J_3')$ can account for the trends observed through the choice of constants b and c. An exception is found in the function containing constant d because J_3' changes sign with stress. Consequently, this function leads to a discontinuity at $\mu = 0$.

10.4.3 Principal Biaxial Stress

Consider proportional stressing under combined internal pressure-axial loading of a thin-walled cylinder. Principal stresses σ_1 and σ_2 in the axial and circumferential directions, respectively, increase proportionately in a ratio $R = \sigma_1/\sigma_2$. The Levy-Mises prediction sets the radial stress $\sigma_3 = 0$ in eq(10.19a,b) to give a strain ratio

$$\frac{d\varepsilon_1^P}{d\varepsilon_2^P} = \frac{2R - 1}{2 - R} \tag{10.29}$$

Figure 10.9 compares the prediction from eq(10.29) with experimental data for En 24 [17], stainless steel [27] and alloys of lead [28] and zirconium [29]. A similar conclusion can be drawn in relation to the observed deviations from eq(10.29). That is, predictions from the alternative homogenous functions $f(J_2', J_3')$, given in Fig. 10.9, usually match the observed behaviour in a given alloy. For example, Drucker's prediction to the strain ratio is found from putting $\sigma_3 = 0$ in eq(10.21b):

$$\frac{d\varepsilon_1^P}{d\varepsilon_2^P} = \frac{3^4(1 + R^2 - R)^2(2R - 1) - 2c(2R^2 - 1 - 2R)\left[(2 - R)^3 + (2R - 1)^3 - (1 + R)^3\right]}{3^4(1 + R^2 - R)^2(2 - R) - 2c(2 - R^2 - 2R)\left[(2 - R)^3 + (2R - 1)^3 - (1 + R)^3\right]} \tag{10.30}$$

Here, however, the influence of initial anisotropy in 'as received' lead and zirconium alloys is more likely for the discrepancy found with isotropic predictions. The initial condition reported for each alloy [28, 29] suggests that a strain ratio, based upon an orthotropic yield function, would be a more appropriate choice than eq(10.30) when explaining these results (see chapter 12, § 12.6).

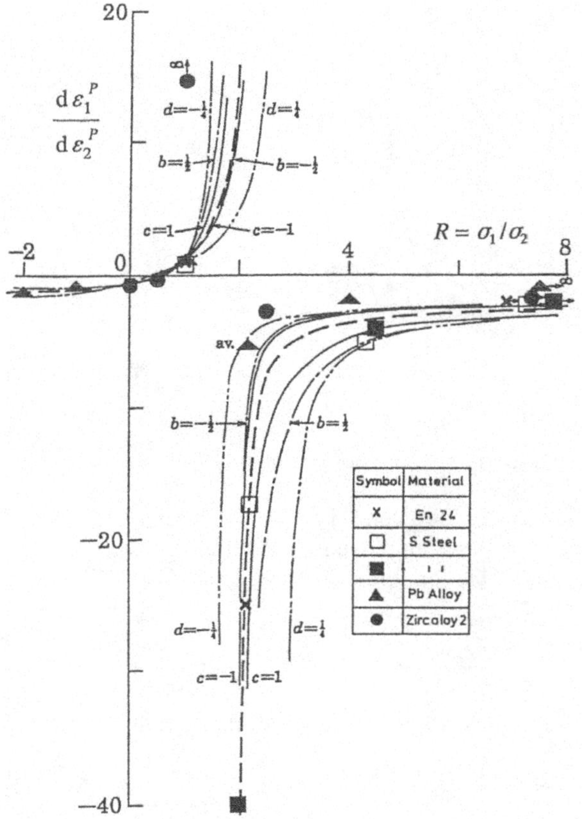

Figure 10.9 Dependence of strain increment ratio upon principal stress

10.4.4 Flow Equivalence

Consider again proportional loading of aluminium under combined tension-torsion. Figure 10.5 shows linear plastic strain paths, between shear and axial components, resulting from increasing the corresponding stress components under constant ratios. The measurement of hoop strain under simple tension confirmed eq(10.24b), indicating that an isotropic yield function applies. A slight deviation from the same gradient, $d\varepsilon_{22}^{P} / d\varepsilon_{11}^{P} = -\frac{1}{2}$, occurred with tension and torsion combined. Moreover, the observed slopes are less than the von Mises prediction $d\gamma_{12}^{P} / d\varepsilon_{11}^{P} = 3R$, from eq(10.24a). Taking co-ordinates 1, 2 and 3, in alignment with the axial, circumferential and radial directions, respectively, the von Mises equivalent stress and strain are found from eqs(9.2c) and (9.6c):

$$\bar{\sigma} = \sqrt{\sigma_{11}^2 + 3\sigma_{12}^2} \tag{10.31a}$$

$$d\bar{\varepsilon}^P = \sqrt{(d\varepsilon_{11}^P)^2 + (1/3)(d\gamma_{12}^P)^2} = d\varepsilon_{11}^P\sqrt{1 + (1/3)(d\gamma_{12}^P/d\varepsilon_{11}^P)^2} \tag{10.31b}$$

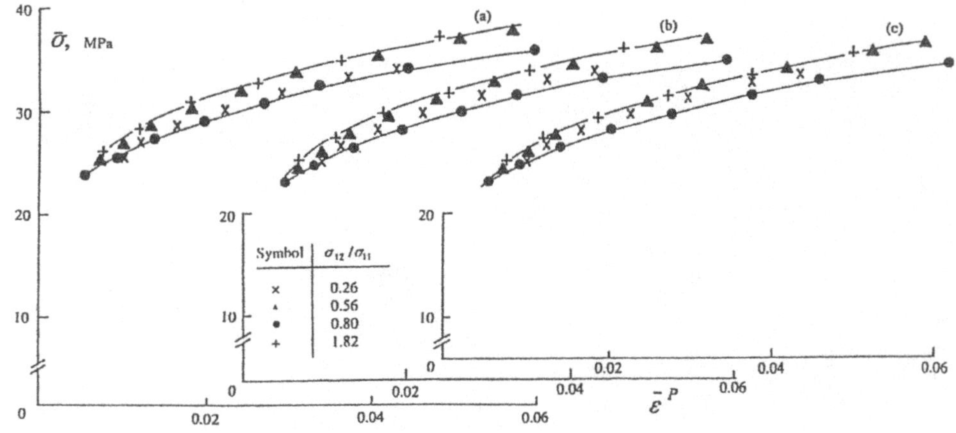

Figure 10.10 Equivalence correlations under combined tension-torsion

Equations (10.31a,b) provide a von Mises correlation in Fig. 10.10a, this being independent of the hardening hypotheses employed for the four radial tests.

Using Drucker's function, the stress invariants J_2' and J_3' in eq(9.5a), reduce to

$$J_2' = \tfrac{1}{2}\,\sigma_{ij}'\,\sigma_{ij}' = \sigma_{12}^{\,2} + \sigma_{11}^{\,2}/\,3, \tag{10.32a}$$

$$J_3' = \tfrac{1}{3}\,\sigma_{ij}'\,\sigma_{jk}'\,\sigma_{ki}' = 2\sigma_{11}^{\,3}/\,27 + \sigma_{11}\,\sigma_{12}^{\,2}/3 \tag{10.32b}$$

Substituting eqs(10.32a,b) into eq(9.4a) gives the equivalent stress expression

$$\bar{\sigma} = \left[\frac{(\sigma_{12}^2 + \sigma_{11}^2/3)^3 - c(2\sigma_{11}^3/3^3 + \sigma_{11}\sigma_{12}^2/3)^2}{1/3^3 - 2^2c/3^6}\right]^{\frac{1}{6}} \tag{10.33}$$

By including J_3' in the yield function, the expression for $\mathbf{d}\bar{\boldsymbol{\varepsilon}}^P$ will depend upon the choice between the work and strain hardening hypothesis.

(a) Strain Hypothesis

When the strain hardening hypothesis defines $\mathbf{d}\bar{\boldsymbol{\varepsilon}}^P$, the invariants in eq(9.5b) become

$$I_2 = \frac{1}{2}\,\mathrm{d}\varepsilon_{ij}^{\,P}\,\mathrm{d}\varepsilon_{ij}^{\,P} = \frac{1}{2}\,[(\mathrm{d}\varepsilon_{11}^{\,P})^2 + (\mathrm{d}\varepsilon_{12}^{\,P})^2 + (\mathrm{d}\varepsilon_{21}^{\,P})^2 + (\mathrm{d}\varepsilon_{22}^{\,P})^2 + (\mathrm{d}\varepsilon_{33}^{\,P})^2]$$

$$I_3 = \frac{1}{3}\,\mathrm{d}\varepsilon_{ij}^{\,P}\,\mathrm{d}\varepsilon_{jk}^{\,P}\,\mathrm{d}\varepsilon_{ki}^{\,P} = \frac{1}{3}\,[(\mathrm{d}\varepsilon_{11}^{\,P})^3 + \mathrm{d}\varepsilon_{11}^{\,P}(\mathrm{d}\varepsilon_{12}^{\,P})^2 + \mathrm{d}\varepsilon_{11}^{\,P}(\mathrm{d}\varepsilon_{21}^{\,P})^2$$
$$+\,\mathrm{d}\varepsilon_{22}^{\,P}(\mathrm{d}\varepsilon_{21}^{\,P})^2 + \mathrm{d}\varepsilon_{11}^{\,P}(\mathrm{d}\varepsilon_{12}^{\,P})^2 + \mathrm{d}\varepsilon_{22}^{\,P}(\mathrm{d}\varepsilon_{21}^{\,P})^2 + \mathrm{d}\varepsilon_{22}^{\,P}(\mathrm{d}\varepsilon_{12}^{\,P})^2 + (\mathrm{d}\varepsilon_{22}^{\,P})^3 + (\mathrm{d}\varepsilon_{33}^{\,P})^3]$$

Substituting $\mathrm{d}\varepsilon_{11}^{\,P} = \mathbf{d}\bar{\boldsymbol{\varepsilon}}^P$, with $\mathrm{d}\varepsilon_{22}^{\,P} = \mathrm{d}\varepsilon_{33}^{\,P} = -\tfrac{1}{2}\,\mathbf{d}\bar{\boldsymbol{\varepsilon}}^P$ and $\mathrm{d}\varepsilon_{12}^{\,P} = \mathrm{d}\varepsilon_{21}^{\,P} = \tfrac{1}{2}\mathrm{d}\gamma_{12}^{\,P}$, simplifies these expressions to

$$I_2 = \frac{1}{4}\left[(3\,\mathrm{d}\varepsilon_{11}^P)^2 + (\mathrm{d}\gamma_{12}^P)^2\right] \tag{10.34a}$$

$$I_3 = \frac{1}{4}\mathrm{d}\varepsilon_{11}^P\left[(\mathrm{d}\varepsilon_{11}^P)^2 + (\mathrm{d}\gamma_{12}^P)^2/2\right] \tag{10.34b}$$

Substituting eqs(10.34a,b) into eq(9.7a) gives

$$\mathrm{d}\bar{\varepsilon}^P = \left\{\frac{(1/2^2)\left[(3\,\mathrm{d}\varepsilon_{11}^P)^2 + (\mathrm{d}\gamma_{12}^P)^2\right]^3 - c\left[(\mathrm{d}\varepsilon_{11}^P)^3 + (1/2)\mathrm{d}\varepsilon_{11}^P(\mathrm{d}\gamma_{12}^P)^2\right]^2}{3^3/2^2 - c}\right\}^{\frac{1}{6}} \tag{10.35}$$

The constant c in eq(10.35) is found from matching observed $\mathrm{d}\gamma_{12}^P/\mathrm{d}\varepsilon_{11}^P$ ratios in Fig. 10.5 with the corresponding prediction. Derivatives of Drucker's function $f(J_2', J_3') = J_2'^3 - cJ_3'^2$ are $\partial f/\partial J_2' = 3J_2'^2$ and $\partial f/\partial J_3' = -2cJ_3'$, so that eq(10.21b) gives the axial, circumferential and shear strain increments, respectively:

$$\mathrm{d}\varepsilon_{11}^P = \mathrm{d}\lambda\,[3(\sigma_{11}^2/3 + \sigma_{12}^2)^2(2\sigma_{11}/3) - 2c(2\sigma_{11}^3/27 + \sigma_{11}\sigma_{12}^2/3)\,t_{11}'] \tag{10.36a}$$

$$\mathrm{d}\varepsilon_{22}^P = \mathrm{d}\lambda\,[3(\sigma_{11}^2/3 + \sigma_{12}^2)^2(-\sigma_{11}/3) - 2c(2\sigma_{11}^3/27 + \sigma_{11}\sigma_{12}^2/3)\,t_{22}'] \tag{10.36b}$$

$$\mathrm{d}\varepsilon_{12}^P = \mathrm{d}\gamma_{12}^P/2 = \mathrm{d}\lambda\,[3(\sigma_{11}^2/3 + \sigma_{12}^2)^2\sigma_{12} - 2c\,(2\sigma_{11}^3/27 + \sigma_{11}\sigma_{12}^2/3)\,t_{12}'] \tag{10.36c}$$

where $t_{11}' = (2\sigma_{11}^2 + 3\sigma_{12}^2)/9$, $t_{22}' = (-\sigma_{11}^2 + 3\sigma_{12}^2)/9$ and $t_{12}' = \sigma_{11}\sigma_{12}/3$. Dividing eqs(10.36b,c) by eq(10.36a) leads to the theoretical plastic strain increment ratios in eqs(10.25a,b). Identifying eqs(10.25a,b) with the gradients in Fig. 10.5, $c = -2$ is chosen within its limiting range, $-27/8 \le c \le 9/4$, for a convex yield function. Equations (10.33), (10.35) and (10.25a) provide theoretical equivalence relationships from strain hardening but the corresponding plot in Fig. 10.10b appears to be no better than that found from the simplified von Mises J_2' correlation (see Fig. 10.10b).

(b) Work Hypothesis

Here $\bar{\sigma}$ remains as defined in eq(10.33) but $\mathrm{d}\bar{\varepsilon}^P$ is re-defined from the work hypothesis (see eq 9.52c):

$$\bar{\sigma}\,\mathrm{d}\bar{\varepsilon}^P = \sigma_{ij}\,\mathrm{d}\varepsilon_{ij}^P = \sigma_{11}\,\mathrm{d}\varepsilon_{11}^P + \sigma_{12}\,\mathrm{d}\gamma_{12}^P$$
$$\mathrm{d}\bar{\varepsilon}^P = (\sigma_{11}/\bar{\sigma})(\mathrm{d}\varepsilon_{11}^P + R\,\mathrm{d}\gamma_{12}^P) \tag{10.37a}$$

Substituting eq(10.33) into eq(10.37a) gives

$$\mathrm{d}\bar{\varepsilon}^P = \left[\frac{1/3^3 - (2^2/3^6)c}{(R^2 + 1/3)^3 - c(R^2/3 + 2/3^3)^2}\right]^{\frac{1}{6}}(\mathrm{d}\varepsilon_{11}^P + R\,\mathrm{d}\gamma_{12}^P) \tag{10.37b}$$

where eq(10.25a) expresses the dependence of R upon the ratio $\mathrm{d}\gamma_{12}^P/\mathrm{d}\varepsilon_{11}^P$. Using the observed ratios for R and $\mathrm{d}\gamma_{12}^P/\mathrm{d}\varepsilon_{11}^P$ in eqs(10.25a) and (10.37b), Fig. 10.10c shows that the work hypothesis provides a slightly improved correlation compared to Figs 10.10a,b. The point to note here is that the associated flow rule will match the strain ratios independently of the work hardening hypothesis.

10.5 Non Associated Flow Rules

Up to now, only isotropic yield functions $f(J_2', J_3')$ have been considered. This implies that the chosen expressions for $\bar{\sigma}$ and $\mathrm{d}\bar{\varepsilon}^P$ (i.e. by von Mises and Drucker) are restricted to material whose flow behaviour is independent of direction. Here, a non-coincidence of flow curves, when based upon the conventional von Mises definitions of equivalent stress and strain, has been attributed to the influence of the third invariant. Strictly, when a material is initially anisotropic to a measurable degree, isotropic invariants should not be used to define its equivalence. Instead, a potential is required that characterises the nature of anisotropy, e.g. a normal or orthotropic form typical of, say, rolled sheet (see chapter 12).

The literature reveals instances where the flow rule has not been associated with the yield criterion. This allows the function employed within the flow rule (the flow potential) to differ from that used to define the yield criterion. This more flexible approach has improved predictions to observed behaviour when compared to making the usual association between the yield criterion and flow potential. Using both the Tresca and von Mises functions will serve to illustrate a non-associated approach with unrelated equivalence and flow. For a Tresca yield function, the equivalent stress is simply

$$\bar{\sigma} = \sigma_1 - \sigma_3 \tag{10.38a}$$

where σ_1 and σ_3 are the greatest and least principal stresses, i.e. the roots to the characteristic eq(1.24a). Tresca's equivalent strain is found from eq(10.38a), corresponding to a uniaxial work expression (9.51b):

$$d\bar{\varepsilon}^P = \frac{2}{3}\left(d\varepsilon_1^P - d\varepsilon_3^P\right) \tag{10.38b}$$

While eqs(10.38a,b) reduce correctly to $\bar{\sigma} = \sigma_1$ and $d\bar{\varepsilon}^P = d\varepsilon_1^P$ for uniaxial deformation, they will otherwise differ from a von Mises equivalence. For example, consider pure shear, where the stresses are $\sigma_1 = \tau$, $\sigma_2 = 0$ and $\sigma_3 = -\tau$ and the strain increments are $d\varepsilon_1^P = d\gamma^P/2$, $d\varepsilon_2^P = 0$ and $d\varepsilon_3^P = -d\gamma^P/2$. Substituting these into eqs(9.2d) and (9.6d), von Mises gives $\bar{\sigma} = \sqrt{3}\,\tau$ and $d\bar{\varepsilon}^P = d\gamma^P/\sqrt{3}$. With similar substitutions into eqs(10.38a,b), Tresca gives $\bar{\sigma} = 2\tau$ and $d\bar{\varepsilon}^P = 2d\gamma^P/3$. The incremental work of pure shear is

$$dW^P = \bar{\sigma}\, d\bar{\varepsilon}^P = \sigma_1\, d\varepsilon_1^P + \sigma_3\, d\varepsilon_3^P$$

which is consistent with von Mises equivalence definitions only. A similar plastic work violation would be found from $d\bar{\varepsilon}^P$ expressions derived earlier from $f(J_2', J_3')$ and the strain hypothesis (see chapter 9). However, this has not precluded the use of a Tresca equivalent stress in the non-associated approach. For example, Marin [30] and Soderberg [31] employed a von Mises potential in a flow rule for rate dependent plasticity (creep). Equation (8.17a) gave this as

$$\dot{\varepsilon}_{ij}^P = \dot{\lambda}\, \partial(J_2')/\partial\sigma_{ij} = \dot{\lambda}\, \sigma_{ij}' \tag{10.39a}$$

where $\dot{\lambda}$ is defined in terms of $\bar{\sigma}$ and $\dot{\bar{\varepsilon}}^P$ for simple tension:

$$\dot{\lambda} = \frac{3\dot{\bar{\varepsilon}}^P}{2\bar{\sigma}} \tag{10.39b}$$

For their equivalence, the authors defined $\bar{\sigma}$ and $\dot{\bar{\varepsilon}}^P$ in eq(10.39b) by both von Mises (associated) and Tresca (non associated). For the latter, $\bar{\sigma}$ is given by eq(10.38a) and $\dot{\bar{\varepsilon}}^P$ is the time derivative of eq(10.38b):

$$\dot{\bar{\varepsilon}}^P = \frac{\mathrm{d}\bar{\varepsilon}^P}{\mathrm{d}t} = \frac{2}{3}\left(\dot{\varepsilon}_1^P - \dot{\varepsilon}_3^P\right) \tag{10.39c}$$

Their experiments revealed less spread in Tresca equivalent stress–strain rate plots compared to von Mises. Consequently, greater accuracy was found from non-associated predictions to creep rates in rotating discs, and tubes under internal pressure combined with axial tension [32–34]. Combining von Mises equivalence with a Tresca flow rule is less realistic practically but has been used for its mathematical simplicity.

10.6 Prandtl-Reuss Flow Theory

In deforming beyond the elastic limit an elastic component of strain continues to accompany the plastic strain. When the latter is large and elastic strain is small enough to be ignored, the Levy-Mises theory estimates the total deformation in an elastic–plastic solid. Such predictions will be unacceptable where elastic and plastic strain components are comparable in their magnitudes. Here, both elastic and plastic incremental strains arising from a loading condition $\mathrm{d}\sigma_{ij} > 0$ are required. The incremental elastic component of strain $\mathrm{d}\varepsilon_{ij}^{\,e}$, follows from Hooke's law and the plastic component $\mathrm{d}\varepsilon_{ij}^{\,P}$ from the Levy-Mises eq(10.14).

10.6.1 Tensor Subscript Notation

The Prandtl-Reuss total incremental strain theory [35, 36] sums the elastic and plastic incremental strains. The theory was originally presented in the tensor subscript notation. Since $\sigma_{kk}' = 0$, five independent deviatoric stress components may act simultaneously to give six total strain increments:

$$\begin{aligned} \mathrm{d}\varepsilon_{ij}^{t} &= \mathrm{d}\varepsilon_{ij}^{P} + \mathrm{d}\varepsilon_{ij}^{e} \\ \mathrm{d}\varepsilon_{ij}^{t} &= \mathrm{d}\lambda\, \sigma_{ij}' + \frac{\mathrm{d}\sigma_{ij}'}{2G} + \frac{(1-2\nu)}{3E}\,\delta_{ij}\,\mathrm{d}\sigma_{kk} \end{aligned} \tag{10.40}$$

The expanded form of eq(10.40) represents three direct and three shear strain increments:

$$\mathrm{d}\varepsilon_{11}^{P} = \frac{2\,\mathrm{d}\lambda}{3}\left[\sigma_{11} - \frac{1}{2}(\sigma_{22} + \sigma_{33})\right] + \frac{1}{E}\left[\mathrm{d}\sigma_{11} - \nu\,(\mathrm{d}\sigma_{22} + \mathrm{d}\sigma_{33})\right], \quad \text{etc}$$

$$\mathrm{d}\varepsilon_{12}^{P} = \mathrm{d}\gamma_{12}^{P}/2 = \mathrm{d}\lambda\sigma_{12} + \mathrm{d}\sigma_{12}/2G\,, \quad \text{etc}$$

Integration of eq(10.40) shows that the permanent, plastic component of total strain depends upon the history of stress, while the recoverable elastic component depends upon the current stress. This applies to the low-temperature, small-strain regime of many polycrystalline materials where they remain insensitive to the rate of straining.

10.6.2 Verification of Prandtl-Reuss Theory

The Prandtl-Reuss theory has matched experimental observations on metal plasticity with both linear and parabolic hardening under a variety of non-radial loading paths [37]. Using Ludwik's hardening eq(9.43a), the constants A and $n < 1$ may be found from reported tension or torsion tests. Alternatively, they may be found from within a loading path that begins with either of these modes. Values of A/Y and n so found are given in Fig. 9.19 for many engineering alloys, some of which will be converted to eq(9.43b) and used here. An exponent $n = 1$ applies to ideal linear hardening material. Near uniform stress states are achieved from loading thin-walled tubular testpieces with combinations of tension, compression, torsion and internal pressure.

(a) Stepped Path Under Tension-Torsion

Equations (10.18a) and (10.40) provide the total incremental shear and axial strains under uniform axial and shear stress components σ_{11} and σ_{12} as follows

$$\mathrm{d}\gamma_{12}^{t} = \frac{3\,\mathrm{d}\bar{\sigma}}{H'\bar{\sigma}}\sigma_{12} + \frac{\mathrm{d}\sigma_{12}}{G} \tag{10.41a}$$

$$\mathrm{d}\varepsilon_{11}^{t} = \frac{\mathrm{d}\bar{\sigma}}{H'\bar{\sigma}}\sigma_{11} + \frac{\mathrm{d}\sigma_{11}}{E} \tag{10.41b}$$

$$\bar{\sigma}^2 = \sigma_{11}^2 + 3\sigma_{12}^2 \tag{10.41c}$$

Let the tensile stress σ_{11} be first applied to exceed the initial yield stress σ_o then held constant ($\sigma_{11} = \sigma_p$) while σ_{12} increases within the plastic range. Equations (9.43b), (10.16c) and (10.41c) are employed as follows:

$$H' = \mathrm{d}\bar{\sigma}/\mathrm{d}\bar{\varepsilon}^P = \left(n\sigma_o/\varepsilon_o\right)\left(\bar{\varepsilon}^P/\varepsilon_o\right)^{n-1} = \left(n\sigma_o/\varepsilon_o\right)\left(\bar{\sigma}/\sigma_o - 1\right)^{\frac{n-1}{n}} \tag{10.42a}$$

$$\bar{\sigma}^2 = \sigma_p^2 + 3\sigma_{12}^2 \quad \Rightarrow \quad \bar{\sigma}\,\mathrm{d}\bar{\sigma} = 3\sigma_{12}\,\mathrm{d}\sigma_{12} \tag{10.42b,c}$$

The total incremental strains under an increasing σ_{12} are found from eqs(10.41) and (10.42):

$$\mathrm{d}\gamma_{12}^{t} = \frac{9\varepsilon_o}{n}\frac{\left[\left(S_p^2 + 3T^2\right)^{\frac{1}{2}} - 1\right]^{\frac{1-n}{n}}}{\left[3 + (S_p/T)^2\right]}\mathrm{d}T + \frac{\sigma_o}{G}\mathrm{d}T \tag{10.43a}$$

$$\mathrm{d}\varepsilon_{11}^{t} = \frac{3\varepsilon_o}{n}\frac{\left[\left(S_p^2 + 3T^2\right)^{\frac{1}{2}} - 1\right]^{\frac{1-n}{n}}}{\left(S_p/T + 3T/S_p\right)}\mathrm{d}T \tag{10.43b}$$

where the dimensionless stresses are $S_p = \sigma_p/\sigma_o$ and $T = \sigma_{12}/\sigma_o$. Normally, numerical integration of eqs(10.43a,b) is required unless $n = ⅓$, when closed solutions are found. For example, Ivey [38] conducted an experiment of this kind on a 24S-T4 aluminium alloy where, from Fig. 9.19, $n \approx ⅓$ and $\varepsilon_o = (Y/A)^3 = 2.24^{-3} = 0.089$. Here, $S_p = 1.46$ exceeded the tensile yield stress and therefore the lower limit of integration for eqs(10.43a,b) is $T = 0$ (see the inset Fig. 10.11a). Under the shear stress branch, eqs(10.43a,b) become

$$\gamma_{12}^{P} = 27\,\varepsilon_o \int_0^T \left[T^2 - \frac{2T^2}{(S_p^2 + 3T^2)^{1/2}} + \frac{T^2}{(S_p^2 + 3T^2)} \right] \mathrm{d}T + \frac{\sigma_o}{G}\int_0^T \mathrm{d}T$$

$$\varepsilon_{11}^{t} = \varepsilon_{11}^{P} = 9 S_p \varepsilon_o \int_0^T \left[T - \frac{2T}{(S_p^2 + 3T^2)^{1/2}} + \frac{T}{(S_p^2 + 3T^2)} \right] \mathrm{d}T \tag{10.43c,d}$$

Equations (10.43c,d) are composed of standard integrals that lead to total strains

$$\gamma_{12}^{t} = 9\,\varepsilon_o \left[T^3 - T(S_p^2 + 3T^2)^{1/2} + \frac{S_p^2}{\sqrt{3}} \ln \left\{ \left[1 + 3\left(\frac{T}{S_p} \right)^2 \right]^{\frac{1}{2}} + \frac{\sqrt{3}T}{S_p} \right\} + T - \frac{S_p}{\sqrt{3}} \tan^{-1}\left(\frac{\sqrt{3}T_p}{S_p} \right) \right] + \frac{\sigma_o T}{G} \tag{10.44a}$$

$$\varepsilon_{11}^{t} = \varepsilon_{11}^{P} = 9 S_p \varepsilon_o \left\{ \frac{T^2}{2} - \frac{2}{3}(S_p^2 + 3T^2)^{1/2} + \frac{1}{6} \ln \left[1 + 3\left(\frac{T}{S_p} \right)^2 \right] + \frac{2S_p}{3} \right\} \tag{10.44b}$$

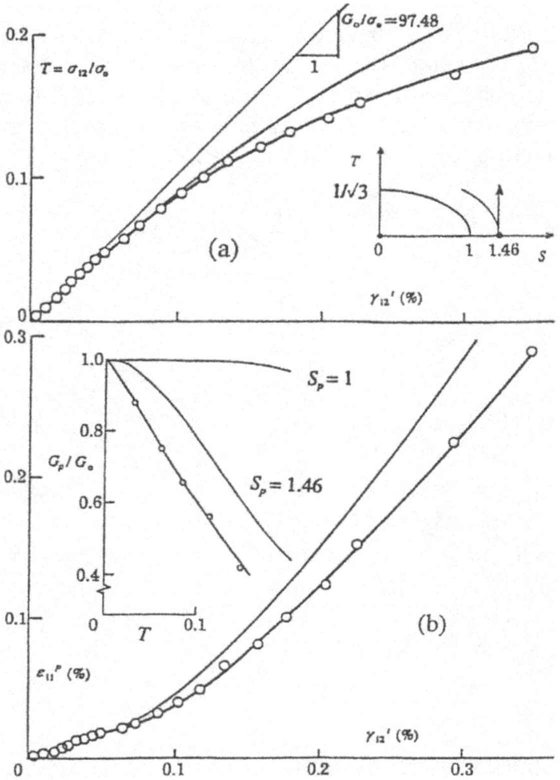

Figure 10.11 Stepped tension-torsion loading of 24S-T4 aluminium alloy

This integration shows that, under the increasing shear stress, (i) the final term in eqs(10.43a) and (10.44a) is the elastic shear strain and (ii) plastic axial strain is present despite the absence of elastic axial strain. Figures 10.11a,b show that the predictions from eqs(10.44a,b) match the trends observed in plastic strain paths and flow stresses. The variation in shear modulus under the shear stress path is shown inset in Fig. 10.11b. It is known that the minimum surface model (see Fig. 8.19) and Hencky's total strain theory both predict an initial shear modulus $G_p = d\sigma_{12} / d\gamma_{12}^t$ of plastically prestrained material different from the elastic shear modulus G_o. Taking an isotropic expansion of the yield surface to the contain prestress, the differentiation of eq(10.44a) provides a modulus ratio

$$\frac{G_p}{G_o} = \left\{ 1 + \frac{27\,\varepsilon_o T^2 G_o}{\sigma_o} \left[1 - \frac{2}{(S_p^2 + 3T^2)^{1/2}} + \frac{1}{(S_p^2 + 3T^2)} \right] \right\}^{-1} \tag{10.45}$$

Equation (10.45) shows that $G_p = G_o$ when $T = 0$, irrespective of S_p. Further confirmation of this was found for similar experiments conducted on steel [39], aluminium [40] and various aluminium alloys [39–43]. As T increases, the fall in G_p/G_o, observed in the inset diagram, is replicated with eq(10.45). *Note*: When $S_p = 1$, the fall in G_p/G_o is so gradual under increasing σ_{12} that one could be deceived into expecting a region of elasticity from Prandtl-Reuss though, strictly, the condition for this is $S_p < 1$.

(b) Stepped Path Under Tension and Internal Pressure

One method of achieving stepped stressed paths in principal biaxial stress space is to subject thin-walled cylinders to combined internal pressure and axial load. Radial stress (σ_3) is negligible and the respective axial and circumferential stress components σ_1 and σ_2 may be assumed to be uniform. Equations(10.12c) and (10.40) supply

$$d\varepsilon_1^t = \frac{d\bar{\sigma}}{H'\bar{\sigma}}\left(\sigma_1 - \frac{\sigma_2}{2}\right) + \frac{1}{E}(d\sigma_1 - \nu d\sigma_2) \tag{10.46a}$$

$$d\varepsilon_2^t = \frac{d\bar{\sigma}}{H'\bar{\sigma}}\left(\sigma_2 - \frac{\sigma_1}{2}\right) + \frac{1}{E}(d\sigma_2 - \nu\, d\sigma_1) \tag{10.46b}$$

$$\bar{\sigma}^2 = \sigma_1^2 - \sigma_1\sigma_2 + \sigma_2^2 \quad \Rightarrow \quad 2\bar{\sigma}\, d\bar{\sigma} = (2\sigma_1 - \sigma_{2p})\, d\sigma_1 \tag{10.46c,d}$$

Equation (10.46d) applies where σ_1 increases in the presence of a constant σ_2 (i.e. σ_{2p}). Substituting eqs(10.42a) and (10.46d) into eqs(10.46a,b) and setting $S_1 = \sigma_1/\sigma_o$, $S_{2p} = \sigma_{2p}/\sigma_o$ and $\bar{S} = \bar{\sigma}/\sigma_o = (S_1^{\,2} - S_1 S_{2p} + S_{2p}^{\,2})^{1/2}$ leads to total, principal strains under increasing σ_1:

$$\varepsilon_1^t = \frac{\varepsilon_o}{4n}\int_{S_{1p}}^{S_1} \frac{(2S_1 - S_{2p})^2(\bar{S} - 1)^{\frac{1-n}{n}}}{\bar{S}^2}\, dS_1 + \frac{\sigma_o}{E}\int_0^{S_1} dS_1$$

$$\varepsilon_2^t = \frac{\varepsilon_o}{4n}\int_{S_{1p}}^{S_1} \frac{(2S_1 - S_{1p})(2S_{2p} - S_1)(\bar{S} - 1)^{\frac{1-n}{n}}}{\bar{S}^2}\, dS_1 - \frac{\nu\sigma_o}{E}\int_0^{S_1} dS_1 \tag{10.47a,b}$$

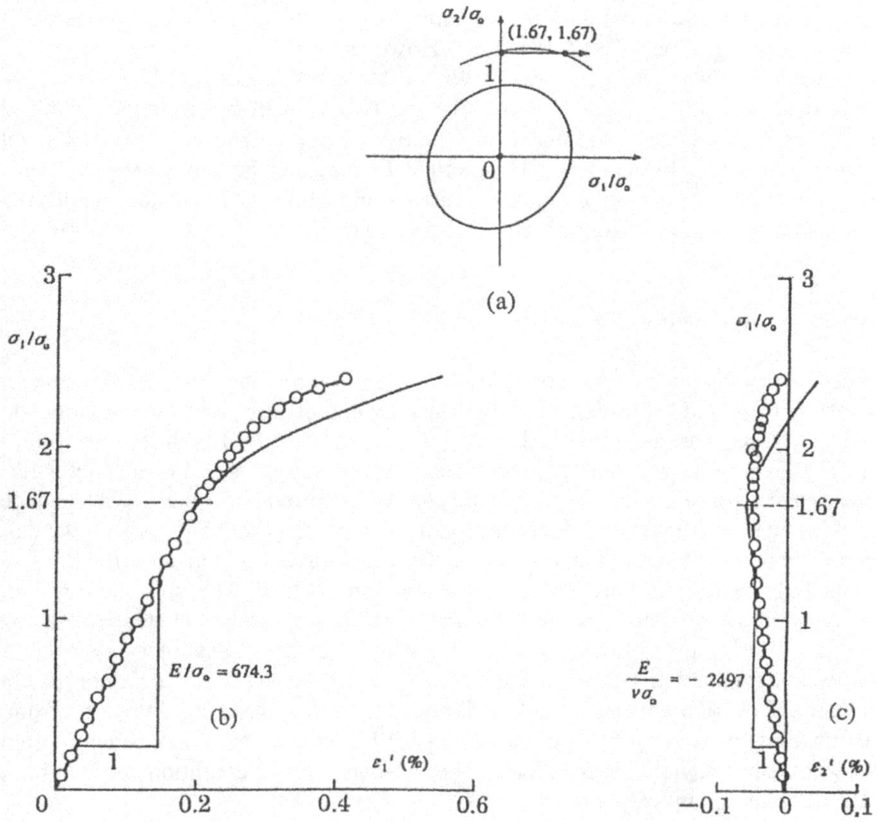

Figure 10.12 Stepped tension-tension loading of SAE 1020 steel

In the experiment of Marin and Hu [44], a mild steel testpiece was prestressed to $S_{2p} = 1.67$. Thereafter, S_1 was increased, as shown in Fig. 10.12a. Figures 10.12b,c show that elastic stress–strain responses prevail until $S_{1p} = 1.67$, this being the lower limit for the first integral in eqs(10.47a,b). Taking from Fig. 9.19 $n \approx ⅓$, $\varepsilon_o = 6.54^{-3} = 3.575 \times 10^{-3}$ for this material, eqs(10.47a,b) become

$$\varepsilon_1^t = \frac{3\,\varepsilon_o}{4}\int_{S_{1p}}^{S_1}(2S_1 - S_{2p})^2\left[1 - \frac{2}{(S_1^2 - S_{2p}S_1 + S_{2p}^2)^{1/2}} + \frac{1}{(S_1^2 - S_{2p}S_1 + S_{2p}^2)}\right]\mathrm{d}S_1 + \frac{\sigma_o}{E}\int_0^{S_1}\mathrm{d}S_1 \tag{10.47c}$$

$$\varepsilon_2^t = \frac{3\,\varepsilon_o}{4}\int_{S_{1p}}^{S_1}(2S_1 - S_{1p})(2S_{2p} - S_1)\left[1 - \frac{2}{(S_1^2 - S_{2p}S_1 + S_{2p}^2)^{1/2}} + \frac{1}{(S_1^2 - S_{2p}S_1 + S_{2p}^2)}\right]\mathrm{d}S_1 - \frac{\nu\sigma_o}{E}\int_0^{S_1}\mathrm{d}S \tag{10.47d}$$

where $\sigma_o/E = (674.3)^{-1}$ and $\nu = 0.27$ from within Figs 10.12b,c. Closed solutions can be found from the integration of eqs(10.47c,d). However, for the predictions shown, it is simpler to apply Simpson's rule repeatedly, with a step length $\Delta S_1 = 0.2$. These reproduce the trends observed and suggest that the deviation found can be attributed to plastic anisotropy in the cold-drawn tube material. The elastic response observed is predictable from isotropic hardening in its range and magnitude. In this test, where the elastic and plastic strain magnitudes are similar, an omission of elastic strain would lead to unacceptable error. Here it is seen that the Prandtl-Reuss theory is essential to give the full strain response to S_1.

10.7 Kinematic Hardening

It has been seen that the rule of isotropic hardening is attractive from a mathematical standpoint when applied to plasticity arising from outward loading paths. Though intended for initially isotropic material, a similar hardening rule can be applied to initially anisotropic material with an appropriate change to the yield function/flow potential (see chapter 12, § 12.2). Moreover, a similar rule adopted in chapter 11 can provide for the rate dependence of stress where a strain rate vector lies normal to an expanding potential for visco-plasticity and creep. The potential may again be associated with the yield function but in the flow rule the incremental plastic strains are replaced by viscous strain rates. The isotropic hardening rule is only useful for outward loading paths and works best for radial, or stepped loadings in which plastic strain history does not exert a strong influence. In chapter 8 (§ 8.9.1 – § 8.9.3) it was explained that the isotropic hardening rule could not account for the Bauschinger effect. Moreover, said rule will not account for plastic strain arising from a 'neutral' loading condition [45] where a stress path follows the boundary of an isotropic expansion of the initial yield locus. Clearly, the subsequent yield condition needs to be re-defined under more complex paths.

10.7.1 Bauschinger Effect

The laboratory testing referred to in chapter 9 was conducted mostly under increasing loading and did not involve a reversal of plasticity. Correspondingly, the descriptions of flow curves given apply to forward deformation. One of the critical validation tests for any theory is whether it can represent, with sufficient accuracy, the stress required to reverse the direction of plastic strain. Bauschinger [46] first observed a common feature of polycrystalline materials in which a reduction in flow stress accompanied a reversal to the plastic strain (see Fig. 8.21b). There is also a marked increase in the reversed hardening rate under the lower flow stress. Physically, a lesser stress is required to activate slip processes following a reversal than the stress required to continue with forward slip on active slip planes. To explain this, Kadashevitch and Novozhilov [47] identified a field of residual micro-stresses within a given slipped state. Therein, the Bauschinger effect was identified with the sense of this micro-stress distribution: opposing forward flow but assisting reversed flow. A lower applied stress is therefore required to promote reversed yielding consistently.

10.7.2 Translation Rules

Prager [48] modelled the Bauschinger effect with a rigid translation of the initial yield surface away from its origin. He assumed that the direction of translation followed the incremental plastic strain vector $d\varepsilon_{ij}^{P}$, as shown in Fig. 10.13a.

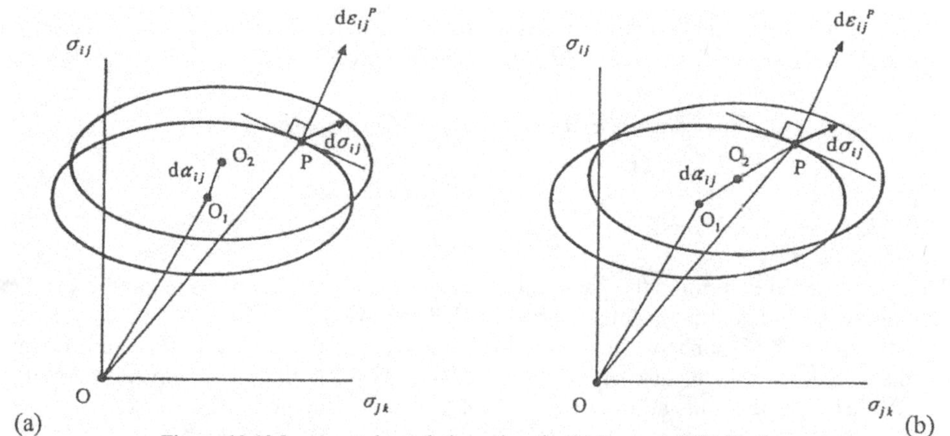

Figure 10.13 Incremental translations given by (a) Prager and (b) Ziegler

The co-ordinates of a rigid, incremental translation, from O_1 to O_2, are

$$d\alpha_{ij} = c\,d\varepsilon_{ij}^{P} \tag{10.48}$$

When eq(10.48) is used to describe the translation of an initial von Mises surface, the subsequent function is referred to the co-ordinates for O_2 as

$$f = \frac{1}{2}\left(\sigma_{ij}' - c\int d\varepsilon_{ij}^{P}\right)^2 = k^2 \tag{10.49}$$

where k is the shear yield stress. Applying eq(10.49) to simple tension:

$$\frac{1}{2}\left[(\sigma_{11}' - c\,\varepsilon_{11}^{P})^2 + (\sigma_{22}' - c\,\varepsilon_{22}^{P})^2 + (\sigma_{33}' - c\,\varepsilon_{33}^{P})2\right] = k^2$$

Substituting $\sigma_{11}' = ⅔\sigma_{11}$, $\sigma_{22}' = -⅓\sigma_{11}$ and $\sigma_{33}' = -⅓\sigma_{11}$, with $\varepsilon_{22}^{P} = \varepsilon_{22}^{P} = -½\varepsilon_{11}^{P}$, leads to the uniaxial hardening law:

$$\sigma_{11} = \sigma_o + (3c/2)\,\varepsilon_{11}^{P} \tag{10.50}$$

where $\sigma_o = \sqrt{3}k$ is the tensile yield stress. Equation (10.50) shows that Prager's translation describes linear hardeming at a rate proportional to c. A useful consequence of this is that the path of the centre translation is proportional to the plastic strain path.

The reader should note the difference between the translations shown in Figs 10.13a,b. The latter corresponds to an alternative rigid translation rule by Shield and Ziegler [49]:

$$d\alpha_{ij} = d\mu(\sigma_{ij} - \alpha_{ij}) \tag{10.51}$$

where scalar $d\mu = d\bar{\sigma}/\sigma_o = (c/\sigma_o)d\bar{\varepsilon}^{P}$ is defined from the linear hardening rate c in simple tension. Equation (10.51) ensures a translation along the radial path in the manner of Fig 3.7 and this may be identified with a linear tensile hardening material, as shown in Fig. 8.22. Of the two rules, Prager's has received most attention, particularly with its modification to non-linear hardening [50]:

$$d\alpha_{ij} = c_1\,d\varepsilon_{ij}^{P} - (c_2/\sigma_o)\,\alpha_{ij}\,d\bar{\varepsilon}^{P} \tag{10.52a}$$

where c_1 and c_2 are functions of the plastic strain invariants I_2 and I_3 (see eqs 9.5a,b). A similar, non-linear modification to Ziegler's law [51] gives

$$d\alpha_{ij} = \left[(c_1/\sigma_o)(\sigma_{ij} - \alpha_{ij}) - (c_2/\sigma_o)\alpha_{ij}\right]d\bar{\varepsilon}^P \tag{10.52b}$$

10.7.3 Reversed Yield Stress

The usual way to determine the parameters c_1 and c_2 in eqs(10.52a,b) is to combine either equation with a standard non-linear hardening law, e.g. Ludwik, Swift, Vocé etc., and then to fit the pair of equations to a flow curve with a reversal. In this case, both expansion and translation of the yield surface are permitted (Fig. 8.21b). For example, integrating eq(10.52a) gives the translation $\alpha_f = \alpha\,(\varepsilon^P)$ for OA under uniaxial deformation

$$\frac{\alpha_f}{\sigma_o} = \frac{c_1}{c_2}\left(1 - e^{-c_2\varepsilon^P/\sigma_o}\right) \tag{10.53}$$

The forward flow stress becomes the sum of two parts $\sigma_f = \alpha_f + \mathfrak{R}\,(\varepsilon^P)$ where $\mathfrak{R}\,(\varepsilon^P)$ is a 'radius' of the expanded yield surface (see Fig. 10.14). Taking σ_f to be given by Ludwik's eq(9.43b), with α_f from eq(10.53), the radius becomes

$$\mathfrak{R}\left(\varepsilon^P\right) = \sigma_f - \alpha_f = \sigma_o\left[1 + (\varepsilon^P/\varepsilon_o)^n\right] - \frac{\sigma_o c_1}{c_2}\left(1 - e^{-c_2\varepsilon^P/\sigma_o}\right) \tag{10.54a}$$

The reversed yield stress σ_P at point P follows:

$$\sigma_P = \alpha_f - \mathfrak{R}\,(\varepsilon^P) \tag{10.54b}$$

$$\frac{\sigma_P}{\sigma_o} = \frac{2\,c_1}{c_2}\left(1 - e^{-c_2\varepsilon^P/\sigma_o}\right) - \left[1 + \left(\varepsilon^P/\varepsilon_o\right)^n\right] \tag{10.54c}$$

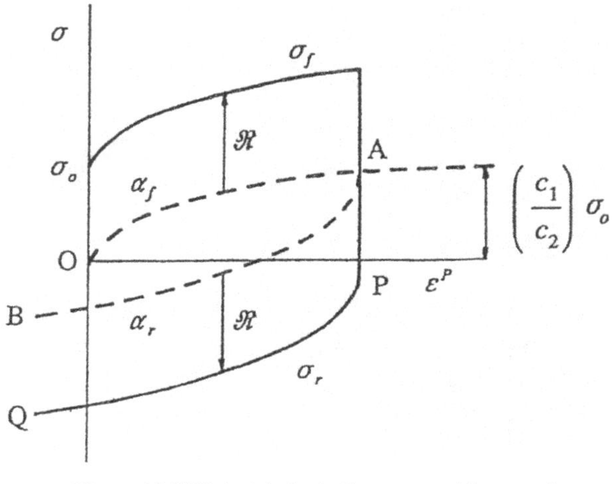

Figure 10.14 Uniaxial plastic flow curve with reversal

10.7.4 Reversed Flow Curve

Equations (10.53) and (10.54c) give normalised stresses for the centre translation OA and the reversed yield point P, respectively, in Fig. 10.14. To describe the reversed flow curve PQ, it is necessary to find the corresponding centre translation AB. Strain reversal requires a change in the sign of the second term in eq(10.52a) [52]. Applying this with lower limits of integration for point A (ε_A^P, α_A) gives

$$\int_{\alpha_A}^{\alpha} \frac{d\alpha}{c_1 + c_2\alpha/\sigma_o} = \int_{\varepsilon_A^P}^{\varepsilon^P} d\varepsilon^P$$

$$\frac{\alpha}{\sigma_o} = \left(\frac{c_1}{c_2} + \frac{\alpha_A}{\sigma_o}\right) e^{c_2(\varepsilon^P - \varepsilon_A^P)/\sigma_o} - \frac{c_1}{c_2} \qquad (10.55a)$$

Substituting α_A/σ_o from eq(10.53), with $\varepsilon^P = \varepsilon_A^P$, into eq(10.55a) leads to α_r for AB:

$$\frac{\alpha_r}{\sigma_o} = \frac{c_1}{c_2}\left(2e^{c_2(\varepsilon^P - \varepsilon_A^P)/\sigma_o} - e^{c_2(\varepsilon^P - 2\varepsilon_A^P)/\sigma_o} - 1\right) \qquad (10.55b)$$

Now the reversed flow curve PQ is $\sigma_r = \alpha_r - \mathfrak{R}$ where $\mathfrak{R} = \sigma_f - \alpha_f$ is based upon the forward flow curve. That is, $\mathfrak{R}$ (ε^P) in eq(10.54a) defines both σ_f and σ_r curves at a given ε^P, as shown in Fig. 10.14:

$$\frac{\sigma_r}{\sigma_o} = \frac{c_1}{c_2}\left(2e^{c_2(\varepsilon^P - \varepsilon_A^P)/\sigma_o} - e^{c_2(\varepsilon^P - 2\varepsilon_A^P)/\sigma_o} - e^{-c_2\varepsilon^P/\sigma_o}\right) - \frac{\sigma_f}{\sigma_o} \qquad (10.56)$$

where from eq(10.53) c_1/c_2 is an asymptote for α/σ_o at large strain The above analysis is based upon a reversal between tension and compression but this can be difficult to achieve in practice. Consequently, torsion of thin walled tubes and bending of thin strips may be used, in which unwanted friction is removed and stress gradients have been reduced. For torsion, the translation OA is equivalent to writing eq(10.53) as

$$\frac{\beta}{k} = \frac{c_1}{c_2'}\left(1 - e^{-c_2'\gamma^P/k}\right) \qquad (10.57)$$

and eq(10.56) is re-written for the reversed flow curve PQ as

$$\frac{\tau_r}{k} = \frac{c_1}{c_2'}\left(2e^{c_2'(\gamma^P - \gamma_A^P)/k} - e^{c_2'(\gamma^P - 2\gamma_A^P)/k} - e^{-c_2'\gamma^P/k}\right) - \frac{\tau_f}{k} \qquad (10.58)$$

where $c_2' = c_2/3$ arises from the use of engineering shear strain. Figure 10.15 shows reversed torsional flow curves for annealed En3B tubes with inner diameter 12.7 mm, outer diameter 16.1 mm and a gauge length of 80 mm.

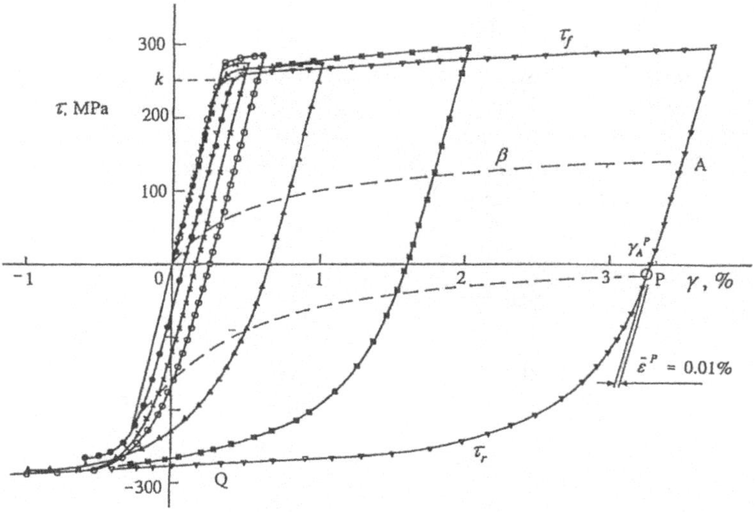

Figure 10.15 Reversed flow curves under torsion

At each γ_A^P value, the reversed yield point P is determined for 0.01 % von Mises proof strain. The mid-point translation locus OA reveals an asymptote in the region of $c_1/c_2' = 0.74$. When eq(10.58) is fitted to the reversed flow curve PQ, having subtracted elastic strain appropriate to each stress level, then c_1/k will vary with γ^P in the manner of Fig. 10.16. This identifies the function $c_1(I_2)$ with a complex polynomial in which $I_2 = (\gamma^P)^2/4$.

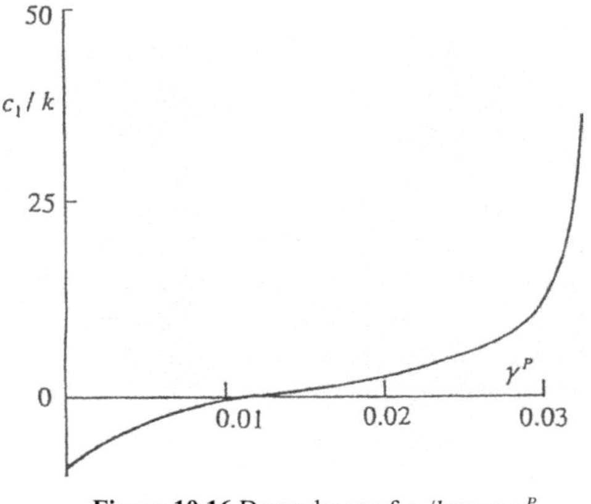

Figure 10.16 Dependence of c_1/k upon γ^P

For multiaxial stress states, a subsequent combined hardening yield function is required. That function is written to combine a translation and an expansion of an initial von Mises yield surface:

$$\frac{3}{2}\left(\sigma_{ij}' - \alpha_{ij}'\right)\left(\sigma_{ij}' - \alpha_{ij}'\right) = \left[\Re\left(\bar{\varepsilon}^P\right)\right]^2$$

in which $\alpha_{ij}' = \alpha_{ij} - \frac{1}{3}\delta_{ij}\alpha_{kk}$ and eq(10.54a) will again define the right-hand side when an equivalent von Mises plastic strain $\bar{\varepsilon}^P$ identifies with the tensile strain ε^P.

10.8 Concluding Remarks

The rule of isotropic hardening is suitable for radial outward loading paths. The size of the subsequent yield surface may be expressed from a hardening hypothesis. Here, it appears that either the work or strain hardening hypotheses may be applied to an $f(J_2', J_3')$ isotropic material. The strain hardening hypothesis may be preferred, for simplicity, since the chosen equivalent strain expression identifies directly with an empirical hardening function (e.g. Hollomon, Swift etc.). However, the work hypothesis, being more exacting, will ensure the correct amount of plastic work for non radial loading. Stress reversal involving a Bauschinger effect is matched better by allowing a shift in the origin of the yield locus coupled to its inflation. This is Hodge's combined hardening scheme [53], which has been adopted by many for a description of subsequent yielding. That scheme includes also a combination that appears to be needed most often in its description of reversed plasticity experiments. Referring to the reversed torsion experiments in Fig. 10.15, it appears that deflation and translation are required to describe the diminishing elastic unloading range preceding a 'growing' Bauschinger effect.

References

1. Hill R. *The Mathematical Theory of Plasticity*, 1989, Oxford University Press.
2. Drucker D. C. *Proc. 1st US Nat. Congr. of Applied Mechanics*, 1952, 487, ASME.
3. Prager W. *Jl. Appl. Phy.*, 1949, **20**, 235.
4. Ford H. and Alexander J. M. *Advanced Mechanics of Materials*, 1963, Longmans.
5. Zyczkowski M. *Combined Loadings in the Theory of Plasticity*, 1981, Polish Sci Pub.
6. Naghdi P. *Plasticity, Proc. 2nd Symp. on Naval Struct. Mech.*, (eds Lee E. H. and Symonds P. S.) 1960, 121, Pergamon Press.
7. Hollomon J. H. *Am. Inst. Mech. Eng.*, 1945, **162**, 268.
8. Ludwick P. *Elemente der Technologischen Mechanik*, 1909, Springer, Berlin.
9. Swift H. W. *Jl. Mech. Phys. Solids*, 1952, **1**, 1
10 Rees D. W. A. *Proc. Roy. Soc.*, 1982, **A383**, 333.
11. Levy M. *Jl. Math. Pures. Appl.*, 1871, **16**, 369.
12. von Mises R. *Ges. Wiss. Gottingen*, 1913, 582.
13. Drucker D. C. *Trans. ASME, Jl. Appl. Mech.*, 1949, **16**, 349.
14. Rees D. W. A. and Mathur S. B. *Proc.: Non Linear Problems in Stress Analysis*, (ed Stanley P.) 1978, 185, Applied Science.
15. Phillips A. and Ricciuti M. *Int. Jl. Solids and Struct.*, 1976, **12**, 159.
16. Phillips A. and Lee C-W. *Int. Jl. Solids and Struct.*, 1979, **15**, 715.
17. Shahabi S. N. and Shelton A. *Jl. Mech. Eng. Sci.*, 1975, **17**, 93.
18. Johnson A. E., Frost N. E. and Henderson J. *The Engineer*, 1955, **199**, 366, 402, 457.
19. Johnson A. E. and Frost N. E. *The Engineer*, 1952, **194**, 713.
20. Khan A. S. and Parikh Y. *Eng. Fract. Mech.*, 1985, **21**, 697.
21. Freudental A. M. and Gou P. F. *Acta Mech.*, 1969, **8**, 34.
22. Rogan J. Ph.D. Thesis, University of London, 1966.
23. Taylor G. I. and Quinney H. *Phil. Trans. Roy. Soc.*, 1931, **A230**, 323, London.
24. Lode W. *Z. Phys.*, 1926, **36**, 913.
25. Baraya G. L. and Parker J. *Int. Jl. Mech. Sci.*, 1963, **5**, 353.
26. Ellington P. *Jl. Mech. Phys. Solids*, 1958, **6**, 276.
27. Dillamore I., Hazel R. J., Watson T. W. and Hadden P. *Int. Jl. Mech. Sci.*, 1971, **13**, 1049.

28. Fessler H. and Hyde T. *Proc.: Non Linear Problems in Stress Analysis*, (ed. Stanley P.) 1978, 233, Applied Science.
29. Mehan R. L. *Trans. ASME, Jl. Basic Engng.*, 1961, **83**, 499.
30. Marin J. *Jl. Appl. Mech.*, 1937, **4**, 55.
31. Soderberg C. R. *Trans ASME*, 1936, **58**, 733.
32. Drucker D. C. and Stockton F. D. *Proc Soc Exp Stress Anal.*, 1953, **10**, 127.
33. Mair W. M. *Jl Strain Anal.*, 1967, **2**, 188.
34. Paul B., Chen W. and Lee L. *Proc. 4th US Nat. Congr. Appl. Mech. ASME*, 1962, 1013.
35. Prandtl L. *ZAMM*, 1928, **8**, 85.
36. Reuss A. *ZAMM*, 1930, **10**, 266.
37. Rees D. W. A. *Proc. R. Soc.*, 1987, **A410**, 443.
38. Ivey H. J. *Jl. Mech. Engng. Sci.*, 1961, **3**, 15.
39. Morrison J. L. M. and Shepherd W. M. *Proc. I. Mech. E.*, 1950, **163**, 1.
40. Daneshi G. H. and Hawkyard J. B. *Int. Jl. Mech. Sci.*, 1976, **18**, 195.
41. Peters R. W., Dow N. F. and Batdorf S. B. *Proc. Soc. Exp. Stress Anal.*, 1950, **7**, 127.
42. Feigen M. *Proc.: 2nd US Nat. Cong. Appl. Mech.*, 1955, 469, Pergamon.
43. Naghdi P. M., Essenberg F. and Koff W. *Jl. Appl. Mech.*, 1958, **25**, 201.
44. Marin J. and Hu L. W. *Trans. ASME*, 1956, **78**, 499.
45. Rees D. W. A. *Int. Jl. Plasticity*, 1988, **4**, 91.
46. Bauschinger J. *Civilingenieur*, 1881, 289.
47. Kadashevitch and Novozhilov V. V. *Prikl. Mat. Mekh.*, 1958, **22**, 78.
48. Prager W. *Jl. Appl. Mech.*, 1956, **23**, 483.
49. Shield R. T. and Ziegler H. *ZAMP*, 1958, **9a**, 260.
50. Armstrong P. J. and Frederick, C. O. *CEGB Rpt.* RD/B/N, 731, 1966.
51. Brunet M., Morestin F. and Walter H. *Int. Jl. Forming Proc.,* 2002, **5**, 225.
52. Chaboche J. L. *Int. Jl. Plasticity*, 1986, **2**, 149.
53. Hodge Jr P. G. *Jl. Appl. Mech.*, 1957, **24**, 482.

Exercises

10.1 Examine whether it is possible to derive the function F in the work hardening hypothesis for the flow curves described by Hollomon, modified Prager and Vocé (see Figs 9.18a, c and f, respectively). *See solution*

10.2 Derive the strain and work hardening derivatives F' and H', respectively, for the non-dimensional form of the Hollomon and Swift hardening laws, eqs(9.39b) and (9.44b).

10.3 Show that the Levy-Mises flow rule may also be written, using the von Mises equivalent stress and equivalent plastic strain increment definitions, as

$$\mathrm{d}\varepsilon_{ij}^{P} = \left(\frac{\mathrm{d}\bar{\varepsilon}^{P}}{\bar{\sigma}}\right)\frac{\partial\bar{\sigma}}{\partial\sigma_{ij}}$$

10.4 Is the strain hardening hypothesis, which defines $\mathrm{d}\bar{\varepsilon}^{P}$ in eq(10.35), consistent with a meaningful work expression? Examine from a consideration of deformation under simple shear.

10.5 Show that $\delta\lambda$ in eqs(10.18a,b) may be derived from eq(10.5) and the von Mises potential eq(10.6).

10.6 Show that $\delta\lambda$ in eqs(10.23a,b) may be derived from eq(10.5) and the Drucker potential (10.20).

10.7 Derive, for a thin-walled cylinder, the Prandtl-Reuss predictions to total strain components $\varepsilon_{zz}{}^t$ and $\gamma_{z\theta}{}^t$ that arise from increasing torque under a constant, initially elastic tension. Assume that in the wall, the uniform shear stress increases in the presence of a constant, initially elastic axial stress.

10.8 A thin walled cylinder with closed ends is constructed from a von Mises hardening material. It is subjected to a constant, principal stress ratio $\sigma_\theta/\sigma_z = 4/3$, where σ_z and σ_θ are axial and circumferential stresses due to independent axial load and internal pressure. Show that the total principal strains in the axial (z), circumferential (θ) and radial (r) directions are, respectively:

$$\varepsilon_z{}^t = \ln\left(\frac{\sigma_\theta}{p}\right)^{\frac{1}{2}} + \frac{\sigma_\theta}{4K}$$

$$\varepsilon_\theta{}^t = \ln\left(\frac{\sigma_\theta}{p}\right)^{\frac{1}{4}} + \frac{\sigma_\theta}{4K}$$

$$\varepsilon_r{}^t = \ln\left(\frac{p}{\sigma_\theta}\right)^{\frac{3}{4}} + \frac{\sigma_\theta}{4K}$$

where p is the internal pressure and K is the bulk modulus of the cylinder material. Determine the angle of inclination between the vectors of equivalent stress and equivalent plastic strain increment on the principal stress plane.
See solution: 15.07°

10.9 A thin-walled closed tube with an initial mean diameter d_o and thickness t_o is subjected to an internal pressure p. Neglecting elasticity and assuming a vonMises material with a Hollomon hardening law $\bar{\sigma}/\sigma_o = (\bar{\varepsilon}^P/\varepsilon_o)^n$, show that the circumferential plastic strain is

$$\frac{\varepsilon_\theta^P}{\varepsilon_o} = 2\left(\frac{\sqrt{3}}{4}\right)^{\frac{n+1}{n}}\left(\frac{pd}{\sigma_o t}\right)^{\frac{1}{n}}$$

Confirm that the relationship between pressure and current geometry that preserves volume is

$$\left(\frac{pd^2}{d_o t_o}\right)\left(\frac{\varepsilon_o^n}{\sigma_o}\right) = 2\left(\frac{2}{\sqrt{3}}\right)^{1+n} \ln\left(\frac{d}{d_o}\right)^n$$

10.10 A thin walled cylinder with closed ends is constructed from a von Mises hardening material. It is subjected to constant stress ratios $\sigma_{\theta\theta}/\sigma_{z\theta} = \sigma_{\theta\theta}/\sigma_{zz} = 2$, where σ_{zz}, $\sigma_{z\theta}$ and $\sigma_{\theta\theta}$ are axial, shear and circumferential stresses due to applied torsion and internal pressure p. Show that the total normal and shear strain components in the axial (z), circumferential (θ) and radial (r) directions are

Normal strains:

$$\varepsilon_{zz}^t = \frac{\sigma_{\theta\theta}}{6K}$$

$$\varepsilon_{\theta\theta}^t = \ln\left(\frac{\sigma_{\theta\theta}}{p}\right)^{\frac{1}{2}} + \frac{\sigma_{\theta\theta}}{6K}$$

$$\varepsilon_{rr}^t = \ln\left(\frac{p}{\sigma_{\theta\theta}}\right)^{\frac{1}{2}} + \frac{\sigma_{\theta\theta}}{6K}$$

Shear strain:

$$\varepsilon_{z\theta}^t = \frac{\gamma_{z\theta}^t}{2} = \ln\left(\frac{\sigma_{\theta\theta}}{p}\right)^{\frac{1}{2}}$$

where K is the bulk modulus of the cylinder material. Show that the stress vector and the plastic strain increment vector are co-incident in the $\sigma_{z\theta}$ versus $\sigma_{\theta\theta}$ plane.
See solution

10.11 In a torsion test, a thin-walled tube is twisted into the plastic range to a plastic shear strain γ^P. The torque is then reversed to twist the tube to its reversed shear yield point. Show that the normalised, shear yield stress for this reversal is expressed as

$$\frac{\tau_r}{k} = \frac{3c_1}{c_2}\left(e^{\frac{c_2 \gamma^P}{3k}} - 1\right) - \left(1 + \frac{\gamma^P}{\gamma_o}\right)^n$$

Assume a von Mises material ($\bar{\sigma} = \sqrt{3}\tau$, $\bar{\varepsilon}^P = \gamma^P/\sqrt{3}$), a modified Prager translation, eq(10.52a) and a Swift forward hardening law $\tau_r/k = (1 + \gamma^P/\gamma_o)^n$, where k is the initial shear yield stress and γ_o and n are constants.
See solution

CHAPTER 11

VISCOUS PLASTICITY

11.1 Introduction

The time-independent plasticity theory, covered previously in chapters 4–7, would normally be applied to metals at low strain rates ($\dot{\varepsilon} < 10^{-5}\ s^{-1}$) . Conditions are regarded as quasi-static when the strain rate does not influence the strength and ductility of polycrystalline materials appreciably. Many investigations have established a flow stress dependency upon more rapid rates of straining within the range $10^{-5} < \dot{\varepsilon} < 10^{3}\ s^{-1}$. Uniaxial experiments, conducted at strain rates lying within this range, show that metallic materials behave in a quasi-dynamic manner where the flow stress increases and the fracture strain decreases. True dynamic behaviour, observed for even higher strain rates under impact loading, accentuates these trends, resulting in a dynamic stress-strain curve lying well above the quasi-static curve. The difference in ordinates between these two curves is referred to as the excess stress or the overstress. It will be seen in this chapter how a rate-dependent overstress appears in the development of a viscous plasticity theory. Also reviewed are further developments that have been made to this theory. For example, predictions to uniaxial rate-dependency are those which appear as reduction to multi-axial theory. The latter fall into two categories: (i) those that extend the principles of incremental plasticity and (ii) those that embody thermodynamic principles. The former adopt a time-dependent plasticity theory, similar to the classical approach outlined in the previous chapters 9 and 10. These do not discriminate between time independent plastic strain and time dependent creep strain. Instead, a single inelastic strain rate term is employed as their sum. The application of the laws of thermodynamics to a deforming solid apply to a dynamic loading with large strain. Here, it will be seen how elastic constitutive relations are coupled to thermodynamic internal state variables. With a similar coupling applied to viscous plasticity, it is found that the constitutive relation has not required a yield function or a flow rule. For large strains produced at rapid rates under impact loading a constitutive relation employs finite stress and strain tensors defined in chapter 2.

Many theoretical developments to viscous plasticity theory have been made relatively recently and may not yet be fully accepted. Indeed, there has been controversy over whether thermodynamic internal state variables should, in principle, displace certain elements of a classical theory, e.g. as to whether the yield function should ever become irrelevant. This chapter contains an extended reference list upon the recent state of viscous plasticity theory in which the opposing views have been published in a learned exchange [1, 2].

11.2 Creep Curves

The influence of time upon plasticity involves: (i) the stress response to rapid rates of straining and (ii) how creep extends the deformation under slower rates of straining, for both short and longer time intervals under load. A period of incubation has been observed for low stresses in complex dispersion hardened structures that are specifically designed to resist dislocation creep, e.g. as in nickel-base alloys [3]. In most metal polycrystals, incubation is absent due to an initial, rapid generation of dislocations. Here, following the initial application of load, the instantaneous strain ε_o may be either all elastic ε^E or a sum of elastic and plastic strains, $\varepsilon^E + \varepsilon^P$. Thereafter, if the load is held constant over long periods, typically 1000 h at elevated temperature, the time-dependent stages of creep ensue in their respective order: primary, secondary and tertiary (see Fig. 11.1). At ambient temperature, only primary creep need be considered within a viscous plastic response to a slow continuous loading rate or a discontinuous incremental loading where time intervals between loads are short (<1000 s). The term 'transient creep' refers to a rate of creep strain that diminishes within this interval.

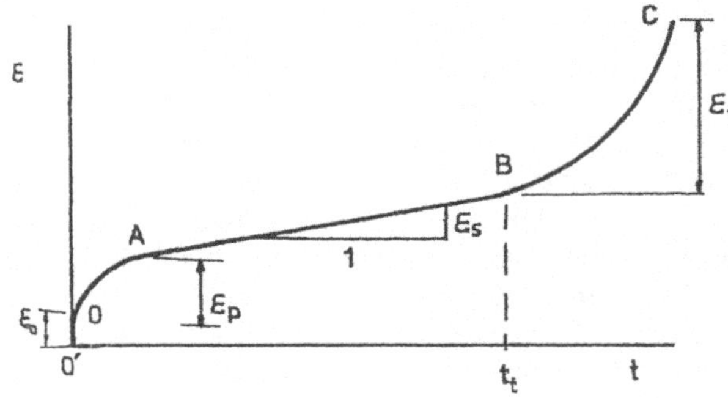

Figure 11.1 Three stage, constant load creep curve

11.2.1 Primary Creep

The region OA, shown in Fig. 11.1, is characterised by a creep rate $d\varepsilon_c/dt$ that decreases with time. In the lower range of homologous temperatures, $T/T_m < 0.5$ (T_m is the absolute melt temperature), primary creep strain dominates the creep curve. The primary region is known alternatively as: (i) transient creep, where the falling rate is due to a reduced mobility of dislocations within their increasing density [4] and (ii) parabolic creep, commonly arising from the kinetics of work hardening and diffusion controlled recovery occurring simultaneously and in opposition [5].

11.2.2 Time Dependence

A number of empirical functions $\varepsilon_c = \varepsilon_c(t)$ have been proposed to describe primary creep within a specified σ and T range. When the stress is low and a limited amount of transient creep is observed, the decreasing strain rate is derived from an exhaustion of diffusion creep processes [4]. This leads to a logarithmic power function in time:

$$\varepsilon_c = \alpha_1 \left(\ln \alpha_2 t \right)^{2/3} \tag{11.1}$$

where α_1 expresses the stress and temperature dependence of lattice parameters and α_2 is the vibrational frequency of a dislocation loop. Equation (11.1) was partly confirmed for a 3 % Au-Cu alloy at 27°C and −183°C but α_2 was found to be six orders of magnitude lower than its theoretical value. The simplified exhaustion theories of Smith [6] and Cottrell [7] both lead to a linear dependence between the creep strain and the logarithm of time

$$\varepsilon_c = \alpha \ln t \tag{11.2a}$$

The theoretical expressions for the temperature dependence of α, by either theory, are known to be unreliable. However, by employing an empirical function $\alpha = \alpha(\sigma, T)$, eq(11.2a) was shown [8] to describe short-time, room-temperature creep in ferritic-steel and copper at lower stresses and temperatures. Olds [9] showed that logarithmic creep appears in a precipitation hardened copper-silver alloy at 100°C. Wyatt [10], Rogan and Alexander [11] and Rees [12] showed that eq(11.2a) described creep in the 15 minute interval ($\approx 10^3$ s) allowed between load steps for carbon steel, annealed OFHC copper and aluminium at room-temperature. Such short-time 'cold creep' strain often appears in plasticity studies that adopt an incremental loading technique with time allowed between increments. Equation (11.2a) may be modified to avoid an unrealistic infinite creep rate for $t = 0$:

$$\varepsilon_c = \alpha_1 \ln\left(1 + \alpha_2 t\right) \tag{11.2b}$$

which gives $\dot{\varepsilon}_c = \alpha_1 \alpha_2$ for $t = 0$. In the derivation of eq(11.2b), the creep strain rate is associated with the thermal activation of dislocations past barriers over a period of time. In pure metals, the dislocations themselves form the barriers. Solute atoms and precipitates provide the barriers to dislocation motion in alloys [13]. Equation (11.2b) follows from the balance between the energy supplied by thermal activation and the work done by a passing dislocation. This theory appears to have been developed initially by Phillips [14] to account for creep in glass and rubber. Later it was shown to agree well with the creep behaviour observed for single crystals of magnesium [15] and tin [16].

As the stress and temperature were increased, the aforementioned authors [10–12] found that primary creep observed was described better by the Bailey [17] power law

$$\varepsilon_c = \alpha t^m \tag{11.3a}$$

where $0 < m < 1$ for each material. Bhattacharya et al. [18] showed that eq(11.3a) applied to the room-temperature creep behaviour of copper, zinc, tin, cadmium, lead and aluminium for times approaching 600 min. Though m values were not specified in this work, wide variations in m have been observed elsewhere [19, 20]. There, eq(11.3a) was used to describe the parabolic creep strains numerically with good accuracy. In particular, $m = ⅓$ has been identified with the mechanistic processes of work hardening [5, 21] and appears within Andrade's law [22] of combined primary and secondary creep:

$$\varepsilon_c = \left(1 + \alpha t^{1/3}\right) \mathrm{e}^{\beta t} - 1 \approx \alpha t^{1/3} \tag{11.3b}$$

The approximation in eq(11.3b) identifies Andrade's primary 'α-strain' component when $\exp(\beta t) \to 1$ as $t \to 0$. In an intermediate region of stress and temperature, Wyatt [11] assumed that primary creep in copper was due to a combination of work hardening and exhaustion within an equation of the form

$$\varepsilon_c = \alpha_1 \ln t + \alpha_2 t^{1/3} \tag{11.4}$$

Such *transient creep* behaviour is produced by a variety of dislocation sources, for which an alternative, exponential description is often preferred [13, 23]

$$\varepsilon_c = \varepsilon_P\left(1 - \mathrm{e}^{-mt}\right) \tag{11.5}$$

for which $\varepsilon_c \to \varepsilon_P$ asymptotically as $t \to \infty$. In contrast, there are no asymptotes from eqs(11.3a,b) and (11.4). Equation (11.5) is often employed to represent the region of transient creep [26] that precedes the steady state in association with a recovery creep theory. The latter predicts that the primary creep curve arises from competing work hardening and recovery processes acting simultaneously. Work hardening is responsible for the decreasing strain rate that dominates shorter times. Consequently, the material becomes harder to deform as the internal stress increases with the dislocation density. The recovery process begins when the internal energy of the structure has risen sufficiently to release the barriers to dislocation motion. In particular, a self-diffusion activates the climb and cross slip of previously locked dislocations. The balance between work hardening and recovery identifies macroscopically with the secondary creep region AB in Fig. 11.1 where the strain rate is constant. The duration of AB depends upon composition in which a prolonged period applies to metals and solid solutions under constant load and moderate temperature ($T/T_m < 0.5$). In contrast, the time within AB is shortened by the onset of void damage in dispersion hardened alloys bearing load at high temperature. Usually, point B identifies where damage begins before growing under an accelerating strain rate in the tertiary region BC of the creep curve. However, the preceding primary creep strain can damage these alloys. Viscous plasticity within primary and secondary regions is considered in § 11.4.2 and § 11.9.3. The presence of primary creep in plasticity follows.

11.2.3 Cold Creep

At ambient temperature, 'cold' creep often contributes to metal plasticity. Though short-lived, the contribution from a cold creep strain to total strain should not be ignored. Such transient strain exhibits a time-dependence that is typically logarithmic or parabolic. Either form of creep strain is seen to follow an instantaneous strain ε_o, with the latter being taken, generally, as the sum of the elastic and plastic components ε_e and ε_P. The total strain is then

$$\varepsilon_t = \varepsilon_o + \alpha \ln t \tag{11.6a}$$

$$\varepsilon_t = \varepsilon_o + \alpha t^m \tag{11.6b}$$

where $\varepsilon_o = \varepsilon_e + \varepsilon_P$ (also written as $\varepsilon_o = \varepsilon^E + \varepsilon^P$). Creep has influenced plasticity studies made under deadweight incremental loading of polycrystalline materials. One such study includes the experimental determination of initial and subsequent yield loci [27, 28] discussed in chapters 3 and 4. Regarding its nature, the cold creep behaviour of commercially pure aluminium (E1A) at 21°C [12] is illustrated in Fig. 11.2. Periods of 1000 s were allowed between successive tensile stress increments of 2–4 MPa as shown inset. At low stress levels the creep strain is linear in the logarithmic time axis of Fig. 11.2, conforming to eq(11.2a). With a further increase in total stress, larger creep strains occur that appear linear initially. However, the corresponding variation in strain rate, lying within the range $10^{-3} - 10^{-7}$ s^{-1}, is greater than is given by the logarithmic law, eq(11.6a). Figure 11.3 shows that this creep strain behaviour is more closely parabolic. In fact, eq(11.6b) applies to $t \geq 60$ s, where the time exponent m lies in the range ⅓ – ½ and $\alpha = \alpha(\sigma)$ is stress dependent. For $t \leq 60$ s, the double logarithmic plot is non-linear, this being sensitive to the 'instantaneous' time chosen (taken here as 2 s). In general, where cold creep occurs the corresponding instantaneous strain ε_o ($= \varepsilon_{(2)}$) is composed of elastic and plastic components, their dependencies upon stress being linear and parabolic respectively. That is, $\varepsilon_o = \sigma/E + A\sigma^n$, where E is Young's modulus and A and n are inverted Hollomon constants.

Note: At the lowest stress levels where little to no cold creep appears for soft aluminium, the instantaneous is all elastic.

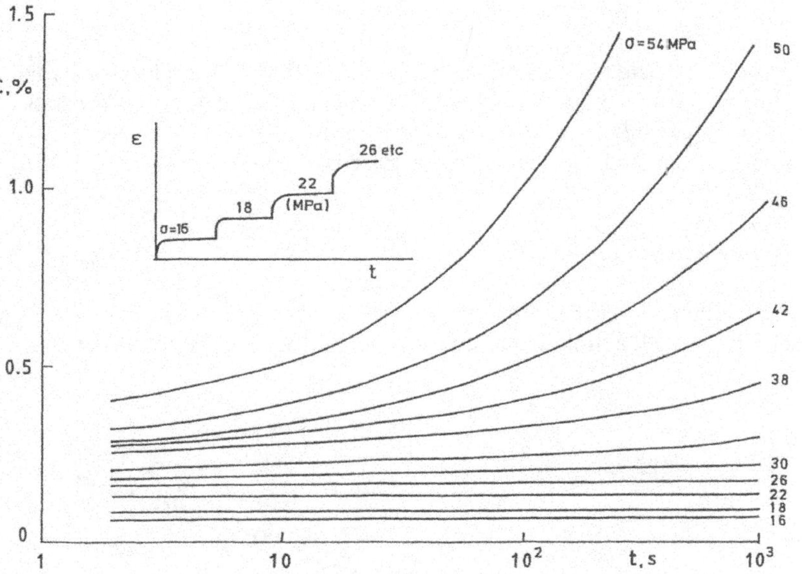

Figure 11.2 Logarithmic and parabolic 'cold' creep for E1A aluminium

Similar cold-creep behaviour was observed for radial loading of aluminium under combined tension–torsion [25]. Both logarithmic and parabolic stress-dependent creep occurred in the axial and shear strains between proportional load increments. Logarithmic creep is prevalent in copper and steel over a far greater range of stress than for aluminium [10, 11]. Wyatt [10] employed eq(11.3b) to describe the parabolic creep that he observed consistently at 27°C in aluminium with differing grain size and purity. In contrast, eq(11.2a) was found to apply to all conditions of aluminium at − 196°C.

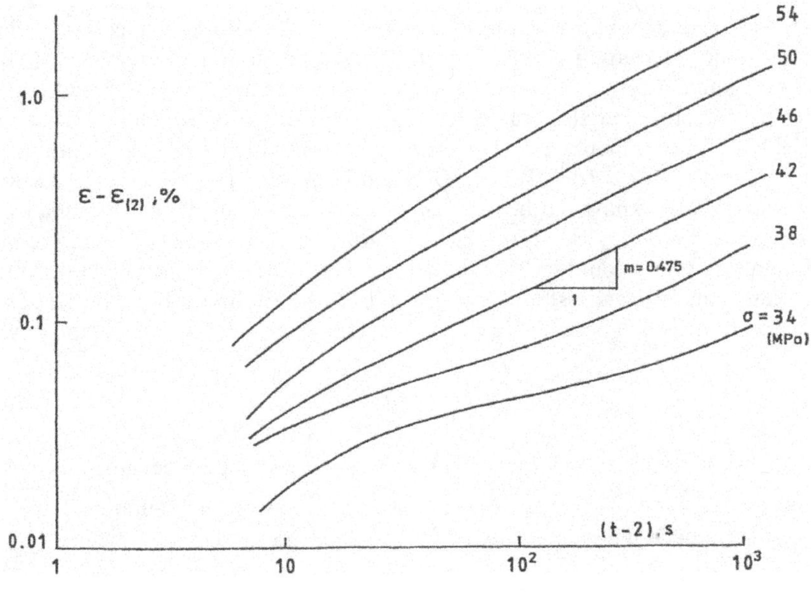

Figure 11.3 Stress dependence of coefficient α in parabolic creep law (11.6b)

11.3 Strain Rate Effects

Increasing the strain rate to lie within the range $10^{-5} - 10^{-1}\ s^{-1}$ introduces a dynamic effect in which plastic flow becomes rate (or time)-dependent. This raises the flow stress required to attain a given plastic strain. Experiments upon metallic materials have shown that the extent to which a uniaxial flow stress can be raised depends upon the crystal structure.

11.3.1 Uniaxial Behaviour

A typical variation in uniaxial flow stress over the range of strain rates stated above is illustrated for annealed aluminium of commercial purity (99.8 %) in Fig. 11.4.

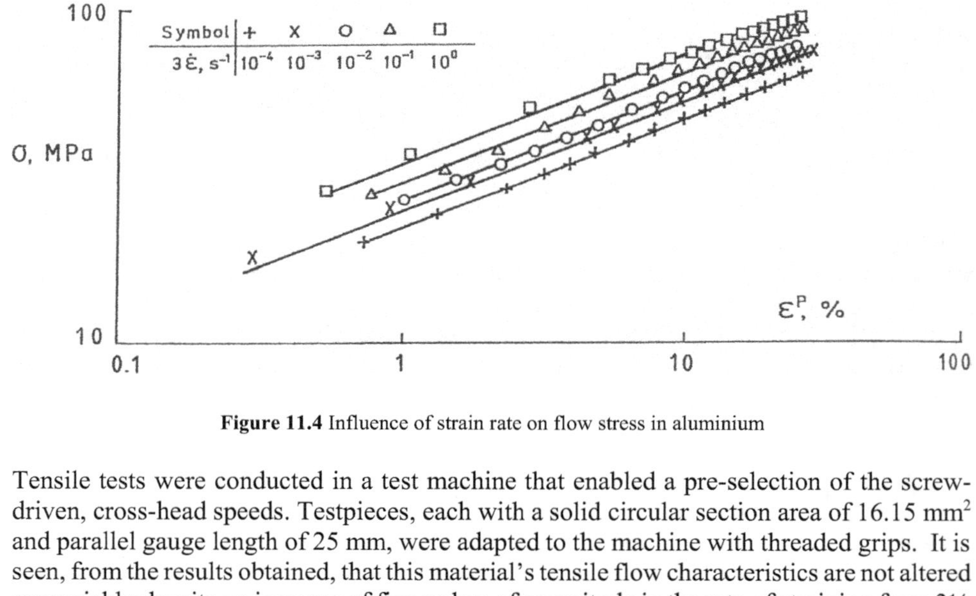

Figure 11.4 Influence of strain rate on flow stress in aluminium

Tensile tests were conducted in a test machine that enabled a pre-selection of the screw-driven, cross-head speeds. Testpieces, each with a solid circular section area of 16.15 mm^2 and parallel gauge length of 25 mm, were adapted to the machine with threaded grips. It is seen, from the results obtained, that this material's tensile flow characteristics are not altered appreciably despite an increase of five orders of magnitude in the rate of straining from 3⅓ × 10^{-5} to 3⅓ × $10^{-1}\ s^{-1}$ (inset). The behaviour is typical of a f.c.c. structure (e.g. Al, Cu, Ni and brass) that is known to become more strain-rate sensitive at the higher end of an extended range $10^{-1} - 10^{4}\ s^{-1}$ [26, 27]. In contrast, the flow stress of b.c.c. structures (Fe, Mo, W and steel) may double within the lower range. The double logarithmic plots of true stress versus natural plastic strain, given in Fig. 11.4, are linear and parallel sensibly for strains exceeding 1 %. These indicate a constant exponent n in the Hollomon power law:

$$\sigma = K(\varepsilon^P)^n \tag{11.7a}$$

Figure 11.5 shows that another power law connects the strength coefficient $K = K(\dot{\varepsilon}^P)$ to the plastic strain rate $\dot{\varepsilon}^P$ across the five decades. Thus, the Hollomon law (11.7a) is partitioned to include the stress dependence upon plastic strain and its rate:

$$\sigma = A(\dot{\varepsilon}^P)^m (\varepsilon^P)^n \tag{11.7b}$$

For pure aluminium, the material constants are: $A \simeq 2$, $m = 0.043$ and $n \simeq ⅓$ (in units of MPa for σ and s^{-1} for $\dot{\varepsilon}^P$).

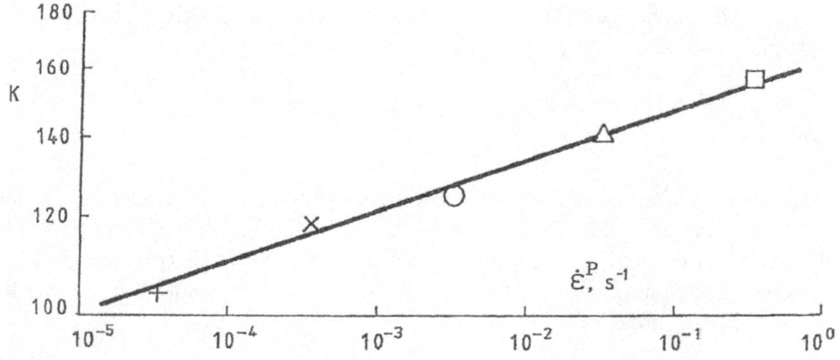

Figure 11.5 Influence of strain rate on the strength coefficient for pure aluminium

11.3.2 Rate Decomposition

Figure 11.6 shows that the test machine's constant crosshead speed does not equate to a constant rate of straining within the testpiece.

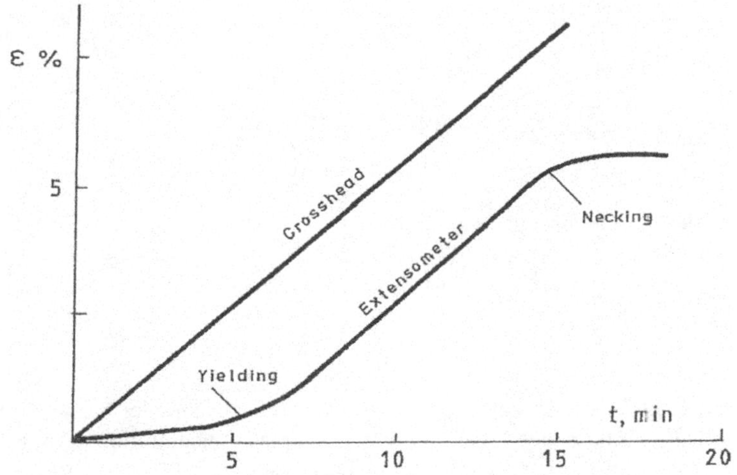

Figure 11.6 Elastic and elastic–plastic rates of strain

Using an extensometer or an axially aligned strain gauge with the testpiece enables its lower rate of elastic strain $\dot{\varepsilon}^E$ to be separated from the total elastic–plastic rate $\dot{\varepsilon}^T$. The latter is proportional to the constant cross-head displacement rate. Hence, the rate of plastic straining $\dot{\varepsilon}^P$ follows from re-arranging an additive decomposition of the total infinitesimal strain rate:

$$\dot{\varepsilon}^P = \dot{\varepsilon}^T - \dot{\varepsilon}^E \tag{11.7c}$$

Figure 11.6 shows an apparent fall in the total engineering strain rate (gradient) beyond the point of necking. In fact, the true (equivalent) strain rate increases as the neck develops under a stress state that becomes increasingly triaxial (see § 9.3). *Note*: Here upper-case superscripts E and P denote elastic and plastic contributions to plasticity, instead of lower-case subscripts e and p.

11.3.3 Superplasticity

Superplastic materials have relatively high m values within eq(11.7b) meaning that they are capable of high strain before fracture. Neck formation is inhibited as the strain rate is more uniformly spread throughout the length of testpiece. Assume that a constant plastic strain applies at given applied load W irrespective of the strain rate. Then, eq(11.7b) shows the rate sensitivity of the flow stress as

$$\sigma = B\dot{\varepsilon}^{m} = \frac{W}{A} \tag{11.8a}$$

For a current length and area of testpiece l and A, deforming at constant volume V:

$$V = AL = \text{constant} \quad \Rightarrow \quad \frac{\delta V}{V} = 0 = \frac{\delta A}{A} + \frac{\delta l}{l}$$

$$\therefore \ \dot{\varepsilon} = \frac{1}{l} \times \frac{\mathrm{d}l}{\mathrm{d}t} = -\frac{1}{A} \times \frac{\mathrm{d}A}{\mathrm{d}t} \tag{11.8b}$$

Combining eqs(11.8a,b):

$$A^{\frac{1-m}{m}} \times \frac{\mathrm{d}A}{\mathrm{d}t} = -\left(\frac{W}{B}\right)^{\frac{1}{m}} = C$$

for which C is constant at a given load

$$\frac{\mathrm{d}A}{\mathrm{d}t} = \frac{C}{A^{\frac{1-m}{m}}} \tag{11.8c}$$

which shows that for most metals, where $m = 0.02 - 0.2$, the smaller the section area, the more rapidly that area is reduced. Consequently, a stress concentration arising from poor surface finish or an imperfection in composition places a site of reduced area under deformation. These raise the strain rate locally to develop a reduced area and restricted length of a neck in which strain is concentrated. In contrast, for 'superplastic' alloys, where the rate exponent $m = 0.8 - 1.0$, eq(11.8c) shows that the current area is reduced by plasticity almost at a constant rate. Any local irregularity in section, which normally provides a site for strain concentration, is then preserved without developing into a neck. Thus, a high m offsets a local neck from forming allowing a high strain to be attained in which the testpiece length can more than double. However, such strain rate sensitivity may only appear within a narrow range of strain rate and temperature, typically at $0.4T_m$ for aluminium-magnesium alloy within the mid range of rates typical of those given in Fig. 11.5.
Note: See data sheets in first entry of bibliography.

11.3.4 Time-Dependent Plasticity

Chapter 3 showed that there were practical difficulties associated with yield stress determination (see § 3.3). The subsequent yield stress for a pre-stressed material is often found from a partial unloading followed by a re-loading to its limit of proportionality. It will be shown that the yield stress does not equate to the pre-stress where there is hysteresis between loading and unloading. However, Phillips and Lee [28] showed that increasing the time under the prestress, before applying an unloading–reloading cycle, served to equalise the pre-stress and the yield stress. This demonstrates that transient strain (cold creep) has an influence on yielding. Where sufficient time under a pre-stress allows the full contribution to strain from creep, the pre-stress point will lie on the yield surface. It follows that the 'true' yield surface, which will contains the pre-stress, is associated with zero creep rate. That rate is reached at low temperatures, when eventually an exhaustion of transient creep strain occurs. For example, under an incremental loading of aluminium at $T/T_m \simeq 0.3$ in Fig. 11.2, the creep strain rate falls from $\simeq 10^{-3}\ s^{-1}$ at $t = 10$ s to $t \simeq 10^{-5}\ s^{-1}$ in $t = 1000$ s. The sequential creep curves, shown schematically in Fig. 11.7a, enable the corresponding stress versus inelastic strain response at each rate to be constructed, in the manner of Fig. 11.7b.

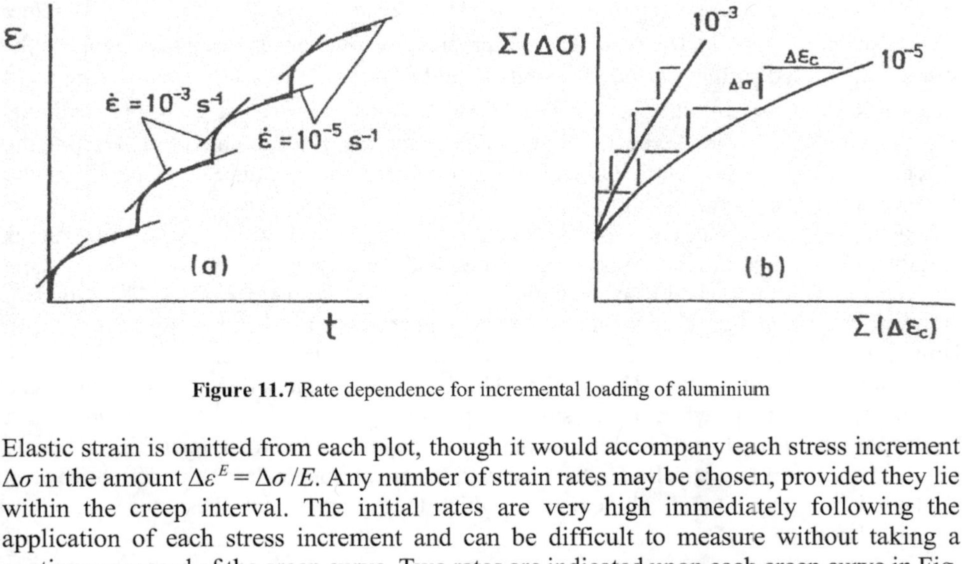

Figure 11.7 Rate dependence for incremental loading of aluminium

Elastic strain is omitted from each plot, though it would accompany each stress increment $\Delta\sigma$ in the amount $\Delta\varepsilon^E = \Delta\sigma / E$. Any number of strain rates may be chosen, provided they lie within the creep interval. The initial rates are very high immediately following the application of each stress increment and can be difficult to measure without taking a continuous record of the creep curve. Two rates are indicated upon each creep curve in Fig. 11.7a. The creep strain increment $\Delta\varepsilon_c$, corresponding to each stress increment $\Delta\sigma$, is greater at a slower rate. A creep equation of state implies that the plots of total stress $\sum(\Delta\sigma)$ versus accumulated strain $\sum(\Delta\varepsilon_c)$ in Fig. 11.7b are equivalent to those that would be obtained from a monotonic loading at each constant strain rate (see Fig. 11.4). The creep contribution to inelastic strain and its effect upon the flow stress becomes apparent from Fig. 11.7b. When allowing for time-dependent strain, the relation between the pre-stress and the yield stress may be interpreted in the manner of Figs 11.8a,b. When the pre-stress point P in Fig. 11.8a lies on the yield surface, the inelastic strain rate $\dot{\varepsilon}^P$ is zero and unloading from P is linearly elastic. Figure 11.8b shows further that an unloading from σ_P, when the strain rate under load is finite ($\dot{\varepsilon}^P \neq 0$), applies to P′ on another upper flow curve. Here a stress reduction in creep (the dip test) has shown that a material may: (i) continue to creep at a lower rate, (ii) attain a zero rate and (iii) reverse creep with negative rate.

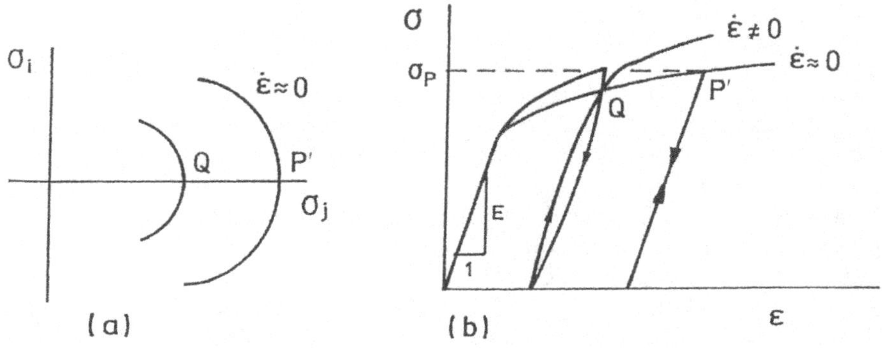

Figure 11.8 Relationship between pre-stress and yield stress

With a continuous unloading from a pre-stress σ_p at P′, a compounding of the three rates in (i), (ii) and (iii) is manifested in a non-linear, unloading curve in Fig. 11.8b. The initial, elastic response to a reloading completes the closed hysteresis cycle at the point Q, on the lower flow curve for which $\dot{\varepsilon}^P = 0$. Thereafter, creep recommences at a positive rate as the stress is increased further. Thus, the points P′ and Q in Fig. 11.8a are taken as true yield stresses, since they lie on the curve in Fig. 11.8b for which $\dot{\varepsilon}^P = 0$. The stress at P is unique while the lower stress at Q will depend upon the creep rate prior to unloading. Despite the absence of a specified creep rate most yield loci given in the literature correspond to yield stresses at the intersection point Q.

It follows that in the region of stress between P′ and Q in Fig. 11.8a, the strain rate is variable. Here, within the division between distinct regions of elastic and inelastic deformation, the rate variation occurs with the time allowed for creep to occur. The boundary is set at P for which the full exhaustion of creep strain has occurred under a given pre-stress. However, a zero creep rate will not be reached under a pre-stress when a recovery process occurs for higher temperatures ($T/T_m \geq 0.5$). The flow law for time-independent plasticity can be adapted to creep when a 'dissipation potential' replaces the 'plastic potential'. Correspondingly, a flow rule for creep supplies the magnitude and direction of the creep rate vector by the normality rule. So far, the observations on rate-dependent behaviour have been explained from experiment. The extent to which theory can match these observations is examined next.

11.3.5 Temperature-Dependent Plasticity

It is known that increasing temperature has the opposite influence upon the flow curve to increasing the strain rate. It follows from this that time under stress allows more strain to accumulate by the process of creep. Creep strain and its dependence upon stress, time and temperature is discussed in detail elsewhere [29]. In particular, for temperature dependent plasticity, schematic diagrams follow when loading at a constant displacement (nominal strain) rate and at a constant stress rate.

(a) Constant Strain Rate

Consider tension tests each conducted each at a different temperature all under a constant displacement (nominal strain) rate. The order of flow curves appears in Fig. 11.9a with the highest temperature plot lying in the lowest position. The constant rate input to the material (see Fig. 11.9b) controls its stress rate and the stiffness in the manner shown in Figs 11.9c,d.

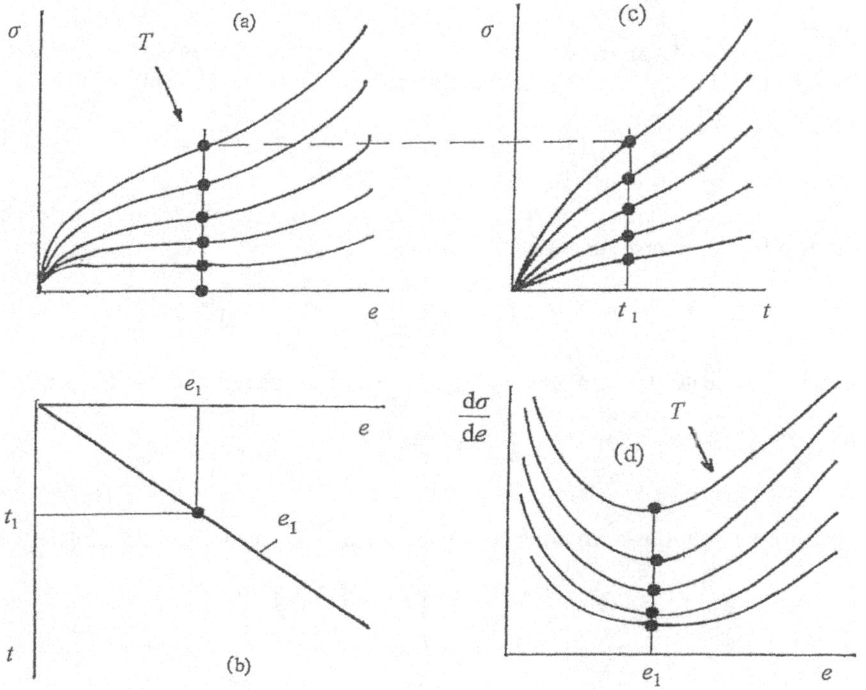

Figure 11.9 Temperature influence upon flow curve for a common displacement rate

Each response is derived from the flow curve function involving either the material stiffness or its compliance in their respective forms $\sigma = \sigma(e, T)$ and $e = e(\sigma, T)$. For simplicity, the variables are separated within a polynomial fit to each flow curve [4] and the former function becomes

$$\sigma = \sigma(e,T) = s_1(T)\,e + s_2(T)\,e^2 + \ldots s_n(T)\,e^n \tag{11.9a}$$

in which $s_i(T)$ indicates the temperature dependence of the stiffness coefficients as

$$S_e = \frac{\partial \sigma}{\partial e} = s_1(T) + 2s_2(T)e + \ldots ns_n(T)e^{n-1} \tag{11.9b}$$

The stress rate follows from the change in stress $\delta\sigma$ occurring in a time interval δt:

$$\frac{\delta\sigma}{\delta t} = \left(\frac{\partial\sigma}{\partial e}\right)\left(\frac{\delta e}{\delta t}\right) + \left(\frac{\partial\sigma}{\partial T}\right)\left(\frac{\delta T}{\delta t}\right) \tag{11.10a}$$

Since $\delta T/\delta t = 0$ for an isothermal test, eq(11.10a) provides the limiting stress rate as $\delta t \rightarrow 0$:

$$\frac{\mathrm{d}\sigma}{\mathrm{d}t} = \left(\frac{\partial\sigma}{\partial e}\right)\left(\frac{\mathrm{d}e}{\mathrm{d}t}\right) \tag{11.10b}$$

Combining eqs(11.9b) and (11.10b) gives

$$\dot{\sigma} = S_e \dot{e} = \left[s_1(T) + 2s_2(T)e + \ldots ns_n(T)e^{n-1} \right] \dot{e} \tag{11.11a}$$

in which the bracketed quantity fits the variation in tangential stiffness shown in Fig. 11.9c. Integrating eq(11.11a) with respect to t, the σ versus t plot follows simply

$$\sigma = t\left[s_1(T) + 2s_2(T)e + \ldots ns_n(T)e^{n-1} \right] \dot{e} \tag{11.11b}$$

and with $\dot{e}$ constant, the plot mirrors that of σ versus e. Alternatively, should the flow curve be described by the inverse function

$$e = e(\sigma, T) = c_1(T)\,\sigma + c_2(T)\,\sigma^2 + \ldots c_n(T)\,\sigma^n \tag{11.12a}$$

in which $c_i(T)$ indicates the temperature dependence of the compliance coefficients as

$$C_\sigma = \frac{\partial e}{\partial \sigma} = c_1(T) + 2c_2(T)\sigma + \ldots nc_n(T)\sigma^{n-1} \tag{11.12b}$$

Here, the stress rate follows from the change in strain δe occurring in a time interval δt:

$$\frac{\delta e}{\delta t} = \left(\frac{\partial e}{\partial \sigma} \right)\left(\frac{\delta \sigma}{\delta t} \right) + \left(\frac{\partial e}{\partial T} \right)\left(\frac{\delta T}{\delta t} \right) \tag{11.13a}$$

and since $\delta T/\delta t = 0$ for an isothermal test the limiting stress rate for $\delta t \to 0$ appears within

$$\frac{\mathrm{d}e}{\mathrm{d}t} = \left(\frac{\partial e}{\partial \sigma} \right)\left(\frac{\mathrm{d}\sigma}{\mathrm{d}t} \right) \quad \text{or} \quad \dot{e} = C_\sigma \dot{\sigma} \tag{11.13b,c}$$

Substituting eq(11.12b):

$$\dot{\sigma} = C_\sigma^{-1} \dot{e} = \left[c_1(T) + 2c_2(T)\sigma + \ldots nc_n(T)\sigma^{n-1} \right]^{-1} \dot{e} \tag{11.14a}$$

and integrating for $\dot{e}$ constant provides the σ versus t response equivalent to eq(11.12a) as the 'mirror' of the σ versus e relationship.

$$\sigma = t\left[c_1(T) + 2c_2(T)\sigma + \ldots nc_n(T)\sigma^{n-1} \right]^{-1} \dot{e} \tag{11.14b}$$

(b) Constant Stress Rate

Consider each test conducted at a different temperature all under a constant stress (load) rate. The same order of flow curves appears in Fig. 11.10a with the highest temperature plot lying in the lowest position. The constant stress rate input to the material controls its strain rate in the manner shown in Figs 11.10b,c. Each e versus t response is derived from the flow curve function, again written in terms of the material stiffness or its compliance in their respective forms: $\sigma = \sigma(e, T)$ and $e = e(\sigma, T)$. For simplicity, the variables are separated within a polynomial fit to each flow curve from the former function $\sigma = \sigma(e, T)$.

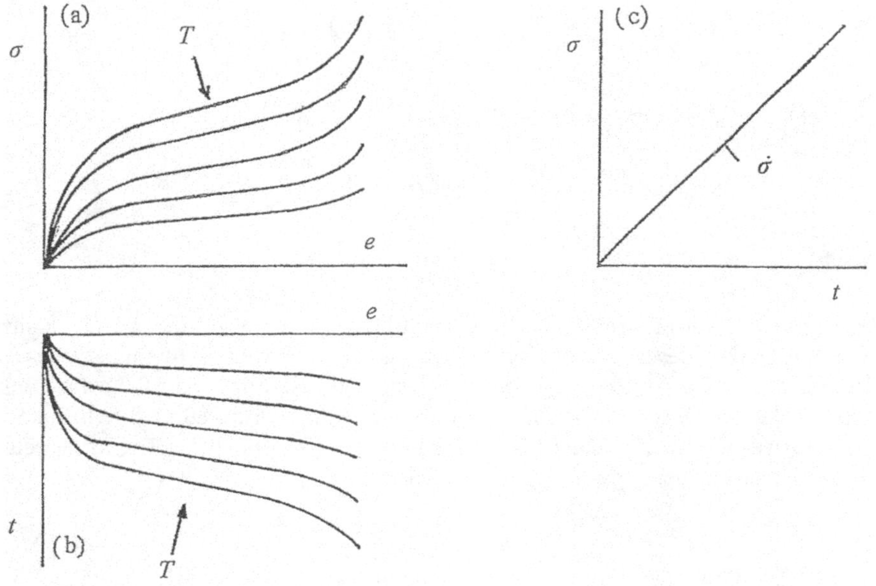

Figure 11.10 Temperature influence upon flow curve for a constant loading (nominal stress) rate

The stress rate appears with the stiffness derivative and the strain rate, i.e. the change in strain δe over time interval δt:

$$\frac{\delta\sigma}{\delta t} = \left(\frac{\partial\sigma}{\partial e}\right)\left(\frac{\delta e}{\delta t}\right) + \left(\frac{\partial\sigma}{\partial T}\right)\left(\frac{\delta T}{\delta t}\right) \tag{11.15a}$$

Now as $\delta T/\delta t = 0$ for an isothermal test eq(11.15a) gives the limiting stress rate for $\delta t \to 0$:

$$\frac{d\sigma}{dt} = \left(\frac{\partial\sigma}{\partial e}\right)\left(\frac{de}{dt}\right) \tag{11.15b}$$

Equation (11.15b) shows a rate for the polynomial eq(11.14a):

$$\dot{\sigma} = S_e\dot{e} = \left[s_1(T) + 2s_2(T)e + \ldots ns_n(T)e^{n-1}\right]\dot{e} \tag{11.16a}$$

and with $\dot{\sigma}$ constant the strain versus time response follows:

$$e = \dot{\sigma}t\left[s_1(T) + 2s_2(T)e + \ldots ns_n(T)e^{n-1}\right]^{-1} \tag{11.16b}$$

which mirrors the flow curve σ versus e as shown. Alternatively, starting with $e = e(\sigma, T)$:

$$\frac{\delta e}{\delta t} = \left(\frac{\partial e}{\partial\sigma}\right)\left(\frac{\delta\sigma}{\delta t}\right) + \left(\frac{\partial e}{\partial T}\right)\left(\frac{\delta T}{\delta t}\right) \tag{11.17a}$$

and taking $\delta T/\delta t = 0$ the stress rate required appears from eq(11.17a) as

$$\frac{de}{dt} = \left(\frac{\partial e}{\partial \sigma}\right)\left(\frac{d\sigma}{dt}\right) \quad \text{or} \quad \dot{e} = C_\sigma \dot{\sigma} \tag{11.17b,c}$$

Equation (11.17c) shows the polynomial forms

$$\dot{e} = C_\sigma \dot{\sigma} = \left[c_1(T) + 2c_2(T)\sigma + \ldots nc_n(T)\sigma^{n-1}\right]\dot{\sigma} \tag{11.18a}$$

$$e = t\dot{\sigma}\left[c_1(T) + 2c_2(T)\sigma + \ldots nc_n(T)\sigma^{n-1}\right] \tag{11.18b}$$

In Fig. 7.7b, eq(11.18b) is seen to be equivalent to eq(11.16b) as a 'mirror' of σ versus e.

The alternative descriptions given in (a) and (b) above, involving stiffness and compliance coefficients (s_q and c_q), provide equivalent descriptions of the flow curve. Therefore, either may be chosen according to whether the numerical curve fit is to cut off the σ versus e curves in Figs 11.9a and 11.10a at a common stress or strain. The former cut-off appears by either manner shown in Figs 11.11a,b

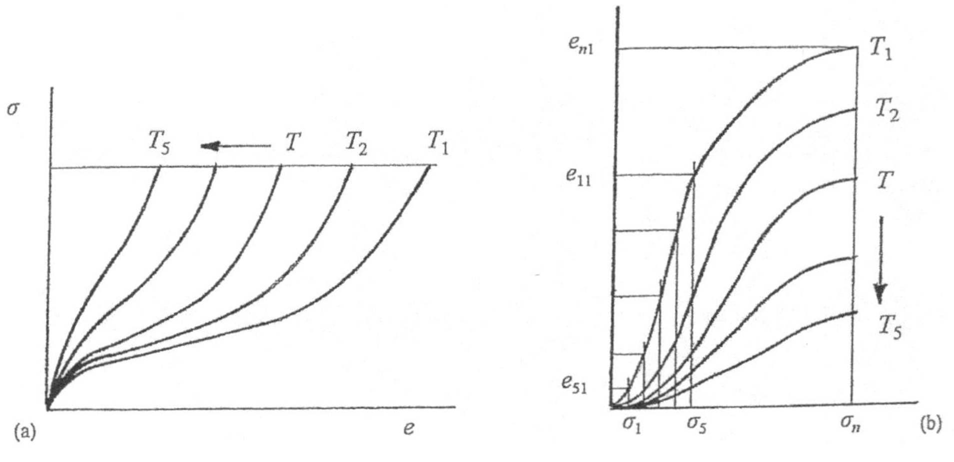

Figure 11.11 Alternative stress cut-offs for temperature dependent flow curves

Here, the coefficients c_q and s_q tied to each flow curve are identified with a specific temperature. It follows that an empirical form $c_q(T)$ and $s_q(T)$ of each coefficient may be established in the temperature range of interest.

(c) Varying Temperature

The assumption made of a constant temperature for a testpiece in an 'isothermal' test in paragraphs (a) and (b) above would be in error should its initial temperature be influenced by either the strain rate or the stress rate. For example, if a temperature change occurs under the fast strain rate f in Figs 11.12a–c both terms in the total stress rate differential must be retained:

$$\frac{\delta\sigma}{\delta t} = \left(\frac{\partial\sigma}{\partial e}\right)\left(\frac{\delta e}{\delta t}\right) + \left(\frac{\partial\sigma}{\partial T}\right)\left(\frac{\delta T}{\delta t}\right) \tag{11.19a}$$

Equation (11.19a) is written as

$$\dot{\sigma} = S_e \dot{e} + S_T \dot{T} \tag{11.19b}$$

where from eq(11.9b):

$$S_e = \frac{\partial \sigma}{\partial e} = s_1(T) + 2s_2(T)e + \ldots ns_n(T)e^{n-1} \tag{11.20a}$$

$$S_T = \frac{\partial \sigma}{\partial T} = \frac{\partial s_1(T)}{\partial T}e + \frac{\partial s_2(T)}{\partial T}e^2 + \ldots \frac{\partial s_n(T)}{\partial T}e^n \tag{11.20b}$$

in which the temperature dependence of the stiffness coefficients s_i may be established from σ versus e plots each at a constant test temperature T_o for a common, low strain rate test as in Fig. 11.9b. Re-arranging eq(11.19b) gives

$$\dot{T} = \left(\dot{\sigma} - S_e \dot{e}\right)/S_T \tag{11.20c}$$

within which eqs(11.20a,b) appear, enabling an integration giving the testpiece temperature variation for the test duration. If the stress rates at each temperature are also taken from Fig. 11.9c then, with no change in test temperature, eq(11.19b) shows that the stress rate becomes $\dot{\sigma} = S_e \dot{e}$. However, a high strain rate will alter the disposition of the flow curves under this isothermal condition.

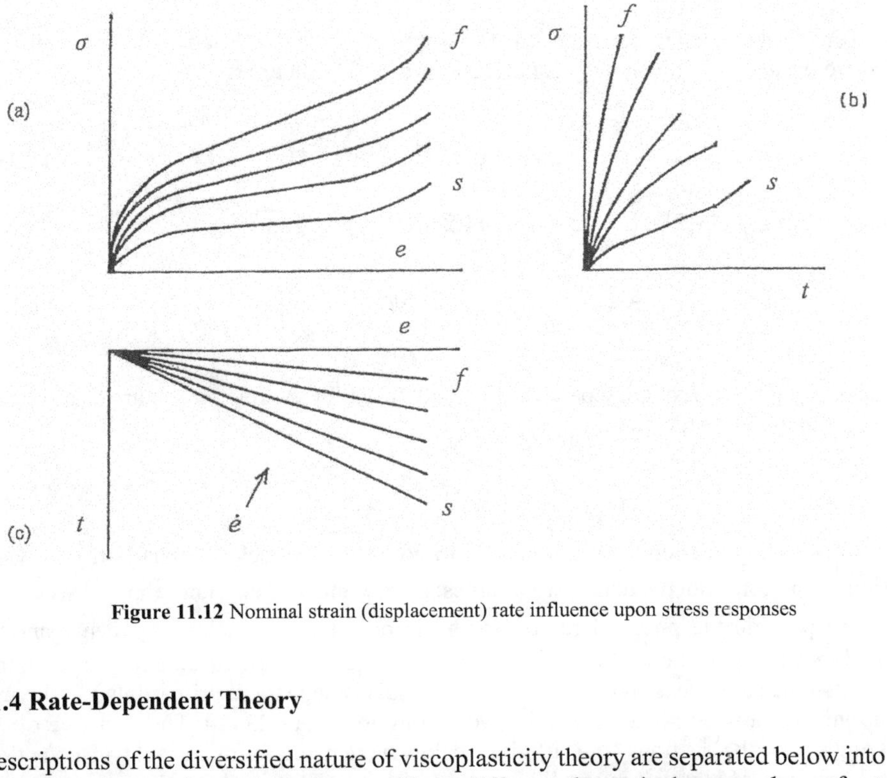

Figure 11.12 Nominal strain (displacement) rate influence upon stress responses

11.4 Rate-Dependent Theory

Descriptions of the diversified nature of viscoplasticity theory are separated below into four distinct accounts of rate-dependence. These different theoretical approaches refer to the micro-structure, micro-mechanics, the overstress definition and a heredity integral.

11.4.1 Microstructure

Perzyna [30] employed micro-structural parameters to characterise the dynamic dislocation arrangements produced by viscous plastic deformation. For example, the rate of uniaxial straining was connected to four parameters as follows:

$$\dot{\varepsilon} = \alpha N b v_m \tag{11.21}$$

where $b = |\mathbf{b}|$ is the magnitude of the Burgers vector, i.e. the displacement per dislocation line, N is the mean dislocation density, α is an orientation factor and v_m is the mean dislocation velocity. The shear stress τ, required to overcome short-range obstacles that impede dislocation motion, is assisted by thermal activation within the dislocated structure. The response function φ expresses this:

$$U = \varphi\left[(\tau - \tau^*)Lb\right] \tag{11.22a}$$

where U is the activation energy, τ^* is a back stress and L is the mean distance between dislocation arrest points. The mean velocity v_m is then

$$v_m = A\nu \mathrm{e}^{-\frac{U}{kT}} \tag{11.22b}$$

where T is the absolute temperature, A is the swept area, k is Boltzmann's constant and ν is the Debye frequency. Substituting eq(11.22b) into eq(11.21) gives

$$\dot{\varepsilon} = \alpha N b A \nu \mathrm{e}^{-\frac{U}{kT}} = \nu^* \mathrm{e}^{-\frac{U}{kT}} \tag{11.23a}$$

and then, from eqs (11.21), (11.22a) and (11.23a), τ is given as

$$\tau = \tau^* + \frac{1}{Lb}\varphi^{-1}\left[kT \ln\left(\frac{\nu^*}{\dot{\varepsilon}}\right)\right] \tag{11.23b}$$

In eqs(11.23a,b), the frequency parameter ν^* is defined for the microstructure as a product of the five included parameters:

$$\nu^* = \alpha N A b \nu \tag{11.23c}$$

Macroscopically, ν^* depends only upon T. In contrast, the back (internal) stress τ^* is a function of the four independent test variables: ε, $\dot{\varepsilon}$, τ and T. Common terms between eqs (11.23a,b) provide the physical reason for a viscoplastic connection between τ and $\dot{\varepsilon}$, assumed in the macro-theory (see § 11.2). Perzyna's [30] review of uniaxial experiments showed that the stress sensitivity of metallic materials is dependent primarily upon thermal activation, over a wide range of ε and T, according to eqs(11.23a,b). The first five of the following seven observations (i)–(vii) given below were made for monotonic, viscous flow in single crystals of aluminium and lithium flouride and polycrystals of aluminium, α-iron, carbon-steel, lead and copper:

(i) The mean dislocation velocity v_m increased with both stress τ and temperature T.
(ii) The activation volume Av_m decreased with strain when $\tau > \tau^*$, at a given T.
(iii) The strain rate $\dot{\varepsilon}$ of the metal polycrystals displayed both logarithmic and power law dependencies upon τ for a given strain ε and temperature T.
(iv) The flow stress τ increased with ε for a given T.
(v) The flow stress τ decreased with T for a given ε and $\dot{\varepsilon}$.
(vi) The creep strain that occurred within the time intervals between increments of loading conformed to a logarithmic or power time law.
(vii) The strain rates were non-linear with increasing applied stress for a given T.

It has been seen earlier in § 11.2 that (vi) and (vii) apply to creep in copper and aluminium under an incremental loading. Within a restricted range of ε and T, Seeger's linear approximation to eq(11.23b) shows that the stress dependency upon the viscous plastic strain rate is thermally activated [31]:

$$\tau = \left(\tau^* + \frac{U^*}{V^*}\right) + \frac{kT}{V^*}\ln\left(\frac{\dot{\varepsilon}}{\nu^*}\right) \tag{11.24}$$

where U^* and V^* are, respectively, the activation energy and the volume corresponding to the back stress τ^*. Lindholm [32] showed that eq(11.24) may be extended to multi-axial stress states when the stress invariant $\surd J_2'$ plotted linearly with the logarithm of the strain invariant $\surd I_2$ for a given rate of the latter ($\surd \dot{I}_2$). The test results were found to violate the assumption of isotropic hardening, this observation being common to the results of many quasi-static investigations on the shape of the subsequent yield surface (see § 10.3 and § 10.7).

11.4.2 Micro-Mechanics

(a) Uniaxial

Miller [33] expressed the changing strain rates within viscous plasticity arising: (i) under monotonic and cyclic loading and (ii) during hardening and softening in creep. Consideration of the physical mechanisms involved led to a common, inelastic strain rate function, incorporating respective isotropic and kinematic hardening parameters, D and R, as follows:

$$\dot{\varepsilon} = f\left(\frac{\sigma - R}{D}\right) \tag{11.25a}$$

Specifically, D identified a drag stress, i.e. the measure of a structure's resistance to dislocation motion provided by its grain boundaries, precipitates, solute atoms, sub-grains and sub-boundaries [34]. The rest stress R (also known as the internal or back stress) develops with the build-up of dislocations pinned and entangled by these 'obstacles'. The particular deformation process to be described govern the explicit forms of D and R. For example, during secondary creep both D and R should remain constant but they should change continuously to match the preceding transient creep phase. Hence, a continuous function f in eq(11.25a) is required for both creep regions whilst admitting these variations in D and R. The Norton law meets these requirements when expressed as

$$\dot{\varepsilon} = A\left[f\left(\frac{\sigma - R}{D}\right)\right]^n \tag{11.25b}$$

where $A = A'\exp(-Q/RT)$ applies to high temperatures ($T \geq 0.6T_m$). For $T/T_m \leq 0.6$, a different function, $A = A(T)$, was employed to account for the dependence of the activation energy Q upon temperature T. Alternatively, exponential or hyperbolic sine forms may replace a power function in (11.25b) to match the rate dependence upon the effective stress more accurately within a narrower range. In eq(11.25b), both D and R vary with primary creep strain. Since the latter is dependent upon previous deformation, so too are D and R. When $\dot{\varepsilon}$ becomes constant in secondary creep, eq(11.25b) must employ stationary values of D and R to ensure a constant effective stress. Thus, D and R must satisfy the condition

$$\dot{\varepsilon} = f\left(\frac{\sigma - R}{D}\right), \quad \Rightarrow \quad \dot{\varepsilon} = f(\sigma - \sigma_i) \qquad (11.26a,b)$$

Here, the secondary creep rate becomes a function of the effective stress only with σ_i as the back stress. Otherwise, the primary creep rate given in eq(11.25a) requires that the effective stress is normalised by D. For eqs(11.26a,b) to hold, the variations in D and R, corresponding to primary and secondary creep deformation, are as shown in Figs 11.13a,b. The figures show how D and R attain their constant values when primary straining is completed. In fact, the variations that occur in D and R during primary creep reflect the opposing work hardening and thermal softening mechanisms within recovery creep theory [29].

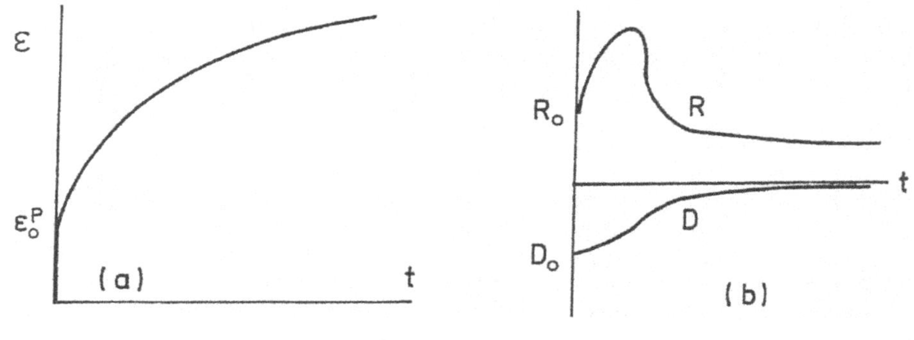

Figure 11.13 Variation in D and R with primary and secondary creep time

Given the respective rates of hardening and recovery, $h = \mathrm{d}\sigma/\mathrm{d}\varepsilon$ and $r = -\mathrm{d}\sigma/\mathrm{d}t$ respectively, the rates of change to each internal stress, D and R, produced by a strain increment $\mathrm{d}\varepsilon$ in a time interval $\mathrm{d}t$ at absolute temperature T become

$$\dot{R} = h_1\,\dot{\varepsilon} - r_1\,(R, T) \qquad (11.27a)$$

$$\dot{D} = h_2\,\dot{\varepsilon} - r_2\,(D, T) \qquad (11.27b)$$

Within eqs(11.27a,b), the Bailey-Orowan recovery theory shows that the two hardening moduli, $h_i = \mathrm{d}\sigma/\mathrm{d}\varepsilon$, and the two softening rates, $r_i = -\mathrm{d}\sigma/\mathrm{d}t = r(R,T)$, for $i = 1, 2$ become constant in secondary creep. With $\dot{R} = \dot{D} = 0$, as both R and D become constant, it follows from eqs(11.27a,b) that the constant secondary creep rate becomes $\dot{\varepsilon} = r_1/h_1 = r_2/h_2$. The transient primary creep behaviour is referred to the initial values D_o and R_o that arise from the instantaneous plastic strain ε_o^P, shown in Fig. 11.13a. Taking the difference between the rates of hardening and recovery r_1 and h_1 (i.e. between the ordinates r and h in Fig. 11.14) results in an R–variation having the maximum shown in Fig.11.13b. Also, the variation in D can be deduced from Fig. 11.14, where it shows the time-dependencies of h_2 and r_2 given that the fall in r_2 is more gradual than that in r_1.

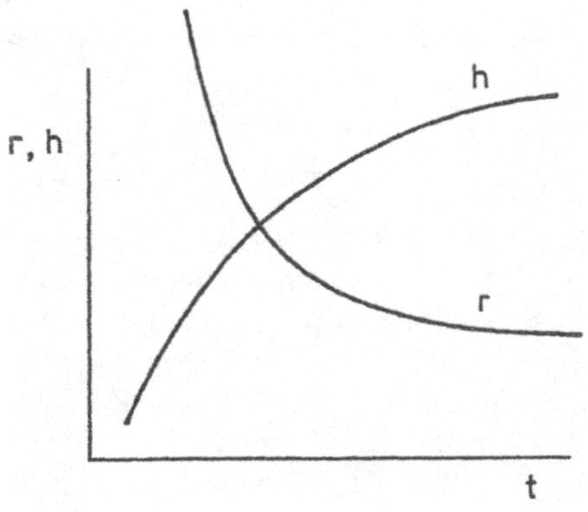

Figure 11.14 Rates of hardening and recovery in secondary creep

The explicit forms of D and R in eqs(11.27a,b) apply to all transient behaviour in which the strain rate changes. These include creep, warm working and annealing. Consider the derivation of one particular form of recovery function for a Norton steady state condition with $T/T_m \geq 0.6$. Setting $\dot{D} = 0$ in eq(11.27b) gives

$$r_2(D, T) = h_2 \dot{\varepsilon} \tag{11.28a}$$

Substituting $\dot{\varepsilon}_s \propto \sigma^n$ and $\dot{\varepsilon}_s \propto \exp(-Q/RT)$ into eq(11.28a) leads to

$$r_2(D, T) = \left[A' e^{-\frac{Q}{RT}} \sigma^n \right] h_2 \tag{11.28b}$$

in which A', n, Q and R are constants. In the case of transient creep, eq(11.28a) is re-written to admit the effect of the changing strain rate upon D:

$$r_2(D, T) = A' e^{-\frac{Q}{RT}} \left[f\left(\frac{\sigma - R}{D} \right) \right]^n h_2 \tag{11.29a}$$

where f needs to be established experimentally from within Fig. 11.6b. Under steady conditions, the room-temperature flow stress Y of previously warm-worked 3 % Si-steel confirmed isotropic hardening and hence the proportionality relationship $Y \propto D$. Further experimental data [33] were used to establish a relationship between D and the pre-stress as $D \propto \sigma^{1/3}$. This relationship led to an explicit form of eq(11.25b): $\dot{\varepsilon} \propto [(\sigma - R)/D]^{3/2}$. Given that R is low at steady state, an approximation applies: $\dot{\varepsilon} \simeq (\sigma/D)^{3/2}$. Substituting $D \propto \sigma^{1/3}$ into eq(11.28b) gives

$$r_2(D, T) = A' D^{n'} e^{-\frac{Q}{RT}} h_2 \tag{11.29b}$$

where $n' = 3n$ is also assumed to hold during transient creep. Finally, substituting eq(11.29b) into (11.27b), leads to

$$\dot{D} = h_2 \left(\dot{\varepsilon} - A' D^{n'} \mathrm{e}^{-\frac{Q}{RT}} \right) \tag{11.30a}$$

which, evidently, can be applied to a primary regime, which terminates at the steady state. Correspondingly, the recovery function (11.27a) must provide for the steady state:

$$\dot{R} = h_1 \left(\dot{\varepsilon} - A' R^{n'} \mathrm{e}^{-\frac{Q}{RT}} \right) \tag{11.30b}$$

Miller [33] developed functions for $\dot{D}$ and $\dot{R}$ in a similar manner extended over wider stress and temperature ranges. To do this, he employed an empirical temperature function and Garofalo's hyperbolic sine stress relationship [30]. Equations (11.25a,b) and (11.30a,b) provide the inelastic component of strain for all transient phenomena, e.g. annealing, pre-straining, short-time plasticity and long-time creep. To account for hardening and softening behaviour under strain cycling, D increases and decreases respectively with time while R cycles through zero consistently. This theory is restricted to initially isotropic materials that harden appreciably such as pure aluminium and stainless steel. Particular applications of eqs(11.30a,b) were made [35] for balanced strain cycling of annealed 304 stainless steel at 593°C, with varying periods of relaxation at peak strain. In the absence of a relaxation hold, a theoretical approximation to the hysteresis loop reproduced cyclic hardening and the Bauschinger effect. Provision for a greater decrease in D was made where peak stress softening resulted from dynamic recovery under longer hold times.

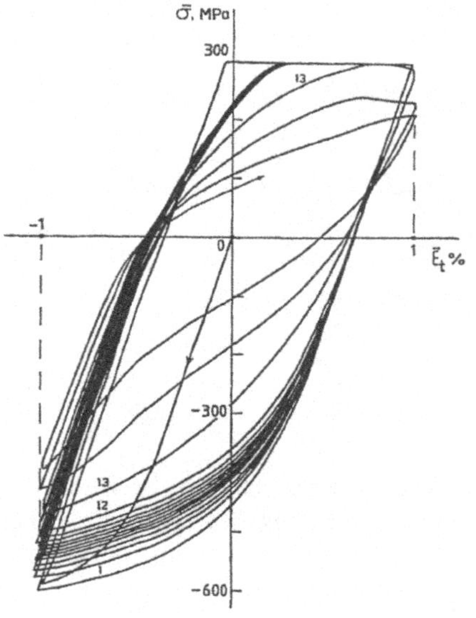

Figure 11.15 Peak stress softening with creep dwell

Other materials were shown to exhibit peak stress softening due to cavity nucleation and growth. For example, Fig. 11.15 shows such softening under strain-limited, torsional cycling for the Ni-base alloy (IN 597) at 850°C [36]. The creep component of each creep–plasticity

(CP) cycle allows a period of creep dwell under $\bar{\sigma} = \sqrt{3}\tau = 300$ MPa to attain a limiting equivalent strain $\bar{\varepsilon}^T = \gamma^T/\sqrt{3} = 1$ %. When this deformation is reversed by rapid plasticity to a peak strain $\bar{\varepsilon}^T = -1$ %, a successive reduction in the magnitude of the reversed peak stress $\bar{\sigma}_p$ occurs, leading to failure under relatively few cycles as shown.

In a further test [36], $\bar{\sigma}_p$ was allowed to relax under the limiting reversed strain $\bar{\varepsilon}^T = -1$ % (see Fig. 11.16). The continued cyclic softening behaviour observed for IN597 shows that stress reversal does not fully 'heal' the intergranular void growth that occurs during periods of forward creep dwell. The cycles show that a relaxation of the peak stress has not altered the deleterious effect of reversed plasticity upon the creep resistance of this Ni-base alloy.

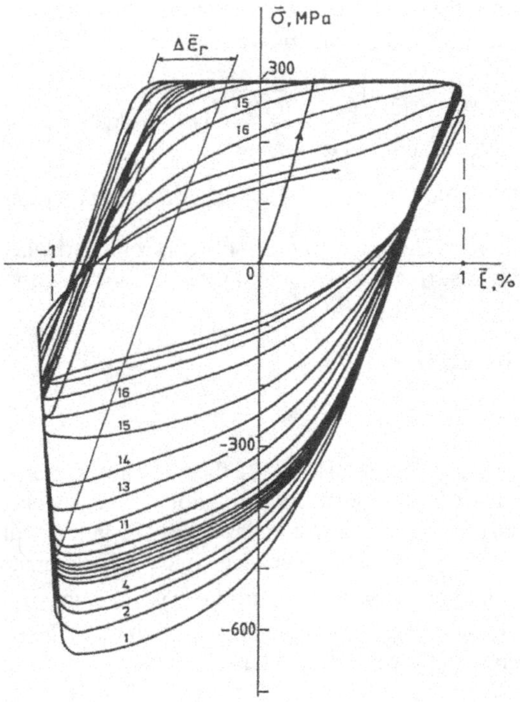

Figure 11.16 Peak stress softening with creep and relaxation dwell

(b) Multiaxial Deformation

Miller [33] assumed isotropic hardening in extending his theory to multi-axial deformation. This adopts time derivatives of the strain components given by an isotropic flow rule, thereby leading to the particular Levy-Mises form in § 10.3.1. Thus, the elastic strain rate was separated from the total strain rate according to eq(11.8). The Levy-Mises flow rule (10.14a) provides the viscoplastic strain rate tensor in a form similar to that used for creep:

$$d\varepsilon^P = d\lambda \sigma_{ij}' \quad \rightarrow \quad \dot{\varepsilon}_{ij} = \dot{\lambda} \sigma_{ij}' \tag{11.31a}$$

Evidently, the comparison must apply to secondary creep in the absence of a time function.

The scalar $\dot{\lambda}$ in eq(11.31a) is given as

$$\dot{\lambda} = \frac{3}{2\bar{\sigma}} f\left(\frac{\bar{\sigma} - \bar{R}}{\bar{D}}\right) \tag{11.31b}$$

where the bar refers to a von Mises equivalence between multi-axial and uniaxial stressing. If, instead of isotropic hardening, Hodge's combined isotropic-kinematic hardening rule [29] is employed, then the yield function contains both the rest and drag stresses as follows:

$$\frac{3}{2}(\sigma_{ij}' - R_{ij}')(\sigma_{ij}' - R_{ij}') = \bar{\sigma}^2 = (K\bar{D})^2 \tag{11.32a}$$

where K is the constant of proportionality, expressing the current radius of the yield surface as $\bar{\sigma} = K\bar{D}$. Applying the flow rule to eq(11.32a) shows how the components of the back stress tensor R_{ij}' re-appear within the constitutive relation

$$\dot{\varepsilon}_{ij} = \frac{3f\left[(\bar{\sigma} - \bar{R})/\bar{D}\right]}{2(\bar{\sigma} - \bar{R})}(\sigma_{ij}' - R_{ij}') \tag{11.32b}$$

where $\bar{R} = \sqrt{(\frac{3}{2} R_{ij}' R_{ij}')}$ is the centre of a yield surface that translates and expands simultaneously. The kinematic hardening contribution is removed when R_{ij}' is set to zero in eq(11.32a) so that the yield surface can expand only in an isotropic manner.

11.4.3 Overstress and Total Strain

(a) Uniaxial

In a theory, developed by Liu and Krempl [37–39], the rate effects upon hardening, creep and relaxation appear within a single constitutive relation. Their theory assumes a smooth transition from linear elasticity to non-linear inelasticity. The latter is taken from non-linear viscoelasticity theory in the form of a 'piecewise' linear approximation employing total strain. For a given dynamic stress σ, the strain rate $\dot{\varepsilon}$ is the sum of elastic and inelastic components. The latter depends upon an overstress σ_v, which is the stress in excess of an equilibrium value $g = g[\varepsilon]$, as shown in Fig. 11.17.

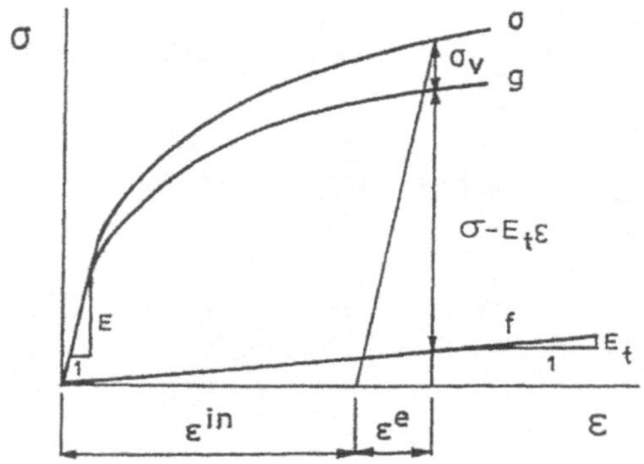

Figure 11.17 Equilibrium and dynamic flow curves showing overstress

The overstress follows as

$$\sigma_v = \sigma - g[\varepsilon] \tag{11.33a}$$

The corresponding total, engineering strain rate (omitting superscript *T*) is given from eq(11.7b) as

$$\dot{\varepsilon} = \dot{\varepsilon}^e + \dot{\varepsilon}^{in} = \frac{\dot{\sigma}}{E} + \frac{\sigma_v}{E\,k[\sigma_v]} \tag{11.33b}$$

from which

$$\dot{\varepsilon}\left[1 - \frac{1}{E}\left(\frac{\mathrm{d}\sigma}{\mathrm{d}\varepsilon}\right)\right] = \frac{\sigma_v}{E\,k[\sigma_v]} \tag{11.34a}$$

where $k[\sigma_v]$ is a viscosity function with a dimension of time. For metal plasticity, the right-hand side of eq(11.33b) may be defined from the parabolic hardening function given, say, by the Ramberg-Osgood eq(9.40). It can be seen that the left-hand side of eq(11.34a) will become constant when $\mathrm{d}\sigma/\mathrm{d}\varepsilon$ is identified with the asymptotic tangent modulus E_t under a constant strain rate. For the intermediate, inelastic strain range, the gradient $\mathrm{d}\sigma/\mathrm{d}\varepsilon$ lies in the range $E >> \mathrm{d}\sigma/\mathrm{d}\varepsilon > E_t$. Ignoring the elastic contribution in eq(11.33b), the inelastic strain rate may be assumed constant at

$$\dot{\varepsilon} \simeq \dot{\varepsilon}^{in} = \frac{\sigma_v}{E\,k[\sigma_v]} \tag{11.34b}$$

This implies that σ_v will become constant for a dynamic curve under a given total strain rate. It follows that separate dynamic curves, pertaining to different strain rates, will eventually become parallel, separated by a constant viscous stress, though they are neither linear nor spaced in proportion to $\dot{\varepsilon}$. This was demonstrated for a Ti-alloy [37] and 304 stainless steel [38] for strain rates $\dot{\varepsilon}$ lying in the range $10^{-6} - 10^3\ \mathrm{s}^{-1}$. The time-integration of eq(11.33b) identifies the elastic and inelastic components of strain shown in Fig. 11.17. Combining eqs(11.33a) and (11.34a):

$$E k[\sigma_v]\,\dot{\varepsilon} + g[\varepsilon] = \sigma + k[\sigma_v]\,\dot{\sigma} \tag{11.35a}$$

and putting $m[\sigma_v] = E\,k[\sigma_v]$, eq(11.35a) appears as

$$m[\sigma_v]\,\dot{\varepsilon} + g[\varepsilon] = \sigma + k[\sigma_v]\dot{\sigma} \tag{11.35b}$$

Equation (11.35b) gives the stress rate during relaxation, when $\dot{\varepsilon} = 0$ under a given total strain ε_o, as

$$\dot{\sigma} = \frac{g[\varepsilon_o] - \sigma}{k[\sigma_v]} \tag{11.36a}$$

Equation (11.36a) shows that $\dot{\sigma} = 0$ when $\sigma = g[\varepsilon_o]$. Assuming $g[\varepsilon]$ is known, the function $k[\sigma_v]$ may be determined from a relaxation curve. For example, the room-temperature relaxation behaviour of AISI type 304 stainless steel [38, 39] conforms to

$$\ln \sigma_v = \ln r_o + r_1\,\mathrm{e}^{-|\sigma_v/r_2|^{r_3}} \tag{11.36b}$$

where r_0, r_1, r_2 and r_3 are material constants. The rate dependent behaviour of this material may be predicted from the numerical solution to eq(11.36b). This reveals that unequally spaced stress–strain curves apply to the total strain rates 10^{-6}, 10^{-5} . . . 10^{-1} s^{-1}, which was confirmed by experiment.

Equation (11.35a) gives a strain rate for creep under a constant stress σ_o $(\dot{\sigma} = 0)$ as

$$Ek[\sigma_v]\,\dot{\varepsilon} + g[\varepsilon] = \sigma_o \tag{11.36c}$$

The requirement here is that both primary and secondary creep should appear in a numerical solution $\varepsilon = \varepsilon(t)$ to eq(11.36c) for an appropriate choice of $k[\sigma_v]$ and $g[\varepsilon]$.

(b) Multiaxial

Krempl et al [40–42] have developed further multiaxial, viscoplasticity theories, based upon overstress, for initially isotropic [40, 41] and orthotropic [42] materials. These theories decompose total engineering strain components ε_{ij} into the sum of their elastic and inelastic components. The inelastic strain rate does not distinguish further between its plastic, creep and relaxation components. For the isotropic theory, components of the total, deviatoric strain rate are used:

$$\dot{e}_{ij} = \dot{\varepsilon}_{ij} - \tfrac{1}{3}\,\delta_{ij}\,\dot{\varepsilon}^{e}_{kk}, \quad \text{where} \quad \dot{\varepsilon}^{e}_{kk} = \frac{(1-2\nu)\,\dot{\sigma}_{kk}}{E} \tag{11.37a,b}$$

along with the components of the deviatoric stress:

$$\sigma_{ij}' = \sigma_{ij} - \delta_{ij}\,\sigma_m \quad \text{where} \quad \sigma_m = \tfrac{1}{3}\sigma_{kk} = \tfrac{1}{3}(\sigma_{11} + \sigma_{22} + \sigma_{33}) \tag{11.38a,b}$$

These deviatoric components of stress and strain rate are related through a multiaxial counterpart of eq(11.33b):

$$\dot{e}_{ij} = \dot{e}^{e}_{ij} + \dot{e}^{in}_{ij} = \frac{(1+\nu)}{E}\,\dot{\sigma}_{ij}' + \frac{(1+\nu)}{E}\,\frac{\sigma^{v\,\prime}_{ij}}{k[\bar{\sigma}_v]} \tag{11.39a}$$

where Poisson's ratio ν is the constant ratio between the elastic strain rate(s) for the width(s) (directions 2 and 3) and the length (direction 1) in rectangular-section bar under uniaxial tension:

$$\frac{\dot{\varepsilon}_2}{\dot{\varepsilon}_1} = \frac{\dot{\varepsilon}_3}{\dot{\varepsilon}_1} = -\nu \tag{11.39b}$$

The axial, deviatoric strain rate for the length is, from eq(11.37a):

$$\dot{e}_1 = \dot{\varepsilon}_1 - \tfrac{1}{3}(\dot{\varepsilon}_1 + \dot{\varepsilon}_2 + \dot{\varepsilon}_3) \tag{11.40a}$$

Combining eqs(11.39b) and (11.40a) gives

$$\dot{e}_1 = \tfrac{2}{3}(1+\nu)\,\dot{\varepsilon}_1 \tag{11.40b}$$

Now from eqs(11.38a,b) the axial deviatoric stress is $\sigma_1' = \tfrac{2}{3}\sigma_1$. Taking this with the $(1 + \nu)$ factor in eq(11.40b), the correct reduction to eq(11.39a) is ensured for simple tension. The overstress components are given by

$$\sigma_{ij}^{v\,\prime} = \sigma_{ij}{}' - g_{ij}{}' \qquad (11.41a)$$

where $g_{ij}{}'$ are the components of an equilibrium deviatoric stress tensor

$$g_{ij}{}' = g_{ij} - \tfrac{1}{3}\delta_{ij}\, g_{kk} \qquad (11.41b)$$

The development of an expression for the rate of growth in the components of g_{ij} can be understood from a consideration of simple tensile behaviour, given in Fig. 11.17. This shows

$$\dot{g} = \psi[\sigma_v]\,\dot{\varepsilon} - \frac{(g - f)}{b[\sigma_v]}\,\dot{\varepsilon}^{in} \qquad (11.42a)$$

in which σ_v is the axial viscous stress, $\dot{\varepsilon}$ and $\dot{\varepsilon}^{in}$ are the total and inelastic axial strain rates. Figure 11.7 also shows that both $g - f$ and $\sigma - g$ will attain constant, asymptotic values. Therein, $b[\sigma_v]$ in eq(11.42a) is a scaling factor:

$$b[\sigma_v] = \frac{A}{\psi[\sigma_v] - E_t} \qquad (11.42b)$$

which applies to lesser ordinates than the asymptotic value $g - f = A$. The multiaxial counterpart of eq(11.42a) is written as

$$\dot{g}_{ij} = \psi\,[\bar{\sigma}_v]\, T_{ijkl}\,\dot{\varepsilon}_{kl} - \frac{(g_{ij} - f_{ij})}{b[\bar{\sigma}_v]}\,\dot{\bar{\varepsilon}}^{in} \qquad (11.42c)$$

where $f_{ij} = E_t\, T_{ijkl}\, \varepsilon_{kl}$ in which E_t is the tangent modulus and T_{ijkl} is a modified form of the isotropic tensor $I_{ijkl} = \tfrac{1}{2}(\delta_{ik}\delta_{jl} + \delta_{il}\delta_{jk}) - \tfrac{1}{3}\delta_{ij}\delta_{kl}$, admitting the correct uniaxial reduction:

$$T_{ijkl} = \frac{1}{(1+\nu)}\left[\frac{1}{2}(\delta_{ik}\delta_{jl} + \delta_{il}\delta_{jk}) + \frac{\nu}{(1-2\nu)}\delta_{ij}\delta_{kl}\right] \qquad (11.43a)$$

The equivalent viscous stress and strain rate appearing in eq(11.42c) are given by the scalars

$$\bar{\sigma}_v = \sqrt{\frac{3}{2}\,(\sigma_{ij}{}' - g_{ij}{}')(\sigma_{ij}{}' - g_{ij}{}')}\,, \qquad \dot{\bar{\varepsilon}}^{in} = \frac{\left(\dot{\varepsilon}_{ij}^{in}\,\dot{\varepsilon}_{ij}^{in}\right)^{1/2}}{\sqrt{(1 + 2\nu^2)}} \qquad (11.43b,c)$$

In the correspondence between eqs(11.42a) and (11.42c), eqs(11.43b,c) reduce to a uniaxial, 'viscoplastic' stress and strain rate:

$$\bar{\sigma}^{v} = \sigma - g_1 \quad \text{and} \quad \dot{\bar{\varepsilon}}^{in} = \dot{\varepsilon}^{in}$$

Further reductions to eq(11.42c) for plane stress states enable comparisons with corresponding experiments. For example, the room-temperature viscous behaviour of T6 aluminium alloy under radial and non-radial cyclic loadings [40, 41] was reproduced from assuming incompressible elasticity, i.e. from taking $v = ½$. The numerical integration of eqs(11.39a) and (11.42a) requires four material constants, v, E, E_t and A, and the two overstress functions, $k[\bar{\sigma}_v]$ and $b[\bar{\sigma}_v]$, to be established from monotonic tension [43]. Chaboche [44] has pointed out the similarity between eq(11.42b) and the non-linear translation rule in eqs(10.52a,b). Both contain driving and recall terms that are proportional respectively to the strain rate tensor $(\dot{\varepsilon}_{ij})$ and an equivalent strain rate

$$\dot{\bar{\varepsilon}} \text{ or } \dot{\bar{\varepsilon}}^{in} = \left(\frac{2}{3} \dot{\varepsilon}_{ij}^{in} \dot{\varepsilon}_{ij}^{in} \right)^{1/2}$$

However, while total strain rate appears in eq(11.42c), plastic strain is employed with eq(10.52a) for which $\nu = ½$. This is because the viscoplastic model requires a gradual transition into non-linearity within the function $\psi[\bar{\sigma}_v]$, while an elastic–viscoplastic model retains a quasi-static yield surface to distinguish between elastic and plastic behaviour.

11.4.4 Integral Constitutive Equations

Krempl [45] also developed non-linear integral equations for the history of plasticity and ageing found in metals. The stress and strain at time t under smooth loadings are adaptations of the superposition principle [46]:

$$\boldsymbol{\varepsilon}(t) = \int_0^t \mathbf{S}(\boldsymbol{\sigma}, t - \tau) \frac{\mathrm{d}\boldsymbol{\sigma}}{\mathrm{d}\tau} \mathrm{d}\tau \quad \text{or} \quad \varepsilon_{ij}(t) = \int_0^t S_{ijkl}(\sigma_{ij}, t - \tau) \frac{\mathrm{d}\sigma_{kl}}{\mathrm{d}\tau} \mathrm{d}\tau \qquad (11.44a)$$

$$\boldsymbol{\sigma}(t) = \int_0^t \mathbf{R}(\boldsymbol{\varepsilon}, t - \tau) \frac{\mathrm{d}\boldsymbol{\varepsilon}}{\mathrm{d}\tau} \mathrm{d}\tau \quad \text{or} \quad \sigma_{ij}(t) = \int_0^t R_{ijkl}(\varepsilon_{ij}, t - \tau) \frac{\mathrm{d}\varepsilon_{kl}}{\mathrm{d}\tau} \mathrm{d}\tau \qquad (11.44b)$$

where **S** and **R** are *response kernals*, t is the current time and τ represents past times within the interval $0 - t$. Uniaxial, creep, relaxation and recovery appear within eqs(11.44a,b) from applying the respective conditions:

(i) Creep $\sigma = \sigma_o$ is constant for $0 \leq \tau \leq t$ and

$$\frac{\mathrm{d}\varepsilon}{\mathrm{d}\tau} \geq 0, \quad \frac{\mathrm{d}^2\varepsilon}{\mathrm{d}\tau^2} \leq 0$$

(ii) Relaxation $\varepsilon = \varepsilon_o$ is constant for $0 \leq \tau \leq t$ and

$$\frac{\mathrm{d}\sigma}{\mathrm{d}\tau} \leq 0, \quad \frac{\mathrm{d}^2\sigma}{\mathrm{d}\tau^2} \geq 0$$

(iii) Recovery $\varepsilon > 0$ for $\sigma = 0$ at $t \geq 0$ and

$$\frac{d\varepsilon}{d\tau} \leq 0, \quad \frac{d^2\varepsilon}{d\tau^2} \geq 0$$

The limit of each differential in (i), (ii) and (iii) is zero for times $\tau = t \rightarrow 0$. Under monotonic loading the rates $d\sigma/d\tau$ and $d\varepsilon/d\tau$ are usually constant, obeying $d\sigma/d\tau > 0$ and $d\varepsilon/d\tau > 0$. The initial, elastic response is assumed to be independent of history and the elastic–plastic transition is taken to be smooth.

The fading memory of metals under certain conditions refers to their inability to 'remember' previous deformation paths in their entirety. For example, at higher temperatures, recovery processes will soften a previously hardened material and may restore its original structure completely. To account for this, structure tensors $\mathbf{M}^{\circ}$and $\mathbf{M}^{\alpha}$ were introduced within the kernals **S** and **R** in eqs(11.20a,b). $\mathbf{M}^{\circ}$ defines the initial structure, which, if isotropic, is given as $\mathbf{M}^{\circ} = p\,\mathbf{I}$, where p is a constant and **I** is the unit tensor. $\mathbf{M}^{\alpha}$ characterises the deformed structure where, depending upon the memory, none, some, or all of the past deformation may be 'remembered' within a plastic work function. Further refinements were necessary to account for the deformation-induced anisotropy. Here, the integral equation (11.44a) was shown to account for a Bauschinger effect, cross-hardening expansion and the effects of history under cyclic creep for a 1045 steel.

11.5 Quasi-Dynamic Viscoplasticity

The quasi-static plasticity theory may be extended to account for the small increases in flow stress due to moderate changes in the rate of straining. The time-independent hypotheses of strain hardening or work hardening, as in § 9.9.1 and § 9.9.2, are taken to apply at a given strain rate. Here, an incremental plasticity theory accounts for viscous flow with a choice to be made between isotropic and anisotropic hardening, depending upon the loading path.

11.5.1 Isotropic Hardening

(a) Small Strain Theory

The strain hardening hypothesis states that the current stress depends upon the plastic strain accumulated. From eq(9.48):

$$\bar{\sigma} = H\left(\int d\bar{\varepsilon}^{P}\right) \tag{11.45a}$$

where H is the hardening function. Combining eq(11.45a) with the rate-independent Levy-Mises flow rule, eqs(10.14) and (10.15) show

$$d\varepsilon_{ij}^{P} = \frac{3\,d\bar{\varepsilon}^{P}}{2\bar{\sigma}}\,\sigma_{ij}' = \frac{3\,d\bar{\sigma}}{2H'\bar{\sigma}}\,\sigma_{ij}' \tag{11.45b}$$

where $H' = d\bar{\sigma}/d\bar{\varepsilon}^{P}$ is the gradient to the plastic flow curve (the hardening modulus). Defining H' within eq(11.45b) from the equivalent form of Hollomon's eq(9.39a) gives the plastic strain increment

$$\bar{\sigma} = A(\dot{\bar{\varepsilon}})^{m}\,(\bar{\varepsilon}^{P})^{n}$$

$$\mathrm{d}\varepsilon_{ij}^{P} = \frac{3\bar{\sigma}^{\frac{1-2n}{n}}\,\mathrm{d}\bar{\sigma}}{2nA^{1/n}\,(\dot{\bar{\varepsilon}})^{m/n}}\,\sigma_{ij}' \tag{11.45c}$$

where the plastic rate equals the total rate approximately. Increasing the strain rate increases the flow stress and the interior range of elasticity within the inflation of a subsequent yield surface. The elastic component of the total strain is calculated from Hooke's law (see eq(10.40)) in which elastic constants E, G and v are not altered by the rate of straining.

Using the equivalent stress and strain, the work hypothesis is given by eqs(9.51a,b):

$$\bar{\sigma} = F\left(\int \mathrm{d}w^{P}\right) = F\left(\int \bar{\sigma}\,\mathrm{d}\bar{\varepsilon}^{P}\right) \tag{11.46a}$$

Here, the current equivalent stress depends upon prior plastic work. Combining eq(11.46a) with the Levy-Mises flow rule:

$$\mathrm{d}\varepsilon_{ij}^{P} = \frac{3\,\mathrm{d}\bar{\sigma}}{2F'\bar{\sigma}^{2}}\,\sigma_{ij}' \tag{11.46b}$$

where
$$F' = \frac{\mathrm{d}\bar{\sigma}}{\mathrm{d}w^{P}} = \frac{1}{\bar{\sigma}} \times \frac{\mathrm{d}\bar{\sigma}}{\mathrm{d}\bar{\varepsilon}^{P}} \tag{11.46c}$$

in which $\mathrm{d}\bar{\sigma}/\mathrm{d}\bar{\varepsilon}^{P} = F'\bar{\sigma}$ from eq(11.46a). Equation (11.46c) allows the modulus F' to be derived from a uniaxial curve at a given strain rate as described by eq(11.46a).

(b) Large Strain Theory

Bell [47] accounted for large strains accompanying moderately rapid rates of straining (up to $10^{2}\ \mathrm{s}^{-1}$) in ductile metals. An engineering, or nominal, effective stress and strain were re-defined as

$$T = \sqrt{(s_{ij}'\,s_{ij}')}, \qquad \mu = \sqrt{(e_{ij}^{P}\,e_{ij}^{P})} \tag{11.47a,b}$$

Simple, tensile or torsional loading connects T and μ in eqs(11.47a,b), appearing as a universal square law:

$$T = \beta\sqrt{(\mu - \mu_b)} \quad \Rightarrow \quad T^{2} = \beta^{2}\mu - \beta^{2}\mu_b \tag{11.48a,b}$$

It follows from eq(11.48b) that at a given absolute temperature (symbol is θ here), β^2 appears as the gradient and $-\beta^{2}\mu_b$ is the intercept value for the linear plot of T^2 versus μ. The following rate-independent expression for β was developed [48] from the physics of crystalline deformation:

$$\beta = \left(\frac{2}{3}\right)^{\frac{r}{2}} G(0)\,B_o\left(1 - \frac{\theta}{\theta_m}\right) \tag{11.48c}$$

where θ/θ_m is the homologous temperature and the $G(0)$ is the shear modulus at absolute zero temperature. In eq(11.48c), r is a deformation mode index and $B_o = 0.0207$, common to 50 polycrystalline solids, including non-metals. Annealed polycrystalline copper [49, 50] was shown to conform to eq(11.48c), for a single universal index $r = 4$, at a rate of $10^{-4}\ \mathrm{s}^{-1}$ for engineering strains to 15 %. The hardening of annealed aluminuium, on the other hand, required r to increase with strain, in the range $2 \le r \le 7$, for similar test conditions [51, 52]. Similarly, for 1020 steel, r varied within the range $2 \le r \le 9$ [53]. For quasi-dynamic straining, the index r was taken to be constant over a limited range of strain, so that a piecewise linear approximation to a curve of T^2 versus μ applied. The range of r decreases with increasing strain rate, to the extent that under impact loading of these materials, a single r–value applies. A relation, very similar to eq(11.48a), relates the critical resolved shear stress to strain for stage III deformation in a single crystal [48]. Thus, eqs(11.48a–c) serve to correlate the hardening behaviour of single crystals with polycrystals [26] in a given lattice structure (see Figs 8.8a,b).

The usual assumption of normality leads to an incremental constitutive relation for plasticity at large engineering (or nominal) strain [54]. Retaining Bell's notation, this gives

$$\mathrm{d}e_{ij}^{P} = \mathrm{d}\lambda\, s_{ij}' = \left(\frac{\mathrm{d}\mu}{T}\right) s_{ij}' \tag{11.49a}$$

where the nominal stress deviator $s_{ij}' = s_{ij} - ⅓\delta_{ij}s_{kk}$, applies to $T > T_Y$, in which T_Y is a limiting equivalent elastic stress (i.e. the yield stress). Now from eq(11.48b):

$$\frac{\mathrm{d}\mu}{T} = \left(\frac{2}{\beta^2}\right)\mathrm{d}T \tag{11.49b}$$

Substituting eq(11.49b) into eq(11.49a) gives the nominal plastic strain tensor:

$$\mathrm{d}e_{ij}^{P} = \left(\frac{2}{\beta^2}\right)\mathrm{d}T\, s_{ij}' \tag{11.49c}$$

In particular, for combined tension-torsion, eq(11.49c) supplies the respective axial and shear strain components

$$\mathrm{d}e_{11}^{P} = \left(\frac{4}{3\beta^2}\right) s_{11}\,\mathrm{d}T, \qquad 2\,\mathrm{d}e_{12}^{P} = \left(\frac{4}{\beta^2}\right) s_{12}\,\mathrm{d}T \tag{11.50a,b}$$

where

$$\mathrm{d}T = \left(\tfrac{2}{3}s_{11}\mathrm{d}s_{11} + 2s_{12}\mathrm{d}s_{12}\right)/T$$

Equations (11.50a,b) were shown to provide an account of large deformation behaviour in copper [49, 50, 54] and aluminium [51, 52] under radial loading (where s_{12}/s_{11} = constant) and under various non proportional loading paths comprising linear segments. The maximum strains attained for each material were, typically, $e_{11} \simeq 15$ % and $2e_{12} \simeq 30$ %. The essential differences between Bell's approach and the Levy-Mises incremental theory ($\mathrm{d}\varepsilon_{ij}^{P} = \mathrm{d}\lambda\sigma_{ij}'$) lies in Bell's use of engineering definitions for stress and strain, connected with a universal hardening law, with which to define the scalar $\mathrm{d}\lambda$. His piecewise representation of an engineering stress–strain curve provides a close approximation over a wider range of strain. The Levy-Mises theory adopts the continuous expressions of Hollomon, Ludwik or Swift (eqs(9.39a,b), (9.43a,b) and (9.44a,b) respectively) to define $\mathrm{d}\lambda$. These are able to relate derived measures: the true (Cauchy) stress to the true natural strain for a comparable range of strain. Therein, greater deviations arise between the two approaches as the corresponding true and engineering stress and strain measures diverge with increasing strain.

(c) Combined Work and Strain Hardening

Bodner and Pardom [55] employed a combined hardening function $\bar{\varepsilon}^P = \bar{\varepsilon}^P(\bar{\sigma}, w^P)$ for rate-dependent viscoplasticity of commercially pure titanium in the range $\dot{\varepsilon} \simeq 10^{-3}$–$10^{-5}\text{s}^{-1}$. They expressed $\bar{\varepsilon}^P$ as decreasing exponential functions of w^P and $\bar{\sigma}$ in separate forms:

$$\bar{\varepsilon}^P = \varepsilon_o(w^P)\,\mathrm{e}^{-k/\bar{\sigma}^n} \quad \text{and} \quad \bar{\varepsilon}^P = \varepsilon_o\,\mathrm{e}^{-k(w^P)/\bar{\sigma}} \tag{11.51a,b}$$

where n is a constant and ε_o and k may be chosen either as material constants or work expressions. Combining the work and strain hardening hypotheses (11.45a) and (11.46a) gives the equivalent stress in a more general form:

$$\bar{\sigma} = \bar{\sigma}\left(w^P, \bar{\varepsilon}^P\right) \tag{11.52a}$$

for which the equivalent stress differential is

$$\mathrm{d}\bar{\sigma} = \frac{\partial \bar{\sigma}}{\partial w^P}\,\mathrm{d}w^P + \frac{\partial \bar{\sigma}}{\partial \bar{\varepsilon}^P}\,\mathrm{d}\bar{\varepsilon}^P \tag{11.52b}$$

Substituting from eqs (11.45a) and (11.46a):

$$\mathrm{d}\bar{\sigma} = F'\bar{\sigma}\,\mathrm{d}\bar{\varepsilon}^P + H'\mathrm{d}\bar{\varepsilon}^P \tag{11.53a}$$

Correspondingly, the equivalent plastic strain increment follows as

$$\mathrm{d}\bar{\varepsilon}^P = \frac{\mathrm{d}\bar{\sigma}}{F'\bar{\sigma} + H'} \tag{11.53b}$$

Equations (11.51a,b) may be regarded as two integrated forms of eq(11.53b) that provide appropriate descriptions of the stress–strain curve for this material. Here, F' and H' are identified with each partial derivative in eq(11.52b). Combining eq(11.53b) with eq(10.15) leads to the scalar $\mathrm{d}\lambda$

$$\mathrm{d}\lambda = \frac{3\,\mathrm{d}\bar{\varepsilon}^P}{2\bar{\sigma}} = \frac{3\,\mathrm{d}\bar{\sigma}}{2\bar{\sigma}\,(F'\bar{\sigma} + H')} \tag{11.53c}$$

Equation (11.53c) may then be employed with the Levy-Mises flow rule (10.14) to provide the strain increment tensor or the strain rate tensor for viscoplasticity.

11.5.2 Bauschinger Effect

The Bauschinger effect is known to reduce the range of elasticity when the direction of deformation is reversed (see § 8.8). This conflicts with an isotropic hardening rule. To admit a Bauschinger effect, an anisotropic hardening model is required to describe a translated domain of elasticity (the subsequent yield surface), surrounded by an elastic–plastic exterior region. The subsequent yield surface *f* must account for the influence that an increasing strain rate has upon raising the flow stress. Here, eq(11.7b) is combined with a description of a translated yield surface, appropriate to non-linearly hardening material (see § 10.7.3).

One example is the Baltov-Sawczuk function *f* [56, 57]:

$$f = \frac{1}{2}(I_{ijkl} + a\alpha_{ij}'\alpha_{kl}')(\sigma_{ij}' - \alpha_{ij}')(\sigma_{kl}' - \alpha_{kl}') = k^2 = \frac{A^2}{3}\left[(\dot{\bar{\varepsilon}}^P)^m(\bar{\varepsilon}^P)^n\right]^2 \tag{11.54}$$

in which the coefficient a and the shear yield stress k ($k^2 = \bar{\sigma}^2/3$) admit the effects of strain history within quasi-dynamic rate sensitivity. Where a history of prior straining occurs, within a time interval $t_o \le t \le t_1$, the deformation is characterised within two controlling parameters appearing as arguments of α and k:

$$\gamma_h = \frac{1}{\sqrt{2}}\int_{t_o}^{t_1}\left(\dot{\varepsilon}_{ij}'\,\dot{\varepsilon}_{ij}'\right)^{\frac{1}{2}}\,\mathrm{d}t \quad \text{and} \quad \beta_h = \frac{\gamma h}{(t_1 - t_o)} \tag{11.55a,b}$$

Equation (11.55a) uses the total strain deviator $\varepsilon_{ij}' = \varepsilon_{ij} - ⅓\delta_{ij}\varepsilon_{kk}$ for defining the strain rate. Further parameters: the equivalent deviatoric plastic strain and the total deviatoric strain rate were defined respectively for the subsequent deformation:

$$\gamma = \frac{1}{\sqrt{2}}\int\left(\mathrm{d}\varepsilon_{ij}^{P\,\prime}\,\mathrm{d}\varepsilon_{ij}^{P\,\prime}\right)^{\frac{1}{2}} \quad \text{and} \quad \dot{\beta} = \frac{1}{\sqrt{2}}\left(\dot{\varepsilon}_{ij}'\,\dot{\varepsilon}_{ij}'\right)^{\frac{1}{2}} \tag{11.56a,b}$$

Since $\varepsilon_{ij}^{P\prime} = \varepsilon_{ij}^P - ⅓\delta_{ij}\varepsilon_{kk}^P$, eqs(11.56a,b) can provide an account of compressible deformation, when $\varepsilon_{kk}^P \neq 0$. This applies when plastic strain causes damage to reduce volume contiguity. Therein, γ is found by imposing $\varepsilon_{ij}^{P\,\prime} \neq \varepsilon_{ij}^P$ within the integration of eq(11.56a). Also, compressible plasticity requires that a and k in eq(11.54) be replaced with the functions $a = a(\gamma_h, \gamma, \beta_h, \dot{\beta})$ and $k^2 = k^2(\gamma_h, \gamma, \beta_h, \dot{\beta})$. Equation (11.54) can then describe a limit surface for the initiation and nucleation of microcracks under a critical internal micro-stress distribution α_{ij}'. In contrast, time-dependent viscoplasticity under constant volume is considered in § 11.6 and § 11.7.

11.6 Isotropic Viscoplasticity

The presence of creep and viscous strain prevents a clear distinction from being made between regions of pure elasticity and time-independent elastic–plasticity. Thus, the classical viscoplasticity theory connects the static yield surface to a dynamic loading surface. The latter becomes the potential function from which the viscoplastic strain rate is defined in a similar manner to creep. The normality rule is assumed to apply for higher strain rates in the range: $10^{-1} - 10^{-3}\,\mathrm{s}^{-1}$, where the deformation is both time- and path-dependent. The dynamic loading surface F contains the current stress point. The exterior normal to F at that point gives the direction of the viscoplastic strain rate vector with components $\dot{\varepsilon}_{ij}^P$ (Fig. 11.18a). The dynamic loading surface F encloses the subsequent quasi-static yield surface f, whose interior defines the region of linear elasticity observed in metals [58]. Originally, F was assumed to be an isotropic expansion of f. Under the rule of isotropic hardening, it follows that the two surfaces, F and f, remain concentric with the initial yield surface f_o, as shown in Fig. 11.18a. Figure 11.18b shows how the separation in f and F relates to equivalent static and dynamic flow curves. Surface F applies to any rapid strain rate while f refers to quasi-static loading at a slow rate. Previously it was shown that F and f can coincide only in the absence of time-dependent strain, i.e. when P′ coincides with Q in Fig. 11.8a.

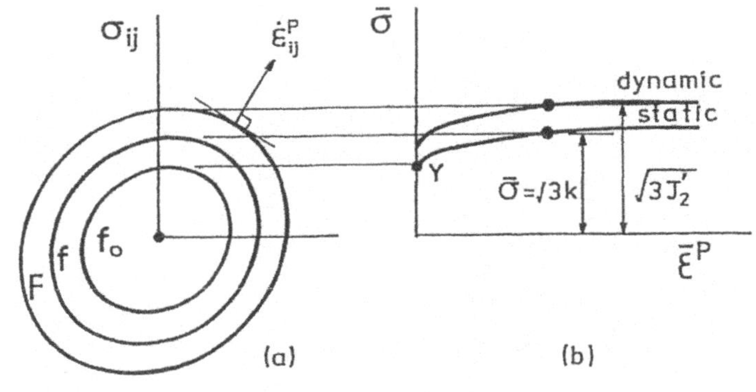

Figure 11.18 Concentric dynamic and quasi-static flow potentials F and f

With f bounding a region of subsequent elasticity, the net strain rate for isothermal, infinitesimal deformation becomes $\dot{\varepsilon}_{ij} = \dot{\varepsilon}_{ij}^{e} + \dot{\varepsilon}_{ij}^{P}$. The elastic strain rate component $\dot{\varepsilon}_{ij}^{e}$ is the time-derivative of the elastic strain increment term in eq(10.40) and is independent of the strain rate. The viscoplastic rates $\dot{\varepsilon}_{ij}^{P}$ are provided by the following form of the flow rule:

$$\dot{\varepsilon}_{ij}^{P} = K\langle \varphi(f) \rangle \frac{\partial F}{\partial \sigma_{ij}} \tag{11.57}$$

where K is a viscosity constant and F is a function of the viscous stress in excess of that defined from f, at the limit of elasticity. The latter identifies the quasi-static surface, which is expressed in the form $f = 0$. For stress states giving $f > 0$, the Macaulay brackets imply $\langle\varphi(f)\rangle = \varphi(f)$ and $\dot{\varepsilon}_{ij}^{P} > 0$. Otherwise, when the stress state lies at or below the limit of elasticity, then $f \le 0$, $\langle\varphi(f)\rangle = 0$ and $\dot{\varepsilon}_{ij}^{P} = 0$. *Note*: Here all viscoplastic strain is taken to be time-dependent, there being no separation into plastic and creep strain components.

11.6.1 Rate Sensitivity

The time-dependence of $\dot{\varepsilon}_{ij}^{P}$ does not appear explicitly in eq(11.57), but as a time derivative of the dynamic flow potential F. This implies that a linear, viscoplastic strain versus time relationship applies, approximating to the viscoplastic strain observed over a short time interval, say, under the very rapid rates of impact loading. Taken with incompressible viscoplasticity ($\dot{\varepsilon}_{kk}^{P} = 0$) and an initially isotropic condition, a general isotropic function $F(J_2', J_3')$, applies to eq(11.57). For example, taking F to be a von Mises function, the viscoplastic potential is defined as

$$F(\sigma_{ij}) = \sqrt{J_2'} = \varkappa(\psi) \tag{11.58a}$$

where $\varkappa(\psi)$ is an isotropic hardening function. With k as the current shear yield stress, the quasi-static yield function becomes

$$f(\sigma_{ij}) = \sqrt{J_2'} - k = 0 \tag{11.58b}$$

It follows from eqs(11.58a,b) that f and F are related:

$$f(\sigma_{ij}) = \frac{F(\sigma_{ij})}{\varkappa(\psi)} - 1 \tag{11.58c}$$

in which $k = \varkappa(\psi)$. It is instructive to derive the viscoplastic response to tension and torsion as a reference for the prediction of viscoplastic strain rates under more complex stress states. Setting $F = \sqrt{J_2'} = \sqrt{(\tfrac{1}{2}\,\sigma_{ij}'\sigma_{ij}')}$, eq(11.57) becomes

$$\dot{\varepsilon}_{ij}^{P} = \frac{K}{2}\langle\varphi(f)\rangle \frac{\sigma_{ij}'}{\sqrt{J_2'}} = \frac{K}{2}\left\langle\varphi\left|\frac{\sqrt{J_2'}}{\varkappa(\psi)} - 1\right|\right\rangle \frac{\sigma_{ij}'}{\sqrt{J_2'}} \tag{11.59a}$$

Assume that ψ is given by the strain hardening hypothesis in eq(11.45a). For tension, where $\psi = \varepsilon^P$, the axial deviatoric stress is $\sigma_1' = 2\sigma/3$ and the invariant is $\sqrt{J_2'} = \sigma/\sqrt{3}$. Then, from eq(11.59a), the axial, viscoplastic strain rate $\dot{\varepsilon}_1^P = \dot{\varepsilon}^P$ becomes

$$\dot{\varepsilon}^P = \frac{K}{\sqrt{3}}\left\langle\varphi\left|\frac{\sigma}{\sqrt{3}\,\varkappa(\varepsilon^P)} - 1\right|\right\rangle \tag{11.59b}$$

which contributes to the total strain rate as $\dot{\sigma}/E + \dot{\varepsilon}^P$. The inverse form of eq(11.59b) supplies the viscoplastic stress

$$\sigma = \sqrt{3}\varkappa(\varepsilon^P)\left[1 + \varphi^{-1}\left(\frac{\sqrt{3}\,\dot{\varepsilon}^P}{K}\right)\right] \tag{11.60a}$$

For torsional equivalence, σ and ε^P are replaced by $\bar{\sigma} = \sqrt{3}\tau$ and $\bar{\varepsilon}^P = \gamma^P/\sqrt{3}$ respectively, when eq(11.60a) becomes

$$\tau = \varkappa\left(\frac{\gamma^P}{\sqrt{3}}\right)\left[1 + \varphi^{-1}\left(\frac{\dot{\gamma}^P}{K}\right)\right] \tag{11.60b}$$

in which the inverse function φ^{-1} will depend upon the strain rate sensitivity. A geometrical interpretation of eqs(11.60a,b) is given in Fig. 11.19a. As the strain rates decrease and φ^{-1} approaches zero, the function $\varkappa(\psi)$ must express the corresponding quasi-static stress versus plastic strain curve. Equation (11.59b) will describe the strain rate sensitivity for a non-hardening material when the term $\sigma/\sqrt{3}\varkappa(\varepsilon^P)$ is replaced with the constant tensile flow stress. The function φ^{-1} in eqs(11.60a,b) is required to express the stress dependence upon visco-plastic strain rate for the dynamic curve. Here Fig. 11.5 hows how φ^{-1} is applied to constant rate, uni-axial flow curves (i.e. $\dot{\varepsilon}^P$= constant), with or without hardening. More generally, where the equivalent plastic strain rate $\dot{\bar{\varepsilon}}^P$ under a multiaxial stress state depends upon the equivalent plastic strain $\bar{\varepsilon}^P$, the flow curves appear in the manner of Figs 11.19a,b.

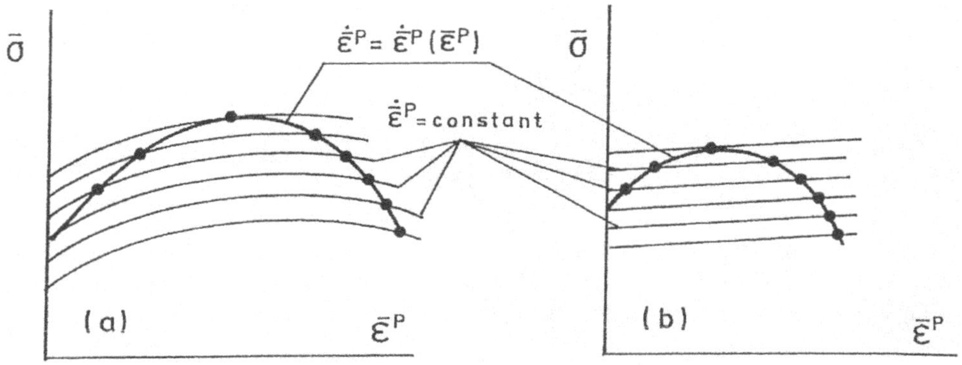

Figure 11.19 Dependence of flow stress upon variable strain rates

Equations (11.60a,b) provide the variations in flow stress with strain rate under uniaxial and torsional impact behaviour. For simplicity, a non-hardening behaviour is often assumed, where the tensile, compressive and shear strain rates are constant. Power and exponential functions are found to be in closest agreement with rate-sensitive behaviour [59] when eq(11.59b) appears in these respective forms:

$$\dot{\varepsilon}^P = K'\left(\frac{\sigma}{Y} - 1\right)^{\delta} \tag{11.61a}$$

$$\dot{\varepsilon}^P = K'\left[\mathrm{e}^{(\sigma/Y - 1)} - 1\right] \tag{11.61b}$$

where δ is a constant, Y is the quasi-static yield stress and $K' = K/\sqrt{3}$ is again assumed constant for the infinitesimal time scale implied.

Campbell et al. [60] also included hardening and rate sensitivity for copper under torsional impacts within the following form of eq(11.60b):

$$\tau = A\left(\gamma^P\right)^n\left[1 + m\ln\left(1 + \frac{\dot{\gamma}^P}{B}\right)\right] \tag{11.61c}$$

where A, n, m and B are experimental constants.

Malvern [61] expressed the total strain rate within a uniaxial wave propagation from the addition of an elastic rate of strain to the plastic strain rate. This gives

$$\dot{\varepsilon} = \frac{\dot{\sigma}}{E} + g(\sigma, \varepsilon^P) \tag{11.62a}$$

where $g(\sigma, \varepsilon^P)$ is one of two rate-dependent hardening functions consistent with eq(11.62a):

$$g(\sigma, \varepsilon^P) = \frac{K\sigma_o}{E}\left(\frac{\sigma}{\sigma_o} - 1\right), \quad g(\sigma, \varepsilon^P) = \frac{1}{Q}\left(\frac{\sigma}{\sigma_o} - 1\right)^{\frac{1}{M}} \tag{11.62b,c}$$

in which E, K, M and Q are material constants and $\sigma_o = \sigma_o(\varepsilon^P)$ is the static flow stress (slow rate) for a given ε^P expressed, typically, by Hollomon's or Ludwik's eqs(9.39a,b) and (9.43a,b), respectively. Setting $\dot{\varepsilon}^P = g(\sigma, \varepsilon^P)$ and re-arranging eqs(11.62b,c) describes, in either form, the dynamic, tensile stress–strain curve for a given plastic strain rate [62]:

$$\sigma = \sigma_o(\varepsilon^P) + E\dot{\varepsilon}^P/K, \quad \sigma = \sigma_o(\varepsilon^P)\left[1 + (Q\dot{\varepsilon}^P)^M\right] \qquad (11.63a,b)$$

Equation (11.59a) has allowed further solutions to the propagation of elastic–plastic stress waves under very high strain rates arising from the impact loading of tubes under combined tension-torsion [63, 64]. The solutions appear as stress wave trajectories that emanate radially from the boundary of a yield surface f_o, enclosing the stress-space (σ, τ), to the point of failure. It was found that trajectories, under simultaneous compressive and torsional impacts of initially unstressed tubes, were more linear than those for statically pre-stressed tubes. Only the radial trajectories were consistent with an assumption of isotropic hardening. Non-radial trajectories required an anisotropic potential, which is described later.

The classical theory thus assumes that the material is linearly elastic within the interior of f_o. While the total strain rate may be decomposed into elastic and inelastic rates, no distinction is made between plasticity and creep. Naghdi and Murch [65], Cernocky [66, 67] and Haupt [68] have dealt with another class of rate-sensitive behaviour, namely: that for which an initial viscoelastic material becomes plastic. Because viscous behaviour occurs in both regions, the yield and dynamic loading surfaces are 'non-inviscid', i.e. they are both time- and stress-dependent. Consequently, this material behaviour is harder to model than a simplified elastic–viscoplastic solid, due to the difficulty of establishing a function for a time-dependent yield condition. Krempl's approach of combining elasticity with piecewise, non-linear viscoelasticity (see § 11.4.3) overcomes this problem. This model enables the yield surface to be discarded for a material model that responds to stress with a smooth transition from linear elasticity/non-linear viscoelasticity to non-linear inelasticity.

11.6.2 Creep

A non-linear strain versus time relationship is required where a strain rate sensitivity appears over much longer durations, typical of a creep test. Here, in a generalisation of eq(11.57), the constant K is replaced with a time function $K(t)$ so that one interpretation in Fig. 11.20a lies in the construction of isochronous stress–strain curves for creep [69]. These are derived from a family of primary creep curves at given times t_1, t_2, t_3 etc., in the manner of Fig. 11.20b.

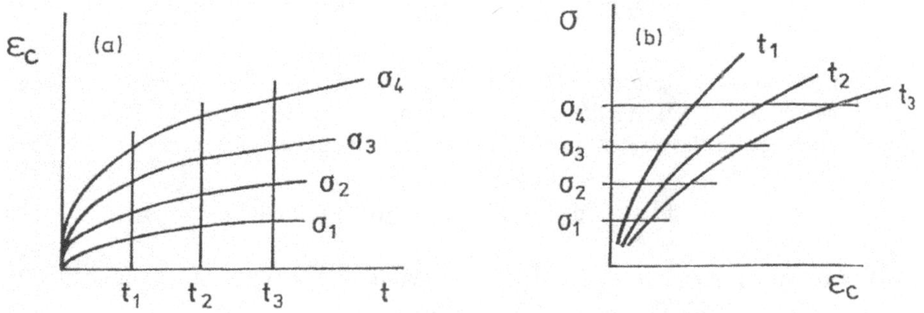

Figure 11.20 Construction of isochronous stress–strain curves

If the creep curves in Fig. 11.20a are given by the law $\varepsilon_c = a\sigma^n t^m$, this can be applied at each time in the isochronous plots of Fig. 11.20b. Moreover, with the creep rates expressed as a time-hardening derivative $\dot{\varepsilon}_c = ma\sigma^n t^{m-1}$, the two functions φ and $\varkappa$ appearing in eq(11.59b) may be established. *Note*: For a material with both temperature and strain rate sensitivity, the quantities K, $\varkappa$ and φ in eq(11.59a) become temperature-dependent.

It is instructive to adapt isotropic viscoplastic theory to high-temperature, long-time creep by recovering the classical multiaxial creep theory. K is replaced by the constant $3A/2$ for secondary creep deformation. Primary creep rates will appear when $K(t) \rightarrow Qt^{m-1}$, in which $Q = 3mA/2$. Here, F in eq(11.57) is identified with a von Mises dissipation potential J_2'. Furthermore, $\varphi(f)$ is identified with the equivalent stress function $\bar{\sigma}^{n-1}$ from Norton's law. The Macaulay brackets imply that an effective stress for creep $\bar{\sigma} - \bar{\sigma}_i$ must surpass an internal back stress $\bar{\sigma}_i$ [70]. That is: $\langle \varphi(f) \rangle \rightarrow (\bar{\sigma} - \bar{\sigma}_i)^{n-1}$, meaning that the creep rates $\dot{\varepsilon}_{ij}^{c} = 0$ for $\bar{\sigma} \leq \bar{\sigma}_i$ and $\dot{\varepsilon}_{ij}^{c} > 0$ for $\bar{\sigma} > \bar{\sigma}_i$. Thus, at a given temperature, the association between eq(11.32) and primary creep rates is made in reversed order as follows:

$$\dot{\varepsilon}_{ij}^{c} = \left[\frac{3}{2}(mA)\,t^{m-1}\right]\bar{\sigma}^{n-1}\sigma_{ij}' \Rightarrow \left(Q\,t^{m-1}\right)\langle(\bar{\sigma} - \bar{\sigma}_i)^{n-1}\rangle\left(\frac{\partial J_2'}{\partial \sigma_{ij}}\right) \Rightarrow K(t)\,\langle\varphi(f)\rangle\left(\frac{\partial F}{\partial \sigma_{ij}}\right)$$

11.7 Anisotropic Viscoplasticity

Earlier discussions of time-independent plasticity showed that the rule of isotropic hardening did not represent the elastic domain of pre-strained material realistically. It follows that a similar conclusion will apply where the dynamic surface F and the quasi-static surface f each appear as isotropic expansions of the initial yield surface f_o (see Fig. 11.18a). Despite this, a concentric arrangement between f and F remains reliable for outward, radial loadings. Later versions of viscoplasticity theory allow anisotropic hardening to appear within the quasi-static yield surface f. This modification is particularly appropriate to non-radial loading and load reversal. Goel and Malvern [64] examined impact loadings under these conditions. They adopted Hodge's combined hardening model for viscoplasticity in which f involved both isotropic and kinematic hardening motion (see § 8.9.3). That is, f translates in the direction of the plastic strain vector (Prager's rule) as it expands with the accumulation of plastic strain. The description of this quasi-static surface provides the required flow potential f:

$$f = \frac{3}{2}(\sigma_{ij}' - c\,\varepsilon_{ij}^{P})(\sigma_{ij}' - c\,\varepsilon_{ij}^{P}) - \left[H\left(\int \mathrm{d}\bar{\varepsilon}^{P}\right)\right]^2 = 0 \tag{11.64a}$$

Viscoplastic strain was assumed to lie in the direction of the outward normal to this surface under all stress states. Within the strain hardening function H in eq(11.64a), the authors attributed fractions of the tangent modulus to translation and expansion.

$$\mathrm{d}\bar{\sigma} = H'\mathrm{d}\bar{\varepsilon}^{P} = H'[(1 - m)\mathrm{d}\bar{\varepsilon}^{P} + m\mathrm{d}\bar{\varepsilon}^{P}] \tag{11.64b}$$

Later, the surface expansion, in the absence of rotation, is associated with a change to the radius of the translated hypersphere that f represents.

Comparing eq(11.64a) with eq(11.54) shows that the latter also allows a contraction and a rotation to accompany a translation in surface f. Taking particular values of β or γ in the arguments of a and k^2 in eqs(11.56a,b) results in a family of nested surfaces, lying between f and F. In the absence of pre-strain, say under creep conditions, a single valued parameter, i.e. the equivalent creep strain rate or the rate of the specific work dissipation, is sufficient to define each of these surfaces as follows [71, 72]

$$\dot{\bar{\varepsilon}}^{in} = \left[\frac{2}{3}\,\dot{\varepsilon}_{ij}^{in}\,\dot{\varepsilon}_{ij}^{in}\right]^{\frac{1}{2}} \quad \text{and} \quad \dot{w}^{in} = \sigma_{ij}\,\dot{\varepsilon}_{ij}^{in} \qquad (11.65a,b)$$

11.7.1 Experimental Observations

Drucker [71] commented upon the difficulties that arise when determining creep potential surfaces experimentally. These surfaces may, however, be constructed from physical principles. Brown [72] employed the normality rule with Rice's continuum slip theory [73] to construct 2D loci for a constant work dissipation rate $\dot{w}^P$. They were produced for combined tension-torsion to account for the short-time creep of aluminium at 250°C. Figures 11.21a–c shows these loci for $\dot{w}^{in} = 3 \times 10^4$ lbfin/in^{3}s (206.9 MJ/m^{3}s) corresponding to initial radial stress paths under: (a) tension, (b) torsion and (c) tension-torsion, at respective creep times of 280, 250 and 75 s. The translations in *f* are seen to be directed along each radial stress vector and are accompanied by expansion, contraction and rotation during creep. Here the amount of distortion appears small enough to be ignored. The comparison was made with isotropic expansions, associated with a 'zero current state', in the absence of lattice slip. With increasing creep time under constant stress, further $\dot{w}^{in}$ loci of type (a), (b) and (c) were found to be concentric, 'centred' at each position *A* shown. The co-ordinates $A(\alpha_{ij}) = (\sigma_A, \tau_A)$ show that generally the back stress α_{ij} for creep depends upon the load path. These observations suggest that the flow potential *f* will appear from eqs(11.54) and (11.64a) when an isotropic hardening function *F* for creep is based upon the dissipation work rate in eq(11.65b).

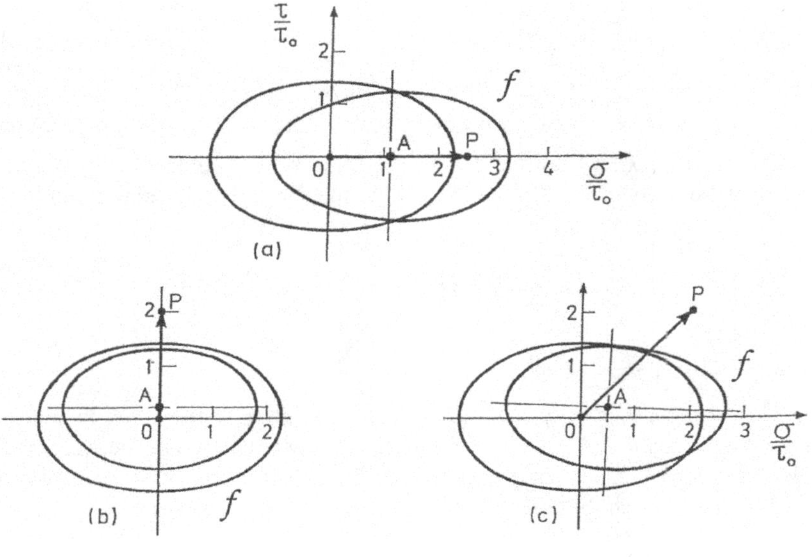

Figure 11.21 Creep potentials under radial loading

This association between incompressible creep and viscoplasticity provides a singe flow potential *f* coincident with the loading surface *F*:

$$f = \frac{3}{2}(\sigma_{ij}' - \alpha_{ij}')(\sigma_{ij}' - \alpha_{ij}') - F\left(\int_0^t \bar{\sigma}\,\dot{\bar{\varepsilon}}^{in}\,\mathrm{d}t\right) = 0 \tag{11.66}$$

Equation (11.66) shows no overstress with $\bar{\sigma}$ a constant in creep, where rotations and unsymmetrical distortion do not appear in f. However, isotropic hardening applies to the changing work rate with increasing creep time. *Note*: With its use of the stress deviators σ_{ij}' and α_{ij}', eq(11.66) assumes incompressible deformation. The integration is simplified, when short periods of transient 'cold creep' strain obeying Andrade's t^m law ($m \approx 1/3$) appear during a slow loading. With extended time after loading, the strain rate provided by that law would have made its transition to a constant secondary creep rate under the constant load.

11.7.2 Concentric Potentials f and F

(a) Isotropic Hardening

Assume that, for rapid loading, separate flow potentials f and F remain centred at the origin, as shown in Fig. 11.18a. The overstress $\bar{\sigma}_v$ is defined from eqs(11.58a,b) in one of the following three forms:

$$\bar{\sigma}_v = \sqrt{3}\,[(\varkappa(\psi) - k] \tag{11.67a}$$

$$\bar{\sigma}_v = \sqrt{\tfrac{3}{2}\sigma_{ij}'\sigma_{ij}'} - \bar{\sigma} = \sqrt{3J_2'} - \sqrt{3}k \tag{11.67b,c}$$

In eq(11.67c), the equivalent, quasi-static stress $\bar{\sigma} = \sqrt{3}k$ and the dynamic stress $\sqrt{(3J_2')}$ are measured at the same $\psi (= \bar{\varepsilon}^P)$, as shown in Fig. 11.18b. Equations (11.67b,c) are the multiaxial extensions to Perzyna's isotropic hardening theory [33], where $\dot{\varepsilon}_{ij}^P$ are found from the normal to surface F at the dynamic stress point σ_{ij}. Equation (11.67a) provides for a uniaxial overstress when $\varkappa(\psi)$ is equated to either the strain or work hardening function:

$$\varkappa(\psi) = H\left(\int \dot{\bar{\varepsilon}}^P \mathrm{d}t\right) \quad \text{or} \quad \varkappa(\psi) = F\left(\int \dot{w}^P \mathrm{d}t\right)$$

(b) Anisotropic Hardening

Brown's [72] observations on the motion of the dissipation potential surface during creep suggest another version of anisotropic viscoplasticity theory. The concentric bounding potentials, f and F in Fig. 11.22a, are each displaced to O′ from the stress origin by α_{ij} and remain connected through the following modification to eq(11.58c):

$$f(\sigma_{ij} - \alpha_{ij}) = \frac{F(\sigma_{ij} - \alpha_{ij})}{\varkappa(\psi)} - 1 \tag{11.68a}$$

The equivalent overstress $\bar{\sigma}_v$ becomes a scalar stress measure of the distance PQ between the current quasi-static yield surface f and the stress point σ_{ij} on F, as shown in Fig. 11.22a.

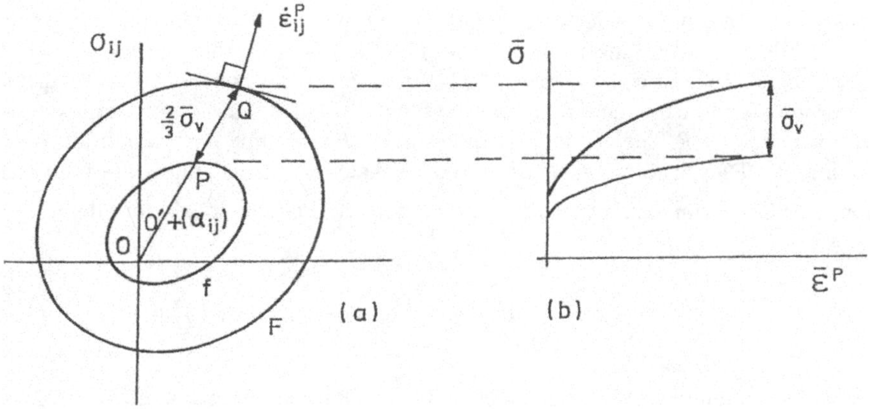

Figure 11.22 Concentric, translated viscoplastic potentials

Under the radial loading path OPQ shown, let the centre α_{ij} of f lie at O′, along OPQ, to ensure that $\bar{\sigma}_v$ is a scalar measure of the radial distance PQ, between f and F. In fact, the magnitude of the overstress $\bar{\sigma}_v$ is proportional to PQ. It will be seen that true equivalent stress magnitudes can only appear on the deviatoric plane. For uniaxial tension, the overstress is identified directly with the difference between the ordinates for the static and dynamic flow curves, as shown in Fig. 11.22b. The function $F(\bar{\sigma}_v)$ allows viscoplastic strain rates $\dot{\varepsilon}_{ij}^P$ to obey the normality condition

$$\dot{\varepsilon}_{ij}^P = \dot{\lambda}\ \frac{\partial F(\bar{\sigma}_v)}{\partial \sigma_{ij}} \tag{11.68b}$$

where, from eq(11.57), $\dot{\lambda} = K\langle \varphi(f)\rangle$ expresses a known, or an assumed, dependence of $\dot{\bar{\varepsilon}}^P$ upon $\bar{\sigma}_v$ for when $f > 0$ under uniaxial stressing. This simplified interpretation is examined further on the deviatoric plane in Fig. 11.23a.

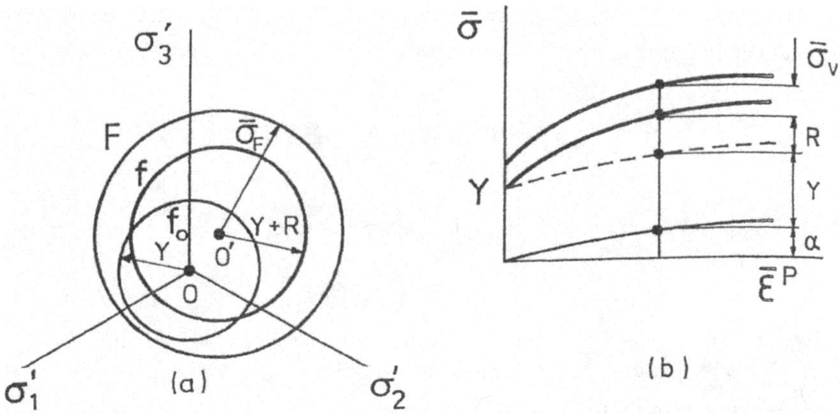

Figure 11.23 Concentric translations in f and F on the deviatoric plane

The translation in an initial, von Mises yield circle f_o, of radius Y, combines isotropic and kinematic hardening, according to eqs(11.64a) and (11.66). In the absence of distortion, within a principal deviatoric stress space, both f and F appear as concentric circles displaced from the origin. Let R be the increase in radius of the subsequent quasi-static surface f, so that its radius becomes $Y + R$. The dynamic loading surface F is assumed to be an inflation of f, to a new radius $\bar{\sigma}_F$. Thus, a common centre O′ (co-ordinates α_{ij}') is maintained for both f and F. Equation (11.68a) permits an overstress $\bar{\sigma}_v$ to be defined unambiguously for all stress states lying on F as

$$\bar{\sigma}_v = \sqrt{\left[\frac{3}{2}(\sigma_{ij}' - \alpha_{ij}')(\sigma_{ij}' - \alpha_{ij}')\right]} - [Y + R(\psi)] \qquad (11.69a)$$

The first and second terms on the right-hand side of eq(11.46a) are, respectively, radii of the dynamic and quasi-static surfaces as shown. When α_{ij}' is zero, eq(11.69a) reduces to the isotropic overstress in eq(11.67b,c) with $\bar{\sigma} = Y + R(\psi)$. When R is zero, the translation in f is purely kinematic. The radii of f and F may be identified with the equivalent stress in tension, for equal quasi-static and viscoplastic equivalent strains ($\psi = \bar{\varepsilon}^P$), as shown in Fig. 11.23b. Equation (11.69a) becomes

$$\bar{\sigma}_v = \bar{\sigma}_F - [Y + R(\psi)] \qquad (11.69b)$$

where the equivalent, tensile quasi-static flow stress appears as

$$\bar{\sigma} = Y + R(\psi) + \alpha(\psi) \qquad (11.69c)$$

This theory requires, therefore, that the translation function $\alpha(\psi)$ and the expansion or contraction function $R(\psi)$ be specified for the quasi-static surface. For example, with $\psi = \bar{\varepsilon}^P$ in eq(11.69c), the translation rate given by eq(10.52a):

$$\dot{\alpha}(\bar{\varepsilon}^P) = c_1 \dot{\bar{\varepsilon}}^P - c_2 \alpha |\dot{\bar{\varepsilon}}^P| = (c_1 - c_2 \alpha) \frac{\mathrm{d}\bar{\varepsilon}^P}{\mathrm{d}t} \qquad (11.70a)$$

Equation (11.70a) refers to a monotonic loading under uniaxial tension. Integration provides the time-independent translation in closed-form for an initial condition $\bar{\varepsilon}^P = \alpha = 0$:

$$\alpha(\bar{\varepsilon}^P) = \frac{c_1}{c_2}\left(1 - \mathrm{e}^{-c_2 \bar{\varepsilon}^P}\right) \qquad (11.70b)$$

Chaboche et al. [74–76] employed the radial expansion rate:

$$\dot{R}(\bar{\varepsilon}^P) = b(Q - R)\,\dot{\bar{\varepsilon}}^P \qquad (11.71a)$$

where b is a coefficient and Q is an asymptotic value of R, giving a radius $Y + Q$ for the limiting surface f. Equation (11.71a) integrates to a time-independent, closed form, when $R = 0$ for $\bar{\varepsilon}^P = 0$, under proportional loading:

$$R(\bar{\varepsilon}^P) = Q\left(1 - \mathrm{e}^{-b\bar{\varepsilon}^P}\right) \qquad (11.71b)$$

Finally, the function $F(\bar{\sigma}_v)$ in eq(11.68) is specified with the condition $\bar{\sigma}_v > 0$ for viscous flow. The viscoplastic potential for creep $F(\bar{\sigma}_v)$ assumes a power function in the equivalent, viscous overstress [77, 78]. This enables eq(11.68b) to be written in the following normalised form for secondary creep rates:

$$\frac{\dot{\varepsilon}_{ij}^{P}}{\dot{\bar{\varepsilon}}_o} = \frac{1}{(n+1)\,\bar{\sigma}_o^{\,n}} \times \frac{\partial(\bar{\sigma}_v^{\,n+1})}{\partial\sigma_{ij}} = \left(\frac{\bar{\sigma}_v}{\bar{\sigma}_o}\right)^n \frac{\partial\bar{\sigma}_v}{\partial\sigma_{ij}} \tag{11.72a}$$

where the reference state obeys a power function $\dot{\bar{\varepsilon}}_o = A\bar{\sigma}_o^{\,n}$ and the partial derivative follows from eq(11.69a). Equation (11.72a) then provides the normalised rates

$$\frac{\dot{\varepsilon}_{ij}^{P}}{\dot{\bar{\varepsilon}}_o} = \left(\frac{\bar{\sigma}_v}{\bar{\sigma}_o}\right)^n \frac{3}{2\bar{\sigma}_F}\,(\sigma_{ij}' - \alpha_{ij}') \tag{11.72b}$$

$$\dot{\varepsilon}_{ij}^{P} = \frac{3}{2}A\bar{\sigma}_v^{\,n-1}\,(\sigma_{ij}' - \alpha_{ij}') \tag{11.72c}$$

For creep under constant viscous stress, eq(11.72c) assumes $\bar{\sigma}_F = \bar{\sigma}_v$, i.e. $Y + R(\psi) = 0$. This gives an effective stress for creep that is the applied stress less the back stress: $\bar{\sigma}_v = \bar{\sigma}_F - \bar{\alpha}$. Equation (11.72c) gives the three principal creep strain rates:

$$\dot{\varepsilon}_1^{P} = A\bar{\sigma}_v^{\,n-1}\left[S_1 - \tfrac{1}{2}(S_2 + S_3)\right] \tag{11.73a}$$

$$\dot{\varepsilon}_2^{P} = A\bar{\sigma}_v^{\,n-1}\left[S_2 - \tfrac{1}{2}(S_3 + S_1)\right] \tag{11.73b}$$

$$\dot{\varepsilon}_3^{P} = A\bar{\sigma}_v^{\,n-1}\left[S_3 - \tfrac{1}{2}(S_1 + S_2)\right] \tag{11.73c}$$

where $S_1 = \sigma_1 - \alpha_1$, $S_2 = \sigma_2 - \alpha_2$ and $S_3 = \sigma_3 - \alpha_3$. Equations (11.73a,b,c) were shown to describe the inelastic strain response to monotonic and cyclic proportional loadings for 316 stainless steel at room-temperature. Translations in *f*, given by the non-linear, kinematic hardening rule in eq(11.70b), matched the cyclically stable hysteresis loops under tension-compression and forward-reversed torsion.

A constitutive relation, similar to eq(11.72c), was employed by Mroz et al. [79] for the prediction of cyclic creep in pure copper at 300°C and in 304 stainless steel at 600°C (see §11.8.1). Within this prediction, a single term for the current, quasi-static yield stress replaced the sum $Y + R(\psi)$ in eq(11.69b), though this stress was found to be negligible for high temperatures. It is instructive to compare principal rates in eqs(11.73a–c) to those from the original Marin-Soderberg isotropic creep theory. Therein, S_1, S_2 and S_3 are absolute principal stresses when creep is taken to occur, with an expansion but not a translation in *f*. Also, the reference state becomes Norton's secondary creep law. In eqs(11.73a–c), that law is modified with an internal stress $\bar{\sigma}_i$ (= back stress $\bar{\alpha}$) as follows:

$$\dot{\bar{\varepsilon}}_o = A\bar{\sigma}_o^{\,n} \quad \Rightarrow \quad \dot{\bar{\varepsilon}}_o = A(\bar{\sigma}_o - \bar{\sigma}_i)^n$$

which shows that creep strain is restricted to stress magnitudes greater than $\bar{\sigma}_i$, as in § 11.4.2. Figure 11.23b shows that a uniaxial flow stress may be partitioned as $\sigma = \alpha + Y + R + \sigma_v$. This gives a viscous stress contribution $\sigma_v = \sigma - (Y + R + \alpha)$. Clearly different interpretations are placed upon the stress producing time dependent flow when the rate is slow (creep) and when it is rapid (viscous). Here, a principal rate correspondence $\sigma_i = \alpha$ is made in the absence of an elastic region $(Y + R)$ for creep at higher temperatures.

Exponential functions and power series for $F(\bar{\sigma}_v)$ have been chosen to cover a wider range of strain rates [74, 75, 80]. These two forms have been applied to uniaxial and multiaxial cyclic loadings including cyclic and non proportional loading of 316 stainless steel at ambient and elevated temperatures. In each application, the influence of creep enhanced the rate of deformation to lie within the viscous regime. Further refinements to eqs(11.70a,b) and (11.71a,b), i.e. the contributions from kinematic and isotropic hardening, were required to account for influences of strain history under non-radial loadings.

(c) Initial Anisotropy

An account of an initial structural anisotropy appears with the introduction of fourth order tensor components C_{ijkl} into eq(11.69a). This gives the equivalent viscous overstress as

$$\bar{\sigma}_v = \sqrt{\tfrac{3}{2} C_{ijkl} (\sigma_{ij}' - \alpha_{ij}')(\sigma_{kl}' - \alpha_{kl}')]} - [(Y + R(\psi)] \qquad (11.74)$$

The simplest form of eq(11.74) identifies C_{ijkl} with those 6 constants, F, G, H etc., that Hill originally used to characterise an orthotropic material (see § 12.2.1). Lee and Zaverl [80] assumed anisotropic viscous flow for all stress levels lying in a medium temperature range. Their effective stress was defined from the first term in eq(11.74) as

$$\bar{\sigma} = \sqrt{\tfrac{3}{2} C_{ijkl} (\sigma_{ij}' - \alpha_{ij}')(\sigma_{kl}' - \alpha_{kl}')]} \qquad (11.75a)$$

Constant strain rates were found from eq(11.72a) as

$$\dot{\varepsilon}_{ij} = \dot{\bar{\varepsilon}} \left(\frac{\partial \bar{\sigma}}{\partial \sigma_{ij}} \right) = \frac{\dot{\bar{\varepsilon}}}{2\bar{\sigma}} \times \frac{3}{2} C_{ijkl} (\sigma_{kl}' - \alpha_{kl}')$$

$$\dot{\varepsilon}_{ij} = \frac{\dot{\bar{\varepsilon}}}{2\bar{\sigma}} C_{ijkl} \, (\sigma_{kl} - \alpha_{kl}) \qquad (11.75b)$$

The equivalent strain rate was taken to obey the function $\dot{\bar{\varepsilon}} = \dot{\bar{\varepsilon}}\,(\sigma, \tau, T)$, where T is absolute temperature, σ is the climb stress and τ is the shear (glide) stress resistance. Explicit forms of $\dot{\bar{\varepsilon}}$ apply to: (i) viscous dislocation glide and (ii) diffusion controlled climb and viscous flow (subscripts g and d), respectively as

$$\dot{\bar{\varepsilon}}_g = \dot{\bar{\varepsilon}}_{og} \left(\frac{\sigma}{\tau} \right)^{m_g} e^{-\frac{\Delta H}{RT}} \qquad (11.76a)$$

$$\bar{\dot{\varepsilon}}_d = \bar{\dot{\varepsilon}}_{od} \left(\frac{\sigma}{\tau} \right)^{m_d} e^{-\frac{D}{kT}} \tag{11.76b}$$

The thermal activation energy ΔH, from short-range obstacles, was assumed to depend upon the climb/glide stress ratio σ/τ in the form $\Delta H = B[1 - (\sigma/\tau)^p]$, where $p \simeq ½$ and B is the component of thermal energy that dislocations require to overcome obstacles when unassisted by stress. Further constants D, $\bar{\dot{\varepsilon}}_o$ and m appearing in eqs(11.76a,b) are appropriate to each mode of deformation, i.e. $1 \le m_d \le 3$ for viscous flow and $m_d = 3$ for climb. For $T \le 0.4T_m$, the contribution from grain boundary sliding is ignored and the net strain rate becomes the sum of eqs(11.76a) and (11.76b).

$$\bar{\dot{\varepsilon}}_{net} = \bar{\dot{\varepsilon}}_g + \bar{\dot{\varepsilon}}_d \tag{11.76c}$$

An equivalent back stress $\bar{\alpha}$ provides the resistance to the net equivalent strain rate: This condition enables uniaxial flow as a model with the three-element dashpot [79], shown in Fig. 11.24a.

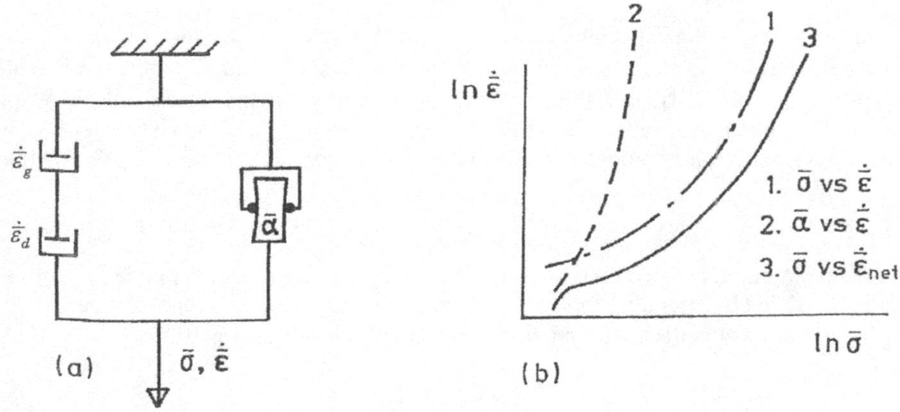

Figure 11.24 Three-element dashpot model

The sum in eq(11.76c) is simplified upon setting $\bar{\dot{\varepsilon}}_o = \bar{\dot{\varepsilon}}_{og} = \bar{\dot{\varepsilon}}_{od}$ to give

$$\bar{\dot{\varepsilon}}_{net} = \bar{\dot{\varepsilon}}_o \left(\frac{\sigma}{\tau} \right)^m \left\{ e^{-\frac{B}{RT}\left[1 - \left(\frac{\sigma}{\tau}\right)^p\right]} + e^{-\frac{D}{kT}} \right\} \tag{11.76c}$$

in which it was assumed that $m = m_g = m_d$ for low values of σ. Equation (11.76c) is represented graphically in the applied stress versus net strain rate curve 1 in Fig. 11.24b. The gradient alters according to the more dominant contributor: low rates are diffusion controlled, high rates are glide controlled. Curve 3 is the 'sum' of curve 1 and the back stress curve 2. The latter curve 2 may be determined from the stress-dip test, referred to in § 11.3.4. When the ratio σ/τ is negligibly small, under constant-temperature creep conditions, the exponential terms { } in eq(11.76c) become constant. Upon setting $S = \sigma - \alpha$ for isotropic material, eqs(11.73a–c) are restored and a constant, universal exponent $n = m = 4$ in eq(11.7b) applies. The theory was applied to a wider range of test conditions in which the plastic strain rate varied continuously under load relaxation (not to be confused with stress relaxation in which the total strain remains constant). In this test, a tensile load was increased before the cross-head of a screw-driven machine was brought to rest. Thereafter, the fall in load and displacement were recorded continuously. Equation (11.76c) was shown to fit the dependence of $\dot{\bar{\varepsilon}}$ upon σ, observed in two materials: (i) anisotropic, Zircaloy-2 at 450°C and (ii) isotropic, 304 stainless steel at 25°C. The corresponding material parameters, C_{ijkl} and α_{ij} in eqs(11.75a,b), were established from monotonic loading and reversed loading tests performed in tension and compression at comparable rates.

Note: The reduction to a multiaxial theory to explain the results found from uniaxial experiments, may not provide, fully, a conclusive test of that theory. This is a consequence of there being neither a requirement to define the equivalent strain rate nor to specify a hypothesis of hardening for the uniaxial mode. Each requirement arises when correlating the strain rates arising under a multiaxial mode. Indeed, one of the standard procedures is to show that flow data from multiaxial and uniaxial tests will lie on a single curve, with $\bar{\sigma}$ and $\dot{\bar{\varepsilon}}$ defined by work or strain hardening. Chapter 12 shows that, with $\bar{\sigma}$ and $\mathrm{d}\bar{\varepsilon}^P$ defined by eqs(12.4b) and (12.14) respectively, an equivalence in plastic flow applies to an initially anisotropic material that hardens 'isotropically' (i.e. the initial anisotropy present is preserved). Here, however, little attention has been paid to defining an equivalent strain rate for viscoplasticity when hardening is anisotropic. It appears that the viscoplastic theories described here retain the von Mises isotropic definition of $\dot{\bar{\varepsilon}}^P$, despite employing an overstress expression that is kinematic in character. While it is convenient to adopt this definition for uniaxial applications, it should be recognised that in extending this approach to multiaxial conditions, the flow rule becomes one of a mixed combination within a non associated flow rule (see § 10.5):

$$\dot{\varepsilon}_{ij} = KM^n \frac{\partial D}{\partial \sigma_{ij}} \tag{11.77a}$$

where D and M are different equivalent stresses applied to multiaxial flow and its correlation. For example, with M by Tresca and D by von Mises for an initially isotropic condition, the principal stress forms are ($\sigma_1 > \sigma_2 > \sigma_3$):

$$M = \sigma_T = \sigma_1 - \sigma_3 \tag{11.77b}$$

$$D = \bar{\sigma}_V = \frac{1}{2}[(\sigma_1 - \sigma_2)^2 + (\sigma_2 - \sigma_3)^2 + (\sigma_3 - \sigma_1)^2]^{\frac{1}{2}} \tag{11.77c}$$

in which the principal strain rates follow from eq(11.77a) (see exercise 11.9)

11.7.3 Rheological Models

The use of rheological models for a description of the inelastic strain rates found in viscoelastic, polymeric materials is well established [46]. For example, the simple Maxwell model, with a spring and dashpot connected in series, admits creep and relaxation but not recovery. On the other hand, the Kelvin model, with a spring and a dashpot in parallel, gives creep and recovery but not relaxation. Creep, relaxation and recovery phenomena all appear within the type-1 standard linear solid, i.e. with a further a spring placed in series with the Kelvin element [29]. In § 11.7.2 it was demonstrated how an alternative model accounted for the dependence of diffusion and dislocation controlled strain rates upon stress. Hart [81] proposed another rheological model to describe high-temperature, rate-dependence in metals. This consists of a series connection between the two models shown in Figs 11.25a,b for the matrix and the grain respectively.

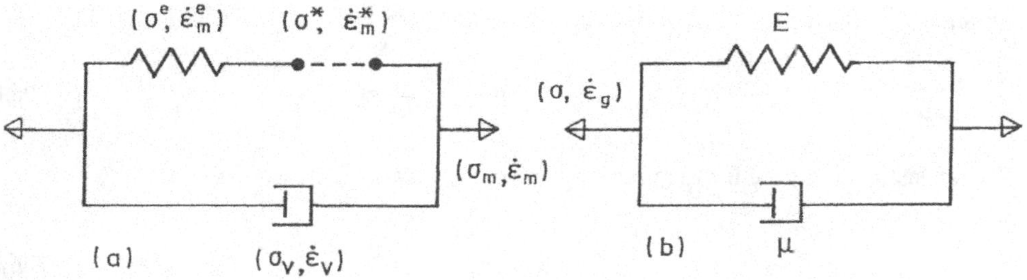

Figure 11.25 Models for inelasticity and anelasticity

In Fig. 11.25a, the viscoelastic flow within the grain matrix is provided by connecting a viscous dashpot in parallel with two, series-connected, elastic springs. The springs provide for a strain summation from elastic lattice distortion (superscript *e*) and anelastic creep (superscript *). A discussion of anelastic and recoverable creep strain is given in [29]. The net, uniaxial stress and strain rate for the matrix (subscript *m*) are correspondingly

$$\sigma_m = \sigma^* + \sigma_v = \sigma^e + \sigma_v \tag{11.78a}$$

$$\dot{\varepsilon}_m = \dot{\varepsilon}^e_m + \dot{\varepsilon}^*_m = \dot{\varepsilon}_v \tag{11.78b}$$

where σ_v is the stress component producing non-linear viscous flow. The strain rate sensitivity of the viscous stress σ_v is assumed to be less temperature-dependent than that of σ^e (= σ^*). In Fig. 11.25b, a Kelvin model represents the amount of grain boundary shear sliding (subscript *g*) that is required to accommodate matrix flow. Thus, sliding and anelasticity between grains is identified with the respective creep and recovery responses from the Kelvin element. This shows [29] forward grain boundary sliding creep strain [80] to be exponential in time:

$$\varepsilon_g = \frac{\sigma}{E}\left(1 - e^{-\frac{Et}{\mu}}\right) \tag{11.79a}$$

where E is the elasticity modulus for the Hookean spring and μ is the coefficient of viscosity for the Newtonian dashpot in Fig. 11.25b. The rate of forward grain boundary sliding is the derivative of eq(11.79a)

$$\dot{\varepsilon}_g = \frac{\sigma}{\mu} e^{-\frac{Et}{\mu}} \tag{11.79b}$$

Following a stress removal at time $t = t_1$, the recovery curve is given as

$$\varepsilon_r = \frac{\sigma}{E} e^{-\frac{Et}{\mu}} \left(e^{-\frac{Et_1}{\mu}} - 1 \right) \tag{11.79c}$$

One difficulty lies in identifying anelastic strain for $t = \infty$ when, from eq(11.79c), all strain will recover at a rate $\dot{\varepsilon}_r = - E\varepsilon_r/\mu$ when $t_1 < t < \infty$. The net creep strain rate $\dot{\varepsilon}$ is composed of matrix flow and grain boundary sliding. Each component is partly recoverable, leading to the net anelastic strain rate $\dot{\varepsilon}^*$. At finite time these rates are

$$\dot{\varepsilon} = \dot{\varepsilon}_m + \dot{\varepsilon}_g \quad \text{and} \quad \dot{\varepsilon}^* = \dot{\varepsilon}_m^* + \dot{\varepsilon}_g^* \tag{11.80a,b}$$

The anelastic strain rate for the grain, $\dot{\varepsilon}_g^*$ in eq(11.80b), is expressed as

$$\dot{\varepsilon}_g^* = \dot{\varepsilon}_o \left(\frac{\sigma}{M_g} - \varepsilon_g^* \right) \tag{11.80c}$$

where $\dot{\varepsilon}_o$ is given by an Arrhenius temperature function and M_g is an anelastic modulus. The relative contribution from sliding within eqs(11.80a,b,c) depends upon the test conditions. Matrix deformation results from high strain rates under high stress and temperature. Sliding is associated with brittle intergranular shear cracking under lower stresses and longer durations.

Hart's model was formulated with stress and strain rate tensors though its application was restricted to uniaxial behaviour. The simplification $\dot{\varepsilon}^* \simeq 0$ was made for a steady-state condition. This was confirmed from room-temperature, load-relaxation tests in high purity aluminium. Pure viscoplasticity within a grain requires the removal of the Hookean spring in Fig. 11.25a, as shown in Fig. 11.26a. A coupling between Fig. 11.26a and the spring-dashpot, parallel combination in Fig. 11.25b maintains an isotropic shear sliding behaviour between grains when pure viscoplasticity occurs within each grain.

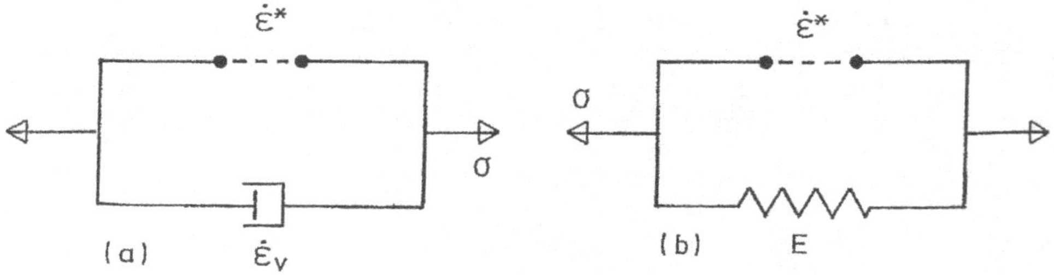

Figure 11.26 Models for isotropic and kinematic hardening

Replacing the dashpot in Fig. 11.25b with a linear spring, as in Fig. 11.26b, enables kinematic hardening to be reproduced. That is, the model then becomes a parallel combination of elastic and anelastic spring components in Fig. 11.26b connected in series to Fig. 11.26a. This model is intended to represent a surface containing an interior elastic domain, translated from the stress free origin, surrounded by a viscoplastic region. It is left to exercises 11.10 and 11.11 for the reader to examine the modified sliding and recovery responses from Figs 11.26a,b.

11.7.4 Dissimilar Potentials f and F

Complex features of anisotropic hardening, including distortion and rotation, appear in the quasi-static surface from loading under multi-axial stress states. The formulation of appropriate potential functions f and F to match these features for viscoplastic deformation are more capable than attempting the same within a rheological model. The potential method adopts an alternative overstress definition for the exterior region of viscoplasticity. Up to now, the potential theory has assumed that F is an inflation of f, with their respective surfaces having the same centre. Now, however, the overstress is also required to admit these complex features of anisotropic hardening within the region of viscoplasticity between f and F. It is known from experiment how anisotropic hardening manifests itself within the quasi-static yield surface f. In particular, it sharpens the curvature of the yield surface at the load point which influences the prediction of plastic strain behaviour under a non-radial load path. Thus, in a more general account of a viscous anisotropic hardening, f and F no longer conform to the relation given in eq(11.58c). Consequently, a new definition for $\bar{\sigma}_v$ is required. Here, Phillips and Wu [82] and Eisenberg and Yen [83] allowed for relative translations between f and F, as shown in the respective Figs 11.27a,b.

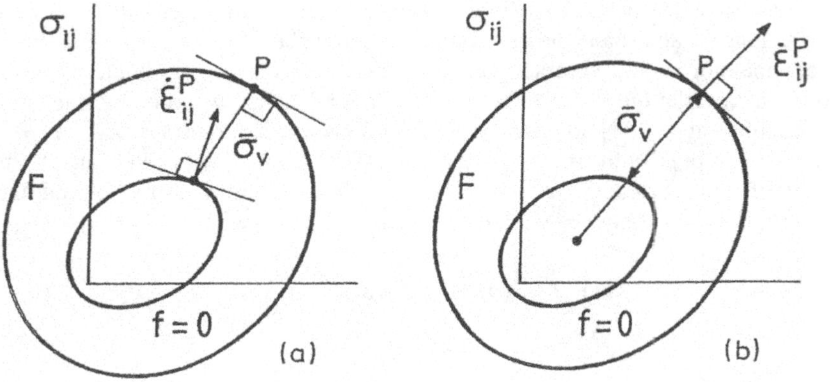

Figure 11.27 Definitions of overstress within non-concentric potentials

The motion of the dynamic surface F is taken to be similar to that of the quasi-static yield surface f in that rotation, expansion or contraction and cross hardening may all accompany their translations, relative to the initial yield surface f_o (not shown). The essential difference between the two theories lies in the definition given to the overstress (also called the excess or viscous stress) and to the direction assumed for $\dot{\varepsilon}_{ij}^P$. One measure of $\bar{\sigma}_v$ and the location of vector $\dot{\varepsilon}_{ij}^P$ are given in Fig. 11.27a. At load point P, $\bar{\sigma}_v$ is the minimum perpendicular distance from F to f and $\dot{\varepsilon}_{ij}^P$ lies normal to the quasi-static yield surface f. Alternatively, in Fig. 11.27b, $\dot{\varepsilon}_{ij}^P$ lies normal to F and $\bar{\sigma}_v$ lies between f and F on a line joining the dynamic stress point P to the centre of f. When the viscoplastic strain rate $\dot{\varepsilon}_{ij}^{P}$ is to be found from eq(11.68b), the latter measure of $\bar{\sigma}_v$ appears to be more convenient geometrically for defining an anisotropic, viscoplastic potential $F(\bar{\sigma}_v)$.

A relative rotation between the dynamic and quasi-static surfaces is achieved with F as an anisotropic hardening function which describes the subsequent dynamic surface. For example, modifying the general, quadratic, stress deviator form for $\bar{\sigma}_v$ in eq(11.74):

$$\bar{\sigma}_v = \sqrt{\tfrac{3}{2}\, H_{ijkl}\,(\sigma_{ij}' - \alpha_{ij}')(\sigma_{kl}' - \alpha_{kl}')]} - [(Y + R(\psi))] \qquad (11.81a)$$

where now H_{ijkl} is taken to be composed of isotropic I_{ijkl} and anisotropic A_{ijkl} tensors, as in eq(11.54). The particular form for A_{ijkl} is chosen to match the loading path. The geometrical interpretation of eq(11.81a) is similar to Fig. 11.27a, when the rotation between f and F is referred to a common centre. For simplicity, it is assumed that there is no relative rotation between f and the initial yield surface f_o. Hence, f is derived from f_o by translation and expansion. This implies that the directions of $d\varepsilon_{ij}^{P}$ and $\dot{\varepsilon}_{ij}^{P}$, lying normal to f_o and f respectively, will differ for the alterative definition of $\bar{\sigma}_v$, given in Fig. 11.27b. However, this configuration is consistent with the preservation of constant normal directions to f_o and f at the intermediate load points, i.e. where linear inelastic strain paths are observed from a radial loading. Substituting eq(11.81a) into eq(11.68b) leads to similar viscoplastic strain rate behaviour, where $\bar{\sigma}_v$ is modified by the term A_{ijkl}. In writing eq(11.81a) as $\bar{\sigma}_v = F - f$ then, for each geometrical interpretation in Figs 11.27a,b, A_{ijkl} accounts for a rotation in surface F while $R(\psi)$ accounts for an expansion, or a contraction, in surface f. Where $R(\psi) = 0$, the quasi-static surface f does not expand, contract or cross-harden (i.e. widen perpendicular to the stress path). Therein, the translation in f_o to f obeys a non-linear, kinematic hardening rule.

Nouailhas et al. [84] adopted Baltov and Sawczuk's definition: $H_{ijkl} = I_{ijkl} + A\varepsilon_{ij}^{P}\varepsilon_{kl}^{P}$ in eq(11.81a) to represent the pronounced cross-effect observed with a 90° out-of-phase, cyclic straining of stainless steel. However, to account for the deformation observed under cyclic, non-proportional loadings, $\bar{\sigma}_v$ was later 'strain-modified' (see § 11.8.2):

$$\bar{\sigma}_v = \sqrt{\left[\frac{(3/2)\, H_{ijkl}\,(\sigma_{ij}' - \alpha_{ij}')(\sigma_{kl}' - \alpha_{kl}')]}{1 + (3/2)\,B(\bar{\varepsilon}^{P})^2}\right]} - [(Y + R(\psi))] \qquad (11.81b)$$

in which $B \neq 0$ provided for a rotation in F. Also, $R(\psi) \neq 0$ in permitting an expansion or a contraction in f. It is seen that eqs(11.81a,b) are equivalent under uniaxial loading and under multiaxial loadings when $B = 0$.

Other two-surface models of plasticity considered earlier in § 8.9 could also be adapted to provide for a suitable positioning between viscoplastic potentials f and F. Variations between them will arise from the overstress definition and choice of translation rule. Distortion of viscoplastic potential surfaces is likely to appear under rapid rates of straining, but overstress expressions have not included distortion specifically. The relative rotations allowed between f and F, in Figs 11.27a,b may offset the need for this. *Note*: An equi-strain potential, admitting distortion under plane stress (see § 11.8) has accounted for interactions between plasticity and creep in stainless steel [85, 86].

Walker [87] based a viscoplastic potential on effective stress, not overstress, for when anisotropic deformation occurs within the high-temperature creep regime (see § 11.6.2).

Consequently, with no region of elasticity, the quasi-static surface disappears, along with the second, flow stress term $Y + R(\psi)$ in eq(11.81a). Then, creep strain rates from eq(11.68b) become the derivatives to a Norton power function $F(\bar{\sigma}_v) = P\bar{\sigma}_v^{\,q}$, in which the viscous creep stress becomes the effective stress for creep, i.e. the difference between the absolute stress and the back stress (see exercise 11.4).

11.8 Strain-Space Theory

It is also possible to formulate a flow potential for viscoplasticity in strain space. This model may be again be interpreted geometrically within the division of viscous strain into its permanent and recoverable components, as with creep strain [29]. The creep potential g defines a range of recovered (anelastic) strain at time t. Figure 11.28 shows that this range is time-dependent, being bounded by the curves of creep strain ε_c and permanent strain ε_p.

Note: Here ε_p is the time-dependent, permanent component of the forward creep strain, not to be confused with the plastic strain ε^P in time-independent plasticity.

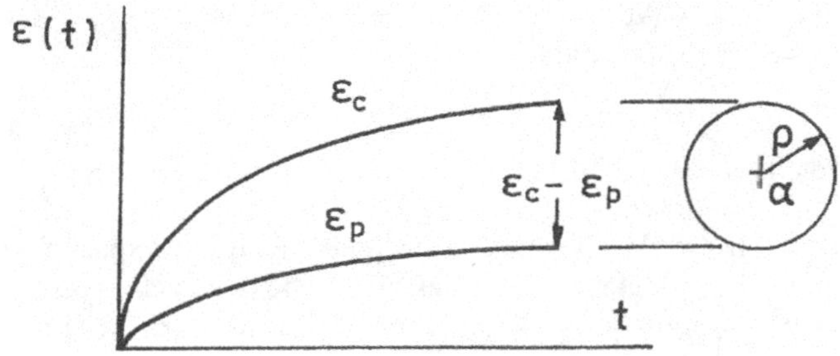

Figure 11.28 Creep strain potential

The stress-dip test involves negative decrements in contrast to § 11.3.4 for positive increments. The dip test reveals that for negative creep rates (i.e. recovery) to be possible, the stress decrement $\Delta\sigma$ should exceed the internal stress σ_i (see Fig. 11.29). Where the net stress $\sigma - \Delta\sigma$ is positive and $\Delta\sigma > \sigma_i$, Fig. 11.29 shows that the amount of recovered strain remains within the anelastic range ($\varepsilon_a^{\,*} = \varepsilon_c - \varepsilon_p$), and ε_p remains unaffected. A greater amount of recovered strain occurs when the stress is fully removed ($\Delta\sigma = \sigma$) at time t_1. Recall that $\varepsilon_a^{\,*}$ becomes the strain recovered following an infinite time $t - t_1$ allowed for recovery. With a stress reversal, the net stress becomes negative ($\Delta\sigma > \sigma$). Here, Fig. 11.29 shows that ε_p will only remain unchanged in the time interval $t_2 - t_1$, i.e. where the lowest reversed creep curve passes through the anelastic range.

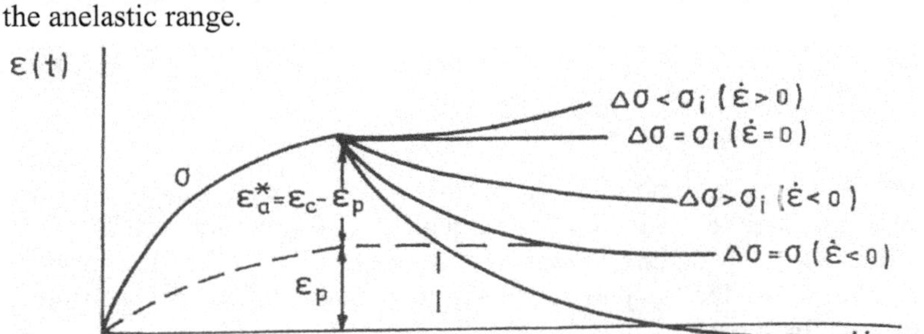

Figure 11.29 Stress reductions within the anelastic range

For $t > t_2$, the plastic strain is altered by some of the previously pinned dislocations regaining their mobility under the negative stress. Consequently, the range of anelastic strain will increase with time. This is because at time $t_3 > t_2$ in Fig. 11.30, a further component of anelastic strain $\Delta\varepsilon_a^*$ derives from the recovery curve BC when the applied stress is reversed. Thus, at time t_3 in Fig. 11.30, the ordinate CF defines the new anelastic range.

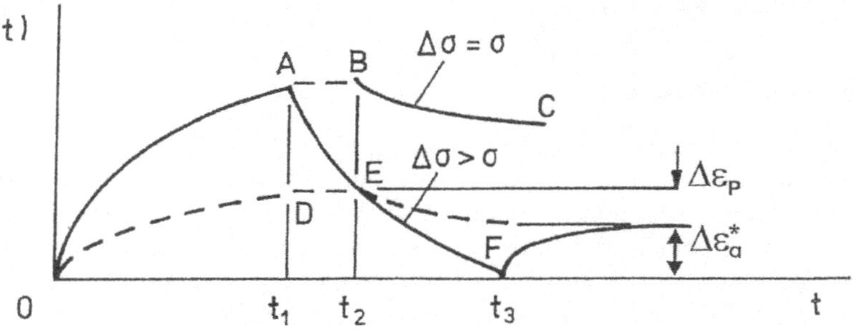

Figure 11.30 Bounds to the range of anelastic strain.

It follows that the curves OABC and ODEF define the respective upper and lower bounds to the uniaxial range of anelastic strain.

11.8.1 Creep Potential

Murakami and Ohno [88, 89] formulated a flow potential in strain space for multi-axial creep. Their extension to the uniaxial scheme described above is achieved by representing the anelastic range with a sphere of radius ρ and centre co-ordinates α_{ij} (see Fig. 11.28). This leads to a viscous potential of the form:

$$g = \frac{2}{3}(\varepsilon_{ij} - \alpha_{ij})(\varepsilon_{ij} - \alpha_{ij}) - \rho^2 \leq 0 \tag{11.82}$$

where ε_{ij} are the components of the anelastic (recoverable) strain range. The factor ⅔ ensures the correct reduction to uniaxial tension, where $\varepsilon_c = \rho + \alpha$ in Fig. 11.28. For recovered strain lying within the bounds of eq(11.82), we have $g < 0$ and $d\rho = d\alpha_{ij} = 0$, indicating that the surface remains stationary. As the forward creep strain increases the spherical surface ($g = 0$) expands and translates between the bounding curves shown, according to the current values of ρ and α_{ij}. Their incremental changes (or rates with respect to time) are given as

$$d\rho = \frac{2}{3}\lambda\, d\varepsilon_{ij}\, n_{ij} \tag{11.83a}$$

$$d\alpha_{ij} = (1 - \lambda)\, d\varepsilon_{kl}\, n_{kl}\, n_{ij} \tag{11.83b}$$

where n_{ij} is the outward unit normal to $g = 0$ and λ is a material constant that separates the isotropic and kinematic contributions to hardening. The translation is directed along the outward normal and the bounding segments, OA and EF in Fig. 11.30, are controlled by strain hardening. This allows the Levy-Mises flow rule to be combined with the strain hardening equation of state for creep as follows:

$$\dot{\varepsilon}_c = \dot{\varepsilon}_c(\sigma, T, \varepsilon_c)$$

$$\frac{\dot{\varepsilon}_{ijc}}{\dot{\bar{\varepsilon}}} = \frac{3\sigma_{ij}'}{2\bar{\sigma}} \tag{11.84a}$$

Eliminating time t between the uniaxial primary creep strain equation and its rate equation

$$\varepsilon_c = A\sigma^n t^m \;\Rightarrow\; \dot{\varepsilon}_c = mA\sigma^n t^{m-1} = mA^{1/m}\sigma^{n/m}\varepsilon_c^{(m-1)/m}$$

Equivalently, (see § 11.6.2) with rearrangement of eq(11.84a) the multiaxial creep rates become

$$\dot{\varepsilon}_{ijc} = \frac{3}{2}\, m A^{\frac{1}{m}}\, \dot{\bar{\varepsilon}}^{\frac{m-1}{m}}\, \bar{\sigma}^{\frac{n-m}{m}}\, \sigma_{ij}' \tag{11.84b}$$

where

$$\dot{\bar{\varepsilon}} = \int \sqrt{\tfrac{2}{3}\,\dot{\varepsilon}_{ij}\,\dot{\varepsilon}_{ij}}\;\mathrm{d}t \quad \text{and} \quad \bar{\sigma} = \sqrt{\tfrac{3}{2}\,\sigma_{ij}'\,\sigma_{ij}'} \tag{11.84c,d}$$

Creep strain rates, found from eq(11.84a–d), would account for the plastic contribution where there is stress reversal. The separation of creep strain into its plastic and anelastic strain components is not essential when using the function $g = 0$ as the creep potential to predict multiaxial creep rates. Particular applications made in [88, 89] include the predictions to the creep strain arising in 304 stainless steel at 650°C under repeated and reversed stress applications in tension, torsion and combined tension-torsion. Where certain results from the combined loading did not conform to eq(11.84a) refinements to the above theory were found to be necessary.

11.8.2 Memory Function

Nouailhas et al. [84, 90] identified eqs(11.83a,b) with the respective radius and centre of a memory surface in strain space as follows:

$$F = \sqrt{\tfrac{2}{3}\,(\varepsilon_{ij} - \alpha_{ij})(\varepsilon_{ij} - \alpha_{ij})} - \rho \le 0 \tag{11.85a}$$

The maximum value of ρ is associated with an asymptotic value of the plastic strain range attained under isotropic hardening. This asymptote derives from a stable dislocation structure and is analogous to the parameter Q, employed with eq(11.71a) in stress space. The rates of Q and ρ, for the preceding deformation, may be related through eqs(11.71a) and (11.83a) to enable a continuous interchange between the viscoplastic stress and strain spaces. This procedure has been found advantageous for memorising a previous history of plastic strain under multiaxial, cyclic loadings. Initial applications were made with equal weighting from isotropic and kinematic hardening, i.e. for $\lambda = ½$ in eqs(11.83a,b). However, Ohno [91] found that the influence of mean strain in the cycle required a lesser value of λ.

Note: Equations (11.82) and (11.85a) are similar in form to the strain potential $g(\varepsilon_{ij})$ for an elastic–plastic strain space, originally proposed by Besseling [92, 93]:

$$g(\varepsilon_{ij}) = \left(\varepsilon_{ij} - \alpha_{ij}\right)\left(\varepsilon_{ij} - \alpha_{ij}\right) = \rho^2 \tag{11.85b}$$

In this model ρ describes the differing yield limits for 'sub-elements' (i.e. grains) of material that are microscopically inhomogenous. These grains comprise an aggregate whose initial yield behaviour is homogenous. When all grains experience a similar inelastic strain, in excess of their limiting elastic values, then ρ in eq(11.85b) is altered by hardening. As a consequence of hardening, the aggregate strain become anisotropic and time dependent. Besseling derived each deformation process, mathematically, from its associated non-uniform rate of energy dissipation.

11.9 Comparison Between Theories

The literature reveals various theoretical predictions to the results from reversed torsional creep tests for stainless steel [88, 89, 94, 95, 96] at 600°C–700°C, and for copper [78, 96] at 300°C. In these experiments, a shear stress was applied to a thin-walled tube for a given time before a stress reversal enabled negative viscous flow for a similar time interval. Alternatively, the total shear strain may provide the limit to the forward and reversed deformation. Figures 11.31 and 11.32 give respective examples of creep curves that result from these time- and strain-controlled test conditions. They apply to a nickel-base alloy (IN597) at 850°C under an alternating shear stress of ± 144.3 MPa ($\bar{\sigma}$ = 250 MPa) [97]. Figure 11.31 shows how the equivalent creep strain has increased (ratchet strain) at the end of each 24 hr creep period. Figure 11.32 shows how the successive time intervals decrease when total strain limits of ± 2 % are imposed.

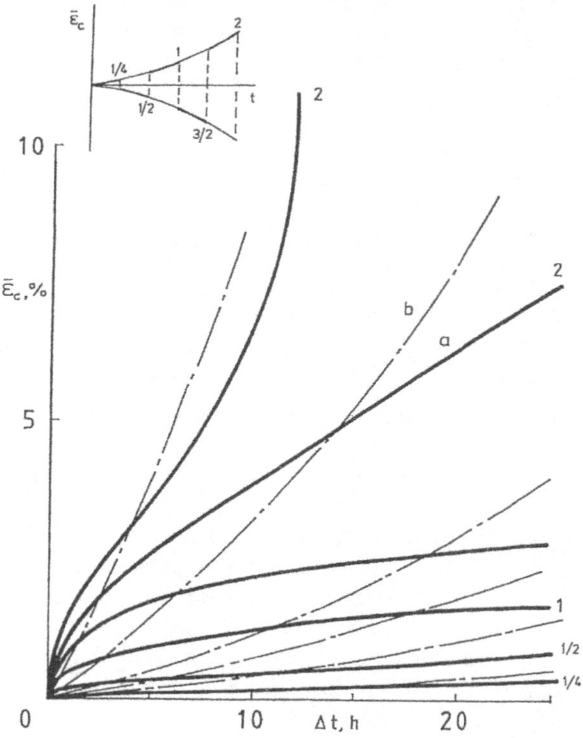

Figure 11.31 Reversed torsional creep under time control
Key: (a) - experiment; (b) - eq(11.86a)

The theoretical predictions, given in Figs 11.31 and 11.32, are those supplied by the time and strain hardening equations of state [46]. Their applications amount to a simple partitioning of a monotonic creep curve in the manner shown within the inset curves. The latter are given from integrating the following two descriptions of the increasing creep rate within tertiary creep, the dominant stage for this alloy:

$$\dot{\bar{\varepsilon}}_c = \dot{\bar{\varepsilon}}_s(1 + m\bar{\varepsilon}_c), \qquad \dot{\bar{\varepsilon}}_c = \dot{\bar{\varepsilon}}_s\, e^{C\bar{\varepsilon}_c} \tag{11.86a,b}$$

With m and C as constants in eqs(11.86a,b), the tertiary rates are taken to increase from the initial secondary rate, linearly and exponentially respectively with tertiary creep strain. Whilst the effects described above for each cycle are reproduced it is clear that state equation predictions found from eqs(11.86a,b) do not match the cyclic rates observed for this Ni-base alloy. In fact, the reversals appear to promote a primary stage in cyclic creep curves that was never present in the monotonic creep curve!

Greenstreet [98] employed the state equation approach in a similar manner with acceptable accuracy for tension-compression cycling in 304 stainless steel at 593°C. Descriptions of two alternative predictions to these results now follow.

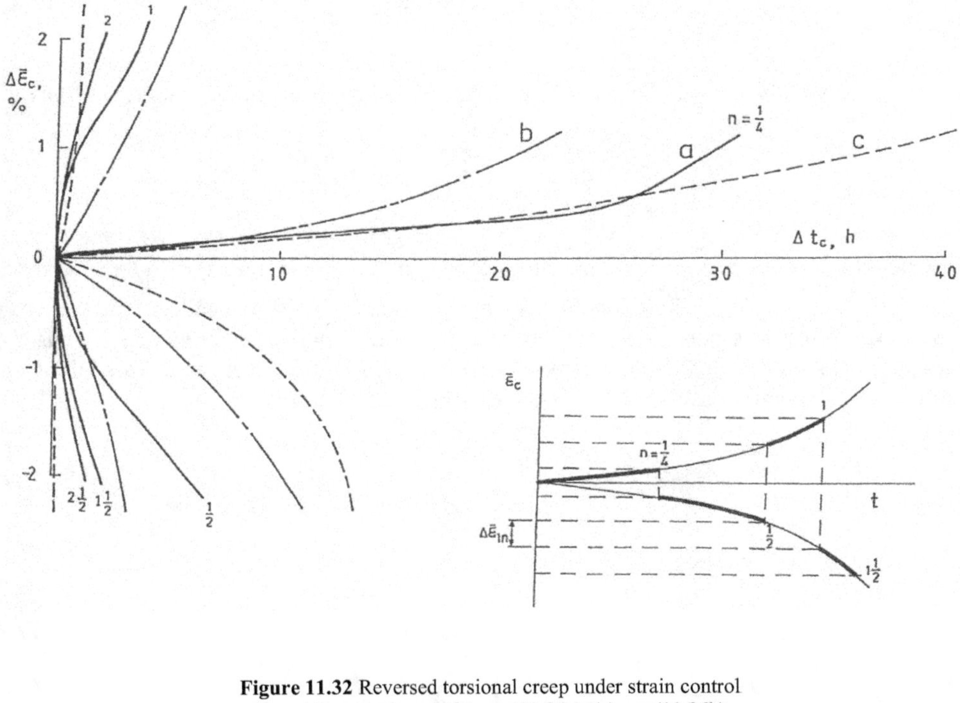

Figure 11.32 Reversed torsional creep under strain control
Key: (a) - experiment, (b) - eq(11.86a), (c) - eq(11.86b)

11.9.1 Stress Potential

Mroz et al. [79, 96] used eq(11.72c) to predict cyclic torsional creep curves for copper at 300°C and 304 stainless steel at 600°C. The Norton law took the form

$$\frac{\dot{\bar{\varepsilon}}}{\dot{\bar{\varepsilon}}_o} = \left(\frac{\bar{\sigma} - \bar{\alpha}_f}{\bar{\sigma}_o}\right)^n \tag{11.87a}$$

where the back stress $\bar{\alpha}_f = \mu\bar{\sigma}$ (with $\mu \simeq 0.35$ for copper and $\mu = 0.6$ for steel). This modification will supply equivalent creep rates identical to those from the basic Norton law

$$\frac{\dot{\bar{\varepsilon}}}{\dot{\bar{\varepsilon}}_o} = \left(\frac{\bar{\sigma}_N}{\bar{\sigma}_o}\right)^n \tag{11.87b}$$

for $\bar{\sigma}_N = \bar{\sigma}(1 - \mu)$ with both $\dot{\bar{\varepsilon}}_o$ and n as constants in this law. Equations (11.87a,b) are combined within the flow rule

$$\frac{\dot{\varepsilon}_{ijc}}{\dot{\bar{\varepsilon}}} = \frac{3(\sigma_{ij}' - \alpha_{ij}')}{2\bar{\sigma}_N} \tag{11.88a}$$

such that for simple torsion, the normalised shear creep rate is given by eqs(11.87a,b) as

$$\frac{\dot{\gamma}_c}{\dot{\bar{\varepsilon}}_o} = 3\left(\frac{\bar{\sigma} - \bar{\alpha}_f}{\bar{\sigma}_o}\right)^{n-1}\left(\frac{\tau - \beta}{\bar{\sigma}_o}\right) \tag{11.88b}$$

in which the von Mises equivalence relationships apply: $\bar{\sigma} = \sqrt{3}\tau$ and $\bar{\alpha}_f = \sqrt{3}\beta$.

Note: A separate time function does not appear in eq(11.88b). Instead, the back stress β is assumed to increase during creep, from its initial value to a maximum value β_m. Thus, for a given shear stress at P, on the loading surface, $F = 0$ in Fig. 11.33, β grows radially along OP such that, at a specific time s, $\sqrt{3}\beta = |\mathbf{OA}|$ defines the centre of F.

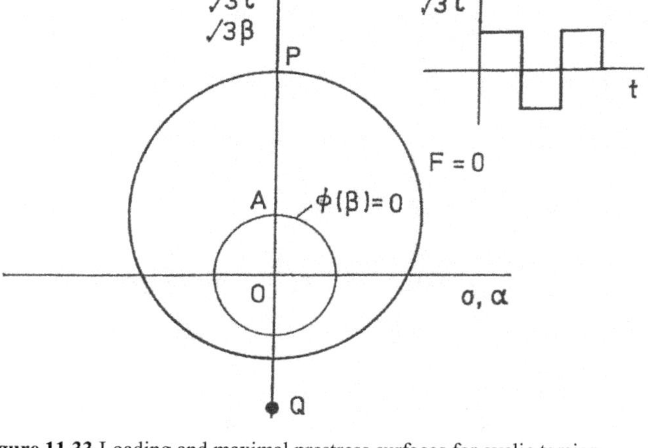

Figure 11.33 Loading and maximal prestress surfaces for cyclic torsion

The 'maximal prestress' surface $\varphi(\beta) = 0$ passes through A, being given by $\varphi(\beta) = \beta - \beta_m = 0$, where $\beta_m = \sup\beta\ (t - s)$ for $0 < s < t$. This surface provides a memory for the history of deformation, admitting a current steady-state, when β_m tends to its asymptotic value β_f. When the shear stress is fully reversed to point Q, β reverses from A and grows in the direction of AQ during the creep period. Figure 11.34 shows the configuration for when cyclic torsion is superimposed upon steady tension.

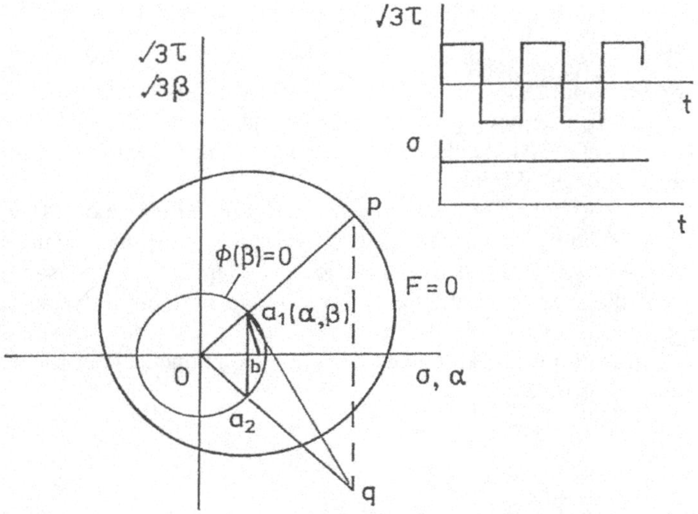

Figure 11.34 Loading and maximal prestress surfaces for cyclic torsion with static tension

For an initial radial loading to point p, the intersection point a_1, between Op and the pre-stress surface $\varphi(\beta) = 0$, locates the back stress co-ordinates $a_1(\alpha, \beta)$ for the loading surface $F = 0$. Stress reversal, from position p to q in Fig. 11.34, requires that two components of the back stress vector $\mathbf{a_1b}$ are directed along a_1q and a_1a_2. When the back stress is tracked in this manner, the shear creep rates for points p and q are again given by eq(11.88b) and the companion axial creep rate is

$$\frac{\dot{\varepsilon}_c}{\dot{\varepsilon}_o} = \left(\frac{\bar{\sigma} - \bar{\alpha}_f}{\bar{\sigma}_o}\right)^{n-1} \left(\frac{\sigma - \alpha}{\bar{\sigma}_o}\right) \tag{11.88c}$$

where $\bar{\sigma} = \sqrt{(\sigma^2 + 3\tau^2)}$ and $\bar{\alpha}_f = \mu\bar{\sigma}$. In Mroz's theory, a plasticity–creep interaction appears within the rate summation $\dot{\varepsilon}_{in} = \dot{\varepsilon}_P + \dot{\varepsilon}_c$. Instantaneous plastic strain ε_P is found from taking F as the plastic potential surface in Figs 11.33 and 11.34. These figures show that F is centred at the back stress with a radius equal to an effective stress $\bar{\sigma}_Y$. That is:

$$F = \sqrt{\tfrac{3}{2}\,(\sigma_{ij}' - \alpha_{ij}')(\sigma_{ij}' - \alpha_{ij}')} - \bar{\sigma}_Y = 0 \tag{11.89a}$$

In taking the relationship between the potentials for creep and plasticity (i.e. the yield surface) to be $F = f - \bar{\sigma}_Y$, it follows from eq(11.89a) that the viscous (creep) potential function f is

$$f = \sqrt{\tfrac{3}{2}\,(\sigma_{ij}' - \alpha_{ij}')(\sigma_{ij}' - \alpha_{ij}')} \tag{11.89b}$$

Equation (11.89b) implies that creep occurs under all stresses, i.e. $\dot{\varepsilon}_{ijc} > 0$ in eq(11.88a). If an elastic domain exists, with radius $\bar{\sigma}_e$, then eq(11.88a) gives $\dot{\varepsilon}_{ijc} > 0$ for $f - \bar{\sigma}_e > 0$ and $\dot{\varepsilon}_{ijc} < 0$ for $f - \bar{\sigma}_e < 0$. Correspondingly, eq(11.89b) is modified to

$$f = \sqrt{\frac{3}{2}(\sigma_{ij}' - \alpha_{ij}')(\sigma_{ij}' - \alpha_{ij}')} - \bar{\sigma}_e = 0 \tag{11.89c}$$

Equations (11.88b,c) reproduced the shear and axial rates observed in experimental creep curves for each material, after having removed the instantaneous loading strain. However, for balanced, cyclic torsion of 304 stainless steel, the theory did not account for changes to the peak strains following each 24 h interval.

Walker [87] and Hartmann [99] employed Walker's and Miller's models (see § 11.4.2 and § 11.7.4)) to predict the viscous deformation observed under high-temperature, strain-controlled cycling of 347 stainless steel and Hastelly-X. Here the peak stress may increase or decrease initially as the material hardens or softens, before attaining its stable state. Walker's model was more descriptive of the initial asymmetric cycles but both models required modification to their isotropic and kinematic hardening components in order to reproduce the cyclic stability found.

11.9.2 Strain Potential

Figure 11.35 applies the creep hardening surface of Murakami et al. [88, 89, 95], to a cyclic torsion creep test (the instantaneous strain having been removed).

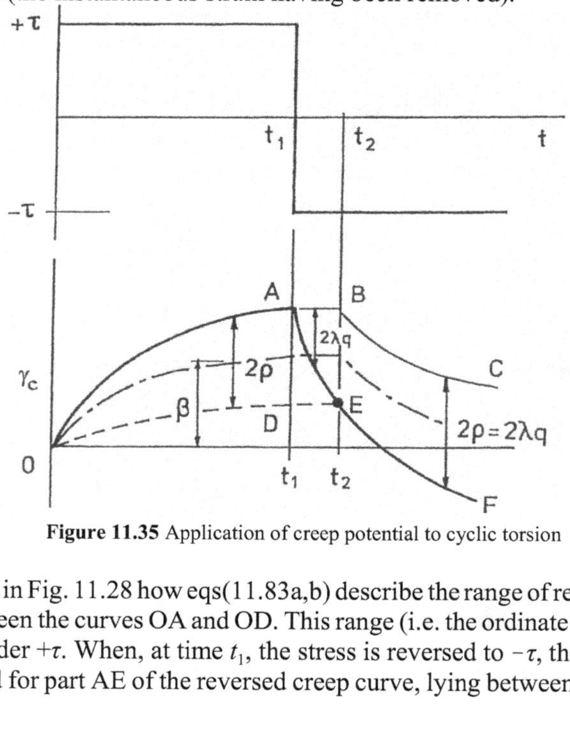

Figure 11.35 Application of creep potential to cyclic torsion

It has been shown in Fig. 11.28 how eqs(11.83a,b) describe the range of recoverable (anelastic) strain, lying between the curves OA and OD. This range (i.e. the ordinate 2ρ), increases during forward creep under $+\tau$. When, at time t_1, the stress is reversed to $-\tau$, the anelastic range AD remains unaltered for part AE of the reversed creep curve, lying between bounds AB and DE, for $t_1 \le t < t_2$.

Thereafter, for a creep time $t > t_2$ the ordinate 2ρ, between the bounds BC and EF, will re-define the anelastic strain range. Equation (11.82) bounds the shear strain as $\gamma_c = \beta \pm \rho$. The rates of increase in ρ and β during hardening partition the creep curve as follows:

$$\gamma_c = \rho + \beta = \lambda\gamma_c + (1 - \lambda)\gamma_c \tag{11.90a}$$

so that

$$\rho = \lambda\gamma_c, \qquad \beta = (1 - \lambda)\gamma_c \tag{11.90b}$$

In eqs(11.90a,b), λ is a material constant which follows from setting $2\rho = 2\lambda q$ in Fig. 11.35. Setting $\lambda = 0$ will restore the strain hardening equation of state. For the interval $0 \le t \le t_1$, $q = |\gamma_c| = \rho/\lambda$, implying that curves OA and OD are geometrically similar. When the stress is reversed at time t_1, q falls to zero but increases within the interval $t_1 \le t < t_2$, despite the absence of hardening, so that $q_E = q_B = \rho/\lambda$. Now, as q is proportional to the ordinate between the constant bounds AB and DE, this gives, for the interval $t_1 < t < t_2$:

$$2\lambda q = \beta + \rho - \gamma_c, \quad \Rightarrow \quad q = \frac{\beta + \rho - \gamma_c}{2\lambda} \tag{11.91a,b}$$

where γ_c is the creep strain ordinate for curve AE. For the reversed hardening time interval ($t \ge t_2$), a geometric similarity between curves BC and EF is assumed. Again, this gives a range of anelasticity: $2\lambda q = 2\rho$, for $q = \rho/\lambda$. The discontinuity in q, for $t = t_1$, may be avoided by setting $\lambda = ½$ in eqs(11.83a,b). Then, q expresses the alternating ranges of anelasticity as follows [100]:

$$q = |\gamma_c - (\beta - \rho)| \quad \text{for} \quad \tau \ge 0 \tag{11.92a}$$

$$q = |\gamma_c - (\beta + \rho)| \quad \text{for} \quad \tau < 0 \tag{11.92b}$$

In the application of eqs(11.91a,b) and (11.92a,b) to cyclic torsion of 304 stainless steel at 650°C [88, 89], γ_c was given by the strain hardening equation of state (see eq(11.84b)). Here the constants A, m and n are determined from the fit to the first forward creep curve, with constant $\lambda = ½$ being assumed. This approach led to predictions comparable with the Mroz theory. However, to predict cyclic peak strains that decrease successively, a further fit to the first reversed creep curve ($n = ½$) was required. Later studies [95] on 304 stainless steel at 600°C showed that the prior ageing time affected its peak stress and transient creep behaviour. An ageing time of 500 hr resulted in an almost steady range of shear strain under subsequent torsional cycling with $\sqrt{3}\tau = \pm$ 125 and $\pm$ 140 MPa for Δt_c = 96 hr. A shorter ageing time of 20 hr, under similar test conditions, led to considerable peak stress softening before a cyclically stable state was attained. To account for the precipitation of the carbide $M_{23}C_6$, the components of hardening due to ageing and to creep were separated. Ohashi, Ohno and Kawai [94] compared the results for the fully aged material with further predictions from Miller's combined hardening theory [33] and from Malinin and Khadjinsky's non-linear kinematic hardening theory [76]. Of all the predictions those from the modified creep potential theory, as outlined here, were more realistic, though no one theory could match the creep curves precisely.

One can draw a similar conclusion for the test results given in Figs. 11.31 and 11.32, but, undoubtedly, further refinements would be necessary to account for the pronounced softening observed throughout life in a cavity-forming alloy, where the assumption of incompressible flow is violated. It was seen that the 'softening' is manifested in: (i) ratchet strain under time-limited cycles (Fig. 11.31) and (ii) successive reductions to the time interval under strain-limited cycling (Fig. 11.32).

11.9.3 Damage Mechanics

It is appropriate to conclude this study with reference to certain viscous damage studies. The tertiary region, BC in Fig. 11.1, results from various metallurgical instabilities including necking, cracking, cavitation and grain coarsening. From a physical standpoint, tertiary creep is essentially a region of micro-structural damage. The concept of a creep damage parameter to measure loss of section from these damage sources was first advanced by Katchanov [57]. A similar concept was developed later by Rabotnov [58], specifically for brittle creep rupture where the parameter is intended to account for damage under low stresses over long periods. Consequently, this parameter will predict a dominant tertiary stage resulting from early grain boundary cavitation and cracking. The loss of material continuity, arising from these instabilities, is expressed in a structural damage parameter ω, with limiting values; $\omega = 0$, at the start of tertiary creep ($t = 0$ for simplicity) and $\omega = 1$ at fracture ($t = t_f$). For a given temperature, the chosen damage function must remain consistent with each limiting ω–value. During the tertiary creep period $0 \le t \le t_f$, both the strain rate $\dot{\varepsilon}_c$ and the damage rate $\dot{\omega}$ are assumed to depend upon σ, T and ω. Hence, the theory implies that the strain rate is coupled to the damage rate in their respective functions:

$$\dot{\varepsilon} = f(\omega, \sigma, T) \quad \Rightarrow \quad \dot{\varepsilon} = A\sigma^n (1 - \omega)^{-q} \qquad (11.93a,b)$$

$$\dot{\omega} = g(\omega, \sigma, T) \quad \Rightarrow \quad \dot{\omega} = B\sigma^k (1 - \omega)^{-r} \qquad (11.94a,b)$$

where A and B are temperature dependent and k, n, q and r are positive material constants. It follows from eq(11.93b) that damage does not influence $\dot{\varepsilon}$ when $q = 0$. When $q = n$, in eq(11.93b), it is seen that $\dot{\varepsilon}$ depends upon the current state of true stress; $\sigma/(1-\omega)$, within the Norton power law. Here $(1 - \omega)$ is often associated with the remaining cohesive area of the section. The Rabotnov-Katchanov theory, given in eqs(11.93a,b) and (11.94a,b), implies a change to the magnitude of a creep stress for a given value of the damage parameter, ω. This is more likely to be true when the direction of deformation remains unchanged. Otherwise, a stress reversal results in crack closure, rendering some damaged areas inactive. Dyson and McLean [101] developed cavity nucleation and growth laws from tensile and torsional creep tests for a Ni-base alloy (80A) at 750°C. For a given time, within the alloy's dominant tertiary region, the equivalent creep rate appeared as

$$\dot{\bar{\varepsilon}} = A\bar{\sigma}^4 V^{4/9} \qquad (11.95)$$

where A is a constant and V is the total void volume fraction. Measurements [102] of the growth of the number of grain boundary voids n, nucleated within a given volume, showed

$$n = B\left(\frac{\sigma_1}{\bar{\sigma}}\right)^2 \bar{\varepsilon}^{1/2} \quad \Rightarrow \quad \frac{dn}{dt} = \frac{B}{2}\left(\frac{\sigma_1}{\bar{\sigma}}\right)^2 \frac{\dot{\bar{\varepsilon}}}{\bar{\varepsilon}^{1/2}} \qquad (11.96a,b)$$

which enabled the void growth rate dv/dt for individual voids within a unit volume of material to be expressed as

$$\frac{dv}{dt} = C\left(\frac{\sigma_1}{\bar{\sigma}}\right)^{0.7} \dot{\bar{\varepsilon}} \qquad (11.96c)$$

In eqs(11.96a–c) B and C are constants and $\bar{\sigma}$ and $\dot{\bar{\varepsilon}}$ take the usual von Mises forms (see eqs 11.84c,d). Nucleation and growth display: (i) linear dependencies upon $\bar{\varepsilon}$ and (ii) strong and weak dependencies, respectively, upon the normalised principal stress state $(\sigma_1/\bar{\sigma})$. Equations (11.95) and (11.96a,b) also show that, while the nucleation rate is infinite initially, the subsequent creep rate is controlled only by those voids that grow under the major principal tensile stress σ_1.

A further relationship holds between the total volume of voids V and their simultaneous rates of nucleation and growth, dn/dt and dv/dt respectively, for a given creep time t. Let a nucleation rate d$n(\tau)$/dτ apply to an earlier time $\tau < t$, so that in time dτ, the number of voids formed is dn = [d$n(\tau)$/dτ]dτ. At the later time t, the volume growth rate for dn voids is given by dv/dt = d$v(t, \tau)$/dt. Hence, in the time interval $t - \tau$, the voids formed at time τ will occupy a volume

$$V(t) = \int_0^{\tau} \frac{\mathrm{d}n(\tau)}{\mathrm{d}\tau}\,\mathrm{d}\tau \int_{\tau}^{t} \frac{\mathrm{d}v(t,\tau)}{\mathrm{d}t}\,\mathrm{d}t \tag{11.97a}$$

Taking the lower limit as $\tau = 0$ in eq(11.97a) gives the total cavity volume within a unit volume of material, with a critical value for $t = t_f$, at fracture. Under a constant, uni-axial stress, the substitution of eqs(11.96a,b) into eq(11.97a) leads to a rupture condition [101]:

$$V = D\left(\frac{\sigma_1}{\bar{\sigma}}\right)^{2.7} \bar{\varepsilon}^{\,3/2} \tag{11.97b}$$

where D is a constant. Hayhurst et al. [102, 103] identified all constants A, B, C and D in eqs(11.95)–(11.97a,b) with the values of $\bar{\sigma}$, $\bar{\varepsilon}$, v, V and n at rupture within the 80A alloy. Predictions to creep curves from reversed torsion tests, based upon these equations, were found to depend upon the directional nature of damage assumed. The best prediction indicated that the initial nucleation rate n was scalar while the subsequent growth rate v and the total void volume V were vectorial in character. Both v and V were associated with cavitated grain boundaries lying normal to σ_1. This approach admitted further the strain-ratcheting observed in 24 hr time-controlled torsional cycling, similar to that shown in Fig. 11.31. However, it was noted previously from this figure that nimonic alloy regenerates primary creep following each stress reversal under torsion.

In contrast to eqs(11.95) and (11.96b,c) for 80A, the corresponding rate expressions for copper at 500°C were given as [104]:

$$\dot{\bar{\varepsilon}} = A'\bar{\sigma}^{5}\, \mathrm{e}^{-\frac{\Delta H}{RT}} \tag{11.98}$$

$$\frac{\mathrm{d}n}{\mathrm{d}t} = B'\sigma_1^{2}\,\dot{\bar{\varepsilon}} \tag{11.99a}$$

$$\frac{\mathrm{d}v}{\mathrm{d}t} = C'\sigma_1\, \mathrm{e}^{-\frac{\Delta H_v}{RT}} \tag{11.99b}$$

where A', B' and C' in eqs(11.98) and (11.99a,b) are constants. Also, within the Arrhenius exponential term in eqs(11.98) and (11.99b), ΔH is the activation energy for self diffusion and $R = 1.98$ cal/mole is the characteristic gas constant. This exponential term is constant at a given temperature. Equations (11.98)–(11.99a,b) show again for copper that a uniaxial tensile stress σ_1 controls void nucleation and growth rates arising from diffusion. The voids lie along boundaries normal to σ_1 under uniaxial and multiaxial stress states when, for the latter, σ_1 is the major principal tensile stress.

Perzyna [105] and Eftis and Nemes [106] introduced an internal damage variable to the isotropic visco-plastic theory, outlined in § 11.6 above. The theory, as applied to ductile void growth, modified eq(11.58c) to become

$$f\left(\sigma_{ij}, \varepsilon^P, V\right) = \frac{F(J_1, J_2', V)}{\varkappa(\varepsilon^P, V)} - 1 \tag{11.100}$$

where V is the void volume fraction and $J_1 = \sigma_{kk}$ accounts for the influence of hydrostatic stress on plastic compressibility through void formation. Correspondingly, in eq(11.100), the strain hardening function $\varkappa\,(\varepsilon^P, V)$ should admit softening. Typical functions follow for ε^P as a uniaxial plastic strain:

$$\varkappa(\varepsilon^P, V) = \left[(\varkappa_o + H\varepsilon^P)(1 - n\sqrt{V})\right]^2 \tag{11.101a}$$

$$\varkappa(\varepsilon^P, V) = \left\{\left[q + (\varkappa_o - q)\, e^{-\beta\varepsilon^P}\right](1 - n\sqrt{V})\right\}^2 \tag{11.101b}$$

where H, β, q are material parameters. A constant ($\varkappa_o$) is found from the initial yield stress and the number of voids n comprising the void volume fraction at fracture ($n \sim 1/10$ for copper). The rate at which V increases, from a typical initial value of 0.1 %, becomes a sum of those contributions from nucleation and growth:

$$\dot{V} = \dot{V}_n\left(V, |\sigma - \sigma_n|, T, k\right) + \frac{1}{\mu}\dot{V}_g\left(V, V_o, \varkappa_o, |\sigma - \sigma_g|\right) \tag{11.102}$$

where | | are threshold axial stress levels necessary for each process to commence, T is the absolute temperature and k is Boltman's constant. Exponential functions for $\dot{V}_n$ and $\dot{V}_g$ in eq(11.102) were applied to lower temperatures without diffusion. It was shown [106] that the softening, which accompanied void growth in copper, became dominant for higher V. The micro-viscosity coefficient μ, in eq(11.102), accounted for viscous flow within localized hot spots at the nucleation sites. This affected the strain-dependence of both the uniaxial flow stress and the damage accumulation.

11.10 Internal State Parameters

Various internal, or hidden, state parameters have appeared in formulating alternative, energy-based theories of viscoplasticity. These parameters aim to describe the path dependence of a polycrystalline material, i.e. the extent to which viscous flow is influenced by strain history. It has been seen that classical theories achieve this through an incremental strain formulation, embodying the normality rule with an identity placed between the yield function and the plastic potential. The latter is retained as a common feature in theories that employ internal parameters but the manner in which they are introduced into the potential function varies. Internal state parameters fall broadly into two categories: those that are purely mechanical and those that are coupled to the thermodynamics of elasto–plasticity.

Thermodynamics becomes important to consider, particularly where significant temperature changes accompany finite deformations [107]. The internal parameters appear in a free-energy response function in order to facilitate plastic irreversibility. One such approach, developed in § 11.11.3, is a recent endochronic theory, which applies to metals and non-metals. In contrast, the classical, infinitesimal plasticity theory (see chapter 10) is restricted to metallic materials and is not derived from thermodynamics.

11.10.1 Mechanical

One mechanical internal state parameter is that adopted by Katchanov for the evolution of brittle creep damage. Another internal parameter is the scalar measure of hardening ψ used in eqs(11.58a,c), to characterise a uniform expansion of the yield surface. While classical plasticity theory equates ψ to the accumulated plastic strain or work, other scalars could equally well be used, provided they increase monotonically to provide for an isotropic hardening. Furthermore, for kinematic hardening, the parameter locating the centre of a translated yield surface, known as the internal or back stress, may also be interpreted as an internal variable. Chaboche [107] developed this further within the anisotropic hardening potential (see § 11.7), in which the translation tensor α_{ij} was identified with a residual micro-stress. An expression for α_{ij} is the multi-axial counterpart of a steady state uniaxial condition in which hardening and recovery oppose one another:

$$d\sigma_{ij} = \frac{\partial \sigma_{ij}}{\partial \varepsilon_{ij}} \times d\varepsilon_{ij} - \frac{\partial \sigma_{ij}}{\partial t} \times dt \qquad (11.103a)$$

The opposition between these terms becomes an incremental micro-stress $d\alpha_{ij}$, for an incremental plastic strain $d\varepsilon_{ij}^P$ within a time interval dt, as

$$d\alpha_{ij} = F(\sigma_{ij}, \alpha_{ij}, \varepsilon_{ij}^P)\, d\varepsilon_{ij}^P - G(\sigma_{ij}, \alpha_{ij}, \varepsilon_{ij}^P)\, dt \qquad (11.103b)$$

In eq(11.103b), the hardening function F increases the translation components α_{ij}, due to strain hardening. The softening function G, accounts for a reduction in α_{ij} due to recovery and relaxation. Here the arguments of F and G, i.e. σ_{ij}, α_{ij} and ε_{ij}^P become the *internal state parameters*. Thus, isotropic and kinematic hardening potentials employ internal parameters both explicitly and implicitly.

Kratchovil et al. [109, 110] included in an isotropic potential function $f = J_2'$ the change in specific internal energy u^P of plastic deformation. The corresponding argument for f appeared as a derivative of the plastic component of the specific internal energy $u^P = u^P(a_{ij})$, where a_{ij} is a structure tensor. That is, $A_{ij} = \partial u^P / \partial a_{ij}$. Taken with the flow rule, the plastic strain rate tensor becomes

$$\dot{\varepsilon}_{ij}^P = \frac{\partial f(\sigma_{ij}, A_{ij})}{\partial \sigma_{ij}} \qquad (11.104a)$$

in which the decreasing rate of change in α_{ij} is given by

$$\dot{\alpha}_{ij} = -\frac{\partial f(\sigma_{ij}, A_{ij})}{\partial A_{ij}} \qquad (11.104b)$$

Separating σ_{ij} and A_{ij} into deviatoric (σ_{ij}' and A_{ij}') and hydrostatic components, the response (potential) function f, involving eqs(11.104a,b), becomes a function of the three invariants of σ_{ij}' and A_{ij}', taken alone and in combination, as follows:

$$f = f\left[A_{kk}, \left(\sigma_{ij}'\sigma_{ij}'\right), \left(A_{ij}'A_{ij}'\right), \left(\sigma_{ij}'A_{ij}'\right), \left(\sigma_{ij}'\sigma_{jk}'\sigma_{ki}'\right), \left(A_{ij}'A_{jk}'A_{ki}'\right), \left(\sigma_{ij}'A_{jk}'A_{ki}'\right), \left(\sigma_{ij}'\sigma_{jk}'A_{ki}'\right)\right] \tag{11.105a}$$

where u^P is a function of the invariants of a_{ij}':

$$u^P = u^P\left[a_{kk}, \left(a_{ij}'a_{ij}'\right), \left(a_{ij}'a_{jk}'a_{ki}'\right)\right] \tag{11.105b}$$

In eqs(11.105a,b), hydrostatic components A_{kk} and a_{kk} appear along with σ_{kk}, which do not influence f. When ε_{ij}^P replaces A_{ij} in eq(11.104a), and deformation is incompressible, then $A_{kk} = \varepsilon_{kk}^P = 0$. This leads to the general invariant form for f

$$f = f(J_2', I_2, K_2, J_3', I_3, K_3, K_4) \tag{11.105c}$$

in which the respective invariants (J_2', J_2'), (I_2, I_3) and (K_2, K_3, K_4) of the stress deviator, of plastic strain and of their products appear as

$$J_2' = \frac{1}{2}\sigma_{ij}'\sigma_{ij}', \quad J_3' = \frac{1}{3}\sigma_{ij}'\sigma_{jk}'\sigma_{ki}',$$

$$I_2 = \frac{1}{2}\varepsilon_{ij}^P\varepsilon_{ij}^P, \quad I_3 = \frac{1}{3}\varepsilon_{ij}^P\varepsilon_{jk}^P\varepsilon_{ki}^P,$$

$$K_2' = \frac{1}{2}\sigma_{ij}'\varepsilon_{ij}^P, \quad K_3' = \frac{1}{3}\sigma_{ij}'\sigma_{jk}'\varepsilon_{ki}^P, \quad K_4' = \frac{1}{3}\sigma_{ij}'\varepsilon_{jk}^P\varepsilon_{ki}^P$$

Note: Equation (11.105c) is equivalent to eq(11.64b) which describes combined isotropic and kinematic hardening for when an expansion and translation appear in the subsequent yield surface (see exercise 11.12). Equations(11.104a,b) and (11.105a,b) are simplified when applied to radial loading. Then, reduced forms of eqs(11.105a,b) express a rate dependency

$$f = f\left(A_{kk}, \sigma_{ij}'\sigma_{ij}'\right), \quad u^P = u^P\left(a_{kk}\right) \tag{11.106a,b}$$

where $A_{kk} = \partial u^P/\partial a_{kk}$ accounts for compressibility. In eq(11.106b), a_{kk} was interpreted from within a family of constant hardening curves given by Hart [111, 112].

Other internal parameters of importance are both micro-structurally based and thermodynamic in nature. In particular, parameters of the former kind have been employed to express: (i) the motion, arrangement and interaction of dislocations [113] for plasticity and (ii) diffusion, phase changes and micro-void formation [114] for creep. It would appear that such micro-structural parameters, which serve the transition from the micro- to macro-scale, are easier to formulate for metals than for engineering alloys [114] including the added complexity of solution treatment and age hardening through the formation of time dependent second phase particles. One advantage of employing thermodynamic internal state parameters, as proposed by Coleman and Noll [115], is that they do not require this distinction from micro– to macro–scale to be made. Here it will first be necessary to outline the thermodynamics of elastic and plastic deformation to show how internal parameters are more convenient to apply within a viscoplastic potential.

11.10.2 Thermodynamic State

A number of internal thermodynamic state parameters are drawn from the following laws of thermodynamics and applied to a deforming solid.

(a) Laws of Thermodynamics
The *first law of thermodynamics* relates a closed system undergoing a non-cyclic process where incremental heat and work transfers, dq and dw respectively, change the internal energy by an amount du. Firstly, take the closed system to be a deforming elastic solid. Conventionally, positive transfers dq and dw are taken to and from the solid respectively, i.e. for the directions shown in Fig. 11.36.

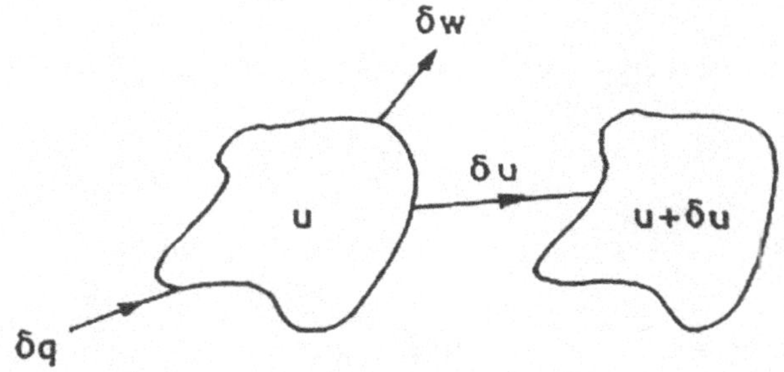

Figure 11.36 Elastic solid undergoing a non-cyclic process

The first law is written in terms of specific quantities (per unit mass) as

$$dq - dw = du \tag{11.107a}$$

Here the simplest 'system' is taken as a volume V of solid, deforming elastically under a uniaxial tensile stress σ. With no heat transfer, dq = 0 and the work done equates directly to the strain energy stored. This gives the specific work density (per unit volume) as

$$\rho_o dw = -\sigma d\varepsilon^e \tag{11.107b}$$

where ρ_o is the initial density of the undeformed material. The density decreases with the tension according $\delta\rho \simeq -\sigma\rho_o/3K$, where K is the bulk modulus. Equation (11.107b) shows that negative work is done when deforming the solid. Substituting eq(11.107b) into eq(11.107a) gives the first law for when a heat transfer occurs to a unit mass

$$du = Tds + \frac{1}{\rho_o}\sigma d\varepsilon^e \tag{11.107c}$$

where dq = Tds, in which T is the absolute temperature and ds is the change in entropy of the system. The *second law of thermodynamics* states that ds cannot be negative. That is:

$$ds = \frac{dq}{T} \geq 0 \tag{11.108}$$

showing that if dq is negative in Fig. 11.36 then T is negative, which is not possible on an absolute temperature scale. In the context of a deforming solid, the equality ds = 0 applies to reversible elasticity. The inequality ds > 0 applies generally to all inelastic processes.

(b) Elastic Deformation

The thermodynamic state of an elastically deformed material is defined by any two of the variables: stress, strain and temperature, appearing in the single-valued, thermal state equation [116, 117]:

$$g(\sigma,\ \varepsilon^e,\ T) = 0 \tag{11.109}$$

where stress, strain and temperature appear as *thermodynamic, equilibrium state properties*. The remaining state properties internal energy u, entropy s and enthalpy h, (per unit mass) do not appear within eq(11.109). Therefore, it is necessary to couple to eq(11.109) a response function whose partial derivatives deliver the remaining state variables. Any one of the following response functions is chosen for this purpose:

(i) The specific internal energy u, as a function of elastic strain ε^e and specific entropy s:

$$u = u(\varepsilon^e,\ s) \tag{11.110a}$$

From eq(11.107c):

$$u = u(\varepsilon^e, s) = T\int \mathrm{d}s + \frac{1}{\rho_o}\int \sigma\,\mathrm{d}\varepsilon^e \tag{11.110b}$$

so that by maintaining s and ε^e as respective constants:

$$\sigma = \rho_o\left(\frac{\partial u}{\partial \varepsilon^e}\right),\quad T = \frac{\partial u}{\partial s} \tag{11.110c,d}$$

(ii) The specific enthalpy h, as a function of the applied stress σ and the specific entropy s:

$$h = h(\sigma,\ s) \tag{11.111a}$$

where, by definition:

$$h = \int \mathrm{d}u + \int p\,\mathrm{d}V \tag{11.111b}$$

From eqs(11.107b,c):

$$h(\sigma,\ s) = \int \mathrm{d}u - \frac{1}{\rho_o}\int \sigma\,\mathrm{d}\varepsilon^e = T\int \mathrm{d}s \tag{11.111c}$$

so that:

$$\sigma = -\rho_o\frac{\partial h}{\partial \varepsilon^e} = -\rho_o E\frac{\partial h}{\partial \sigma},\quad T = \frac{\partial h}{\partial s} \tag{11.111d,e,f}$$

Hooke's law $\sigma = E\varepsilon^e$ applies to linear elasticity, connecting eqs(11.111d,e) with the elastic modulus:

$$E = \frac{\mathrm{d}\sigma}{\mathrm{d}\varepsilon^e} \tag{11.111g}$$

(iii) Helmholtz's free-energy function H, of temperature T and elastic strain ε^e:

$$H = H(T,\ \varepsilon^e) \tag{11.112a}$$

where, by definition:

$$H = \int \mathrm{d}u - \int T\,\mathrm{d}s \tag{11.112b}$$

Hence, from eqs(11.107c):

$$H(T, \varepsilon^e) = \int \mathrm{d}u - \int T\,\mathrm{d}s = \frac{1}{\rho_o}\int \sigma\,\mathrm{d}\varepsilon^e \tag{11.112c}$$

and with H identifying the work transferred from a unit mass:

$$\sigma = \rho_o \frac{\partial H}{\partial \varepsilon^e}, \quad T = -\frac{\partial H}{\partial s}, \quad s = -\frac{\partial H}{\partial T} \tag{11.112d,e,f}$$

Equation (11.112f) has converted eq(11.112e) to the argument T given for H in eq(11.112a). This assumes that

$$\mathrm{d}q = T\mathrm{d}s = s\,\mathrm{d}T \tag{11.112g}$$

(iv) Gibb's free enthalpy function G, of temperature T and stress σ, defined as

$$G = G(T, \sigma) \tag{11.113a}$$

where, by definition:

$$G = \int \mathrm{d}u + \int p\,\mathrm{d}V - \int T\mathrm{d}s = 0 \tag{11.113b}$$

From eq(11.107b):

$$G(T,\sigma) = \int \mathrm{d}u - \frac{1}{\rho_o}\int \sigma\,\mathrm{d}\varepsilon^e - \int T\mathrm{d}s = 0 \tag{11.113c}$$

giving

$$\sigma = -\rho_o \frac{\partial G}{\partial \varepsilon^e}, \quad T = -\frac{\partial G}{\partial s} \tag{11.113d,e}$$

It is now required to convert eqs(11.113d,e) to the arguments T and σ, given for G in eq(11.113a). Using eqs(11.111g) and (11.112g):

$$\varepsilon^e = -\rho_o \frac{\partial G}{\partial \sigma}, \quad s = -\frac{\partial G}{\partial T} \tag{11.113f,g}$$

Equation (11.109), when taken with any pair of partial derivatives (11.110c,d), (11.111e,f), (11.112d,f) or (11.113f,g), is sufficient to describe infinitesimal, uniaxial elasticity. Green and Naghdi [118, 119] and Phillips [120] considered the extension to eqs(11.112d,f) and (11.113f,g) for linear elasticity under multiaxial stress. The latter only modifies the work term within the first law of thermodynamics so eq(11.107c) appears in its multiaxial form

$$\mathrm{d}u = T\mathrm{d}s + \frac{1}{\rho_o}\sigma_{ij}\,\mathrm{d}\varepsilon_{ij}^e \tag{11.114}$$

where $\mathrm{d}s > 0$. Combining eq(11.114) with the Helmholtz eq(11.112b) gives

$$H(T, \varepsilon_{ij}^e) = \int \mathrm{d}u - \int T\mathrm{d}s = \frac{1}{\rho_o}\int \sigma_{ij}\,\mathrm{d}\varepsilon_{ij}^e \tag{11.115a}$$

from which

$$\sigma_{ij} = \rho_o \frac{\partial H}{\partial \varepsilon_{ij}^e}, \quad T = -\frac{\partial H}{\partial s}, \quad s = -\frac{\partial H}{\partial T} \tag{11.115b,c}$$

Alternatively, Gibb's potential (11.113c) becomes

$$G(T,\sigma_{ij}) = \int \mathrm{d}u - \frac{1}{\rho_o}\int \sigma_{ij}\,\mathrm{d}\varepsilon_{ij}^e - \int T\mathrm{d}s = 0 \tag{11.116a}$$

from which

$$\sigma_{ij} = -\rho_o \frac{\partial G}{\partial \varepsilon_{ij}^e}, \quad T = -\frac{\partial G}{\partial s} \tag{11.116b,c}$$

Other response functions (or thermodynamic potentials) may be extended to multiaxial elasticity in a similar manner. Clearly, the derivative of H in eq(11.115b) appears with respect to its elastic strain tensor argument and is therefore the more convenient form. Equations(11.116b,c) require further elastic constitutive relations for their conversion to the partial derivatives of G required with respect to its arguments T and σ_{ij} (see exercise 11.16).

Consider the application of the first law to uniaxial deformation within the closed, elastic cycle 010, given in Fig. 11.37a. Equation (11.107a) gives $\mathrm{d}u = 0$.

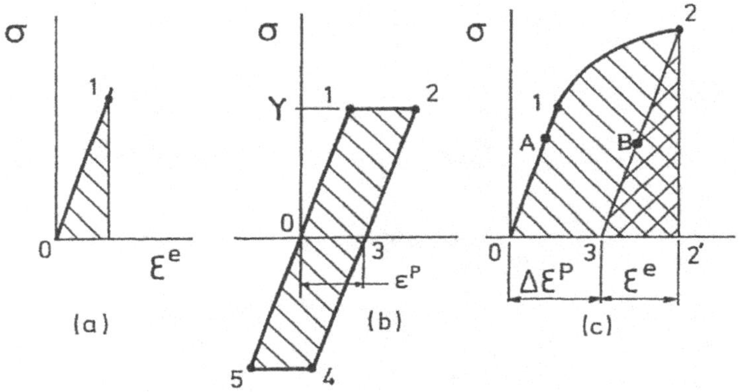

Figure 11.37 Elastic and elastic–plastic deformation paths

The resulting specific heat and specific work transfers, q and w respectively, conform to

$$\oint \mathrm{d}q - \oint \mathrm{d}w = 0 \tag{11.117a}$$

Equation (11.117a) applies to quasi-static, reversible elasticity, both linear and non-linear. With linear elasticity there is no hysteresis between a slow elastic loading–unloading cycle. Consequently, with there being no area enclosed, eq(11.117a) shows

$$\oint \mathrm{d}q = \oint \mathrm{d}w = \frac{1}{\rho_o}\oint \sigma\,\mathrm{d}\varepsilon^e = 0 \tag{11.117b}$$

In the case of non-linear elasticity, where there is an area enclosed within the cycle, it follows from eq(11.117b) that $\oint \mathrm{d}q = \oint \mathrm{d}w \neq 0$. Under the forward elastic loading path 01, the two laws of thermodynamics state that $\mathrm{d}u \neq 0$ and that the rate of entropy production is positive. From eq(11.108):

$$\frac{\mathrm{d}s}{\mathrm{d}t} = \frac{1}{T}\frac{\mathrm{d}q}{\mathrm{d}t} > 0 \tag{11.117c}$$

For a *unit mass* of material, eqs(11.107a) and (11.108) lead to

$$\int dq = \int du + \int dw = \int T ds \tag{11.118a}$$

From eq(11.107b), with the negative work done upon a *unit volume* of material, eq(11.118a) gives

$$\rho_o \int T ds = \rho_o \int du - \int \sigma d\varepsilon^e \tag{11.118b}$$

where $\int \sigma d\varepsilon^e$ is the shaded area in Fig. 11.37a. Equation (11.118b) shows that this work differs from the dissipated heat by an amount equal to the change in the internal energy.

(c) Elastic–Plastic Deformation

Equation (11.117b) will remain valid when a closed cycle is applied to an elastic–perfectly plastic solid at constant temperature, e.g. cycle 0123450 in Fig. 11.37b. Here, two state parameters, stress and temperature, are constants for the paths 12 and 45. Equation (11.109) will remain single-valued for each path, provided the strain is identified with the constant, elastic component of the total strain. It follows that all the preceding thermodynamic potentials should employ the elastic strain when applied within the rigid plastic region. Thus, the thermodynamic state (σ, T, ε^e) will only change during elastic loading and unloading, i.e. along paths 501 and 234. Now, from eqs(11.107b) and (11.117a), the reversible work for the closed cycle becomes

$$\oint dw = -\frac{1}{\rho_o}\oint \sigma d\varepsilon^e = \oint dq \tag{11.119}$$

where both σ and ε^e are constants for paths 12 and 45 but are variables for paths 01, 234 and 50. Here the work path integral $\oint dw = 2Y\varepsilon^P$ corresponds to the enclosed, shaded area in Fig. 11.37b, when eq(11.119) reveals that this is also the heat dissipation $\oint dq$.

If the material hardens during the irreversible paths 12 and 45, both σ and ε^e change continuously within the thermal state equation $g(\sigma, \varepsilon^e, T) = 0$. Corresponding to a given state upon each path, ε^e is identified with the elastic component of strain that recovers upon unloading from that state.

The elastic–plastic, isothermal path 0123, in Fig. 11.37c, is represented by two sequential thermodynamic processes between the states existing at 0, 2 and 3, as shown in Fig. 11.38.

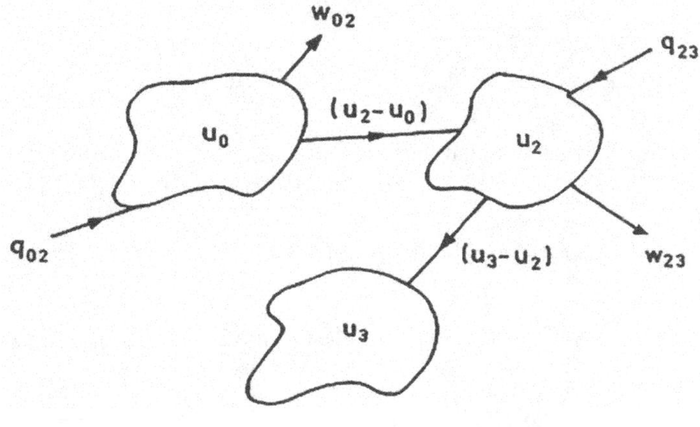

Figure 11.38 Elastic–plastic solid undergoing an open cyclic process

Applying the first law to each process 0–2 and 2–3 allows the corresponding work and heat transfers per unit volume to be identified with the shaded areas, shown in Fig. 11.37c. Thus, for the loading process between states at 0 and 2:

$$q_{02} - w_{02} = (u_2 - u_0) \tag{11.120a}$$

Area 0122′ is the total (negative) work done *on* the solid in reaching state point 2:

$$w_{02} = -\frac{1}{\rho_o}\int \sigma\,\mathrm{d}\varepsilon \tag{11.120b}$$

With an unloading, from state point 2 to 3:

$$q_{23} - w_{23} = (u_3 - u_2) \tag{11.121a}$$

Area 22′3 is the reversible (positive) elastic work done *by* the solid:

$$w_{23} = \frac{1}{\rho_o}\int \sigma\,\mathrm{d}\varepsilon^e \tag{11.121b}$$

For the complete cycle 0123 in Fig. 11.38, adding eqs(11.120a) and (11.121a):

$$\sum q - \sum w = \sum(\Delta u) \tag{11.122a}$$

In full:

$$(q_{02} + q_{23}) - (w_{02} + w_{23}) = (u_2 - u_0) + (u_3 - u_2) = u_3 - u_0 \tag{11.122b}$$

Points A and B, and all other horizontal intercepts at constant stress on the elastic loading and unloading lines, have equal elastic strain. Since the process is isothermal, it follows that σ, ε^e and T are the same for A and B. That is, the elastic equilibrium state A is equivalent to the partially unloaded, plastically deformed state at B. It follows that states 0 and 3 are identical with similar $\sigma\,(=0)$, $\varepsilon^e(=0)$ and T. Consequently, these two states have an identical internal energy ($u_3 = u_0$), so that eqs(11.120b)–(11.122b) lead to

$$\sum q = (w_{02} + w_{23}) = -\frac{1}{\rho_o}\left[\int \sigma\left(\mathrm{d}\varepsilon - \mathrm{d}\varepsilon^e\right)\right] \tag{11.123a}$$

or,

$$\rho_o \sum q = -\int \sigma\,\mathrm{d}\varepsilon^P \tag{11.123b}$$

Equation (11.123b) shows again that the dissipated heat is the work of plastic deformation (area 0123). Equations (11.123a,b) may also be applied to the path A12B in Fig. 11.37c:

$$\rho_o \int \mathrm{d}q = \rho_o\left(\int_{A2}\mathrm{d}q + \int_{2B}\mathrm{d}q\right) = -\int \sigma\,\mathrm{d}\varepsilon^P \tag{11.124a}$$

where from eq(11.108) $\mathrm{d}q = T\mathrm{d}s$ applies to the elastic path 2B. Substituting into eq(11.124a):

$$\rho_o\left(\int_{A2}\mathrm{d}q + \int_{2B}T\mathrm{d}s\right) = -\int \sigma\,\mathrm{d}\varepsilon^P \tag{11.124b}$$

Hence eq(11.124b) gives the heat dissipated by a unit volume of material deformed under the elastic–plastic path A12 as

$$\rho_o \int_{A2} \mathrm{d}q \; = \; - \, \rho_o \int_{2B} T \mathrm{d}s \; - \int \sigma \, \mathrm{d}\varepsilon^P \tag{11.125a}$$

Now let $\Delta\sigma = \sigma_2 - \sigma_A$ be an incremental excursion from the elastic to the plastic region under isothermal conditions. This gives an entropy change

$$\Delta s \; = \; \frac{\partial s}{\partial \sigma} \, \Delta\sigma \; = \; \left(\frac{\alpha}{\rho_o} \right) \Delta\sigma \tag{11.125b}$$

where α is the linear, thermal expansion coefficient. It follows that eqs(11.125a,b) provide the heat transfer for path A12 as

$$- \rho_o \Delta q \; \approx \; (T\alpha \; + \; \Delta\varepsilon^P)\,\Delta\sigma, \qquad \therefore \; - \rho_o \Delta q \; - \; \Delta\varepsilon^P \Delta\sigma \; = \; T\alpha\Delta\sigma \tag{11.125c,d}$$

where from Drucker's postulate, $\Delta\sigma\,\Delta\varepsilon^P > 0$ (see eq(10.8c) in § 10.2.3). Equation (11.125c) reveals the very small amount by which the dissipated heat exceeds the increment in plastic work for this irreversible path. The difference is the amount by which the internal energy changes. Horrie [117] attributed this to changes in the energies of internal lattice distortion and dislocation re-arrangements. He decomposed du, in eq(11.107a), into the sum of a recoverable elastic component, associated with lattice distortion, and a plastic component, associated with a change to the dislocation density. The dislocations themselves act as heat sinks, which release their energy into the lattice when annihilated. This shows that, unlike a perfect gas, it is possible to have a change in internal energy for a solid without a change to its temperature.

In practice, a permanent change to du may never amount to more than a few degrees rise in temperature for metallic conductors placed under small strains applied at low rates. On the other hand, with the inability of insulators to dissipate heat, small strains can result in a significant temperature rise. This may lead to appreciable softening, particularly when polymers are subjected to repeated loading [121]. When straining metals at higher rates, the temperature increases with strain and is greatest in the necked region, where local deformation rates are accentuated. For example, tensile testing of steel at a rate $\approx 10^{-2}\,\mathrm{s}^{-1}$ resulted in a temperature rise of $\approx$40°C in the neck at 40–50 % engineering strain [122]. This shows especially why it becomes important to consider a thermo–mechanical coupling for large deformation in metals. The coupling is achieved [29] within free energy expressions, similar to those in eqs(11.112a) and (11.113a), when stress and strain appear in their various finite forms, given in chapters 1 and 2.

11.10.3 Endochronic Theory

In the small-strain, endochronic theory of multi-axial plasticity, the Helmholtz eq(11.112a) is divided into elastic and plastic components. This provides the following free-energy potential for multiaxial, elastic–plasticity [119, 120]:

$$H(\varepsilon_{ij}^e,\, \varepsilon_{ij}^P,\, T) \; \equiv \; H^e(\varepsilon_{ij}^e,\, T) \; + \; H^P(\varepsilon_{ij}^P) \tag{11.126a}$$

Equations (11.112d,f) have shown how the stress and entropy follow from the elastic contribution only to the free energy (H^e):

$$\sigma_{ij} = \rho_o \frac{\partial H^e}{\partial \varepsilon_{ij}^e}, \qquad s = -\frac{\partial H^e}{\partial T} \tag{11.126b,c}$$

Equation (11.126a) supplies, further, a thermodynamic reference stress for plasticity:

$$\sigma_{ij\,o} = \rho_o \frac{\partial H^P}{\partial \varepsilon_{ij}^P} \tag{11.126d}$$

which lies on, or within, the yield surface expressed by the function $f(\sigma_{ij}, \varepsilon_{ij}^P, T)$. In order to account for irreversibility, arising from plasticity and creep, it has been found necessary to include further internal state variables as arguments in the thermodynamic potential eq(11.103a). Tracing the development of this approach, Biot [123], Coleman and Noll [115] and Leigh [124] considered firstly the thermodynamics of materials with memory. Important to Biot's work on thermo-elasticity and viscoelasticity were: (i) the use of 'hidden variables', in addition to state variables, within the free-energy function and (ii) the development of linear constitutive laws, using Onsager's reciprocity relation [125]. Schapery [126, 127] adopted (i) and (ii) to account for non-linear, viscoelasticity for which the constitutive relations were reduced to linear forms through the use of reduced variables. A similar approach was adopted by Valanis [128] for the development of an endochronic theory of viscoplasticity. Here, all the non-linearity due to strain history was expressed within a single state variable called the *intrinsic time*. Under non-equilibrium inelastic states, internal state variables q_α ($\alpha = 1, 2, \ldots n$) defined the entropy generation. These variables appear, along with the observed variables temperature T and total strain ε_{ij}, in Helmholtz's free-energy eq(11.112b) as

$$H(\varepsilon_{ij}, T, q_\alpha) = u - \int T\,\mathrm{d}s = \frac{1}{\rho_o} \int \sigma_{ij}\,\mathrm{d}\varepsilon_{ij} \tag{11.127a}$$

Note: q_α apply only to inelastic deformation. Otherwise, q_α are set to zero for elastic equilibrium states. Thus, for any elastic process, defined uniquely from the observed state variables, $q_\alpha = 0$. When eq(11.127a) is referred to a unit volume, eqs(11.115b,c) give the following thermic and caloric equations for any equilibrium state:

$$\sigma_{ij} = \rho_o \frac{\partial H}{\partial \varepsilon_{ij}}, \qquad s = -\frac{\partial H}{\partial T} \tag{11.127b,c}$$

It has been seen from eq(11.115a) that strain is a true state variable when it is elastic. It is therefore questionable whether the total strain tensor ε_{ij} should appear in eqs(11.127a,b). Despite this, the endochronic theory adopts total strain as a state variable for its irreversible thermodynamics of both compressible and incompressible viscoplasticity. The role of q_α in eq(11.127a) is to express the rate of entropy generation in the absence of thermodynamic equilibrium during loading (q_α are not to be confused with heat flow q). This condition is expressed, from eqs(11.108) and (11.127a), as

$$T\frac{\mathrm{d}s}{\mathrm{d}t} = -\left(\frac{\partial H}{\partial q_\alpha}\right)\left(\frac{\mathrm{d}q_\alpha}{\mathrm{d}t}\right) \geq 0 \tag{11.128}$$

Valanis showed, from eq(11.128), that because the product of the thermodynamic 'force' $-(\partial H/\partial q_\alpha)$ and the inelastic 'flux' $\mathrm{d}q_\alpha/\mathrm{d}t$ gives positive entropy production, each may be connected, linearly, through Onsager's principle [125]. This gives a system of equations, abbreviated as follows:

$$\frac{\partial H}{\partial q_\alpha} + b_{\alpha\beta}'\left(\frac{\partial q_\beta}{\partial t}\right) = 0 \tag{11.129a}$$

in which the second term represents a summation for β (= 1, 2, 3 ... n). Now, as the matrix $b_{\alpha\beta}'$ in eq(11.129a) appears non-linear in state variables, including q_α, it cannot provide closed solutions for q_α. Schapery [126, 127] was confronted with a differential equation, similar to eq(11.129a), for non-linear, viscoelasticity. He linearised the equation with the introduction of a non-linear, differential parameter dx, appearing in

$$b_{\alpha\beta} = \left(\frac{\mathrm{d}x}{\mathrm{d}t}\right) b_{\alpha\beta}' \tag{11.129b}$$

where x denotes *reduced time* and t is real time. Equation (11.129b) ensures a linear response for $b_{\alpha\beta}$ in the neighbourhood of an equilibrium state, where $q_\alpha = 0$. In the endochronic theory, an *intrinsic time scale* z replaces x in eq(11.129b), so that eq(11.129a) appears in its linearised form

$$\frac{\partial H}{\partial q_\alpha} + b_{\alpha\beta}\left(\frac{\mathrm{d}q_\beta}{\mathrm{d}z}\right) = 0 \tag{11.129c}$$

Equation (11.129c) allows the measure of intrinsic time to be taken from its relationship to a strain hardening function f. This gives

$$\mathrm{d}z = \frac{\mathrm{d}\zeta}{f(\zeta)} \tag{11.130a}$$

where $f(\zeta)$ increases monotonically with ζ during hardening and decreases with ζ during softening. The function $f(\zeta)$ otherwise remains constant in the absence of hardening. Linear and asymptotic expressions were adopted by Wu et al. [129, 130] for isotropic and saturation hardening respectively. These are

$$f(\zeta) = 1 + a \tag{11.130b}$$

$$f(\zeta) = a + (1 - a)\,\mathrm{e}^{-b} \tag{11.130c}$$

where a and b are constants and dζ in eq(11.130a) defines an incremental length of the strain path in a generalised multiaxial form:

$$\mathrm{d}\zeta = \sqrt{p_{ijkl}\,\mathrm{d}\varepsilon_{ij}\,\mathrm{d}\varepsilon_{kl}} \tag{11.131a}$$

The coefficients p_{ijkl} may express an initially anisotropic condition, though they were actually taken to have an isotropic form:

$$p_{ijkl} = k_1\,\delta_{ij}\,\delta_{kl} + k_2\,\delta_{ik}\,\delta_{jl} \tag{11.131b}$$

where k_1 and k_2 are positive constants. Alternatively, p_{ijkl} in eq(11.131a) may be replaced with unity. This provides an isotropic account of strain history in which $d\zeta$ is proportional to an equivalent total strain ($d\zeta = \sqrt{3/2}\, d\bar{\varepsilon}$). Correspondingly, the solution to q_α, found from combining eqs(11.127b) and (11.129c) will supply explicit forms of constitutive laws depending upon the nature of inelastic deformation. For example, a Duhamel heredity integral, eq(11.44b), is applied to isothermal, rate-dependent, deformation [46]:

$$\sigma_{ij} = \delta_{ij} \int_{z'} \lambda(z - z') \left(\frac{\partial \varepsilon_{kk}}{\partial z'} \right) dz' + 2 \int_{z'} \mu(z - z') \left(\frac{\partial \varepsilon_{ij}}{\partial z'} \right) dz' \qquad (11.132a)$$

The separated deviatoric and hydrostatic integral terms employ their respective infinitesimal, total strain tensors with shear and bulk moduli given by the Dirichlet series:

$$\lambda(z) = \lambda_\infty + \sum_r \lambda_r\, e^{-\rho_r z} \qquad (11.132b)$$

$$\mu(z) = \mu_\infty + \sum_r \mu_r\, e^{-\alpha_r z} \qquad (11.132c)$$

in which λ_∞, λ_r, μ_∞, μ_r, ρ_r and α_r are all positive constants, determined experimentally. The initial elastic modulus are restored from eqs(11.132b,c), for $z = 0$.

The endochronic theory of plasticity is fundamentally different from the classical approach where a yield function and associated flow rule are employed. Instead, the stress response is found from integrating the history of total strain. Wu et al. [129, 130] and Valanis et al. [131–134] employed eq(11.132a) to predict viscoplastic behaviour that were not well represented from the isotropic hardening rule of plasticity. Among these were: (i) the effect of a torsional, plastic pre-strain on the lateral versus axial plastic strain path under tension, (ii) the Bauschinger effect, (iii) the strain path resulting from increasing tension under constant torsion and (iv) relaxation and cyclic creep, each with and without prior deformation. Bazant [135] applied eq(11.132a), together with a non-associated plasticity theory, to cyclic loading. The endochronic theory was found to describe the cyclic behaviour of geological materials, including soil, rock and concrete, more consistently than applying the normality rule. These materials all exhibit strain softening, pressure sensitivity and inelastic dilatation, effects which arise from frictional slip and micro-cracking. To provide an account of this damage further state variables were introduced into the endochronic time scale.

Rivlin [1] and Sandler [136] noted the inability of the endochronic theory to represent the hysterisis observed in metals under uniaxial, cyclic straining. In reply, Valanis et al. [137, 138] modified their theory to match the successive yielding behaviour observed from load cycling. It was found necessary, either to present a scalar barrier to each internal variable q_α, appearing in eq(11.127a) [137], or to replace the Helmholtz potential with Gibb's free energy potential in eqs(11.116a,b,c) [138]. Wu and Aboutorabi [139] also employed the Gibb's potential for their application of the endochronic theory to porous and granular materials.

Further modifications to the endochronic theory have been made for its application to non-proportional, cyclic loadings. These include the need to partition strain and to admit a yield surface [140]. Thus, when ε_{ij}^P replaces ε_{ij} in eq(11.132a), the first compressibility term disappears. The kernal in the remaining integral term can embody the concept of an expanding and translating yield surface. To show this, eq(11.132a) is modified to provide the deviatoric stress tensor as follows:

$$\sigma_{ij}' = 2\mu_o\rho_o \int_z \frac{d\varepsilon_{ij}^P}{dz}\, dz + 2\mu_o \int_{z'} \rho_1 (z - z') \frac{d\varepsilon_{ij}^P}{dz'}\, dz' \tag{11.133a}$$

where μ_o is the elastic shear modulus and $dz = d\zeta / f(\zeta)$ becomes a function of the equivalent plastic strain increment tensor. That is, $d\varepsilon_{ij}$ is replaced by $d\varepsilon_{ij}^P$ in eq(11.131a) and p_{ijkl} is made unity to give a scalar measure for $d\zeta$ proportional to the equivalent plastic strain

$$d\zeta = \sqrt{d\varepsilon_{ij}^P\, d\varepsilon_{ij}^P} \tag{11.133b}$$

Wu and Yang [130] applied eq(11.133a) to a non-radial, tension-torsion loading. Their intrinsic time was derived from the hardening function (11.130c), with axial and shear plastic strains appearing from eq(11.131a) as

$$d\zeta = \sqrt{(k_1\, d\varepsilon^P)^2 + (k_2\, d\gamma^P)^2} \tag{11.133c}$$

When the second term in eq(11.133a) is identified with the centre of a yield surface, in the form of the back stress tensor α_{ij}', the following flow rule is found from eq(11.130a):

$$d\varepsilon_{ij}^P = \frac{1}{2\mu_o\rho_o} \left(\sigma_{ij}' - \alpha_{ij}'\right) \frac{d\zeta}{f(\zeta)} \tag{11.134a}$$

Combining eq(11.134a) with eq(11.133b) gives

$$(d\zeta)^2 = d\varepsilon_{ij}^P\, d\varepsilon_{ij}^P = \frac{1}{(2\mu_o\rho_o)^2} (\sigma_{ij}' - \alpha_{ij}')(\sigma_{ij}' - \alpha_{ij}') \left[\frac{d\zeta}{f(\zeta)}\right]^2$$

which leads to

$$\frac{1}{2}(\sigma_{ij}' - \alpha_{ij}')(\sigma_{ij}' - \alpha_{ij}') = 2(\mu_o\rho_o)^2 [f(\zeta)]^2 \tag{11.134b}$$

Referring to eq(11.66), it is seen that eq(11.134b) combines isotropic and kinematic hardening of an initial von Mises yield function through the respective terms $f(\zeta)$ and α_{ij}'. It has been shown [44] that the back stress becomes identical to eq(10.52a) with the kernal $\rho_1(z)$ in eq(11.133a) expressed as an exponential series:

$$\rho_1(z) = \sum_i \rho_i\, e^{-\gamma_i z} \tag{11.134c}$$

Equation (11.134c) allows for the decomposition of the heredity integral into differential forms that are convenient to apply, e.g. when predicting strains under cyclic loading [44]. Alternatively, a power function $\rho_1(z)$ enables integration of eq(11.133a) explicitly under cyclic loading, but the integral must be re-evaluated with each new reversal. *Note*: The modified theory remains independent of real time, with its 'intrinsic' time z having been defined in eq(11.130a). Watanabe and Atluri [140] re-defined the intrinsic time z further to account for the known real-time dependence of viscoplastic strain in metals. The form chosen for this z scale was

$$(\mathrm{d}z)^2 = \frac{(\mathrm{d}\zeta)^2}{f^2(\zeta)} + \frac{(\mathrm{d}t)^2}{g^2} \tag{11.135a}$$

in which $\mathrm{d}\zeta$ refers to all accumulated inelastic strain due to plasticity and creep, t is real time and $g = g(\bar{\sigma}, \zeta)$ is a scalar, having a power law dependence upon the driving stress magnitude. The latter provides an effective stress for viscoplasticity:

$$g(\bar{\sigma}, \zeta) = g\left(|\sigma_{ij}' - \alpha_{ij}'|\right), \quad \bar{\sigma} = \sqrt{\frac{3}{2}(\sigma_{ij}' - \alpha_{ij}')(\sigma_{ij}' - \alpha_{ij}')} \tag{11.135b,c}$$

On combining eqs(11.135a–c), the endochronic theory matched the primary and secondary creep strains that accumulated in 1000 h when 304 stainless steel was subjected to a constant tensile stress at 550°C. *Note*: It appears that this and other internal state variable formulations of viscoplasticity have not been adapted further to the higher-temperature creep regime, where thermal activated recovery acts in opposition to work hardening.

11.11 Concluding Remarks

Viscoplasticity theory applies to higher rates of straining where all inelastic deformation becomes dependent upon time. This theory can account for the effects that strain rate have on the mechanical behaviour, including the stress–strain curve and the plastic strain distribution. Theories examined here are either purely mechanical or thermo-mechanical. The models of deformation behaviour under rapid, infinitesimal straining employ either (i) the concept of an overstress within an anisotropic hardening flow potential, or (ii) an internal variable, known as the intrinsic time. The choice between the isotropic and anisotropic versions of (i) and (ii) depends upon the history associated with the given load path, as with the time-independent theory given in chapter 10. The laws of thermodynamics become more important to rate dependent behaviour at finite strain. They provide the necessary free-energy response potentials for elasticity and can account for plastic irreversibility when used in conjunction with the usual elements of classical plasticity theory, namely the yield function, an associated flow rule and a rule of hardening. However, a free-energy approach does make it possible to dispense with such classical requirements as with the development of a heredity integral, connecting stress to strain, within an endochronic theory. Whilst the irreversibility of metal plasticity and its total strain decomposition remain an issue here, the laws of thermodynamics are certainly required to understand rate-dependent behaviour at finite strain. Therein, it has been shown that the thermodynamics of finite plasticity theory employs a decomposition of the total deformation gradient [29]. For this various energy response functions and their derivatives have been formulated from appropriate definitions of finite stress and strain that were outlined in chapters 1 and 2.

References

1. Rivlin R. S. *Intl. Jl. Solids and Structures*, 1981, **17**, 231.
2. Valanis K. C. *Intl. Jl. Solids and Structures*, 1981, **17**, 249, 267.
3. Webster G. *Phil. Mag.*, 1966, **14**, 775.
4. Mott N. F. and Nabarro F. R. N. *Rpt. on the Strength of Solids*, 1948, *Phys. Soc.*, U.K.
5. Orowan E. *Jl. West Scot. Iron and Steel Inst.*, 1946-7, **54**, 45.
6. Smith C. L. *Proc. Phys. Soc.*, 1948, **61**B, 201.

7. Cottrell A. H. *Jl. Mech. Phys. Solids*, 1952, **1**, 53.
8. Josefsson A. and Lagerberg G. *Jerkontorets Annaler*, 1958, **142**, 570.
9. Olds G. C. E. *Proc. Phys. Soc.*, 1954, **67**B, 832.
10. Wyatt O. H. *Proc. Phys. Soc.*, 1953, **66**B, 459.
11. Rogan J. and Alexander J. M. *Jl. Inst. Metals*, 1966, **94**, 61.
12. Rees D. W. A. *Proc. Roy. Irish Acad.*, 1981, **81**A, 167.
13. Garofalo F. *Fundamentals of Creep and Creep Rupture in Metals*, 1965, MacMillan.
14. Phillips F. *Phil. Mag.*, 1905, **9**, 513.
15. Conrad H. *Acta Met.*, 1958, **6**, 339.
16. Chalmers B. *Jl. Inst. Metals*, 1937, **61**, 103.
17. Bailey R. W. *Proc. I. Mech. E.*, 1935, **131**, 131.
18. Bhattacharya S., Congreve W. K. A. and Thompson F. C. *J. Inst. Met.*, 1952-3, **81**, 83.
19. Conway J. B. *Numerical Methods for Creep and Rupture Analysis*, 1969, Gordon and Breach.
20. Crussard C. *Rev. de Metall.*, 1946, **43**, 310.
21. Mott N. F. *Phil. Mag.*, 1953, **44**, 742.
22. Andrade E. N. da C. *Proc. Roy Soc.*, 1910, **84**A, 1.
23. McVetty P. G. *Mech. Eng.*, 1934, **56**, 149.
24. Lagneborg R. *Int. Met. Reviews*, 1972, **17**, 130.
25. Rees D. W. A. *Biaxial Creep and Plastic Flow of Anisotropic Aluminium*, Ph.D. Thesis, 1976, Kingston University, EthOS, Research Repository CC BY-NC-ND.
26. Hecker S. S. *Constitutive Equations in Viscoplasticity*, (ed. Stricklin J. A.) 1976, ASME, AMD-20, 1.
27. Harding J. *Materials at High Strain Rates*, (ed. Blazynski T. Z.) 1987, 133, Elsevier.
28. Phillips A. and Lee, C. W. *Intl. Jl. Solids and Structures*, 1979, **15**, 715.
29. Rees D. W. A. *Mechanics of Deformable Solids*, 2010, Studium Press, Texas
30. Perzyna P. *Advances in Applied Mechanics*, (ed. Chia-shun Y.) 1971, **11**, 313, Academic Press, New York.
31. Seeger A. *Phil. Mag.*, 1955, **48**, 1194.
32. Lindholm U. S. *Behaviour of Materials Under Dynamic Loading*, 1965, 42, ASME.
33. Miller A. *Jl. Eng. Mat. and Tech.*, Trans ASME, 1976, **98**, 97.
34. Gilman J. J. *Proc. 5th US Nat. Congr. Appl. Mech.*, 1966, 385.
35. Miller A. *Jl. Eng. Mat. and Tech.*, Trans ASME, 1976, **98**, 106.
36. Rees D. W. A. and Dyson B. F. *Intl. Jl. of Plasticity*, 1986, **2**, 1.
37. Krempl E. *Jl. Mech. Phy. Solids*, 1979, **27**, 363.
38. Liu M. C. M and Krempl E. *Jl. Mech. Phy. Solids*, 1979, **27**, 377.
39. Liu M. C. M, Krempl E. and Nairn D. C. *Trans ASME*, 1976, **98**, 322.
40. Krempl E. *Mechanics of Engineering Materials* (eds Desai C. S. and Gallagher R. H.) 1984, Ch. 19, 369, J. Wiley and Sons.
41. Yao D. and Krempl E. *Intl. Jl. of Plasticity*, 1985, **1**, 259.
42. Sutcu M. and Krempl E. *Intl. Jl. of Plasticity*, 1990, **6**, 247.
43. Kempl E. and Lu H. *Mech. of Materials*, 1983, **2**, 183.
44. Chaboche J. L. *Intl. Jl. of Plasticity*, 1989, **5**, 247.
45. Krempl E. *Acta Mech.*, 1975, **22**, 53.
46. Rees D. W. A. *Mechanics of Elastic Solids*, 2019, World Scientific Europe.
47. Bell J. F. *Arch. Rat. Mech. Anal.*, 1981, **75**, 103.
48. Bell J. F. *The Physics of Large Deformation of Crystalline Solids*, 1968, Springer-Verlag, New York.
49. Khan A. S. and Parikh Y. *Eng. Fract. Mech.*, 1985, **21**, 697.
50. Bell J. F. and Khan A. S. *Intl. Jl. Solids and Structures*, 1980, **16**, 683.

51. Mittal R. K. *Jl. Materials*, 1971, **6**, 67.
52. Moon H. *Acta. Mech.*, 1975, **23**, 49.
53. Bell J. F. *Intl. Jl. of Plasticity*, 1985, **1**, 1.
54. Khan A. S. and Parikh Y. *Intl. Jl. of Plasticity*, 1986, **2**, 379.
55. Bodner S. R. and Pardom Y. *Jl. Appl. Mech.*, Trans ASME, 1975, **42**, 385.
56. Baltov A. and Bontcheva N. Colloque Euomech 115: *Comportement Mecanique des Solides Anisotropes*, (ed. Boehler, J. P.) 1982, 243, Martinus Nijhoff.
57. Bontcheva, N. and Baltov, A. Proc. IUTAM/ICM Symp.: *Yielding, Damage and Failure of Anisotropic Solids*, (ed. Boehler J. P.) 1990, 199, MEP Ltd, U.K.
58. Perzyna P. *Quart. Appl. Math.*, 1963, **20**, 321.
59. Perzyna P. *Advances in Mechanics*, 1966, **9**, 243, Academic Press.
60. Campbell J. D., Eleiche A. M. and Tsao M. C. C. *Fundamental Aspects of Structural Alloy Design*, 1977, 545, Plenum Publ Corp, New York.
61. Malvern L. E. *Jl. Appl. Mech.*, 1951, **18**, 203.
62. Khan A. S. and Huang S. *Intl. Jl. Plasticity*, 1989, **5**, 445.
63. Campbell J. D. *Mat. Sci. Engng.*, 1973, **12**, 3.
64. Goel R. P. and Malvern L. E. *Jl. Appl. Mech.*, Trans ASME, 1970, **37**, 1100.
65. Naghdi P. M. and Murch S. A. *Jl. Appl. Mech.*, Trans ASME, 1963, **30**, 321.
66. Cernocky E. P. *Intl. Jl. Non-linear Mech.*, 1982, **17**, 255.
67. Cernocky E. P. *Intl. Jl. Non-linear Mech.*, 1982, **18**, 989.
68. Haupt P. *Ingenieur Archiv.*, 1953, **51**, 1.
69. Schaefer A. O. *The Generation of Isochronous Stress-Strain Curves*, 1972, ASME, New York.
70. Oytana C., Delobelle P. and Mermet A. *Jl. Eng. Mat. and Tech.*, Trans ASME, 1982, **104**, 1.
71. Drucker D. C. *Recent Advances in Applied Mechanics*, (eds Broberg H. et al.) 1967, 209, Almqvist and Wiksell, Stockholm.
72. Brown G. M. *Jl. Mech. Phy. Solids*, 1970, **18**, 367.
73. Rice J. R. *Jl. Appl. Mech.*, Trans ASME, 1970, **37**, 728.
74. Chaboche J. L. *Int. Jl. Plasticity*, 1986, **2**, 149.
75. Chaboche J. L, Dang-van K. and Cordier G. Trans SmiRT 5, Vol L, 1979.
76. Benallal A., Cailletaud G., Chaboche J. L., Marquis D., Nouailhas D. and Rousset M. Proc. EGF3: *Biaxial and Multiaxial Fatigue*, (eds Brown M. W. and Miller K. J.) MEP Ltd., 1989, 107.
77. Malanin N. N. and Khadjinsky, G. M. *Int. Jl. Mech. Sci.*, 1972, **14**, 235.
78. Delobelle P. and Oytana C. *Nuclear Engng. and Design*, 1984, **83**, 333.
79. Mroz Z. and Trampczynski W. A. *Intl. Jl. Solids and Struct.*, 1984, **20**, 467.
80. Lee D. and Zaverl F. *Metallurgica*, 1978, **26**, 1771.
81. Hart E. W. *Jl. Eng. Mat. and Tech.*, Trans ASME, 1976, **98**, 193.
82. Phillips A. and Wu H. C. *Intl. Jl. Solids and Structures*, 1973, **9**, 15.
83. Eisenberg M. A. and Yen C. F. *Jl. Appl. Mech.*, Trans ASME, 1981, **48**, 276.
84. Nouailhas D., Cailletaud G., Policella H., Marquis D., Dufailly J., Lieurade H. P., Ribes A. and Bollinger E. *Eng. Fract. Mech.*, 1985, **21**, 609.
85. Ikegami K. and Niitsu Y. *Eng. Fract. Mech.*, 1985, **21**, 887.
86. Ikegami K. and Niitsu Y. *Intl. Jl. Plasticity*, 1985, **1**, 331.
87. Walker K. P. NASA PWA-5700-50, CR 165533, 1981.
88. Murakami S. and Ohno N. *Intl. Jl. Solids and Structures*, 1982, **18**, 597.
89. Murakami S. and Ohno N. Trans SMiRT 6, Vol L, 1981, L7/3.
90. Nouailhas D., Chaboche J. L., Savalle S. and Cailletaud G. *Intl. Jl. Solids and Structures*, 1985, **1**, 317.
91. Ohno N. *Jl. Appl. Mech.*, Trans ASME, 1982, **49**, 722.

92. Besseling J. F. *A Theory of Plastic Flow for Anisotropic Hardening in Plastic Deformation of Initially Isotropic Materials*, Rpt. S410, 1953, Nat. Aero. Res. Inst., Amsterdam (NLL).
93. Besseling J. F. *Jl. Appl. Mech.*, Trans ASME, 1958, **28**, 529.
94. Ohashi Y., Ohno N. and Kawai M. *Jl. Engng. Mat. and Tech.*, Trans ASME, 1982, **104**, 159.
95. Ohno N., Murakami S. and Kawabata M. EGF5 Proc.: *Yielding, Damage and Failure of Anisotropic Solids*, (ed. Boehler J-P.) 1990, 207, MEP Ltd., UK.
96. Mroz Z., Trampczynski W. and Hayhurst D. *Int. Jl. of Plasticity*, 1988, **4**, 279.
97. Rees D. W. A. Proc.: *Applied Solid Mechanics*, (eds Allison I. M. and Ruiz C.) 1989, **3**, 395, Elsevier.
98. Greenstreet W. L. *Nuc. Engng. and Design*, 1977, **41**, 375.
99. Hartmann G. *Int. Jl. of Plasticity*, 1990, **6**, 189.
100. Corum J. M., Greenstreet W. L., Lin K. C., Pugh C. E. and Swindeman R. W. Oak Ridge Nat., Lab Rpt.: ORNL-5014, 1974.
101. Dyson B. F. and McLean D. *Met. Sci.*, 1977, **2**, 37.
102. Hayhurst D. R., Trampczynski W. A. and Leckie F. A. *Acta Met.*, 1980, **28**, 1171.
103. Hayhurst D. R. *Engineering Approaches to High Temperature Design*, (eds Wilshire B. and Owen D. R. J.) 1983, **2**, 85, Pineridge Press, Swansea.
104. Greenwood G. Proc. 3rd Int. Conf.: *Strength of Metals and Alloys*: *Microstructure and the Design of Alloys*, 1973, 2, 91, Inst. Metals/Iron and Steel Inst.
105. Perzyna P. *Int. Jl. Solids and Structures*, 1986, **22**, 797.
106. Eftis J. and Nemes J. A. *Int. Jl. of Plasticity*, 1991, **7**, 275.
107. Boley B. A. IUTAM Symp.: *Thermoinelasticity*, 1970, Springer-Verlag, Vienna.
108. Chaboche J. L. *Bull. de l'Acad. des Sci.*, Ser. Sci. Tech., 1977, **25**, 33.
109. Kratchovil J. and Dillon O. W. *Jl. Appl. Phys.*, 1970, **41**, 1470.
110. Kratchovil J. *Letters in Appl. Eng. Sci.*, 1978, **16**, 403.
111. Hart E. W. *Acta Met.*, 1967, **15**, 1545.
112. Hart E. W. *Acta Met.*, 1970, **18**, 599.
113. Li J. C. M. and Mukherjee A. K. Symp.: *Rate Processes in Plastic Deformation of Metals*, (ed. Dorn J. E.) 1975, 4, ASM.
114. Argon A. S. (ed.) *Constitutive Equations in Plasticity*, 1975, MIT Press, Cambridge.
115. Coleman B. D. and Noll W. *Arch. Rat. Mech. Anal.*, 1963, **13**, 167.
116. Kestin J. Proc. IUTAM Symp.: *Irreversible Aspects of Continuum Mechanics*, (eds Parkus H. and Sedov L. I.) 1968, 177, Springer-Verlag, Berlin.
117. Horrie Y. *Foundations of Plasticity*, (ed. Sawczuk A.) 1973, 219, Noodrhoff, Leyden.
118. Naghdi P. Proc. 2nd Symp. on Naval Struct. Mech.: *Plasticity*, (eds Lee E. H. and Symonds P. S.) 1960, 121, Pergamon Press.
119. Green A. E. and Naghdi P. *Arch. Rat. Mech. Anal.*, 1965, **18**, 251.
120. Phillips A. IUTAM Symp.: *Thermoinelasticity*, (ed. Boley B. A.) 1970, 241, Springer-Verlag, Vienna.
121. Borduas H. F., Culver L. F. and Burns D. J. *Jl. Strain Anal.*, 1968, **3**, 193.
122. Raghavan K. S. and Wagoner R. H. *Int. Jl. Plasticity*, 1987, **3**, 33.
123. Biot M. A. Proc. 3rd US Nat. Congr. Appl. Mech., 1958, 1, ASME.
124. Leigh D. C. Proc. 5th US Nat. Congr. Appl. Mech., 1966, 481, ASME.
125. Onsager L. *Phys. Rev.*, 1931, **37**, 405; **38**, 2256.
126. Schapery R. A. *Int. Jl. Solids and Struct.*, 1966, **2**, 407.
127. Schapery R. A. Proc. IUTAM Symp.: *Thermoinelasticity*, (ed. Boley B. A.) 1970, 259, Springer-Verlag, Vienna.
128. Valanis K. C. *Arch. Mech. Stows.*, 1971, **23**, 517, 535.
129. Wu H. C. and Yip M. C. *Jl. Eng. Mat. Tech.*, Trans ASME, 1981, **103**, 211.

130. Wu H. C. and Yang R. J. *Int. Jl. Non-Linear Mech.*, 1983, **181**, 395.
131. Valanis K. C. Proc.: *Foundations of Plasticity*, (ed. Sawczuk A.) 1973, 235, Noodrhoff, Leyden.
132. Valanis K. C. *Jl. Appl. Mech.*, Trans ASME, 1974, **41**, 441.
133. Valanis K. C. and Wu H. C. *Jl. Appl. Mech.*, Trans ASME, 1975, **42**, 67.
134. Valanis K. C. and Lee C. F. Proc.: *Mechanics of Engineering Materials*, (eds Desai C. S. and Gallagher R. H.) 1984, Ch. 30, 591, J. Wiley and Sons.
135. Bazant Z. P. *Int. Jl. of Solids and Structures*, 1978, **14**, 691.
136. Sandler I. S. *Jl. Appl. Mech.*, Trans ASME, 1978, **45**, 263.
137. Valanis K. C. Trans SMiRT 4, Vol L (eds Jaeger T. A. and Boley B. A.) 1977, L1/2, North-Holland, Amsterdam.
138. Wu H. C., Valanis K. C. and Yao R. F. *Letters in Appl. Sci. and Engng.*, 1976, **4**, 127.
139. Wu H. C. and Aboutorabi M. R. *Int. Jl. of Plasticity*, 1988, **4**, 163.
140. Watanabe O. and Atluri S. N. *Int. Jl. of Plasticity*, 1986, **2**, 107.

Bibliography

aalco® Aluminium Alloy 5083 SPF, Aalco Metals Ltd, Cobham, Surrey, U. K.
Billington E. W. and Tate A. *The Physics of Deformation and Flow*, 1981, McGraw-Hill.
Conway J. B. *Numerical Methods for Creep and Creep Rupture Analyses*, 1966, Gordon and Breach.
Cristescu N. and Suliciu I. *Viscoplasticity*, 1982, Martinus Nijhoff.
Khan A. S. and Huang S. *Continuum Theory of Plasticity*, 1998, Wiley.
Kraus A. S. and Krausz K. (eds) *Unified Constitutive Laws of Plastic Deformation*, 1996, Academic Press.
Mase G. T. and Mase G. E. *Continuum Mechanics for Engineers*, 1998, CRC Press.
Miller A. K. *Unified Constitutive Relations for Creep and Plasticity*, 1982, Elsevier.
Phan-Thien N. *Understanding Viscoelasticity*, 2002, Springer-Verlag.
Rees D. W. A. *Mechanics of Solids and Structures*, 2nd Ed., 2016, I.C Press, World Scientific.

Exercises

11.1 Determine an explicit function $F' = \mathrm{d}\bar{\sigma}/\mathrm{d}W^P$ from the work-hardening hypothesis when rate-dependent, stress–strain curves conform to eq(11.7b).

11.2 Derive the two forms of the Levy-Mises flow rule $\mathrm{d}\varepsilon_{ij}^P = \mathrm{d}\lambda\, \sigma_{ij}'$ corresponding to the combined hardening work and strain laws in eqs(11.51a,b) when $\mathrm{d}\lambda$ is given by eq(11.53c).

11.3 Show how the total strain rate tensor in eq(11.39a) reduces to eq(11.35a) for a uniaxial tension.

11.4 Using eq(11.68b), compare the axial and shear strain rates under combined tension and torsion when $F(\bar{\sigma}_v) = P\bar{\sigma}_v^{\,q}$, in which the overstress $\bar{\sigma}_v$ is defined by eqs(11.81a) and (11.81b). What is the advantage in using eq(11.81b)?

11.5 Referring to eqs(11.36a–c), predict the uniaxial responses under creep, relaxation and recovery conditions from eq(11.35b) when the hardening and viscosity function are taken from the Hollomon and universal Norton laws respectively:

$$g[\varepsilon] = K\sigma^n \quad \text{and} \quad k[\sigma_v] = A\sigma^4$$

11.6 Using eq(11.45a) and (11.46a), derive the hardening moduli H' and F' for a uniaxial flow curve in which eq(11.7b) applies to a given strain rate.

11.7 Confirm that the strain rate tensor is derived from eq(11.68b) as

$$\varepsilon_{ij}^{P} = \frac{\frac{3}{2}\,p\,(\sigma_{ij}' - \alpha_{ij}')}{\frac{1}{2}\,(\sigma_{ij}' - \alpha_{ij}')^{2}} \quad \text{for} \quad p = \left(\frac{\bar{\sigma}_v}{K}\right)^{n} \exp[\alpha\,(\bar{\sigma}_v/K)^{n+1}]$$

when $F(\bar{\sigma}_v)$ is given by a single-term exponential expression in which $\bar{\sigma}_v$ is defined by eq(11.69a):

$$F(\bar{\sigma}_v) = \frac{K}{\alpha\,(1+n)} \exp[(\bar{\sigma}_v/K)^{n+1}]$$

11.8 Confirm that the strain rate tensor is derived from eq(11.68b) as

$$\varepsilon_{ij}^{P} = \frac{\frac{3}{2}\,p\,(\sigma_{ij}' - \alpha_{ij}')}{\frac{1}{2}\,(\sigma_{ij}' - \alpha_{ij}')^{2}} \quad \text{for} \quad p = \left(\frac{\bar{\sigma}_v}{K_1}\right)^{n_1} + \left(\frac{\bar{\sigma}_v}{K_2}\right)^{n_2}$$

when $F(\bar{\sigma}_v)$ is given by the following, two-term exponential expression in which $\bar{\sigma}_v$ is defined by eq(11.69a):

$$F(\bar{\sigma}_v) = \frac{K_1}{1+n_1}\left(\frac{\bar{\sigma}_v}{K_1}\right)^{1+n_1} + \frac{K_2}{1+n_2}\left(\frac{\bar{\sigma}_v}{K_2}\right)^{1+n_2}$$

11.9 Show that the following major principal creep strain rate is found from a non associated flow rule, eqs(11.77a–c), in which M and D are given by Tresca and von Mises equivalent stress:

$$\dot{\varepsilon}_1 = \frac{K(\bar{\sigma}_T)^n}{\bar{\sigma}_M}\left[\sigma_1 - \frac{1}{2}(\sigma_2 + \sigma_3)\right]$$

where K and n are constants in the Norton law for Tresca equivalence.

11.10 The parallel combination of elastic and anelastic springs, shown in Fig. 11.26b, can represent a surface containing an interior elastic domain, translated from the stress free origin, surrounded by a viscoplastic region. Examine the creep, relaxation and recovery predictions from Fig. 11.26b. *Hint*: Use, as a guide, the solutions given in eqs(11.79a–c) for the Kelvin-Voigt model in Fig. 11.25b, where a dashpot connects with an anelastic spring.

11.11 Confirm that the parallel combination of a dashpot and an anelastic spring, shown Fig. 11.26a, can represent pure viscoplasticity in which hardening is isotropic.

11.12 Show that eq(11.105c) embodies both isotropic and kinematic hardening from expanding the subsequent yield surface function of eq(11.64a).

11.13 Reduce the endochronic theory, eqs(11.132a–c), for an application to combined tension-torsion. Examine the strain path prediction for loading paths: (i) in which the axial and shear stress components, σ and τ, increase proportionately and (ii) where σ is held constant at the yield stress while τ increases.

11.14 Show that when the integral in eq(11.119) is applied to the closed path in Fig. 11.37b, the heat dissipation is given by the shaded area. Confirm this result by applying the first law to the paths 01, 12, 234, 45 and 50 separately.

11.15 Find the heat dissipated when loading a solid bar in tension along an elastic-plastic path 012, in Fig. 11.37c. Take the stress and strain for this path to increase according to the Ramberg-Osgood law $\varepsilon = \sigma/E + (\sigma/E_o)^m$, where E, E_o and m are constants.

11.16 Using the relationships between the four elastic constants: E, G, K and v, i.e. $E = 2G(1 + v) = 3K(1 - 2v)$, convert eq(11.116b), given as a strain derivative of Gibb's potential $\partial[G(T, \sigma_{ij})]/\partial\varepsilon_{ij}$, to one of a stress derivative $\partial[G(T, \sigma_{ij})]/\partial\sigma_{ij}$ where

$$G(T,\sigma_{ij}) = \int du - \frac{1}{\rho_o}\int \sigma_{ij}\,d\varepsilon_{ij}^e - \int T ds = 0$$

Note: The appropriate multiaxial elastic constitutive relation adopts deviatoric and mean stress components as follows:

$$\varepsilon_{ij}^e = \frac{\sigma_{ij}'}{2G} + \frac{\delta_{ij}\,\sigma_m}{3K}$$

$$\sigma_{ij}' = \sigma_{ij} - \delta_{ij}\,\sigma_m \quad \text{for} \quad \sigma_m = \frac{1}{3}\sigma_{kk}$$

in which constant G, the shear modulus, is not to be confused with the energy potential $G(T, \sigma_{ij})$.

11.17 Show how the replacement of the dashpot in Fig. 11.26a with a linear spring, as in Fig. 11.26b, enables a uniaxial kinematic hardening to be reproduced when two of Fig. 11.26b are connected in series. Examine whether Fig. 11.26b alone, having a parallel combination of elastic and anelastic springs, can model viscous plasticity.

11.18 Extend exercise 11.17 in which a model is required for a multi-axial viscoplastic theory giving: (i) a three-dimensional surface containing an interior elastic domain, (ii) a translation in that surface from the stress free origin and (iii) having an exterior domain of viscoplastic deformation.

11.19 Explain why it becomes necessary to modify the fourth order isotropic tensor T_{ijkl} with Poisson's ratio v in eq(11.43a) for use with a multiaxial viscoplastic overstress theory.

11.20 Derive expressions for the nominal strain, strain rate and stress for a tension test conducted at constant velocity v upon a solid circular section bar of original area A_o and length l_o. Convert each nominal measure respectively into true measures of strain, strain rate and stress, showing that no one of these is constant in that test. Present each of the true variations in a plot with a common axis of nominal strain, taking the origin at the intercept on that axis made by the tangent to the true stress plot at the point of tensile instability. State and justify any assumption made.
See solution

CHAPTER 12

ORTHOTROPIC PLASTICITY

12.1 Introduction

Flow curves from different tests upon the same material are not always brought into line by isotropic definitions of equivalent stress and strain. Chapter 9 showed flow data from different tests, plotted in isotropic axes of $\bar{\sigma}$ versus $\bar{\varepsilon}^P$, too widely spread to determine a unique hardening law. These flow curves did not correlate well for material in an initially anisotropic condition, suggesting that re-definitions of $\bar{\sigma}$ and $\bar{\varepsilon}^P$ are necessary. To achieve this the following analysis assumes that orthogonal axes of orthotropy lie within both bar and sheet metal stock. In this way, the anisotropy is defined within three orthogonal planes as they arise from the manufacturing processes, including the production of rolled and drawn sheet, extruded and forged billets. The residual stress resulting from each process may be relieved by post heat treatment but the stock retains an orthotropic character within the grain structure. Klinger and Sachs [1] were among the first to show this for a heat treated 24ST aluminium alloy plate. They found for tensile testpieces, aligned with orthogonal directions, that incremental plastic strain ratios did not conform to the isotropic ratio $d\varepsilon_{22}{}^P/d\varepsilon_{11}{}^P = -\frac{1}{2}$. This behaviour was attributed to the partial retention of the original preferred grain orientation imparted by rolling. Therein lies a crystallographic anisotropy, originating from the highly anisotropic single crystals that comprise the polycrystalline matrix. This is distinct from laminar inclusions, cavities and various arrangements of alloy phases that can produce similar deviations from isotropic plasticity.

12.2 Orthotropic Flow Potential

Hill's quadratic yield function eq(3.27) accounts for directionally dependent yield stresses in orthotropic, metallic materials. To describe the plastic flow behaviour a similar orthotropic function is taken as a plastic potential in the flow rule, eq(3.22). The theory is simplified when: (i) Cartesian stresses align with each material's orthogonal axes and (ii) an isotropic hardening rule (see § 10.3) defines the subsequent yield surface. The latter is expressed in a generalised quadratic form of eq(3.29). Combining this with eq(10.10a) describes an 'isotropic hardening' for an initially anisotropic, incompressible material:

$$\frac{1}{2} H_{ijkl}\sigma_{ij}'\sigma_{kl}' = \chi(\psi) \tag{12.1}$$

For an orthotropic condition and with plastic incompressibility imposed by the stress deviators, it was shown in § 3.4.1 that the fourth order tensor H_{ijkl} is equivalent to the six anisotropy parameters F, G, H, L, M and N in eq(3.27). Equation (12.1) ensures that an initial orthotropy remains unchanged given constant ratios F/G, G/H etc., during hardening. For this to apply, it is sufficient that the argument ψ be identified with an appropriate anisotropic form of equivalent plastic strain. That is, $d\bar{\varepsilon}^P$ may be derived from either the work or strain hardening hypothesis so that the choice between them remains within the argument $\psi = \bar{\varepsilon}^P$ in eq(12.1). Furthermore, because the left-hand side of eq(12.1) identifies with the square of an orthotropic equivalent stress expression, the hardening rule is simplified to $\bar{\sigma} = \bar{\sigma}(\bar{\varepsilon}^P)$. In fact, most equivalent stress–plastic strain flow curve correlations for anisotropic materials [2–4] have been derived from this hardening function.

The flow rule in eq(3.22) provides associated othotropic, incremental constitutive relations. These relations must be used with appropriate definitions of $\bar{\sigma}$ and $d\bar{\varepsilon}^P$ to establish the flow rule's scalar multiplier $d\lambda$. The following theories of othotropic plasticity are similar to those proposed by Jackson, Smith and Lankford [5], Dorn [6], Prager [7] and Fisher [8]. Certain modifications to these quadratic stress potential theories were proposed by Hu [9] and Jones and Gillis [10] to provide for a wider range of anisotropic behaviour.

12.2.1 Equivalent Stress

Firstly, it is necessary to derive expressions the equivalent stress and plastic strain increment for an orthotropic material in terms of its anisotropy parameters H_{ijkl} within eq(12.1). Hill [11] employed the reduced form with six coefficients F, G, H, L, M and N (see eq(3.27)). When the six stress components σ_{ij} align with principal material (orthotropic) axes 1, 2 and 3, the plastic potential becomes

$$2f(\sigma_{ij}) = F(\sigma_{22} - \sigma_{33})^2 + G(\sigma_{33} - \sigma_{11})^2 + H(\sigma_{11} - \sigma_{22})^2 + 2L\sigma_{23}{}^2 + 2M\sigma_{13}{}^2 + 2N\sigma_{12}{}^2 = 1 \quad (12.2)$$

The unity value on the right hand side is preserved during hardening. It follows that F, G, H etc., (with units stress^{-2}) must diminish as the stress components σ_{11}, σ_{22} etc., increase. For uniaxial flow along each axis, eq(12.2) shows

$$(G + H)\,\sigma_1{}^2 = 1 \quad (12.3a)$$

$$(F + H)\,\sigma_2{}^2 = 1 \quad (12.3b)$$

$$(F + G)\,\sigma_3{}^2 = 1 \quad (12.3c)$$

When the flow stresses σ_1, σ_2 and σ_3 are constrained to increase proportionately with hardening along 1, 2 and 3, Hill connected an equivalent stress to function $f(\sigma_{ij})$ in eq(12.2)

$$(F + G + H)\,\bar{\sigma}^2 = 3\,f(\sigma_{ij}) \quad (12.4a)$$

Substituting eq(12.2) into eq(12.4a), gives an equivalent stress expression

$$\bar{\sigma} = \sqrt{\frac{3}{2}\left[\frac{F(\sigma_{22} - \sigma_{33})^2 + G(\sigma_{33} - \sigma_{11})^2 + H(\sigma_{11} - \sigma_{22})^2 + 2L\sigma_{23}^2 + 2M\sigma_{13}^2 + 2N\sigma_{12}^2}{(F + G + H)}\right]} \quad (12.4b)$$

Setting $F = G = H$ and $L = M = N = 3F$ in eq(12.4b) restores a von Mises definition, eq(9.2c).

Finnie [12] defined an alternative form of equivalent stress for creep of orthotropic material, simply by taking F, G, H etc., as constants:

$$2\bar{\sigma}^2 = F(\sigma_{22} - \sigma_{33})^2 + G(\sigma_{33} - \sigma_{11})^2 + H(\sigma_{11} - \sigma_{22})^2 + 6L\sigma_{23}{}^2 + 6M\sigma_{13}{}^2 + 6N\sigma_{12}{}^2 \quad (12.5a)$$

Here a conversion of eq(12.5a) to a von Mises isotropic form is achieved from setting all constants equal to unity, corresponding to $f(\sigma_{ij}) = J_2' = \bar{\sigma}^2/3$. Consequently, $\bar{\sigma}$ in eq(12.5a) identifies with its orthotropic potential:

$$6f(\sigma_{ij}) = 2\bar{\sigma}^2 \tag{12.5b}$$

12.2.2 Equivalent Flow Rule

The flow rule provides plastic strain increments when the plastic potential is that given in either eq(12.4a) or (12.5b). Alternatively, the flow rule, eq(10.4b), can be based upon each definition of $\bar{\sigma}$ given by Hill or Finnie as follows:

$$\mathrm{d}\varepsilon_{ij}^{P} = \mathrm{d}\lambda \frac{\partial f}{\partial \sigma_{ij}} = \mathrm{d}\lambda \frac{\partial f}{\partial \bar{\sigma}} \times \frac{\partial \bar{\sigma}}{\partial \sigma_{ij}} \tag{12.6}$$

Clearly, the relationship between $f(\sigma_{ij})$ and $\bar{\sigma}$ will differ between eqs(12.4a) and (12.5b). From Hill, we substitute eq(12.4a) into eq(12.6):

$$\mathrm{d}\varepsilon_{ij}^{P} = \frac{2}{3}\mathrm{d}\lambda\,(F+G+H)\,\bar{\sigma}\left(\frac{\partial \bar{\sigma}}{\partial \sigma_{ij}}\right) = \left(\frac{\mathrm{d}\lambda}{\bar{\sigma}}\right)\left(\frac{\partial \bar{\sigma}}{\partial \sigma_{ij}}\right) \tag{12.7a}$$

and from Finnie, substitute eq(12.5b) into eq(12.6):

$$\mathrm{d}\varepsilon_{ij}^{P} = \mathrm{d}\lambda\left(\frac{2\bar{\sigma}}{3}\right)\left(\frac{\partial \bar{\sigma}}{\partial \sigma_{ij}}\right) \tag{12.7b}$$

Since function $f(\sigma_{ij})$ is homogenous, of degree $n = 2$ in stress, Euler's theorem states

$$nf = 2f = \sigma_{ij}\left(\frac{\partial f}{\partial \sigma_{ij}}\right) \tag{12.8a}$$

Combining eq(12.8a) with the work hypothesis and the flow rule gives

$$\mathrm{d}W^P = \bar{\sigma}\,\mathrm{d}\bar{\varepsilon}^P = \sigma_{ij}\mathrm{d}\varepsilon_{ij}^P = \mathrm{d}\lambda\,\sigma_{ij}\left(\frac{\partial f}{\partial \sigma_{ij}}\right) = 2f\,\mathrm{d}\lambda \tag{12.8b}$$

when, from eq(12.8b), it follows

$$\mathrm{d}\lambda = \frac{\bar{\sigma}\,\mathrm{d}\bar{\varepsilon}^P}{2f} \tag{12.8c}$$

Correspondingly, substitute into eq(12.8c): (i) $2f = 1$ from eq(12.2) and (ii) $2f = 2\bar{\sigma}^2/3$, from eq(12.5b). Thus, $\mathrm{d}\lambda$ is defined by Hill and Finnie respectively:

$$\mathrm{d}\lambda = \bar{\sigma}\,\mathrm{d}\bar{\varepsilon}^P \quad \text{and} \quad \mathrm{d}\lambda = \frac{3\,\mathrm{d}\bar{\varepsilon}^P}{2\bar{\sigma}} \tag{12.9a,b}$$

These scalars ensure that substituting from eqs(12.9a,b) into eqs(12.7a,b) leads to an identical *equivalent rule of flow*:

$$\mathrm{d}\varepsilon_{ij}^P = \mathrm{d}\bar{\varepsilon}^P\left(\frac{\partial\bar{\sigma}}{\partial\sigma_{ij}}\right) \tag{12.10}$$

12.2.3 Incremental Constitutive Relations

Substituting eq(12.2) into eq(12.6), leads to the following relations between the six independent components of stress and incremental plastic strains

$$\mathrm{d}\varepsilon_{11}{}^P = \mathrm{d}\lambda\,[H(\sigma_{11} - \sigma_{22}) - G(\sigma_{33} - \sigma_{11})] \tag{12.11a}$$

$$\mathrm{d}\varepsilon_{22}{}^P = \mathrm{d}\lambda\,[F(\sigma_{22} - \sigma_{33}) - H(\sigma_{11} - \sigma_{22})] \tag{12.11b}$$

$$\mathrm{d}\varepsilon_{33}{}^P = \mathrm{d}\lambda\,[G(\sigma_{33} - \sigma_{11}) - F(\sigma_{22} - \sigma_{33})] \tag{12.11c}$$

$$\mathrm{d}\gamma_{12}{}^P = 2\mathrm{d}\lambda\, N\,\sigma_{12},\ \mathrm{d}\gamma_{23}{}^P = 2\mathrm{d}\lambda\, L\,\sigma_{23}\ \text{and}\ \mathrm{d}\gamma_{13}{}^P = 2\mathrm{d}\lambda\, M\,\sigma_{13} \tag{12.11d,e,f}$$

Summing eqs(12.11a–c) confirms an incompressible plastic flow (i.e. $\mathrm{d}\varepsilon_{ii}{}^P = 0$). Finnie's potential (12.5b) gives identical engineering shear strains (12.11d–f) but the coefficient $\mathrm{d}\lambda$ in eqs(12.11a–c) is replaced by $\mathrm{d}\lambda/3$. Consequently, one must select $\mathrm{d}\lambda$ appropriately from eqs(12.9a,b) to ensure that the normal strain components from either potential agree.

12.2.4 Equivalent Plastic Strain

To determine Hill's equivalent plastic strain increment $\mathrm{d}\bar{\varepsilon},^P$ we write from eqs(12.9a) and (12.11a–c):

$$F\,\mathrm{d}\varepsilon_{11}{}^P - G\,\mathrm{d}\varepsilon_{22}{}^P = (FG + GH + HF)(\sigma_{11} - \sigma_{22})(\bar{\sigma}\,\mathrm{d}\bar{\varepsilon}^P) \tag{12.12a}$$

$$G\,\mathrm{d}\varepsilon_{22}{}^P - H\,\mathrm{d}\varepsilon_{33}{}^P = (FG + GH + HF)(\sigma_{22} - \sigma_{33})(\bar{\sigma}\,\mathrm{d}\bar{\varepsilon}^P) \tag{12.12b}$$

$$H\,\mathrm{d}\varepsilon_{33}{}^P - F\,\mathrm{d}\varepsilon_{11}{}^P = (FG + GH + HF)(\sigma_{33} - \sigma_{11})(\bar{\sigma}\,\mathrm{d}\bar{\varepsilon}^P) \tag{12.12c}$$

Combining eq(12.4b) with eqs(12.12a–c) gives the equivalent plastic strain increment:

$$\mathrm{d}\bar{\varepsilon}^P = \sqrt{\frac{2(F+G+H)}{3}}\left[F\left(\frac{G\mathrm{d}\varepsilon_{22}^P - H\mathrm{d}\varepsilon_{33}^P}{FG+GH+HF}\right)^2 + G\left(\frac{H\mathrm{d}\varepsilon_{33}^P - F\mathrm{d}\varepsilon_{11}^P}{FG+GH+HF}\right)^2 + H\left(\frac{F\mathrm{d}\varepsilon_{11}^P - G\mathrm{d}\varepsilon_{22}^P}{FG+GH+HF}\right)^2 + \frac{(\mathrm{d}\gamma_{23}^P)^2}{2L} + \frac{(\mathrm{d}\gamma_{13}^P)^2}{2M} + \frac{(\mathrm{d}\gamma_{12}^P)^2}{2N}\right]^{\frac{1}{2}} \tag{12.13}$$

Alternatively, from Finnie's eqs(12.5a) and (12.10):

$$d\bar{\varepsilon}^P = \sqrt{2}\left[F\left(\frac{G\,d\varepsilon_{22}^P - H\,d\varepsilon_{33}^P}{FG + GH + HF}\right)^2 + G\left(\frac{H\,d\varepsilon_{33}^P - F\,d\varepsilon_{11}^P}{FG + GH + HF}\right)^2\right.$$

$$\left. + H\left(\frac{F\,d\varepsilon_{11}^P - G\,d\varepsilon_{22}^P}{FG + GH + HF}\right)^2 + \frac{(d\gamma_{23}^P)^2}{6L} + \frac{(d\gamma_{13}^P)^2}{6M} + \frac{(d\gamma_{12}^P)^2}{6N}\right]^{\frac{1}{2}} \qquad (12.14)$$

Equation (12.14) reduces to an isotropic definition (9.6b) for *constants* of unity. To achieve a similar reduction from eq(12.13) the *coefficients* obey $F = G = H$ and $L = M = N = 3F$.

12.2.5 Anisotropy Parameters

The parameters F, G, H, L, M and N can be expressed either in terms of the yield stresses or the plastic strain increment ratios obtained from tension tests. Identifying σ_{1y}, σ_{2y} and σ_{3y} with the uniaxial yield stresses in principal material directions and σ_{12y}, σ_{23y} and σ_{13y} with the shear yield stresses in each plane, Hill's equivalent stress (12.4b) gives coefficients F, G, H etc. (units of $(\text{MPa})^{-2}$):

$$F = \tfrac{1}{2}\,(1/\sigma_{3y}^{\,2} + 1/\sigma_{2y}^{\,2} - 1/\sigma_{1y}^{\,2})$$

$$G = \tfrac{1}{2}\,(1/\sigma_{1y}^{\,2} + 1/\sigma_{3y}^{\,2} - 1/\sigma_{2y}^{\,2}) \qquad (12.15a\text{–}f)$$

$$H = \tfrac{1}{2}\,(1/\sigma_{2y}^{\,2} + 1/\sigma_{1y}^{\,2} - 1/\sigma_{3y}^{\,2})$$

$$N = 1/(2\sigma_{12y}^{\,2}),\ L = 1/(2\sigma_{23y}^{\,2})\ \text{and}\ M = 1/(2\sigma_{13y}^{\,2})$$

Finnie's equivalent stress (12.5a) gives F, G, H etc. as dimensionless constants:

$$F = (1/\sigma_{3y}^{\,2} + 1/\sigma_{2y}^{\,2} - 1/\sigma_{1y}^{\,2})\,\bar{\sigma}^2$$

$$G = (1/\sigma_{1y}^{\,2} + 1/\sigma_{3y}^{\,2} - 1/\sigma_{2y}^{\,2})\,\bar{\sigma}^2 \qquad (12.16a\text{–}f)$$

$$H = (1/\sigma_{2y}^{\,2} + 1/\sigma_{1y}^{\,2} - 1/\sigma_{3y}^{\,2})\,\bar{\sigma}^2$$

$$N = \bar{\sigma}^2/(3\sigma_{12y}^{\,2}),\ L = \bar{\sigma}^2/(3\sigma_{23y}^{\,2})\ \text{and}\ M = \bar{\sigma}^2/(3\sigma_{13y}^{\,2})$$

where $\bar{\sigma}^2 = (\sigma_{1y}^{\,2} + \sigma_{2y}^{\,2} + \sigma_{3y}^{\,2})$. Absolute values of F, G, H etc., are unnecessary as it is more common to identify the ratios between them with plastic strain increment ratios. For example, let $r = d\varepsilon_w^P/d\varepsilon_t^P$ be the gradient to a plot between width and thickness plastic strains from tensile testing sheet material along its orthotropic axes. The tensile stress lies in the plane of the sheet so that in one test σ_1 is aligned with 1 and in another test σ_2 is aligned with 2. A through-thickness compressive stress σ_3 aligns with direction 3. Equations (12.11a–c) give corresponding r–values

$$r_1 = d\varepsilon_2^{\,P}/\,d\varepsilon_3^{\,P} = H/G \qquad (12.17a)$$

$$r_2 = d\varepsilon_1^{\,P}/\,d\varepsilon_3^{\,P} = H/F \qquad (12.17b)$$

$$r_3 = d\varepsilon_2^{\,P}/\,d\varepsilon_1^{\,P} = F/G \qquad (12.17c)$$

in which $r_3 = r_1/r_2$. The departure from unity in each r–value provides: (i) a sensitive measure of degree of orthotropy and (ii) a useful measure of formability in sheet metals.

12.3 Orthotropic Flow Curves

12.3.1 Work Hardening Hypothesis

Hill's theory, which is based upon the work hardening hypothesis, is employed for the following analysis. When in-plane stresses σ_{11}, σ_{22} and σ_{12} are applied to a rolled sheet material, eq(12.4b) reduces to

$$\bar{\sigma} = \sqrt{\frac{3}{2}}\left[\frac{F\sigma_{22}^2 + G\sigma_{11}^2 + H(\sigma_{11} - \sigma_{22})^2 + 2N\sigma_{12}^2}{F + G + H}\right]^{\frac{1}{2}} \tag{12.18}$$

in which direction 1 is aligned with the rolling direction. Reducing the stress components in eq(12.18) and substituting from eqs(12.17a,b) gives equivalent, uni-axial flow stresses for the rolling and transverse directions respectively:

$$\bar{\sigma} = \sqrt{\frac{3r_2(1 + r_1)}{2(r_1 + r_2 + r_1 r_2)}}\,\sigma_1 \tag{12.19a}$$

$$\bar{\sigma} = \sqrt{\frac{3r_1(1 + r_2)}{2(r_1 + r_2 + r_1 r_2)}}\,\sigma_2 \tag{12.19b}$$

The equivalent stresses are the same when, from eqs(12.19a,b):

$$\sigma_2 = \sqrt{\frac{r_2(1 + r_1)}{r_1(1 + r_2)}}\,\sigma_1 \tag{12.20}$$

Given $d\varepsilon_1^P$ as the plastic strain increment under σ_1, the two lateral strain increments become

$$d\varepsilon_3^P = -\frac{d\varepsilon_1^P}{1 + r_1} \quad \text{and} \quad d\varepsilon_2^P = -\frac{r_1\, d\varepsilon_1^P}{1 + r_1} \tag{12.21a,b}$$

in which eq(12.17a) has been combined with eqs(12.11a–c). Similarly, for a strain increment $d\varepsilon_2^P$ arising from the tensile stress σ_2, the lateral strains become

$$d\varepsilon_1^P = -\frac{r_2\, d\varepsilon_2^P}{1 + r_2} \quad \text{and} \quad d\varepsilon_3^P = -\frac{d\varepsilon_2^P}{1 + r_2} \tag{11.22a,b}$$

Note: $d\varepsilon_1^P + d\varepsilon_2^P + d\varepsilon_3^P = 0$ applies to each test. Substituting eqs(12.21a,b) and (12.22a,b) into eq(12.13) and omitting shear strain terms provides equivalent plastic strain increments for each direction:

$$d\bar{\varepsilon}^P = \sqrt{\frac{2(r_1 + r_2 + r_1 r_2)}{3r_2(1 + r_1)}}\, d\varepsilon_1^P \tag{12.23a}$$

$$d\bar{\varepsilon}^P = \sqrt{\frac{2(r_1 + r_2 + r_1 r_2)}{3r_1(1 + r_2)}}\, d\varepsilon_2^P \tag{12.23b}$$

The equivalent plastic strain increments are the same, when from eqs(12.23a,b):

$$d\varepsilon_2^P = \sqrt{\frac{r_1(1 + r_2)}{r_2(1 + r_1)}}\, d\varepsilon_1^P \tag{12.24}$$

Equations (12.20) and (12.24) enable a transverse flow curve (2–direction) to be predicted from the axial flow curve (1–direction). The validity of the theory may be checked from the experimental measurement of the two flow curves.

It is also possible to predict compressive flow in the thickness (3–direction) for a thin sheet given its tensile flow curve for the 1–direction. Equations (12.4c) and (12.19a) provide the following stress relationships:

$$\bar{\sigma} = \sqrt{\frac{3(r_1 + r_2)}{2(r_1 + r_1 r_2 + r_2)}}\, \sigma_3 \tag{12.25}$$

$$\sigma_3 = \sqrt{\frac{r_2(1 + r_1)}{(r_1 + r_2)}}\, \sigma_1 \tag{12.26}$$

Taking the thickness strain $d\varepsilon_3^P$ to be positive, the lateral strains are found from the incompressibility condition

$$d\varepsilon_1^P = -\frac{r_3\, d\varepsilon_3^P}{1 + r_3} = -\frac{r_2\, d\varepsilon_3^P}{r_1 + r_2} \tag{12.27a}$$

$$d\varepsilon_2^P = -\frac{d\varepsilon_3^P}{1 + r_3} = -\frac{r_1\, d\varepsilon_3^P}{r_1 + r_2} \tag{12.27b}$$

where eq(12.17c) defines r_3. Substituting eqs(12.27a,b) into eq(12.13) gives the equivalent thickness strain

$$d\bar{\varepsilon}^P = \sqrt{\frac{2(r_1 + r_2 + r_1 r_2)}{3(r_1 + r_2)}}\, d\varepsilon_3^P \tag{12.28}$$

Equating (12.23a) and (12.28) relates incremental strains for the 1– and 3–directions

$$d\varepsilon_3^P = \sqrt{\frac{(r_1 + r_2)}{r_2(1 + r_1)}}\, d\varepsilon_1^P \tag{12.29}$$

When it is not practical to compress a thin sheet, the hydraulic bulge test (see § 9.7) will provide the required equivalence between in-plane and thickness strains. To show this, put $\sigma_1 = \sigma_2$ and $\sigma_3 = 0$ in eq(12.4b). This gives a similar equivalent stress to eq(12.25) when σ_1 replaces σ_3. From eq(12.11a–c), the principal strain increments become

$$d\varepsilon_1{}^P = d\lambda G\sigma_1\,,\;\; d\varepsilon_2{}^P = d\lambda F\sigma_1 \;\;\text{and}\;\; d\varepsilon_3{}^P = -\,(G + F)\,d\lambda\sigma_1 \qquad (12.30a\text{–}c)$$

Expressing $d\varepsilon_1{}^P$ and $d\varepsilon_2{}^P$ in terms of r_1, r_2 and $d\varepsilon_3{}^P$ confirms strain increments (12.27a,b) and therefore eq(12.28).

12.3.2 Strain Hardening Hypothesis

Chakrabarty [13] defined an in-plane equivalent stress as

$$\bar{\sigma} = \sqrt{\frac{(G + H)\sigma_{11}^2 - 2H\sigma_{11}\sigma_{22} + (H + F)\sigma_{22}^2 + 2N\sigma_{12}^2}{G + H}} \qquad (12.31)$$

Equation (12.31) gives $\bar{\sigma} = \sigma_{11}$ for uniaxial flow in the 1–direction, which differs from eq(12.18). In taking $f = \bar{\sigma}$, the flow rule (12.10) gave the equivalent strain increment as

$$d\bar{\varepsilon}^P = (G + H)\sqrt{\frac{(d\varepsilon_{11}^P)^2 + (d\varepsilon_{22}^P)^2 + (d\varepsilon_{33}^P)^2 + 2(d\varepsilon_{12}^P)^2}{2(G^2 + GH + H^2)}} \qquad (12.32)$$

The following principal stress and strain relationships, between uniaxial flow in the 1– and 2–directions, are found from eqs(12.31) and (12.32):

$$\sigma_2 = \sqrt{\frac{r_2(1 + r_1)}{r_1(1 + r_2)}}\,\sigma_1 \qquad (12.33)$$

$$d\varepsilon_2^P = \frac{(1 + r_2)}{(1 + r_1)}\sqrt{\frac{1 + r_1 + r_1^2}{1 + r_2 + r_2^2}}\,d\varepsilon_1^P \qquad (12.34)$$

Equations (12.20) and (12.33) agree, but eqs(12.24) and (12.34) show that the work and strain hypotheses connect the two in-plane strains differently.

Thickness compression is assumed identical to in-plane, equi-biaxial tension. Hence the 1– and 3– flow relationships follow from substituting $\sigma_{11} = \sigma_{22} = \sigma_3$ and $\sigma_{12} = 0$ in eq(12.31)

$$\bar{\sigma} = \sigma_1 = \sqrt{\frac{(r_1 + r_2)}{r_2(1 + r_1)}}\,\sigma_3 \qquad (12.35a)$$

$$\sigma_3 = \sqrt{\frac{r_2(1 + r_1)}{(r_1 + r_2)}}\,\sigma_1 \qquad (12.35b)$$

Equation (12.35b) agrees with eq(12.26). The reduction to $d\bar{\varepsilon}^P$ is found from substituting eqs(12.27a,b) into eq(12.32), with $d\bar{\varepsilon}^P = d\varepsilon_1^P$:

$$d\bar{\varepsilon}^P = \frac{(1 + r_1)}{(r_1 + r_2)} \sqrt{\frac{r_1^2 + r_2^2 + r_1 r_2}{1 + r_1 + r_1^2}}\, d\varepsilon_3^P \qquad (11.36a)$$

$$d\varepsilon_3^P = \frac{(r_1 + r_2)}{(1 + r_1)} \sqrt{\frac{1 + r_1 + r_1^2}{r_1^2 + r_2^2 + r_1 r_2}}\, d\varepsilon_1^P \qquad (11.36b)$$

12.3.3 Comparisons with Experiment

In Fig. 12.1, a comparison is made between the longitudinal and transverse flow in as-rolled, 6.35 mm (¼″) thick, copper sheet. The specification C101 (B.S. 2870) refers to an electrolytic, tough-pitch copper of 99.9 % purity.

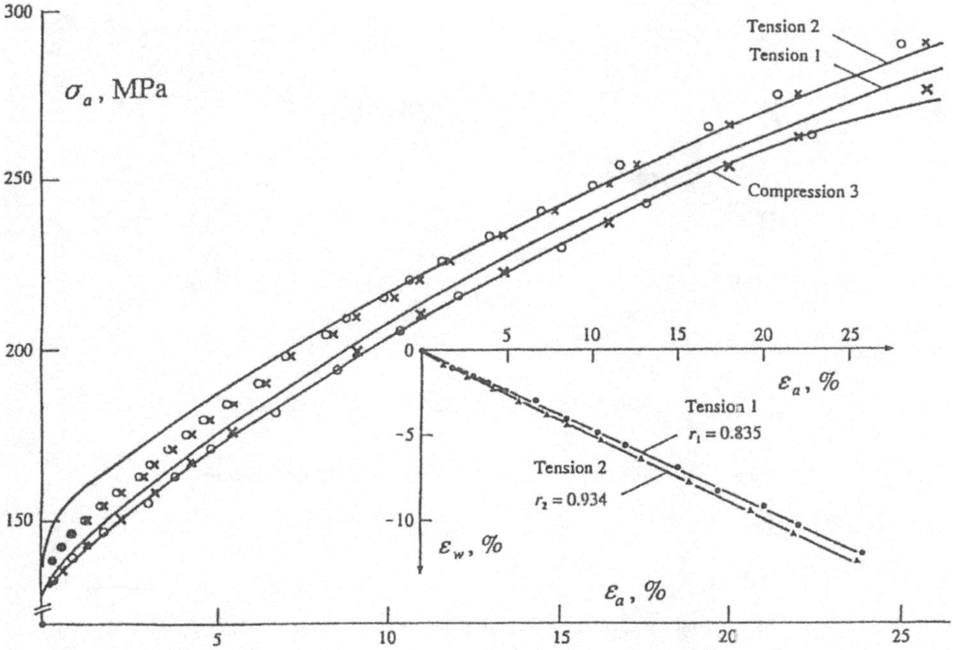

Figure 12.1 Longitudinal, transverse and thickness flow in copper sheet

(**Key**: 1, 2 and 3 are experimental flow curves; ○ work hardening predictions to curves 2 and 3 as calculated from curve 1; × strain hardening predictions to curves 2 and 3 calculated from curve 1. Inset figure shows axial versus width plastic strain paths for 1- and 2- directions, from which r_1 and r_2 are calculated from their gradients.)

Waisted testpieces were machined from the 1– and 2–directions with their parallel gauge dimensions 55 mm long by 12.5 mm wide, according to B.S. 18. Incremental tension tests were performed in a tensile test machine. Permanent length and width changes were made with the load removed. The continuous lines 1 and 2 in Fig. 12.1 are the experimental true stress versus natural axial plastic strain curves for the 1– and 2–directions. Shown inset are plots between the width strain and axial plastic strains. Using $d\varepsilon_{ii}^{P} = 0$ enables the calculation of $r_1 = 0.835$ and $r_2 = 0.934$ from the slope to each path. Linear strain paths show that ratios between F, G and H remain constant in plastic flow. This fact enables the curve for the 2–direction to be derived from the 1–direction curve in the manner outlined above. Work hardening eqs(12.20) and (12.24) give $\sigma_2 = 1.03\sigma_1$ and $\varepsilon_2^{P} = 0.971\varepsilon_1^{P}$. Strain hardening eqs(12.33) and (12.34) give $\sigma_2 = 1.03\sigma_1$ and $\varepsilon_2^{P} = 1.0012\varepsilon_1^{P}$. Applying these relations to curve 1 provides the calculated points from work and strain hardening. Their coincidence with curve 2 is good beyond 10 % strain, where strain hardening appears slightly superior. The thickness flow curve 3 in Fig. 12.1 applies to lubricated uniaxial compression. To predict compressive flow (i.e. σ_3 versus ε_3^{P}) from tensile flow, substitute $r_1 = 0.835$ and $r_2 = 0.934$ in eqs(12.26) and (12.29). Hence $\sigma_3 = 0.984\sigma_1$ and $\varepsilon_3^{P} = 1.016\varepsilon_1^{P}$ for work hardening. Making similar substitutions, into eqs(12.35b) and (12.36b), gives equally good predictions, $\sigma_3 = 0.984\sigma_1$ and $\varepsilon_3^{P} = 1.001\varepsilon_1^{P}$, from strain hardening.

Figure 12.2 confirms that each hypothesis provides an equivalent account of the spread in flow behaviour between the rolling and thickness directions. Points predicted from work and strain hardening are indistinguishable, these all being consistent with the slightly lower flow stresses observed under compression. With r_1 and r_2 values marginally less than unity, the sheet anisotropy was not severe.

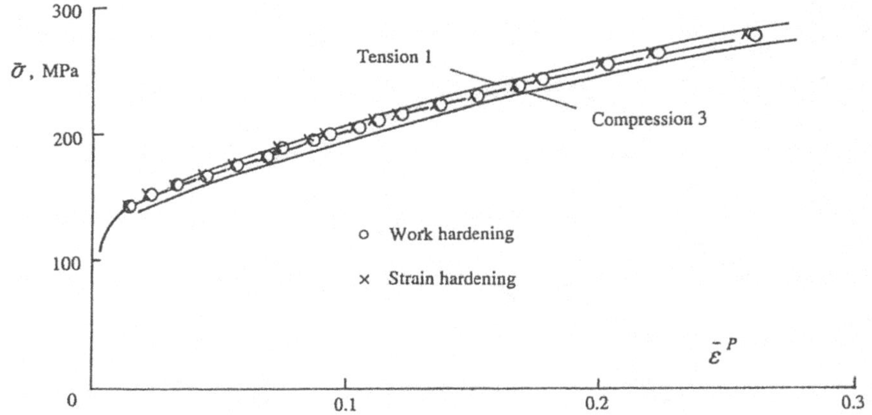

Figure 12.2 Equivalent tensile and compressive flow in copper sheet

12.4 Planar Isotropy

When shear stress is absent along material directions 1 and 2, eq(12.18) reduces to a principal, plane stress form:

$$\tfrac{2}{3}(F + G + H)\bar{\sigma}^2 = (F + H)\sigma_2^2 + (G + H)\sigma_1^2 - 2H\sigma_1\sigma_2 = 1 \qquad (12.37a)$$

Dividing throughout by F, eq(12.37a) beomes

$$r_1(1 + r_2)\sigma_2^2 + r_2(1 + r_1)\sigma_1^2 - 2r_1 r_2\sigma_1\sigma_2 = \text{constant} \qquad (12.37b)$$

Setting $\sigma_2 = 0$ and $\sigma_1 = \sigma_{1y}$ in eq(12.37b) gives $r_2(1+r_1)\sigma_{1y}^2 = \text{constant}$, where σ_{1y} is the yield strength for the 1–direction. Equation (12.37b) becomes

$$r_1(1+r_2)\sigma_2^2 + r_2(1+r_1)\sigma_1^2 - 2r_1 r_2 \sigma_1 \sigma_2 = r_2(1+r_1)\sigma_{1y}^2 \tag{12.37c}$$

where from eqs(12.17a,b), $r_1 = H/G$, $r_2 = H/F$. An r_3–value is defined from a thickness compression test as either $r_3 = G/F$ or $r_3 = F/G$, depending upon how the width and thickness is arranged (see Fig. 12.3). It follows that either one of the following relationships holds:

$$r_1 r_3 = r_2 \quad \text{or} \quad r_2 r_3 = r_1 \tag{12.38a,b}$$

Consider any metal stock production processes, i.e. extrusion, cogging, hot and cold rolling with inter-stage anneals. Assume that a given process imparts a grain structure in one of three directional forms shown in Figs 12.3a–c. Each schematic block arrangement of grains indicates a preferred orientation. These show a planar isotropic character for the material when the preferred direction aligns with each orthogonal axes 1, 2 and 3.

(a) Flow Behaviour Independent of Orientation in the 1–2 Plane (Fig. 12.3a)

When rolled grains are compressed uniformly in the 1–2 plane, $r_1 = r_2 = r_{12}$ and $\sigma_{1y} = \sigma_{2y}$. For uniaxial flow in the 3–direction, e.g. from a through-thickness compression test, both eqs(12.38a,b) show $r_3 = 1$. The yield function (12.37c) becomes

$$\sigma_1^2 + \sigma_2^2 - \frac{2r_{12}}{1+r_{12}}\sigma_1\sigma_2 = \sigma_{1y}^2 \tag{12.39a}$$

$$x^2 + y^2 - \frac{2r_{12}}{1+r_{12}}xy = 1 \tag{12.39b}$$

where $x = \sigma_1/\sigma_{1y}$ and $y = \sigma_2/\sigma_{2y}$. Equation (12.39b) describes a yield criterion for sheet metal with *normal anisotropy*. Here the thickness flow stress differs from the in-plane flow stress, the latter being invariant to its orientation (see also § 12.8). In practice, an r–variation often appears in the 1–2 plane of rolled sheet metal (see § 12.5). If the variation is slight, a single, weighted $\bar{r}$ value replaces r_{12} in eq(12.39b), i.e. $\bar{r} = \frac{1}{4}(r_0 + 2r_{45} + r_{90})$ in which 0°, 45° and 90° are orientations to the roll.

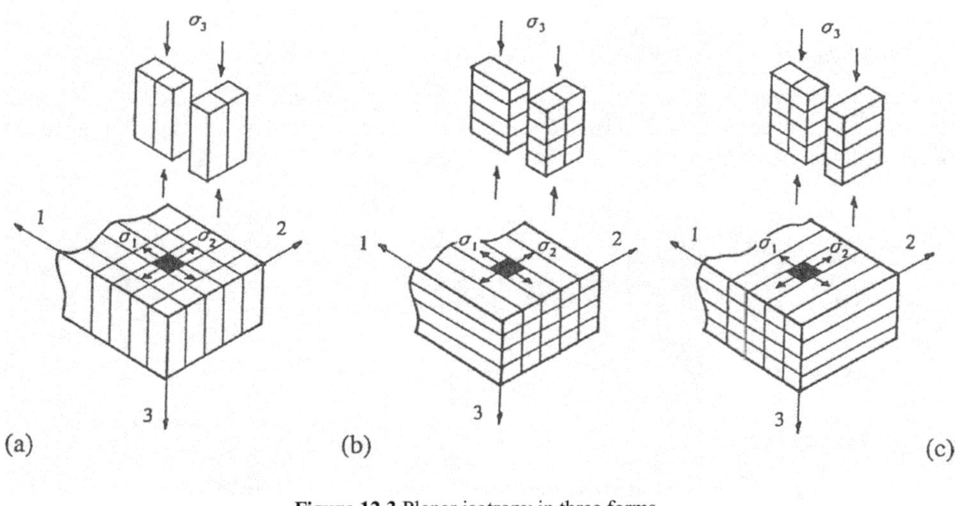

Figure 12.3 Planar isotropy in three forms

(b) Flow Behaviour Independent of Orientation in the 2–3 Plane (Fig. 12.3b)

Grains with a preferential orientation aligned with 1 are taken to be transversely isotropic in their 2–3 plane. This gives $r_2 = r_3 = r_{23}$ and $\sigma_{2y} = \sigma_{3y}$. Equation (12.38a) gives $r_1 = 1$ so that eq(12.37c) becomes

$$\sigma_1^2 + \frac{1 + r_{23}}{2r_{23}}\sigma_2^2 - \sigma_1\sigma_2 = \sigma_{1y}^2 \tag{12.40a}$$

Equation (12.40a) is normalised in the following manner:

$$\left(\frac{\sigma_1}{\sigma_{1y}}\right)^2 + \frac{(1 + r_{23})}{2r_{23}}\left(\frac{\sigma_2}{\sigma_{2y}}\right)^2\left(\frac{\sigma_{2y}}{\sigma_{1y}}\right)^2 - \left(\frac{\sigma_1}{\sigma_{1y}}\right)\left(\frac{\sigma_2}{\sigma_{2y}}\right)\left(\frac{\sigma_{2y}}{\sigma_{1y}}\right) = 1 \tag{12.40b}$$

where, from eq(12.37b):

$$\left(\frac{\sigma_{2y}}{\sigma_{1y}}\right)^2 = \frac{r_2(1 + r_1)}{r_1(1 + r_2)} = \frac{2r_{23}}{1 + r_{23}} \tag{12.40c}$$

Substituting eq(12.40c) into eq(12.40b):

$$x^2 + y^2 - \left(\frac{2r_{23}}{1 + r_{23}}\right)^{\frac{1}{2}} xy = 1 \tag{12.41a}$$

Equation (12.38b) also gives $r_1 = r_{23}^{\,2}$, which leads to an alternative yield criterion:

$$x^2 + y^2 - \left(\frac{4r_{23}^3}{(1 + r_{23})(1 + r_{23}^2)}\right)^{\frac{1}{2}} xy = 1 \tag{12.41b}$$

The two yield criteria (12.41a,b) together define the inner elastic bound of the yield locus.

(c) Flow Behaviour Independent of Orientation in the 1–3 Plane

Grains aligned with 2 are transversely isotropic in the 1–3 plane. This gives $\sigma_{1y} = \sigma_{3y}$ with $r_1 = r_3 = r_{13}$. Equations (12.38a,b) show that $r_2 = r_{13}^{\,2}$ or $r_2 = 1$. Two corresponding forms of normalised yield criteria follow from eq(12.37c):

$$x^2 + y^2 - \left[\frac{4r_{13}^3}{(1 + r_{13})(1 + r_{13}^2)}\right]^{\frac{1}{2}} xy = 1 \tag{12.42a}$$

and

$$x^2 + y^2 - \left(\frac{2r_{13}}{1 + r_{13}}\right)^{\frac{1}{2}} xy = 1 \tag{12.42b}$$

For a given r_{13} value, eqs(12.42a,b) identify an inner elastic bound for the yield locus. In Fig. 12.4, a comparison is made between eqs(12.39b), (12.41a) and (12.42b).

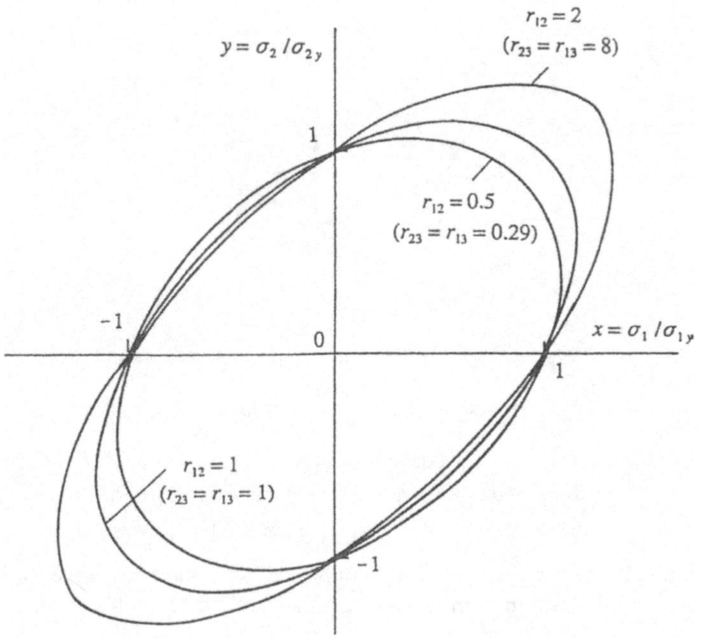

Figure 12.4 Yield loci for transversely isotropic material

Recall that, for the three transversely isotropic planes 1–2, 2–3 and 1–3, identified within Figs 12.3a–c, a single r-value applies to each plane identified, respectively, as r_{12}, r_{23} and r_{13}. Therefore, the loci from eqs(12.41a) and (12.42b) will coincide for similar r_{23} and r_{13} values. However, these loci can only coincide with the locus defined by eq(12.39b) for r–values of unity in a fully isotropic material. Otherwise, it is seen that the r_{12}, r_{23} and r_{23} values in the plane of isotropy must differ if the three loci are to coincide. Within the first quadrant of Fig. 12.4, it is seen that the resistance to yielding increases as r exceeds unity. The yield stresses are more sensitive to changes in r_{12} than to changes in r_{13} or r_{23}.

12.5 Rolled Sheet Metals

Heavily rolled sheet metals are orthotropic in character, having their principal directions aligned with the rolling and transverse directions. Assume that the strength and, therefore, the flow behaviour (i.e. the stress–strain curve) varies with orientation in the plane of the sheet but that it shows transverse isotropy as in Fig. 12.3b. The r–values for orthotropic plasticity are again used to characterise the dependence of flow upon in-plane orientation.

12.5.1 In-Plane r-Variation

The off-axis tensile test will quantify in-plane sheet anisotropy when the precise processing history is not known. Conducting a number of tensile tests upon testpieces machined at different orientations θ to the roll direction (see Fig. 12.5) reveals the variation in r with θ. Here the respective axes 1 and 2 refer to the roll and transverse material directions in the sheet while axes 1′ and 2′ refer to the testpiece axial and width directions.

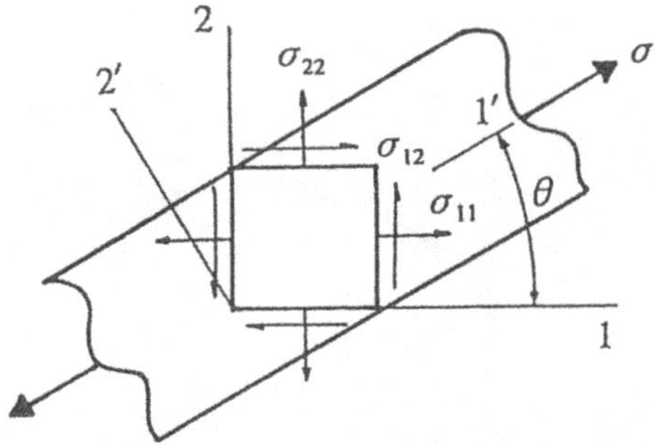

Figure 12.5 Off-axis tensile testpiece in the 1–2 plane

The plane strain transformation eqs(2.9a–c) give off-axis strain increments:

$$d\varepsilon'^{P}_{11} = d\varepsilon^{P}_{11}\cos^2\theta + d\varepsilon^{P}_{22}\sin^2\theta + d\varepsilon^{P}_{12}\sin 2\theta \qquad (12.43a)$$
$$d\varepsilon'^{P}_{22} = d\varepsilon^{P}_{11}\sin^2\theta + d\varepsilon^{P}_{22}\cos^2\theta - d\varepsilon^{P}_{12}\sin 2\theta \qquad (12.43b)$$
$$d\varepsilon'^{P}_{12} = -\tfrac{1}{2}(d\varepsilon^{P}_{11} - d\varepsilon^{P}_{22})\sin 2\theta + d\varepsilon^{P}_{12}\cos 2\theta \qquad (12.43c)$$

The in-plane stress state σ_{11}, σ_{22} and σ_{12} is aligned with material directions 1 and 2, as shown. The material strain increment eqs(12.11a–d) are reduced to

$$d\varepsilon^{P}_{11} = d\lambda\,[(H+G)\sigma_{11} - H\sigma_{22}] \qquad (12.44a)$$
$$d\varepsilon^{P}_{22} = d\lambda\,[(F+H)\sigma_{22} - H\sigma_{11}] \qquad (12.44b)$$
$$d\varepsilon^{P}_{33} = -\,d\lambda\,(G\sigma_{11} + F\sigma_{22}) \qquad (12.44c)$$
$$d\varepsilon^{P}_{12} = d\lambda\,N\sigma_{12} \qquad (12.44d)$$

A tensile stress σ, aligned with the testpiece 1′– axis, transforms to give stress components along the material axis as

$$\sigma_{11} = \sigma\cos^2\theta \qquad (12.45a)$$
$$\sigma_{22} = \sigma\sin^2\theta \qquad (12.45b)$$
$$\sigma_{12} = \tfrac{1}{2}\sigma\sin 2\theta \qquad (12.45c)$$

Substituting eqs(12.45a–c) and (12.44a,b,d) into eqs(12.43a–c) leads to

$$d\varepsilon'^{P}_{11} = d\lambda\sigma[(H+G)\cos^4\theta - 2H\sin^2\theta\cos^2\theta + (F+H)\sin^4\theta + (N/2)\sin^2 2\theta] \qquad (12.46a)$$
$$d\varepsilon'^{P}_{22} = d\lambda\sigma[(2H+F+G)\sin^2\theta\cos^2\theta - H(\sin^4\theta + \cos^4\theta) - (N/2)\sin^2 2\theta] \qquad (12.46b)$$
$$d\varepsilon'^{P}_{12} = d\lambda\sigma\{\tfrac{1}{2}\sin 2\theta\,[(F+H)\sin^2\theta - (H+G)\cos^2\theta] + (H/2)(\sin^2\theta\sin 2\theta - \cos^2\theta\sin 2\theta) + (N/2)\sin 2\theta\cos 2\theta\} \qquad (12.46c)$$
$$d\varepsilon^{P}_{33} = d\varepsilon^{P}_{3} = -\,d\lambda\sigma\,(G\cos^2\theta + F\sin^2\theta) \qquad (12.46d)$$

where ε_3^{P} is the principal thickness strain. Dividing eqs(12.46b,d) gives the r–θ variation

$$r_\theta = \frac{d\varepsilon'^{P}_{22}}{d\varepsilon^{P}_{3}} = -\frac{(2H+F+G)\sin^2\theta\cos^2\theta - H(\sin^4\theta + \cos^4\theta) - (N/2)\sin^2 2\theta}{G\cos^2\theta + F\sin^2\theta} \qquad (12.47a)$$

The coefficients F, G, H and N in eq(12.47a) may be found from measured r_θ–values for testpieces with orientations $\theta = 0°$, 45° and 90°. Equation(12.47a) shows

$$r_1 = r_{0°} = H/G,\quad r_{45°} = (2N - G - F)/[2(G+F)] \text{ and } r_2 = r_{90°} = H/F \qquad (12.47b)$$

conforming to eqs(12.17a,b). Equation (12.47a) is re-written for these three r_θ-values:

$$r_\theta = \frac{2 r_0 r_{90} (\sin^4\theta + \cos^4\theta) - [r_0 r_{90} - r_{45}(r_0 + r_{90})] \sin^2 2\theta}{2(r_0 \sin^2\theta + r_{90} \cos^2\theta)} \tag{12.47c}$$

12.5.2 Comparison with Experiment

Consider the off-axis tensile plasticity of four rolled sheet materials: C101 annealed copper, C638 copper-alloy (3 % Al, 2 % Si, 0.46 % Co), each 6 mm thick, and two plated steels 1 mm and ¼ mm, used in the automotive and canning industries. Tensile testpieces were machined with parallel gauge dimensions: 12 mm wide × 62 mm long. End widths were enlarged to 25 mm for mounting in wedge grips. Axial, width and thickness changes were all measured following unloading in steps from the plastic range. An axial extensometer, with 50 mm gauge length, was mounted upon each testpiece. The greater ductility of the copper and steel allowed micrometer measurement of current width and thickness to be made with good accuracy.

(a) Copper

The linear, natural plastic strain paths found for annealed C101 copper sheet (Fig. 12.6) confirm that the anisotropy parameters remain constant during deformation. The linearity applies to 20 % plastic axial strain, i.e. $\varepsilon'^{P}_{11} = \ln(l/l_o) = 0.2$.

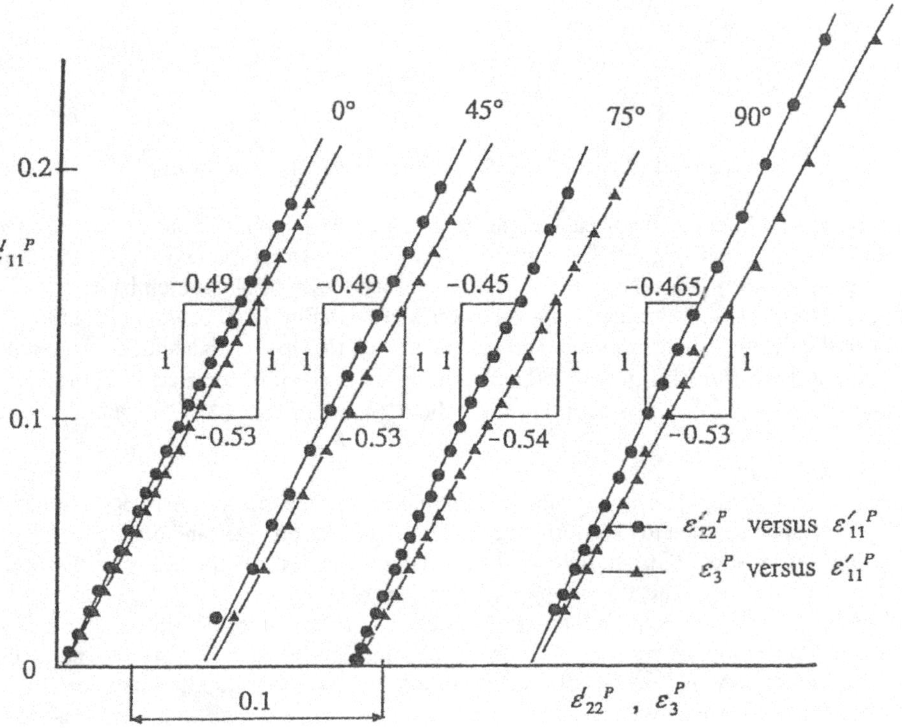

Figure 12.6 Plastic strain paths for off-axis tension in copper sheet

Note: Engineering strain axes would not produce a linear plot. Natural width and thickness plastic strains are similarly defined: $\varepsilon'_{22}{}^P = \ln(w/w_o)$ and $\varepsilon'_3{}^P = \ln(t/t_o)$. Each gradient in Fig. 12.6 conforms to the incompressibility condition

$$\frac{d\varepsilon'^{P}_{22}}{d\varepsilon'^{P}_{11}} + \frac{d\varepsilon^{P}_{3}}{d\varepsilon'^{P}_{11}} = -1 \tag{12.48}$$

Applying eq(12.48) to the 0°, 45° and 90° strain gradients gives $r_0 = 0.924$, $r_{90} = 0.877$ and $r_{45} = 0.924$. These enable eq(12.47c) to predict the observed r–variation for all intermediate orientations, as shown in upper curve of Fig. 12.7.

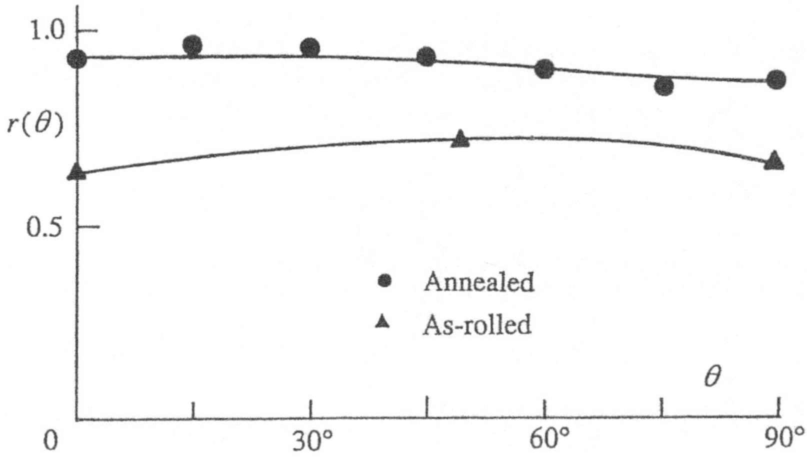

Figure 12.7 Variation in strain ratios with orientation for C101 copper

The agreement between theory and experiment is acceptable but this sheet does not depart greatly from an isotropic condition ($r_o = r_{45} = r_{90} = 1$). A second, as-rolled C101 copper sheet gave lower r–values, $r_0 = 0.631$, $r_{90} = 0.658$ and $r_{45} = 0.705$, which have again been used with eq(12.47c) to predict an r_θ–variation, shown in Fig. 12.7. It would be appropriate to describe the near planar isotropic condition of each material with a single $\bar{r}$–value for a measure of their formability. For example, the $\bar{r}$–value is often correlated with the largest diameter blank that can be drawn through a die without tearing [17].

(b) Copper-Alloy

A thickness reduction gauge was mounted upon the testpiece in addition to post-yield strain gauges, each being used to determine the smaller thickness changes and plastic width strain in copper-alloy sheet (C638). The strain paths so measured for this alloy revealed a far greater degree of in-plane anisotropy with $r_0 = 0.6$, $r_{45} = 1.15$ and $r_{90} = 2.03$. The r–value for the 90° orientation is more than 100 % greater than the isotropic, unity value within a range of 1 % axial plastic strain. The orthotropy and limited ductility is a consequence of a 52 % thickness reduction by cold-rolling in three passes with no inter-stage or after roll subsequent heat treatment. Figure 12.8 again shows acceptable predictions to the r–variation as provided from eq(12.47c). Within its width/thickness strain ratio, r_θ provides a more sensitive measure of in-plane anisotropy than would be given by a width to axial strain ratio.

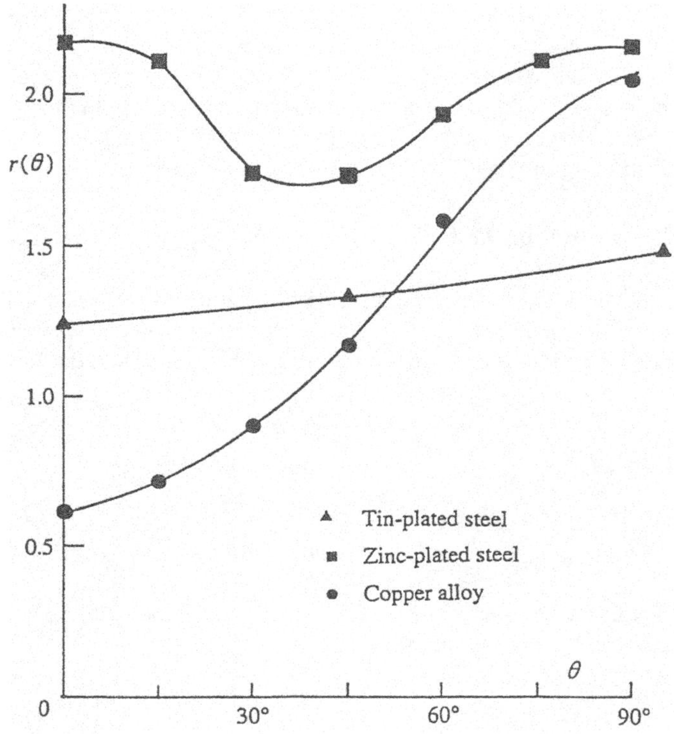

Figure 12.8 Variation in r–value with orientation for copper alloy and coated steel sheets

(c) Plated Steels

In practice, consistently high r–values are desirable if a sheet is to resist thinning when being formed at moderately high strain, particularly under in-plane, biaxial tension [14–16]. Figure 12.8 shows that $r > 1$ for all orientations θ in a 0.85 mm zinc-plated, automotive steel (zintec) and a 0.3 mm can body steel (tin plate). It was seen from eq(12.47c) that a homogenous, quadratic yield function expressed the full r_θ variation in terms of three known r–values for $\theta = 0°$, 45° and 90°. Consequently, this prediction may not fit intermediate r_θ–values precisely. An improved fit is provided by a higher order yield function. For example, a derivation from a homogenous, cubic yield function $f(\sigma_{ij}) = C_{ijklmn}\sigma_{ij}\sigma_{kl}\sigma_{mn}$ gives r_θ as

$$r_\theta = \frac{d\varepsilon'^{P}_{22}}{d\varepsilon^{P}_{3}} = \frac{C_1\cos^6\theta + C_2\sin^6\theta + C_3\cos^4\theta\sin^2\theta + C_4\cos^2\theta\sin^4\theta}{C_5\cos^4\theta + C_6\sin^4\theta + \cos^2\sin^4\theta} \tag{12.49}$$

The six parameters, $C_1, C_2 \ldots C_6$, in eq(12.49) allow a good fit to most observed variations in r_θ [18]. Moreover, a cubic function can describe earing in deep drawn cups [19]. There appears to be no single yield function available to match all manifestations of plastic anisotropy across a range of engineering alloys. The choice of function (see Table 3.3, § 3.4.1) will reflect the match to an r_θ variation with minimum mathematical complexity.

12.5.3 Equivalence Correlation

To establish an equivalence between the individual off-axis flow curves, the parameters F, G, H and N in eqs(12.4b) and (12.18) are required. For example, with the r–values quoted above for the rolled C101 sheet, eqs(12.17) and (12.47b) give $G = 1.583H$, $F = 1.52H$ and $N = 3.74H$. Substituting into eq(12.13) leads to an equivalent plastic strain increment

$$(\mathrm{d}\bar{\varepsilon}^P)^2 = 1.25(\mathrm{d}\varepsilon_{11}{}^P)^2 + 0.993(\mathrm{d}\varepsilon_{11}{}^P)(\mathrm{d}\varepsilon_{22}{}^P) + 1.282(\mathrm{d}\varepsilon_{22}{}^P)^2 + 0.366(\mathrm{d}\varepsilon_{12}{}^P)^2 \quad (12.50)$$

where $\mathrm{d}\varepsilon_{11}{}^P$, $\mathrm{d}\varepsilon_{22}{}^P$ and $\mathrm{d}\varepsilon_{12}{}^P$ refer to increments of plastic strain aligned with the axes of anisotropy (see Fig. 12.5). In the off-axis test, the strain components $\mathrm{d}\varepsilon'_{11}{}^P$ and $\mathrm{d}\varepsilon'_{22}{}^P$ were measured. An inverse strain transformation $\mathbf{E} = \mathbf{M}^{\mathrm{T}}\mathbf{E}'\mathbf{M}$, taken from eq(2.5a), gives

$$\mathrm{d}\varepsilon_{11}{}^P = \mathrm{d}\varepsilon'_{11}{}^P \cos^2\theta + \mathrm{d}\varepsilon'_{22}{}^P \sin^2\theta - 2\mathrm{d}\varepsilon'_{12}{}^P \sin\theta \cos\theta \quad (12.51\mathrm{a})$$

$$\mathrm{d}\varepsilon_{22}{}^P = \mathrm{d}\varepsilon'_{11}{}^P \sin^2\theta + \mathrm{d}\varepsilon'_{22}{}^P \cos^2\theta + 2\mathrm{d}\varepsilon'_{12}{}^P \sin\theta \cos\theta \quad (12.51\mathrm{b})$$

$$\mathrm{d}\varepsilon_{12}{}^P = \mathrm{d}\varepsilon_{21}{}^P = \tfrac{1}{2}(\mathrm{d}\varepsilon'_{11}{}^P - \mathrm{d}\varepsilon'_{22}{}^P)\sin 2\theta + \mathrm{d}\varepsilon'_{12}{}^P(\cos^2\theta - \sin^2\theta) \quad (12.51\mathrm{c})$$

$$\mathrm{d}\gamma_{12}{}^P = (\mathrm{d}\varepsilon_{12}{}^P + \mathrm{d}\varepsilon_{21}{}^P) = (\mathrm{d}\varepsilon'_{11}{}^P - \mathrm{d}\varepsilon'_{22}{}^P)\sin 2\theta + \mathrm{d}\gamma'_{12}{}^P \cos 2\theta \quad (12.51\mathrm{d})$$

Substituting eqs(12.51a,b,d) into eq(12.50) provides the equivalent plastic strain increment in terms of incremental plastic strain ratios:

$$\begin{aligned}(\mathrm{d}\bar{\varepsilon}^P)^2 = 1.25\Bigg\{&\left[\cos^2\theta + \left(\frac{\mathrm{d}\varepsilon'^P_{22}}{\mathrm{d}\varepsilon'^P_{11}}\right)\sin^2\theta - \frac{1}{2}\left(\frac{\mathrm{d}\gamma'^P_{12}}{\mathrm{d}\varepsilon'^P_{11}}\right)\sin 2\theta\right]^2 \\ &+ 0.993\left[\cos^2\theta + \left(\frac{\mathrm{d}\varepsilon'^P_{22}}{\mathrm{d}\varepsilon'^P_{11}}\right)\sin^2\theta - \frac{1}{2}\left(\frac{\mathrm{d}\gamma'^P_{12}}{\mathrm{d}\varepsilon'^P_{11}}\right)\sin 2\theta\right] \\ &\times\left[\sin^2\theta + \left(\frac{\mathrm{d}\varepsilon'^P_{22}}{\mathrm{d}\varepsilon'^P_{11}}\right)\cos^2\theta + \frac{1}{2}\left(\frac{\mathrm{d}\gamma'^P_{12}}{\mathrm{d}\varepsilon'^P_{11}}\right)\sin 2\theta\right] \\ &+ 1.282\left[\sin^2\theta + \left(\frac{\mathrm{d}\varepsilon'^P_{22}}{\mathrm{d}\varepsilon'^P_{11}}\right)\cos^2\theta + \frac{1}{2}\left(\frac{\mathrm{d}\gamma'^P_{12}}{\mathrm{d}\varepsilon'^P_{11}}\right)\sin 2\theta\right]^2 \\ &+ 0.366\left[1 - \left(\frac{\mathrm{d}\varepsilon'^P_{22}}{\mathrm{d}\varepsilon'^P_{11}}\right)\sin 2\theta + \left(\frac{\mathrm{d}\gamma'^P_{12}}{\mathrm{d}\varepsilon'^P_{11}}\right)\cos 2\theta\right]^2\Bigg\}(\mathrm{d}\varepsilon'^P_{11})^2\end{aligned} \quad (12.52)$$

Equation (12.52) accepts either the measured ratios $\mathrm{d}\varepsilon'_{22}{}^P/\mathrm{d}\varepsilon'_{11}{}^P$ and $\mathrm{d}\varepsilon'_{12}{}^P/\mathrm{d}\varepsilon'_{11}{}^P$ for each orientation or theoretical ratios derived from eqs(12.46a–d). Since no attempt was made to measure the small order of shear strain $\mathrm{d}\gamma'_{12}{}^P$, in the off-axis tests, it is necessary to employ the theoretical expression eq(12.46c) with eq(12.52). For example, $\theta = 45°$ shows a strain ratio $\mathrm{d}\gamma'_{12}{}^P/\mathrm{d}\varepsilon'_{11}{}^P = -0.0237$. Substituting this and $\mathrm{d}\varepsilon'_{22}{}^P/\mathrm{d}\varepsilon'_{11}{}^P = -0.414$ into eq(12.52) gives $\mathrm{d}\bar{\varepsilon}^P = 1.0232\,\mathrm{d}\varepsilon'_{11}{}^P$. Hazlett, Robinson and Dorn [20] demonstrated the appearance of shear strain in off-axis tension of an anisotropic magnesium-aluminium alloy (AZ31X).

Transverse lines, marked on the surface for all but the 0° and 90° testpieces, were rotated by applying perpendicular tension. *Note*: Engineering shear strain $d\gamma_{1'2'}^P$ is the sum of the rotations of transverse and longitudinal fibres, but gripping constrains rotation of the latter. The equivalent stress corresponding to eq(12.52) is found from eq(12.18):

$$\bar{\sigma}^2 = 0.944\,\sigma_{11}^2 - 0.731\,\sigma_{11}\sigma_{22} + 0.922\,\sigma_{22}^2 + 2.735\,\sigma_{12}^2 \qquad (12.53a)$$

Substituting eqs(12.45a–c) into eq(12.53a):

$$\bar{\sigma} = \surd[0.944\cos^4\theta - 0.731\cos^2\theta\,\sin^2\theta + 0.922\,\sin^4\theta + 0.684\,\sin^2 2\theta] \times \sigma \quad (12.53b)$$

where σ is the stress aligned with each orientation θ. For example, with $\theta = 45°$, eq(12.53b) gives $\bar{\sigma} = 0.982\sigma$. Figure 12.9 gives the plot between eqs(12.52) and (12.53b).

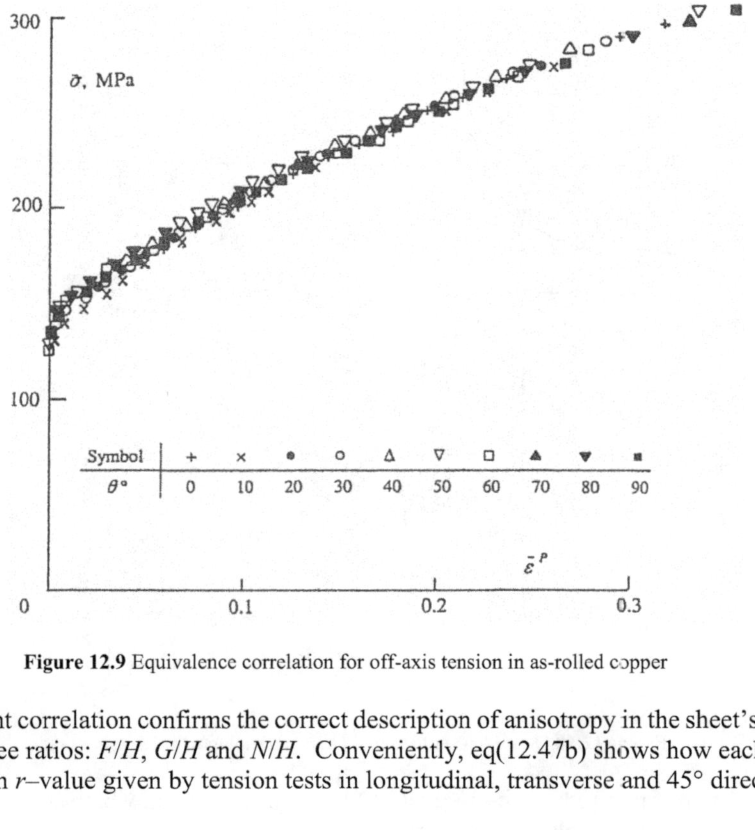

Figure 12.9 Equivalence correlation for off-axis tension in as-rolled copper

The excellent correlation confirms the correct description of anisotropy in the sheet's plane from the three ratios: *F*/*H*, *G*/*H* and *N*/*H*. Conveniently, eq(12.47b) shows how each ratio equates to an *r*–value given by tension tests in longitudinal, transverse and 45° directions.

12.6 Extruded Tubes

To account for anisotropy in a tube let direction 1 be aligned with the tube's extruded direction. The principal material directions 1, 2 and 3 align with the tube's polar co-ordinates: the axial, circumferential and radial directions, respectively. Consider the plasticity of a thin-walled tube subjected to various combined loadings that produce plane stress states aligned with the coincident co-ordinate and material directions.

12.6.1 Biaxial Tension

The combination of axial force and internal pressure gives axial and hoop (i.e. principal) stresses, σ_1 and σ_2 respectively, in a ratio $R = \sigma_1/\sigma_2$. Provided the wall thickness is less than one tenth of the tube's mean diameter then radial stress σ_3 in the wall is judged small enough to be ignored.

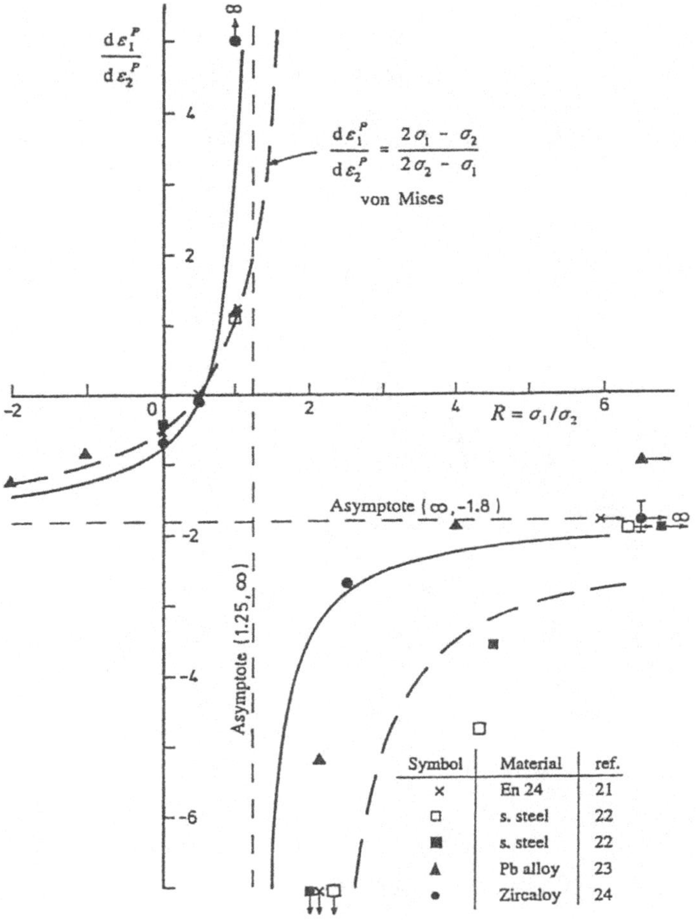

Figure 12.10 Dependence of strain increment ratio upon biaxial stress ratio

Making appropriate reductions to eqs(12.11a,b) gives the ratio between the axial and circumferential (hoop) plastic strain increments:

$$\frac{d\varepsilon_1^P}{d\varepsilon_2^P} = \frac{H(R-1) + GR}{F - H(R-1)} = \frac{(H/G)(R-1) + R}{(F/H)(H/G) - (H/G)(R-1)}$$

$$= \frac{r_1(R-1) + R}{(r_1/r_2) - r_1(R-1)} = \frac{r_1 r_2(R-1) + R r_2}{r_1 - r_1 r_2(R-1)} \qquad (12.54)$$

where r_1 and r_2 in eqs(12.17a,b) are found from separate axial and circumferential tension tests. Figure 12.10 shows that eq(12.54) accounts for the manner in which the strain increment ratio depends upon the stress ratio for five materials [21–24]. With the exception of En 24 steel, which conforms to a von Mises condition, it can be seen elsewhere how materials with r–values different from unity in eq(12.54) depart from the said condition. For example, the asymptotes given for a zirconium alloy (Zircaloy) correspond approximately to $r_1 = 1.3$ and $r_2 = 0.7$.

12.6.2 Tension-Torsion

(a) Strain Ratios

Combining an axial force with torsion produces two independent stress components: σ_{11} and σ_{12}, respectively. Dividing eqs(12.11a,b,d) gives the following ratios between increments of shear, axial, diametral and thickness plastic strains, $d\gamma_{12}^{P}$, $d\varepsilon_{11}^{P}$, $d\varepsilon_{22}^{P}$ and $d\varepsilon_{33}^{P}$, respectively:

$$\frac{d\gamma_{12}^{P}}{d\varepsilon_{11}^{P}} = \frac{2d\varepsilon_{12}^{P}}{d\varepsilon_{11}^{P}} = \frac{2N\sigma_{12}}{(H + G)\sigma_{11}} = \frac{(r_1 + r_2)(1 + 2r_{45})\sigma_{12}}{r_2(1 + r_1)\sigma_{11}} \tag{12.55a}$$

$$\frac{d\varepsilon_{22}^{P}}{d\varepsilon_{11}^{P}} = -\frac{H}{(H + G)} = -\frac{r_1}{(1 + r_1)} \tag{12.55b}$$

$$\frac{d\varepsilon_{33}^{P}}{d\varepsilon_{11}^{P}} = -\frac{G}{(H + G)} = -\frac{1}{(1 + r_1)} \tag{12.55c}$$

in which eq(12.47b) has been applied. Either of eqs(12.55b,c) provides r_1 when diametral or thickness changes are measured along with twist and extension. A circumferential tension provides r_2 while r_{45} is calculated from eq(12.55a) for a given stress ratio $R = \sigma_{12}/\sigma_{11}$. Alternatively, r–values can be found from miniature tensile testpieces machined from the tube wall. Equation (12.55a) predicts a strain increment ratio linearly dependent upon R. Figure 12.11 presents evidence of this behaviour for 5 materials [3, 21, 25, 26], previously considered with isotropic potentials (see Fig. 10.6). When anisotropy is present, eq(12.55a) shows how r–values explain a gradient different from 3 (by Levy-Mises). Test temperatures 300°C and 400°C apply to the 0.17 % carbon steels (m.s.). One steel conforms to a von Mises isotropic condition at the higher temperature. For the other materials at ambient temperature, it is seen how r–values different from unity explains their departure from a von Mises condition.

(b) Equivalence Correlation

The work hypothesis gives an equivalent stress in eq(12.18) as

$$\frac{\bar{\sigma}}{\sigma_{11}} = \frac{3}{2}\sqrt{\frac{r_2(1 + r_1) + (r_1 + r_2)(1 + 2r_{45})R^2}{r_1 + r_1 r_2 + r_2}} \tag{12.56a}$$

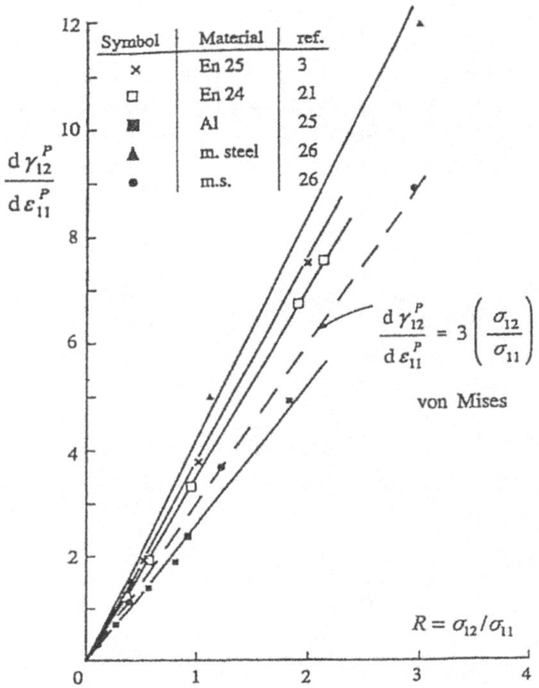

Figure 12.11 Dependence of strain increment ratio upon combined tension-torsion stress ratio

Setting $g = d\gamma_{12}^P / d\varepsilon_{11}^P$ in eq(12.13), the equivalent plastic strain appears as

$$\frac{d\bar{\varepsilon}^P}{d\varepsilon_{11}^P} = \frac{2}{3}\sqrt{\frac{(r_1 + r_1 r_2 + r_2)[1 + r_2(1 + r_1)g^2]}{r_2(1 + r_1)(1 + 2r_{45})(r_1 + r_2)}} \tag{12.56b}$$

In the hypothesis of strain hardening, eqs(12.31) and (12.32) are reduced correspondingly:

$$\frac{\bar{\sigma}}{\sigma_{11}} = \sqrt{1 + \frac{(r_1 + r_2)(1 + 2r_{45})g^2}{r_2(1 + r_1)}} \tag{12.57a}$$

$$\frac{d\bar{\varepsilon}^P}{d\varepsilon_{11}^P} = \sqrt{1 + \frac{(1 + r_1)^2 g^2}{4(1 + r_1 + r_1^2)}} \tag{12.57b}$$

where eqs(12.55b,c) have been used. To compare correlations of $\bar{\sigma}$ versus $\int d\bar{\varepsilon}^P$ from each hypothesis, four radial loading tests with $R = \sigma_{12}/\sigma_{11} = 0.39$, 0.92, 1.77 and 7.1, were conducted on tubes machined from as-received, 99.8 % pure aluminium bar. The testpiece diameters were 25.4 mm i.d. and 28.4 mm o.d. Of the 75 mm parallel length, 50 mm defined a gauge length for attachment of an extensometer. The latter housed displacement transducers for the measurement of extension and angular twist simultaneously. Reductions to the outer diameter were measured with a micrometer. Force and torque were applied to the testpiece through threads and a square register machined into each of its enlarged ends [26]. Elastic strains were removed from total strains using elastic constants determined from initial elastic incremental loading, i.e. $E = 68$ GPa, $G = 25.5$ GPa and $\nu = E/(2G) - 1 = 0.33$. The resulting inelastic strains enabled the strain paths to be plotted for plasticity under incremental loading and, thereafter, creep under a constant load, as shown in Fig. 12.12.

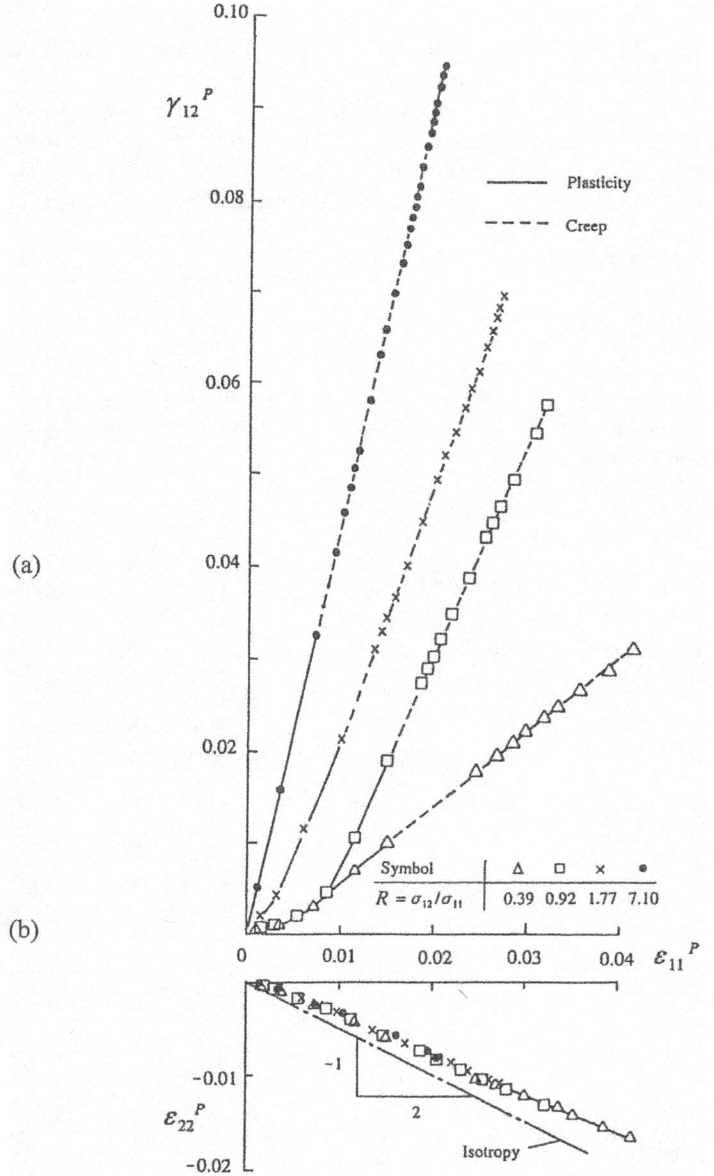

Figure 12.12 Anisotropic plastic strain paths under combined tension-torsion

The axes in each figure correspond to the removal of elastic strain from total natural strain:

$$\gamma_{12}^{P} = \gamma_{12} - \frac{\sigma_{12}}{G}$$

$$\varepsilon_{11}^{P} = \varepsilon_{11} - \frac{\sigma_{11}}{E}$$

$$\varepsilon_{22}^{P} = \varepsilon_{22} - \frac{\nu\sigma_{11}}{E}$$

Note: Paths become linear with increasing plastic strain. The non-linearity, which is a characteristic of pre-strain history, is considered in the following section § 12.7.

The ratios between the incremental strains in eqs(12.55a,b) are identified with the slopes of linear paths in Figs 12.12a,b. The *r*–values were also found from miniature tensile testpieces, machined from the bar in longitudinal and transverse directions. Equations (12.56a,b) give the equivalent stress and equivalent plastic strain from the work hardening hypothesis

$$\bar{\sigma} = 0.907\,(1 + 2.05R^2)^{1/2}\,\sigma_{11} \tag{12.58a}$$

$$\bar{\varepsilon}^{P} = 1.103\,[\,1 + 0.488\,g^2\,]^{1/2}\,\varepsilon_{11}^{\,P} \tag{12.58b}$$

in which $R = \sigma_{12}/\sigma_{11}$ is the stress ratio and $g = d\gamma_{12}^{\,P}/d\varepsilon_{11}^{\,P}$ is the plastic strain increment ratio, i.e. the constant slope to a linear plastic strain path. For the strain-hardening hypothesis, an equivalence is expressed through eqs(12.57a,b)

$$\bar{\sigma} = (1 + 2.05R^2)^{1/2}\,\sigma_{11} \tag{12.59a}$$

$$\bar{\varepsilon}^{P} = [\,1 + 0.33g^2\,]^{1/2}\,\varepsilon_{11}^{\,P} \tag{12.59b}$$

Substituting the observed values of *g* for each *R* enables equivalence plots in Figs. 12.13a,b to be constructed for each hypothesis.

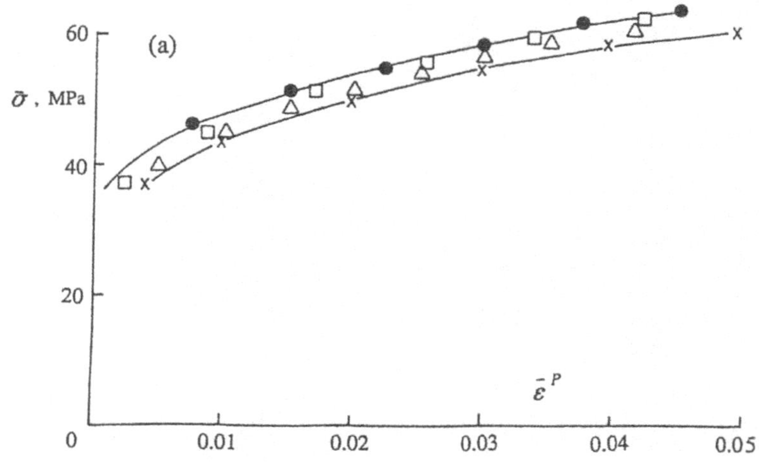

Figure 12.13a Equivalence based upon work hardening

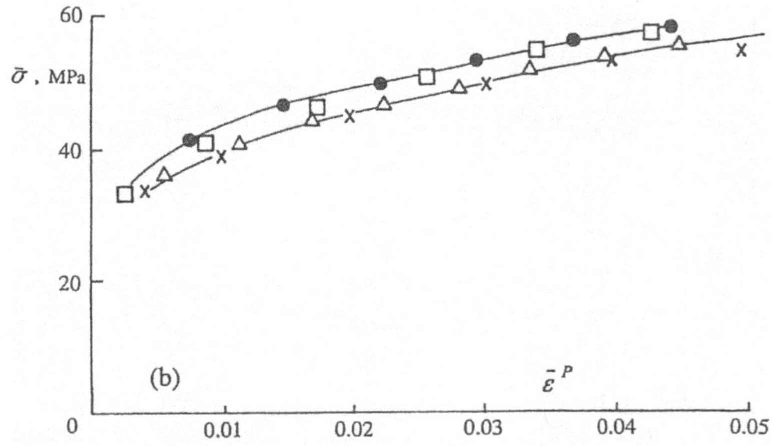

Figure 12.13b Equivalence based upon strain hardening

Clearly, these experiments are unable to reveal which is the better hypothesis. Narrow scatter bands result from the observed strain paths and confirm that orthotropic plasticity theory may be employed equally well with either hypothesis.

12.7 Non-Linear Strain Paths

The foregoing equivalence correlations assume linear plastic strain paths. Consequently, the constant ratios between the anisotropy parameters *F*, *G*, *H*, *L*, *M* and *N* describe an unchanging residual anisotropy. For this condition, Hill [27] showed that the parameters decrease as yield stresses, in eq(12.2), increase in strict proportion to the amount of cold work. In Finnie's alternative formulation (12.5a) *F*, *G*, *H*, etc., are dimensionless constants. The literature reports linear plastic strain paths for orthotropic sheets of aluminium [28], aluminium alloy [1, 20, 25] and mild steel [20, 28]. In contrast, non linear plastic strain paths were found for alloys of magnesium [20, 29], titanium [30–32] and aluminium 3S [33]. Here, the gradient, i.e. the incremental plastic strain ratio, changes with increasing plastic strain. An example of non-linear behaviour of aluminium alloy strip (HE 30TF) under tension is given in Figs 12.14a,b. The stress–strain plots in Fig. 12.14a apply to a testpiece with gauge dimensions: 12.7 mm wide × 50.8 mm long, machined from 9.5 mm thick, rolled sheet. Testpiece axes were aligned with the roll direction, 1, and the transverse direction, 2. Total strains were measured with bonded, post-yield, strain gauges which also provided elastic constants: $E = 78.3$ GPa and $\nu = 0.30$ for the 1–direction; $E = 77.4$ GPa and $\nu = 0.336$ for the 2–direction. Appropriate E and ν were used in Fig. 12.14b to remove elastic strains from the total axial and lateral strains:

$$\varepsilon'^{P}_{11} = \varepsilon'_{11} - \sigma / E$$
$$\varepsilon'^{P}_{22} = \varepsilon'_{22} - \nu\sigma / E$$

where 1' and 2' denote testpiece length and width co-ordinates, i.e. in Fig. 12.5, 1′ coincides with 1 and 2′ coincides with 2 in each testpiece. The quadratic potential supplies an incremental strain ratio $d\varepsilon'^{P}_{22} / d\varepsilon'^{P}_{11}$ from eqs(12.46a,b) with $\theta = 0°$ and $90°$. Using the normality rule and preservation of volume, it follows that this potential can match a non-linear plastic strain path only when *F*, *G*, *H*, *L*, *M* and *N* become functions of plastic strain.

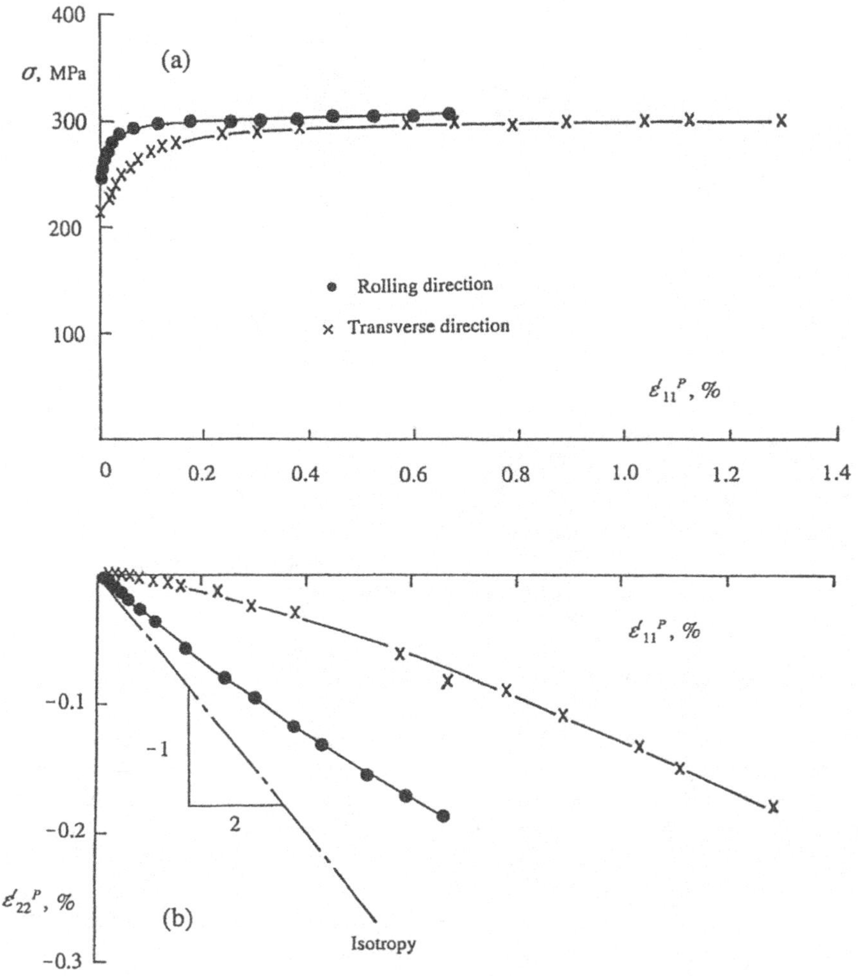

Figure 12.14 Non-linear tensile flow behaviour of aluminium alloy

Alternatively [34], the initial, constant values of the anisotropy parameters are retained within a combined linear-quadratic theory in eq(3.30). A reduction to plane stress gives

$$f(\sigma_{ij}') = H_{ij}\sigma_{ij}' + \tfrac{1}{2}H_{ijkl}\,\sigma_{ij}'\sigma_{kl}' \tag{12.60a}$$

$$= L_1\sigma_{11} + L_2\sigma_{22} + L_3\sigma_{12} + Q_1\sigma_{11}^{\,2} + Q_2\sigma_{22}^{\,2} + Q_3\sigma_{11}\sigma_{22} + Q_4\sigma_{12}^{\,2} + Q_5\sigma_{11}\sigma_{12} + Q_6\sigma_{22}\sigma_{12} \tag{12.60b}$$

when, from eq(10.4b), the incremental stress–strain relations become

$$d\varepsilon_{11}^{P} = d\lambda\,(\partial f/\partial\sigma_{11}) = d\lambda\,(L_1 + 2Q_1\sigma_{11} + Q_3\sigma_{22} + Q_5\sigma_{12}) \tag{12.61a}$$

$$d\varepsilon_{22}^{P} = d\lambda\,(\partial f/\partial\sigma_{22}) = d\lambda\,(L_2 + 2Q_2\sigma_{22} + Q_3\sigma_{11} + Q_6\sigma_{12}) \tag{12.61b}$$

$$d\varepsilon_{12}^{P} = d\lambda\,(\partial f/\partial\sigma_{12}) = d\lambda\,(L_3 + 2Q_4\sigma_{12} + Q_5\sigma_{11} + Q_6\sigma_{22}) \tag{12.61c}$$

$$d\varepsilon_{3}^{P} = -\,(d\varepsilon_{11}^{P} + d\varepsilon_{22}^{P})$$

$$= -\,d\lambda[(L_1 + L_2) + (2Q_1 + Q_3)\sigma_{11} + (Q_3 + 2Q_2)\sigma_{22} + (Q_5 + Q_6)\sigma_{12} \tag{12.61d}$$

Applying the stress and strain transformation eqs(12.45a–c) and (12.43a–c) to an off-axis testpiece, the axial and transverse strains become

$$\begin{aligned} d\varepsilon'^{P}_{11} = d\lambda\{&[L_1 + 2Q_1(\sigma\cos^2\theta) + Q_3(\sigma\sin^2\theta) + Q_5(\tfrac{1}{2}\sigma\sin 2\theta)]\cos^2\theta \\ &+ [L_2 + 2Q_2(\sigma\sin^2\theta) + Q_3(\sigma\cos^2\theta) + Q_6(\tfrac{1}{2}\sin 2\theta)]\sin^2\theta \\ &+ [L_3 + 2Q_4(\tfrac{1}{2}\sigma\sin 2\theta) + Q_5(\sigma\cos^2\theta) + Q_6(\sigma\sin^2\theta)]\sin 2\theta\} \end{aligned} \quad (12.62a)$$

$$\begin{aligned} d\varepsilon'^{P}_{22} = d\lambda\{&[L_1 + 2Q_1(\sigma\cos^2\theta) + Q_3(\sigma\sin^2\theta) + Q_5(\tfrac{1}{2}\sigma\sin 2\theta)]\sin^2\theta \\ &+ [L_2 + 2Q_2(\sigma\sin^2\theta) + Q_3(\sigma\cos^2\theta) + Q_6(\tfrac{1}{2}\sin 2\theta)]\cos^2\theta \\ &- [L_3 + 2Q_4(\tfrac{1}{2}\sigma\sin 2\theta) + Q_5(\sigma\cos^2\theta) + Q_6(\sigma\sin^2\theta)]\sin 2\theta\} \end{aligned} \quad (12.62b)$$

$$\begin{aligned} d\varepsilon'^{P}_{12} = d\lambda\{&-[L_1 + 2Q_1(\sigma\cos^2\theta) + Q_3(\sigma\sin^2\theta) + Q_5(\tfrac{1}{2}\sigma\sin 2\theta)]\tfrac{1}{2}\sin 2\theta \\ &+ [L_2 + 2Q_2(\sigma\sin^2\theta) + Q_3(\sigma\cos^2\theta) + Q_6(\tfrac{1}{2}\sin 2\theta)]\tfrac{1}{2}\sin 2\theta \\ &+ [L_3 + 2Q_4(\tfrac{1}{2}\sigma\sin 2\theta) + Q_5(\sigma\cos^2\theta) + Q_6(\sigma\sin^2\theta)]\cos 2\theta\} \end{aligned} \quad (12.62c)$$

$$d\varepsilon_3^{P} = -d\lambda[(L_1 + L_2) + (2Q_1 + Q_3)\sigma\cos^2\theta + (Q_3 + 2Q_2)\sigma\sin^2\theta + (Q_5 + Q_6)\tfrac{1}{2}\sin 2\theta] \quad (12.62d)$$

Taking the ratios between the strain increments shows that the applied stress σ is retained within the resulting expression. For example, setting $\theta = 0°$ and $90°$ in eqs(12.62a–d) gives two plastic strain increment ratios corresponding to each orientation:

$$r_{0°} = \frac{d\varepsilon'^{P}_{22}}{d\varepsilon^{P}_{3}} = \frac{-(L_2 + Q_3\sigma)}{(L_1 + L_2) + (2Q_1 + Q_3)\sigma} \quad (12.63a)$$

$$\frac{d\varepsilon'^{P}_{22}}{d\varepsilon'^{P}_{11}} = \frac{L_2 + Q_3\sigma}{L_1 + 2Q_1\sigma} \quad (12.63b)$$

$$r_{90°} = \frac{d\varepsilon'^{P}_{22}}{d\varepsilon^{P}_{3}} = \frac{-(L_1 + Q_3\sigma)}{(L_1 + L_2) + (2Q_2 + Q_3)\sigma} \quad (12.63c)$$

$$\frac{d\varepsilon'^{P}_{22}}{d\varepsilon'^{P}_{11}} = \frac{L_1 + Q_3\sigma}{L_2 + 2Q_2\sigma} \quad (12.63d)$$

It follows that when eqs(12.63b,d) are applied to Fig. 12.14b, with L_1, L_2, Q_1, Q_2 and Q_3 assumed constant, the applied stress σ governs the ratio. Thus, taking $\sigma = \sigma(\varepsilon'^{P}_{11})$ from Fig. 12.14a enables each strain path in Fig. 12.14b to be found by numerical integration.

The yield function, eq(12.60b), assumes different tensile and compressive yield stresses, as seen in Fig. 3.12. *Note*: When using an associated flow rule, normality is implied between the plastic strain increment vector and the yield surface. The stress-dependent, plastic strain increment ratios, given in eqs(12.63a–d), suggest that the curvature of the yield surface changes continuously in the vicinity of an advancing stress. Clearly, this contravenes the assumption that the yield surface expands uniformly as for an isotropic hardening rule.

12.8 Alternative Yield Criteria

There have been inconsistencies reported between the predicted and observed equi-biaxial yield stress in materials with r–values less than unity [35]. To overcome this, further generalisations have been made to Hill's criterion, including modifications made by that author [36] (see also Table 3.3). The reduction of these criteria to a plane, principal stress state is the most common form and these will now be examined.

Identifying anisotropy parameters F, G and H with the r–values in the usual way and omitting shear stress terms, the principal stress form of Hill's yield function (12.2) becomes

$$r_1(\sigma_2 - \sigma_3)^2 + r_2(\sigma_1 - \sigma_3)^2 + r_1 r_2(\sigma_1 - \sigma_2)^2 = r_2(1 + r_1)\sigma_{1y}^{\,2} \tag{12.64}$$

where σ_{1y} is the yield stress for the material's 1–direction. Here principal stress aligns with the material axes. In a modification proposed by Hosford [35], eq(12.64) was written as

$$r_1|\sigma_2 - \sigma_3|^a + r_2|\sigma_1 - \sigma_3|^a + r_1 r_2|\sigma_1 - \sigma_2|^a = r_2(1 + r_1)\sigma_{1y}^{\,a} \tag{12.65}$$

where $a > 2$ is an integer and each modulus $|\;|$ requires a positive stress difference. Values of a were suggested lying in a range from 6 to 8 for pencil glide and from 8 to 10 for bcc materials in general.

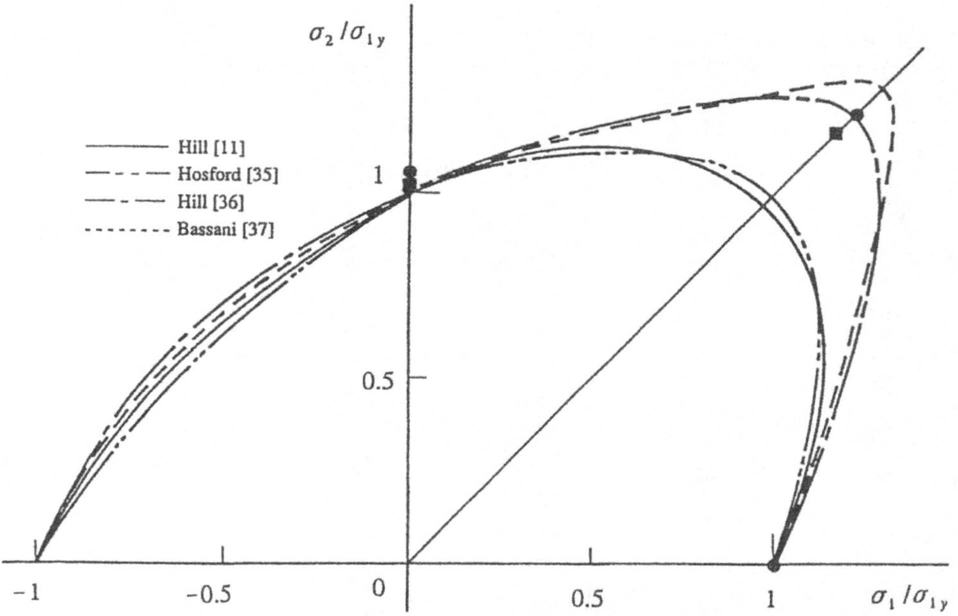

Figure 12.15 Comparison between planar-isotropic yield loci

For a planar isotropic yield criterion set $\sigma_3 = 0$ and $r = r_1 = r_2$ in eq(12.65) to give

$$\sigma_1^{\,a} + \sigma_2^{\,a} + r|\sigma_1 - \sigma_2|^a = (1 + r)\sigma_{1y}^{\,a} \tag{12.66a}$$

If $a = 2$ in eq(12.66a), the corresponding quadratic form becomes

$$(\sigma_1 + \sigma_2)^2 + (1 + 2r)(\sigma_1 - \sigma_2)^2 = 2(1 + r)\sigma_{1y}^{\,2} \tag{12.66b}$$

which can be written as

$$\sigma_1^2 + \sigma_2^2 - \frac{2r}{1+r}\sigma_1\sigma_2 = \sigma_{1y}^2 \tag{12.66c}$$

This is the planar isotropic form previously identified in eq(12.39a). In Hill's 1979 modification to eq(12.66b), a non-quadratic potential was proposed [36]:

$$|\sigma_1 + \sigma_2|^m + (1+2r)\,|\sigma_1 - \sigma_2|^m = 2(1+r)\sigma_{1y}{}^m \tag{12.66d}$$

where $2 \ge m \ge 1$. Bassani [37] employed different exponents in a non-quadratic yield criterion

$$|\sigma_1 + \sigma_2|^q + (q/p)(1+2r)\sigma_{1y}{}^{q-p}\,|\sigma_1 - \sigma_2|^p = [1 + (q/p)(1+2r)]\,\sigma_{1y}{}^q$$

where $p \ge 1$, $q \ge 1$ are integer exponents and $r \equiv \bar{r}$. A comparison between eqs(12.66a–d) and flow data for two ¼ mm can-end steels is given in Fig. 12.15. Taking $m = 1.5$, $a = 6$, $p = 1$ and $q = 2$, the respective criteria demonstrate the diverge for a material with $\bar{r} = 0.874$. Here, the equi-biaxial yield stresses, found from bulge forming, exceed the tensile yield stress in each $\bar{r} < 1$ material [38]. The loci are sensitive to their stress exponents, which control the curvature (and hence the strain path) in a region of biaxial tension. It is seen that Hill's modification in eq(12.66d) describes this. *Note*: The small difference observed between tensile yield stresses in the rolling and transverse directions, i.e. $\sigma_{2y}/\sigma_{1y} \neq 1$, does not appear within these symmetrical loci. Hill [39] corrected for this by combining cubic and quadratic non-dimensional stress terms in a criterion:

$$\frac{\sigma_1^2}{\sigma_{1y}^2} - \frac{c\,\sigma_1\sigma_2}{\sigma_{1y}\sigma_{2y}} + \frac{\sigma_2^2}{\sigma_{2y}^2} + \left[(p+q) - \frac{(p\sigma_1 + q\sigma_2)}{\sigma_{by}}\right]\frac{\sigma_1\sigma_2}{\sigma_{1y}\sigma_{2y}} = 1$$

where σ_{by} is the biaxial yield stress, p and q are constants and c combines the three yield points in Fig. 12.15:

$$\frac{c}{\sigma_{1y}\sigma_{2y}} = \frac{1}{\sigma_{1y}^2} + \frac{1}{\sigma_{2y}^2} + \frac{1}{\sigma_{by}^2}$$

Other forms of yield criteria that account for initial anisotropy, listed in Table 3.3, can be adapted to the planar isotropic condition as required. They reduce conveniently when a principal biaxial stress state aligns with the material directions, 1 and 2, say. Shear stress terms, along with σ_3, are dropped allowing the table to be reconstructed for this specific case (see exercise 12.15). These reductions pre-suppose initial anisotropy in a metallic material has an orthotropic character that has arisen from the manner of its production, say, from a prior rolling or extrusion process. The form remains even with post heat treatment provided the rolled grains retain preferred orientation. However, there is another form of anisotropy arising from subsequent straining that has been referred to earlier in § 10.7 that lies within a translation in the subsequent yield surface. Writing its function generally as $f(\sigma_{ij}' - \alpha_{ij}')$ for incompressible plasticity, the literature reveals a number of specific forms that differ in their description of a translation, as given in Table 12.1.

Table 12.1 Anisotropic Yield Criteria

ID	Ref.	$f(\sigma_{ij}' - \alpha_{ij}') =$
(a)	40	$\tfrac{1}{2}C_{ijkl}\,(\sigma_{ij}' - m\varepsilon_{ij}^P)(\sigma_{kl}' - m\varepsilon_{kl}^P)$
(b)	41	$\tfrac{1}{2}\,\sigma_{ij}'\sigma_{ij}' - B\varepsilon_{ij}^P\sigma_{ij}'$
(c)	42	$\tfrac{1}{2}C_{ijkl}\,\sigma_{ij}'\sigma_{kl}' - B\varepsilon_{ij}^P\sigma_{ij}'$
(d)	43	$\tfrac{1}{2}\,[(\sigma_{ij}' - \alpha_{ij}')(\sigma_{ij}' - \alpha_{ij}')]^{3/2} + h\sqrt{I_2}\,(\sigma_{ij}' - \alpha_{ij}')(\sigma_{jk}' - \alpha_{jk}')(\sigma_{ki}' - \alpha_{ki}')$
(e)	44	$\tfrac{1}{2}C_{ijkl}\,(\sigma_{ij}' - m\varepsilon_{ij}^P)(\sigma_{kl}' - m\varepsilon_{kl}^P) + \tfrac{1}{3}p(\sigma_{ij}' - \alpha_{ij}')(\sigma_{jk}' - \alpha_{jk}')(\sigma_{ki}' - \alpha_{ki}')$
(f)	45	$C_{ijkl}\,(\sigma_{ij}' - m\varepsilon_{ij}^P)(\sigma_{kl}' - m\varepsilon_{kl}^P) + C_{ijklmn}(\sigma_{ij}' - m\varepsilon_{ij}^P)(\sigma_{jk}' - m\varepsilon_{jk}^P)(\sigma_{lm}' - m\varepsilon_{lm}^P)$

In Table 12.1, non rigid kinematic translations appear. The hardening is expressed by generalised second and third stress deviator invariants J_2' and J_3'. These appear with C_{ijkl} and C_{ijklmn} in (a), (c), (e) and (f) to account for any initial anisotropy and distortion present in the initial yield surface. Subsequent translation of that surface have been described by the prior plastic work $\varepsilon_{ij}^P\sigma_{ij}'$ in (b) and (c) or elsewhere in (a), (e) and (f) by Prager's rule and α_{ij} in (d) by Ziegler rule, eqs(10.48) and (10.51) respectively. Non incremental formulations suggest that proportional loadings apply to each description. Otherwise, a path dependent (differential) form would be required wherever integrated plastic strains ε_{ij}^P are required to appear (see exercise 12.16). For example, the second invariant of plastic strain I_2, appearing as a coefficient in (d), is proportional to the measure of equivalent plastic strain given in eq(10.12a). The specific work done by a similar path is given by eq(10.12b), where it will be seen that either the absolute or deviatoric stress may appear, i.e. $w^P = \sigma_{ij}\varepsilon_{ij}^P = \sigma_{ij}'\varepsilon_{ij}^P$.

12.9 Concluding Remarks

In plasticity theory the choice of yield function (i.e. the plastic potential) depends upon the initial condition of a material. Generally, an isotropic potential will predict the plasticity in heat treated material. Otherwise, an account needs to be made of crystallographic anisotropy resulting from processing, particularly in rolled sheet and extruded bars/tubes. Hill's 1948 homogenous, quadratic yield function, with its associated flow rule, provides a good account of plasticity under radial loading of an orthotropic material, e.g. a rolled sheet metal. However, certain anomalies have arisen with a quadratic stress potential. Among these are the absence of a Bauschinger effect and differences between equi-biaxial and uni-axial yield stresses. These have prompted modified potentials including the addition of linear and/or cubic stress terms or in adopting non integer stress exponents. Increasing flexibility by these means allows the anisotropy parameters to match observed behaviour more closely. As with isotropic material there appears to be no particular advantage in using the work or strain hypotheses of hardening for an orthotropic solid. Therein, equivalent stress versus equivalent plastic strain correlations from each hypothesis would appear to be equally suitable for use with a normality rule of flow.

References

1. Klinger L. J. and Sachs G. *Jl. Aero. Sci.*, 1948, **15**, 599.
2. Cunninghan D. M., Thompsen E. G. and Dorn J. E. *Proc. ASTM,* 1947, **47**, 546.
3. Rogan J. and Shelton A. *Jl. Strain Analysis*, 1969, **4**, 127.
4. Shahabi S. N. and Shelton A. *Jl. Mech. Eng. Sci.*, 1975, **17**, 93.
5. Jackson L. R., Smith K. F. and Lankford W. T. *Jl. Metals*, 1949, **323**, 1.
6. Dorn J. E. *Jl. Appl. Phys.*, 1949, **20**, 15.
7. Prager W. *Jl. Appl. Phys.*, 1949, **20**, 234.
8. Fisher J. C. *Trans ASME*, 1949, **71**, 349.
9. Hu L. W. *Trans ASME, Jl. Appl. Mech.*, 1956, **23**, 444.
10. Jones S. E. and Gillis P. P. *Met. Trans*, 1984, **15**A, 129.
11. Hill R. *Proc. Roy. Soc. A*, 1948, **A193**, 281.
12. Finnie I. and Heller W. R. *Creep of Engineering Materials*, 1959, McGraw-Hill, N.Y.
13. Chakrabarty J. *Int. Jl. Mech. Sci.*, 1970, **12**, 169.
14. Whiteley R. L. *Trans ASM*, 1960, **52**, 154.
15. Bramley A. N. and Mellor P. B. *Int. Jl. Mech. Sci.*, 1966, **8**, 101.
16. Bramley A. N. and Mellor P. B. *Int. Jl. Mech. Sci.*, 1968, **10**, 211.
17. Atkinson M. *Sht. Met. Ind.*, 1967, **44**, 167.
18. Li K. Y. *A Study of Certain Classes of Constitutive Equations for Anisotropic Solid Materials*, Ph.D. thesis, Trinity College, Dublin, 1985.
19. Bourne L. and Hill R. *Phil. Mag.*, 1950, **41**, 671.
20. Hazlett T. H., Robinson A. T. and Dorn J. E. *Trans ASM,* 1950, **42**, 1326.
21. Shahabi S. N. and Shelton A. *Jl. Mech. Eng. Sci.*, 1975, **17**, 82.
22 Dillamore I. L, Hazel R. J, Watson T.W and Hadden P. *Int. Jl. Mech. Sci.*, 1971, **13**, 1049.
23. Fessler H. and Hyde T. H. *Proc.: Non Linear Problems in Stress Analysis*, p. 233, (ed. Stanley P.) Inst. of Physics, 1977.
24. Mehan R. L. *Trans ASME, J. Basic Engng.*, 1961, **83**, 499.
25. Rees D. W. A. and Mathur S. B. *Proc.: Non Linear Problems in, Stress Analysis*, 1977, p. 185, (ed. Stanley P.) Inst. of Physics, UK.
26. Johnson A. E., Frost N. E. and Henderson J. *The Engineer*, 1955, 199, 366, 402, 457.
27. Hill R. *The Mathematical Theory of Plasticity*, 1989, Clarendon Press, Oxford.
28. Rees D. W. A. *Proc. Roy. Soc.*, 1982, **A383**, 333.
29. Avery D. H., Hosford W. F and Backofen W. A. *Trans. Met. Soc., AIME*, 1965, **223**, 71.
30. Lee D. and Backofen W. A. *Trans. Met. Soc. AIME*, 1966, **236**, 1077.
31. Mellor P. B. *Mechanics of Solids*, (eds Hopkins H. G. and Sewell M. J.) 1982, 383, Pergamon Press.
32. Larson F. R. *Trans. ASM* 1964, **27**, 321.
33. McEvily A. J. and Hughes P. J. *NACA Tech. Note*, 1954, 3248.
34. Rees D. W. A. *Acta Mech.*, 1982, **43**, 223.
35. Hosford W. F. *Proc: 7th Nth Amer Metal Working Res Conf*, 1979, p. 191, SME, Dearbon, Michigan.
36. Hill R. *Math. Proc. Camb. Phil. Soc.*, 1979, **85**, 179.
37. Bassani J. L. *Int. Jl. Mech. Sci.*, 1977, **19**, 651.
38. Rees D. W. A. *Jl. de Phy. IV*, 2003, **105**, 19. See also Procedings: *Non-Linear Mechanics of Anisotropic Materials*, EMMC 6, 2002, p. 33, (eds Cescotto S. et al.) Euromech-Mecamat, Univesity of Liége.

39. Hill R. *Int. Jl. Mech. Sci.*, 1993, **35**, 19.
40. Williams J. F. and Svennson N. L. *Jl. Strain Analysis*, 1970, **5**, 128: 1971, **6**, 263.
41. Kishi T. and Tanabe T. *Jl. Mech. Phy. Sols.*, 1973, **21**, 303.
42. Yoshimura Y. and Takenaka Y. *Proc.: Second Order Effects in Elasticity, Plasticity and Fluid Dynamics*, IUTAM, 1964, p. 729, (eds Reiner M. and Abir D.) Pergamon.
43. Freudenthal A. M. and Gou P. F. *Acta Mech.*, 1969, **8**, 34.
44. Shrivastava H. P., Mroz Z. and Dubey R. N. *ZAMM*, 1973, **53**, 626.
45. Rees D. W. A. ZAMM, 1983, **197C**, 31.

Exercises

12.1 Derive the relationships between the components of H_{ijkl} in eq(12.1) and the six anisotropy coefficients employed with Hill's orthotropic yield function in eq(12.2).

12.2 Figure 12.16 shows an 'off-axis' cylinder subjected to a combined tension-torsion test. Axial and shear stress components (σ_{11}', σ_{12}') are produced in the wall of the thin-cylinder by the combined action of tension and torque W and T as shown. The material's orthotropy axes 1, 2 are inclined to tube axis (the inclination θ lies between 1 and 1′), while axis 3 remains radial, as shown.

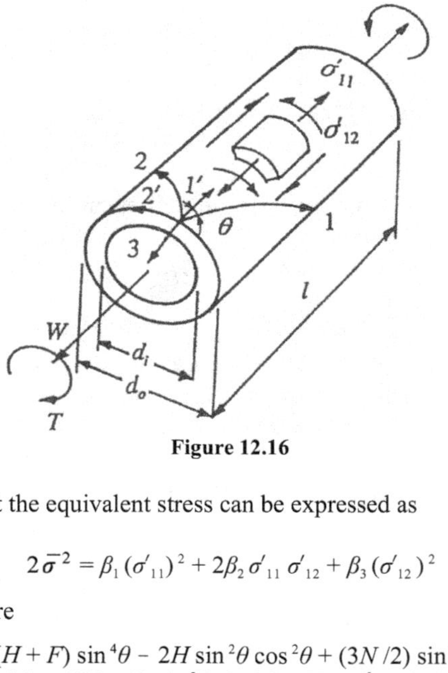

Figure 12.16

Show, from eq(12.4b), that the equivalent stress can be expressed as

$$2\bar{\sigma}^2 = \beta_1(\sigma'_{11})^2 + 2\beta_2\sigma'_{11}\sigma'_{12} + \beta_3(\sigma'_{12})^2$$

where the β–coefficients are

$$\beta_1 = (H+G)\cos^4\theta + (H+F)\sin^4\theta - 2H\sin^2\theta\cos^2\theta + (3N/2)\sin^2 2\theta$$
$$\beta_2 = (H+G)\cos 2\theta\sin 2\theta - (2H+F)\sin^2\theta\sin 2\theta + H\cos^2\theta\ \sin 2\theta - 3N\cos 2\theta\sin 2\theta$$
$$\beta_3 = (4H+G+F)\sin^2 2\theta + 6N\cos^2 2\theta$$

See solution

12.3 Show that the plastic strain increment ratios for the cylinder in Fig. 12.16 are given by

$$\frac{d\gamma'^{P}_{12}}{d\varepsilon'^{P}_{11}} = \frac{\beta_2\sigma'_{11} + \beta_3\sigma'_{12}}{\beta_1\sigma'_{11} + \beta_2\sigma'_{12}} \quad \text{and} \quad \frac{d\varepsilon'^{P}_{22}}{d\varepsilon'^{P}_{11}} = \frac{\beta_4\sigma'_{11} + \beta_5\sigma'_{12}}{\beta_1\sigma'_{11} + \beta_2\sigma'_{12}}$$

where coefficients β_1, β_2 and β_3 are given in exercise 12.2 and

$$\beta_4 = (2H + G - 6N)\sin^2\theta\cos^2\theta - H(\sin^4\theta + \cos^4\theta)$$

$$\beta_5 = (2H + G)\sin 2\theta\sin^2\theta - (2H + F)\sin 2\theta\cos^2\theta + 3N\sin 2\theta\cos 2\theta$$

12.4 If the stresses for the tube in Fig. 12.16 are applied in a constant ratio $R = \sigma'_{12}/\sigma'_{11}$, show that the principal axes of stress coincide with the principal material axes when $\theta = \frac{1}{2}\tan^{-1}(2R)$. What then is the incremental plastic strain ratio $d\gamma_{12}{}^P/d\varepsilon_{11}{}^P$? Employ eqs(12.11a–c) to show that the principal plastic strain increments for this condition become

$$d\varepsilon_1^P = \frac{\delta\lambda\,\sigma_{11}'}{3}\left[\frac{G}{2} + \left(H + \frac{G}{2}\right)\sqrt{(1 + 4R^2)}\right],\quad d\varepsilon_2^P = \frac{\delta\lambda\,\sigma_{11}'}{3}\left[\frac{F}{2} - \left(H - \frac{F}{2}\right)\sqrt{(1 + 4R^2)}\right]$$

$$d\varepsilon_3^P = -\frac{\delta\lambda\,\sigma_{11}'}{6}\left[(F + G) + (G - F)\sqrt{(1 + 4R^2)}\right]$$

Hint: The principal stress directions w.r.t.1′ are found from $\tan 2\alpha = 2\sigma'_{12}/\sigma'_{11} = 2R$
See solution

12.5 If the off-axis tube in Fig. 12.16 is subjected to a separate tensile force W and a torque T, determine the corresponding $d\gamma'_{12}{}^P/d\varepsilon'_{11}{}^P$ ratios. Show that: (i) in the absence of torsion, the axial plastic strain increment becomes

$$d\varepsilon'^P_{11} = \frac{d\bar{\varepsilon}^P}{2\bar{\sigma}}\left(\beta_2\,\sigma'_{11}\right)$$

and (ii) in the absence of tension, the plastic shear strain increment becomes

$$d\gamma'^P_{12} = \frac{d\bar{\varepsilon}^P}{2\bar{\sigma}}\left(\beta_3\,\sigma'_{12}\right)$$

For what testpiece orientations, with θ lying between 1 and 1′, do $d\varepsilon'_{11}{}^P$ and $d\gamma'_{12}{}^P$ attain their maxima?

12.6 Show that when the thin-walled tubular specimen in Fig. 12.16 is subjected to torsion only, the equivalent stress expression for the orientation θ given in exercise 12.2 reduces to

$$\bar{\sigma} = (\sigma_{12\,y}/\sqrt{2})\,[\,(4H + G + F)\sin^2 2\theta + 6N\cos^2 2\theta\,]^{1/2}$$

where $\sigma_{12\,y}$ is the shear flow stress aligned with the material co-ordinates.

12.7 Confirm, from first principles, the $\bar{\sigma}$ expression given in exercise 12.6 when, under pure torsion, θ is the inclination between the material's 1–axis and the testpiece axis 1′. *Hint*: Take the two axes to be aligned initially.

12.8 Show (a) that the following ratio between the shear yield stress σ_{12y} and the tensile yield stress σ_{1y} for an orthotropic material is given by eq(12.2) as

$$\frac{\sigma_{12y}}{\sigma_{1y}} = \left(\frac{H + G}{6N}\right)^{\frac{1}{2}}$$

What are the corresponding ratios between yield stresses: (b) σ_{12y} and σ_{2y} and (c) σ_{12y} and σ_{3y}?
See solution: (b) $\sigma_{12\,y}/\sigma_{2\,y} = [(H + F)/(6N)]^{1/2}$, c) $\sigma_{12\,y}/\sigma_{3\,y} = [(F + G)/(6N)]^{1/2}$}

12.9 Establish, from eq(12.2), further relationships between the following shear and tensile yield stresses for orthotropic material: (a) $\sigma_{23\,y}$ and $\sigma_{1\,y}$ (b) $\sigma_{23\,y}$ and $\sigma_{2\,y}$, (c) $\sigma_{23\,y}$ and $\sigma_{3\,y}$, (d) $\sigma_{31\,y}$ and $\sigma_{1\,y}$, (e) $\sigma_{31\,y}$ and $\sigma_{2\,y}$ and (f) $\sigma_{31\,y}$ and $\sigma_{3\,y}$.

Ans: (a) $\sigma_{23\,y}/\sigma_{1\,y} = [(H+G)/(6L)]^{1/2}$, (b) $\sigma_{23\,y}/\sigma_{2y} = [(H+F)/(6L)]^{1/2}$, (c) $\sigma_{23\,y}/\sigma_{3\,y} = [(F+G)/(6L)]^{1/2}$, (d) $\sigma_{31\,y}/\sigma_{1y} = [(H+G)/(6M)]^{1/2}$, (e) $\sigma_{31\,y}/\sigma_{2\,y} = [(H+F)/(6M)]^{½}$, (f) $\sigma_{31\,y}/\sigma_{3\,y} = [(F+G)/(6M)]^{1/2}]\}$

12.10 A sheet is transversely isotropic within the 1–2 plane ($r = r_{0°} = r_{90°}$). Show that plane strain compression and simple tension are related through the following relationships:

$$p = \frac{(1+r)\sigma_1}{\sqrt{(1+2r)}} \quad \text{and} \quad \frac{dt}{t} = \frac{\sqrt{(1+2r)}\,\varepsilon_1^P}{(1+r)}$$

where σ_1, $\varepsilon_1^{\,P}$ are the stress and plastic strain for a tension test. In plane compression p is the pressure, dt/t is an incremental through-thickness strain and t is the current thickness (see Fig. 9.13).

12.11 Show that the variation in flow stress for tensile testpieces lying in the 1–2 plane with their axes inclined at θ to the 1–direction is given by

$$\sigma = \sqrt{2}\,\bar{\sigma}\,[\,(H+G)\cos^4\theta + (H+F)\sin^4\theta + ½\,(3N - H)\sin^2 2\theta\,]^{-1/2}$$

when $\bar{\sigma}$ is defined from Finnie's eq(12.5a). Hence, show that the condition for a maximum or minimum yield stress is given by

$$\tan^2\theta = \frac{2H + G - 3N}{2H + F - 3N}$$

See solution

12.12 Repeat exercise 12.11 for testpieces lying in the 1–3 and 2–3 planes. That is, show that the yield stress variations are respectively

$$\sigma = \sqrt{2}\,\bar{\sigma}\,[\,(H+G)\cos^4\theta + (G+F)\sin^4\theta + ½\,(3M - G)\sin^2 2\theta\,]^{-1/2}$$

$$\sigma = \sqrt{2}\,\bar{\sigma}\,[\,(H+F)\cos^4\theta + (G+F)\sin^4\theta + ½\,(3L - F)\sin^2 2\theta\,]^{-1/2}$$

for which the maximum or minimum correspond to

$$\tan 2\theta = \frac{2G + H - 3M}{2G + F - 3M}, \qquad \tan 2\theta = \frac{2F + H - 3L}{2F + G - 3L}$$

12.13 What is the condition of instability when the off-axis tensile equivalence correlation for copper, shown in Fig. 12.9, obeys the Swift law in eq(9.44b)? Find the particular instability condition for a testpiece orientation of $\theta = 45°$.

12.14 A worked planar isotropic material resembles that given in Fig. 12.3a. A subsequent plane strain deformation, which allows plasticity to occur only in the 2– and 3–directions, converts the structure to that shown in Fig. 12.3b. Assuming that the r–value for each isotropic plane is similar, i.e. $r_{12} = r_{23}$, compare the yield loci for when the principal plane stress states arise as shown within the 1–2 plane of each worked material.

12.15 Reduce the yield criteria (a)–(h) in Table 3.3 to a principal biaxial stress state σ_1, σ_2 in alignment with an orthotropic metal's 'material directions', 1 and 2. Check that the reduction to criterion (d) takes the form given in eq(12.66d).

12.16 How should an integrated strain describing a path dependent translation appear for functions (a), (e) and (f) in Table 12.1 when they required to describe a subsequent yield surface following a non radial loading?
Ans: $m = m(I_2, I_3)$

CHAPTER 13

PLASTIC INSTABILITY

13.1 Introduction

There are many examples where the influence of plasticity upon the load carrying capacity of a structure must be considered. A purely elastic analysis will supply the critical loading at which plasticity first appears but the analysis will not extend into the plastic regime without modifications. These can take the form of a change to the modulus of a material to reflect the loss in stiffness of a strut, the introduction of empirically-based plasticity reduction factors for plate buckling and the derivation of an appropriate sub-tangent under multiaxial stressing. Such problems adopt the plasticity theory appropriate to the initial condition of the material. The Levy-Mises theory is adequate for an isotropic condition but in orthotropic sheets such problems should employ anisotropy parameters, e.g. the r-values as discussed in the previous chapter. For example, r-values arise in the determination of the limiting strains in sheet metal forming. The theory may be coupled to either a diffuse or local instability criterion for the sheet material to determine the combination of limiting, in-plane plastic strains. These strains are used in the construction of a forming limit diagram.

13.2 Inelastic Buckling of Struts

13.2.1 Tangent Modulus

Engesser's modification [1] to Euler's buckling theory accounts for inelastic buckling simply by replacing the elastic modulus with a plastic tangent modulus. With this reduction in stiffness, the section stress is given by

$$\sigma = \frac{P}{A} = \frac{\pi^2 E_t}{\left(L_e/k\right)^2} \tag{13.1a}$$

where L_e is an effective length that accounts for the particular rotational restraint exerted by the end fixing. Within the range of slenderness ratios $50 \le L_e/k \le 100$, the effect of end constraint on the plastic buckling load P is less than that of an elastic strut with similar end fixings. For example, we should not assume that the plastic buckling load of a pinned-end strut, lying in the same range, will be doubled by fixing its ends. However, for $L_e/k \le 50$ an elastic constraint is assumed and an equivalent elastic length $L_e = cL$ may be employed, particularly when a safety factor is used. This gives: $L_e = L$ for pinned ends, $L_e = L/2$ for fixed ends, $L_e = L/\sqrt{2}$ for pinned-encastre end fixings and $L_e = 2L$ for fixed-free ends fixings. The tangent modulus $E_t = \mathrm{d}\sigma/\mathrm{d}\varepsilon$ in eq(13.1a) is the gradient of the tangent to the uniaxial compressive stress–strain curve within its plastic range (see Fig. 13.1). It follows from this definition of E_t that the buckling stress σ in eq(13.1a) must satisfy the following condition

$$\sigma = \frac{\pi^2(\mathrm{d}\sigma/\mathrm{d}\varepsilon)}{\left(L_e/k\right)^2} \tag{13.1b}$$

Since σ appears on both sides of eq(13.1b) it may be solved by trial. That is, a plastic buckling stress σ is selected to be greater than the yield stress σ_o. Assuming that a stress-strain curve is available, $E_t = \mathrm{d}\sigma/\mathrm{d}\varepsilon$ is found and eq(13.1b) is solved for L_e/k. The solution is correct only when L_e/k matches that for the given strut. The procedure is aided when all such solutions to eq(13.1b) appear as points on a plot of σ versus L_e/k. Alternatively, a suitable empirical representation to the σ versus ε curve may be employed, the simplest being the Hollomon law (9.39b), which is represented graphically in Fig. 13.1.

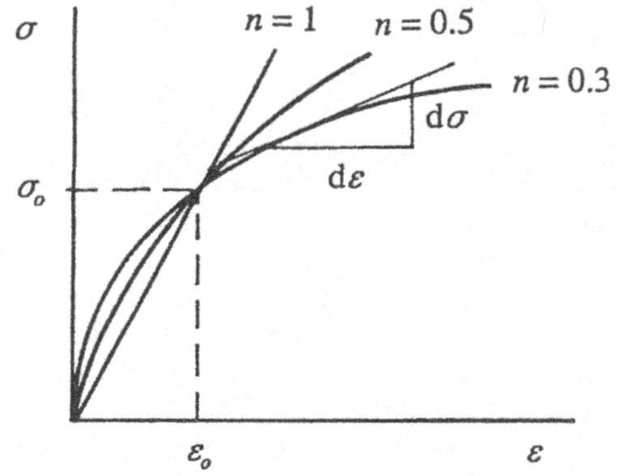

Figure 13.1 Hollomon curve with tangent modulus

Given that the stress–strain curve is described by

$$\frac{\sigma}{\sigma_o} = \left(\frac{\varepsilon}{\varepsilon_o}\right)^n \tag{13.2a}$$

the gradient of its tangent is

$$E_t = \frac{\mathrm{d}\sigma}{\mathrm{d}\varepsilon} = n\left(\frac{\sigma_o}{\varepsilon_o}\right)\left(\frac{\sigma}{\sigma_o}\right)\left(\frac{\sigma_o}{\sigma}\right)^{1/n} \tag{13.2b}$$

where ε_o and n are material constants. Substituting eq(13.2b) into eq(13.1b) results in an equation for σ/σ_o:

$$\frac{\sigma}{\sigma_o} = \left(\frac{n}{\varepsilon_o}\right)^n\left(\frac{\pi}{L_e/k}\right)^{2n} \tag{13.3}$$

Equation (13.3) defines the Engesser's curve 1 in Fig. 13.2, this curve being valid for a net section stress $\sigma > \sigma_o$. In contrast, Euler's elastic curve 2 in this figure applies to section stresses $\sigma < \sigma_o$ and is expressed as

$$\sigma = \frac{P}{A} = \frac{\pi^2 E}{\left(L_e/k\right)^2} \tag{13.4}$$

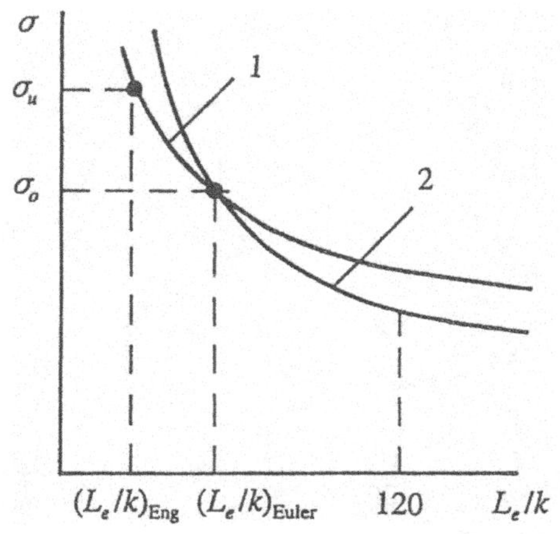

Figure 13.2 Buckling curves: 1–Engesser, 2–Euler

To ensure an intersection between the two curves at $\sigma = \sigma_o$, first take Euler's critical slenderness ratio from eq(13.4):

$$\left(\frac{L_e}{k}\right)_{Eul} = \sqrt{\left(\frac{\pi^2 E}{\sigma_o}\right)} \tag{13.5}$$

Since $(L_e/k)_{\text{Euler}}$ defines an intersection co-ordinate, substitution of eq(13.5) into eq(13.3) leads to the condition

$$n\sigma_o = E\varepsilon_o \tag{13.6}$$

Equation (13.6) shows that Euler's theory becomes a special case of eq(13.3) when $n = 1$. Thus, a curve of $0 < n < 1$ value will pass through this common point $[(L_e/k)_{Eul}, \sigma_o]$. Correspondingly, an intersection occurs at a common yield point $(\varepsilon_o, \sigma_o)$ for Hollomon curves with different n–values (see Fig. 13.1).

Contrary to the Hollomon law, the stress in a plastic strut cannot be increased indefinitely. The cut-off ordinate in Fig. 13.2 occurs at the ultimate compressive strength σ_u as shown. Thus, the critical value of L_e/k for an Engesser strut follows from eqs(13.3) and (13.6) as

$$\left(\frac{L_e}{k}\right)_{Eng} = \sqrt{\frac{\pi^2 E}{\sigma_o\left(\sigma_u/\sigma_o\right)^{1/n}}} \tag{13.7}$$

13.2.2 Empirical Formulae

Empirical 'plastic strut' formulae extend L_e/k partly into the elastic range accounting for a range of short struts with L_e/k initially less than the critical value from Engesser's eq(13.7). Figures 13.3a,b show the range of interest in which section stresses may decrease from the ultimate value σ_u to less than the yield stress σ_o. In each diagram the Engesser and Euler curves, 1 and 2 respectively, intersect at point A. The empirical approach replaces the two segments of each curve with a continuous curve that plots to the safe side and terminates at σ_u on the σ axis. The following strut formulae are capable of this, in which the choice between them depends upon the strut material.

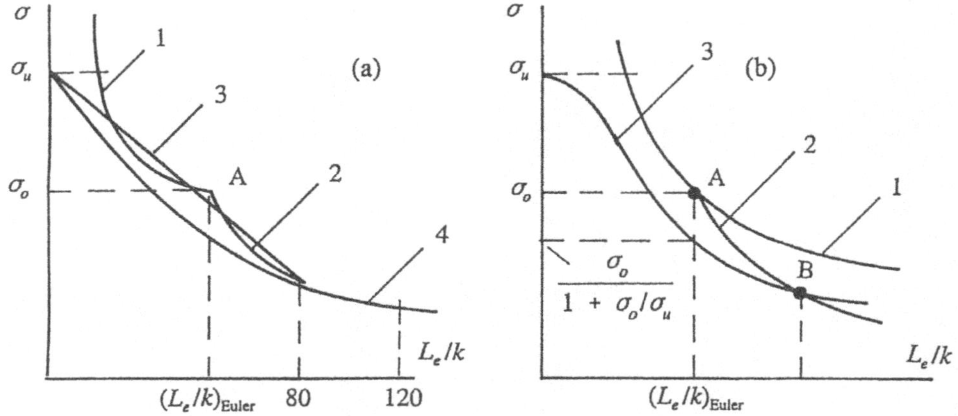

Figure 13.3 Empirical approximations (3) and (4) compared to Euler (1) and Engesser (2)

(a) Straight Line

The straight line 3 in Fig. 13.3a is written simply as

$$\sigma = \sigma_u [\, 1 - q\,(L_e/\,k)\,] \tag{13.8}$$

where q is a material constant. For aluminium alloy struts, eq(13.8) is arranged to intersect the Euler curve at $L_e/k = 80$ while for other non-ferrous metals an intersection at $L_e/k = 120$ may be preferred. The Euler stress ordinate at each intersection is found from eq(13.4). Substituting this into eq(13.8), q is found. Typically, for aluminium alloy with $E = 75$ GPa and $\sigma_u = 320$ MPa, $q = 8 \times 10^{-3}$.

(b) Parabola

The parabola 4 in Fig. 13.3a provides a safer prediction for steel struts with effective slenderness ratios lying in the range $L_e/k \le 120$. This is written as

$$\sigma = \sigma_u [\, 1 - b\,(L_e/\,k)^2\,] \qquad (13.9]$$

where b is a material constant found from known co-ordinates (σ, L_e/k) at the intersection with the Euler curve. Values of b for steels lie in the range $(40 - 50) \times 10^{-6}$ for ultimate strengths between 400 and 500 MPa and with a modulus $E = 210$ GPa.

(c) Rankine-Gordon

Their critical plastic buckling load $P_{R\text{-}G}$ is found from the condition

$$\frac{1}{P_{R\text{-}G}} = \frac{1}{P_{Eng}} + \frac{1}{P_u} \tag{13.10a}$$

where $P_u = A\sigma_u$ is the ultimate load and $P_{Eng} = A\sigma$. Re-arranging eq(13.10a) for P_{RG} gives

$$P_{R\text{-}G} = \frac{P_u}{1 + P_u/P_{Eng}} \tag{13.10b}$$

Substituting P_{Eng} separately from eqs(13.1a) and (13.3) with $P_o = \sigma_o A$ gives two alternative forms for $P_{R\text{-}G}$:

$$P_{R\text{-}G} = \frac{A\sigma_u}{1 + \dfrac{\sigma_u}{\pi^2 E_t}\left(\dfrac{L_e}{k}\right)^2} = \frac{A\sigma_u}{1 + \dfrac{\sigma_u}{\sigma_o}\left(\dfrac{\sigma_o}{\pi^2 E}\right)^n \left(\dfrac{L_e}{k}\right)^{2n}} \qquad (13.11a,b)$$

Equation (13.11b) is a more useful for combining with eq(13.6). This reveals material constants σ_o, σ_u, E and n within the section stress

$$\sigma = \frac{\sigma_u}{1 + \dfrac{\sigma_u}{\sigma_o}\left(\dfrac{\sigma_o}{\pi^2 E}\right)^n \left(\dfrac{L_e}{k}\right)^{2n}} \qquad (13.12)$$

Equation(13.12) appears as curve 3 in Fig. 13.3b for which the stress is $\sigma = \sigma_o/(1 + \sigma_o/\sigma_u)$ at $(L_e/k)_{Eul}$ and lies on the safe side of the Euler curve 2. The slenderness ratio at an intersection point B with curve 2 is found from equating eqs(13.4) and (13.12).

Example 13.1 An 800 mm long steel strut has a thin-walled, elliptical cross-section, shown in Fig. 13.4a. The mean lengths of the major and minor axes are 80 and 30 mm respectively and the wall thickness is 3 mm. At its end fixings, Fig. 13.4a shows that the strut is free to rotate about a pin aligned with its y–axis but it is prevented from rotating about its x–axis by the rigid walls shown. Compare the allowable compressive plastic lcads according to the Engesser, parabolic and Rankine-Gordon formulae, using a safety factor of 1.5. For steel take $\sigma_o = 300$ MPa, $\sigma_u = 450$ MPa, $E = 210$ GPa and $n = ⅓$.

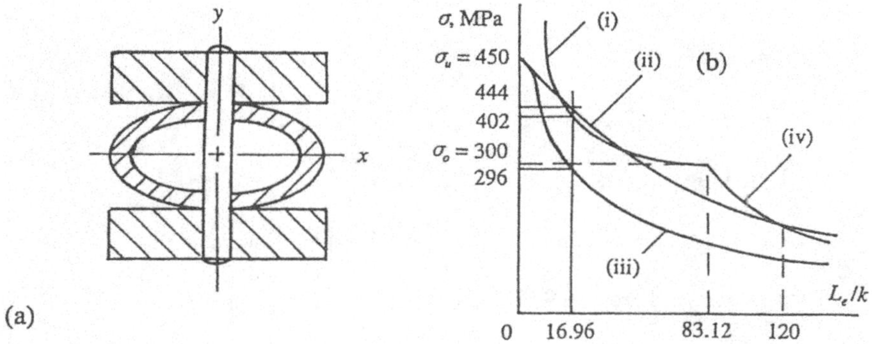

Figure 13.4 Elliptical section showing plastic buckling stress

The properties required of the cross-section are

$$A = \pi ab = \pi[(83 \times 35) - (77 \times 27)] = 2073.45 \text{ mm}^2$$
$$I_x = \pi ab^3/4 = (\pi/4)[83(33)^3 - 77(27)^3] = 1.1523 \times 10^6 \text{ mm}^4$$
$$k_x = \sqrt{(I_x/A)} = 23.583 \text{ mm}$$
$$I_y = \pi ba^3/4 = (\pi/4)[33(83)^3 - 27(77)^3] = 5.1385 \times 10^6 \text{ mm}^4$$
$$k_y = \sqrt{(I_y/A)} = 49.782 \text{ mm}$$

The strut will buckle about the axis for which L_e/k has the greater value (this gives the lower load). For the x–axis the ends are fixed, so $L_e = 400$ mm and

$$L_e / k_x = 400 / 23.583 = 16.96$$

For the y–axis the ends are pinned so that $L_e = 800$ mm giving

$$L_e / k_y = 800 / 49.782 = 16.07$$

Having established that buckling is more likely to occur about the x–axis, the following safe loads can be found.

(i) Engesser

Applying eqs(13.3) and (13.6), Engesser's plastic buckling stress is

$$\sigma / \sigma_o = (E/\sigma_o)^n [\pi / (L_e / k)]^{2n}$$
$$= (210 \times 10^3/300)^{1/3} (\pi / 16.96)^{2/3} = 1.339$$
$$\sigma = 1.339 \times 300 = 401.76 \text{ MPa}$$

Hence the safe stress is 401.76 /1.5 = 267.84 MPa, from which the safe load is 267.84 × 2073.45 = 555.35 kN.

(ii) Parabola

Taking the intersection with the Euler curve to occur at $L_e/k = 120$, the corresponding stress in eq(13.4) is

$$\sigma = \pi^2 E / (L_e / k)^2 = 143.93 \text{ MPa}$$

and b is found from eq(13.9) as

$$b = (1 - 143.93 / 450) / (120)^2 = 47.23 \times 10^{-6}$$

Hence the plastic bucking stress is

$$\sigma = 450[1 - (47.23 \times 10^{-6})(16.96)^2] = 443.89 \text{ MPa}$$

Thus, the safe stress and load are 295.92 MPa and 613.58 kN respectively.

(iii) Rankine-Gordon

The buckling stress follows from eq(13.12), in units of N and mm, as:

$$\sigma = 450 \Bigg/ \left[1 + \frac{450}{300} \left(\frac{300}{\pi^2 \times 210000} \right)^{\frac{1}{3}} (16.96)^{\frac{2}{3}} \right] = 296.1 \text{ MPa}$$

Thus, the safe stress and load are 197.38 MPa and 409.27 kN respectively. Figure 13.4b compares the three buckling stress predictions graphically. It is seen that Engesser gives a valid stress ($\sigma < \sigma_u$) and that Rankine Gordon ($\sigma < \sigma_o$) is rather conservative. The intersection between curve (i) and Euler's curve (iv) is found from eq(13.5) as $L_e/k = 83.12$.

13.2.3 Wide Strut

Consider the compression of a wide, thin plate, thickness t, with similar length and width dimensions, a and b respectively (see Fig. 13.5a). With unsupported sides, the strut deflects laterally with uniform curvature while the cross-section $b \times t$ remains rectangular during bending. Consequently, biaxial, in-plane stresses, σ_x and σ_y, exist in the body of the plate.

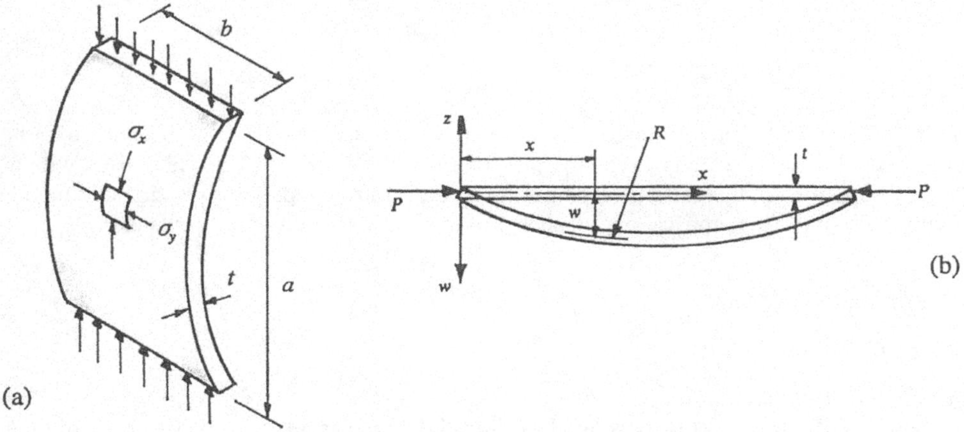

Figure 13.5 Buckling of a wide plate strut

With no dimensional change in the y–direction:

$$\varepsilon_y = 0 = (1/E)(\sigma_y - \nu\,\sigma_x)$$

$$\varepsilon_x = (1/E)(\sigma_x - \nu\,\sigma_y) = (1 - \nu^2)\sigma_x/E \qquad (13.13)$$

When b is small compared to a, the absence of ε_y results in *anticlastic curvature*, i.e. the cross-section does not remain rectangular. The distortion arises from opposing lateral strains induced within the tensile and compressive surfaces (the Poisson effect). Here $\sigma_y = 0$ and the strains along the length and width are simply $\varepsilon_x = \sigma_x/E$ and $\varepsilon_y = -\nu\,\sigma_x/E$. Comparison with eq(13.13) reveals a difference in ε_x between narrow and wide struts, i.e. it is necessary to modify the flexure equation for a wide strut by the factor $(1 - \nu^2)$. From Fig. 13.5b:

$$\frac{d^2w}{dx^2} = \frac{1}{R} = \frac{\varepsilon_x}{z}$$

Substituting from eq(13.13), with $M/I = \sigma_x/z$, it follows that

$$\frac{d^2w}{dx^2} = \frac{(1 - \nu^2)\sigma_x}{Ez} = (1 - \nu^2)\frac{M}{EI}$$

With $M = -Pw$, the solution to the critical buckling load for a pinned-end plate is

$$P_{cr} = \frac{\pi^2 EI}{(1 - \nu^2)a^2}$$

Substituting $P_{cr} = \sigma_{cr} bt$ and $I = Ak^2$ where $k^2 = t^2/12$ leads to the section stress

$$\sigma_{cr} = \frac{\pi^2 E}{(1 - \nu^2)(a/k)^2} \qquad (13.14a)$$

Setting $L_e = a\surd(1 - \nu^2)$ for simply supported ends, other edge fixings may be accounted for by re-writing eq(13.14a) as

$$\sigma_{cr} = \frac{\pi^2 E}{(L_e/k)^2} \tag{13.14b}$$

in which

$L_e = (a/2)\,\surd(1 - \nu^2)$ for fixed ends

$L_e = (a/\surd 2)\,\surd(1 - \nu^2)$ for pinned-fixed ends

$L_e = 2a(1 - \nu^2)$ for fixed-free ends

Engesser's inelastic theory applies to these equivalent lengths when E in eq(13.14b) is replaced by E_t. For example, $(L_e/k)_{Eng} = (a/k)\surd(1 - \nu^2)$ in Fig. 13.2 is identified with a simply supported plate. All other relationships, derived previously for plastic buckling of struts, including the empirical approaches, can then be applied to wide struts using L_e.

13.3 Buckling of Plates

13.3.1 Uniaxial Compression

(a) Simple Supports

Under a uni-directional compression, the critical buckling stress σ_x, for a plate with all four sides simply supported, is given as [2]:

$$(\sigma_x)_{cr} = (D\pi^2/t)\,[(m/a)^2 + (n/b)^2]^2 (a/m)^2$$

$$(\sigma_x)_{cr} = (D\pi^2/t\,b^2)\,[mb/a + n^2a/mb]^2 \tag{13.15}$$

where $D = Et^3/[12(1 - \nu^2)]$ is the flexural stiffness, t, a and b are the plate thickness, length and breadth respectively. The number of half-waves in the x– and y–directions are denoted by m and n respectively. For simple supports, the interior buckles with a half-wave ($n = 1$) in the y–direction, as shown in Fig. 13.6.

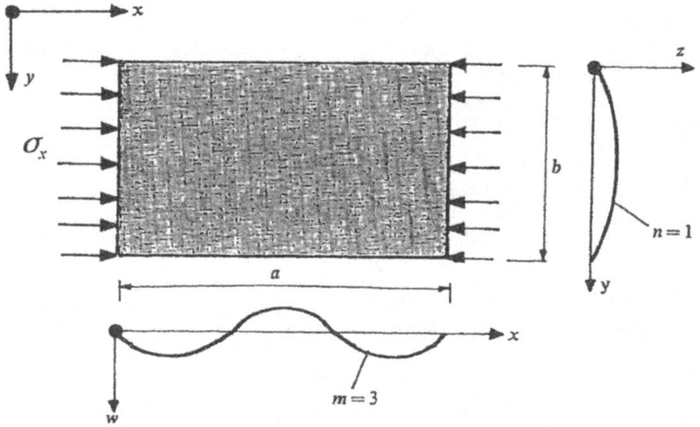

Figure 13.6 Buckling of a thin, simply-supported plate under uniaxial compression

From eq(13.15), the buckling stress becomes

$$(\sigma_x)_{cr} = (D\pi^2/t\,b^2)[m/r + r/m]^2 \tag{13.16}$$

where $r = a/b$. Differentiating for m-values that minimise eq(13.16) with integral r–values:

$$\begin{aligned} &\mathrm{d}(\sigma_x)_{cr}/\mathrm{d}m = 2(m/r + r/m)(1/r - r/m^2) = 0 \\ &m/r^2 + 1/m - 1/m - r^2/m^3 = 0 \\ &m^4 - r^4 = 0 \\ &(m - r)(m + r)(m^2 + r^2) = 0 \end{aligned}$$

The condition $m = r$ implies that the plate interior will buckle into an integral number of square cells $a \times a$ each under the same stress. That is, from eq(13.16):

$$(\sigma_x)_{cr} = \frac{4D\pi^2}{tb^2} = \frac{\pi^2 E}{3(1 - \nu^2)}\left(\frac{t}{b}\right)^2 \qquad (13.17)$$

The graph in Fig. 13.7 provides the buckling stress when non-integer values of r are taken with particular values of m in eq(13.16).

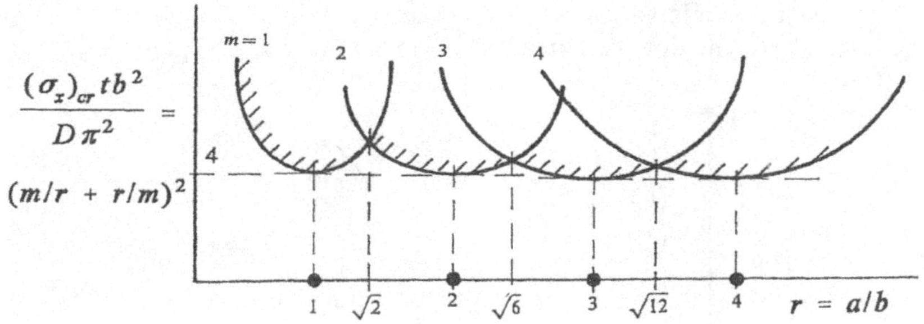

Figure 13.7 Effect of r and m on uniaxial buckling stress

The trough in each curve corresponds to eq(13.17) and at the intersections of these curves where $r = \sqrt{2}, \sqrt{6}, \sqrt{12}$ etc., m has been increased by one. This occurs when, from eq(13.16):

$$[\, m/r + r/m \,] = [\, (m + 1)/r + r/(m + 1) \,]$$

$$\therefore \quad r = \sqrt{m(m + 1)}$$

from which: $r = \sqrt{2}$ for $m = 1$, $r = \sqrt{6}$ for $m = 2$, $r = \sqrt{12}$ for $m = 3$ etc. Equation (13.17) should not be confused with the buckling stress for a thin, wide plate acting as a strut with its longer, parallel sides unsupported (see § 13.2.3).

(b) Other Edge Fixings

A number of approaches have been proposed for constrained edges. The simplest of these employs the expression

$$(\sigma_x)_{cr} = \frac{\pi^2 K_r D}{tb^2} \qquad (13.18a)$$

where the restraint coefficient K_r applies to any edge fixings when

$$K_r = (m/r)^2 + p + q\,(r/m)^2 \qquad (13.18b)$$

Clearly, with restraint factors $p = 2$ and $q = 1$, eqs(13.18a,b) contain eq(13.16) in the special case of simple supports. The dependence of K_r upon the two rotational edge restraint factors

(p and q), plate dimensions ($r = a/b$) and buckling mode m has been established experimentally in certain cases. Table 13.1 applies to the case of fixed sides.

Table 13.1 Restraint coefficients for a plate with fixed edges

$r = a/b$	0.75	1.0	1.5	2.0	2.5	3.0
K_r	11.69	10.07	8.33	7.88	7.57	7.37

As r increases, the effect of edge restraint lessens and K_r approaches the minimum value of 4 as found from eq(13.17) for a plate with simply supported edges. An alternative graphical approach [3] employs design curves derived from the restraint coefficients. The latter supply the ratio between the critical elastic buckling stress $(\sigma_{cr})_e$ for a plate with a given edge fixing to that of a simply supported plate with a similar geometry: $r = a/b$ in eq(13.16). Figure 13.8 shows how this stress ratio varies with r for clamped and various mixed-edge fixings. A similar graphical presentation has extended this to biaxially loaded plates [3].

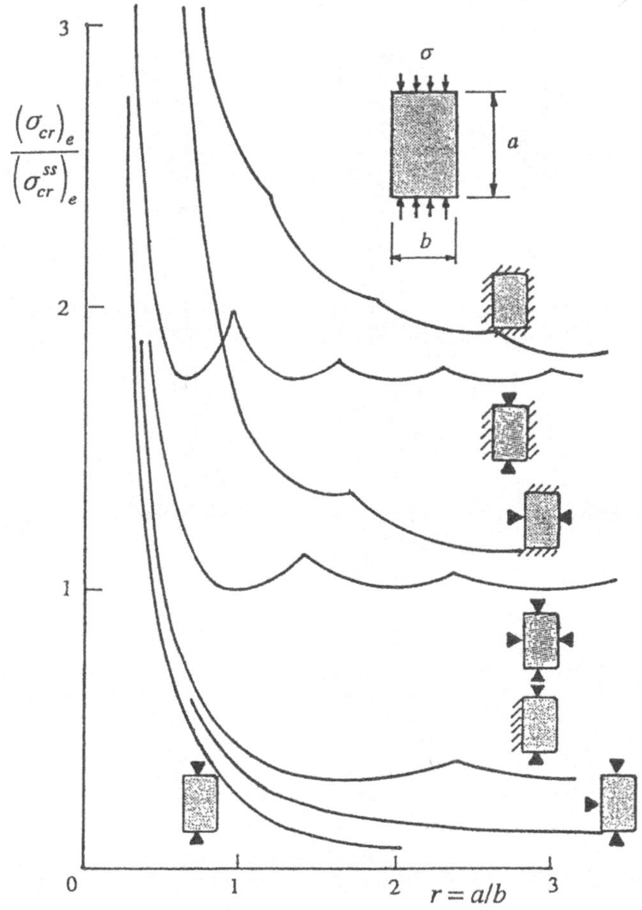

Figure 13.8 Plate buckling under compression

(c) Inelastic Buckling

In thicker plates the critical elastic stress $(\sigma_{cr})_e$, as calculated from eq(13.17), can exceed the yield stress Y of the plate material. The solution will be invalid because, with the use of E and ν in eq(13.17), linear elasticity is assumed at the critical stress level. Figure 13.9a shows that where plasticity has occurred it reduces the buckling stress to a lower level $(\sigma_{cr})_p$. Figure 13.9b shows how tangent and secant moduli, E_t and E_s respectively, are used to account for a plastic stress level.

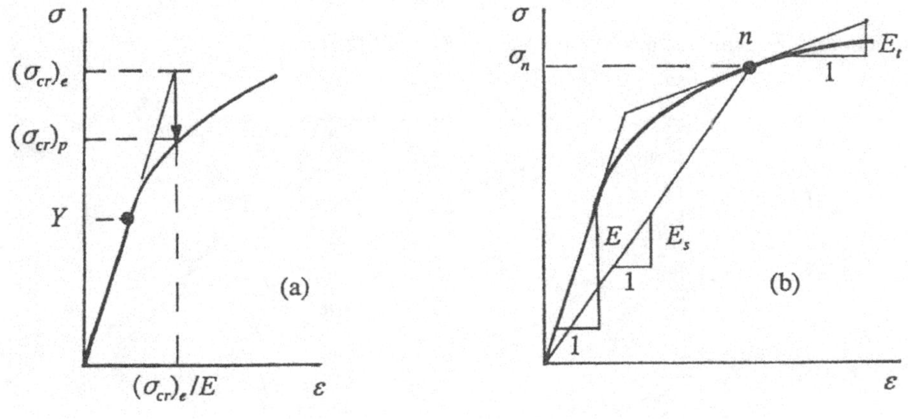

Figure 13.9 Tangent and secant modulus

The secant modulus gives the total strain at a reference point n (see Fig. 13.9b) as

$$\varepsilon_n = \sigma_n / E_s \tag{13.19}$$

The Ramberg-Osgood description [4] of a stress–strain curve gives the total strain under a plastic stress:

$$\varepsilon = \sigma / E + \alpha (\sigma / E)^m \tag{13.20}$$

Combining eqs(13.19) and (13.20) for the reference stress σ_n gives

$$E/E_s - 1 = \alpha (\sigma_n / E)^{m-1} \tag{13.21a}$$

Differentiating eq(13.20) gives the gradient $E_t = \mathrm{d}\sigma/\mathrm{d}\varepsilon$ at point n:

$$(1/m)(E/E_t - 1) = \alpha (\sigma / E)^{m-1} \tag{13.21b}$$

Let σ_n be the stress level at which $E_t = E/2$. Equation (13.21b) gives

$$\alpha = (1/m)(\sigma_n / E)^{1-m} \tag{13.21c}$$

Substituting eq(13.21c) into eq(13.20) and multiplying through by E/σ_n leads to a normalised stress–total strain relationship

$$\varepsilon = \frac{\sigma_n}{E}\left[\frac{\sigma}{\sigma_n} + \frac{1}{m}\left(\frac{\sigma}{\sigma_n}\right)^m\right] \tag{13.22}$$

where σ_n and m are material properties found from fitting eq(13.22) to a stress–strain curve.

Typical m–values lie in a range 16–29 for aluminium alloys and in a range 5–17 for steels. The m–value depends upon whether the material is in sheet or bar form. Figure 13.9a shows that the elastic strain under $(\sigma_{cr})_e$ is $\varepsilon = (\sigma_{cr})_e/E$. Substituting this value for ε in eq(13.22) allows $\sigma = (\sigma_{cr})_p$ to be found. Design data [5] employ a *plasticity reduction factor* $\mu < 1$ with a graphical solution for $(\sigma_{cr})_p$:

$$\mu = \frac{(\sigma_{cr})_p}{(\sigma_{cr})_e} \tag{13.23}$$

Both m and σ_n influence μ in the manner shown in Fig. 13.10.

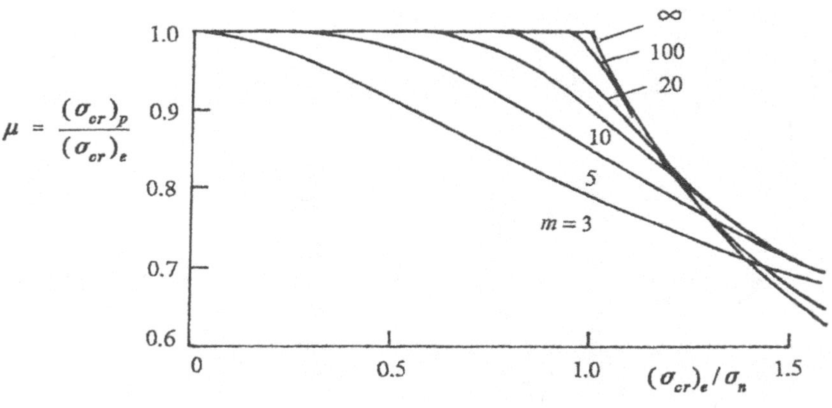

Figure 13.10 Plasticity reduction factor

Firstly, a critical elastic stress is found from eq(13.17). This determines the ratio $(\sigma_{cr})_e/\sigma_n$ from which a μ–value is found from Fig. 13.10. Equation (13.23) is then employed to find $(\sigma_{cr})_p$ as in the following example.

Example 13.2 A 320 mm square steel plate is 7 mm thick. It is is simply supported along all sides and carries a uni-axial compressive stress. Determine the critical elastic and plastic buckling stresses. What is the influence of clamping the unloaded sides upon the plastic buckling stress? Take: $E = 210$ GPa, $Y = 310$ MPa, $\sigma_n = 450$ MPa, $\nu = 0.27$ and $m = 5$.

The plate aspect ratio is an integral number $r = 1$. Equation (13.17) provides the critical elastic buckling stress for simply-supported sides

$$(\sigma_{cr})_e = \frac{\pi^2 \times 210 \times 10^3}{3(1 - 0.27^2)}\left(\frac{7}{320}\right)^2 = 356.59 \text{ MPa}$$

Since this stress exceeds the compressive yield stress $Y = 310$ MPa, a correction for plasticity becomes necessary. Thus, $(\sigma_{cr})_e/\sigma_n = 356.59/450 = 0.792$, when Fig. 13.10 gives $\mu = 0.916$ for $m = 5$. Hence, we find from eq(13.23) a critical plastic buckling stress:

$$(\sigma_{cr})_p = 0.916 \times 356.59 = 326.64 \text{ MPa}$$

With the loaded sides clamped, Fig. 13.8 gives $(\sigma_{cr})_e = 1.7 \times 356.59 = 606.2$ MPa. Correcting for plasticity as before: $(\sigma_{cr})_e/\sigma_n = 606.2/450 = 1.347$. Figure 13.10 gives a reduction factor $\mu = 0.753$ when, from eq(13.23), the plastic buckling stress has been increased to $(\sigma_{cr})_p = 0.753 \times 606.2 = 456.47$ MPa.

13.3.2 Biaxial Compression

Consider a simply-supported, thin rectangular plate $a \times b$ with thickness t. Let uniform compressive stresses σ_x act normal to $b \times t$ and σ_y act normal to $a \times t$, as shown in Fig. 13.11.

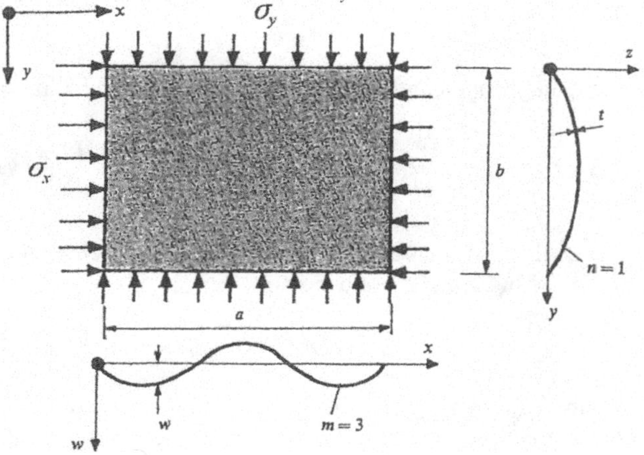

Figure 13.11 Buckling of a thin plate under biaxial stress

When the stresses increase proportionately in the ratio $\beta = \sigma_y/\sigma_x$, the actual number of half waves of buckling are those which minimise σ_x. In general, this stress is expressed as [2]

$$(\sigma_x)_{cr} = \frac{D\pi^2[(m/a)^2 + (n/b)^2]^2}{t[(m/a)^2 + \beta(n/b)^2]} \tag{13.24}$$

where D is the flexural stiffness and m and n are the respective number of half waves m and n for buckling in the x and y directions. In Fig. 13.11, for example, $m = 3$ and $n = 1$ is shown. For a square plate, where $a = b$, buckling occurs with one half wave in each direction and $m = n = 1$. Equation (13.24) reduces to

$$(\sigma_x)_{cr} = \frac{\pi^2 E}{3(1 - \nu^2)(1 + \beta)}\left(\frac{t}{a}\right)^2 \tag{13.25a}$$

and in the case of equi-biaxial compression $\beta = 1$, when eq(13.25a) gives

$$(\sigma_x)_{cr} = \frac{\pi^2 E}{6(1 - \nu^2)}\left(\frac{t}{a}\right)^2 \tag{13.25b}$$

Another useful buckling stress $(\sigma_x)_{cr}$ applies to $b \ll a$. Firstly, find the number of half wavelengths of the buckled shape for the x–direction. Taking $n = 1$ for the much smaller b dimension, eq(13.24) becomes

$$(\sigma_x)_{cr} = \frac{D\pi^2(m^2 + r^2)^2}{a^2 t(m^2 + \beta r^2)} = \frac{D\pi^2(m/r + r/m)^2}{b^2 t[1 + \beta(r/m)^2]} \tag{13.26}$$

and m is found from the condition that $(\sigma_x)_{cr}$ is a minimum. This is

$$d(\sigma_x)_{cr}/dm = [1 + \beta\,(r/m)^2]\,2(m/r + r/m)(1/r - r/m^2) - (m/r + r/m)^2 \times 2\beta\,(r/m)(-\,r/m^2) = 0$$

$$[1 - (r/m)^2][1 + \beta\,(r/m)^2] + \beta\,(r/m)^2[1 + (r/m)^2] = 0$$

$$\therefore\ (r/m)^2(1 - 2\beta) = 1$$

This gives

$$m = r\,(1 - 2\beta)^{1/2} \tag{13.27}$$

with the corresponding half wavelength $a/m \approx r/m = (1 - 2\beta)^{-1/2}$. Substituting eq(13.27) into eq(13.26):

$$(\sigma_x)_{cr} = \frac{D\pi^2[(1 - 2\beta) + 2 + 1/(1 - 2\beta)]}{bt^2[1 + \beta/(1 - 2\beta)]} = \frac{4D\pi^2}{b^2 t}(1 - \beta) \tag{13.28a}$$

It is apparent from eq(13.28a) that as $\beta \to ½$, the half wavelength in the x–direction approaches infinity as the buckling stress becomes

$$(\sigma_x)_{cr} \to (2D\pi^2)\,/\,(b^2 t) \tag{13.28b}$$

The critical lateral stress follows from each of these solutions as $(\sigma_y)_{cr} = \beta\,(\sigma_x)_{cr}$. These stresses remain elastic provided they satisfy a von Mises yield criterion:

$$\sigma_x^2 - \sigma_x\sigma_y + \sigma_y^2 \le Y^2$$

which gives

$$(\sigma_x)_{cr} \le Y / \sqrt{(1 - \beta + \beta^2)}$$

If $(\sigma_x)_{cr}$ exceeds this limiting elastic value a correction for plasticity may be made with a reduction factor as before. When applying Fig. 13.10 to one component of a biaxial stress it becomes necessary to re-define the abscissa as $(\sigma_x)_{cr}/\sigma_n'$ where $\sigma_n' = \sigma_n / (1 - \beta - \beta^2)^{½}$.

13.3.3 Local Buckling of Plate Sections

The straight, thin walls of an open section may distort without translation or rotation. Localised stress concentrations at corners may exceed the yield stress and cause either a local crippling failure or a reduction in the resistance to buckling by other modes. Strut sections, consisting of straight sides such as I, ⊤, ⏋, ⎣ and ⊔, as well as thin-walled closed tubes, are prone to local buckling at sharp corners. When the strut length is at least four times the section depth h, each limb may be treated as a plate with simple side support from one or both its neighbouring limbs. A supported side will restrain an unsupported side. The solution to the local elastic compressive buckling stress takes the common form

$$\sigma_{be} = KE\left(\frac{t}{h}\right)^2 \tag{13.29}$$

where the thickness t of the sides $t < d/5$, where d is the semi-flange length. The buckling coefficient K depends upon the d/h ratio and the neighbouring restraints within the four sections as shown in Fig. 13.12. When $\nu \neq 0.3$, K should be corrected by a multiplying factor $(1 - 0.3^2)/(1 - \nu^2)$. If σ_{be} from eq(13.29) is found to exceed the yield stress of the strut material it again becomes necessary to employ the plasticity reduction factor $\sigma_b = \mu\sigma_{be}$ with a suitable description to the stress–strain curve, as in the previous example.

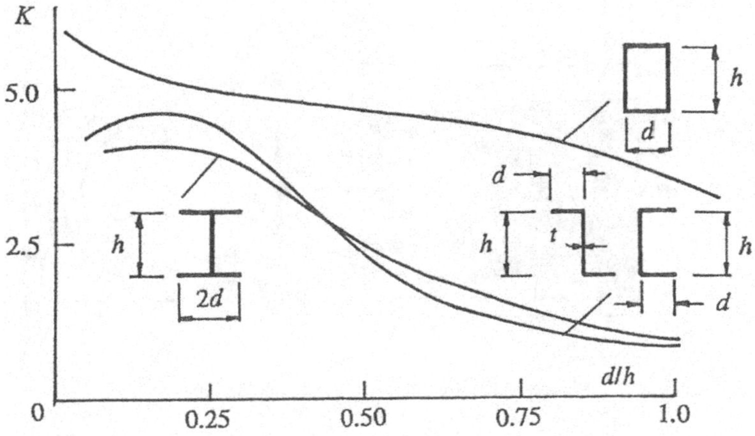

Figure 13.12 Local buckling coefficient for uniform thin sections

In medium length struts, local instability results in a loss of stiffness but without complete failure. However, the influence of local instability upon global buckling arising from flexure and torsion is important. If the strut is very long the global buckling stresses are attained well before the onset of local buckling. On the other hand, a short strut may carry more compression beyond that producing local buckling. Final failure is estimated to occur under a crippling stress σ_c, found from

$$\sigma_c = \sqrt{\sigma_b \times \sigma_{0.1}} \tag{13.30}$$

where $\sigma_{0.1}$ refers to the 0.1 % compressive proof stress for the strut material. Test data shows that eq(13.30) is accurate to within 10 %.

Example 13.3 Find the buckling and crippling stress for a short strut of angle section (⏋), given that its dimensions are: $h = 125$ mm, $d = 25$ mm and $t = 4$ mm (see Fig. 13.12). Take $E = 74$ GPa, $\nu = 0.3$, $m = 10$, $\sigma_{0.1} = 325$ MPa and $\sigma_n = 280$ MPa.

From Fig. 13.12, we find $K = 4$ for $d/h = 0.2$. Equation (13.29) gives the local elastic buckling stress

$$\sigma_{be} = 4 \times 74 \times 10^3 (4 / 125)^2 = 303.1 \text{ MPa}$$

This exceeds the yield stress because $\sigma_n = 280$ MPa corresponds to $E_t = 2E$ as in Fig. 13.9b. The ratio $\sigma_{be}/\sigma_n = 303.1/280 = 1.083$ is used with Fig. 13.10 to give a plasticity reduction factor $\mu = 0.877$. Equation (13.23) provides the local buckling stress

$$\sigma_b = \mu\, \sigma_{be} = 0.877 \times 303.1 = 265.82 \text{ MPa}$$

and using eq(13.30), a crippling stress is estimated:

$$\sigma_c = \sqrt{(265.82 \times 325)} = 293.92 \text{ MPa}$$

13.3.4 Post-Buckling of Flat Plates

When a plate buckles the load may be increased further as the axial compressive stress increases in the material along the side supports (see Fig. 13.13).

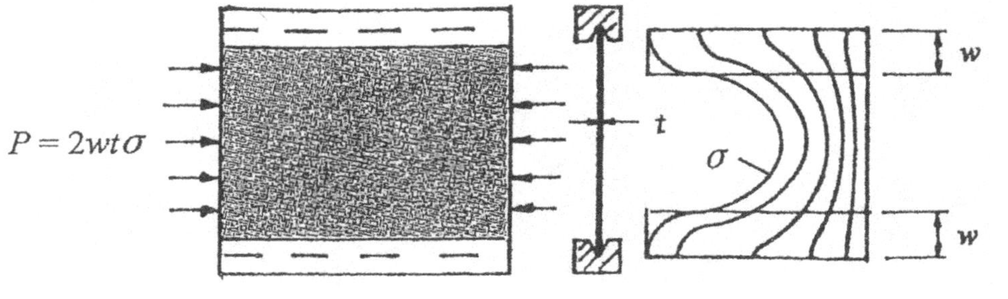

Figure 13.13 Stress distribution in a buckled plate

Only a slight increase in axial stress occurs in the central buckled material. Assuming that the whole load P is carried by two edge strips of effective width $2w$, over which σ is assumed constant, the load supported becomes

$$P = 2wt\sigma$$

With the edges of our equivalent elastic plate all simply supported, the critical buckling stress is [2]

$$\left(\sigma_x\right)_{cr} = \frac{\pi^2 E}{3(1 - \nu^2)}\left(\frac{t}{2w}\right)^2 \tag{13.31a}$$

from which w may be found once $(\sigma_x)_{cr}$ attains the yield stress Y:

$$w = \frac{\pi t}{2}\sqrt{\frac{E}{3Y(1 - \nu^2)}} \tag{13.31b}$$

Taking $\nu = 0.3$ in eq(13.31b) gives

$$w = 0.95\, t\sqrt{E/Y}$$

Experiment shows that the coefficient is nearer 0.85. With other edge fixings along the long sides, the asymptotic values of the critical elastic buckling stress ratio applies. These appear in Fig. 13.8 in which eq(13.31a) defines the denominator. For example, when one side is simply supported and the other is free, Fig. 13.8 gives $\sigma_{cr}/\sigma_b = 0.106$ and this modifies the semi-effective width:

$$w = 0.31\, t\sqrt{E/Y}$$

13.3.5 Buckling of Plates in Shear

When the sides of a thin plate $a \times b \times t$ are subjected to shear stress τ, the principal stress state (σ_1, σ_2) within the plate becomes one of diagonal tension and compression. The principal stresses are of equal magnitude: $\sigma_1 = \tau$ and $\sigma_2 = -\tau$ (see Fig. 13.14a). Hence, the principal, biaxial stress ratio $\beta = -1$. Shear buckling occurs with parallel wrinkles lying perpendicular to the compressive stress σ_2 (see Fig. 13.14b).

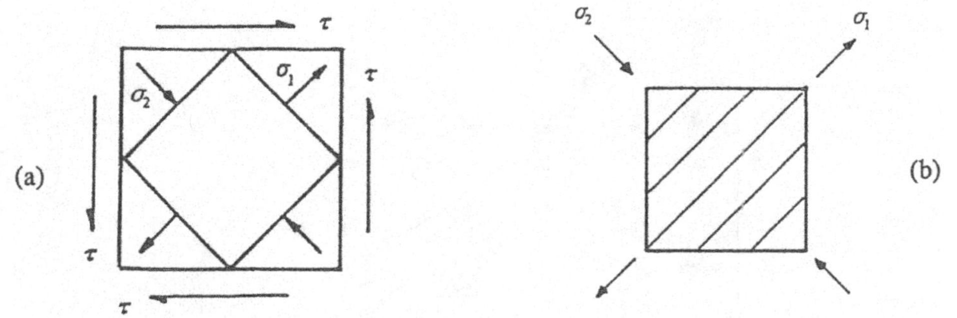

Figure 13.14 Plate in shear showing principal stress directions

Under this condition the plate cannot sustain a further increase in diagonal compression though an increase in diagonal tension is possible. Therefore, a compressive buckling mode of flat plates arises under shear. The critical elastic shear stress is given by [6]

$$(\tau_{cr})_e = K_s E\,(t/b)^2 \tag{13.32}$$

where b is the lesser side length. The shear buckling coefficient K_s depends upon the edge fixing in the manner of Fig. 13.15 when $\nu = 0.3$.

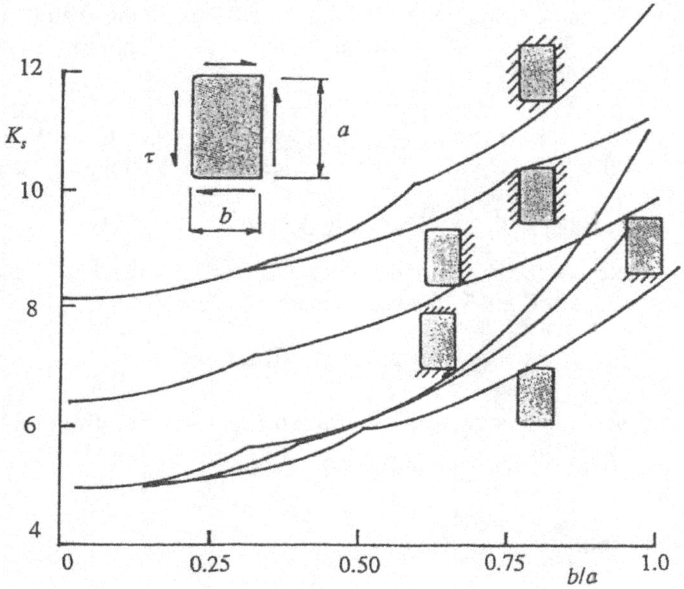

Figure 13.15 Dependence of shear buckling coefficient upon geometry and edge-fixing

If $\nu \neq 0.3$, the K_s value is factored by $0.91/(1 - \nu^2)$. Equation (13.32) supplies $(\tau_{cr})_e$ but this critical elastic stress will need further correction if it exceeds the shear yield stress k of the plate material [7]. The plasticity reduction factor μ is obtained from Fig. 13.16 knowing σ_n and m for the material. These give

$$(\tau_{cr})_p = \mu\,(\tau_{cr})_e \tag{13.33}$$

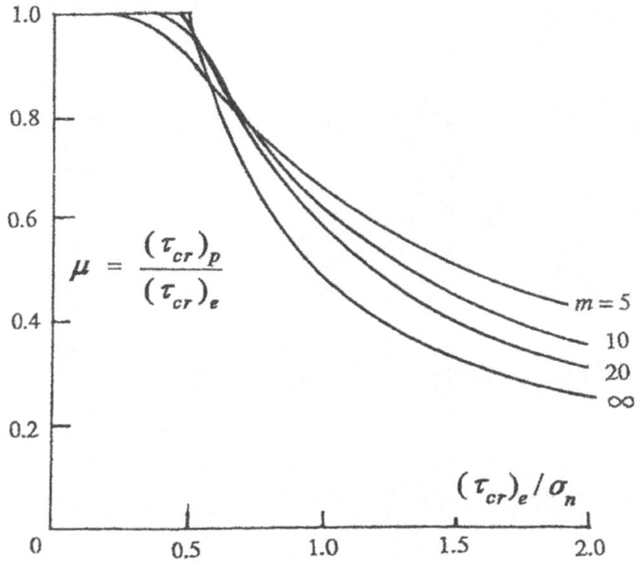

Figure 13.16 Plasticity reduction factor in shear

Example 13.4 Find the critical buckling stress in shear loading an aluminium plate 2.5 mm thick, 100 mm wide and 200 mm long. The longer sides are clamped and the shorter sides are simply supported. Take the plate material properties as: $k = 200$ MPa, $\sigma_n = 350$ MPa, $m = 15$, $E = 70$ GPa and $\nu = 0.33$.

For $b/a = 0.5$, the appropriate curve in Fig. 13.15 gives $K_s = 9.0$. Correcting for Poisson's ratio gives $K_s = 9.0 \times 0.91/(1 - 0.33^2) = 9.19$. The elastic buckling stress is found from eq(13.32):

$$(\tau_{cr})_e = 9.19 \times 70 \times 10^3 (2.5 / 100)^2 = 402.1 \text{ MPa}$$

Since this stress exceeds the shear yield stress $k = 200$ MPa, a plasticity reduction factor is required. The abscissa in Fig. 13.16 takes a value:

$$(\tau_{cr})_e / \sigma_n = 402.1 / 350 = 1.149$$

Interpolating for $m = 15$ gives an ordinate of $\mu = 0.52$. Therefore, from eq(13.33):

$$(\tau_{cr})_p = 0.52 \times 402.1 = 209.1 \text{ MPa}$$

13.4 Tensile Instability

13.4.1 Uniaxial Tension

In chapter 9 it was shown that the maximum load condition provided the derivative relation, eq(9.13a), between nominal stress and strain at the inception of necking in a tension test. The various Considére expressions (9.14a–c) for this condition employ nominal or true stress, engineering or natural (logarithmic) strain according to axes chosen to present the stress–strain curve. Empirical equations used to describe flow curves (see Figs 9.18a–f) employ true stress and natural strain to be used with eq(9.14b), as in the following example.

Example 13.5 Determine the true stress and true strain at the point of tensile instability for the Hollomon and Ludwik functions.

From eq(9.39b):
$$\frac{d\sigma}{d\varepsilon} = n\left(\frac{\sigma_o}{\varepsilon_o}\right)\left(\frac{\varepsilon}{\varepsilon_o}\right)^{n-1}$$

and from eq(9.14b), at the point of instability $d\sigma/d\varepsilon = \sigma$:

$$n\left(\frac{\sigma_o}{\varepsilon_o}\right)\left(\frac{\varepsilon}{\varepsilon_o}\right)^{n-1} = \sigma_o\left(\frac{\varepsilon}{\varepsilon_o}\right)^{n}$$

which shows that

$$\varepsilon = n \quad \text{and} \quad \sigma = \sigma_o\left(\frac{n}{\varepsilon_o}\right)^{n}$$

That is, the strain at which necking begins equals the work hardening exponent for a given material. The Ludwik law (9.43a) gives

$$\frac{d\sigma}{d\varepsilon^P} = nA(\varepsilon^P)^{n-1}$$

Combining this with eqs(9.14c) and (9.43a) we find the instability strain

$$\varepsilon^P \simeq n\left(1 - \frac{Y}{\sigma}\right)$$

giving a quadratic in the instability stress

$$\sigma^2 - (Y + An^n)\sigma + An^{n+1}Y = 0$$

13.4.2 Orthotropic Sheet

The previous example assumes an isotropic condition, i.e. the analysis is independent of the testpiece orientation. In rolled sheet metal both the initial condition and the orientation of the testpiece needs to be considered. Recall that an equivalent flow curve expresses a current stress and strain state aligned with the material's orthogonal directions. Using eqs(12.2) and (12.4a), a plane stress yield function *f* appears in terms of an anisotropy parameters *F*, *G*, *H* and the equivalent stress $\bar{\sigma}$ as

$$2f = (2/3)(F + G + H)\bar{\sigma}^2 = (G+H)\sigma_1^2 - 2H\sigma_1\sigma_2 + (F + H)\sigma_2^2 = 1 \qquad (13.34)$$

Re-writing eq(13.34) with the *r*–values $r_1 = H/G$ and $r_2 = H/F$ leads to an equivalent plane stress expression:

$$\bar{\sigma}^2 = \frac{r_2(1 + r_1)\sigma_1^2 - 2r_1r_2\sigma_1\sigma_2 + r_1(1 + r_2)\sigma_2^2}{2(r_1 + r_1r_2 + r_2)/3} \qquad (13.35)$$

The equivalent plastic strain increment is found from the plastic work expression (12.8b):

$$\bar{\sigma}\,d\bar{\varepsilon}^P = \sigma_{ij}\,d\varepsilon_{ij}^P$$

From eq(13.35):

$$\left(\mathrm{d}\bar{\varepsilon}^{P}\right)^2 = \frac{2(r_1 + r_1 r_2 + r_2)\left[(1/r_2)(\mathrm{d}\varepsilon_1^P)^2 + (1/r_1)(\mathrm{d}\varepsilon_2^P)^2 + (\mathrm{d}\varepsilon_3^P)^2\right]}{3(1 + r_1 + r_2)} \tag{13.36}$$

When a tensile testpiece is aligned with the sheet's rolling direction, the equivalent stress follows from eq(13.35) as

$$\bar{\sigma}^2 = \frac{3r_2(1 + r_1)\sigma_1^2}{2(r_1 + r_1 r_2 + r_2)} \tag{13.37}$$

where σ_1 is the tensile stress. The lateral strain increments are

$$\mathrm{d}\varepsilon_2^P = -\frac{r_1\,\mathrm{d}\varepsilon_1^P}{1 + r_1} \quad \text{and} \quad \mathrm{d}\varepsilon_3^P = -\frac{\mathrm{d}\varepsilon_1^P}{1 + r_1} \tag{13.38a,b}$$

Substituting eqs(13.38a,b) into eq(13.36) gives the total equivalent plastic strain

$$\left(\bar{\varepsilon}^P\right)^2 = \frac{2(r_1 + r_1 r_2 + r_2)}{3r_2(1 + r_1)}\left(\varepsilon_1^P\right)^2 \tag{13.39}$$

Writing eq(13.37) as $\bar{\sigma} = K\sigma_1$, then eq(13.39) becomes $\bar{\varepsilon}^P = (\varepsilon_1^P)/K$ where K is the square root of the coefficient containing r_1 and r_2. It follows that the instability condition, eq(9.14b), applies to the true stress–strain curve for the 1–direction as

$$\frac{\mathrm{d}\sigma_1}{\mathrm{d}\varepsilon_1^P} = \sigma_1 \tag{13.40a}$$

Equation (13.40a) is written as

$$\frac{\mathrm{d}\sigma_1}{\mathrm{d}\bar{\sigma}} \times \frac{\mathrm{d}\bar{\varepsilon}^P}{\mathrm{d}\varepsilon_1} \times \frac{\mathrm{d}\bar{\sigma}}{\mathrm{d}\bar{\varepsilon}^P} = \frac{\bar{\sigma}}{K}$$

giving

$$\frac{\mathrm{d}\bar{\sigma}}{\mathrm{d}\bar{\varepsilon}^P} = K\bar{\sigma} \tag{13.40b}$$

so that the reciprocal of K becomes the strain intercept (i.e. z in Fig. 13.17) made by the tangent to the $\bar{\sigma}$ versus $\bar{\varepsilon}^P$ curve at the point of instability.

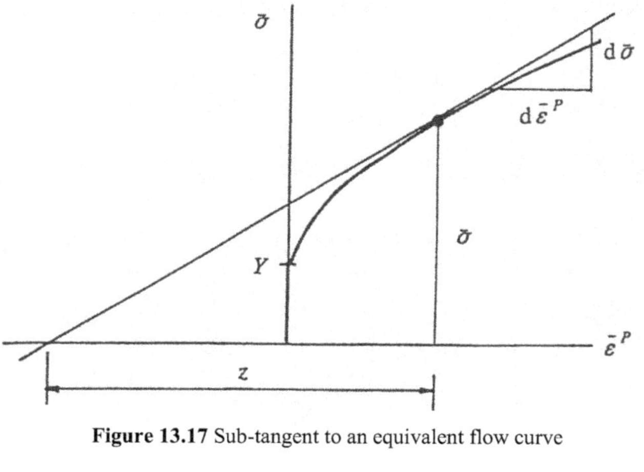

Figure 13.17 Sub-tangent to an equivalent flow curve

If we approximate this curve with the Hollomon law (9.39b), this gives $\mathrm{d}\bar{\sigma}/\mathrm{d}\bar{\varepsilon}^P = n\bar{\sigma}/\bar{\varepsilon}^P$ from which $K = n/\bar{\varepsilon}^P$. Therefore, the equivalent strain at the point of instability is $\bar{\varepsilon}^P = n/K$. For tension upon an orthotropic sheet, discussed above:

$$z = \frac{1}{K} = \sqrt{\frac{2(r_1 + r_2 + r_1 r_2)}{3r_2(1 + r_1)}}$$

It is also possible to predict this strain in a similar manner for off-axis testpieces using a single flow curve equation and the strain transformation equations [8].

13.4.3 Biaxial Subtangent, Pressure Vessels

We may extend the previous instability analysis to problems involving more than one stress with the use of equivalent stress and plastic strain. Equation (13.40b) shows that this will modify Considére's condition to

$$\frac{\mathrm{d}\bar{\sigma}}{\mathrm{d}\bar{\varepsilon}^P} = \frac{\bar{\sigma}}{z} \tag{13.40c}$$

where z is the sub-tangent that defines the particular problem. Figure 13.17 gives a geometrical interpretation to eq(13.40c) in which z is the intercept along the strain axis made with the gradient $\mathrm{d}\bar{\sigma}/\mathrm{d}\bar{\varepsilon}^P$ of the tangent and the ordinate $\bar{\sigma}$ as shown. Clearly, for a tension test upon isotropic material $z = 1$ since $\bar{\sigma}$ and $\bar{\varepsilon}^P$ are, by definition, the axial stress and axial plastic strain in that test. Pressure vessels require z to be found from combining a maximum pressure instability criterion with constitutive relations defining the material condition. The conversions will enable $\bar{\sigma}$ and $\bar{\varepsilon}^P$ to be written in terms of a single dependent variable, allowing the application of eq(13.40c) to the vessels that follow.

(a) Spherical Pressure Vessel

Let the pressure vessel's isotropic steel have an equivalent stress–strain curve represented by a Swift law:

$$\bar{\sigma} = \sigma_o\left(1 + \bar{\varepsilon}^P/\varepsilon_o\right)^n$$

It is required to find expressions for: (i) the internal pressure at which a bi-axial tensile instability occurs, (ii) the equivalent strain and (iii) the maximum hoop stress and strain reached at the point of instability. The hoop and meridional stresses are $\sigma_\theta = \sigma_\varphi = pr/2t$, from which

$$\ln \sigma_\theta = \ln (p/2) + \ln r - \ln t$$

$$\mathrm{d}\sigma_\theta/\sigma_\theta = \mathrm{d}p/p + \mathrm{d}r/r - \mathrm{d}t/t$$

Since $\mathrm{d}p/p = 0$ at the point of instability and $\mathrm{d}\varepsilon_\theta^{\,P} = \mathrm{d}r/r$, $\mathrm{d}\varepsilon_r^{\,P} = \mathrm{d}t/t$ and then

$$\mathrm{d}\sigma_\theta/\sigma_\theta = \mathrm{d}\varepsilon_\theta^{\,P} - \mathrm{d}\varepsilon_r^{\,P} \tag{13.41}$$

For the von Mises equivalent stress (9.2d) we replace subscripts 1, 2 and 3 by θ, φ and r to identify the stress components: $\sigma_1 = \sigma_\theta$, $\sigma_2 = \sigma_\varphi = \sigma_\theta$ and $\sigma_3 = \sigma_r = 0$. Thus:

$$\bar{\sigma} = \frac{1}{\sqrt{2}}\sqrt{0 + \sigma_\varphi^{\,2} + \sigma_\theta^{\,2}} = \sigma_\theta \tag{13.42}$$

The Levy-Mises equation (9.29a–c) gives

$$\mathrm{d}\varepsilon_\theta^{\;P} = (2\,\mathrm{d}\lambda/3)\sigma_\theta/2$$

$$\mathrm{d}\varepsilon_\varphi^{\;P} = (2\,\mathrm{d}\lambda/3)\sigma_\theta/2 = \mathrm{d}\varepsilon_\theta^{\;P}$$

$$\mathrm{d}\varepsilon_r^{\;P} = -\,(\mathrm{d}\varepsilon_\theta^{\;P} + \mathrm{d}\varepsilon_\varphi^{\;P}) = -\,2\,\mathrm{d}\varepsilon_\theta^{\;P}$$

$$\therefore\ \mathrm{d}\varepsilon_\theta^{\;P} = \mathrm{d}\varepsilon_\varphi^{\;P} = -\,\mathrm{d}\varepsilon_r^{\;P}/2 \tag{13.43}$$

when from eq(9.6d) the equivalent plastic strain increment is

$$\mathrm{d}\bar{\varepsilon}^{P} = \sqrt{(2/3)}\,\sqrt{[(\mathrm{d}\varepsilon_\theta^{\;P})^2 + (\mathrm{d}\varepsilon_\theta^{\;P})^2 + (-\,2\mathrm{d}\varepsilon_\theta^{\;P})^2]} = 2\,\mathrm{d}\varepsilon_\theta^{\;P} \tag{13.44}$$

From eqs(13.42) and (13.44):

$$\mathrm{d}\bar{\sigma}/\mathrm{d}\bar{\varepsilon}^{P} = 1/2\,(\mathrm{d}\sigma_\theta/\mathrm{d}\varepsilon_\theta^{P})$$

where, from eqs(13.41) and (13.43):

$$\frac{\mathrm{d}\sigma_\theta}{\mathrm{d}\varepsilon_\theta^{P}} = \sigma_\theta\left(1 - \frac{\mathrm{d}\varepsilon_r^{P}}{\mathrm{d}\varepsilon_\theta^{P}}\right) = 3\sigma_\theta$$

$$\therefore\ \frac{\mathrm{d}\bar{\sigma}}{\mathrm{d}\bar{\varepsilon}^{P}} = \frac{3\bar{\sigma}}{2}$$

Thus, in the equivalent plot of Fig. 13.17, the sub-tangent value is $z = 2/3$. The equivalent plastic instability strain is found from the Swift law:

$$\frac{\mathrm{d}\bar{\sigma}}{\mathrm{d}\bar{\varepsilon}^{P}} = \frac{n\bar{\sigma}}{[\varepsilon_o(1 + \bar{\varepsilon}^{P}/\varepsilon_o)]} = \frac{3\bar{\sigma}}{2}$$

$$\therefore\ \bar{\varepsilon}^{P} = 2n/3 - \varepsilon_o$$

from which the maximum hoop strain is $\varepsilon_\theta^{P} = ½\,\bar{\varepsilon}^{P}$ and the equivalent stress at instability is $\bar{\sigma} = \sigma_o[2n/(3\varepsilon_o)]^n$. Since $\bar{\sigma} = \sigma_\theta = pr/2t$, this gives the critical pressure as

$$p = \frac{2t\sigma_o}{r}\left(\frac{2n}{3\varepsilon_o}\right)^n \tag{13.45}$$

(b) Cylindrical Pressure Vessel

For a thin-walled, closed cylinder the stresses are $\sigma_r = 0$ and $\sigma_\theta = pr/t = 2\sigma_z$. Hence, eq(13.41) again applies to the point of instability. The equivalent stress eq(9.2d) gives

$$\bar{\sigma} = (1/\surd 2)\,\surd[(\sigma_\theta - \sigma_z)^2 + \sigma_\theta^{\,2} + \sigma_z^{\,2}]$$

$$= (1/\surd 2)\,\surd[\tfrac{1}{4}(\sigma_\theta)^2 + \sigma_\theta^{\,2} + \tfrac{1}{4}(\sigma_\theta)^2] = \surd 3\sigma_\theta/2 \qquad (13.46)$$

and from eq(9.29a–c), it follows that $d\varepsilon_z^{\,P} = 0$ and

$$d\varepsilon_\theta^{\,P} = (2\,d\lambda/3)[\sigma_\theta - \tfrac{1}{2}(0 + \tfrac{1}{2}\sigma_\theta)] = (d\lambda/2)\sigma_\theta$$

$$d\varepsilon_r^{\,P} = (2\,d\lambda/3)[0 - \tfrac{1}{2}(\sigma_\theta + \tfrac{1}{2}\sigma_\theta)] = -\,(d\lambda/2)\sigma_\theta$$

$$\therefore\ d\varepsilon_r^{\,P} = -\,d\varepsilon_\theta^{\,P} \qquad (13.47)$$

The equivalent plastic strain is, from eq(9.6e)

$$d\bar{\varepsilon}^{P} = (\surd 2/3)\surd[(d\varepsilon_\theta^{\,P})^2 + (-d\varepsilon_\theta^{\,P})^2 + (0)^2] = (2/\surd 3)\,d\varepsilon_\theta^{\,P} \qquad (13.48)$$

Combining eqs(13.46) and (13.48):

$$d\bar{\sigma}/d\bar{\varepsilon}^{P} = (3d\sigma_\theta)\,/\,(4d\varepsilon_\theta^{\,P})$$

where, from eqs(13.41) and (13.46):

$$d\bar{\sigma}/d\bar{\varepsilon}^{P} = \tfrac{3}{4}\sigma_\theta(d\varepsilon_\theta^{\,P} - d\varepsilon_r^{\,P})/(d\varepsilon_\theta^{\,P}) = 3\sigma_\theta/2$$

$$\therefore\quad \frac{d\bar{\sigma}}{d\bar{\varepsilon}^{P}} = \surd 3\,\bar{\sigma}$$

i.e. a sub-tangent of $1/\surd 3$ in this case. Hence from the given law:

$$d\bar{\sigma}/d\bar{\varepsilon}^{P} = n\bar{\sigma}/(\varepsilon_o + \bar{\varepsilon}^{P}) = \surd 3\bar{\sigma}$$

$$\therefore\quad \bar{\varepsilon}^{P} = n/\surd 3 - \varepsilon_o$$

The corresponding equivalent stress is $\bar{\sigma} = \sigma_o(n/\surd 3\,\varepsilon_o)^n$, giving a critical pressure as

$$p = \frac{2t\sigma_o}{\surd 3 r}\left(\frac{n}{\surd 3\,\varepsilon_o}\right)^n$$

13.5 Circular Bulge Instability

The attainment of maximum pressure in a bulge test is similar to that for a spherical vessel. However, eq(13.45) is inappropriate because the bulge forms continuously out of the plane of the sheet. Consider the section of the bulge shown in Fig. 9.16 in which an instability arises from bulging isotropic material in a circular die. The radius of curvature at an instant where the current height is h, is given by

$$R = (a^2 + h^2)\,/\,(2h) \qquad (13.49)$$

where $a = D/2$ is the radius of the circular die and h is the bulge height. The in-plane, equi-biaxial stress at the pole consists of hoop and meridional components (see eqs(9.33a,b)), as with a spherical vessel:

$$\sigma_\theta = \sigma_\varphi = pR/2t \tag{13.50}$$

Differentiating eq(13.50):

$$\frac{\delta\sigma_\theta}{\sigma_\theta} = \frac{\delta p}{p} + \frac{\delta R}{R} - \frac{\delta t}{t} \tag{13.51a}$$

where $\delta\varepsilon_r^P = \delta t/t$ defines the pole thickness strain. Setting $\delta p/p = 0$ in eq(13.51a) gives an instability condition

$$\left(\frac{1}{\sigma_\theta}\right)\left(\frac{\delta\sigma_\theta}{\delta\varepsilon_r^P}\right) = -1 + \left(\frac{1}{R}\right)\left(\frac{\delta R}{\delta\varepsilon_r^P}\right) \tag{13.51b}$$

The hoop and meridional pole strain increments are each equal to $\delta h/R$ so it follows from the incompressibility condition that $\delta\varepsilon_r^P = -2\delta h/R$ giving $\delta h/\delta\varepsilon_r^P = -\tfrac{1}{2}R$. Also, from differentiating eq(13.49), $\delta R/\delta h = 1 - R/h$ and, therefore

$$\frac{1}{R} \times \frac{\delta R}{\delta\varepsilon_r^P} = \frac{1}{R} \times \frac{\delta R}{\delta h} \times \frac{\delta h}{\delta\varepsilon_r^P} = -\frac{1}{2}\left(1 - \frac{R}{h}\right) \tag{13.52}$$

Substituting eq(13.52) into eq(13.51b) gives

$$\frac{1}{\sigma_\theta}\,\frac{\delta\sigma_\theta}{\delta\varepsilon_r^P} = \frac{1}{\bar\sigma}\,\frac{\delta\bar\sigma}{\delta\bar\varepsilon^P} = -\frac{3}{2} + \frac{R}{2h} \tag{13.53}$$

We have converted the left-hand side of eq(13.53) into equivalent stress and strain readily since $\bar\sigma = \sigma_\varphi = \sigma_\theta$ and $\mathrm{d}\bar\varepsilon^P = \mathrm{d}\varepsilon_r^P$ at the pole of a bulge (see eqs(9.33b) and (9.36a)). The term $R/2h$ within the right-hand side of eq(13.53) appears from an integration involving eq(13.49) and a series expansion as follows:

$$\varepsilon_r^P = -2\int \frac{\mathrm{d}h}{R} = -2\int_0^h \frac{2h\mathrm{d}h}{(a^2 + h^2)} = -2\ln\left[\frac{2Rh}{(2Rh - h^2)}\right]$$

$$\begin{aligned}
1 - h/2R &= \exp(\varepsilon_r^P/2) = 1 + \varepsilon_r^P/2 + (\varepsilon_r^P)^2/8 + \ldots \\
h/2R &= -\tfrac{1}{2} \times \varepsilon_r^P(1 + \varepsilon_r^P/4) \\
R/2h &= -[1/(2\varepsilon_r^P)](1 + \varepsilon_r^P/4)^{-1} \\
&= -[1/(2\varepsilon_r^P)][1 - \varepsilon_r^P/4 + (\varepsilon_r^P)^2/16 \ldots] \\
&\simeq -1/(2\varepsilon_r^P) + 1/8
\end{aligned} \tag{13.54}$$

in which squared strain is neglected. *Note*: Radial strain ε_r^P is negative since the pole material thins but, by taking $\bar\varepsilon^P = \varepsilon_r^P$ to be positive, it follows from eqs(13.53) and (13.54) that

$$\frac{\mathrm{d}\bar\sigma}{\mathrm{d}\bar\varepsilon^P} = \left(\frac{11}{8} - \frac{1}{2\bar\varepsilon^P}\right)\bar\sigma \tag{13.55}$$

Comparing eq(13.55) with eq(13.40b), the sub-tangent is written as [9]

$$\frac{1}{z} = \frac{11}{8} - \frac{1}{2\bar{\varepsilon}_I^P} \tag{13.56}$$

where $\bar{\varepsilon}_I^P$ is the equivalent instability strain. Combined with Swift law (9.44b), this sub-tangent becomes

$$\frac{1}{z} = \frac{n}{\bar{\varepsilon}_I^P + \varepsilon_o} \tag{13.57}$$

When eqs(13.56) and (13.57) are combined $\bar{\varepsilon}_I^P$ becomes one root to the quadratic

$$(\bar{\varepsilon}_I^P)^2 + [\varepsilon_o - (4/11)(2n+1)]\,\bar{\varepsilon}_I^P - 4\varepsilon_o/11 = 0 \tag{13.58a}$$

In the case of an annealed material, where $\varepsilon_o \simeq 0$, eq(13.58a) provides

$$\bar{\varepsilon}_I^P = (4/11)(2n+1) \tag{13.58b}$$

Equation (13.58b) will also apply for a Hollomon hardening law (9.39b). Alternatively, using his own hardening law, Swift [10] identified the critical pole strain as a root to the cubic equation in $\bar{\varepsilon}_I^P$. If we use a Hollomon's hardening law instead, Swift's analysis results in the following quadratic equation:

$$(\bar{\varepsilon}_I^P)^2 - (n + 33/10)\,\bar{\varepsilon}_I^P + (9/10)[1 + \surd(363/50)] = 0 \tag{13.59}$$

For example, with a hardening exponent $n = 0.6$ for annealed stainless steel, $\bar{\varepsilon}_I^P = 0.8$ and 1.26 from eqs(13.58b) and (13.59). These compare with $\bar{\varepsilon}_I^P = 0.77$ from eq(13.58a) on taking $\varepsilon_o = 0.048$ for this material.

13.6 Ellipsoidal Bulging, Orthotropic Sheet

The circular bulge analysis may be extended to a more general analysis of ellipsoidal bulging of orthotropic sheet metal. The problem is simplified when the rolling direction of the sheet is aligned with the minor axis of the die aperture (i.e. the major principal strain direction), as shown in Fig. 13.18. The principal radii of curvature are

$$R_1 = (h^2 + b^2)/2h \quad \text{and} \quad R_2 = (h^2 + a^2)/2h \tag{13.60a,b}$$

where h is the bulge height and a, b are the semi-lengths of the major and minor axes of the die aperture respectively. Figure 13.18 shows the principal stress state at the pole, from which the corresponding strain increments are

$$\mathrm{d}\varepsilon_1^{\,P} = \mathrm{d}h/R_1 = \mathrm{d}\lambda\,[H(\sigma_1 - \sigma_2) + G\sigma_1] \tag{13.61a}$$

$$\mathrm{d}\varepsilon_2^{\,P} = \mathrm{d}h/R_2 = \mathrm{d}\lambda\,[F\sigma_2 - H(\sigma_1 - \sigma_2)] \tag{13.61b}$$

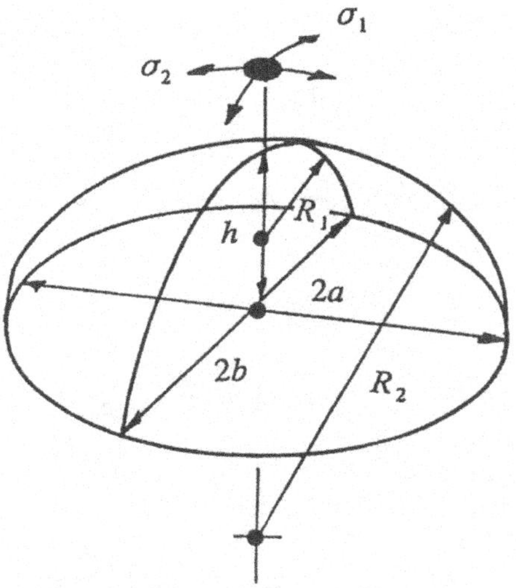

Figure 13.18 Ellipsoidal bulge geometry

Dividing eqs(13.61a,b) allows the stress ratio to be found:

$$Q = \frac{\sigma_2}{\sigma_1} = \frac{[R_2 + R_1(1 + 1/r_1)]}{[R_1 + R_2(1 + 1/r_2)]} \tag{13.62}$$

Given a lateral pressure p and current thickness t, the membrane equilibrium equation for the pole's in plane principal stresses is [9]

$$\frac{\sigma_1}{R_1} + \frac{\sigma_2}{R_2} = \frac{p}{t} \tag{13.63a}$$

which is written as

$$\sigma_1 = \frac{pR}{t} \quad \text{where} \quad R = \frac{R_1 R_2}{(R_2 + QR_1)} \tag{13.63b,c}$$

13.6.1 Instability Strains

Differentiating eq(13.63b) and setting $\delta p/p = 0$ gives the condition of instability as

$$\frac{\delta\sigma_1}{\sigma_1} = \frac{\delta R}{R} - \delta\varepsilon_3^P \tag{13.64}$$

where
$$\frac{\delta R}{R} = \frac{\partial R}{\partial R_1} \times \frac{\delta R_1}{R} + \frac{\partial R}{\partial R_2} \times \frac{\delta R_2}{R} \tag{13.65a}$$

Now from eqs(13.60a,b) and (13.63c), eq(13.65a) becomes

$$\frac{\delta R}{R} = \Phi \frac{\delta R_1}{R_1} + \Omega \frac{\delta R_2}{R_2} \tag{13.65b}$$

where

$$\Phi = \frac{2 + (1 + 1/r_2)[2q - (1 + 1/r_1) + q^2(1 + 1/r_2)]}{[2 + q(1 + 1/r_2) + (1 + 1/r_1)/q][1 + (1 + 1/r_2)q]} \quad (13.66a)$$

$$\Omega = \frac{(1 + 1/r_2) + 2(1 + 1/r_1)(1 + 1/r_2)/q + (1 + 1/r_1)/q^2]}{[2 + q(1 + 1/r_2) + (1 + 1/r_1)/q][1/q + (1 + 1/r_2)]} \quad (13.66b)$$

in which $q = R_2/R_1$. Dividing eq(13.64) by $\delta\varepsilon_3^{\,P}$ and introducing eqs(13.65b) leads to

$$-\frac{1}{\sigma_1}\frac{\delta\sigma_1}{\delta\varepsilon_3^P} = -1 + \frac{\Phi}{R_1}\frac{\delta R_1}{\delta\varepsilon_3^P} + \frac{\Omega}{R_2}\frac{\delta R_2}{\delta\varepsilon_3^P} \quad (13.67a)$$

Introducing, from eqs(13.60a,b), the following relationships

$$\frac{\delta R_1}{\delta h} = 1 - \frac{R_1}{h}, \quad \frac{\delta R_2}{\delta h} = \frac{1 - R_2}{h} \quad \text{and} \quad \frac{\delta h}{\delta\varepsilon_3^P} = -\frac{R_1 R_2}{R_1 + R_2}$$

where $\delta\varepsilon_3^{\,P} = -(\delta\varepsilon_1^{\,P} + \delta\varepsilon_2^{\,P})$, eq(13.67a) becomes

$$-\frac{1}{\sigma_1}\frac{\delta\sigma_1}{\delta\varepsilon_3^P} = -1 - \Phi\frac{(1 - R_1/h)}{(1 + R_1/R_2)} - \Omega\frac{(1 - R_2/h)}{(1 + R_2/R_1)} \quad (13.67b)$$

Equation (13.67b) will supply the critical sub-tangent at instability when σ_1 and $\delta\varepsilon_3^{\,P}$ are converted to equivalent stress and strain. To do this we use reduce eqs(11.4b) and (11.13) to one dependent variable as follows:

$$\bar{\sigma} = X\sigma_1, \quad \mathrm{d}\bar{\varepsilon}^P = Z\,\mathrm{d}\varepsilon_1^P \quad (13.68a,b)$$

where

$$X = \sqrt{\frac{3\left[r_2(1 + r_1) - 2Qr_1r_2 + Q^2 r_1(1 + r_2)\right]}{2(r_1 + r_2 + r_1 r_2)}} \quad (13.69a)$$

$$Z = \sqrt{a + b\left(\mathrm{d}\varepsilon_2^P/\mathrm{d}\varepsilon_1^P\right)^2 + c\left(\mathrm{d}\varepsilon_2^P/\mathrm{d}\varepsilon_1^P\right)} = \sqrt{a + b\left(R_1/R_2\right)^2 + c\left(R_1/R_2\right)} \quad (13.69b)$$

The coefficients a, b and c in eq(13.69b) are

$$a = [2\,(r_1 + r_2 + r_1 r_2)(1 + r_2)]/\,[3\,r_2\,(1 + r_1 + r_2)]$$
$$b = [2\,(r_1 + r_2 + r_1 r_2)(1 + r_1)]/\,[3\,r_1\,(1 + r_1 + r_2)]$$
$$c = [4\,(r_1 + r_2 + r_1 r_2)]/\,[3\,(1 + r_1 + r_2)]$$

The thickness strain follows from the incompressibility condition as

$$\delta\varepsilon_3^{\,P} = -(\delta\varepsilon_1^{\,P} + \delta\varepsilon_2^{\,P}) = -(1 + R_1/R_2)\delta\varepsilon_1^{\,P} \quad (13.69c)$$

Combining eqs(13.67b), (13.68a,b) and (13.69c) and applying eq(13.40b) leads to the sub-tangent expression

$$\frac{1}{z} = \frac{(R_1 + R_2)}{ZR_2}\left\{1 + \frac{\Phi[(1 + 1/r_1) - Q]}{(1/r_1 + Q/r_2)}\left(\frac{3}{4} - \frac{Z}{2\bar{\varepsilon}^P}\right)\right.$$

$$\left. + \frac{\Omega\,[Q(1 + 1/r_2) - 1]}{(1/r_1 + Q/r_2)}\left(\frac{3}{4} - \frac{ZR_2}{2R_1\bar{\varepsilon}^P}\right)\right\} \qquad (13.70)$$

Taking $1/z = n/\bar{\varepsilon}^P$ from Hollomon's law enables eq(13.70) to be solved for the equivalent strain at instability. The pole strain follows from eqs(13.69a,b,c) as

$$\varepsilon_1^P = \frac{\bar{\varepsilon}^P}{Z}, \quad \varepsilon_2^P = \frac{R_1}{R_2}\varepsilon_1^P \quad \text{and} \quad \varepsilon_3^P = -(\varepsilon_1^P + \varepsilon_2^P)$$

For example, with a circular bulge setting $R_1 = R_2$, eq(13.70) gives

$$\bar{\varepsilon}^P = \frac{Z\left\{n(1/r_1 + Q/r_2) + \Phi[(1 + 1/r_1) - Q] + \Omega[Q(1 + 1/r_2) - 1]\right\}}{2\left\{(1/r_1 + Q/r_2) + (3/4)\left\{\Phi[(1 + 1/r_1) - Q] + \Omega[Q(1 + 1/r_2) - 1]\right\}\right\}}$$

Also, setting $r_1 = r_2 = 1$ for an isotropic sheet gives $\Phi = 1/3$, $\Omega = 2/3$, $Q = 1$ and $Z = 2$, so recovering eq(13.56) from eq(13.70).

13.6.2 Pressure Versus Height

So far the analysis has dealt with the pole instability condition only. One ready measure of prior pole deformation is the pressure versus height curve, which may be predicted from eq(13.63b) and (13.68a) as

$$p = \frac{\bar{\sigma}t}{RX} \qquad (13.71)$$

The current thickness t follows from the thickness strain ε_3^P as

$$\varepsilon_3^P = \ln(t/t_o) \;\Rightarrow\; t = t_o \exp(\varepsilon_3^P) \qquad (13.72a,b)$$

Now from eqs(13.60a,b):

$$1 + \frac{R_1}{R_2} = \frac{2h^2 + a^2 + b^2}{h^2 + a^2} \qquad (13.73)$$

Substituting eq(13.73) into eq(13.69c) and integrating:

$$\varepsilon_3^P = -\frac{2h^2 + a^2 + b^2}{h^2 + a^2}\varepsilon_1^P \qquad (13.74)$$

where

$$\varepsilon_1^P = \int \frac{dh}{R_1} = 2\int_0^h \frac{h\,dh}{h^2 + b^2} = \ln\left(1 + \frac{h^2}{b^2}\right) \tag{13.75a}$$

$$\therefore \quad \bar{\varepsilon}^P = Z\varepsilon_1^P = Z\ln\left(1 + \frac{h^2}{b^2}\right) \tag{13.75b}$$

Thus, for a given h–value, find $\bar{\varepsilon}^P$ from eq(13.75b) when $\bar{\sigma}$ follows from the flow law chosen for the material. Combining eqs(13.74) and (13.75a) and substituting into eq(13.72b) gives an equation in the current pole thickness t that appears in eq(13.71):

$$\frac{t_o}{t} = \left(1 + \frac{h^2}{b^2}\right)^{\frac{2h^2 + a^2 + b^2}{h^2 + a^2}} \tag{13.76}$$

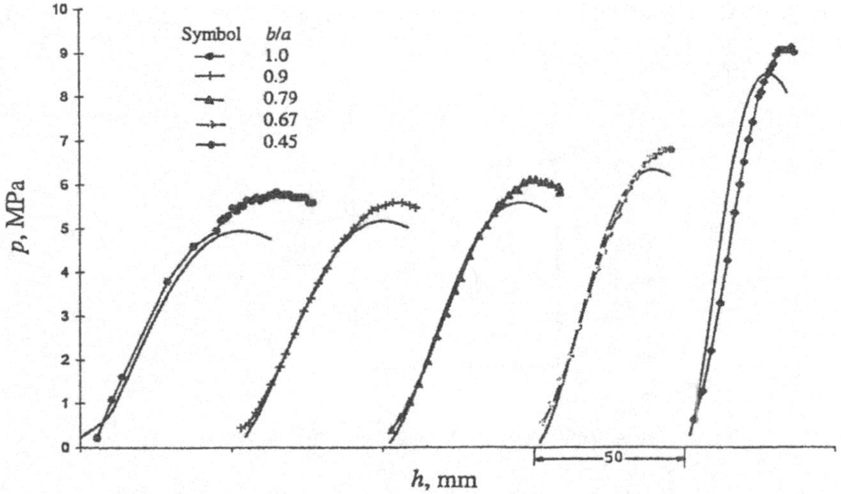

Figure 13.19 Pressure versus height for ellipsoidal bulging through 5 dies

The predicted pressure follows from eqs(13.71) and (13.76). For example, this procedure has been used [11] with Hollomon's law to predict the p, h curves for elliptical bulging of an automotive steel (where A = 570, n = 0.385). In each of 5 dies oil is pumped to the underside of the disc to form a circular or elliptical bulge depending upon the die aperture. The major axis of each ellipse was constant at 180 mm while the minor axis varied with the die aspect ratios b/a = 1, 0.90, 0.79, 0.67 and 0.45. Sealing was achieved with a rubber O-ring in the lower ring and by a circumferential bead in the blank which formed when the top and bottom rings were bolted together. A pressure transducer was connected to the oil feed. A dial gauge measured the pole height h above the die surface. In all the present tests the rolling direction was aligned with the ellipse minor axis which is the major principal axis of stress and strain. Though there is not complete agreement with the experimental curves, Fig. 13.19 shows that the S-shaped curve and a pressure maximum at instability are reproduced satisfactorily.

13.7 Plate Stretching

There are a number of theoretical predictions to the instability arising from in-plane, biaxial stressing of thin sheets. Those reviewed in [12–14] all appear as particular sub-tangents z, within the Considére expression (13.40b). This section will examine the possibility of both diffuse and local instability conditions promoting necking in a thin sheet. Since either condition places a theoretical limit on the strain available for forming, the two predictions may be used to construct a material's forming limit diagram (FLD).

13.7.1 Diffuse Instability

In the bulge test, where both in-plane strains are positive, a visible neck does not form instantly with the attainment of maximum pressure. Instead, thinning spreads around the pole at this pressure showing that there still remains useful strain for forming before a local neck finally forms into a crack. A similar diffuse instability condition arises in thin flat sheets under in-plane biaxial tension. Swift [10] and Moore and Wallace [15] identified this condition with the simultaneous attainment of maxima in applied perpendicular forces. Figure 13.20 shows a sheet with rolling direction (1) aligned to the major principal stress.

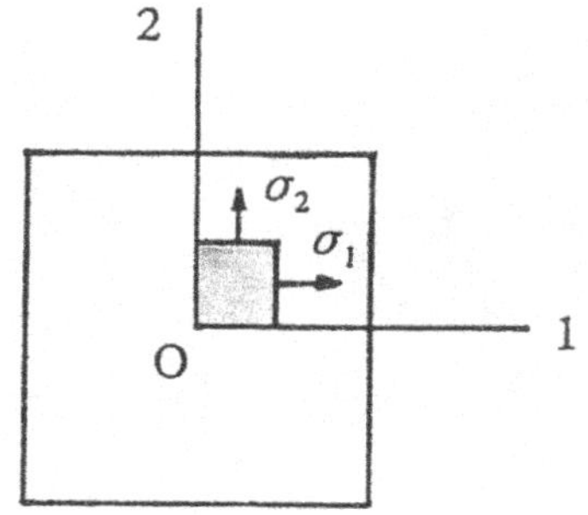

Figure 13.20 Sheet orientation and co-ordinates

The following plastic strain increment ratios are found from eqs(13.61a,b):

$$w = \frac{\delta\varepsilon_2^P}{\delta\varepsilon_1^P} = \frac{r_1[Q - r_2(1 - Q)]}{r_2[r_1(1 - Q) + 1]} \tag{13.77a}$$

$$r = \frac{\delta\varepsilon_2^P}{\delta\varepsilon_3^P} = \frac{r_1[r_2(1 - Q) - Q]}{Qr_1 + r_2} \tag{13.77b}$$

where $-1 < (Q = \sigma_2/\sigma_1) < 1$. The equivalent stress is again given by eqs(13.68a), i.e. $\bar{\sigma} = X\sigma_1$, where X has been defined in eq(13.69a). The equivalent strain is written from eq(13.68b) as $\bar{\varepsilon}^P = Z\varepsilon_1^P$ where Z now follows from eqs(13.36) and (13.77a) as

$$Z = \sqrt{(a + bw^2 + cw)} \tag{13.78}$$

in which a, b and c appear with eq(13.69b). Now, at any instant in the deformation

$$\delta\bar{\sigma} = \frac{\partial\bar{\sigma}}{\partial\sigma_1}\delta\sigma_1 + \frac{\partial\bar{\sigma}}{\partial\sigma_2}\delta\sigma_2 = \left[\frac{\partial\bar{\sigma}}{\partial\sigma_1} + \left(\frac{\partial\bar{\sigma}}{\partial\sigma_2} \times \frac{\delta\sigma_2}{\delta\sigma_1}\right)\right]\delta\sigma_1 \qquad (13.79)$$

where, from eq(13.35):

$$\frac{\partial\bar{\sigma}}{\partial\sigma_1} = \frac{3\sigma_1\left[r_2(1 + r_1) - Qr_1r_2\right]}{2\bar{\sigma}\left(r_1 + r_2 + r_1r_2\right)} \qquad (13.80a)$$

$$\frac{\partial\bar{\sigma}}{\partial\sigma_2} = \frac{3\sigma_1\left[Qr_1(1 + r_2) - r_1r_2\right]}{2\bar{\sigma}\left(r_1 + r_2 + r_1r_2\right)} \qquad (13.80b)$$

Assuming homogenous deformation and that a neck band forms perpendicular to the major principal stress direction, the simultaneous instability conditions apply:

$$\frac{\delta\sigma_1}{\delta\varepsilon_1^P} = \sigma_1 \quad \text{and} \quad \frac{\delta\sigma_2}{\delta\varepsilon_2^P} = \sigma_2 \qquad (13.81a,b)$$

From eqs(13.81a,b) and (13.77a):

$$\frac{\delta\sigma_2}{\delta\sigma_1} = \frac{\sigma_2\delta\varepsilon_2^P}{\sigma_1\delta\varepsilon_1^P} = Qw \qquad (13.82)$$

Substituting eqs(13.80a,b) and (13.82) into eq(13.79):

$$\bar{\sigma}\,\delta\bar{\sigma} = K\sigma_1\,\delta\sigma_1 \qquad (13.83a)$$

where K depends upon r_1, r_2 and Q as follows:

$$K = \frac{3\left\{ r_2[r_2(1 + r_1) - Qr_1r_2][r_1(1 - Q) + 1] + Qr_1[Qr_1(1 + r_2) - r_1r_2][Q - r_2(1 - Q)] \right\}}{2r_2(r_1 + r_2 + r_1r_2)[r_1(1 - Q) + 1]} \qquad (13.83b)$$

The sub-tangent z is identified from combining eqs(13.68a,b) and (13.83a):

$$\frac{\delta\bar{\sigma}}{\delta\bar{\varepsilon}^P} = \frac{K\bar{\sigma}}{X^2Z} \qquad (13.84)$$

Comparing eq(13.84) with eq(13.40c) shows $z = X^2Z/K$ and we may substitute for X, Z and K from eqs(13.69a), (13.78) and (13.83b) to give

$$z = \frac{\left[\frac{2}{3}(1 + r_1 + r_2)\right]^{\frac{1}{2}}\left[(1 + 1/r_1) - 2Q + Q^2(1 + 1/r_2)\right]^{\frac{3}{2}}}{(1 + 1/r_1)^2 - Q(1 + 2/r_1) - Q^2(1 + 2/r_2) + Q^3(1 + 1/r_2)^2} \qquad (13.85a)$$

Knowing z allows $\bar{\varepsilon}^P$ to be found, from which the component strains are $\varepsilon_1^P = \bar{\varepsilon}^P / Z$ and $\varepsilon_2^P = w\varepsilon_1^P$. An alternative expression [10] leading to the diffuse instability sub-tangent z_D in eq(13.85a) employs the flow rule with the yield criterion f in eq(13.34):

$$\frac{1}{z_D} = \frac{\sigma_1(\partial f/\partial\sigma_1)^2 + \sigma_2(\partial f/\partial\sigma_2)^2}{[\sigma_1(\partial f/\partial\sigma_1) + \sigma_2(\partial f/\partial\sigma_2)]\mathrm{d}f/\mathrm{d}\bar{\sigma}} \qquad (13.85b)$$

13.7.2 Local Instability

For simple tension of a thin strip, the transverse strain is negative and a local neck can occur obliquely to the applied stress. Similarly, under in-plane, biaxial plate stressing a negative transverse strain is crucial to local neck band formation. As with the onset of necking in a tensile test, it has been suggested [16] that local neck formation in a plate requires plane strain deformation, irrespective of the original straining path. This means that for a groove to form strain along its length is prohibited as widening and thinning continue. The condition for the initiation of local necking is that the maximum force per unit width of plate reaches a maximum. Thus, in Fig. 13.20 when $Q = \sigma_2/\sigma_1$ is constant, a condition of local instability occurs when force $T_1 = \sigma_1 t$ aligned with the sheet's rolling direction reaches its maximum. This gives

$$\frac{\delta T_1}{T_1} = \frac{\delta\sigma_1}{\sigma_1} + \frac{\delta t}{t} = 0 \qquad (13.86)$$

Now from eqs(13.68a,b) and (13.77a):

$$\delta\varepsilon_3^P = \frac{\delta t}{t} = -\frac{(1+w)}{Z}\delta\bar{\varepsilon}^P \quad \text{and} \quad \bar{\sigma} = X\sigma_1 \qquad (13.87a,b)$$

Substituting eqs(13.87a,b) into eq(13.86) gives

$$\frac{\delta\bar{\sigma}}{\delta\bar{\varepsilon}^P} = \frac{(1+w)\bar{\sigma}}{Z} \qquad (13.88)$$

from which the subtangent of local instability is

$$z = \frac{Z}{(1+w)} \qquad (13.89)$$

Substituting for Z, w, a, b and c from eqs(13.69b) and (13.77a):

$$1 + w = \frac{r_2 + Qr_1}{r_2[r_1(1-Q)+1]} \qquad (13.90a)$$

$$Z = \sqrt{a + bw^2 + cw}$$

$$= \frac{\sqrt{(2/3)(r_1 + r_2 + r_1r_2)(1 + r_1 + r_2)\,[Q^2r_1 + r_2 + r_1r_2(1-Q)^2]}}{r_2\,[r_1(1-Q)+1]} \qquad (13.90b)$$

From eq(13.89), divide eq(13.90b) by eq(13.90a) to give z:

$$z = \frac{\sqrt{(2/3)(1 + 1/r_1 + 1/r_2)[(1 + 1/r_1) - 2Q + Q^2(1 + 1/r_2)]}}{1/r_1 + Q/r_2} \tag{13.91a}$$

An alternative expression leading to the sub-tangent (z_L) for local necking uses the flow rule and yield function f in eq(13.34) as follows [17]:

$$\frac{1}{z_L} = \frac{\partial f/\partial \sigma_1 + \partial f/\partial \sigma_2}{\mathrm{d}f/\mathrm{d}\bar{\sigma}} \tag{13.91b}$$

in which $\bar{\sigma}$ appears in eq(13.35). The equivalent instability strain follows from Hollomon as $\bar{\varepsilon}_I^P = nz$ or from Swift (13.57), as $\bar{\varepsilon}_I^P = nz - \varepsilon_o$. The components of $\bar{\varepsilon}_I^P$ are $\varepsilon_1^P = \bar{\varepsilon}_I^P/Z$ and $\varepsilon_2^P = w\varepsilon_1^P$. Equation (13.91a) is important when ε_2^P is negative for $-1 \le Q \le 0.5$ since a local groove will form at a fixed orientation to the stress axes 1 and 2 (see Figs 13.21a,b).

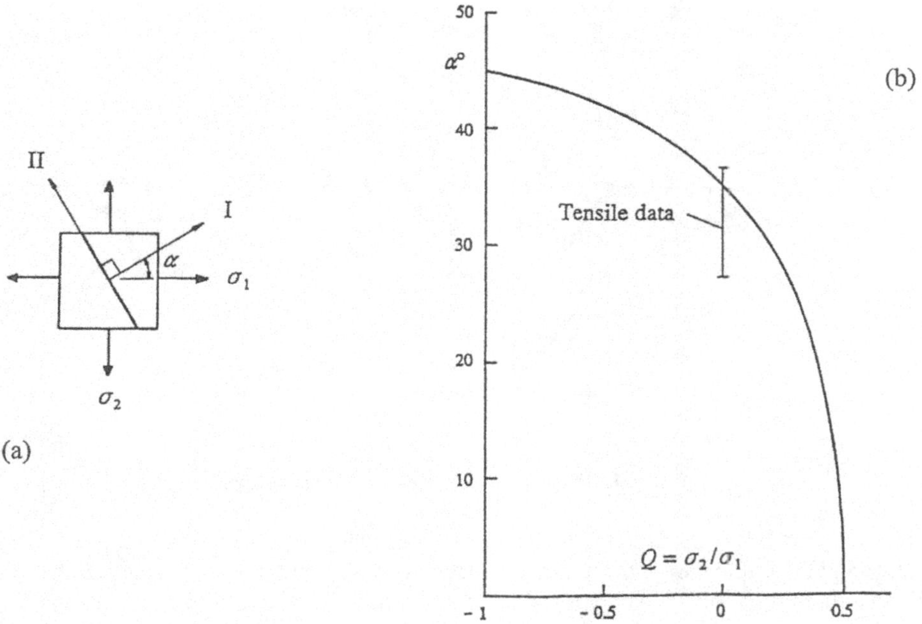

Figure 13.21 Groove formation under local instability

Here the axes of stress coincide with the material axes (with 1 parallel to the roll and 2 transverse to the roll). Let an axis I lie normal to the groove at an orientation α to 1. Axis II lies parallel to the groove as shown. Transforming the principal strain increments $\mathrm{d}\varepsilon_1^P$ and $\mathrm{d}\varepsilon_2^P$ along the axis II gives the normal strain component aligned with the groove as

$$\mathrm{d}\varepsilon_{II}^P = \mathrm{d}\varepsilon_1^P \sin^2\alpha + \mathrm{d}\varepsilon_2^P \cos^2\alpha$$

Setting $\mathrm{d}\varepsilon_{II}^P = 0$ leads to

$$\cos 2\alpha = (1 + w)/(1 - w) \tag{13.92a}$$

where $w = d\varepsilon_2^P/d\varepsilon_1^P$ depends upon the stress ratio $Q = \sigma_2/\sigma_1$, as in eq(13.77a). Substituting eq(13.77a) into eq(13.92a) gives the groove orientation in terms of Q and the two r–values:

$$\cos 2\alpha = \frac{r_2 + Qr_1}{r_2(1 + 2r_1) - Qr_1(1 + 2r_2)} \qquad (13.92b)$$

In the case of a simple tension test under an axial stress σ_1 with $\sigma_2 = 0$ (see Fig. 13.22a), we set $Q = 0$ in eq(13.92b) to give $\cos 2\alpha = 1/(1 + 2r_1)$. If the material is isotropic, then $r_1 = 1$ and we find $\alpha = \frac{1}{2}\cos^{-1}(\frac{1}{3}) = 35.26°$ (i.e. the groove lies at 90° to this). The Mohr's strain circle (see Fig. 13.22b) provides a geometrical interpretation of this solution.

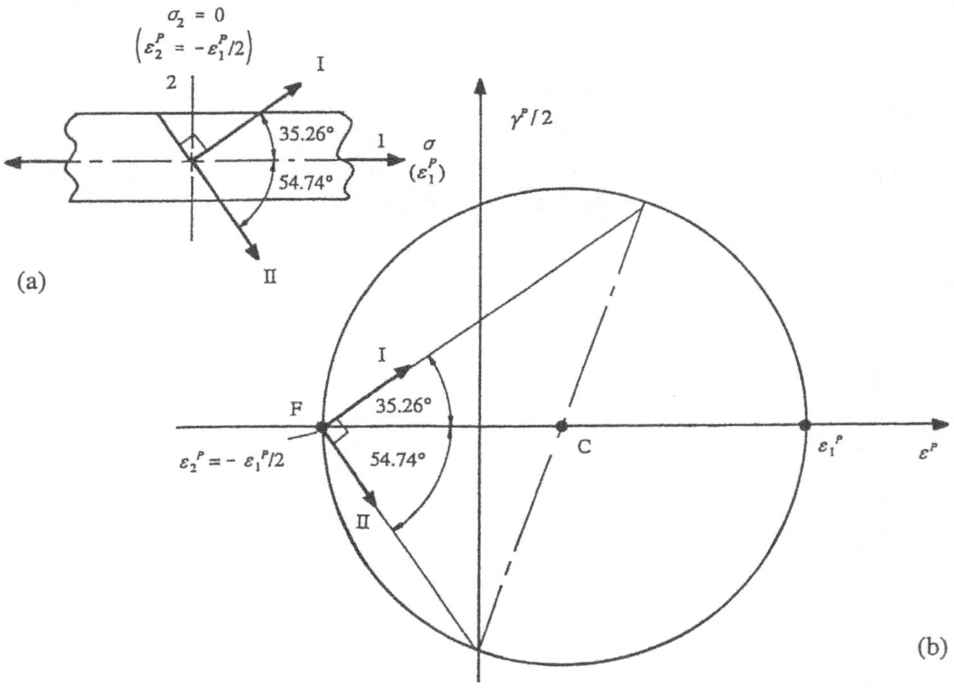

Figure 13.22 Mohr's circle showing groove orientation

Setting ε_1^P and $\varepsilon_2^P = -\frac{1}{2}\varepsilon_1^P$ along the axis of plastic strain ε^P, the centre C bisects these points and the circle passes through them. The groove direction II passes though the focus F and a point of zero ε^P as shown. For other stress ratios, applied to isotropic material, the groove orientation follows from eq(13.92b) as

$$\cos 2\alpha = (1 + Q)/[3(1 - Q)] \qquad (13.92c)$$

As Q increases from zero the centre C moves to the right along its ε^P axis. With $Q = \frac{1}{2}$, under plane strain, the circle becomes tangential to the vertical, semi-shear strain axis $\gamma^P/2$. Correspondingly, $\alpha = 0°$ and the groove becomes vertically aligned . For $Q > \frac{1}{2}$, it follows that the circle will lie to the right of the $\gamma^P/2$ axis, when no further inclined grooves can form. Figure 13.21b shows the dependence of groove inclination α upon stress ratio Q in the range $-1 \le Q \le \frac{1}{2}$. Superimposed tensile data at $Q = 0$ reveal that rolled sheets of aluminium alloy and tin-plate form grooves across a range of inclinations deviating from the isotropic prediction. For stress ratios $Q > \frac{1}{2}$, failure may be diffuse or a groove may form perpendicular to σ_1 by another mechanism, as outlined in the following section.

13.7.3 Forming Limit Diagram

The local and diffuse theories of instability given above provide a combination of critical principal strains (ε_1^P, ε_2^P) depending upon the stress ratio. When the full range of stress ratios $-1 \le Q \le 1$ are considered across crucial instability regimes, the critical strains plot to define a forming limit diagram (FLD) for the material. Different instability criteria result in different FLD's in co-ordinates of ε_1^P and ε_2^P, i.e. the criteria control both the shape and position of the FLD within a given quadrant of strain. The question of which prediction supplies the most appropriate limiting strains for forming operations may be answered from making a comparison with experimental data. Figure 13.23 compares results from laboratory tests upon an automotive steel with predictions using sub-tangents from: (i) maximum pressure, (ii) diffuse and (iii) local instability [18]. The sub-tangents appear in eqs(13.70), (13.85a) and (13.91a) respectively.

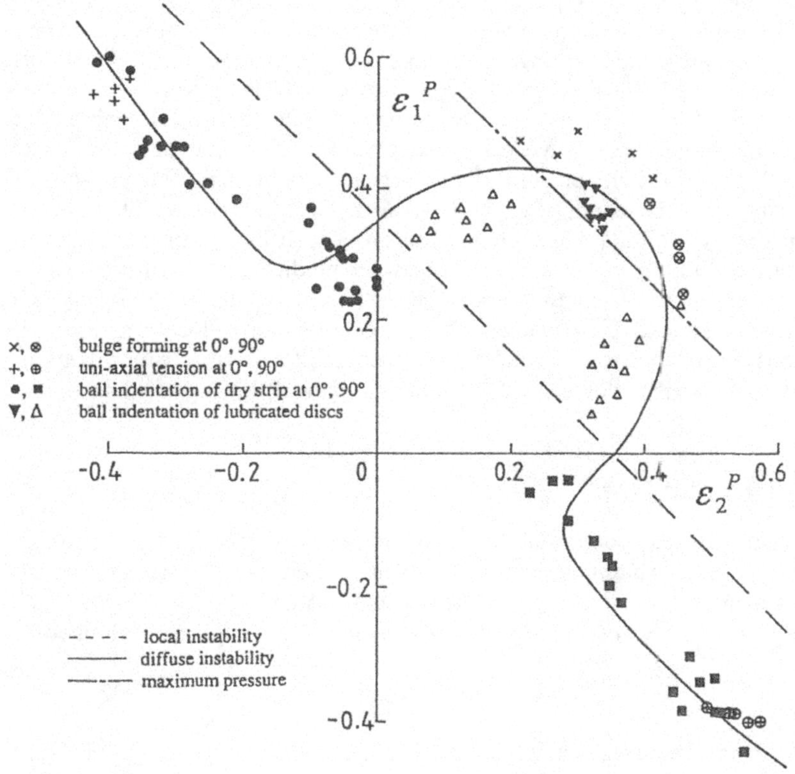

Figure 13.23 Forming limit diagram in natural strain axes

The sub-tangents were calculated with material properties $r_1 = 2.31$ and $r_2 = 2.54$ for $n = 0.37$, $\varepsilon_o = 1/25$ and $\sigma_o = 200$ MPa in Swift's eq(9.44b). Stress ratios Q cover three strain quadrants. For example, $Q = 0$ for tension in the 1–direction, $Q = \infty$ for tension in the 2–direction and $1 < Q < 2$ for elliptical bulging, depending upon a/b for the die and $r_{1,2}$ for the material. The conversion to the equivalent instability strain uses $\bar{\varepsilon}_I^P = nz - \varepsilon_o$. The component strains ε_1^P and ε_2^P are derived from $\bar{\varepsilon}_I^P$ in the manner outlined above. Natural strains apply to these calculations and to the plot in Fig. 13.23 though they may be converted to engineering strains for a convenient working FLD using the relationship $e = \exp\varepsilon - 1$. In Fig. 13.23 the sheet's othotropic axes are taken as the co-ordinates. Thus, data in quadrants 2 and 4 apply to 0° and 90° orientations respectively for tension testpieces and for punch indentation of strips with different widths. The experimental data lying in the first quadrant applies to bulge forming, with 0° and 90° orientations to the roll and to hemispherical punch indentation of discs. The symmetry of the latter test reflects all its data about a 45° line in this quadrant. The strains in all tests were measured from the deformation to a circular grid pattern in the vicinity of a failure. Of the three predictions shown, that based upon diffuse instability appears the most representative of the limit strains measured. Veerman and Neve [19] postulated that a true FLD applies in the absence of strain gradients at the instant localised necking begins. They were able to detect this from the plot of strains across adjacent circles within the necked region of spherically punched testpieces. Hence, one would expect local instability to provide the true fracture strains in quadrants 2 and 4. The two predictions agree only at points of intersection with the axes, i.e. the plane strain limit. The local instability prediction is not appropriate to quadrant 1. The maximum pressure line is adequate for providing limiting strains under near bi-axial tension in bulge forming.

Figure 13.23 shows that instability predictions are likely to provide a conservative estimate to the forming limits in a low carbon, ductile sheet steel. This is because the forming limit strains are sensitive to the practical definition of that limit. Strains found from the distortion to grid circles or from thickness measurements in the fracture zone can exceed the predictions by the extent of diffuse straining. The latter is most likely in the positive strain quadrant of an FLD which predicts the onset of diffuse straining as a lower forming limit. Consequently, a precise agreement between predicted and measured fracture strains should not expected. Alternative approaches to FLD construction: (i) rely solely upon experimental data to give a shape resembling a combination of local and diffuse instability predictions between respective quadrants 1 and 2 [20, 21] and (ii) assume that a groove will form from a pre-existing defect [22], as outlined in the following section.

13.7.4 FLD, Groove Formation

A modification to the local instability prediction is clearly required for quadrant 1. In the original hypothesis Marciniak et al. [22] assumed a pre-existing defect produced a local reduction in sheet thickness between the two zones A and B in Fig. 13.24a With original thicknesses t_{Bo} and t_{Ao} in each respective zone, the *defect size factor* is defined:

$$f_o = t_{\mathrm{Bo}} / t_{\mathrm{Ao}} \tag{13.93a}$$

Under in-plane, biaxial tension, the current thickness reduction factor is modified as follows:

$$f = \frac{t_{\mathrm{B}}}{t_{\mathrm{A}}} = \frac{t_{\mathrm{Bo}}}{t_{\mathrm{Ao}}} \times \frac{t_{\mathrm{B}}}{t_{\mathrm{Bo}}} \times \frac{t_{\mathrm{Ao}}}{t_{\mathrm{A}}} \tag{13.93b}$$

Using the following thickness strain definitions

$$\varepsilon_{3\mathrm{A}} = \ln (t_{\mathrm{A}} / t_{\mathrm{Ao}}) \text{ and } \varepsilon_{3\mathrm{B}} = \ln (t_{\mathrm{B}} / t_{\mathrm{Bo}}) \tag{13.94a,b}$$

it follows from eqs(13.93b) and (13.94a,b) that

$$f = f_o \exp\left(\varepsilon_{3\mathrm{B}} - \varepsilon_{3\mathrm{A}}\right) \tag{13.95}$$

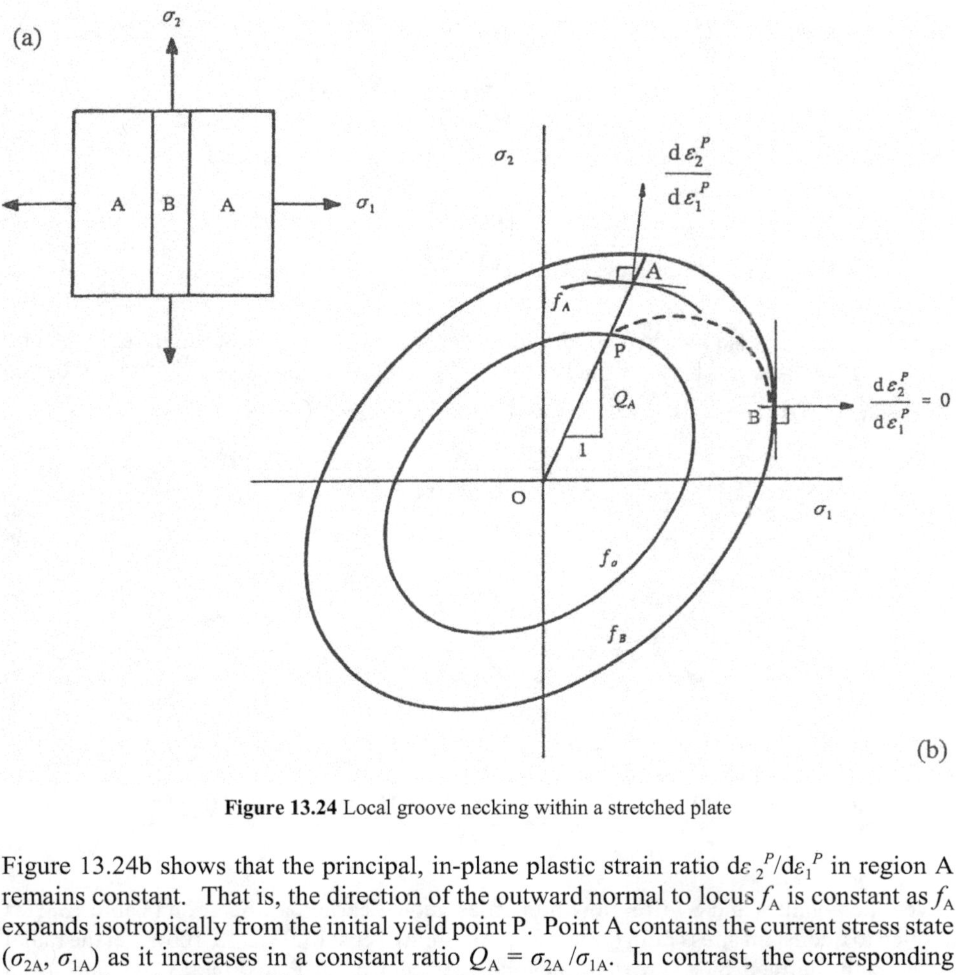

Figure 13.24 Local groove necking within a stretched plate

Figure 13.24b shows that the principal, in-plane plastic strain ratio $d\varepsilon_2^P/d\varepsilon_1^P$ in region A remains constant. That is, the direction of the outward normal to locus f_A is constant as f_A expands isotropically from the initial yield point P. Point A contains the current stress state $(\sigma_{2A}, \sigma_{1A})$ as it increases in a constant ratio $Q_A = \sigma_{2A}/\sigma_{1A}$. In contrast, the corresponding stress and strain ratios within groove B continuously change to approach the plane strain condition $d\varepsilon_2^P/d\varepsilon_1^P = 0$, as shown. At this point ε_{1B}^P increases rapidly to failure when the strain state in the adjacent region A is taken to constitute the forming limit. To find the limiting strain in A first note that the major principal stress σ_{1B} in the groove leads that within A to satisfy force equilibrium

$$\sigma_{1A}\, t_A = \sigma_{1B}\, t_B \tag{13.96a}$$

from which

$$\sigma_{1B} = \sigma_{1A}/f \tag{13.96b}$$

where $f < 1$. Also, a minor principal strain compatibility condition applies:

$$d\varepsilon_{2A}^P = d\varepsilon_{2B}^P \tag{13.97a}$$

In an orthotropic sheet the principal axes of stress and plastic strain coincide when the stress axes align with material axes. This permits a substitution from eq(12.11b) into eq(13.97a)

$$[F(\sigma_2 - \sigma_3) - H(\sigma_1 - \sigma_2)]_A = [F(\sigma_2 - \sigma_3) - H(\sigma_1 - \sigma_2)]_B \tag{13.97b}$$

Setting $\sigma_3 = 0$ and $r_2 = H/F$ leads to the minor principal stress within the groove:

$$\sigma_{2B} = \sigma_{2A} - (\sigma_{1A} - \sigma_{1B}) r_1 r_2/(1 + r_2) \tag{13.98}$$

Dividing eqs(11.11a,b) connects the stress and strain ratios within each region as

$$\frac{d\varepsilon_2^P}{d\varepsilon_1^P} = \frac{r_1[(Q - 1)r_2 + Q]}{r_2[(1 - Q)r_1 + 1]} \tag{13.99}$$

where $Q = \sigma_2/\sigma_1$. Also required are the equivalent stress and plastic strain in region A and an appropriate relation between them, e.g. Hollomon, Swift etc. We may write these as $\bar{\sigma} = X\sigma_1$ and $d\bar{\varepsilon}^P = Z d\varepsilon_1^P$ where X and Z follow from eqs(13.34) and (13.36) as

$$X = \sqrt{\frac{r_2(1 + r_1) - 2Qr_1r_2 + Q^2 r_1(1 + r_2)}{(2/3)(r_1 + r_2 + r_1r_2)}} \tag{13.100a}$$

in which $r_1 = H/G$ and

$$Z = \sqrt{a + b\left(d\varepsilon_2^P/d\varepsilon_1^P\right)^2 + c\left(d\varepsilon_2^P/d\varepsilon_1^P\right)} \tag{13.100b}$$

where, from eq(13.69b), a, b and c depend upon the two r–values:

$$a = \frac{2(r_1 + r_2 + r_1r_2)(1 + r_2)}{3r_2(1 + r_1 + r_2)}, \quad b = \frac{2(r_1 + r_2 + r_1r_2)(1 + r_1)}{3r_1(1 + r_1 + r_2)}, \quad c = \frac{4(r_1 + r_2 + r_1r_2)}{3(1 + r_1 + r_2)}$$

Given the r-values and Swift's flow properties quoted earlier, an iteration (Table 13.2) is applied to a constant stress ratio Q_A within the system of eqs(13.93)–(13.100). Let the radial stress path, defined by Q_A, extend beyond the yield point P in Fig. 13.24b due to an increment $d\bar{\varepsilon}^P$. A repeated calculation of the strain increment ratio $w = d\varepsilon_2^P/d\varepsilon_1^P$ is made from following the steps given. The iteration stops as $w \to 0$ within stage 8 so that the strain sums formed from stage 4 constitute the forming limits. These limiting strains can be found for any stress ratio but more often the analysis is confined to quadrant 1. In quadrant 2, the limit strains would coincide with those found from the sub-tangent for an inclined local neck, considered earlier.

Figure 13.25 presents the FLD in axes of engineering strain, which is more often applied in practice. Conversions from the natural plastic principal strains within Table 13.2 are applied as follows:

$$\varepsilon_1^P = \ln(1 + e_1^P), \quad \varepsilon_2^P = \ln(1 + e_2^P) \quad \text{and} \quad \varepsilon_3^P = -(\varepsilon_1^P + \varepsilon_2^P)$$

$$d\varepsilon_1^P = \frac{de_1^P}{1 + e_1^P}, \quad d\varepsilon_2^P = \frac{de_2^P}{1 + e_2^P} \quad \text{and} \quad d\varepsilon_3^P = -(d\varepsilon_1^P + d\varepsilon_2^P)$$

In quadrant 1 of this FLD the limit line from Table 13.2 lies above the diffuse instability limit, thereby extending the bi-axial stretch strain limits for an automotive steel at the chosen defect ratio f_o value.

Table 13.2 Iteration for a groove plane strain failure

1) Assume or find from measurement an initial f_o value, normally $0.95 < f_o < 1$

2) Apply an increment of equivalent plastic strain to region A, i.e. $d\bar{\varepsilon}_A^P$

3) Find the three principal plastic strain increments:

$$d\varepsilon_{1A}{}^P = d\bar{\varepsilon}_A^P / Z,\quad d\varepsilon_{2A}{}^P = w_A\, d\varepsilon_{1A}{}^P \quad \text{and} \quad d\varepsilon_{3A}{}^P = -\,(d\varepsilon_{1A}{}^P + d\varepsilon_{2A}{}^P)$$

4) Form sums $\sum d\varepsilon_{1A}{}^P$ and $\sum d\varepsilon_{1A}{}^P$ over successive iterations

5) Find $\bar{\sigma}_A$ from $\sum d\bar{\varepsilon}_A^P$ using Swift's law: $\bar{\sigma} = \sigma_o\left(1 + \bar{\varepsilon}^P / \varepsilon_o\right)^n$

6) Find the principal stresses in region A:

$$\sigma_{1A} = \bar{\sigma}_A / X \text{ and } \sigma_{2A} = Q_A\, \sigma_{1A}$$

7) Determine the principal stresses in region B:

$$\sigma_{1B} = \sigma_{1A} / f_o \text{ (initially) or } \sigma_{1B} = \sigma_{1A} / f \text{ (subsequently) with } \sigma_{2B} \text{ from eq(13.98)}$$

8) For $Q_B = \sigma_{2B} / \sigma_{1B}$, find w_B from eq(13.99). Stop when $w_B \to 0$

9) Find the principal plastic strain increments within the groove

$$d\varepsilon_{1B}{}^P = d\varepsilon_{2A}{}^P / w_B \text{ and } d\varepsilon_{3B}{}^P = -\,(d\varepsilon_{1B}{}^P + d\varepsilon_{2B}{}^P) \text{ where } d\varepsilon_{2A}{}^P = d\varepsilon_{2B}{}^P$$

10) Sum the thickness strains increments

$$\varepsilon_{3A}{}^P = \sum d\varepsilon_{3A}{}^P \text{ and } \varepsilon_{3B}{}^P = \sum d\varepsilon_{3B}{}^P$$

within successive iterations and re-define f from eq(13.95).

11) Return to step 2.

Either prediction given in quadrant 1, when combined with the local instability line in quadrant 2, provides an FLD of in-plane limit strains under any stress ratio Q. For example, in axes of engineering strain, Fig. 13.25 shows that the strain path is linear for equi-biaxial tension ($Q = 1$) but is non-linear for uni-axial tension ($Q = 0$) and pure shear ($Q = -1$). Limit strains are provided by their intersections with the chosen FLD. Note, that plane strain constitutes the lowest strain limit and is responsible for most product failures in the region of bends and folds.

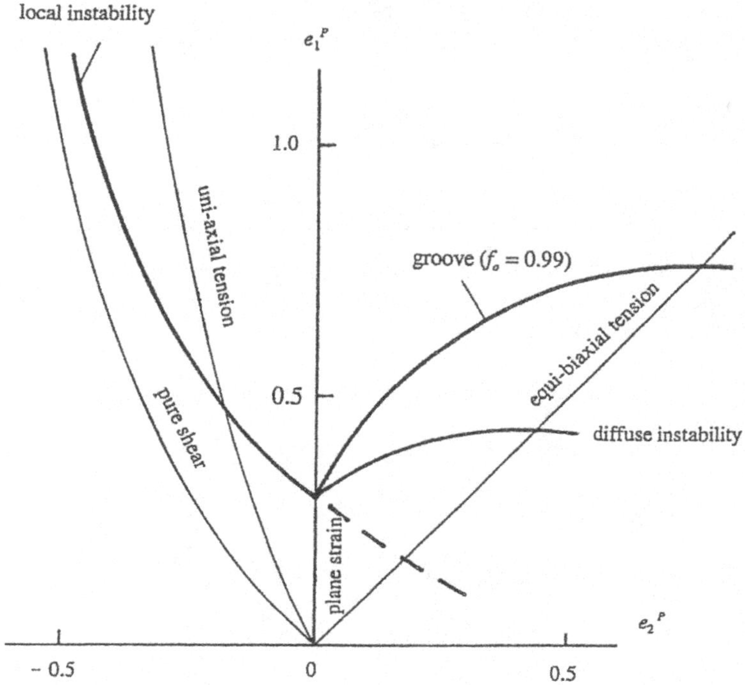

Figure 13.25 Comparison of FLD's from diffuse, local and groove formation

13.8 Concluding Remarks

Plastic instability appears in many forms: from the buckling of long struts and thin plates at loads beyond the elastic limit to the material instabilities that control fracture. Among the analytical methods used are extensions to elastic analyses using a suitable plastic modulus and the realisation of pressure and load maxima co-incident with the onset of failure, e.g. as with the Considére condition in simple tension. The extension to biaxial stress states may involve local or diffuse instability depending upon the applied stress ratio. In practice, where biaxial stress states occur in pressure vessels and in sheet metal forming, the initial orthotropy of the rolled sheet may become an important factor in these analyses. In the most common plane strain failure mode, anisotropy is almost always present along with an initial imperfection. Each is admitted when constructing the forming limit diagram based upon failure from a narrow, inclined local neck.

References

1. Shanley F. R. *Jl. Aero. Sci.*, 1947, **14**, 261.
2. Vinson J. R. *Structural Mechanics*, 1974, Wiley.
3. ESDU 72019 *Buckling of Flat Isotropic Plates Under Uniaxial and Biaxial Loading*, August 1972.
4. Ramberg W. and Osgood W. R. *Nat. Adv. Comm. for Aero*., 1943, TN902.
5. ESDU 83044 *Plasticity Correction Factors for Plate Buckling*, December 1983.
6. ESDU 71005 *Buckling of Flat Plates in Shear*, February 1971.

7. Gerard G. *Jl. Appl. Mech.*, 1948, **15**(1), 7.
8. Rees D. W. A. *Jl. Strain Anal.*, 1995, **30**, 305.
9. Hill R. *Phil. Mag.*, 1950, **41**, 1133.
10. Swift H. W. *Jl. Mech. Phys. Solids*, 1952, **1**, 1.
11. Rees D. W. A. *Jl. Mats. Proc. Tech.*, 1999, **92-93**, 508.
12. Lee S. H. and Kobayashi S. *Proc.: 3rd North. Amer. Metal Working Res. Conf.*, 1975, p. 277, Carnegie Press, Pittsburgh.
13. Hasek V. V. *Proc.: ICF4 Fracture*, 1977, **2**, p. 475, Pergamon Press, Waterloo, Canada.
14. Korhonen A. S. *Trans ASME, Jl. Eng. Mat. and Tech.*, 1978, **100**, 303.
15. Moore G. G. and Wallace J. F. *Jl. Inst. Met.*, 1964-65, **93**, 33.
16. Marciniak Z. and Kuczynski K. *Int. Jl. Mech. Sci.*, 1967, **9**, 609.
17. Hill R. *Jl. Mech. Phys. Solids*, 1952, **1**, 19.
18. Rees D. W. A. *Jl. Mats. Proc. Tech.*, 1995, **55**, 146.
19. Veerman C and Neve P. V. *Sht. Met. Ind.*, 1972, **49**, 421.
20. Keeler S. *Sht. Met. Ind.*, 1965, **42**, 683.
21. Pearce R. *Sht. Met. Ind.*, 1971, **48**, 943.
22. Marciniak Z., Duncan J. and Hu S. J. *Mechanics of Sheet Metal Forming*, 2002, Butterworth-Heinemann.

Exercises

13.1 It is required to find the maximum compressive load that a brass strut with an elliptical thin-walled section can support without plastic buckling. If the strut length is 825 mm, the major and minor axes are 80 mm and 30 mm respectively and the wall thickness is 3 mm, estimate the buckling load for the minor axis where the ends may be assumed pinned. Take the Hollomon stress–strain law for brass as $\sigma/\sigma_c = 7.83\varepsilon^{1/3}$ where the compressive yield stress $\sigma_c = 144$ MPa.
See solution

13.2 A pinned-end strut 1.5 m long has the cross-section in Fig. 13.26. Find Engesser's plastic buckling load given that Hollomon's law $\sigma/\sigma_c = 5\varepsilon^{1/3}$ describes the compressive stress–strain curve beyond a yield stress of $\sigma_c = 300$ MPa. Compare Engesser's load with predictions from the straight line and parabolic formulae using mid-range constants for steel given in text example 13.1.
Ans: 817.2 kN

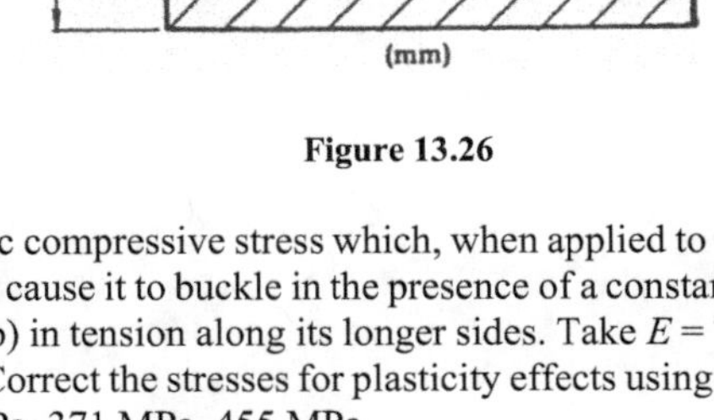

Figure 13.26

13.3 Determine the elastic compressive stress which, when applied to the shorter sides of a thin plate 390 × 200 × 7.5 mm, will cause it to buckle in the presence of a constant stress of 80 MPa which acts: (a) in compression and (b) in tension along its longer sides. Take $E = 72.4$ GPa and $v = 0.33$ with all sides simply supported. Correct the stresses for plasticity effects using Fig. 13.10. Take $\sigma_n = 405$ MPa and $m = 16$. *Ans*: 265 MPa, 371 MPa, 455 MPa

13.4 A plate $a \times b \times t$ is subjected to a normal tensile stress $\sigma_y = \sigma$ on its $a \times t$ sides and a compressive stress $\sigma_x = -\sigma$ on its $b \times t$ sides. If all edges are simply supported show that the minimum value of critical buckling stress is given by $(\sigma_x)_{cr} = (8\pi^2 D)/(tb^2)$ for one half-wave in the y–direction. To what values of $r = a/b$ does this minimum apply?
See solution : $1/\surd 3$, $2/\surd 3$ etc.

13.5 Find the shear stress for which a rectangular plate 200 mm wide × 260 mm long × 4.5 mm thick buckles when the short sides are clamped and the long sides are simply supported. Using Figs 13.15 and 13.16, correct for plasticity given $\sigma_n = 340$ MPa, $m = 16$, $E = 73$ GPa and $\nu = 0.3$. *Ans*: 198 MPa
See solution

13.6 Determine the equivalent stress and strain at the point of instability for a thin-walled, open-ended pressurized cylinder of mean diameter d and thickness t, if the material hardens according to the law $\bar{\sigma} = \sigma_o(1 + B\bar{\varepsilon}^P)^n$. What is the pressure that causes this instability?
See solution: $\bar{\sigma} = \sigma_o(2nB/3)^n$, $\bar{\varepsilon}_u^P = (1/B)(2nB/3 - 1)$, $p = (2t\sigma_o/d)(2nB/3)^n$

13.7 Determine the true fracture strain of a ductile material in terms of: (a) the % elongation at fracture and (b) the % reduction of area at fracture in a tensile test.
See solution

13.8 The following table of results applies to a tension test on a cylindrical testpiece 13 mm diameter and 50 mm long:

Force, kN	10	20	30	40	52.29	61.5	67.2	72	74.8
	76.6	77	78	78.3	78.53	78.72	78.79	78.73	78.6
	78.37	78.16	77.82	77.50	77.12	76.71			
Extension, mm	0.02	0.04	0.05	0.07	0.50	1.20	2.03	3.45	4.3
	5.0	6.2	7.37	8.21	9.01	9.90	11.23	12.20	13.0
	14.2	15.0	16.1	16.9	17.5	18.1			

Determine graphically, from the Considére condition, the true stress and natural strain at the point of instability. Determine the strength coefficient, hardening exponent and instability strain from a Hollomon description of the given data. Compare with Considére's instability strain.
See solution.

13.9 Construct an FLD for conditions of simple local and diffuse instability within quadrants 1 and 3 respectively. The sheet metal has a yield stress 150 MPa with a Hollomon exponent $n = 0.25$. The in-plane anisotropy is characterised with $r_1 = 1.7$ and $r_2 = 2.0$. *Hint*: Determine the sub-tangents z from eqs(13.85) and (13.91) for a range of stress ratios. Convert each z to an equivalent instability strain which decomposes into the two principal in-plane engineering strains.

CHAPTER 14

STRESS WAVES IN BARS

14.1 Introduction

This introduction to elastic and plastic wave generation within round bars applies to slow heavy strikers and lighter struck members, as in a forging operation, and faster impacts from projectiles upon heavier bodies. By equating external work to the strain energy stored, the dynamic stresses and deflections may be found, usually from assuming that the static and dynamic force versus deflection curves are the same. This static approach can lead to a good approximate solution though it provides little information on the mechanics of loading. Where a hard, fast moving striker is of similar weight to the struck member, as from an explosive charge, the stress waves generated apply to relatively short times after impact. The full solution for longer times after impact would require an examination of normal vibration modes using the exact equations of elasticity. Approximate dynamic solutions considered here are more dependable than those found from a static stress analysis and may be used to describe certain observed effects, including spalting and scabbing. The analysis begins with a geometrical interpretation of the elastic impact wave equations from which bar stresses and particle velocities are found. This is followed with an analysis of plastic waves using a simple bi-linear approximation to the stress–strain curve. The theory is applied to elastic and elastic-plastic impacts between metal bars and to the drop weight test where experimental data is available for comparison. For a more expansive discussion the reader is referred to dedicated works [1–10] upon this topic.

14.2 The Wave Equation

Let the longitudinal wave propagation in a bar be c_o m/s and the particle velocity in the bar be v m/s. The latter is intimately related to the stress in the bar and is readily converted to an absolute velocity from knowing the bar velocity after impact. A given cross-section in the bar, distance x from the left end origin, displaces by an amount u where the compressive stress is σ. At a further distance δx away from this cross section, the displacement and stress increase by the respective amounts $(\partial u/\partial x)\delta x$ and $(\partial \sigma/\partial x)\delta x$ (see Fig. 14.1).

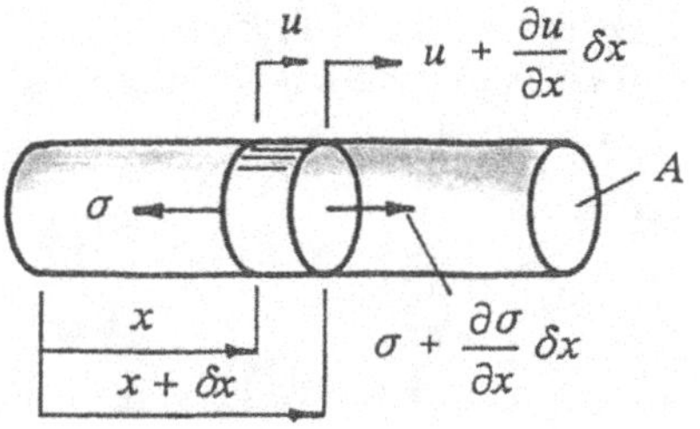

Figure 14.1 Elastic stress wave in a bar

Hooke's law gives the elastic stress as

$$\sigma = E\varepsilon = E\frac{\partial u}{\partial x}$$

so that its rate of change with respect to x becomes

$$\frac{\partial \sigma}{\partial x} = E\left(\frac{\partial^2 u}{\partial x^2}\right) \tag{14.1}$$

where E is the elastic modulus of the bar. Newton's second law of motion connects the axial force due to the stress change, with the force required to accelerate the plane through δx in time δt. Writing the mass $m = \rho A \delta x$, where ρ is the density and A is the bar's section area, this becomes

$$F = \rho A \delta x \times \frac{\partial^2 u}{\partial t^2} = A\frac{\partial \sigma}{\partial x}\delta x$$

Substituting from eq(14.1):

$$\rho A\frac{\partial^2 u}{\partial t^2}\delta x = AE\frac{\partial^2 u}{\partial x^2}\delta x$$

This results in a one-dimensional wave equation:

$$\frac{\partial^2 u}{\partial t^2} = c_o^2\frac{\partial^2 u}{\partial x^2} \tag{14.2}$$

where $c_o = \sqrt{(E/\rho)}$. The general solution [1] to eq(14.2) is

$$u = f_1(c_o t - x) + f_2(c_o t + x) \tag{14.3a}$$

in which the two functions f_1 and f_2 may be interpreted from a simple harmonic motion of the displacement in the form $u = a \sin \omega t$. With a period T, the wave frequency is $\omega = 2\pi/T$ and the wavelength is $\lambda = c_o T$. Then

$$u = a\sin\left(\frac{2\pi c_o t}{\lambda}\right) \tag{14.3b}$$

To match eq(14.3b) with the first function f_1 in eq(14.3a), the former equation must be modified with a time delay x/c_o for position x as follows:

$$u = a\sin\left[\frac{2\pi}{\lambda}(c_o t - x)\right] \tag{14.3c}$$

This reveals a wave propagating in direction x–positive. Similarly, within the second function f_2 in eq(14.3a), a wave propagates in direction x–negative. Thus, the solution given in eq(14.3c) reveals two waves in both travelling at the propogation velocity $c_o = \sqrt{(E/\rho)}$ but in opposite directions.

14.3 Particle Velocity

Consider the bar in Fig. 14.2 with a compressive stress maintained at its left-hand end. The velocity v of particles, due to the effect of the stress wave produced, may be found from equating impulse to the change of linear momentum.

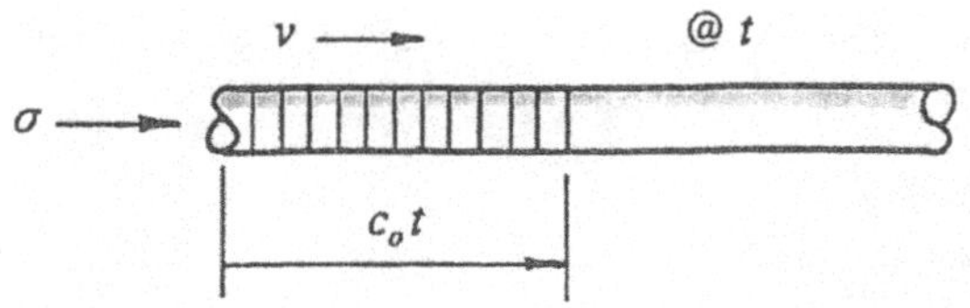

Figure 14.2 Impulse and particle velocity due to a stress application

That is, at time t in Fig. 14.2:

$$A\sigma t = A\rho(c_o t)v$$

from which

$$\sigma = \rho c_o v \quad \text{or} \quad v = \frac{\sigma}{\rho c_o} \tag{14.4a,b}$$

in which ρc_o is the acoustic impedance. Equations (14.4a,b) show that both σ and v are independent of the bar's dimensions. The two waves identified above may now be associated with the *sense* of the wave and particle velocities, c_o and v respectively, as follows:

- A *compressive wave* has both c_o and v in direction x–positive (same sense).
- A *tensile wave* has c_o in direction x–positive and v in direction x–negative (opposite sense).

14.3.1 Reflection at a Boundary

A compressive wave will be reflected at the free-end of a rod to become a tensile wave, i.e. compression returns as tension and vice versa (see Figs 14.3a,b).

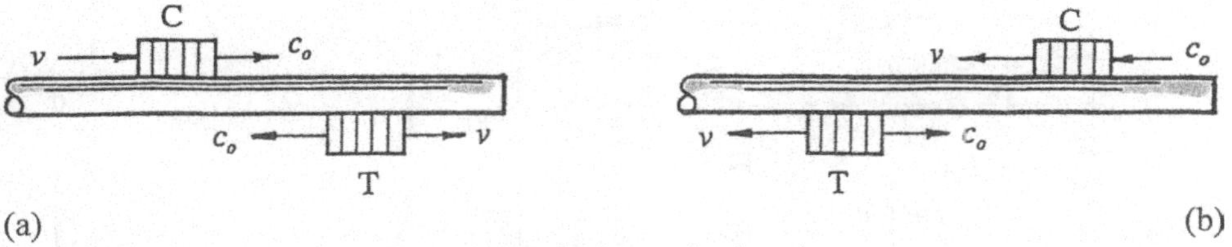

Figure 14.3 Reflection at a free-boundary

The stress reversal must produce a condition of zero stress at the free-end if an identical wave of opposite particle sign is reflected by the boundary. As a consequence both the velocity and displacement are doubled at the free end. In contrast, at a fixed boundary the end displacement must remain zero so that the reflected wave remains in the same sense as the incident wave, i.e. it remains either in tension or compression (see Figs 14.4a,b).

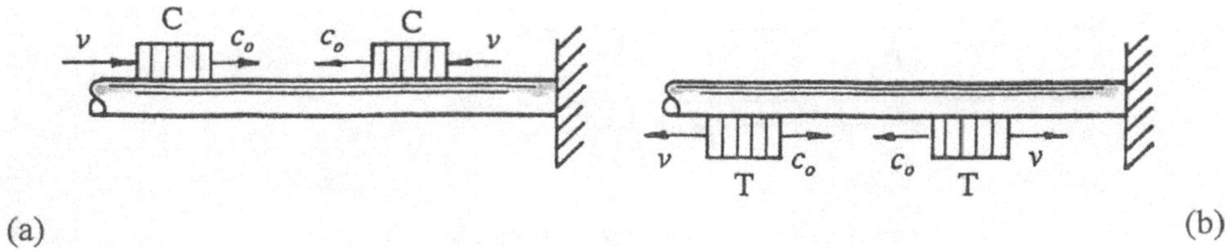

Figure 14.4 Reflections at a fixed boundary

It follows that upon reflection at a fixed boundary the stress is doubled but the velocity and hence the displacement must cancel to zero.

14.3.2 Displacement–Time Diagram

The position of both compressive and tensile elastic wave fronts at their propagation and particle velocities $c_o = \sqrt{(E/\rho)}$ and $v = \sigma/(\rho c_o)$ can appear with similar stress levels but opposing directions on an x–t diagram. In any region of this diagram the force at all points must be in equilibrium if additional stress waves due to discontinuities in section area are to be avoided. Moreover, a compatibility condition is ensured when all points within this region have the same absolute velocity.

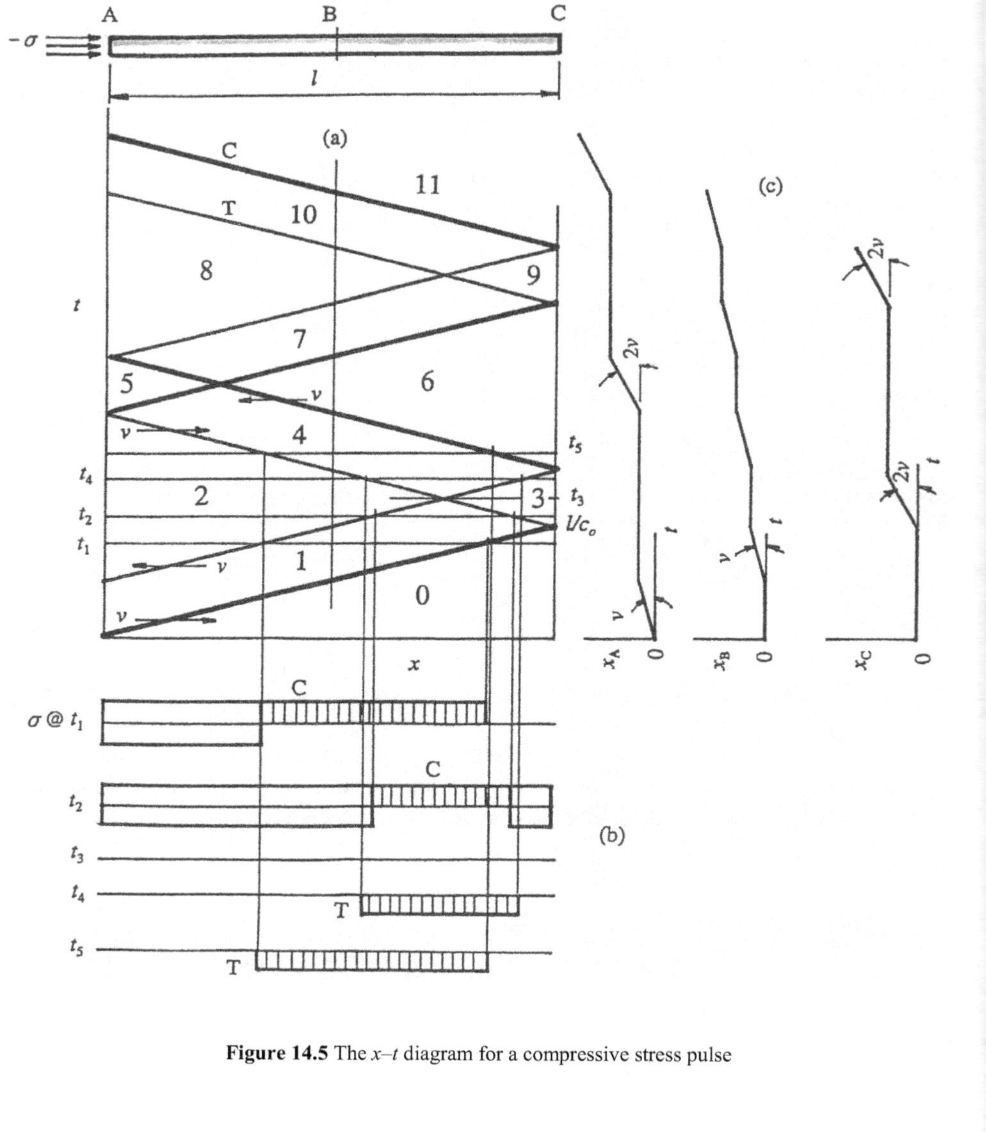

Figure 14.5 The x–t diagram for a compressive stress pulse

Finally, to conserve momentum at a given time, the momentum change must equal the impulse applied up to that time. Example 14.1 embodies these principles within the construction the *x–t* diagram in Fig. 14.5 when a stress pulse is applied to one end of a bar.

Example 14.1 A uniform circular bar of length l is subjected to a constant compressive stress at its left-hand end for a pulse time $t = l/(2c_o)$. Construct the *x–t* diagram showing the tensile and compressive waves. Use this diagram to show how the displacements of the bar's ends and centre vary with time.

The *x–t* diagram consists of a repeated pattern of wave reflections with alternating sense as shown in Fig. 14.5a. The given pulse is equivalent to applying a compressive wave followed by a tensile wave separated by the pulse time (see Fig. 14.5b). The time for this wave to reach the right-hand end is l/c_o. Hence, the compressive wave reaches the bar centre before the tensile wave begins. Within the regions 0, 1, 2 etc. in Fig. 14.5a, the net stress and absolute velocities within Table 14.1 apply.

Table 14.1 Stress and velocity within given regions of the *x–t* diagram

region	0	1	2	3	4	5	6	7	8	9	10	11
stress	0	$-\sigma$	0	0	σ	0	0	$-\sigma$	0	0	σ	0
velocity	0	v	0	$2v$	v	$2v$	0	v	0	$2v$	v	0

The time-dependent displacements of the bar ends and centre, A, C and B respectively, may be constructed from the wave intersections at these positions, as shown in Fig. 14.5c. The gradients dx/dt of the sloping parts of these diagrams define the particle velocities v and $2v$.

14.4 Longitudinal Impact of Bars

14.4.1 Equal Section Areas

Impact between circular bars with similar section areas but different lengths must satisfy a dynamic equilibrium and a compatibility condition. Let the velocity of bar 1 be v_1 and that of bar 2 be v_2, where $v_1 > v_2$ (see Fig. 14.6a). At the instant of longitudinal impact each bar experiences a compressive wave of identical magnitude σ at a particle velocity v, found from

$$v_1 - v = v_2 + v \qquad \therefore\ v = \tfrac{1}{2}(v_1 - v_2) \qquad (14.5a,b)$$

Substituting eq(14.5b) into either side of eq(14.5a) shows that the absolute velocity of each bar becomes $\tfrac{1}{2}(v_1 + v_2)$. The *x–t* diagram in Fig. 14.6b applies to this condition. From this, the time at separation point S is found as $t_s = 6l/c_o$. The conditions of stress and velocity across each region are given in Table 14.2.

Table 14.2 Stress and velocity regions

Region	0	1	2	3	4	5	6
stress	0	0	$-\sigma$	0	0	σ	0
abs vel	v_2	v_1	$\tfrac{1}{2}(v_1 + v_2)$	v_2	v_1	$\tfrac{1}{2}(v_1+v_2)$	v_1

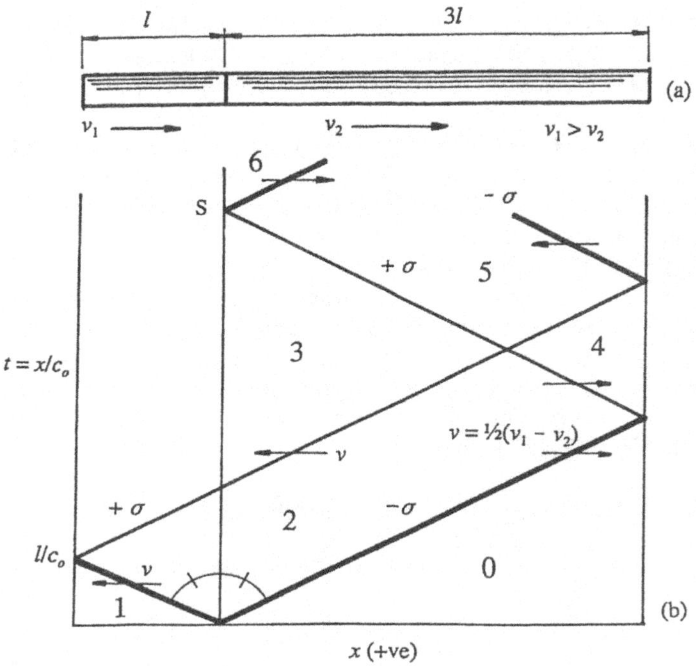

Figure 14.6 Longitudinal impact between two bars

14.4.2 Unequal Section Areas

When a bar or striker impacts longitudinally another bar with larger section area, a wave reflection occurs at the change in section (see Fig. 14.7a). Firstly, let the incident wave *a* exist in the smaller section. Waves *b* and *c* reflect and transmit respectively at the section change. Their stress levels σ_a, σ_b and σ_c, where $\sigma_a > \sigma_b$, are denoted positive when compressive, as shown in Fig. 14.7b.

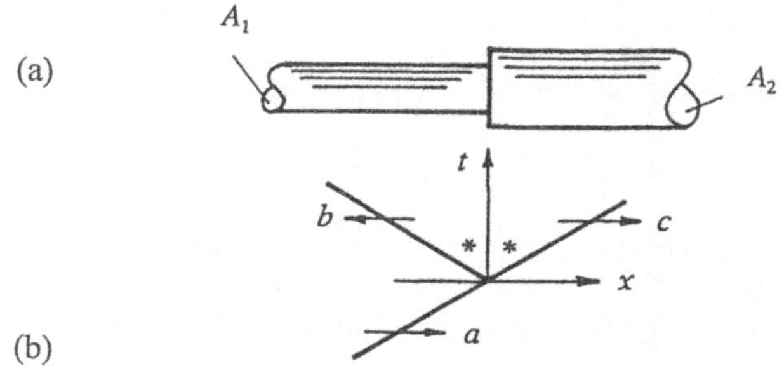

Figure 14.7 Reflection with a step-up in area

By assuming compression within the three waves *a*, *b* and *c*, equilibrium and compatibility conditions provide simultaneous equations:

$$(\sigma_a + \sigma_b)A_1 = \sigma_c A_2 \tag{14.6a}$$

$$\frac{(\sigma_a - \sigma_b)}{\rho c_o} = \frac{\sigma_c}{\rho c_o} \tag{14.6b}$$

Given $A_2 = 2A_1$, for example, the solution to eqs(14.6a,b) shows $\sigma_b = \sigma_a/3$ and $\sigma_c = 2\sigma_a/3$. However, if A_2 is very large, the effect is similar to incidence upon a fixed-end. Now let the incident wave exist in the larger section (see Fig. 14.8a).

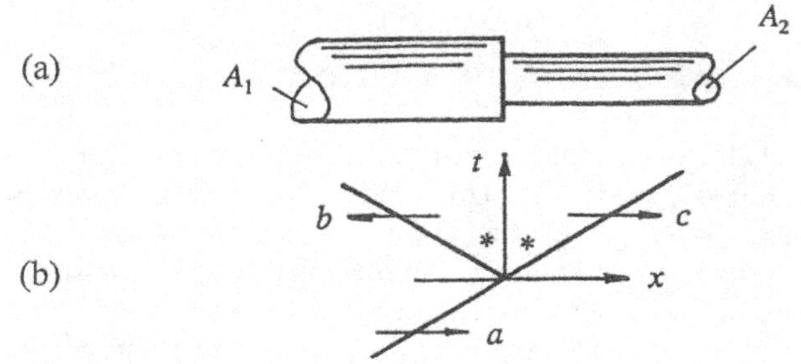

Figure 14.8 Reflection from a step-down in area

The two conditions in eqs(14.6a,b) are simplified:

$$(\sigma_a + \sigma_b)A_1 = \sigma_c A_2 \tag{14.7a}$$

$$\sigma_a - \sigma_b = \sigma_c \tag{14.7b}$$

and, if $A_1 = 2A_2$, then from eqs(14.7a,b) the stress levels are $\sigma_b = -\sigma_a/3$ and $\sigma_c = 4\sigma_a/3$, showing that the reflected wave b changes sign, i.e. it becomes tensile.

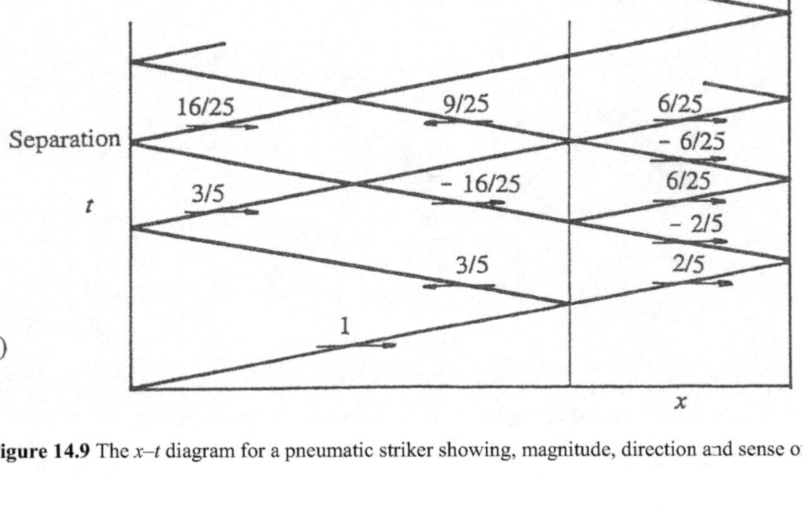

Figure 14.9 The x–t diagram for a pneumatic striker showing, magnitude, direction and sense of stress waves

Example 14.2 Figure 14.9a shows the stepped steel striker of a pneumatic tool which impacts upon a fixed target at its left end. Plot the *x*–*t* diagram and find the highest impact velocity if the dynamic elastic stress limit is 675 MPa. The initial stress may be taken as unity by setting $\sigma_o = \rho c_o v = 1$. Take $E = 210$ GPa and $\rho = 7890$ kg/m^3.

With zero net velocity at the target end $v_1 - v = 0$. From eq(14.4b), at the stress limit:

$$v_1 = \frac{\sigma}{\rho c_o} = \frac{\sigma}{\sqrt{E\rho}} = \frac{(675 \times 10^6)}{\sqrt{(210 \times 10^9) \times 7890}} = 16.58 \text{ m/s}$$

Applying eqs(14.6a,b) across the stepped increase in the cross-section, where $A_2 = 4A_1$, gives $\sigma_b = 3\sigma_a/5$ and $\sigma_c = 2\sigma_a/5$. Applying eqs(14.7a,b) to the step-down section change, where $A_2 = A_1/4$, gives $\sigma_b = -3\sigma_a/5$ and $\sigma_c = 8\sigma_a/5$ and. The *x*–*t* diagram (see Fig. 14.9b) is constructed from the repeated application of these stress multiplication factors to transmission and reflection of incident waves as they meet the section change from either side. Recall that at the fixed, target end, the reflected wave retains the sense of the stress within the incident wave but the particle velocity is reversed.

Example 14.3 A striker travels lengthwise with a velocity $3v_o/2$ to impact the left end of a bar that is fixed rigidly at its right end (see Fig. 14.10a). Assuming elastic impact conditions, plot the *x*–*t* diagram, showing the separation point on this diagram and from it produce stress–time and velocity–time plots for locations X, Y and Z. Assume that the striker and bar material is the same.

Table 14.3 Regional wave stresses and velocities

Region	Stress	Absolute velocity
0	0	0
1	-2σ	$v = 3v_o/4$
2	$(-2\sigma + 4\sigma/3) = -2\sigma/3$	$-v/3 = -v_o/4$
3	$(-2\sigma - 2\sigma) = -4\sigma$	0
4	$(-4\sigma + 4\sigma/3) = -8\sigma/3$	$-v/3 - v = -4v/3 = -v_o$
5	$(-8\sigma/3 + 2\sigma) = -2\sigma/3$	$-4v/3 - v = -7v/3 = -7v_o/4$
6	$(-8\sigma/3 + 4\sigma/3) = -4\sigma/3$	$-4v/3 + 4v/3 = 0$
7	$(-2\sigma/3 + 4\sigma/3) = 2\sigma/3$	$-v = -3v_o/4$
8	$(+2\sigma/3 + 2\sigma) = 8\sigma/3$	0
L	0	$2v = 3v_o/2$
M	$-\sigma$	$3v_o/2 - v = (3v_o/2 - 3v_o/4) = 3v_o/4$
N	0	$3v_o/4 - v = (3v_o/4 - 3v_o/4) = 0$
O	$-\sigma/3$	$-v/3 = -v_o/4$
P	0	$-v/3 - v/3 = -v_o/2$
Q	$-\sigma/3$	$-2v/3 - v/3 = -v = -3v_o/4$
R	0	$-v - v/3 = -4v/3 = -v_o$

Force balance gives $\sigma_o A = \sigma_1 A / 2$, which shows that the bar stress is twice that of the striker. The absolute velocities of particles at the contact surfaces must be the same:

$$3v_o/2 - v = v + v_1$$

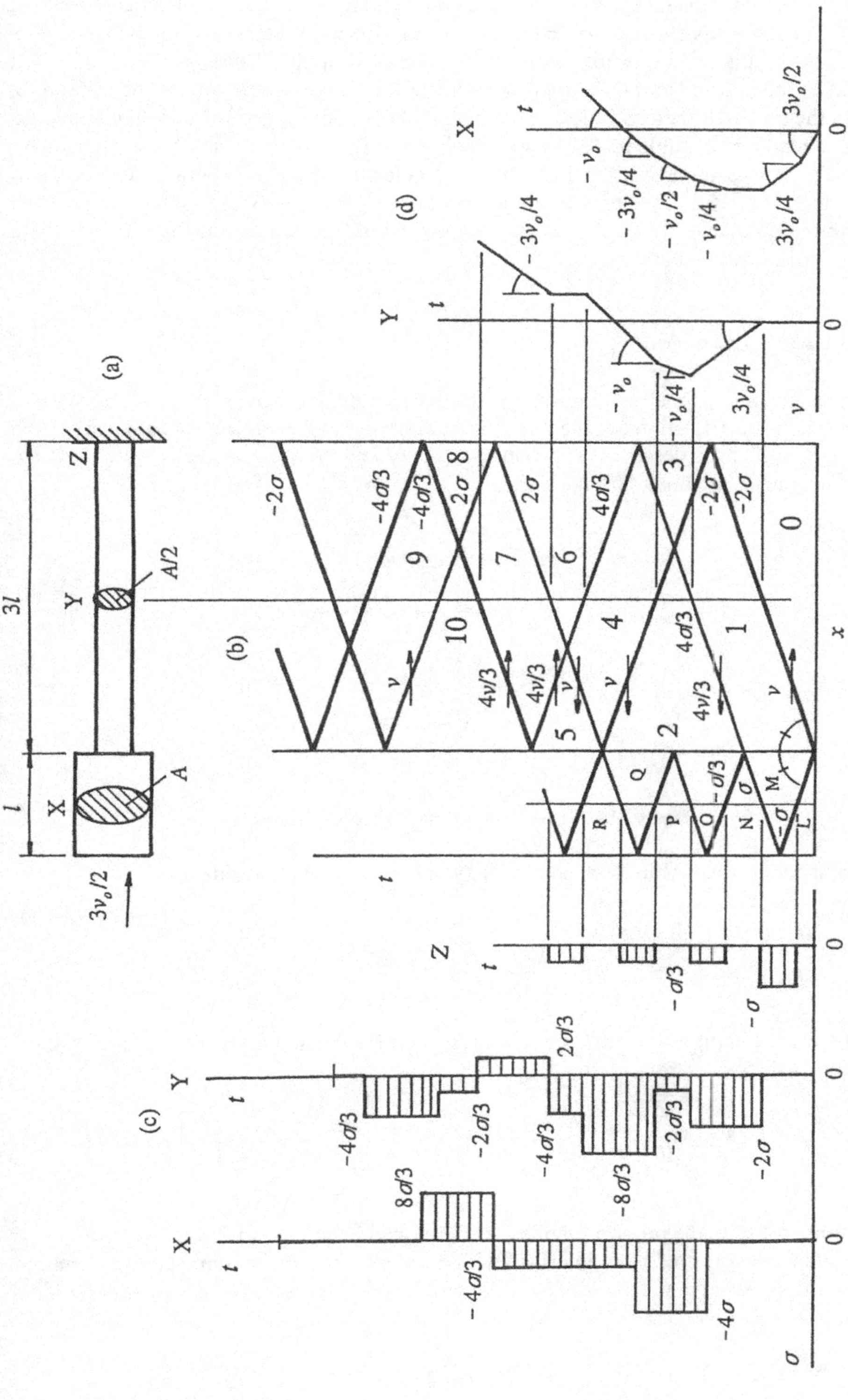

Figure 14.10 (a) Striker and bar showing (b) the x - t, (c) σ - t and (d) v - t diagrams; Key —— tension, —— compression

Since $v_1 = 0$, this shows that $v = 3v_o/4$ is the particle velocity of two identical compressive waves directed away from the contacting surfaces. Because the area of the bar section is halved, the multiplication factors in eqs(14.6) and (14.7) apply. These provide a sufficient portion of the *x–t* diagram (Fig. 14.10b) to reveal that a bar separation occurs at point 2. Within the enclosed regions 0, 1, 2 . . . and L, M, N . . . , the stress and absolute velocities are calculated within Table 14.3. The stress column in Table 14.3 enables the accompanying stress plots for positions X, Y and Z to be constructed to the left side of the *x–t* diagram, as shown in Fig. 14.10c. The absolute velocity column in Table 14.3 allows the displacement plots for points X and Y to be constructed to the right of the *x–t* plot (see Fig. 14.10d), there being no displacement at Z.

14.4.3 Change of Material

Where a change of material occurs in the bar length, or between impacting bars, a transmission and reflection occur at the interface, similar to the effect of a change in section area. Let E, ρ and c describe the modulus, density and propagation velocity in each bar material 1 and 2, as shown in Fig. 14.11a.

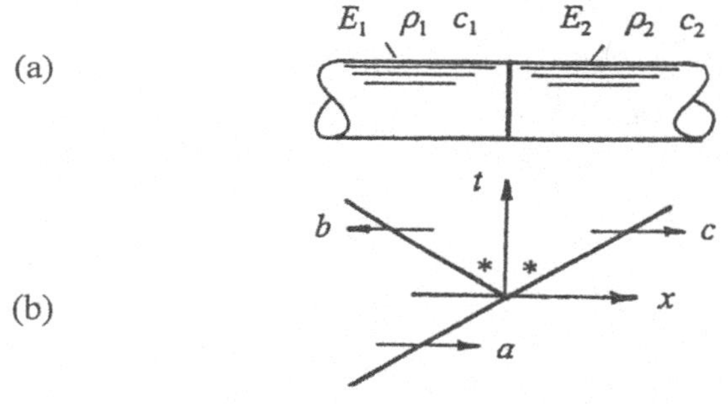

Figure 14.11 Reflection at the interface of a composite bar

For an incident wave a the force and velocity balance equations yield

$$\sigma_a + \sigma_b = \sigma_c \tag{14.8a}$$

$$\frac{(\sigma_a - \sigma_b)}{\rho_1 c_1} = \frac{\sigma_c}{\rho_2 c_2} \tag{14.8b}$$

Solving eqs(14.8a,b) leads to the stress within the reflected and transmitted waves b and c (see Fig. 14.11b):

$$\sigma_b = \left[\frac{\sqrt{E_2\rho_2} - \sqrt{E_1\rho_1}}{\sqrt{E_2\rho_2} + \sqrt{E_1\rho_1}}\right]\sigma_a\,, \qquad \sigma_c = \left[\frac{2\sqrt{E_2\rho_2}}{\sqrt{E_2\rho_2} + \sqrt{E_1\rho_1}}\right]\sigma_a \tag{14.9a,b}$$

Example 14.4 Find the proportion of stress for an incident wave within epoxy resin that is transmitted and reflected at an interface with glass. Repeat for the same arrangement but with an incident wave within the glass. For epoxy: $E = 3.24$ GPa and $\rho = 1245$ kg/m^3 and for glass: $E = 62$ GPa and $\rho = 2215$ kg/m^3.

For a transmission from epoxy (subscript 1) to glass (subscript 2):

$$\sqrt{(E_1\rho_1)} = \sqrt{(3.24 \times 10^9 \times 1245)} = 2.008 \times 10^6 \text{ Nm/s}^2$$

$$\sqrt{(E_2\rho_2)} = \sqrt{(62 \times 10^9 \times 2215)} = 11.718 \times 10^6 \text{ Nm/s}^2$$

Substituting into eq(14.9a,b):

$$\sigma_b = \frac{(11.718 - 2.008)\sigma_a}{(11.718 + 2.008)} = 0.707\sigma_a, \qquad \sigma_c = \frac{(2 \times 11.718)\sigma_a}{(11.718 + 2.008)} = 1.707\sigma_a$$

Interchanging subscripts 1 and 2 for glass to epoxy, eq(14.9a,b) gives

$$\sigma_b = \frac{(2.008 - 11.718)}{(2.008 + 11.718)} = -0.707\sigma_a, \qquad \sigma_c = \frac{(2 \times 2.008)}{(2.008 + 11.718)} = 0.293\sigma_a$$

Note: For air $\sqrt{(E\rho)}$ is small enough to be neglected compared to glass (say). So, for a glass to air interface eqs(14.9a,b) show $\sigma_b \approx -\sigma_a$ and $\sigma_c \approx 0$. That is, the wave is fully reflected with a change in sign, as was shown earlier for a free-end condition.

14.5 Plastic Waves

Previously, when the end of a long rod is struck in such a way that a constant elastic stress is maintained at its end, then a wave of compression with constant amplitude will travel along the rod at a speed of $c_o = \sqrt{(E/\rho)}$. Where the stress level in a longitudinal wave exceeds the yield stress of the bar material, i.e. $\sigma > \sigma_o$, two wave fronts arise: (i) an elastic wave of amplitude σ_o travelling at c_o and (ii) a plastic wave of amplitude σ which follows the elastic wave with a lower velocity c (see Fig. 14.12a). The distances x travelled by each wave at time t are respectively $c_o t$ and ct. Since $c < c_o$, at the plastic wave front, there will be a step up in the stress from σ_o to σ, where σ remains constant between the plastic wave front and the left struck end of the bar.

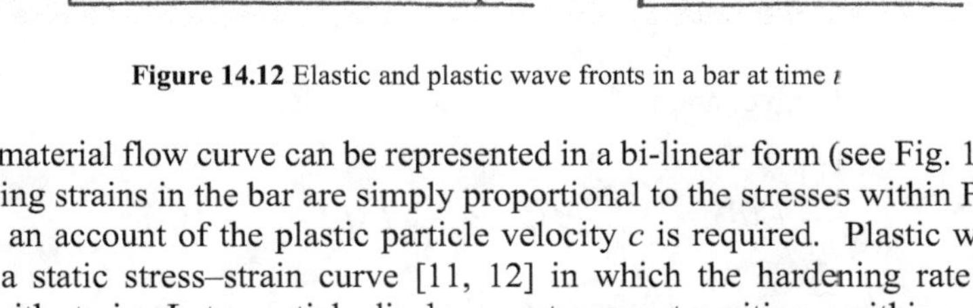

(a) (b)

Figure 14.12 Elastic and plastic wave fronts in a bar at time t

Where the material flow curve can be represented in a bi-linear form (see Fig. 14.12b), the corresponding strains in the bar are simply proportional to the stresses within Fig. 14.12a. Otherwise, an account of the plastic particle velocity c is required. Plastic wave theory applies to a static stress–strain curve [11, 12] in which the hardening rate $S = d\sigma/d\varepsilon$ decreases with strain. Let a particle displacement occur at position x within a circular bar of area A, due to a change in stress over the infinitesimal length δx, as shown previously in Fig. 14.1. Replacing E with S in eq(14.2) gives

$$\frac{\partial^2 u}{\partial t^2} = \left(\frac{S}{\rho}\right)\frac{\partial^2 u}{\partial x^2} \qquad (14.10)$$

where $S = S(\varepsilon)$ is a variable. The rigorous solution to eq(14.10) is due to von Karman [1]. Taylor's simplified solution $x/t = \sqrt{(S/\rho)}$ [11] employs the following boundary conditions:

(i) for time t at $x = 0$: $u\,(0, t\,) = -\,v_o t$

(ii) for time t at $x = \infty$, $u\,(\infty, t\,) = 0$

Since the strain aligned with x is found from $\varepsilon = \partial u/\partial x$, the displacement u follows as

$$u = \int_{\infty}^{x} \varepsilon\,\mathrm{d}x \tag{14.11a}$$

Set $\beta = x/t$ so that the strain $\varepsilon = f(\beta)$ for a given time $t = \mathrm{d}x/\mathrm{d}\beta$. Equation (14.11a) becomes

$$u = t\int_{\infty}^{\beta} f(\beta)\,\mathrm{d}\beta \tag{14.11b}$$

Using condition (ii) above, the substitution of the lower limit must give zero displacement. Equation eq(14.11b) integrates to

$$u = t\,F(\beta) \tag{14.12}$$

Now if t varies, successive differentiation of eq(14.12) reveals

$$\frac{\partial u}{\partial t} = F(\beta) + t\,F'(\beta)\frac{\partial \beta}{\partial t}$$

$$\frac{\partial^2 u}{\partial t^2} = 2\,F'(\beta)\frac{\partial \beta}{\partial t} + t\,F''(\beta)\left(\frac{\partial \beta}{\partial t}\right)^2 + t\,F'(\beta)\frac{\partial^2 \beta}{\partial t^2} \tag{14.13}$$

where $F'(\beta) = f(\beta)$ and $F''(\beta) = f'(\beta)$. With $\beta = x/t$, the following partial derivatives apply to the right hand side of eq(14.13):

$$\frac{\partial \beta}{\partial t} = -\frac{x}{t^2}, \quad \left(\frac{\partial \beta}{\partial t}\right)^2 = \frac{x^2}{t^4}, \quad \frac{\partial^2 \beta}{\partial t^2} = \frac{2x}{t^3} \tag{14.14a,b,c}$$

The left side of eq(14.13) is identified with the right side of eq(14.10), so that

$$\frac{\partial^2 u}{\partial x^2} = \frac{\partial \varepsilon}{\partial x} = \frac{\partial f(\beta)}{\partial x} = \frac{1}{t}\,f'(\beta) \tag{14.15}$$

Substituting eqs(14.14a,b,c) and (14.15) into (14.13):

$$\frac{S}{\rho t}\,f'(\beta) = -\,\frac{2x}{t^2}\,f(\beta) + \frac{x^2}{t^3}\,f'(\beta) + \frac{2x}{t^2}\,f(\beta)$$

$$\frac{1}{t}\,f'(\beta)\left[\left(\frac{x^2}{t^2}\right) - \frac{S}{\rho}\right] = 0$$

$$\frac{1}{t}\,f'(\beta)\left(\beta^2 - \frac{S}{\rho}\right) = 0 \tag{14.16}$$

Equation (14.16) is satisfied either by

$$f'(\beta) = \frac{\partial f(\varepsilon)}{\partial \beta} = 0 \quad \Rightarrow \quad \varepsilon = \text{constant (w.r.t. } \beta) \tag{14.17a}$$

or

$$\beta = \sqrt{S/\rho} \tag{14.17b}$$

Figure 14.13 provides a geometrical interpretation to the solutions (14.17a,b) within a region of work hardening between the yield strain ε_o and a plastic strain ε_1 (see Fig. 14.13a). The derived curve in Fig. 14.13b shows that only the second solution, eq(14.17b), is valid.

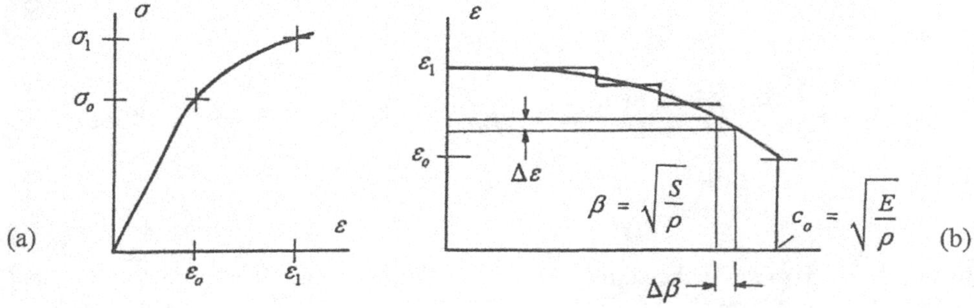

Figure 14.13 Variation in β within the plastic range

An incremental solution enables curve (b) to be approximated with a series of small steps in which β is taken to be constant for an increment of inelastic strain. This is equivalent to approximating the plastic region of the stress–strain curve with linear segments.

14.5.1 Plastic Particle Velocity

Combining condition (i) above with eq(14.11b) gives

$$-\,v_o = \int_{\infty}^{0} f(\beta)\,\mathrm{d}\beta = -\int_{0}^{\infty} \varepsilon\,\mathrm{d}\beta \tag{14.18a}$$

The integral in eq(14.18a) may be replaced by its equivalent strip area $\varepsilon \times \mathrm{d}\beta = \beta \times \mathrm{d}\varepsilon$, as shown in Fig. 14.13b.

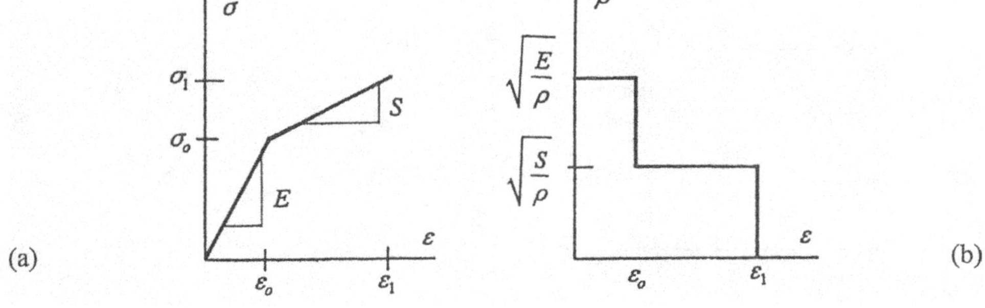

Figure 14.14 Bi-linear stress–strain curve

Transforming eq(14.18a): $$v_o = \int_0^{\varepsilon_1} \beta \, d\varepsilon \qquad (14.18b)$$

With the simplest bi-linear approximation to a stress–strain curve (see Fig. 14.14a), the modulus within the elastic and plastic regions remain constant as E and S respectively. Hence, eq(14.18b) may be integrated as follows:

$$\begin{aligned} v_o &= \int_0^{\varepsilon_o} \sqrt{E/\rho}\, d\varepsilon + \int_{\varepsilon_o}^{\varepsilon_1} \sqrt{S/\rho}\, d\varepsilon \\ &= \sqrt{E/\rho}\,\varepsilon_o + \sqrt{S/\rho}\,(\varepsilon_1 - \varepsilon_o) \end{aligned} \qquad (14.19a)$$

Substituting $\varepsilon_o = \dfrac{\sigma_o}{E}$, $\varepsilon_1 - \varepsilon_o = \dfrac{(\sigma_1 - \sigma_o)}{S}$, $\sqrt{E\rho} = \rho c_o$ and $\sqrt{S\rho} = \rho c$ into eq(14.19a):

$$v_o = \frac{\sigma_o}{\rho c_o} + \frac{(\sigma_1 - \sigma_o)}{\rho c_o}\left(\frac{c_o}{c}\right) \qquad (14.19b)$$

in which the first term is the particle velocity arising from an elastic impact at the yield stress. The second term gives the plastic particle velocity when the stress level exceeds the yield stress. Here the identity $c = \beta = \sqrt{(S/\rho)}$ defines the plastic wave propagation velocity, which appears in an inverse ratio with the elastic propagation velocity c_o:

$$\frac{c_o}{c} = \frac{\sqrt{E/\rho}}{\sqrt{S/\rho}} = \sqrt{\frac{E}{S}} \qquad (14.20)$$

Figure 14.14b suggests a simple extension for a multi-linear plastic stress–strain curve in which further plastic terms may be added to eq(14.19b) for each plastic stress increment.

14.5.2 Reflection at a Fixed Boundary

Recall that the stress within an elastic wave retains its sense and magnitude while the particle velocity changes direction upon reflection at a fixed boundary. Consequently, the stress in the material must double at the fixed end of the bar in Fig. 14.15.

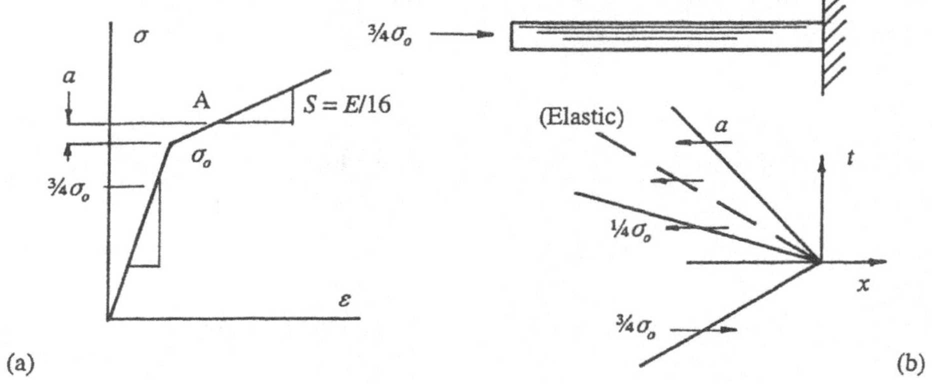

Figure 14.15 Elastic and plastic reflections at a fixed boundary

It follows that plasticity effects must be considered for fixed boundary reflections of incident elastic waves with stress magnitudes greater than $\frac{1}{2}\sigma_o$. Under these conditions both the elastic and plastic waves, as expressed by respective terms within eq(14.19b), will reflect to ensure zero displacement. To show this, let the stress within an incident elastic wave be $\frac{3}{4}\sigma_o$ in the material model of Fig. 14.15a. Firstly, assume that an elastic stress reflection $\frac{1}{4}\sigma_o$ (Fig. 14.15b) gives a net stress with a magnitude of the yield stress σ_o. Let a stress increment a, beyond the yield stress σ_o, also exist within a reflected plastic wave in Fig. 14.15b. From eq(14.20), $c_o/c = 4$ and setting $v = 0$ after reflection gives

$$\frac{3}{4}\sigma_o - \left(\frac{1}{4}\sigma_o + 4a\right) = 0$$

from which $a = \sigma_o/8$. *Note*: If the incident stress had equalled σ_o, then only a plastic wave can reflect. As an alternative to a bi-linear model in Fig. 14.15a, let the plastic gradients in Fig. 14.16a give $c_o/c = 2$, 4, 8 and 16 etc., these being separated as far as possible by stress increments $\sigma_o/16$.

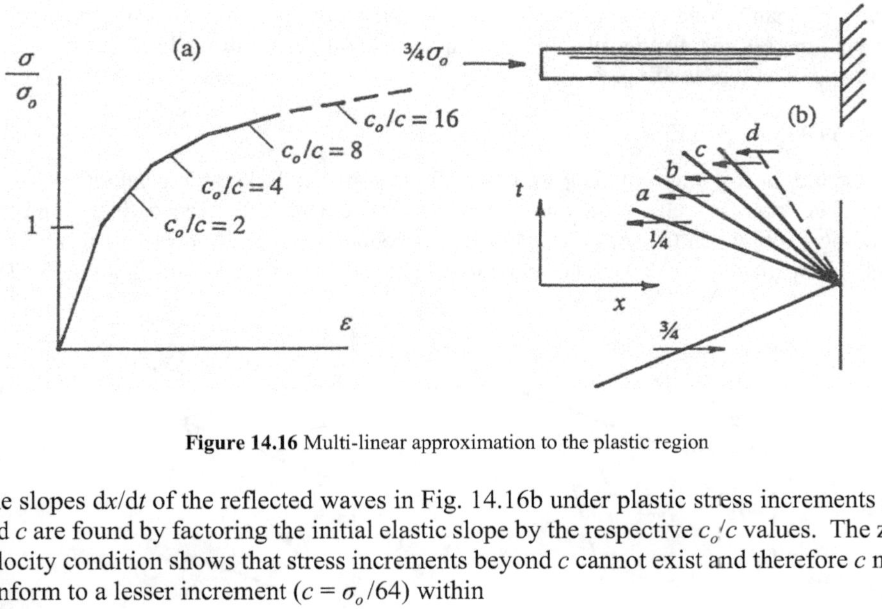

Figure 14.16 Multi-linear approximation to the plastic region

The slopes dx/dt of the reflected waves in Fig. 14.16b under plastic stress increments a, b and c are found by factoring the initial elastic slope by the respective c_o/c values. The zero velocity condition shows that stress increments beyond c cannot exist and therefore c must conform to a lesser increment ($c = \sigma_o/64$) within

$$\frac{3\sigma_o}{4} - \frac{\sigma_o}{4} - \left(2 \times \frac{\sigma_o}{16}\right) - \left(4 \times \frac{\sigma_o}{16}\right) - 8c = 0$$

14.5.3 Unloading Waves

Consider loading a bar at time $t = 0$ to a plastic stress level σ_1 at its left-hand end. At a time $t = t_1$, this stress is released. With a bi-linear approximation to the bar's stress–strain curve, loading is represented by two waves: (i) elastic under the yield stress σ_o and (ii) plastic under the stress increment $\sigma_1 - \sigma_o$ (Figs 14.17a,b). The elastic unloading wave is parallel to that for loading under σ_o but offset by time t_1. Consequently, given its faster particle velocity, the unloading wave will intersect the plastic increment loading wave at point A in Fig. 14.17b.

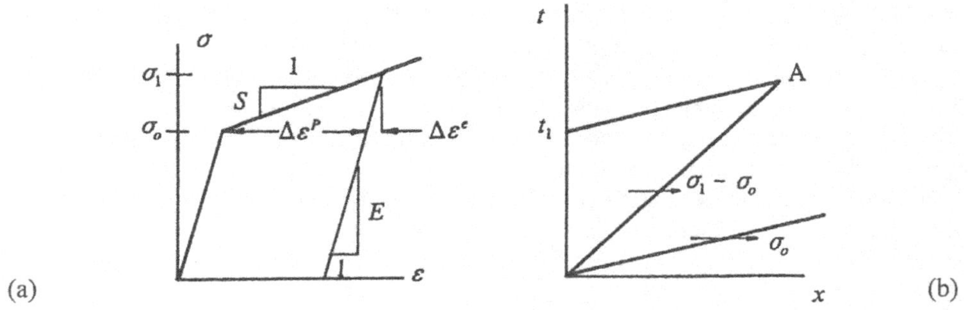

Figure 14.17 Loading and unloading waves

What happens at A will depend upon the magnitude of the stress increment $\sigma_1 - \sigma_o$. While the stress within the unloading wave will always exceed that in the plastic wave, the respective particle velocities of the two waves will not necessarily be in the proportion expressed by eq(14.19b). The plastic wave velocity will depend upon whether both stress and velocity can be overcome by the interaction. Depending upon the stress increment the following may occur: further plastic straining, a general elastic unloading into the elastic region and elastic re-loading.

(I) *Large* $(\sigma_1 - \sigma_o)$

The σ–ε and the x–t diagrams in Figs 14.18a,b are now interlinked with state points B, C, D etc. The particle velocity within the plastic increment will exceed that within the unloading wave when $(\sigma_1 - \sigma_o)c_o/c_1 > \sigma_1$. Consequently, a reduced plastic wave under stress increment $b = \sigma_2 - \sigma_o$ will continue to propagate to give state point E in Fig. 14.18b.

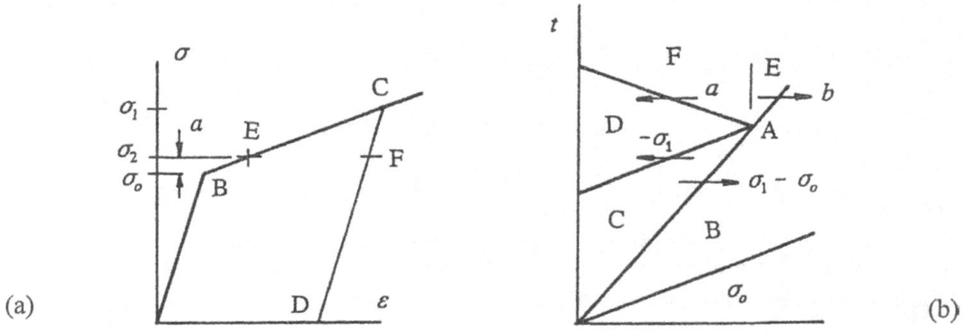

Figure 14.18 Wave interactions at point A following unloading

This figure also shows that when the left-hand end is unloaded to D, following the wave interaction at A, a further elastic re-loading wave under stress a travels back from A giving state point F. The stress levels a and b must satisfy both the equilibrium and compatibility conditions. These conditions are applied to the left and right of the vertical section through A, using the signs and directions shown

$$(\sigma_1 - \sigma_o) - \sigma_1 + a = b \tag{14.21a}$$

$$(\sigma_1 - \sigma_o)\frac{c_o}{c} - \sigma_1 - a = \left(b \times \frac{c_o}{c}\right) \tag{14.21b}$$

(II) *Small* $(\sigma_1 - \sigma_o)$

The particle velocity within a plastic wave of small stress increment may be overcome by the velocity of the elastic unloading wave. The latter will then have sufficient energy to continue unloading beyond the interaction point A (i.e. within region E) under a reduced stress level $-b = \sigma_o - \sigma_2$, as shown in Figs 14.19a,b.

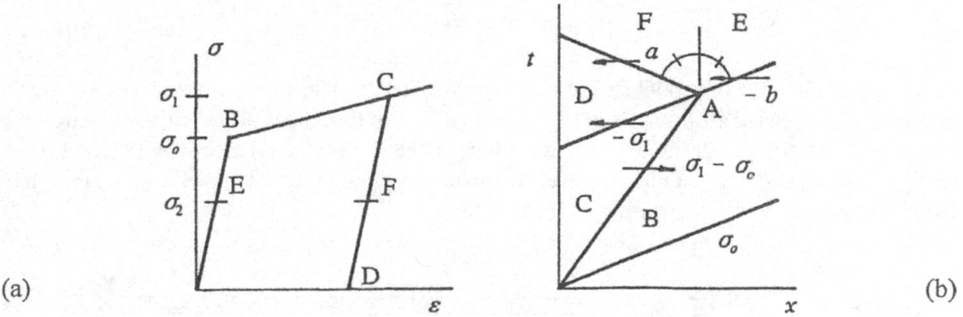

Figure 14.19 Interaction at A for a small plastic stress increment

Following the interaction, a re-loading wave under stress a must also travel back to the left hand end (state F). Both stress levels a and b must preserve equilibrium and compatibility:

$$(\sigma_1 - \sigma_o) - \sigma_1 + a = -b \tag{14.22a}$$

$$(\sigma_1 - \sigma_o)\frac{c_o}{c} - \sigma_1 - a = -b \tag{14.22b}$$

(III) *Special Case*

At a critical stress difference $\sigma_1 - \sigma_o$, lying between cases I and II above, no further plasticity can occur and nor can a continued unloading. To find the corresponding stress level σ_1, refer to Figs 14.20a,b.

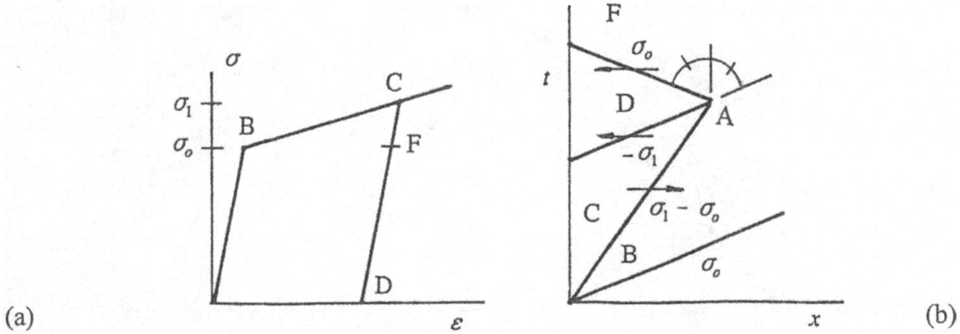

Figure 14.20 Interaction at A without plasticity or unloading

Figure 14.20b shows that when a re-loading wave returns to the free-end, its stress level must be σ_o to satisfy equilibrium:

$$(\sigma_1 - \sigma_o) - \sigma_1 + \sigma_o = 0 \tag{14.23a}$$

Using eq(14.19b) to ensure velocity compatibility:

$$(\sigma_1 - \sigma_o)\frac{c_o}{c} - \sigma_1 - \sigma_o = 0 \tag{14.23b}$$

Solving eqs(14.23a,b):
$$\sigma_1 = \left(\frac{c_o/c + 1}{c_o/c - 1}\right)\sigma_o \tag{14.23c}$$

where particle velocities for elastic unloading and re-loading are $\sigma_o/(\rho c_o)$ and $\sigma_1/(\rho c_o)$.

Example 14.5 A bar of model material is compressed at its left end to a stress level $3\sigma_o/2$ before being released at time t_1 (see Fig. 14.21a). Given that the gradient of the plastic line is $S = E/81$, determine all the wave fronts when the bar length is: (i) infinite and (ii) chosen to allow the elastic wave under σ_o to reach its fixed right end at time t_1. Show the residual stress and plastic strain distribution for each case.

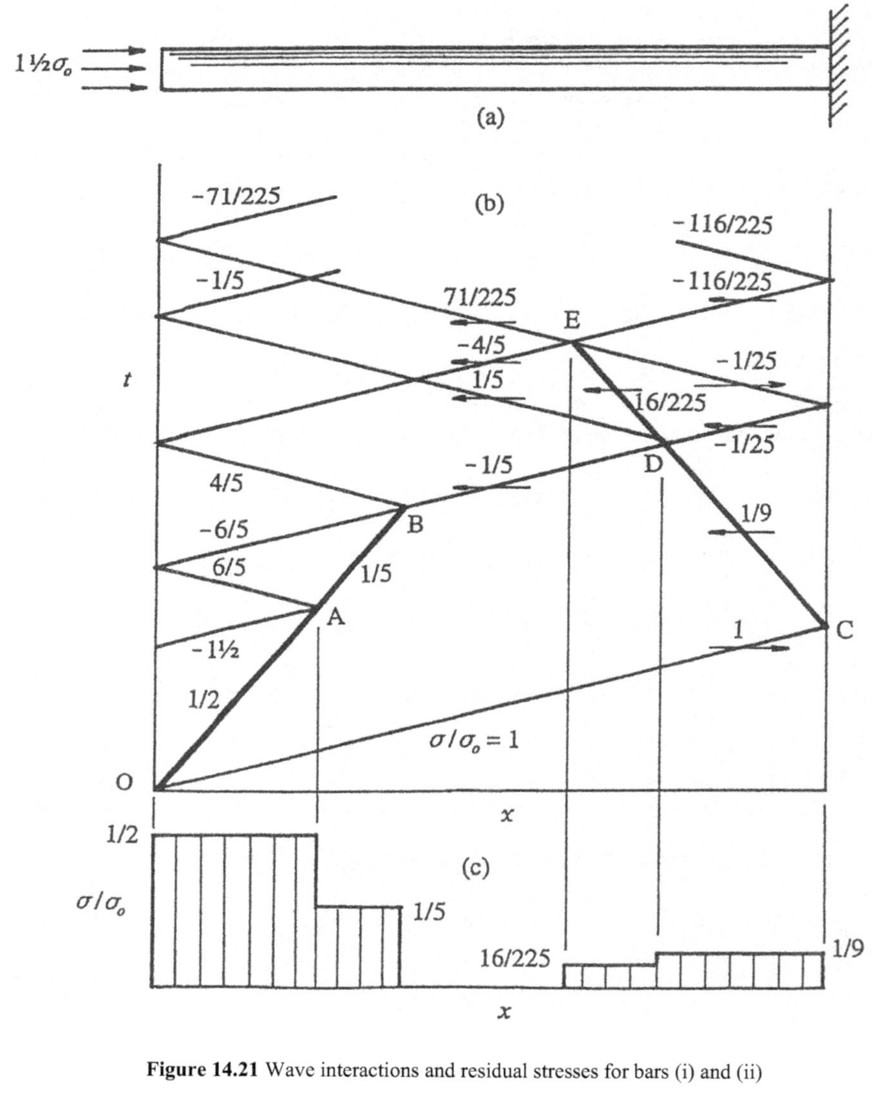

Figure 14.21 Wave interactions and residual stresses for bars (i) and (ii)

Figure 14.21b gives the a combined wave solution for bars (i) and (ii). For bar (i) interactions occur at A and B while the waves transmitted beyond these points do not return (see Figs 14.22a–c). For bar (ii) the wave returns at $t_1 = l/c_o$ from the fixed end at C resulting in additional interactions at D and E (see Figs 14.23a–c).

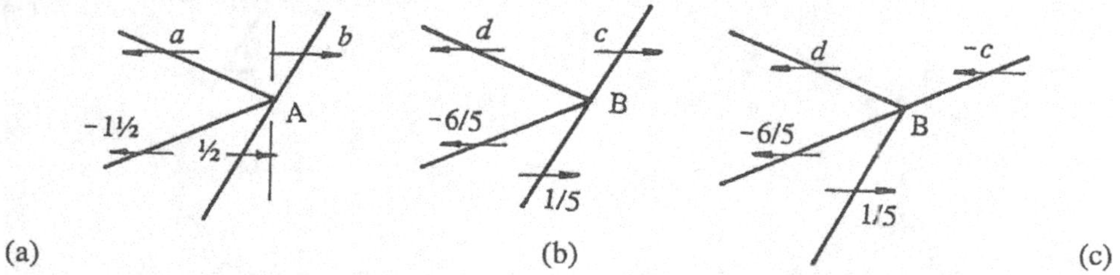

Figure 14.22 Possible wave interactions at A and B showing plastic stress levels

To determine the wave stresses levels *a* and *b* for point A it is necessary to decide which of the three cases above applies. First, assume case I where a weaker plastic wave continues under stress *b* and re-loads elastically under stress *a* (see Fig. 14.22a). The equilibrium and compatibility requirements are expressed from eq(14.20) and eqs(14.21a,b):

$$\frac{1}{2} - \frac{3}{2} + a = b$$

$$\left(\frac{1}{2} \times 9\right) - \frac{3}{2} - a = (b \times 9)$$

from which $a = 6/5$ and $b = 1/5$. Similarly, at B, by the same assumption (see Fig. 14.22b):

$$\frac{1}{5} - \frac{6}{5} + d = c$$

$$\left(\frac{1}{5} \times 9\right) - \frac{6}{5} - d = (c \times 9)$$

These equations show that the plastic stress increment c is negative, so invalidating case I. Assuming that case II applies to B (see Fig. 14.22c), the force and velocity balance eqs(14.22a,b) yield

$$\frac{1}{5} - \frac{6}{5} + d = -c$$

$$\left(\frac{1}{5} \times 9\right) - \frac{6}{5} - d = -c$$

which gives the valid result: $c = 1/5$ and $d = 4/5$. This may be applied to the x–t diagram in Fig. 14.21b to reveal that unloading and re-loading waves follow each other along an infinite length without further interactions. Figure 14.21c shows the stress increment levels $\Delta\sigma$ within the plastic wave OAB. Referring to Fig. 14.17a, $\Delta\sigma$ may be converted to plastic strain as follows:

$$\Delta\varepsilon^P = \frac{\Delta\sigma}{S} - \frac{\Delta\sigma}{E} \tag{14.24}$$

For bar (ii), the plastic wave reflection at the fixed-end point C (see Fig. 14.23a) occurs with zero net velocity. From eq(14.19b):

$$\frac{1}{\rho c_o} - \frac{a}{\rho c_o} \times \left(\frac{c_o}{c}\right) = 0$$

where $a = 1/9$ given $c_o/c = 9$.

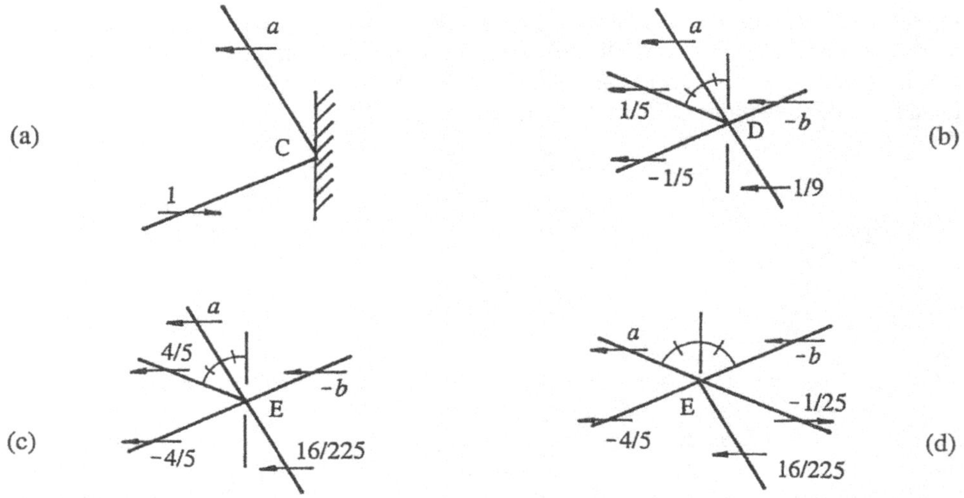

Figure 14.23 Wave interactions at C, D and E

The interaction at point D in Fig. 14.23b is tenable physically since the force and velocity balance equations

$$\frac{1}{5} - \frac{1}{5} + a = \frac{1}{9} - b$$

$$\frac{1}{5} + \frac{1}{5} + (a \times 9) = b + \left(\frac{1}{9} \times 9\right)$$

show, correctly, that $a = 16/225$ and $b = 1/25$ are positive. However, if a similar interaction is assumed for point E (see Fig. 14.23c), then

$$-\frac{4}{5} + \frac{4}{5} + a = \frac{16}{225} - b$$

$$\frac{4}{5} + \frac{4}{5} + (a \times 9) = \left(\frac{16}{225} \times 9\right) + b$$

It follows, unacceptably, that a is negative, since it is impossible to have a tensile plastic increment beyond a compressive yield stress! Hence, it is deduced that at E the plastic increment is overcome as in Fig. 14.23d. A force and velocity balance then give

$$-\frac{4}{5} + a = -b - \frac{1}{25} + \frac{16}{225}$$

$$\frac{4}{5} + a = b - \frac{1}{25} + \left(\frac{16}{225} \times 9\right)$$

from which $a = 71/225$ and $b = 116/225$ are acceptable physically. All waves that follow remain elastic. Equation (14.24) again provides the conversion from the stress levels within plastic wave CDE to residual plastic strain (see Fig. 14.21c).

Example 14.6 A free bar of length l is subjected to a sudden compressive stress $\sigma_1 = 4\sigma_o$ at its left end for a time $T = l/c_o$ and thereafter removed. Taking a bi-linear stress-strain curve, with $S = E/9$, draw the x–t diagram for a period $t = 0$ to $t = 5T/2$ s. Indicate the stress levels and particle velocities within each region of an x–t diagram.

The x–t diagram (Fig. 14.24) may be constructed from assuming that a plastic wave continues beyond the first interaction point A since $\sigma_1 - \sigma_o = 3$ is large. Also intersecting at this point is the tensile wave following its reflection from the right free-end as shown.

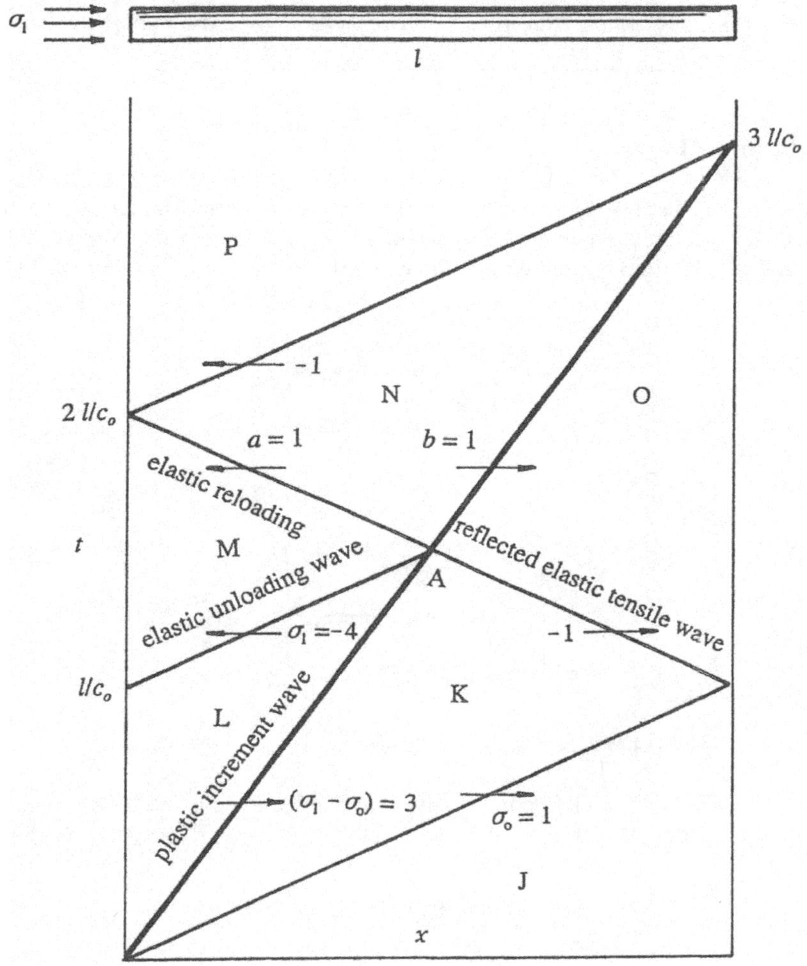

Figure 14.24 The x–t diagram for a free-end bar

Thus, with $c_o/c = \sqrt{(E/S)} = 3$, and with the stress signs and velocity directions given, the force and velocity balance equations for point A become

$$a + 3 - 4 = b - 1$$

$$(3 \times 3) - 4 - a = 3b + 1$$

from which $a = b = 1$ are acceptable. The re-loading compressive wave a is reflected in tension at the left end and meets the transmitted compressive plastic wave b at the right end at a time $3l/c_o = 3T$. Thereafter, only the elastic wave reflects. Within the regions J, K, L etc., for this diagram, the net stress and absolute velocities are given in Table 14.4.

Table 14.4 Normalised wave stresses and velocities for bar in Fig. 14.24

Region	J	K	L	M	N	O	P
σ / σ_o	0	1	4	0	1	0	0
$v \times \rho\, c_o$	0	1	10	6	5	8	4

14.6 Plastic Stress Levels

In practice it is necessary to know how to determine the stress under plastic impact conditions. For example, when a plastic impact occurs (i.e. $\sigma_1 > \sigma_o$) between a bar with known speed v and a rigid target, reflections within the bar are as shown in Figs 14.25a,b.

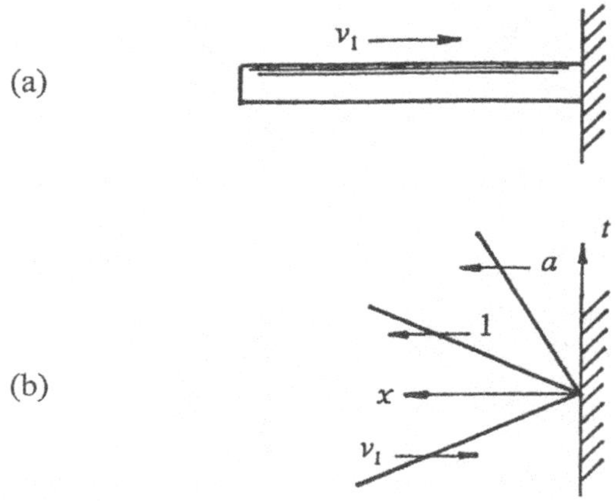

Figure 14.25 Reflections from an impact between bar and rigid target

The impact stress is split into two compressive wave reflections: an elastic wave under σ_o and a plastic increment wave under $a = \sigma_1 - \sigma_o$. Within Fig. 14.25b, the velocities balance:

$$v - \frac{\sigma_o}{\rho c_o} - \frac{(\sigma_1 - \sigma_o)}{\rho c_o}\left(\frac{c_o}{c}\right) = 0 \tag{14.25}$$

Hence, from knowing: v_1, σ_o, ρ, E and S, it follows that σ_1 and a can be calculated.

A plastic impact can occur between bars of different sizes. Let the stationary bar 1 be impacted by a larger bar 2 with an initial velocity v. The plastic stress levels upon impact become σ_1 and σ_2 respectively, as shown in Fig. 14.26a. Both elastic and plastic compressive waves reflect into each bar from the interface under the stress levels indicated in Fig. 14.26b. The force equilibrium condition and a common particle velocity at impact ensure that

$$A_1 \sigma_1 = A_2 \sigma_2 \tag{14.26a}$$

$$v - \frac{\sigma_o}{\rho c_o} - \frac{(\sigma_2 - \sigma_o)}{\rho c_o}\left(\frac{c_o}{c}\right) = \frac{\sigma_o}{\rho c_o} + \frac{(\sigma_1 - \sigma_o)}{\rho c_o}\left(\frac{c_o}{c}\right) \tag{14.26b}$$

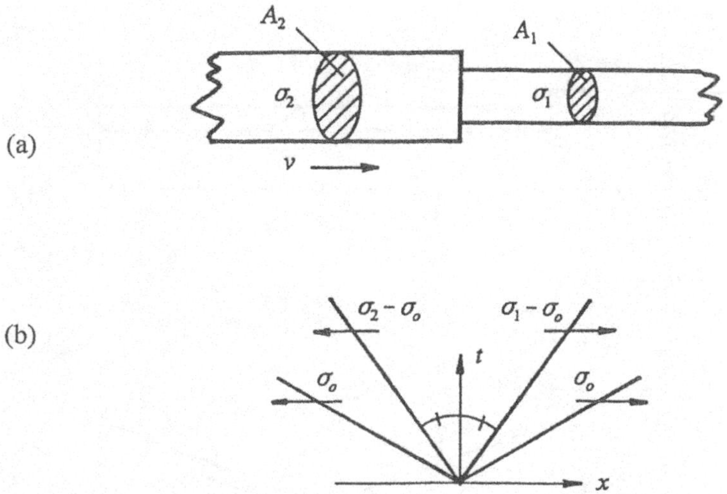

Figure 14.26 Plastic impact between bars of different areas

Equations (14.26a,b) allow for the calculation of σ_1 and σ_2, as the following example shows.

Example 14.7 A striker of length l, moving with a velocity 20 m/s, strikes a stationary rod half its diameter and twice its length (see Fig. 14.27a). Both striker and rod are of steel with a modulus $E = 207$ GPa and density $\rho = 7900$ kg/m^3. Using a a bi-linear stress–strain curve in which $\sigma_o = 465$ MPa and $S = E/16$, draw the x–t diagram from the moment of impact to a later time $t = 7l/c_o$. Plot the corresponding residual stress distribution present at this time.

Since $A_1/A_2 = (d_1/d_2)^2 = 1/4$, it follows from eq(14.26a) that $\sigma_2 = \sigma_1/4$. Assume, firstly, that both bars become plastic by substituting into eq(14.26b). This leads to

$$\frac{\sigma_1}{\sigma_o} = \frac{1}{5}\left(6 + \frac{v\sqrt{E\rho}}{\sigma_o}\right)$$

Using the material constants gives $\sigma_1/\sigma_o = 1.55$, from which $\sigma_2/\sigma_o = 0.387$. These show that bar 2 remains elastic while bar 1 is plastic. Since this contravenes the impact assumption it is necessary to re-calculate stress ratios under these conditions. Equation (14.26b) is modified to become

$$v - \frac{\sigma_2}{\rho c_o} = \frac{\sigma_o}{\rho c_o} + \frac{(\sigma_1 - \sigma_o)}{\rho c_o}\left(\frac{c_o}{c}\right)$$

Re-arranging with eq(14.26a) as before:

$$\frac{\sigma_1}{\sigma_o} = \frac{4}{17}\left(3 + \frac{v\sqrt{E\rho}}{\sigma_o}\right)$$

for which $\sigma_1/\sigma_o = 1.115$ and $\sigma_2/\sigma_o = 0.278$. Figure 14.27b shows the x–t diagram based upon an elastic, elastic–plastic impact. The full solution provides the wave stress components at each intersection between reflections from the ends and from the contacting interface. A sample of the interactions involved appears within Figs 14.28a–d. Only those solutions possible physically for each intersection are given, i.e. no alternative interactions can apply.

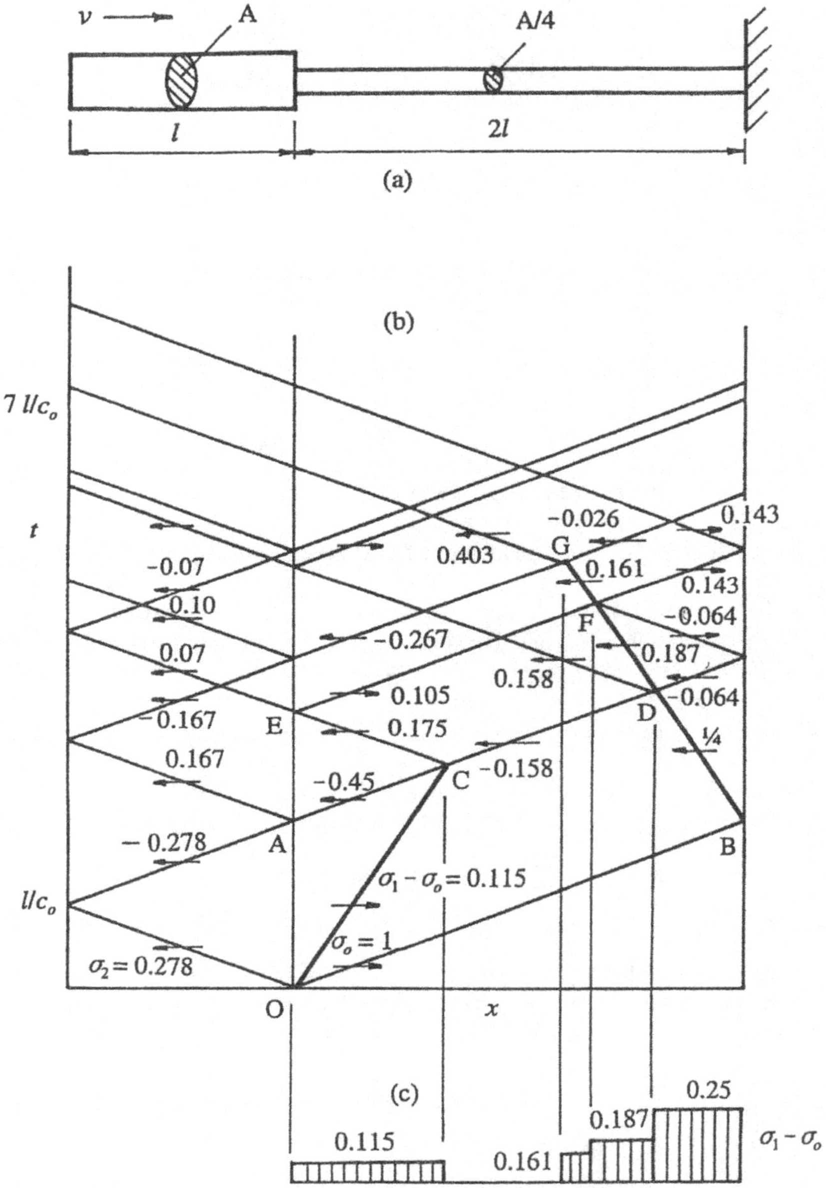

Figure 14.27 The x–t diagram for an elastic–plastic impact between bars

<u>At A</u>: (see Fig. 14.28a). The reflected tensile wave is split elastically at the interface.

Applying eqs(14.22a,b):

$$-(0.278 + b)A = -a \times \frac{A}{4}$$

$$-0.278 + b = -a$$

from which $a = 0.45$ and $b = -0.167$ (i.e. wave b becomes tensile).

At B: Applying the condition that $v_B = 0$ in Fig. 14.28b gives

$$\frac{1}{\rho c_o} - \frac{a}{\rho c_o}\left(\frac{c_o}{c}\right) = 0$$

from which $a = c/c_o = 1/4$.

At C: Assume that the elastic wave is overcome, as in Fig. 14.28c:

$$b + 0.115 - 0.450 = -a$$

$$-(0.115 \times 4) + 0.450 + b = a$$

from which $a = 0.158$, $b = 0.175$.

At D: Let Fig. 14.28d be the wave interaction. This reveals

$$-0.158 + 0.158 + a = 0.25 - b$$

$$0.158 + 0.158 + 4a = b + \left(4 \times \frac{1}{4}\right)$$

from which $a = 0.187$ and $b = 0.064$.

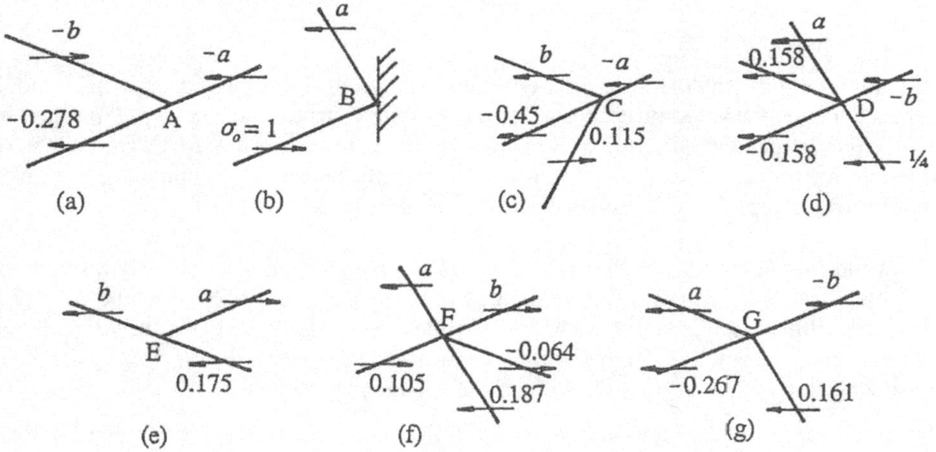

Figure 14.28 Wave stresses at interaction points A–G within Fig. 14.27

At E: Assume a reflected wave b with the same sign (see Fig. 14.28e):

$$b \times A = (a + 0.175)\frac{A}{4}$$

$$b = 0.175 - a$$

from which $a = 0.105$ and $b = 0.07$.

At F: Assume the wave interaction shown in Fig. 14.28f:

$$a + 0.105 = b + 0.187 - 0.064$$

$$0.105 - 4a = b + 0.064 - (4 \times 0.187)$$

from which $a = 0.161$ and $b = 0.143$.

At G: The interaction given in Fig. 14.28f is assumed:

$$a - 0.267 = 0.161 - b$$

$$a + 0.267 = (4 \times 0.161) + b$$

from which $a = 0.403$ and $b = 0.026$. This shows that no plastic wave is transmitted beyond G and only elastic waves will follow after a time $7l/c_o$. The projection, Fig. 14.27c, shows the residual plastic stress distribution, derived from the plastic limbs OC and BDFG. The incremental proportions of the yield stress σ_o shown are converted to strain readily using eq(14.24) with the given modulus.

14.7 Drop Weight Test

The solution to stress and velocity in a suspended small diameter rod or wire impacted at its free end is considered. Firstly, calculations are illustrated with a worked example and then the results from a similar experiment are discussed. Thereafter, the impact theory outlined within this chapter is compared with an energy exchange. It is shown that both approaches are required for a complete theory of the drop weight test. Energy conservation provides the impact velocity. Wave equilibrium and compatibility requirements reveal the impact stress.

14.7.1 Solution Strategy

The following example illustrates how the calculations for a drop weight test should proceed. Given the velocity at impact between a tubular striker and a solid rod, the stresses within them depend upon nature of deformation that occurs. From the stresses found, the *x–t* diagram is constructed for both tube and rod to reveal the respective compressive and tensile wave propagation in each length both elastic and plastic appropriately.

Example 14.8 A circular section rod of area A and length $3L$ is fixed to a rigid support at one end. The other end is free and flanged as shown in Fig. 14.29a. A tube of similar material, with section area $2A$ and length L, slides freely along the bar to strike the flange with an impact velocity 15 m/s. Draw the *x–t* diagram from $t = 0$ to $t = 6l/c_o$ given the following: $E = 207$ GPa, $\sigma_o = 275$ MPa, $S = E/16$ and $\rho = 7890$ kg/m^3.

Stress at Impact: The three assumptions that follow are made to establish an impact stress, the last of which provides a realistic solution.

(i) Zero velocity after impact in which the bar becomes plastic
This is an unlikely result but it will be considered as a possibility. The condition requires eq(14.25) be set to zero showing that the rod does become plastic (i.e. $\sigma_1 > \sigma_o$):

$$v_o - \frac{\sigma_o}{\rho c_o} - \frac{(\sigma_1 - \sigma_o)}{\rho c_o} \times \frac{c_o}{c} = 0$$

$$\frac{\sigma_1}{\sigma_o} = 1 + \left(\frac{v_o \rho c_o}{\sigma_o} - 1\right)\frac{c}{c_o} = 1 + \left(\frac{v_o \sqrt{E\rho}}{\sigma_o} - 1\right)\sqrt{\frac{S}{E}}$$

$$= 1 + \left(\frac{15 \times \sqrt{(207 \times 10^9) \times 7890}}{(275 \times 10^6)} - 1\right)\frac{1}{4} = 1.30$$

(a)

(b)

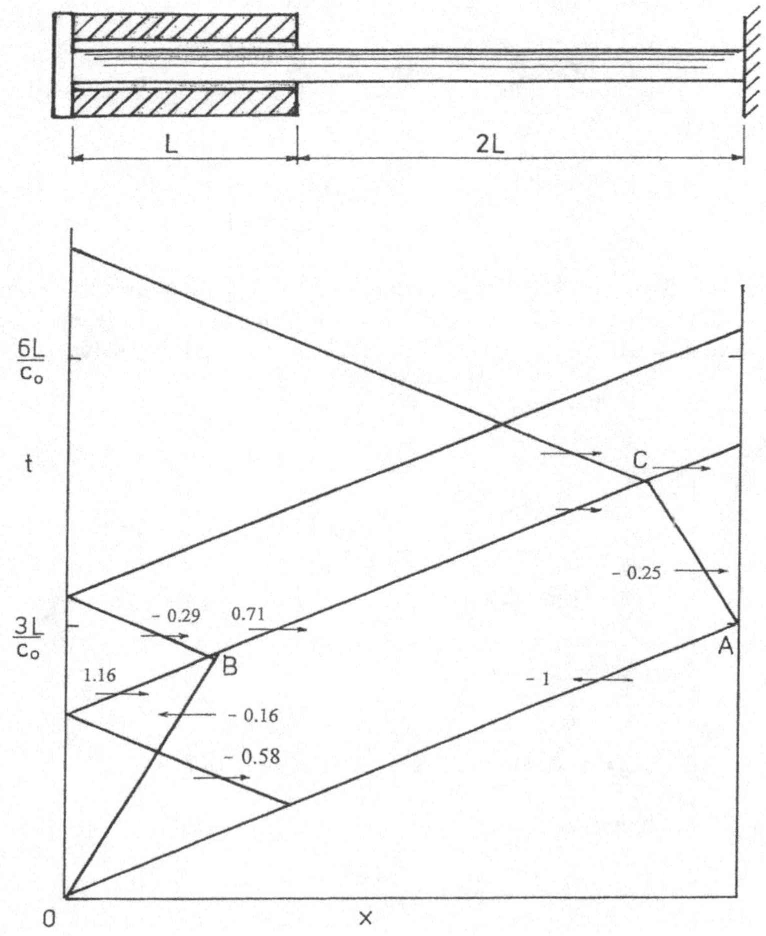

Figure 14.29 Wave propagation *x–t* diagram for an elastic–plastic impact of striker upon a flanged rod

(ii) Equal velocities after impact with rod and striker plastic

$$v_o - \frac{\sigma_o}{\rho c_o} - \frac{(\sigma_2 - \sigma_o)}{\rho c_o} \times \frac{c_o}{c} = \frac{(\sigma_1 - \sigma_o)}{\rho c_o} \times \frac{c_o}{c} + \frac{\sigma_o}{\rho c_o}$$

and with $\sigma_2 = \sigma_1/2$, $c_o/c = 4$:

$$\frac{\sigma_1}{\sigma_o} = 1 + \frac{v_o\sqrt{E\rho}}{6\sigma_o}$$

$$= 1 + \frac{15\sqrt{(207 \times 10^9) \times 7890}}{6 \times (275 \times 10^6)} = 1.367$$

giving $\sigma_1 > \sigma_o$. But with $\sigma_2 < \sigma_o$, this solution violates the assumption made in (ii).

(iii) Equal velocities after impact with rod plastic and striker elastic

$$v_o - \frac{\sigma_2}{\rho c_o} = \frac{(\sigma_1 - \sigma_o)}{\rho c_o} \times \frac{c_o}{c} + \frac{\sigma_o}{\rho c_o}$$

and with $\sigma_2 = \sigma_1/2$, $c_o/c = 4$:

$$\frac{\sigma_1}{\sigma_o} = \frac{2}{9}\left(3 + \frac{v_o\sqrt{E\rho}}{\sigma_o}\right) = \frac{2}{9}\left(3 + \frac{15\sqrt{(207 \times 10^9) \times 7890}}{(275 \times 10^6)}\right) = 1.16$$

giving $\sigma_1 > \sigma_o$ and with $\sigma_2 < \sigma_o$, This solution validates the assumption made in (iii). Compared to solution (i) this lower solution to $\sigma_1 > \sigma_o$ is the more tenable, upon which the following wave interactions are applied to points A, B and C in Figs 14.30a–c.

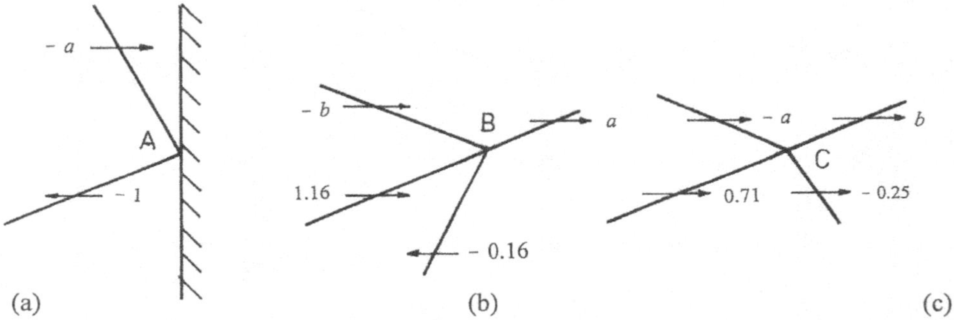

Figure 14.30 Wave interaction stress at interaction points A, B and C

<u>(a) At A</u> (Rigid end fixing):

$v_A = 0$:
$$\frac{1}{\rho c_o} - \frac{a}{\rho c_o}\left(\frac{c_o}{c}\right) = 0$$

$$1 - 4a = 0, \quad \therefore a = 1/4$$

<u>(b) At B</u> (Elastic–plastic wave interaction):

Equilibrium: $1.16 - b - 0.16 = a \quad \Rightarrow \quad a + b = 1$

Compatibility: $1.16 + b - (4 \times 0.16) = a \quad \Rightarrow \quad a - b = 0.42$

$$\therefore a = 0.71, \; b = 0.29$$

<u>(c) At C</u> (Elastic–plastic wave interaction):

Equilibrium: $0.71 - a = b - 0.25 \quad \Rightarrow \quad a + b = 0.96$

Compatibility: $0.71 + a = b + 1 \quad \Rightarrow \quad a - b = 0.29$

$$\therefore a = 0.625, \; b = 0.335$$

These wave interactions are applied to the *x–t* diagram in Fig. 14.29b in which it appears that a separation between rod and striker occurs at point D with the unloading assumed.

14.7.2 Rig Design

In this drop weight test (see Fig. 14.31), a weight is dropped vertically from a height to strike the buffer end of various non-ferrous metal alloy wires 1 m long and ≈1 mm diameter. The bottom end of each hanging wire was elastically pre-tensioned by the mass of the buffer and before its release the weight was in contact with pre-compressed springs. Alignment of the guided drop was ensured with frictionless bearings around two polished supporting columns. Resistance heating was employed to raise temperature to 300°C. A thermocouple lead was threaded from the bottom to centre of the wire within a surrounding sleeve of mineral insulation running the full length of drop. A tubular load cell was attached to the top end of the wire. A disc with a hole at its centre accepted the wire and its clamp. The disc was screwed to the bottom flange of the tensile cell bonded with outer strain gauges connected into a four-arm bridge. A compressive load cell of similar inverted design was mounted concentrically in the buffer at the wire bottom. Within the interval of the drop the wire is stretched as the end clamps transmit the impact forces to the cell at each end of the wire simultaneously. A time-base recording was made of each force. Correspondingly, a displacement transducer in contact with the buffer recorded the wire displacement through to the rig's full stretch or to a failure depending upon the wire ductility. Thus, each test was quantified within the load and temperature versus displacement/time plots so measured.

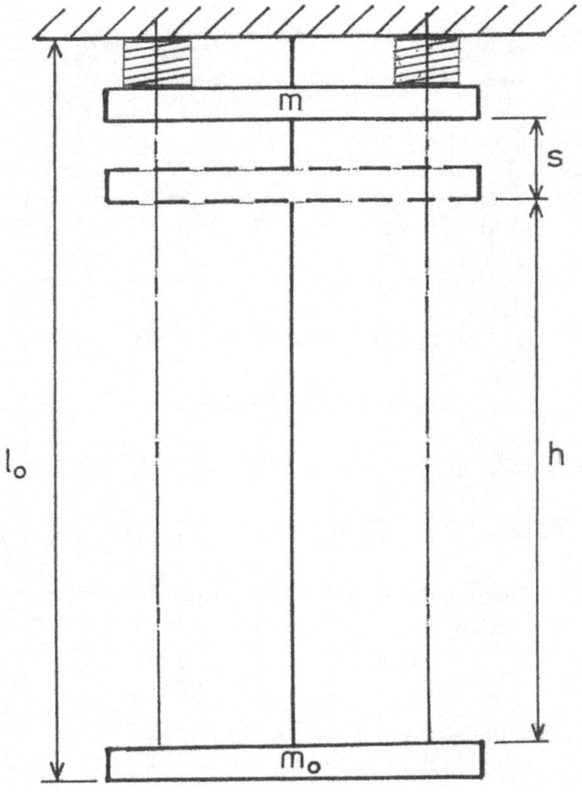

Figure 14.31 Drop weight test for 1 m length wire (see notebook)

14.7.3 Experimental Analyses

It is seen from example 14.8 that the impact velocity v_o of the 'striker' is required for the construction of the *x–t* diagram. An energy conservation exchange was applied to a drop weight with mass m starting from rest in which the gravitational acceleration is boosted by the release of compressed parallel springs, each of stiffness k displaced by an amount s (see Fig. 14.31). The mass of the buffer is m_o and, without the spring pre-compression, the free height of the fall is h.

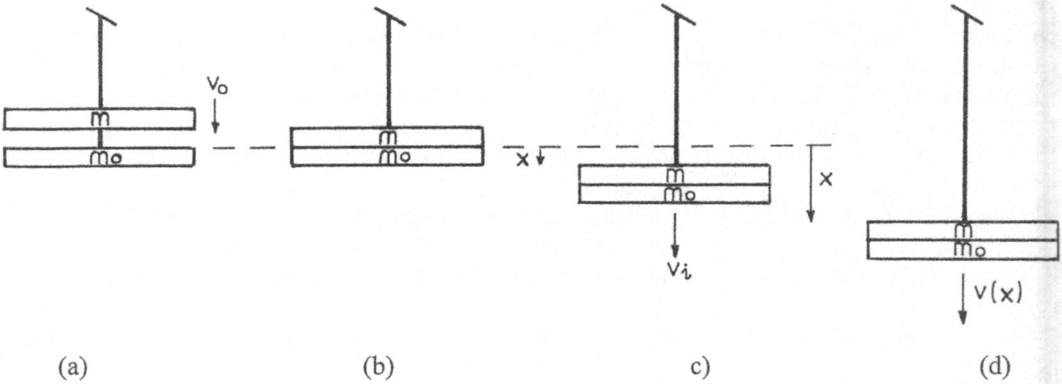

Figure 14.32 Drop weight stages pre- and post-impact

(a) Velocity and Strain Rate at Impact

For the total energy of the unreleased mass m, its potential energy is added to the stored energy in the two springs, each of equal stiffness k and pre-compression s. Upon its release, the full conversion to the kinetic energy applies to the drop h at the point of impact. This provides an initial estimate of a pre-impact velocity v_o (see Fig. 14.32a):

$$2\left(\frac{1}{2}ks^2\right) + mg(h+s) = \frac{1}{2}mv_o^2$$

$$v_o = \sqrt{2ks^2/m + 2g(h+s)} \tag{14.27}$$

For $k = 5.25$ N/mm, $s = 50$ mm, $h = 1$ m and $m = 15$ kg, eq(14.27) reveals (in N, m units)

$$v_o = \sqrt{2 \times (5.25 \times 10^3) \times 0.05^2/15 + 2 \times 9.81(1.0 + 0.05)} = 4.73 \text{ m/s}$$

which is independent of wire diameter and its properties. The conservation of momentum is applied to the impacting masses m and m_o that move together thereafter with initial velocity v_i (see Fig. 14.32b):

$$m_o u_o + m v_o = (m_o + m)v_i$$

and with $u_o = 0$ for the stationary buffer (of mass $m_o = 7$ kg):

$$v_i = \frac{v_o}{(1 + m_o/m)} \quad \text{and} \quad \dot{e} = \frac{v_i}{l_o} \tag{14.28a,b}$$

With $v_o = 4.73$ m/s:

$$v_i = \frac{4.73}{1 + 7/15} = 3.23 \text{ m/s} \quad \text{and} \quad \dot{e}_i = \frac{3.23}{1} = 3.23 \text{ s}^{-1}$$

which is seen to provide an initial nominal strain rate $\dot{e}_i$ for an original wire length $l_o = 1$ m. The true strain rate $\dot{\varepsilon}$ is based upon the deforming length as follows:

$$\varepsilon = \ln(1 + e), \quad \therefore \dot{\varepsilon} = \frac{\dot{e}}{1 + e} = \frac{v}{l_o + x} \tag{14.28c}$$

(b) Instantaneous Deformation

The energy conservation principle is applied to the instant before and immediately after the impact as shown in Figs 14.32a,b respectively. That is, the kinetic and potential energies of the falling mass m with velocity v_o is equal to the kinetic energy of the combined mass $m_o + m$ with initial velocity v_i plus the work done U within an instantaneous deformation of the wire, for which x is the corresponding displacement at time $t = 0$:

$$\frac{mv_o^2}{2} + (m + m_o)gx = \frac{(m + m_o)v_i^2}{2} + U + N \tag{14.29a}$$

where N is the energy lost to collision noise. Substituting for v_o from eq(14.28a) and rearranging eq(14.29a) provides U for an elastic–plastic deformation under velocity v_o:

$$U = \int F\mathrm{d}x = \frac{m\,v_o^2}{2}\left(\frac{m_o}{m + m_o}\right) + (m + m_o)gx - N \tag{14.29b}$$

If N is ignored, then the impact force $F = \mathrm{d}U/\mathrm{d}x = (m + m_o)g$ is the total weight. However, the displacement x is undefined and so the energy balance between Figs 14.32b,c should refer to velocity v and displacement x at time t for which

$$\frac{(m + m_o)v_1^2}{2} + (m + m_o)gx = \frac{(m + m_o)v^2}{2} + \int_0^x F\mathrm{d}x \tag{14.30a}$$

Differentiating eq(14.30a) w. r. t. x for the impact force:

$$F = (m + m_o)g - (m + m_o)v\frac{\mathrm{d}v}{\mathrm{d}x} \tag{14.30b}$$

The condition for a maximum F is, from eq(14.30b):

$$\frac{\mathrm{d}F}{\mathrm{d}x} = -(m + m_o)\left[v\frac{\mathrm{d}^2v}{\mathrm{d}x^2} + \left(\frac{\mathrm{d}v}{\mathrm{d}x}\right)^2\right] = 0 \tag{14.30c}$$

The simplest polynomial $v = v(x)$ to satisfy eq(14.30c) is quadratic with constants a and b:

$$v(x) = \frac{a^2}{4b} - ax + bx^2 \tag{14.31a}$$

Equation (14.31a) has a single root $x = a/2b$ for $v = 0$, corresponding to its minimum, i.e. this quadratic function shows that when the impact force is at its maximum the velocity is zero.

Therein, eq(14.30a) shows for velocity $v = 0$ that the maximum force becomes the total dynamic weight applied:

$$F_{max} = mg + m_o g \tag{14.31b}$$

Figures 14.33a,b show the relationship between F, v_1 and x.

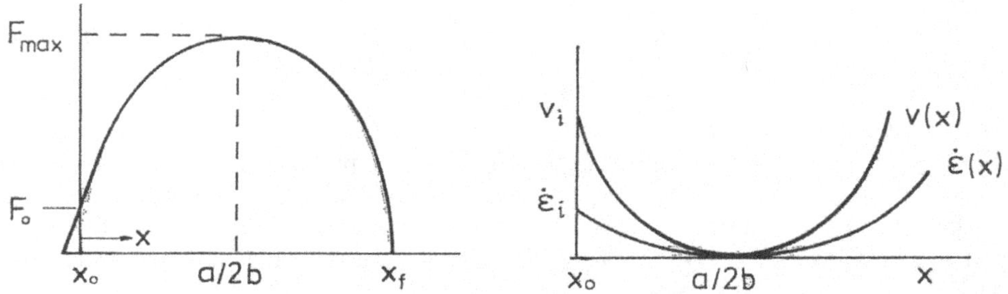

(a) (b)

Figure 14.33 Impact force, velocity and strain rate variations with displacement

Constants a and b may be found from an appropriate choice of origin. Here the initial elastic displacement under the buffer pre-load $F_o = m_o g$ is $x_o = m_o gl/AE$. This sets the origin for x due to the impact shown in Fig. 14.33a, when, from eq(14.30b)

$$g - \left(\frac{a^2}{4b} - ax_o + bx_o^2\right)(-a + 2bx_o) = \frac{m_o g}{m + m_o} \tag{14.32a}$$

In the fit to one further point take the co-ordinates at the point of failure as $x = x_f$, $F = 0$. Equations (14.30b) and (14.31a) provide for this condition when

$$\left(\frac{a^2}{4b} - ax_f + bx_f^2\right)(-a + 2bx_f) = g \tag{14.32b}$$

Equations (14.32a,b) may be solved for the ratio a/b, one half of which equals the deflection under F_{max}. An increased number of coefficients within a higher order polynomial solution to eq(14.30c) should provide a closer fit to the F–x response observed in a drop weight test. The nominal and true strain rate dependency upon x follows from eqs(14.28b,c) and (14.31a):

$$\dot{e} = \frac{v}{l_o} = \frac{a^2/4b - ax + bx^2}{l_o} \tag{14.33a}$$

$$\dot{\varepsilon} = \frac{v}{l_o + x} = \frac{a^2/4b - ax + bx^2}{l_o + x} \tag{14.33b}$$

Figure 14.33b shows that $\dot{e}(x)$ is directly proportional to $v(x)$ but the true rate $\dot{\varepsilon}$ rate falls from an initial value $a^2/4bl_o$ at $x = 0$ to zero at $x = a/2b$ and then rises to $\approx b/2l_o$ for $x = l_o$. Once the impact stress has been estimated realistically, to follow in paragraphs (c) and (d) below, a comparison between these and other strain rate predictions with experimental behaviour can be made, as in (a), (f) and (g) to follow.

(c) Impact Stresses from Energy Conservation

Without spring compression an impact stress may be attempted using an equivalent static load P_E, which gives a deflection Δ equal to that from the free falling weight. This enables the potential energy lost in the fall to be equated to the work done by P_E in deforming the wire:

$$mg(h + \Delta) = \int_0^{\Delta} P_E \times \delta\Delta \tag{14.34a}$$

An elastic solution to eq(14.34a) appears as a quadratic in P_E with Δ as an elastic deflection and work done equated to strain energy stored:

$$mg\left(h + \frac{P_E L}{AE}\right) = \frac{P_E^2 L}{2AE}$$

$$P_E^2 - (2mg)P_E - (2mg)\frac{AEh}{L} = 0 \tag{14.34b}$$

For a Ni-Cr wire: $A = 0.322$ mm^2 (diameter = 0.64 mm), $m = 15$ kg, $h = L = 1$ m and $E = 200$ GPa. Equation (14.34b) becomes

$$P_E^2 - (2 \times 15 \times 9.81)P_E - (2 \times 15 \times 9.81)(0.322 \times 10^{-6})(200 \times 10^9) = 0$$

$$P_E^2 - 294.3P_E - (18.953 \times 10^6) = 0$$

giving an unrealistic solution: $P_E = 4.5$ kN. An elastic–plastic impact solution, which is more appropriate, adopts a power law relationship between the static load and the deflection $\Delta = B'(P_E)^n$ for the wire. This leads to

$$mg(h + B'P_E^n) = \frac{nB'}{n+1} P_E^{n+1}$$

$$P_E^{n+1} - \left(1 + \frac{1}{n}\right)(mg)P_E^n - \left(1 + \frac{1}{n}\right)(mg)\frac{h}{B'} = 0 \tag{14.34c}$$

in which $n < 1$ is the hardening exponent in Hollomon's stress–strain law $\varepsilon = B\sigma^n$. The constant B' in eq(14.34c) connects with the strength coefficient B as $B' = BL/A^n$ in the region where engineering definitions of stress and strain hold. Then L and A may be taken respectively as the original length and cross-sectional area of the wire. For example, a law of the form $\varepsilon = (8 \times 10^{-6})\,\sigma^{0.25}$, with σ in Pa, leads to

$$P_E^{1.25} - 5(15 \times 9.81)P_E^{0.25} - 5(15 \times 9.81)(0.322 \times 10^{-6})^{0.25}/(8 \times 10^{-6}) = 0$$

$$P_E^{1.25} - 735.75P_E^{0.25} - (2.191 \times 10^6) = 0$$

Again, a trial solution does not provide a tenable load level for the material. Exercise 14.13 provides examples of where an energy conservation approach is appropriate for a larger diameter bar under an elastic and an elastic–plastic impact. For a smaller wire diameter the wave analysis that follows is the solution recommended to estimate the impact stress.

(d) Impact Stresses from Wave Theory

Guided by example 14.8 it is assumed that equal velocities apply between an elastic–plastic Ni-Cr wire and a hardened steel elastic striker. The dimensions of the latter apply to its annular area within 30 mm outer and 9.5 mm inner diameters (i.e. A_2 = 636 mm^2) for a parallel length of 60 mm. The respective densities for the wire and striker are ρ_1 = 8900 kg/m^3 and ρ_2 = 7860 kg/m^3. The elastic modulus and yield strength for Cu-Ni are E_1 = 200 GPa and σ_o = 60 MPa. Equilibrium and compatibility are satisfied when

$$v_o - \frac{\sigma_2}{\rho_2 c_o} = \frac{(\sigma_1 - \sigma_o)}{\rho_1 c_o} \times \frac{c_o}{c} + \frac{\sigma_o}{\rho_1 c_o} \tag{14.35}$$

$$\frac{\sigma_1}{\sigma_o}\left[\frac{1}{r} + \left(\frac{\rho_2}{\rho_1}\right)\left(\frac{c_o}{c}\right)\right] = \frac{v_o(\rho_2 c_o)}{\sigma_o} + \frac{\rho_2}{\rho_1}\left(\frac{c_o}{c} - 1\right)$$

With $r = A_2/A_1 = 1977$, $\rho_1 c_o = \sqrt{(E_1\rho_1)}$ and $c_o/c = 4$, the solution to eq(14.35) is (sensibly)

$$\frac{\sigma_1}{\sigma_o} = \frac{\rho_1 c_o v_o}{\sigma_o}\left(\frac{c}{c_o}\right) + \left(1 - \frac{c}{c_o}\right)$$

$$= \frac{\sqrt{(200 \times 10^9) \times 8900} \times 4.73}{60 \times 10^6 \times 4} + \left(1 - \frac{1}{4}\right) = 1.58$$

giving $\sigma_1 > \sigma_o$. With $\sigma_2 = \sigma_1 \times A_1/A_2 << \sigma_o$, it is seen that the yield stress in the wire is exceeded while the stress in the striker is a negligible elastic value.

(e) Dynamic Stress versus Strain Curve

The load versus deflection response of a wire receiving a dropped weight is taken from the fixed-end load cell and the free-end displacement. Figure 14.34 is typical of the response of a 0.64 mm Ni-Cr wire at different temperatures. Each plot shows a rapid rise in load to a maximum and thereafter an extended displacement occurs under a falling load. Peak loads do not identify local necking since percentage elongations lay consistently in the range 15–20 % accompanied by 5–10 % reductions to the area that were found to be uniform over the length. It appears that the curve shape, which is characteristic of this material at the rate and temperature of testing, defies direct description by any empirical static stress–strain law.

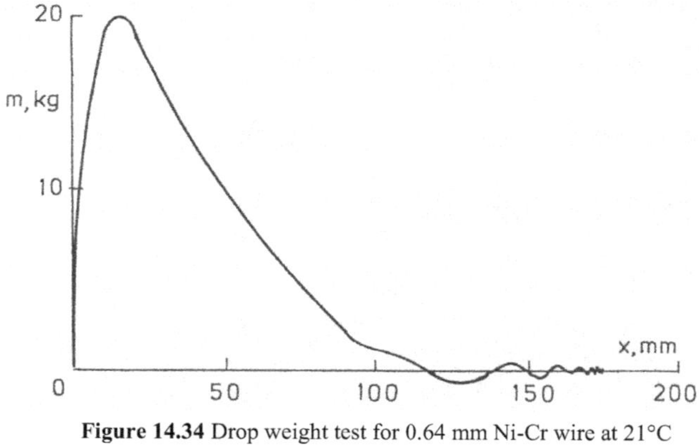

Figure 14.34 Drop weight test for 0.64 mm Ni-Cr wire at 21°C

Chapter 11 refers to one viscoplastic form appropriate to rate-dependent plasticity under a changing strain rate. Here an explicit form of eq(11.37a) is required to describe the flow curve derived within the scheme illustrated in Figs 11.9a,b. The forms adopted by Malvern [14] are convenient extensions to descriptions of a quasi-static flow curve:

$$\sigma = B(\varepsilon^P)^n(\dot{\varepsilon}^P)^m \quad \text{and} \quad \sigma = \sigma_o(\varepsilon^P)\left[1 + (Q\dot{\varepsilon}^P)^M\right] \qquad (14.36a,b)$$

in which the rate term is consistent with rising, non-linear flow curves at constant increasing strain rates in the manner of Fig 11.19a. In eq(14.36b), $\sigma_o(\varepsilon^P)$ is the static description of a flow curve that originates at the yield stress given by Ludwik or Swift. Hollomon's approximation describes a curve that originates from the origin and is useful given that the strain exponent n identifies with the instability strain in a nominal stress versus strain plot. When $\dot{\varepsilon}^P$ is constant, eqs(14.36a,b) describe a family of rate dependent flow curves (i = 1, 2, 3 . . . N):

$$\sigma_i = B_i(\varepsilon^P)^n(\dot{\varepsilon}_i^P)^m, \qquad \sigma_i = B_i(\varepsilon^P)^n\left[1 + (Q\dot{\varepsilon}_i^P)^M\right] \qquad (14.36c,d)$$

which apply to increasing plastic strain rates $(\dot{\varepsilon}_1^P < \dot{\varepsilon}_2^P < \dot{\varepsilon}_3^P \;...\; \dot{\varepsilon}_N^P)$, each a constant within a single non-linear flow curve in Fig. 11.19a. For example, Hollomon describes flow curves 1, 2, 3 . . N rising from the origin, simply, from a change to the strength coefficient B_i in eq(14.36c) as follows:

$$\sigma_1 = B_1(\varepsilon^P)^n, \quad \sigma_2 = B_2(\varepsilon^P)^n, \quad \sigma_3 = B_3(\varepsilon^P)^n \quad \text{etc.}$$

where $B_1 < B_2 < B_3 \ldots B_N$. Where the relationship between strain and strain rate is known for a drop weight test, the flow curve may be interpolated in the manner shown. The theoretical relationship follows from eqs(14.28b,c) as

$$\dot{\varepsilon}^P = \dot{\varepsilon}\exp(-\varepsilon^P) = \frac{v(x)}{l_o}\exp\left[\frac{-v(x)}{(l_o + x)}\right] \qquad (14.37)$$

It would appear from the following experimental analysis that when $v(x)$ in eq(14.37) is given by the quadratic eq(14.31a) it does not represent the gradual fall in strain rate observed in the drop weight test. *Note*: An exponential solution $y = A\exp(-mx)$ will not satisfy eq(14.30c). Other quadratic solutions to eq(14.30c) apply when the two roots are real and imaginary (see exercise 14.15). Of these it would appear that a quadratic $v(x) = a_o - ax + bx^2$ can be made to fit the velocity data given in Table 14.5 more closely when the roots are imaginary. Alternatively, the greater flexibility provided by an a higher order polynomial should provide a closer fit to the velocity variation calculated below.

(f) Velocity and Strain Rate

The question remains as to how to express the changing strain rate experimentally within the drop test. To answer this it is assumed that an energy conversion applies between the kinetic and potential energies from the total moving mass $m_t = m_o + m$ and the work done upon the deforming wire. The origin for plastic strain applies to the initial velocity v_1 at time $t = 0$. Let a change to the velocity δv m/s accompany a displacement δx and an incremental plastic strain $\delta\varepsilon^P$ within the wire at time $t > 0$. The energy exchange is expressed as

$$m_t g\delta x + m_t v\delta v = \sigma\delta\varepsilon^P\delta V \qquad (14.38a)$$

Integrating eq(14.38a) for plastic incompressibility, i.e. for a constant volume V_o of wire:

$$m_t g \int_0^x \mathrm{d}x + m_t \int_{v_1}^{v} v\,\mathrm{d}v = V_o \int_0^{\varepsilon^P} \sigma\,\mathrm{d}\varepsilon^P$$

$$m_t g x + m_t(v^2 - v_1^2)/2 = u V_o \tag{14.38b}$$

where $u = U/V_o$, the work density, is the area beneath the true stress versus natural strain diagram. Equation (14.38b) provides the velocity:

$$v = \sqrt{v_1^2 - 2(gx - U/m_t)} \tag{14.39a}$$

Conveniently, U may be found from the area beneath a nominal stress–strain plot which connects to the area beneath a load–extension plot (W versus x) as follows:

$$U = \int_0^x W\mathrm{d}x = \int_0^{e^P} (\sigma A_o)\mathrm{d}(l_o e^P) = V_o \int_0^{e^P} \sigma\,\mathrm{d}e^P \tag{14.39b}$$

where $V_o = A_o l_o$. Finally, eqs(14.28b,c) and are employed for the nominal and true strain rates. Starting with $v_1 = 3.23$ m/s, $l_o = 1$ m and $n = 1$, divide Fig. 14.34 into fifteen strips (say) of equal width giving a constant displacement increment Δx. Allowing for elastic pre-strain in the wire under the buffer mass m_o, the initial deflection x_o and elastic work U_o are

$$x_o = \frac{(m_o g) l_o}{A_o E} = \frac{7 \times 9.81 \times 1}{[(\pi/4)(0.64)^2 \times 10^{-6}] \times (200 \times 10^9)} = 1.067 \times 10^{-3}\ \text{m}$$

$$U_o = \frac{(m_o g) x_o}{2} = \frac{(7 \times 9.81) \times (1.067 \times 10^{-3})}{2} = 36.65 \times 10^{-3}\ \text{Nm}$$

or

$$U_o = \frac{(m_o g)^2 l_o}{2 A_o E} = \frac{(7 \times 9.81)^2 \times 1}{2 \times [(\pi/4)(0.64)^2 \times 10^{-6}] \times (200 \times 10^9)} = 36.65 \times 10^{-3}\ \text{Nm}\ \checkmark$$

The experimental determination of the true strain rate follows in sequence:

(i) Calculate the displacement from $x = x_o + n\Delta x$ (m) within Fig. 14.34

(ii) Calculate the engineering strain from $e = x/l_o$

(iii) Determine the strip area: $\Delta U_n = \int_0^{\Delta x} W\mathrm{d}x$ (Nm), within Fig. 14.34

(iv) Determine the work done:

$$U_n = U_o + \sum_{i=1}^{n} \Delta U_i + n(m_o g)\Delta x = \sum_{i=1}^{n} \Delta U_i + m_o g(x_o/2 + n\Delta x)$$

(v) Calculated the velocity: $v = \sqrt{v_1^2 - 2(gx - U_n/m_t)}$, (m/s)

(vi) Find the true strain rate: $\dot{\varepsilon} = \dfrac{v}{l_o(1+e)} = \dfrac{v}{1+x}$ (s^{-1}, for $l_o = 1$ m and x in m)

(vii) Set $n = n + 1$ and return to step (i).

Table 14.5 Velocity and strain rate variations at room temperature

n	1	2	3	4	5	6	7	8
$x/10^{-3}$ m	8.12	16.0	23.82	31.67	39.52	47.37	55.27	63.07
U_n, Nm	0.84	2.19	3.18	4.34	5.34	6.19	6.91	7.51
$gx - U_n/m_t$	0.042	0.057	0.089	0.113	0.145	0.183	0.228	0.277
v, m/s	3.217	3.212	3.202	3.195	3.185	3.173	3.159	3.143
$e/10^{-3}$	8.12	16.0	23.82	31.67	39.52	47.37	55.27	63.07
$\dot{\varepsilon}$, s^{-1}	3.217	3.197	3.182	3.165	3.150	3.138	3.127	3.118

n	9	10	11	12	13	14	15
$x/10^{-3}$ m	70.93	78.78	86.63	94.48	102.38	110.28	118.08
U_n, Nm	7.97	8.33	8.59	8.75	8.84	8.90	8.92
$gx - U_n/m_t$	0.334	0.394	0.459	0.529	0.603	0.678	0.753
v, m/s	3.125	3.106	3.085	3.062	3.038	3.013	2.988
$e/10^{-3}$	70.93	78.78	86.63	94.48	102.38	110.28	118.08
$\dot{\varepsilon}$, s^{-1}	3.089	3.105	3.10	3.097	3.095	3.092	3.090

(g) Plastic Work

In a refinement to step (iv) in the above sequence the plastic work is extracted from U by removing the elastic strain energy under the maximum load at each step. With A and l as the current length and area of wire and A_o and l_o its initial values, an additional calculation is required for U^P following step (iv):

$$U_n = \sum_{i=1}^{n} \Delta U_i + m_o g(x_o/2 + n\Delta x)$$

$$U_n^P = U_n - \frac{[(m+m_o)g]^2 l}{2AE} = U_n - \frac{[(m+m_o)g]^2 l_o(1+e)^2}{2A_o E}$$

$$= U_n - \frac{[(15+7)9.81]^2 \times 1 \times (1+e)^2}{2[(\pi/4)(0.64)^2 \times 10^{-6}] \times (200 \times 10^9)} = U_n - (362 \times 10^{-3})(1+e)^2$$

Substituting the pre-calculated U_o and x_o:

$$U_n^P = \sum_{i=1}^{n} \Delta U_i + m_o g(x_o/2 + n\Delta x) - (362\times 10^{-3})(1 + e_n)^2$$

$$U_n^P = U_o + \sum_{i=1}^{n} \Delta U_i + n(m_o g)\Delta x - (362\times 10^{-3})(1 + e_n)^2$$

$$U_n^P = (36.65 \times 10^{-3}) + \sum_{i=1}^{n} \Delta U_i + n(m_o g)\Delta x - (362\times 10^{-3})(1 + e_n)^2$$

Correspondingly, in calculating v in step (v), U^P replaces U. Also, x^P should replace x. Then, with v re-calculated, step (vi) provides the true plastic strain rate $\dot{\varepsilon}^P$. However, the difference $gx^P - U^P/m_t$ is not altered significantly from the $gx - U/m_t$ values given in table 14.5 and therefore the velocities given are acceptable.

(h) The x–t Diagram
Here x refers to the position in the wire length with origin at the left impacted end in Figure 14.35 This shows the elastic–plastic wave reflections within the wire in isolation where there is no interaction with a negligible elastic stress wave for the striker. The tensile wave is split into its elastic component under a stress level σ_o at speed c_o and the plastic component under a stress level $\sigma_1 - \sigma_o = 0.58$ at speed c, where $c/c_o = 4$.

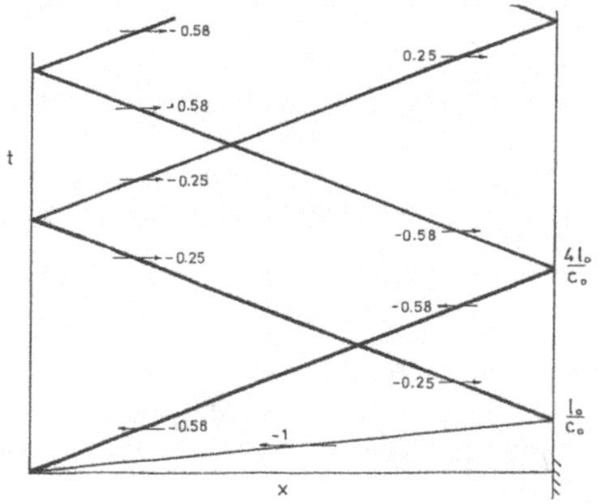

Figure 14.35 The x–t diagram for the drop weight test

14.8 Concluding Remarks

It has been seen that the approximate solution to the wave equation lends itself to a useful graphical interpretation of stress waves within impacting model materials. Stress, strain and velocity, within travelling and intersecting waves, are provided by this method. Stress levels in excess of the elastic limit are contained most conveniently within a single plastic modulus. More accurately, over a wide plastic stress range, the approach permits a more accurate muli-linear representation of the flow curve. The worked examples and exercises show the nature of the solution furnished by this approach when dealing with the longitudinal impact of a bar with both a rigid solid and another bar of different sectional area and material.

References

1. Kolsky H. *Stress Waves in Solids*, 1953, Clarendon, Oxford.
2. Goldsmith W. *Impact*, 1960, Edward Arnold.
3. Batchelor J. and Davies R. *Surveys in Mechanics*, 1956, Cambridge, pp. 64-138.
4. Abramson H. N., Plass H. J. and Ripperger E. A. *Advances in Appl. Mech.*, 1958, **5**, Academic Press, New York.
5. Cottrell A. H. *Chartered Mech. Eng.*, Nov. 1957, I. Mech. E., London.
6. Rinehart J. S. and Pearson J. *Behaviour of Metals under Impulsive Loads*, 1965, Dover Pubs, New York.
7. Lee E. H. and Symonds P. S. *Plasticity, Proc. 2nd Symposium Naval Structural Mechs.*, 1960, Brown University, Pergamon.
8. Redwood M. *Mechanical Waveguides*, 1960, Pergamon Press, London.
9. Kolsky H. and Douch J. *Jl. Mech. Phys. Sols.*, 1962, **10**, 195.
10. Johnson W. *Impact Strength of Materials*, 1972, Arnold.
11. Taylor G. I. *Proc. Roy. Soc.*, 1948, **A194**, 289.
12. Bell J. F. *The Physics of Large Deformation of Crystalline Solids*, 1968, Springer-Verlag, New York.
13. Clark and Duwez, *Proc. ASTM*, 1947, **47**.
14. Malvern L. E. *Jl. Appl. Mech.*, 1951, **18**, 203.

Exercises

14.1 A steel bar of length 510 mm and sectional area 645 mm^2 travels with a velocity 3.65 m/s to collide axially with a similar stationary bar of length 127.5 mm. Determine the time at which the impacting surfaces separate and find the final translation velocity of each bar. Determine the events subsequent to impact if the bar material is known to fracture under a tensile impact speed of 1.83 m/s.

14.2 An explosive charge detonated on one end of a free uniform section concrete beam causes five scabs to be thrown off at the other end. The scab lengths are (from 1 to 5): 237.5 mm, 162.5 mm, 187.5 mm, 325 mm and 612.5 mm respectively. Assuming the pulse to be steep fronted, construct a graph showing the variation of stress in the applied pulse against distance travelled in the concrete. What is the possible error in the stress ordinate for this plot? Take the tensile strength as 2 MPa.

14.3 The compressive stress pulse in Fig. 14.36a is applied at time $t = 0$ to the left-hand end of the concrete and steel composite bar, shown in Fig. 14.36b.

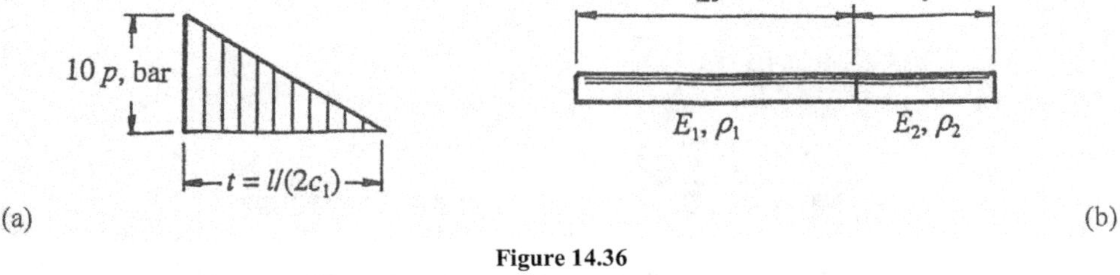

Figure 14.36

Find the sequence of events during time $0 < t < (2l/c_1 + 2l/c_2)$ where subscripts 1 and 2 refer to steel and concrete respectively and c_1 and c_2 are the propagation velocities within these materials. Illustrate the answer with sketches of the transmitted pulses at specific times within this interval. Take the maximum allowable tensile stress in the concrete as $-p$ MPa.

14.4 A tension impact specimen of length l is held by two loading bars, as shown in Fig. 14.37. The end X is pulled suddenly and maintained at a constant velocity v_o. The resulting, square-fronted, wave of tensile stress amplitude σ_o crosses section PP at $t = 0$. Derive stress–time curves for the sections PP, QQ and RR from $t = 0$ to $t = 4l/c_o$. Assume the same bar material for which the yield stress is not exceeded. Take the ends to be sufficiently remote for no reflections to occur during the time interval.
See solution

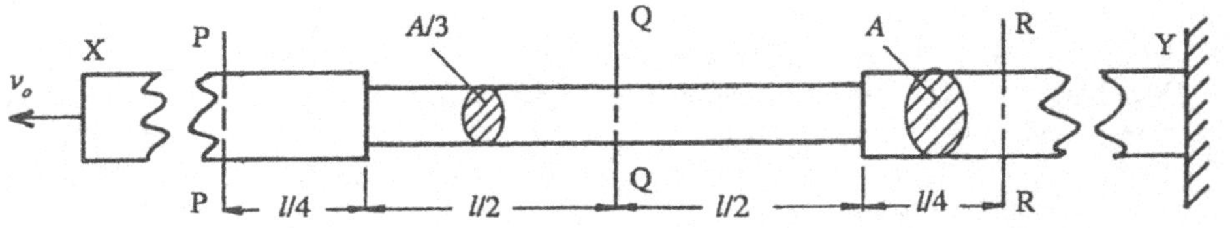

Figure 14.37

14.5 Derive all the elastic–plastic wavefronts resulting from the interactions shown in Figs 14.38a–d. The given numbers/fractions refer to stress ratios σ/σ_0. Positive and negative signs indicate compression and tension respectively. Take the velocity ratio $c_o/c = 4$ in all cases.
See solution

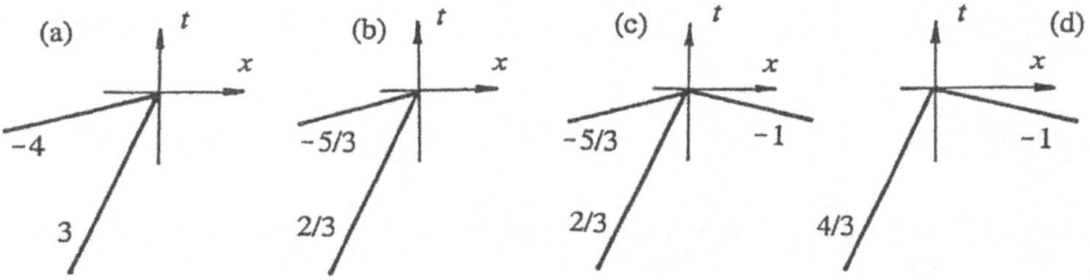

Figure 14.38

14.6 A thin rod of uniform cross-section and length l is rigidly clamped at one end. A tensile stress $25\sigma_o/12$ is suddenly applied at the other end for an interval l/c_o and then removed. Given the plastic modulus $S = E/16$, plot on an x–t diagram the resulting stress increments for time $t = 0$ to $t = (5l)/(2c_o)$. The yield stress of the rod is σ_o.

14.7 A bar of area A and length $4L$ is fixed to a rigid support at one end. The other end is free and flanged as shown in Fig. 14.39. A tube of similar material, with section area $3A$ and length L, slides freely along the bar to strike the flange with an impact velocity 150 m/s. Draw the x–t diagram from $t = 0$ to $t = 8l/c_o$ given the following: $E = 207$ GPa, $\sigma_o = 275$ MPa, $S = E/16$ and $\rho = 7890$ kg/m^3.

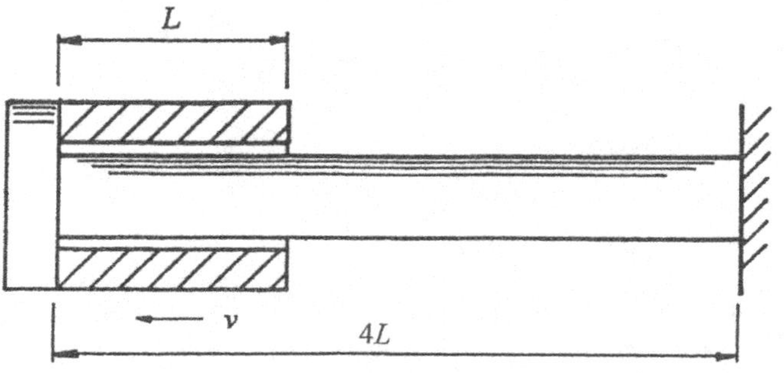

Figure 14.39

14.8 The large diameter end of a light alloy flagpole (see Fig. 14.40) is dropped 2.75 m on to a solid rigid base plate. Construct the x–t diagram from the time of impact $t = 0$ to $t = 7l/c_o$. Show that the residual plastic strain distribution following this impact duration is then completed. Take $g = 9.81$ m/s^2 with the following material properties for the bar : $E = 124$ GPa, $S = E/16$, $\sigma_o = 198$ MPa, and $\rho = 3320$ kg/m^3.

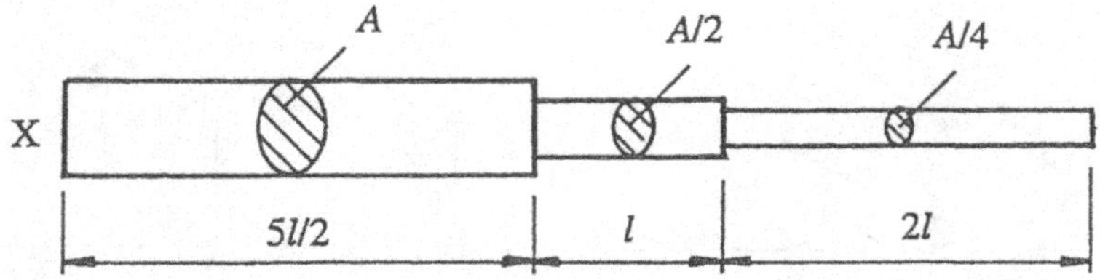

Figure 14.40

14.9 A uniform section bar, 255 mm long, travelling at 76.5 m/s, strikes a rigid perpendicular surface. Plot the x–t diagram for the period in which plasticity occurs and sketch the residual plastic strain distribution. Assume a bi-linear stress–strain curve for the bar in which $\sigma_o = 138$ MPa, $E = 6.9$ GPa, $S = E/64$ and $\rho = 8304$ kg/m^3.
See solution

14.10 A uniform bar 255 mm long is projected axially at constant speed to strike a rigid perpendicular target. If the initial impact stress within the striking end of the bar is 69 MPa, find the striking velocity and plot the x–t diagram, showing the magnitude of each wave front for the duration in which plasticity occurs. Sketch the residual plastic strain distribution after this period. Assume the following: $E = 13.8$ GPa, $\sigma_o = 55$ MPa, $S = E/100$ and $\rho = 7200$ kg/m^3.
Ans: $v_o = 19.56$ m/s.

14.11 A uniform bar of length l and area A is rigidly clamped at one end. A second bar of the same material with length $l/3$ and area $2A$ impacts the clamped bar at its free-end with velocity of 14.75 m/s. Plot the x–t diagram showing the stress magnitudes in both bars from time $t = 0$ to $t = 5l/c_o$. Discuss whether any further plastic wave fronts are generated after this time interval for a material with the following properties: $E = 207$ GPa, $\sigma_o = 345$ MPa, $S = E/16$ and $\rho = 7890$ kg/m^3.
See solution

14.12 A stepped circular bar consists of two lengths L and $3L$ with respective areas $2A$ and A. The smaller end strikes a rigid perpendicular target at 12.7 m/s. Plot the location-time diagram showing the magnitudes and direction of the wavefronts for as long as plasticity continues to occur. Plot the residual plastic strain distribution. Take the following material properties for the bar: $E = 207$ GPa, $\sigma_o = 345$ MPa, $S = E/16$ and $\rho = 7890$ kg/m^3.

14.13 Find the equivalent static load and the instantaneous deflection when a weight of mass 15 kg falls freely through a height of 100 mm to make an impact with a straight steel bar of length: (a) 5 m and (b) 1 m, each of diameter 8 mm. Take $E = 207$ GPa with a yield stress of 300 MPa, an ultimate strength of 600 MPa and a 25 % elongation at failure.
See solution

14.14 Show that the velocity function $v = v(x)$ for the maximum force condition, eq(14.30c), in a drop weight test cannot be satisfied by: (a) positive and negative exponentials $v = A\exp(\pm mx)$ where A and m are constants and (b) simple harmonic motion with an acceleration $a = -K^2x$, where $x = R\cos Kt$ in which t is time and K and R are constants.

14.15 Other quadratic solutions to eq(14.30b,c) apply when the two roots are real and imaginary compared to when they are equal as given in eqs(14.31a,b). Confirm the diagrammatic solutions given in Fig. 14.41a,b for a quadratic $v(x) = a_o - ax + bx^2$.

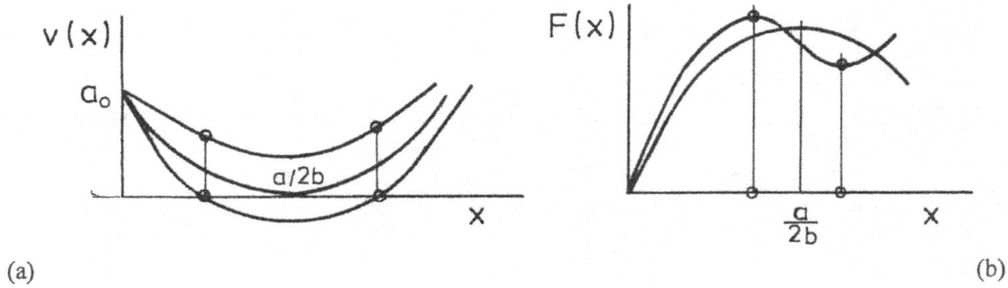

Figure 14.41

CHAPTER 15

PRODUCTION PROCESSES

15.1 Introduction

In the earlier chapters 6 and 7 on slip lines and limit loading, analyses were made of extrusion and rolling. These processes are among a number of large scale bulk forming processes where pressure, temperature, friction and strain rate are controlling parameters. These also include forging, cogging, drawing and machining. Alternative analyses of each of these shaping methods given in this chapter employ work conservation and slab equilibrium principles. Within these analyses some previous results from SLF and bounding theory will be employed for accuracy checks by comparison. The introductory treatment given here will serve to extend the range of methods available for the analyses of forming and shaping of metals within the plastic range.

15.2 Hot Forging

In the case of hot forging, a billet of material is compressed between flat, parallel dies increasingly so that thickness is reduced by the required amount. Uniform thickness bars and rings may be produced by a cogging method, either by applying rapid, repeated hammering or by applying a slower hydraulic pressure successively in steps along the length. More complex shapes, e.g. connecting rods and propellor blades, may be produced by forcing material into die cavities. Working material into such shapes is inhibited by its loss in temperature and an increase in surface friction, both of which increase the compressive force required.

15.2.1 Geometrical Constraints

Figure 15.1 shows the process of deforming a volume of material beneath dies that reduce the thickness from a cogging process. Depending upon the final dimensions required, there will usually be some degree of spreading and bulging in the width. That is, $w_1 > w_o$, but for each cogging step the material preserves its initial volume:

$$w_o b\, t_o = w_1 b_1 t_1 \tag{15.1a}$$

in which b_1 is the formed length of material beneath the die breadth b. Equation (15.1a) is written as

$$\ln(w_1 / w_o) + \ln(b_1 / b) + \ln(t_1 / t_o) = 0 \tag{15.1b}$$

A *coefficient of spread s* is defined as

$$s = \frac{\ln(w_1/w_o)}{\ln(t_o/t_1)} \tag{15.2a}$$

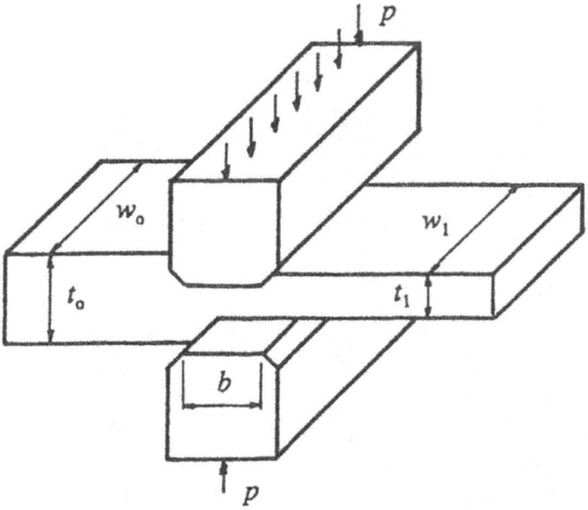

Figure 15.1 Forging between flat dies

A *coefficient of elongation* follows from eqs(15.1b) and (15.2a) as

$$1 - s = \frac{\ln(b_1/b)}{\ln(t_o/t_1)} \tag{15.2b}$$

The two coefficients in eqs(15.2a,b) give width and breadth ratios:

$$\frac{w_1}{w_o} = \left(\frac{t_o}{t_1}\right)^s \quad \text{and} \quad \frac{b_1}{b} = \left(\frac{t_o}{t_1}\right)^{1-s} \tag{15.3a,b}$$

where s is regarded as an empirical constant that depends upon the bite ratio b/w_o. If $b = w_o$, eqs(15.2a,b) give $s = 0.5$. Where w_o is large, giving $b/w_o \approx 0$, then $s = 0$ to give no spread under plane strain. Conversely, if w_o is small then $s = 1$. A practical formula, that embodies these three cases, defines s as

$$s = \frac{(b/w)_o}{1 + (b/w_o)} \tag{15.4}$$

Equations (15.1–4) apply to a forging pass in which the squeeze ratio t_o/t_1 is constant. The following recommendations [1] are made to avoid a central region of unworked material, excessive distortion and overlapping at the die edges:

$$\frac{t_o}{t_1} \le 1.3, \quad \frac{b}{t_o} \ge \frac{1}{3} \quad \text{and} \quad \frac{t_o}{w_o} \le \frac{3}{2}$$

15.2.2 Die Pressure and Loading

As a first approximation it is assumed that the material beneath the die (see Fig. 15.1) deforms in plane strain. Thus, when the width w remains unaltered, the die pressure will be given by the SLF solutions discussed in Section 6.7. These solutions are combined within Fig. 6.50 to give a normalised die pressure appropriate to the b/t ratio in lubricated dies. However, in the presence of friction, SLF theory underestimates die pressure. The accuracy of the latter is improved when it is assumed that the shear yield stress k of the material has been reached at the interface, as shown in Fig. 15.2.

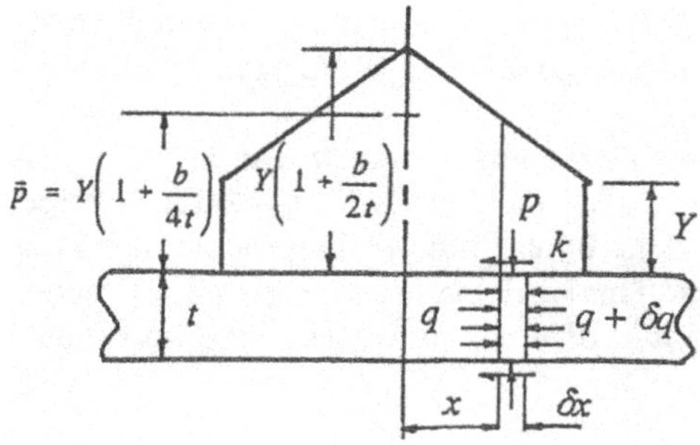

Figure 15.2 Forging of strip showing a material element and the friction hill

Plane strain conditions are assumed so that horizontal equilibrium of a strip with a unit width ($w = 1$) reveals

$$(2k \times \delta x \times 1) + (t \times \delta q \times 1) = 0 \tag{15.5}$$

Within the strip, the two compressive principal stresses p and q conform to the Tresca yield criterion

$$p - q = Y = 2k \tag{15.6}$$

Note: The von Mises yield criterion for plane strain, where $\sigma_3 = ½(\sigma_1 + \sigma_2)$, is $p - q = 2Y/\surd 3$, which also equals $2k$ by that criterion since $k = Y/\surd 3$. Combining eqs(15.5) and (15.6) leads to the pressure gradient for the die half:

$$\mathrm{d}p/\mathrm{d}x = \mathrm{d}q/\mathrm{d}x = -\,2k\,/\,t \tag{15.7a}$$

Integrating gives the die pressure

$$p = -\,Yx\,/\,t + C \tag{15.7b}$$

in which the constant $C = Y + Yb/2t$ is found from the condition that $p = Y$ for $x = b/2$ at the die edge. Hence, the die pressure is distributed as

$$p = Y\left(1 + \frac{b}{2t} - \frac{x}{t}\right) \tag{15.8a}$$

Equation (15.8a) defines a linear pressure distribution over one half of the die face, as shown in Fig. 15.2. The so-called *friction hill* estimates the forces required to forge both square and circular prisms [2]. The three-dimensional pressure required is distributed within a pyramid and a cone respectively over the die face. To describe a simple cogging operation, with $b/t > 1$, the mean pressure is adequate. Equation (15.8a) gives this as

$$\bar{p} = Y(1 + b/4t) \tag{15.8b}$$

The following empirical expressions [1] provide a closer estimate for each range of die breadth to thickness ratio:

$$p = Y(\,0.797 + 0.203t/b\,) \text{ for } b/t < 1 \tag{15.9a}$$

$$p = Y\,(0.750 + 0.250b/t\,) \text{ for } b/t > 1 \tag{15.9b}$$

Example 15.1 A block, 60 mm square and 300 mm long, is to be hot forged into a thinner strip, 20 mm thick, 60 mm wide by 900 mm long, by repeated cogging steps between dies whose breadth b is to minimise the plastic work required. Given that the b/t ratios involved with the reduction in thickness from 60 mm to 20 mm involve die pressures expressed by both eqs(15.9a,b), find the required value of b. Neglect work hardening and spread.

Since $60 < b < 20$ mm, eq(15.9a) applies initially with $t_o = 60$ mm for a reduction to $t = b$ (i.e. $b/t \le 1$). Thereafter, eq(15.9b) will apply for $t_o = b$ for final reductions to $t_1 = 20$ mm (i.e. $b/t \ge 1$). The work done by each die is the product of the force F applied and the displacement y it produces. This work may be referred to a current strip thickness t with an appropriate change to the limits of integration. Hence, the work done by the die pair per step becomes

$$W = 2\int_0^{(t_o - t)/2} F\,\mathrm{d}y = \int_t^{t_o} F\,\mathrm{d}t$$

Substituting from eqs(15.9a,b), with $F = pwb$ and multiplying by the number of cogging steps l_o/b (l_o is the initial length), gives the total work done

$$\sum W = Ywl_o\left[\int_b^{t_o}\left(0.797 + 0.203\frac{t}{b}\right)\mathrm{d}t + \int_{t_1}^{b}\left(0.750 + 0.250\frac{b}{t}\right)\mathrm{d}t\right]$$

$$= Ywl_o\left[0.797(t_o - b) + \frac{0.203}{2b}(t_o^2 - b^2) + 0.75(b - t_1) + 0.25\,b\ln\left(\frac{b}{t_1}\right)\right]$$

The die breadth which minimises the total work follows from

$$\frac{\mathrm{d}\sum W}{\mathrm{d}b} = 0 = -\,0.797 - \frac{0.203\,t_o^2}{2b^2} - \frac{0.203}{2} + 0.75 + 0.25\left[1 + \ln\left(\frac{b}{t_1}\right)\right]$$

$$0 = 0.1015\left(1 - \frac{t_o^2}{b^2}\right) + 0.25\ln\left(\frac{b}{t_1}\right) = 0$$

Substituting $t_o = 60$ mm and $t_1 = 20$ mm provides the required die width: $b = 37.56$ mm.

15.2.3 Sliding Friction

With contacting surfaces not being perfectly flat, asperities lie between them, reducing the area of contact to α at the junctions. The latter are subject to yielding under normal and tangential stress at their respective yield stresses Y and k in compression and shear. The corresponding normal and frictional forces distribute over the junctions as $N = \alpha Y$ and $F = \alpha k$, as shown in Fig. 15.3. It follows that the ratio between forces F and R defines the coefficient of friction independently of the contact area α

$$\mu = \frac{F}{N} = \frac{\alpha k}{\alpha Y} = \frac{k}{Y}$$

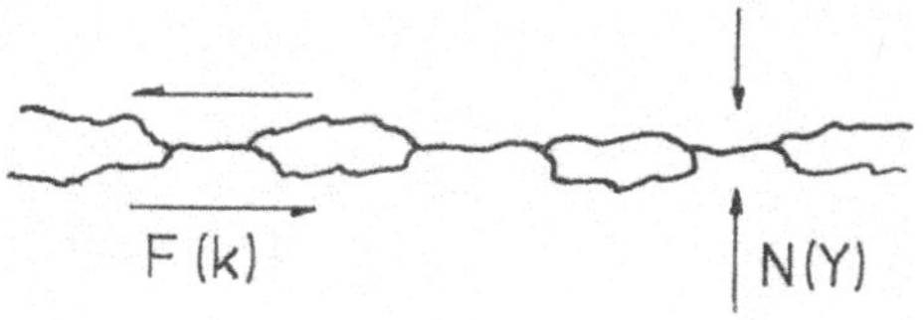

Figure 15.3 Forces between contacting asperity linkages

When $Y = 2k$, then $\mu = 1/2$, corresponding to the adhesion of contacting asperities. Experiment shows that this ratio lies in the region $1/\surd 3 - 1/2$, confirming the respective ratios between uniaxial and shear yield stresses according to the von Mises and Tresca yield criteria. A penetrating contact among some junctions is not uniaxial and involves a threefold increase in the normal force for the material to yield, as with a pyramid indenting hardness test. In this case the friction coefficient is lowered to $\mu \approx 1/6$. Here an effective lubricant can serve to reduce μ further. However, the more likely combination of sliding and penetrative contacts from ploughing behaviour suggests a range for μ between 1/6 and 1/2. Here it is the friction force that is lowered for a given normal force. Hence, the friction stress is controlled by yielding in shear which, in the presence of strain hardening, can never exceed the current flow stress in shear. Writing the latter as $\bar{\tau}$, the friction stress becomes $m\bar{\tau} = m\bar{\sigma}/\surd 3$, where $m < 1$. A von Mises shear to compressive flow stress conversion is applied to corresponding equivalent plastic strains, namely $\gamma^P = \surd 3\bar{\varepsilon}^P$. When $m = 1$ and $\bar{\tau} = \bar{\sigma}/\surd 3$, sticking friction applies to the interface shearing, given the high normal stress that exists between contacting surfaces in metal working. *Note*: Increasing forming forces are required for continuity where slipping is interrupted continually. Slip-stick behaviour does not conform to the quantitative descriptions of Coulomb friction outlined above.

15.3 Cold Forging

Small parts may be cold forged to improve their surface finish and hardness. The process is usually conducted at rapid rates under a drop hammer using dies to produce the complicated shapes required, e.g. in crankshaft and coin production. Forging may be employed for non ferrous alloys and carbon steels when the interface frictional stress is less than the shear yield stress k in the contact zone. Assuming Coulomb friction $\tau = \mu p \le k$ implies that the material beneath the dies slides under a constant frictional coefficient μ. The strip equilibrium equation (15.7a) becomes

$$\frac{dp}{dx} = -\frac{2\mu p}{t} \tag{15.10a}$$

Integrating eq(15.10a) leads to

$$\ln p = -2\mu x / t + C \tag{15.10b}$$

Along the edges of the die ($x = \pm b/2$) friction is absent and $p = 2k$ is reached. This gives $C = \mu b/t + \ln 2k$ and eq(15.10b) becomes

$$p = 2ke^{\frac{\mu b}{t}\left(1 - \frac{2x}{b}\right)} \tag{15.10c}$$

Equation (15.10c) describes the friction hill distribution shown in Fig. 15.4a.

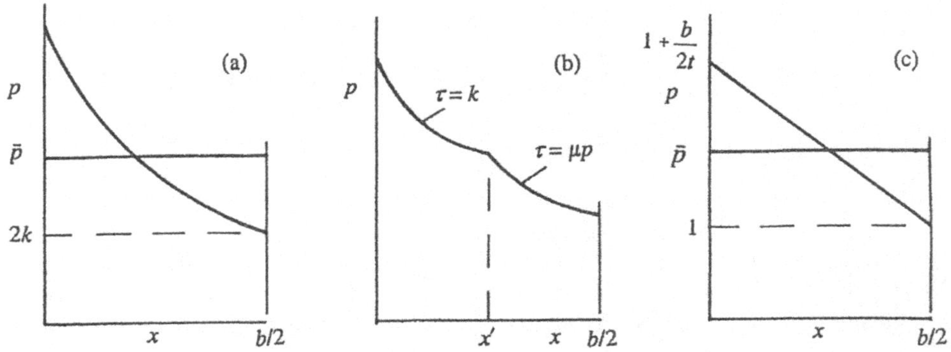

Figure 15.4 Friction hills for cold forging

From this, the mean pressure $\bar{p}$ in Fig. 15.4a may be found:

$$\bar{p} \times \frac{b}{2} = \int_0^{b/2} p\,\mathrm{d}x = 2k \int_0^{b/2} e^{\frac{\mu b}{t}\left(1 - \frac{2x}{b}\right)} \mathrm{d}x$$

$$\frac{\bar{p}}{2k} = \frac{t}{\mu b}\left(e^{\mu b/t} - 1\right) \tag{15.11}$$

The peak pressure for the hill in Fig. 15.4a is restricted to where $\mu p = k$ and so, with a widening of the die, the pressure in the central region becomes $p = k/\mu$. If μ remains constant in this region, the friction hill is modified with a flat plateau. Figure 15.4b shows a more likely pressure distribution under a varying μ. Equation (15.10b) gives the corresponding position $x = x'$ at which edge sliding under $\tau = \mu p$ becomes central sticking under $\tau = k$:

$$x' = \frac{b}{2}\left[1 - \frac{t}{\mu b}\ln\left(\frac{1}{2\mu}\right)\right] \tag{15.12}$$

If x' is to lie within the die's half-width ($0 < x' < b/2$), it follows that $\ln(1/2\mu) < \mu b/t$, from eq(15.12). When $x' < b/2$, the pressure within the central zone ($0 \le x \le x'$) is found from integrating eq(15.7a):

$$p = C - \frac{2kx}{t}$$

where C follows from setting p to eq(15.10b) with $x = x'$ from eq(15.12). This leads to

$$p = \frac{2k}{t}\left(\frac{b}{2} - x\right) + \frac{k}{\mu}\left(1 - \ln\frac{1}{2\mu}\right) \tag{15.13}$$

At the friction limit, $\mu = ½$ and eq(15.12) gives $x' = b/2$. Correspondingly, eq(15.13) shows that the pressure then becomes distributed linearly over the half-width (see Fig. 15.4c) as

$$p = \frac{2k}{t}\left(t + \frac{b}{2} - x\right)$$

for which the mean value is

$$\frac{\bar{p}}{2k} = 1 + \frac{b}{4t}$$

In the case of Fig. 15.4b, the mean pressure is

$$\frac{\bar{p}}{2k} = \frac{t}{\mu b}\left\{\frac{1}{4\mu}\left[(1 + \alpha)^2 + 1\right] - 1\right\}$$

where $\alpha = \mu b/t - \ln[1/(2\mu)]$. The present theory is regarded as acceptable where $b/t < 3$ and $\mu < 0.3$. Where $b/t > 3$ and $\mu > 0.3$ a thin strip analysis follows but slip-line field solutions (see § 6.7) will provide more reliable estimates of die pressure particularly where μ is high.

(a) Thin Strip Forging

A thin strip of thickness t and $b/t > 3$, when cold forged with dies overhanging its breadth b is analysed with a similar friction hill to Fig. 15.4a. The central peak pressure is found for the centre line for $x = 0$ in eq(15.10c):

$$p_{max} = 2k\mathrm{e}^{\frac{\mu b}{t}}$$

where k is the shear yield stress and μ is the Coulomb friction coefficient. The edge pressure $p = 2k$ applies when $x = b/2$ in eq(15.10c). The die loading upon a strip width w is taken as the area beneath the exponential distribution either side of the centre line to which these pressure limits apply:

$$F = 2w\int_0^{b/2} p\,\mathrm{d}x = 4wk\int_0^{b/2} \mathrm{e}^{\frac{\mu b}{t}\left(1 - \frac{2x}{b}\right)}\mathrm{d}x$$

$$= -\frac{4wkt}{2\mu}\left|\mathrm{e}^{\frac{\mu b}{t}\left(1 - \frac{2x}{b}\right)}\right|_0^{\frac{b}{2}} = \frac{2kwt}{\mu}\left(\mathrm{e}^{\frac{\mu b}{t}} - 1\right)$$

For example, if $b = 80$ mm, $t = 20$ mm with $2k = 250$ MPa and $\mu = 0.35$, then the forging force per mm of width becomes

$$\frac{F}{w} = \frac{250 \times 20}{0.35}\left(\mathrm{e}^{\frac{0.35 \times 80}{20}} - 1\right) = 43.65 \text{ kN/mm}$$

Note: The tensile yield stress equals $2k$ by Tresca's yield criterion.

(b) Friction Hill

This theory, which applies to a perfectly plastic material, allows forging to the depth required under the constant forging force calculated for the strip width. In the presence of hardening, incremental increases in the forces calculated above are required to reach the final thickness successively. There appears to be no theory available to control this trial procedure though the plane strain compression test provides the appropriate von Mises equivalent stress and equivalent plastic strain increment expressions for isotropic strip (see § 9.6.2):

$$\bar{\sigma} = \frac{\sqrt{3}}{2}\sigma_1, \quad \delta\bar{\varepsilon}^P = \frac{2}{\sqrt{3}}\delta\varepsilon_1^P$$

In Fig. 6.38, co-ordinate 1 is aligned with the compressive force F allowing σ_1 to identify with the friction hill normal pressure:

$$\sigma_1 = \frac{F}{wb} = \frac{2kt}{\mu b}\left(e^{\frac{\mu b}{t}} - 1\right) \tag{15.14a}$$

which applies across the width beneath the die centre line, i.e. beneath the line of F shown in Fig. 6.38. Correspondingly, the normalised equivalent stress and strain increment are

$$\frac{\bar{\sigma}}{2k} = \frac{\sqrt{3}\,t}{2\mu b}\left(e^{\frac{\mu b}{t}} - 1\right) \tag{15.14b}$$

$$\delta\bar{\varepsilon}^P = \frac{2}{\sqrt{3}}\frac{\delta t}{t} \Rightarrow \bar{\varepsilon}^P = \frac{2}{\sqrt{3}}\ln\left(\frac{t}{t_o}\right) \tag{15.14c}$$

With an isotropic (von Mises) definition for` $\bar{\sigma}$ in eq(15.14b), the initial yield stress is $\bar{\sigma}/\sqrt{3}k = 1$ for $t = t_o$. This corresponds to a series approximation applied to eq(15.14a):

$$e^{\mu b/t_o} \approx 1 + \mu b/t_o$$

which restricts the μ value and the ratio b/t_o to low values given additional terms in the exponential series expansion have been ignored. Therein, the following modification to eq(15.14b) is suggested for higher μ and b/t_o values:

$$\frac{\bar{\sigma}}{2k} = \frac{\sqrt{3}}{2}\left[\frac{\mu b}{t} + \frac{1}{2!}\left(\frac{\mu b}{t}\right)^2 + \frac{1}{3!}\left(\frac{\mu b}{t}\right)^3 + \ldots\right]^{-1}\left(e^{\frac{\mu b}{t}} - 1\right)$$

which does satisfy $\bar{\sigma}/\sqrt{3}k = 1$ for $t = t_o$ within a convergent series for which $b/t < 1/\mu$. This inequality can only accept higher μ with reduced b/t.

15.4 Extrusion

SLF solutions to various extrusion geometries and friction conditions were presented in chapter 6. Therein, making an assumption of plane strain restricts SLF applications to rectangular dies and billets. Consequently, the axi-symmetric extrusion of bars and tubes remains to be analysed. These products may be produced by direct and indirect methods. In conventional hot extrusion (see Fig. 6.24), the ram must exert a force of several hundred tonnes to overcome friction at the container walls. This method is favoured for the production of various bar sections in non-ferrous alloys where glass has been used as a lubricant. With even higher temperatures required to extrude steels, the molten glass solidifies upon the surface of the extrusion. A less convenient, inverted extrusion process (Fig. 15.5a) reduces the forces required, since the billet remains stationary in the container under a moving die.

Direct extrusion with a ram and mandrel is used for tube production by either method shown in Figs 15.5b,c. In the preferred design, the mandrel remains fixed in position by webbing to the port hole die (see Fig. 15.5c). In providing better support of the extruded tube the webbing also serves improve the concentricity between tube diameters.

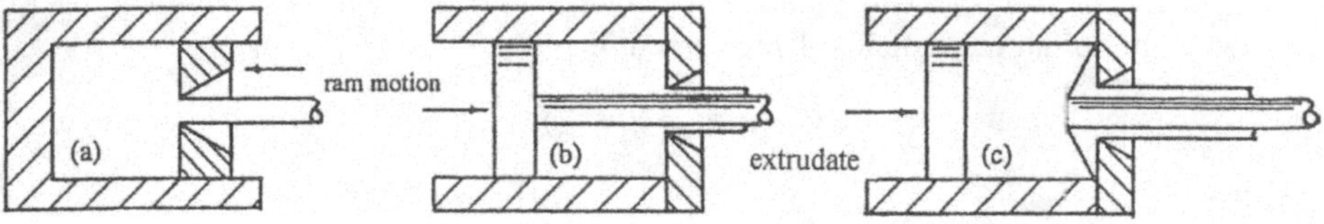

Figure 15.5 Extrusion processes

Taking an annular strip between mandrel and container wall from Fig. 15.5b allows a common analysis for axi-symmetric extrusion.

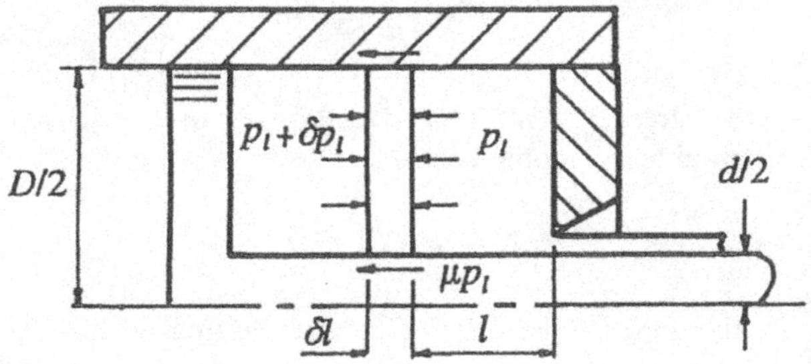

Figure 15.5b Analysis of extruded tube half-section

At the distance l from the die face, the equilibrium of this strip element is expressed as

$$\frac{\pi}{4}(D^2 - d^2) \times \delta p_l = \pi (D + d)\,\delta l \times \mu p_l \tag{15.15}$$

Separating the variables p and l within eq(15.15) allows an integration to

$$\frac{p_l}{p} = \exp\left(\frac{4\mu l}{D - d}\right) \tag{15.16a}$$

where p is taken as a mean extrusion pressure, revealed by SLF analyses:

$$\frac{p}{2k} = a + b \ln R \tag{15.16b}$$

where $R = H/h$ is the extrusion ratio and a and b are constants, which have been found [3] for this process as $a = 0.47$ and $b = 1.2$. Combining eqs(15.16a,b) gives

$$\frac{p_l}{Y} = (0.47 + 1.2 \ln R) \exp\left(\frac{4\mu l}{D - d}\right) \tag{15.17}$$

When the mean extrusion pressure is found for two billets of lengths l_1 and l_2, eq(15.17) may be re-arranged to provide the friction coefficient

$$\mu = \frac{(D - d)}{4(l_1 - l_2)} \ln\left(\frac{p_{l_1}}{p_{l_2}}\right) \tag{15.18}$$

A numerical value $\mu \approx 0.03$ was found from eq(15.18) for a range of extruded lengths [3]. The work done for an incremental displacement is

$$\delta W = p_l \times A \times \delta l$$

Substituting p_l from eq(15.17) and integrating for the total work done:

$$W = AY_m(0.47 + 1.2\ln R)\int_0^l \exp\left(\frac{4\mu l}{D - d}\right) dl$$

$$= AY_m(0.47 + 1.2\ln R)\frac{(D - d)}{4\mu}\exp\left(\frac{4\mu l}{D - d} - 1\right) \qquad (15.19)$$

In eq(15.19), a mean yield stress Y_m is introduced to allow for the influence of a temperature rise from the work required to overcome friction. This is

$$W = mC_V(T - T_o) \qquad (15.20)$$

where m is the mass of the pre-extruded billet, C_V is its specific heat at constant volume and T_o is its initial, pre-heated temperature. Equating (15.19) and (15.20) gives the temperature rise $\Delta T = T - T_o$:

$$\Delta T = \frac{AY_m}{\rho AlC}(0.47 + 1.2\ln R)\frac{4\mu}{(D - d)}\exp\left(\frac{4\mu l}{D - d} - 1\right) \qquad (15.21)$$

where ρ is the material density and l is the non extruded length. Equation (15.21) allows the maximum extrusion ratio $R \approx D/d$ to be found. To determine Y_m, the mean strain rate at the extrusion temperature must be estimated. For example, in a conventional extrusion, assume that a ring of dead metal exists in the corner of the vessel. The deformation then becomes confined to the shaded conical volume in Fig. 15.6a.

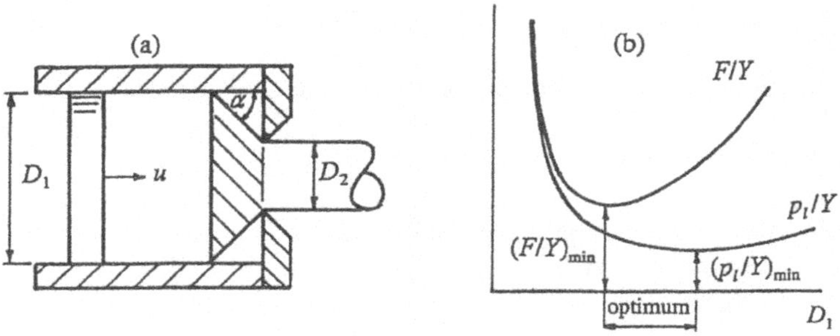

Figure 15.6 Deforming volume in conventional extrusion

Let the ram force a solid cylindrical billet of diameter D_1 into the 45° cone at a velocity of u. The time t for this to occur is found from

$$\frac{\pi}{4}D_1^2 u \times t = \frac{\pi}{24}(D_1^3 - D_2^3) \qquad (15.22a)$$

Taking the mean strain in the cone as $\varepsilon_m = \ln R$, where $R = D_1/D_2$, the mean strain rate in time t is approximated from eq(15.22a):

$$\dot{\varepsilon}_m = \frac{\varepsilon_m}{t} = \frac{6D_1^2 u}{D_1^3 - D_2^3}\ln R \approx \frac{6u}{D_1}\ln R \qquad (15.22b)$$

Equation (15.17) may be used to minimise the extrusion pressure and force with respect to the container diameter D_1 [3]. For a solid circular extrusion, the minimum force and pressure are not coincident (see Fig. 15.6b). The curves show that it would be advisable to select a value of D_1 lying between these minima. When the above theory is applied to an extrusion within conical dies some modifications are necessary. Firstly, for an inverted extrusion process, the coefficients a and b in eq(15.16) depend upon the semi-cone angle

$$a = a' - a''\cot\alpha \quad \text{and} \quad b = b' + b''\cot\alpha \qquad (15.23a,b)$$

where a', a'', b' and b'' are constants. Secondly, for direct extrusion through conical dies, the mean pressure is further raised to overcome additional wall friction

$$\frac{p}{Y} = \frac{(a + b\ln R)}{(1 - 4\mu/D_1)}$$

where a and b are given by eqs(15.23a,b).

Example 15.2 Show that for simple homogenous, cylindrical extrusion the force acting upon the ram in Fig. 15.5b is given by $F = YA_r \ln(A_r/A_e)$, where A_r is the area of the ram, A_e the area of the extruded bar and Y is the yield stress for rigid-plastic material. In a two-stage, extrusion process a 20 mm diameter billet is first produced from a 10 % reduction to its original area. Thereafter, the billet is extruded to 18 mm diameter. Given that the true stress-natural strain law for the annealed material is expressed by Hollomon as $\sigma = 980\varepsilon^{0.19}$, determine the force necessary for extrusion accounting for work hardening.

Let the ram force $F = YA_r$ move through a distance δl to give the work increment

$$\delta W = YA_r\delta l = \frac{YV}{l}\delta l \qquad \text{(i)}$$

where V is the volume displaced in a length l. With a homogenous deformation of rigid plastic material, occurring between an initial length l_r and a final length l_e, the work/unit volume becomes

$$\frac{W}{V} = Y\int_{l_r}^{l_e}\frac{\mathrm{d}l}{l} = Y\ln\left(\frac{l_e}{l_r}\right) = Y\ln\left(\frac{A_r}{A_e}\right) \qquad \text{(ii)}$$

where $A_r l_r = A_e l_e$ provides the reduction in area ratio. On the ram side, the specific work is

$$\frac{W}{V} = \frac{F\,l_r}{A_r l_r} = \frac{F}{A_r} \qquad \text{(iii)}$$

Equating (ii) and (iii) gives the ram force

$$F = YA_r \ln\left(\frac{A_r}{A_e}\right) \tag{iv}$$

In the first stage, the area reduction is

$$\frac{A_o - A_1}{A_o} = 0.1 \quad \Rightarrow \quad \frac{A_o}{A_1} = 1.11$$

and from this, the true strain is

$$\varepsilon = \ln\left(\frac{l_1}{l_o}\right) = \ln\left(\frac{A_o}{A_1}\right) = \ln(1.11) = 0.105$$

From the Hollomon law, the corresponding flow stress is $\sigma = 638.6$ MPa. In the second stage the strain increment is similarly found:

$$\Delta\varepsilon = \ln\left(\frac{l_e}{l_r}\right) = \ln\left(\frac{A_r}{A_e}\right) = \ln\left(\frac{20}{18}\right)^2 = 0.211$$

The total strain after stage 2 is $\varepsilon = 0.105 + 0.211 = 0.316$, for which the stress is 787.3 MPa. The average stress for stage 2 is $\sigma_m = 712.97$ MPa and from this the extrusion force in eq(iv) may be modified to account for work hardening:

$$F = \sigma_m A_r \ln\left(\frac{A_r}{A_e}\right) = 712.97 \times \frac{\pi}{4}(20)^2 \ln\left(\frac{20}{18}\right)^2 = 47.2\,\text{kN}$$

Example 15.3 The uniform 625 mm^2 cross section of a metal bar is reduced by 30 % in an extrusion process. Find the extrusion force given that the material conforms to a Ludwik law $\sigma = Y + B\varepsilon^n$, with $Y = 200$ MPa, $B = 250$ MPa and $n = 0.25$. Find the temperature increase arising from the process given that the specific heat and density for constant volume deformation are $C_V = 0.50$ kJ/kgK and $\rho = 7200$ kg/m^3 respectively.

In Ludwik's account $\sigma = Y + B\varepsilon^n$ of the influence of hardening upon cylindrical extrusion of bar, the hardening term $B\varepsilon^n$ is included within the specific work integral $w = W/V$:

$$w = \int \sigma\,\mathrm{d}\varepsilon = \int_{\varepsilon_r = 0}^{\varepsilon_e} (Y + B\varepsilon^n)\,\mathrm{d}\varepsilon \tag{i}$$

The extrusion force $F = A_r w$ follows from eq(i) as

$$F = A_r\left[Y\varepsilon_e + \frac{B\varepsilon_e^{n+1}}{n+1}\right] = A_r Y \ln\left(\frac{A_r}{A_e}\right) + \frac{A_r B}{n+1}\left[\ln\left(\frac{A_r}{A_e}\right)\right]^{n+1} \tag{ii}$$

These analyses assume simple frictionless homogenous deformation in the absence of redundant work. The latter refers to the work consumed as the die forces the material to flow within its taper. It is the shearing deformation, in addition to the extension, which occurs within the taper that accounts for the redundant work. The analyses given in examples 15.2 and 15.3 above are based upon the diameter reduction within tapered dies involving tensile extension only.

A homogenous deformation provides the temperature change from entry to exit in that the work of extruding changes the internal energy and the exit temperature of the extrusion:

$$W = mC_V \Delta T = \rho V C_V \Delta T \;\Rightarrow\; w = W/V = F/A_r = \rho C_V \Delta T$$

$$\therefore \Delta T = w/(\rho C_V) = F/(A_r \rho C_V) \qquad \text{(iii)}$$

where F is given by eq(ii). The area reduction, which applies to *any* uniform cross-section, provides the ratio between areas at entry and exit for a single reduction (where $A_r = A_o$):

$$(A_o - A_e)/A_o = 0.3 \;\Rightarrow\; A_o/A_e = 1.43$$

This gives $\ln(A_o/A_e) = \ln(1.43) = 0.357$ and $[\ln(A_o/A_e)]^{n+1} = 0.357^{1.25} = 0.276$

The extrusion force is found from eq(ii). In units of N and m:

$$F = (625 \times 10^{-6})[(200 \times 10^6 \times 0.357) + (250 \times 10^6/1.25) \times 0.276] = 79125 \text{ N}$$

Equation (iii) reveals the accompanying temperature increase during adiabatic extrusion

$$\Delta T = 79125/(625 \times 10^{-6} \times 7200 \times 0.5 \times 10^3) = 35.2 \text{ K}$$

which applies irrespective of the initial section area when all the heat generated serves to raise the temperature.

15.5 Hot Rolling

This continuous cogging process may be performed hot or cold between rolling mills arranged as 2-high, 4-high, clustered, planetary and pendulum (see Figs 15.7a–e).

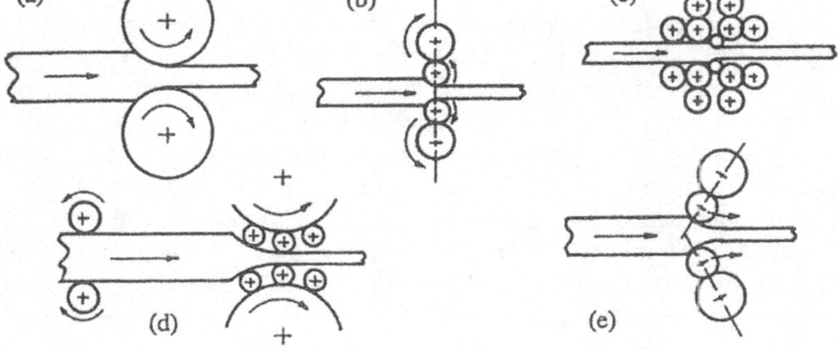

Figure 15.7 Rolling processes

The aim of all rolling is to reduce sheet thickness in a continuous process while maintaining uniform thickness and width throughout [4]. The roll force and torque will dictate the reduction, which may be increased with more sophisticated mills, involving a number of smaller rollers in contact with the sheet.

15.5.1 Wide Sheet

The theory of hot rolling assumes plane strain in which no spread occurs, i.e. transverse cross-sections remain plane and no roll deformation occurs. Hence, the theory is similar to that for cold rolling strip (see § 7.7). In Fig. 15.8a, a 2-high mill transmits the roll radii R to the sheet within an arc of contact. At the entry and exit planes the thicknesses are h_1 and h_2 and axial stresses on these planes are zero when front and back tensions are absent.

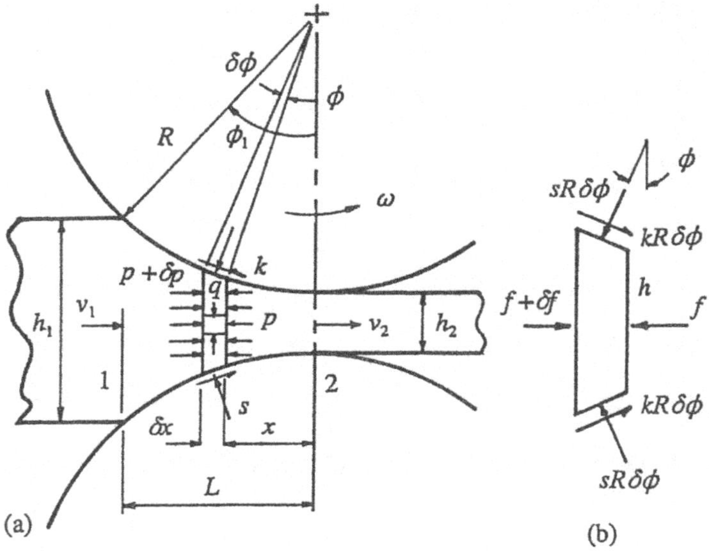

Figure 15.8 Geometry and forces within simple hot rolling

The volume rate remains constant, so for sections at entry, exit and another between these

$$b_1 h_1 v_1 = bhv = b_2 h_2 v_2$$

where v_1, v_2 and v are the respective velocities at these positions. Since $b_1 \approx b \approx b_2$, it follows that $v_2/v_1 = h_1/h_2 > 1$. The tangential roll velocity $v_R = \omega R$ lies between v_1 and v_2, i.e. $v_R > v_1$ at entry and $v_R < v_2$ at exit. Where v_R equals the strip speed there is no relative motion and at this *neutral point*, the frictional force between contacting surfaces changes direction. In a hot rolling process the shear yield stress k is attained by the friction present. The direction of k at the intermediate position shown applies to where the roll speed is greater than the strip. This is coupled with a normal pressure s along the arc interface and a distribution in the axial compression from p to $p + \delta p$. The forces acting upon the elemental strip, distance x from the vertical symmetry line, are shown in Fig. 15.8b. The balance between horizontal forces from pressure ($f = ph$) and those from the roll is expressed for a unit of width as

$$(p + \delta p)(h + \delta h) + 2kR\delta\varphi \cos\varphi = ph + 2sR\delta\varphi \sin\varphi \qquad (15.24a)$$

Expanding (15.24a) and neglecting the product ($\delta p \times \delta h$) leads to

$$\frac{\mathrm{d}}{\mathrm{d}\varphi}(ph) = 2R(s \sin \varphi \pm k \cos \varphi) \tag{15.24b}$$

in which $-k \cos\varphi$ applies to the horizontal friction force component at the entry side of the neutral point and $+k \cos\varphi$ applies to the exit side of this point. Strictly, eq(15.24b) requires a numerical solution to p but a good approximate closed solution follows from assuming a rolling reduction between flat, parallel dies. This gives $\varphi = 0$, $h_1 = h_2$ and $\delta x = R\delta\varphi$ and eq(15.24b) becomes

$$h_2 \, \mathrm{d}p / \mathrm{d}x = \pm 2k \tag{15.25a}$$

Both p and s are compressive and assuming $s \approx q$ for the strip's central element shown, the plane strain yield criterion, eq(15.6), is simply

$$s - p = 2k \tag{15.25b}$$

Combining eqs(15.25a,b) gives

$$h_2 \frac{\mathrm{d}s}{\mathrm{d}x} = \pm 2k \tag{15.25c}$$

Integrating eq(15.25c) at exit:

$$s = s_2 + \frac{2k}{h_2} x \tag{15.26a}$$

where s_2 applies to the exit plane ($x = 0$) where $p = 0$. Hence, eq(15.25b) gives $s_2 = 2k$ and eq(15.26a) becomes

$$s = 2k\left(C + \frac{x}{h_2}\right) \tag{15.26b}$$

(a) Lower Bound

A simple homogenous compression throughout the reduction ignores friction. Taking a mean yield stress $Y_m = 2k$ for the pass and a flat 'friction hill' with $s = 2k$ in Fig. 15.8 provides a first approximation to the roll force P for a sheet width of w:

$$P = 2k \times L \times w = Y_m w \sqrt{R\delta} \tag{15.27a}$$

where
$$\delta = h_1 - h_2 = 2R(1 - \cos\varphi_1) \approx R\varphi_1^{\,2} \tag{15.27b}$$

and
$$L \approx R\varphi_1 = R \times \surd(\delta/R) = \surd(R\delta) \tag{15.27c}$$

Equations (15.27a–c) serve to guide the roll force for thickness reductions of a hardening material using multiple passes as the following example shows.

Example 15.4 The thickness of an annealed sheet, 150 mm wide, was reduced by 30 % to 3 mm before a second pass reduced the thickness further to 2 mm. Calculate the roll force required for each pass with a 175 mm roll radius allowing for strain hardening given that Ludwik's law $\sigma = Y + B(\varepsilon^P)^n$ for annealed material applies with $Y = 180$ MPa, $B = 390$ MPa and $n = 0.6$.

With a 30 % reduction within the first pass the area ratio between entry and exit is found:

$$\frac{(A_0 - A_1)}{A_0} = 0.3, \quad \therefore \frac{A_1}{A_0} = 0.7$$

The original thickness follows from constant volume rolling under a plane strain condition, i.e. there is no change in width w as an original length changes from l_o to l_1:

$$A_0 l_0 = A_1 l_1, \quad \therefore l_1/l_0 = A_0/A_1 = 1.429$$

$$w h_0 l_0 = w h_1 l_1, \quad \therefore h_0 = h_1 l_1/l_0 = 3 \times 1.429 = 4.287 \text{ mm}$$

The length ratio from the first pass provides the natural axial strain and the amount by which the yield stress of the annealed material has been raised (Y_0 to Y_1) by Ludwik's law:

$$\varepsilon_{0-1} = \ln(l_1/l_0) = \ln(1.429) = 0.357$$

$$Y_1 = Y_0 + B(\varepsilon_1)^n = 180 + 390(0.357)^{0.6} = 390.2 \text{ MPa}$$

where $Y_m = (180 + 390.2)/2 = 285.1$ MPa, $\Delta h = 1.287$ mm. Equation(15.27a) gives

$$P_1 = Y_m w \sqrt{R \Delta h} = 285.1 \times 150 \sqrt{175 \times 1.287} = 641.8 \text{ kN}$$

For the second pass:

$$w h_1 l_1 = w h_2 l_2, \quad \therefore l_2/l_1 = h_1/h_2 = 3/2$$

$$\varepsilon_{1-2} = \ln(l_2/l_1) = \ln(1.5) = 0.406$$

$$\varepsilon_2 = \varepsilon_{0-1} + \varepsilon_{1-2} = 0.357 + 0.406 = 0.763$$

$$Y_2 = Y_1 + B(\varepsilon_2)^n = 180 + 390(0.763)^{0.6} = 551.6 \text{ MPa}$$

where $Y_m = (390.2 + 551.6)/2 = 470.9$ MPa, $\Delta h = 1.0$ mm. Equation(15.27a) gives

$$P_1 = Y_m w \sqrt{R \Delta h} = 470.9 \times 150 \sqrt{175 \times 1.0} = 934.4 \text{ kN}$$

(b) Friction Hill

An account of the extra redundant work is needed over and above that for a pure homogenous compression in (a) above. The friction hill is modified from experiment, where C in eq(15.26b) is taken as $\pi/4 \approx 0.8$ and not unity [5]. Equation (15.26b) defines the *friction hill* for hot rolling shown in Fig. 15.9.

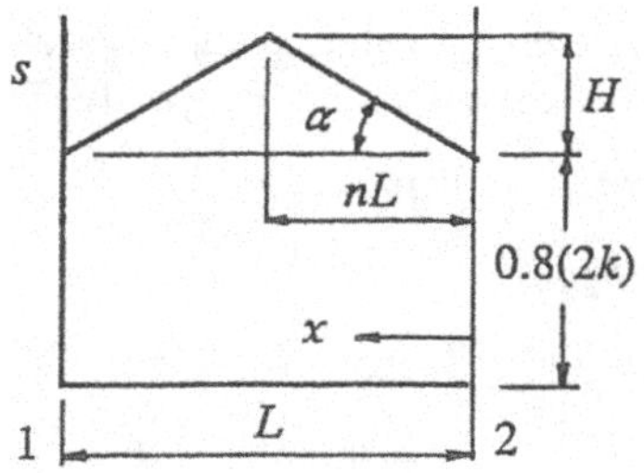

Figure 15.9 Friction hill for hot rolling

Equation (15.26b) gives the gradient, $\tan\alpha = H/(nL)$, of the hill as

$$\frac{ds}{dx} = \frac{2k}{h_2}$$

in which the neutral point is assumed to bisect the projected arc length L. The area beneath the hill gives the roll force/unit width

$$P = \int s\,dx = 2 \times 2k \int_0^{L/2} \left(C + \frac{x}{h_2}\right) dx = 2kL\left(C + \frac{L}{4h_2}\right) \qquad (15.28a)$$

which is comparable with eqs(15.9a,b) for a cogging operation. Each roll torque/unit width may be approximated from eq(15.28a) as

$$T = P \times L/2 \qquad (15.28b)$$

Now, from the geometry in Fig. 15.8:

$$L \approx R\varphi_1 \text{ and } x \approx R\varphi \qquad (15.29a)$$

$$h_1 - h_2 = 2R(1 - \cos\varphi_1) \approx R\varphi_1^2 \qquad (15.29b)$$

$$h - h_2 = 2R(1 - \cos\varphi) \approx R\varphi^2 \qquad (15.29c)$$

Setting the 'draft' $\delta = h_1 - h_2$, eqs(15.29a,b) give the geometry of the contact arc:

$$\varphi_1 = \sqrt{\delta / R} \quad \text{and} \quad L = \sqrt{R\delta} \qquad (15.29d,e)$$

Thus, knowing R, δ, h_2 and k, the roll force and torque follow from eqs(15.28a,b). Sims [6] modified the yield criterion (15.25b) to correspond with $C \approx \pi/4$:

$$p = s - \frac{\pi}{4}(2k) \qquad (15.30a)$$

and set $\sin\varphi \approx \varphi$ and $\cos\varphi \approx 1$ in eq(15.24b) for rolling mills with a small arc of contact. Substituting from eqs(15.29c) and (15.30a), the equilibrium eq(15.24b) becomes

$$\frac{d}{d\varphi}\left(\frac{s}{2k} - \frac{\pi}{4}\right) = \frac{\pi R\varphi}{2(h_2 + R\varphi^2)} \pm \frac{R}{(h_2 + R\varphi^2)} \qquad (15.30b)$$

Equation (15.30b) may be integrated, using the appropriate limits for φ between exit and entry, where $s = (\pi/4) \times (2k)$ at each position. Respective closed solutions are found:

$$\frac{s^+}{2k} = \frac{\pi}{4}\ln\left(\frac{h}{h_2}\right) + \frac{\pi}{4} + \sqrt{\frac{R}{h_2}}\tan^{-1}\left(\sqrt{\frac{R}{h_2}}\varphi\right) \quad \text{(exit)} \tag{15.31a}$$

$$\frac{s^-}{2k} = \frac{\pi}{4}\ln\left(\frac{h}{h_1}\right) + \frac{\pi}{4} + \sqrt{\frac{R}{h_2}}\tan^{-1}\left(\sqrt{\frac{R}{h_2}}\varphi_1\right) - \sqrt{\frac{R}{h_2}}\tan^{-1}\left(\sqrt{\frac{R}{h_2}}\varphi\right) \quad \text{(entry)} \tag{15.31b}$$

The neutral point $\varphi = \varphi_n$ is found from equating (15.31a,b) for $h = h_n = h_2 + R\varphi_n^2$:

$$\frac{\pi}{4}\ln\left(\frac{h_2}{h_1}\right) = 2\sqrt{\frac{R}{h_2}}\tan^{-1}\sqrt{\frac{R}{h_2}}\varphi_n - \sqrt{\frac{R}{h_2}}\tan^{-1}\sqrt{\frac{h_1 - h_2}{h_2}} \tag{15.32}$$

The roll force, acting normal to a unit width of roll, is found from eq(15.32) as follows:

$$P = R\int_0^{\varphi_1} s\,\mathrm{d}\varphi = R\int_0^{\varphi_n} s\,\mathrm{d}\varphi + R\int_{\varphi_n}^{\varphi_1} s\,\mathrm{d}\varphi$$

$$= 2kR\left[\frac{\pi}{2}\sqrt{\frac{h_2}{R}}\tan^{-1}\sqrt{\frac{h_1 - h_2}{h_2}} - \frac{\pi\varphi_1}{4} - \ln\left(\frac{h_n}{h_2}\right) + \frac{1}{2}\ln\left(\frac{h_1}{h_2}\right)\right] \tag{15.33}$$

where $\sqrt{[(h_1 - h_2)/h_2]} = \sqrt{(R/h_2)}\varphi_1$. Sims [6] interpreted eq(15.33) graphically by writing $P = 2kRQ_P$, where Q_P depends upon R/h_2 and h_1/h_2. The roll torque/unit width may be estimated from eq(15.28b) simply as $T = PL/2$ where $L = \sqrt{(R\delta)}$ or, more accurately, from k as

$$T = 2\int_{\varphi_n}^{\varphi_1/2} kR^2\mathrm{d}\varphi = 2kR^2(\varphi_1/2 - \varphi_n) \tag{15.34a}$$

The limits of integration apply to the range of φ over which the direction of k is not opposed within the adjacent region. Since φ_n appears in eq(15.32) and $\varphi_1 = \sqrt{[(h_1 - h_2)/R]}$, eq(15.34a) may be re-written as

$$T = 2kR^2 Q_T \tag{15.34b}$$

The dependence of Q_T upon R/h_2 and h_1/h_2 has appeared in graphical form [6]. The shear yield stress k in eqs(15.33) and (15.34b) must apply to the mean strain rate for the process. Then $2k$ becomes the mean value of the plane strain yield stress for rolling. For this a stress–strain curve is required at the mean strain rate for rolling (see § 15.5.3). The curve allows a mean stress $2\bar{k}$ to be read or, where work-hardening is evident (e.g. see Fig. 15.10), as the ordinate of a rectangle with an equal enclosed area:

$$2\bar{k}\,\varepsilon_2 = \int_0^{\varepsilon_2} 2k\,\mathrm{d}\varepsilon \tag{15.34c}$$

where $\varepsilon_2 = \ln(h_1/h_2)$ is the exit strain simulated, say, from a plane strain compression test (see § 9.6.2).

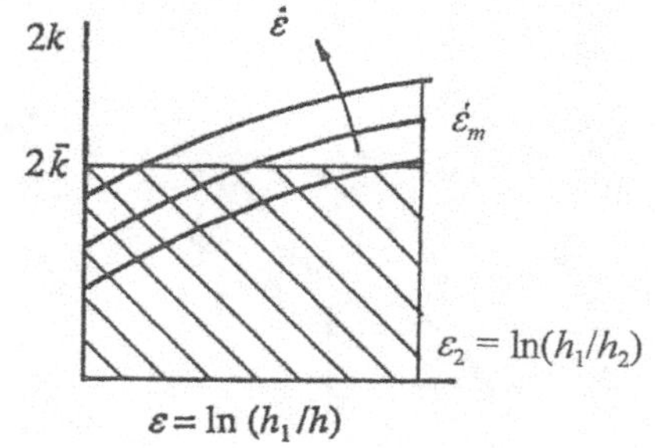

Figure 15.10 Mean stress and strain in plane strain

Example 15.5 Estimate the roll force and torque required to roll a 300 mm wide, annealed sheet from 4 mm to 3.5 mm thick between two rolls each of radius 175 mm. How is the force altered when the original area had been attained by a previous reduction of 10 %? The plane strain, true stress versus natural strain curve is given by a Swift law, eq(9.44b), in which $\sigma_o = 2k = 200$ MPa, $\varepsilon_o = 0.05$ and $n = 0.25$.

The compressive strain involved in a reduction from 4 to 3.5 mm is

$$\varepsilon_2 = \int_{h_2}^{h_1} \frac{dh}{h} = \ln\left(\frac{h_1}{h_2}\right) = \ln\left(\frac{4}{3.5}\right) = 0.134$$

Applying eq(15.34c) gives a mean flow stress $2\bar{k}$ for the strain range $0 \le \varepsilon \le 0.134$ (compression implied) as

$$2\bar{k} \times 0.134 = 200 \int_0^{0.134} (1 + 20\,\varepsilon)^{0.25}\, d\varepsilon \approx 32.78$$

from which $2\bar{k} = 244.6$ MPa. Equation (15.29e) gives the arc length as

$$L = \sqrt{R(h_1 - h_2)} = \sqrt{175(4 - 3.5)} = 9.35 \text{ mm}$$

from which eqs(15.27a,b) supply each roll force and the torque as

$$P = 2\bar{k}Lw\left(\frac{\pi}{4} + \frac{L}{4h_2}\right) = 244.6 \times 9.35 \times 300 \times 10^{-3}\left(0.7854 + \frac{9.35}{4 \times 3.5}\right) = 997.1 \text{ kN}$$

$$T = PL/2 = 997.1 \times (9.35 \times 10^{-3})/2 = 4.66 \text{ kNm}$$

In the case of a 10 % pre-reduction in area ($A_o/A_1 = l_1/l_o = 1.11$), the lower limit of integration in eq(15.34c) is $\ln(1.11) = 0.105$ and the upper limit is $0.105 + 0.134 = 0.239$.

15.5.2 Narrow Strip

Rolling narrow strip modifies the friction hill (Fig. 15.9) between the planes of entry and exit. The hill top slopes at β to sheet edges but extends toward the centre of the rolls, retaining its constant peak, to give the *friction roof* in Fig. 15 11.

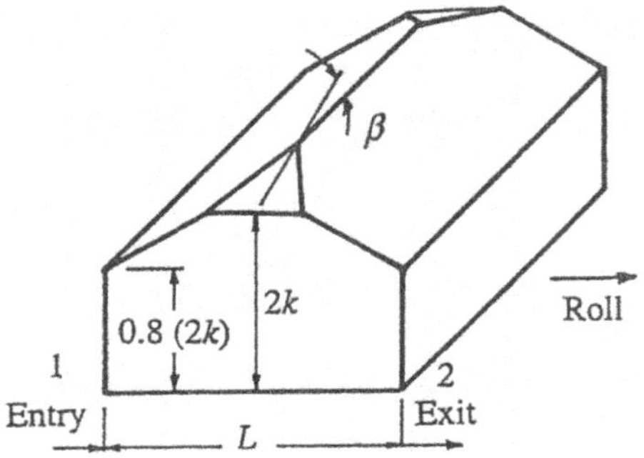

Figure 15.11 Friction roof for rolling narrow strip

The total roll force is found from the volume beneath the roof [7]:

$$P_T = 2k\bar{b}L\left[C + \frac{g}{2} + \frac{\bar{h}}{3\bar{b}g}(g - 0.2)^3\right] \tag{15.35}$$

where $g \approx L/2h_2$ when the neutral point is central. The mean thickness and breadth, $\bar{h}$ and $\bar{b}$ respectively, are based upon a parabola describing the shape of the contact arc and the spread between entry and exit. These give:

$$\tan\beta = \frac{2k}{\bar{h}}, \quad \bar{h} = \frac{1}{3}(2h_2 + h_1) \quad \text{and} \quad \bar{b} = \frac{1}{3}(2b_2 + b_1)$$

As with hot forging, a *spread factor* S_r and a *coefficient of spread* s_r (see eq(15.4)) can be defined for strip rolling:

$$S_r = \frac{b_2}{b_1} \quad \text{and} \quad s_r = \frac{L/b_1}{1 + L/b_1} \approx \frac{1}{1 + b_1/\sqrt{R\delta}}$$

In a two-roll mill, eqs(15.28b) and (15.29e) approximate the total roll torque as

$$T_T = 2\left(\frac{P_T L}{2}\right) = P_T\sqrt{R\delta}$$

15.5.3 Strain Rate

A consideration of the strain rate in hot rolling becomes important when estimating the appropriate shear yield stress k in eqs(15.27a) and (15.35). For rolling a wide strip, the through-thickness strain increment is $\delta\varepsilon = \delta h/h$, giving its rate as

$$\dot{\varepsilon} = \frac{1}{h}\frac{\mathrm{d}h}{\mathrm{d}t} = \frac{1}{h}\left(\frac{\mathrm{d}h}{\mathrm{d}\varphi} \times \frac{\mathrm{d}\varphi}{\mathrm{d}x} \times \frac{\mathrm{d}x}{\mathrm{d}t}\right) \tag{15.36}$$

where, from the geometry in Fig. 15.8

$$x = R\sin\varphi \quad \Rightarrow \quad \frac{\mathrm{d}x}{\mathrm{d}\varphi} = R\cos\varphi \tag{15.37a}$$

$$h = h_2 + 2R(1 - \cos\varphi) \quad \Rightarrow \quad \frac{\mathrm{d}h}{\mathrm{d}\varphi} = 2R\sin\varphi \tag{15.37b}$$

$$\frac{\mathrm{d}x}{\mathrm{d}t} = -\,v = -\frac{V}{bh} \tag{15.37c}$$

where $V = bhv$ is the volume of material rolled per second. Substituting eqs(15.37a–c) into eq(15.36) gives

$$\dot{\varepsilon} = -\frac{2V}{bh^2}\tan\varphi \tag{15.38}$$

Equation (15.29c) approximates the thickness at the x–plane to

$$h = h_2 + 2R\,(1 - \cos\varphi\,) \approx h_2 + R\varphi^2 \tag{15.39a}$$

from which we may write an approximation for $\tan\varphi$:

$$\tan\varphi \approx \varphi = \sqrt{\frac{h - h_2}{R}} \tag{15.39b}$$

Substituting eq(15.39b) into (15.38) gives

$$\dot{\varepsilon} = -\frac{2V}{bh^2}\sqrt{\frac{h - h_2}{R}} = -\frac{2V}{b}\sqrt{\frac{h_2}{R}}\left[\frac{\sqrt{(h/h_2) - 1}}{h^2}\right] \tag{15.40}$$

Equation (15.40) allows the strain rate distribution within the pass to be determined. This shows a near maximum rate at the entry plane $h = h_1$, falling to zero at the exit plane $h = h_2$. A mean strain rate $\dot{\varepsilon}_m$ may be identified with the ordinate of a rectangle with the same area as that beneath the curve of $\dot{\varepsilon}$ versus h:

$$(h_1 - h_2)\,\dot{\varepsilon}_m = \int_{h_1}^{h_2} \dot{\varepsilon}\,\mathrm{d}h$$

Under plane strain conditions, the flow in the transverse direction is zero. Hence $\dot{\varepsilon}_m$ applies to the mean rate of thinning, this equalling the mean rate of axial extension beneath the rolls.

15.6 Cold Rolling

Dimensional stability, hardness and improved surface finish may be achieved with a final cold rolling of pre-tensioned sheet. This ensures that the yield stress for the sheet material is far greater than that for all previous hot passes. Cold rolling raises the normal pressure s in Fig. 15.8 and flattens the roll itself elasticity by altering its radius from R to R'. As with cold forging, the interfacial shear stress is lowered to μs and this cannot exceed the shear yield stress k for the material. Thus, the horizontal force equilibrium condition for a wide sheet modifies eq(15.24b) to

$$\frac{\mathrm{d}}{\mathrm{d}\varphi}(ph) = 2R's(\sin\varphi \pm \mu\cos\varphi)\,{}^{+\,exit}_{-\,entry} \tag{15.41}$$

Bland and Ford's [8] simplifying assumptions $s \approx q$, $\sin\varphi \approx \varphi$ and $\cos\varphi \approx 1$ allowed the plane strain yield criterion $q - p = 2k$ to be applied to a plastic hardening material. Here, the plane strain yield stress $2k$ depends upon the plastic strain level, and eq(15.41) becomes

$$\frac{\mathrm{d}}{\mathrm{d}\varphi}[(s - 2k)h] = 2R's(\varphi \pm \mu)\,{}^{+\,exit}_{-\,entry}$$

$$2kh\frac{\mathrm{d}}{\mathrm{d}\varphi}\left(\frac{s}{2k} - 1\right) + 2\left(\frac{s}{2k} - 1\right)\frac{\mathrm{d}}{\mathrm{d}\varphi}(kh) = 2R's(\varphi \pm \mu)$$

$$2kh\frac{\mathrm{d}}{\mathrm{d}\varphi}\left(\frac{s}{2k}\right) + 2\left(\frac{s}{2k} - 1\right)\frac{\mathrm{d}}{\mathrm{d}\varphi}(kh) = 2R's(\varphi \pm \mu) \tag{15.42}$$

With the exception of cold rolling an annealed material [8], the second term on the left-hand side of eq(15.42) is negligible, i.e. the product kh is almost a constant and the bracket term is small. Thus, with the substitution from eq(15.39a), the integration follows as

$$2k(h_2 + R'\varphi^2)\frac{\mathrm{d}}{\mathrm{d}\varphi}\left(\frac{s}{2k}\right) = 2R's(\varphi \pm \mu)$$

$$\int\frac{\mathrm{d}(s/2k)}{(s/2k)} = \int\frac{2R'\varphi\,\mathrm{d}\varphi}{h_2 + R'\varphi^2} \pm \int\frac{2R'\mu\,\mathrm{d}\varphi}{h_2 + R'\varphi^2}$$

$$\ln\left(\frac{s}{2k}\right) = \ln\left(h_2 + R'\varphi^2\right) \pm 2\mu\sqrt{\frac{R'}{h_2}}\tan^{-1}\sqrt{\frac{R'}{h_2}}\,\varphi + \ln C$$

$$\ln\left(\frac{s}{2khC}\right) = \pm\mu Q$$

$$s = 2khC\,e^{\pm\mu Q} \tag{15.43}$$

in which

$$Q = 2\sqrt{\frac{R'}{h_2}}\tan^{-1}\left(\sqrt{\frac{R'}{h_2}}\,\varphi\right)$$

Constant C is found from the boundary conditions for a pre-tensioned strip. Let these be t_1 ($= -p_1$) and t_2 ($= -p_2$) at the entry and exit position, where $\varphi = \varphi_1$ and $\varphi_2 = 0$ respectively. Applying the plane strain yield criterion to each position gives $s_1 = 2k_1 - t_1$ and $s_2 = 2k_2 - t_2$ correspondingly. With these substitutions into eq(15.43), the roll pressures become

$$s^+ = \frac{2kh}{h_2}\left(1 - \frac{t_2}{2k_2}\right)e^{\mu Q} \tag{15.44a}$$

$$s^- = \frac{2kh}{h_1}\left(1 - \frac{t_1}{2k_1}\right) e^{\mu(Q_1 - Q)} \qquad (15.44b)$$

where Q_1 applies to the entry condition $\varphi = \varphi_1$. The exponential terms in eqs(15.44a,b) describe friction hills with both front and back tensions, t_1 and t_2 respectively, in Fig. 15.12. In the absence of t_1 and t_2, the friction hill (shaded) is superimposed upon the plane strain yield stress, the latter increasing from $2k_1$ to $2k_2$ in the contact zone.

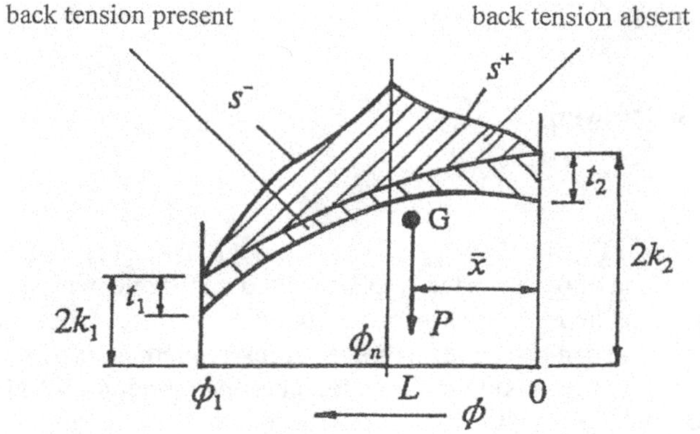

Figure 15.12 Roll pressure showing the influence of strip tensions

The yield stress variation may be replaced with a mean value, found from eq(15.34c). When present, pre-tensions reduce both the roll pressure and the roll force for a given reduction. When $t_1 > t_2$ these tensions shift the neutral point φ_n toward the exit plane and, conversely, toward the entry plane when $t_2 > t_1$. The position φ_n follows from equating (15.44a,b):

$$\varphi_n = \sqrt{\frac{h_2}{R'}} \tan\left(\frac{Q_n}{2}\sqrt{\frac{h_2}{R'}}\right)$$

where

$$Q_n = \frac{Q_1}{2} - \frac{1}{2\mu} \ln\left\{\frac{h_1[1 - (t_2/2k_2)]}{h_2[1 - (t_1/2k_1)]}\right\}$$

The roll force/unit width of strip is given by the enclosed area in Fig. 15.12:

$$P = R'\left(\int_0^{\varphi_n} s^+ \mathrm{d}\varphi + \int_{\varphi_n}^{\varphi_1} s^- \mathrm{d}\varphi\right) \qquad (15.45)$$

where s^+ and s^- are given by eqs(15.44a,b). Each roll torque/unit width may be found [9] from the sum of the torques due to s, t_1 and t_2:

$$T = \mu R R' \int_0^{\varphi_1} s\, \mathrm{d}\varphi + \frac{R}{2}(t_1 h_1 - t_2 h_2) \qquad (15.46)$$

This may be written as $T = P\bar{x}$. Taking the centroid $\bar{x}$ for the pressure distribution as approximately one half the contact length, i.e. $\bar{x} \approx L/2$ in Fig. 15.12, gives a reasonable estimate of the torque upon a single roll. The deformed roll radius in eqs(15.45) and (15.46) may be estimated by replacing the friction hill with an elliptical distribution of pressure along the arc of contact [10]:

$$R' = R(1 + KP/\delta) \qquad (15.47)$$

where, typically, $K = 10.8 \times 10^{-6}$ m^2/MN for hardened steel rolls. The error involved in estimating R' from eq(15.47) is likely to be less than the error arising from neglecting elastic strain when cold rolling thin, hard strip [9].

15.7 Wire and Strip Drawing

15.7.1 Frictionless Drawing

The simplest analysis of cold drawing a wire through frictionless dies assumes an homogenous deformation of rigid, perfectly-plastic, incompressible material. For example, Fig. 15.13a shows the reduction in cross-sectional area from A_o to A_1 for drawing wire in conical dies. This method can be used to achieve successive reductions from a 10–12 mm stock diameter, usually at the maximum reduction possible (about 60 %) in a single pass. Multiple passes are employed to obtain greater reductions and, with an intermediate anneal, wire as fine as 0.02 mm diameter may be cold drawn.

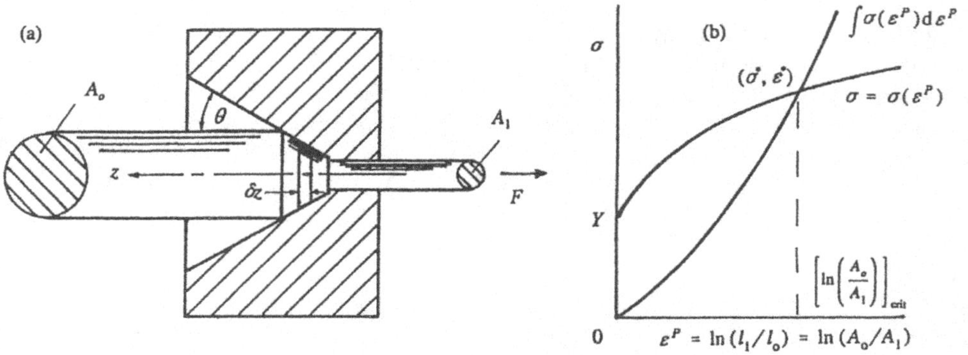

Figure 15.13 Wire drawing showing (a) conical dies and (b) the reduction limit

The drawing force within a given reduction is found from the specific work done:

$$W/V = \int_{\varepsilon^P} \sigma \, d\varepsilon^P = Y \int_{\varepsilon^P} d\varepsilon^P \qquad (15.48a)$$

where Y is the constant flow stress and V is a given volume of material for which the corresponding lengths before and after reduction are l_o and l_1. Since $V = A_o l_o = A_1 l_1$, eq(15.48a) gives

$$W/V = Y \int_{l_o}^{l_1} dl/l = Y \ln(l_1/l_o) = Y \ln(A_o/A_1) \qquad (15.48b)$$

Also, the work done in drawing a length l_1 under a force F is $W = Fl_1$. Hence

$$W/V = (F\,l_1)/(A_1\,l_1) = F/A_1 \tag{15.49}$$

Equating (15.48b) and (15.49) gives the wire drawing force

$$F = A_1\,Y\ln\left(A_o/A_1\right) \tag{15.50a}$$

The drawing stress follows from eq(15.50a) as

$$\sigma_w = F/A_1 = Y\ln\left(A_o/A_1\right) = Y\ln\left(l_1/l_o\right) \tag{15.50b}$$

Now, σ_w should be less than Y if the wire is not to break. Setting $\sigma_w = Y$ in eq(15.50b) gives $A_o/A_1 = 2.718$, from which the maximum percentage reduction in area is 63.2 %. For a hardening material, $\sigma = \sigma(\varepsilon^P)$ in eq(15.48a), giving the drawing stress as $\sigma_w = \int \sigma(\varepsilon^P)\mathrm{d}\varepsilon^P$. Since the integral defines the area beneath the flow curve, it permits the geometrical interpretation of the failure condition shown in Fig. 15.13b. The point of intersection gives the maximum possible reduction. For example, using Swift's law (9.44b), with intersection co-ordinates (σ^*, ε^*), gives

$$\sigma_o\left(1 + \frac{\varepsilon^*}{\varepsilon_o}\right)^n = \sigma_o\int_0^{\varepsilon^*}\left(1 + \frac{\varepsilon}{\varepsilon_o}\right)^n \mathrm{d}\varepsilon = \frac{\sigma_o\,\varepsilon_o}{(n+1)}\left[\left(1 + \frac{\varepsilon^*}{\varepsilon_o}\right)^{n+1} - 1\right] \tag{15.51a}$$

Setting $\sigma^*/\sigma_o = (1 + \varepsilon^*/\varepsilon_o)^n$ in eq(15.51a) provides an equation in the normalised stress σ^*/σ_o:

$$\left(\frac{\sigma^*}{\sigma_o}\right)^{\frac{n+1}{n}} - \frac{(n+1)}{\varepsilon_o}\left(\frac{\sigma^*}{\sigma_o}\right) - 1 = 0 \tag{15.51b}$$

which depends upon the hardening exponent n and, more strongly, upon the strain history term ε_o. Alternatively, we may use an average yield stress for the process within eq(15.50a) to account more simply for a history of hardening, as the following example shows.

Example 15.6 The true stress–strain relationship for an annealed material is given by the Swift law $\sigma = 75(1 + 30\varepsilon)^{0.65}$. Determine the draw force required to reduce a 3 mm diameter wire to 2.75 mm, given that the 3 mm wire was produced originally from a 20 % area reduction. What is the greatest reduction from 3 mm in a single pass?

Let the sequence of processes be 0–1–2 where 0 is for the annealed material, 1 applies to 3 mm wire with a previous 20 % area reduction on entry to the die and 2 applies to the wire at exit, following a further area reduction of $100(3^2 - 2.75^2)/3^2 = 15.9$ %. The natural strains associated with the 20 % and 15.9 % reductions are respectively

$$\varepsilon_{0\text{-}1} = \ln(l_1/l_0) = \ln(A_0/A_1) = \ln(1/0.80) = 0.223$$

$$\varepsilon_{1\text{-}2} = \ln(l_2/l_1) = \ln(A_1/A_2) = \ln(1/0.84) = 0.174$$

Thus, the natural strains at entry to and exit from the die are 0.223 and 0.397. The Swift law gives the corresponding stresses as 282.4 MPa and 395.5 MPa. Using a mean yield stress

$Y_m = 339$ MPa for the draw process, eq(15.50a) is written as

$$F = A_2 Y_m \ln(A_1/A_2) = 5.94 \times 339 \times \ln(7.069/5.94) = 350.4 \text{ N}$$

The greatest reduction, based upon the flow stress in the drawn material, is found from

$$350.4 = 395.5 \ln(A_1/A_2)$$

giving $A_1/A_2 = 2.425$ and $d_2 = 1.926$ mm.

15.7.2 Friction and Die Angle

There have been many refinements to the simple wire drawing analysis, notable among these are the accounts of friction, geometry and redundant work, reviewed in [11]. A theoretical account of the first two takes a disc of perfectly-plastic material within the contact zone (see Fig. 15.13a) to reveal the normal pressure p and friction stress μp around the conical interface. These are equilibrated by a change to the axial stress $\delta\sigma_z$ over thickness δz. Let the position z increase with D, as shown in Fig. 15.14a.

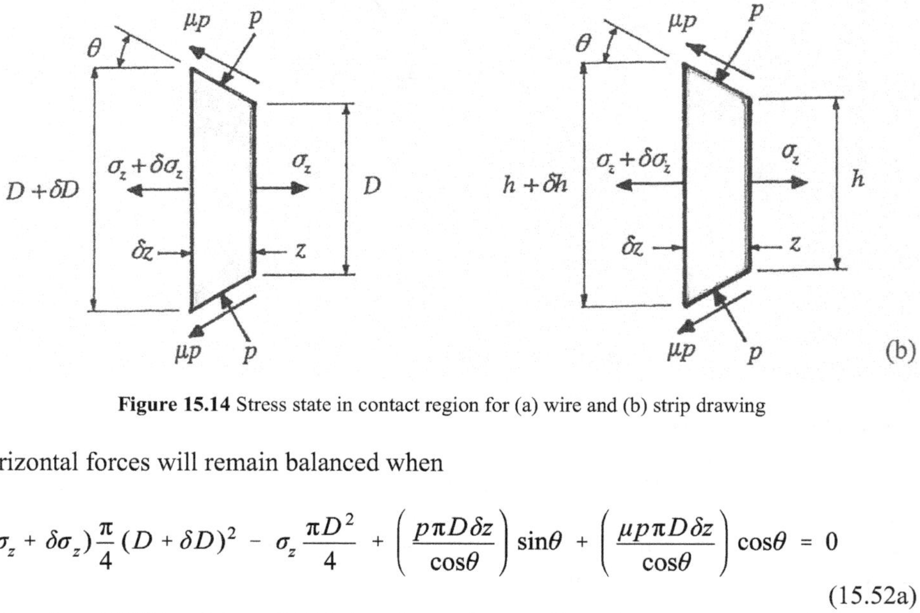

Figure 15.14 Stress state in contact region for (a) wire and (b) strip drawing

Horizontal forces will remain balanced when

$$(\sigma_z + \delta\sigma_z)\frac{\pi}{4}(D + \delta D)^2 - \sigma_z\frac{\pi D^2}{4} + \left(\frac{p\pi D\delta z}{\cos\theta}\right)\sin\theta + \left(\frac{\mu p\pi D\delta z}{\cos\theta}\right)\cos\theta = 0 \quad (15.52a)$$

Setting $\delta z = \delta D/(2\tan\theta)$ in eq(15.52a) and ignoring small products upon expansion:

$$D\delta\sigma_z + 2\sigma_z\delta D + 2p(1 + \beta)\delta D = 0 \quad (15.52b)$$

where $\beta = \mu/\tan\theta$. The radial and circumferential strain increments within the cone are the same since $\delta\varepsilon_\theta^P = (\pi\delta D)/(\pi D) = \delta\varepsilon_r^P$. Substituting this condition into the Levy-Mises eqs(6.1a–c) shows that $\sigma_\theta = \sigma_r$ for a cylindrical reduction. As a consequence both the Tresca and von Mises yield criteria reduce to $\sigma_z - \sigma_r = Y$ when the frictional stress μp is ignored. Moreover, with a small semi-cone angle θ, the radial stress $\sigma_r \approx -p$ to give

$$p = Y - \sigma_z \quad (15.53)$$

Combining eqs(15.52b) and (15.53):

$$\frac{2\delta D}{D} = \frac{\delta\sigma_z}{\beta\sigma_z - (1 + \beta)Y} \tag{15.54a}$$

Given $D_o > D_1$, eq(15.54a) is integrated between exit position 1 and entry position 0:

$$2\int_{D_1}^{D_0} \frac{\mathrm{d}D}{D} = \int_{\sigma_{z1}}^{\sigma_{z0}} \frac{\mathrm{d}\sigma_z}{\beta\sigma_z - (1 + \beta)Y}$$

$$\left(\frac{D_1}{D_0}\right)^{2\beta} = \frac{\beta\sigma_{z1} - (1 + \beta)Y}{\beta\sigma_{z0} - (1 + \beta)Y} \tag{15.54b}$$

Re-arranging eq(15.54b) gives a wire draw stress $\sigma_w = \sigma_{z1}$ in the presence of back tension σ_{z0}

$$\sigma_w = \frac{Y(1 + \beta)}{\beta}\left[1 - \left(\frac{D_1}{D_0}\right)^{2\beta}\right] + \sigma_{z0}\left(\frac{D_1}{D_0}\right)^{2\beta} \tag{15.54c}$$

In the absence of σ_{z0} the drawing stress and force required to give a reduction in area ratio

$$r = (A_o - A_1)/A_o = 1 - (D_1/D_o)^2 \tag{15.55a}$$

are respectively

$$\sigma_w = Y(1 + 1/\beta)\left[1 - (1 - r)^\beta\right], \quad F = A_1\,\sigma_w \tag{15.55b,c}$$

The greatest reduction ratio $r = r_{max}$ applies when $\sigma_w = Y$ in eq(15.55b):

$$r_{max} = 1 - 1/(1 + \beta)^{1/\beta} \tag{15.56}$$

Equation (15.53) shows that $\sigma_z = Y - p$ and $\delta\sigma_z = -\delta p$. Substituting into eq(15.54a) gives the pressure distribution with an integration from the exit position 1:

$$2\int_{D_1}^{D} \frac{\mathrm{d}D}{D} = \int_{p_1}^{p} \frac{\mathrm{d}p}{Y + \beta p}$$

from which

$$p = (p_1 + Y/\beta)(D/D_1)^{2\beta} - Y/\beta \tag{15.57}$$

where $p_1 = Y - \sigma_1$ and $\sigma_1 = \sigma_w$ is given by the draw stress eq(15.55b).

In the case of a strip drawing analysis (see Fig. 15.14b), the strip thickness h replaces the wire diameter D. The horizontal force equilibrium equation becomes

$$(\sigma_z + \delta\sigma_z)(h + \delta h)\,w - \sigma_z h w + \left(\frac{2pw\delta z}{\cos\theta}\right)\sin\theta + \left(\frac{2\mu p w\delta z}{\cos\theta}\right)\cos\theta = 0 \tag{15.58a}$$

where w is the strip width. With $\delta z = \delta h/(2\tan\theta)$, eq(15.58a) is simplified:

$$h\delta\sigma_z + \sigma_z\delta h + p(1+\beta)\delta h = 0 \tag{15.58b}$$

where $\beta = \mu/\tan\theta$. As the width strain is zero, eq(6.1c) gives the width stress as $(\sigma_z - p)/2$. Both Tresca and von Mises provide a plane strain yield criterion: $\sigma_z + p = Y$. Hence, the integration of eq(15.58b) leads to a draw stress without back pull and a maximum reduction identical to eqs(15.55b) and (15.56). With a reduction in thickness from h_o to h_1, the reduction ratio is written as $r = 1 - h_1/h_o$.

Example 15.7 What is the maximum reduction that can be achieved in strip drawing when $\mu = 0.07$ and $\alpha = 15°$? Given an initial thickness of 5 mm, what is the drawing force per unit width under these conditions?

$$\beta = \mu/\tan\alpha = 0.07/\tan 15° = 0.261$$

$$r_{max} = 1 - 1/(1+\beta)^{1/\beta} = 1 - 1/(1+0.261)^{3.83} = 0.589$$

$$r_{max} = 1 - h_1/h_o \quad \Rightarrow \quad h_1 = h_o(1 - r_{max}) = 5(1 - 0.589) = 2.06 \text{ mm}$$

$$\sigma_w = Y(1 + 1/\beta)\left[1 - (1 - r)^{\beta}\right], \quad F = A_1\sigma_w$$

$$\sigma_w = Y \quad \Rightarrow \quad F = A_1 Y \quad \therefore \; F/w = h_1 Y = 2.06 \times 200 = 412 \text{ N/mm}$$

In Fig. 15.13a an alternative, semi-empirical account [12] of work lost to friction and shear employs the mean radial pressure upon the conical interface. This is determined when friction and back pull are absent. Thus, in Fig. 15.15a, a horizontal force balance reveals

$$p_m = \frac{A_1\sigma_w^*}{A_o - A_1} \tag{15.59a}$$

It is assumed that the same p_m exists when friction of low μ is present. As the draw stress is altered to σ_w (see Fig. 15.15b), the new equilibrium equation becomes

$$\mu p_m\left(\frac{A_o - A_1}{\sin\theta}\right)\cos\theta + p_m\left(\frac{A_o - A_1}{\sin\theta}\right)\sin\theta = \sigma_w A_1 \tag{15.59b}$$

in which the bracket term is the sloping surface area. Combining eqs(15.59a,b) leads to

$$\sigma_w = \sigma_w^*(1 + \mu\cot\theta) \tag{15.60a}$$

When σ_w^* is defined by eq(15.50b) and a redundant shear work factor $\varphi > 1$ is introduced, the drawing stress eq(15.60a) is modified to

$$\sigma_w = (1 + \mu\cot\theta)\varphi Y_m \ln\left(A_o/A_1\right) \tag{15.60b}$$

where, for a hardening material, Y_m is the mean flow stress in the given reduction. As φ increases with increasing die angle θ, the term $(1 + \mu\cot\theta)$ in eq(15.60b) decreases when μ is constant. This agrees with experimental data [12] showing that an optimum die angle minimises σ_w.

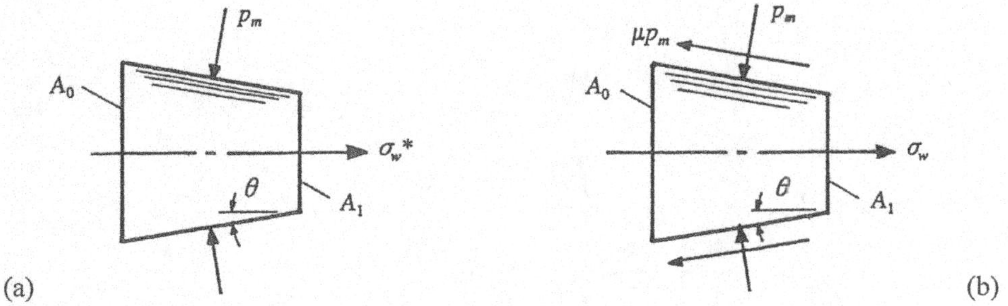

Figure 15.15 Conical contact area (a) without friction and (b) with friction

15.8 Orthogonal Machining

In an over- simplified analysis of machining, the cutting edge of the tool face is taken to lie perpendicular to the workpiece but the force, friction and heat involved shorten tool life and impair accuracy. In practice, after setting the *depth of cut b*, the inclined rake face of the hardened tool removes workpiece material under a constant tool velocity, i.e. the *feed v*. The workpiece remains stationary during planing, but it rotates at a constant rotational speed when turning. The *rake angle* α is crucial to the type of chip produced and the *clearance angle* δ avoids excessive rubbing. Depending upon the material and tool geometry, the chip formation may be discontinuous or continuous, both of with can form a *built-up edge* upon the tool (see Figs 15.16a–c). In brittle workpiece materials, a periodic rupture occurs ahead of the tool to produce the discontinuous segments in Fig. 15.16a. For ductile materials the continuous chip occurs by shear (Fig. 15.16b) whilst the building of an edge between chemically similar materials in contact is due to friction welding (Fig. 15.16c).

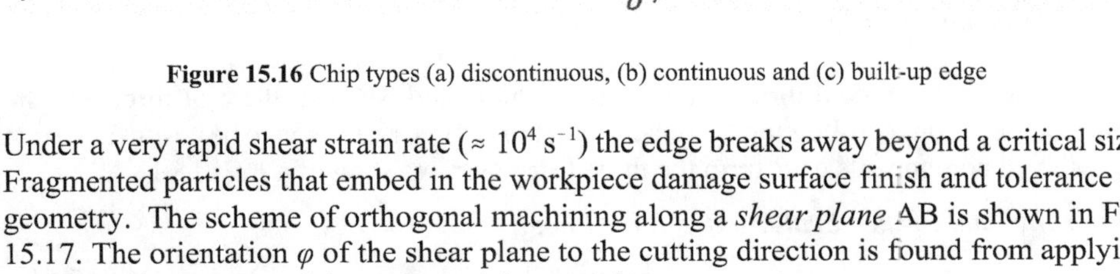

Figure 15.16 Chip types (a) discontinuous, (b) continuous and (c) built-up edge

Under a very rapid shear strain rate ($\approx 10^4$ s^{-1}) the edge breaks away beyond a critical size. Fragmented particles that embed in the workpiece damage surface finish and tolerance on geometry. The scheme of orthogonal machining along a *shear plane* AB is shown in Fig. 15.17. The orientation φ of the shear plane to the cutting direction is found from applying a constant volume condition to material removal:

$$l\,w\,b = l'w'b' \tag{15.61a}$$

Plane strain involves no side flow and so $w = w'$. This allows the chip's machined length l' and thickness b' to be combined with original dimensions l and b in a single ratio:

$$r_t = l'/\,l = b\,/\,b' \tag{15.61b}$$

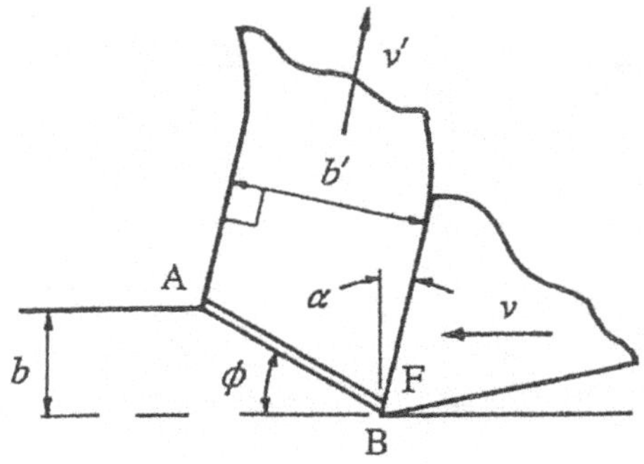

Figure 15.17 Orthogonal cutting geometry

The length of the shear plane AB is found from the geometry shown in Fig. 15.17:

$$AB = \frac{b}{\sin\varphi} = \frac{b'}{\cos(\varphi - \alpha)} \tag{15.62}$$

Combining eqs(15.61b) and (15.62) gives

$$\tan\varphi = \frac{r_t \cos\alpha}{1 - r_t \sin\alpha} \tag{15.63a}$$

The most accurate way to find r_t is from chip weight measurement. From eqs(15.61a,b):

$$r_t = \frac{\rho l' b w}{\rho l b w} = \frac{\rho\, b w}{\text{chip weight/unit length}} \tag{15.63b}$$

where ρ is the workpiece density, b defines the depth of cut and w is the tool width.

15.8.1 Merchant's Circle

During machining an equilibrium condition is maintained between the tool force and the workpiece reaction R. Following Merchant [13], these forces are taken to be co-linear so that the resolved components of R form the three force systems shown in Figs 15.18.

(a) Parallel and Perpendicular to the Rake Face:

$$F = R\sin\beta \tag{15.64a}$$

$$N = R\cos\beta \tag{15.64b}$$

(b) Parallel and Perpendicular to the Shear Plane:

$$F_S = R\cos(\varphi + \beta - \alpha) \tag{15.65a}$$

$$F_N = R\sin(\varphi + \beta - \alpha) \tag{15.65b}$$

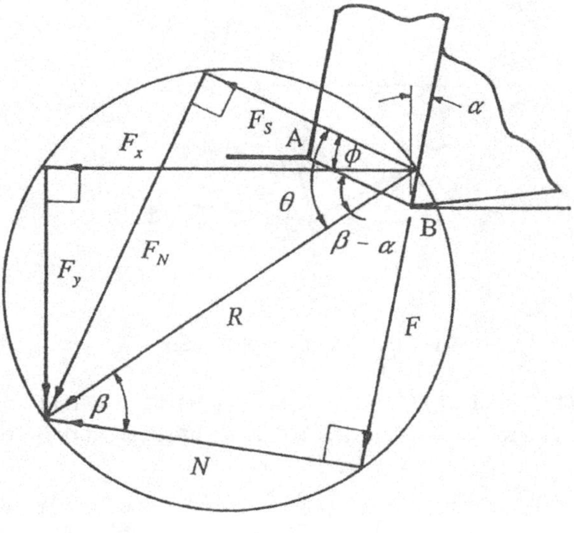

Figure 15.18 Resolved reaction forces showing Merchant's circle

(c) Parallel and Perpendicular to the Tool Motion:

$$F_x = R\cos(\beta - \alpha) \tag{15.66a}$$

$$F_y = R\sin(\beta - \alpha) \tag{15.66b}$$

Forces in eqs(15.66a,b) are measurable with a *dynanometer* and may be used to determine R and the other force components in eqs(15.64a,b) and (15.65a,b). Since R is common to each expression it is identified with the diameter of a *Merchant's circle* that superimposes the three force systems (a), (b) and (c), as shown in Fig. 15.18. The conversion of F_S and F_N to shear plane stress components τ_s and σ_s, respectively, requires the division of these forces by the surface area of the shear plane, i.e. $wb/\sin\varphi$. This gives

$$\begin{aligned}\tau_s &= \frac{F_S}{wb/\sin\varphi} = \frac{R}{wb}\cos(\varphi + \beta - \alpha)\sin\varphi \\ &= \frac{F_x}{wb}\sec(\beta - \alpha)\cos(\varphi + \beta - \alpha)\sin\varphi\end{aligned} \tag{15.67a}$$

$$\begin{aligned}\sigma_s &= \frac{F_N}{wb/\sin\varphi} = \frac{R}{wb}\sin(\varphi + \beta - \alpha)\sin\varphi \\ &= \frac{F_x}{wb}\sec(\beta - \alpha)\sin(\varphi + \beta - \alpha)\sin\varphi\end{aligned} \tag{15.67b}$$

in which R was eliminated from eqs(15.66a,b). At the tool tip, the shear zone has a finite thickness, i.e. length FG, within the enlarged scale of Fig. 15.19. Let a particle displace from F to F′ with shear distortion as it crosses this plane. This defines the shear strain γ as

$$\gamma = \frac{BF'}{FG} = \frac{BG}{FG} + \frac{GF'}{FG} = \tan(\varphi - \alpha) + \cot\varphi \tag{15.68}$$

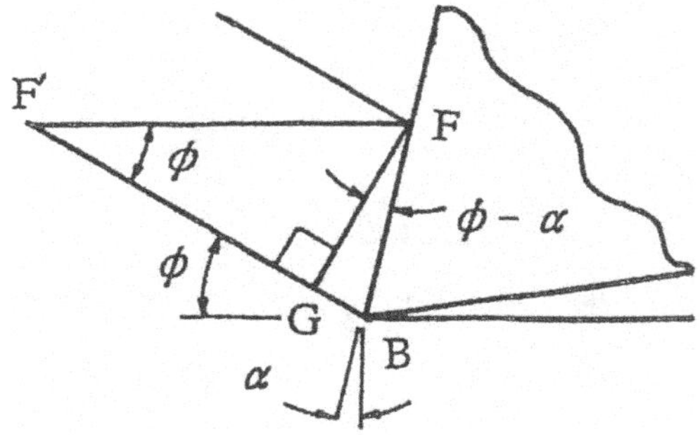

Figure 15.19 Shear zone displacement

The product of eqs(15.67a) and (15.68) provides the work done/unit volume of sheared material. With a cutting tool velocity v (m/s), the rate of shear work becomes

$$\dot{W}_s = \tau_S \gamma \times wbv = \frac{F_S}{(wb/\sin\varphi)} \gamma \times wbv = F_S \gamma v \sin\varphi \tag{15.69a}$$

In addition, work is done to overcome friction along the rake face. In force system (a) above, frictional work is the product of the force F and the chip velocity v'

$$\dot{W}_f = F \times v' = F \times r_t v \tag{15.69b}$$

where r_t in eq(15.61b) also expresses the ratio of cut to uncut lengths in a given time, i.e. the velocity ratio $r_t = b/b' = v'/v$ (Fig. 15.17). Adding eqs(15.69a,b) gives the total work rate

$$\dot{W} = \dot{W}_s + \dot{W}_f = (F_S \gamma \sin\varphi + F r_t) v \tag{15.70}$$

15.8.2 Shear Plane Orientation

(a) Maximum Shear Stress

The previous analyses assume that the shear angle φ is known. Equations (15.63a,b) allow φ to be found experimentally. Alternatively, a purely theoretical estimate for φ may be found from assuming that τ_s is the maximum shear stress within the shear band. From eq(15.67a), the condition that τ_s is a maximum is given by

$$\begin{aligned} d\tau_s/d\varphi &= \cos(\varphi + \beta - \alpha)\cos\varphi - \sin(\varphi + \beta - \alpha)\sin\varphi = 0 \\ \tan\varphi &= \cot(\varphi + \beta - \alpha) = \tan\left[\pi/2 - (\varphi + \beta - \alpha)\right] \\ \therefore \quad \varphi &= \pi/4 - (\beta - \alpha)/2 \end{aligned} \tag{15.71a}$$

So, for a given β, the effect of increasing α and φ will lower both γ and $\dot{W}_s$. Conversely, for a given α and increasing β, φ is lowered while both γ and $\dot{W}_s$ are raised. Equation (15.71a) implies: (i) the shear plane is not material dependent and (ii) a friction coefficient μ applies to the force system (a) above, as follows:

$$\mu = \frac{F}{N} = \tan\beta \Rightarrow \beta = \tan^{-1}\mu = \tan^{-1}\frac{F}{N} \tag{15.71b}$$

Equation (15.71a) describes a line with a gradient – ½ and an intercept of $\pi/4$ in axes of φ versus $\beta - \alpha$ (see Fig. 15.20).

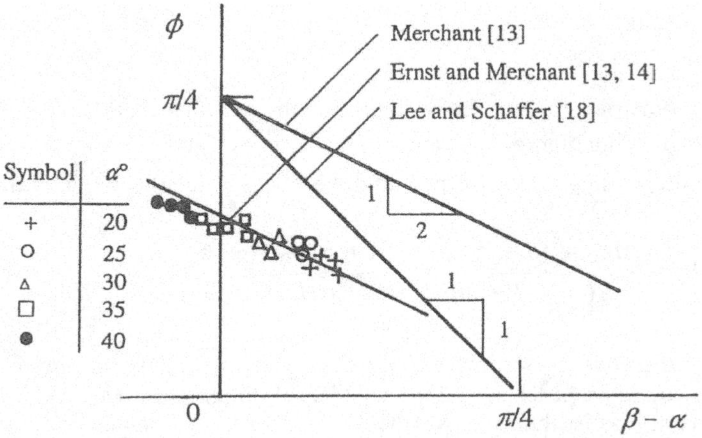

Figure 15.20 Theoretical predictions to shear angle φ. Data from Eggleston et al. [15]

Because experiment has not confirmed the prediction of φ from eq(15.71a), the theory has been modified in various ways. Ernst [14] and Merchant [13] postulated that τ_s and σ_s within eqs(15.67a,b) were dependent linearly:

$$\tau_s = \tau_o + K\sigma_s \tag{15.72a}$$

where K is a constant and $\tau_s = \tau_o$ for $\sigma_s = 0$ within eqs(15.67a,b). Substituting this condition into eq(15.72a) and differentiating for a maximum shear stress as before leads to a modified straight-line relationship

$$\varphi = \frac{1}{2}\cot^{-1}K - \frac{1}{2}(\beta - \alpha) \tag{15.72b}$$

from which the intercept $\frac{1}{2}\cot^{-1}K$ can match any observed value different from $\pi/4$. Data for machining brass with a range of rake angles from 20°–40° [15] is given as an example in Fig. 15.20. These conform quite well to eq(15.72b) when $K = 37°$. However, experiment [16] has shown that the shear yield stress is not as sensitive to a superimposed hydrostatic stress as implied by eqs(15.72a,b). Pugh [17] preferred to express observed data empirically through a linear equation

$$\varphi + s(\beta - \alpha) = c$$

where s and c were found to be dependent upon the workpiece material, e.g. for an aluminium alloy, $s = 1.5$ and $c = 25°$ were found.

(b) Minimum Work

An alternative approach to locating φ minimises the total work done in machining. This requires that the two work rate components in eq(15.70) appear in terms of φ. Thus, forces F and F_S are written in terms of F_x from eqs(15.64a) and (15.66a) then substituted with r_t and γ from eqs(15.63b) and (15.68). This leads to a normalised work expression

$$w = \frac{\dot{W}}{F_x v} = \left[\sec(\beta - \alpha)\cos(\varphi + \beta - \alpha)\sin\varphi\right]\left[\cot\varphi + \tan(\varphi - \alpha)\right] + \sin\beta\sec(\beta - \alpha)\frac{\sin\varphi}{\cos(\varphi - \alpha)} \tag{15.73a}$$

In this purely geometric approach the two terms on the right-hand side of eq(15.73a) sum to unity for any combination of α, β and γ. Hence, $w = 1$ and a minimum $\dot{W}$ (= $F_x v$) corresponds to minimising F_x with respect to φ. From eqs(15.65a) and (15.66a):

$$F_x = \frac{F_S\cos(\beta - \alpha)}{\cos(\varphi + \beta - \alpha)} = \frac{\tau_s w b\cos(\beta - \alpha)}{\sin\varphi\cos(\varphi + \beta - \alpha)} \tag{15.73b}$$

Taking τ_s to be the constant shear yield strength for the material, the differentiation of eq(15.73b), i.e. $\mathrm{d}F_x/\mathrm{d}\varphi = 0$, leads again to eq(15.71a). In practice, there are non uniform shear and normal stress distributions upon the flank face, due to a combination of sliding and sticking. Here a minimum work condition reveals a different φ value (see example 15.8).

15.8.3 Slip-Line Field

Making the usual assumptions of a rigid plastic workpiece, the existence of a shear plane and a stress-free chip, Lee and Schaffer [18] produced two slip line fields for orthogonal machining, given in Figs 15.21a,b.

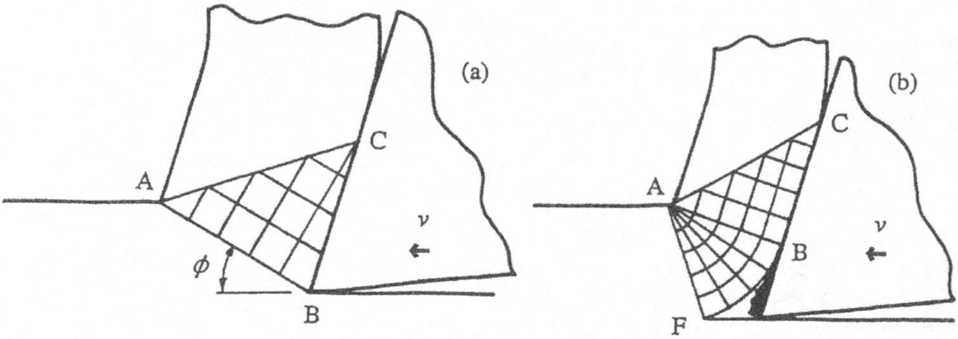

Figure 15.21 Slip-line fields for machining

In the simpler of these (Fig. 15.21a), the stress free line AC limits the field so that the slip plane AB is one of a number of orthogonal slip lines within the deformation zone that all meet AC at 45°. The zone between AB and AC moves as a rigid block. The shear plane orientation φ, revealed from a Mohr's circle construction, applies to the stress state at point B (see Fig. 15.28b):

$$\varphi = \pi/4 - (\beta - \alpha) \tag{15.74a}$$

Equation (15.74a) describes the line in Fig. 15.20 with a similar intercept to Merchant but with a gradient of − 1, this often being in better agreement with experimental data. The shear and normal stress on AB are each k, these corresponding to force components, from eqs(15.65a,b) and (15.66a,b), as follows:

$$F_S = F_N = \frac{kbw}{\sin\varphi}$$

$$F_{\substack{x(+)\\y(-)}} = kbw\,(\cot\varphi \pm 1) \tag{15.74b,c}$$

where k is the shear yield stress. Equations (15.74b,c) are invalid when $\beta - \alpha = \pi/4$ (since $\cot 0 = \infty$), a condition associated with the formation of a built-up edge. In the corresponding slip line field solution (see Fig. 15.21b) two shear planes AB and AF contain a continuous deformation zone ABF ahead of the rigid motion of block ABC. While this field matches conditions for an edge to form, strictly, it is not an appropriate field for materials that violate perfect plastic behaviour under very rapid rates of machining.

15.8.4 Work Hardening

Palmer and Oxley [19] added further integral tems $\int(\partial k/\partial\beta)\mathrm{d}\alpha$ and $\int(\partial k/\partial\alpha)\mathrm{d}\beta$ respectively to Hencky's eqs(6.8a,b) to allow for the influence of hardening along an orthogonal pair of α, β slip lines. These *modified Hencky equations* confirm a fan of slip lines within the shear zone and a consequent chip curvature. To simplify the analysis they extended the single shear zone either side of AB using parallel lines CD and EF, as shown in Fig. 15.22a.

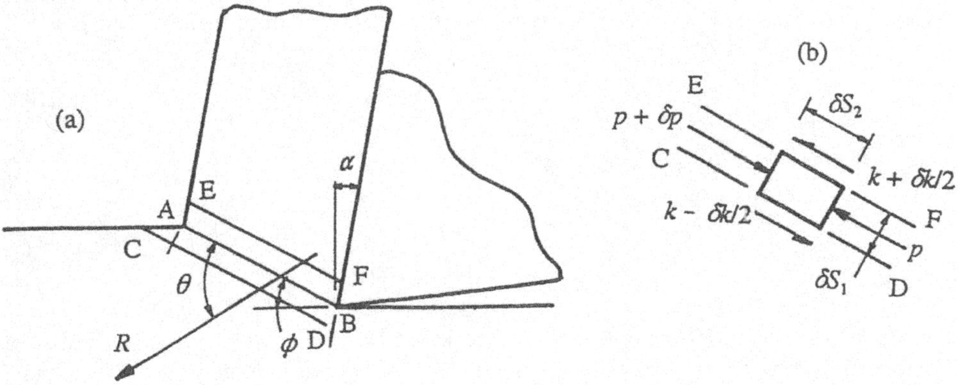

Figure 15.22 Simple approximation to a shear zone

Across an element $\delta S_1 \times \delta S_2$ within this zone (Fig. 15.22b), the shear and hydrostatic stress increase by amounts δk and δp. The following equilibrium equation applies along AB:

$$(p + \delta p)\delta S_1 + (k - \delta k/2)\delta S_2 = p\delta S_1 + (k + \delta k/2)\delta S_2$$

$$\delta p = \frac{\delta k}{\delta S_1}\delta S_2 \quad \Rightarrow \quad p = \int \frac{\partial k}{\partial S_1}\mathrm{d}S_2 + C_{S_2} \tag{15.75a,b}$$

This hydrostatic pressure p also lies normal to AB. C_{S_2} is an integration constant pertaining to a known pressure along S_2. Taking the partial derivative to be constant between parallel lines CD and EF, eq(15.75b) shows, for an integration with respect to S_2, that p increases in proportion to the length of AB. Since the area of shear plane AB is $bw/\sin\varphi$, the average normal and shear forces follow from stresses p and k:

$$F_N = \frac{(p_A + p_B) b w}{2 \sin \varphi}$$
$$F_S = \frac{k b w}{\sin \varphi} \tag{15.76a,b}$$

The inclination θ of the resultant force R to the shear plane in Fig. 15.22a follows from eqs(15.76a,b):

$$\tan \theta = \frac{F_N}{F_S} = \frac{p_A + p_B}{2k} \tag{15.77a}$$

where θ and φ are related to α and β in Fig. 15.18:

$$\theta = \varphi + \beta - \alpha \tag{15.77b}$$

Further relationships involving θ, α and φ are required, as in eq(15.76b), to enable each of these orientations to be separated.

The Lee and Schaffer solution [18] corresponds to setting $p_A = p_B$ with $\theta = 45°$. This implies that AB lies in the direction of maximum shear so that R aligns with the maximum principal stress direction. Otherwise, the pressure at B is known from applying the modified Hencky equations to the rotation $\pi/4 - \varphi$ at point C. Knowing that one principal stress is zero at C, the shear stress must meet the free surface at 45° (see Fig. 15.23). This gives

$$p_A = k \left[1 + 2 \left(\frac{\pi}{4} - \varphi \right) \right] \tag{15.78a}$$

Applying eq(15.75a), the pressure at B becomes

$$p_B = p_A - \left(\frac{\partial k}{\partial S_1} \right) \frac{b}{\sin \varphi} \tag{15.78b}$$

and, since $\partial k/\partial S_1$ is a constant between parallel lines CD and EF, the variation in p between A and B becomes linear. Figure 15.23 shows that two hydrostatic pressure distributions along AB are possible when the pressure ordinates at A and B are defined by eqs(15.78a,b).

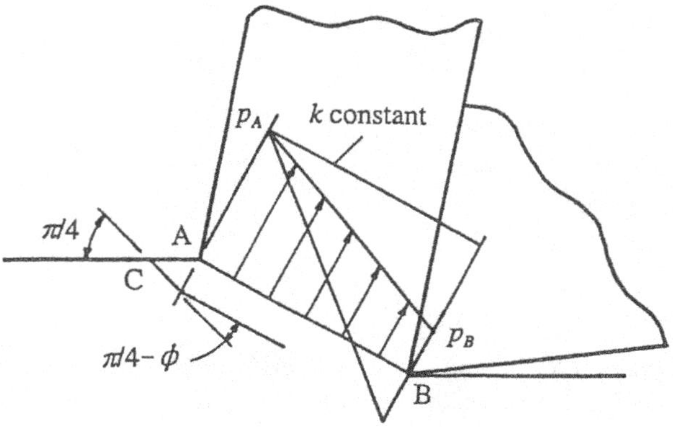

Figure 15.23 Hydrostatic pressure distribution along AB

At B, the pressure may become tensile if the rate of hardening $\partial k/\partial S$ across the strip is sufficiently large. Here, a discontinuous chip is likely when θ and φ are low within eq(15.77b), given that $\beta - \alpha$ is reasonably constant. The amount of hardening will depend upon the material, the shear rate and the temperature arising from interface friction.

15.8.5 Shear Strain and Strain Rate

The shear strain across a narrow deformation band for machining has been previously defined in eq(15.68). Figure 15.24 shows a hodograph for orthogonal machining. The orientations α and φ appear in the physical plane (Fig. 15.24a) for which v is the cutting speed and v' is the chip velocity.

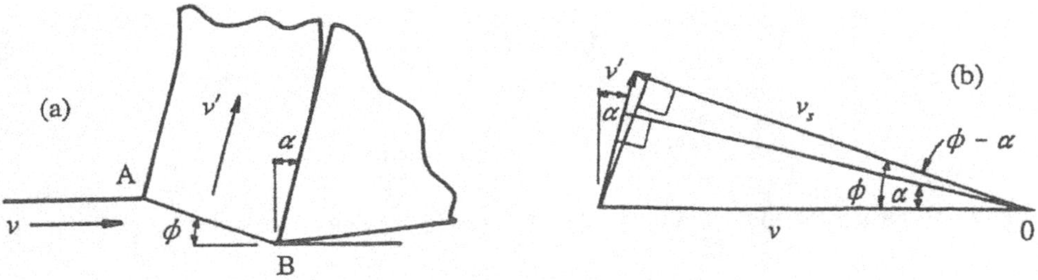

Figure 15.24 Hodograph for orthogonal machining

In the hodograph (Fig. 15.24b), v_s is the velocity along the shear plane AB. This velocity is found by ensuring continuity of normal and tangential velocities across the rake face:

$$v \cos \alpha = v_s \cos (\varphi - \alpha) \tag{15.79a}$$

$$v \sin \varphi = v' \cos (\varphi - \alpha) \tag{15.79b}$$

Equations (15.79a,b) supply the chip velocity and shear strain. Figure 15.24b expresses γ in terms of the velocities within eq(15.79a) as

$$\gamma = \frac{v_s}{v \sin\varphi} = \frac{\cos\alpha}{\sin\varphi \cos(\varphi - \alpha)} \tag{15.80a}$$

The reader should confirm, from the previous derivation of shear strain for machining, that eqs(15.80a) and (15.68) are identical. The mean shear strain rate for AB is defined as

$$\dot{\gamma} = \frac{v_s}{\delta S_1} = \frac{v}{\delta S_1} \times \frac{\cos\alpha}{\cos(\varphi - \alpha)} \tag{15.80b}$$

where δS_1 is the perpendicular distance between EF and CD (see Fig. 15.22a).

Example 15.8 The conditions for cutting a work hardening material are: rake angle $\alpha = 33°$, cutting speed $v = 3.75$ m/s, depth of cut $b = 0.1$ mm and width of cut $w = 4.3$ mm. Assume an experimental value for the shear angle $\varphi = 25.5°$ for a rectangular shear zone of aspect ratio 10:1. Calculate the shear strain, the shear strain rate, the friction angle β and the cutting forces F_x and F_y. The initial shear yield stress k and the linear hardening gradient $m = dk/d\gamma$ depend upon the mean shear strain rate in the manner of Fig. 15.25.

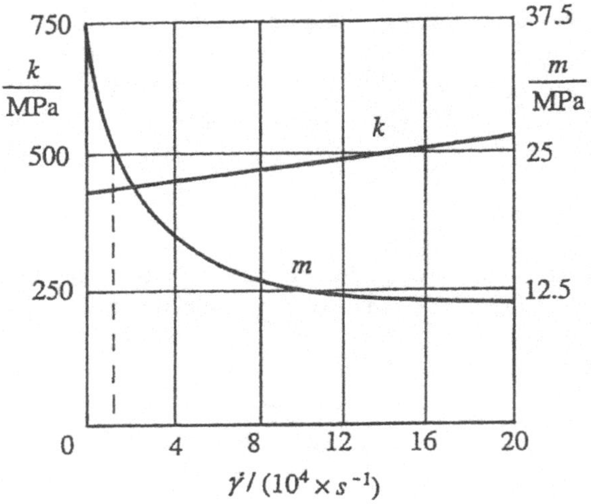

Figure 15.25 Flow rate dependent properties k and m

Firstly, the shear zone thickness is estimated from Fig. 15.22a as

$$\frac{b}{\delta S_1 \sin\varphi} = 10, \;\Rightarrow\; \delta S_1 = \frac{b}{10\sin\varphi} = \frac{0.2}{10 \sin 25.5^\circ} = 0.023 \text{ mm}$$

Equations (15.80a,b) provide the shear strain and shear strain rate:

$$\gamma = \frac{\cos\alpha}{\sin\varphi \cos(\varphi - \alpha)} = \frac{\cos 33^\circ}{\sin 25.5^\circ \cos(25.5^\circ - 33^\circ)} = 1.965$$

$$\dot{\gamma} = \frac{v\cos\alpha}{\delta S_1 \cos(\varphi - \alpha)} = \frac{(3.75 \times 10^3)\cos 33^\circ}{0.023 \cos(25.5^\circ - 33^\circ)} = 138 \times 10^3 \, s^{-1}$$

From Fig. 15.25, the initial yield stress on plane CD (Fig. 15.22b) and the work hardening gradient are

$$k - \frac{\delta k}{2} = 435.4 \text{ MPa} \quad \text{and} \quad m = 25.2 \text{ MPa}$$

Since $\delta k = m\gamma$, the shear stress k upon plane AB is found from

$$k - \frac{m\gamma}{2} = 435.4 \;\Rightarrow\; k = 435.4 + \frac{(25.2 \times 1.965)}{2} = 460.2 \text{ MPa}$$

The normal pressures upon the slip plane at end points A and B follow from eqs(15.78a,b):

$$\frac{p_A}{k} = 1 + 2\left(\frac{\pi}{4} - \varphi\right) = 1 + 2\left(\frac{\pi}{4} - \frac{(25.5 \times \pi)}{180}\right) = 1.681$$

$$\frac{p_B}{k} = \frac{p_A}{k} - \frac{\delta k}{k} \times \frac{b}{\delta S_1 \sin\varphi} = 1.681 - \frac{10 \times 49.6}{460.2} = 0.603$$

Then, from eqs(15.77a,b):

$$\tan\theta = \frac{p_A + p_B}{2k} = \frac{1.681 + 0.603}{2} = 1.142 \quad \Rightarrow \quad \theta = 48.79°$$

$$\beta - \alpha = \theta - \varphi = 48.79 - 25.50 = 23.29°$$

from which the friction angle $\beta = 56.29°$ and the shear angle $\varphi = 25.5°$ is confirmed. Finally, the cutting forces F_x and F_y follow from eliminating R within Merchant's two force systems, eqs(15.65a,b) and (15.66a,b). Setting $F_S = kbw/\sin\varphi$, these give

$$F_x = \frac{kwb\cos(\beta - \alpha)}{\sin\varphi\cos(\varphi + \beta - \alpha)} = \frac{460.2 \times 4.3 \times 0.1\cos(56.29° - 33°)}{\sin 25.5°\cos(25.5° + 56.29° - 33°)} = 640.8 \text{ N}$$

$$F_y = \frac{kwb\sin(\beta - \alpha)}{\sin\varphi\cos(\varphi + \beta - \alpha)} = \frac{460.2 \times 4.3 \times 0.1\sin(56.29° - 33°)}{\sin 25.5°\cos(25.5° + 56.29° - 33°)} = 275.9 \text{ N}$$

Note: From this example, if b is reduced so too is ΔS_1. Consequently, $\dot{\gamma}$ and k are raised and with them p_A and p_B are increased. This implies that a *size-effect* exists in machining.

15.8.6 Friction

In an account of friction upon the rake face, eq(15.71b) shows that we may either employ the friction coefficient $\mu = F/N$, i.e. the ratio between shear and normal forces, or the friction angle $\beta = \tan^{-1}(F/N)$. It is assumed that μ, or β, is independent of (i) the contacting area between the rake face and the chip and (ii) the chip speed v'. Both forces F and N are the resultant of the respective shear and normal stress distributions (τ, σ) along the rake face.

(a) Empirical Approach

The normal stress σ reaches a maximum at B and falls to zero at the free surface. Between these Zorev [20] assumed a parabolic σ distribution. The shear force was assumed to be spread over two zones: (i) sticking, where $\tau = \tau_B =$ constant, within the nose region and (ii) sliding, where μ is constant but τ decreases to zero at the free surface (see Fig. 15.26).

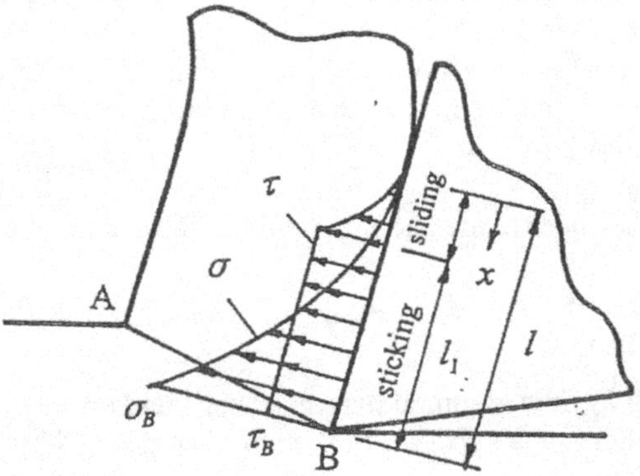

Figure 15.26 Rake face normal and shear stress distributions

With the x–co-ordinate as shown, the following stress ratio holds for $l \le x \le 0$:

$$\sigma = p x^{q}, \qquad \sigma_B = p l^{q}, \qquad \frac{\sigma}{\sigma_B} = \left(\frac{x}{l}\right)^{q} \tag{15.81a,b,c}$$

where p and q are constants. For $0 \le x \le (l - l_1)$, the sliding shear stress becomes

$$\tau = \mu\sigma = \mu\sigma_B\left(\frac{x}{l}\right)^{q} \tag{15.82a}$$

For $(l - l_1) \le x \le l$, the constant, sticking shear stress at B is found from eq(15.82a):

$$\tau = \tau_B = \mu\sigma_B\left(\frac{l - l_1}{l}\right)^{q} \tag{15.82b}$$

where μ in eqs(15.82a,b) is the constant friction coefficient for the sliding zone. Integrating eqs(15.81c) and (15.82a,b) with respect to flank area $w\delta x$ gives N and F:

$$N = w\int_0^l \sigma\,\mathrm{d}x = \frac{\sigma_B l w}{1 + q}$$

$$F = w\int_0^l \tau\,\mathrm{d}x = \tau_B w\left[l_1 + \frac{l - l_1}{1 + q}\right] \tag{15.83a,b}$$

Setting $\tau_B = k$, where k is the shear yield strength, then dividing eqs(15.83a,b) leads to a simplified friction angle

$$\beta = \tan^{-1}\left[\frac{k}{\sigma_B}\left(1 + \frac{q\,l_1}{l}\right)\right] \tag{15.84a}$$

With no sliding set $l_1 = l$ in eq(15.84a) and assume k remains constant in sticking to give

$$\beta = \tan^{-1}\left[\frac{k(1 + q)}{\sigma_B}\right] = \tan^{-1}\left(\frac{k}{\sigma_m}\right) \tag{15.84b}$$

where, from eq(15.83a), the mean normal stress is $\sigma_m = N/(wl)$. Under these conditions eq(15.84b) shows that the friction angle depends solely upon the mean normal stress.

Example 15.9 Using Zorev's analysis of the flank face stresses, determine the orientation of the shear plane, based upon a minimum work condition in a non-hardening material, when: $\alpha = 5°$, $l_1/l = 1/2$, $q = 2$ for $l/b = 1, 2, 3 \ldots 6$.

Let subscript s refer to the shear plane and f to the flank face. The total work rate is simply

$$\dot{W}_t = F_s v_s + F_f v_f \tag{i}$$

To find the forces F_s and F_f, it is assumed that the shear yield stress k remains constant all along the shear plane and along a portion of the flank face defined by l_1 in Fig. 15.26. The two velocity components v_s and $v_f = v'$ are found from the hodograph in Fig. 15.24b. Thus, the two work terms in eq(i) appear as

$$F_s v_s = \frac{kwb}{\sin\varphi} \times \frac{v\cos\alpha}{\cos(\varphi - \alpha)} \tag{ii}$$

$$F_f v_f = \left[kwl_1 + \frac{kw}{(1+q)}(l - l_1) \right] \frac{v\sin\varphi}{\cos(\varphi - \alpha)} \tag{iii}$$

Substituting eqs(ii) and (iii) into eq(i) leads to the normalised work rate expression

$$\frac{\dot{W}_t}{kwbv} = \frac{\cos\alpha}{\sin\varphi\cos(\varphi - \alpha)} + \frac{l}{b}\left[\frac{l_1}{b} + \frac{(1 - l_1/l)}{(1+q)} \right] \frac{\sin\varphi}{\cos(\varphi - \alpha)} \tag{iv}$$

Setting $\alpha = 5°$ and $q = 2$ with $l_1/l = 1/2$ for $l/b = 1, 2, 3 \ldots 6$ enables plots from eq(iv) that show work rate versus φ, as shown in Fig. 15.27.

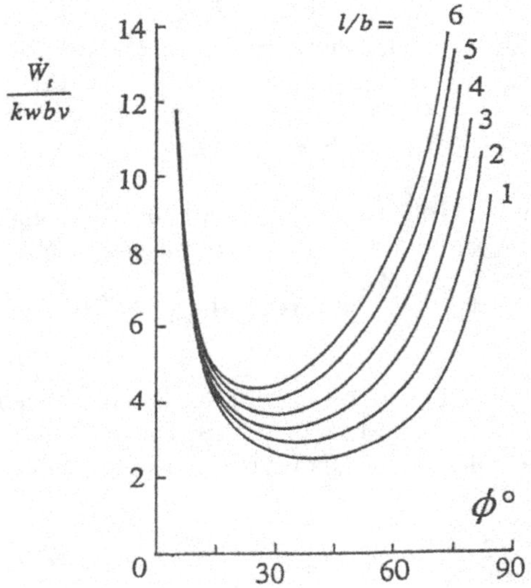

Figure 15.27 Minimum work condition

With an increasing orientation φ of the shear plane, the work rate decreases within the first term but increases for the flank face in the second term. The result is that their sum, i.e. the normalised total work rate, attains a minimum value under these cutting conditions. As the chip length to thickness ratio increases from 1 to 6, the shear angle φ, corresponding to the minimum work condition, diminishes successively from 40° to 25° as shown. This analysis is in qualitative agreement with the many factors known to influence the work rate under practical cutting conditions [21].

(b) Analytical Approach

Following [19], the stress states (σ, τ) for the rake face and the shear plane are related by the equations of stress transformation at the tool tip. Figure 15.28a shows the rake face stress state at point B as σ_B and τ_B. Correspondingly, upon the shear plane AB, the normal pressure is p_B and the shear stress is k.

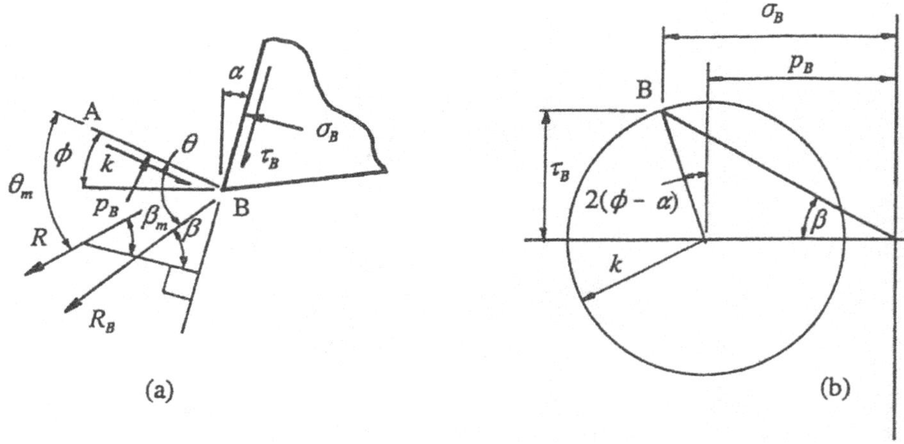

Figure 15.28 Mohr's circle for point B at tool tip

For orientations α and φ of the two planes shown in Fig. 15.28a, the Mohr's circle is given in Fig. 15.28b. The geometry of this circle reveals

$$[\,p_B + k\sin 2(\varphi - \alpha)\,]\tan\beta_B = k\cos 2(\varphi - \alpha) \tag{15.85a}$$

Combining eqs(15.77a) and (15.78a) and eliminating p_B from eq(15.85a) gives the orientation θ of the resultant force R_B at B relative to the shear plane AB (see Fig. 15.28a)

$$\theta = \tan^{-1}\left[\frac{1}{2} + \frac{\pi}{4} - \varphi + \frac{\cos 2(\varphi - \alpha)}{2\tan\beta} - \frac{\sin 2(\varphi - \alpha)}{2}\right] \tag{15.85b}$$

where β applies to point B only. Equation (15.85b) needs a modification to include the mean values θ_m and β_m across the whole shear plane AB. The distributions in σ and τ along the rake face are taken to be similar to those shown in Fig. 15.26. These provide the average normal and shear forces:

$$N = \sigma_B l\,w/2$$

$$F = \tau_B l_1 w + \frac{\tau_B w}{2}\left(l - l_1\right) = \frac{\tau_B w}{2}\left(l + l_1\right) \tag{15.86a,b}$$

Dividing eqs(15.86a,b) gives the mean friction angle for the flank face:

$$\tan\beta_m = \frac{F}{N} = \frac{\tau_B}{\sigma_B}\left(1 + \frac{l_1}{l}\right) = \tan\beta\left(1 + \frac{l_1}{l}\right) \tag{15.87a}$$

Substituting $\tan\beta$ from eq(15.87a) into eq(15.85b) allows the orientation θ_m of the resultant force $R = \sqrt{(N^2 + F^2)}$ to be found:

$$\theta_m = \tan^{-1}\left[\frac{1}{2} + \frac{\pi}{4} - \varphi + \frac{(1 + l_1/l)\cos 2(\varphi - \alpha)}{2\tan\beta_m} - \frac{\sin 2(\varphi - \alpha)}{2}\right] \tag{15.87b}$$

where F and N are measured forces. Taken with eq(15.77b), the shear angle φ may be calculated from eq(15.87b) for a given rake angle α and an assumed ratio $l_1/l < 1$.

15.9 Concluding Remarks

Various analyses of metal shaping processes are available. The simplest work conservation formulae, as used for extrusion and wire drawing, apply only to homogenous deformation. They ignore contributions to work done from friction and redundant shearing. Consequently, the applied forces required to extrude and draw are underestimated. The strip equilibrium method assumes a constant friction coefficient in non-hardening material. This theory allows corrections based upon mean rates and stresses to be made when a process work hardens a material. The convenient closed-form solutions afforded by these analyses allow rapid estimates of the forces involved compared, say, to the slip line field method. The latter is exact in the case of plane strain plasticity involving a plastic rigid material. The library of SLF solutions (see chapter 6) applies mainly to forming by extrusion and cogging (repeated indentation). Orthogonal machining is predominantly a plastic shearing process but is complicated by rapid variations in strain rates and temperature gradients. Simplifying assumptions allow for estimates of the cutting forces, the shear angle, work hardening and friction. In practice, these calculations are used to offset adverse cutting conditions that lead to unacceptable rates of wear and poor surface finish.

References

1. Wistreich J. G. and Shutt A. *Jl. Iron and Steel Inst.*, 1959, **6**, 132.
2. Bishop J. F. W. *Jl. Mech. Phys. and Solids*, 1958, **6**, 132-144.
3. Hirst S. and Ursell D. H. Proc.: *Tech. of Eng. Manuf.*, 1958, p. 58, I. Mech. E., London.
4. Larke E. C. *The Rolling of Strip, Sheet and Plate*, 1957, Chapman and Hall, London.
5. Orowan E. *Proc. I. Mech. E.,* 1943, **150**, 140.
6. Sims R. B. *Proc. I. Mech. E.*, 1954, **168**, 191.
7. Orowan E. and Pascoe K. J. *Iron and Steel Inst. Rpt.*, 1946, **34**, 124.
8. Bland D. R. and Ford H. *Proc. I. Mech. E.*, 1948, **159**, 144.
9. Bland D. R. and Ford H. *Jl. Iron and Steel Inst.*, 1952, **171**, 245.
10. Hitchcock J. *Roll Neck Bearings*, 1935, ASME Res Pubs.
11. Avitzur B. *Metal Forming: Processes and Analysis*, 1968, McGraw-Hill.
12. Wistreich J. G. *Proc. I. Mech. E.*, 1955, **169**, 659.
13. Merchant M. E. *Jl. Appl. Physics*, 1945, **16**, 267.
14. Ernst H. *Annals New York Acad. Sci.*, 1951, **53**, 936.
15. Eggleston D. M., Herzog R. P. and Thomsen E. G. *Trans ASME* (*B*), 1959, **81**, 263.
16. Crossland B. *Proc. I. Mech. E.*, 1954, **168**, 935.
17. Pugh H. Ll. D. *Jl. Mech. Eng. Sci.*, 1964, **6**, 4.
18. Lee E. H. and Schaffer B. W. *Jl. Appl. Mech.*, 1951, **18**, 405.
19. Palmer W. B. and Oxley P. L. B. *Proc. I. Mech. E.*, 1959, **173**, 623.
20. Zorev N. N. Proc.: *Int. Conf. Prod. Eng. Res.*, 1963, p. 43, Pittsburg.
21. Rowe G. W. and Spick P. T. *Trans ASME*, 1967, **89B**, 530.

Exercises

15.1 Tresca's yield criterion simplifies the 'friction hill' pressure distribution, required for plane strain forging of thin strip. The distribution is expressed by eq(15.10c) and Fig. 15.4a as

$$p = 2k\mathrm{e}^{\frac{\mu b}{t}\left(1 - \frac{2x}{b}\right)}$$

where $Y = 2k$ is the constant plane strain yield stress, μ is the frictional coefficient, b is the breadth and t is the thickness beneath the dies. Determine the forging force given $\mu = 0.2$ and the forged strip dimensions: $b = 80$ mm and $t = 20$ mm for a strip width $w = 300$ mm lying perpendicular to b.
See solution

15.2 A piece of lead 25 mm × 25 mm × 150 mm, with a yield stress 7 MPa, is forged between flat dies to a final approximate size: 6 mm × 104 mm × 150 mm. Assume that a friction coefficient $\mu = 0.25$ prevails with sliding between the lead and the full width of the dies in the contact region. Determine the pressure distribution that the dies exert upon the lead and the total forging force.
See solution

15.3 Confirm that the mean pressure for the friction hill shown in Fig. 15.4b is

$$\frac{\bar{p}}{2k} = \frac{t}{\mu b}\left\{\frac{1}{4\mu}\left[(1 + \alpha)^2 + 1\right] - 1\right\}$$

where $\alpha = \mu b/t - \ln(1/2\mu)$.

15.4 Compare the work required to hot forge a 0.75 m long billet, with 150 mm square cross-section, into a strip 50 mm thick, 150 mm wide and 2.25 m long, using dies of breadth (a) 50 mm, (b) 100 mm and (c) 150 mm. Hence, show that the minimum work is done when the die breadth lies between (a) and (b). Neglect work hardening and spread and assume that small reductions are made so that the existence of the transitional shape at the edges of the die may be neglected. Take the load L at any thickness to be given by $L = YAC$ in which Y is the yield stress. The contact area is $A = wb$, where b is the die width and w the billet width. For this cogging process take the constant C as
See solution

$$C = 0.797 + 0.203t/b \text{ for } b/t < 1$$

$$C = 0.75 + 0.25b/t \text{ for } b/t > 1$$

15.5 In forging the ring shown in Fig. 15.29 a plastic hinge is formed opposite the forging platens. The shear yield stress of the ring is constant at k, its mean radius is R and the other dimensions are as shown. Prove that the additional pressure on the platens required to form the plastic hinge is approximately $p' = kt/4R$, given that the plastic hinge moment is $kwt^2/2$. The final dimensions of the ring are to be: $t = 125$ mm, $R = 1.075$ m and $w = 175$ mm for $k = 40$ MPa. Given that a maximum force of 1.5 MN is available, determine the maximum internal diameter of the ring at which forging can begin. Assume plane strain conditions and take the total pressure upon the platens to be

$$p = 2k[C + t/(8R)] \text{ where } C = 0.797 + 0.203t/b.$$

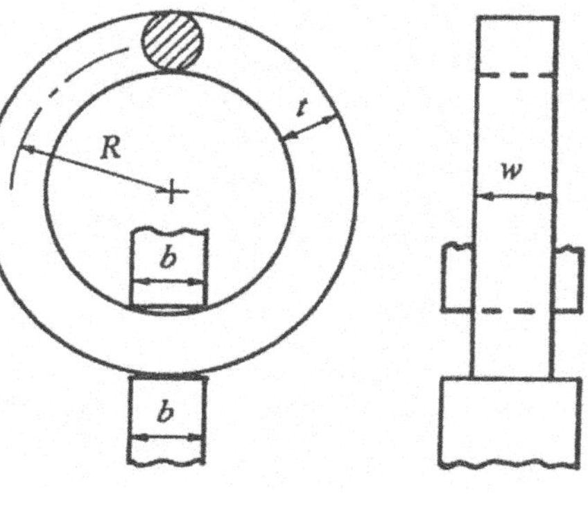

Figure 15.29

15.6 What drawing stress is required to reduce a rod from 100 mm to 70 mm in a conical die with 30° included angle? Assume that the friction coefficient is 0.05 and that the material is rigid-perfectly plastic with a tensile yield stress of 160 MPa. What is the maximum reduction that can be achieved under these conditions?
Ans: 127 MPa

15.7 A cylindrical component of original height $h_o = 50$ mm is forged between flat platens to a height $h_1 = 40$ mm. The lower platen is rigid whilst the upper one descends with a velocity $v_o = 3$ m/s. Show that the true mean strain rate for the process is given by the following expression and from this calculate its magnitude.

$$\dot{\varepsilon} = -\frac{v_o \ln(h_o/h_1)}{2(h_o - h_1)}$$

15.8 Determine the pressure on a ram required to reduce a billet from 40 mm diameter to 37 mm diameter in a homogenous extrusion process. The 40 mm diameter billet is the result of a previous extrusion where the bar section area was reduced by 15 % for annealed material obeying a Hollomon flow law $\sigma = 670\varepsilon^{0.5}$. Assume that the ram force for simple extrusion is given by
See solution

$$F = Y_m A_r \ln\left(\frac{A_r}{A_e}\right)$$

where Y_m is the mean compressive yield stress, A_r and A_e are the cross sectional areas for the ram and the extrusion respectively.

15.9 In an extrusion operation it is required, within a second reduction, to reduce the diameter of a bar from 150 mm to 140 mm. This bar's section area at a diameter of 150 mm had already been reduced by 20 % in a previous reduction. The stress versus strain curve for the original annealed material is provided by a Ludwik law $\sigma = 200 + 182(\varepsilon^P)^{0.25}$, where σ is the true stress in MPa and ε^P is the natural plastic strain. Determine the extrusion forces required for the first and second reductions assuming homogeneous deformation.

15.10 The force required to reduce the section area of a bar in each pass of a drawing operation is

$$F = Y_m A_r \ln\left(\frac{A_0}{A_r}\right)$$

where A_r is the reduced area from each pass, A_o is the original area before each pass and Y_m is the mean yield stress for each pass. The first pass of a drawing operation upon a rod had reduced its diameter from 10 to 8.5 mm. For the second pass it is required to reduce the section area of the 8.5 mm rod by a further 30 %. Before any drawing the annealed material's stress versus strain curve is approximated by a Hollomon law $\sigma = 300\varepsilon^{0.5}$, where σ is the true stress and ε is the natural strain. Determine the force required for the second pass with 30 % reduction. What diameter applies to the maximum reduction available for the second pass? If the material behaved in a perfectly plastic manner what would the smallest diameter possible in a single pass?
Ans: 3125 N, 4.75 mm, 5.16 mm
See solution

15.11 Show that the maximum possible change in section area for wire drawing a rigid, perfectly plastic material is 63 %. If 3 mm diameter wire is to be drawn at maximum reduction in each of three sequential processes, what is the smallest diameter possible after the third process? Examine whether an optimum die angle exists from a consideration of friction and contact length assuming a homogenous deformation within the tapered die.

15.12 In a homogenous wire drawing operation upon hardening material, the drawing force F may be estimated as

$$F = Y_m A_1 \ln\left(\frac{A_0}{A_1}\right)$$

where A_0 is the original cross-sectional area, A_1 is area of the drawn wire and Y_m is the average yield stress for material hardened by the process. Using an average yield stress between entry and exit, determine the drawing force required to reduce a wire from 3 mm to 2.75 mm, given that a 20 % reduction in area had previously been applied in reaching the 3 mm wire size. These reductions should be referred to the annealed state for which the Hollomon flow law is $\sigma = 627\varepsilon^{0.5}$, where σ and ε are the true stress and natural strain respectively.

15.13 What drawing stress is required to draw a 500 mm wide, annealed steel strip from 2.5 to 2.4 mm thick in dies with 15° included angle and a friction coefficient 0.1? What would be the influence of a previous 20 % reduction in area arising from rolling the annealed material to 2.5 mm given that its true stress versus natural strain curve conforms to a Ludwik law: $\sigma = 200 + 100\varepsilon^{0.8}$?
See solution

15.14 What drawing stress is required to draw a 150 mm wide × 6 mm thick strip with a 20 % reduction in area? This strip had been reduced in thickness by 30 % in a previous process from its annealed condition. What drawing force would be required for a further 20 % reduction in area? The die angle is 15° and the friction coefficient is 0.05. The true stress versus strain behaviour of the annealed material shows natural (logarithmic) strains of 0.356, 0.580 and 0.803 for stress levels 350, 430 and 490 MPa respectively.
Ans: 73 kN, 69 kN

15.15 Assuming homogenous deformation under plane strain, show the roll force in sheet rolling of plastic rigid material may be approximated as

$$F = Yw\sqrt{R\Delta h}$$

where Y is the compressive yield stress, w is the sheet width, R is the roll radius and Δh is the reduction in sheet thickness. What further assumption has been made concerning the relative magnitudes of R and Δh?

How should the roll force expression given above be modified so that it may be applied to a work hardening material? The compressive flow curve for an annealed material is described by the Ludwik hardening law $\sigma = Y + B(\varepsilon^P)^n$ for $Y = 200$ MPa, $B = 310$ MPa and $n = 0.68$. Determine the roll force required to reduce a 250 mm wide sheet from 4 mm thick to 3.44 mm in two rolls each of radius 175 mm. This sheet had previously received a 10 % reduction in its cross sectional area with a pre-roll.
Ans: 729.8 kN
See solution

15.16 Orowan [5] proposed a solution to the problem of hot rolling narrow flat strip with a width/thickness ratio for which the decrease in roll pressure at the edges is not negligible (Fig. 15.11). Assuming that the neutral point lies midway between exit and entry, show that the total roll force is given by

$$P = kbL\left[0.8 + \frac{L}{4h_o} - \frac{2hh_o}{3bL}\left(\frac{l}{2h_o} - 0.2\right)^3\right]$$

where k is the plane strain yield stress, b is the mean strip width, h is the mean thickness of strip within the roll gap, h_o is the final thickness of the strip and L is the projected length over the arc of contact.
See solution

15.17 What do you understand by the term friction hill in forging and rolling operations? Outline the various frictional conditions which can occur in hot and cold working and sketch the associated friction hills. Describe the geometrical factors which affect the friction hill in such operations, particularly spread. Define the term plane strain and explain the phrase 'plane sections remain plane'. Prove, when forging under plane strain a strip of thickness t and breadth b of the flat platens less than

$$b = \frac{t}{\mu} \ln\left(\frac{1}{2\mu}\right),$$

the frictional shear stress between the strip and platens will never attain the shear yield stress of the strip material. The coefficient of friction in the contact zone is μ.

15.18 In an analysis of cold rolling thin strip both rolls and strip remain elastic. Show that the distribution of roll pressure is approximately parabolic, being given by

$$s = \frac{E_s(a^2 - x^2)}{(1 - \nu_s^2)R'h}$$

where a is the half-length of the constant arc, E_s and ν_s are the strip's elastic constants, R' is the deformed roll radius and x is the distance from the axis of symmetry. The value of R' may be derived from Hitchcock's analysis [10], namely

$$\frac{1}{R'} = \frac{1}{R} - \frac{4(1 - \nu_r^2)P}{E_r \pi a^2}$$

where R is the undeformed roll radius, P is the total roll force and E_r and ν_r are the elastic constants for the roll material. It may be assumed that at the centre of the strip, on the axis of symmetry, a longitudinal compressive stress p is induced due to the combined effect of frictional stresses acting on the surface of the strip and the mean of the back and front strip tensions $\bar{t}$ from

$$p = \frac{\mu P}{h} - \bar{t}$$

where μ is the Coulomb friction coefficient. Assuming a yield criterion $s - p = mY$, where Y is the tensile yield stress and m is a factor lying between 1 and $2/\sqrt{3}$, show that the minimum thickness which could be rolled plastically is approximately

$$h_{min} = (mY - \bar{t})\left[\frac{14.22}{E_s}\mu^2 R(1 - \nu_s^2) + \frac{9.05}{E_r}\mu R(1 - \nu_r^2)\right]$$

15.19 Metal cutting is basically a process of plastic deformation. Discuss an orthogonal cutting operation on a practical work material emphasising the role of the important geometrical, physical and metallurgical factors involved and their inter-relationships. What are the parameters of particular importance to the conduct and control of machining?

15.20 Explain the unconstrained geometry of plastic deformation in orthogonal machining. Discuss the limitations of the friction angle expression $\tan\beta = F/N$ when applied to work/tool interface friction in machining. Given that the shear strain γ depends upon the shear angle φ and the rake angle α in the form $\gamma = \cot\varphi + \tan(\varphi - \alpha)$, show that there will be a greater tendency for the chip to break away when the rake angle is negative rather than positive.

15.21 Discuss briefly the formulation and implementation of two shear plane models of the orthogonal metal cutting process indicating their limitations. Explain how a model which admits work hardening of the workpiece provides a more realistic explanation of machining behaviour in practice.

15.22 The conditions when orthogonal machining a work hardening material are: rake angle $\alpha = 10°$, cutting speed $v = 0.5$ m/s and depth of cut $b = 0.2$ mm. Assume an experimental value $\varphi = 25°$ for the shear angle with a rectangular shear zone of aspect ratio 10:1. Calculate the shear strain, the shear strain rate and the inclinations of the resultant force R to the flank face and to the shear plane. Use the flow properties for the material given in Fig. 15.25.

See solution

15.23 A work hardening material is machined with rake angle 8°, cutting speed 1.8 m/s, depth of cut 0.25 mm, width of cut 6.5 mm and chip thickness 0.4 mm. Assume a rectangular shear zone of aspect ratio 10:1. Calculate the shear strain γ, the shear strain rate $\dot{\gamma}$, the friction angle β and the two cutting forces F_x and F_y. The dependence of the initial shear yield stress k and the work hardening gradient $m = dk/d\gamma$ upon the shear strain rate are as given in Fig. 15.25.

CHAPTER 16

APPLICATIONS OF FINITE ELEMENTS

16.1 Introduction

Up to now the analyses given of plasticity theory together with its applications have resulted, largely, in closed solutions across a wide range of topics. The reader will no doubt be aware of the availability of commercial finite element (FE) packages that provide numerical solutions to many of these topics involving bulk plasticity and sheet metal forming. Consequently it is pertinent to end this book with an overview and a few applications of the FE technique. Firstly, the theory behind the FE method is explained as it applies to elastic deformation. This is intended to aid an understanding of how the linear elastic theory contributes to total strain within an incremental elastic–plastic FE analysis. Specific details can be found in many of the dedicated FE texts [1–4]. These reveal that the elastic–plastic FE theory calls upon: the yield criterion, the constitutive relations, the flow and hardening rules, which can all be found in our earlier chapters on plasticity theory. Outlined here is the method by which each is transcribed into its matrix form and then assembled into a final stiffness matrix. The assemblage ensures efficient programming within a solution to displacement, strain and stress from that matrix. Various FE codes are applied here to sheet metal forming problems. The particular package employed is described and its outputs are selected for comparison with available experimental data. In the first application, an example is given of hydrostatic bulge forming within circular and elliptical dies. This particular test provides FE users with a validation check in which the predicted plot between pressure and displacement is overlaid with experimental data. The agreement found between the two plots inspires a trust in an FE simulation showing how deformation evolves when forming more complex shapes to which an analytical solutions are unavailable. In a second application, the Erichsen test is simulated in order to provide strain distributions under a ball indentation. Again, the predictions are overlaid with experimental data from this test. Finally, it is shown how FE is used as a tool for preserving the integrity of a pressed sheet metal product at the die try-out stage of manufacture. Among the more sophisticated predictions that FE can make are of strain contours and signatures. As measured, these show the proximity of the strain state in the pressing to the sheet material's forming limit.

16.2 Elastic Stiffness Matrix

Elastic FE applies the theory of elasticity, within the sub-division of a body into smaller elements inter-connected throughout at nodal points. This technique of discretization enables an assembly of elements to describe the initial unstrained shape of the body. Thus, the behaviour of the entire body under stress may be computed from the known elastic behaviour of its elements. FE does this for a co-ordinate frame lying within the continuum, providing for equilibrium and compatibility of its internal stress and strain distributions and matching external boundary conditions, usually, of known force and displacement.

In the *stiffness or displacement method* of finite elements, the displacements at the nodal points are the unknowns. These displacements are solved numerically from their relation to the nodal forces. This relation is established by assembling a stiffness matrix to connect the 'vectors' of nodal point forces and displacements for each element. An overall stiffness matrix **K** is then assembled, where for a given mesh $\mathbf{f} = \mathbf{K}\,\boldsymbol{\delta}$, in which **f** and $\boldsymbol{\delta}$ represent forces and displacements at its many nodes as column matrices. The displacements are found numerically from a matrix inversion $\boldsymbol{\delta} = \mathbf{K}^{-1}\mathbf{f}$. Strain follows from a strain–displacement relation and the stress from its stress–strain (constitutive) relation. While such relations are well established for elasticity, the procedure must assume a function expressing the displacements of a node upon its co-ordinates. Thus, from applying the displacement function across all elements, the stiffness method provides numerical solutions to the stress and strain distributions within a loaded body. In the *force or flexibility method* of finite elements, the displacements are found from an assumed stress distribution. Though less popular generally, this method has proven a better choice for statically indeterminate, framed-structures [1] but it will not be explained here.

In essence, the stiffness FE method aims to compile and invert the matrix, connecting force to displacement, for the whole assembly of elements in a computationally efficient manner. There are many different types of 2D and 3D finite elements. Those that are plane and with axial symmetry include: triangular, rectangular, bar, shell and volume elements. These elements appear commonly within many well documented applications of FE including: beam and plate bending, torsion of prismatic bars, plates and bars under plane stress and strain, pressure vessels and rotating discs.

In static FE codes, the common approach is to derive the element stiffness matrix $\mathbf{K}^e$, connecting its nodal point force and displacement vectors $\mathbf{f}^e$ and $\boldsymbol{\delta}^e$ respectively:

$$\mathbf{f}^e = \mathbf{K}^e\,\boldsymbol{\delta}^e \tag{16.1}$$

The method for doing this will be shown briefly in selected examples that follow: simple tension, torsion and bending, before considering an element used for plane stress and strain in more detail. Bold upper-case Roman letters **A**, **B**, **C**, **K** denote matrices and bold lower-case Roman and Greek letters $\mathbf{f}$, $\boldsymbol{\alpha}$, $\boldsymbol{\varepsilon}$, $\boldsymbol{\delta}$, $\boldsymbol{\sigma}$ denote column matrices or column vectors. The word 'vector' is reserved strictly for quantities that take a physical meaning, as with force **f** and displacement $\boldsymbol{\delta}$. The FE analysis begins with a derivation of the matrix $\mathbf{K}^e$ in eq(16.1) for simple bar elements in tension, torsion and bending.

16.2.1 Tension

For simple tension of a bar element of length L, with uniform cross-sectional area A, connecting nodal points 1 and 2, in Fig. 16.1.

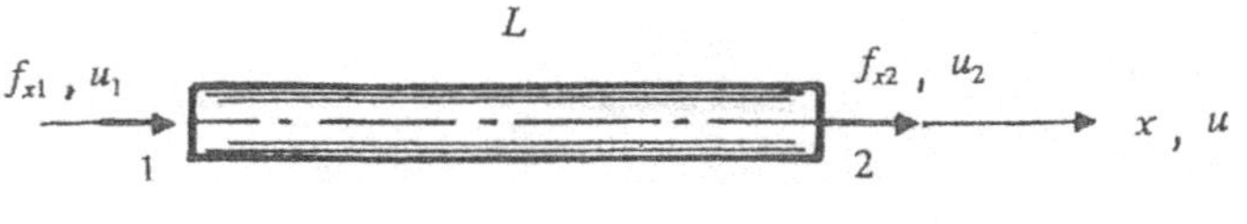

Figure 16.1 Tension bar element

The axis of the bar is aligned with the x–direction. The nodal forces f_{x1} and f_{x2} are taken to act at nodes 1 and node 2 respectively in the positive x–direction. These forces produce nodal displacements u_1 and u_2 aligned with positive x. Here, the stiffness matrix $\mathbf{K}^e$ is obvious from the uni-axial relationship between force and displacement: $f = (AE/L)u$. Written in the matrix form of eq(16.1), this becomes

$$\begin{bmatrix} f_{x_1} \\ f_{x_2} \end{bmatrix} = \frac{AE}{L}\begin{bmatrix} 1 & -1 \\ -1 & 1 \end{bmatrix}\begin{bmatrix} u_1 \\ u_2 \end{bmatrix} \tag{16.2}$$

where $\mathbf{f}^e = [\, f_{x1} \;\; f_{x2} \,]^T$ (in which $f_{x2} = -f_{x2}$) and $\boldsymbol{\delta}^e = [\, u_1 \;\; u_2 \,]^T$ are force and displacement vectors. Equation (16.2) reveals a symmetrical, 2×2 element stiffness matrix $\mathbf{K}^e$, as would be expected when each of the two nodes has a single degree of freedom. The element's stresses and strains follow applying the relationships $\sigma = E\varepsilon = Eu/L$ and $\varepsilon = \partial u/\partial x = \sigma/E$, given a displacement function $u = u(x)$.

16.2.2 Torsion

In Fig. 16.2, the uniform, solid circular shaft element of length L and radius r is subjected to nodal torques t_1 and t_2, as shown, which twist nodes 1 and 2 by amounts θ_1 and θ_2.

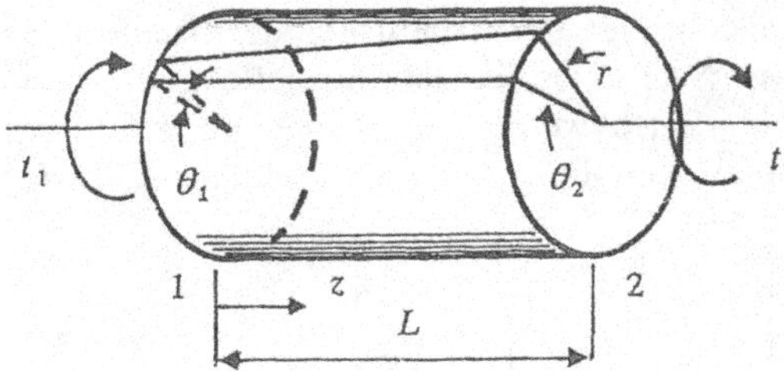

Figure 16.2 Circular bar element under axial torque

The components of the torsional stiffness matrix are defined from $\delta t/\delta\theta = JG/L$ where $J = \pi r^4/2$ is the polar second moment of area and G is the rigidity modulus. As with a tension bar, we may deduce that the element stiffness matrix for simple torsion is 2×2, there being two nodes with a single degree of freedom at each node. Thus, $\mathbf{K}^e$ follows directly from the stiffness relation as

$$\begin{bmatrix} t_1 \\ t_2 \end{bmatrix} = \frac{JG}{L}\begin{bmatrix} 1 & -1 \\ -1 & 1 \end{bmatrix}\begin{bmatrix} \theta_1 \\ \theta_2 \end{bmatrix} \tag{16.3a}$$

Comparing eq(16.3a) with eq(16.1) identifies $\mathbf{f}^e = [t_1 \;\; t_2]^T$ and $\boldsymbol{\delta}^e = [\theta_1 \;\; \theta_2]^T$ as the torque and twist vectors. Once the nodal twists have been found from a function $\theta = \theta(z)$, the element's nodal shear strains $\boldsymbol{\gamma}^e = [\gamma_1 \;\; \gamma_2]^T$ and stresses $\boldsymbol{\tau}^e = [\tau_1 \;\; \tau_2]^T$ follow proportionately from the elastic relation $\gamma = \tau/G = r\theta/L$ as

$$\boldsymbol{\gamma}^e = (r/L)\,\boldsymbol{\delta}^e, \quad \boldsymbol{\tau}^e = (Gr/L)\,\boldsymbol{\delta}^e \tag{16.3b,c}$$

16.2.3 Beam Bending

The element shown in Fig. 16.3a is used to determine the displacements, strains and stresses for a beam under bending. Shear force q must accompany bending moment m, each of which differ between nodes 1 and 2. Such actions produce both a vertical deflection v and a rotation θ in the positive directions shown.

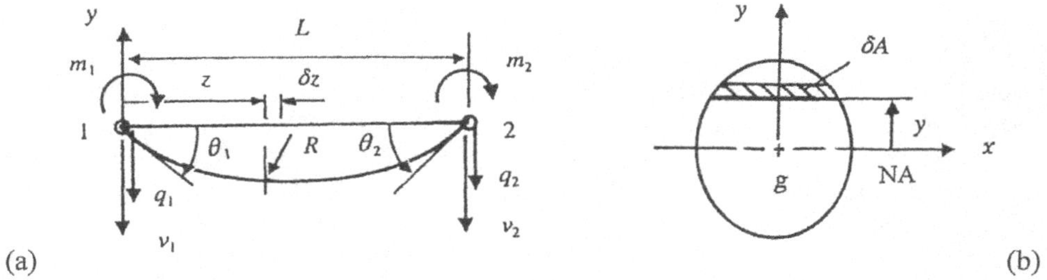

Figure 16.3 Beam element, showing arbitrary cross-section

The nodal force and displacement vectors include shear and bending: $\mathbf{f} = [\,q_1\ m_1\ q_2\ m_2]^{\mathrm{T}}$ and $\boldsymbol{\delta} = [v_1\ \theta_1\ v_2\ \delta_2]^{\mathrm{T}}$. The beam cross-section, which may be of any shape, is uniform along the length. Co-ordinate axes, x and y in the section and z in the length, pass through the centroid g of the section (see Fig. 16.3b). Under bending, the longitudinal axis z becomes the neutral axis (NA) of the section; this deflects to a radius of curvature R but suffers no stress or strain. Since two degrees of freedom exist at each node, the stiffness matrix $\mathbf{K}^{\mathrm{e}}$, connecting **f** to $\boldsymbol{\delta}$, has dimensions 4 × 4 with the following symmetry [5]:

$$\begin{bmatrix} q_1 \\ m_1 \\ q_2 \\ m_2 \end{bmatrix} = \frac{EI}{L^3}\begin{bmatrix} 12 & 6L & -12 & 6L \\ 6L & 4L^2 & -6L & 2L^2 \\ -12 & -6L & 12 & -6L \\ 6L & 2L^2 & -6L & 4L^2 \end{bmatrix}\begin{bmatrix} v_1 \\ \theta_1 \\ v_2 \\ \theta_2 \end{bmatrix} \tag{16.4}$$

Once the beam's nodal displacements v and $\theta = \mathrm{d}v/\mathrm{d}z$ have been found from a displacement function $v = v(z)$, the strain $\boldsymbol{\varepsilon}^{\mathrm{e}} = [\varepsilon_1\ \varepsilon_2]^{\mathrm{T}}$ and stress $\boldsymbol{\sigma}^{\mathrm{e}} = [\sigma_1\ \sigma_2]^{\mathrm{T}}$ follow from bending theory $\varepsilon = y/R = y(\mathrm{d}^2v/\mathrm{d}z^2)$ and $\sigma = E\varepsilon$.

16.3 Energy Methods

The forgoing review has shown that the FE stiffness method resolves into finding the stiffness matrix $\mathbf{K}^{\mathrm{e}}$ for a given element. Clearly, bar and beam elements allow their stiffness matrices, eqs(16.2)–(16.4), to be expressed quite simply. However, a derivation of each stiffness component K_{ij}^{e} of matrix $\mathbf{K}^{\mathrm{e}}$ is required for most other elements. The basis for this is to employ a convenient energy method. Either the principle of virtual work (PVW) or the principal of stationary potential energy (SPE) may be used [5]. These principles show that K_{ij}^{e} can be expressed in terms of the product of matrices in the nodal point co-ordinates and the elastic constants for the element material.

16.3.1 Principle of Virtual Work

The stiffness method of FE imposes virtual displacements at the nodes of a deformable element where real forces are applied. Conversely, virtual forces and real displacements are used with the flexibility method of FE. For the former method, the system of real forces, f_k ($k = 1, 2, 3 \ldots$) is in equilibrium with its internal stress state σ_{ij}. When the real forces experience virtual, in-line displacements $\delta_k^{\,v}$, the corresponding virtual strains are $\varepsilon_{ij}^{\,v}$. Superscript v denotes that displacements and strains are virtual. The principal of virtual work states that no net work can be done by imposing a displacement upon an equilibrium force system. When displacement due to self-weight is neglected, we can express this principle in two useful, equivalent ways:

$$f_k\delta_k^{\,v} - \int \sigma_{ij}\varepsilon_{ij}^{\,v}\,\mathrm{dV} = 0 \quad \Rightarrow \quad (\mathbf{f}^{\mathrm{e}})^{\mathrm{T}}\,\boldsymbol{\delta}^{\mathrm{ev}} - \int \boldsymbol{\sigma}^{\mathrm{T}}\boldsymbol{\varepsilon}^{\mathrm{v}}\,\mathrm{d}V = 0 \qquad (16.5\mathrm{a,b})$$

$$\delta_k^{\,v} f_k - \int \varepsilon_{ij}^{\,v}\sigma_{ij}\,\mathrm{dV} = 0 \quad \Rightarrow \quad (\boldsymbol{\delta}^{\mathrm{ev}})^{\mathrm{T}}\,\mathbf{f}^{\mathrm{e}} - \int (\boldsymbol{\varepsilon}^{\mathrm{v}})^{\mathrm{T}}\boldsymbol{\sigma}\,\mathrm{d}V = 0 \qquad (16.5\mathrm{c,d})$$

In the matrix notation, eq(16.5b) expresses $\mathbf{f}$, $\boldsymbol{\delta}$, $\boldsymbol{\sigma}$ and $\boldsymbol{\varepsilon}$ as column matrices. The reversal to the order of their multiplication within eq(16.5d) governs which matrix to transpose. The resulting scalar products are the same: $\mathbf{f}^{\mathrm{T}}\boldsymbol{\delta} = \boldsymbol{\delta}^{\mathrm{T}}\mathbf{f}$ and $\boldsymbol{\varepsilon}^{\mathrm{T}}\boldsymbol{\sigma} = \boldsymbol{\sigma}^{\mathrm{T}}\boldsymbol{\varepsilon}$. The additional superscript e in eqs(16.5b) and (16.5d) applies the principle specifically to an element's nodal forces and displacements. Components of the nodal force vector $\mathbf{f}^{\mathrm{e}}$ and the internal stress matrix $\boldsymbol{\sigma}$ are real and in equilibrium. Components of the nodal displacements vector $\boldsymbol{\delta}^{\mathrm{ev}}$ and the internal strain matrix $\boldsymbol{\varepsilon}^{\mathrm{v}}$ are compatible and virtual, i.e. independent of the real force-stress system. Hence, we can integrate the strain independently of stress over a volume V, within which the stress and strain can vary. Actually, eq(16.5d) is more convenient for deriving $\mathbf{K}^{\mathrm{e}}$ in the FE analysis that follows. Here, $(\boldsymbol{\delta}^{\mathrm{ev}})^{\mathrm{T}}$ is the transpose of the element's virtual, nodal displacement vector $\boldsymbol{\delta}^{\mathrm{ev}}$. Here, the word 'displacement' refers generally to deflection, rotation and twist at nodal points. Correspondingly, the real 'force' vector $\mathbf{f}^{\mathrm{e}}$ contains the appropriate nodal forces, moments and torques. Should forces be distributed, the virtual work principle allows $\mathbf{f}^{\mathrm{e}}$ to be compiled from an equivalent system of concentrated nodal forces [6].

16.3.2 Stationary Potential Energy

With both stress and strain being real, a stationary value of the total potential energy P will apply to an equilibrium condition. Now P is the sum of the internal strain energy stored U within a deformable element and the negative work of its external forces $-V$. Hence, a stationary P will lie in the condition

$$\mathrm{d}\left[\int\int \boldsymbol{\sigma}^{\mathrm{T}}\,\mathrm{d}\boldsymbol{\varepsilon}\,\mathrm{d}V - (\mathbf{f}^{\mathrm{e}})^{\mathrm{T}}\,\boldsymbol{\delta}^{\mathrm{e}}\right] = 0 \qquad (16.6\mathrm{a})$$

With Hooke's law providing for proportionality between stress and strain, an integration over strain reduces eq(16.6a) to

$$\mathrm{d}\left[\tfrac{1}{2}\int \boldsymbol{\sigma}^{\mathrm{T}}\boldsymbol{\varepsilon}\,\mathrm{d}V - (\mathbf{f}^{\mathrm{e}})^{\mathrm{T}}\,\boldsymbol{\delta}^{\mathrm{e}}\right] = 0 \qquad (16.6\mathrm{b})$$

Equation (16.6b) is applied as a partial differentiation with respect to the element's displacement vector $\boldsymbol{\delta}^{\mathrm{e}}$. Using the relationship $\boldsymbol{\sigma}^{\mathrm{T}}\boldsymbol{\varepsilon} = \boldsymbol{\varepsilon}^{\mathrm{T}}\boldsymbol{\sigma}$, stationary PE condition becomes

$$\frac{\partial P}{\partial \boldsymbol{\delta}^{e}} = \frac{1}{2}\int_V \left(\boldsymbol{\sigma}^{T}\frac{\partial \boldsymbol{\varepsilon}}{\partial \boldsymbol{\delta}^{e}} + \boldsymbol{\varepsilon}^{T}\frac{\partial \boldsymbol{\sigma}}{\partial \boldsymbol{\delta}^{e}} \right) dV - (\mathbf{f}^{e})^{T} = 0 \qquad (16.6c)$$

16.4 Plane Triangular Element

Let us illustrate the application of each energy method (PVW and SPE) to find the stiffness matrix for a plane triangular element in Fig. 16.4a. This element applies to bodies in plane stress and plane strain, i.e. those loaded in the plane x–y, where, respectively, the z–dimension can be both small (through a plate thickness) and large (along a bar axis).

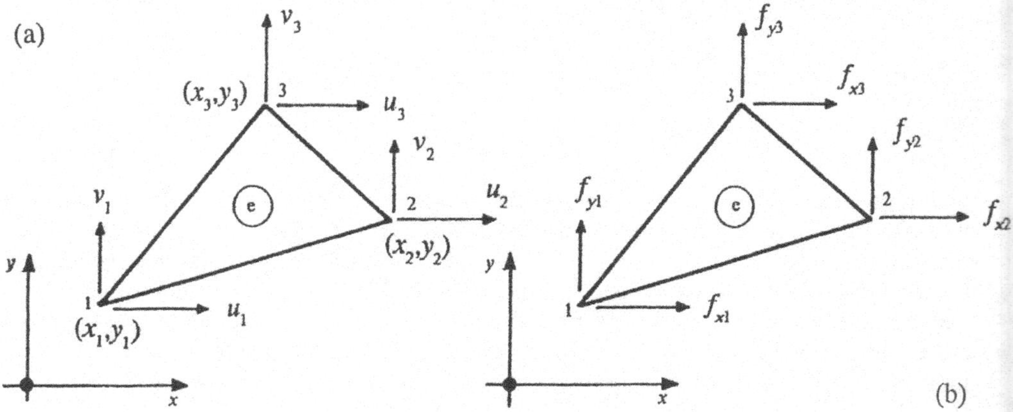

Figure 16.4 Forces and displacements at the nodes of a plane triangular element

16.4.1 Nodal Displacements

Firstly, the displacements (u, v) for any point (x, y) in the element are assumed to obey linear displacement functions:

$$u(x, y) = \alpha_1 + \alpha_2 x + \alpha_3 y \qquad (16.7a)$$

$$v(x, y) = \alpha_4 + \alpha_5 x + \alpha_6 y \qquad (16.7b)$$

where coefficients $\alpha_1, \alpha_2 \ldots \alpha_6$ are to be found. In matrix form, eqs(16.7a,b) appear as a displacement vector $\boldsymbol{\delta} = [u \quad v]^T$:

$$\boldsymbol{\delta} = \begin{bmatrix} u \\ v \end{bmatrix} = \begin{bmatrix} 1 & x & y & 0 & 0 & 0 \\ 0 & 0 & 0 & 1 & x & y \end{bmatrix} \begin{bmatrix} \alpha_1 \\ \alpha_2 \\ \alpha_3 \\ \alpha_4 \\ \alpha_5 \\ \alpha_6 \end{bmatrix} \qquad \boldsymbol{\delta} = \mathbf{A}\,\boldsymbol{\alpha} \qquad (16.7c,d)$$

where $\boldsymbol{\alpha} = [\alpha_1 \ \alpha_2 \ \alpha_3 \ \alpha_4 \ \alpha_5 \ \alpha_6]^T$ is a column matrix of six components, matching the number of degrees of freedom, and **A** is a 6 × 2 matrix in co-ordinates (x, y). The displacements for nodes 1, 2 and 3 may be found from substituting the co-ordinates (x_1, y_1), (x_2, y_2) and (x_3, y_3) into eq(16.7a,b). This gives the three nodal point displacement vectors $\boldsymbol{\delta}_1 = [u_1 \ v_1]^T$, $\boldsymbol{\delta}_2 = [u_2 \ v_2]^T$ and $\boldsymbol{\delta}_3 = [u_3 \ v_3]^T$, which re-appear in the element's nodal point displacement vector $\boldsymbol{\delta}^e$ as

$$\boldsymbol{\delta}^e = \begin{bmatrix} \boldsymbol{\delta}_1 \\ \boldsymbol{\delta}_2 \\ \boldsymbol{\delta}_3 \end{bmatrix} = \begin{bmatrix} u_1 \\ v_1 \\ u_2 \\ v_2 \\ u_3 \\ v_3 \end{bmatrix} = \begin{bmatrix} 1 & x_1 & y_1 & 0 & 0 & 0 \\ 0 & 0 & 0 & 1 & x_1 & y_1 \\ 1 & x_2 & y_2 & 0 & 0 & 0 \\ 0 & 0 & 0 & 1 & x_2 & y_2 \\ 1 & x_3 & y_3 & 0 & 0 & 0 \\ 0 & 0 & 0 & 1 & x_3 & y_3 \end{bmatrix} \begin{bmatrix} \alpha_1 \\ \alpha_2 \\ \alpha_3 \\ \alpha_4 \\ \alpha_5 \\ \alpha_6 \end{bmatrix} \qquad \boldsymbol{\delta}^e = \mathbf{A}^e \boldsymbol{\alpha} \qquad (16.8a,b)$$

where $\boldsymbol{\delta}^e = [\boldsymbol{\delta}_1 \ \boldsymbol{\delta}_2 \ \boldsymbol{\delta}_3]^T$ and $\mathbf{A}^e$ is a 6 × 6 matrix of nodal co-ordinates. It follows that the coefficients $\alpha_1, \alpha_2 \ldots \alpha_6$ are found from inverting matrix $\mathbf{A}^e$ in eq(16.8b):

$$\boldsymbol{\alpha} = (\mathbf{A}^e)^{-1} \boldsymbol{\delta}^e \qquad (16.9a)$$

This gives

$$(\mathbf{A}^e)^{-1} = \frac{1}{2\Delta} \begin{bmatrix} x_2y_3 - x_3y_2 & 0 & x_3y_1 - x_1y_3 & 0 & x_1y_2 - x_2y_1 & 0 \\ y_2 - y_3 & 0 & y_3 - y_1 & 0 & y_1 - y_2 & 0 \\ x_3 - x_2 & 0 & x_1 - x_3 & 0 & x_2 - x_1 & 0 \\ 0 & x_2y_3 - x_3y_2 & 0 & x_3y_1 - x_1y_3 & 0 & x_1y_2 - x_2y_1 \\ 0 & y_2 - y_3 & 0 & y_3 - y_1 & 0 & y_1 - y_2 \\ 0 & x_3 - x_2 & 0 & x_1 - x_3 & 0 & x_2 - x_1 \end{bmatrix} \qquad (16.9b)$$

The square matrix relation applies $\mathbf{A}^e (\mathbf{A}^e)^{-1} = \mathbf{I}$, where **I** is a unit matrix, and Δ gives the area of the element as $\Delta = \frac{1}{2} [(x_1 - x_2)(y_2 - y_3) - (x_2 - x_3)(y_1 - y_2)]$. Combining eqs(16.7b) and (16.9a) expresses the element's general displacement vector $\boldsymbol{\delta} = [u \ \ v]^T$ in terms of its three nodal displacement vectors $\boldsymbol{\delta}^e = [\boldsymbol{\delta}_1 \ \boldsymbol{\delta}_2 \ \boldsymbol{\delta}_3]^T$:

$$\boldsymbol{\delta} = \mathbf{A} (\mathbf{A}^e)^{-1} \boldsymbol{\delta}^e = \mathbf{N} \boldsymbol{\delta}^e \qquad (16.10)$$

where the matrix product $\mathbf{A} (\mathbf{A}^e)^{-1}$ provides the element's *shape function* (see exercise 16.1).

16.4.2 Strain-Displacement Matrix

Taking the displacement derivatives from eqs(16.7a,b) provides compatible, infinitesimal strains in the x, y plane:

$$\varepsilon_x = \partial u/\partial x = \alpha_2, \ \varepsilon_y = \partial v/\partial y = \alpha_6 \text{ and } \gamma_{xy} = \partial u/\partial y + \partial v/\partial x = \alpha_3 + \alpha_5 \qquad (16.11a\text{–}c)$$

Writing eqs(16.11a–c) in matrix form gives

$$\begin{bmatrix} \varepsilon_x \\ \varepsilon_y \\ \gamma_{xy} \end{bmatrix} = \begin{bmatrix} 0 & 1 & 0 & 0 & 0 & 0 \\ 0 & 0 & 0 & 0 & 0 & 1 \\ 0 & 0 & 1 & 0 & 1 & 0 \end{bmatrix} \begin{bmatrix} \alpha_1 \\ \alpha_2 \\ \alpha_3 \\ \alpha_4 \\ \alpha_5 \\ \alpha_6 \end{bmatrix} \qquad \boldsymbol{\varepsilon} = \mathbf{C}\boldsymbol{\alpha} \qquad (16.11\text{d,e})$$

The symbolic notation of eq(16.11e) shows that the 6 × 3 matrix **C** connects the column matrices $\boldsymbol{\varepsilon} = [\varepsilon_x \ \varepsilon_y \ \gamma_{xy}]^T$ and $\boldsymbol{\alpha} = [\alpha_1 \ \alpha_2 \ \alpha_3 \ \alpha_4 \ \alpha_5 \ \alpha_6]^T$. Substituting eq(16.9a) into eq(16.11e) gives strain $\boldsymbol{\varepsilon} = [\varepsilon_x \ \varepsilon_y \ \gamma_{xy}]^T$ at point (x, y) from the nodal point displacements $\boldsymbol{\delta}^e$:

$$\boldsymbol{\varepsilon} = \mathbf{C}(\mathbf{A}^e)^{-1}\boldsymbol{\delta}^e = \mathbf{B}\,\boldsymbol{\delta}^e \qquad (16.12\text{a})$$

where the matrix multiplication $\mathbf{B} = \mathbf{C}(\mathbf{A}^e)^{-1}$ employs eqs(16.9b) and (16.11d):

$$\mathbf{B} = \frac{1}{2\Delta}\begin{bmatrix} y_2 - y_3 & 0 & y_3 - y_1 & 0 & y_1 - y_2 & 0 \\ 0 & x_3 - x_2 & 0 & x_1 - x_3 & 0 & x_2 - x_1 \\ x_3 - x_2 & y_2 - y_3 & x_1 - x_3 & y_3 - y_1 & x_2 - x_1 & y_1 - y_2 \end{bmatrix} \qquad (16.12\text{b})$$

We see that **B** in eq(16.12b) appears only in terms of the nodal point co-ordinates, thereby revealing a 'constant strain' triangle. This arises from the linear displacement functions assumed in eqs(16.7a,b), leading to matrix **C** in eq(16.11d) being independent of x and y.

16.4.3 Constitutive Relation

Plane stress–strain relations may be expressed in symbolic notation as $\boldsymbol{\varepsilon} = \mathbf{P}\,\boldsymbol{\sigma}$, where **P** is a (3 × 3) elasticity matrix, connecting column matrices of stress $\boldsymbol{\sigma} = [\sigma_x \ \sigma_y \ \tau_{xy}]^T$ and strain $\boldsymbol{\varepsilon} = [\varepsilon_x \ \varepsilon_y \ \gamma_{xy}]^T$. Alternatively, since $\boldsymbol{\sigma} = \mathbf{P}^{-1}\boldsymbol{\varepsilon}$, then eq(16.12a) may be written as

$$\boldsymbol{\sigma} = \mathbf{D}\,\boldsymbol{\varepsilon} = \mathbf{D}\,\mathbf{B}\,\boldsymbol{\delta}^e \qquad (16.13\text{a})$$

where $\mathbf{D} = \mathbf{P}^{-1}$ is a 3 × 3 inverse elasticity matrix, whose plane stress components are

$$\mathbf{D} = \frac{E}{(1 - \nu^2)}\begin{bmatrix} 1 & \nu & 0 \\ \nu & 1 & 0 \\ 0 & 0 & \tfrac{1}{2}(1 - \nu) \end{bmatrix} \qquad (16.13\text{b})$$

and whose plane strain components are

$$\mathbf{D} = \frac{E(1-\nu)}{(1+\nu)(1-2\nu)}\begin{bmatrix} 1 & \nu/(1-\nu) & 0 \\ \nu/(1-\nu) & 1 & 0 \\ 0 & 0 & (1-2\nu)/[2(1-\nu)] \end{bmatrix} \tag{16.13c}$$

As with the strain components, the three stress components σ_x, σ_y, τ_{xy} are also constant throughout a triangular element for given nodal displacement vector $\boldsymbol{\delta}^e$. When plotting a triangular element's stress and strain distributions their magnitudes would be assigned to the triangle's centroid or to a mid-position of one side.

16.4.4 Element Stiffness Matrix

Finally, the element stiffness matrix $\mathbf{K}^e$ may be derived. Here a matrix of dimensions 6 × 6 must connect our nodal point force and displacement vectors:

$$\mathbf{f}^e = \mathbf{K}^e\,\boldsymbol{\delta}^e \tag{16.14a}$$

Here the element's force vector $\mathbf{f}^e = [\,f_{x1}\;\; f_{y1}\;\; f_{x2}\;\; f_{y2}\;\; f_{x3}\;\; f_{y3}\,]^T$ may be shortened to

$$\mathbf{f}^e = [\mathbf{f}_1\;\; \mathbf{f}_2\;\; \mathbf{f}_3\,]^T \tag{16.14b}$$

in which the nodal forces are $\mathbf{f}_1 = [\,f_{x1}\;\; f_{y1}\,]^T$, $\mathbf{f}_2 = [\,f_{x2}\;\; f_{y2}\,]^T$ and $\mathbf{f}_3 = [\,f_{x3}\;\; f_{y3}\,]^T$ (see Fig. 16.4b). Correspondingly, the element's displacement vector $\boldsymbol{\delta}^e$ is shortened:

$$\boldsymbol{\delta}^e = [\boldsymbol{\delta}_1\;\; \boldsymbol{\delta}_2\;\; \boldsymbol{\delta}_3\,]^T \tag{16.14c}$$

where the nodal displacements vectors are $\boldsymbol{\delta}_1 = [u_1\;\; v_1\,]^T$, $\boldsymbol{\delta}_2 = [u_2\;\; v_2\,]^T$ and $\boldsymbol{\delta}_3 = [u_3\;\; v_3\,]^T$ (see Fig. 16.4a). Here $\boldsymbol{\delta}^e$ follows from inverting $\mathbf{K}^e$ in eq(16.14a):

$$\boldsymbol{\delta}^e = (\mathbf{K}^e)^{-1}\,\mathbf{f}^e \tag{16.14d}$$

The components K_{ij} of $\mathbf{K}^e$ in eq(16.14a) follow from nodal point co-ordinates (x_i, y_i, for i = 1, 2, 3) and the elastic constants E and ν, i.e. within the components of matrices **D** and **B**, given in eqs(16.12b) and (16.13b,c). To find the relation between $\mathbf{K}^e$, **D** and **B** we may use either the principal of virtual work or stationary potential energy.

(a) Virtual Work

On substituting eqs(16.12a) and (16.13a) into eq(16.5d), the PVW gives

$$(\boldsymbol{\delta}^{ev})^T\mathbf{f}^e = \int(\mathbf{B}\boldsymbol{\delta}^{ev})^T(\mathbf{D}\mathbf{B}\boldsymbol{\delta}^e)\,\mathrm{d}V = (\boldsymbol{\delta}^{ev})^T(\mathbf{B}^T\mathbf{D}\mathbf{B}\boldsymbol{\delta}^e)\int\mathrm{d}V \tag{16.15a}$$

where neither **B** nor **D** depend upon x and y. Noting that $(\mathbf{B}\,\boldsymbol{\delta}^{ev})^T = (\boldsymbol{\delta}^{ev})^T\,\mathbf{B}^T$, one may cancel $(\boldsymbol{\delta}^{ev})^T$ to give

$$\mathbf{f}^e = \mathbf{B}^T\,\mathbf{D}\,\mathbf{B}\,\boldsymbol{\delta}^e\,V \tag{16.15b}$$

where $\int dV = V = At$, where A and t are, respectively, the area and thickness of the triangle. It follows from eqs(16.14a) and (16.15b) that this element's stiffness matrix becomes

$$\mathbf{K}^e = \mathbf{B}^T\,\mathbf{D}\,\mathbf{B}\,V \tag{16.15c}$$

where $\mathbf{K}^e$ is a (6×6) symmetrical matrix.

(b) Stationary Potential Energy

Alternatively, eq(16.15c) may be derived from substituting eqs(16.12a) and (16.13a) into eq(16.6c):

$$\frac{1}{2}\int_V \left[(\mathbf{D}\,\boldsymbol{\varepsilon})^T \frac{\partial}{\partial \boldsymbol{\delta}^e}(\mathbf{B}\,\boldsymbol{\delta}^e) + \boldsymbol{\varepsilon}^T \frac{\partial}{\partial \boldsymbol{\delta}^e}(\mathbf{D}\,\mathbf{B}\,\boldsymbol{\delta}^e)\right] dV - (\mathbf{f}^e)^T = 0$$

$$\frac{1}{2}\int_V [(\mathbf{D}\,\boldsymbol{\varepsilon})^T\,\mathbf{B} + \boldsymbol{\varepsilon}^T\,\mathbf{D}\,\mathbf{B}]\,dV - (\mathbf{f}^e)^T = 0$$

Using the relation $\mathbf{D} = \mathbf{D}^T$:

$$(\mathbf{f}^e)^T = \int_V \boldsymbol{\varepsilon}^T\,\mathbf{D}\,\mathbf{B}\,dV = \int_V (\mathbf{B}\,\boldsymbol{\delta}^e)^T\,\mathbf{D}\,\mathbf{B}\,dV$$

which can be written as

$$(\mathbf{f}^e)^T = (\boldsymbol{\delta}^e)^T\,(\mathbf{B}^T\,\mathbf{D}\,\mathbf{B})\,V$$

Taking the transpose of both sides:

$$\mathbf{f}^e = [(\boldsymbol{\delta}^e)^T\,(\mathbf{B}^T\mathbf{D}\,\mathbf{B})]^T\,V = (\mathbf{B}^T\mathbf{D}\,\mathbf{B})^T\,\boldsymbol{\delta}^e\,V = (\mathbf{D}\,\mathbf{B})^T\,\mathbf{B}\,\boldsymbol{\delta}^e\,V = (\mathbf{B}^T\,\mathbf{D}^T\mathbf{B}\,V)\,\boldsymbol{\delta}^e$$

when eq(16.15c) follows.

Either stiffness matrix derivation given in (a) and (b) above allows expressions for the components K_{ij} of $\mathbf{K}^e$ to be found from a matrix multiplication $\mathbf{B}^T\mathbf{D}\,\mathbf{B}$. For example, the diagonal and off-diagonal components $K_{11}{}^e$ and $K_{12}{}^e$ appear as

$$K_{11}{}^e = [t/(4\Delta)][D_{11}\,(y_2 - y_3)^2 + D_{33}\,(x_3 - x_2)^2]$$

$$K_{12}{}^e = K_{21}{}^e = [t/(4\Delta)][D_{12}\,(x_3 - x_2)(y_2 - y_3) + D_{33}\,(x_3 - x_2)(y_2 - y_3)]$$

where D_{11}, D_{33} and $D_{12} = D_{21}$ depend upon the plane problem, as defined in eqs(16.13b,c).

16.4.5 Overall Stiffness Matrix

Finally, the overall stiffness matrix $\mathbf{K}$ is assembled from the individual element stiffness matrices $\mathbf{K}^e$. The dimension of $\mathbf{K}$ may be established in advance. For example, consider the four-element, plane stress cantilever, shown in Fig. 16.5. With 6 nodes, each with 2 degrees of freedom, $\mathbf{K}$ becomes a 12×12 matrix.

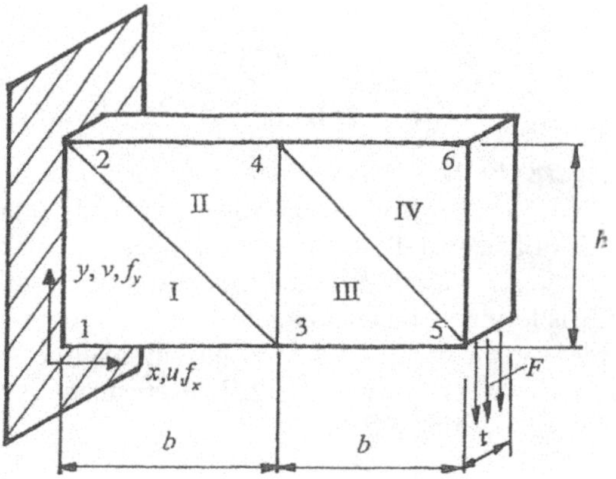

Figure 16.5 Plane stress cantilever composed of four triangular elements

The dimension may be reduced to (6 × 6) using sub-matrices $\mathbf{M}_{ij}^{e}$ with the shortened force and displacement notation adopted within eqs(16.14b,c). Writing the element number e = I, II, III and IV gives **K** more compactly as

$$\begin{bmatrix} \mathbf{f}_1 \\ \mathbf{f}_2 \\ \mathbf{f}_3 \\ \mathbf{f}_4 \\ \mathbf{f}_5 \\ \mathbf{f}_6 \end{bmatrix} = \begin{bmatrix} \mathbf{M}_{11}^{I} & \mathbf{M}_{12}^{I} & \mathbf{M}_{13}^{I} & \cdot & \cdot & \cdot \\ \mathbf{M}_{21}^{I} & \mathbf{M}_{22}^{I}+\mathbf{M}_{22}^{II} & \mathbf{M}_{23}^{I}+\mathbf{M}_{23}^{II} & \mathbf{M}_{24}^{II} & \cdot & \cdot \\ \mathbf{M}_{31}^{I} & \mathbf{M}_{32}^{I}+\mathbf{M}_{32}^{II} & \mathbf{M}_{33}^{I}+\mathbf{M}_{33}^{II}+\mathbf{M}_{33}^{III} & \mathbf{M}_{34}^{II}+\mathbf{M}_{34}^{III} & \mathbf{M}_{35}^{III} & \cdot \\ \cdot & \mathbf{M}_{42}^{II} & \mathbf{M}_{43}^{II}+\mathbf{M}_{43}^{III} & \mathbf{M}_{44}^{II}+\mathbf{M}_{44}^{III}+\mathbf{M}_{44}^{IV} & \mathbf{M}_{45}^{III}+\mathbf{M}_{45}^{IV} & \mathbf{M}_{46}^{IV} \\ \cdot & \cdot & \mathbf{M}_{53}^{III} & \mathbf{M}_{54}^{III}+\mathbf{M}_{54}^{IV} & \mathbf{M}_{55}^{III}+\mathbf{M}_{55}^{IV} & \mathbf{M}_{56}^{IV} \\ \cdot & \cdot & \cdot & \mathbf{M}_{64}^{IV} & \mathbf{M}_{65}^{IV} & \mathbf{M}_{66}^{IV} \end{bmatrix} \begin{bmatrix} \boldsymbol{\delta}_1 \\ \boldsymbol{\delta}_2 \\ \boldsymbol{\delta}_3 \\ \boldsymbol{\delta}_4 \\ \boldsymbol{\delta}_5 \\ \boldsymbol{\delta}_6 \end{bmatrix} \qquad (16.16)$$

Individual stiffness components appear within the sub-matrix addition. For example, the diagonal element $\mathbf{M}_{33}^{I} + \mathbf{M}_{33}^{II} + \mathbf{M}_{33}^{III}$ in eq(16.16) represents

$$\mathbf{M}_{33}^{I} + \mathbf{M}_{33}^{II} + \mathbf{M}_{33}^{III} = \begin{bmatrix} K_{33}^{I} & K_{34}^{I} \\ K_{43}^{I} & K_{44}^{I} \end{bmatrix} + \begin{bmatrix} K_{33}^{II} & K_{34}^{II} \\ K_{43}^{II} & K_{44}^{II} \end{bmatrix} + \begin{bmatrix} K_{33}^{III} & K_{34}^{III} \\ K_{43}^{III} & K_{44}^{III} \end{bmatrix}$$

The final step is to ensure that the elements provide zero displacement at support points, e.g. $u_1 = v_1 = u_2 = v_2 = 0$ at the cantilever fixing. Numerically, this is achieved by multiplying the appropriate stiffness components by a large number, such that with the inversion the deflection is eliminated at the required position.

Following this assembly of elements, the resulting stress, strain and displacement relations are applied in the same sense as the element node numbering. The nodal displacement vector can then be found from the following inversion process:

$$\mathbf{f} = \mathbf{K}\,\boldsymbol{\delta}, \quad \Rightarrow \quad \boldsymbol{\delta} = \mathbf{K}^{-1}\,\mathbf{f} \qquad (16.17a,b)$$

Thus, knowing $\boldsymbol{\delta}^{e}$, the three, constant strain and stress components are referred to the centroid of each element. They follow from the matrix multiplications:

$$\boldsymbol{\varepsilon}^{e} = \mathbf{B}^{e}\,\boldsymbol{\delta}^{e} \tag{16.18a}$$

$$\boldsymbol{\sigma}^{e} = \mathbf{D}\,\mathbf{B}^{e}\,\boldsymbol{\delta}^{e} = \mathbf{H}^{e}\,\boldsymbol{\delta}^{e} \tag{16.18b}$$

In eqs(16.12b) and (16.18a,b), $\mathbf{B}^e = \mathbf{B}$ as it depends only upon the nodal point co-ordinates. Also, $\mathbf{H}^e = \mathbf{D}\,\mathbf{B}^e$, from eq(16.18b), depends upon the elastic constants, according to the plane condition. The assembly of an overall stiffness matrix for several hundred elements requires a computer. With efficient node-numbering, we see from the cantilever example that the symmetrical matrix **K** is formed with its non-zero elements lying within a narrow diagonal band. Such nodal order and the partitioning of **K** into sub-matrices, as indicated, both serve to the reduce the computer memory required [7]. Both features would be incorporated within modern automated meshing practice.

16.5 Elastic–Plastic Stiffness Matrix

16.5.1 Vector Representation

The matrix form of the Prandtl-Reuss theory appeared in eq(4.3b) as

$$\mathrm{d}\mathbf{E} = \mathrm{d}\lambda\,\mathbf{T}' + \frac{(1-2\nu)}{3E}\,\mathbf{I}\,\mathrm{tr}\,(\mathrm{d}\mathbf{T}) + \frac{1}{2G}\,\mathrm{d}\mathbf{T}' \tag{16.19}$$

where **E**, **T**′ and **I** are, respectively, the total strain, the deviatoric stress and unit (3 × 3) matrices. Alternatively, column matrices (vectors) of stress and strain are often more convenient to use within the application of finite elements to incremental plasticity. Correspondingly, in the vector conversion of eq(16.19) the following column matrix representations of deviatoric stress and plastic strain increment tensor components are used:

$$\boldsymbol{\sigma}' = [\sigma_{11}'\;\; \sigma_{22}'\;\; \sigma_{33}'\;\; \surd 2\,\sigma_{12}'\;\; \surd 2\,\sigma_{23}'\;\; \surd 2\sigma_{13}']^{\mathrm{T}} \tag{16.20a}$$

$$\mathrm{d}\boldsymbol{\varepsilon}^{P} = [\mathrm{d}\varepsilon_{11}^{\,P}\;\; \mathrm{d}\varepsilon_{22}^{\,P}\;\; \mathrm{d}\varepsilon_{33}^{\,P}\;\; \surd 2\mathrm{d}\varepsilon_{12}^{\,P}\;\; \surd 2\mathrm{d}\varepsilon_{23}^{\,P}\;\; \surd 2\mathrm{d}\varepsilon_{13}^{\,P}]^{\mathrm{T}} \tag{16.20b}$$

where $\mathrm{d}\varepsilon_{12}^{\,P}$, $\mathrm{d}\varepsilon_{23}^{\,P}$ and $\mathrm{d}\varepsilon_{13}^{\,P}$ tensor shear strains. If engineering shear strains are preferred, where $\mathrm{d}\varepsilon_{12}^{\,P} = \mathrm{d}\gamma_{12}^{\,P}/2$, $\mathrm{d}\varepsilon_{13}^{\,P} = \mathrm{d}\gamma_{13}^{\,P}/2$ etc., then eq(16.20b) is re-written:

$$\mathrm{d}\boldsymbol{\varepsilon}^{P} = [\mathrm{d}\varepsilon_{11}^{\,P}\;\; \mathrm{d}\varepsilon_{22}^{\,P}\;\; \mathrm{d}\varepsilon_{33}^{\,P}\;\; \mathrm{d}\gamma_{12}^{\,P}/\surd 2\;\; \mathrm{d}\gamma_{23}^{\,P}/\surd 2\;\; \mathrm{d}\gamma_{13}^{\,P}/\surd 2]^{\mathrm{T}} \tag{16.20c}$$

Others [8] make use of column matrices of engineering stress and strain components, as in eq(16.13a), but these do not always convert correctly into the quantities derived earlier in our tensor notation of plasticity. It follows from eqs(16.20a,b) that the respective expressions, given in § 9.2.1, § 9.2.2 and § 9.9.2, for: (i) equivalent stress, (ii) equivalent plastic strain increment and (iii) plastic work increment convert correctly between the two notations.

(i)
$$\bar{\sigma}^2 = \frac{3}{2}\,\sigma_{ij}'\sigma_{ij}' = \frac{3}{2}\,\boldsymbol{\sigma}'^{\,\mathrm{T}}\boldsymbol{\sigma}' \tag{16.21a}$$

(ii)
$$\left(\mathrm{d}\bar{\varepsilon}^{P}\right)^2 = \frac{2}{3}\,\mathrm{d}\varepsilon_{ij}^{P}\,\mathrm{d}\varepsilon_{ij}^{P} = \frac{2}{3}\left(\mathrm{d}\boldsymbol{\varepsilon}^{P}\right)^{\mathrm{T}}\mathrm{d}\boldsymbol{\varepsilon}^{P} \tag{16.21b}$$

(iii)
$$\mathrm{d}W^P = \sigma_{ij}\,\mathrm{d}\varepsilon_{ij}^P = \boldsymbol{\sigma}^{\mathrm{T}}\,\mathrm{d}\boldsymbol{\varepsilon}^P = (\mathrm{d}\boldsymbol{\varepsilon}^P)^{\mathrm{T}}\boldsymbol{\sigma}$$
$$\mathrm{d}W^P = \sigma_{ij}'\,\mathrm{d}\varepsilon_{ij}^P = \boldsymbol{\sigma}'^{\mathrm{T}}\,\mathrm{d}\boldsymbol{\varepsilon}^P = (\mathrm{d}\boldsymbol{\varepsilon}^P)^{\mathrm{T}}\boldsymbol{\sigma}' \qquad (16.21\mathrm{c,d})$$

For, example, writing eq(16.21a) in full, the matrix multiplication gives the correct scalar quantity for equivalent stress:

$$\bar{\sigma}^2 = \frac{3}{2}\begin{bmatrix}\sigma_{11}' & \sigma_{22}' & \sigma_{33}' & \sqrt{2}\sigma_{12}' & \sqrt{2}\sigma_{23}' & \sqrt{2}\sigma_{13}'\end{bmatrix}\begin{bmatrix}\sigma_{11}' \\ \sigma_{22}' \\ \sigma_{33}' \\ \sqrt{2}\sigma_{12}' \\ \sqrt{2}\sigma_{23}' \\ \sqrt{2}\sigma_{13}'\end{bmatrix} \qquad (16.22\mathrm{a})$$

$$\bar{\sigma}^2 = \frac{3}{2}\left\{(\sigma_{11}')^2 + (\sigma_{22}')^2 + (\sigma_{33}')^2 + 2\,[(\sigma_{12}')^2 + (\sigma_{13}')^2 + (\sigma_{23}')^2\,]\right\} \qquad (16.22\mathrm{b})$$

agreeing with the summation convention employed in the tensor notation:

$$\bar{\sigma}^2 = \frac{3}{2}\sigma_{ij}'\sigma_{ij}' = \frac{3}{2}[\sigma_{1j}'\sigma_{1j}' + \sigma_{2j}'\sigma_{2j}' + \sigma_{3j}'\sigma_{3j}'] \text{ for } i = 1,2,3$$

$$= \frac{3}{2}\left[(\sigma_{11}'\sigma_{11}' + \sigma_{12}'\sigma_{12}' + \sigma_{13}'\sigma_{13}') + (\sigma_{21}'\sigma_{21}' + \sigma_{22}'\sigma_{22}' + \sigma_{23}'\sigma_{23}') + (\sigma_{31}'\sigma_{31}' + \sigma_{32}'\sigma_{32}' + \sigma_{33}'\sigma_{33}')\right]_{j\,=\,1,2,3}$$

in which the reduction to tension is provided by its non zero stress deviators:

$$\sigma_{11}' = \frac{2}{3}\bar{\sigma}, \quad \sigma_{22}' = \sigma_{33}' = -\frac{1}{3}\bar{\sigma}$$

16.5.2 Deviatoric Stress

The conversion from absolute to deviatoric stress in the vector notation appears as

$$\begin{bmatrix}\sigma_{11}' \\ \sigma_{22}' \\ \sigma_{33}' \\ \sqrt{2}\sigma_{12}' \\ \sqrt{2}\sigma_{23}' \\ \sqrt{2}\sigma_{13}'\end{bmatrix} = \begin{bmatrix}1&0&0&0&0&0\\0&1&0&0&0&0\\0&0&1&0&0&0\\0&0&0&1&0&0\\0&0&0&0&1&0\\0&0&0&0&0&1\end{bmatrix}\begin{bmatrix}\sigma_{11} \\ \sigma_{22} \\ \sigma_{33} \\ \sqrt{2}\sigma_{12} \\ \sqrt{2}\sigma_{23} \\ \sqrt{2}\sigma_{13}\end{bmatrix} - \frac{1}{3}\begin{bmatrix}1\\1\\1\\0\\0\\0\end{bmatrix}\begin{bmatrix}1&1&1&0&0&0\end{bmatrix}\begin{bmatrix}\sigma_{11} \\ \sigma_{22} \\ \sigma_{33} \\ \sqrt{2}\sigma_{12} \\ \sqrt{2}\sigma_{23} \\ \sqrt{2}\sigma_{13}\end{bmatrix} \qquad (16.23\mathrm{a})$$

Matrix multiplication in eq(16.23a) gives $\sigma_{11}' = \sigma_{11} - ⅓(\sigma_{11} + \sigma_{22} + \sigma_{33})$ etc. and $\sigma_{12}' = \sigma_{12}$ etc., consistent with conversions within the tensor notation $\sigma_{ij}' = \sigma_{ij} - ⅓\delta_{ij}\sigma_{kk}$. It is seen that eq(16.23a) may be re-written as

$$\boldsymbol{\sigma}' = \mathbf{I}\,\boldsymbol{\sigma} - \tfrac{1}{3}\,\mathbf{a}\,\mathbf{a}^{\mathrm{T}}\boldsymbol{\sigma} = [\mathbf{I} - \tfrac{1}{3}\,\mathbf{a}\,\mathbf{a}^{\mathrm{T}}]\,\boldsymbol{\sigma} = \boldsymbol{\mu}\,\boldsymbol{\sigma} \tag{16.23b}$$

where $\mathbf{a} = [1\ 1\ 1\ 0\ 0\ 0]^{\mathrm{T}}$. Also, the matrix $\boldsymbol{\mu}$ follows from eqs(16.23a,b) as

$$\boldsymbol{\mu} = \mathbf{I} - \tfrac{1}{3}\,\mathbf{a}\,\mathbf{a}^{\mathrm{T}} \tag{16.24a}$$

$$\boldsymbol{\mu} = \begin{bmatrix} 2/3 & -1/3 & -1/3 & 0 & 0 & 0 \\ -1/3 & 2/3 & -1/3 & 0 & 0 & 0 \\ -1/3 & -1/3 & 2/3 & 0 & 0 & 0 \\ 0 & 0 & 0 & 1 & 0 & 0 \\ 0 & 0 & 0 & 0 & 1 & 0 \\ 0 & 0 & 0 & 0 & 0 & 1 \end{bmatrix} \tag{16.24b}$$

Since $\boldsymbol{\mu} = \boldsymbol{\mu}^{\mathrm{T}}$, the following relationships hold:

$$\boldsymbol{\mu}\boldsymbol{\mu} = \boldsymbol{\mu}\boldsymbol{\mu}^{\mathrm{T}} = \boldsymbol{\mu}^{\mathrm{T}}\boldsymbol{\mu} = \boldsymbol{\mu}^{\mathrm{T}}\boldsymbol{\mu}^{\mathrm{T}} \tag{16.25a}$$

This product also appears from eq(16.24a) as

$$\begin{aligned} \boldsymbol{\mu}\boldsymbol{\mu} &= [\mathbf{I} - \tfrac{1}{3}\,\mathbf{a}\,\mathbf{a}^{\mathrm{T}}]\,[\mathbf{I} - \tfrac{1}{3}\,\mathbf{a}\,\mathbf{a}^{\mathrm{T}}] \\ &= \mathbf{I}\,[\mathbf{I} - \tfrac{1}{3}\,\mathbf{a}\,\mathbf{a}^{\mathrm{T}}] - \tfrac{1}{3}\,\mathbf{a}\,\mathbf{a}^{\mathrm{T}}[\mathbf{I} - \tfrac{1}{3}\,\mathbf{a}\,\mathbf{a}^{\mathrm{T}}] \\ &= [\mathbf{I} - \tfrac{1}{3}\,\mathbf{a}\,\mathbf{a}^{\mathrm{T}}] - \tfrac{1}{3}\,\mathbf{a}\,\mathbf{a}^{\mathrm{T}} + \tfrac{1}{3}\,\mathbf{a}\,(\tfrac{1}{3}\,\mathbf{a}^{\mathrm{T}}\mathbf{a})\,\mathbf{a}^{\mathrm{T}} \\ &= \mathbf{I} - \tfrac{1}{3}\,\mathbf{a}\,\mathbf{a}^{\mathrm{T}} = \boldsymbol{\mu} \end{aligned} \tag{16.25b}$$

which employs the fact that $\mathbf{a}^{\mathrm{T}}\mathbf{a} = 3$.

16.5.3 Matrix Constitutive Relations

In converting the Prandtl-Reuss theory to this vector notation, use is made of the fact that the sum of the final two terms in eq(4.3a) make up the elastic increment of strain. This sum is equivalent to the usual 3D elastic (superscript E) constitutive relations (see § 10.6.1):

$$\mathrm{d}\varepsilon_{11}{}^{E} = (1/E)[\mathrm{d}\sigma_{11} + \nu(\mathrm{d}\sigma_{22} + \mathrm{d}\sigma_{33})]$$

$$\mathrm{d}\varepsilon_{22}{}^{E} = (1/E)[\mathrm{d}\sigma_{22} + \nu(\mathrm{d}\sigma_{11} + \mathrm{d}\sigma_{33})]$$

$$\mathrm{d}\varepsilon_{33}{}^{E} = (1/E)[\mathrm{d}\sigma_{33} + \nu(\mathrm{d}\sigma_{11} + \mathrm{d}\sigma_{22})]$$

$$\mathrm{d}\gamma_{12}{}^{E} = \mathrm{d}\tau_{12}/G,\ \mathrm{d}\gamma_{23}{}^{E} = \mathrm{d}\tau_{23}/G,\ \mathrm{d}\gamma_{13}{}^{E} = \mathrm{d}\tau_{13}/G$$

where E, G and ν are the elastic constants. These constants define a (6×6) elastic flexibility matrix that links the column matrices of strain increment to stress increments according to the above relations, as follows:

$$\begin{bmatrix} \mathrm{d}\varepsilon_{11}^E \\ \mathrm{d}\varepsilon_{22}^E \\ \mathrm{d}\varepsilon_{33}^E \\ \sqrt{2}\,\mathrm{d}\varepsilon_{12}^E \\ \sqrt{2}\,\mathrm{d}\varepsilon_{23}^E \\ \sqrt{2}\,\mathrm{d}\varepsilon_{13}^E \end{bmatrix} = \begin{bmatrix} 1/E & -\nu & -\nu & 0 & 0 & 0 \\ -\nu & 1/E & -\nu & 0 & 0 & 0 \\ -\nu & -\nu & 1/E & 0 & 0 & 0 \\ 0 & 0 & 0 & 1/(2G) & 0 & 0 \\ 0 & 0 & 0 & 0 & 1/(2G) & 0 \\ 0 & 0 & 0 & 0 & 0 & 1/(2G) \end{bmatrix} \begin{bmatrix} \mathrm{d}\sigma_{11} \\ \mathrm{d}\sigma_{22} \\ \mathrm{d}\sigma_{33} \\ \sqrt{2}\,\mathrm{d}\sigma_{12} \\ \sqrt{2}\,\mathrm{d}\sigma_{23} \\ \sqrt{2}\,\mathrm{d}\sigma_{13} \end{bmatrix} \tag{16.26a}$$

which is simplified in the symbolic notation

$$\mathrm{d}\boldsymbol{\varepsilon}^{\mathrm{E}} = \mathbf{P}^{\mathrm{E}}\,\mathrm{d}\boldsymbol{\sigma} \tag{16.26b}$$

An incremental plastic component of strain (superscript P) is, from eqs(10.14) and (10.18):

$$\mathrm{d}\varepsilon_{ij}^{P} = \frac{3\,\bar{\sigma}\,\mathrm{d}\bar{\sigma}\,\sigma_{ij}'}{2E^{P}\bar{\sigma}^{2}} \tag{16.27a}$$

where $E^P = \mathrm{d}\bar{\sigma}/\mathrm{d}\bar{\varepsilon}^P$ has now replaced H' as the plastic tangent modulus, i.e. the gradient to the flow curve with the elastic components of strain removed. If a Ludwik or a Swift law describes the plastic flow curve from a tension test, then E^P becomes the derivative of eq(9.43a) or (9.44a). Alternatively, using total strain, the gradient E^T of the tangent to a uni-axial stress–strain curve may be written as

$$E^{T} = \frac{\mathrm{d}\bar{\sigma}}{\mathrm{d}\varepsilon^{T}} = \frac{\mathrm{d}\bar{\sigma}}{\mathrm{d}\bar{\varepsilon}^{P} + \mathrm{d}\varepsilon^{E}} \tag{16.27b}$$

Substituting $\mathrm{d}\bar{\varepsilon}^P = \mathrm{d}\bar{\sigma}/E^P$ and $\mathrm{d}\varepsilon^E = \mathrm{d}\bar{\sigma}/E$ into eq(16.27b) relates the elastic modulus E and plastic modulus E^{P} to the tangent modulus E^T as

$$E^{P} = \frac{E^{T}}{1 - E^{T}/E} \tag{16.27c}$$

The product $\bar{\sigma}\,\mathrm{d}\bar{\sigma}$ in eq(16.27a) is found from eq(16.21) as

$$\bar{\sigma}\,\mathrm{d}\bar{\sigma} = \frac{3}{2}\sigma_{ij}'\,\mathrm{d}\sigma_{ij}' = \frac{3}{2}\boldsymbol{\sigma}'^{\mathrm{T}}\,\mathrm{d}\boldsymbol{\sigma}' \tag{16.28a}$$

The matrix form of eq(16.27a) follows from eqs(16.20a,b), (16.21a) and (16.28a) as

$$\mathrm{d}\boldsymbol{\varepsilon}^{\mathrm{P}} = \frac{3\,\boldsymbol{\sigma}'^{\mathrm{T}}\,\mathrm{d}\boldsymbol{\sigma}'\,\boldsymbol{\sigma}'}{2E^{P}(\boldsymbol{\sigma}'^{\mathrm{T}}\,\boldsymbol{\sigma}')} \tag{16.28b}$$

Since $\boldsymbol{\sigma}'^{\mathrm{T}} \mathbf{d}\boldsymbol{\sigma}'$ is a scalar, the product term in the numerator of eq(16.28b) is written as

$$(\boldsymbol{\sigma}'^{\mathrm{T}}\, \mathbf{d}\boldsymbol{\sigma}')\, \boldsymbol{\sigma}' = \boldsymbol{\sigma}'(\boldsymbol{\sigma}'^{\mathrm{T}}\, \mathbf{d}\boldsymbol{\sigma}') = (\boldsymbol{\sigma}'\boldsymbol{\sigma}'^{\mathrm{T}})\, \mathbf{d}\boldsymbol{\sigma}'$$

Substituting from eq(16.23b) gives the absolute stress form

$$(\boldsymbol{\mu\sigma})\,(\boldsymbol{\mu\sigma})^{\mathrm{T}}\mathrm{d}(\boldsymbol{\mu\sigma}) = \boldsymbol{\mu\sigma\sigma}^{\mathrm{T}}(\boldsymbol{\mu}^{\mathrm{T}}\boldsymbol{\mu})\mathrm{d}\boldsymbol{\sigma} = \boldsymbol{\mu\sigma\sigma}^{\mathrm{T}}\boldsymbol{\mu}\mathrm{d}\boldsymbol{\sigma} = (\boldsymbol{\mu\sigma})(\boldsymbol{\mu\sigma})^{\mathrm{T}}\mathrm{d}\boldsymbol{\sigma} = \boldsymbol{\sigma}'\boldsymbol{\sigma}'^{\mathrm{T}}\mathrm{d}\boldsymbol{\sigma}$$

from which eq(16.28b) is written as

$$\mathbf{d}\boldsymbol{\varepsilon}^{\mathrm{P}} = \frac{3\,\boldsymbol{\sigma}'(\boldsymbol{\sigma}')^{\mathrm{T}}\,\mathbf{d}\boldsymbol{\sigma}}{2E^{P}(\boldsymbol{\sigma}')^{\mathrm{T}}\boldsymbol{\sigma}'} = \mathbf{P}^{\mathrm{P}}\,\mathbf{d}\boldsymbol{\sigma} \qquad (16.28\mathrm{c})$$

Equation (16.28c) is added to eq(16.26b) to give the total incremental strain (superscript T) in a symbolic matrix form, corresponding to the Prandtl-Reuss theory:

$$\mathbf{d}\boldsymbol{\varepsilon}^{\mathrm{T}} = \left(\mathbf{P}^{\mathrm{E}} + \mathbf{P}^{\mathrm{P}}\right)\mathbf{d}\boldsymbol{\sigma} \qquad (16.28\mathrm{d})$$

The symmetrical plasticity matrix $\mathbf{P}^{\mathrm{P}}$ in eq(16.28c) forms from a matrix product $\boldsymbol{\sigma}'(\boldsymbol{\sigma}')^{\mathrm{T}}$ (where $(\boldsymbol{\sigma}')^{\mathrm{T}}$ is the transpose of $\boldsymbol{\sigma}'$). *Note*: From eq(16.21a): $(\boldsymbol{\sigma}')^{\mathrm{T}}\boldsymbol{\sigma}' = (2/3)\,\bar{\sigma}^{2}$.

$$\boldsymbol{\sigma}'(\boldsymbol{\sigma}')^{\mathrm{T}} = \begin{bmatrix} \sigma_{11}'^{2} & \sigma_{11}'\sigma_{22}' & \sigma_{11}'\sigma_{33}' & \surd 2\sigma_{11}'\sigma_{12}' & \surd 2\sigma_{11}'\sigma_{13}' & \surd 2\sigma_{11}'\sigma_{23}' \\ \sigma_{22}'\sigma_{11}' & \sigma_{22}'^{2} & \sigma_{22}'\sigma_{33}' & \surd 2\sigma_{22}'\sigma_{12}' & \surd 2\sigma_{22}'\sigma_{13}' & \surd 2\sigma_{22}'\sigma_{23}' \\ \sigma_{33}'\sigma_{11}' & \sigma_{33}'\sigma_{22}' & \sigma_{33}'^{2} & \surd 2\sigma_{33}'\sigma_{12}' & \surd 2\sigma_{33}'\sigma_{13}' & \surd 2\sigma_{33}'\sigma_{23}' \\ \surd 2\sigma_{12}'\sigma_{11}' & \surd 2\sigma_{12}'\sigma_{22}' & \surd 2\sigma_{12}'\sigma_{33}' & 2\sigma_{12}'^{2} & 2\sigma_{12}'\sigma_{13}' & 2\sigma_{12}'\sigma_{23}' \\ \surd 2\sigma_{13}'\sigma_{11}' & \surd 2\sigma_{13}'\sigma_{22}' & \surd 2\sigma_{13}'\sigma_{33}' & 2\sigma_{13}'\sigma_{12}' & 2\sigma_{13}'^{2} & 2\sigma_{13}'\sigma_{23}' \\ \surd 2\sigma_{23}'\sigma_{11}' & \surd 2\sigma_{23}'\sigma_{22}' & \surd 2\sigma_{23}'\sigma_{33}' & 2\sigma_{23}'\sigma_{12}' & 2\sigma_{23}'\sigma_{13}' & 2\sigma_{23}'^{2} \end{bmatrix} \qquad (16.29\mathrm{a})$$

An absolute stress form for $\mathbf{P}^{\mathrm{P}}$ follows from eq(16.28c) as

$$\mathbf{P}^{\mathrm{P}} = \frac{3\boldsymbol{\mu\sigma}(\boldsymbol{\mu\sigma})^{\mathrm{T}}}{2E^{P}(\boldsymbol{\mu\sigma})^{\mathrm{T}}(\boldsymbol{\mu\sigma})} = \frac{3\,\boldsymbol{\mu\sigma}(\boldsymbol{\sigma})^{\mathrm{T}}\boldsymbol{\mu}}{2E^{P}(\boldsymbol{\sigma})^{\mathrm{T}}\boldsymbol{\mu\sigma}} \qquad (16.29\mathrm{b})$$

Matrix multiplication within eq(16.28c) leads to the plastic strain increments. For example, taking one direct and one shear strain component:

$$d\varepsilon_{11}^{P} = \frac{\begin{array}{c}[(2\sigma_{11} - \sigma_{22} - \sigma_{33})\,\mathrm{d}\sigma_{11} + (2\sigma_{22} - \sigma_{11} - \sigma_{33})\,\mathrm{d}\sigma_{22} + (2\sigma_{33} - \sigma_{11} - \sigma_{22})\mathrm{d}\sigma_{33} \\ + 6(\sigma_{12}\mathrm{d}\sigma_{12} + \sigma_{13}\mathrm{d}\sigma_{13} + \sigma_{23}\mathrm{d}\sigma_{23})]\left[\sigma_{11} - \tfrac{1}{2}(\sigma_{22} + \sigma_{33})\right]\end{array}}{E^{P}[(\sigma_{11} - \sigma_{22})^{2} + (\sigma_{22} - \sigma_{33})^{2} + (\sigma_{11} - \sigma_{33})^{2} + 6(\sigma_{12}^{2} + \sigma_{13}^{2} + \sigma_{23}^{2})]}$$

$$d\varepsilon_{12}^{P} = \frac{\begin{array}{c}[(2\sigma_{11} - \sigma_{22} - \sigma_{33})\,\mathrm{d}\sigma_{11} + (2\sigma_{22} - \sigma_{11} - \sigma_{33})\,\mathrm{d}\sigma_{22} + (2\sigma_{33} - \sigma_{11} - \sigma_{22})\mathrm{d}\sigma_{33} \\ + 6(\sigma_{12}\mathrm{d}\sigma_{12} + \sigma_{13}\mathrm{d}\sigma_{13} + \sigma_{23}\mathrm{d}\sigma_{23})](3/\surd 2)\sigma_{12}\end{array}}{E^{P}[(\sigma_{11} - \sigma_{22})^{2} + (\sigma_{22} - \sigma_{33})^{2} + (\sigma_{11} - \sigma_{33})^{2} + 6(\sigma_{12}^{2} + \sigma_{13}^{2} + \sigma_{23}^{2})]}$$

to which are added the incremental elastic strains from $\mathbf{P}^{E}$ in eq(16.26a). Setting $E^{P} = \mathrm{d}\bar{\sigma}/\mathrm{d}\bar{\varepsilon}^{P}$ and, with substitutions from eqs(16.22b) and (16.28a), the two incremental plastic strains above agree with the Levy-Mises equivalent forms:

$$\mathrm{d}\varepsilon_{11}^{P} = \frac{\mathrm{d}\bar{\sigma}}{E^{P}\bar{\sigma}}\left[\sigma_{11} - \frac{1}{2}\left(\sigma_{22} + \sigma_{33}\right)\right]$$

$$\mathrm{d}\varepsilon_{12}^{P} = \frac{\mathrm{d}\bar{\sigma}}{E^{P}\bar{\sigma}}\left(\frac{3}{2}\sigma_{12}\right)$$

16.5.4 Virtual Work

Inverting eq(16.28d) gives incremental stress $\mathrm{d}\boldsymbol{\sigma}$ from total incremental strain $\mathrm{d}\boldsymbol{\varepsilon}^{T}$:

$$\mathrm{d}\boldsymbol{\sigma} = \mathbf{D}^{EP}\mathrm{d}\boldsymbol{\varepsilon}^{T}, \quad \mathbf{D}^{EP} = (\mathbf{P}^{E} + \mathbf{P}^{P})^{-1}$$

Applying virtual work in its incremental form provides a tangential stiffness matrix for elastic–plastic deformation. Omitting the superscript (e – for element) this becomes

$$(\Delta\boldsymbol{\delta}^{v})^{T}\Delta\mathbf{f} - \int_{V}(\Delta\boldsymbol{\varepsilon}^{v})^{T}\Delta\boldsymbol{\sigma}\,\mathrm{d}V = 0 \qquad (16.30a)$$

where

$$\Delta\boldsymbol{\sigma} = (\mathbf{P}^{E} + \mathbf{P}^{P})^{-1}\Delta\boldsymbol{\varepsilon}^{T} = \mathbf{D}^{EP}\Delta\boldsymbol{\varepsilon}^{T} \qquad (16.30b)$$

giving

$$\Delta\mathbf{f} = \mathbf{K}\,\Delta\boldsymbol{\delta} \qquad (16.30c)$$

in which $\mathbf{K}$ is variable stiffness, corresponding to the variable E^{P}. Combining eqs(16.30a,b) with eq(16.12a):

$$(\Delta\boldsymbol{\delta}^{v})^{T}\Delta\mathbf{f} - \int_{V}(\mathbf{B}\,\Delta\boldsymbol{\delta}^{v})^{T}(\mathbf{D}^{EP}\,\mathbf{B}\,\Delta\boldsymbol{\delta})\,\mathrm{d}V = 0$$

Cancelling $(\Delta\boldsymbol{\delta}^{v})^{T}$ and removing $\Delta\boldsymbol{\delta}$ from the integrand:

$$\Delta\mathbf{f} = \left\{\int_{V}(\mathbf{B})^{T}\mathbf{D}^{EP}\,\mathbf{B}\,\mathrm{d}V\right\}\Delta\boldsymbol{\delta} \qquad (16.31a)$$

Finally, identifying eq(16.31a) with eq(16.30c) defines the tangent stiffness matrix:

$$\mathbf{K} = \int_{V}(\mathbf{B})^{T}\mathbf{D}^{EP}\mathbf{B}\,\mathrm{d}V \qquad (16.31b)$$

In general, for most plane and axi-symmetric elements, the co-ordinates x, y appear within matrix $\mathbf{B}$ and so the matrix product in eqs(16.31a,b) must be integrated over the volume. This, and the fact that the elements K_{ij} must be updated continually to form a tangent stiffness matrix for each increment of plastic strain, are the essential differences compared to the linear elastic case, previously described.

16.5.5 FE Codes

The analysis above shows that the elastic–plastic and elastic FE program structures are essentially the same, but solving $\Delta\boldsymbol{\delta} = \mathbf{K}^{-1}\Delta\mathbf{f}$, with each non-linear strain increment, is equivalent to a complete linear analysis, repeatedly applied. Consequently, the cost of running an elastic–plastic analysis will be greater where the stiffness matrix needs to be inverted and convergence criteria applied to each new increment of strain. Such repetitions, administered within static codes (as described above), tend to be slow, though alternative, faster FE simulations of sheet metal forming are now available commercially. For example, the matrix inversion is avoided within an incremental, explicit dynamic code. Moreover, the incremental procedure would be avoided altogether in an approximate, one-step FE method, based upon Hencky's total strain constitutive relations (see § 4.2). The user needs to choose between these codes when using brick or shell elements to simulate the required shape. Mesh refinement in regions of high strain is now automated and the choice offered to the user between alternative yield criteria for orthotropic sheet metals obviates any need to write specific user subroutines. The more recent 'explicit dynamic' code offsets the need to invert **K**, thereby reducing storage and avoiding the slower, convergent-dependent solutions found with an 'implicit static' code [9]. Though developed for plasticity under dynamic impacts, these explicit codes are also efficient for analyses of non-linear quasi-static processes including sheet metal forming, extrusion and rolling. The explicit code requires an increase to the real speeds of each process followed by a check that the outputs have remained unaffected, i.e. they remain within the steady state solution pertaining to a range of speeds. In particular, the accuracy of spring-back predictions in sheet metal is especially important to monitor, where it may be necessary to revert to the static code [10]. A rigid punch/die combination is usually assumed, and with a geometry that is often axi-symmetric, the meshing need only be applied to one quadrant of an initially flat sheet. This imposes certain boundary conditions that must be met, i.e. the constraining of points that do not displace during the forming operation. Where the sheet makes contact with punch and die a friction coefficient is required to attest the interfacial shear arising from forming forces. With the metal-to-metal contact, the punch/die geometry is also required. Normally, a rigid geometry is assumed given that the sheet forms in an elastic-plastic manner under distributed pressure. Plastic incompressibility limits the element choice. Solid elements will require fewer nodes to enforce volume constancy, say, at each corner node when simulating forging with brick elements. Shell elements, which define the full thickness, require fewer integration points. Inputs include the sheet's elastic constants and its flow curve, expressed in co-ordinates of true stress and natural strain, to allow for either a piecewise linear approximation or a closer curve fit. Codes which admit anisotropy require the r– and n–values, the strength coefficient and the pre-strain, as in Swift's law, eq(9.44a,b).

Essentially there are three incremental FE formulations of non-linear plasticity, which depend upon the severity of deformation [4]: (i) small displacements and small strains, (ii) large displacements with small strains and (iii) large displacements with finite strains. Within (i), the elastic and plastic strains are of similar magnitudes and the FE analysis used is that described in § 16.5 above. In (ii) and (iii), more generally, the appropriate finite stress and strain definitions replace engineering measures. In (ii), the true stress and natural strain are adequate to provide for the deformation found, typically, in the sheets formed from large displacements. Here, the natural strain is converted to engineering strain when applying the FLD. The particular finite measures chosen for combination (iii) are the second Piola-Kirchhoff stress and the Lagrange (Green's) strain, which were defined in chapter 1. Here the yield criterion and the hardening rule for finite deformation may retain the same form as in the small strain theory. Outputs from FE are usually displayed as contours of equivalent stress and strain and the displacements found provide an image of the distortion. In sheet metal forming, the safeness of forming zones is assessed within (i) coloured regions in the pressing and (ii) from the imposition of high strain states and strain signature peaks upon the FLD, examples of which follow.

16.6 FE Simulations

In the three applications of FE that follow, consideration is made of stretching thin sheet by hydraulic bulging, ball indentation and the pressing of a panel into a flanged box shape. Three different, available FE codes are applied to provide the simulations: the first two are incremental, implicit codes, as described above, and the third is a rapid, one-step solution used at the die try-out stage for new product designs. The elements employed are respectively: shell, 2-layered brick and 8-node, single-layered rectangles. In every simulation of sheet metal forming, a criterion of fracture is essential to limit the strain in the sheet to beneath its fracture value. Invariably, the forming limit diagram (FLD) is employed for this purpose (see Fig. 13.25). The simulated in-plane strains are placed at their appropriate positions within the FLD's principal, engineering strain axes. This locates the proximity of forming strain to each branch of the FLD. The branches are usually taken as the lower lines in a band of scatter to provide a margin of safety for working strain levels in practice. Points for strain combinations lying beneath the FLD are safe while points found to lie on or above the FLD are unsafe. The simulation thus locates the most likely site of failure, providing the critical load and displacement correspondingly. The many different strain combinations that arise during a forming operation produce a spread in points across the two quadrants of strain that continually changes. Consequently, the monitoring of strains in this way is integrated within the simulation, given that the FLD for the sheet material is available. Normally, a FLD is found from laboratory tests that ensure representative strains paths.

16.6.1 Hydraulic Bulging

The bulge test has the advantage of eliminating contact friction, where oil under pressure is applied to the underside of a clamped and sealed disc, forcing it to bulge through a top die (see Fig. 9.16). Consequently, the bulging in a circular die may be used as a convenient check upon FE simulation of this stretch forming processes. Given that an analytical bulge theory is available for both isotropic and anisotropic sheet (see § 9.7 and § 13.6), the theoretical relation between the pressure and pole displacement provides the required check upon an FE prediction. The theoretical plot (Fig. 16.6a) between these two parameters shows a rising curve that reaches a pressure maximum at the inception of a pole instability and thereafter pole failure under diffuse straining. Such behaviour is confirmed by experiment and reproduced by FE, indicating that other outputs from the simulation can be relied upon.

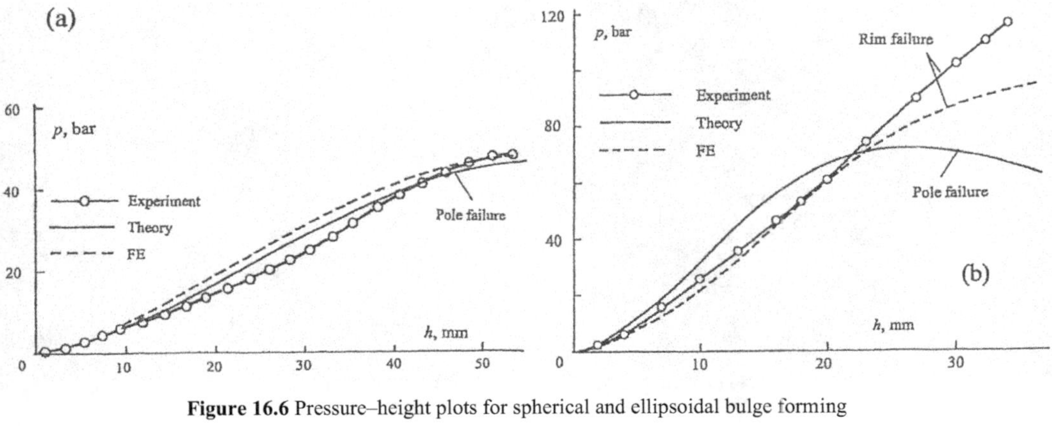

Figure 16.6 Pressure–height plots for spherical and ellipsoidal bulge forming

Actually, an experimental pressure versus height plot provides a better check upon FE because the theory assumes that bulge sections are circular arcs and that pole deformation is equi-biaxial. In practice, some drawing of rim material occurs over the locking bead [11] with a sharp bending of the sheet at this position. Moreover, for a severely rolled sheet, the pole curvature may oblate [12, 13]. In the FE simulation of the bulge test, shell elements were used within an implicit, static code. The simulation also included bulging with elliptical dies in which rim failures have occurred [14]. Not surprisingly here, the pressure–height plot is reproduced better by FE than the pole failure theory (see Fig. 16.6b).

(a)

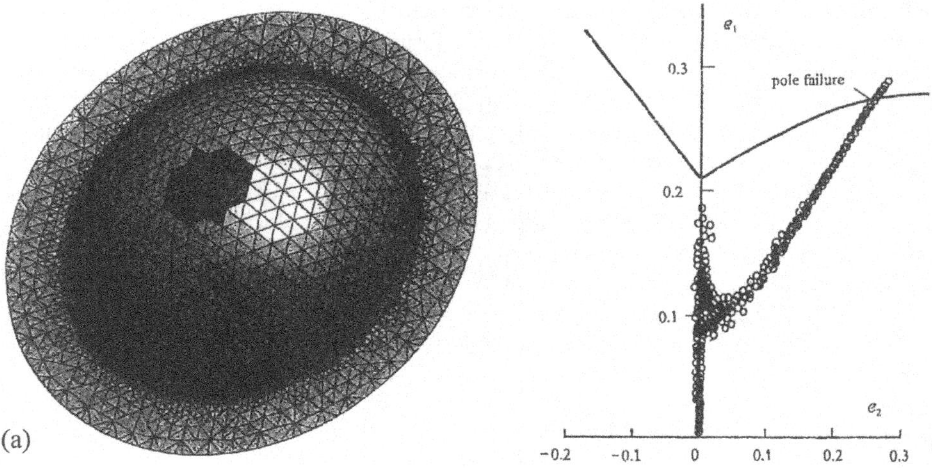

(b)

Figure 16.7 FE simulation of circular bulge strains

The full field strain predictions (Figs. 16.7a and 16.8a) reveal the most highly strained regions and their proximity to the forming limit. They show that, while a pole failure is expected for a circular bulge, a rim failure occurs in a narrow ellipsoidal die with axis ratio 0.42. The accompanying FLD's (see Figs 16.7b and 16.8b) reveal that these failures correspond to the equi-biaxial and plane strain paths respectively. Limiting plane strain corresponds to a major engineering strain e_1 lying at the base of the 'V' in the FLD, coincident with a constraint of zero minor strain e_2. This point defines the plane strain condition, known to be responsible for many service failures.

(a)

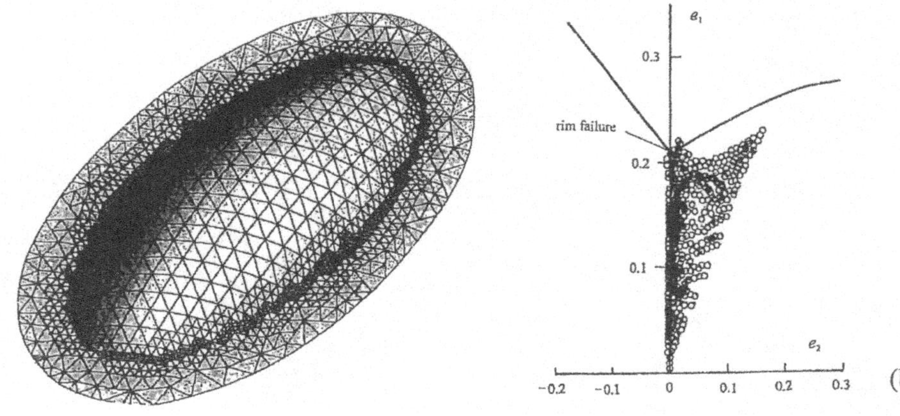

(b)

Figure 16.8 FE simulation of ellipsoidal bulge strains

Such predictions have been confirmed by experiment [14] in the case of a 1.22 mm aluminium alloy (6016). Strains for circular bulging lie within the first quadrant of the FLD at positions dependent upon the position in the bulge wall. The pole of a circular bulge approximates to a state of high, equi-biaxial strain. It is seen from Fig. 16.7b that a pole failure is predicted when, of all the strains at different locations from rim to pole, the pole's equi-biaxial strain path is the first to meet the FLD. On the other hand, for the narrow elliptical die shown in Fig. 16.8, it is the plane strain path, at the indicated position on the rim, that is the first to reach the FLD, this locating the rim failures position.

The strain predicted in the bulge wall can offer an alternative basis for the confirmation of FE output, provided they can be measured accurately (see Figs 16.9a,b). These strains are calculated from after-test optical measurements of the distortions to a small circular grid pattern etched to the sheet surface. Typically, the hoop ε_θ and meridional natural strain ε_r distributions along principal axes, from rim to pole of a bulge, provide a basis for comparison. Modern methods employ full-field, digital imaging of the deforming grid to confirm the strain maps provided by an FE simulation [11].

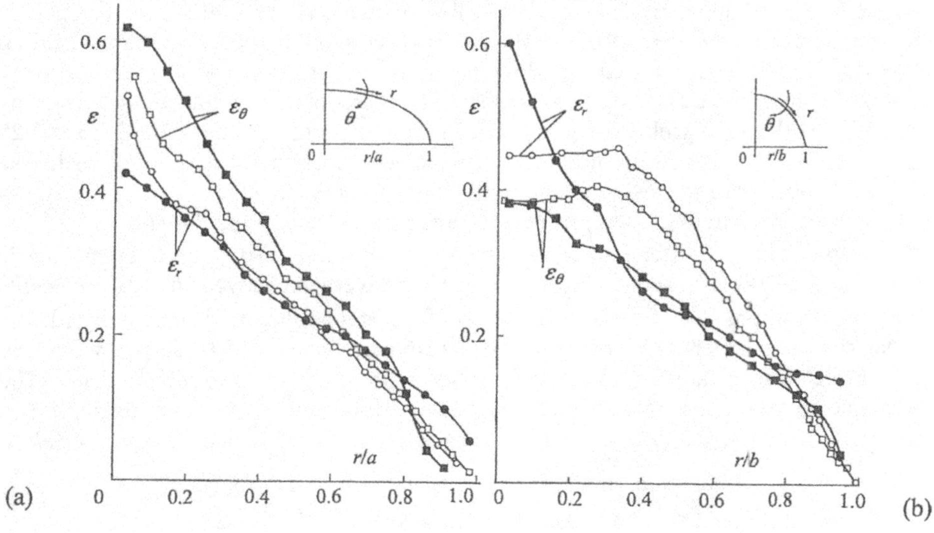

Figure 16.9 Predictions to principal strains along (a) major, (b) minor axes for an ellipsoidal bulge test

16.6.2 Ball Indentation

In the Erichsen test (see Fig. 16.10) no die is used as this allows the material in contact with the punch to stretch and thin under a tensile, biaxial stress state. The Erichsen number is simply the maximum height of the indentation at the initiation of failure. The comparative ductility measure provided by this number is particularly useful when assessing the suitability of different sheet metals for stretch forming, recognised universally with their now being international standards for this test's conditions [15–19].

In the simulation of an Erichsen test, as the punch velocity does not increase beyond 10 mm/s, the slower, static code is adequate. This test provides its measure of formability from test conditions (applied here) in which a 20 mm ball indents the centre of a 65 mm diameter disc, clamped at 15.3 bar around its periphery. With an axial symmetry present, only one quarter of the initially flat disc need be meshed.

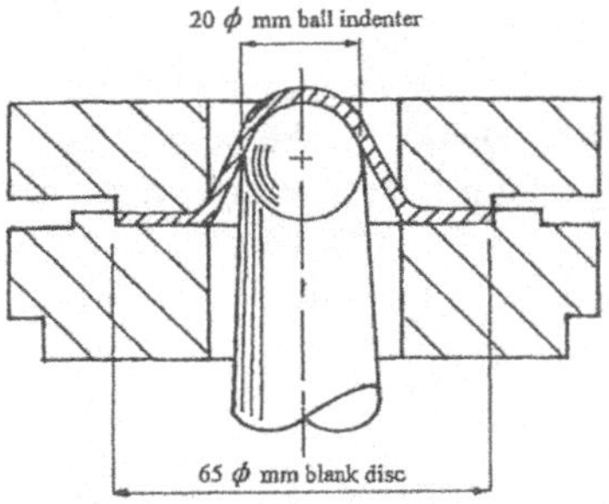

Figure 16.10 Erichsen indentation test geometry

An implicit Lagrangian code is used to provide for in-plane natural strains (< 80 %) from large punch displacements. This code employs two layers of 8-noded, solid elements to build the sheet thickness [20]. The in-plane, principal strain distributions, as predicted for the indented disc, will apply to any meridional line drawn from pole to rim if the sheet is initially isotropic. However, with the rolled steel sheet tested here (1.24 mm thick, En 3B steel), the principal strains varied with direction to the roll; this being a consequence of the different *r*–values in the plane of the sheet. The FE predictions allow for plastic anisotropy by adopting Hill's 1948 yield criterion, eq(12.2). This requires *r*–values 0.95, 1.1 and 1.28 respectively and the flow curves for the 0°, 45° and 90° directions. The flow curves for the rolled steel sheet were expressed from Swift's law eq(9.44b) in which the constants σ_o, n and ε_o connect true stress to natural strain (average values used were 245 **MPa**, 0.17 and 0.0057 respectively). Elastic anisotropy is negligible by comparison, thus allowing elastic moduli E and ν (200 GPa and 0.27 respectively) to be averaged between tensile test results in the three directions. The Coulomb friction coefficient for either lubricated or dry contact between sheet and punch is also required. Here, respective values of 0.05 and 0.2 were used in which the former value applies to lubrication with a thin film of polyethylene. The natural radial and hoop strains were calculated from after-test radial displacement measurements (Δr) separating concentric circles, whose diameters had changed by Δd:

$$\varepsilon_r = \ln(1 + \Delta r/r), \qquad \varepsilon_\theta = \ln(1 + \Delta d/d)$$

Figure 16.11a shows these principal strains at position s from the pole along the roll meridian for a lubricated test. The comparison suggests that measured strain distributions should be predicted reasonably well by FE. Inevitably, errors arise in the measurements of indentation distortion to a grid of initially concentric circles applied to the small flat discs. Within the lubricated contact zone ($s < 10$ mm) strains are approximately equi-biaxial, reaching their maxima at the pole. Greater differences occur in these strains for unlubricated contact (see Fig. 16.11b). In the non-contact region ($s > 10$ mm) the smaller strains differ, falling toward zero at the clamp end. Unsupported material, lying between the tangent point upon the ball to the edge of the rim clamp, is formed into a conical shape. In this region, friction is absent and a near plane strain condition prevails in which the clamp constrains the circumferential strain. In the annular region of flat material lying under the rim clamp ($s >$ 18 mm), Coulomb friction is again present to oppose the sliding motion under the normal clamping pressure. Some drawing-in is evident where ε_θ has become compressive. Where these three distinct regions receive separate analyses [21, 22], it is essential that they should marry properly in transition. Here, an FE simulation of the Erichsen test can assist, given that the simulation has been validated for certain test parameters, of which the most obvious is the force versus displacement relation.

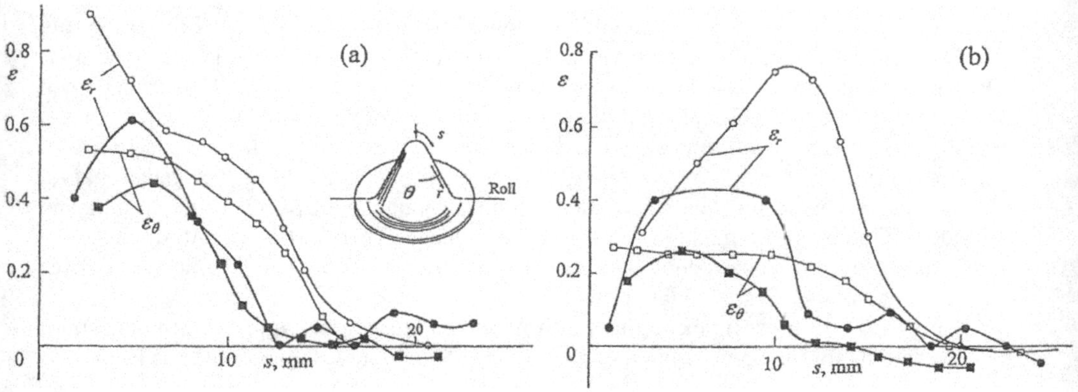

Figure 16.11 Predictions to principal strains in an Erichsen test: (a) lubricated, (b) dry

16.6.3 Pressed Panel

In simulating the deformation within a pressed panel, say for a car body, the FE analysis can begin with the final shape required. The 8-node rectangular mesh is applied to the panel design and an edge tension introduced to avoid wrinkling [23]. The approximate, one-step FE solution simply flattens the panel to ascertain its strains and the sheet size required. Of course, as the one-step method does not track strain history, it is prone to error for positions in the panel that underwent non-radial strain paths.

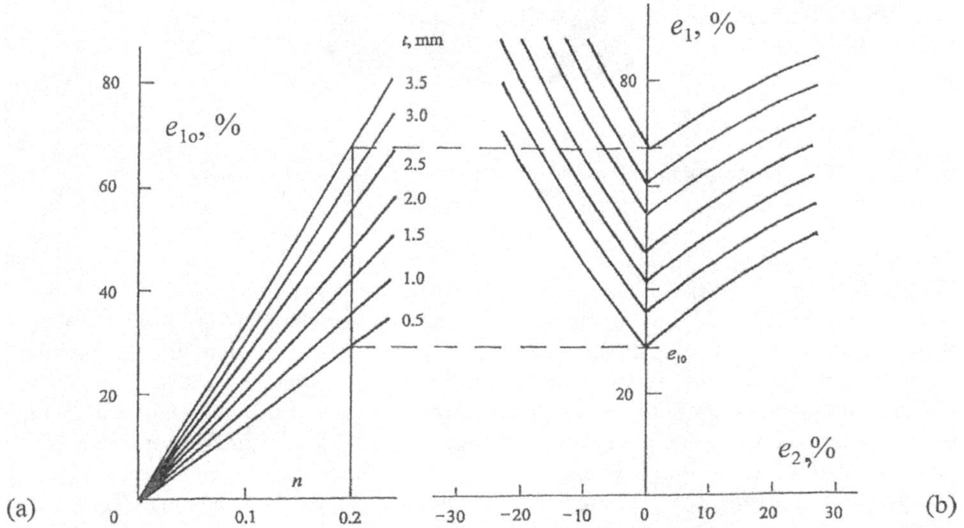

Figure 16.12 FLD dependencies upon thickness and n-value

There has been much debate upon the influence of non-radial paths upon the FLD, including experimental work that shows a translation in the position of this diagram [24]. The strains from simpler radial paths are superimposed upon a fixed FLD to examine their spread and safety. Ideally, all the material should be strained without violating the FLD. If the true FLD is not available, use may be made of a construction in which geometrically similar FLDs are assumed. Each major plane strain intercept e_1 is derived from the known influences of the strain hardening exponent and thickness of the material [25, 26], as shown in Fig. 16.12. The rapid, one-step simulation assists with die try-out, the selection of a material grade, blank sizing and assessing the feasibility of manufacturing new panel shapes. To achieve a worked sheet with acceptable strain, a change to the panel design, tooling and material choice can be made at the die 'try-out' stage. To facilitate a re-design, the FE outputs consist of coloured contour plots showing forming zones, safety zones, von Mises equivalent strains and strain signatures.

Figure 16.13 gives an example of a forming zone display in which all in-plane strain combinations for the corner region of a box-shaped panel lie safely beneath the FLD.

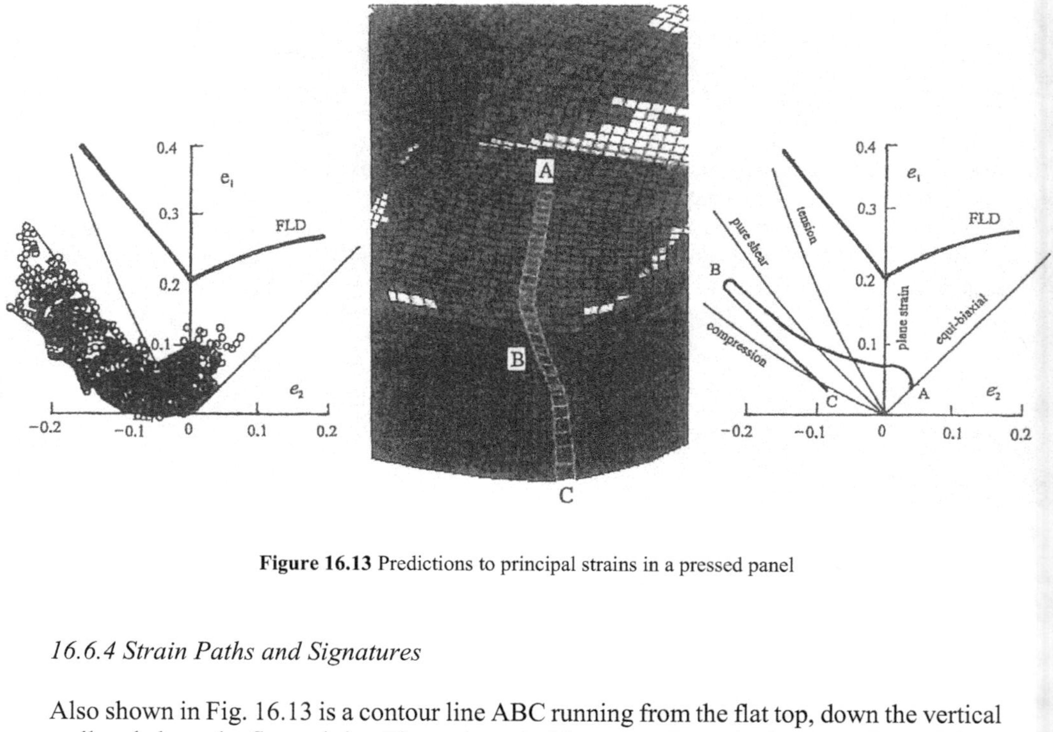

Figure 16.13 Predictions to principal strains in a pressed panel

16.6.4 Strain Paths and Signatures

Also shown in Fig. 16.13 is a contour line ABC running from the flat top, down the vertical wall and along the flanged rim. Shown inset is this contour's strain signature, formed from connecting the principal strain combinations from elements lying along ABC. This reveals that the full gamut of strain paths arise when forming a corner: from biaxial tension at the top through plane strain and pure shear in the wall to a simple compression at the rim. The signature may be verified experimentally with an economical circle grid analysis, i.e. by applying a grid to regions predicted to suffer high strain levels. It has been seen that a pressed product can contain all the common engineering strain paths appearing on the FLD: (i) equi-biaxial tension, (ii) plane strain, (iii) tension, (iv) pure shear and (v) compression. Typically, these paths apply to respective operations: (i) stretching over the punch face, (ii) forming a bend, (iii) axial stretching over a punch, (iv) deep drawing of a wall and (v) flanging a rim. The strain signature is a useful method of tracking these modes within highly strained contours. The signature refers to the plot between the two, in-plane engineering strains calculated for points (i.e. from deformed circles) lying around the contour. The signatures taken from parallel sections and contours provide a useful measure of the distribution of strain around and across the pressing. They show where die modifications are necessary to improve the uniformity in the spread of strain. Normally, the strain signature for each contour is superimposed upon the FLD. A useful isolated strain signature applied to a section of symmetry can capture all strain combinations across the two regions of the FLD. The overlay establishes the closest proximity of the signature to the forming limit curve. For example, Figs 16.14a,b show the signatures for radial section contours for forming a cup with (ABCD) and without (ABC) a flange.

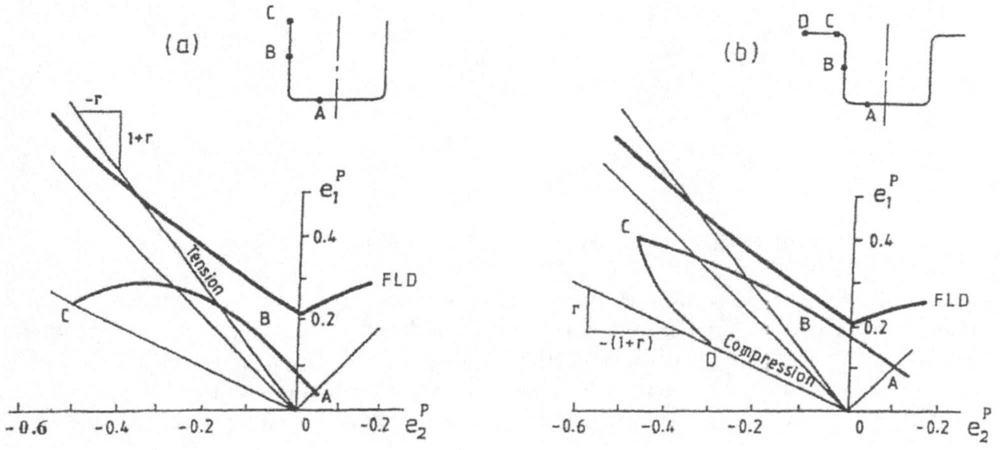

Figure 16.14 Strain signatures for simple axi-symmetric products

Signatures, taken for a radial section within each axi-symmetric, cylindrical product, traverse the full range of strain combinations. That is, from biaxial tension at A in the base through plane strain, tension and pure shear beyond B to uniaxial compression at C for the rim. Figure 16.15a gives further signatures for perpendicular radii AB and AC under hemispherical punch indentation without lubrication. Here, the extent of the sheet's in-plane anisotropy is revealed for the flanged cup in Fig. 16.15a where signatures differ for the radial contours AB and AC. In Fig. 16.15b, the signature ABCD for one corner of a flanged box is given. An isolated contour is taken from within the flat base at A, along the vertical corner BC to the rim at D. The five radial lines, including the minor strain axis shown are identified, in an anti-clockwise direction, with the deformation modes (i)–(v) stated above. It is seen how similar signatures, when taken from base to rim, will contain some or all of these strain states to varying degree depending upon the product. If the clearance between the signature's peak and the forming limit curve is deemed too great, then this may be reduced by changing to a lower grade, thinner material with a reduced yield stress. On the other hand, where the signature encroaches upon the forming limit's scatter band, indicating a risk of failure, two remedial measures may be applied. Firstly, the selection of a higher grade material may serve to increase the clearance to the degree of safety desired. Secondly, an equivalent safety margin can be achieved with a die re-design that serves to offset the undesirable peak strain without a material change. In applying either measure, trial tests upon a sheet with grid are undertaken before a full production run.

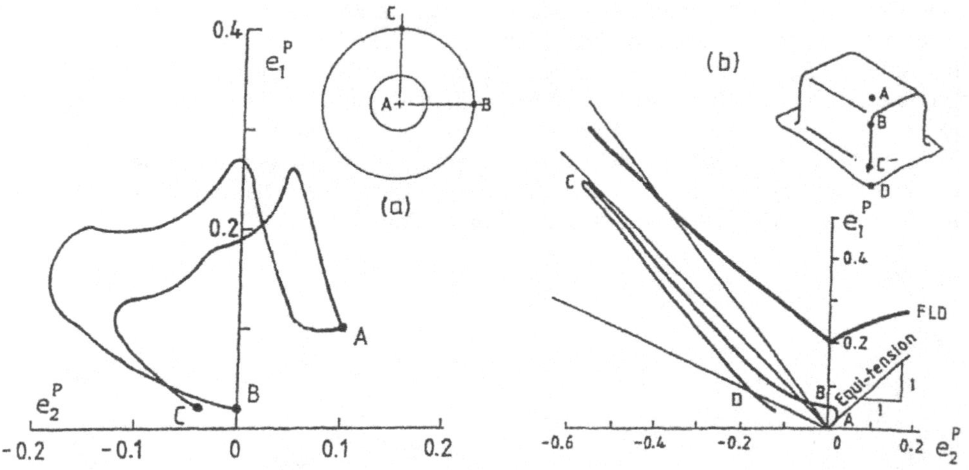

Figure 16.15 Strain signature for (a) punch indentation and (b) a box pressing

At the die tryout stage, strains are calculated from after-test optical measurements of the distortions to a small, circular grid pattern etched to the sheet's surface. A single signature may be established with an economical circular grid application, applied to a selected region known to suffer high strain levels. For example, when applied to contain the box corner ABCD in Fig. 16.15b, a grid provides the principal strain distribution from rim to base. Modern methods of strain measurement employ a full-field, digital imaging system for recording the deforming grid strains, these being used as a means of checking the strain coloured contour maps provided by an FE simulation.

16.7 Concluding Remarks

It has been shown how FE is applied incrementally to the regime of elastic-plastic deformation. The examples cited have applied various static FE codes to quasi-static sheet metal forming processes. Commercial FE codes adopt both implicit and explicit formulations, depending upon the nature of the plasticity. For gradual rates of loading the implicit (static) method fulfills equilibrium requirements in meeting its convergence criteria with each numerical time step. The latter involves a new strain increment for a quasi-static process and this increment may be made large to reduce the computing time, say, where hardening is relatively linear. On the other hand, implicit codes involve a far greater computation time to attain large strain with non-linear hardening, where accuracy demands smaller increments. In this case the use of an explicit code obviates the need for matrix inversion at each time step through its use of the central difference solution to the equation of motion. Where the straining process is rapid, the time step in a simulation becomes physically meaningful. The explicit codes were developed for dynamic plasticity including slower metal forming processes. If the time step is to be increased to reduce computation time it should not exceed a stability limit. To avoid an instability the forming process time can be reduced, say from simulating an increase in punch speed, provided this is known not to alter the material response.

References

1. Zeinkiewicz O. C. *The Finite Element Method*, 1977, McGraw-Hill.
2. Owen D. R. J. and Hinton E. *Finite Elements in Plasticity*, 1980, Pineridge Press.
3. Belytschko T., Liu W. K. and Moran B. *Non-linear Finite Elements for Continua and Structures*, 2000, Wiley.
4. Bathe K-J. *Finite Element Procedures*, 1996, Prentice-Hall.
5. Rees D. W. A. *Mechanics of Solids and Structures*, 2000, I.C. Press.
6. Spencer W. J. *Fundamental Structural Analysis*, 1988, Macmillan.
7. Fenner R. T. *Finite Element Methods for Engineers*, 1975, Macmillan.
8. Rowe G. W., Sturgess C. E. N., Hartley P. and Pillinger I. *Finite-Element Plasticity and Metalforming Analysis*, 1991, Cambridge University Press.
9. Wenner M. L. Proceedings: *Numisheet 2005*, CP778A, pp 3-7 (eds Smith L. M. et al.).
10. Wang C.T. Proceedings: *Numisheet 2002*, pp 13-24 (eds Yang D. Y. et al.).
11. Hu C. Q. and Rees D. W. A. Proceedings: *SAE 2005*, 06M-311.
12. Storakers B. *Int. Jl. Mech. Sci.*, 1996, **8**, 619.
13. Shang H. M. and Hsu T. C. *Trans ASME, Jl. Eng. for Industry*, 1979, **101**, 341.
14. Rees D. W. A. *Jl. Strain Analysis*, 2000, **35**(2), 109.
15. BS 3855: 1965, 1983, *Modified Erichsen Cupping Test for Sheet and Strip Metal.*
16. DIN 50 101 / 50 102 *Olsen Testing.*
17. ISO 8490, R 149-60.
18. ASTM 643-84.
19. Euronorm 14-58/67.
20. Rees D. W. A. and Solen T. Proceedings: *7th ESAFORM Conference*, Glasgow, 2005.
21. Kaftanoglu B. and Alexander J. M. *Jl. Inst. Metals*, 1961-2, **90**, 457.
22. Marciniak Z., Duncan J. L. and Hu S. J. *Mechanics of Sheet Metal Forming*, 2002, Butterworth-Heinemann.
23. Rees D. W. A. and Power R. *Jl. Mats. Proc. Tech.*, 1998, **77**, 134–144.
24. Laukonis J. A. and Gosh K. *Met. Trans.*, 1978, **A9**, 1849.
25. Hiam J. and Lee A. *Sheet Metal Industries*, May 1978.
26. Rees D. W. A. Proceedings: *4th ESAFORM Conference*, Leige, 2001, Vol. 2, 293.

Exercises

16.1 Using the principle of superposition between the two stiffness matrices (16.2) and (16.3a), find the stiffness matrix for a circular bar element subjected to combined axial tension and torsion.
See solution

16.2 Using eq(16.4), construct the stiffness matrix for two adjacent beam elements where each of their three nodes has two degrees of freedom.

16.3 Examine ways to invert the elastic–plastic matrix sum given in eqs(16.28d) and § 16.5.4 to a combined matrix $\mathbf{D}^{EP}$ in the following form: $d\boldsymbol{\sigma} = \mathbf{D}^{EP} d\boldsymbol{\varepsilon}$. *Hint*: See references [2, 8].

16.4 The load applied to a single beam element, length l and width w, is distributed uniformly from node 1 to node 2 as q/unit area of the top surface. Examine ways in which to construct the element's equivalent nodal force vector. *Note*: For the exact method, based upon the shape factor, the equivalent nodal force vector includes both shear forces and bending moments as $\mathbf{f}^e = (qlw/12)[6 \; l \; 6 \; -l]$.

16.5 Assemble the overall stiffness matrix for the 4-element cantilever beam shown in Fig. 16.5, for $F = 10$ kN. Show that the vertical deflection beneath the load is 0.127 mm. Take $b = 120$ mm, $h = 80$ mm and $t = 10$ mm respectively, with elastic modulus $E = 200$ GPa and $\nu = 0.3$.

16.6 Show that, for a plane triangular element, the displacements u and v may be expressed as

$$u = N_i\, u_i = N_1\, u_1 + N_2\, u_2 + N_3\, u_3 \text{ and } v = N_i\, v_i = N_1\, v_1 + N_2\, v_2 + N_3\, u_3$$

in which $N_i = a_i + b_i x + c_i y$ (i = 1, 2, 3) represent the shape functions

$$N_1 = a_1 + b_1 x + c_1 y,\ N_2 = a_2 + b_2 x + c_2 y \text{ and } N_3 = a_3 + b_3 x + c_3 y$$

Using the nodal displacements $u = u_1$, $v = v_1$ for $x = x_1$ and $y = y_1$ etc., show that coefficients a_i, b_i and c_i appear in terms of nodal point co-ordinates and the element's area Δ, as follows:

$$\Delta = \tfrac{1}{2}(x_1 - x_2)(y_3 - y_3) - \tfrac{1}{2}(x_2 - x_3)(y_1 - y_2)$$
$$a_1 = (x_2 y_3 - x_3 y_2)/(2\Delta),\ b_1 = (y_2 - y_3)/(2\Delta),\ c_1 = (x_3 - x_2)/(2\Delta)$$
$$a_2 = (x_3 y_1 - x_1 y_3)/(2\Delta),\ b_2 = (y_3 - y_1)/(2\Delta),\ c_2 = (x_1 - x_3)/(2\Delta)$$
$$a_3 = (x_1 y_2 - x_2 y_1)/(2\Delta),\ b_3 = (y_1 - y_2)/(2\Delta),\ c_3 = (x_2 - x_1)/(2\Delta)$$

See solution

16.7 When a uniformly distributed normal pressure is applied to the side 23 of the triangular element in Fig. 16.4b, the principle of virtual displacements gives an equivalent nodal force vector as

$$\mathbf{f}^{e} = t\left(\mathbf{A}^{e-1}\right)^{T} \int \mathbf{A}^{T}\mathbf{p}\, dl$$

where $\mathbf{p} = [\,p_x \quad p_y\,]^T$ and $\mathbf{f}^e$, $\mathbf{A}^e$ and $\mathbf{A}$ are given in eqs(16.14b), (16.8a) and (16.7c) respectively. Show, from matrix multiplication and integration, that the equivalent nodal force components satisfy equilibrium: $f_{x1} + f_{x2} + f_{x3} = p_x l_{23} t$ and $f_{y1} + f_{y2} + f_{y3} = p_y l_{23} t$. Take $dl = l_{23}\, dx/(x_2 - x_3)$, in which the side length $l_{23} = \sqrt{[(x_2 - x_3)^2 + (y_3 - y_2)^2]}$
See solution

16.8 Show, for an element, that when real nodal body forces exist (vector **b**) in addition to real external nodal forces $\mathbf{f}^e$ then the principal of virtual elastic displacements modifies eqs(16.5a–d) into the following forms:

$$f_k \delta_k^{v} + b_k \delta_k^{v} - \int_V \sigma_{ij} \varepsilon_{ij}^{v}\, dV = 0 \ \Rightarrow\ (\mathbf{f}^e)^T \boldsymbol{\delta}^{ev} + (\mathbf{b})^T \boldsymbol{\delta}^{ev} - \int_V (\boldsymbol{\sigma})^T \boldsymbol{\varepsilon}^{v}\, dV = 0$$

$$\delta_k^{v} f_k + \delta_k^{v} b_k - \int_V \varepsilon_{ij}^{v} \sigma_{ij}\, dV = 0 \ \Rightarrow\ (\boldsymbol{\delta}^{ev})^T \mathbf{f}^e + (\boldsymbol{\delta}^{ev})^T \mathbf{b} - \int_V (\boldsymbol{\varepsilon}^{v})^T \boldsymbol{\sigma}\, dV = 0$$

16.9 Show that the principal of virtual displacements can be modified as follows in the case of incremental plasticity with body forces distributed as self weight/unit volume.

$$(\Delta\boldsymbol{\delta}^{ev})^T \mathbf{f}^e + \int_V [\,(\Delta\boldsymbol{\delta}^{v})^T \mathbf{b} - (\Delta\boldsymbol{\varepsilon}^{v})^T \boldsymbol{\sigma}\,]\, dV = 0$$

16.10 At the die try-out stage the strain signature, taken across the corner of a pressed steel box, is found to lie in the danger region of the FLD (see Fig. 16.16). List the most likely modifications that could be made to make the pressing safe. Given that the original sheet thickness is 1.5 mm and the n-value is 0.2, employ the construction in Fig. 16.12 to recommend possible material changes.
See solution

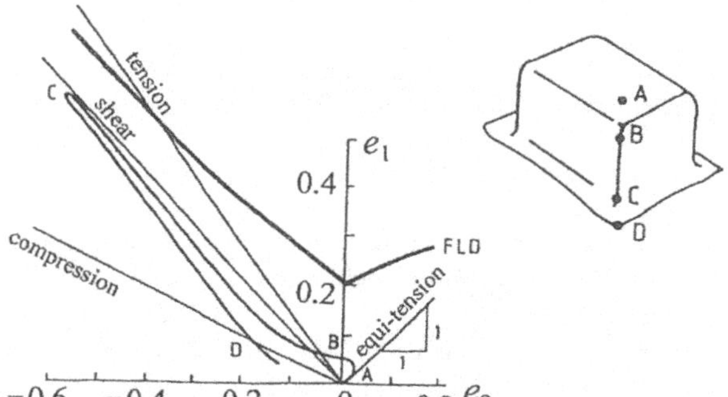

Figure 16.16

EXERCISES

WORKED SOLUTIONS

Solutions to selected exercises are given here. They are indicated by the same number combination that is used in the book, e.g. **10.3** refers to exercise 3 following chapter 10. The user of these selected solutions will find that reference is made to certain figures and equations appearing in the book in a similar manner (e.g. Fig. 3.4, eq(8.7) etc.). However, within the solutions given here, additional equations and figures are employed. These are referred to differently. Equations appear as lower case Roman numerals: (i), (ii) etc., and these are repeated within each self-contained solution. Diagrams are numbered to correspond with the exercise number, e.g. Diag. 2.13 appears in the solution to exercise 2.13 and is referred to as such. Thus, the term 'Fig.' is used for figures within the book and 'Diag.' for figures within the solutions. The user will find suggestions for alternative solutions where they exist. The worked solutions given below should equip the user with the necessary techniques to provide their own solutions to those exercises remaining. More solutions can be added depending upon demand but for the present the exercises chosen here should be sufficient for those planning to teach the subject through lectures and worked examples.

CHAPTER 1

Ex. 1.1 From eq(1.25a):

$$J_1 = \sigma_{11} + \sigma_{22} + \sigma_{33}$$

$$\sigma_{33} = J_1 - (\sigma_{11} + \sigma_{22}) = 5 - (2 + 2) = 1 \text{ MPa}$$

The principal stresses follow from the determinant, eq(1.23c):

$$\begin{vmatrix} 2 - \sigma & 3 & 1 \\ 3 & 2 - \sigma & 3 \\ 1 & 3 & 1 - \sigma \end{vmatrix} = 0$$

Expanding leads to the principal stress cubic (*the characteristic equation*):

$$\sigma^3 - 5\sigma^2 - 11\sigma + 7 = 0$$

The cubic roots, which may be solved by trial, by Newtons's approximation or graphically, give the major, intermediate and minor values ($\sigma_1 > \sigma_2 > \sigma_3$). Then τ_{max} from eq(1.33b):

$$\sigma_1 = 6.523,\ \sigma_2 = 0.523,\ \sigma_3 = -2.045 \text{ (MPa)}$$

$$\tau_{max} = ½(\sigma_1 - \sigma_3) = 4.284 \text{ MPa}$$

Ex. 1.3 With principal stresses as co-ordinates, eqs(1.31a,b) for the normal and shear stresses upon any oblique plane are simplified. In fact, this problem lends itself to a graphical solution (Fig. 1.17) but unless the Mohr's circles are drawn suitably to large scales then slight errors are inevitable (as in the answers given). The analytical solutions provide the correct answers. Check firstly that $l^2 + m^2 + n^2 = 1$, then substitute $l = 0.732$, $m = 0.521$ and $n = 0.439$ into eq(1.31a) to give the normal stress upon the oblique plane:

$$\sigma = \sigma_1 l^2 + \sigma_2 m^2 + \sigma_3 n^2$$

$$= 15.4\,(0.732)^2 + 12.65\,(0.521)^2 + 6.8\,(0.439)^2$$

$$\sigma = 12.995 \text{ MPa}$$

The components of the resultant force upon this plane are

$$r_1 = \sigma_1 l = 15.40 \times 0.732 = 11.273$$

$$r_2 = \sigma_2 m = 12.65 \times 0.521 = 6.591$$

$$r_3 = \sigma_3 n = 6.80 \times 0.439 = 2.985$$

when from eqs(1.12c) and (1.34a,b) the resultant shear and octahedral stresses are

$$\tau^2 = r^2 - \sigma^2 = (r_1^2 + r_2^2 + r_3^2) - \sigma^2$$

$$= (11.273^2 + 6.591^2 + 2.985^2) - 12.995^2 = 10.43, \therefore \tau = 3.23 \text{ MPa}$$

$$\sigma_o = ⅓(\sigma_1 + \sigma_2 + \sigma_3) = 11.62 \text{ MPa},$$

$$\tau_o = ⅓[(\sigma_1 - \sigma_2)^2 - (\sigma_2 - \sigma_3)^2 + (\sigma_3 - \sigma_2)^2]^{½} = 3.59 \text{ MPa}$$

Ex. 1.4 Check that $l^2 + m^2 + n^2 = 1$ then substitute $l = 2/\sqrt{14}$, $m = -1/\sqrt{14}$ and $n = 3/\sqrt{14}$ into eq(1.12b) to give the normal stress upon the plane. In the mixed notation given:

$$\sigma = \sigma_{11} l^2 + \sigma_{22} m^2 + \sigma_{33} n^2 + 2(lm\sigma_{12} + mn\sigma_{23} + ln\sigma_{13})$$

$$= (14 \times 4/14) + (10 \times 1/14) + (35 \times 9/14) + 2\,[\,-(7 \times 2/14) - (0 \times 3/14) - (7 \times 6/14)\,]$$

$$\sigma = 19.214 \text{ MPa}$$

To find the shear stress upon this plane first find the components of the resultant normal force r from eqs(1.11a–c):

$$r_{11} = [(2 \times 14) - (1 \times 7) - (3 \times 7)\,] / \sqrt{14} = 0$$

$$r_{22} = [-(1 \times 10) + (3 \times 0) + (2 \times 7)\,] / \sqrt{14} = 4/\sqrt{14}$$

$$r_{33} = [(3 \times 35) - (2 \times 7) - (1 \times 0)\,] / \sqrt{14} = 91/\sqrt{14}$$

when from eq(1.12c):

$$\tau^2 = r^2 - \sigma^2 = (r_{11}^2 + r_{22}^2 + r_{33}^2) - \sigma^2$$

$$= [0^2 + (4/\sqrt{14})^2 + (91/\sqrt{14})^2] - 19.214^2 = 223.465$$

$$\tau = 14.949 \text{ MPa}$$

Ex. 1.7 Check, firstly, the following relationships:

$$l_{11}^{\,2} + l_{12}^{\,2} + l_{13}^{\,2} = 1$$
$$l_{21}^{\,2} + l_{22}^{\,2} + l_{23}^{\,2} = 1$$
$$l_{31}^{\,2} + l_{32}^{\,2} + l_{33}^{\,2} = 1$$

Eq(1.22b) provides transformed stress components for the primed axes 1′, 2′ and 3′:

$$\begin{bmatrix} \sigma_{11}' & \sigma_{12}' & \sigma_{13}' \\ \sigma_{21}' & \sigma_{22}' & \sigma_{23}' \\ \sigma_{31}' & \sigma_{32}' & \sigma_{33}' \end{bmatrix} = \begin{bmatrix} 0.6 & 0 & -0.8 \\ 0 & 1 & 0 \\ 0.8 & 0 & 0.6 \end{bmatrix} \begin{bmatrix} 15 & -10 & 0 \\ -10 & 5 & 0 \\ 0 & 0 & 20 \end{bmatrix} \begin{bmatrix} 0.6 & 0 & 0.8 \\ 0 & 1 & 0 \\ -0.8 & 0 & 0.6 \end{bmatrix}$$

Matrix multiplication provides the transformed components in which the expected symmetry $\sigma_{ij}' = \sigma_{ji}'$ is confirmed:

$$\begin{bmatrix} \sigma_{11}' & \sigma_{12}' & \sigma_{13}' \\ \sigma_{21}' & \sigma_{22}' & \sigma_{23}' \\ \sigma_{31}' & \sigma_{32}' & \sigma_{33}' \end{bmatrix} = \begin{bmatrix} 18.2 & -6 & -2.4 \\ -6 & 5 & -8 \\ -2.4 & -8 & 16.8 \end{bmatrix}$$

Ex. 1.10 Firstly, check that the three vectors given are orthogonal, i.e. their dot products are zero: $\mathbf{a} \cdot \mathbf{b} = 0$, $\mathbf{b} \cdot \mathbf{c} = 0$ and $\mathbf{a} \cdot \mathbf{c} = 0$. For example:

$$\mathbf{a} \cdot \mathbf{b} = (1 \times 1) + (2 \times 1) - (3 \times 1) = 0$$

The unit vector equation for the axes x_1', x_2' and x_3' are respectively

$$\mathbf{u}_1' = \frac{\mathbf{a}}{|\mathbf{a}|} = \frac{1}{\sqrt{14}}\mathbf{u}_1 + \frac{2}{\sqrt{14}}\mathbf{u}_2 + \frac{3}{\sqrt{14}}\mathbf{u}_3$$

$$\mathbf{u}_2' = \frac{\mathbf{b}}{|\mathbf{b}|} = \frac{1}{\sqrt{3}}\mathbf{u}_1 + \frac{1}{\sqrt{3}}\mathbf{u}_2 - \frac{1}{\sqrt{3}}\mathbf{u}_3$$

$$\mathbf{u}_3' = \frac{\mathbf{c}}{|\mathbf{c}|} = -\frac{5}{\sqrt{42}}\mathbf{u}_1 + \frac{4}{\sqrt{42}}\mathbf{u}_2 - \frac{1}{\sqrt{42}}\mathbf{u}_3$$

which identify the direction cosines $l_{11} = 1/\sqrt{14}$, $l_{12} = 2/\sqrt{14}$, $l_{13} = 3/\sqrt{14}$ etc., so that eq(1.22b) provides the required transformation:

$$\begin{bmatrix} \sigma_{11}' & \sigma_{12}' & \sigma_{13}' \\ \sigma_{21}' & \sigma_{22}' & \sigma_{23}' \\ \sigma_{31}' & \sigma_{32}' & \sigma_{33}' \end{bmatrix} = \begin{bmatrix} 0.267 & 0.535 & 0.802 \\ 0.577 & 0.577 & -0.577 \\ -0.772 & 0.617 & -0.154 \end{bmatrix} \begin{bmatrix} 1 & 5 & -5 \\ 5 & 0 & 0 \\ -5 & 0 & -1 \end{bmatrix} \begin{bmatrix} 0.267 & 0.577 & -0.772 \\ 0.535 & 0.577 & 0.617 \\ 0.802 & -0.577 & -0.154 \end{bmatrix}$$

The matrix multiplication provides the transformed components:

$$\begin{bmatrix} \sigma_{11}' & \sigma_{12}' & \sigma_{13}' \\ \sigma_{21}' & \sigma_{22}' & \sigma_{23}' \\ \sigma_{31}' & \sigma_{32}' & \sigma_{33}' \end{bmatrix} = \begin{bmatrix} -1.286 & 1.389 & 1.980 \\ 1.389 & 6.667 & -2.762 \\ 1.980 & -2.762 & -5.381 \end{bmatrix}$$

CHAPTER 2

Ex. 2.1 The infinitesimal strain components ε_{ij} are given by the partial derivatives in eq(2.3):

$$\varepsilon_{11} = ½(\partial u_1/\partial x_1 + \partial u_1/\partial x_1) = 2(k_1 x_1 - k_3 x_3) \times k_1$$

$$\varepsilon_{12} = ½(\partial u_1/\partial x_2 + \partial u_2/\partial x_1) = ½\,(0 + 0) = 0$$

$$\varepsilon_{13} = ½(\partial u_1/\partial x_3 + \partial u_3/\partial x_1) = ½\,[2(k_1 x_1 - k_3 x_3) \times (-k_3) + k_1 k_2 x_2]$$

$$\varepsilon_{21} = ½(\partial u_2/\partial x_1 + \partial u_1/\partial x_2) = ½\,(0 + 0) = 0 = \varepsilon_{12}$$

$$\varepsilon_{22} = ½(\partial u_2/\partial x_2 + \partial u_2/\partial x_2) = 2(k_2 x_2 + k_3 x_3) \times k_2$$

$$\varepsilon_{23} = ½(\partial u_2/\partial x_3 + \partial u_3/\partial x_2) = ½\,[2(k_2 x_2 + k_3 x_3)\, k_3 + k_1 k_2 x_1]$$

$$\varepsilon_{31} = ½(\partial u_3/\partial x_1 + \partial u_1/\partial x_3) = ½\,[k_1 k_2 x_2 + 2(k_1 x_1 - k_3 x_3) \times (-k_3)] = \varepsilon_{31}$$

$$\varepsilon_{32} = ½(\partial u_3/\partial x_2 + \partial u_2/\partial x_3) = ½\,[k_1 k_2 x_1 + 2(k_2 x_2 + k_3 x_3)\, k_3] = \omega_{23}$$

$$\varepsilon_{33} = ½(\partial u_3/\partial x_3 + \partial u_3/\partial x_3) = 0$$

The rigid body rotation components ω_{ij} are given by the partial derivatives in eq(2.4):

$$\omega_{11} = ½(\partial u_1/\partial x_1 - \partial u_1/\partial x_1) = 0$$

$$\omega_{12} = ½(\partial u_1/\partial x_2 - \partial u_2/\partial x_1) = 0$$

$$\omega_{13} = ½(\partial u_1/\partial x_3 - \partial u_3/\partial x_1) = ½\,[2(k_1 x_1 - k_3 x_3) \times (-k_3) - k_1 k_2 x_2]$$

$$\omega_{21} = ½(\partial u_2/\partial x_1 - \partial u_1/\partial x_2) = 0$$

$$\omega_{22} = ½(\partial u_2/\partial x_2 - \partial u_2/\partial x_2) = 0$$

$$\omega_{23} = ½(\partial u_2/\partial x_3 - \partial u_3/\partial x_2) = ½\,[2(k_2 x_2 + k_3 x_3)k_3 - k_1 k_2 x_1]$$

$$\omega_{31} = ½(\partial u_3/\partial x_1 - \partial u_1/\partial x_3) = ½\,[k_1 k_2 x_2 - 2(k_1 x_1 - k_3 x_3) \times (-k_3)] = -\,\omega_{13}$$

$$\omega_{32} = ½(\partial u_3/\partial x_2 - \partial u_2/\partial x_3) = ½\,[k_1 k_2 x_1 - 2(k_2 x_2 + k_3 x_3)\, k_3] = -\,\omega_{23}$$

$$\omega_{33} = ½(\partial u_3/\partial x_3 + \partial u_3/\partial x_3) = 0$$

At a point P(0, 1, 2), with $k_1 = k_2 = 1$ and $k_3 = 2$, (each $k \times 10^{-6}$) the micro-strain components ε_{ij} are

$$\varepsilon_{11} = -\,8,\;\; \varepsilon_{12} = 0,\;\; \varepsilon_{13} = 17/2$$

$$\varepsilon_{21} = 0,\;\; \varepsilon_{22} = 10,\;\; \varepsilon_{23} = 10$$

$$\varepsilon_{31} = 17/2,\;\; \varepsilon_{32} = 10,\;\; \varepsilon_{33} = 0$$

The tensor shear strain between two perpendicular lines passing through P requires a transformation (2.8a) upon these matrix ε_{ij} components; where l_1, m_1 and n_1 are the direction cosines of one line and l_2, m_2 and n_2 are the direction cosines of the other. Thus:

$$\frac{\gamma}{2} = \begin{bmatrix} 1/\sqrt{3} & 1/\sqrt{3} & -1/\sqrt{3} \end{bmatrix} \begin{bmatrix} -8 & 0 & 17/2 \\ 0 & 10 & 10 \\ 17/2 & 10 & 0 \end{bmatrix} \begin{bmatrix} 1/\sqrt{14} \\ 2/\sqrt{14} \\ 3/\sqrt{14} \end{bmatrix} = \frac{36}{\sqrt{42}}$$

Hence, the engineering shear strain is $\gamma = 11.11 \times 10^{-6}$, this being the radian measure of the angular change to the right angle. *Note*: The dot product for the two lines is zero initially.

Ex. 2.3 The direction cosines for the x_2' direction $l_{21} = 1/\sqrt{3}$, $l_{22} = 1/\sqrt{3}$ and $l_{23} = -1/\sqrt{3}$ ensure consistency with the following relationships:

$l_{21}^2 + l_{22}^2 + l_{23}^2 = 1$ and $l_{11}\, l_{21} + l_{12}\, l_{22} + l_{13}\, l_{23} = 0$

The direction cosines for the third orthogonal direction x_3' follows from the cross product between unit vectors aligned with 1′ and 2′: $\mathbf{u}_3' = \mathbf{u}_1' \times \mathbf{u}_2'$ (see text example 2.2):

$$\mathbf{u}_3' = \begin{vmatrix} \mathbf{u}_1 & \mathbf{u}_2 & \mathbf{u}_3 \\ 1/\sqrt{14} & 2/\sqrt{14} & 3/\sqrt{14} \\ 1/\sqrt{3} & 1/\sqrt{3} & -1/\sqrt{3} \end{vmatrix} = -\frac{5}{\sqrt{42}}\mathbf{u}_1 + \frac{4}{\sqrt{42}}\mathbf{u}_2 - \frac{1}{\sqrt{42}}\mathbf{u}_3$$

from which $l_{31} = -5/\sqrt{42}$, $l_{32} = 4/\sqrt{42}$ and $l_{33} = -1/\sqrt{42}$. One may now use eq(2.5b) to transform the strains:

$$\frac{\varepsilon_{ij}}{10^{-4}} = \begin{bmatrix} \frac{1}{\sqrt{14}} & \frac{2}{\sqrt{14}} & \frac{3}{\sqrt{14}} \\ \frac{1}{\sqrt{3}} & \frac{1}{\sqrt{3}} & -\frac{1}{\sqrt{3}} \\ -\frac{5}{\sqrt{42}} & \frac{4}{\sqrt{42}} & -\frac{1}{\sqrt{42}} \end{bmatrix} \begin{bmatrix} 1 & 5 & -5 \\ 5 & 0 & 0 \\ -5 & 0 & -1 \end{bmatrix} \begin{bmatrix} \frac{1}{\sqrt{14}} & \frac{1}{\sqrt{3}} & \frac{-5}{\sqrt{42}} \\ \frac{2}{\sqrt{14}} & \frac{1}{\sqrt{3}} & \frac{4}{\sqrt{42}} \\ \frac{3}{\sqrt{14}} & -\frac{1}{\sqrt{3}} & -\frac{1}{\sqrt{42}} \end{bmatrix} = \begin{bmatrix} \frac{-9}{7} & \frac{9}{\sqrt{42}} & \frac{24}{7\sqrt{3}} \\ \frac{9}{\sqrt{42}} & \frac{20}{3} & \frac{-31}{3\sqrt{14}} \\ \frac{24}{7\sqrt{3}} & \frac{-31}{3\sqrt{14}} & \frac{-226}{42} \end{bmatrix}$$

Ex. 2.7 It is often necessary to convert between the spacial and material components describing finite deformation (as in Ex. 2.13 below). This amounts to solving three simultaneous equations to give components X_j from a spacial description $x_i = x_i(X_j, t)$ or to give components x_j from a material description $X_i = X_i(x_j, t)$. Here, the former problem is simplified with the equality between one component in each reference frame ($x_3 = X_3$). Hence, the two equations to be solved are written as

$$x_1(\mathrm{e}^{-t} - K) = X_1(\mathrm{e}^{-t} - K) + X_2(\mathrm{e}^{t} - K)(\mathrm{e}^{-t} - K) \qquad \text{(i)}$$

$$x_2 = X_1(\mathrm{e}^{-t} - K) + X_2 \qquad \text{(ii)}$$

Subtracting eq(ii) from eq(i):

$x_1(e^{-t} - K) - x_2 = X_2 [(e^{t} - K)(e^{-t} - K) - 1]$

from which:

$$X_2 = \frac{x_1(e^{-t} - K) - x_2}{K(K - e^{t} - e^{-t})} \tag{iii}$$

Substituting eq(iii) into eq(ii), gives, after much manipulation:

$$X_1 = \frac{x_2 - X_2}{e^{-t} - K} = \frac{-x_1 + x_2(e^{t} - K)}{K(K - e^{t} - e^{-t})}$$

Ex. 2.10 The displacement vector **u** (or **U**) between the two reference frames x_i and X_i in Fig. 2.6 must be identical because they describe the same motion Pp. It is more common to use **u** only, so that here one writes $u_i = U_i$, giving the material displacement gradients $\partial u_i/\partial X_j$ as

$$u_1 = X_1 X_3^2 \Rightarrow \partial u_1/\partial X_1 = X_3^2, \quad \partial u_1/\partial X_2 = 0, \quad \partial u_1/\partial X_3 = 2X_1X_3$$

$$u_2 = X_1^2 X_2 \Rightarrow \partial u_2/\partial X_1 = 2X_1X_2, \quad \partial u_2/\partial X_2 = X_1^2, \quad \partial u_2/\partial X_3 = 0$$

$$u_3 = X_2^2 X_3 \Rightarrow \partial u_3/\partial X_1 = 0, \quad \partial u_3/\partial X_2 = 2X_2X_3, \quad \partial u_3/\partial X_3 = X_2^2$$

Now from eq(2.25a):

$x_i = U_i + X_i = u_i + X_i$

Knowing u_i, the components x_i become

$$x_1 = u_1 + X_1 = X_1X_3^2 + X_1$$

$$x_2 = u_2 + X_2 = X_1^2X_2 + X_2$$

$$x_3 = u_3 + X_3 = X_2^2X_3 + X_3$$

from which the deformation gradients $F_{ij} = \partial x_i/\partial X_i$ are

$$F_{11} = X_3^2 + 1, \quad F_{12} = 0, \quad F_{13} = 2X_1X_3$$

$$F_{21} = 2X_1X_2, \quad F_{22} = X_1^2 + 1, \quad F_{23} = 0$$

$$F_{31} = 0, \quad F_{32} = 2X_2X_3, \quad F_{33} = X_2^2 + 1$$

in which the relationship $F_{ij} = \partial u_i/\partial X_i + \delta_{ij}$ holds.

Ex. 2.13 In calculating the extension ratio from eq(2.30c), a unit vector $\mathbf{m} = [0 \;\; 1 \;\; 0]^T$ is aligned with the X_2–axis (see Diag. 2.13) and $\mathbf{G} = \mathbf{F}^T\mathbf{F}$ is calculated from the following components $F_{ij} = \partial x_i/\partial X_j$ of the deformation gradient matrix **F**:

$$x_1 = X_1, \Rightarrow F_{11} = 1,\ F_{12} = F_{13} = 0$$

$$x_2 = X_2 + KX_3, \Rightarrow F_{21} = 0,\ F_{22} = 1,\ F_{23} = K$$

$$x_3 = X_3 + KX_2, \Rightarrow F_{31} = 0,\ F_{32} = K,\ F_{33} = 1$$

Hence, the right Cauchy-Green deformation tensor **G** is formed as

$$\mathbf{G} = \begin{bmatrix} 1 & 0 & 0 \\ 0 & 1 & K \\ 0 & K & 1 \end{bmatrix}\begin{bmatrix} 1 & 0 & 0 \\ 0 & 1 & K \\ 0 & K & 1 \end{bmatrix} = \begin{bmatrix} 1 & 0 & 0 \\ 0 & 1+K^2 & 2K \\ 0 & 2K & 1+K^2 \end{bmatrix}$$

from which the required stretch ratio $\lambda_{\mathbf{X}} = \lambda_{\mathbf{m}}$ is found from

$$\lambda_{\mathbf{m}}^2 = \mathbf{m}^{\mathrm{T}}\mathbf{G}\mathbf{m} = \begin{bmatrix} 0 & 1 & 0 \end{bmatrix}\begin{bmatrix} 1 & 0 & 0 \\ 0 & 1+K^2 & 2K \\ 0 & 2K & 1+K^2 \end{bmatrix}\begin{bmatrix} 0 \\ 1 \\ 0 \end{bmatrix} = 1 + K^2$$

If the line element OP has unit length then the origin O ($X_i = 0, 0, 0$) does not displace, i.e. $x_1 = x_2 = x_3 = 0$ but the other end P ($X_i = 0, 1, 0$) does displace to position p ($x_1 = 0$ $x_2 = 1$ and $x_3 = K$). Thus, in Diag. 2.13 the extension ratio to which $\lambda_{\mathbf{m}} = \sqrt{(1 + K^2)}$ refers is apparent.

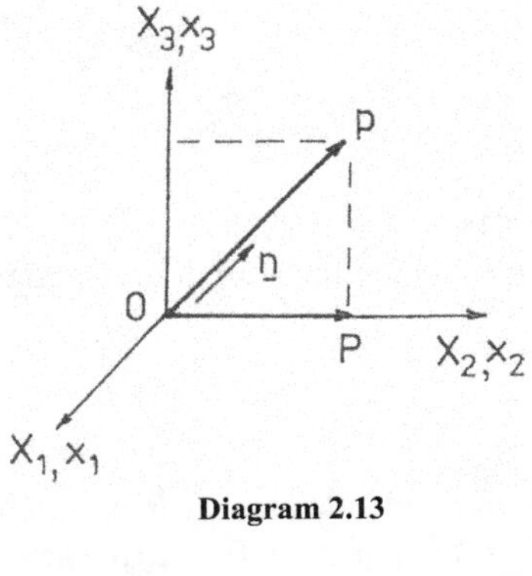

Diagram 2.13

Referring to Diag. 2.13, when calculating the extension ratio from eq(2.30c), a unit vector $\mathbf{n} = [0 \quad (1 + K^2)^{-½} \quad K/(1 + K^2)^{-½}]$ is aligned with Op. The Cauchy deformation tensor **C** = $\mathbf{H}^{\mathrm{T}}\mathbf{H}$ is derived from the inverse relationships $X_i = X_i(x_j, t)$:

$$X_1 = x_1$$

$$X_2 = -x_2/(K^2 - 1) + Kx_3/(K^2 - 1)$$

$$X_3 = (Kx_2 - x_3)/(K^2 - 1)$$

These give the components $H_{ij} = \partial X_i/\partial x_j$ of the spacial deformation gradient **H** as

$H_{11} = 1,\ \ H_{12} = 0,\ \ H_{13} = 0$

$H_{21} = 0,\ \ H_{22} = -1/(K^2 - 1),\ \ H_{23} = K/(K^2 - 1)$

$H_{31} = 0,\ \ H_{32} = K/(K^2 - 1),\ \ H_{33} = -\ 1/(K^2 - 1)$

and hence **C** is formed:

$$\mathbf{C} = \begin{bmatrix} 1 & 0 & 0 \\ 0 & -1/(K^2-1) & K/(K^2-1) \\ 0 & K/(K^2-1) & -1/(K^2-1) \end{bmatrix} \begin{bmatrix} 1 & 0 & 0 \\ 0 & -1/(K^2-1) & K/(K^2-1) \\ 0 & K/(K^2-1) & -1/(K^2-1) \end{bmatrix}$$

$$= \begin{bmatrix} 1 & 0 & 0 \\ 0 & (K^2+1)/(K^2-1)^2 & -2K/(K^2-1)^2 \\ 0 & -2K/(K^2-1)^2 & (K^2+1)/(K^2-1)^2 \end{bmatrix}$$

from which the required stretch ratio $\lambda_{\mathbf{x}} = \lambda_{\mathbf{n}}$ is found from

$$\frac{1}{\lambda_{\mathbf{n}}^2} = \mathbf{n}^{\mathrm{T}}\mathbf{C}\mathbf{n} = \begin{bmatrix} 0 & \dfrac{1}{\sqrt{1+K^2}} & \dfrac{K}{\sqrt{1+K^2}} \end{bmatrix} \begin{bmatrix} 1 & 0 & 0 \\ 0 & \dfrac{K^2+1}{(K^2-1)^2} & \dfrac{-2K}{(K^2-1)^2} \\ 0 & \dfrac{-2K}{(K^2-1)^2} & \dfrac{(K^2+1)}{(K^2-1)^2} \end{bmatrix} \begin{bmatrix} 0 \\ \dfrac{1}{\sqrt{1+K^2}} \\ \dfrac{K}{\sqrt{1+K^2}} \end{bmatrix}$$

$$= \frac{1}{(1+K)(1\ +\ K^2)}$$

giving $\lambda_{\mathbf{n}} = \surd[(1 + K)(1 + K^2)]$. This differs from $\lambda_{\mathbf{m}} = \surd(1 + K^2)$ as a consequence of the different directions **n** and **m** to which the stretch ratio refers.

CHAPTER 3

Ex. 3.6 In eqs(3.15a–c), the right-hand sides are expressed, conveniently, in terms of the shear yield stress k, where $J_2' = k^2$ and $J_3' = 0$. At the tensile yield stress Y, the two deviatoric invariants, eqs(3.10b,c), become

$$J_2' = \frac{Y^2}{3},\ \ J_3' = \frac{2Y^3}{27}$$

Substituting into eq(3.15a), Drucker's function gives the following relationship between the two yield stresses:

$$\left(\frac{Y^2}{3}\right)^3 - c\left(\frac{2Y^3}{27}\right)^2 = k^6, \quad \Rightarrow \quad Y = \frac{k}{(1/27\ -\ 4c/9^3)^{1/6}}$$

Similarly, the relationships between Y and k from the remaining two homogenous functions (3.15b,c) are

$$\frac{Y^2}{3} - b\left(\frac{2Y^3}{27} \times \frac{3}{Y^2}\right)^2 = k^2, \quad \Rightarrow \quad Y = \frac{9k}{(27 - 4b)^{1/2}}$$

$$\left(\frac{Y^2}{3}\right)^{3/2} - d\left(\frac{2Y^3}{27}\right) = k^3, \quad \Rightarrow \quad Y = \frac{k}{[(1/3)^{3/2} - 2d/27]^{1/3}}$$

Hence, the constants b, c and d may be determined from known yield stresses.

Ex. 3.10 The reduction to eq(c) (in Table 3.3) to tensile yielding requires similar substitutions for J_2' and J_3', given in ex. 3.6 above, together with the reduction in the term $A_{ijkl}\sigma_{ij}'\sigma_{kl}'$ by successive summations that retain the normal stress deviators as follows:

$$A_{ijkl}\sigma_{ij}'\sigma_{kl}' = A_{1jkl}\sigma_{1j}'\sigma_{kl}' + A_{2jkl}\sigma_{2j}'\sigma_{kl}' + A_{3jkl}\sigma_{3j}'\sigma_{kl}' \quad \text{for } i = 1, 2, 3$$

$$= A_{11kl}\sigma_{11}'\sigma_{kl}' + A_{22kl}\sigma_{22}'\sigma_{kl}' + A_{33kl}\sigma_{3j}'\sigma_{kl}' \quad \text{for } j = 1, 2, 3$$

$$\begin{aligned} &= A_{111l}\sigma_{11}'\sigma_{1l}' + A_{112l}\sigma_{11}'\sigma_{2l}' + A_{113l}\sigma_{11}'\sigma_{3l}' \\ &+ A_{221l}\sigma_{22}'\sigma_{1l}' + A_{222l}\sigma_{22}'\sigma_{2l}' + A_{223l}\sigma_{22}'\sigma_{3l}' \\ &+ A_{331l}\sigma_{33}'\sigma_{1l}' + A_{332l}\sigma_{22}'\sigma_{2l}' + A_{333l}\sigma_{33}'\sigma_{3l}' \quad \text{for } k = 1, 2, 3 \end{aligned}$$

$$\begin{aligned} &= A_{1111}\sigma_{11}'\sigma_{11}' + A_{1122}\sigma_{11}'\sigma_{22}' + A_{1133}\sigma_{11}'\sigma_{33}' \\ &+ A_{2211}\sigma_{22}'\sigma_{11}' + A_{2222}\sigma_{22}'\sigma_{22}' + A_{2233}\sigma_{22}'\sigma_{33}' \\ &+ A_{3311}\sigma_{33}'\sigma_{11}' + A_{3322}\sigma_{33}'\sigma_{22}' + A_{3333}\sigma_{33}'\sigma_{33}' \quad \text{for } l = 1, 2, 3 \end{aligned}$$

Substituting $\sigma_{11}' = ⅔\sigma_{1y}$, $\sigma_{22}' = \sigma_{33}' = -⅓\sigma_{1y}$ with $A_{1122} = A_{2211}$ etc. gives

$$A_{ijkl}\sigma_{ij}'\sigma_{kl}' = (1/9)[4(A_{1111} - A_{1122} - A_{1133}) + 2A_{2233} + A_{2222} + A_{3333}]\sigma_{1y}^2 \quad \text{(i)}$$

Where the retained tensor components A_{ijkl} are individual constants, one may replace eq(i) with a single constant $A_{ijkl}\sigma_{ij}'\sigma_{kl}' = A\,\sigma_{1y}^2$ so that the tensile yield function becomes

$$f = \left(\frac{\sigma_{1y}^2}{3}\right)^2\left(\frac{\sigma_{1y}^2}{3} + A\sigma_{1y}^2\right) - C\left(\frac{2\sigma_{1y}^3}{27}\right)^2 = \left(1 + 3A - \frac{4C}{27}\right)\left(\frac{\sigma_{1y}}{\sqrt{3}}\right)^6$$

The reduction to eq(d) in Table 3.3 for tension becomes

$$f = g\sigma_{1y}^m + h\sigma_{1y}^m + a(2\sigma_{1y})^m + b(-\sigma_{1y})^m + c(-\sigma_{1y})^m \quad \text{for } 1 \le m \le 2 \quad \text{(ii)}$$

When $m = 2$ or $m = 1$, eq(ii) becomes

$$f = (g + h + a2^m \pm b \pm c)\sigma_{1y}^m = \sigma^m \quad \text{(iii)}$$

When $1 < m < 2$ one should interpret the term $(-\sigma_{1y})^m$ as meaning $-|\sigma_{1y}|^m$. A further reduction in eq(iii) applies to planar isotropy when $a = b = g = 0$, giving $\sigma^m = (h \pm c)\sigma_{1y}^{\ m}$.

Ex. 3.12 The flow rule eq(3.22) requires a partial derivative of each function f in eqs(3.15a–c) with respect to principal stress σ_i ($i = 1, 2, 3$). These are respectively

$$f = J_2'^3 - cJ_3'^2, \quad \Rightarrow \quad \frac{\partial f}{\partial \sigma_i} = 3J_2'^2\left(\frac{\partial J_2'}{\partial \sigma_i}\right) - 2cJ_3'\left(\frac{\partial J_3'}{\partial \sigma_i}\right) \tag{i}$$

$$f = J_2' - b\left(\frac{J_3'}{J_2'}\right)^2, \quad \Rightarrow \quad \begin{aligned} \frac{\partial f}{\partial \sigma_i} &= \frac{\partial J_2'}{\partial \sigma_i} - 2b\left(\frac{J_3'}{J_2'}\right)\frac{\partial}{\partial \sigma_i}\left(\frac{J_3'}{J_2'}\right) \\ &= \frac{\partial J_2'}{\partial \sigma_i} - \frac{2bJ_3'}{J_2'^3}\left[J_2'\left(\frac{\partial J_3'}{\partial \sigma_i}\right) - J_3'\left(\frac{\partial J_2'}{\partial \sigma_i}\right)\right] \end{aligned} \tag{ii}$$

$$f = J_2'^{3/2} - dJ_3', \quad \Rightarrow \quad \frac{\partial f}{\partial \sigma_i} = \frac{3}{2}J_2'^{1/2}\left(\frac{\partial J_2'}{\partial \sigma_i}\right) - d\left(\frac{\partial J_3'}{\partial \sigma_i}\right) \tag{iii}$$

where J_2' and J_3' are given in eqs(3.10b,c), from which, with $i = 1$, for example:

$$\frac{\partial J_2'}{\partial \sigma_1} = \frac{1}{3}(2\sigma_1 - \sigma_2 - \sigma_3)$$

$$\frac{\partial J_3'}{\partial \sigma_1} = \frac{1}{3^3}\left[(2\sigma_1 - \sigma_2 - \sigma_3)^2 - (2\sigma_2 - \sigma_1 - \sigma_3)^2 - (2\sigma_3 - \sigma_1 - \sigma_2)^2\right]$$

The derivatives for $i = 2$ and 3 are found by rotating the subscripts. Then substituting the derivatives from eqs (i)–(iii) into eq(3.22) provides the three principal strain increments $d\varepsilon_i^{\ P}$.

Ex. 3.15 Equation (3.27) applies when the stress axes are aligned with the material axes x, y and z. (a) Aligned principal stresses (σ_1, σ_2) replace σ_x with σ_1 and σ_y with σ_2 for all other stress components set to zero. A plane, orthotropic yield function applies:

$$2f = F\sigma_2^{\ 2} + G\sigma_1^{\ 2} + H(\sigma_1 - \sigma_2)^2 = 1 \tag{i}$$

(b) The tensile yield stresses Y and Z are equal for their respective y and z directions, or 2 and 3 in the alternative notation of eq(12.2). Equation (3.27) shows

$$2f = (F + H)Y^2 = (F + G)Z^2 = 1$$

giving $H = G$, when eq(i) reduces to

$$2f = F\sigma_2^{\ 2} + G\sigma_1^{\ 2} + G(\sigma_1 - \sigma_2)^2 = 1$$

Ex. 3.18 (a) When $\theta = 0°$ in Fig. 3.15, tension σ becomes aligned with x and complementary shear τ with both x and y. Within eq(3.27), one may identify σ_x with σ and τ_{xy} with τ directly and set all other stress components to zero, giving

$$2f = (G + H)\sigma^2 + 2N\tau^2 = 1 \qquad \text{(i)}$$

Knowing the tensile yield stress X, eq(i) becomes

$$(G + H)\sigma^2 + 2N\tau^2 = (G + H)X^2$$

from which the yield criterion becomes

$$\sigma^2 + \left(\frac{2N}{G + H}\right)\tau^2 = X^2$$

(b) With an inclination $\theta°$ between the axes, as shown in Diag. 3.18, it becomes necessary to resolve the applied stresses σ and τ into the material axes x and y.

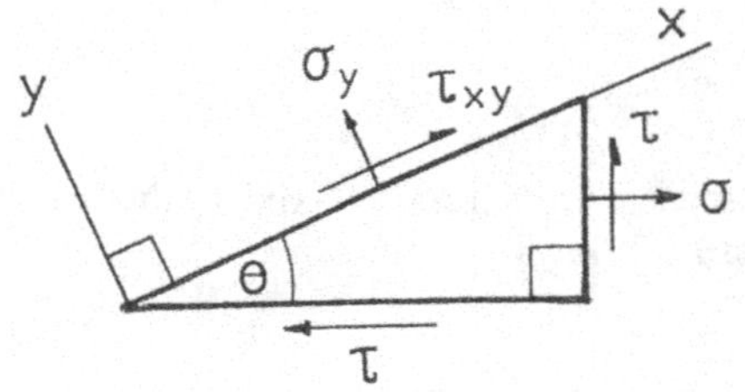

Diagram 3.18

Standard plane stress transformations may be applied, or better, resolve forces acting upon the right triangle. For the applied stress directions shown, with tension and clockwise shear (σ_y, τ_{xy}) assumed to act upon the plane of x, the latter method gives

$$\sigma_{yy} = \sigma \sin^2\theta - \tau \sin 2\theta \qquad \text{(i)}$$
$$\tau_{xy} = \tau \cos 2\theta - \tfrac{1}{2}\sigma \sin 2\theta \qquad \text{(ii)}$$

Setting $\theta = \theta + 90°$ in eqs (i) and (ii) provides the stress state (σ_{xx}, τ_{xy}) outward upon the plane of y:

$$\sigma_{xx} = \sigma \cos^2\theta + \tau \sin 2\theta \qquad \text{(iii)}$$
$$\tau_{yx} = -\tau \cos 2\theta + \tfrac{1}{2}\sigma \sin 2\theta \qquad \text{(iv)}$$

That is, eq(iii) and (iv) involve a sign change to ensure complementary shear $\tau_{xy} = \tau_{yx}$. It is the rectilinear element set in the plane of x–y for which σ_{xx}, σ_{yy} τ_{xy} appear in eq(3.27) (with all other stress components absent). This gives

$$2f = F\sigma_{yy}^2 + G\sigma_{xx}^2 + H(\sigma_{xx} - \sigma_{yy})^2 + 2N\tau_{xy}^2 = (G + H)X^2$$

and substituting from eqs(i)–(iv):

$$2f = F(\sigma \sin^2\theta - \tau \sin 2\theta)^2 + G(\sigma \cos^2\theta + \tau \sin 2\theta)^2 + H[\sigma(\cos^2\theta - \sin^2\theta) + 2\tau \sin 2\theta]^2 + 2N(\tau \cos 2\theta - \tfrac{1}{2}\sigma \sin 2\theta)^2 = (G + H)X^2$$

Hence, the stress combination (σ, τ) at yield may be found for any orientation θ.

CHAPTER 4

Ex. 4.1 A uniform stress state is assumed in a cylinder with a thin wall. This means that the shear stress gradient from outer to inner radius may be ignored. Here, the shear strain at yield γ_o, following a radial stress path $\sigma/\tau = 2$, is held constant while further tension is applied. The stress state (σ_o, τ_o) at yield conforms to

$$\sigma_o^2 + 3\,\tau_o^2 = Y^2 \tag{i}$$

and since $\sigma_o/\tau_o = 2$, the initial yield co-ordinates are $\sigma_o = 2Y/\surd 7$ and $\tau_o = Y/\surd 7$. Because the cylinder material is non hardening, the stress combination (σ, τ) must continually adjust to follow the yield locus, which in non-dimensional form becomes

$$(\sigma/Y)^2 + 3\,(\tau/Y)^2 = 1 \quad \Rightarrow \quad S^2 + 3T^2 = 1 \tag{ii, iii}$$

The manner of this adjustment with increasing axial strain can be found from the Hencky and Prandtl-Reuss theories as follows.

(a) Hencky

Equation (4.10a) describes the ensuing total axial strain where $\gamma_o = \tau_o/G = Y/\surd 7G$. This leads to

$$\varepsilon^t = \frac{\sigma Y}{\surd 21\, G \surd(Y^2 - \sigma^2)} + \frac{(1 - 2\nu)\sigma}{3E}$$

Substituting $E = 2G(1 + \nu)$, eq(iv) may be written in non-dimensional form:

$$\frac{6\,\varepsilon^t G}{Y} = \frac{6\,S}{\surd 21\,\sqrt{(1 - S^2)}} + \left(\frac{1 - 2\nu}{1 + \nu}\right) S \tag{iv}$$

(b) Prandtl-Reuss

The integration given in eqs(4.14a,b) applies to this problem. Substituting $\sigma_o = 2Y/\surd 7$:

$$\varepsilon^t = \frac{Y}{6G} \ln\left[\frac{(1 + \sigma/Y)(1 - 2/\surd 7)}{(1 + 2/\surd 7)(1 - \sigma/Y)}\right] + Y\left(\frac{1}{E} - \frac{1}{3G}\right)\left(\frac{\sigma}{Y} - \frac{2}{\surd 7}\right)$$

With $E = 2G(1 + \nu)$, the non-dimensional form becomes

$$\frac{6\,\varepsilon^t G}{Y} = \ln\left[\frac{(1 + S)(\surd 7 - 2)}{(\surd 7 + 2)(1 - S)}\right] + \frac{(1 - 2\nu)(\surd 7 S - 2)}{\surd 7(1 + \nu)} \tag{v}$$

When the given values of G, v and Y are substituted into eqs(iv) and (v), and taken with eq(iii), the variations in Figs 4.2a–c are found. *Note*: Only v is needed when $\varepsilon^t G/Y$ is plotted against S and T.

Ex 4.4 This is an example of constrained deformation under a principal biaxial stress state. It can be achieved from subjecting the thin tube to internal pressure and axial loading. The load would require a continual adjustment under a rising internal pressure if the axial strain at yield is to be maintained constant. Using z and θ for the axial and circumferential directions respectively, the Prandtl-Reuss, total strain increments under a rising circumferential tension are

$$d\varepsilon_z{}^t = d\lambda[\sigma_z - ⅓(\sigma_\theta + \sigma_z)] + (1/E)(d\sigma_z - v\, d\sigma_\theta) \quad \text{(i)}$$

$$d\varepsilon_\theta{}^t = d\lambda[\sigma_\theta - ⅓(\sigma_\theta + \sigma_z)] + (1/E)(d\sigma_\theta - v\, d\sigma_z) \quad \text{(ii)}$$

So, when the axial strain is held constant, $d\varepsilon_Z{}^t = 0$ and eq(i) gives $d\lambda$ as

$$d\lambda = \frac{-3(d\sigma_z - v d\sigma_\theta)}{E(2\sigma_z - \sigma_\theta)} \quad \text{(iii)}$$

Implicit differentiation of the yield criterion eq(4.32a) gives

$$\frac{d\sigma_z}{d\sigma_\theta} = \frac{-(2\sigma_\theta - \sigma_z)}{(2\sigma_z - \sigma_\theta)} \quad \text{(iv)}$$

Substituting eqs(iii) and (iv) into (ii) leads to

$$d\varepsilon_\theta^t = \frac{d\sigma_\theta}{E(2\sigma_z - \sigma_\theta)^2}\left[2(2 - v)Y^2 + (1 - 2v)(Y^2 - 3\sigma_\theta\sigma_z)\right] \quad \text{(v)}$$

in which eq(4.32a) allows σ_z to be expressed in terms of σ_θ:

$$(2\sigma_z - \sigma_\theta)^2 = 4Y^2 - 3\sigma_\theta{}^2 \quad \text{(vi)}$$

giving

$$\sigma_z = \frac{1}{2}\left[\sigma_\theta^2 \pm \sigma_\theta(4Y^2 - 3\sigma_\theta{}^2)^{1/2}\right] \quad \text{(vii)}$$

Substituting eqs(vi) and (vii) (with the positive root) into eq(v) leads to $\varepsilon_Z{}^t$ as the sum of three standard integrals:

$$\varepsilon_\theta^t = \frac{Y^2(5 - 4v)}{E}\int_0^{\sigma_\theta}\frac{d\sigma_\theta}{(4Y^2 - 3\sigma_\theta^2)} - \frac{3(1 - 2v)}{2E}\int_0^{\sigma_\theta}\frac{\sigma_\theta^2\, d\sigma_\theta}{(4Y^2 - 3\sigma_\theta^2)} - \frac{3(1 - 2v)}{2E}\int_0^{\sigma_\theta}\frac{\sigma_\theta\, d\sigma_\theta}{(4Y^2 - 3\sigma_\theta^2)^{1/2}}$$

leading to

$$\varepsilon_\theta^t = \frac{\sqrt{3}\,Y}{4E}\ln\left(\frac{Y + \sqrt{3}\,\sigma_\theta/2}{Y - \sqrt{3}\,\sigma_\theta/2}\right) - \frac{(1 - 2v)}{E}\left(Y - \frac{\sigma_\theta}{2}\right) + \frac{(1 - 2v)}{E}\sqrt{Y^2 - \frac{\sqrt{3}\,\sigma_\theta^2}{2}}$$

this being equivalent to the non-dimensional form given. The corresponding plots are those given in Fig. 4.6a–c, for which stress asymptotes of 2/√3 and 1/√3 appear in Fig. 4.6(c).

Ex. 4.7 (a) Following the incompressible Hencky theory for radial loading a solid circular bar, the bar first becomes plastic at its outer radius $\rho_{ep} = 1$ so that eqs(4.50a,b) apply:

$$n^2 + m^2 = 1 \quad \Rightarrow \quad n^2 [1 + (m/n)^2] = 1 \tag{i}$$

in which n and m are non-dimensional force and torque in eqs(4.42a,b), which divide as

$$\frac{m}{n} = \frac{2\surd 3}{r_o} \times \frac{C}{P} = \frac{2\surd 3}{7.375} \times \frac{1}{0.375} = 1.253$$

giving, from eq(i), $n = 0.624$ and $m = 0.782$. The corresponding loading is

$$P = nP_y = n\pi Y r_o^{\,2} = 0.624 \times \pi \times 283 \times 7.375^2 = 30.18 \times 10^3 \text{ N} = 30.18 \text{ kN}$$

$$C = mC_y = m\pi Y r_o^{\,3}/(2\surd 3) = 0.782 \times \pi \times 283 \times 7.375^3/(2\surd 3) = 80.51 \times 10^3 \text{ Nmm} = 80.51 \text{ Nm}$$

in which P_y and C_y are the yield loads under simple tension and simple torsion.

(b) By increasing P and C in the same proportion, beyond those found in (a), a plastic zone penetrates inward. The elastic–plastic interface coincides with the mean radius of the bar $\rho_{ep} = 0.5$. Equation (4.49) gives the following relationship between the normalised strains:

$$e^2 + 0.25g^2 = 1 \tag{ii}$$

Here, the assumption of incompressibility is invoked to provide n and m from eqs(4.52a,b):

$$n = 0.25e + (2e/g^2)(e^2 + g^2)^{1/2} - (2e/g^2)(e^2 + 0.25g^2)^{1/2} \tag{iii}$$

$$m = 0.0625g + (4/g)(e^2 + g^2)^{1/2} - (8e/3g^3)(e^2 + 0.25g^2)^{3/2} - (1.0/g)(e^2 + 0.25g^2)^{1/2} + (8/3g^3)(e^2 + 0.25g^2)^{3/2} \tag{iv}$$

Since $n/m = 0.798$, we may combine eqs(iii) and (iv) with eq(ii) to give an equation in g:

$$0.25(1 - 0.25g^2)^{1/2} + \frac{2}{g^2}(1 - 0.25g^2)^{1/2}(1 + 0.75g^2)^{1/2} - \frac{2}{g^2}(1 - 0.25g^2)^{1/2}$$
$$= 0.05g + \frac{3.19}{g}(1 + 0.75g^2)^{1/2} - \frac{2.13}{g^3}(1 + 0.75g^2)^{3/2} - \frac{0.798}{g} + \frac{2.13}{g^3}$$

for which an approximate solution $g = 1.12$ is found by trial. Hence, from eq(ii), $e = 0.83$ when $n = 0.73$ and $m = 0.915$. Then, as in (a): $P = nP_y = 35.3$ kN and $C = mC_y = 94.1$ Nm. *Note*: Here, e and g are normalised outer diameter strains accompanying this penetration.

(c) By increasing P and C in the same proportion to full plasticity, the plastic radius is zero ($\rho_{ep} = 0$) and from eq(4.49), $e = 1$. Following the route in (b) above leads to the required equations in n and m:

$$n = (2/g^2)(1 + g^2)^{1/2} - (2/g^2) \tag{v}$$

$$m = (4/g)(1 + g^2)^{1/2} - (8/3g^3)(1 + g^2)^{3/2} + 8/(3g^3) \tag{vi}$$

which are solved with $n = 0.798m$ to provide $g = 1.38$, giving $n = 0.74$ and $m = 0.93$. Equation (4.54a) provides a check on the load limits for which $P = nP_y = 35.8$ kN and $C = mC_y = 95.7$ Nm.

Ex. 4.9 Here, a 12.8 mm diameter sold bar is subjected to increasing tension under a constant torque. Hencky's incompressible theory may again be used to simplify the loading calculations involving plasticity (see § 4.4.3). Firstly, check that the torque is initially elastic, i.e. $m < 1$. In units of N and mm:

$$m = \frac{C}{C_y} = \frac{2\surd 3C}{\pi r_o^3 Y} = \frac{2\surd 3 \times 60.7 \times 10^3}{\pi \times (6.4)^3 \times 305} = 0.837$$

(a) In the 'fully–elastic' bar the yield stress has been reached at the outer radius from the added tension. This is found from eq(4.50b):

$$n = \surd(1 - m^2) = \surd[1 - (0.837)^2] = 0.547$$

giving the force as

$$P = nP_y = n\pi Y r_o^2 = 0.547 \times \pi \times 305 \times 6.4^2 = 21.47 \times 10^3 \text{ N} = 21.47 \text{ kN}$$

(b) By increasing P beyond 21.47 kN, with C constant, a plastic zone penetrates inward. When the elastic–plastic interface coincides with the mean radius of the bar, $\rho_{ep} = 0.5$. Equation (4.49) gives the following relationship between the normalised strains:

$$e^2 = 1 - 0.25g^2$$

when from eqs(4.52a,b):

$$n = 0.25(1 - 0.25g^2)^{1/2} + \frac{2}{g^2}(1 - 0.25g^2)^{1/2}(1 + 0.75g^2)^{1/2} - \frac{2}{g^2}(1 - 0.25g^2)^{1/2}$$

$$m = 0.837 = 0.0625g + \frac{4}{g}(1 + 0.75g^2)^{1/2} - \frac{8}{3g^3}(1 + 0.75g^2)^{3/2} - \frac{1}{g} + \frac{8}{3g^3}$$

The solution from the second equation is $g = 0.98$ (approx.) when $e = 0.872$ and $n = 0.784$ from the first equation. Hence the force has been increased to

$$P = n\,P_y = 30.8 \text{ kN}$$

(c) By increasing P to full plasticity ($\rho_{ep} = 0$), we have $e = 1$ from eq(4.49). Equations (4.52a,b) become

$$n = (2/g^2)(1 + g^2)^{1/2} - (2/g^2)$$

$$m = 0.837 = (4/g)(1 + g^2)^{1/2} - (8/3g^3)(1 + g^2)^{3/2} + 8/(3g^3)$$

for which $g \approx 1.11$, giving $n = 0.802$ and $P = 31.5$ kN.

Exs. 4.8 and **4.10** The non-dimensional stress distributions accompanying each penetration in exercises 4.7 and 4.9 above follow from eqs(4.51a,b). These produce plots typified by Figs 4.8a,b and 4.11a,b respectively. Normalised residual stresses follow from eqs(4.74b) and (4.75b). The corresponding plots (Figs 4.12a,c) are typical of those provided by Hencky with assumed elastic incompressibility.

CHAPTER 5

Ex. 5.1 Moments:

$$M_E = \frac{bd^2Y}{6} = \frac{50 \times 100^2 \times 320}{6} = 26.\dot{6} \text{ kNm}$$

$$M_P = \frac{bd^2Y}{4} = \frac{50 \times 100^2 \times 320}{4} = 40 \text{ kNm}$$

Shape factor $\frac{M_P}{M_E} = \frac{3}{2}$ is a parameter giving the moment capacity for the cross-section.

$$\frac{M_{ep}}{M_E} = 1 + \frac{2h}{d}\left(1 - \frac{h}{d}\right) = 1 + \frac{2 \times 25}{100}\left(1 - \frac{25}{100}\right) = 1.375$$

$$M_{ep} = 1.375 \times 26.\dot{6} = 36.\dot{6} \text{ kNm}$$

Elastic–plastic central hinge $0 \le x \le L_p/2$:

$$M_x = -\frac{W}{2}\left(\frac{L}{2} - x\right) \quad \therefore \frac{M_x}{M_{ep}} = 1 - \frac{2x}{L}, \quad M_{ep} = -\frac{WL}{4}$$

$$0 \le x \le L_p/2: (M_{ep} \ge M_x \ge M_E) \Rightarrow 1.375 \ge M_x/M_E \ge 1$$

$$\frac{M_x}{M_E} = \frac{M_{ep}}{M_E}\left(1 - \frac{2x}{L}\right) = \left[1 + \frac{2h}{d}\left(1 - \frac{h}{d}\right)\right]\left(1 - \frac{2x}{L}\right)$$

and with $h/d = 25/100$ for $x = 0$:

$$\frac{M_x}{M_E} = 1.375\left(1 - \frac{2x}{L}\right)$$

$$x = 0,\ M_x = M_{ep},\ \frac{M_{ep}}{M_E} = 1.375$$

$$x = \frac{L_p}{2},\ M_x = M_E;\ 1 = 1.375\left(1 - \frac{L_P}{L}\right),\ \therefore \frac{L_P}{L} = 0.273$$

Alternatively, from moment diagram geometry (see Diag. 5.1(a))

$$\frac{M_{ep} - M_x}{x} = \frac{M_{ep} - M_E}{L_p/2} = \frac{M_{ep}}{L/2}$$

$$\frac{L_P}{L} = 1 - \frac{M_E}{M_{ep}} = 1 - \frac{1}{1.375} = 0.273$$

$$\therefore L_p = 0.272L = 0.273 \times 2.5 = 0.682 \text{ m}$$

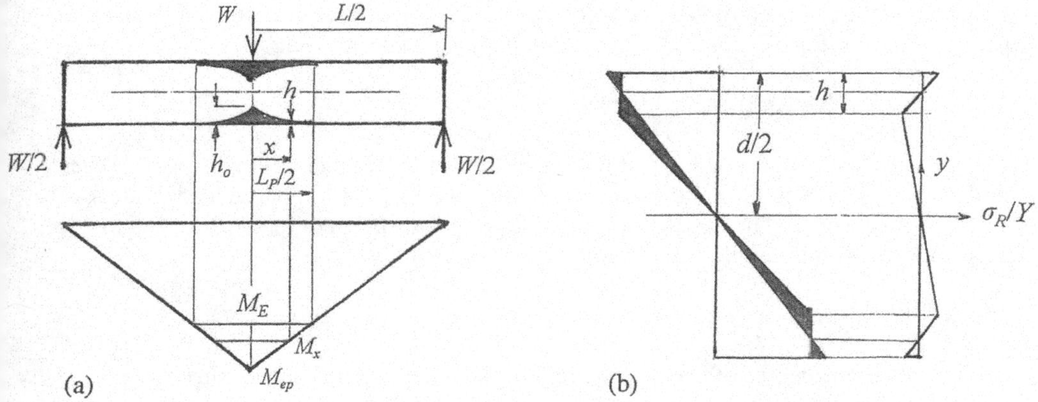

Diagram 5.1 (a) Moment diagram showing central hinge (b) residual stress distribution

Residual Stress (see Diag. 5.1(b)):

$0 \leq y \leq (d/2 - h)$

$$\frac{\sigma_R}{Y} = \frac{2y}{d}\left\{\frac{1}{(1 - 2h/d)} - \left[1 + \frac{2h}{d}\left(1 - \frac{h}{d}\right)\right]\right\}$$

$$= \frac{2y}{100}\left\{\frac{1}{(1 - 50/100)} - \left[1 + \frac{50}{100}\left(1 - \frac{25}{100}\right)\right]\right\} = 0.0125y$$

$y = 25$ mm, $\sigma_R/Y = 0.3125$, $\sigma_R = 100$ MPa

$(d/2 - h) \leq y \leq d/2$

$$\frac{\sigma_R}{Y} = 1 - \frac{2y}{d}\left[1 + \frac{2h}{d}\left(1 - \frac{h}{d}\right)\right]$$

$$= 1 - \frac{2y}{100}\left[1 + \frac{(2 \times 25)}{100}\left(1 - \frac{25}{100}\right)\right] = 1 - 0.0275y$$

$y = 25$ mm, $\sigma_R/Y = 0.3125$, $\sigma_R = 100$ MPa

$y = 50$ mm, $\sigma_R/Y = -\,0.375$, $\sigma_R = -\,120$ MPa

Ex. 5.4(a) The fully plastic moment for a beam depends upon its section dimensions only, as for the rectangular section $b \times d$ in Diag. 5.1a. When plasticity has penetrated from the outer edges inward to the depth of the neutral axis, the stress distribution diagram (see Fig. 5.1d) is composed of two rectangles. As the material does not harden, the stress ordinate is constant at Y. Hence, an ultimate force $F_P = Y \times bd/2$ may be taken to act leftward and rightward at the centroids of the areas above and below the neutral axis (NA). The distance between the two forces s_P is the moment arm of the collapse moment:

$$M_P = F_P \times s_P = (Y \times bd/2) \times d/2 = bd^2Y/4 \quad \text{(i)}$$

The collapse load will depend upon the manner of loading and support in the beam. When uniformly spread over the length of a simply supported beam, the collapse load produces a single, central hinge. The maximum central moment for a beam of length l, loaded in this way, follows from the applied loading as

$$M = wl^2/8 \quad \text{(ii)}$$

irrespective of the degree of plastic penetration. At collapse, one may equate (i) and (ii) to give the collapse loading:

$$wl^2/8 = bd^2Y/4, \quad \Rightarrow \quad w = 2bd^2Y / l^2$$

As the hinge penetrates the depth it also spreads in the length. To find the spread $l_P/2$ either side of the centre we follow a procedure similar to finding $l_P/2$ for a concentrated loading (see Fig. 5.7). Here, however, two parabolas appear as moment diagrams, one within the other, for the fully plastic and fully elastic sections respectively (see Diag. 5.4a).

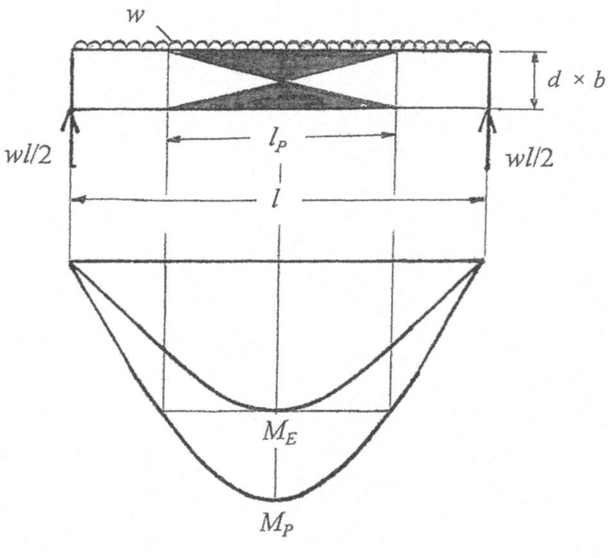

Diagram 5.4a Central hinge spread at collapse under udl

The elastic diagram projects its maximum to the position $l_P/2$ upon the fully plastic diagram. Obviously there are no similar triangles here so we find the moment expression due to the loading at the position $l_P/2$ (where the section is fully elastic). Taking moments at this position for that portion of the beam to the right, with the hogging moment positive and the sagging moment negative, respectively:

$$M_E = w\left(\frac{l}{2} - \frac{l_P}{2}\right) \times \frac{1}{2}\left(\frac{l}{2} - \frac{l_P}{2}\right) - \frac{wl}{2}\left(\frac{l}{2} - \frac{l_P}{2}\right)$$

which may be written as

$$M_E = -\frac{wl^2}{8}\left[1 - \left(\frac{l_P}{l}\right)^2\right] = M_P\left[1 - \left(\frac{l_P}{l}\right)^2\right]$$

Now from eq(5.6a), the shape factor $Q = M_P/M_E$ and so

$$l_p/l = \sqrt{1 - 1/Q}$$

giving $l_p = l/\sqrt{3}$ when $Q = 3/2$ for a rectangular section.

Ex. 5.4(b) Given the beam symmetry, let x lie in the fully formed hinge right of a centre origin in Diag. 5.4b. The following analysis applies to the beam half length: $0 \le x \le L/2$.

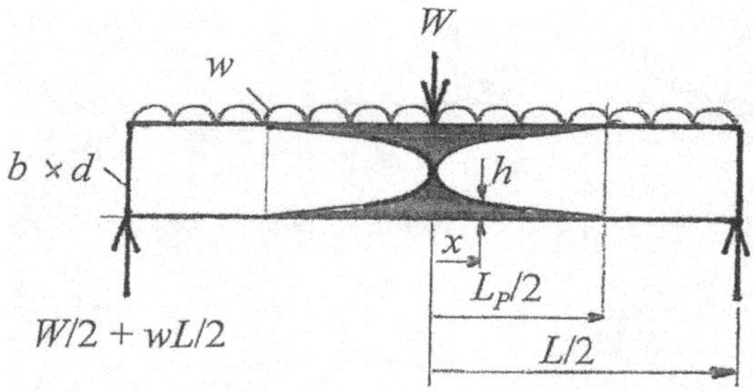

Diag. 5.4b Beam half-length loading

$$M_x = -\frac{1}{2}\left(W + wL\right)\left(\frac{L}{2} - x\right) + \frac{w}{2}\left(\frac{L}{2} - x\right)^2$$

$$x = 0, \quad M_x = M_P = -\frac{WL}{4} - \frac{wL^2}{8}$$

$$x = L/2, \; M_x = 0$$

Within the hinge half length $0 \le x \le L_P/2$:

$$\frac{M_x}{M_P} = \frac{(1 - 2x/L)(W + wL) - (1 - 2x/L)^2 wL/2}{W + wL/2} = \frac{2}{3}\left[1 + \frac{2h}{d}\left(1 - \frac{h}{d}\right)\right]$$

$$x = 0, \quad h/d = 1/2; \quad M_x = M_P$$

$$x = \frac{L_P}{2}, \; \frac{h}{d} = 0, \; M_x = M_E; \quad \frac{M_E}{M_P} = \frac{2}{3} = \frac{(1 - L_P/L)(W + wL) - (1 - L_P/L)^2 wL/2}{W + wL/2}$$

from which a quadratic in L_P/L gives the hinge length under combined loading.

Ex. 5.10 Elastic–plastic torque check:

$$T_E = \frac{kJ}{r_o} = \frac{k \times \pi r_o^4}{2 r_o} = \frac{\pi k r_o^3}{2}$$

$$= \frac{\pi \times 125 \times 50^3}{2} = 24.55 \times 10^6 \text{ Nmm} = 24.55 \text{ kNm}$$

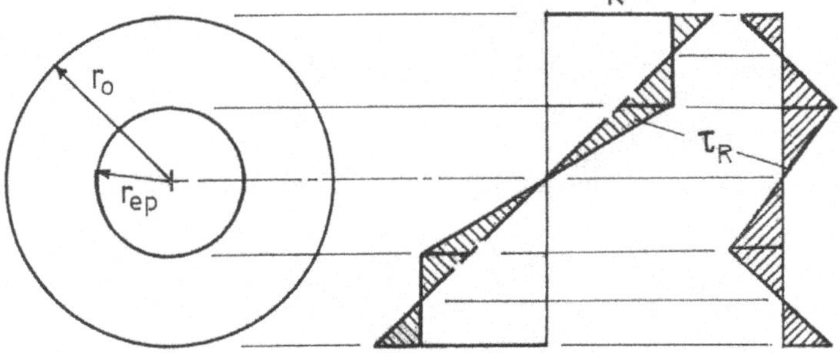

Diag. 5.10 (a) Elastic–plastic torsion bar; (b) residual stress distribution

$$T_P = \frac{2\pi k r_o^3}{3} = \frac{4T_E}{3} = 32.73 \text{ kNm}$$

$$T_E \le T_{ep} \le T_P \checkmark$$

Elastic–plastic radius (see Diag. 5.10(a)):

$$T_{ep} = \frac{\pi k r_{ep}^3}{2} + \frac{2\pi k}{3}(r_o^3 - r_{ep}^3)$$

$$r_{ep}^3 = 4r_o^3 - \frac{6T_{ep}}{\pi k} = 4 \times (50)^3 - \frac{6 \times (30 \times 10^6)}{\pi \times 125}$$

$$r_{ep} = 34.66 \text{ mm}$$

Angular twist:

$$\frac{T_e}{J_e} = \frac{G\theta}{l} = \frac{k}{r_{ep}}$$

$$\theta = \frac{T_e l}{GJ_e} = \frac{kl}{G r_{ep}}$$

$$= \frac{125 \times (3 \times 10^3)}{(82 \times 10^3) \times 34.66} = 0.132^c = 7.56°$$

Residual stress (see Diag. 5.10(b)):

$$r_o \le r \le r_{ep}; \quad \tau_R = k - \frac{T_{ep} r}{J}$$

$$r = r_o; \; \tau_R = 125 - \frac{(30 \times 10^6) \times 50}{\pi \times 50^4/2} = -27.8 \text{ MPa}$$

$$0 \le r \le r_{ep}; \quad \tau_R = \frac{kr}{r_{ep}} - \frac{T_{ep} r}{J}$$

$$r = r_{ep}; \; \tau_R = 125 - \frac{(30 \times 10^6) \times 34.66}{\pi \times 50^4/2} = 19.10 \text{ MPa}$$

Ex. 5.11 The dimensions of the hollow section are $r_i = d/4$ and $r_o = d/2$ and of the solid section $r_o = d/2$. When the shear yield stress k is reached at the outer radius of the solid bar, the fully elastic torque is given by eq(5.29) as

$$T_E = \frac{\pi (d/2)^3 k}{2} = \frac{\pi d^3 k}{16} \tag{i}$$

This same torque must be carried by the tubular section of the bar but in an elastic–plastic manner. This torque is given by eqs(5.29) and (5.32) as

$$T_{ep} = \frac{\pi k r_o^3}{2} \left\{ \frac{4}{3}\left[1 - \frac{1}{4}\left(\frac{r_{ep}}{r_o}\right)^3\right] - \frac{r_i}{r_{ep}}\left(\frac{r_i}{r_o}\right)^3 \right\} \tag{ii}$$

Setting $r_i/r_o = ½$ and $r_o = d/2$, eq(ii) becomes

$$T_{ep} = \frac{\pi d^3 k}{16} \left\{ \frac{4}{3}\left[1 - \frac{1}{4}\left(\frac{d_{ep}}{d_o}\right)^3\right] - \frac{1}{16}\left(\frac{d_o}{d_{ep}}\right) \right\} \tag{iii}$$

Equating (i) and (ii) to arrive at the *correct* equation in d_{ep}:

$$\left(\frac{d_{ep}}{d}\right)^4 - \left(\frac{d_{ep}}{d}\right) + \frac{3}{16} = 0, \quad \Rightarrow \quad d_{ep}^4 - d^3 d_{ep} + \frac{3}{16} d^4 = 0$$

The twist ratio between the two sections apply to the outer radii of their inner elastic zones. Equation (5.33) applies to the solid fully elastic bar with $r_{ep} = r_o$:

$$\theta_s = \frac{kl}{G r_o}$$

Equation (5.48a) applies to the elastic–plastic radius in the hollow bar:

$$\theta_h = \frac{kl}{G r_{ep}}$$

Hence, $\theta_h/\theta_s = r_o/r_{ep} = d/d_{ep}$.

Ex. 5.15 Nadai's ingenious, parametric theory of annular discs provides an internal pressure to produce a fully plastic disc, given by eq(5.72c). The parameter $-\pi/3 < \theta_i < \pi/6$ appearing in this equation connects to the inner radius r_i with an accompanying equation:

$$\left(\frac{r_o}{r_i}\right)^2 = \frac{2}{\sqrt{3}}\cos\theta_i \, e^{\sqrt{3}(\pi/6 - \theta_i)} \qquad \text{(i)}$$

from which a maximum radius ratio follows from differentiation:

$$2\left(\frac{r_o}{r_i}\right)\frac{d}{d\theta_i}\left(\frac{r_o}{r_i}\right) = \frac{2}{\sqrt{3}}\left[-\sqrt{3}\cos\theta_i \, e^{\sqrt{3}(\pi/6 - \theta_i)} - \sin\theta_i \, e^{\sqrt{3}(\pi/6 - \theta_i)}\right] = 0$$

giving $\tan\theta_i = -\sqrt{3}$ and $\theta_i = -\pi/3$. Substituting into eq(i):

$$\left(\frac{r_o}{r_i}\right)^2_{max} = \frac{2}{\sqrt{3}}\cos\left(-\frac{\pi}{3}\right) e^{\sqrt{3}(\pi/6 + \pi/3)} = 8.779, \quad \Rightarrow \quad \left(\frac{r_o}{r_i}\right)_{max} = 2.963$$

with a corresponding maximum pressure from eq(5.72c):

$$(p_i)_{max} = \frac{2Y}{\sqrt{3}}\sin\left[(\pi/6) - (-\pi/3)\right] = \frac{2Y}{\sqrt{3}}$$

Discs with radii ratios greater than 2.963 will not yield fully and no pressure greater than p_{max} is required to burst a disc of non-hardening material at a lesser radius ratio.

Ex. 5.18 Yielding in a solid, spinning disc commences at the centre where both the radial and hoop stresses, σ_r and σ_θ respectively, attain the value of the yield stress Y when the axial stress σ_z is elastic. For this principal, triaxial stress state, Tresca's yield criterion may be stated as

$$\textit{Greatest p. stress} - \textit{Least p. stress} = Y, \quad \Rightarrow \quad \sigma_\theta \,(\text{or } \sigma_r) - \sigma_z = Y \qquad \text{(i)}$$

If the disc is thin in the z–direction it may be assumed that $\sigma_z = 0$ when eq(i) becomes

$$\sigma_\theta = Y \qquad \text{(ii)}$$

where σ_θ is a limiting elastic stress at the centre. Setting $r = 0$ in eq(5.78b):

$$\sigma_\theta = \frac{(3 + \nu)}{8}\rho\omega_Y^2 r_o^2 \qquad \text{(iii)}$$

Equating (ii) and (iii), the angular speed ω_Y to initiate yielding is found to be

$$\omega_Y = \sqrt{\frac{8Y}{(3 + \nu)\rho r_o^2}} \qquad \text{(iv)}$$

The von Mises yield criterion is given by eq(5.88a):

$$\sigma_\theta^2 - \sigma_\theta\sigma_r + \sigma_r^2 = Y^2 \qquad \text{(v)}$$

where, with $\sigma_r = \sigma_\theta$, it follows that eq(iii) again applies. For a hollow disc the elastic radial stress distribution is

$$\sigma_r = \frac{(3 + \nu)\rho\omega^2}{8}\left(r_i^2 + r_o^2 - r^2 - \frac{r_i^2 r_o^2}{r^2} \right)$$

which shows that $\sigma_r = 0$ at the inner radius $r = r_i$. The elastic hoop stress distribution is

$$\sigma_\theta = \frac{\rho\omega^2}{8}\left[(3 + \nu)\left(r_i^2 + r_o^2 - \frac{r_i^2 r_o^2}{r^2} \right) - (1 + 3\nu)r^2 \right]$$

where at the inner radius $r = r_i$:

$$\sigma_\theta = \frac{\rho\omega^2}{4}\left[(1 - \nu)r_i^2 + (3 + \nu)r_o^2 \right] \qquad \text{(vi)}$$

and eq(ii) is again provided by either yield criterion eqs(i) or (v). Equating (ii) and (vi), the angular speed ω_Y to initiate yielding at r_i is found as

$$\omega_Y = \sqrt{\frac{4Y}{\rho\,[r_i^2(1 - \nu) + r_o^2(3 + \nu)]}}$$

CHAPTER 6

Ex. 6.11 The equivalent stress and plastic strain conversions from load and thickness measurements, W and t, employ the following formulae and are given in the tables below.

$$\bar{\sigma} = \frac{\sqrt{3}}{2}\left(\frac{W}{bw} \right), \quad \bar{\varepsilon}^P = \frac{2}{\sqrt{3}}\ln\left(\frac{t_o}{t} \right)$$

$b = 4$ mm, $t_o = 2.48$ mm

$\bar{\sigma}$, MPa	169.5	198.0	217.2	231.5	245.7	256.4	267.1	274.9
$\bar{\varepsilon}^P$	0.014	0.038	0.062	0.097	0.138	0.181	0.237	0.296

$b = 3.2$ mm, $t_o = 1.92$ mm

$\bar{\sigma}$, MPa	264.4	277.8	284.9	290.2	293.8	298.2	302.7
$\bar{\varepsilon}^P$	0.018	0.062	0.120	0.175	0.247	0.308	0.381
$\sum\bar{\varepsilon}^P$	0.314	0.358	0.416	0.471	0.543	0.604	0.677

$b = 2.5$ mm, $t_o = 1.38$ mm

$\bar{\sigma}$, MPa	302.0	310.0	313.4	315.6	319.1	323.6	327.0
$\bar{\varepsilon}^P$	0.025	0.096	0.181	0.262	0.372	0.481	0.630
$\sum \bar{\varepsilon}^P$	0.702	0.773	0.858	0.939	1.049	1.158	1.307

$b = 1.28$ mm, $t_o = 0.8$ mm

$\bar{\sigma}$, MPa	333.8	336.1	338.3	340.5	345.0	349.4	358.3	371.7	387.3	405.1	456.3
$\bar{\varepsilon}^P$	0.085	0.179	0.306	0.435	0.597	0.764	0.951	1.168	1.415	1.678	2.013
$\sum \bar{\varepsilon}^P$	1.392	1.486	1.613	1.742	1.904	2.071	2.258	2.475	2.722	2.985	3.320

The last table entry shows an impressive accumulated strain in excess of 300 % but this should be checked within a plot of equivalent stress versus equivalent plastic strain to ensure the correct shape. A concave curvature indicates that an interaction has occurred between opposing die faces beyond a critical thickness reduction. With such large strains, elasticity is negligible, allowing thicknesses to be measured under load. *Note*: The maximum elastic strain to be removed is $453.6/(70\times10^3) = 0.0065$. Diagram 6.11 shows that an interaction does occur and therefore Hollomon's law fits the acceptable data well within the reduced strain range $0.05 < \bar{\varepsilon}^P < 2$. That is, with $\bar{\sigma}$ in MPa:

$$\bar{\sigma} = 316.5\,(\varepsilon^P)^{0.13} \quad \text{or} \quad \frac{\bar{\sigma}}{\sigma_o} = \left(\frac{\bar{\varepsilon}^P}{\varepsilon_o}\right)^n \quad \text{for} \quad \sigma_o = 150 \text{ MPa}, \; \varepsilon_o = 3.2\times10^{-3}$$

In contrast, a tensile test limits strain to $n = 0.13$ at which a local neck develops.

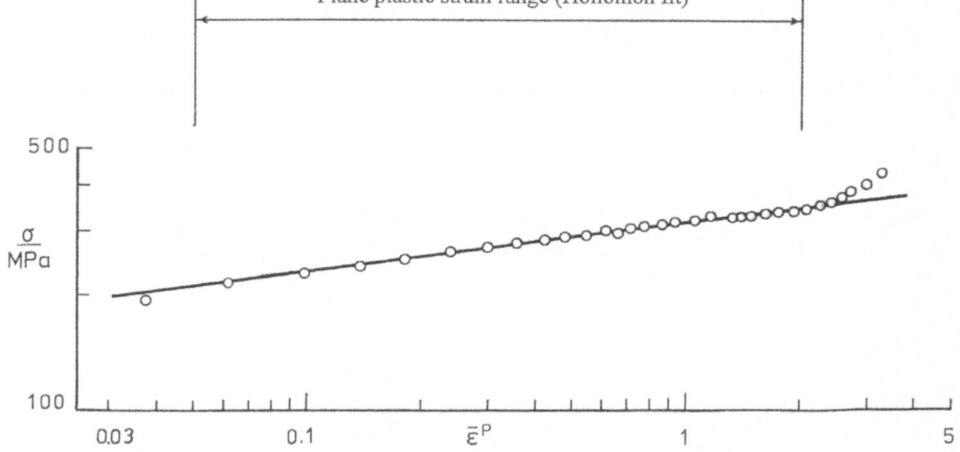

Diagram 6.11 Hollomon log-log plot: $\bar{\sigma}$ versus $\sum \bar{\varepsilon}^P$

Alternative empirical laws of Ludwik and Swift fit these data with a change to their matched constants. Respectively:

$$\bar{\sigma} = 150 + 162.6\left(\bar{\varepsilon}^P\right)^{0.282} \quad \text{or} \quad \frac{\bar{\sigma}}{\sigma_o} = 1 + \left(\frac{\bar{\varepsilon}^P}{\varepsilon_o}\right)^n \quad \text{for } n = 0.282, \; \varepsilon_o = 0.751$$

$$\bar{\sigma} = 317.7\left(3.25 \times 10^{-3} + \bar{\varepsilon}^P\right)^{0.131} \quad \text{or} \quad \frac{\bar{\sigma}}{\sigma_o} = \left(1 + \frac{\bar{\varepsilon}^P}{\varepsilon_o}\right)^n, \quad \text{for } n = 0.13, \; \varepsilon_o = 3.2 \times 10^{-3}$$

Note: Equations are normalised with the same yield stress σ_o but ε_o differs, appearing as the yield strain and the intercept strain in Hollomon and Swift respectively. For Ludwik, ε_o is a curve fitting parameter.

CHAPTER 7

Ex. 7.1 The lower bound moment is found from the fully plastic stress fields in the ligaments above and below the triangular cut-out (see Diag. 7.1a).

$M = 2Fd$ (i)

For a beam with unit thickness and yield stress $Y = 2k$:

$F = 2k \times a/2 \times 1$ (ii)

and with $d = a/2$ we may combine eqs(i) and (ii):

$$M_L = 2\left(2k \times \frac{a}{2} \times 1\right)\frac{a}{2} = ka^2$$

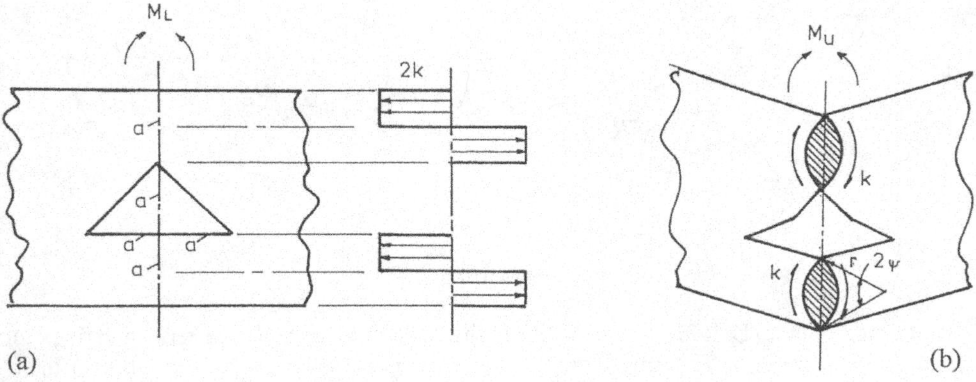

Diagrams 7.1a,b

In the upper bound solution (see Diag. 7.1b), the work done W is given by

$W = M_U \times 2\delta\theta = 2M_U\delta\theta$ (i)

The energy dissipated E within the two plastic hinges is given by

$E = \text{Force} \times \text{Displacement} = 2 \times 2\,[k\,r(2\psi) \times 1 \times r\delta\theta]$ (ii)

where the second multiplication factor 2 applies to the formation of two sides/hinge. Since $W = E$, we may equate (i) and (ii):

$$M_U = 4kr^2\psi$$

The hinge radius r is eliminated using the hinge geometry relation $a = 2r\sin\psi$, giving

$$M_U = \frac{ka^2\psi}{\sin^2\psi} \qquad \text{(iii)}$$

and minimising M_U for ψ:

$$\frac{\mathrm{d}M_U}{\mathrm{d}\psi} = \frac{(ka^2)\sin^2\psi - (ka^2\psi)2\sin\psi\cos\psi}{\sin^4\psi} = 0$$

which gives $\tan\psi = 2\psi, \;\; \Rightarrow \;\; \psi = 66°47'$

Substituting into eq(iii): $M_U = 1.38ka^2$

Ex. 7.3 Label the spaces between the stress discontinuities A, B, C and D (Diag. 7.3a) and identify, from the geometry, the lengths h and H and angles α and β, as shown.

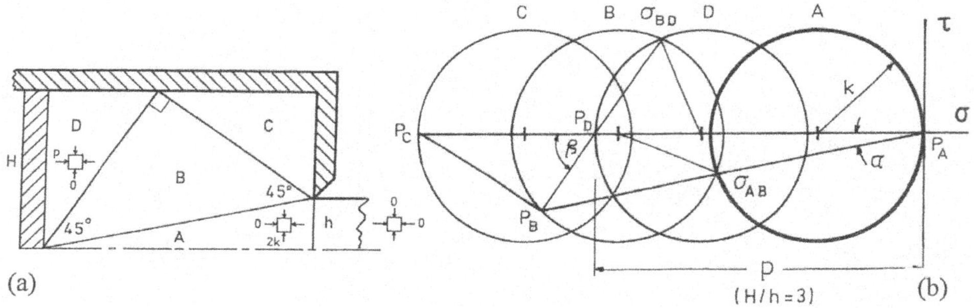

Diagrams 7.3a,b

For the stress plane (Diag. 7.3b) we start from the unstressed, extruded material and fix pole point P_A for circle A. The stress states from within regions A and B, parallel to their common discontinuity, are identical. In particular, the normal stress σ_{AB} is common to circles A and B which allows pole point P_B to be found. Similarly, (i) the normal stress σ_{BD} is common to circles B and D which allows pole point P_D to be found and (ii) the normal stress σ_{BC} is common to circles B and C which allows pole point P_C to be found. All circles have a common radius k. The correspondence between α and β within the physical plane and the stress plane enables the normalised pressure on the die face $p_C/2k$ to be expressed in terms of the extrusion ratio $R = H/h$. Using a horizontal force balance for unit width

$$p_C(H - h) = pH$$

leads to the lower bound extrusion pressure $p = p_L$:

$$\frac{p_L}{2k} = 2\left[1 - \frac{2R}{(2R - 1)^2 + 1}\right]$$

Alternatively, make the scaled construction for $R = 3$ (say) and show that $p/2k = 1.54$, in agreement with the given expression. Compare with the SLF solution eq(6.22b): $p/2k = 1.714$.

Ex. 7.6 This question refers to the collapse condition for an end-propped cantilever, where 2 plastic hinges must form: one at the end fixing and the other beneath the force F applied at distance x from that fixing (see Diag. 7.6a).

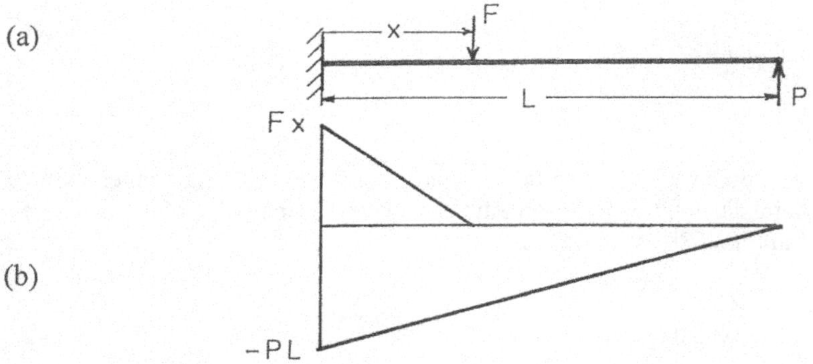

Diagram 7.6

The plastic collapse moment M_P attained at these positions are equal, depending upon the section and not the length, i.e. $M_P = a^3Y/4$. Hence, F and its position x are the variables that ensure a minimum collapse condition. Diagram 7.6b shows that the two moments have opposite sense (hogging + and sagging − are shown) so one can equate their equal magnitudes $|M_P|$ in the following manner:

$$|M_P| = Fx - PL = P(L - x)$$

from which the prop reaction P is

$$P = \frac{Fx}{2L - x}$$

This gives each plastic moment as

$$|M_P| = P(L - x) = \frac{Fx(L - x)}{(2L - x)} \qquad \text{(i)}$$

Re-arranging eq(i) for the variable F:

$$\frac{F}{|M_P|} = \frac{(2L - x)}{x(L - x)} = \frac{1}{(L - x)} + \frac{2}{x} \qquad \text{(ii)}$$

Differentiating eq(ii) for a minimum F:

$$\frac{1}{|M_P|}\frac{\mathrm{d}F}{\mathrm{d}x} = \frac{1}{(L - x)^2} - \frac{2}{x^2} = 0$$

which leads to the quadratic equation:

$$x^2 - 4Lx + 2L^2 = 0$$

The valid root $0 < x < L$ applies:

$$x = (2 - \surd 2)L = \frac{\surd 2 L}{(1 + \surd 2)}$$

That is, $x = 0.585L$ is the position minimising F in eq(ii):

$$F = 5.828\frac{M_P}{L} = 1.457\frac{a^3 Y}{L}$$

Ex. 7.14 (a) Figures 7.8b,c show the two failure modes for the portal frame under a single vertical load W. Using the notation indicated, the virtual work principal, eq(7.6), is applied to the three hinges in Fig. 7.8b as follows:

$$M_{PA}\theta_A + M_{PA}(2\theta_A) + M_{PA}\theta_A = W\delta$$

Substituting $\theta_A = 2\delta/L$ gives the collapse load

$$W = 8M_{PA}/L$$

and for Fig. 7.8c:

$$M_{PB}\theta_B + M_{PA}(2\theta_A) + M_{PB}\theta_B = W\delta$$

Substituting $\theta_A = \theta_B = 2\delta/L$ gives the collapse load

$$W = 4\,(M_{PA} + M_{PB})/L$$

(b) Figures 7.8d,e show the two failure modes for the portal frame under a single horizontal load P. Let the side sway under P be Δ. The application of eq(7.6) to the four hinges gives:

$$M_{PB}\theta_B + M_{PA}\theta_A + M_{PA}\theta_A + M_{PB}\theta_B = P\Delta$$

Substituting $\theta_A = \theta_B = \Delta/H$ gives the collapse load

$$P = 2\,(M_{PA} + M_{PB})\,/H$$

and for Fig. 7.6e:

$$4\,M_{PB}\theta_B = P\Delta$$

Substituting $\theta_B = \Delta/H$ gives the collapse load

$$P = 4M_{PB}/H$$

(c) Figures 7.8f,g show the two possible failure modes for the portal frame under combined vertical and horizontal loads W and P respectively. Let the side sway under P be Δ and the deflection under W be Δ. Applying eq(7.6) to the four hinges in Fig.7.6f:

$$M_{PB}\theta_B + M_{PA}(2\theta_A) + M_{PA}(2\theta_A) + M_{PB}\theta_B = W\delta + P\Delta$$

$$2(2M_{PA}\theta_A + M_{PB}\theta_B) = W\delta + P\Delta$$

Substituting $\theta_A = \theta_B = \Delta/H = 2\delta/L$ gives the collapse loading equation

$$4(2M_{PA} + M_{PB}) = WL + 2PH$$

and for Fig. 7.6f:

$$M_{PB}\theta_B + M_{PA}(2\theta_A) + M_{PB}(2\theta_B) + M_{PB}\theta_B = W\delta + P\Delta$$

$$2(M_{PA}\theta_A + 4M_{PB}\theta_B) = W\delta + P\Delta$$

Substituting $\theta_A = \theta_B = \Delta/H = 2\delta/L$ gives the collapse loading equation

$$4(M_{PA} + 2M_{PB}) = WL + 2PH$$

These equations may be simplified when the ratios H/L and P/W are known. The loads H and P follow from the hinge moments M_{PB} and M_{PB}, which depend upon when the section dimensions and the yield stress, e.g. for a rectangular section eq(5.5), apply. For a uniform section $M_{PA} = M_{PB}$, when the collapse loading for (a), (b) and (c) simplify to various graphical solutions in axes of M_p/WL versus P/W (or H/L) for given H/L (or P/W):

$$\frac{M_P}{WL} = \frac{1}{8}, \quad \frac{M_P}{WL} = \frac{1}{4}\left(\frac{P}{W}\right), \quad \frac{M_P}{WL} = \frac{1}{12}\left[1 + 2\left(\frac{P}{W}\right)\left(\frac{H}{L}\right)\right]$$

CHAPTER 8

Ex. 8.1 The shear stress τ acting in the slip plane and aligned with the slip direction is called the resolved shear stress. It is expressed in terms of two angles and a nominal applied tensile stress according to eq(8.1). The angles φ and λ define the orientation of the plane's normal and slip direction to the stress axis. In seeking a maximum τ, it is convenient to fix φ within $0° \le \varphi \le 90°$ and allow λ to vary within $0° \le \lambda \le 90°$. Thus, for $\varphi = 0°$ then $\tau = \sigma \cos\lambda$, but with the normal parallel to the stress axis λ is restricted to 90° and so $\tau = 0$. For $\varphi = 90°$ it is seen that $\tau = 0$ for all λ.

Plots of τ/σ versus λ, lying in a restricted range $0° < \lambda \le 90°$, for constant, intermediate φ values ($0° < \varphi < 90°$) can be examined in a similar manner. For example, with $\varphi = 45°$ then $\tau/\sigma = (1/\sqrt{2})\cos\lambda$, which gives, apparently, a variation from $\tau/\sigma = (1/\sqrt{2})$ for $\lambda = 0°$ to $\tau/\sigma = 0$ for $\lambda = 90°$. However, a glance at Fig. 8.1a shows that λ is now restricted to $45° \le \lambda \le 90°$ and the maximum that can be found in plotting over this range is $\tau/\sigma = 1/2$ for $\lambda = 45°$. It can be shown from similar reasoning that $\tau/\sigma = ½$ is not exceeded for all other slip plane orientations.

Alternatively, when the normal, the slip plane direction and the stress axis all lie in the same plane, then $\lambda + \varphi = 90°$ and Diag. 8.1 applies. The resolved shear stress appears as τ_φ, which follows from force resolution parallel to the slip plane. Taking the wedge to have unit thickness: τ_φ AC = σ AB sin φ. Substituting AB = AC cos φ gives: $\tau_\varphi = \sigma \sin\varphi \cos\varphi = \tfrac{1}{2}\sigma \sin 2\varphi$, showing again that $\tau_\varphi = \tfrac{1}{2}\sigma$ is at its maximum for $\varphi = 45°$. *Note*: The normal stress σ_φ has no bearing upon slip but is important to crack propagation.

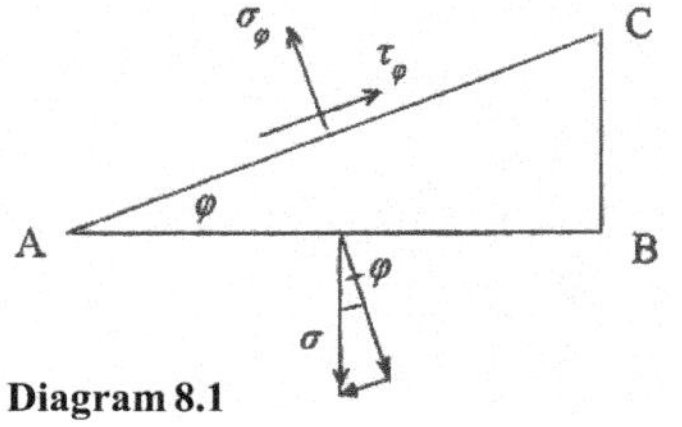

Diagram 8.1

Ex. 8.3 In appropriate places throughout this book equivalent expressions between the tensor subscript and matrix notations have been given. The earlier developments to plasticity theory from Prandtl, Reuss, von Mises etc., grew with the use of the tensor notation. Now, with increasing use of computers for providing numerical solutions to the non-linear equations presented by plasticity theory, matrix methods have been adopted. For example, in chapter 16 it is shown that the essence of the FE method lies with the formation of a stiffness matrix for a discretised structure. Here, an example is given on how the incremental measure of plastic work appears in the two notations at both the micro- and macro-levels. In Einstein's tensor subscript notation, repeated subscripts imply summation, so that the left hand side of eq(8.11a) expands across i and j = 1, 2 and 3 as

$$
\begin{aligned}
\sigma_{ij}\,\delta\varepsilon_{ij}{}^{P} &= \sigma_{1j}\,\delta\varepsilon_{1j}{}^{P} + \sigma_{2j}\,\delta\varepsilon_{2j}{}^{P} + \sigma_{3j}\,\delta\varepsilon_{3j}{}^{P} \quad (\text{for } i = 1, 2 \text{ and } 3)\\
&= \sigma_{11}\,\delta\varepsilon_{11}{}^{P} + \sigma_{12}\,\delta\varepsilon_{12}{}^{P} + \sigma_{13}\,\delta\varepsilon_{13}{}^{P} \quad (\text{for } i = 1 \text{ and } j = 2 \text{ and } 3)\\
&+ \sigma_{21}\,\delta\varepsilon_{21}{}^{P} + \sigma_{22}\,\delta\varepsilon_{22}{}^{P} + \sigma_{23}\,\delta\varepsilon_{23}{}^{P} \quad (\text{for } i = 2 \text{ and } j = 2 \text{ and } 3)\\
&+ \sigma_{31}\,\delta\varepsilon_{31}{}^{P} + \sigma_{32}\,\delta\varepsilon_{32}{}^{P} + \sigma_{33}\,\delta\varepsilon_{33}{}^{P} \quad (\text{for } i = 3 \text{ and } j = 2 \text{ and } 3)
\end{aligned}
$$

In the equivalent matrix notation, $\mathbf{T}\delta\mathbf{E}^{\mathrm{P}}$ denotes a matrix multiplication:

$$
\mathbf{T}\delta\mathbf{E}^{\mathrm{P}} = \begin{bmatrix} \sigma_{11} & \sigma_{12} & \sigma_{13} \\ \sigma_{21} & \sigma_{22} & \sigma_{23} \\ \sigma_{31} & \sigma_{32} & \sigma_{33} \end{bmatrix} \begin{bmatrix} \delta\varepsilon_{11}^{P} & \delta\varepsilon_{12}^{P} & \delta\varepsilon_{13}^{P} \\ \delta\varepsilon_{21}^{P} & \delta\varepsilon_{22}^{P} & \delta\varepsilon_{23}^{P} \\ \delta\varepsilon_{31}^{P} & \delta\varepsilon_{32}^{P} & \delta\varepsilon_{33}^{P} \end{bmatrix}
$$

and tr($\mathbf{T}\delta\mathbf{E}^{\mathrm{P}}$) refers to the sum of the three diagonal term (the trace) within the 3 × 3 matrix resulting from this multiplication:

$$
\mathrm{tr}(\mathbf{T}\delta\mathbf{E}^{\mathrm{P}}) = \begin{bmatrix} \sigma_{11} & \sigma_{12} & \sigma_{13} \end{bmatrix} \begin{bmatrix} \delta\varepsilon_{11}^{P} \\ \delta\varepsilon_{12}^{P} \\ \delta\varepsilon_{13}^{P} \end{bmatrix} + \begin{bmatrix} \sigma_{21} & \sigma_{22} & \sigma_{23} \end{bmatrix} \begin{bmatrix} \delta\varepsilon_{21}^{P} \\ \delta\varepsilon_{22}^{P} \\ \delta\varepsilon_{23}^{P} \end{bmatrix} + \begin{bmatrix} \sigma_{31} & \sigma_{32} & \sigma_{33} \end{bmatrix} \begin{bmatrix} \delta\varepsilon_{31}^{P} \\ \delta\varepsilon_{32}^{P} \\ \delta\varepsilon_{33}^{P} \end{bmatrix}
$$

which agrees with the last three lines given above by the summation convention.

Ex. 8.4 In the absence of the lattice imperfections, the critical resolved shear stress in a perfect crystal can attain a value as high as $\tau_{cr}^* = G/(2\pi)$. One may take this value of the elastic constant G from the polycrystalline aggregate since it is based upon elastic lattice distortion and not slip. Here, G is calculated for the three materials using their values of E and ν given using the relation $E = 2G(1 + \nu)$. These values appear in the table below.

Material	Silver	Aluminium	Iron	Steel
G, GPa	25.55	26.49	79.85	82.7
$\tau_{cr}^* = G/(2\pi)$, GPa	4.07	4.22	12.71	13.15
$\tau_{cr}^* = G/(2\pi)$, N/mm^2	4.07×10^3	4.22×10^3	12.71×10^3	13.15×10^3
τ_{cr}, dyne/cm^2	6×10^6	4×10^6	2×10^6	20×10^6
τ_{cr}, N/mm^2	0.6	0.4	0.2	20

To make comparisons with τ_{cr} values given for real metals, a common unit of N/mm^2 (MPa) has been used. Conversions to MPa require a multiplication of GPa by 10^3 and a division of dyne/cm^2 by 10^7. Comparing row three with row five shows that the strength τ_{cr} is less than τ_{cr}^* by a factor of ~10^4. *Note*: The theoretical values τ_{cr}^* apply to single crystals. Elsewhere this book shows that the shear yield strength of polycrystalline steel is nearer to 150 MPa. The multiplication factor of 7.5 indicates the level to which the τ_{cr} should be raised to produce slip in grains with least favourable orientations, as shown in Fig. 8.9.

Ex. 8.5 A multitude of imperfections arise in a real crystal lattice to explain why its theoretical strength is never reached. They include the following: (a) point, (b) line, (c) plane and (d) volume imperfections.

(a) Point Defects
Vacancies - atoms missing from point in a crystal lattice. They arise in quenching, excessive cold work and radiation. The Schottky defect refers to the vacancy created by an atom diffusing to a free surface.

Interstitial atoms - intruding atoms lying within those at normal lattice sites. Interstitials are normally 'foreign' atoms from alloying elements, e.g. carbon in iron. However, the Frankel defect refers to the vacancy created in a lattice where an atom of the same metal has moved to a neighbouring interstitial site.

Impurity atoms - either substitution or interstitial depending upon the size of the impurity atom and the space available between parent atoms at normal lattice sites.

(b) Line Defects
Edge dislocations - arise through the extra half plane of atoms within an imperfect lattice. They promote slip between neighbouring atomic planes. The slip, which occurs in a direction perpendicular to the half plane and of magnitude equal to that of the Burgers vector, is activated by a shear stress parallel to the slip direction.

Screw dislocations - arise from a similar extra half plane of atoms. However, slip now occurs in a direction parallel to the half plane with magnitude equal to that of the Burgers vector. This dislocation is activated by complementary shear stress (perpendicular to that which activates an edge dislocation).

c) *Plane Defects*
Grain boundaries - regions of misfit between the regular arrangement of atoms within neighbouring grains. The arise from the change in lattice orientation between grains and have a major influence upon deformation behaviour, regarded as the main reason for the loss in theoretical strength. The shape of the grains that form on solidification from the liquid serve to minimise grain boundary area

Twin boundaries - a mirrored reflection between neighbouring lattice orientations (see Fig. 8.6). The twin plane has no width, unlike a grain boundary. Annealing twins arises with the grain boundary migration and stress relief that occurs at high temperatures, particularly in lattices with face and hexagonal atom packing. Plastic deformation can produce mechanical twins in all lattices (b.c.c., f.c.c. and h.c.p.). These twins arise with slip between the most closely packed atomic planes, e.g. (111) for a f.c.c. lattice in Fig. 8.5a.

(d) *Volume Defects*
Stacking faults - refer to a loss in symmetry between the normal layers of atomic packing within ideal closely packed f.c.c. and h.c.p. lattices. While the layers A, B, C in an ideal f.c.c. lattice appear as ABC ABC etc., those in an ideal h.c.p appears as AB AB AB etc. Thus the stacking ABC ABAB ABC is a mixture of f.c.c. and h.c.p. As the accompanying energy change is relatively low, stacking faults are quite common.

Inter-phase boundaries - are the boundaries that arise between different lattices in alloys. With differences between their atomic spacings, the respective lattices can meet with little or no match; said to be coherent, semi- or non-coherent. Where an atomic plane of one element ends at the interface with the other element the non-coherency results in an edge dislocation.

CHAPTER 9

Ex. 9.1 Differentiating eq(9.44b):

$$\frac{d\sigma}{d\varepsilon^P} = \frac{n\sigma_o}{\varepsilon_o}\left(1 + \frac{\varepsilon^P}{\varepsilon_o}\right)^{n-1} = \frac{n\sigma_o}{\varepsilon_o}\left(\frac{\sigma}{\sigma_o}\right)^{\frac{n-1}{n}}$$

and equating to eq(9.14c) at the point (ε_u^P, σ_u) of tensile instability:

$$\frac{\sigma_u}{\left(1 - \sigma_u/E\right)} = \frac{n\sigma_o}{\varepsilon_o}\left(\frac{\sigma_u}{\sigma_o}\right)^{\frac{n-1}{n}}$$

which may be re-arranged to give:

$$\left(\frac{\sigma_u}{\sigma_o}\right)^{\frac{1}{n}} = \frac{n}{\varepsilon_o}\left(1 - \frac{\sigma_o}{E}\right)$$

from which we can confirm the approximations given in eqs(9.44c,d).

Ex. 9.6 Use eqs(9.30) and (9.32b) to convert force (F) and thickness displacement (Δt) data given into equivalent stress and strain. *Note*: In these equations F (= $-W$) and $t = t_o - \Delta t$. The minus signs which accompanies compression conventionally are omitted.

$\bar{\sigma}$	14.35	21.59	28.78	35.97	43.16	50.34	57.56	64.75	71.94
(MPa)	79.13	86.31	100.72	115.10	129.50	143.83	158.25	172.66	187.03

$\bar{\varepsilon}^P$	5.9	6.3	7.1	8.0	11.7	15.4	18.8	23.6	30.4
(%)	40.6	56.0	86.0	113.2	142.3	160.0	181.0	256.0	367.0

When the equivalent stress and plastic strain are plotted together in logarithmic co-ordinates Hollomon's law fits their pairing for an $\bar{\varepsilon}^P$ range from 10 % up to an impressive 300 %. Within this range that law applies with strength coefficient and hardening exponent as

$$\bar{\sigma} = 115.5\left(\bar{\varepsilon}^P\right)^{0.45}$$

Ex. 9.7 Firstly, plot T versus θ and draw a smooth curve. Then draw the tangents to this curve at suitable positions along it, say, for 50 Nm torque steps. The tangents establish AC and AB (in Nm units) within Fig. 9.6 as follows:

T = AC	300	350	400	450	500	550	600	650	700
$\theta°$	25	45	70	100	147	220	303	500	695
AB	56.7	106.7	126.7	130.0	141.3	153.3	110.0	128.0	113.3

Equations (9.24a) and (9.27b) (θ in radians) are applied to find the shear strain and stress at the outer radius. Here, eqs(9.25a,b) are used to provide the equivalent stress and strain.

τ_o, MPa	146.3	176.8	202.8	226.3	250.9	275.7	292.0	317.7	338.4
γ_o, %	2.2	3.9	6.1	8.7	12.8	19.2	26.4	43.5	60.5
$\bar{\sigma}$, MPa	253.4	306.2	351.2	392.0	434.6	477.5	505.7	550.2	586.0
$\bar{\varepsilon}$, %	1.3	2.3	3.5	5.0	7.4	11.1	15.2	25.1	34.9
$\bar{\varepsilon}^P$, %	1.19	2.17	3.35	4.8	7.2	10.9	15.0	24.9	34.7

Note, The removal of elastic shear strain ($\gamma^e = \tau/G$) from each total outer diameter shear strain is almost negligible. For example, at most $\gamma^e = 338.4/(80 \times 10^3) = 0.423$ %. The equivalent plastic strains are calculated for the removal of γ^e as shown.

Ex. 9.9 Equations(9.9a,b) are employed to convert load and extension data as given into nominal stress (**MPa**) and % engineering strain for loading and unloading as follows:

Loading:

σ 25.20	56.33	76.80	95.20	115.6	136.0	156.4	175.8	197.2
219.6	259.4	298.2	330.3	358.4	380.9	397.3	411.8	423.5
433.3	442.9	449.8	462.4	483.8	487.7	493.5	497.4	501.3
504.1	507.1	509.0	511.9					

ε	0.01	0.02	0.03	0.04	0.05	0.06	0.07	0.08	0.09
	0.10	0.12	0.14	0.16	0.179	0.20	0.22	0.24	0.26
	0.279	0.299	0.350	0.40	0.50	0.549	0.60	0.65	0.70
	0.75	0.799	0.850	0.90					

Unloading:

σ	400.3	312.7	221.5	131.7	49.6	0
ε	0.85	0.79 9	0.750	0.70	0.65	0.62

The plot between σ and ε shows a slight zero error, which emphasises the need to find E from the *gradient* to the initial elastic line:

$$E = \frac{\Delta\sigma}{\Delta\varepsilon} = \frac{390}{(0.19 \times 10^{-2})} = (205 \times 10^3) \text{ MPa} = 205 \text{ GPa}$$

Drawing lines with gradient $E = 205$ GPa, offset by strains 0.1 % and 0.5 %, intersect the stress–strain curve at the corresponding proof stresses: $\sigma_{0.1} = 445$ MPa and $\sigma_{0.5} = 503$ MPa.

The unloading plot reveals some non-linear hysteresis as σ approaches zero. Despite this, an E for unloading may be calculated from the initial region of linearity observed:

$$E = \frac{\Delta\sigma}{\Delta\varepsilon} = \frac{525}{(0.29 \times 10^{-2})} = (181 \times 10^3) \text{ MPa} = 181 \text{ GPa}$$

Classical plasticity theory assumes that E remains unchanged by prior plastic strain. Experiment shows some loss of stiffness from plastic pre-strain.

Ex. 9.10 The Bridgeman analysis of tensile necking requires, firstly, that a mean stress is calculated at each current neck diameter:

$$\sigma_m = \frac{W}{(\pi d^2/4)} = 1273.2\,\frac{W}{d^2}$$

where W, in kN, converts the load data given to mean stress in MPa:

σ_m	274.4	304.8	362.8	384.1	426.8	451.7	488.2	510.3	535.5
	539.5	550.5	559.4	571.8	579.3	588.9	602.3	608.7	618.1
	635.8	646.0	657.5	685.9	692.2	711.3	718.2	730.9	730.4
	751.8	774.3	782.0	805.9	798.4	815.4	827.1	826.6	828.6

The corresponding radius ratios $a/r_o = d/d_o$ are calculated and applied to Fig. 9.21 to give the ordinates values $\bar{\sigma}/\sigma_m$. Multiplying these ordinates by σ_m given in the table above provides an equivalent stress for the neck stress in MPa:

$\bar{\sigma}$	274.4	304.8	362.8	384.1	426.8	451.7	488.2	510.3	535.5
(MPa)	537.3	546.7	553.8	563.2	568.9	575.4	586.1	590.5	597.7
	611.0	616.9	624.6	646.8	650.7	662.9	666.5	675.4	673.4
	688.7	702.3	707.7	722.8	713.0	725.7	732.0	727.4	727.5

The corresponding equivalent plastic strains follow from

$$\bar{\varepsilon}^P = 2\ln(d_o/d)$$

This strain is assumed to be plastic. Strictly, the current neck diameter should be measured following an unloading from each of the listed loads, allowing elasticity to recover. However, the influence of lateral elastic displacement upon d is expected to be small when larger plastic strain does not alter volume. Thus, equivalent strains found from d measured under each load given below can be taken to be reliable for strains in excess of 10 %. Strains less than 10 % do present some error from this source:

$\bar{\varepsilon}^P$	1.42	2.1	3.2	4.0	5.5	6.9	9.4	11.1	12.4
(%)	14.1	15.8	17.1	19.3	20.5	22.4	24.7	26.0	27.9
	31.6	33.9	36.3	40.7	43.1	46.7	49.2	51.8	53.3
	57.0	61.3	63.5	69.1	70.8	73.1	77.1	80.7	81.9

The plot between $\bar{\sigma}$ and $\bar{\varepsilon}^P$ provides a true flow curve for the material. The curve is not terminated by the necking that begins when the maximum load is reached, as would be the case when a mounted extensometer is used to measure axial displacement instead of the diameter change. Here, the table extends the flow curve to 80 % strain, revealing true stress well beyond the ultimate strength as taken from the usual nominal stress–strain plot.

Ex. 9.11 The data supplied should be plotted in a manner similar to Fig. 9.11 for the three d_o/h_o ratios. An abscissa of engineering plastic strain is intended $\bar{\varepsilon}^P = \Delta h/h_o$. While high load data show similar curvature to Fig. 9.11, the low load data reverses the curvature. Thus, a linear extrapolation should not be attempted. The graphical extrapolations to $d_o/h_o = 0$ give the following $\%\,\bar{\varepsilon}^P$ for each load. In Fig. 9.12, these engineering strains have been plotted against the nominal stress $\bar{\sigma} = W/A_o$. Where a true stress and true plastic strain plot is required it may be derived from

$$\bar{\sigma}_t = \bar{\sigma}(1 - \bar{\varepsilon}^P), \quad \bar{\varepsilon}_t^{\,P} = \ln(1 - \bar{\varepsilon}^P)$$

The following table converts selected nominal values to true values (% and MPa):

$\%\,\bar{\varepsilon}^P$	0	0.84	2.90	5.30	7.80	10.76	14.26	17.60	22.0	25.84	29.60	33.70
$\bar{\sigma}_t$	39.3	77.93	114.5	148.9	181.2	210.4	235.9	259.0	275.9	291.4	304.3	312.6
$\bar{\varepsilon}_t^{\,P}$	0	0.84	2.9	5.40	8.1	11.4	15.4	19.4	24.9	30.0	35.1	41.1

$\%\,\bar{\varepsilon}^P$	37.4	41.1	44.8	47.8	50.5	53.0	55.4	57.5	60.5	61.9	64.6
$\bar{\sigma}_t$	319.8	324	325.4	328.2	330.7	332.5	333	334	341.5	344.4	347.8
$\bar{\varepsilon}_t^{\,P}$	46.8	52.9	59.4	65.0	70.3	75.5	80.7	85.6	92.9	96.5	103.8

Note: As with plane strain compression, Cook and Larke provide more than 100 % strain.

CHAPTER 10

Ex. 10.1 Normally equations which describe the stress–strain curve, e.g. Hollomon, Swift etc., appear in the strain hardening form of eq(9.48). A less convenient form, the work hardening hypothesis eq(9.51a), may be a more natural choice but is less convenient when converting data from a tensile test. Certain functions allow an interchange between the two hypotheses, as with Swift's law for example in Fig. 9.20. Following a similar approach to that given for Swift, Hollomon's law (Fig. 9.18a) converts as follows:

$$W^P = \int \bar{\sigma}\, d\bar{\varepsilon}^P = \frac{\sigma_o}{\varepsilon_o^n} \int (\bar{\varepsilon}^P)^n\, d\bar{\varepsilon}^P = \frac{\sigma_o}{\varepsilon_o^n} \frac{(\bar{\varepsilon}^P)^{\frac{(1+n)}{n}}}{(1+n)} = \frac{\sigma_o \varepsilon_o}{(1+n)} \left(\frac{\bar{\sigma}}{\sigma_o} \right)^{\frac{(1+n)}{n}}$$

from which a work hardening equivalent is found:

$$\bar{\sigma} = F(W^P) = \sigma_o \left[\frac{(1+n) W^P}{\sigma_o \varepsilon_o} \right]^{\frac{n}{(1+n)}}$$

Prager's modified law (Fig. 9.18c) does not lend itself to a similar conversion:

$$W^P = \int \bar{\sigma}\, d\bar{\varepsilon}^P = \sigma_o \int_0^{\bar{\varepsilon}^P} \left[1 + K \tanh\left(\frac{\bar{\varepsilon}^P}{\varepsilon_o} \right) \right] d\bar{\varepsilon}^P = \sigma_o \left\{ \bar{\varepsilon}^P + K \left[\ln\left(e^{\bar{\varepsilon}^P/\varepsilon_o} + e^{-\bar{\varepsilon}^P/\varepsilon_o} \right) - 2 \right] \right\}$$

Hence the work calculated from the formula above may be tabulated with the co-ordinates $(\bar{\sigma}, d\bar{\varepsilon}^P)$ from the law and a suitable hardening function F sought.

The Vocé equation (9.45b) provides the area beneath its flow curves (Fig. 9.18f) as

$$W^P = \int_0^{\bar{\varepsilon}^P} \left[\sigma_\infty - (\sigma_\infty - \sigma_o)\, e^{\frac{\bar{\varepsilon}^P}{\varepsilon_o}} \right] d\bar{\varepsilon}^P = \sigma_\infty \bar{\varepsilon}^P + \varepsilon_o (\sigma_\infty - \sigma_o) \left(e^{\frac{-\bar{\varepsilon}^P}{\varepsilon_o}} - 1 \right)$$

The second term may be re-written in terms of stress, and so too the first term at small plastic strain using an approximation:

$$\frac{\bar{\varepsilon}^P}{\varepsilon_o} \approx 1 - e^{\frac{-\bar{\varepsilon}^P}{\varepsilon_o}} = \frac{(\bar{\sigma} - \sigma_o)}{(\sigma_\infty - \sigma_o)}$$

A conversion to wok hardening then becomes possible:

$$\bar{\sigma} = F(W^P) = \sigma_o + \frac{(\sigma_\infty - \sigma_o)}{\sigma_o \varepsilon_o} W^P$$

which, interestingly, has a linear form.

Ex. 10.8 Let subscripts $i, j = 1, 2, 3$ denote the stress and strain components to the axial, circumferential and radial directions respectively. The radial stress σ_{33} in a thin walled cylinder may be ignored. The three total strain increments appear from eq(10.40) with single subscripts denoting principal values as

$$d\varepsilon_1^t = d\lambda\,\sigma_1' + d\sigma_1'/2G + (\sigma_1 + \sigma_2)/9K$$
$$d\varepsilon_2^t = d\lambda\,\sigma_2' + d\sigma_2'/2G + (\sigma_1 + \sigma_2)/9K \qquad \text{(i–iii)}$$
$$d\varepsilon_3^t = d\lambda\,\sigma_3' + d\sigma_3'/2G + (\sigma_1 + \sigma_2)/9K$$

where $\sigma_1 = (\sigma_1)_p + (\sigma_1)_F$ is the total axial stress due to both the pressure and independent axial force. The former arises from the pressure acting upon the closed ends. Working in terms of the hoop stress σ_2 and with the given ratio $\sigma_2/(\sigma_1)_F = 4/3$, this becomes

$$\sigma_1 = ½ \times \sigma_2 + ¾ \times \sigma_2 = 5\sigma_2/4$$

The deviatoric stresses become

$$\sigma_1' = \sigma_1 - ⅓(\sigma_1 + \sigma_2) = 5\sigma_2/4 - ⅓(5\sigma_2/4 + \sigma_2) = \sigma_2/2$$

$$\sigma_2' = \sigma_2 - ⅓(\sigma_1 + \sigma_2) = \sigma_2 - ⅓(5\sigma_2/4 + \sigma_2) = \sigma_2/4$$

$$\sigma_3' = -⅓(\sigma_1 + \sigma_2) = -⅓(5\sigma_2/4 + \sigma_2) = -3\sigma_2/4$$

and substituting into eqs(i)–(iii):

$$d\varepsilon_1^t = d\lambda\sigma_2/2 + d\sigma_2/4G + d\sigma_2/4K \qquad \text{(i)}$$

$$d\varepsilon_2^t = d\lambda\sigma_2/4 + d\sigma_2/8G + d\sigma_2/4K \qquad \text{(ii)}$$

$$d\varepsilon_3^t = -3d\lambda\sigma_2/4 - 3d\sigma_2/8G + d\sigma_2/4K \qquad \text{(iii)}$$

Now as $\sigma_2 = pr/t$, where r and t are the cylinder's inner radius and wall thickness, this gives the differential relationship

$$d\sigma_2/\sigma_2 = dp/p + dr/r - dt/t \qquad \text{(iv)}$$

which connects to the strain increments through

$$d\varepsilon_2^t = dr/r \text{ and } d\varepsilon_3^t = dt/t \qquad \text{(v)}$$

Eliminating $d\lambda$ from combining eqs(ii) and (iii) with eqs(iv) and (v) will eventually lead to

$$d\varepsilon_1^t = ½(d\sigma_2/\sigma_2 - dp/p) + d\sigma_2/4K$$

$$d\varepsilon_2^t = ¼(d\sigma_2/\sigma_2 - dp/p) + d\sigma_2/4K$$

$$d\varepsilon_3^t = -¾(d\sigma_2/\sigma_2 - dp/p) + d\sigma_2/K$$

which integrate to

- $\varepsilon_1^t = \ln(\sigma_2/p)^{1/2} + \sigma_2/4K, \quad \varepsilon_2^t = \ln(\sigma_2/p)^{1/4} + \sigma_2/4K, \quad \varepsilon_3^t = \ln(p/\sigma_2)^{3/4} + \sigma_2/4K$

Equation (10.46c) gives the von Mises yield criterion in terms of the two principal stress components. The gradient of the normal defines the direction of the equivalent plastic strain vector, i.e. normal to the yield locus. In the stress plane σ_2 versus σ_1 (y versus x), the gradient to the normal is

$$\frac{d\sigma_2}{d\sigma_{12}} = \frac{2\sigma_2 - \sigma_1}{2\sigma_1 - \sigma_2} = \frac{1}{2} \quad \left(\text{for } \frac{\sigma_2}{\sigma_1} = \frac{4}{5}\right)$$

With the gradient of the equivalent stress vector as 4/5, the angle between the two vectors is $\tan^{-1}(4/5) - \tan^{-1}(1/2) = 12.1°$. For a radial path $\sigma_2/\sigma_1 = 4/3$, a similar analysis gives $\tan^{-1}(5/2) - \tan^{-1}(4/3) = 15.07°$.

Ex. 10.10 The Prandtl-Reuss, total incremental strains follow from eq(10.40) as

$$d\varepsilon_{ij}^{t} = d\lambda\, \sigma_{ij}' + \frac{d\sigma_{ij}'}{2G} + \frac{1}{9K}\delta_{ij}\,\sigma_{kk} \tag{i}$$

where $E = 3K(1 - 2v)$. Subscripts $i, j = 1, 2, 3$ refer stress and strain components to the co-ordinates directions: axial, circumferential and radial respectively. Double subscripts k in the final hydrostatic stress term denotes a summation. The stress state components (σ_{11}, σ_{22}, σ_{12}) are increased proportionately in two similar ratios: $\sigma_{22}/\sigma_{11} = 2$ from a rising pressure and $\sigma_{22}/\sigma_{12} = 2$ from increasing torsion applied independently. The radial stress σ_{33} in a thin walled cylinder may be ignored. Working in terms of the hoop stress σ_{22}, the deviatoric stress components σ_{ij}' (see eq(10.14)) become

$$\sigma_{11}' = \sigma_{11} - ⅓(\sigma_{11} + \sigma_{22}) = ½\sigma_{22} - ⅓(½\sigma_{22} + \sigma_{22}) = 0$$

$$\sigma_{22}' = \sigma_{22} - ⅓(\sigma_{11} + \sigma_{22}) = \sigma_{22} - ⅓(½\sigma_{22} + \sigma_{22}) = ½\sigma_{22}$$

$$\sigma_{33}' = -⅓(\sigma_{11} + \sigma_{22}) = -⅓(½\sigma_{22} + \sigma_{22}) = -½\sigma_{22}$$

$$\sigma_{12} = \sigma_{12}' = ½\sigma_{22}$$

With $\sigma_{kk} = \sigma_{11} + \sigma_{22} + \sigma_{33} = ½\sigma_{22} + \sigma_{22} + 0 = 3\sigma_{22}/2$, eq(i) gives the strain increments:

$$d\varepsilon_{11}^{t} = d\sigma_{22}/6K \tag{i}$$

$$d\varepsilon_{22}^{t} = d\lambda\sigma_{22}/2 + d\sigma_{22}/4G + \sigma_{22}/6K \tag{ii}$$

$$d\varepsilon_{33}^{t} = -d\lambda\sigma_{22}/2 - d\sigma_{22}/4G + \sigma_{22}/6K \tag{iii}$$

$$d\varepsilon_{12}^{t} = d\lambda\sigma_{22}/2 + d\sigma_{22}/4G \tag{iv}$$

Equation (i) integrates directly to give

- $\varepsilon_{11}^{t} = \sigma_{22}/6K$

Now as $\sigma_{22} = pr/t$, where r and t are the cylinder's inner radius and wall thickness, this gives the differential relationship

$$d\sigma_{22}/\sigma_{22} = dp/p + dr/r - dt/t \qquad \text{(v)}$$

which connects to the strain increments through

$$d\varepsilon_{22}{}^{t} = dr/r \text{ and } d\varepsilon_{33}{}^{t} = dt/t \qquad \text{(vi)}$$

Adding eqs(ii) and (iii) to eliminate $d\lambda$, then substituting the result in eqs(v) and (vi), gives

$$d\varepsilon_{22}{}^{t} = \tfrac{1}{2}(d\sigma_{22}/\sigma_{22} - dp/p) + d\sigma_{22}/6K$$

which integrates to give

- $\varepsilon_{22}^{t} = \ln\left(\sigma_{22}/p\right)^{1/2} + \sigma_{22}/6K$

Further manipulation between eqs(i)–(iv) gives the thickness strain

- $\varepsilon_{33}^{t} = \ln\left(p/\sigma_{22}\right)^{1/2} + \sigma_{22}/6K$

and the (tensor) shear strain

- $\varepsilon_{12}^{t} = \ln\left(\sigma_{22}/p\right)^{1/2}$

Equation (4.24a) gives the von Mises yield criterion in terms of the three stress components. With its reduction to the stress plane σ_{12} versus σ_{22} (y versus x), this criterion becomes

$$\frac{3}{4}\sigma_{22}^{2} + 3\sigma_{22}^{2} = Y^{2}$$

The plastic strain vector is coincident with a stress vector that is the normal to this locus (i.e. the normality rule). The gradient of the normal is

$$\frac{d\sigma_{12}}{d\sigma_{22}} = 4\left(\frac{\sigma_{12}}{\sigma_{22}}\right) = 2 \quad \left(\text{for } \frac{\sigma_{12}}{\sigma_{22}} = \frac{1}{2}\right)$$

Note: The normal gradient is *not* coincident with gradient to the stress path $\sigma_{22}/\sigma_{12} = 2$, with the angle between them being $\tan^{-1}2 - \tan^{-1}(1/2) = 36.87°$.

Ex. 10.11 Under torsion, the kinematic translation of the yield surface in Fig. 10.4a occurs along the τ axis, in contrast to the isotropic expansions shown in this figure. The position β of the current centre of the translated locus along this axis is found from integrating the incremental translation eq(10.52a). The relevant component from this equation is

$$d\alpha_{12} = c_1 d\varepsilon_{12}^{P} - \frac{c_2}{\sigma_o}\alpha_{12} d\bar{\varepsilon}^{P} \qquad \text{(i)}$$

in which both c_1 and c_2 take units of stress. In torsion, the following substitutions are made:

$$d\alpha_{12} = d\beta, \quad \sigma_o = \sqrt{3}k, \quad d\varepsilon_{12} = \frac{d\gamma^P}{2}, \quad d\bar{\varepsilon}^P = \frac{d\gamma^P}{\sqrt{3}}$$

and eq(i) then gives the centre translation increment

$$d\beta = \frac{c_1}{2} d\gamma^P - \frac{c_2}{3k} \beta \, d\gamma^P$$

Separating the variables, and integrating for β:

$$\int_0^\beta \frac{2 d\beta}{c_1 - (2c_2\beta)/(3k)} = \int_0^{\gamma^P} d\gamma^P$$

which gives

$$\frac{\beta}{k} = \frac{3c_1}{2c_2}\left(1 - e^{-c_2\gamma^P/3k}\right) \tag{ii}$$

and since for large strain $\beta/k \to 3c_1/2c_2$ asymptotically, this should be taken as the *correct* expression for the first reversal. Referring to Fig. 10.14, with a Swift hardening law providing the forward shear stress τ_f, the 'radius' or semi-range of elasticity is given as $\mathfrak{R} = \tau_f - \beta$. The reversed yield stress at point P follows as

$$\tau_r = \tau_f - 2\mathfrak{R} = \tau_f - 2(\tau_f - \beta) = 2\beta - \tau_f \tag{iii}$$

Substituting eq(ii) with Swift's law for torsion into eq(iii) gives the required expression for the reversed yield stress from forward shear strain γ^P:

$$\frac{\tau_r}{k} = \frac{3c_1}{c_2}\left(1 - e^{-\frac{c_2\gamma^P}{3k}}\right) - \left(1 + \frac{\gamma^P}{\gamma_o}\right)^n \tag{iii}$$

Note: Care should be taken with the signs that alternate within the second term of eq(i) when integrating with origin at A for the reversed flow curve AB (see § 10.7.4).

CHAPTER 11

Ex. 11.20 Deformation imposed upon a bar at a constant velocity v gives a displacement $x = vt$, at time t, from which the *nominal* strain e and strain rate $\dot{e}$ in an *original* length l_o are

$$e = \frac{x}{l_o} = \frac{vt}{l_o}, \quad \dot{e} = \frac{v}{l_o}$$

A *true* incremental total strain $\delta\varepsilon$ and its strain rate $\dot{\varepsilon} = \delta\varepsilon/\delta t$ are referred to a change δl in the *current* length l as follows:

$$\delta\varepsilon = \frac{\delta l}{l}, \qquad \frac{\delta\varepsilon}{\delta t} = \frac{\delta}{\delta t}\left(\int \frac{\delta l}{l}\right)$$

The integration imposes the initial and current lengths in the limit of each sum as $\delta l \to 0$:

$$\varepsilon = \int_{l_o}^{l} \frac{dl}{l} = \ln\left(\frac{l}{l_o}\right) = \ln\left(\frac{l_o + x}{l_o}\right) = \ln(1 + e)$$

$$\dot{\varepsilon} = \frac{d}{dt}\left[\ln(1 + e)\right] = \frac{\dot{e}}{(1 + e)}$$

A *nominal* stress s divides the applied load W by the original section area A_o of the bar, this giving s as an applied load per unit original area:

$$s = W / A_o \qquad \text{(i)}$$

A *true* stress σ divides the applied load W by the current section area A of the bar, this giving σ as an applied load per unit of current area. The latter follows from an assumption of constant volume of the deforming bar that neglects compressible elasticity:

$$A_o l_o = A l \qquad \text{(ii)}$$

Combining eqs(i) and (ii) provides an estimate of the true stress:

$$\sigma = \frac{W}{A} = \frac{Wl}{A_o l_o} = \frac{W(l_o + x)}{A_o l_o} = s(1 + e)$$

These relationships lie within a sequence of calculations for true strain, strain rate and stress at time t, given a load W versus extension x plot, conducted at constant velocity, v:

$t \to x$	time → displacement
$x \to e, \dot{e}$	displacement → nominal strain & strain rate
$e \to \varepsilon$	nominal strain → true strain
$\dot{e} \to \dot{\varepsilon}$	nominal strain rate → true strain rate
$s \to \sigma$	nominal stress → true stress

These show, with the exception of the nominal strain rate $\dot{e}$, that *all* quantities so calculated will vary with co-ordinate points (x, W) taken from the load versus extension plot. Variations are shown in respect of an origin for e coincident with the intercept shown. This intercept gives the *sub-tangent* as $1 + e$ for a true stress versus nominal strain plot (σ vs e) at the point of instability, i.e. corresponding to $\delta W/\delta x = 0$, as shown. That is:

$$\frac{\delta W}{\delta x} = \frac{\delta(sA_o)}{\delta(el_o)} = \frac{A_o}{l_o}\left(\frac{\delta s}{\delta e}\right) = \frac{A_o}{l_o} \times \frac{\delta}{\delta e}\left[\frac{\sigma}{(1 + e)}\right] = 0$$

In the limit as $\delta x \to 0$, then $\delta e \to 0$, giving

$$\frac{dW}{dx} = \frac{\frac{A_o}{l_o}\left[(1 + e)\frac{d\sigma}{de} - \sigma\right]}{(1 + e)^2} = 0 \quad \Rightarrow \quad \frac{d\sigma}{de} = \frac{\sigma}{(1 + e_I)}$$

where e_I is the nominal instability strain shown in Diag.11.20.

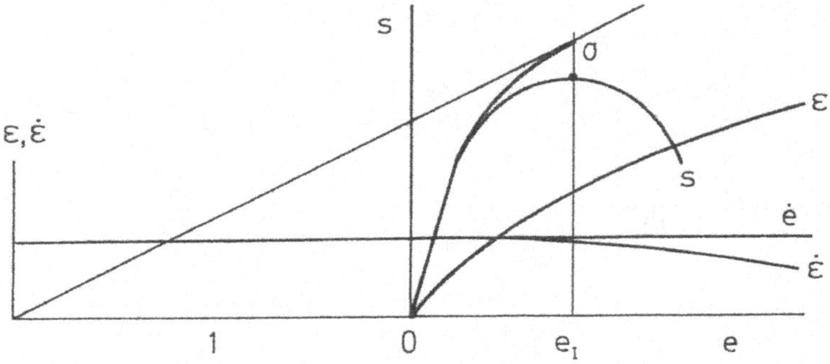

Diagram 11.20

The constant volume assumption applies strictly to a plastically deforming volume. This requires the removal of compressible elastic volume change from the total deforming volume V. Let the volume change be δV:

$$\delta V = \delta V_P + \delta V_E$$

Given $\delta V_P = 0$, then $\delta V = \delta V_E$ follows from the bulk modulus K and the mean stress σ_m:

$$K = \frac{\sigma_m}{(\delta V/V)_E}$$

Setting $\sigma_m = \sigma/3$ for uniaxial tension with V_E as the initial volume V_o gives

$$\delta V_E = \frac{\sigma V_o}{3K}$$

which is recoverable. Hence, the permanently deforming volume follows from

$$V_P = A_P l_P = V_o - \frac{\sigma V_o}{3K} = V_o\left(1 - \frac{\sigma}{3K}\right)$$

With $K > 100$ GPa for engineering alloys, it is seen that the difference between V_P and V_o is negligible.

CHAPTER 12

Note: Some expressions derived here have been based upon Finnie's (subscsript F) simplified, orthotropic equivalent stress and strain expressions, eqs(12.5a,b) and (12.14). If Hill's (subscript H) theory is preferred, then the following conversions can be applied:

$$L_F \to \frac{L_H}{3}, \quad M_F \to \frac{M_H}{3}, \quad N_F \to \frac{N_H}{3}$$

$$\bar{\sigma}_F^2 \rightarrow \frac{1}{3}(F + G + H)\bar{\sigma}_H^2$$

$$(\mathrm{d}\bar{\varepsilon}_F^P)^2 \rightarrow \frac{3\,(\mathrm{d}\bar{\varepsilon}_H^P)^2}{(F + G + H)}$$

$$\mathrm{d}\lambda_F \rightarrow 3\,\mathrm{d}\lambda_H$$

Ex. 12.2 The off-axis stress state (σ_{11}', σ_{12}') within the cylinder wall in Fig. 12.16 produces stress components σ_{11}, σ_{22} and σ_{12} aligned with the material's orthotropic axes 1, 2. Given an inclination between 1′ and 1 of θ, the plane stress transformation eqs(12.45a–c) apply:

$$\sigma_{11} = \sigma_{11}'\cos^2\theta + \sigma_{12}'\sin 2\theta \quad \text{(i)}$$

$$\sigma_{22} = \sigma\sin^2\theta - \sigma_{12}'\sin 2\theta \quad \text{(ii)}$$

$$\sigma_{12} = \sigma_{12}'\cos 2\theta - ½\sigma_{11}'\sin 2\theta \quad \text{(iii)}$$

With all other stress components absent, Finnie's equivalent expression (12.5a) reduces to

$$2\bar{\sigma}^2 = (G + H)\sigma_{11}^2 + (F + H)\sigma_{22}^2 - 2H\sigma_{11}\sigma_{22} + 6N\sigma_{12}^2 \quad \text{(iv)}$$

Substituting eqs(i)–(iii) into (iv) provides the equivalent stress expression

$$2\bar{\sigma}^2 = (G + H)(\sigma_{11}'\cos^2\theta + \sigma_{12}'\sin 2\theta)^2 + (F + H)(\sigma\sin^2\theta - \sigma_{12}'\sin 2\theta)^2$$
$$-2H(\sigma_{11}'\cos^2\theta + \sigma_{12}'\sin 2\theta)(\sigma\sin^2\theta - \sigma_{12}'\sin 2\theta) + 6N(\sigma_{12}'\cos 2\theta - ½\sigma_{11}'\sin 2\theta)^2$$

which is expanded and simplified into the following form:

$$2\bar{\sigma}^2 = \beta_1(\sigma_{11}')^2 + 2\beta_2\sigma_{11}'\sigma_{22}' + \beta_1(\sigma_{22}')^2$$

where $\beta_i = \beta_i(\theta)$ (i = 1, 2, 3) are the orientation coefficients given.

Ex. 12.4 The orientation α of the major principal axis of stress to the tube axis 1′ is not influenced by the orthotropic condition of the material. It follows from equilibrium that

$$\alpha = \tan^{-1}\left(\frac{2\sigma_{12}'}{\sigma_{11}'}\right) = \tan^{-1}(2R)$$

Therefore, if the orientation θ of the material axis 1 is made to coincide with α, the principal axes of stress (major σ_1 and minor σ_2) will lie long the material axes 1 and 2.

The principal plastic strain increments follow from eq(12.5a), with $\sigma_3 = 0$, as

$$\mathrm{d}\varepsilon_1^P = ⅓\,\mathrm{d}\lambda[H(\sigma_1 - \sigma_2) + G\sigma_1] \quad \text{(i)}$$

$$\mathrm{d}\varepsilon_2^P = ⅓\,\mathrm{d}\lambda[F\sigma_2 - H(\sigma_1 - \sigma_2)] \quad \text{(ii)}$$

$$\mathrm{d}\varepsilon_3^P = -⅓\,\mathrm{d}\lambda(G\sigma_1 + F\sigma_2) \quad \text{(iii)}$$

where the principal stresses are given by

$$\sigma_1 = ½\,\sigma_{11}' + ½\,\sqrt{[(\sigma_{11}')^2 + 4\,(\sigma_{12}')^2]} = ½\,\sigma_{11}'\,[1 + \sqrt{(1 + 4R^2)}] \tag{iv}$$

$$\sigma_2 = ½\,\sigma_{11}' - ½\,\sqrt{[(\sigma_{11}')^2 + 4\,(\sigma_{12}')^2]} = ½\,\sigma_{11}'\,[1 - \sqrt{(1 + 4R^2)}] \tag{v}$$

Substituting eqs(iv) and (v) into eqs(i)–(iii) leads to the *intended* forms by Finnie or Hill:

$$d\varepsilon_1^P = ⅓\,d\lambda\,\sigma_{11}'\,[½G + (H + ½G)\sqrt{(1 + 4R^2)}]$$

$$d\varepsilon_2^P = ⅓\,d\lambda\,\sigma_{11}'\,[½F - (H + ½F)\sqrt{(1 + 4R^2)}]$$

$$d\varepsilon_3^P = -1/6\,d\lambda\,\sigma_{11}'\,[F + G + (G - F)\sqrt{(1 + 4R^2)}]$$

where σ_{11}' is the applied axial stress and $R = \sigma_{12}'/\sigma_{11}'$.

Ex. 12.8 The solution given here is based upon Finnie's equivalent stress expression (12.5a,b). For the given yield function *f* this relates the shear yield stress to

(a) the tensile yield stress for the 1–direction as

$$6f = G\sigma_{1y}^2 + H\sigma_{1y}^2 = 6N\sigma_{12y}^2, \quad \Rightarrow \quad \frac{\sigma_{12y}}{\sigma_{1y}} = \left(\frac{H + G}{6N}\right)^{1/2}$$

(b) the tensile yield stress for the 2–direction as

$$6f = F\sigma_{2y}^2 + H\sigma_{2y}^2 = 6N\sigma_{12y}^2, \quad \Rightarrow \quad \frac{\sigma_{12y}}{\sigma_{2y}} = \left(\frac{H + F}{6N}\right)^{1/2}$$

(c) the tensile yield stress for the 3–direction as

$$6f = G\sigma_{3y}^2 + H\sigma_{3y}^2 = 6N\sigma_{12y}^2, \quad \Rightarrow \quad \frac{\sigma_{12y}}{\sigma_{3y}} = \left(\frac{F + G}{6N}\right)^{1/2}$$

Note: For Hill's potential, eq(12.2), 3 replaces 6 in the denominator of each expression.

Ex. 12.11 The off-axis tensile stress σ produces non-zero stress components σ_{11}, σ_{22} and σ_{12} aligned with the material's orthotropic axes, which are found from the stress transformation eqs(12.45a–c):

$$\sigma_{11} = \sigma\cos^2\theta \tag{i}$$

$$\sigma_{22} = \sigma\sin^2\theta \tag{ii}$$

$$\sigma_{12} = ½\sin 2\theta \tag{iii}$$

With all other stress components absent, Hill's equivalent expression (12.4b) reduces to

$$\frac{2}{3}(F + G + H)\bar{\sigma}^2 = (G + H)\sigma_{11}^2 + (F + H)\sigma_{22}^2 - 2H\sigma_{11}\sigma_{22} + 2N\sigma_{12}^2 \tag{iv}$$

Substituting eqs(i)–(iii) into (iv) gives one possible form:

$$\frac{2}{3}(F + G + H)\bar{\sigma}^2 = \sigma^2\left[(G + H)\cos^4\theta + (F + H)\sin^4\theta + \frac{1}{2}(N - H)\sin^2 2\theta\right]$$

Finnie's function (12.5a) provides the simpler result:

$$2\bar{\sigma}^2 = \sigma^2\left[(G + H)\cos^4\theta + (F + H)\sin^4\theta + \frac{1}{2}(3N - H)\sin^2 2\theta\right]$$

so that the yield stress variation becomes

$$\sigma = \sqrt{2}\bar{\sigma}\left[(G + H)\cos^4\theta + (F + H)\sin^4\theta + \frac{1}{2}(3N - H)\sin^2 2\theta\right]^{-\frac{1}{2}} \qquad \text{(v)}$$

Given that $\bar{\sigma}$ is constant at yield for all orientations, the maximum (or minimum) σ in eq(v) is found from the condition $d\sigma/d\theta = 0$. Differentiating eq(v) leads to the equation

$$(H + F)\sin^3\theta\cos\theta + \frac{1}{2}(3N - H)\sin 2\theta\cos 2\theta - (H + G)\cos^3\theta\sin\theta = 0$$

from which:

$$\tan^2\theta = \frac{(2H + G - 3N)}{(2H + F - 3N)} = \frac{[3N - (G + 2H)]}{[3N - (F + 2H)]}$$

which reveal maxima or minima when $3N > F + 2H$ and $3N > G + 2H$, making $\tan^2\theta$ positive, and the opposite when $3N > F + 2H$ and $3N < G + 2H$, making $\tan^2\theta$ negative.

CHAPTER 13

Ex. 13.1 The elliptical tube is said to buckle about its minor axis. Taking the mean semi-lengths of the major and minor axis as $a = 80$ and $b = 30$ mm (as intended in example 13.1), we find the second moment of area:

$$I_y = \frac{\pi}{4}\left(b_o a_o^3 - b_i a_i^3\right) = \frac{\pi}{4}\left(31.5 \times 81.5^3 - 28.5 \times 78.5^3\right) = 2.565 \times 10^6 \text{ mm}^4$$

The section area is

$$A = \pi\,(a_o b_o - a_i b_i) = \pi\,(31.5 \times 81.5 - 28.5 \times 78.5) = 1036.76 \text{ mm}^2$$

Hence, the radius of gyration is

$$k_y = \sqrt{\frac{I_y}{A}} = \sqrt{\frac{2.565 \times 10^6}{1036.76}} = 49.74$$

from which: $\dfrac{L_e}{k_y} = \dfrac{825}{49.74} = 16.59$

In order to apply eq(13.3), identify firstly $\varepsilon_o = 2.083 \times 10^{-3}$ within Hollomon's eq(13.2a). Then, with $\sigma_o = 144$ MPa and $n = 1/3$, eq(13.3) gives

$$\frac{\sigma}{\sigma_o} = \left(\frac{1}{3 \times 2.083 \times 10^{-3}}\right)^{\frac{1}{3}} \left(\frac{\pi}{16.59}\right)^{\frac{2}{3}} = 1.79$$

from which: $\sigma = 1.79 \times 144 = 257.8$ **MPa** and $P_{crit} = \sigma A = 257.8 \times 1036.76 = 267.3$ kN.

Ex. 13.4 When the direction of σ_y in Fig. 13.11 is reversed and its magnitude equals that of σ_x, the plate is placed under pure shear (as in Fig. 13.14). Setting $\beta = -1$, $n = 1$ and $r = a/b >> 1$ in eq(13.24) gives the critical elastic value of σ_x:

$$(\sigma_x)_{cr} = \frac{D\pi^2}{a^2 t} \frac{(m^2 + r^2)^2}{(m^2 - r^2)} \tag{i}$$

where m, the number of half waves in the x–direction, is found when $(\sigma_x)_{cr}$ is a minimum:

$$\frac{\mathrm{d}(\sigma_x)_{cr}}{\mathrm{d}m} = (m^2 - r^2) \times 2(m^2 + r^2) \times 2m - (m^2 + r^2)^2 \times 2m = 0$$

which gives $m = \sqrt{3}r$. Therefore, from eq(i):

$$(\sigma_x)_{cr} = \frac{8\pi^2 D}{tb^2} = \frac{2\pi^2 E}{3(1 - \nu^2)} \left(\frac{t}{b}\right)^2 \tag{ii}$$

Equation (ii) is seen to take the form of eq(13.32) upon substituting $D = Et^3/[12(1 - \nu^2)]$. Taking integral values of $m = 1, 2, 3, \ldots$ within the length a, corresponding values of the ratio $r = 1/\sqrt{3}, 2/\sqrt{3}, 3/\sqrt{3}$ apply to the minima.

Ex. 13.5 With $b/a = 200/260 = 0.77$, the shear buckling coefficient for a plate, with both simple and clamped supports, is read from Fig. 13.15 as $K_s = 8.9$. The elastic buckling shear stress follows from eq(13.32) as

$$(\tau_{cr})_e = 8.9 \times 73 \times 10^3 \left(\frac{4.5}{200}\right)^2 = 329 \text{ MPa}$$

This gives the ratio $\dfrac{(\tau_{cr})_e}{\sigma_n} = \dfrac{329}{340} = 0.97$ where, from Fig. 13.16, it is apparent that a plasticity reduction factor needs to be applied. Interpolating from this figure for $m = 15$ gives ital $\eta = 0.6$ when $(\tau_{cr})_p = 0.6 \times 329 = 197.4$ MPa.

Ex. 13.6 In an 'open' cylinder the internal pressure acts against sliding plugs and does not stress the cylinder wall axially. The pressure results in a circumferential wall tension for which the equivalent stress and strain are by definition

$$\bar{\sigma} = \sigma_\theta = \frac{pr}{t}, \qquad d\bar{\varepsilon}^P = d\varepsilon_\theta^P \tag{i, ii}$$

Differentiating eq(i) gives

$$\frac{d\sigma_\theta}{\sigma_\theta} = \frac{dp}{p} + \frac{dr}{r} - \frac{dt}{t} = d\varepsilon_\theta^P - d\varepsilon_r^P \tag{iii}$$

when the condition for instability is $dp/p = 0$. Now, as $d\varepsilon_r^P = -½\, d\varepsilon_\theta^P$, eq(iii) becomes:

$$\frac{d\sigma_\theta}{\sigma_\theta} = \frac{3}{2}\, d\varepsilon_\theta^P \tag{iv}$$

Combining eqs(i), (ii) and (iv) defines the point of instability on the equivalent flow curve:

$$\frac{d\bar{\sigma}}{d\bar{\varepsilon}^P} = \frac{3}{2}\bar{\sigma} \tag{v}$$

Differentiating the Swift expression given for this flow curve:

$$\frac{d\bar{\sigma}}{d\bar{\varepsilon}^P} = B\sigma_o n(1 + B\bar{\varepsilon}^P)^{n-1} = B\sigma_o n\left(\frac{\bar{\sigma}}{\sigma_o}\right)^{\frac{n-1}{n}} \tag{vi}$$

Equating (v) and (vi) gives the co-ordinates to the instability point and the critical pressure:

$$\bar{\sigma} = \left(\frac{2Bn}{3}\right)^n \sigma_o, \quad \bar{\varepsilon}^P = \frac{2n}{3} - \frac{1}{B}, \quad p_{crit} = \frac{t\bar{\sigma}}{r} = \frac{2t\sigma_o}{d}\left(\frac{2Bn}{3}\right)^n$$

Ex. 13.7 Let E_L be the % elongation and R_A be the % reduction in the area of a fractured tensile testpiece. Each will supply an estimate of the true fracture strain when defined as

$$\varepsilon_f = \ln(1 + e_f) = \ln(l_f/l_o) \tag{i}$$

where $e_f = (l_f - l_o)/l_o$ is the engineering fracture strain. Since $E_L = 100e$, it follows from (i):

$$\varepsilon_f = \ln(1 + E_L/100)$$

Now, by definition:

$$\%R_A = 100(A_o - A_f)/A_o \tag{ii}$$

and with no change in volume:

$$A_o l_o = A_f l_f \tag{iii}$$

For uniform straining to failure, the combination of eqs(i)–(iii) leads to:

$$\varepsilon_f = \ln\left(\frac{100}{100 - R_A}\right)$$

Ex. 13.8 To find the point of tensile instability, the Consideré tangent condition applies to a true stress versus true strain curve with a sub-tangent of unity, as in Fig. 9.2a:

$$\frac{d\sigma}{d\varepsilon} = \sigma \tag{i}$$

Conveniently, the instability co-ordinates are found by differentiating the equation to the flow curve, if it is known. Here, Hollomon's law is assumed to apply to the data supplied:

$$\sigma = A\varepsilon^n \tag{ii}$$

for which constants A and n are to be found from the force (W) and extension (Δl) data. These data are converted to true stress and natural strain as follows:

$$\sigma = \frac{W}{A_o}\left(1 + \frac{\Delta l}{l}\right) = \frac{4W}{\pi d_o^2}\left(1 + \frac{\Delta l}{l_o}\right), \qquad \varepsilon = \ln\left(1 + \frac{\Delta l}{l_o}\right)$$

σ	75	150.8	226.2	301.8	397.8	474.4	526.9	579.8	612
(MPa)	643.8	652.0	674.2	686.8	698.2	710.5	726.9	737.8	746.2
	758.1	765.4	775.1	781.3	784.4	787.1			

% ε	0.04	0.08	0.10	0.14	1.0	2.4	3.98	6.67	8.25
	9.53	11.7	13.75	15.2	16.57	18.07	20.26	21.83	23.11
	25.0	26.24	27.92	29.12	30.0	30.9			

Plotted in double logarithmic co-ordinates, σ and ε fall on a straight line (for $\varepsilon > 1$ %), thereby conforming to a Hollomon law:

$$\sigma = 10^3 \varepsilon^{0.2} \tag{iii}$$

This plot ignores the initial elasticity involved in rising from zero stress to meet the data in the range beyond 1 % strain, where the law is found to be accurate. Since elastic strain is small within that range, sensibly, the instability co-ordinates (i.e. those at the UTS) follow from combining eqs(i) and (ii) as

$$nA\varepsilon_u^{n-1} = A\varepsilon_u^n \quad \therefore \ \varepsilon_u = n \quad \text{and} \quad \sigma_u = A\varepsilon_u^n$$

$$\varepsilon_u = n = 0.2 \quad \text{and} \quad \sigma_u = An^n = 10^3(0.2)^{0.2} = 724.8 \text{ MPa}$$

CHAPTER 14

Ex. 14.4 This is an example of a purely elastic wave interaction between the three section areas. Referring to the left-hand change in section (see also Figs 14.8a,b), when a wave crosses from the large to the small section, eqs(14.7a,b) become

Force: $A(a + b) = cA/3$
Velocity: $a - b = c$

from which $c = 3a/2$ and $b = -a/2$, the latter showing that the reflected wave changes sign.

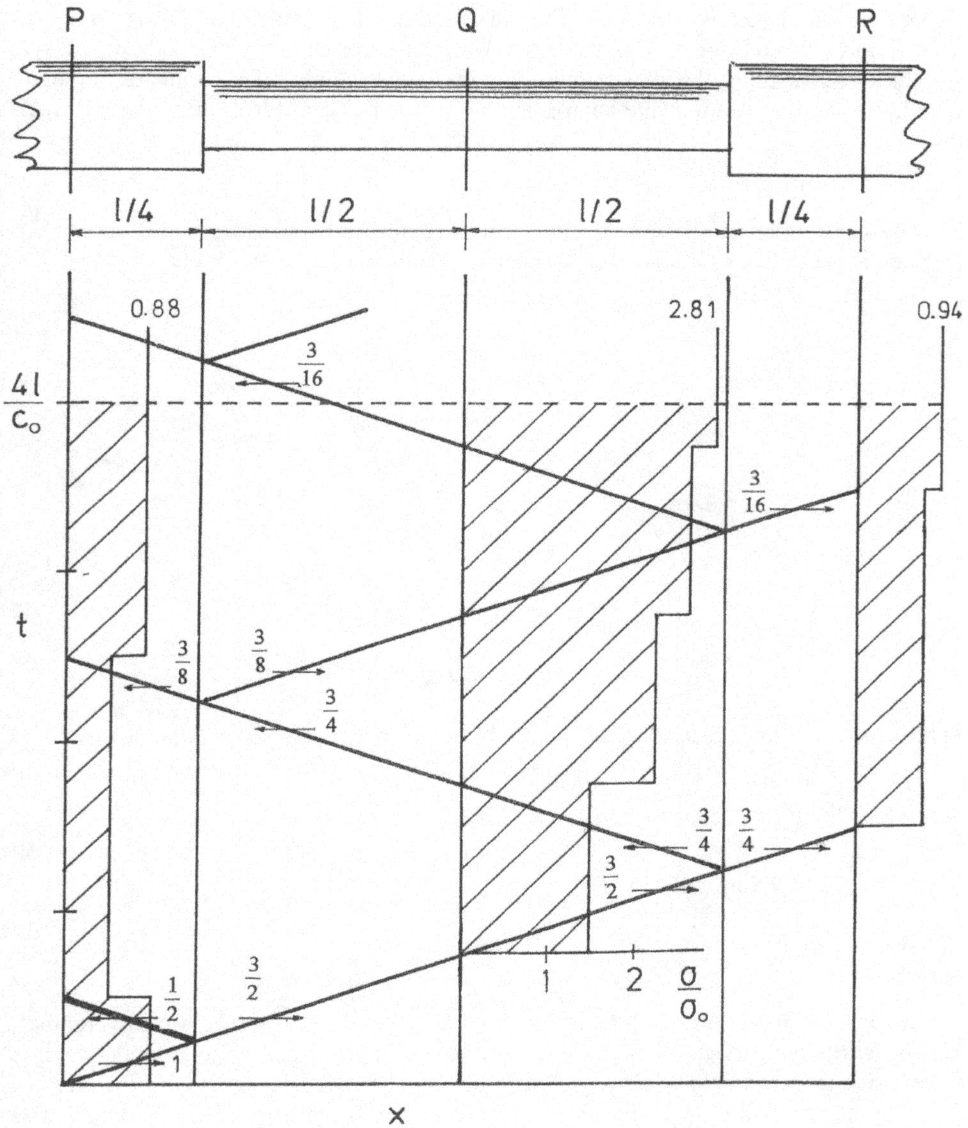

Diagram 14.4

Referring to the right-hand change in section (see Fig. 14.7a,b), when a wave crosses from the small to the large section, eqs(14.6a,b) become

Force: $(a + b)A/3 = cA$
Velocity: $a - b = c$

from which $c = a/2$ and $b = a/2$. Transmitted waves b do not reflect in the time interval. We may use these results for repeated interactions of those waves reflected from the larger ends within the smaller diameter bar. This leads to the x–t Diag. 14.4, where the bold line applies to the single compressive waves and finer lines for tensile waves. From this, the σ–t plots at sections P, Q and R are derived as shown within the interval $0 \leq t \leq 4l/c_o$. They show that for all three sections the tensile stress increase with time. The ordinates in these distributions are the sum of the appropriate fractions in the x–t diagram to two decimals places.

Ex. 14.5 (a) Referring to Diag. 14.5a, it is assumed in the first of the interactions a continued propagation of the plastic wave with a reloading from compressive wave a:

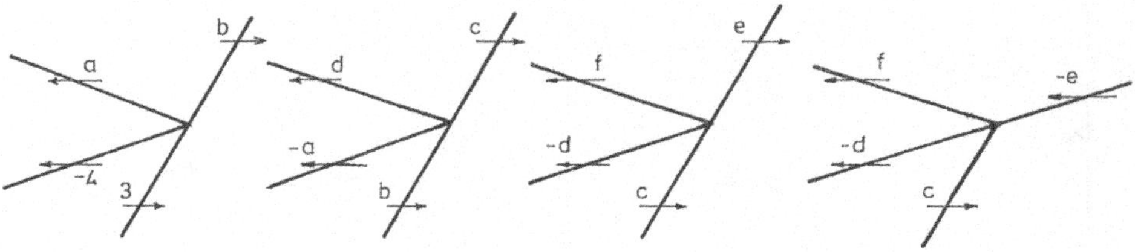

Diagram 14.5a

Firstly, assume Case I (see also Fig. 14.18a,b) in which $\sigma_1 - \sigma_o$ (= 3) is large.

Force (i.e. signs only):

$3 - 4 + a = b$

$-1 + a = b$ (i)

Velocity (i.e. directions only):

$(\sigma_1 - \sigma_o)(c_o/c) - \sigma_1 - a = b(c_o/c)$

$(3 \times 4) - 4 - a = b \times 4$

$8 - a = 4b$ (ii)

Solving (i) and (ii) gives $a = 12/5$ and $b = 7/5$, indicating that the plastic wave overcomes. This is a sufficient answer, but if on the side $-x$ waves are reflected at a free boundary, while the $+x$ side is infinite (no reflections), we could encounter further successive interactions given in the remaining three figures. In the second diagram, the wave a is reflected in tension to interact with the reduced, compressive plastic wave b.

Force: $b + d - a = c'$

$7/5 + d - 12/5 = c$

$-1 + d = c'$ (iii)

Velocity: $b(c_o/c) - a - d = c'(c_o/c)$

$(7/5 \times 4) - 12/5 - d = 4c'$

$16/5 - d = 4c'$ (iv)

Solving (iii) and (iv) gives $c' = 11/25$ and $d = 36/25$, i.e. plastic wave still overcomes.

In the third diagram, a tensile wave d is reflected in to meet wave c'. The interaction assumes that a weaker plastic wave e is propagated. Then:

Force: $c' + f - d = e$
$11/25 + f - 36/25 = e$
$-1 + f = e$ (v)

Velocity: $c'(c_o/c) - d - f = e(c_o/c)$
$(11/25 \times 4) - 36/25 - f = 4e$
$8/25 - f = 4e$ (vi)

Solving (v) and (vi) gives $e = -17/125, f = 108/125$. The negative value indicates that e has become a tensile wave, which is not possible.

Hence, the fourth diagram 14.5a examines Case II (see Figs 14.19a,b). With $\sigma_1 - \sigma_o \equiv c'$ having become small, this provides an interaction with d in which the elastic wave overcomes the plastic wave.

Force: $c' - d + f = -e$
$11/25 - 36/25 + f = -e$
$-1 + f = -e$ (vii)

Velocity: $c' - d - f = -e$
$(11/25 \times 4) - 36/25 - f = -e$
$8/25 - f = -e$ (viii)

Solving eqs(vii) and (viii) gives $e = 17/50$ and $f = 33/50$. This is the correct solution as no further plasticity is confirmed from these positive e and f values.

(b) This is an example of a special Case III (see Figs 14.20a,b), where the reloading wave shown in Diag. 14.5b satisfies eq(14.23c):

$$\sigma_1 = \frac{(c_o/c + 1)\sigma_o}{(c_o/c - 1)} = \frac{(4 + 1)\sigma_o}{(4 - 1)} = \frac{5\sigma_o}{3}$$

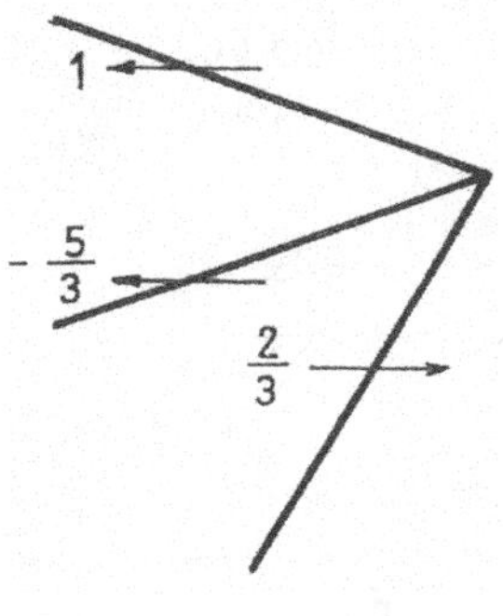

Diagram 14.5b

Alternatively, using the magnitudes shown, it can be seen that both the forces and the velocities balance:

$2/3 - 5/3 + 1 = 0$
$4(2/3) - 5/3 - 1 = 0$

Hence, there is no further plasticity or an unloading wave in this case.

(c) Applying the force and velocity equations with the directions shown in Fig. 14.38c reveals that a wave cancellation occurs from this interaction, i.e. there is no further propagation of waves, either elastic or plastic.

Force: $-5/3 + 2/3 = -1$

Velocity: $-5/3 + 4(2/3) = 1$

(d) Assume the two possible wave interactions in Diag. 14.5d.

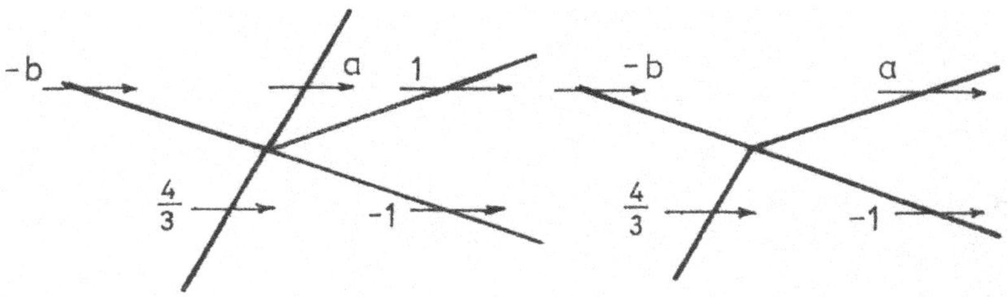

Diagram 14.5d

In the first interaction, a weaker plastic wave a propagates (Case I).

Force: $4/3 - b = a + 1 - 1$
$a + b = 4/3$ (i)

Velocity: $(4/3 \times 4) + b = (a \times 4) + 1 + 1$
$4a - b = 10/3$ (ii)

Solving (i) and (ii) gives $a = 14/15$ and $b = 2/5$. The positive stresses suggest that the assumed interaction is correct. To check that there is not an alternative where the plastic wave is overcome, refer to the second interaction Diag. 14.5d in which Case II is assumed (see also Fig. 14.19b).

Force: $4/3 - b = a - 1$
$a + b = 7/3$ (iii)

Velocity: $(4/3 \times 4) + b = 1 + a$
$a - b = 13/3$ (iv)

Solving (iii) and (iv) gives $a = 13/3$ and $b = -1$, of which a sign change is not possible. Hence, we may continue with the reduced plastic wave until the elastic wave overcomes, as for (a) above.

Ex. 14.9 Firstly, find the stress σ_1 at impact from taking zero velocity of the bar at impact assuming that it becomes plastic. From eq(14.25):

$$\text{Velocity: } v_o - \frac{\sigma_o}{\rho c_o} - \frac{(\sigma_1 - \sigma_o)}{\rho c_o}\left(\frac{c_o}{c}\right) = 0$$

Setting $c_o = \surd(E/\rho)$ and solving for σ_1/σ_o gives

$$\frac{\sigma_1}{\sigma_o} = 1 + \left(\frac{v_o\sqrt{E\rho}}{\sigma_o} - 1\right)\frac{c}{c_o}$$

Setting $c_o/c = \surd(E/S) = \surd 64 = 8$, $E = 6.9 \times 10^9$ Pa, $\rho = 8304$ kg/m^3, $\sigma_o = 138 \times 10^6$ Pa and v_o = 76.5 m/s gives $\sigma_1/\sigma_o = 1.4$.

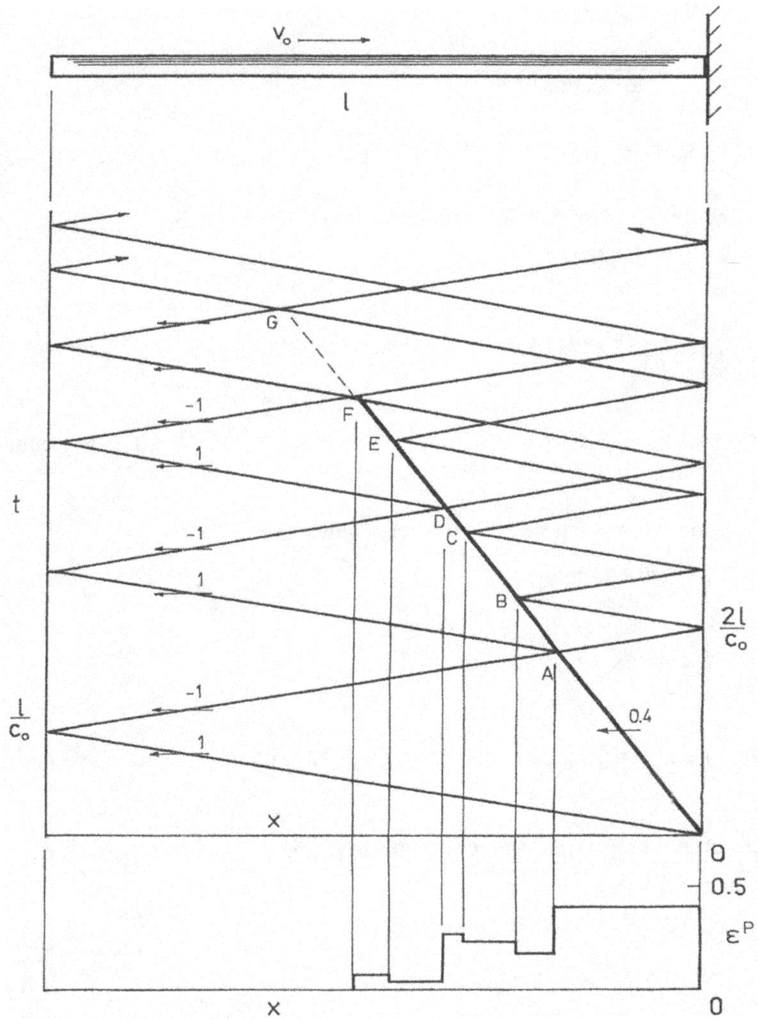

Diagram 14.9

Let positive stress be that of the initial wave, i.e. compression. The x–t plot in Diag. 14.9 is constructed from accompanying plastic interaction calculations at A, B, C, D, E, F and G, as follows:

Point A – Assuming the plastic wave propagates as a (as in Fig. 14.23b):

Force: $a + b = 0.4$ (i)

Velocity: $1 + 1 + 8a = (0.4 \times 8) + b$
$8a - b = 1.2$ (ii)

Solving eqs (i) and (ii) yields $a = 0.178$ and $b = 0.222$, indicating that they are physically possible.

Point B – Assuming the plastic wave propagates as b:

Force: $b = -a + 0.178 + 0.222$
$a + b = 0.4$ (iii)

Velocity: $8b = a + 0.222 + (8 \times 0.178)$
$-a + 8b = 1.646$ (iv)

Solving eqs (iii) and (iv) yields $a = 0.173$, $b = 0.227$, i.e. the plastic wave is strengthened.

Point C – Assuming the plastic wave continues to propagate as b:

Force: $b = -a + 0.173 + 0.227$
$a + b = 0.4$ (v)

Velocity: $8b = a + 0.173 + (8 \times 0.227)$
$8b - a = 1.989$ (vi)

Solving eqs(v) and (vi) yields $a = 0.135$, $b = 0.265$, i.e. the plastic wave continues to propagate.

Point D – Assuming the plastic wave now propagates as a:

Force: $a + 1 - 1 = -b + 0.265$
$a + b = 0.265$ (vii)

Velocity: $8a + 2 = b + (8 \times 0.265)$
$8a - b = 0.120$ (viii)

Solving eqs(vii) and (viii) yields $a = 0.043$, $b = 0.222$, i.e. the plastic wave continues to propagate.

Point E – Assuming the plastic wave continues to propagate as a:

Force: $a = 0.043 + 0.135 - b$
$a + b = 0.178$ (ix)

Velocity: $8a = b + 0.135 + (8 \times 0.043)$
$8a - b = 0.479$ (x)

Solving eqs(ix) and (x) yields a = 0.073, b = 0.105, i.e. the plastic wave continues to propagate but of small enough magnitude to assume the elastic wave overcomes it at F.

Point F – Assuming that a weaker plastic wave propagates as a:

Force: $a = 0.073 + 0.222 - b$
$a + b = 0.295$ (xi)

Velocity: $8a = b + 0.22 + (8 \times 0.073)$
$a - 0.125b = 0.101$ (xii)

Solving eqs(xi) and (xii) yields a = 0.122, b = 0.173, i.e. the plastic wave continues to propagate but of small enough magnitude to assume elastic wave overcomes it at G.

Point G – An elastic wave interaction is assumed beyond this point:

Force: $b - 1 = - a + 0.122$
$a + b = 1.122$ (xiii)

Velocity: $b + 1 = a + (8 \times 0.122)$
$a - b = 0.0213$ (xiv)

Solving eqs(xiii) and (xiv) yields $a = 0.571$, $b = 0.55$, which confirms the assumption made.

The projection of the residual stress distribution for the plastic limb OABCDEFG is shown in Diag. 14.9. By virtue of eq(14.24), the residual strain are proportional to this. That is:

$$\varepsilon^P = \frac{(\sigma_1 - \sigma_o)}{S} - \frac{(\sigma_1 - \sigma_o)}{E} = \frac{(\sigma_1 - \sigma_o)}{\sigma_o} \times \frac{\sigma_o}{E}\left(\frac{E}{S} - 1\right)$$

showing that a strain conversion amounts to multiplying the stress ordinate by a factor of (138 × 63)/69000 = 1.26, thereby indicating maximum residual strain of order 50 %.

Ex. 14.11 Firstly, find the stress σ_1 at impact by taking equal velocities of the bars at impact, assuming that they each become plastic. From eqs(14.26a,b):

Force Equation: $A_1\sigma_1 = A_2\sigma_2$ (i)

Velocity Equation: $$v_o - \frac{\sigma_o}{\rho c_o} - \frac{(\sigma_2 - \sigma_o)}{\rho c_o}\left(\frac{c_o}{c}\right) = \frac{(\sigma_1 - \sigma_o)}{\rho c_o}\left(\frac{c_o}{c}\right) + \frac{\sigma_o}{\rho c_o}$$ (ii)

By setting $c_o/c = \sqrt{(E/S)} = \sqrt{16} = 4$ and $A_2/A_1 = 2$ in eqs (i) and (ii), they may be solved for σ_1/σ_o:

$$\frac{\sigma_1}{\sigma_o} = \frac{v_o\sqrt{E\rho}}{6\sigma_o} + 1$$

Substituting $E = 207 \times 10^9$ Pa, $\rho = 7890$ kg/m^3, $\sigma_o = 345 \times 10^6$ Pa and $v_o = 14.75$ m/s gives $\sigma_1 / \sigma_o = 1.287$. *Note*: From eq(i), $\sigma_2 = \sigma_1/2 = 0.644\sigma_o$, which is elastic. Hence, the initial assumption was wrong and we must modify eq(ii) bar 1 plastic, bar 2 elastic:

$$v_o - \frac{\sigma_2}{\rho c_o} = \frac{(\sigma_1 - \sigma_o)}{\rho c_o}\left(\frac{c_o}{c}\right) + \frac{\sigma_o}{\rho c_o} \tag{iii}$$

Solving eq(iii) with eq(i):

$$\frac{\sigma_1}{\sigma_o} = \frac{2 v_o \sqrt{E\rho}}{9\sigma_o} + \frac{2}{3}$$

giving $\sigma_1 / \sigma_o = 1.05$ with $\sigma_2 / \sigma_o = 0.53$, this being consistent with our second assumption. The x–t plot in Diag. 14.11 can then be constructed from each of the accompanying plastic interaction calculations at A, B, C, D, E and F as follows:

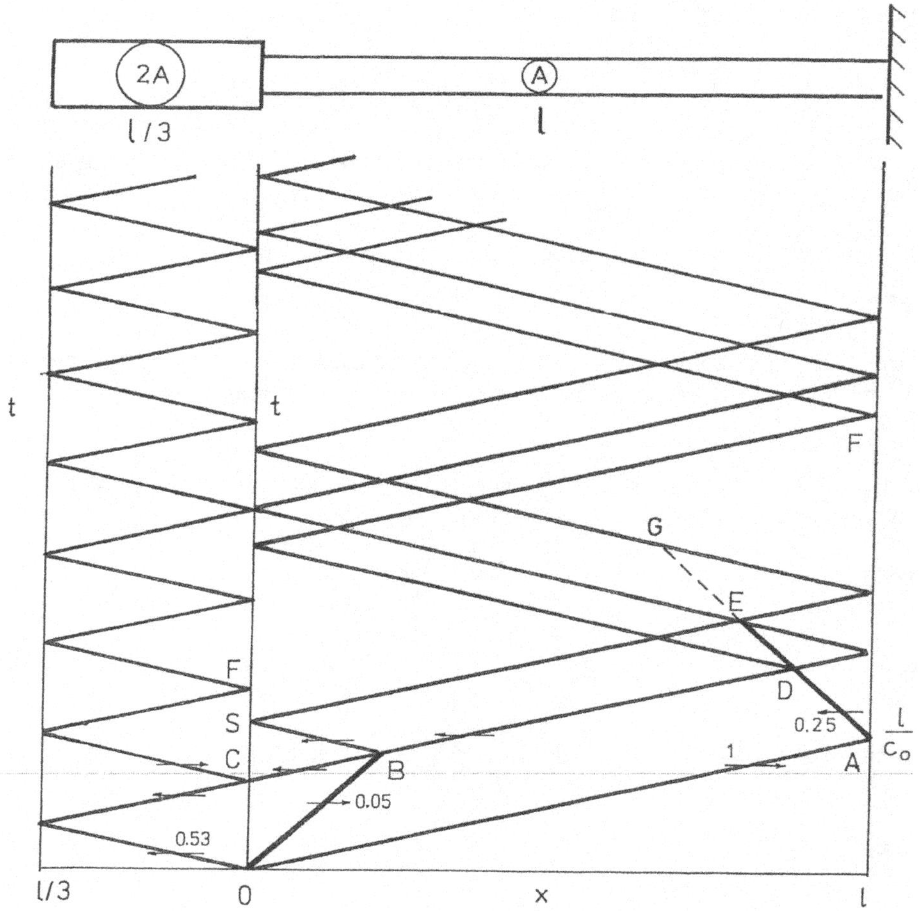

Diagram 14.11

Point A – An elastic wave reaches the fixed end where the particle velocity is zero:

$1 - 4a = 0$
giving $a = 0.25$ for the plastic reflected wave.

Point B – Assuming waves have similar signs:

Force: $(-b - 0.53)2A = -aA$
$2b - a = -1.06$ (iv)

Velocity: $b - 0.53 = -a$
$b + a = 0.53$ (v)

Solving eqs(iv) and (v) yields $a = 0.707$ and $b = -0.177$ (the negative value indicates that a compressive wave is reflected).

Point C – Assuming that the reflected wave overcomes the plastic wave:

Force: $b + 0.05 - 0.707 = -a$ (vi)
$b + a = 0.657$

Velocity: $(0.05 \times 4) - 0.707 - b = -a$
$a - b = 0.507$ (vii)

Solving eqs(vi) and (vii) yields $a = 0.582$ and $b = 0.075$ (the positive value indicates that they are possible).

Point D – Assuming that a weaker plastic wave b propagates:

Force: $b = +0.25 - a$
$a + b = 0.25$ (viii)

Velocity: $0.582 + 0.582 + 4b = a + 1$
$a - 4b = 0.164$ (ix)

Solving eqs(viii) and (ix) yields $a = 0.233$ and $b = 0.0172$ (the positive value confirms the assumption).

Point E – The compressive elastic wave is both transmitted and reflected at the enlarged section, as in Fig. 14.7:

Force: $2aA = (c + b)A$
$2a = 0.075 + b$ (x)

Velocity: $a = c - b$
$a = 0.075 - b$ (xi)

Solving eqs(x) and (xi) yields $a = 0.05$, $b = 0.025$.

Point F – The tensile elastic wave is both transmitted in tension but reflected in compression at the reduced section, as in Fig. 14.8:

Force: $(-b - 0.177)2A = -aA$
$2b - a = -0.354$ (xii)

Velocity: $b - 0.177 = -a$
$b + a = 0.177$ (xiii)

Solving eqs(xii) and (xiii) yields $a = 0.236$, $b = -0.059$.

Point G – Assuming plastic wave is overcome, i.e. only elastic interactions follow:

Force: $0.025 - b = a - 0.233 + 0.0172$

$$a + b = 0.241 \qquad \text{(xiv)}$$

Velocity: $0.025 + b = a + 0.233 - (4 \times 0.0172)$

$$a - b = -0.14 \qquad \text{(xv)}$$

Solving eqs(xiv) and (xv) yields $a = 0.051$ and $b = 0.191$ (+ confirming the assumption).

The plastic interactions ADG and OC, which occur at both ends of the longer bar, are completed within a time $2l/c_o$. No further plasticity occurs for times greater than this where elastic waves follow. The returning elastic waves in the smaller diameter bar are shown in Diagram 14.11 to change sense at the free end and retain their sense at the fixed end. Reflected waves also change sense at both ends of the larger diameter bar but the situation is complicated with the transmission of waves from the larger to the smaller bar for as long as they remain in contact. The transmission of waves from the smaller bar to the larger also occurs, where a separation has been judged to occur at S where the wave remains tensile at the interface.

Ex. 14.13(a) Assuming an elastic impact the equivalent static load is found from eq(14.34b):

$$P_E^2 - (2mg)P_E - (2mg)\frac{AEh}{L} = 0$$

$$P_E^2 - (2 \times 15 \times 9.81)P_E - (2 \times 15 \times 9.81) \times (16\pi \times 10^{-6}) \times (207 \times 10^9) \times 0.1/5 = 0$$

$$P_E^2 - 294.3P_E - (61.25 \times 10^6) = 0$$

giving $P_E = 7975$ N and the bar's elastic stress as 158.6 MPa, with a deflection 3.83 mm.

(b) A similar elastic analysis applied to the shorter length $L = 1$ m gives $P_E = 17648$ N. This shows the bar's stress as 351.1 MPa with a deflection 1.7 mm. However, since the yield stress is exceeded in this case, applying the Hollomon stress–strain law $\varepsilon = B\sigma^n$ would be more appropriate to the elastic-plastic impact. Taking a hardening exponent $n = 0.25$ from the % elongation given, let the strength coefficient B be defined by the yield stress co-ordinates with σ_y in Pa and ε_y without dimension:

$$\sigma_y = 300 \times 10^6 \text{ Pa}, \quad \varepsilon_y = \frac{\sigma_y}{E} = \frac{300 \times 10^6}{207 \times 10^9} = 1.45 \times 10^{-3}$$

$$\varepsilon_y = B\sigma_y^n \quad \therefore B = \frac{\varepsilon_y}{\sigma_y^n} = \frac{1.45 \times 10^{-3}}{(300 \times 10^6)^{0.25}} = 11 \times 10^{-6}$$

From eq(14.24c) the equivalent static load P_E is found from the solution to eq(4.34c):

$$P_E^{n+1} - \left(1 + \frac{1}{n}\right)(mg)P_E^n - \left(1 + \frac{1}{n}\right)(mg)\frac{hA^n}{BL} = 0$$

where in units of N and m:

$$\frac{hA^n}{BL} = \frac{0.1 \times (16\pi \times 10^{-6})^{0.25}}{(11 \times 10^{-6}) \times 1} = 765.5$$

$$P_E^{1.25} - 5(15 \times 9.81)P_E^{0.25} - 5(15 \times 9.81) \times 765.5 = 0$$

$$P_E^{1.25} - 735.75P_E^{0.25} - (563.22 \times 10^3) = 0$$

from which $P_E \approx 40.5 \times 10^3$ N and $\sigma = 806$ MPa in excess of the UTS = 600 MPa, indicating that the rod would fail!

CHAPTER 15

Ex. 15.1 Taking the origin of co-ordinates to lie at the centre of the strip section gives the mean pressure/unit width as

$$\bar{p} \times \frac{b}{2} = 2k \int_0^{b/2} e^{\frac{\mu b}{t}(1 - \frac{2x}{b})} dx = \frac{kt}{\mu}(e^{\frac{\mu b}{t}} - 1)$$

$$F = \bar{p} \times b \times w = 2k \times \frac{wt}{\mu}\left(e^{\frac{\mu b}{t}} - 1\right)$$

$$= \frac{250 \times 300 \times 20}{0.2}\left(e^{\frac{0.2 \times 80}{20}} - 1\right) = 9.192 \times 10^6 \text{ N} = 9.192 \text{ MN}$$

Ex. 15.2 As the 150 mm dimension remains unchanged, a plane strain compression may be assumed with $b = 25$ mm, $t_o = 25$ mm and $w_o = 150$ mm (as in Fig. 15.1). The single step cogging process produces $b_1 = 104$ mm, $t_1 = 6$ mm and $w_1 = 150$ mm, for which a constant flow stress $2k = 7$ MPa applies to lead during the cold forging process. The pressure distributions are shown in Figs 15.4a–c and from the first of these an initial estimate to the mean pressure is given by eq(15.11):

$$\frac{\bar{p}}{2k} = \frac{t}{\mu b}\left(e^{\mu b/t} - 1\right) = \frac{6}{(0.25 \times 25)}\left(e^{0.25 \times 25/6} - 1\right) = 1.761$$

Correspondingly, the forging force is

$$F = 1.761 \times 2k \times bw_o = 1.761 \times 7 \times 25 \times 150 = 43.23 \times 10^3 \text{ N} = 43.23 \text{ kN}$$

Given that the b/t ratio varies from 1 initially to exceed 4 at the end of the process, it is appropriate to examine the 'rough die' SLF solutions in § 6.8.1 and 6.8.2. These show a mean normalised pressure (Fig. 6.46) variation from 1 to 1.75.

Ex. 15.4 As the 150 mm dimension remains unchanged, a plane strain compression may be assumed for $t_o = 150$ mm and $w_o = 150$ mm with each of three dies $b = 50$, 100 and 150 mm (see Fig. 15.1). Multiple cogging steps extend the length from $l_o = 0.75$ m to $l_f = 2.25$ m when the thickness is reduced from 150 mm to 50 mm. The load expression $L = YAC$ employs two forms for the coefficient C. The first applies to $b = 50$ mm ($b/t_o < 1$) where the work done per step is given by

$$W = \int L\,\mathrm{d}t = YA\int_{t_f}^{t_o}\left(0.797 + 0.203\frac{t}{b}\right)\mathrm{d}t = YA\left[0.797(t_o - t_f) + \frac{0.203}{2b}(t_o^2 - t_f^2)\right]$$

Substituting $A = bw_o$ and multiplying by the number of cogging steps l_o/b gives the total work:

$$W = Yw_o l_o\left[0.797(t_o - t_f) + \frac{0.203}{2b}(t_o^2 - t_f^2)\right] \qquad \text{(i)}$$

The second coefficient applies to $b = 150$ mm ($b/t_o > 1$), where

$$W = \int L\,\mathrm{d}t = YA\int_{t_f}^{t_o}\left(0.75 + 0.25\frac{b}{t}\right)\mathrm{d}t = YA\left[0.75(t_o - t_f) + 0.25\,b\ln\left(\frac{t_o}{t_f}\right)\right]$$

which gives the total work done

$$W = Yw_o l_o\left[0.75(t_o - t_f) + 0.25\,b\ln\left(\frac{t_o}{t_f}\right)\right] \qquad \text{(ii)}$$

As the intermediate die ($b = 100$ mm) violates the condition ($b/t_o < 1$) when the thickness becomes less than 100 mm, both coefficients C are required. Let $t_i = 100$ mm be the thickness for which a die change is made. We may add the work expression (i) and (ii) as follows:

$$W = Yw_o l_o\left[0.797(t_o - t_i) + \frac{0.203}{2b}(t_o^2 - t_i^2)\right] + Yw_o l_o\left[0.75(t_i - t_f) + 0.25\,b\ln\left(\frac{t_i}{t_f}\right)\right] \qquad \text{(iii)}$$

The reader should show that eq(iii) gives an intermediate amount of work and then find a die breadth that minimises the work done (see example 15.1).

Ex. 15.8 The mean pressure for homogenous, cylindrical extrusion appears in example 15.2, where Y is the average flow stress for a hardening material. The initial flow stress must first be found for a material pre-strained with a 15 % area reduction. Using constant volume

$$\frac{A_o - A_r}{A_o} = 0.15, \quad \Rightarrow \quad \frac{A_o}{A_r} = \frac{l_r}{l_o} = 1.176$$

Hence, the natural pre-strain and initial flow stress appear within Hollomon's law:

$$\varepsilon_{o-r} = \ln(l_r/l_o) = \ln(1.176) = 0.163, \quad \Rightarrow \quad \sigma_r = 670\,(0.163)^{0.5} = 270.1 \text{ MPa}$$

With a further reduction in diameter from 40 mm to 37 mm between the ram and extrusion, the additional natural strain imparted to the material is

$$\frac{l_e}{l_r} = \frac{A_r}{A_e} = \left(\frac{d_r}{d_e}\right)^2 = \left(\frac{40}{37}\right)^2 = 1.169, \quad \Rightarrow \quad \varepsilon_{r-e} = 0.156$$

The final flow stress reached, following this natural process strain appears within Hollomon's law:

$$\varepsilon_{o-e} = \varepsilon_{o-r} + \varepsilon_{r-e} = 0.163 + 0.156 = 0.319, \quad \Rightarrow \quad \sigma_r = 670\,(0.319)^{0.5} = 378.4 \text{ MPa}$$

Thus, the mean flow stress for the extrusion process is $Y_m = \frac{1}{2}(270.1 + 378.4) = 324.3$ MPa, from which the mean ram pressure is

$$p_m = F/A_r = Y_m \ln(A_r/A_e) = 2Y_m \ln(d_r/d_e) = 2 \times 324.3 \times \ln(40/37) = 50.57 \text{ MPa}$$

Alternatively, the mean pressure identifies with specific work done, i.e. the area beneath the stress–strain curve between strain limits of 0.163 and 0.319:

$$p_m = \frac{W}{V} = \int_{\varepsilon_{o-r}}^{\varepsilon_{o-e}} \sigma\,d\varepsilon = 670 \int_{0.163}^{0.319} \varepsilon^{0.5}\,d\varepsilon = \frac{2 \times 670}{3}\left[0.319^{1.5} - 0.163^{1.5}\right] = 51.08 \text{ MPa}$$

Ex. 15.10 For process 0 – 1, 10 mm – 8.5 mm diameter reduction at constant volume:

$$A_0 l_0 = A_1 l_1 \quad \therefore \quad \frac{l_1}{l_0} = \frac{A_0}{A_1} = \left(\frac{d_0}{d_1}\right)^2$$

$$\varepsilon_{0-1} = \ln\left(\frac{l_1}{l_o}\right) = 2\ln\left(\frac{d_0}{d_1}\right) = 2\ln\left(\frac{10}{8.5}\right) = 0.325$$

$$\sigma_1 = 300\,(0.325)^{0.5} = 171 \text{ MPa}$$

Process 1–2, 30 % area reduction at constant volume:

$$\frac{A_1 - A_2}{A_1} = 0.3 \quad \therefore \quad \frac{A_1}{A_2} = 1.429 = \frac{l_2}{l_1}$$

$$\varepsilon_{1-2} = \ln\left(\frac{l_2}{l_1}\right) = \ln 1.429 = 0.357$$

$$\varepsilon_{0-2} = \varepsilon_{0-1} + \varepsilon_{1-2} = 0.325 + 0.357 = 0.682$$

$$\sigma_2 = 300\,(0.682)^{0.5} = 247.75 \text{ MPa}$$

$$\sigma_m = \frac{1}{2}(\sigma_1 + \sigma_2) = \frac{1}{2}(171 + 247.8) = 209.4 \text{ MPa}$$

$$F_2 = \sigma_m A_2 \ln\left(\frac{A_1}{A_2}\right) = 209.4 \times \frac{\pi \times (8.5)^2}{4 \times 1.429} \times \ln(1.429) = 2968.4 \text{ N}$$

Minimum d_2 with hardening:

$$\sigma_2 = \frac{F_2}{A_2} = \sigma_m \ln\left(\frac{A_1}{A_2}\right) \quad \therefore \quad \frac{A_1}{A_2} = \left(\frac{d_1}{d_2}\right)^2 = \exp\left(\frac{\sigma_2}{\sigma_m}\right)$$

$$\left(\frac{8.5}{d_2}\right)^2 = \exp\left(\frac{247.75}{209.4}\right) = 3.2646$$

$$\therefore\ d_2 = \frac{8.5}{\sqrt{3.2646}} = 4.704 \text{ mm}$$

Minimum d_2 without hardening:

$$\sigma_2 = Y = Y\ln\left(\frac{A_1}{A_2}\right) \quad \therefore \quad \frac{A_1}{A_2} = \left(\frac{d_1}{d_2}\right)^2 = \mathrm{e}$$

$$\left(\frac{8.5}{d_2}\right)^2 = \mathrm{e} = 2.7183$$

$$\therefore\ d_2 = \frac{8.5}{\sqrt{2.7183}} = 5.156 \text{ mm}$$

Ex. 15.13 The strip draw stress is given by eq(15.55b):

$$\sigma_s = Y_m(1 + 1/\beta)[1 - (1 - r)^\beta] \tag{i}$$

where Y_m is the mean stress between entry 0 and exit 1 to the conical die (see Fig. 15.13a). At entry, Y_0 = 200 MPa for annealed material. At exit, the natural strain follows from deformation at constant volume, where the width remains constant:

$$\frac{l_1}{l_0} = \frac{A_0}{A_1} = \frac{h_0}{h_1}, \quad \Rightarrow \quad \varepsilon = \ln\left(\frac{h_0}{h_1}\right) = \ln\left(\frac{2.5}{2.4}\right) = 0.0408$$

Then Y_1 follows from Ludwik's law:

$Y_1 = 200 + 100\,(0.0408)^{0.8} = 207.7$ MPa

which gives

$Y_m = ½\,(200 + 207.7) = 203.9$ MPa

Also within eq(i):

$$\beta = \frac{\mu}{\tan\theta} = \frac{0.1}{\tan 7.5°} = 0.76, \quad r = \frac{A_o - A_1}{A_o} = 1 - \frac{t_1}{t_o} = 1 - \frac{2.4}{2.5} = 0.04$$

Substituting Y_m, β and r into eq(i) gives the strip draw stress $\sigma_s = 14.42$ MPa.
With a previous 20 % reduction from rolling, the work hardened material's flow stress at entry position 0 is found from the pre-strain:

$$\frac{A_0 - A_1}{A_0} = 0.2, \quad \frac{l_1}{l_0} = \frac{A_0}{A_1} = 1.25, \quad \Rightarrow \quad \varepsilon = \ln\left(\frac{l_1}{l_0}\right) = \ln 1.25 = 0.223$$

and from Ludwik's law: $Y_0 = 200 + 100\,(0.223)^{0.8} = 230.1$ MPa.

Again, the process reduction upon this pre-strained material strains it further by 0.0408, giving a total natural strain at exit position 1 of (0.0408 + 0.223) = 0.2638. Correspondingly, the flow stress at exit has risen to

$Y_1 = 200 + 100\,(0.2638)^{0.8} = 234.4$ MPa.

Hence, the mean stress between entry and exit becomes

$Y_m = ½\,(230.1 + 234.4) = 232.3$ MPa

when, from eq(i), the strip draw stress is raised to 14.42 × 232.3/203.9 = 16.43 MPa.

Ex. 15.15 Here: $R >> \Delta h$, $Y \rightarrow \sigma_m$ for $\sigma = Y + B(\varepsilon^P)^n$

Process 0 – 1 involves 10 % reduction with constant volume:

$$\frac{A_0 - A_1}{A_0} = 0.1 \quad \therefore \quad \frac{A_0}{A_1} = 1.\dot{1} = \frac{l_1}{l_0}$$

$$\varepsilon_{0\text{–}1} = \ln\left(\frac{l_1}{l_o}\right) = \ln 1.\dot{1} = 0.105$$

$$\sigma_1 = 200 + 310(0.105)^{0.68} = 267 \text{ MPa}$$

Process 1–2:

$$\frac{l_2}{l_1} = \frac{A_1}{A_2} = \frac{b \times h_1}{b \times h_2} = \frac{h_1}{h_2} = \frac{4}{3.44} = 1.163$$

$$\varepsilon_{1-2} = \ln\left(\frac{l_2}{l_1}\right) = \ln 1.163 = 0.151$$

$$\varepsilon_{0-2} = \varepsilon_{0-1} + \varepsilon_{1-2} = 0.105 + 0.151 = 0.256$$

$$\sigma_2 = 200 + 310(0.256)^{0.68} = 322.7 \text{ MPa}$$

$$\sigma_m = \frac{1}{2}(\sigma_1 + \sigma_2) = \frac{1}{2}(267 + 322.7) = 294.9 \text{ MPa}$$

$$F = \sigma_m w \sqrt{R\Delta h} = 294.9 \times 250\sqrt{175 \times (4 - 3.44)} = 729.8 \text{ kN}$$

Ex. 15.16 In the hot rolling of narrow strip, the edge spread alters the plane strain condition slightly such that a greater edge pressure $2k$ is present. So if we look at the roll end-on, the constant peak pressure within the central region falls to $2k$ at the edges, in the manner of Fig. 15.11. The effect is to cut off the top corners of the usual 'friction roof' that would apply across the full width when looking squarely at the rolling of a wide strip. For the latter, the roll force/unit width P in (eq 15.27a) was identified with the area beneath the roof. Hence, the full roll force becomes the volume beneath the roof:

$$P = 2kbL\left(0.8 + \frac{L}{4h_o}\right) \tag{i}$$

Thus, in rolling narrow strip, the two corner volumes that have been removed must be subtracted from eq(i). These are identical right pyramids: the base being an isosceles triangle with equal angles α (see Diag. 15.12). The hypotenuse is of length

$$q = 2\left(L/2 - 0.2h_o\right) \tag{ii}$$

and from the hypotenuse mid-position, the perpendicular height of the base triangle is

$$r = \frac{q}{2}\tan\alpha = \left(\frac{L}{2} - 0.2h_o\right)\tan\alpha \tag{iii}$$

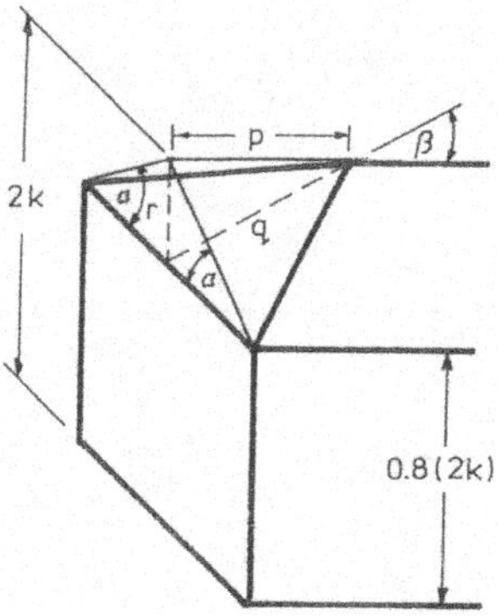

Diagram 15.12

The perpendicular height of the pyramid is

$$p = \frac{q}{2}\frac{\tan\alpha}{\tan\beta} = \left(\frac{L}{2} - 0.2h_o\right)\frac{\tan\alpha}{\tan\beta} \tag{iv}$$

where, in eqs(iii) and (iv), the slopes of the roof are

$$\tan\alpha = \frac{2k}{h_o}, \quad \tan\beta = \frac{2k}{h} \tag{v}$$

Given that the volume of a right pyramid may be expressed as $qrp/6$, the volume to be subtracted from eq(i) is

$$\frac{qrp}{3} = \frac{q^3}{12} \times \frac{\tan^2\alpha}{\tan\beta} = \frac{q^3}{12}\left(\frac{2k}{h_o}\right)^2\left(\frac{h}{2k}\right) = \frac{4kh}{3h_o^2}\left(\frac{q}{2}\right)^3 = \frac{4kh}{3h_o^2}\left(\frac{L}{2} - 0.2h_o\right)^3 \tag{vi}$$

Subtracting eq(vi) from eq(i):

$$P = 2kbL\left(0.8 + \frac{L}{4h_o}\right) - \frac{4kh}{3h_o^2}\left(\frac{L}{2} - 0.2h_o\right)^3 = 2kbL\left[0.8 + \frac{L}{4h_o} - \frac{2hh_o}{3bL}\left(\frac{L}{2h_o} - 0.2\right)^3\right]$$

this being the correct expression irrespective of whether the lengths b and h are simple or weighted means (as in eq(15.35)).

Ex. 15.22 The shear zone thickness δS_1 (see Fig. 15.22b) is found from the aspect ratio 10:1, the depth of cut and the shear angle $\varphi = 25°$:

$$b = AB\sin\varphi, \quad \frac{AB}{\delta S_1} = 10 \quad \Rightarrow \quad \delta S_1 = \frac{AB}{10} = \frac{b}{10\sin 25°} = 0.047 \text{ mm}$$

Other machining conditions specified allow for the calculation of the shear strain:

$$\gamma = \frac{\cos\alpha}{\sin\varphi\cos(\varphi - \alpha)} = \frac{\cos 10^\circ}{\sin 25^\circ \cos(25^\circ - 10^\circ)} = 2.413$$

and the shear strain rate:

$$\dot{\gamma} = \frac{v\cos\alpha}{\delta S_1 \sin\varphi\cos(\varphi - \alpha)} = \frac{0.5 \times 10^3 \cos 10^\circ}{0.047 \sin 25^\circ \cos(25^\circ - 10^\circ)} = 10.85 \times 10^3\ s^{-1}$$

Reading from Fig. 15.25, the mean shear stress along plane AB and the hardening exponent:

$k - \delta k/2 = 448$ MPa and $m = \delta k/\delta\gamma = 27.5$ MPa

These combine to give

$k = 448 + m\gamma/2 = 448 + 27.5 \times 2.413/2 = 481.2$ MPa

The normal pressures at A and B are found from eqs(15.78a,b):

$$p_a = 1 + 2\left(\frac{\pi}{4} - \frac{25 \times \pi}{180}\right) = 1.698, \quad p_b = 1.698 - 10\frac{\delta k}{k} = 0.319$$

Hence, the inclination of R to the shear plane AB (see Fig. 15.22a) is

$$\theta = \tan^{-1}\left(\frac{p_A + p_B}{2k}\right) = 45.2^\circ$$

and the inclination of R to flank face BF (Fig. 15.22) is

$\alpha + 90^\circ - \varphi + \theta = 10^\circ + 90^\circ - 25^\circ + 45.2^\circ = 120.24^\circ$

as it enters and produces $b_1 = 104$ mm, $t_1 = 6$ mm and $w_1 = 150$ mm, for which a constant flow stress $2k = 7$ MPa applies to lead during the cold forging process.

CHAPTER 16

Ex. 16.1 In an isotropic material it may be assumed that there is no interaction between the displacements from the combined action of an axial force and a torque. So for a circular section bar the separate forces and displacements shown in Figs 16.1 and 16.2 may be superimposed. Consequently, we may combine the stiffness matrices eqs(16.2) and (16.3a) in the most convenient manner:

$$\begin{bmatrix} f_{x1} \\ t_1 \\ f_{x2} \\ t_2 \end{bmatrix} = \begin{bmatrix} AE/L & 0 & -AE/L & 0 \\ 0 & JG/L & 0 & -JG/L \\ -AE/L & 0 & AE/L & 0 \\ 0 & -JG/L & 0 & JG/L \end{bmatrix} \begin{bmatrix} u_1 \\ \theta_1 \\ u_2 \\ \theta_2 \end{bmatrix}$$

Modification is necessary to allow for anisotropy, when, for example, a bar changes its length and twists under a torque.

Ex. 16.6 The general displacement vector $\boldsymbol{\delta} = [u \quad v]^T$ in eq(16.10) applies to point (x, y) in the triangular element. To prove the relationships given, it is necessary to form the shape function matrix $\mathbf{N} = \mathbf{A}\,(\mathbf{A}^e)^{-1}$ where $\mathbf{A}$ and $(\mathbf{A}^e)^{-1}$ appear in eqs (16.7c) and (16.9b). This gives

$$(2\Delta)\,\mathbf{N} = \begin{bmatrix} 1 & x & y & 0 & 0 & 0 \\ 0 & 0 & 0 & 1 & x & y \end{bmatrix} \begin{bmatrix} x_2y_3 - x_3y_2 & 0 & x_3y_1 - x_1y_3 & 0 & x_1y_2 - x_2y_1 & 0 \\ y_2 - y_3 & 0 & y_3 - y_1 & 0 & y_1 - y_2 & 0 \\ x_3 - x_2 & 0 & x_1 - x_3 & 0 & x_2 - x_1 & 0 \\ 0 & x_2y_3 - x_3y_2 & 0 & x_3y_1 - x_1y_3 & 0 & x_1y_2 - x_2y_1 \\ 0 & y_2 - y_3 & 0 & y_3 - y_1 & 0 & y_1 - y_2 \\ 0 & x_3 - x_2 & 0 & x_1 - x_3 & 0 & x_2 - x_1 \end{bmatrix}$$

Expanding shows that $\mathbf{N}$ is a 6×2 matrix, containing both the general and nodal point co-ordinates. When in turn $\mathbf{N}$ is multiplied by $\boldsymbol{\delta}^e = [\,u_1 \quad v_1 \quad u_2 \quad v_2 \quad u_3 \quad v_3]^T$ as in eq(16.10), the general displacement matrix $\boldsymbol{\delta} = [u \quad v]^T$ is found. Correspondingly, the expressions for the general displacements u and v may be reduced as follows:

$$u = (a_1 + b_1x + c_1y)u_1 + (a_2 + b_2x + c_2y)u_2 + (a_3 + b_3x + c_3y)u_3 \qquad \text{(i)}$$

$$= N_1u_1 + N_2u_2 + N_3u_3 = N_i u_i \text{ (for } i = 1, 2, 3)$$

Similarly:

$$v = (a_1 + b_1x + c_1y)v_1 + (a_2 + b_2x + c_2y)v_2 + (a_3 + b_3x + c_3y)v_3 \qquad \text{(ii)}$$

$$= N_1v_1 + N_2v_2 + N_3v_3 = N_i v_i \text{ (for } i = 1, 2, 3)$$

in which $N_i = a_i + b_ix + c_iy$ are common to the two displacements. To find the coefficients a_i, b_i and c_i, eqs(i) and (ii) must match nodal point displacements: $u = u_1$ and $v = v_1$ for $x = x_1$ and $y = y_1$ etc.

Ex. 16.7 When a normal pressure p is applied to side 23 in Fig. 16.4, the nodal force vector $\mathbf{f}$ in (16.14b) requires that equivalent nodal point forces f_{x1}, f_{y1} etc. be found. Employing the principle of virtual displacements allows the work done by the pressure to be equated to the virtual work done by the equivalent forces:

$$(\boldsymbol{\delta}^{ev})^T \mathbf{f}^e = \int (\boldsymbol{\delta}^v)^T \mathbf{p}\, t\, dl \qquad \text{(i)}$$

where vector $\boldsymbol{\delta}^{ev}$ gives virtual nodal point displacements, $\boldsymbol{\delta}$ is the general displacement vector, $\mathbf{p} = [p_x \quad p_y\,]^T$ is the pressure component vector and $t\,dl$ is an element of area along

23 upon which p acts normally. If we now substitute the relationship between $\boldsymbol{\delta}$ and $\boldsymbol{\delta}^e$ from (16.10) into eq(i):

$$(\boldsymbol{\delta}^{ev})^T \mathbf{f}^e = \int [\mathbf{A}\,(\mathbf{A}^e)^{-1}\,\boldsymbol{\delta}^{ev}]^T \mathbf{p}\; t\; \mathrm{d}l$$

$$= t \int (\boldsymbol{\delta}^{ev})^T [\mathbf{A}\,(\mathbf{A}^e)^{-1}]^T\, \mathbf{p}\; \mathrm{d}l$$

$$= t\, (\boldsymbol{\delta}^{ev})^T [(\mathbf{A}^e)^{-1}]^T \int \mathbf{A}^T \mathbf{p}\; \mathrm{d}l \qquad \text{(ii)}$$

Upon cancelling $(\boldsymbol{\delta}^{ev})^T$, eq(ii) provides the equivalent nodal force vector

$$\mathbf{f}^e = [(\mathbf{A}^e)^{-1}]^T \mathbf{p}^e$$

where $(\mathbf{A}^e)^{-1}$ appears in eq(16.9b) and $\mathbf{p}^e$ must be found from the integral

$$\mathbf{p}^e = t \int \mathbf{A}^T \mathbf{p}\; \mathrm{d}l \qquad \text{(iii)}$$

Substituting $\mathbf{A}$ from eq(16.7c), the integral becomes

$$\mathbf{A}^T\mathbf{p} = \begin{bmatrix} 1 & 0 \\ x & 0 \\ y & 0 \\ 0 & 1 \\ 0 & x \\ 0 & y \end{bmatrix} \begin{bmatrix} p_x \\ p_y \end{bmatrix} = \begin{bmatrix} p_x \\ xp_x \\ yp_x \\ p_y \\ xp_y \\ yp_y \end{bmatrix} \qquad \text{(iv)}$$

The equation of the side 23 is

$$y - y_2 = \frac{(y_3 - y_2)}{(x_3 - x_2)}(x - x_2) \qquad \text{(v)}$$

and by proportion:

$$\frac{\mathrm{d}x}{\mathrm{d}l} = \frac{(x_3 - x_2)}{l_{23}} \qquad \text{(vi)}$$

Substituting eq(iv) into eq(iii), with y and $\mathrm{d}l$ from eqs(v) and (vi), allows each term in $\mathbf{A}^T\mathbf{p}$ to be integrated separately. This gives expressions for each component of the equivalent nodal force vector $\mathbf{p}^e = [f_{x1}\;\; f_{y1}\;\; f_{x2}\;\; f_{y2}\;\; f_{x3}\;\; f_{y3}]^T$, for which the following summations can be seen to apply:

$$f_{x1} + f_{x2} + f_{x3} = p_x l_{23}\, t$$

$$f_{y1} + f_{y2} + f_{y3} = p_y l_{23}\, t$$

Ex. 16.10 The 'strain signature' taken along the corner of a box shape could be produced from an FE simulation of the strains arising in pressing a panel to the required shape (see Fig. 16.13). The cost involved here is that of the FE software and the computing facilities. A far greater cost is incurred if 'safe' predictions are found to be inaccurate in a production run, resulting in much scrap metal. For this reason a die-tryout should also be employed. Here, a grid is etched to the surface of the flat sheet and the strains are found from the distortion that occurs to that grid after the pressing is made. A grid of tangential or overlapping circles allows the principal engineering strains e_1 and e_2 to be calculated from the change to their diameters. Because the detection of strain gradients requires small circles ($\approx$5 mm), the errors in measurement are accentuated, though modern optical and digital imaging methods have improved the accuracy. The ideal safe design is one where the pressing is deemed not to have failed from splitting, necking and wrinkling through the material suffering too greater strain. Moreover, the material should not be under-strained as this can result in unwanted areas of loose metal, as in a car roof, for example. In practice, a number of signatures would be taken within high strain regions and superimposed upon the FLD in the manner of Fig. 16.15b. Typically, the signature ABCD shows that the material has been subjected to stretch at base A, draw along the corner BC and compression in the flange rim at C. Here, the principal strain combination (e_1, e_2) at D lies closest to the FLC. Elsewhere, at A and D especially, it appears that the strain limits presented by the FLD are excessive and therefore the material has not been used to its full extent. This information that this signature provides would need to be considered with that from other signatures to decide whether a panel redesign could remove any risky strain peaks. The alternative is retain the panel shape but upgrade the material. A change in material raises the FLD to take the peak strain out of the danger region. There are many factors which influence the FLD. Those which concern sheet selection are

yield stress - low value preferred
strain rate sensitivity - raises the flow stress
strain hardening exponent n - high value preferred
anisotropy - certain sheet orientations may allow more thinning
thickness t - governs press capacity

Those which arise from the process are

friction - contact forces 'add' to forming forces
high temperature - lowers flow stress
lubrication - reduces friction
poor die condition - unwanted stress raisers

The effects of sheet thickness and the n–value appear together in Fig. 16.12. They are connected through the lowest limiting, plane strain intercept value e_{10}, which appears as an ordinate in (a) and defines the 'V' of the FLD in (b). So in a sheet of given thickness, for material of known n–value, e_{10} is read from (a), then used to locate the FLD in (b). The arms of the FLD rely on geometric similarity and on one FLD being available. Thus, with an increase in thickness, the FLD may be shifted vertically to the e_{10} intercept provided by (a). The figure demonstrates this for $n = 0.2$ with t ranging from 0.5 to 3.5 mm. The classical, incompressible plasticity theory cannot predict this thickness effect, at least not when sheet instability limits its in-plane strain. Of course, strains are limited as a consequence of material breakdown; witness the cup and cone fracture in a simple tension test, for example. The greater strain reached in a local neck are too large for a practical FLD; one would normally measure grid strain away from the neck in setting a safe limit to work to. Moreover, in producing the one FLD required for each material when applying the scheme in Fig. 16.12, the lower limit is taken from a band of strain (scatter) that arises from the laboratory tests. The latter are numerous. In addition to simple tension, Figs 16.8 and 16.10 show hydraulic bulging and punch indentation, all of which apply to different positions on the FLD (see Fig. 13.23). Whatever testing is performed should serve to span the two quadrants of strain, similar to when pressing a panel. While the simple laboratory tests reach their limiting strains via a linear path, this is unlikely to be the case in practice. In this regard, we shall finish on a contentious note. Despite the concept of an FLD having been accepted as a material property for some 30 years, the effect of strain history upon the FLD is uncertain. One could draw a parallel with initial and subsequent yield surfaces linked by a path dependent strain history.

INDEX

Key to abbreviations: FE – Finite element, FLD – Forming Limit Diagram, H-I – Hencky-Illyushin, P-R – Prandtl-Reuss, PVD – Principle of Virtual Displacements, PVW– Principle of Virtual Work, SLF – Slip Line Field, SMF – Sheet Metal Forming, VD – Velocity Discontinuity